Selected Titles in This Series

53 **Andreas Kriegl and Peter W. Michor,** The convenient setting of global analysis, 1997
52 **V. A. Kozlov, V. G. Maz'ya, and J. Rossmann,** Elliptic boundary value problems in domains with point singularities, 1997
51 **Jan Malý and William P. Ziemer,** Fine regularity of solutions of elliptic partial differential equations, 1997
50 **Jon Aaronson,** An introduction to infinite ergodic theory, 1997
49 **R. E. Showalter,** Monotone operators in Banach space and nonlinear partial differential equations, 1997
48 **Paul-Jean Cahen and Jean-Luc Chabert,** Integer-valued polynomials, 1997
47 **A. D. Elmendorf, I. Kriz, M. A. Mandell, and J. P. May (with an appendix by M. Cole),** Rings, modules, and algebras in stable homotopy theory, 1997
46 **Stephen Lipscomb,** Symmetric inverse semigroups, 1996
45 **George M. Bergman and Adam O. Hausknecht,** Cogroups and co-rings in categories of associative rings, 1996
44 **J. Amorós, M. Burger, K. Corlette, D. Kotschick, and D. Toledo,** Fundamental groups of compact Kähler manifolds, 1996
43 **James E. Humphreys,** Conjugacy classes in semisimple algebraic groups, 1995
42 **Ralph Freese, Jaroslav Ježek, and J. B. Nation,** Free lattices, 1995
41 **Hal L. Smith,** Monotone dynamical systems: an introduction to the theory of competitive and cooperative systems, 1995
40.2 **Daniel Gorenstein, Richard Lyons, and Ronald Solomon,** The classification of the finite simple groups, number 2, 1995
40.1 **Daniel Gorenstein, Richard Lyons, and Ronald Solomon,** The classification of the finite simple groups, number 1, 1994
39 **Sigurdur Helgason,** Geometric analysis on symmetric spaces, 1994
38 **Guy David and Stephen Semmes,** Analysis of and on uniformly rectifiable sets, 1993
37 **Leonard Lewin, Editor,** Structural properties of polylogarithms, 1991
36 **John B. Conway,** The theory of subnormal operators, 1991
35 **Shreeram S. Abhyankar,** Algebraic geometry for scientists and engineers, 1990
34 **Victor Isakov,** Inverse source problems, 1990
33 **Vladimir G. Berkovich,** Spectral theory and analytic geometry over non-Archimedean fields, 1990
32 **Howard Jacobowitz,** An introduction to CR structures, 1990
31 **Paul J. Sally, Jr. and David A. Vogan, Jr., Editors,** Representation theory and harmonic analysis on semisimple Lie groups, 1989
30 **Thomas W. Cusick and Mary E. Flahive,** The Markoff and Lagrange spectra, 1989
29 **Alan L. T. Paterson,** Amenability, 1988
28 **Richard Beals, Percy Deift, and Carlos Tomei,** Direct and inverse scattering on the line, 1988
27 **Nathan J. Fine,** Basic hypergeometric series and applications, 1988
26 **Hari Bercovici,** Operator theory and arithmetic in H^∞, 1988
25 **Jack K. Hale,** Asymptotic behavior of dissipative systems, 1988
24 **Lance W. Small, Editor,** Noetherian rings and their applications, 1987
23 **E. H. Rothe,** Introduction to various aspects of degree theory in Banach spaces, 1986
22 **Michael E. Taylor,** Noncommutative harmonic analysis, 1986
21 **Albert Baernstein, David Drasin, Peter Duren, and Albert Marden, Editors,** The Bieberbach conjecture: Proceedings of the symposium on the occasion of the proof, 1986
20 **Kenneth R. Goodearl,** Partially ordered abelian groups with interpolation, 1986
19 **Gregory V. Chudnovsky,** Contributions to the theory of transcendental numbers, 1984
18 **Frank B. Knight,** Essentials of Brownian motion and diffusion, 1981

(Continued in the back of this publication)

The Convenient Setting of Global Analysis

Mathematical
Surveys
and
Monographs

Volume 53

The Convenient Setting of Global Analysis

Andreas Kriegl
Peter W. Michor

American Mathematical Society

1991 *Mathematics Subject Classification.* Primary 22E65, 26E15, 26E20, 46A17, 46G05, 46G20, 46E25, 46E50, 58B10, 58B12, 58B20, 58B25, 58C20, 46E50, 58D05, 58D10, 58D15, 58D17, 58D19, 58F25; Secondary 22E45, 58C40, 22E67, 46A16, 57N20, 58B05, 58D07, 58D25, 58D27, 58F05, 58F06, 58F07.

ABSTRACT. The aim of this book is to lay foundations of differential calculus in infinite dimensions and to discuss those applications in infinite dimensional differential geometry and global analysis which do not involve Sobolev completions and fixed point theory. The approach is very simple: A mapping is called smooth if it maps smooth curves to smooth curves. All other properties are proved results and not assumptions: Like chain rule, existence and linearity of derivatives, powerful smooth uniformly boundedness theorems are available. Up to Fréchet spaces this notion of smoothness coincides with all known reasonable concepts. In the same spirit calculus of holomorphic mappings (including Hartogs' theorem and holomorphic uniform boundedness theorems) and calculus of real analytic mappings are developed. Existence of smooth partitions of unity, the foundations of manifold theory in infinite dimensions, the relation between tangent vectors and derivations, and differential forms are discussed thoroughly. Special emphasis is given to the notion of regular infinite dimensional Lie groups. Many applications of this theory are included: manifolds of smooth mappings, groups of diffeomorphisms, geodesics on spaces of Riemannian metrics, direct limit manifolds, perturbation theory of operators, and differentiability questions of infinite dimensional representations.

Corrections and complements to this book will be posted on the internet at the URL
`http://www.mat.univie.ac.at/~michor/apbook.ps`

Library of Congress Cataloging-in-Publication Data

Kriegl, Andreas.
The convenient setting of global analysis / Andreas Kriegl, Peter W. Michor.
p. cm. — (Mathematical surveys and monographs, ISSN 0076-5376 ; v. 53)
Includes bibliographical references (p. –) and index.
ISBN 0-8218-0780-3 (alk. paper)
1. Global analysis (Mathematics) I. Michor, Peter W., 1949– . II. Title. III. Series: Mathematical surveys and monographs ; no. 53.
QA614.K75 1997
514′.74—dc21 97-25857
CIP

---

10 9 8 7 6 5 4 3 2 1 02 01 00 99 98 97

To Elli, who made working on this book into a culinary experience.

Table of Contents

Introduction

At the very conception of the notion of manifolds, in the Habilitationsschrift [Riemann, 1854], infinite dimensional manifolds were mentioned explicitly:

> "Es giebt indess auch Mannigfaltigkeiten, in welchen die Ortsbestimmung nicht eine endliche Zahl, sondern entweder eine unendliche Reihe oder eine stetige Mannigfaltigkeit von Grössenbestimmungen erfordert. Solche Mannigfaltigkeiten bilden z.B. die möglichen Bestimmungen einer Function für ein gegebenes Gebiet, die möglichen Gestalten einer räumlichen Figur u.s.w."

The purpose of this book is to lay the foundations of infinite dimensional differential geometry. The book [Palais, 1968] and review article [Eells, 1966] have similar titles and treat global analysis mainly on manifolds modeled on Banach spaces. Indeed classical calculus works quite well up to and including Banach spaces: Existence and uniqueness hold for solutions of smooth ordinary differential equations (even Lipschitz ones), but not existence for all continuous ordinary differential equations. The inverse function theorem works well, but the theorem of constant rank presents problems, and the implicit function theorem requires additional assumptions about existence of complementary subspaces. There are also problems with partitions of unity, with the Whitney extension theorem, and with Morse theory and transversality.

Further development has shown that Banach manifolds are not suitable for many questions of Global Analysis, as shown by the following result, which is due to [Omori, de la Harpe, 1972], see also [Omori, 1978b]: If a Banach Lie group acts effectively on a finite dimensional smooth manifold it must be finite dimensional itself. The study of Banach manifolds per se is not very interesting, since they turn out to be open subsets of the modeling space for many modeling spaces, see [Eells, Elworthy, 1970].

Our aim in this book is to treat manifolds which are modeled on locally convex spaces, and which are smooth, holomorphic, or real analytic in an appropriate sense. To do this we start with a careful exposition of smooth, holomorphic, and real analytic calculus in infinite dimensions. Differential calculus in infinite dimensions has already quite a long history; in fact it goes back to Bernoulli and Euler, to the beginnings of variational calculus. During the 20-th century the urge to differentiate in spaces which are more general than Banach spaces became stronger, and many different approaches and definitions were attempted. The main difficulty encountered was that composition of (continuous) linear mappings ceases to be a jointly continuous operation exactly at the level of Banach spaces, for any suitable topology on spaces of linear mappings. This can easily be explained in a somewhat simpler example:

Consider the evaluation $\mathrm{ev} : E \times E^* \to \mathbb{R}$, where E is a locally convex space and E^* is its dual of continuous linear functionals equipped with any locally convex topology. Let us assume that the evaluation is jointly continuous. Then there are neighborhoods $U \subseteq E$ and $V \subseteq E^*$ of zero such that $\mathrm{ev}(U \times V) \subseteq [-1, 1]$. But then U is contained in the polar of V, so it is bounded in E, and so E admits a bounded neighborhood and is thus normable.

The difficulty described here was the original motivation for the development of a whole new field within general topology, convergence spaces. Fortunately it is no longer necessary to delve into this, because [Frölicher, 1981] and [Kriegl, 1982], [Kriegl, 1983] presented independently the solution to the question for the right differential calculus in infinite dimensions, see the monograph [Frölicher, Kriegl, 1988]. The smooth calculus which we present here is the same as in this book, but our exposition is based on functional analysis rather than on category theory.

Let us try to describe the basic ideas of smooth calculus: One can say that it is a (more or less unique) consequence of taking variational calculus seriously. We start by looking at the space of smooth curves $C^\infty(\mathbb{R}, E)$ with values in a locally convex space E and note that it does not depend on the topology of E, only on the underlying system of bounded sets. This is due to the fact, that for a smooth curve difference quotients converge to the derivative much better than arbitrary converging nets or filters. Smooth curves have integrals in E if and only if a weak completeness condition is satisfied: it appeared as 'bornologically complete' or 'locally complete' in the literature; we call it c^∞-complete. Surprisingly, this is equivalent to the condition that scalarwise smooth curves are smooth. All calculus in this book will be done on convenient vector spaces. These are locally convex vector spaces which are c^∞-complete. Note that the locally convex topology on a convenient vector space can vary in some range – only the system of bounded set must remain the same. The next steps are then easy: a mapping between convenient vector spaces is called smooth if it maps smooth curves to smooth curves, and everything else is a theorem – existence, smoothness, and linearity of derivatives, the chain rule, and also the most important feature, cartesian closedness

$$C^\infty(E \times F, G) \cong C^\infty(E, C^\infty(F, G)) \tag{1}$$

holds without any restriction, for a natural convenient vector space structure on $C^\infty(F, G)$: So the old dream of variational calculus becomes true in a concise way. If one wants (1) and some other mild properties of calculus, then smooth calculus as described here is unique. Let us point out that on some convenient vector spaces there are smooth functions which are not continuous for the locally convex topology. This is not so horrible as it sounds, and is unavoidable if we want the chain rule, since $\mathrm{ev} : E \times E^* \to \mathbb{R}$ is always smooth but continuous only if E is normable, by the discussion above. This just tells us that locally convex topology is not appropriate for non-linear questions in infinite dimensions. We will, however, introduce the c^∞-topology on any convenient vector space, which survives as the fittest for non-linear questions.

An eminent mathematician once said that for infinite dimensional calculus each serious application needs its own foundation. By a serious application one obviously means some application of a hard inverse function theorem. These theorems can be proved, if by assuming enough a priori estimates one creates enough Banach space situation for some modified iteration procedure to converge. Many authors try to build their platonic idea of an a priori estimate into their differential calculus. We think that this makes the calculus inapplicable and hides the origin of the a priori estimates. We believe that the calculus itself should be as easy to use as possible, and that all further assumptions (which most often come from ellipticity of some nonlinear partial differential equation of geometric origin) should be treated separately, in a setting depending on the specific problem. We are sure that in this sense the setting presented here (and the setting in [Frölicher, Kriegl, 1988]) is useful for most applications. To give a basis to this statement we present also the hard implicit function theorem of Nash and Moser, in the approach of [Hamilton, 1982] adapted to convenient calculus, but we give none of its serious applications.

A surprising and very satisfying feature of the notion of convenient vector spaces is that it is also the right setting for holomorphic calculus as shown in [Kriegl, Nel, 1985], for real analytic calculus as shown by [Kriegl, Michor, 1990], and also for multilinear algebra.

In chapter III we investigate the existence of smooth bump functions and smooth partitions of unity. This is tied intimately to special properties of the locally convex spaces in question. There is also a section on differentiability of finite order, based on Lipschitz conditions, whereas the rest of the book is devoted to differentiability of infinite order. Chapter IV answers the question whether real valued algebra homomorphisms on algebras of smooth functions are point evaluations. Germs, extension results like (22.17), and liftings are the topic of chapter V. Here we also treat Frölicher spaces (i.e. spaces with a fairly general smooth structure) and free convenient vector spaces over them.

Chapters VI to VIII are devoted to the theory of infinite dimensional manifolds and Lie groups and some of their applications. We treat here only manifolds described by charts although this limits cartesian closedness of the category of manifolds drastically, see (42.14) and section (23) for more thorough discussions. Then we investigate tangent vectors seen as derivations or kinematically (via curves): these concepts differ, and there are some surprises even on Hilbert spaces, see (28.4). Accordingly, we have different kinds of tangent bundles, vector fields, differential forms, which we list in a somewhat systematic manner. The theorem of De Rham is proved, and a (small) version of the Frölicher-Nijenhuis bracket in infinite dimensions is treated. Finally, we discuss Weil functors (certain product preserving functors of manifolds) as generalized tangent bundles. The theory of infinite dimensional Lie groups can be pushed surprisingly far: Exponential mappings are unique if they exist. A stronger requirement (leading to regular Lie groups) is that one assumes that smooth curves in the Lie algebra integrate to smooth curves in the group in a smooth way (an 'evolution operator' exists). This is due to [Milnor, 1984] who weakened the concept of [Omori, Maeda, Yoshioka, 1982]. It turns out that regular Lie groups have strong permanence properties. Up to now (April 1997) no non-regular Lie group is known. Connections on smooth principal bundles with a regular Lie group as structure group have parallel transport (39.1), and for flat connections the horizontal distribution is integrable (39.2). So some (equivariant) partial differential equations in infinite dimensions are very well behaved, although in general there are counter-examples in every possible direction. As consequence we obtain in (40.3) that a bounded homomorphism from the Lie algebra of simply connected Lie group into the Lie algebra of a regular Lie group integrates to a smooth homomorphism of Lie groups.

The rest of the book describes applications: In chapter IX we treat manifolds of mappings between finite dimensional manifolds. We show that the group of all diffeomorphisms of a finite dimensional manifold is a regular Lie group, also the group of all real analytic diffeomorphisms, and some subgroups of diffeomorphism groups, namely those consisting of symplectic diffeomorphisms, volume preserving diffeomorphism, and contact diffeomorphisms. Then we treat principal bundles with structure group a diffeomorphism group. The first example is the space of all embeddings between two manifolds, a sort of nonlinear Grassmann manifold, which leads to a smooth manifold which is a classifying space for the diffeomorphism group of a compact manifold. Another example is the nonlinear frame bundle of a fiber bundle with compact fiber, for which we investigate the action of the gauge group on the space of generalized connections and show that there are no slices. In section (45) we compute explicitly all geodesics for some natural (pseudo) Riemannian metrics on the space of all Riemannian metrics. Section (46) is devoted to the Korteweg–De Vrieß equation which is shown to be the geodesic equation of a certain right invariant Riemannian metric on the Virasoro group.

Chapter X start with section (47) on direct limit manifolds like the sphere S^∞ or the Grassmannian $G(k,\infty)$ and shows that they are real analytic regular Lie groups or associated homogeneous spaces. This put some constructions of algebraic topology directly into differential geometry. Section (48) is devoted to weak symplectic manifolds (where the symplectic form is injective but not surjective as a mapping from the tangent bundle into the cotangent bundle). Here we describe precisely the space of smooth functions for which the Poisson bracket makes sense. In section (49) on representation theory we show how easily the spaces of smooth (real analytic) vectors can be treated with the help of the calculus developed in this book. The results (49.3) – (49.5) and their real analytic analogues (49.8) – (49.10) should convince the reader who has seen the classical proofs that convenient analysis is worthwhile to use. We included also some material on the moment mapping for unitary representations. This mapping is defined on the space of smooth (real analytic) vectors. Section (50) is devoted to the preparations and the proof of theorem (50.16) which says that a smooth curve of unbounded selfadjoint operators on Hilbert space with compact resolvent admits smooth parameterizations of its eigenvalues and eigenvectors, under some condition. The real analytic version of this is due to [Rellich, 1940]; we also give a new and simpler proof of this result. In our view, the best advantage of our approach is the natural and easy way to express what a smooth or real analytic curve of unbounded operators really is.

Hints for the reader. The numbering of subsections is done extensively and consecutively, the number valid at the bottom of each page can be found in the running head, opposite to the page number. Concepts which are not central are usually defined after the formulation of the result, before the proof, and sometimes even in the proof. So please look ahead rather than behind (which is advisable in everyday life also). Related materials from the literature are listed under the name **Result** if we include them without proofs. Appendix (52) collects some background material from functional analysis in compressed form, and appendix (53) contains a tool for analyzing non-separable Banach spaces which is used in sections (16) and (19). A **list of symbols** has been worked into the index.

Reading map for the cross reader. Most of chapter I is essential. Chapter II is for readers who also want to know the holomorphic and real analytic calculus, others may leave it for a second reading. Chapters III–V treat special material which can be looked up later whenever properties like smooth partitions of unity in infinite dimensions are asked for. In chapter VI section (35) can be skipped, in chapter VII one may omit some proofs in sections (33) and (35). Chapter VIII contains Lie theory and bundle theory, and is necessary for chapter IX and parts of chapter X.

Thanks. The work on this book was done from 1989 onwards, most of the material was presented in our joint seminar and elsewhere several times, which led to a lot of improvement. We want to thank all participants, who devoted a lot of attention and energy, in particular our (former) students who presented talks on that subject, also those who helped with proofreading or gave good advise: Eva Adam, Dmitri Alekseevsky, Andreas Cap, Stefan Haller, Ann and Bertram Kostant, Grigori Litvinov, Mark Losik, Josef Mattes, Martin Neuwirther, Tudor Ratiu, Konstanze Rietsch, Hermann Schichl, Erhard Siegl, Josef Teichmann, Klaus Wegenkittl. The second author acknowledges the support of 'Fonds zur Förderung der wissenschaftlichen Forschung, Projekt P 10037 PHY'.

Chapter I
Calculus of Smooth Mappings

This chapter is devoted to calculus of smooth mappings in infinite dimensions. The leading idea of our approach is to base everything on smooth curves in locally convex spaces, which is a notion without problems, and a mapping between locally convex spaces will be called smooth if it maps smooth curves to smooth curves.

We start by looking at the set of smooth curves $C^\infty(\mathbb{R}, E)$ with values in a locally convex space E, and note that it does not depend on the topology of E, only on the underlying system of bounded sets, its bornology. This is due to the fact, that for a smooth curve difference quotients converge to the derivative much better (2.1) than arbitrary converging nets or filters: we may multiply it by some unbounded sequences of scalars without disturbing convergence (or, even better, boundedness).

Then the basic results are proved, like existence, smoothness, and linearity of derivatives, the chain rule (3.18), and also the most important feature, the 'exponential law' (3.12) and (3.13): We have

$$C^\infty(E \times F, G) \cong C^\infty(E, C^\infty(F, G)),$$

without any restriction, for a natural structure on $C^\infty(F, G)$.

Smooth curves have integrals in E if and only if a weak completeness condition is satisfied: it appeared as bornological completeness, Mackey completeness, or local completeness in the literature, we call it c^∞-complete. This is equivalent to the condition that weakly smooth curves are smooth (2.14). All calculus in later chapters in this book will be done on convenient vector spaces: These are locally convex vector spaces which are c^∞-complete; note that the locally convex topology on a convenient vector space can vary in some range, only the system of bounded sets must remain the same.

Linear or more generally multilinear mappings are smooth if and only if they are bounded (5.5), and one has corresponding exponential laws (5.2) for them as well.

Furthermore, there is an appropriate tensor product, the bornological tensor product (5.7), satisfying

$$L(E \otimes_\beta F, G) \cong L(E, F; G) \cong L(E, L(F, G)).$$

An important tool for convenient vector spaces are uniform boundedness principles as given in (5.18), (5.24) and (5.26).

It is very natural to consider on E the final topology with respect to all smooth curves, which we call the c^∞-topology, since all smooth mappings are continuous for it: the vector space E, equipped with this topology is denoted by $c^\infty E$, with lower case c in analogy to kE for the Kelley-fication and in order to avoid any confusion with any space of smooth functions or sections. The special curve lemma (2.8) shows that the c^∞-topology coincides with the usual Mackey closure topology. The space $c^\infty E$ is not a topological vector space in general. This is related to the fact that the evaluation $E \times E' \to \mathbb{R}$ is jointly continuous only for normable E, but it is always smooth and hence continuous on $c^\infty(E \times E')$. The c^∞-open subsets are the natural domains of definitions of locally defined functions. For nice spaces (e.g. Fréchet and strong duals of Fréchet-Schwartz spaces, see (4.11)) the c^∞-topology coincides with the given locally convex topology. In general, the c^∞-topology is finer than any locally convex topology with the same bounded sets.

In the last section of this chapter we discuss the structure of spaces of smooth functions on finite dimensional manifolds and, more generally, of smooth sections of finite dimensional vector bundles. They will become important in chapter IX as modeling spaces for manifolds of mappings. Furthermore, we give a short account of reflexivity of convenient vector spaces and on (various) approximation properties for them.

1. Smooth Curves

1.1. Notation. Since we want to have unique derivatives all locally convex spaces E will be assumed Hausdorff. The family of all bounded sets in E plays an important rôle. It is called the *bornology* of E. A linear mapping is called *bounded*, sometimes also called bornological, if it maps bounded sets to bounded sets. A bounded linear bijection with bounded inverse is called *bornological isomorphism*. The *space of all continuous linear functionals on E* will be denoted by E^* and the *space of all bounded linear functionals on E* by E'. The *adjoint* or *dual mapping* of a linear mapping ℓ, however, will be always denoted by ℓ^*, because of differentiation.

See also the appendix (52) for some background on functional analysis.

1.2. Differentiable curves. The concept of a smooth curve with values in a locally convex vector space is easy and without problems. Let E be a locally convex vector space. A curve $c : \mathbb{R} \to E$ is called *differentiable* if the *derivative*

$c'(t) := \lim_{s\to 0} \frac{1}{s}(c(t+s) - c(t))$ at t exists for all t. A curve $c : \mathbb{R} \to E$ is called *smooth* or C^∞ if all iterated derivatives exist. It is called C^n for some finite n if its iterated derivatives up to order n exist and are continuous.

Likewise, a mapping $f : \mathbb{R}^n \to E$ is called smooth if all iterated partial derivatives $\partial_{i_1,\dots,i_p} f := \frac{\partial}{\partial x^{i_1}} \dots \frac{\partial}{\partial x^{i_p}} f$ exist for all $i_1, \dots, i_p \in \{1, \dots, n\}$.

A curve $c : \mathbb{R} \to E$ is called *locally Lipschitzian* if every point $r \in \mathbb{R}$ has a neighborhood U such that the *Lipschitz condition* is satisfied on U, i.e., the set

$$\left\{\tfrac{1}{t-s}\Big(c(t) - c(s)\Big) : t \neq s; t, s \in U\right\}$$

is bounded. Note that this implies that the curve satisfies the Lipschitz condition on each bounded interval, since for (t_i) increasing

$$\frac{c(t_n) - c(t_0)}{t_n - t_0} = \sum \frac{t_{i+1} - t_i}{t_n - t_0} \, \frac{c(t_{i+1}) - c(t_i)}{t_{i+1} - t_i}$$

is in the absolutely convex hull of a finite union of bounded sets.

A curve $c : \mathbb{R} \to E$ is called $\mathcal{L}ip^k$ or $C^{(k+1)-}$ if all derivatives up to order k exist and are locally Lipschitzian. For these properties we have the following implications:

$$C^{n+1} \implies \mathcal{L}\text{ip}^n \implies C^n,$$
$$\text{differentiable} \implies C.$$

In fact, continuity of the derivative implies locally its boundedness, and since this can be tested by continuous linear functionals (see (52.19)) we conclude from the one dimensional mean value theorem the boundedness of the difference quotient. See also the lemma (1.3) below.

1.3. Lemma. Continuous linear mappings are smooth. *A continuous linear mapping $\ell : E \to F$ between locally convex vector spaces maps $\mathcal{L}\text{ip}^k$-curves in E to $\mathcal{L}\text{ip}^k$-curves in F, for all $0 \le k \le \infty$, and for $k > 0$ one has $(\ell \circ c)'(t) = \ell(c'(t))$.*

Proof. As a linear map ℓ commutes with difference quotients, hence the image of a Lipschitz curve is Lipschitz since ℓ is bounded. As a continuous map it commutes with the formation of the respective limits. Hence $(\ell \circ c)'(t) = \ell(c'(t))$.

Note that a differentiable curve is continuous, and that a continuously differentiable curve is locally Lipschitz: For $\ell \in E^*$ we have

$$\ell\left(\frac{c(t) - c(s)}{t - s}\right) = \frac{(\ell \circ c)(t) - (\ell \circ c)(s)}{t - s} = \int_0^1 (\ell \circ c)'(s + (t - s)r)dr,$$

which is bounded, since $(\ell \circ c)'$ is locally bounded.

Now the rest follows by induction. □

1.4. The mean value theorem. In classical analysis the basic tool for using the derivative to get statements on the original curve is the mean value theorem. So we try to generalize it to infinite dimensions. For this let $c : \mathbb{R} \to E$ be a differentiable curve. If $E = \mathbb{R}$ the classical mean value theorem states, that the difference quotient $(c(a)-c(b))/(a-b)$ equals some intermediate value of c'. Already if E is two dimensional this is no longer true. Take for example a parameterization of the circle by arclength. However, we will show that $(c(a) - c(b))/(a - b)$ lies still in the closed convex hull of $\{c'(r) : r\}$. Having weakened the conclusion, we can try to weaken the assumption. And in fact c may be not differentiable in at most countably many points. Recall however, that there exist strictly monotone functions $f : \mathbb{R} \to \mathbb{R}$, which have vanishing derivative outside a Cantor set (which is uncountable, but has still measure 0).

Sometimes one uses in one dimensional analysis a generalized version of the mean value theorem: For an additional differentiable function h with non-vanishing derivative the quotient $(c(a)-c(b))/(h(a)-h(b))$ equals some intermediate value of c'/h'. A version for vector valued c (for real valued h) is that $(c(a) - c(b))/(h(a) - h(b))$ lies in the closed convex hull of $\{c'(r)/h'(r) : r\}$. One can replace the assumption that h' vanishes nowhere by the assumption that h' has constant sign, or, more generally, that h is monotone. But then we cannot form the quotients, so we should assume that $c'(t) \in h'(t) \cdot A$, where A is some closed convex set, and we should be able to conclude that $c(b) - c(a) \in (h(b) - h(a)) \cdot A$. This is the version of the *mean value theorem* that we are going to prove now. However, we will make use of it only in the case where $h = \mathrm{Id}$ and c is everywhere differentiable in the interior.

Proposition. Mean value theorem. *Let $c : [a, b] =: I \to E$ be a continuous curve, which is differentiable except at points in a countable subset $D \subseteq I$. Let h be a continuous monotone function $h : I \to \mathbb{R}$, which is differentiable on $I \setminus D$. Let A be a convex closed subset of E, such that $c'(t) \in h'(t) \cdot A$ for all $t \notin D$.*

Then $c(b) - c(a) \in (h(b) - h(a)) \cdot A$.

Proof. Assume that this is not the case. By the theorem of Hahn Banach (52.16) there exists a continuous linear functional ℓ with $\ell(c(b)-c(a)) \notin \overline{\ell((h(b) - h(a)) \cdot A)}$. But then $\ell \circ c$ and $\overline{\ell(A)}$ satisfy the same assumptions as c and A, and hence we may assume that c is real valued and A is just a closed interval $[\alpha, \beta]$. We may furthermore assume that h is monotonely increasing. Then $h'(t) \geq 0$, and $h(b) - h(a) \geq 0$. Thus the assumption says that $\alpha h'(t) \leq c'(t) \leq \beta h'(t)$, and we want to conclude that $\alpha(h(b) - h(a)) \leq c(b) - c(a) \leq \beta(h(b) - h(a))$. If we replace c by $c - \beta h$ or by $\alpha h - c$ it is enough to show that $c'(t) \leq 0$ implies that $c(b) - c(a) \leq 0$. For given $\varepsilon > 0$ we will show that $c(b) - c(a) \leq \varepsilon(b - a + 1)$. For this let J be the set $\{t \in [a, b] : c(s) - c(a) \leq \varepsilon\left((s - a) + \sum_{t_n<s} 2^{-n}\right) \text{ for } a \leq s < t\}$, where $D =: \{t_n : n \in \mathbb{N}\}$. Obviously, J is a closed interval containing a, say $[a, b']$. By continuity of c we obtain that $c(b') - c(a) \leq \varepsilon\left((b' - a) + \sum_{t_n<b'} 2^{-n}\right)$. Suppose $b' < b$. If $b' \notin D$, then there exists a subinterval $[b', b' + \delta]$ of $[a, b]$ such that for $b' \leq s < b' + \delta$ we have $c(s) - c(b') - c'(b')(s - b') \leq \varepsilon(s - b')$. Hence we get

$$c(s) - c(b') \leq c'(b')(s - b') + \varepsilon(s - b') \leq \varepsilon(s - b'),$$

and consequently

$$\begin{aligned} c(s)-c(a) &\le c(s)-c(b')+c(b')-c(a) \\ &\le \varepsilon(s-b')+\varepsilon\Big(b'-a+\sum_{t_n<b'}2^{-n}\Big)\le \varepsilon\Big(s-a+\sum_{t_n<s}2^{-n}\Big). \end{aligned}$$

On the other hand if $b' \in D$, i.e., $b' = t_m$ for some m, then by continuity of c we can find an interval $[b', b'+\delta]$ contained in $[a,b]$ such that for all $b' \le s < b'+\delta$ we have

$$c(s)-c(b') \le \varepsilon 2^{-m}.$$

Again we deduce that

$$c(s)-c(a) \le \varepsilon 2^{-m}+\varepsilon\Big(b'-a+\sum_{t_n<b'}2^{-n}\Big)\le \varepsilon\Big(s-a+\sum_{t_n<s}2^{-n}\Big).$$

So we reach in both cases a contradiction to the maximality of b'. □

Warning: One cannot drop the monotonicity assumption. In fact take $h(t) := t^2$, $c(t) := t^3$ and $[a,b] = [-1,1]$. Then $c'(t) \in h'(t)[-2,2]$, but $c(1)-c(-1) = 2 \notin \{0\} = (h(1)-h(-1))[-2,2]$.

1.5. Testing with functionals. Recall that in classical analysis vector valued curves $c : \mathbb{R} \to \mathbb{R}^n$ are often treated by considering their components $c_k := \operatorname{pr}_k \circ c$, where $\operatorname{pr}_k : \mathbb{R}^n \to \mathbb{R}$ denotes the canonical projection onto the k-th factor $\mathbb{R}$. Since in general locally convex spaces do not have appropriate bases, we use all continuous linear functionals instead of the projections pr_k. We will say that a property of a curve $c : \mathbb{R} \to E$ is *scalarly true*, if $\ell \circ c : \mathbb{R} \to E \to \mathbb{R}$ has this property for all continuous linear functionals ℓ on E.

We want to compare scalar differentiability with differentiability. For finite dimensional spaces we know the trivial fact that these to notions coincide. For infinite dimensions we first consider $\mathcal{L}$ip-curves $c : \mathbb{R} \to E$. Since by (52.19) boundedness can be tested by the continuous linear functionals we see, that c is $\mathcal{L}$ip if and only if $\ell \circ c : \mathbb{R} \to \mathbb{R}$ is $\mathcal{L}$ip for all $\ell \in E^*$. Moreover, if for a bounded interval $J \subset \mathbb{R}$ we take B as the absolutely convex hull of the bounded set $c(J) \cup \{\frac{c(t)-c(s)}{t-s} : t \ne s; t, s \in J\}$, then we see that $c|_J : J \to E_B$ is a well defined $\mathcal{L}$ip-curve into E_B. We denote by E_B the linear span of B in E, equipped with the *Minkowski functional* $p_B(v) := \inf\{\lambda > 0 : v \in \lambda.B\}$. This is a normed space. Thus we have the following equivalent characterizations of $\mathcal{L}$ip-curves:

(1) locally c factors over a $\mathcal{L}$ip-curve into some E_B;
(2) c is $\mathcal{L}$ip;
(3) $\ell \circ c$ is $\mathcal{L}$ip for all $\ell \in E^*$.

For continuous instead of Lipschitz curves we obviously have the analogous implications ($1 \Rightarrow 2 \Rightarrow 3$). However, if we take a non-convergent sequence $(x_n)_n$, which converges weakly (e.g. take an orthonormal base in a separable Hilbert space), and consider an infinite polygon c through these points x_n, say with $c(\frac{1}{n}) = x_n$ and

$c(0) = 0$. Then this curve is obviously not continuous but $\ell \circ c$ is continuous for all $\ell \in E^*$.

Furthermore, the "worst" continuous curve - i.e. $c : \mathbb{R} \to \prod_{C(\mathbb{R},\mathbb{R})} \mathbb{R} =: E$ given by $(c(t))_f := f(t)$ for all $t \in \mathbb{R}$ and $f \in C(\mathbb{R},\mathbb{R})$ - cannot be factored locally as a continuous curve over some E_B. Otherwise, $c(t_n)$ would converge into some E_B to $c(0)$, where t_n is a given sequence converging to 0, say $t_n := \frac{1}{n}$. So $c(t_n)$ would converge *Mackey* to $c(0)$, i.e., there have to be $\mu_n \to \infty$ with $\{\mu_n(c(t_n) - c(0)) : n \in \mathbb{N}\}$ bounded in E. Since a set is bounded in the product if and only if its coordinates are bounded, we conclude that for all $f \in C(\mathbb{R},\mathbb{R})$ the sequence $\mu_n(f(t_n) - f(0))$ has to be bounded. But we can choose a continuous function f with $f(0) = 0$ and $f(t_n) = \frac{1}{\sqrt{\mu_n}}$ and conclude that $\mu_n(f(t_n) - f(0)) = \sqrt{\mu_n}$ is unbounded.

Similarly, one shows that the reverse implications do not hold for differentiable curves, for C^1-curves and for C^n-curves. However, if we put instead some Lipschitz condition on the derivatives, there should be some chance, since this is a bornological concept. In order to obtain this result, we should study convergence of sequences in E_B.

1.6. Lemma. Mackey-convergence. *Let B be a bounded and absolutely convex subset of E and let $(x_\gamma)_{\gamma\in\Gamma}$ be a net in E_B. Then the following two conditions are equivalent:*

(1) *x_γ converges to 0 in the normed space E_B;*
(2) *There exists a net $\mu_\gamma \to 0$ in $\mathbb{R}$, such that $x_\gamma \in \mu_\gamma \cdot B$.*

In (2) we may assume that $\mu_\gamma \geq 0$ and is decreasing with respect to γ, at least for large γ. In the particular case of a sequence (or where we have a cofinal countable subset of Γ) we can choose all $\mu_n > 0$ and hence we may divide.

A net (x_γ) for which a bounded absolutely convex $B \subseteq E$ exists, such that x_γ converges to x in E_B is called Mackey convergent to x or short *M-convergent.*

Proof. ($\Uparrow$) Let $x_\gamma = \mu_\gamma \cdot b_\gamma$ with $b_\gamma \in B$ and $\mu_\gamma \to 0$. Then $p_B(x_\gamma) = |\mu_\gamma|\, p_B(b_\gamma) \leq |\mu_\gamma| \to 0$, i.e. $x_\gamma \to x$ in E_B.

($\Downarrow$) Let $\delta > 1$, and set $\mu_\gamma := \delta\, p_B(x_\gamma)$. By assumption, $\mu_\gamma \to 0$ and $x_\gamma = \mu_\gamma \frac{x_\gamma}{\mu_\gamma}$, where $\frac{x_\gamma}{\mu_\gamma} := 0$ if $\mu_\gamma = 0$. Since $p_B(\frac{x_\gamma}{\mu_\gamma}) = \frac{1}{\delta} < 1$ or is 0, we conclude that $\frac{x_\gamma}{\mu_\gamma} \in B$.

For the final assertions, choose γ_1 such that $|\mu_\gamma| \leq 1$ for $\gamma \geq \gamma_1$, and for those γ we replace μ_γ by $\sup\{|\mu_{\gamma'}| : \gamma' \geq \gamma\}$. Thus we may choose $\mu \geq 0$ and decreasing with respect to γ.

If we have a sequence $(\gamma_n)_{n\in\mathbb{N}}$ which is cofinal in Γ, i.e. for every $\gamma \in \Gamma$ there exists an $n \in \mathbb{N}$ with $\gamma \leq \gamma_n$, then we may replace μ_γ by

$$\max(\{\mu_\gamma\} \cup \{\mu_{\gamma_m} : \gamma_m \geq \gamma\} \cup \{\frac{1}{m} : \gamma_m \geq \gamma\})$$

to conclude that $\mu_\gamma \neq 0$ for all γ. □

If Γ is the ordered set of all countable ordinals, then it is not possible to find a net $(\mu_\gamma)_{\gamma\in\Gamma}$, which is positive everywhere and converges to 0, since a converging net is finally constant.

1.7. The difference quotient converges Mackey. Now we show how to describe the quality of convergence of the difference quotient.

Corollary. *Let $c:\mathbb{R}\to E$ be a $\mathcal{L}\mathrm{ip}^1$-curve. Then the curve $t\mapsto \frac{1}{t}(\frac{1}{t}(c(t)-c(0))-c'(0))$ is bounded on bounded subsets of $\mathbb{R}\setminus\{0\}$.*

Proof. We apply (1.4) to c and obtain:

$$\begin{aligned}\frac{c(t)-c(0)}{t}-c'(0)&\in\Big\langle c'(r):0<|r|<|t|\Big\rangle_{\text{closed, convex}}-c'(0)\\ &=\Big\langle c'(r)-c'(0):0<|r|<|t|\Big\rangle_{\text{closed, convex}}\\ &=\Big\langle r\,\frac{c'(r)-c'(0)}{r}:0<|r|<|t|\Big\rangle_{\text{closed, convex}}\end{aligned}$$

Let $a>0$. Since $\{\frac{c'(r)-c'(0)}{r}:0<|r|<a\}$ is bounded and hence contained in a closed absolutely convex and bounded set B, we can conclude that

$$\frac{1}{t}\left(\frac{c(t)-c(0)}{t}-c'(0)\right)\in\Big\langle\frac{r}{t}\,\frac{c'(r)-c'(0)}{r}:0<|r|<|t|\Big\rangle_{\text{closed, convex}}\subseteq B.\quad\square$$

1.8. Corollary. Smoothness of curves is a bornological concept. *For $0\le k<\infty$ a curve c in a locally convex vector space E is $\mathcal{L}\mathrm{ip}^k$ if and only if for each bounded open interval $J\subset\mathbb{R}$ there exists an absolutely convex bounded set $B\subseteq E$ such that $c|_J$ is a $\mathcal{L}\mathrm{ip}^k$-curve in the normed space E_B.*

Attention: A smooth curve factors locally into some E_B as a $\mathcal{L}\mathrm{ip}^k$-curve for each finite k only, in general. Take the “worst” smooth curve $c:\mathbb{R}\to\prod_{C^\infty(\mathbb{R},\mathbb{R})}\mathbb{R}$, analogously to (1.5), and, using Borel's theorem, deduce from $c^{(k)}(0)\in E_B$ for all $k\in\mathbb{N}$ a contradiction.

Proof. For $k=0$ this was shown before. For $k\ge 1$ take a closed absolutely convex bounded set $B\subseteq E$ containing all derivatives $c^{(i)}$ on J up to order k as well as their difference quotients on $\{(t,s):t\ne s,t,s\in J\}$. We show first that c is differentiable, say at 0, with derivative $c'(0)$. By the proof of the previous corollary (1.7) we have that the expression $\frac{1}{t}(\frac{c(t)-c(0)}{t}-c'(0))$ lies in B. So $\frac{c(t)-c(0)}{t}-c'(0)$ converges to 0 in E_B. For the higher order derivatives we can now proceed by induction.

The converse follows from lemma (1.3). $\square$

A consequence of this is, that smoothness does not depend on the topology but only on the dual (so all topologies with the same dual have the same smooth curves), and in fact it depends only on the bounded sets, i.e. the bornology. Since on $L(E,F)$ there is essentially only one bornology (by the uniform boundedness principle, see (52.25)) there is only one notion of $\mathcal{L}\mathrm{ip}^n$-curves into $L(E,F)$. Furthermore, the class of $\mathcal{L}\mathrm{ip}^n$-curves doesn't change if we pass from a given locally convex topology to its *bornologification*, see (4.2), which by definition is the finest locally convex topology having the same bounded sets.

Let us now return to scalar differentiability. Corollary (1.7) gives us $\mathcal{L}\mathrm{ip}^n$-ness provided we have appropriate candidates for the derivatives.

1.9. Corollary. Scalar testing of curves. *Let $c^k : \mathbb{R} \to E$ for $k < n+1$ be curves such that $\ell \circ c^0$ is $\mathcal{L}\mathrm{ip}^n$ and $(\ell \circ c^0)^{(k)} = \ell \circ c^k$ for all $k < n+1$ and all $\ell \in E^*$. Then c^0 is $\mathcal{L}\mathrm{ip}^n$ and $(c^0)^{(k)} = c^k$.*

Proof. For $n = 0$ this was shown in (1.5). For $n \geq 1$, by (1.7) applied to $\ell \circ c$ we have that

$$\ell\left(\frac{1}{t}\left(\frac{c^0(t) - c^0(0)}{t} - c^1(0)\right)\right)$$

is locally bounded, and hence by (52.19) the set

$$\left\{\frac{1}{t}\left(\frac{c^0(t) - c^0(0)}{t} - c^1(0)\right) : t \in I\right\}$$

is bounded. Thus $\frac{c^0(t)-c^0(0)}{t}$ converges even Mackey to $c^1(0)$. Now the general statement follows by induction. □

2. Completeness

Do we really need the knowledge of a candidate for the derivative, as in (1.9)? In finite dimensional analysis one often uses the Cauchy condition to prove convergence. Here we will replace the Cauchy condition again by a stronger condition, which provides information about the quality of being Cauchy:

A net $(x_\gamma)_{\gamma\in\Gamma}$ in E is called *Mackey-Cauchy* provided that there exist a bounded (absolutely convex) set B and a net $(\mu_{\gamma,\gamma'})_{(\gamma,\gamma')\in\Gamma\times\Gamma}$ in $\mathbb{R}$ converging to 0, such that $x_\gamma - x_{\gamma'} \in \mu_{\gamma,\gamma'} B$. As in (1.6) one shows that for a net x_γ in E_B this is equivalent to the condition that x_γ is Cauchy in the normed space E_B.

2.1. Lemma. The difference quotient is Mackey-Cauchy. *Let $c : \mathbb{R} \to E$ be scalarly a $\mathcal{L}\mathrm{ip}^1$-curve. Then $t \mapsto \frac{c(t)-c(0)}{t}$ is a Mackey-Cauchy net for $t \to 0$.*

Proof. For $\mathcal{L}\mathrm{ip}^1$-curves this is a immediate consequence of (1.7) but we only assume it to be scalarly $\mathcal{L}\mathrm{ip}^1$. It is enough to show that $\frac{1}{t-s}\left(\frac{c(t)-c(0)}{t} - \frac{c(s)-c(0)}{s}\right)$ is bounded on bounded subsets in $\mathbb{R} \setminus \{0\}$. We may test this with continuous linear functionals, and hence may assume that $E = \mathbb{R}$. Then by the fundamental theorem of calculus we have

$$\begin{aligned}\frac{1}{t-s}\left(\frac{c(t)-c(0)}{t} - \frac{c(s)-c(0)}{s}\right) &= \int_0^1 \frac{c'(tr) - c'(sr)}{t-s}\, dr \\ &= \int_0^1 \frac{c'(tr) - c'(sr)}{tr - sr}\, r\, dr.\end{aligned}$$

Since $\frac{c'(tr)-c'(sr)}{tr-sr}$ is locally bounded by assumption, the same is true for the integral, and we are done. □

2.2. Lemma. Mackey Completeness. *For a space E the following conditions are equivalent:*

(1) *Every Mackey-Cauchy net converges in E;*
(2) *Every Mackey-Cauchy sequence converges in E;*
(3) *For every absolutely convex closed bounded set B the space E_B is complete;*
(4) *For every bounded set B there exists an absolutely convex bounded set $B' \supseteq B$ such that $E_{B'}$ is complete.*

A space satisfying the equivalent conditions is called Mackey complete. Note that a sequentially complete space is Mackey complete.

Proof. (1) $\Rightarrow$ (2), and (3) $\Rightarrow$ (4) are trivial.

(2) $\Rightarrow$ (3) Since E_B is normed, it is enough to show sequential completeness. So let (x_n) be a Cauchy sequence in E_B. Then (x_n) is Mackey-Cauchy in E and hence converges in E to some point x. Since $p_B(x_n - x_m) \to 0$ there exists for every $\varepsilon > 0$ an $N \in \mathbb{N}$ such that for all $n, m \geq N$ we have $p_B(x_n - x_m) < \varepsilon$, and hence $x_n - x_m \in \varepsilon B$. Taking the limit for $m \to \infty$, and using closedness of B we conclude that $x_n - x \in \varepsilon B$ for all $n > N$. In particular $x \in E_B$ and $x_n \to x$ in E_B.

(4) $\Rightarrow$ (1) Let $(x_\gamma)_{\gamma\in\Gamma}$ be a Mackey-Cauchy net in E. So there is some net $\mu_{\gamma,\gamma'} \to 0$, such that $x_\gamma - x_{\gamma'} \in \mu_{\gamma,\gamma'} B$ for some bounded set B. Let γ_0 be arbitrary. By (4) we may assume that B is absolutely convex and contains x_{γ_0}, and that E_B is complete. For $\gamma \in \Gamma$ we have that $x_\gamma = x_{\gamma_0} + x_\gamma - x_{\gamma_0} \in x_{\gamma_0} + \mu_{\gamma,\gamma_0} B \in E_B$, and $p_B(x_\gamma - x_{\gamma'}) \leq \mu_{\gamma,\gamma'} \to 0$. So (x_γ) is a Cauchy net in E_B, hence converges in E_B, and thus also in E. □

2.3. Corollary. Scalar testing of differentiable curves. *Let E be Mackey complete and $c : \mathbb{R} \to E$ be a curve for which $\ell \circ c$ is $\mathcal{L}\text{ip}^n$ for all $\ell \in E^*$. Then c is $\mathcal{L}\text{ip}^n$.*

Proof. For $n = 0$ this was shown in (1.5) without using any completeness, so let $n \geq 1$. Since we have shown in (2.1) that the difference quotient is a Mackey-Cauchy net we conclude that the derivative c' exists, and hence $(\ell \circ c)' = \ell \circ c'$. So we may apply the induction hypothesis to conclude that c' is $\mathcal{L}\text{ip}^{n-1}$, and consequently c is $\mathcal{L}\text{ip}^n$. □

Next we turn to integration. For continuous curves $c : [0,1] \to E$ one can show completely analogously to 1-dimensional analysis that the *Riemann sums* $R(c, \mathcal{Z}, \xi)$, defined by $\sum_k (t_k - t_{k-1})c(\xi_k)$, where $0 = t_0 < t_1 < \cdots < t_n = 1$ is a partition $\mathcal{Z}$ of $[0,1]$ and $\xi_k \in [t_{k-1}, t_k]$, form a Cauchy net with respect to the partial strict ordering given by the size of the *mesh* $\max\{|t_k - t_{k-1}| : 0 < k < n\}$. So under the assumption of sequential completeness we have a *Riemann integral* of curves. A second way to see this is the following reduction to the 1-dimensional case.

2.4. Lemma. *Let $L(E^*_{\text{equi}}, \mathbb{R})$ be the space of all linear functionals on E^* which are bounded on equicontinuous sets, equipped with the complete locally convex topology* *of uniform convergence on these sets. There is a natural topological embedding* $\delta : E \to L(E^*_{\text{equi}}, \mathbb{R})$ *given by* $\delta(x)(\ell) := \ell(x)$.

Proof. Let $\mathcal{U}$ be a basis of absolutely convex closed 0-neighborhoods in E. Then the family of *polars* $U^o := \{\ell \in E^* : |\ell(x)| \leq 1 \text{ for all } x \in U\}$, with $U \in \mathcal{U}$ form a basis for the equicontinuous sets, and hence the *bipolars* $U^{oo} := \{\ell^* \in L(E^*_{\text{equi}}, \mathbb{R}) : |\ell^*(\ell)| \leq 1 \text{ for all } \ell \in U^o\}$ form a basis of 0-neighborhoods in $L(E^*_{\text{equi}}, \mathbb{R})$. By the bipolar theorem (52.18) we have $U = \delta^{-1}(U^{oo})$ for all $U \in \mathcal{U}$. This shows that δ is a homeomorphism onto its image. □

2.5. Lemma. Integral of continuous curves. *Let* $c : \mathbb{R} \to E$ *be a continuous curve in a locally convex vector space. Then there is a unique differentiable curve* $\int c : \mathbb{R} \to \widehat{E}$ *in the completion* $\widehat{E}$ *of* E *such that* $(\int c)(0) = 0$ *and* $(\int c)' = c$.

Proof. We show uniqueness first. Let $c_1 : \mathbb{R} \to \widehat{E}$ be a curve with derivative c and $c_1(0) = 0$. For every $\ell \in E^*$ the composite $\ell \circ c_1$ is an anti-derivative of $\ell \circ c$ with initial value 0, so it is uniquely determined, and since E^* separates points c_1 is also uniquely determined.

Now we show the existence. By the previous lemma we have that $\widehat{E}$ is (isomorphic to) the closure of E in the obviously complete space $L(E^*_{\text{equi}}, \mathbb{R})$. We define $(\int c)(t) : E^* \to \mathbb{R}$ by $\ell \mapsto \int_0^t (\ell \circ c)(s)ds$. It is a bounded linear functional on E^*_{equi} since for an equicontinuous subset $\mathcal{E} \subseteq E^*$ the set $\{(\ell \circ c)(s) : \ell \in \mathcal{E}, s \in [0, t]\}$ is bounded. So $\int c : \mathbb{R} \to L(E^*_{\text{equi}}, \mathbb{R})$.

Now we show that $\int c$ is differentiable with derivative $\delta \circ c$.

$$\left(\frac{(\int c)(t + r) - (\int c)(r)}{t} - (\delta \circ c)(r)\right)(\ell) =$$
$$= \frac{1}{t}\left(\int_0^{t+r} (\ell \circ c)(s)ds - \int_0^r (\ell \circ c)(s)ds - t(\ell \circ c)(r)\right) =$$
$$= \frac{1}{t}\int_r^{r+t} \Big((\ell \circ c)(s) - (\ell \circ c)(r)\Big)ds = \int_0^1 \ell\Big(c(r + ts) - c(r)\Big)ds.$$

Let $\mathcal{E} \subseteq E^*$ be equicontinuous, and let $\varepsilon > 0$. Then there exists a neighborhood U of 0 such that $|\ell(U)| < \varepsilon$ for all $\ell \in \mathcal{E}$. For sufficiently small t, all $s \in [0, 1]$ and fixed r we have $c(r + ts) - c(r) \in U$. So $|\int_0^1 \ell(c(r + ts) - c(r))ds| < \varepsilon$. This shows that the difference quotient of $\int c$ at r converges to $\delta(c(r))$ uniformly on equicontinuous subsets.

It remains to show that $(\int c)(t) \in \widehat{E}$. By the mean value theorem (1.4) the difference quotient $\frac{1}{t}((\int c)(t) - (\int c)(0))$ is contained in the closed convex hull in $L(E^*_{\text{equi}}, \mathbb{R})$ of the subset $\{c(s) : 0 < s < t\}$ of E. So it lies in $\widehat{E}$. □

Definition of the integral. For continuous curves $c : \mathbb{R} \to E$ the *definite integral* $\int_a^b c \in \widehat{E}$ is given by $\int_a^b c = (\int c)(b) - (\int c)(a)$.

2.6. Corollary. Basics on the integral. *For a continuous curve $c : \mathbb{R} \to E$ we have:*

(1) $\ell(\int_a^b c) = \int_a^b (\ell \circ c)$ *for all* $\ell \in E^*$.
(2) $\int_a^b c + \int_b^d c = \int_a^d c$.
(3) $\int_a^b (c \circ \varphi)\varphi' = \int_{\varphi(a)}^{\varphi(b)} c$ *for* $\varphi \in C^1(\mathbb{R}, \mathbb{R})$.
(4) $\int_a^b c$ *lies in the closed convex hull in* $\widehat{E}$ *of the set* $\{(b-a)c(t) : a < t < b\}$ *in* E.
(5) $\int_a^b : C(\mathbb{R}, E) \to \widehat{E}$ *is linear.*
(6) *(Fundamental theorem of calculus.) For each* C^1*-curve* $c : \mathbb{R} \to E$ *we have* $c(s) - c(t) = \int_t^s c'$. □

We are mainly interested in smooth curves and we can test for this by applying linear functionals if the space is Mackey complete, see (2.3). So let us try to show that the integral for such curves lies in E if E is Mackey-complete. So let $c : [0,1] \to E$ be a smooth or just a $\mathcal{L}$ip-curve, and take a partition $\mathcal{Z}$ with mesh $\mu(\mathcal{Z})$ at most δ. If we have a second partition, then we can take the common refinement. Let $[a, b]$ be one interval of the original partition with intermediate point t, and let $a = t_0 < t_1 < \cdots < t_n = b$ be the refinement. Note that $|b - a| \le \delta$ and hence $|t - t_k| \le \delta$. Then we can estimate as follows.

$$(b-a)\,c(t) - \sum_k (t_k - t_{k-1})c(t_k) = \sum_k (t_k - t_{k-1})\,(c(t) - c(t_k)) = \sum_k \mu_k b_k,$$

where $b_k := \frac{c(t)-c(t_k)}{\delta}$ is contained in the *absolutely convex Lipschitz bound*

$$B := \left\langle \left\{ \frac{c(t) - c(s)}{t - s} : t, s \in [0,1] \right\} \right\rangle_{abs.conv}$$

of c and $\mu_k := (t_k - t_{k-1})\delta \ge 0$ and satisfies $\sum_k \mu_k = (b-a)\delta$. Hence we have for the Riemann sums with respect to the original partition $\mathcal{Z}_1$ and the refinement $\mathcal{Z}'$ that $R(c, \mathcal{Z}_1) - R(c, \mathcal{Z}')$ lies in $\delta \cdot B$. So $R(c, \mathcal{Z}_1) - R(c, \mathcal{Z}_2) \in 2\delta B$ for any two partitions $\mathcal{Z}_1$ and $\mathcal{Z}_2$ of mesh at most δ, i.e. the Riemann sums form a Mackey-Cauchy net with coefficients $\mu_{\mathcal{Z}_1,\mathcal{Z}_2} := 2\max\{\mu(\mathcal{Z}_1), \mu(\mathcal{Z}_2)\}$ and we have proved:

2.7. Proposition. Integral of Lipschitz curves. *Let $c : [0,1] \to E$ be a Lipschitz curve into a Mackey complete space. Then the Riemann integral exists in E as (Mackey)-limit of the Riemann sums.* □

2.8. Now we have to discuss the relationship between differentiable curves and Mackey convergent sequences. Recall that a sequence (x_n) converges if and only if there exists a continuous curve c (e.g. a reparameterization of the infinite polygon) and $t_n \searrow 0$ with $c(t_n) = x_n$. The corresponding result for smooth curves uses the following notion.

Definition. We say that a sequence x_n in a locally convex space E *converges fast* to x in E, or *falls fast* towards x, if for each $k \in \mathbb{N}$ the sequence $n^k(x_n - x)$ is bounded.

Special curve lemma. *Let x_n be a sequence which converges fast to x in E. Then the infinite polygon through the x_n can be parameterized as a smooth curve $c : \mathbb{R} \to E$ such that $c(\frac{1}{n}) = x_n$ and $c(0) = x$.*

Proof. Let $\varphi : \mathbb{R} \to [0,1]$ be a smooth map, which is 0 on $\{t : t \leq 0\}$ and 1 on $\{t : t \geq 1\}$. The parameterization c is defined as follows:

$$c(t) := \begin{cases} x & \text{for } t \leq 0, \\ x_{n+1} + \varphi\left(\frac{t-\frac{1}{n+1}}{\frac{1}{n}-\frac{1}{n+1}}\right)(x_n - x_{n+1}) & \text{for } \frac{1}{n+1} \leq t \leq \frac{1}{n}, \\ x_1 & \text{for } t \geq 1 \end{cases}.$$

Obviously, c is smooth on $\mathbb{R} \setminus \{0\}$, and the p-th derivative of c for $\frac{1}{n+1} \leq t \leq \frac{1}{n}$ is given by

$$c^{(p)}(t) = \varphi^{(p)}\left(\frac{t - \frac{1}{n+1}}{\frac{1}{n} - \frac{1}{n+1}}\right)(n(n+1))^p(x_n - x_{n+1}).$$

Since x_n converges fast to x, we have that $c^{(p)}(t) \to 0$ for $t \to 0$, because the first factor is bounded and the second goes to zero. Hence c is smooth on $\mathbb{R}$, by the following lemma. □

2.9. Lemma. Differentiable extension to an isolated point. *Let $c : \mathbb{R} \to E$ be continuous and differentiable on $\mathbb{R} \setminus \{0\}$, and assume that the derivative $c' : \mathbb{R} \setminus \{0\} \to E$ has a continuous extension to $\mathbb{R}$. Then c is differentiable at 0 and $c'(0) = \lim_{t\to 0} c'(t)$.*

Proof. Let $a := \lim_{t\to 0} c'(t)$. By the mean value theorem (1.4) we have $\frac{c(t)-c(0)}{t} \in \langle c'(s) : 0 \neq |s| \leq |t| \rangle_{\text{closed, convex}}$. Since c' is assumed to be continuously extendable to 0 we have that for any closed convex 0-neighborhood U there exists a $\delta > 0$ such that $c'(t) \in a + U$ for all $0 < |t| \leq \delta$. Hence $\frac{c(t)-c(0)}{t} - a \in U$, i.e. $c'(0) = a$. □

The next result shows that we can pass through certain sequences $x_n \to x$ even with given velocities $v_n \to 0$.

2.10. Corollary. *If $x_n \to x$ fast and $v_n \to 0$ fast in E, then there are smoothly parameterized polygon $c : \mathbb{R} \to E$ and $t_n \to 0$ in $\mathbb{R}$ such that $c(t_n + t) = x_n + tv_n$ for t in a neighborhood of 0 depending on n.*

Proof. Consider the sequence y_n defined by

$$y_{2n} := x_n + \tfrac{1}{4n(2n+1)} v_n \quad \text{and} \quad y_{2n+1} := x_n - \tfrac{1}{4n(2n+1)} v_n.$$

It is easy to show that y_n converges fast to x, and the parameterization c of the polygon through the y_n (using a function φ which satisfies $\varphi(t) = t$ for t near $1/2$) has the claimed properties, where

$$t_n := \tfrac{4n+1}{4n(2n+1)} = \frac{1}{2}\left(\frac{1}{2n} + \frac{1}{2n+1}\right). \quad □$$

As first application (2.10) we can give the following sharpening of (1.3).

2.11. Corollary. Bounded linear maps. *A linear mapping $\ell : E \to F$ between locally convex vector spaces is bounded (or bornological), i.e. it maps bounded sets to bounded ones, if and only if it maps smooth curves in E to smooth curves in F.*

Proof. As in the proof of (1.3) one shows using (1.7) that a bounded linear map preserves $\mathcal{L}ip^k$-curves. Conversely, assume that a linear map $\ell : E \to F$ carries smooth curves to locally bounded curves. Take a bounded set B, and assume that $f(B)$ is unbounded. Then there is a sequence (b_n) in B and some $\lambda \in F'$ such that $|(\lambda \circ \ell)(b_n)| \geq n^{n+1}$. The sequence $(n^{-n}b_n)$ converges fast to 0, hence lies on some compact part of a smooth curve by (2.8). Consequently, $(\lambda \circ \ell)(n^{-n}b_n) = n^{-n}(\lambda \circ \ell)(b_n)$ is bounded, a contradiction. □

2.12. Definition. The *c^∞-topology* on a locally convex space E is the final topology with respect to all smooth curves $\mathbb{R} \to E$. Its open sets will be called *c^∞-open*. We will treat this topology in more detail in section (4): In general it describes neither a topological vector space (4.20) and (4.26), nor a uniform structure (4.27). However, by (4.4) and (4.6) the finest locally convex topology coarser than the c^∞-topology is the bornologification of the locally convex topology.

Let (μ_n) be a sequence of real numbers converging to ∞. Then a sequence (x_n) in E is called *μ-converging* to x if the sequence $(\mu_n(x_n - x))$ is bounded in E.

2.13. Theorem. c^∞-open subsets. *Let $\mu_n \to \infty$ be a real valued sequence. Then a subset $U \subseteq E$ is open for the c^∞-topology if it satisfies any of the following equivalent conditions:*

(1) *All inverse images under $\mathcal{L}ip^k$-curves are open in $\mathbb{R}$ (for fixed $k \in \mathbb{N}_\infty$).*
(2) *All inverse images under μ-converging sequences are open in $\mathbb{N}_\infty$.*
(3) *The traces to E_B are open in E_B for all absolutely convex bounded subsets $B \subseteq E$.*

Note that for closed subsets an equivalent statement reads as follows: A set A is c^∞-closed if and only if for every sequence $x_n \in A$, which is μ-converging (respectively M-converging, resp. fast falling) towards x, the point x belongs to A.

The topology described in (2) is also called *Mackey-closure topology.* It is not the Mackey topology discussed in duality theory.

Proof. (1) $\Rightarrow$ (2) Suppose (x_n) is μ-converging to $x \in U$, but $x_n \notin U$ for infinitely many n. Then we may choose a subsequence again denoted by (x_n), which is fast falling to x, hence lies on some compact part of a smooth curve c as described in (2.8). Then $c(\frac{1}{n}) = x_n \notin U$ but $c(0) = x \in U$. This is a contradiction.

(2) $\Rightarrow$ (3) A sequence (x_n), which converges in E_B to x with respect to p_B, is Mackey convergent, hence has a μ-converging subsequence. Note that E_B is normed, and hence it is enough to consider sequences.

(3) $\Rightarrow$ (2) Suppose (x_n) is μ-converging to x. Then the absolutely convex hull B of $\{\mu_n(x_n - x) : n \in \mathbb{N}\} \cup \{x\}$ is bounded, and $x_n \to x$ in (E_B, p_B), since $\mu_n(x_n - x)$ is bounded.

(2) $\Rightarrow$ (1) Use that for a converging sequence of parameters t_n the images $x_n := c(t_n)$ under a $\mathcal{L}ip$-curve c are Mackey converging. $\square$

Let us show next that the c^∞-topology and c^∞-completeness are intimately related.

2.14. Theorem. Convenient vector spaces. *Let E be a locally convex vector space. E is said to be* c^∞-complete *or* convenient *if one of the following equivalent (completeness) conditions is satisfied:*

1. *Any Lipschitz curve in E is locally Riemann integrable.*
2. *For any $c_1 \in C^\infty(\mathbb{R}, E)$ there is $c_2 \in C^\infty(\mathbb{R}, E)$ with $c_2' = c_1$ (existence of an* anti-derivative*).*
3. *E is c^∞-closed in any locally convex space.*
4. *If $c : \mathbb{R} \to E$ is a curve such that $\ell \circ c : \mathbb{R} \to \mathbb{R}$ is smooth for all $\ell \in E^*$, then c is smooth.*
5. *Any Mackey-Cauchy sequence converges; i.e. E is Mackey complete, see (2.2).*
6. *If B is bounded closed absolutely convex, then E_B is a Banach space. This property is called* locally complete *in* [Jarchow, 1981, p196].
7. *Any continuous linear mapping from a normed space into E has a continuous extension to the completion of the normed space.*

Condition (4) says that in a convenient vector space one can recognize smooth curves by investigating compositions with continuous linear functionals. Condition (5) says via (2.2.4) that c^∞-completeness is a bornological concept. In [Frölicher, Kriegl, 1988] a convenient vector space is always considered with its bornological topology — an equivalent but not isomorphic category.

Proof. In (2.3) we showed (5) $\Rightarrow$ (4), in (2.7) we got (5) $\Rightarrow$ (1), and in (2.2) we had (5) $\Rightarrow$ (6).

(1) $\Rightarrow$ (2) A smooth curve is Lipschitz, thus locally Riemann integrable. The indefinite Riemann integral equals the "weakly defined" integral of lemma (2.5), hence is an anti-derivative.

(2) $\Rightarrow$ (3) Let E be a topological linear subspace of F. To show that E is c^∞-closed we use (2.13). Let $x_n \to x_\infty$ be fast falling, $x_n \in E$ but $x_\infty \in F$. By (2.8) the polygon c through (x_n) can be smoothly parameterized. Hence c' is smooth and has values in the vector space generated by $\{x_n : n \neq \infty\}$, which is contained in E. Its anti-derivative c_2 is up to a constant equal to c, and by (2) $x_1 - x_\infty = c(1) - c(0) = c_2(1) - c_2(0)$ lies in E. So $x_\infty \in E$.

(3) $\Rightarrow$ (5) Let F be the completion $\widehat{E}$ of E. Any Mackey Cauchy sequence in E has a limit in F, and since E is by assumption c^∞-closed in F the limit lies in E. Hence, the sequence converges in E.

(6) $\Rightarrow$ (7) Let $f : F \to E$ be a continuous mapping on a normed space F. Since the image of the unit ball is bounded, it is a bounded mapping into E_B for some closed absolutely convex B. But into E_B it can be extended to the completion, since E_B is complete.

(7) ⇒ (1) Let $c : \mathbb{R} \to E$ be a Lipschitz curve. Then c is locally a continuous curve into E_B for some absolutely convex bounded set B. The inclusion of E_B into E has a continuous extension to the completion of E_B, and c is Riemann integrable in this Banach space, so also in E.

(4) ⇒ (3) Let E be embedded in some space F. We use again (2.13) in order to show that E is c^∞-closed in F. So let $x_n \to x_\infty$ fast falling, $x_n \in E$ for $n \neq 0$, but $x_\infty \in F$. By (2.8) the polygon c through (x_n) can be smoothly symmetrically parameterized in F, and $c(t) \in E$ for $t \neq 0$. We consider $\tilde{c}(t) := tc(t)$. This is a curve in E which is smooth in F, so it is scalarwise smooth in E, thus smooth in E by (4). Then $x_\infty = \tilde{c}'(0) \in E$. □

2.15. Theorem. Inheritance of c^∞-completeness. *The following constructions preserve c^∞-completeness: limits, direct sums, strict inductive limits of sequences of closed embeddings, as well as formation of $\ell^\infty(X, \quad)$, where X is a set together with a family $\mathcal{B}$ of subsets of X containing the finite ones, which are called bounded and $\ell^\infty(X,F)$ denotes the space of all functions $f : X \to F$, which are bounded on all $B \in \mathcal{B}$, supplied with the topology of uniform convergence on the sets in $\mathcal{B}$.*

Note that the definition of the topology of uniform convergence as initial topology shows, that adding all subsets of finite unions of elements in $\mathcal{B}$ to $\mathcal{B}$ does not change this topology. Hence, we may always assume that $\mathcal{B}$ has this stability property; this is the concept of a *bornology on a set.*

Proof. The projective limit (52.8) of $\mathcal{F}$ is the c^∞-closed linear subspace

$$\Big\{(x_\alpha) \in \prod \mathcal{F}(\alpha) : \mathcal{F}(f)x_\alpha = x_\beta \text{ for all } f : \alpha \to \beta\Big\},$$

hence is c^∞-complete, since the product of c^∞-complete factors is obviously c^∞-complete.

Since the coproduct (52.7) of spaces X_α is the topological direct sum, and has as bounded sets those which are contained and bounded in some finite subproduct, it is c^∞-complete if all factors are.

For colimits this is in general not true. For strict inductive limits of sequences of closed embeddings it is true, since bounded sets are contained and bounded in some step, see (52.8).

For the result on $\ell^\infty(X,F)$ we consider first the case, where X itself is bounded. Then c^∞-completeness can be proved as in (52.4) or reduced to this result. In fact let $\mathcal{B}$ be bounded in $\ell^\infty(X,F)$. Then $\mathcal{B}(X)$ is bounded in F and hence contained in some absolutely convex bounded set B, for which F_B is a Banach space. So we may assume that $\mathcal{B} := \{f \in \ell^\infty(X,F) : f(X) \subseteq B\}$. The space $\ell^\infty(X,F)_\mathcal{B}$ is just the space $\ell^\infty(X,F_B)$ with the supremum norm, which is a Banach space by (52.4).

Let now X and $\mathcal{B}$ be arbitrary. Then the restriction maps $\ell^\infty(X,F) \to \ell^\infty(B,F)$ give an embedding ι of $\ell^\infty(X,F)$ into the product $\prod_{B\in\mathcal{B}} \ell^\infty(B,F)$. Since this product is complete, by what we have shown above, it is enough to show that this embedding has a closed image. So let $f_\alpha|_B$ converge to some f_B in $\ell^\infty(B,F)$. Define $f(x) := f_{\{x\}}(x)$. For any $B \in \mathcal{B}$ containing x we have that $f_B(x) = (\lim_\alpha f_\alpha|_B)(x) = \lim_\alpha(f_\alpha(x)) = \lim_\alpha f_\alpha|_{\{x\}} = f_{\{x\}}(x) = f(x)$, and $f(B)$ is bounded for all $B \in \mathcal{B}$, since $f|_B = f_B \in \ell^\infty(B,F)$. □

Example. In general, a quotient and an inductive limit of c^∞-complete spaces need not be c^∞-complete. In fact, let $E_D := \{x \in \mathbb{R}^\mathbb{N} : \operatorname{supp} x \subseteq D\}$ for any subset $D \subseteq \mathbb{N}$ of *density* $\operatorname{dens} D := \limsup\{\frac{|D\cap[1,n]|}{n}\} = 0$. It can be shown that $E := \bigcup_{\operatorname{dens} D=0} E_D \subset \mathbb{R}^\mathbb{N}$ is the inductive limit of the Fréchet subspaces $E_D \cong \mathbb{R}^D$. It cannot be c^∞-complete, since finite sequences are contained in E and are dense in $\mathbb{R}^\mathbb{N} \supset E$.

3. Smooth Mappings and the Exponential Law

Now let us start proving the exponential law $C^\infty(U \times V, F) \cong C^\infty(U, C^\infty(V,F))$ first for $U = V = F = \mathbb{R}$.

3.1. Proposition. *For a continuous map $f : \mathbb{R} \times [0,1] \to \mathbb{R}$ the partial derivative $\partial_1 f$ exists and is continuous if and only if $f^\vee : \mathbb{R} \to C([0,1],\mathbb{R})$ is continuously differentiable. And in this situation $I((f^\vee)'(t)) = \frac{d}{dt}\int_0^1 f(t,s)\,ds = \int_0^1 \frac{\partial}{\partial t} f(t,s)\,ds$, where $I : C([0,1],\mathbb{R}) \to \mathbb{R}$ is integration.*

Proof. We assume that $\partial_1 f$ exists and is continuous. Hence, $(\partial_1 f)^\vee : \mathbb{R} \to C([0,1],\mathbb{R})$ is continuous. We want to show that $f^\vee : \mathbb{R} \to C([0,1],\mathbb{R})$ is differentiable (say at 0) with this function (at 0) as derivative. So we have to show that the mapping $t \mapsto \frac{f^\vee(t)-f^\vee(0)}{t}$ is continuously extendable to $\mathbb{R}$ by defining its value at 0 as $(\partial_1 f)^\vee(0)$. Or equivalently, by what is obvious for continuous maps, that the map

$$(t,s) \mapsto \begin{cases} \frac{f(t,s)-f(0,s)}{t} & \text{for } t \neq 0 \\ \partial_1 f(0,s) & \text{otherwise} \end{cases}$$

is continuous. This follows immediately from the continuity of $\partial_1 f$ and of integration since it can be written as $\int_0^1 \partial_1 f(r\,t,s)\,dr$ by the fundamental theorem.

So we arrive under this assumption at the conclusion, that $\int_0^1 f(t,s)\,ds$ is differentiable with derivative

$$\frac{d}{dt}\int_0^1 f(t,s)\,ds = I((f^\vee)'(t)) = \int_0^1 \frac{\partial}{\partial t} f(t,s)\,ds.$$

The converse implication is obvious. □

3.2. Theorem. Simplest case of exponential law. *Let $f : \mathbb{R}^2 \to \mathbb{R}$ be an arbitrary mapping. Then all iterated partial derivatives exist and are locally bounded if and only if the associated mapping $f^\vee : \mathbb{R} \to C^\infty(\mathbb{R},\mathbb{R})$ exists as a smooth curve,* *where* $C^\infty(\mathbb{R},\mathbb{R})$ *is considered as the Fréchet space with the topology of uniform convergence of each derivative on compact sets. Furthermore, we have* $(\partial_1 f)^\vee = d(f^\vee)$ *and* $(\partial_2 f)^\vee = d \circ f^\vee = d_*(f^\vee)$.

Proof. We have several possibilities to prove this result. Either we show Mackey convergence of the difference quotients, via the boundedness of $\frac{1}{t}\left(\frac{c(t)-c(0)}{t} - c'(0)\right)$, and then use the trivial exponential law $\ell^\infty(X\times Y,\mathbb{R}) \cong \ell^\infty(X,\ell^\infty(Y,\mathbb{R}))$; or we use the induction step proved in (3.1), namely that $f^\vee : \mathbb{R} \to C(\mathbb{R},\mathbb{R})$ is differentiable if and only if $\partial_1 f$ exists and is continuous $\mathbb{R}^2 \to \mathbb{R}$, together with the exponential law $C(\mathbb{R}^2,\mathbb{R}) \cong C(\mathbb{R},C(\mathbb{R},\mathbb{R}))$. We choose the latter method.

For this we have to note first that if for a function g the partial derivatives $\partial_1 g$ and $\partial_2 g$ exist and are locally bounded then g is continuous:

$$\begin{aligned} g(x,y) - g(0,0) &= g(x,y) - g(x,0) + g(x,0) - g(0,0) \\ &= y\partial_2 g(x, r_2 y) + x\partial_1 g(r_1 x, 0) \end{aligned}$$

for suitable r_1, $r_2 \in [0,1]$, which goes to 0 with (x,y).

Proof of $(\Rightarrow)$ By what we just said, all iterated partial derivatives of f are continuous. First observe that $f^\vee : \mathbb{R} \to C^\infty(\mathbb{R},\mathbb{R})$ makes sense and that for all $t \in \mathbb{R}$ we have

$$d^q(f^\vee(t)) = (\partial_2^q f)^\vee(t). \tag{1}$$

Next we claim that $f^\vee : \mathbb{R} \to C^\infty(\mathbb{R},\mathbb{R})$ is differentiable, with derivative $d(f^\vee) = (\partial_1 f)^\vee$, or equivalently that for all a the curve

$$c : t \mapsto \begin{cases} \frac{f^\vee(t+a) - f^\vee(a)}{t} & \text{for } t \neq 0 \\ (\partial_1 f)^\vee(a) & \text{otherwise} \end{cases}$$

is continuous as curve $\mathbb{R} \to C^\infty(\mathbb{R},\mathbb{R})$. Without loss of generality we may assume that $a = 0$. Since $C^\infty(\mathbb{R},\mathbb{R})$ carries the initial structure with respect to the linear mappings $d^p : C^\infty(\mathbb{R},\mathbb{R}) \to C(\mathbb{R},\mathbb{R})$ we have to show that $d^p \circ c : \mathbb{R} \to C(\mathbb{R},\mathbb{R})$ is continuous, or equivalently by the exponential law for continuous maps, that $(d^p \circ c)^\wedge : \mathbb{R}^2 \to \mathbb{R}$ is continuous. For $t \neq 0$ and $s \in \mathbb{R}$ we have

$$\begin{aligned} (d^p \circ c)^\wedge(t,s) &= d^p(c(t))(s) = d^p\left(\frac{f^\vee(t) - f^\vee(0)}{t}\right)(s) \\ &= \frac{\partial_2^p f(t,s) - \partial_2^p f(0,s)}{t} && \text{by (1)} \\ &= \int_0^1 \partial_1\partial_2^p f(t\,\tau, s)\,d\tau && \text{by the fundamental theorem.} \end{aligned}$$

For $t = 0$ we have

$$\begin{aligned} (d^p \circ c)^\wedge(0,s) &= d^p(c(0))(s) = d^p((\partial_1 f)^\vee(0))(s) \\ &= (\partial_2^p(\partial_1 f))^\vee(0)(s) && \text{by (1)} \\ &= \partial_2^p\partial_1 f(0,s) \\ &= \partial_1\partial_2^p f(0,s) && \text{by the theorem of Schwarz.} \end{aligned}$$

So we see that $(d^p \circ c)^\wedge(t,s) = \int_0^1 \partial_1\partial_2^p f(t\,\tau, s)\,d\tau$ for all (t,s). This function is continuous in (t,s), since $\partial_1\partial_2^p f : \mathbb{R}^2 \to \mathbb{R}$ is continuous, hence $(t,s,\tau) \mapsto \partial_1\partial_2^p f(t\,\tau, s)$ is continuous, and therefore also $(t,s) \mapsto (\tau \mapsto \partial_1\partial_2^p f(t\,\tau, s))$ from $\mathbb{R}^2 \to C([0,1],\mathbb{R})$. Composition with the continuous linear mapping $\int_0^1 : C([0,1],\mathbb{R}) \to \mathbb{R}$ gives the continuity of $(d^p \circ c)^\wedge$.

Now we proceed by induction. By the induction hypothesis applied to $\partial_1 f$, we obtain that $d(f^\vee) = (\partial_1 f)^\vee$ and $(\partial_1 f)^\vee : \mathbb{R} \to C^\infty(\mathbb{R},\mathbb{R})$ is n times differentiable, so $f^\vee$ is $(n+1)$-times differentiable.

Proof of ($\Leftarrow$) First remark that for a smooth map $f : \mathbb{R} \to C^\infty(\mathbb{R},\mathbb{R})$ the associated map $f^\wedge : \mathbb{R}^2 \to \mathbb{R}$ is locally bounded: Since f is smooth $f(I_1)$ is compact, hence bounded in $C^\infty(\mathbb{R},\mathbb{R})$ for all compact intervals I_1. In particular, $f(I_1)(I_2) = f^\wedge(I_1 \times I_2)$ has to be bounded in $\mathbb{R}$ for all compact intervals I_1 and I_2.

Since f is smooth both curves df and $d \circ f = d_* f$ are smooth (use (1.3) and that d is continuous and linear). An easy calculation shows that the partial derivatives of $f^\wedge$ exist and are given by $\partial_1 f^\wedge = (df)^\wedge$ and $\partial_2 f^\wedge = (d \circ f)^\wedge$. So one obtains inductively that all iterated derivatives of $f^\wedge$ exist and are locally bounded, since they are associated to smooth curves $\mathbb{R} \to C^\infty(\mathbb{R},\mathbb{R})$. □

In order to proceed to more general cases of the exponential law we need a definition of C^∞-maps defined on infinite dimensional spaces. This definition should at least guarantee the chain rule, and so one could take the weakest notion that satisfies the chain rule. However, consider the following

3.3. Example. We consider the following 3-fold "singular covering" $f : \mathbb{R}^2 \to \mathbb{R}^2$ given in polar coordinates by $(r,\varphi) \mapsto (r, 3\varphi)$. In cartesian coordinates we obtain the following formula for the values of f:

$$\begin{aligned}(r\cos(3\varphi), r\sin(3\varphi)) &= r\Big((\cos\varphi)^3 - 3\cos\varphi(\sin\varphi)^2, 3\sin\varphi(\cos\varphi)^2 - (\sin\varphi)^3\Big)\\ &= \left(\frac{x^3 - 3xy^2}{x^2+y^2}, \frac{3x^2y - y^3}{x^2+y^2}\right).\end{aligned}$$

Note that the composite from the left with any orthonormal projection is just the composite of the first component of f with a rotation from the right (Use that f intertwines the rotation with angle δ and the rotation with angle 3δ).

Obviously, the map f is smooth on $\mathbb{R}^2 \setminus \{0\}$. It is homogeneous of degree 1, and hence the directional derivative is $f'(0)(v) = \frac{\partial}{\partial t}|_{t=0} f(tv) = f(v)$. However, both components are nonlinear with respect to v and thus are not differentiable at $(0,0)$.

Obviously, $f : \mathbb{R}^2 \to \mathbb{R}^2$ is continuous.

We claim that f is differentiable along differentiable curves, i.e. $(f \circ c)'(0)$ exists, provided $c'(0)$ exists.

Only the case $c(0) = 0$ is not trivial. Since c is differentiable at 0 the curve c_1 defined by $c_1(t) := \frac{c(t)}{t}$ for $t \neq 0$ and $c'(0)$ for $t = 0$ is continuous at 0. Hence $\frac{f(c(t)) - f(c(0))}{t} = \frac{f(t\,c_1(t)) - 0}{t} = f(c_1(t))$. This converges to $f(c_1(0))$, since f is continuous.

Furthermore, if $f'(x)(v)$ denotes the directional derivative, which exists everywhere, then $(f\circ c)'(t) = f'(c(t))(c'(t))$. Indeed for $c(t)\neq 0$ this is clear and for $c(t)=0$ it follows from $f'(0)(v) = f(v)$.

The directional derivative of the 1-homogeneous mapping f is 0-homogeneous: In fact, for $s\neq 0$ we have

$$f'(sx)(v) = \frac{\partial}{\partial t}\Big|_{t=0} f(s\,x+tv) = s\,\frac{\partial}{\partial t}\Big|_{t=0} f(x+\frac{t}{s}v) = s\,f'(x)(\frac{1}{s}v) = f'(x)(v).$$

For any $s\in\mathbb{R}$ we have $f'(s\,v)(v) = \frac{\partial}{\partial t}|_{t=0}f(s\,v+tv) = \frac{\partial}{\partial t}|_{t=s}t\,f(v) = f(v)$.

Using this homogeneity we show next, that it is also continuously differentiable along continuously differentiable curves. So we have to show that $(f\circ c)'(t)\to (f\circ c)'(0)$ for $t\to 0$. Again only the case $c(0)=0$ is interesting. As before we factor c as $c(t) = t\,c_1(t)$. In the case, where $c'(0) = c_1(0)\neq 0$ we have for $t\neq 0$ that

$$\begin{aligned}(f\circ c)'(t) - (f\circ c)'(0) &= f'(t\,c_1(t))(c'(t)) - f'(0)(c_1(0))\\ &= f'(c_1(t))(c'(t)) - f'(c_1(0))(c_1(0))\\ &= f'(c_1(t))(c'(t)) - f'(c_1(0))(c'(0)),\end{aligned}$$

which converges to 0 for $t\to 0$, since $(f')^\wedge$ is continuous (and even smooth) on $(\mathbb{R}^2\setminus\{0\})\times\mathbb{R}^2$.
In the other case, where $c'(0) = c_1(0) = 0$ we consider first the values of t, for which $c(t)=0$. Then

$$\begin{aligned}(f\circ c)'(t) - (f\circ c)'(0) &= f'(0)(c'(t)) - f'(0)(c'(0))\\ &= f(c'(t)) - f(c'(0)) \to 0,\end{aligned}$$

since f is continuous. For the remaining values of t, where $c(t)\neq 0$, we factor $c(t) = |c(t)|\,e(t)$, with $e(t)\in\{x : \|x\| = 1\}$. Then

$$(f\circ c)'(t) - (f\circ c)'(0) = f'(e(t))(c'(t)) - 0 \to 0,$$

since $f'(x)(c'(t))\to 0$ for $t\to 0$ uniformly for $\|x\|=1$, since $c'(t)\to 0$.

Furthermore, $f\circ c$ is smooth for all c which are smooth and nowhere infinitely flat. In fact, a smooth curve c with $c^{(k)}(0) = 0$ for $k<n$ can be factored as $c(t) = t^n c_n(t)$ with smooth c_n, by Taylor's formula with integral remainder. Since $c^{(n)}(0) = n!\,c_n(0)$, we may assume that n is chosen maximal and hence $c_n(0)\neq 0$. But then $(f\circ c)(t) = t^n\cdot(f\circ c_n)(t)$, and $f\circ c_n$ is smooth.

A completely analogous argument shows also that $f\circ c$ is real analytic for all real analytic curves $c:\mathbb{R}\to\mathbb{R}^2$.

However, let us show that $f\circ c$ is not Lipschitz differentiable even for smooth curves c. For $x\neq 0$ we have

$$\begin{aligned}(\partial_2)^2 f(x,0) = \left(\tfrac{\partial}{\partial s}\right)^2|_{s=0}f(x,s) = x\left(\tfrac{\partial}{\partial s}\right)^2|_{s=0}f(1,\tfrac{1}{x}s) =\\ = \tfrac{1}{x}\left(\tfrac{\partial}{\partial s}\right)^2|_{s=0}f(1,s) =: \tfrac{a}{x}\neq 0.\end{aligned}$$

Now we choose a smooth curve c which passes for each n in finite time t_n through $(\frac{1}{n^{2n+1}}, 0)$ with locally constant velocity vector $(0, \frac{1}{n^n})$, by (2.10). Then for small t we get

$$(f \circ c)'(t_n + t) = \partial_1 f(c(t_n + t)) \underbrace{\mathrm{pr}_1(c'(t_n + t))}_{=0} + \partial_2 f(c(t_n + t)) \,\mathrm{pr}_2(c'(t_n + t))$$

$$(f \circ c)''(t_n) = (\partial_2)^2 f(c(t_n))\,(\mathrm{pr}_2(c'(t_n)))^2 + 0 = a\,\frac{n^{2n+1}}{n^{2n}} = n\,a,$$

which is unbounded.

So although preservation of (continuous) differentiability of curves is not enough to ensure differentiability of a function $\mathbb{R}^2 \to \mathbb{R}$, we now prove that smoothness can be tested with smooth curves.

3.4. Boman's theorem. [Boman, 1967] *For a mapping $f : \mathbb{R}^2 \to \mathbb{R}$ the following assertions are equivalent:*

1. *All iterated partial derivatives exist and are continuous.*
2. *All iterated partial derivatives exist and are locally bounded.*
3. *For $v \in \mathbb{R}^2$ the iterated directional derivatives*

$$d_v^n f(x) := (\tfrac{\partial}{\partial t})^n|_{t=0}(f(x + tv))$$

   exist and are locally bounded with respect to x.

4. *For all smooth curves $c : \mathbb{R} \to \mathbb{R}^2$ the composite $f \circ c$ is smooth.*

Proof. (1) $\Rightarrow$ (2) is obvious.

(2) $\Rightarrow$ (1) follows immediately, since the local boundedness of $\partial_1 f$ and $\partial_2 f$ imply the continuity of f (see also the proof of (3.2)):

$$f(t, s) - f(0, 0) = t \int_0^1 \partial_1 f(\tau t, s) d\tau + s \int_0^1 \partial_2 f(0, \sigma s) d\sigma.$$

(1) $\Rightarrow$ (4) is a direct consequence of the chain rule, namely that $(f \circ c)'(t) = \partial_1 f(c(t)) \cdot x'(t) + \partial_2 f(c(t)) \cdot y'(t)$, where $c = (x, y)$.

(4) $\Rightarrow$ (3) Obviously, $d_v^p f(x) := (\frac{d}{dt})^p|_{t=0} f(x + tv)$ exists, since $t \mapsto x + tv$ is a smooth curve. Suppose $d_v^p f$ is not locally bounded. So we may find a sequence x_n which converges fast to x, and such that $|d_v^p f(x_n)| \geq 2^{n^2}$. Let c be a smooth curve with $c(t + t_n) = x_n + \frac{t}{2^n} v$ locally for some sequence $t_n \to 0$, by (2.8). Then $(f \circ c)^{(p)}(t_n) = d_v^p f(x_n) \frac{1}{2^{np}}$ is unbounded, which is a contradiction.

(3) $\Rightarrow$ (1) First we claim that $d_v^p f$ is continuous. We prove this by induction on p: $d_v^p f(\ \ + tv) - d_v^p f(\ \ ) = t \int_0^1 d_v^{p+1} f(\ \ + t\tau v) d\tau \to 0$ for $t \to 0$ uniformly on bounded sets. Suppose now that $|d_v^p f(x_n) - d_v^p f(x)| \geq \varepsilon$ for some sequence $x_n \to x$. Without loss of generality we may assume that $d_v^p f(x_n) - d_v^p f(x) \geq \varepsilon$. Then by the

uniform convergence there exists a $\delta > 0$ such that $d_v^p f(x_n + tv) - d_v^p f(x + tv) \geq \frac{\varepsilon}{2}$ for $|t| \leq \delta$. Integration $\int_0^\delta dt$ yields

$$d_v^{p-1} f(x_n + \delta v) - d_v^{p-1} f(x_n) - \Big(d_v^{p-1} f(x + \delta v) - d_v^{p-1} f(x)\Big) \geq \tfrac{\varepsilon\delta}{2},$$

but by induction hypothesis the left hand side converges towards

$$\Big(d_v^{p-1} f(x + \delta v) - d_v^{p-1} f(x)\Big) - \Big(d_v^{p-1} f(x + \delta v) - d_v^{p-1} f(x)\Big) = 0.$$

To complete the proof we use convolution by an *approximation of unity*. So let $\varphi \in C^\infty(\mathbb{R}^2, \mathbb{R})$ have compact support, $\int \varphi = 1$, and $\varphi(y) \geq 0$ for all y. Define $\varphi_\varepsilon(x) := \frac{1}{\varepsilon^2}\varphi(\frac{1}{\varepsilon}x)$, and let

$$f_\varepsilon(x) := (f \star \varphi_\varepsilon)(x) = \int_{\mathbb{R}^2} f(x - y)\, \varphi_\varepsilon(y)\, dy = \int_{\mathbb{R}^2} f(x - \varepsilon y)\varphi(y) dy.$$

Since the *convolution* $f_\varepsilon := f \star \varphi_\varepsilon$ of a continuous function f with a smooth function φ_ε with compact support is differentiable with directional derivative $d_v(f \star \varphi_\varepsilon)(x) = (f \star d_v\varphi_\varepsilon)(x)$, we obtain that f_ε is smooth. And since $f \star \varphi_\varepsilon \to f$ in $C(\mathbb{R}^2, \mathbb{R})$ for $\varepsilon \to 0$ and any continuous function f, we conclude $d_v^p f_\varepsilon = d_v^p f \star \varphi_\varepsilon \to d_v^p f$ uniformly on compact sets.

We remark now that for a smooth map $g : \mathbb{R}^2 \to \mathbb{R}$ we have by the chain rule

$$d_v g(x + tv) = \frac{d}{dt} g(x + tv) = \partial_1 g(x + tv) \cdot v_1 + \partial_2 g(x + tv) \cdot v_2$$

and by induction that

$$d_v^p g(x) = \sum_{j=0}^{p} \binom{p}{i} v_1^i v_2^{p-i} \partial_1^i \partial_2^{p-i} g(x).$$

Hence, we can calculate the iterated derivatives $\partial_1^i \partial_2^{p-i} g(x)$ for $0 \leq i \leq p$ from $p + 1$ many derivatives $d_{v^j}^p g(x)$ provided the v^j are chosen in such a way, that the *Vandermonde's determinant* $\det((v_1^j)^i (v_2^j)^{p-i})_{ij} \neq 0$. For this choose $v_2 = 1$ and all the v_1 pairwise distinct, then $\det((v_1^j)^i (v_2^j)^{p-i})_{ij} = \prod_{j>k}(v_1^j - v_1^k) \neq 0$.

Hence, the iterated derivatives of f_ε are linear combinations of the derivatives $d_v^p f_\varepsilon$ for $p + 1$ many vectors v, where the coefficients depend only on the v's. So we conclude that the iterated derivatives of f_ε form a Cauchy sequence in $C(\mathbb{R}^2, \mathbb{R})$, and hence converge to continuous functions f^α. Thus, all iterated derivatives $\partial^\alpha f$ of f exist and are equal to these continuous functions f^α, by the following lemma (3.5). □

3.5. Lemma. *Let $f_\varepsilon \to f$ in $C(\mathbb{R}^2, \mathbb{R})$ and $d_v f_\varepsilon \to f_v$ in $C(\mathbb{R}^2, \mathbb{R})$. Then $d_v f$ exists and equals f_v.*

Proof. We have to show that for fixed $x, v \in \mathbb{R}^2$ the curve

$$c : t \mapsto \begin{cases} \frac{f(x+tv) - f(x)}{t} & \text{for } t \neq 0 \\ f_v(x) & \text{otherwise} \end{cases}$$

is continuous from $\mathbb{R} \to \mathbb{R}$. The corresponding curve c_ε for f_ε can be rewritten as $c_\varepsilon(t) = \int_0^1 d_v f_\varepsilon(x + \tau\, t\, v)\, d\tau$, which converges by assumption uniformly for t in compact sets to the continuous curve $t \mapsto \int_0^1 f_v(x + \tau\, t\, v)\, d\tau$. Pointwise it converges to $c(t)$, hence c is continuous. □

For the vector valued case of the exponential law we need a locally convex structure on $C^\infty(\mathbb{R}, E)$.

3.6. Definition. Space of curves. Let $C^\infty(\mathbb{R}, E)$ be the locally convex vector space of all smooth curves in E, with the pointwise vector operations, and with the topology of uniform convergence on compact sets of each derivative separately. This is the initial topology with respect to the linear mappings $C^\infty(\mathbb{R}, E) \xrightarrow{d^k} C^\infty(\mathbb{R}, E) \to \ell^\infty(K, E)$, where k runs through $\mathbb{N}$, where K runs through all compact subsets of $\mathbb{R}$, and where $\ell^\infty(K, E)$ carries the topology of uniform convergence, see also (2.15).

Note that the derivatives $d^k : C^\infty(\mathbb{R}, E) \to C^\infty(\mathbb{R}, E)$, the point evaluations $\mathrm{ev}_t : C^\infty(\mathbb{R}, E) \to E$ and the pull backs $g^* : C^\infty(\mathbb{R}, E) \to C^\infty(\mathbb{R}, E)$ for all $g \in C^\infty(\mathbb{R}, \mathbb{R})$ are continuous and linear.

3.7. Lemma. *A space E is c^∞-complete if and only if $C^\infty(\mathbb{R}, E)$ is.*

Proof. ($\Rightarrow$) The mapping $c \mapsto (c^{(n)})_{n\in\mathbb{N}}$ is by definition an embedding of $C^\infty(\mathbb{R}, E)$ into the c^∞-complete product $\prod_{n\in\mathbb{N}} \ell^\infty(\mathbb{R}, E)$. Its image is a closed subspace, since the previous lemma can be easily generalized to curves $c : \mathbb{R} \to E$.

($\Leftarrow$) Consider the continuous linear mapping $\mathrm{const} : E \to C^\infty(\mathbb{R}, E)$ given by $x \mapsto (t \mapsto x)$. It has as continuous left inverse the evaluation at any point (e.g. $\mathrm{ev}_0 : C^\infty(\mathbb{R}, E) \to E$, $c \mapsto c(0)$). Hence, E can be identified with the closed subspace of $C^\infty(\mathbb{R}, E)$ given by the constant curves, and is thereby itself c^∞-complete. □

3.8. Lemma. Curves into limits. *A curve into a c^∞-closed subspace of a space is smooth if and only if it is smooth into the total space. In particular, a curve is smooth into a projective limit if and only if all its components are smooth.*

Proof. Since the derivative of a smooth curve is the Mackey limit of the difference quotient, the c^∞-closedness implies that this limit belongs to the subspace. Thus, we deduce inductively that all derivatives belong to the subspace, and hence the curve is smooth into the subspace.

The result on projective limits now follows, since obviously a curve is smooth into a product, if all its components are smooth. □

We show now that the bornology, but obviously not the topology, on function spaces can be tested with the linear functionals on the range space.

3.9. Lemma. Bornology of $C^\infty(\mathbb{R}, E)$. *The family*

$$\{\ell_* : C^\infty(\mathbb{R}, E) \to C^\infty(\mathbb{R}, \mathbb{R}) : \ell \in E^*\}$$

generates the bornology of $C^\infty(\mathbb{R},E)$. This also holds for E^* replaced by E'.

A set in $C^\infty(\mathbb{R},E)$ is bounded if and only if each derivative is uniformly bounded on compact subsets.

Proof. A set $\mathcal{B}\subseteq C^\infty(\mathbb{R},E)$ is bounded if and only if the sets $\{d^n c(t) : t\in I, c\in\mathcal{B}\}$ are bounded in E for all $n\in\mathbb{N}$ and compact subsets $I\subset\mathbb{R}$.

This is furthermore equivalent to the condition that the set $\{\ell(d^n c(t)) = d^n(\ell\circ c)(t) : t\in I, c\in\mathcal{B}\}$ is bounded in $\mathbb{R}$ for all $\ell\in E^*$, $n\in\mathbb{N}$, and compact subsets $I\subset\mathbb{R}$ and in turn equivalent to: $\{\ell\circ c : c\in\mathcal{B}\}$ is bounded in $C^\infty(\mathbb{R},\mathbb{R})$.

For E^* replaced by $E'\supseteq E^*$ the statement holds, since ℓ_* is bounded for all $\ell\in E'$ by the explicit description of the bounded sets. □

3.10. Proposition. Vector valued simplest exponential law. *For a mapping $f:\mathbb{R}^2\to E$ into a locally convex space (which need not be c^∞-complete) the following assertions are equivalent:*

(1) *f is smooth along smooth curves.*
(2) *All iterated directional derivatives $d^p_v f$ exist and are locally bounded.*
(3) *All iterated partial derivatives $\partial^\alpha f$ exist and are locally bounded.*
(4) *$f^\vee:\mathbb{R}\to C^\infty(\mathbb{R},E)$ exists as a smooth curve.*

Proof. We prove this result first for c^∞-complete spaces E. Then each of the statements (1-4) are valid if and only if the corresponding statement for $\ell\circ f$ is valid for all $\ell\in E^*$. Only (4) needs some arguments: In fact, $f^\vee(t)\in C^\infty(\mathbb{R},E)$ if and only if $\ell_*(f^\vee(t)) = (\ell\circ f)^\vee(t)\in C^\infty(\mathbb{R},\mathbb{R})$ for all $\ell\in E^*$ by (2.14). Since $C^\infty(\mathbb{R},E)$ is c^∞-complete, its bornologically isomorphic image in $\prod_{\ell\in E^*}C^\infty(\mathbb{R},\mathbb{R})$ is c^∞-closed. So $f^\vee:\mathbb{R}\to C^\infty(\mathbb{R},E)$ is smooth if and only if $\ell_*\circ f^\vee = (\ell\circ f)^\vee : \mathbb{R}\to C^\infty(\mathbb{R},\mathbb{R})$ is smooth for all $\ell\in E^*$. So the proof is reduced to the scalar valid case, which was proved in (3.2) and (3.4).

Now the general case. For the existence of certain derivatives we know by (1.9) that it is enough that we have some candidate in the space, which is the corresponding derivative of the map considered as map into the c^∞-completion (or even some larger space). Since the derivatives required in (1-4) depend linearly on each other, this is true. In more detail this means:

(1) ⇒ (2) is obvious.

(2) ⇒ (3) is the fact that ∂^α is a universal linear combination of $d_v^{|\alpha|}f$.

(3) ⇒ (1) follows from the chain rule which says that $(f\circ c)^{(p)}(t)$ is a universal linear combination of $\partial_{i_1}\dots\partial_{i_q}f(c(t))c_{i_1}^{(p_1)}(t)\dots c_{i_q}^{(p_q)}(t)$ for $i_j\in\{1,2\}$ and $\sum p_j = p$, see also (10.4).

(3) ⇔ (4) holds by (1.9) since $(\partial_1 f)^\vee = d(f^\vee)$ and $(\partial_2 f)^\vee = d\circ f^\vee = d_*(f^\vee)$. □

For the general case of the exponential law we need a notion of smooth mappings and a locally convex topology on the corresponding function spaces. Of course, it would be also handy to have a notion of smoothness for locally defined mappings. Since the idea is to test smoothness with smooth curves, such curves should have locally values in the domains of definition, and hence these domains should be c^∞-open.

3.11. Definition. Smooth mappings and spaces of them. A mapping $f : E \supseteq U \to F$ defined on a c^∞-open subset U is called *smooth* (or C^∞) if it maps smooth curves in U to smooth curves in F.

Let $C^\infty(U,F)$ denote the locally convex *space of all smooth mappings* $U \to F$ with pointwise linear structure and the initial topology with respect to all mappings $c^* : C^\infty(U,F) \to C^\infty(\mathbb{R},F)$ for $c \in C^\infty(\mathbb{R},U)$.

For $U = E = \mathbb{R}$ this coincides with our old definition. Obviously, any composition of smooth mappings is also smooth.

Lemma. *The space $C^\infty(U,F)$ is the (inverse) limit of spaces $C^\infty(\mathbb{R},F)$, one for each $c \in C^\infty(\mathbb{R},U)$, where the connecting mappings are pull backs g^* along reparameterizations $g \in C^\infty(\mathbb{R},\mathbb{R})$.*

Note that this limit is the closed linear subspace in the product

$$\prod_{c\in C^\infty(\mathbb{R},U)} C^\infty(\mathbb{R},F)$$

consisting of all (f_c) with $f_{c\circ g} = f_c \circ g$ for all c and all $g \in C^\infty(\mathbb{R},\mathbb{R})$.

Proof. The mappings $c^* : C^\infty(U,F) \to C^\infty(\mathbb{R},F)$ define a continuous linear embedding $C^\infty(U,F) \to \lim_c\{C^\infty(\mathbb{R},F) \xrightarrow{g^*} C^\infty(\mathbb{R},F)\}$, since $c^*(f)\circ g = f\circ c\circ g = (c\circ g)^*(f)$. It is surjective since for any $(f_c) \in \lim_c C^\infty(\mathbb{R},F)$ one has $f_c = f\circ c$ where f is defined as $x \mapsto f_{\text{const}_x}(0)$. $\square$

3.12. Theorem. Cartesian closedness. *Let $U_i \subseteq E_i$ be c^∞-open subsets in locally convex spaces, which need not be c^∞-complete. Then a mapping $f : U_1 \times U_2 \to F$ is smooth if and only if the canonically associated mapping $f^\vee : U_1 \to C^\infty(U_2,F)$ exists and is smooth.*

Proof. We have the following implications:

$f^\vee : U_1 \to C^\infty(U_2,F)$ is smooth.

$\Leftrightarrow$ $f^\vee \circ c_1 : \mathbb{R} \to C^\infty(U_2,F)$ is smooth for all smooth curves c_1 in U_1, by (3.11).

$\Leftrightarrow$ $c_2^* \circ f^\vee \circ c_1 : \mathbb{R} \to C^\infty(\mathbb{R},F)$ is smooth for all smooth curves c_i in U_i, by (3.11) and (3.8).

$\Leftrightarrow$ $f\circ(c_1\times c_2) = (c_2^*\circ f^\vee\circ c_1)^\wedge : \mathbb{R}^2 \to F$ is smooth for all smooth curves c_i in U_i, by (3.10).

$\Leftrightarrow$ $f : U_1 \times U_2 \to F$ is smooth.

Here the last equivalence is seen as follows: Each curve into $U_1 \times U_2$ is of the form $(c_1,c_2) = (c_1\times c_2)\circ\Delta$, where Δ is the diagonal mapping. Conversely, $f\circ(c_1 \times c_2) : \mathbb{R}^2 \to F$ is smooth for all smooth curves c_i in U_i, since the product and the composite of smooth mappings is smooth by (3.11) (and by (3.4)). $\square$

3.13. Corollary. Consequences of cartesian closedness. *Let E, F, G, etc. be locally convex spaces, and let U, V be c^∞-open subsets of such. Then the following canonical mappings are smooth.*

(1) $\operatorname{ev} : C^\infty(U,F)\times U \to F$, $(f,x)\mapsto f(x)$;
(2) $\operatorname{ins} : E \to C^\infty(F, E\times F)$, $x\mapsto (y\mapsto (x,y))$;
(3) $(\ \ )^\wedge : C^\infty(U, C^\infty(V,G)) \to C^\infty(U\times V, G)$;
(4) $(\ \ )^\vee : C^\infty(U\times V, G) \to C^\infty(U, C^\infty(V,G))$;
(5) $\operatorname{comp} : C^\infty(F,G)\times C^\infty(U,F)\to C^\infty(U,G)$, $(f,g)\mapsto f\circ g$;
(6) $C^\infty(\ \ ,\ \ ) : C^\infty(E_2,E_1)\times C^\infty(F_1,F_2)\to$
$\to C^\infty(C^\infty(E_1,F_1), C^\infty(E_2,F_2))$, $(f,g)\mapsto (h\mapsto g\circ h\circ f)$;
(7) $\prod : \prod C^\infty(E_i,F_i) \to C^\infty(\prod E_i, \prod F_i)$, *for any index set.*

Proof. (1) The mapping associated to *ev* via cartesian closedness is the identity on $C^\infty(U,F)$, which is C^∞, thus *ev* is also C^∞.

(2) The mapping associated to *ins* via cartesian closedness is the identity on $E\times F$, hence *ins* is C^∞.

(3) The mapping associated to $(\ \ )^\wedge$ via cartesian closedness is the smooth composition of evaluations $\operatorname{ev}\circ(\operatorname{ev}\times \operatorname{Id}) : (f;x,y)\mapsto f(x)(y)$.

(4) We apply cartesian closedness twice to get the associated mapping $(f;x;y)\mapsto f(x,y)$, which is just a smooth evaluation mapping.

(5) The mapping associated to *comp* via cartesian closedness is $(f,g;x)\mapsto f(g(x))$, which is the smooth mapping $\operatorname{ev}\circ(\operatorname{Id}\times\operatorname{ev})$.

(6) The mapping associated to the one in question by applying cartesian closed twice is $(f,g;h,x)\mapsto g(h(f(x)))$, which is the C^∞-mapping $\operatorname{ev}\circ(\operatorname{Id}\times\operatorname{ev})\circ(\operatorname{Id}\times\operatorname{Id}\times\operatorname{ev})$.

(7) Up to a flip of factors the mapping associated via cartesian closedness is the product of the evaluation mappings $C^\infty(E_i,F_i)\times E_i\to F_i$. □

Next we generalize (3.4) to finite dimensions.

3.14. Corollary. [Boman, 1967]. *The smooth mappings on open subsets of $\mathbb{R}^n$ in the sense of definition* (3.11) *are exactly the usual smooth mappings.*

Proof. Both conditions are of local nature, so we may assume that the open subset of $\mathbb{R}^n$ is an open box and in turn even $\mathbb{R}^n$ itself.

($\Rightarrow$) If $f:\mathbb{R}^n\to F$ is smooth then by cartesian closedness (3.12), for each coordinate the respective associated mapping $f^{\vee_i}:\mathbb{R}^{n-1}\to C^\infty(\mathbb{R},F)$ is smooth, so again by (3.12) we have $\partial_i f = (d_* f^{\vee_i})^\wedge$, so all first partial derivatives exist and are smooth. Inductively, all iterated partial derivatives exist and are smooth, thus continuous, so f is smooth in the usual sense.

($\Leftarrow$) Obviously, f is smooth along smooth curves by the usual chain rule. □

3.15. Differentiation of an integral. We return to the question of differentiating an integral. So let $f: E\times\mathbb{R}\to F$ be smooth, and let $\widehat{F}$ be the completion of the locally convex space F. Then we may form the function $f_0 : E\to\widehat{F}$ defined by

$x \mapsto \int_0^1 f(x,t)\,dt$. We claim that it is smooth, and that the directional derivative is given by $d_v f_0(x) = \int_0^1 d_v(f(\quad,t))(x)\,dt$. By cartesian closedness (3.12) the associated mapping $f^\vee : E \to C^\infty(\mathbb{R}, F)$ is smooth, so the mapping $\int_0^1 \circ f^\vee : E \to \widehat{F}$ is smooth since integration is a bounded linear operator, and

$$\begin{aligned} d_v f_0(x) &= \tfrac{\partial}{\partial s}\big|_{s=0} f_0(x+sv) = \tfrac{\partial}{\partial s}\big|_{s=0} \int_0^1 f(x+sv,t)dt \\ &= \int_0^1 \tfrac{\partial}{\partial s}\big|_{s=0} f(x+sv,t)dt = \int_0^1 d_v(f(\quad,t))(x)\,dt. \end{aligned}$$

But we want to generalize this to functions f defined only on some c^∞-open subset $U \subseteq E \times \mathbb{R}$, so we have to show that the natural domain $U_0 := \{x \in E : \{x\} \times [0,1] \subseteq U\}$ of f_0 is c^∞-open in E. We will do this in lemma (4.15). From then on the proof runs exactly the same way as for globally defined functions. So we obtain the

Proposition. *Let $f : E \times \mathbb{R} \supseteq U \to F$ be smooth with c^∞-open $U \subseteq E \times \mathbb{R}$. Then $x \mapsto \int_0^1 f(x,t)\,dt$ is smooth on the c^∞-open set $U_0 := \{x \in E : \{x\} \times [0,1] \subseteq U\}$ with values in the completion $\widehat{F}$ and $d_v f_0(x) = \int_0^1 d_v(f(\quad,t))(x)\,dt$ for all $x \in U_0$ and $v \in E$.* □

Now we want to define the derivative of a general smooth map and prove the chain rule for them.

3.16. Corollary. Smoothness of the difference quotient. *For a smooth curve $c : \mathbb{R} \to E$ the difference quotient*

$$(t,s) \mapsto \begin{cases} \dfrac{c(t)-c(s)}{t-s} & \text{for } t \neq s \\ c'(t) & \text{for } t = s \end{cases}$$

is a smooth mapping $\mathbb{R}^2 \to E$.

Proof. By (2.5) we have $f : (t,s) \mapsto \frac{c(t)-c(s)}{t-s} = \int_0^1 c'(s+r(t-s))dr$, and by (3.15) it is smooth $\mathbb{R}^2 \to \widehat{E}$. The left hand side has values in E, and for $t \neq s$ this is also true for all iterated directional derivatives. It remains to consider the derivatives for $t = s$. The iterated directional derivatives are given by (3.15) as

$$\begin{aligned} d_v^p f(t,s) &= d_v^p \int_0^1 c'(s+r(t-s))\,dr \\ &= \int_0^1 d_v^p c'(s+r(t-s))\,dr, \end{aligned}$$

where d_v acts on the (t,s)-variable. The later integrand is for $t = s$ just a linear combination of derivatives of c which are independent of r, hence $d_v^p f(t,s) \in E$. By (3.10) the mapping f is smooth into E. □

3.17. Definition. Spaces of linear mappings. Let $L(E,F)$ denote the *space of all bounded* (equivalently smooth by (2.11)) *linear mappings* from E to F. It is a closed linear subspace of $C^\infty(E,F)$ since f is linear if and only if for all x, $y \in E$ and $\lambda \in \mathbb{R}$ we have $(\mathrm{ev}_x + \lambda\, \mathrm{ev}_y - \mathrm{ev}_{x+\lambda y})f = 0$. We equip it with this topology and linear structure.

So a mapping $f : U \to L(E,F)$ is smooth if and only if the composite mapping $U \xrightarrow{f} L(E,F) \to C^\infty(E,F)$ is smooth.

3.18. Theorem. Chain rule. *Let E and F be locally convex spaces, and let $U \subseteq E$ be c^∞-open. Then the differentiation operator*

$$d : C^\infty(U,F) \to C^\infty(U, L(E,F)),$$
$$df(x)v := \lim_{t\to 0} \frac{f(x+tv)-f(x)}{t},$$

exists, is linear and bounded (smooth). Also the chain rule holds:

$$d(f\circ g)(x)v = df(g(x))dg(x)v.$$

Proof. Since $t \mapsto x+tv$ is a smooth curve we know that $d^{\wedge\wedge} : C^\infty(U,F)\times U \times E \to F$ exists. We want to show that it is smooth, so let $(f,x,v) : \mathbb{R} \to C^\infty(U,F)\times U\times E$ be a smooth curve. Then

$$d^{\wedge\wedge}(f(t),x(t),v(t)) = \lim_{s\to 0}\frac{f(t)(x(t)+sv(t)) - f(t)(x(t))}{s} = \partial_2 h(t,0),$$

which is smooth in t, where the smooth mapping $h : \mathbb{R}^2 \to F$ is given by $(t,s) \mapsto f^{\wedge}(t, x(t)+sv(t))$. By cartesian closedness (3.12) the mapping $d^{\wedge} : C^\infty(U,F)\times U \to C^\infty(E,F)$ is smooth.

Now we show that this mapping has values in the subspace $L(E,F)$: $d^{\wedge}(f,x)$ is obviously homogeneous. It is additive, because we may consider the smooth mapping $(t,s) \mapsto f(x+tv+sw)$ and compute as follows, using (3.14).

$$\begin{aligned} df(x)(v+w) &= \tfrac{\partial}{\partial t}\big|_0 f(x+t(v+w)) \\ &= \tfrac{\partial}{\partial t}\big|_0 f(x+tv+0w) + \tfrac{\partial}{\partial t}\big|_0 f(x+0v+tw) = df(x)v + df(x)w. \end{aligned}$$

So we see that $d^{\wedge} : C^\infty(U,F)\times U \to L(E,F)$ is smooth, and the mapping $d : C^\infty(U,F) \to C^\infty(U,L(E,F))$ is smooth by (3.12) and obviously linear.

We first prove the chain rule for a smooth curve c instead of g. We have to show that the curve

$$t \mapsto \begin{cases} \frac{f(c(t))-f(c(0))}{t} & \text{for } t \neq 0 \\ df(c(0)).c'(0) & \text{for } t = 0 \end{cases}$$

is continuous at 0. It can be rewritten as $t \mapsto \int_0^1 df(c(0)+s(c(t)-c(0))).c_1(t)\,ds$, where c_1 is the smooth curve given by

$$t \mapsto \begin{cases} \frac{c(t)-c(0)}{t} & \text{for } t \neq 0 \\ c'(0) & \text{for } t = 0 \end{cases}.$$

Since $h : \mathbb{R}^2 \to U \times E$ given by

$$(t,s) \mapsto (c(0) + s(c(t) - c(0)), c_1(t))$$

is smooth, the map $t \mapsto \Big(s \mapsto df(c(0) + s(c(t) - c(0))).c_1(t)\Big)$ is smooth $\mathbb{R} \to C^\infty(\mathbb{R}, F)$, and hence $t \mapsto \int_0^1 df(c(0) + s(c(t) - c(0))).c_1(t)\, ds$ is smooth, and hence continuous.

For general g we have

$$\begin{aligned} d(f \circ g)(x)(v) &= \tfrac{\partial}{\partial t}\big|_0 (f \circ g)(x + tv) = (df)(g(x + 0v))(\tfrac{\partial}{\partial t}\big|_0 (g(x + tv))) \\ &= (df)(g(x))(dg(x)(v)). \quad \square \end{aligned}$$

3.19. Lemma. *Two locally convex spaces are locally diffeomorphic if and only if they are linearly diffeomorphic.*

Any smooth and 1-homogeneous mapping is linear.

Proof. By the chain rule the derivatives at corresponding points give the linear diffeomorphisms.

For a 1-homogeneous mapping f one has $df(0)v = \frac{\partial}{\partial t}\big|_0 f(tv) = f(v)$, and this is linear in v. $\square$

4. The c^∞-Topology

4.1. Definition. *A locally convex vector space E is called bornological if and only if the following equivalent conditions are satisfied:*

1. *(1) For any locally convex vector space F any bounded linear mapping $T : E \to F$ is continuous; it is sufficient to know this for all Banach spaces F.*
2. *(2) Every bounded seminorm on E is continuous.*
3. *(3) Every absolutely convex bornivorous subset is a 0-neighborhood.*

A *radial subset* U (i.e. $[0,1]U \subseteq U$) of a locally convex space E is called *bornivorous* if it *absorbs* each bounded set, i.e. for every bounded B there exists $r > 0$ such that $[0,r]U \supseteq B$.

Proof. ($3 \Rightarrow 2$), since for $a > 0$ the inverse images under bounded seminorms of intervals $(-\infty, a)$ are absolutely convex and bornivorous. In fact, let B be bounded and $a > 0$. Then by assumption $p(B)$ is bounded, and so there exists a $C > 0$ with $p(B) \subseteq C \cdot (-\infty, a)$. Hence, $B \subseteq C \cdot p^{-1}(-\infty, a)$.

($2 \Rightarrow 1$), since $p \circ T$ is a bounded seminorm, for every continuous seminorm on F.

($2 \Rightarrow 3$), since the Minkowski-functional p generated by an absolutely convex bornivorous subset is a bounded seminorm.

($1 \Rightarrow 2$) Since the canonical projection $T : E \to E/\ker p$ is bounded, for any bounded seminorm p, it is by assumption continuous. Hence, $p = \tilde{p} \circ T$ is continuous, where $\tilde{p}$ denotes the canonical norm on $E/\ker p$ induced from p. $\square$

4.2. Lemma. Bornologification. *The bornologification E_{born} of a locally convex space can be described in the following equivalent ways:*

(1) *It is the finest locally convex structure having the same bounded sets;*
(2) *It is the final locally convex structure with respect to the inclusions $E_B \to E$, where B runs through all bounded (closed) absolutely convex subsets.*

Moreover, E_{born} is bornological. For any locally convex vector space F the continuous linear mappings $E_{\text{born}} \to F$ are exactly the bounded linear mappings $E \to F$. The continuous seminorms on E_{born} are exactly the bounded seminorms of E. An absolutely convex set is a 0-neighborhood in E_{born} if and only if it is bornivorous, i.e. absorbs bounded sets.

Proof. Let E_{born} be the vector space E supplied with the finest locally convex structure having the same bounded sets as E.

($\Uparrow$) Since all bounded absolutely convex sets B in E are bounded in E_{born}, the inclusions $E_B \to E_{\text{born}}$ are bounded and hence continuous. Thus, the final structure on E induced by the inclusions $E_B \to E$ is finer than the structure of E_{born}.

($\Downarrow$) Since every bounded subset of E is contained in some absolutely convex bounded set $B \subseteq E$ it has to be bounded in the final structure given by all inclusions $E_B \to E$. Hence, this final structure has exactly the same bounded sets as E, and we have equality between the final structure and that of E_{born}.

A seminorm p on E is bounded, if and only if $p(B)$ is bounded for all bounded B, and this is exactly the case if $p|_{E_B}$ is a bounded (=continuous) seminorm on E_B for all B, or equivalently that p is a continuous seminorm for the final structure E_{born} on E induced by the inclusions $E_B \to E$.

As a consequence, all bounded seminorms on E_{born} are continuous, and hence E_{born} is bornological.

An absolutely convex subset U is a 0-neighborhood for the final structure induced by $E_B \to E$ if and only if $U \cap E_B$ is a 0-neighborhood, or equivalently if U absorbs B, for all bounded absolutely convex B, i.e. U is bornivorous. All other assertions follow from (4.1). □

4.3. Corollary. Bounded seminorms. *For a seminorm p and a sequence $\mu_n \to \infty$ the following statements are equivalent:*

(1) *p is bounded;*
(2) *p is bounded on compact sets;*
(3) *p is bounded on M-converging sequences;*
(4) *p is bounded on μ-converging sequences;*
(5) *p is bounded on images of bounded intervals under $\mathcal{L}\text{ip}^k$-curves (for fixed $0 \le k \le \infty$).*

The corresponding statement for subsets of E is the following. For a radial subset $U \subseteq E$ (i.e., $[0,1] \cdot U \subseteq U$) the following properties are equivalent:

(1) *U is bornivorous.*

(2) *For all absolutely convex bounded sets B, the trace $U \cap E_B$ is a 0-neighborhood in E_B.*
(3) *U absorbs all compact subsets in E.*
(4) *U absorbs all Mackey convergent sequences.*
(4') *U absorbs all sequences converging Mackey to 0.*
(5) *U absorbs all μ-convergent sequences (for a fixed μ).*
(5') *U absorbs all sequences which are μ-converging to 0.*
(6) *U absorbs the images of bounded sets under $\mathcal{L}\mathrm{ip}^k$-curves (for a fixed $0 \le k \le \infty$).*

Proof. We prove the statement on radial subsets, for seminorms p it then follows by considering the radial set $U := \{x \in E : p(x) \le 1\}$ and using the equality $K \cdot U = \{x \in E : p(x) \le K\}$.

(1) $\Leftrightarrow$ (2) $\Rightarrow$ (3) $\Rightarrow$ (4) $\Rightarrow$ (5) $\Rightarrow$ (5'), (4) $\Rightarrow$ (4'), (3) $\Rightarrow$ (6), (4') $\Rightarrow$ (5'), are trivial.

(6) $\Rightarrow$ (5') Suppose that (x_n) is μ-converging to x but is not absorbed by U. Then for each $m \in \mathbb{N}$ there is an n_m with $x_{n_m} \notin mU$ and clearly we may suppose that $1/\mu_{n_m}$ is fast falling. The sequence $(x_{n_m})_m$ is then fast falling and lies on some compact part of a smooth curve by the special curve lemma (2.8). The set U absorbs this by (6), a contradiction.

(5') $\Rightarrow$ (1) Suppose U does not absorb some bounded B. Hence, there are $b_n \in B$ with $b_n \notin \mu_n^2 U$. However, $\frac{b_n}{\mu_n}$ is μ-convergent to 0, so it is contained in KU for some $K > 0$. Equivalently, $b_n \in \mu_n KU \subseteq \mu_n^2 U$ for all $\mu_n \ge K$, which gives a contradiction. □

4.4. Corollary. Bornologification as locally convex-ification. *The bornologification of E is the finest locally convex topology with one (hence all) of the following properties:*

(1) *It has the same bounded sets as E.*
(2) *It has the same Mackey converging sequences as E.*
(3) *It has the same μ-converging sequences as E (for some fixed μ).*
(4) *It has the same $\mathcal{L}\mathrm{ip}^k$-curves as E (for some fixed $0 \le k \le \infty$).*
(5) *It has the same bounded linear mappings from E into arbitrary locally convex spaces.*
(6) *It has the same continuous linear mappings from normed spaces into E.*

Proof. Since the bornologification has the same bounded sets as the original topology, the other objects are also the same: they depend only on the bornology – this would not be true for compact sets. Conversely, we consider a topology τ which has for one of the above mentioned types the same objects as the original one. Then τ has by (4.3) the same bornivorous absolutely convex subsets as the original one. Hence, any 0-neighborhood of τ has to be bornivorous for the original topology, and hence is a 0-neighborhood of the bornologification of the original topology. □

4.5. Lemma. *Let E be a bornological locally convex vector space, $U \subseteq E$ a convex subset. Then U is open for the locally convex topology of E if and only if U is open*

for the c^∞-topology.
Furthermore, an absolutely convex subset U of E is a 0-neighborhood for the locally convex topology if and only if it is so for the c^∞-topology.

Proof. ($\Rightarrow$) The c^∞-topology is finer than the locally convex topology, cf. (4.2).

($\Leftarrow$) Let first U be an absolutely convex 0-neighborhood for the c^∞-topology. Hence, U absorbs Mackey-0-sequences. By (4.1.3) we have to show that U is bornivorous, in order to obtain that U is a 0-neighborhood for the locally convex topology. But this follows immediately from (4.3).

Let now U be convex and c^∞-open, let $x \in U$ be arbitrary. We consider the c^∞-open absolutely convex set $W := (U - x) \cap (x - U)$ which is a 0-neighborhood of the locally convex topology by the argument above. Then $x \in W + x \subseteq U$. So U is open in the locally convex topology. □

4.6. Corollary. *The bornologification of a locally convex space E is the finest locally convex topology coarser than the c^∞-topology on E.* □

4.7. In (2.12) we defined the c^∞-topology on an arbitrary locally convex space E as the final topology with respect to the smooth curves $c : \mathbb{R} \to E$. Now we will compare the c^∞-topology with other refinements of a given locally convex topology. We first specify those refinements.

Definition. Let E be a locally convex vector space.

(i) We denote by kE the *Kelley-fication* of the locally convex topology of E, i.e. the vector space E together with the final topology induced by the inclusions of the subsets being compact for the locally convex topology.

(ii) We denote by sE the vector space E with the final topology induced by the curves being continuous for the locally convex topology, or equivalently the sequences $\mathbb{N}_\infty \to E$ converging in the locally convex topology. The equivalence holds since the infinite polygon through a converging sequence can be continuously parameterized by a compact interval.

(iii) We recall that by $c^\infty E$ we denote the vector space E with its c^∞-topology, i.e. the final topology induced by the smooth curves.

Using that smooth curves are continuous and that converging sequences $\mathbb{N}_\infty \to E$ have compact images, the following identities are continuous: $c^\infty E \to sE \to kE \to E$.

If the locally convex topology of E coincides with the topology of $c^\infty E$, resp. sE, resp. kE then we call E smoothly generated, resp. sequentially generated, resp. compactly generated.

4.8. Example. *On $E = \mathbb{R}^J$ all the refinements of the locally convex topology described in (4.7) above are different, i.e. $c^\infty E \neq sE \neq kE \neq E$, provided the cardinality of the index set J is at least that of the continuum.*

Proof. It is enough to show this for J equipotent to the continuum, since $\mathbb{R}^{J_1}$ is a direct summand in $\mathbb{R}^{J_2}$ for $J_1 \subseteq J_2$.

($c^\infty E \neq sE$) We may take as index set J the set c_0 of all real sequences converging to 0. Define a sequence (x^n) in E by $(x^n)_j := j_n$. Since every $j \in J$ is a 0-sequence we conclude that the x^n converge to 0 in the locally convex topology of the product, hence also in sE. Assume now that the x^n converge towards 0 in $c^\infty E$. Then by (1.8) some subsequence converges Mackey to 0. Thus, there exists an unbounded sequence of reals λ_n with $\{\lambda_n x^n : n \in \mathbb{N}\}$ bounded. Let j be a 0-sequence with $\{j_n \lambda_n : n \in \mathbb{N}\}$ unbounded (e.g. $(j_n)^{-2} := 1 + \max\{|\lambda_k| : k \leq n\}$). Then the j-th coordinate $j_n \lambda_n$ of $\lambda_n x^n$ is not bounded with respect to n, contradiction.

($sE \neq kE$) Consider in E the subset

$$A := \Big\{x \in \{0,1\}^J : x_j = 1 \text{ for at most countably many } j \in J\Big\}.$$

It is clearly closed with respect to the converging sequences, hence closed in sE. But it is not closed in kE since it is dense in the compact set $\{0,1\}^J$.

($kE \neq E$) Consider in E the subsets

$$A_n := \big\{x \in E : |x_j| < n \text{ for at most } n \text{ many } j \in J\big\}.$$

Each A_n is closed in E since its complement is the union of the open sets $\{x \in E : |x_j| < n \text{ for all } j \in J_o\}$ where J_o runs through all subsets of J with $n+1$ elements. We show that the union $A := \bigcup_{n\in\mathbb{N}} A_n$ is closed in kE. So let K be a compact subset of E; then $K \subseteq \prod \operatorname{pr}_j(K)$, and each $\operatorname{pr}_j(K)$ is compact, hence bounded in $\mathbb{R}$. Since the family $(\{j \in J : \operatorname{pr}_j(K) \subseteq [-n,n]\})_{n\in\mathbb{N}}$ covers J, there has to exist an $N \in \mathbb{N}$ and infinitely many $j \in J$ with $\operatorname{pr}_j(K) \subseteq [-N,N]$. Thus $K \cap A_n = \emptyset$ for all $n > N$, and hence, $A \cap K = \bigcup_{n\in\mathbb{N}} A_n \cap K$ is closed. Nevertheless, A is not closed in E, since 0 is in $\bar{A}$ but not in A. □

4.9. $\mathbf{c^\infty}$-convergent sequences. By (2.13) every M-convergent sequence gives a continuous mapping $\mathbb{N}_\infty \to c^\infty E$ and hence converges in $c^\infty E$. Conversely, a sequence converging in $c^\infty E$ is not necessarily Mackey convergent, see [Frölicher, Kriegl, 1985]. However, one has the following result.

Lemma. *A sequence (x_n) is convergent to x in the c^∞-topology if and only if every subsequence has a subsequence which is Mackey convergent to x.*

Proof. ($\Leftarrow$) is true for any topological convergence. In fact if x_n would not converge to x, then there would be a neighborhood U of x and a subsequence of x_n which lies outside of U and hence cannot have a subsequence converging to x.

($\Rightarrow$) It is enough to show that (x_n) has a subsequence which converges Mackey to x, since every subsequence of a c^∞-convergent sequence is clearly c^∞-convergent to the same limit. Without loss of generality we may assume that $x \notin A := \{x_n : n \in \mathbb{N}\}$. Hence, A cannot be c^∞-closed, and thus there is a sequence $n_k \in \mathbb{N}$ such that (x_{n_k}) converges Mackey to some point $x' \notin A$. The set $\{n_k : k \in \mathbb{N}\}$ cannot be bounded, and hence we may assume that the n_k are strictly increasing by passing to a subsequence. But then (x_{n_k}) is a subsequence of (x_n) which converges in $c^\infty E$ to x and Mackey to x' hence also in $c^\infty E$. Thus $x' = x$. □

Remark. A consequence of this lemma is, that there is no topology having as convergent sequences exactly the M-convergent ones, since this topology obviously would have to be coarser than the c^∞-topology.

One can use this lemma also to show that the c^∞-topology on a locally convex vector space gives a so called *arc-generated vector space*. See [Frölicher, Kriegl, 1988, 2.3.9 and 2.3.13] for a discussion of this.

Let us now describe several important situations where at least some of these topologies coincide. For the proof we will need the following

4.10. Lemma. [Averbukh, Smolyanov, 1968] *For any locally convex space E the following statements are equivalent:*

(1) *The sequential closure of any subset is formed by all limits of sequences in the subset.*
(2) *For any given double sequence $(x_{n,k})$ in E with $x_{n,k}$ convergent to some x_k for $n \to \infty$ and k fixed and x_k convergent to some x, there are strictly increasing sequences $i \mapsto n(i)$ and $i \mapsto k(i)$ with $x_{n(i),k(i)} \to x$ for $i \to \infty$.*

Proof. ($1\Rightarrow 2$) Take an $a_0 \in E$ different from $k\cdot(x_{n+k,k}-x)$ and from $k\cdot(x_k - x)$ for all k and n. Define $A := \{a_{n,k} := x_{n+k,k} - \frac{1}{k}\cdot a_0 : n,k \in \mathbb{N}\}$. Then x is in the sequential closure of A, since $x_{n+k,k} - \frac{1}{k}\cdot a_0$ converges to $x_k - \frac{1}{k}\cdot a_0$ as $n\to\infty$, and $x_k - \frac{1}{k}\cdot a_0$ converges to $x - 0 = x$ as $k \to \infty$. Hence, by (1) there has to exist a sequence $i \mapsto (n_i, k_i)$ with a_{n_i,k_i} convergent to x. By passing to a subsequence we may suppose that $i\mapsto k_i$ and $i \mapsto n_i$ are increasing. Assume that $i \mapsto k_i$ is bounded, hence finally constant. Then a subsequence $x_{n_i+k_i,k_i} - \frac{1}{k_i}\cdot a_0$ is converging to $x_k - \frac{1}{k}\cdot a_0 \neq x$ if $i\mapsto n_i$ is unbounded, and to $x_{n+k,k} - \frac{1}{k}\cdot a_0 \neq x$ if $i \mapsto n_i$ is bounded, which both yield a contradiction. Thus, $i\mapsto k_i$ can be chosen strictly increasing. But then

$$x_{n_i+k_i,k_i} = a_{n_i,k_i} + \tfrac{1}{k_i}a_0 \to x.$$

(1) $\Leftarrow$ (2) is obvious. □

4.11. Theorem. *For any bornological vector space E the following implications hold:*

(1) *$c^\infty E = E$ provided the closure of subsets in E is formed by all limits of sequences in the subset; hence in particular if E is metrizable.*
(2) *$c^\infty E = E$ provided E is the strong dual of a Fréchet Schwartz space;*
(3) *$c^\infty E = kE$ provided E is the strict inductive limit of a sequence of Fréchet spaces.*
(4) *$c^\infty E = sE$ provided E satisfies the M-convergence condition, i.e. every sequence converging in the locally convex topology is M-convergent.*
(5) *$sE = E$ provided E is the strong dual of a Fréchet Montel space;*

Proof. (1) Using the lemma (4.10) above one obtains that the closure and the sequential closure coincide, hence $sE = E$. It remains to show that $sE \to c^\infty E$ is (sequentially) continuous. So suppose a sequence converging to x is given, and let

(x_n) be an arbitrary subsequence. Then $x_{n,k} := k(x_n - x) \to k \cdot 0 = 0$ for $n \to \infty$, and hence by lemma (4.10) there are subsequences k_i, n_i with $k_i \cdot (x_{n_i} - x) \to 0$, i.e. $i \mapsto x_{n_i}$ is M-convergent to x. Thus, the original sequence converges in $c^\infty E$ by (4.9).

(3) Let E be the strict inductive limit of the Fréchet spaces E_n. By (52.8) every E_n carries the trace topology of E, hence is closed in E, and every bounded subset of E is contained in some E_n. Thus, every compact subset of E is contained as compact subset in some E_n. Since E_n is a Fréchet space such a subset is even compact in $c^\infty E_n$ and hence compact in $c^\infty E$. Thus, the identity $kE \to c^\infty E$ is continuous.

(4) is valid, since the M-closure topology is the final one induced by the M-converging sequences.

(5) Let E be the dual of any Fréchet Montel space F. By (52.29) E is bornological. First we show that $kE = sE$. Let $K \subseteq E = F'$ be compact for the locally convex topology. Then K is bounded, hence equicontinuous since F is barrelled by (52.25). Since F is separable by (52.27) the set K is metrizable in the weak topology $\sigma(E, F)$ by (52.21). By (52.20) this weak topology coincides with the topology of uniform convergence on precompact subsets of F. Since F is a Montel space, this latter topology is the strong one, and even the bornological one, as remarked at the beginning. Thus, the (metrizable) topology on K is the initial one induced by the converging sequences. Hence, the identity $kE \to sE$ is continuous, and therefore $sE = kE$.

It remains to show $kE = E$. Since F is Montel the locally convex topology of the strong dual coincides with the topology of uniform convergence on precompact subsets of F. Since F is metrizable this topology coincides with the so-called equicontinuous weak*-topology, cf. (52.22), which is the final topology induced by the inclusions of the equicontinuous subsets. These subsets are by the Alaoğlu-Bourbaki theorem (52.20) relatively compact in the topology of uniform convergence on precompact subsets. Thus, the locally convex topology of E is compactly generated.

Proof of (2) By (5), and since Fréchet Schwartz spaces are Montel by (52.24), we have $sE = E$ and it remains to show that $c^\infty E = sE$. So let (x_n) be a sequence converging to 0 in E. Then the set $\{x_n : n \in \mathbb{N}\}$ is relatively compact, and by [Frölicher, Kriegl, 1988, 4.4.39] it is relatively compact in some Banach space E_B. Hence, at least a subsequence has to be convergent in E_B. Clearly its Mackey limit has to be 0. This shows that (x_n) converges to 0 in $c^\infty E$, and hence $c^\infty E = sE$. One can even show that E satisfies the Mackey convergence condition, see (52.28). □

4.12. Example. We give now a non-metrizable example to which (4.11.1) applies. Let E denote the subspace of $\mathbb{R}^J$ of all sequences with countable support. Then the closure of subsets of E is given by all limits of sequences in the subset, but for non-countable J the space E is not metrizable. This was proved in [Balanzat, 1960].

4.13. Remark. The conditions (4.11.1) and (4.11.2) are rather disjoint since every locally convex space, that has a countable basis of its bornology and for which the *sequential adherence* of subsets (the set of all limits of sequences in it) is sequentially closed, is normable as the following proposition shows:

Proposition. *Let E be a non-normable bornological locally convex space that has a countable basis of its bornology. Then there exists a subset of E whose sequential adherence is not sequentially closed.*

Proof. Let $\{B_k : k \in \mathbb{N}_0\}$ be an increasing basis of the von Neumann bornology with $B_0 = \{0\}$. Since E is non-normable we may assume that B_k does not absorb B_{k+1} for all k. Now choose $b_{n,k} \in \frac{1}{n}B_{k+1}$ with $b_{n,k} \notin B_k$. We consider the double sequence $\{b_{k,0} - b_{n,k} : n, k \geq 1\}$. For fixed k the sequence $b_{n,k}$ converges by construction (in $E_{B_{k+1}}$) to 0 for $n \to \infty$. Thus, $b_{k,0} - 0$ is the limit of the sequence $b_{k,0} - b_{n,k}$ for $n \to \infty$, and $b_{k,0}$ converges to 0 for $k \to \infty$. Suppose $b_{k(i),0} - b_{n(i),k(i)}$ converges to 0. So it has to be bounded, thus there must be an $N \in \mathbb{N}$ with $B_1 - \{b_{k(i),0} - b_{n(i),k(i)} : i \in \mathbb{N}\} \subseteq B_N$. Hence, $b_{n(i),k(i)} = b_{k(i),0} - (b_{k(i),0} - b_{n(i),k(i)}) \in B_N$, i.e. $k(i) < N$. This contradicts (4.10.2). □

4.14. Lemma. *Let U be a c^∞-open subset of a locally convex space, let $\mu_n \to \infty$ be a real sequence, and let $f : U \to F$ be a mapping which is bounded on each μ-converging sequence in U. Then f is bounded on every bornologically compact subset (i.e. compact in some E_B) of U.*

Proof. By composing with linear functionals we may assume that $F = \mathbb{R}$. Let $K \subseteq E_B \cap U$ be compact in E_B for some bounded absolutely convex set B. Assume that $f(K)$ is not bounded. So there is a sequence (x_n) in K with $|f(x_n)| \to \infty$. Since K is compact in the normed space E_B we may assume that (x_n) converges to $x \in K$. By passing to a subsequence we may even assume that (x_n) is μ-converging. Contradiction. □

4.15. Lemma. *Let U be c^∞-open in $E \times \mathbb{R}$ and $K \subseteq \mathbb{R}$ be compact. Then $U_0 := \{x \in E : \{x\} \times K \subseteq U\}$ is c^∞-open in E.*

Proof. Let $x : \mathbb{R} \to E$ be a smooth curve in E with $x(0) \in U_0$, i.e. $(x(0), t) \in U$ for all $t \in K$. We have to show that $x(s) \in U_0$ for all s near 0. So consider the smooth map $x \times \mathbb{R} : \mathbb{R} \times \mathbb{R} \to E \times \mathbb{R}$. By assumption $(x \times \mathbb{R})^{-1}(U)$ is open in $c^\infty(\mathbb{R}^2) = \mathbb{R}^2$. It contains the compact set $\{0\} \times K$ and hence also a $W \times K$ for some neighborhood W of 0 in $\mathbb{R}$. But this amounts in saying that $x(W) \subseteq U_0$. □

4.16. The c^∞-topology of a product. Consider the product $E \times F$ of two locally convex vector spaces. Since the projections onto the factors are linear and continuous, and hence smooth, we always have that the identity mapping $c^\infty(E \times F) \to c^\infty(E) \times c^\infty(F)$ is continuous. It is not always a homeomorphism: Just take a bounded separately continuous bilinear functional, which is not continuous (like the evaluation map) on a product of spaces where the c^∞-topology is the bornological topology.

However, if one of the factors is finite dimensional the product is well behaved:

Corollary. *For any locally convex space E the c^∞-topology of $E\times\mathbb{R}^n$ is the product topology of the c^∞-topologies of the two factors, so that we have $c^\infty(E\times\mathbb{R}^n) = c^\infty(E)\times\mathbb{R}^n$.*

Proof. This follows recursively from the special case $E\times\mathbb{R}$, for which we can proceed as follows. Take a c^∞-open neighborhood U of some point $(x,t)\in E\times\mathbb{R}$. Since the inclusion map $s\mapsto(x,s)$ from $\mathbb{R}$ into $E\times\mathbb{R}$ is continuous and affine, the inverse image of U in $\mathbb{R}$ is an open neighborhood of t. Let's take a smaller compact neighborhood K of t. Then by the previous lemma $U_0 := \{y\in E : \{y\}\times K\subseteq U\}$ is a c^∞-open neighborhood of x, and hence $U_0\times K^o$ is a neighborhood of (x,t) in $c^\infty(E)\times\mathbb{R}$, what was to be shown. $\square$

4.17. Lemma. *Let U be c^∞-open in a locally convex space and $x\in U$. Then the star $\operatorname{st}_x(U) := \{x+v : x+\lambda v\in U$ for all $|\lambda|\le 1\}$ with center x in U is again c^∞-open.*

Proof. Let $c:\mathbb{R}\to E$ be a smooth curve with $c(0)\in\operatorname{st}_x(U)$. The smooth mapping $f:(t,s)\mapsto(1-s)x+sc(t)$ maps $\{0\}\times\{s:|s|\le 1\}$ into U. So there exists $\delta>0$ with $f(\{(t,s):|t|<\delta,|s|\le 1\})\subseteq U$. Thus, $c(t)\in\operatorname{st}_x(U)$ for $|t|<\delta$. $\square$

4.18. Lemma. *The (absolutely) convex hull of a c^∞-open set is again c^∞-open.*

Proof. Let U be c^∞-open in a locally convex vector space E.
For each $x\in U$ the set

$$U_x := \{x+t(y-x) : t\in[0,1], y\in U\} = U\cup\bigcup_{0<t\le 1}(x+t(U-x))$$

is c^∞-open. The convex hull can be constructed by applying n times the operation $U\mapsto\bigcup_{x\in U}U_x$ and taking the union over all $n\in\mathbb{N}$, which respects c^∞-openness.

The absolutely convex hull can be obtained by forming first $\{\lambda : |\lambda|=1\}.U = \bigcup_{|\lambda|=1}\lambda U$ which is c^∞-open, and then forming the convex hull. $\square$

4.19. Corollary. *Let E be a bornological convenient vector space containing a nonempty c^∞-open subset which is either locally compact or metrizable in the c^∞-topology. Then the c^∞-topology on E is locally convex. In the first case E is finite dimensional, in the second case E is a Fréchet space.*

Proof. Let $U\subseteq E$ be a c^∞-open metrizable subset. We may assume that $0\in U$. Then there exists a countable neighborhood basis of 0 in U consisting of c^∞-open sets. This is also a neighborhood basis of 0 for the c^∞-topology of E. We take the absolutely convex hulls of these open sets, which are again c^∞-open by (4.18), and obtain by (4.5) a countable neighborhood basis for the bornologification of the locally convex topology, so the latter is metrizable and Fréchet, and by (4.11) it equals the c^∞-topology.

If U is locally compact in the c^∞-topology we may find a c^∞-open neighborhood V of 0 with compact closure $\overline{V}$ in the c^∞-topology. By lemma (4.18) the absolutely convex hull of V is also c^∞-open, and by (4.5) it is also open in the bornologification E_{born} of E. The set $\overline{V}$ is then also compact in E_{born}, hence precompact. So the absolutely convex hull of $\overline{V}$ is also precompact by (52.6). Therefore, the absolutely convex hull of V is a precompact neighborhood of 0 in E_{born}, thus E is finite dimensional by (52.5). So $E_{\text{born}} = c^\infty(E)$. □

Now we describe classes of spaces where $c^\infty E \neq E$ or where $c^\infty E$ is not even a topological vector space. Finally, we give an example where the c^∞-topology is not completely regular. We begin with the relationship between the c^∞-topology and the locally convex topology on locally convex vector spaces.

4.20. Proposition. *Let E and F be bornological locally convex vector spaces. If there exists a bilinear smooth mapping $m : E \times F \to \mathbb{R}$ that is not continuous with respect to the locally convex topologies, then $c^\infty(E \times F)$ is not a topological vector space.*

We shall show in lemma (5.5) below that multilinear mappings are smooth if and only if they are bounded.

Proof. Suppose that addition $c^\infty(E\times F)\times c^\infty(E\times F) \to c^\infty(E\times F)$ is continuous with respect to the product topology. Using the continuous inclusions $c^\infty E \to c^\infty(E \times F)$ and $c^\infty F \to c^\infty(E \times F)$ we can write m as composite of continuous maps as follows: $c^\infty E \times c^\infty F \to c^\infty(E \times F) \times c^\infty(E \times F) \xrightarrow{+} c^\infty(E \times F) \xrightarrow{m} \mathbb{R}$. Thus, for every $\varepsilon > 0$ there are 0-neighborhoods U and V with respect to the c^∞-topology such that $m(U \times V) \subseteq (-\varepsilon, \varepsilon)$. Then also $m(\langle U\rangle \times \langle V\rangle) \subseteq (-\varepsilon,\varepsilon)$ where $\langle\ \ \rangle$ denotes the absolutely convex hull. By (4.5) one concludes that m is continuous with respect to the locally convex topology, a contradiction. □

4.21. Corollary. *Let E be a non-normable bornological locally convex space. Then $c^\infty(E \times E')$ is not a topological vector space.*

Proof. By (4.20) it is enough to show that $\text{ev} : E \times E' \to \mathbb{R}$ is not continuous for the bornological topologies on E and E'; if it were so there was be a neighborhood U of 0 in E and a neighborhood U' of 0 in E' such that $\text{ev}(U \times U') \subseteq [-1, 1]$. Since U' is absorbing, U is scalarwise bounded, hence a bounded neighborhood. Thus, E is normable. □

4.22. Remark. In particular, for a Fréchet Schwartz space E and its dual E' we have $c^\infty(E\times E') \neq c^\infty E\times c^\infty E'$, since by (4.11) we have $c^\infty E = E$ and $c^\infty E' = E'$, so equality would contradict corollary (4.21).

In order to get a large variety of spaces where the c^∞-topology is not a topological vector space topology the next three technical lemmas will be useful.

4.23. Lemma. *Let E be a locally convex vector space. Suppose a double sequence $b_{n,k}$ in E exists which satisfies the following two conditions:*

- (b') *For every sequence $k \mapsto n_k$ the sequence $k \mapsto b_{n_k,k}$ has no accumulation point in $c^\infty E$.*
- (b") *For all k the sequence $n \mapsto b_{n,k}$ converges to 0 in $c^\infty E$.*

Suppose furthermore that a double sequence $c_{n,k}$ in E exists that satisfies the following two conditions:

(c') *For every 0-neighborhood U in $c^\infty E$ there exists some k_0 such that $c_{n,k} \in U$ for all $k \geq k_0$ and all n.*

(c") *For all k the sequence $n \mapsto c_{n,k}$ has no accumulation point in $c^\infty E$.*

Then $c^\infty E$ is not a topological vector space.

Proof. Assume that the addition $c^\infty E \times c^\infty E \to c^\infty E$ is continuous. In this proof convergence is meant always with respect to $c^\infty E$. We may without loss of generality assume that $c_{n,k} \neq 0$ for all n, k, since by (c") we may delete all those $c_{n,k}$ which are equal to 0. Then we consider $A := \{b_{n,k} + \varepsilon_{n,k} c_{n,k} : n, k \in \mathbb{N}\}$ where the $\varepsilon_{n,k} \in \{-1, 1\}$ are chosen in such a way that $0 \notin A$.

We first show that A is closed in the sequentially generated topology $c^\infty E$: Let $b_{n_i,k_i} + \varepsilon_{n_i,k_i} c_{n_i,k_i} \to x$, and assume that (k_i) is unbounded. By passing if necessary to a subsequence we may even assume that $i \mapsto k_i$ is strictly increasing. Then $c_{n_i,k_i} \to 0$ by (c'), hence $b_{n_i,k_i} \to x$ by the assumption that addition is continuous, which is a contradiction to (b'). Thus, (k_i) is bounded, and we may assume it to be constant. Now suppose that (n_i) is unbounded. Then $b_{n_i,k} \to 0$ by (b"), and hence $\varepsilon_{n_i,k} c_{n_i,k} \to x$, and for a subsequence where ε is constant one has $c_{n_i,k} \to \pm x$, which is a contradiction to (c"). Thus, n_i is bounded as well, and we may assume it to be constant. Hence, $x = b_{n,k} + \varepsilon_{n,k} c_{n,k} \in A$.

By the assumed continuity of the addition there exists an open and symmetric 0-neighborhood U in $c^\infty E$ with $U + U \subseteq E \setminus A$. For K sufficiently large and n arbitrary one has $c_{n,K} \in U$ by (c'). For such a fixed K and N sufficiently large $b_{N,K} \in U$ by (b'). Thus, $b_{N,K} + \varepsilon_{N,K} c_{N,K} \notin A$, which is a contradiction. □

Let us now show that many spaces have a double sequence $c_{n,k}$ as in the above lemma.

4.24. Lemma. *Let E be an infinite dimensional metrizable locally convex space. Then a double sequence $c_{n,k}$ subject to the conditions (c') and (c") of* (4.23) *exists.*

Proof. If E is normable we choose a sequence (c_n) in the unit ball without accumulation point and define $c_{n,k} := \frac{1}{k} c_n$. If E is not normable we take a countable increasing family of non-equivalent seminorms p_k generating the locally convex topology, and we choose $c_{n,k}$ with $p_k(c_{n,k}) = \frac{1}{k}$ and $p_{k+1}(c_{n,k}) > n$. □

Next we show that many spaces have a double sequence $b_{n,k}$ as in lemma (4.23).

4.25. Lemma. *Let E be a non-normable bornological locally convex space having a countable basis of its bornology. Then a double sequence $b_{n,k}$ subject to the conditions (b') and (b") of* (2.11) *exists.*

Proof. Let B_n ($n \in \mathbb{N}$) be absolutely convex sets forming an increasing basis of the bornology. Since E is not normable the sets B_n can be chosen such that B_n does not absorb B_{n+1}. Now choose $b_{n,k} \in \frac{1}{n} B_{k+1}$ with $b_{n,k} \notin B_k$. □

Using these lemmas one obtains the

4.26. Proposition. *For the following bornological locally convex spaces the c^∞-topology is not a vector space topology:*

(i) *Every bornological locally convex space that contains as c^∞-closed subspaces an infinite dimensional Fréchet space and a space which is non-normable in the bornological topology and having a countable basis of its bornology.*

(ii) *Every strict inductive limit of a strictly increasing sequence of infinite dimensional Fréchet spaces.*

(iii) *Every product for which at least $2^{\aleph_0}$ many factors are non-zero.*

(iv) *Every coproduct for which at least $2^{\aleph_0}$ many summands are non-zero.*

Proof. (i) follows directly from the last 3 lemmas.

(ii) Let E be the strict inductive limit of the spaces E_n $(n \in \mathbb{N})$. Then E contains the infinite dimensional Fréchet space E_1 as subspace. The subspace generated by points $x_n \in E_{n+1} \setminus E_n$ $(n \in \mathbb{N})$ is bornologically isomorphic to $\mathbb{R}^{(\mathbb{N})}$, hence its bornology has a countable basis. Thus, by (i) we are done.

(iii) Such a product E contains the Fréchet space $\mathbb{R}^{\mathbb{N}}$ as complemented subspace. We want to show that $\mathbb{R}^{(\mathbb{N})}$ is also a subspace of E. For this we may assume that the index set J is $\mathbb{R}^{\mathbb{N}}$ and all factors are equal to $\mathbb{R}$. Now consider the linear subspace E_1 of the product generated by the elements $x^n \in E = \mathbb{R}^{\mathbb{N}}$, where $(x^n)_j := j(n)$ for every $j \in J = \mathbb{R}^{\mathbb{N}}$. The linear map $\mathbb{R}^{(\mathbb{N})} \to E_1 \subseteq E$ that maps the n-th unit vector to x^n is injective, since for a given finite linear combination $\sum t_n x^n = 0$ the j-th coordinate for $j(n) := \operatorname{sign}(t_n)$ equals $\sum |t_n|$. It is a morphism since $\mathbb{R}^{(\mathbb{N})}$ carries the finest structure. So it remains to show that it is a bornological embedding. We have to show that any bounded $B \subseteq E_1$ is contained in a subspace generated by finitely many x^n. Otherwise, there would exist a strictly increasing sequence (n_k) and $b^k = \sum_{n \le n_k} t_n^k x^n \in B$ with $t_{n_k}^k \neq 0$. Define an index j recursively by $j(n) := n|t_n^k|^{-1} \cdot \operatorname{sign}\left(\sum_{m<n} t_m^k j(m)\right)$ if $n = n_k$ and $j(n) := 0$ if $n \neq n_k$ for all k. Then the absolute value of the j-th coordinate of b^k evaluates as follows:

$$|(b^k)_j| = \Big|\sum_{n \le n_k} t_n^k j(n)\Big| = \Big|\sum_{n<n_k} t_n^k j(n) + t_{n_k}^k j(n_k)\Big|$$
$$= \Big|\sum_{n<n_k} t_n^k j(n)\Big| + |t_{n_k}^k j(n_k)| \ge |t_{n_k}^k j(n_k)| \ge n_k.$$

Hence, the j-th coordinates of $\{b^k : k \in \mathbb{N}\}$ are unbounded with respect to $k \in \mathbb{N}$, thus B is unbounded.

(iv) We can not apply lemma (4.23) since every double sequence has countable support and hence is contained in the dual $\mathbb{R}^{(A)}$ of a Fréchet Schwartz space $\mathbb{R}^A$ for some countable subset $A \subset J$. It is enough to show (iv) for $\mathbb{R}^{(J)}$ where $J = \mathbb{N} \cup c_0$. Let $A := \{j_n(e_n + e_j) : n \in \mathbb{N}, j \in c_0, j_n \neq 0 \text{ for all } n\}$, where e_n and e_j denote the unit vectors in the corresponding summand. The set A is M-closed, since its intersection with finite subsums is finite. Suppose there exists a symmetric M-open 0-neighborhood U with $U + U \subseteq E \setminus A$. Then for each n there exists a $j_n \neq 0$ with $j_n e_n \in U$. We may assume that $n \mapsto j_n$ converges to 0 and hence defines an element $j \in c_0$. Furthermore, there has to be an $N \in \mathbb{N}$ with $j_N e_j \in U$, thus $j_N(e_N + e_j) \in (U+U) \cap A$, in contradiction to $U + U \subseteq E \setminus A$. □

Remark. A nice and simple example where one either uses (i) or (ii) is $\mathbb{R}^{\mathbb{N}} \oplus \mathbb{R}^{(\mathbb{N})}$. The locally convex topology on both factors coincides with their c^∞-topology (the first being a Fréchet (Schwartz) space, cf. (i) of (4.11), the second as dual of the first, cf. (ii) of (4.11)); but the c^∞-topology on their product is not even a vector space topology.

From (ii) it follows also that each space $C_c^\infty(M, \mathbb{R})$ of smooth functions with compact support on a non-compact separable finite dimensional manifold M has the property, that the c^∞-topology is not a vector space topology.

4.27. Although the c^∞-topology on a convenient vector space is always functionally separated, hence Hausdorff, it is not always completely regular as the following example shows.

Example. The c^∞-topology is not completely regular. *The c^∞-topology of $\mathbb{R}^J$ is not completely regular if the cardinality of J is at least $2^{\aleph_0}$.*

Proof. It is enough to show this for an index set J of cardinality $2^{\aleph_0}$, since the corresponding product is a complemented subspace in every product with larger index set. We prove the theorem by showing that every function $f : \mathbb{R}^J \to \mathbb{R}$ which is continuous for the c^∞-topology is also continuous with respect to the locally convex topology. Hence, the completely regular topology associated to the c^∞-topology is the locally convex topology of E. That these two topologies are different was shown in (4.8). We use the following theorem of [Mazur, 1952]: Let $E_0 := \{x \in \mathbb{R}^J : \operatorname{supp}(x) \text{ is countable}\}$, and let $f : E_0 \to \mathbb{R}$ be sequentially continuous. Then there is some countable subset $A \subset J$ such that $f(x) = f(x_A)$, where in this proof x_A is defined as $x_A(j) := x(j)$ for $j \in A$ and $x_A(j) = 0$ for $j \notin A$. Every sequence which is converging in the locally convex topology of E_0 is contained in a metrizable complemented subspace $\mathbb{R}^A$ for some countable A and therefore is even M-convergent. Thus, this theorem of Mazur remains true if f is assumed to be continuous for the M-closure topology. This generalization follows also from the fact that $c^\infty E_0 = E_0$, cf. (4.12). Now let $f : \mathbb{R}^J \to \mathbb{R}$ be continuous for the c^∞-topology. Then $f|E_0 : E_0 \to \mathbb{R}$ is continuous for the c^∞-topology, and hence there exists a countable set $A_0 \subset J$ such that $f(x) = f(x_{A_0})$ for any $x \in E_0$. We want to show that the same is true for arbitrary $x \in \mathbb{R}^J$. In order to show this we consider for $x \in \mathbb{R}^J$ the map $\varphi_x : 2^J \to \mathbb{R}$ defined by $\varphi_x(A) := f(x_A) - f(x_{A \cap A_0})$ for any $A \subseteq J$, i.e. $A \in 2^J$. For countable A one has $x_A \in E_0$, hence $\varphi_x(A) = 0$. Furthermore, φ_x is sequentially continuous where one considers on 2^J the product topology of the discrete factors $2 = \{0, 1\}$. In order to see this consider a converging sequence of subsets $A_n \to A$, i.e. for every $j \in J$ one has for the characteristic functions $\chi_{A_n}(j) = \chi_A(j)$ for n sufficiently large. Then $\{n(x_{A_n} - x_A) : n \in \mathbb{N}\}$ is bounded in $\mathbb{R}^J$ since for fixed $j \in J$ the j-th coordinate equals 0 for n sufficiently large. Thus, x_{A_n} converges Mackey to x_A, and since f is continuous for the c^∞-topology $\varphi_x(A_n) \to \varphi_x(A)$. Now we can apply another theorem of [Mazur, 1952]:

Any function $f : 2^J \to \mathbb{R}$ that is sequentially continuous and is zero on all countable subsets of J is identically 0, provided the cardinality of J is smaller than the first inaccessible cardinal. Thus, we conclude that $0 = \varphi_x(J) = f(x) - f(x_{A_n})$ for all $x \in \mathbb{R}^J$. Hence, f factors over the metrizable space $\mathbb{R}^{A_0}$ and is therefore continuous for the locally convex topology. □

In general, the trace of the c^∞-topology on a linear subspace is not its c^∞-topology. However, for c^∞-closed subspaces this is true:

4.28. Lemma. Closed embedding lemma. *Let E be a linear c^∞-closed subspace of F. Then the trace of the c^∞-topology of F on E is the c^∞-topology on E*

Proof. Since the inclusion is continuous and hence bounded it is c^∞-continuous. Therefore, it is enough to show that it is closed for the c^∞-topologies. So let $A \subseteq E$ be $c^\infty E$-closed. And let $x_n \in A$ converge Mackey towards x in F. Then $x \in E$, since E is assumed to be c^∞-closed, and hence x_n converges Mackey to x in E. Since A is c^∞-closed in E, we have that $x \in A$. □

We will give an example in (4.33) below which shows that c^∞-closedness of the subspace is essential for this result. Another example will be given in (4.36).

4.29. Theorem. The c^∞-completion. *For any locally convex space E there exists a unique (up to a bounded isomorphism) convenient vector space $\tilde{E}$ and a bounded linear injection $i : E \to \tilde{E}$ with the following universal property:*

> *Each bounded linear mapping $\ell : E \to F$ into a convenient vector space F has a unique bounded extension $\tilde{\ell} : \tilde{E} \to F$ such that $\tilde{\ell} \circ i = \ell$.*

Furthermore, $i(E)$ is dense for the c^∞-topology in $\tilde{E}$.

Proof. Let $\tilde{E}$ be the c^∞-closure of E in the locally convex completion $\widehat{E_{\text{born}}}$ of the bornologification E_{born} of E. The inclusion $i : E \to \tilde{E}$ is bounded (not continuous in general). By (4.28) the c^∞-topology on $\tilde{E}$ is the trace of the c^∞-topology on $\widehat{E_{\text{born}}}$. Hence, $i(E)$ is dense also for the c^∞-topology in $\tilde{E}$.

Using the universal property of the locally convex completion the mapping ℓ has a unique extension $\hat{\ell} : \widehat{E_{\text{born}}} \to \widehat{F}$ into the locally convex completion of F, whose restriction to $\tilde{E}$ has values in F, since F is c^∞-closed in $\widehat{F}$, so it is the desired $\tilde{\ell}$. Uniqueness follows, since $i(E)$ is dense for the c^∞-topology in $\tilde{E}$. □

4.30. Proposition. c^∞-completion via c^∞-dense embeddings. *Let E be c^∞-dense and bornologically embedded into a c^∞-complete locally convex space F. If $E \to F$ has the extension property for bounded linear functionals, then F is bornologically isomorphic to the c^∞-completion of E.*

Proof. We have to show that $E \to F$ has the universal property for extending bounded linear maps T into c^∞-complete locally convex spaces G. Since we are only interested in bounded mappings, we may take the bornologification of G and hence may assume that G is bornological. Consider the following diagram

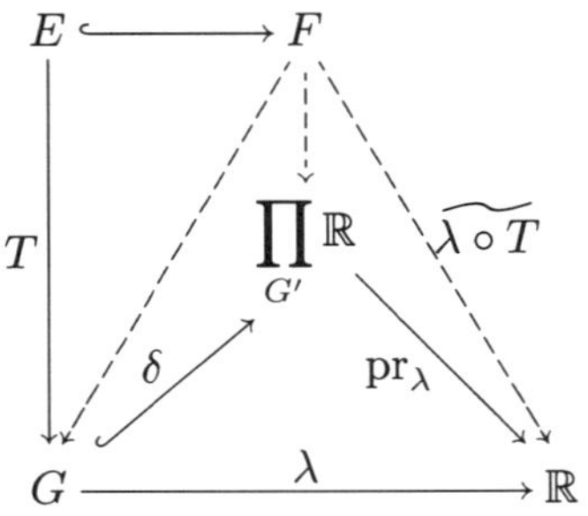

The arrow δ, given by $\delta(x)_\lambda := \lambda(x)$, is a *bornological embedding*, i.e. the image of a set is bounded if and only if the set is bounded, since $B \subseteq G$ is bounded if and only if $\lambda(B) \subseteq \mathbb{R}$ is bounded for all $\lambda \in G'$, i.e. $\delta(B) \subseteq \prod_{G'} \mathbb{R}$ is bounded.

By assumption, the dashed arrow on the right hand side exists, hence by the universal property of the product the dashed vertical arrow (denoted $\tilde{T}$) exists. It remains to show that it has values in the image of δ. Since $\tilde{T}$ is bounded we have

$$\tilde{T}(F) = \tilde{T}(\overline{E}^{c^\infty}) \subseteq \overline{\tilde{T}(E)}^{c^\infty} \subseteq \overline{\delta(G)}^{c^\infty} = \delta(G),$$

since G is c^∞-complete and hence also $\delta(G)$, which is thus c^∞-closed.

The uniqueness follows, since as a bounded linear map $\tilde{T}$ has to be continuous for the c^∞-topology (since it preserves the smooth curves by (2.11) which in turn generate the c^∞-topology), and E lies dense in F with respect to this topology. □

4.31. Proposition. Inductive representation of bornological locally convex spaces. *For a locally convex space E the bornologification E_{born} is the colimit of all the normed spaces E_B for the absolutely convex bounded sets B. The colimit of the respective completions $\tilde{E}_B$ is the linear subspace of the c^∞-completion $\tilde{E}$ consisting of all limits in $\tilde{E}$ of Mackey Cauchy sequences in E.*

Proof. Let $E^{(1)}$ be the *Mackey adherence* of E in the c^∞-completion $\tilde{E}$, by which we mean the limits in $\tilde{E}$ of all sequences in E which converge Mackey in $\tilde{E}$. Then $E^{(1)}$ is a subspace of the locally convex completion $\widehat{E_{\text{born}}}$. For every absolutely convex bounded set B we have the continuous inclusion $E_B \to E_{\text{born}}$, and by passing to the c^∞-completion we get mappings $\widetilde{E_B} = \widehat{E_B} \to \tilde{E}$. These mappings commute with the inclusions $\widehat{E_B} \to \widehat{E_{B'}}$ for $B \subseteq B'$ and have values in the Mackey adherence of E, since every point in $\widehat{E_B}$ is the limit of a sequence in E_B, and hence its image is the limit of this Mackey Cauchy sequence in E.

We claim that the Mackey adherence $E^{(1)}$ together with these mappings has the universal property of the colimit $\varinjlim_B \widehat{E_B}$. In fact, let $T : E^{(1)} \to F$ be a linear mapping, such that $\widehat{E_B} \to E^{(1)} \to F$ is continuous for all B. In particular $T|_E : E \to F$ has to be bounded, and hence $T|_{E_{\text{born}}} : E_{\text{born}} \to F$ is continuous. Thus, it has a unique continuous extension $\widehat{T} : E^{(1)} \to \widehat{F}$, and it remains to show that this extension is T. So take a point $x \in E^{(1)}$. Then there exists a sequence (x_n) in E, which converges Mackey to x. Thus, the x_n form a Cauchy-sequence in some E_B and hence converge to some y in $\widehat{E_B}$. Then $\iota_B(y) = x$, since the mapping $\iota_B : \widehat{E_B} \to E^{(1)}$ is continuous. Since the trace of T to $\widehat{E_B}$ is continuous $T(x_n)$ converges to $T(\iota_B(y)) = T(x)$ and $T(x_n) = \widehat{T}(x_n)$ converges to $\widehat{T}(x)$, i.e. $T(x) = \widehat{T}(x)$. □

In spite of (1) in (4.36) we can use the Mackey adherence to describe the c^∞-closure in the following inductive way:

4.32. Proposition. Mackey adherences. *For ordinal numbers α the Mackey adherence $A^{(\alpha)}$ of order α is defined recursively by:*

$$A^{(\alpha)} := \begin{cases} \text{M-Adh}(A^{(\beta)}) & \text{if } \alpha = \beta + 1 \\ \bigcup_{\beta<\alpha} A^{(\beta)} & \text{if } \alpha \text{ is a limit ordinal number.} \end{cases}$$

Then the closure $\overline{A}$ of A in the c^∞-topology coincides with $A^{(\omega_1)}$, where ω_1 denotes the first uncountable ordinal number, i.e. the set of all countable ordinal numbers.

Proof. Let us first show that $A^{(\omega_1)}$ is c^∞-closed. So take a sequence $x_n \in A^{(\omega_1)} = \bigcup_{\alpha<\omega_1} A^{(\alpha)}$, which converges Mackey to some x. Then there are $\alpha_n < \omega_1$ with $x_n \in A^{(\alpha_n)}$. Let $\alpha := \sup_n \alpha_n$. Then α is a again countable and hence less than ω_1. Thus, $x_n \in A^{(\alpha_n)} \subseteq A^{(\alpha)}$, and therefore $x \in \text{M-Adh}(A^{(\alpha)}) = A^{(\alpha+1)} \subseteq A^{(\omega_1)}$ since $\alpha + 1 \leq \omega_1$.

It remains to show that $A^{(\alpha)}$ is contained in $\overline{A}$ for all α. We prove this by transfinite induction. So assume that for all $\beta < \alpha$ we have $A^{(\beta)} \subseteq \overline{A}$. If α is a limit ordinal number then $A^{(\alpha)} = \bigcup_{\beta<\alpha} A^{(\beta)} \subseteq \overline{A}$. If $\alpha = \beta + 1$ then every point in $A^{(\alpha)} = \text{M-Adh}(A^{(\beta)})$ is the Mackey-limit of some sequence in $A^{(\beta)} \subseteq \overline{A}$, and since $\overline{A}$ is c^∞-closed, this limit has to belong to it. So $A^{(\alpha)} \subseteq \overline{A}$ in all cases. □

4.33. Example. *The trace of the c^∞-topology is not the c^∞-topology, in general.*

Proof. Consider $E = \mathbb{R}^{\mathbb{N}} \times \mathbb{R}^{(\mathbb{N})}$, $A := \{a_{n,k} := (\frac{1}{n}\chi_{\{1,..,k\}}, \frac{1}{k}\chi_{\{n\}}) : n, k \in \mathbb{N}\} \subseteq E$. Let F be the linear subspace of E generated by A. We show that the closure of A with respect to the c^∞-topology of F is strictly smaller than that with respect to the trace topology of the c^∞-topology of E.

A is closed in the c^∞-topology of F: Assume that a sequence (a_{n_j,k_j}) is M-converging to (x, y). Then the second component of a_{n_j,k_j} has to be bounded. Thus, $j \mapsto n_j$ has to be bounded and may be assumed to have constant value n_∞. If $j \mapsto k_j$ were unbounded, then $(x, y) = (\frac{1}{n_\infty}\chi_{\mathbb{N}}, 0)$, which is not an element of F. Thus, $j \mapsto k_j$ has to be bounded too and may be assumed to have constant value k_∞. Thus, $(x, y) = a_{n_\infty,k_\infty} \in A$.

A is not closed in the trace topology since (0,0) is contained in the closure of A with respect to the c^∞-topology of E: For $k \to \infty$ and fixed n the sequence $a_{n,k}$ is M-converging to $(\frac{1}{n}\chi_{\mathbb{N}}, 0)$, and $\frac{1}{n}\chi_{\mathbb{N}}$ is M-converging to 0 for $n \to \infty$. □

4.34. Example. We consider the space $\ell^\infty(X) := \ell^\infty(X, \mathbb{R})$ as defined in (2.15) for a set X together with a family $\mathcal{B}$ of subsets called bounded. We have the subspace $C_c(X) := \{f \in \ell^\infty(X) : \operatorname{supp} f \text{ is finite}\}$. And we want to calculate its c^∞-closure in $\ell^\infty(X)$.
Claim: The c^∞-closure of $C_c(X)$ equals

$$c_0(X) := \{f \in \ell^\infty(X) : f|_B \in c_0(B) \text{ for all } B \in \mathcal{B}\},$$

provided that X is countable.

Proof. The right hand side is just the intersection $c_0(X) := \bigcap_{B \in \mathcal{B}} \iota_B^{-1}(c_0(B))$, where $\iota_B : \ell^\infty(X) \to \ell^\infty(B)$ denotes the restriction map. We use the notation $c_0(X)$, since in the case where X is bounded this is exactly the space $\{f \in \ell^\infty(X) : \{x : |f(x)| \geq \varepsilon\}$ is finite for all $\varepsilon > 0\}$. In particular, this applies to the bounded space $\mathbb{N}$, where $c_0(\mathbb{N}) = c_0$. Since $\ell^\infty(X)$ carries the initial structure with respect to these maps $c_0(X)$ is closed. It remains to show that $C_c(X)$ is c^∞-dense in $c_0(X)$. So take $f \in c_0(X)$. Let $\{x_1, x_2, \dots\} := \{x : f(x) \neq 0\}$.

We consider first the case, where there exists some $\delta > 0$ such that $|f(x_n)| \geq \delta$ for all n. Then we consider the functions $f_n := f \cdot \chi_{x_1,\dots,x_n} \in C_c(X)$. We claim that $n(f - f_n)$ is bounded in $\ell^\infty(X, \mathbb{R})$. In fact, let $B \in \mathcal{B}$. Then $\{n : x_n \in B\} = \{n : x_n \in B \text{ and } |f(x_n)| \geq \delta\}$ is finite. Hence, $\{n(f - f_n)(x) : x \in B\}$ is finite and thus bounded, i.e. f_n converges Mackey to f.

Now the general case. We set $X_n := \{x \in X : |f(x)| \geq \frac{1}{n}\}$ and define $f_n := f \cdot \chi_{X_n}$. Then each f_n satisfies the assumption of the particular case with $\delta = \frac{1}{n}$ and hence is a Mackey limit of a sequence in $C_c(X)$. Furthermore, $n(f - f_n)$ is uniformly bounded by 1, since for $x \in X_n$ it is 0 and otherwise $|n(f - f_n)(x)| = n|f(x)| < 1$. So after forming the Mackey adherence (i.e. adding the limits of all Mackey-convergent sequences contained in the set, see (4.32) for a formal definition) twice, we obtain $c_0(X)$. □

Now we want to show that $c_0(X)$ is in fact the c^∞-completion of $C_c(X)$.

4.35. Example. $\mathbf{c_0(X)}$. We claim that $c_0(X)$ is the c^∞-completion of the subspace $C_c(X)$ in $\ell^\infty(X)$ formed by the finite sequences.
We may assume that the bounded sets of X are formed by those subsets B, for which $f(B)$ is bounded for all $f \in \ell^\infty(X)$. Obviously, any bounded set has this property, and the space $\ell^\infty(X)$ is not changed by adding these sets. Furthermore, the restriction map $\iota_B : \ell^\infty(X) \to \ell^\infty(B)$ is also bounded for such a B, since using the closed graph theorem (52.10) we only have to show that $\mathrm{ev}_b \circ \iota_B = \iota_{\{b\}}$ is bounded for every $b \in B$, which is obviously the case.

By proposition (4.30) it is enough to show the universal property for bounded linear functionals. In analogy to Banach-theory, we only have to show that the dual $C_c(X)'$ is just

$$\ell^1(X) := \{g : X \to \mathbb{R} : \operatorname{supp} g \text{ is bounded and } g \text{ is absolutely summable}\}.$$

In fact, any such g acts even as bounded linear functional on $\ell^\infty(X,\mathbb{R})$ by $\langle g,f\rangle := \sum_x g(x)\,f(x)$, since a subset is bounded in $\ell^\infty(X)$ if and only if it is uniformly bounded on all bounded sets $B\subseteq X$. Conversely, let $\ell: C_c(X)\to\mathbb{R}$ be bounded and linear and define $g: X\to\mathbb{R}$, by $g(x) := \ell(e_x)$, where e_x denotes the function given by $e_x(y) := 1$ for $x=y$ and 0 otherwise. Obviously $\ell(f) = \langle g,f\rangle$ for all $f\in C_c(X)$. Suppose indirectly $\operatorname{supp} g = \{x : \ell(e_x)\neq 0\}$ is not bounded. Then there exists a sequence $x_n\in\operatorname{supp} g$ and a function $f\in\ell^\infty(X)$ such that $|f(x_n)|\geq n$. In particular, the only bounded subsets of $\{x_n : n\in\mathbb{N}\}$ are the finite ones. Hence $\{\frac{n}{|g(x_n)|}e_{x_n}\}$ is bounded in $C_c(X)$, but the image under ℓ is not. Furthermore, g has to be absolutely summable since the set of finite subsums of $\sum_x \operatorname{sign} g(x)\,e_x$ is bounded in $C_c(X)$ and its image under ℓ are the subsums of $\sum_x |g(x)|$.

4.36. Corollary. Counter-examples on c^∞-topology. *The following statements are* **false***:*

1. *The c^∞-closure of a subset (or of a linear subspace) is given by the Mackey adherence, i.e. the set formed by all limits of sequences in this subset which are Mackey convergent in the total space.*
2. *A subset U of E that contains a point x and has the property, that every sequence which M-converges to x belongs to it finally, is a c^∞-neighborhood of x.*
3. *A c^∞-dense subspace of a c^∞-complete space has this space as c^∞-completion.*
4. *If a subspace E is c^∞-dense in the total space, then it is also c^∞-dense in each linear subspace lying in between.*
5. *The c^∞-topology of a linear subspace is the trace of the c^∞-topology of the whole space.*
6. *Every bounded linear functional on a linear subspace can be extended to such a functional on the whole space.*
7. *A linear subspace of a bornological locally convex space is bornological.*
8. *The c^∞-completion preserves embeddings.*

Proof. (1) For this we give an example, where the Mackey adherence of $C_c(X)$ is not all of $c_0(X)$.
Let $X=\mathbb{N}\times\mathbb{N}$, and take as bounded sets all sets of the form $B_\mu := \{(n,k) : n\leq \mu(k)\}$, where μ runs through all functions $\mathbb{N}\to\mathbb{N}$. Let $f: X\to\mathbb{R}$ be defined by $f(n,k) := \frac{1}{k}$. Obviously, $f\in c_0(X)$, since for given $j\in\mathbb{N}$ and function μ the set of points $(n,k)\in B_\mu$ for which $f(n,k)=\frac{1}{k}\geq\frac{1}{j}$ is the finite set $\{(n,k) : k\leq j, n\leq \mu(k)\}$.
Assume there is a sequence $f_n\in C_c(X)$ Mackey convergent to f. By passing to a subsequence we may assume that $n^2(f-f_n)$ is bounded. Now choose $\mu(k)$ to be larger than all of the finitely many n, with $f_k(n,k)\neq 0$. If $k^2(f-f_k)$ is bounded on B_μ, then in particular $\{k^2(f-f_k)(\mu(k),k) : k\in\mathbb{N}\}$ has to be bounded, but $k^2(f-f_k)(\mu(k),k) = k^2\frac{1}{k}-0 = k$.

(2) Let A be a set for which (1) fails, and choose x in the c^∞-closure of A but not in the M-adherence of A. Then $U := E\setminus A$ satisfies the assumptions of (2). In

fact, let x_n be a sequence which converges Mackey to x, and assume that it is not finally in U. So we may assume without loss of generality that $x_n \notin U$ for all n, but then $A \ni x_n \to x$ would imply that x is in the Mackey adherence of A. However, U cannot be a c^∞-neighborhood of x. In fact, such a neighborhood must meet A since x is assumed to be in the c^∞-closure of A.

(3) Let F be a locally convex vector space whose Mackey adherence in its c^∞-completion E is not all of E, e.g. $C_c(X) \subseteq c_0(X)$ as in the previous counter-example. Choose a $y \in E$ that is not contained in the Mackey adherence of F, and let F_1 be the subspace of E generated by $F \cup \{y\}$. We claim that $F_1 \subseteq E$ cannot be the c^∞-completion although F_1 is obviously c^∞-dense in the convenient vector space E. In order to see this we consider the linear map $\ell : F_1 \to \mathbb{R}$ characterized by $\ell(F) = 0$ and $\ell(y) = 1$. Clearly ℓ is well defined.

$\ell : F_1 \to \mathbb{R}$ is bornological: For any bounded $B \subseteq F_1$ there exists an N such that $B \subseteq F + [-N, N]y$. Otherwise, $b_n = x_n + t_n y \in B$ would exist with $t_n \to \infty$ and $x_n \in F$. This would imply that $b_n = t_n(\frac{x_n}{t_n} + y)$, and thus $-\frac{x_n}{t_n}$ would converge Mackey to y; a contradiction.

Now assume that a bornological extension $\bar{\ell}$ to E exists. Then $F \subseteq \ker(\bar{\ell})$ and $\ker(\bar{\ell})$ is c^∞-closed, which is a contradiction to the c^∞-denseness of F in E. So $F_1 \subseteq E$ does not have the universal property of a c^∞-completion.

This shows also that (6) fails.

(4) Furthermore, it follows that F is $c^\infty F_1$-closed in F_1, although F and hence F_1 are c^∞-dense in E.

(5) The trace of the c^∞-topology of E to F_1 cannot be the c^∞-topology of F_1, since for the first one F is obviously dense.

(7) Obviously, the trace topology of the bornological topology on E cannot be bornological on F_1, since otherwise the bounded linear functionals on F_1 would be continuous and hence extendable to E.

(8) Furthermore, the extension of the inclusion $\iota : F \oplus \mathbb{R} \cong F_1 \to E$ to the completion is given by $(x,t) \in E \oplus \mathbb{R} \cong \tilde{F} \oplus \mathbb{R} = \tilde{F_1} \mapsto x + ty \in E$ and has as kernel the linear subspace generated by $(y, -1)$. Hence, the extension of an embedding to the c^∞-completions need not be an embedding anymore, in particular the inclusion functor does not preserve injectivity of morphisms. □

5. Uniform Boundedness Principles and Multilinearity

5.1. The category of locally convex spaces and smooth mappings. The category of all smooth mappings between bornological vector spaces is a subcategory of the category of all smooth mappings between locally convex spaces which is equivalent to it, since a locally convex space and its bornologification (4.4) have the same bounded sets and smoothness depends only on the bornology by (1.8). So it is also cartesian closed, but the topology on $C^\infty(E, F)$ from (3.11) has to be bornologized. For an example showing the necessity see [Kriegl, 1983, p. 297] or [Frölicher, Kriegl, 1988, 5.4.19]: The topology on $C^\infty(\mathbb{R}, \mathbb{R}^{(\mathbb{N})})$ is not bornological.

We will in general, however, work in the category of locally convex spaces and smooth mappings, so function spaces carry the topology of (3.11).

The category of bounded (equivalently continuous) linear mappings between bornological vector spaces is in the same way equivalent to the category of all bounded linear mappings between all locally convex spaces, since a linear mapping is smooth if and only if it is bounded, by (2.11). It is closed under formation of colimits and under quotients (this is an easy consequence of (4.1.1)). The Mackey-Ulam theorem [Jarchow, 1981, 13.5.4] tells us that a product of non trivial bornological vector spaces is bornological if and only if the index set does not admit a Ulam measure, i.e. a non trivial $\{0, 1\}$-valued measure on the whole power set. A cardinal admitting a Ulam measure has to be strongly inaccessible, so we can restrict set theory to exclude measurable cardinals.

Let $L(E_1, \dots, E_n; F)$ denote the *space of all bounded n-linear mappings* from $E_1 \times \dots \times E_n \to F$ with the topology of uniform convergence on bounded sets in $E_1 \times \dots \times E_n$.

5.2. Proposition. Exponential law for L. *There are natural bornological isomorphisms*

$$L(E_1, \dots, E_{n+k}; F) \cong L(E_1, \dots, E_n; L(E_{n+1}, \dots, E_{n+k}; F)).$$

Proof. We proof this for bilinear maps, the general case is completely analogous. We already know that bilinearity translates into linearity into the space of linear functions. Remains to prove boundedness. So let $\mathcal{B} \subseteq L(E_1, E_2; F)$ be given. Then $\mathcal{B}$ is bounded if and only if $\mathcal{B}(B_1 \times B_2) \subseteq F$ is bounded for all bounded $B_i \subseteq E_i$. This however is equivalent to $\mathcal{B}^\vee(B_1)$ is contained and bounded in $L(E_2, F)$ for all bounded $B \subseteq E_1$, i.e. $\mathcal{B}^\vee$ is contained and bounded in $L(E_1, L(E_2, F))$. □

Recall that we have already put a structure on $L(E, F)$ in (3.17), namely the initial one with respect to the inclusion in $C^\infty(E, F)$. Let us now show that bornologically these definitions agree:

5.3. Lemma. Structure on L. *A subset is bounded in* $L(E, F) \subseteq C^\infty(E, F)$ *if and only if it is uniformly bounded on bounded subsets of* E, *i.e.* $L(E, F) \to C^\infty(E, F)$ *is initial.*

Proof. Let $\mathcal{B} \subseteq L(E, F)$ be bounded in $C^\infty(E, F)$, and assume that it is not uniformly bounded on some bounded set $B \subseteq E$. So there are $f_n \in \mathcal{B}$, $b_n \in B$, and $\ell \in F^*$ with $|\ell(f_n(b_n))| \geq n^n$. Then the sequence $n^{1-n} b_n$ converges fast to 0, and hence lies on some compact part of a smooth curve c by the special curve lemma (2.8). So $\mathcal{B}$ cannot be bounded, since otherwise $C^\infty(\ell, c) = \ell_* \circ c^* : C^\infty(E, F) \to C^\infty(\mathbb{R}, \mathbb{R}) \to \ell^\infty(\mathbb{R}, \mathbb{R})$ would have bounded image, i.e. $\{\ell \circ f_n \circ c : n \in \mathbb{N}\}$ would be uniformly bounded on any compact interval.

Conversely, let $\mathcal{B} \subseteq L(E,F)$ be uniformly bounded on bounded sets and hence in particular on compact parts of smooth curves. We have to show that $d^n \circ c^* : L(E,F) \to C^\infty(\mathbb{R},F) \to \ell^\infty(\mathbb{R},F)$ has bounded image. But for linear smooth maps we have by the chain rule (3.18), recursively applied, that $d^n(f \circ c)(t) = f(c^{(n)}(t))$, and since $c^{(n)}$ is still a smooth curve we are done. □

Let us now generalize this result to multilinear mappings. For this we first characterize bounded multilinear mappings in the following two ways:

5.4. Lemma. *A multilinear mapping is bounded if and only if it is bounded on each sequence which converges Mackey to 0.*

Proof. Suppose that $f : E_1 \times \ldots \times E_k \to F$ is not bounded on some bounded set $B \subseteq E_1 \times \ldots \times E_k$. By composing with a linear functional we may assume that $F = \mathbb{R}$. So there are $b_n \in B$ with $\lambda_n^{k+1} := |f(b_n)| \to \infty$. Then $|f(\frac{1}{\lambda_n} b_n)| = \lambda_n \to \infty$, but $(\frac{1}{\lambda_n} b_n)$ is Mackey convergent to 0. □

5.5. Lemma. Bounded multilinear mappings are smooth. *Let $f : E_1 \times \ldots \times E_n \to F$ be a multilinear mapping. Then f is bounded if and only if it is smooth. For the derivative we have the product rule:*

$$df(x_1,\ldots,x_n)(v_1,\ldots,v_n) = \sum_{i=1}^{n} f(x_1,\ldots,x_{i-1},v_i,x_{i+1},\ldots,x_n).$$

In particular, we get for $f : E \supseteq U \to \mathbb{R}$, $g : E \supseteq U \to F$ and $x \in U$, $v \in E$ the Leibniz formula

$$(f \cdot g)'(x)(v) = f'(x)(v) \cdot g(x) + f(x) \cdot g'(x)(v).$$

Proof. We use induction on n. The case $n = 1$ is corollary (2.11). The induction goes as follows:

f is bounded

$\iff$ $f(B_1 \times \ldots \times B_n) = f^\vee(B_1 \times \ldots \times B_{n-1})(B_n)$ is bounded for all bounded sets B_i in E_i;

$\iff$ $f^\vee(B_1 \times \ldots \times B_{n-1}) \subseteq L(E_n,F) \subseteq C^\infty(E_n,F)$ is bounded, by (5.3);

$\iff$ $f^\vee : E_1 \times \ldots \times E_{n-1} \to C^\infty(E_n,F)$ is bounded;

$\iff$ $f^\vee : E_1 \times \ldots \times E_{n-1} \to C^\infty(E_n,F)$ is smooth by the inductive assumption;

$\iff$ $f^\vee : E_1 \times \ldots \times E_n \to F$ is smooth by cartesian closedness (3.13). □

The particular case follows by application to the scalar multiplication $\mathbb{R} \times F \to F$.

Now let us show that also the structures coincide:

5.6. Proposition. Structure on space of multilinear maps. *The injection of $L(E_1,\ldots,E_n;F) \to C^\infty(E_1 \times \ldots \times E_n,F)$ is a bornological embedding.*

Proof. We can show this by induction. In fact, let $\mathcal{B} \subseteq L(E_1,\ldots,E_n;F)$. Then

$\mathcal{B}$ is bounded

$\iff$ $\mathcal{B}(B_1 \times \ldots \times B_n) = \mathcal{B}^\vee(B_1 \times \ldots \times B_{n-1})(B_n)$ is bounded for all bounded B_i in E_i;

$\iff$ $\mathcal{B}^\vee(B_1 \times \ldots \times B_{n-1}) \subseteq L(E_n, F) \subseteq C^\infty(E_n, F)$ is bounded, by (5.3);

$\iff$ $\mathcal{B}^\vee \subseteq C^\infty(E_1 \times \ldots \times E_{n-1}, C^\infty(E_n, F))$ is bounded by the inductive assumption;

$\iff$ $\mathcal{B} \subseteq C^\infty(E_1 \times \ldots \times E_n, F)$ is bounded by cartesian closedness (3.13). □

5.7. Bornological tensor product. It is natural to consider the universal problem of linearizing bounded bilinear mappings. The solution is given by the *bornological tensor product* $E \otimes_\beta F$, i.e. the algebraic tensor product with the finest locally convex topology such that $E \times F \to E \otimes F$ is bounded. A 0-neighborhood basis of this topology is given by those absolutely convex sets, which absorb $B_1 \otimes B_2$ for all bounded $B_1 \subseteq E_1$ and $B_2 \subseteq E_2$. Note that this topology is bornological since it is the finest locally convex topology with given bounded linear mappings on it.

Theorem. *The bornological tensor product is the left adjoint functor to the Hom-functor $L(E, \ \ )$ on the category of bounded linear mappings between locally convex spaces, and one has the following bornological isomorphisms:*

$$\begin{gathered} L(E \otimes_\beta F, G) \cong L(E, F; G) \cong L(E, L(F, G)) \\ E \otimes_\beta \mathbb{R} \cong E \\ E \otimes_\beta F \cong F \otimes_\beta E \\ (E \otimes_\beta F) \otimes_\beta G \cong E \otimes_\beta (F \otimes_\beta G) \end{gathered}$$

Furthermore, the bornological tensor product preserves colimits. It neither preserves embeddings nor countable products.

Proof. We show first that this topology has the universal property for bounded bilinear mappings $f : E_1 \times E_2 \to F$. Let U be an absolutely convex zero neighborhood in F, and let B_1, B_2 be bounded sets. Then $f(B_1 \times B_2)$ is bounded, hence it is absorbed by U. Then $\tilde{f}^{-1}(U)$ absorbs $\otimes(B_1 \times B_2)$, where $\tilde{f} : E_1 \otimes E_2 \to F$ is the canonically associated linear mapping. So $\tilde{f}^{-1}(U)$ is in the zero neighborhood basis of $E_1 \otimes_\beta E_2$ described above. Therefore, $\tilde{f}$ is continuous.

A similar argument for sets of mappings shows that the first isomorphism $L(E \otimes_\beta F, G) \cong L(E, F; G)$ is bornological.

The topology on $E_1 \otimes_\beta E_2$ is finer than the projective tensor product topology, and so it is Hausdorff. The rest of the positive results is clear.

The counter-example for embeddings given for the projective tensor product works also, since all spaces involved are Banach.

Since the bornological tensor-product preserves coproducts it cannot preserve products. In fact $(\mathbb{R} \otimes_\beta \mathbb{R}^{(\mathbb{N})})^\mathbb{N} \cong (\mathbb{R}^{(\mathbb{N})})^\mathbb{N}$ whereas $\mathbb{R}^\mathbb{N} \otimes_\beta \mathbb{R}^{(\mathbb{N})} \cong (\mathbb{R}^\mathbb{N} \otimes_\beta \mathbb{R})^{(\mathbb{N})} \cong (\mathbb{R}^\mathbb{N})^{(\mathbb{N})}$. □

5.8. Proposition. Projective versus bornological tensor product. *If every bounded bilinear mapping on $E \times F$ is continuous then $E \otimes_\pi F = E \otimes_\beta F$. In particular, we have $E \otimes_\pi F = E \otimes_\beta F$ for any two metrizable spaces, and for a normable space F we have $E_{born} \otimes_\pi F = E \otimes_\beta F$.*

Proof. Recall that $E \otimes_\pi F$ carries the finest locally convex topology such that $\otimes : E \times F \to E \otimes F$ is continuous, whereas $E \otimes_\beta F$ carries the finest locally convex topology such that $\otimes : E \times F \to E \otimes F$ is bounded. So we have that $\otimes : E \times F \to E \otimes_\beta F$ is bounded and hence by assumption continuous, and thus the topology of $E \otimes_\pi F$ is finer than that of $E \otimes_\beta F$. Since the converse is true in general, we have equality.

In (52.23) it is shown that in metrizable locally convex spaces the convergent sequences coincide with the Mackey-convergent ones. Now let $T : E \times F \to G$ be bounded and bilinear. We have to show that T is continuous. So let (x_n, y_n) be a convergent sequence in $E \times F$. Without loss of generality we may assume that its limit is $(0,0)$. So there are $\mu_n \to \infty$ such that $\{\mu_n(x_n, y_n) : n \in \mathbb{N}\}$ is bounded and hence also $T\Big(\{\mu_n(x_n, y_n) : n \in \mathbb{N}\}\Big) = \Big\{\mu_n^2 T(x_n, y_n) : n \in \mathbb{N}\Big\}$, i.e. $T(x_n, y_n)$ converges even Mackey to 0.

If F is normable and $T : E_{born} \times F \to G$ is bounded bilinear then $T^\vee : E_{born} \to L(F,G)$ is bounded, and since E_{born} is bornological it is even continuous. Clearly, for normed spaces F the evaluation map $\mathrm{ev} : L(F,G) \times F \to G$ is continuous, and hence $T = \mathrm{ev} \circ (T^\vee \times F) : E_{born} \times F \to G$ is continuous. Thus, $E_{born} \otimes_\pi F = E \otimes_\beta F$. $\square$

Note that the bornological tensor product is invariant under bornologification, i.e. $E_{born} \otimes_\beta F_{born} \cong E \otimes_\beta F$. So it is no loss of generality to assume that both spaces are bornological. Keep however in mind that the corresponding identity for the projective tensor product does not hold. Another possibility to obtain the identity $E \otimes_\pi F = E \otimes_\beta F$ is to assume that E and F are bornological and every separately continuous bilinear mapping on $E \times F$ is continuous. In fact, every bounded bilinear mapping is obviously separately bounded, and since E and F are assumed to be bornological, it has to be separately continuous. We want to find another class beside the Fréchet spaces (see (52.9)) which satisfies these assumptions.

5.9. Corollary. *The following mappings are bounded multilinear.*

(1) $\lim : \mathrm{Nat}(\mathcal{F}, \mathcal{G}) \to L(\lim \mathcal{F}, \lim \mathcal{G})$, *where $\mathcal{F}$ and $\mathcal{G}$ are two functors on the same index category, and where $\mathrm{Nat}(\mathcal{F}, \mathcal{G})$ denotes the space of all natural transformations with the structure induced by the embedding into $\prod_i L(\mathcal{F}(i), \mathcal{G}(i))$.*

(2) $\mathrm{colim} : \mathrm{Nat}(\mathcal{F}, \mathcal{G}) \to L(\mathrm{colim}\, \mathcal{F}, \mathrm{colim}\, \mathcal{G})$.

(3)
$$\begin{aligned} L : L(E_1, F_1) \times \ldots \times L(E_n, F_n) \times L(F, E) &\to \\ \to L(L(F_1, \ldots, F_n; F), &L(E_1, \ldots, E_n; E)) \\ (T_1, \ldots, T_n, T) \mapsto (S \mapsto T \circ S &\circ (T_1 \times \ldots \times T_n)); \end{aligned}$$

(4) $\bigotimes_\beta^n : L(E_1, F_1) \times \ldots \times L(E_n, F_n) \to L(E_1 \otimes_\beta \cdots \otimes_\beta E_n, F_1 \otimes_\beta \cdots \otimes_\beta F_n)$.

(5) $\bigwedge^n : L(E,F) \to L(\bigwedge^n E, \bigwedge^n F)$, *where* $\bigwedge^n E$ *is the linear subspace of all alternating tensors in* $\bigotimes_\beta^n E$. *It is the universal solution of*

$$L\Big(\bigwedge^n E, F\Big) \cong L^n_{\text{alt}}(E;F),$$

where $L^n_{\text{alt}}(E;F)$ *is the space of all bounded* n*-linear alternating mappings* $E \times \ldots \times E \to F$.

(6) $\bigvee^n : L(E,F) \to L(\bigvee^n E, \bigvee^n F)$, *where* $\bigvee^n E$ *is the linear subspace of all symmetric tensors in* $\bigotimes_\beta^n E$. *It is the universal solution of*

$$L\Big(\bigvee^n E, F\Big) \cong L^n_{\text{sym}}(E;F),$$

where $L^n_{\text{sym}}(E;F)$ *is the space of all bounded* n*-linear symmetric mappings* $E \times \ldots \times E \to F$.

(7) $\bigotimes_\beta : L(E,F) \to L(\bigotimes_\beta E, \bigotimes_\beta F)$, *where* $\bigotimes_\beta E := \bigoplus_{n=0}^\infty \overset{n}{\otimes}_\beta E$ *is the tensor algebra of* E. *Note that is has the universal property of prolonging bounded linear mappings with values in locally convex spaces, which are algebras with bounded operations, to continuous algebra homomorphisms:*

$$L(E,F) \cong \operatorname{Alg}(\bigotimes_\beta E, F).$$

(8) $\bigwedge : L(E,F) \to L(\bigwedge E, \bigwedge F)$, *where* $\bigwedge E := \bigoplus_{n=0}^\infty \bigwedge^n E$ *is the exterior algebra. It has the universal property of prolonging bounded linear mappings to continuous algebra homomorphisms into graded-commutative algebras, i.e. algebras in the sense above, which are as vector spaces a coproduct* $\coprod_{n\in\mathbb{N}} E_n$ *and the multiplication maps* $E_k \times E_l \to E_{k+l}$ *and for* $x \in E_k$ *and* $y \in E_l$ *one has* $x \cdot y = (-1)^{kl} y \cdot x$.

(9) $\bigvee : L(E,F) \to L(\bigvee E, \bigvee F)$, *where* $\bigvee E := \bigoplus_{n=0}^\infty \bigvee^n E$ *is the symmetric algebra. It has the universal property of prolonging bounded linear mappings to continuous algebra homomorphisms into commutative algebras.*

Recall that the symmetric product is given as the image of the *symmetrizer* sym : $E \otimes_\beta \cdots \otimes_\beta E \to E \otimes_\beta \cdots \otimes_\beta E$ given by

$$x_1 \otimes \cdots \otimes x_n \to \frac{1}{n!} \sum_{\sigma \in \mathcal{S}_n} x_{\sigma(1)} \otimes \cdots \otimes x_{\sigma(n)}.$$

Similarly the wedge product is given as the image of the *alternator*

$$\operatorname{alt} : E \otimes_\beta \cdots \otimes_\beta E \to E \otimes_\beta \cdots \otimes_\beta E$$

$$\text{given by } x_1 \otimes \cdots \otimes x_n \to \frac{1}{n!} \sum_{\sigma \in \mathcal{S}_n} \operatorname{sign}(\sigma)\, x_{\sigma(1)} \otimes \cdots \otimes x_{\sigma(n)}.$$

Symmetrizer and alternator are bounded projections, so both subspaces are complemented in the tensor product.

Proof. All results follow easily by flipping coordinates until only a composition of products of evaluation maps remains.

That the spaces in (5), and similar in (6), are universal solutions can be seen from the following diagram:

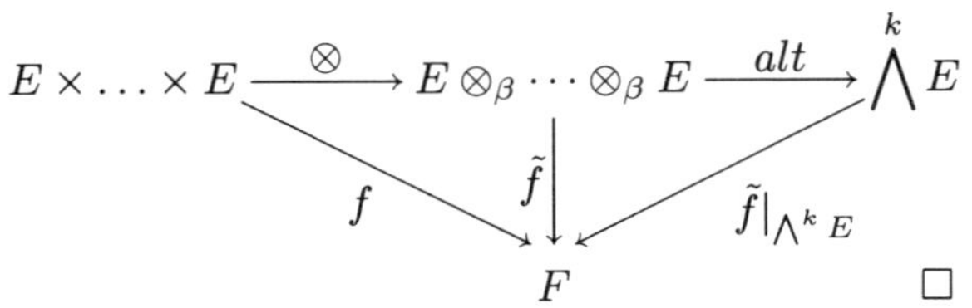

□

5.10. Lemma. *Let E be a convenient vector space. Then $E' \hookrightarrow P_f(E) := \langle E' \rangle_{\text{alg}} \subseteq C^\infty(E, \mathbb{R})$ is the free commutative algebra over the vector space E', i.e. to every linear mapping $f : E' \to A$ in a commmutative algebra, there exists a unique algebra homomorphism $\tilde f : P_f(E) \to A$.*

Elements of the space $P_f(E)$ are called *polynomials of finite type* on E.

Proof. The solution of this universal problem is given by the symmetric algebra $\bigvee E' := \bigoplus_{k=0}^\infty \bigvee^k E'$ described in (5.9.9). In particular we have an algebra homomorphism $\tilde\iota : \bigvee E' \to P_f(E)$, which is onto, since by definition $P_f(E)$ is generated by E'. It remains to show that it is injective. So let $\sum_{k=1}^N \alpha_k \in \bigvee E'$, i.e. $\alpha_k \in \bigvee^k E'$, with $\tilde\iota(\sum_{k=1}^N \alpha_k) = 0$. Thus all derivatives $\iota(\alpha_k)$ at 0 of this mapping in $P_f(E) \subseteq P(E) \subseteq C^\infty(E, \mathbb{R})$ vanish. So it remains to show that $\bigotimes^k E' \to L(E, \ldots, E; \mathbb{R})$ is injective, since then by the polarization identity also $\bigvee^k E' \to P_f(E) \subseteq C^\infty(E, \mathbb{R})$ is injective. Let $\alpha \in E' \otimes F'$ be zero as element on $L(E, F; \mathbb{R})$. We have finitely many $x^k \in E'$ and $y^k \in F'$ with $\alpha = \sum_k x^k \otimes y^k$ and we may assume that the $\{x^k\}$ are linearly independent. So we may choose vectors $x_j \in E$ with $x^k(x_j) = \delta_j^k$. Then $0 = \alpha(x_j, y) = \sum_k x^k(x_j) \cdot y^k(y) = y^j(y)$, so $y^j = 0$ for all j and hence $\alpha = 0$.

Now for the mapping $E_1' \otimes \cdots \otimes E_n' \to L(E_1, \ldots, E_n; \mathbb{R})$. We proceed by induction. Let $\alpha = \sum_k \alpha_k \otimes x^k$, where $\alpha^k \in E_1' \otimes E_{n-1}'$ and $x^k \in E_n'$. We may assume that $(x^k)_k$ is linearly independent. So, as before, we choose $x_j \in E_n$ with $x^k(x_j) = \delta_j^k$ and get $0 = \alpha(y^1, \ldots, y^{n-1}, x_j) = \alpha^j(y^1, \ldots, y^{n-1})$, hence $\alpha^j = 0$ for all j and so $\alpha = 0$. □

5.11. Corollary. Symmetry of higher derivatives. *Let $f : E \supseteq U \to F$ be smooth. The n-th derivative $f^{(n)}(x) = d^n f(x)$, considered as an element of $L^n(E; F)$, is symmetric, so has values in the space $L_{\text{sym}}(E, \ldots, E; F) \cong L(\bigvee^k E; F)$*

Proof. Recall that we can form iterated derivatives as follows:

$$\begin{aligned}
f &: E \supseteq U \to F \\
df &: E \supseteq U \to L(E,F) \\
d(df) &: E \supseteq U \to L(E, L(E,F)) \cong L(E,E;F) \\
&\vdots \\
d(\dots(d(df))\dots) &: E \supseteq U \to L(E,\dots,L(E,F)\dots) \cong L(E,\dots,E;F)
\end{aligned}$$

Thus, the iterated derivative $d^n f(x)(v_1,\dots,v_n)$ is given by

$$\tfrac{\partial}{\partial t_1}|_{t_1=0}\cdots\tfrac{\partial}{\partial t_n}|_{t_n=0} f(x+t_1v_1+\dots+t_nv_n) = \partial_1\dots\partial_n\tilde f(0,\dots,0),$$

where $\tilde f(t_1,\dots,t_n) := f(x+t_1v_1+\dots+t_nv_n)$. The result now follows from the finite dimensional property. □

5.12. Theorem. Taylor formula. *Let $f : U \to F$ be smooth, where U is c^∞-open in E. Then for each segment $[x,x+y] = \{x+ty : 0 \le t \le 1\} \subseteq U$ we have*

$$f(x+y) = \sum_{k=0}^{n} \frac{1}{k!} d^k f(x) y^k + \int_0^1 \frac{(1-t)^n}{n!} d^{n+1} f(x+ty) y^{n+1} dt,$$

where $y^k = (y,\dots,y) \in E^k$.

Proof. This is an assertion on the smooth curve $t \mapsto f(x+ty)$. Using functionals we can reduce it to the scalar valued case, or we proceed directly by induction on n: The first step is (6) in (2.6), and the induction step is partial integration of the remainder integral. □

5.13. Corollary. *The following subspaces are direct summands:*

$$\begin{aligned}
L(E_1,\dots,E_n;F) &\subseteq C^\infty(E_1\times\dots\times E_n, F), \\
L^n_{\text{sym}}(E;F) &\subseteq L^n(E;F) := L(E,\dots,E;F), \\
L^n_{\text{alt}}(E;F) &\subseteq L^n(E;F), \\
L^n_{\text{sym}}(E;F) &\hookrightarrow C^\infty(E,F).
\end{aligned}$$

Note that direct summand is meant in the bornological category, i.e. the embedding admits a left-inverse in the category of bounded linear mappings, or, equivalently, with respect to the bornological topology it is a topological direct summand.

Proof. The projection for $L(E,F) \subseteq C^\infty(E,F)$ is $f \mapsto df(0)$. The statement on L^n follows by induction using cartesian closedness and (5.2). The projections for the next two subspaces are the symmetrizer and alternator, respectively.

The last embedding is given by Δ^*, which is bounded and linear $C^\infty(E\times\dots\times E,F) \to C^\infty(E,F)$. Here $\Delta : E \to E\times\dots\times E$ denotes the diagonal mapping

$x \mapsto (x, \dots, x)$. A bounded linear left inverse $C^\infty(E, F) \to L^k_{\text{sym}}(E; F)$ is given by $f \mapsto \frac{1}{k!} d^k f(0)$. See the following diagram:

$$\begin{array}{ccc} L^k_{\text{sym}}(E;F) & \hookrightarrow & L(E,\dots,E;F) \\ \vdots & & \downarrow \\ C^\infty(E,F) & \xleftarrow[\Delta^*]{} & C^\infty(E\times\dots\times E, F) \end{array} \qquad \square$$

5.14. Remark. We are now going to discuss polynomials between locally convex spaces. Recall that for finite dimensional spaces $E = \mathbb{R}^n$ a polynomial in a locally convex vector space F is just a finite sum

$$\sum_{k\in\mathbb{N}^n} a_k x^k,$$

where $a_k \in F$ and $x^k := \prod_{i=1}^n x_i^{k_i}$. Thus, it is just an element in the algebra generated by the coordinate projections pr_i tensorized with F. Since every (continuous) linear functional on $E = \mathbb{R}^n$ is a finite linear combination of coordinate projections, this algebra is also the algebra generated by E^*. For a general locally convex space E we define the algebra of *finite type polynomials* to be the one generated by E^*.

However, there is also another way to define polynomials, namely as those smooth functions for which some derivative is equal to 0. Take for example the square of the norm $\| \quad \|^2 : E \to \mathbb{R}$ on some infinite dimensional Hilbert space E. Its derivative is given by $x \mapsto (v \mapsto 2\langle x, v\rangle)$, and hence is linear. The second derivative is $x \mapsto ((v, w) \mapsto 2\langle v, w\rangle)$ and hence constant. Thus, the third derivative vanishes. This function is not a finite type polynomial. Otherwise, it would be continuous for the weak topology $\sigma(E, E^*)$. Hence, the unit ball would be a 0-neighborhood for the weak topology, which is not true, since it is compact for it.

Note that for $(x_k) \in \ell^2$ the series $\sum_k x_k^2$ converges pointwise and even uniformly on compact sets. In fact, every compact set is contained in the absolutely convex hull of a 0-sequence x^n. In particular $\mu_k := \sup\{|x^n_k| : n \in \mathbb{N}\} \to 0$ for $k \to \infty$ (otherwise, we can find an $\varepsilon > 0$ and $k_j \to \infty$ and $n_j \in \mathbb{N}$ with $\|x^{n_j}\|_2 \ge |x^{n_j}_{k_j}| \ge \varepsilon$. Since $x^n \in \ell^2 \subseteq c_0$, we conclude that $n_j \to \infty$, which yields a contradiction to $\|x^n\|_2 \to 0$.) Thus

$$K \subseteq \langle x^n : n \in \mathbb{N}\rangle_{\text{absolutely convex}} \subseteq \langle \mu_n e^n\rangle_{\text{absolutely convex}},$$

and hence $\sum_{k\ge n} |x_k| \le \max\{\mu_k : k \ge n\}$ for all $x \in K$.
The series does not converge uniformly on bounded sets. To see this choose $x = e_k$.

5.15. Definition. A smooth mapping $f : E \to F$ is called a *polynomial* if some derivative $d^n f$ vanishes on E. The largest p such that $d^p f \ne 0$ is called the degree of the polynomial. The mapping f is called a *monomial of degree p* if it is of the form $f(x) = \tilde f(x, \dots, x)$ for some $\tilde f \in L^p_{\text{sym}}(E; F)$.

5.16. Lemma. Polynomials versus monomials.

(1) *The smooth p-homogeneous maps are exactly the monomials of degree p.*
(2) *The symmetric multilinear mapping representing a monomial is unique.*
(3) *A smooth mapping is a polynomial of degree $\leq p$ if and only if its restriction to each one dimensional subspace is a polynomial of degree $\leq p$.*
(4) *The polynomials are exactly the finite sums of monomials.*

Proof. (1) Every monomial of degree p is clearly smooth and p-homogeneous. If f is smooth and p-homogeneous, then

$$(d^p f)(0)(x,\dots,x) = (\tfrac{\partial}{\partial t})^p\big|_{t=0} f(tx) = (\tfrac{\partial}{\partial t})^p\big|_{t=0} t^p f(x) = p!f(x).$$

(2) The symmetric multilinear mapping $g \in L^p_{\text{sym}}(E;F)$ representing f is uniquely determined, since we have $(d^p f)(0)(x_1,\dots,x_p) = p!g(x_1,\dots,x_p)$.

(3) & (4) Let the restriction of f to each one dimensional subspace be a polynomial of degree $\leq p$, i.e., we have $\ell(f(tx)) = \sum_{k=0}^p \frac{t^k}{k!}\,(\frac{\partial}{\partial t})^k\big|_{t=0} \ell(f(tx))$ for $x \in E$ and $\ell \in F'$. So $f(x) = \sum_{k=0}^p \frac{1}{k!} d^k f(0.x)(x,\dots,x)$ and hence is a finite sum of monomials. For the derivatives of a monomial q of degree k we have $q^{(j)}(tx)(v_1,\dots,v_j) = k(k-1)\dots(k-j+1)t^{k-j}\tilde{q}(x,\dots,x,v_1,\dots,v_j)$. Hence, any such finite sum is a polynomial in the sense of (5.15).
Finally, any such polynomial has a polynomial as trace on each one dimensional subspace. □

5.17. Lemma. Spaces of polynomials. *The space* $\text{Poly}^p(E,F)$ *of polynomials of degree $\leq p$ is isomorphic to $\sum_{k\leq p} L(\bigvee^k E;F)$ and is a direct summand in $C^\infty(E,F)$ with a complement given by the smooth functions which are p-flat at 0.*

Proof. We have already shown that $L(\bigvee^k E;F)$ embeds into $C^\infty(E,F)$ as a direct summand, where a retraction is given by the derivative of order k at 0. Furthermore, we have shown that the polynomials of degree $\leq p$ are exactly the direct sums of homogeneous terms in $L(\bigvee^k E;F)$. A retraction to the inclusion $\bigoplus_{k\leq p} L(\bigvee^k E;F) \to C^\infty(E,F)$ is hence given by $\bigoplus_{k\leq p} \frac{1}{k!} d^k|_0$. □

Remark. The corresponding statement is false for infinitely flat functions. I.e. the sequence $E \to C^\infty(\mathbb{R},\mathbb{R}) \to \mathbb{R}^{\mathbb{N}}$ does not split, where E denotes the space of smooth functions which are *infinitely flat at 0*. Otherwise, $\mathbb{R}^{\mathbb{N}}$ would be a subspace of $C^\infty([0,1],\mathbb{R})$ (compose the section with the restriction map from $C^\infty(\mathbb{R},\mathbb{R}) \to C^\infty([0,1],\mathbb{R})$) and hence would have a continuous norm. This is however easily seen to be not the case.

5.18. Theorem. Uniform boundedness principle. *If all E_i are convenient vector spaces, and if F is a locally convex space, then the bornology on the space $L(E_1,\dots,E_n;F)$ consists of all pointwise bounded sets.*

So a mapping into $L(E_1,\dots,E_n;F)$ is smooth if and only if all composites with evaluations at points in $E_1 \times \dots \times E_n$ are smooth.

Proof. Let us first consider the case $n = 1$. So let $\mathcal{B} \subseteq L(E,F)$ be a pointwise bounded subset. By lemma (5.3) we have to show that it is uniformly bounded on each bounded subset B of E. We may assume that B is closed absolutely convex, and thus E_B is a Banach space, since E is convenient. By the classical uniform boundedness principle, see (52.25), the set $\mathcal{B}|E_B$ is bounded in $L(E_B, F)$, and thus $\mathcal{B}$ is bounded on B.

The smoothness detection principle: Clearly it suffices to recognize smooth curves. If $c : \mathbb{R} \to L(E,F)$ is such that $\mathrm{ev}_x \circ c : \mathbb{R} \to F$ is smooth for all $x \in E$, then clearly $\mathbb{R} \xrightarrow{c} L(E,F) \xrightarrow{j} \prod_E F$ is smooth. We will show that $(j \circ c)'$ has values in $L(E,F) \subseteq \prod_E F$. Clearly, $(j \circ c)'(s)$ is linear $E \to F$. The family of mappings $\frac{c(s+t)-c(s)}{t} : E \to F$ is pointwise bounded for s fixed and t in a compact interval, so by the first part it is uniformly bounded on bounded subsets of E. It converges pointwise to $(j \circ c)'(s)$, so this is also a bounded linear mapping $E \to F$. By the first part $j : L(E,F) \to \prod_E F$ is a bornological embedding, so c is differentiable into $L(E,F)$. Smoothness follows now by induction on the order of the derivative.

The multilinear case follows from the exponential law (5.2) by induction on n. □

5.19. Theorem. Multilinear mappings on convenient vector spaces. *A multilinear mapping from convenient vector spaces to a locally convex space is bounded if and only if it is separately bounded.*

Proof. Let $f : E_1 \times \ldots \times E_n \to F$ be n-linear and separately bounded, i.e. $x_i \mapsto f(x_1, \ldots, x_n)$ is bounded for each i and all fixed x_j for $j \neq i$. Then $f^\vee : E_1 \times \ldots \times E_{n-1} \to L(E_n, F)$ is $(n-1)$-linear. By (5.18) the bornology on $L(E_n, F)$ consists of the pointwise bounded sets, so $f^\vee$ is separately bounded. By induction on n it is bounded. The bornology on $L(E_n, F)$ consists also of the subsets which are uniformly bounded on bounded sets by lemma (5.3), so f is bounded. □

We will now derive an infinite dimensional version of (3.4), which gives us minimal requirements for a mapping to be smooth.

5.20. Theorem. *Let E be a convenient vector space. An arbitrary mapping $f : E \supseteq U \to F$ is smooth if and only if all unidirectional iterated derivatives $d_v^p f(x) = (\frac{\partial}{\partial t})^p|_0 f(x + tv)$ exist, $x \mapsto d_v^p f(x)$ is bounded on sequences which are Mackey converging in U, and $v \mapsto d_v^p f(x)$ is bounded on fast falling sequences.*

Proof. A smooth mapping obviously satisfies this requirement. Conversely, from (3.4) we see that f is smooth restricted to each finite dimensional subspace, and the iterated directional derivatives $d_{v_1} \ldots d_{v_n} f(x)$ exist and are bounded multilinear mappings in $v_1, \ldots, v_n$ by (5.4), since they are universal linear combinations of the unidirectional iterated derivatives $d_v^p f(x)$, compare with the proof of (3.4). So $d^n f : U \to L^n(E; F)$ is bounded on Mackey converging sequences with respect to the pointwise bornology on $L^n(E; F)$. By the uniform boundedness principle (5.18) together with lemma (4.14) the mapping $d^n f : U \times E^n \to F$ is bounded on sets which are contained in a product of a bornologically compact set in U - i.e. a set in U which is contained and compact in some E_B - and a bounded set in E^n.

Now let $c:\mathbb{R}\to U$ be a smooth curve. We have to show that $\frac{f(c(t))-f(c(0))}{t}$ converges to $f'(c(0))(c'(0))$. It suffices to check that

$$\frac{1}{t}\left(\frac{f(c(t))-f(c(0))}{t}-f'(c(0))(c'(0))\right)$$

is locally bounded with respect to t. Integrating along the segment from $c(0)$ to $c(t)$ we see that this expression equals

$$\begin{aligned}
&\frac{1}{t}\int_0^1\left(f'\Big(c(0)+s(c(t)-c(0))\Big)\left(\frac{c(t)-c(0)}{t}\right)-f'(c(0))(c'(0))\right)ds=\\
&=\int_0^1 f'\Big(c(0)+s(c(t)-c(0))\Big)\left(\frac{\frac{c(t)-c(0)}{t}-c'(0)}{t}\right)ds\\
&\quad+\int_0^1\int_0^1 f''\Big(c(0)+rs(c(t)-c(0))\Big)\left(s\frac{c(t)-c(0)}{t},c'(0)\right)dr\,ds.
\end{aligned}$$

The first integral is bounded since $df:U\times E\to F$ is bounded on the product of the bornologically compact set $\{c(0)+s(c(t)-c(0)):0\le s\le 1, t \text{ near } 0\}$ in U and the bounded set $\{\frac{1}{t}\left(\frac{c(t)-c(0)}{t}-c'(0)\right):t \text{ near } 0\}$ in E (use (1.6)).

The second integral is bounded since $d^2f:U\times E^2\to F$ is bounded on the product of the bornologically compact set $\{c(0)+rs(c(t)-c(0)):0\le r,s\le 1, t \text{ near } 0\}$ in U and the bounded set $\{\left(s\frac{c(t)-c(0)}{t},c'(0)\right):0\le s\le 1, t \text{ near } 0\}$ in E^2.

Thus $f\circ c$ is differentiable in F with derivative $df\circ(c,c')$. Now $df:U\times E\to F$ satisfies again the assumptions of the theorem, so we may iterate. □

5.21. The following result shows that bounded multilinear mappings are the right ones for uses like homological algebra, where multilinear algebra is essential and where one wants a kind of 'continuity'. With continuity itself it does not work. The same results hold for convenient algebras and modules, one just may take c^∞-completions of the tensor products.

So by a *bounded algebra* A we mean a (real or complex) algebra which is also a locally convex vector space, such that the multiplication is a bounded bilinear mapping. Likewise, we consider *bounded modules* over bounded algebras, where the action is bounded bilinear.

Lemma. [Cap et. al., 1993]. *Let A be a bounded algebra, M a bounded right A-module and N a bounded left A-module.*

(1) *There are a locally convex vector space $M\otimes_A N$ and a bounded bilinear map $b:M\times N\to M\otimes_A N$, $(m,n)\mapsto m\otimes_A n$ such that $b(ma,n)=b(m,an)$ for all $a\in A$, $m\in M$ and $n\in N$ which has the following universal property: If E is a locally convex vector space and $f:M\times N\to E$ is a bounded bilinear map such that $f(ma,n)=f(m,an)$ then there is a unique bounded linear map $\tilde f:M\otimes_A N\to E$ with $\tilde f\circ b=f$. The space of all such f is denoted by $L^A(M,N;E)$, a closed linear subspace of $L(M,N;E)$.*

(2) *We have a bornological isomorphism*

$$L^A(M,N;E) \cong L(M \otimes_A N, E).$$

(3) *Let B be another bounded algebra such that N is a bounded right B-module and such that the actions of A and B on N commute. Then $M \otimes_A N$ is in a canonical way a bounded right B-module.*

(4) *If in addition P is a bounded left B-module then there is a natural bibounded isomorphism $M \otimes_A (N \otimes_B P) \cong (M \otimes_A N) \otimes_B P$.*

Proof. We construct $M \otimes_A N$ as follows: Let $M \otimes_\beta N$ be the algebraic tensor product of M and N equipped with the (bornological) topology mentioned in (5.7) and let V be the locally convex closure of the subspace generated by all elements of the form $ma \otimes n - m \otimes an$, and define $M \otimes_A N$ to be $M \otimes_A N := (M \otimes_\beta N)/V$. As $M \otimes_\beta N$ has the universal property that bounded bilinear maps from $M \times N$ into arbitrary locally convex spaces induce bounded and hence continuous linear maps on $M \otimes N$, (1) is clear.

(2) By (1) the bounded linear map $b^* : L(M \otimes_A N, E) \to L^A(M,N;E)$ is a bijection. Thus, it suffices to show that its inverse is bounded, too. From (5.7) we get a bounded linear map $\varphi : L(M,N;E) \to L(M \otimes_\beta N, E)$ which is inverse to the map induced by the canonical bilinear map. Now let $L^{\text{ann}(V)}(M \otimes_\beta N, E)$ be the closed linear subspace of $L(M \otimes_\beta N, E)$ consisting of all maps which annihilate V. Restricting φ to $L^A(M,N;E)$ we get a bounded linear map $\varphi : L^A(M,N;E) \to L^{\text{ann}(V)}(M \otimes_\beta N, E)$.

Let $\psi : M \otimes_\beta N \to M \otimes_A N$ be the the canonical projection. Then ψ induces a well defined linear map $\hat{\psi} : L^{\text{ann}(V)}(M \otimes_\beta N, E) \to L(M \otimes_A N, E)$, and $\hat{\psi} \circ \varphi$ is inverse to b^*. So it suffices to show that $\hat{\psi}$ is bounded.

This is the case if and only if the associated map $L^{\text{ann}(V)}(M \otimes_\beta N, E) \times (M \otimes_A N) \to E$ is bounded. This in turn is equivalent to boundedness of the associated map $M \otimes_A N \to L(L^{\text{ann}(V)}(M \otimes_\beta N, E), E)$ which sends x to the evaluation at x and is clearly bounded.

(3) Let $\rho : B^{op} \to L(N,N)$ be the right action of B on N and let $\Phi : L^A(M,N; M \otimes_A N) \cong L(M \otimes_A N, M \otimes_A N)$ be the isomorphism constructed in (2). We define the right module structure on $M \otimes_A N$ as:

$$B^{op} \xrightarrow{\rho} L(N,N) \xrightarrow{\text{Id} \times} L(M \times N, M \times N) \xrightarrow{b_*}$$
$$\to L^A(M,N; M \otimes_A N) \xrightarrow{\Phi} L(M \otimes_A N, M \otimes_A N).$$

This map is obviously bounded and easily seen to be an algebra homomorphism.

(4) Straightforward computations show that both spaces have the following universal property: For a locally convex vector space E and a trilinear map $f : M \times N \times P \to E$ which satisfies $f(ma,n,p) = f(m,an,p)$ and $f(m,nb,p) = f(m,n,bp)$ there is a unique linear map prolonging f. $\square$

5.22. Lemma. Uniform $\mathcal{S}$-boundedness principle. *Let E be a locally convex space, and let $\mathcal{S}$ be a point separating set of bounded linear mappings with common domain E. Then the following conditions are equivalent.*

(1) *If F is a Banach space (or even a c^∞-complete locally convex space) and $f : F \to E$ is a linear mapping with $\lambda \circ f$ bounded for all $\lambda \in \mathcal{S}$, then f is bounded.*

(2) *If $B \subseteq E$ is absolutely convex such that $\lambda(B)$ is bounded for all $\lambda \in \mathcal{S}$ and the normed space E_B generated by B is complete, then B is bounded in E.*

(3) *Let (b_n) be an unbounded sequence in E with $\lambda(b_n)$ bounded for all $\lambda \in \mathcal{S}$, then there is some $(t_n) \in \ell^1$ such that $\sum t_n b_n$ does not converge in E for the initial locally convex topology induced by $\mathcal{S}$.*

Definition. We say that E satisfies the *uniform $\mathcal{S}$-boundedness principle* if these equivalent conditions are satisfied.

Proof. (1) $\Rightarrow$ (3) : Suppose that (3) is not satisfied. So let (b_n) be an unbounded sequence in E such that $\lambda(b_n)$ is bounded for all $\lambda \in \mathcal{S}$, and such that for all $(t_n) \in \ell^1$ the series $\sum t_n b_n$ converges in E for the initial locally convex topology induced by $\mathcal{S}$. We define a linear mapping $f : \ell^1 \to E$ by $f((t_n)_n) = \sum t_n b_n$, i.e. $f(e_n) = b_n$. It is easily checked that $\lambda \circ f$ is bounded, hence by (1) the image of the closed unit ball, which contains all b_n, is bounded. Contradiction.

(3) $\Rightarrow$ (2): Let $B \subseteq E$ be absolutely convex such that $\lambda(B)$ is bounded for all $\lambda \in \mathcal{S}$ and that the normed space E_B generated by B is complete. Suppose that B is unbounded. Then B contains an unbounded sequence (b_n), so by (3) there is some $(t_n) \in \ell^1$ such that $\sum t_n b_n$ does not converge in E for the weak topology induced by $\mathcal{S}$. But $\sum t_n b_n$ is a Cauchy sequence in E_B, since $\sum_{k=n}^m t_n b_n \in (\sum_{k=n}^m |t_n|) \cdot B$, and thus converges even bornologically, a contradiction.

(2) $\Rightarrow$ (1): Let F be convenient, and let $f : F \to E$ be linear such that $\lambda \circ f$ is bounded for all $\lambda \in \mathcal{S}$. It suffices to show that $f(B)$, the image of an absolutely convex bounded set B in F with F_B complete, is bounded. By assumption, $\lambda(f(B))$ is bounded for all $\lambda \in \mathcal{S}$, the normed space $E_{f(B)}$ is a quotient of the Banach space F_B, hence complete. By (2) the set $f(B)$ is bounded. □

5.23. Lemma. *A convenient vector space E satisfies the uniform $\mathcal{S}$-boundedness principle for each point separating set $\mathcal{S}$ of bounded linear mappings on E if and only if there exists no strictly weaker ultrabornological topology than the bornological topology of E.*

Proof. ($\Rightarrow$) Let τ be an ultrabornological topology on E which is strictly weaker than the natural bornological topology. Since every ultrabornological space is an inductive limit of Banach spaces, cf. (52.31), there exists a Banach space F and a continuous linear mapping $f : F \to (E, \tau)$ which is not continuous into E. Let $\mathcal{S} = \{\mathrm{Id} : E \to (E, \tau)\}$. Now f does not satisfy (5.22.1).

($\Leftarrow$) If $\mathcal{S}$ is a point separating set of bounded linear mappings, the ultrabornological topology given by the inductive limit of the spaces E_B with B satisfying (5.22.2) equals the natural bornological topology of E. Hence, (5.22.2) is satisfied. □

5.24. Theorem. Webbed spaces have the uniform boundedness property. *A locally convex space which is webbed satisfies the uniform $\mathcal{S}$-boundedness principle for any point separating set $\mathcal{S}$ of bounded linear functionals.*

Proof. Since the bornologification of a webbed space is webbed, cf. (52.14), we may assume that E is bornological, and hence that every bounded linear functional is continuous, see (4.1). Now the closed graph principle (52.10) applies to any mapping satisfying the assumptions of (5.22.1). □

5.25. Lemma. Stability of the uniform boundedness principle. *Let $\mathcal{F}$ be a set of bounded linear mappings $f : E \to E_f$ between locally convex spaces, let $\mathcal{S}_f$ be a point separating set of bounded linear mappings on E_f for every $f \in \mathcal{F}$, and let $\mathcal{S} := \bigcup_{f\in\mathcal{F}} f^*(\mathcal{S}_f) = \{g \circ f : f \in \mathcal{F}, g \in \mathcal{S}_f\}$. If $\mathcal{F}$ generates the bornology and E_f satisfies the uniform $\mathcal{S}_f$-boundedness principle for all $f \in \mathcal{F}$, then E satisfies the uniform $\mathcal{S}$-boundedness principle.*

Proof. We check the condition (1) of (5.22). So assume $h : F \to E$ is a linear mapping for which $g \circ f \circ h$ is bounded for all $f \in \mathcal{F}$ and $g \in \mathcal{S}_f$. Then $f \circ h$ is bounded by the uniform $\mathcal{S}_f$- boundedness principle for E_f. Consequently, h is bounded since $\mathcal{F}$ generates the bornology of E. □

5.26. Theorem. Smooth uniform boundedness principle. *Let E and F be convenient vector spaces, and let U be c^∞-open in E. Then $C^\infty(U,F)$ satisfies the uniform $\mathcal{S}$-boundedness principle where $\mathcal{S} := \{\mathrm{ev}_x : x \in U\}$.*

Proof. For $E = F = \mathbb{R}$ this follows from (5.24), since $C^\infty(U,\mathbb{R})$ is a Fréchet space. The general case then follows from (5.25). □

6. Some Spaces of Smooth Functions

6.1. Proposition. *Let M be a smooth finite dimensional paracompact manifold. Then the space $C^\infty(M,\mathbb{R})$ of all smooth functions on M is a convenient vector space in any of the following (bornologically) isomorphic descriptions, and it satisfies the uniform boundedness principle for the point evaluations.*

(1) *The initial structure with respect to the cone*

$$C^\infty(M,\mathbb{R}) \xrightarrow{c^*} C^\infty(\mathbb{R},\mathbb{R})$$

for all $c \in C^\infty(\mathbb{R},M)$.

(2) *The initial structure with respect to the cone*

$$C^\infty(M,\mathbb{R}) \xrightarrow{(u_\alpha^{-1})^*} C^\infty(\mathbb{R}^n,\mathbb{R}),$$

where (U_α,u_α) is a smooth atlas with $u_\alpha(U_\alpha) = \mathbb{R}^n$.

(3) *The initial structure with respect to the cone*

$$C^\infty(M,\mathbb{R}) \xrightarrow{j^k} C(J^k(M,\mathbb{R}))$$

for all $k \in \mathbb{N}$, where $J^k(M,\mathbb{R})$ is the bundle of k-jets of smooth functions on M, where j^k is the jet prolongation, and where all the spaces of continuous sections are equipped with the compact open topology.

It is easy to see that the cones in (2) and (3) induce even the same locally convex topology which is sometimes called the *compact C^∞ topology*, if $C^\infty(\mathbb{R}^n,\mathbb{R})$ is equipped with its usual Fréchet topology. From (2) we see also that with the bornological topology $C^\infty(M,\mathbb{R})$ is nuclear by (52.35), and is a Fréchet space if and only if M is separable.

Proof. For all three descriptions the initial locally convex topology is convenient, since the spaces are closed linear subspaces in the relevant products of the right hand sides. Thus, the uniform boundedness principle for the point evaluations holds for all structures since it holds for all right hand sides (for $C(J^k(M,\mathbb{R}))$ we may reduce to a connected component of M, and we then have a Fréchet space). So the identity is bibounded between all structures. □

6.2. Spaces of smooth functions with compact supports. For a smooth separable finite dimensional Hausdorff manifold M we denote by $C_c^\infty(M,\mathbb{R})$ the vector space of all smooth functions with compact supports in M.

Lemma. *The following convenient structures on the space $C_c^\infty(M,\mathbb{R})$ are all isomorphic:*

(1) *Let $C_K^\infty(M,\mathbb{R})$ be the space of all smooth functions on M with supports contained in the fixed compact subset $K \subseteq M$, a closed linear subspace of $C^\infty(M,\mathbb{R})$. Let us consider the final convenient vector space structure on the space $C_c^\infty(M,\mathbb{R})$ induced by the cone*

$$C_K^\infty(M,\mathbb{R}) \to C_c^\infty(M,\mathbb{R})$$

where K runs through a basis for the compact subsets of M. Then the space $C_c^\infty(M,\mathbb{R})$ is even the strict inductive limit of a sequence of spaces $C_K^\infty(M,\mathbb{R})$.

(2) *We equip $C_c^\infty(M,\mathbb{R})$ with the initial structure with respect to the cone:*

$$C_c^\infty(M,\mathbb{R}) \xrightarrow{e^*} C_c^\infty(\mathbb{R},\mathbb{R}),$$

where $e \in C^\infty_{\text{prop}}(\mathbb{R},M)$ runs through all proper smooth mappings $\mathbb{R} \to M$, and where $C_c^\infty(\mathbb{R},\mathbb{R})$ carries the usual inductive limit topology on the space of test functions, with steps $C_I^\infty(\mathbb{R},\mathbb{R})$ for compact intervals I.

(3) *The initial structure with respect to the cone*

$$C_c^\infty(M,\mathbb{R}) \xrightarrow{j^k} C_c(J^k(M,\mathbb{R}))$$

for all $k \in \mathbb{N}$, where $J^k(M,\mathbb{R})$ is the bundle of k-jets of smooth functions on M, where j^k is the jet prolongation, and where the spaces of continuous sections with compact support are equipped with the inductive limit topology with steps $C_K(J^k(M,\mathbb{R}))$.

The space $C_c^\infty(M,\mathbb{R})$ satisfies the uniform boundedness principle for the point evaluations.

First Proof. We note first that in all descriptions the space $C_c^\infty(M,\mathbb{R})$ is convenient and satisfies the uniform boundedness principle for point evaluations:

In (1) we have $C_c^\infty(M,\mathbb{R}) = \bigoplus C_c^\infty(M_i,\mathbb{R})$ where M_i are the connected components of M, which are separable, so the inductive limit is a strict inductive limit of a sequence of Fréchet spaces, hence each $C_c^\infty(M_i,\mathbb{R})$ is convenient and webbed by (52.13) and (52.12), hence satisfies the uniform boundedness principle by (5.24). So $C_c^\infty(M,\mathbb{R})$ is convenient and satisfies also the uniform boundedness principle for the point evaluations, by [Frölicher, Kriegl, 1988, 3.4.4].

In (2) and (3) the space is a closed subspace of the product of the right hand side spaces, which are convenient and satisfy the uniform boundedness principle, shown as for (1).

Hence, the identity is bibounded for all structures. $\square$

Second Proof. In all three structures the space $C_c^\infty(M,\mathbb{R})$ is the direct sum of the spaces $C_c^\infty(M_\alpha,\mathbb{R})$ for all connected components M_α of M. So we may assume that M is connected and thus separable.

We consider the diagram

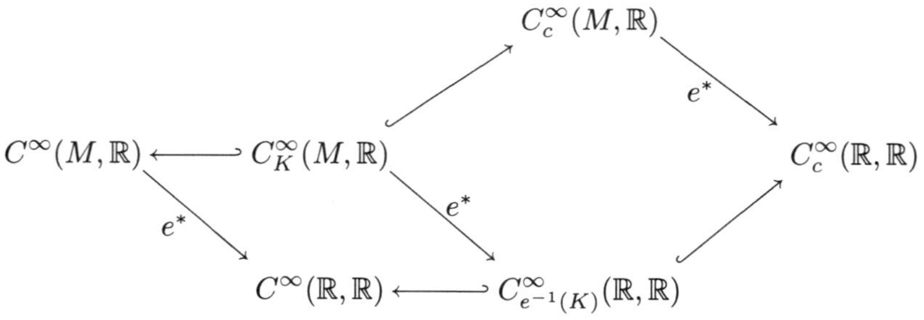

Then obviously the identity on $C_c^\infty(M,\mathbb{R})$ is bounded from the structure (1) to the structure (2).

For the converse we consider a smooth curve $\gamma : \mathbb{R} \to C_c^\infty(M,\mathbb{R})$ in the structure (2). We claim that γ locally factors into some $C_{K_n}^\infty(M,\mathbb{R})$ where (K_n) is an exhaustion of M by compact subsets such that K_n is contained in the interior of K_{n+1}. If not there exist a bounded sequence (t_n) in $\mathbb{R}$ and $x_n \notin K_n$ such that $\gamma(t_n)(x_n) \neq 0$. One may find a proper smooth curve $e : \mathbb{R} \to M$ with $e(n) = x_n$. Then $e^* \circ \gamma$ is a smooth curve into $C_c^\infty(\mathbb{R},\mathbb{R})$. Since the latter space is a strict inductive limit of spaces $C_I^\infty(\mathbb{R},\mathbb{R})$ for compact intervals I, the curve $e^* \circ \gamma$ locally factors into some $C_I^\infty(\mathbb{R},\mathbb{R})$, but $(e^* \circ \gamma)(t_n)(n) = \gamma(t_n)(x_n) \neq 0$, a contradiction. This proves that the curve γ is also smooth into the structure (1), and so the identity on $C_c^\infty(M,\mathbb{R})$ is bounded from the structure (2) to the structure (1).

For the comparison of the structures (3) and (1) we consider the diagram:

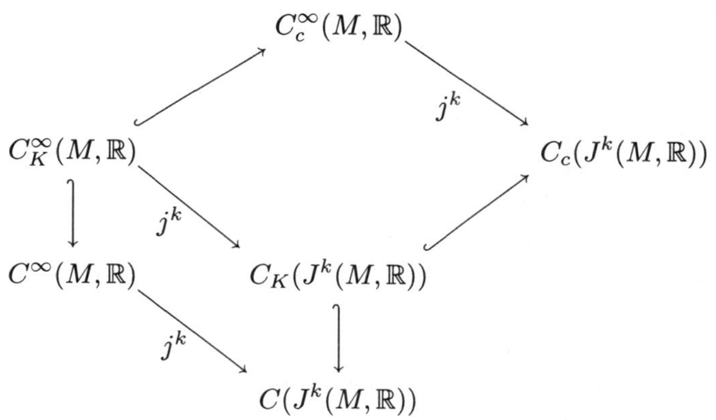

Obviously, the identity on $C_c^\infty(M,\mathbb{R})$ is bounded from the structure (1) into the structure (3).

For the converse direction we consider a smooth curve $\gamma : \mathbb{R} \to C_c^\infty(M,\mathbb{R})$ with structure (3). Then for each k the composition $j^k \circ \gamma$ is a smooth mapping into the strict inductive limit $C_c(J^k(M,\mathbb{R})) = \varinjlim_K C_K(J^k(M,\mathbb{R}))$, thus locally factors into some step $C_K(J^k(M,\mathbb{R}))$ where K chosen for $k=0$ works for any k. Since we have $C_K^\infty(M,\mathbb{R}) = \varprojlim_k C_K(J^k(M,\mathbb{R}))$, the curve γ factors locally into $C_K^\infty(M,\mathbb{R})$ and is thus smooth for the structure (1). For the uniform boundedness principle we refer to the first proof. □

Remark. Note that the locally convex topologies described in (1) and (3) are distinct: The continuous dual of $(C_c^\infty(\mathbb{R},\mathbb{R}),(1))$ is the space of all distributions (generalized functions), whereas the continuous dual of $(C_c^\infty(\mathbb{R},\mathbb{R}),(3))$ are all distributions of finite order, i.e., globally finite derivatives of continuous functions.

6.3. Definition. A convenient vector space E is called *reflexive* if the canonical embedding $E \to E''$ is surjective.

It is then even a bornological isomorphism. Note that reflexivity as defined here is a bornological concept.

Note that this notion is in general stronger than the usual locally convex notion of reflexivity, since the continuous functionals on the strong dual are bounded functionals on E' but not conversely.

6.4. Result. [Frölicher, Kriegl, 1988, 5.4.6]. *For a convenient bornological vector space E the following statements are equivalent.*

(1) *E is reflexive.*
(2) *E is η-reflexive, see* [Jarchow, 1981, p280].
(3) *E is completely reflexive, see* [Hogbe-Nlend, 1977, p. 89].

(4) *E is reflexive in the usual locally convex sense, and the strong dual of E is bornological.*
(5) *The Schwartzening (or nuclearification) of E is a complete locally convex space.*

6.5. Results. [Frölicher, Kriegl, 1988, section 5.4].

(1) *A Fréchet space is reflexive if and only if it is reflexive in the locally convex sense.*
(2) *A convenient vector space with a countable base for its bornology is reflexive if and only if its bornological topology is reflexive in the locally convex sense.*
(3) *A bornological reflexive convenient vector space is complete in the locally convex sense.*
(4) *A closed (in the locally convex sense) linear subspace of a reflexive convenient vector space is reflexive.*
(5) *A convenient vector space is reflexive if and only if its bornological topology is complete and its dual is reflexive.*
(6) *Products and coproducts of reflexive convenient vector spaces are reflexive if the index set is of non-measurable cardinality.*
(7) *If E is a reflexive convenient vector space and M is a finite dimensional separable smooth manifold then $C^\infty(M,E)$ is reflexive.*
(8) *Let U be a c^∞-open subset of a dual of a Fréchet Schwartz space, and let F be a Fréchet Montel space. Then $C^\infty(U,F)$ is a Fréchet Montel space, thus reflexive.*
(9) *Let U be a c^∞-open subset of a dual of a nuclear Fréchet space, and let F be a nuclear Fréchet space. It has been shown by* [Colombeau, Meise, 1981] *that $C^\infty(U,F)$ is not nuclear in general.*

6.6. Definition. Another important additional property for convenient vector spaces E is the approximation property, i.e. the denseness of $E'\otimes E$ in $L(E,E)$. There are at least 3 successively stronger requirements, which have been studied in [Adam, 1995]:

A convenient vector space E is said to have the *bornological approximation property* if $E'\otimes E$ is dense in $L(E,E)$ with respect to the bornological topology. It is said to have the *c^∞-approximation property* if this is true with respect to the c^∞-topology of $L(E,E)$. Finally the *Mackey approximation property* is the requirement, that there is some sequence in $E'\otimes E$ Mackey converging towards Id_E.

Note that although the first condition is the weakest one, it is difficult to check directly, since the bornologification of $L(E,E)$ is hard to describe explicitly.

6.7. Result. [Adam, 1995, 2.2.9] *The natural topology on*

$$L(C^\infty(\mathbb{R},\mathbb{R}),C^\infty(\mathbb{R},\mathbb{R}))$$

of uniform convergence on bounded sets is not bornological.

6.8. Result. [Adam, 1995, 2.5.5] *For any set* Γ *of non-measurable cardinality the space* E *of points in* $\mathbb{R}^\Gamma$ *with countable carrier has the bornological approximation property.*

Note. One first shows that for this space E the topology of uniform convergence on bounded sets is bornological, and the classical approximation property holds for this topology by [Jarchow, 1981, 21.2.2], since E is nuclear.

6.9. Lemma. *Let* E *be a convenient vector space with the bornological (resp.* c^∞*-, resp. Mackey) approximation property. Then for every convenient vector space* F *we have that* $E' \otimes F$ *is dense in the bornological topology of* $L(E,F)$ *(resp. in the* c^∞*-topology, resp. every* $T \in L(E,F)$ *is the limit of a Mackey converging sequence in* $E' \otimes F$*).*

Proof. Let $T \in L(E,F)$ and $T_\alpha \in E' \otimes E$ a net converging to Id_E in the bornological topology of $L(E,F)$ (resp. the c^∞-topology, resp. in the sense of Mackey). Since $T_* : L(E,E) \to L(E,F)$ is bounded and $T \circ T_\alpha \in F' \otimes F$, we get the result in all three cases. □

6.10. Lemma. [Adam, 1995, 2.1.21] *Let* E *be a reflexive convenient vector space. Then* E *has the bornological (resp.* c^∞*-, resp. Mackey) approximation property if and only if* E' *has it.*

Proof. For reflexive convenient vector spaces we have:

$$L(E',E') \cong L^2(E',E;\mathbb{R}) \cong L(E,E'') \cong L(E,E),$$

and $E'' \otimes E$ corresponds to $E' \otimes E$ via this isomorphism. So the result follows. □

6.11. Lemma. [Adam, 1995, 2.4.3] *Let* E *be the product* $\prod_{k\in\mathbb{N}} E_k$ *of a sequence of convenient vector spaces* E_k*. Then* E *has the Mackey (resp.* c^∞*-) approximation property if and only if all* E_k *have it.*

Proof. ($\Rightarrow$) follows since one easily checks that these approximation properties are inherited by direct summands.

($\Leftarrow$) Let $(T_n^k)_n$ be Mackey convergent to T^k in $L(E_k,E_k)$. Then one easily checks the Mackey convergence of $(T_n^k)_k \to (T^k)_n$ in $\prod_k L(E_k,E_k) \subseteq L(E,E)$. So the result follows for the Mackey approximation property.

To obtain it also for the c^∞-topology, one first notes that by the argument given in (6.9) it is enough to approximate the identity. Since the c^∞-closure can be obtained as iterated Mackey-adherence by (4.32) this follows now by transfinite induction. □

6.12. Recall that a set $\mathcal{P} \subseteq \mathbb{R}_+^{\mathbb{N}}$ of sequences is called a *Köthe set* if it is directed upwards with respect to the componentwise partial ordering, see (52.35). To $\mathcal{P}$ we may associate the set

$$\Lambda(\mathcal{P}) := \{x = (x_n)_n \in \mathbb{R}^{\mathbb{N}} : (p_n x_n)_n \in \ell^1 \text{ for all } p \in \mathcal{P}\}.$$

A space $\Lambda(\mathcal{P})$ is said to be a *Köthe sequence space* whenever $\mathcal{P}$ is a Köthe set.

Lemma. *Let $\mathcal{P}$ be a Köthe set for which there exists a sequence μ converging monotonely to $+\infty$ and such that $(\mu_n p_n)_{n\in\mathbb{N}} \in \mathcal{P}$ for each $p \in \mathcal{P}$. Then the Köthe sequence space $\Lambda(\mathcal{P})$ has the Mackey approximation property.*

Proof. The sequence $\left(\sum_{j=1}^n e'_j \otimes e_j\right)_{n\in\mathbb{N}}$ is Mackey convergent in $L(\Lambda(\mathcal{P}), \Lambda(\mathcal{P}))$ to $\mathrm{id}_{\Lambda(\mathcal{P})}$, where e_j and e'_j denote the j-th unit vector in $\Lambda(\mathcal{P})$ and $\Lambda(\mathcal{P})'$ respectively: Indeed, a subset $B \subseteq \Lambda(\mathcal{P})$ is bounded if and only if for each $p \in \mathcal{P}$ there exists $N(p) \in \mathbb{R}$ such that

$$\sum_{k\in\mathbb{N}} p_k |x_k| \le N(p)$$

for all $x = (x_k)_{k\in\mathbb{N}} \in B$. But this implies that

$$\Big\{\mu_{n+1}\Big(\mathrm{Id}_{\Lambda(\mathcal{P})} - \sum_{j=1}^n e'_j \otimes e_j\Big) : n \in \mathbb{N}\Big\} \subseteq L(\Lambda(\mathcal{P}), \Lambda(\mathcal{P}))$$

is bounded. In fact

$$\Big(\Big(\mathrm{Id} - \sum_{j=1}^n e'_j \otimes e_j\Big)(x)\Big)_k = \begin{cases} 0 & \text{for } k \le n, \\ x_k & \text{for } k > n. \end{cases}$$

and hence

$$\sum_k p_k \Big|\mu_{n+1}\Big((\mathrm{Id} - \sum_{j=1}^n e'_j \otimes e_j)(x)\Big)_k\Big| \le \sum_{k>n} p_k |\mu_{n+1} x_k| \le \sum_k p_k \mu_k |x_k| \le N(\mu p) \quad \square$$

Let α be an unbounded increasing sequence of positive real numbers and $\mathcal{P}_\infty := \{(e^{k\alpha_n})_{n\in\mathbb{N}} : k \in \mathbb{N}\}$. Then the associated Köthe sequence space $\Lambda(\mathcal{P}_\infty)$ is called a *power series space of infinite type* (a Fréchet space by [Jarchow, 1981, 3.6.2]).

6.13. Corollary. *Each power series space of infinite type has the Mackey approximation property.* $\square$

6.14. Theorem. *The following convenient vector spaces have the Mackey approximation property:*

(1) *The space $C^\infty(M \leftarrow F)$ of smooth sections of any smooth finite dimensional vector bundle $F \xrightarrow{p} M$ with separable base M, see* (6.1) *and* (30.1).
(2) *The space $C_c^\infty(M \leftarrow F)$ of smooth sections with compact support any smooth finite dimensional vector bundle $F \xrightarrow{p} M$ with separable base M, see* (6.2) *and* (30.4).
(3) *The Fréchet space of holomorphic functions $\mathcal{H}(\mathbb{C}, \mathbb{C})$, see* (8.2).

Proof. The space s of rapidly decreasing sequences coincides with the power series space of infinite type associated to the sequence $(\log(n))_{n\in\mathbb{N}}$. So by (6.13), (6.11) and (6.10) the spaces s, $s^{\mathbb{N}}$ and $s^{(\mathbb{N})} = \big((s')^{\mathbb{N}}\big)'$ have the Mackey approximation property. Now assertions (1) and (2) follow from the isomorphisms $C_c^\infty(M \leftarrow F) =$

$C^\infty(M \leftarrow F) \cong s$ for compact M and $C^\infty(M \leftarrow F) \cong s^{\mathbb{N}}$ for non-compact M (see [Valdivia, 1982] or [Adam, 1995, 1.5.16]) and the isomorphism $C_c^\infty(M \leftarrow F) \cong s^{(\mathbb{N})}$ for non-compact M (see [Valdivia, 1982] or [Adam, 1995, 1.5.16]).

(3) follows since by [Jarchow, 1981, 2.10.11] the space $\mathcal{H}(\mathbb{C}, \mathbb{C})$ is isomorphic to the (complex) power series space of infinite type associated to the sequence $(n)_{n\in\mathbb{N}}$. □

Historical Remarks on Smooth Calculus

Roots in the variational calculus. Soon after the invention of the differential calculus ideas were developed which would later lead to variational calculus. Bernoulli used them to determine the shape of a rope under gravity. It evolved into a 'useful and applicable but highly formal calculus; even Gauss warned of its unreflected application' ([Bemelmans, Hildebrand, von Wahl, 1990, p. 151]). In his Lecture courses Weierstrass gave more reliable foundations to the theory, which was made public by [Kneser, 1900], see also [Bolza, 1909] and [Hadamard, 1910]. Further development concerned mainly the relation between the calculus of variations and the theory of partial differential equations. The use of the basic principle of variational calculus for differential calculus itself appeared only in the search for the exponential law, i.e. a cartesian closed setting for calculus, see below.

The notion of derivative. The first more concise notion of the variational derivative was introduced by [Volterra, 1887], a concept of analysis on infinite dimensional spaces; and this happened even before the modern concept of the total derivative of a function of several variables was born: only partial derivatives were used at that time. The derivative of a function in several variables in finite dimensions was introduced by [Stolz, 1893], [Pierpont, 1905], and finally by [Young, 1910]: A function $f : \mathbb{R}^n \to \mathbb{R}$ is called differentiable if the partial differentials $\frac{\partial f}{\partial x^i}$ exist and

$$f(x^1 + h^1, \dots, x^n + h^n) - f(x^1, \dots, x^n) = \sum_{i=1}^{m} (\frac{\partial f}{\partial x^i} + \varepsilon_i) h^i$$

holds, where $\varepsilon_i \to 0$ for $||h|| \to 0$. The idea that the derivative is an approximation to the function was emphasized frequently by Hadamard. His student [Fréchet, 1911] replaced the remainder term by $\varepsilon.||h||$ with $\varepsilon \to 0$ for $||h|| \to 0$. In [Fréchet, 1937] he writes:

> S.241: "C'est M. Volterra qui a eu le premier l'idée d'étendre le champ d'application du Calcul différentiel à l'Analyse fonctionnelle. [...] Toutefois M. Hadamard a signalé qu'il y aurait grand intérèt á généraliser les définitions de M. Volterra. [...] M. Hadamard a montré le chemin qui devait conduire vers des définitions satisfaisantes en proposant d'imposer à la différentielle d'une fonctionnelle la condition d'être linéaire par rapport à la différentielle de l'argument."

Fréchet derivative. In [Fréchet, 1925a] he defined the derivative of a mapping f between normed spaces as follows: There exists a continuous linear operator A

such that

$$\lim_{||h||\to 0} \frac{f(x+h) - f(x) - A\cdot h}{||h||} = 0.$$

At this time it was, however, not so clear what a normed space should be. Fréchet called his spaces somewhat misleadingly 'vectoriels abstraits distanciés'. Banach spaces were introduced by Stefan Banach in his Dissertation in 1920, with a view also to a non-linear theory, as he wrote in [Banach, 1932]:

S.231: "Ces espaces [complex vector spaces] constituent le point de départ de la théorie des opérations linéaires complexes et d'une classe, encore plus vaste, des opérations analytiques, qui présentent une généralisation des fonctions analytiques ordinaires (cf. p. ex. L. Fantappié, I. funzionali analitici, Citta di Castello 1930). Nous nous proposons d'en exposer la théorie dans un autre volume."

Gâteaux derivative. Another student of Hadamard defined the derivative in [Gâteaux, 1913] with proofs in [Gâteaux, 1922] as follows, see also [Gâteaux, 1922]:

"Considérons $U(z+\lambda t_1)$ (t_1 fonction analogue à z). Supposons que

$$\left[\frac{d}{d\lambda} U(z+\lambda t_1)\right]_{\lambda=0}$$

existe quel que soit t_1. On l'appelle la variation première de U au point z: $\delta U(z, t_1)$. C'est une fonctionnelle de z et de t_1, qu'on suppose habituellement linéaire, en chaque point z, par rapport à t_1."

Several mathematicians gave conditions implying the linearity of the Gâteaux-derivative. In [Daniell, 1919] is was shown that this holds for a Lipschitz function whose Gâteaux-derivative exists locally. Another student of Hadamard assumed linearity in [Lévy, 1922], see again [Fréchet, 1937]:

S.51: "Une fonction abstraite $X = F(x)$ sera dite différentiable au sens de M. Paul Levy pour $x = x_0$, s'il existe une transformation vectorielle linéaire $\Psi(\Delta x)$ de l'accroissement Δx telle que, pour chaque vecteur Δx,

$$\lim_{\lambda\to 0} \frac{\overrightarrow{F(x_0)F(x_0+\lambda\Delta x)}}{\lambda} \text{ existe et } = \Psi(\Delta x)."$$

Hadamard differentiability. In [Hadamard, 1923] a function $f : \mathbb{R}^2 \to \mathbb{R}$ was called differentiable if all compositions with differentiable curves are again differentiable and satisfy the chain rule. He refers to a lecture of Poincaré in 1904. In [Fréchet, 1937] it was shown that Hadamard's notion is equivalent to that of Stolz-Pierpoint-Young:

S.244: "Une fonctionelle $U[f]$ sera dite différentiable pour $f \equiv f_0$ au sens de M. Hadamard généralisé, s'il existe une fonctionnelle $W[df, f_0]$, linéaire par rapport à df, telle que si l'on considère une fonction $f(t,\lambda)$ dérivable par rapport à λ pour $\lambda = 0$, avec $f(t,0) = f_0(t)$, la fonction de λ, $U[f(t,\lambda)]$ soit dérivable en λ pour $\lambda = 0$ et qu'on ait pour $\lambda = 0$

$$\frac{d}{d\lambda} U\Big[f(t,\lambda)\Big] = W\left[\frac{df}{d\lambda}, f_0\right]$$

ou avec les notations des "variations"

$$\delta U[f] = W[\delta f, f_0]."$$

S.245: "la différentielle au sens de M. Hadamard généralisé qui est équivalente à la nôtre dans l'Analyse classique est plus générale dans l'Analyse fonctionnelle."

He also realized the importance of Hadamard's definition:

S.249: "L'intérêt de la définition de M. Hadamard n'est pas épuisé par son utilization en Analyse fonctionnelle. Il est peut-être plus encore dans la possibilité de son extension en Analyse générale.

Dans ce domaine, on peut généraliser la notion de fonctionnelle et considérer des transformations $X = F[x]$ d'un élément abstrait x en un élément abstrait X. Nous avons pu en 1925 [Fréchet, 1925b] étendre notre définition (rappelée plus haut p.241 et 242) de la différentielle d'une fonctionnelle, définir la différentielle de $F[x]$ quand X et x appartiennent à des espaces "vectoriels abstraits distanciés" et en etablir les propriétés les plus importantes.

La définition au sens de M. Hadamard généralisé présente sur notre définition l'avantage de garder un sens pour des espaces abstraits vectoriels non distanciés où notre définition ne s'applique pas. [...]

Il reste à voir si elle conserve les propriétes les plus importantes de la différentielle classique en dehors de la propriété (généralisant le théorème des fonctions composées) qui lui sert de définition. C'est un point sur lequel nous reviendrons ultérieurement."

Hadamard's notion of differentiability was later extended to infinite dimensions by [Michal, 1938] who defined a mapping $f : E \to F$ between topological vector spaces to be differentiable at x if there exists a continuous linear mapping $\ell : E \to F$ such that $f \circ c : \mathbb{R} \to F$ is differentiable at 0 with derivative $(\ell \circ c')(0)$ for each everywhere differentiable curve $c : \mathbb{R} \to E$ with $c(0) = x$.

Independently, a student of Fréchet extended in [Ky Fan, 1942] differentiability in the sense of Hadamard to normed spaces, and proved the basic properties like the chain rule:

S.307: "M. Fréchet a eu l'obligeance de me conseiller d'étudier cette question qu'il avait d'abord l'intention de traiter lui-même."

Hadamard differentiability was further generalized to metrizable vector spaces in [Balanzat, 1949] and to vector spaces with a sequential limit structure in [Long de Foglio, 1960]. Finally, in [Balanzat, 1960] the theory was developed for topological vector spaces. There he proved the chain rule and made the observation that the implication "differentiable implies continuous" is equivalent to the property that the closure of a set coincides with the sequential adherence.

Differentiability via bornology. Here the basic observation is that convergence which appears in questions of differentiability is much better than just topological, cf. (1.7). The relevant notion of convergence was introduced by [Mackey, 1945]. Differentiability based on the von Neumann bornology was first considered in [Sebastião e Silva, 1956a, 1956b, 1957]. In [Sebastião e Silva, 1961] he extended this to bornological vector spaces and referred to Waelbroeck and Fantappié for these spaces:

" ... de généraliser aux espaces localement convexes, réels ou complexes, la notion de fonction différentiable, ainsi que les théorèmes fondamentaux du calcul différentiel et intégral, et de la théorie des fonctions analytiques de plusieurs variables complexes.

Je me suis persuadé que, pour cette généralisation, c'est la notion d'ensemble borné, plutôt que celle de voisinage, qui doit jouer un rôle essentiel."

In [Waelbroeck, 1967a, 1967b] the notion of 'b-space' was introduced, and differentiability in them was discussed. He showed that for Mackey complete spaces a scalar-wise smooth mapping is already smooth, see (2.14.5) $\Rightarrow$ (2.14.4). He refers to [Mikusinski, 1960], [Waelbroeck, 1960], [Marinescu, 1963], and [Buchwalter, 1965]. Bornological vector spaces were developed in full detail in [Hogbe-Nlend, 1970, 1971, 1977], and differential calculus in them was further developed by [Lazet, 1971], and [Colombeau, 1973], see also [Colombeau, 1982]. The importance of differentiability with respect to the bornology generated by the compact subsets was realized in [Sova, 1966b].

An overview on differentiability of first order can be found in [Averbukh, Smolyanov, 1968]. One finds there 25 inequivalent definitions of the first derivative in a single point, and one sees how complicated finite order differentiability really is beyond Banach spaces.

Higher derivatives. In [Maissen, 1963] it was shown that only for normed spaces there exists a topology on $L(E,E)$ such that the evaluation mapping $L(E,E)\times E \to E$ is jointly continuous, and [Keller, 1965] generalized this. We have given the archetypical argument in the introduction.

Thus, a 'satisfactory' calculus seemed to stop at the level of Banach spaces, where an elaborated theory including existence theorems was presented already in the very influential text book [Dieudonné, 1960].

Beyond Banach spaces one had to use convergence structures in order to force the continuity of the composition of linear mappings and the general chain rule. Respective theories based on convergence were presented by [Marinescu, 1963], [Bastiani, 1964], [Frölicher, Bucher, 1966], and by [Binz, 1966]. A review is [Keller, 1974], where the following was shown: Continuity of the derivative implied stronger remainder convergence conditions. So for continuously differentiable mappings the many possible notions collapse to 9 inequivalent ones (fewer for Fréchet spaces). And if one looks for infinitely often differentiable mappings, then one ends up with 6 inequivalent notions (only 3 for Fréchet spaces). Further work in this direction culminated in the two huge volumes [Gähler, 1977, 1978], and in the historically very detailed study [Ver Eecke, 1983] and [Ver Eecke, 1985].

Exponential law. The notion of homotopy makes more sense if it is viewed as a curve $I \to C(X,Y)$. The 'exponential law'

$$Z^{X\times Y} \cong (Z^Y)^X, \quad \text{or } C(X\times Y, Z) \cong C(X, C(Y,Z)),$$

however, is not true in general. It holds only for compactly generated spaces, as was shown by [Brown, 1961], see also [Gabriel, Zisman, 1963/64], or for compactly continuous mappings between arbitrary topological spaces, due to [Brown, 1963] and [Brown, 1964]. Without referring to Brown in the text, [Steenrod, 1967] made this result really popular under the title 'a convenient category of topological spaces', which is the source of the widespread use of 'convenient', also in this book. See also [Vogt, 1971].

Following the advise of A. Frölicher, [Seip, 1972] used compactly generated vector spaces for calculus. In [Seip, 1976] he obtained a cartesian closed category of smooth mappings between compactly generated vector spaces, and in [Seip, 1979] he modified his calculus by assuming both smoothness along curves and compact continuity, for all derivatives. Based on this, he obtained a cartesian closed category of 'smooth manifolds' in [Seip, 1981] by replacing atlas of charts by the set of smooth curves and assuming a kind of (Riemannian) exponential mapping which he called local addition.

Motivated by Seip's work in the thesis [Kriegl, 1980], supervised by Peter Michor, smooth mappings between arbitrary subsets 'Vektormengen' of locally convex spaces were supposed to respect smooth curves and to induce 'tangent mappings' which again should respect smooth curves, and so on. On open subsets of E mappings turned out to be smooth if they were smooth along smooth mappings $\mathbb{R}^n \to E$ for all n. This gave a cartesian closed setting of calculus without any assumptions on compact continuity of derivatives. A combination of this with the result of [Boman, 1967] then quickly lead to [Kriegl, 1982] and [Kriegl, 1983], one of the sources of this book.

Independently, [Frölicher, 1980] considered categories generated by monoids of real valued functions and characterized cartesian closedness in terms of the monoid. [Frölicher, 1981] used the result of [Boman, 1967] to show that on Fréchet spaces usual smoothness is equivalent to smoothness in the sense of the category generated by the monoid $C^\infty(\mathbb{R},\mathbb{R})$. That this category is cartesian closed was shown in the unpublished paper [Lawvere, Schanuel, Zame, 1981].

Already [Boman, 1967] used Lipschitz conditions for his result on finite order differentiability, since it fails to be true for C^n-functions. Motivated by this, finite differentiability based on Lipschitz conditions has then been developed by [Frölicher, Gisin, Kriegl, 1983]. A careful presentation can be found in the monograph [Frölicher, Kriegl, 1988]. Finite differentiability based on Hölder conditions were studied by [Faure, 1989] and [Faure, 1991].

Chapter II
Calculus of Holomorphic and Real Analytic Mappings

This chapter starts with an investigation of holomorphic mappings between infinite dimensional vector spaces along the same lines as we investigated smooth mappings in chapter I. This theory is rather easy if we restrict to convenient vector spaces.

The basic tool is the set of all holomorphic mappings from the unit disk $\mathbb{D} \subset \mathbb{C}$ into a complex convenient vector space E, where all possible definitions of being holomorphic coincide, see (7.4). This replaces the set of all smooth curves in the smooth theory. A mapping between c^∞-open sets of complex convenient vector spaces is then said to be holomorphic if it maps holomorphic curves to holomorphic curves. This can be tested by many equivalent descriptions (see (7.19)), the most important are that f is smooth and $df(x)$ is complex linear for each x (i.e. f satisfies the Cauchy-Riemann differential equation); or that f is holomorphic along each affine complex line and is c^∞-continuous (generalized Hartog's theorem). Again (multi-) linear mappings are holomorphic if and only if they are bounded (7.12).

The space $\mathcal{H}(U, F)$ of all holomorphic mappings from a c^∞-open set $U \subseteq E$ into a convenient vector space F carries a natural structure of a complex convenient vector space (7.21), and satisfies the holomorphic uniform boundedness principle (8.10). Of course our general aim of cartesian closedness (7.22), (7.23) is valid also in this setting: $\mathcal{H}(U, \mathcal{H}(V, F)) \cong \mathcal{H}(U \times V, F)$.

As in the smooth case we have to pay a price for cartesian closedness: holomorphic mappings can be expanded into power series, but these converge only on a c^∞-open subset in general, and not on open subsets.

The second part of this chapter is devoted to real analytic mappings in infinite dimensions. The ideas are similar as in the case of smooth and holomorphic mappings, but our wish to obtain cartesian closedness forces us to some modifications: In (9.1) we shall see that for the real analytic mapping $f : \mathbb{R}^2 \ni (s,t) \mapsto \frac{1}{(st)^2+1} \in \mathbb{R}$ there is no reasonable topology on $C^\omega(\mathbb{R}, \mathbb{R})$, such that the mapping $f^\vee : \mathbb{R} \to C^\omega(\mathbb{R}, \mathbb{R})$ is locally given by its convergent Taylor series, which looks like a counterexample to cartesian closedness. Recall that smoothness (holomorphy) of curves can be tested by applying bounded linear functionals (see (2.14), (7.4)). The example above shows at the same time that this is not true in the real analytic case in general; if E' carries a Baire topology then it is true (9.6).

So we are forced to take as basic tool the space $C^\omega(\mathbb{R}, E)$ of all curves c such that $\ell \circ c : \mathbb{R} \to \mathbb{R}$ is real analytic for each bounded linear functional, and we call these the real analytic curves. In order to proceed we have to show that real analyticity of a curve can be tested with any set of bounded linear functionals which generates the bornology. This is done in (9.4) with the help of an unusual bornological description of real analytic functions $\mathbb{R} \to \mathbb{R}$ (9.3).

Now a mapping $f : U \to F$ is called real analytic if $f \circ c$ is smooth for smooth c and is real analytic for real analytic $c : \mathbb{R} \to U$. The second condition alone is not sufficient, even for $f : \mathbb{R}^2 \to \mathbb{R}$. Then a version of Hartog's theorem is true: f is real analytic if and only if it is smooth and real analytic along each affine line (10.4). In order to get to the aim of cartesian closedness we need a natural structure of a convenient vector space on $C^\omega(U, F)$. We start with $C^\omega(\mathbb{R}, \mathbb{R})$ which we consider as real part of the space of germs along $\mathbb{R}$ of holomorphic functions. The latter spaces of holomorphic germs are investigated in detail in section (8). At this stage of the theory we can prove the real analytic uniform boundedness theorem (11.6) and (11.12), but unlike in the smooth and holomorphic case for the general exponential law (11.18) we still have to investigate mixing of smooth and real analytic variables in (11.17). The rest of the development of section (11) then follows more or less standard (categorical) arguments.

7. Calculus of Holomorphic Mappings

7.1. Basic notions in the complex setting. In this section all locally convex spaces E will be complex ones, which we can view as real ones $E_\mathbb{R}$ together with continuous linear mapping J with $J^2 = -\operatorname{Id}$ (the complex structure). So all concepts for real locally convex spaces from sections (1) to (5) make sense also for complex locally convex spaces.

A set which is absolutely convex in the real sense need not be absolutely convex in the complex sense. However, the $\mathbb{C}$-absolutely convex hull of a bounded subset is still bounded, since there is a neighborhood basis of 0 consisting of $\mathbb{C}$-absolutely convex sets. So in this section absolutely convex will refer always to the complex notion. For absolutely convex bounded sets B the real normed spaces E_B (see (1.5)) inherit the complex structure.

A complex linear functional ℓ on a convex vector space is uniquely determined by its real part $\operatorname{Re} \circ \ell$, by $\ell(x) = (\operatorname{Re} \circ \ell)(x) - \sqrt{-1}(\operatorname{Re} \circ \ell)(Jx)$. So for the respective spaces of bounded linear functionals we have

$$E_\mathbb{R}{}' = L_\mathbb{R}(E_\mathbb{R}, \mathbb{R}) \cong L_\mathbb{C}(E, \mathbb{C}) =: E^*,$$

where the complex structure on the left hand side is given by $\lambda \mapsto \lambda \circ J$.

7.2. Definition. Let $\mathbb{D}$ be the the open unit disk $\{z \in \mathbb{C} : |z| < 1\}$. A mapping $c : \mathbb{D} \to E$ into a locally convex space E is called *complex differentiable*, if

$$c'(z) = \lim_{\mathbb{C} \ni w \to 0} \frac{c(z+w) - c(z)}{w}$$

exists for all $z \in \mathbb{D}$.

7.3. Lemma. *Let E be convenient and $a_n \in E$. Then the following statements are equivalent:*

(1) *$\{r^n a_n : n \in \mathbb{N}\}$ is bounded for all $|r| < 1$.*
(2) *The power series $\sum_{n \geq 0} z^n a_n$ is Mackey convergent in E, uniformly on each compact subset of $\mathbb{D}$, i.e., the Mackey coefficient sequence and the bounded set can be chosen valid in the whole compact subset.*
(3) *The power series converges weakly for all $z \in \mathbb{D}$.*

Proof. (1) $\Rightarrow$ (2) Any compact set is contained in $r\mathbb{D}$ for some $0 < r < 1$, the set $\{R^n a_n : n \in \mathbb{N}\}$ is contained in some absolutely convex bounded B for some $r < R < 1$. So the partial sums of the series form a Mackey Cauchy sequence uniformly on $r\mathbb{D}$ since

$$\frac{1}{(r/R)^N - (r/R)^{M+1}} \sum_{n=N}^{M} z^n a_n \in \frac{1}{1 - (r/R)} B.$$

(2) $\Rightarrow$ (3) is clear.

Proof of (3) $\Rightarrow$ (1) The summands are weakly bounded, thus bounded. $\square$

7.4. Theorem. *If E is convenient then the following statements for a curve $c : \mathbb{D} \to E$ are equivalent:*

(1) *c is complex differentiable.*
(2) *$\ell \circ c : \mathbb{D} \to \mathbb{C}$ is holomorphic for all $\ell \in E^*$*
(3) *c is continuous and $\int_\gamma c = 0$ in the completion of E for all closed smooth ($\mathcal{L}\text{ip}^0$-) curves in $\mathbb{D}$.*
(4) *All $c^{(n)}(0)$ exist and $c(z) = \sum_{n=0}^{\infty} \frac{z^n}{n!} c^{(n)}(0)$ is Mackey convergent, uniformly on each compact subset of $\mathbb{D}$.*
(5) *For each $z \in \mathbb{D}$ all $c^{(n)}(z)$ exist and $c(z+w) = \sum_{n=0}^{\infty} \frac{w^n}{n!} c^{(n)}(z)$ is Mackey convergent, uniformly on each compact set in the largest disk with center z contained in $\mathbb{D}$.*
(6) *$c(z)dz$ is a closed $\mathcal{L}\text{ip}^1$ 1-form with values in $E_{\mathbb{R}}$.*
(7) *c is the complex derivative of some complex curve in E.*
(8) *c is smooth ($\mathcal{L}\text{ip}^1$) with complex linear derivative $dc(z)$ for all z.*

From now on all locally convex spaces will be convenient. A curve $c : \mathbb{D} \to E$ satisfying these equivalent conditions will be called a *holomorphic curve*.

Proof. (2) $\Rightarrow$ (1) By assumption, the difference quotient $\frac{c(z+w)-c(z)}{w}$, composed with a linear functional, extends to a complex valued holomorphic function of w, hence it is locally Lipschitz. So the difference quotient is a Mackey Cauchy net. So it has a limit for $w \to 0$.

Proof of (1) $\Rightarrow$ (2) Suppose that ℓ is bounded. Let $c : \mathbb{D} \to E$ be a complex differentiable curve. Then $c_1 : z \mapsto \frac{1}{z}\left(\frac{c(z)-c(0)}{z} - c'(0)\right)$ is a complex differentiable curve (test with linear functionals), hence

$$(\ell \circ c_1)(z) = \frac{1}{z}\left(\frac{\ell(c(z)) - \ell(c(0))}{z} - \ell(c'(0))\right)$$

is locally bounded in z. So $\ell \circ c$ is complex differentiable with derivative $\ell \circ c'$.

Composition with a complex continuous linear functional translates all statements to one dimensional versions which are all equivalent by complex analysis. Moreover, each statement is equivalent to its weak counterpart, where for (4) and (5) we use lemma (7.3). □

7.5. Remarks. In the holomorphic case the equivalence of (7.4.1) and (7.4.2) does not characterize c^∞-completeness as it does in the smooth case. The complex differentiable curves do not determine the bornology of the space, as do the smooth ones. See [Kriegl, Nel, 1985, 1.4]. For a discussion of the holomorphic analogues of smooth characterizations for c^∞-completeness (see (2.14)) we refer to [Kriegl, Nel, 1985, pp. 2.16].

7.6. Lemma. *Let $c : \mathbb{D} \to E$ be a holomorphic curve in a convenient space. Then locally in $\mathbb{D}$ the curve factors to a holomorphic curve into E_B for some bounded absolutely convex set B.*

First Proof. By the obvious extension of lemma (1.8) for smooth mappings $\mathbb{R}^2 \supset \mathbb{D} \to E$ the curve c factors locally to a $\mathcal{L}ip^1$-curve into some complete E_B. Since it has complex linear derivative, by theorem (7.4) it is holomorphic. □

Second direct proof. Let W be a relatively compact neighborhood of some point in $\mathbb{D}$. Then $c(W)$ is bounded in E. It suffices to show that for the absolutely convex closed hull B of $c(W)$ the Taylor series of c at each $z \in W$ converges in E_B, i.e. that $c|W : W \to E_B$ is holomorphic. This follows from the

Vector valued Cauchy inequalities. If $r > 0$ is smaller than the radius of convergence at z of c then

$$\tfrac{r^k}{k!} c^{(k)}(z) \in B$$

where B is the closed absolutely convex hull of $\{\, c(w) : |w - z| = r\}$. (By the Hahn-Banach theorem this follows directly from the scalar valued case.)

Thus, we get

$$\textstyle\sum_{k=n}^m (\frac{w-z}{r})^k \cdot \frac{r^k}{k!} c^{(k)}(z) \in \sum_{k=n}^m (\frac{w-z}{r})^k \cdot B$$

and so $\sum_k \frac{c^{(k)}(z)}{k!}(w-z)^k$ is convergent in E_B for $|w - z| < r$. □

This proof also shows that holomorphic curves with values in complex convenient vector spaces are topologically and bornologically holomorphic in the sense analogous to (9.4).

7.7. Lemma. *Let E be a regular (i.e. every bounded set is contained and bounded in some step E_α) inductive limit of complex locally convex spaces $E_\alpha \subseteq E$, let $c : \mathbb{C} \supseteq U \to E$ be a holomorphic mapping, and let $W \subseteq \mathbb{C}$ be open and such that the closure $\overline{W}$ is compact and contained in U. Then there exists some α, such that $c|W : W \to E_\alpha$ is well defined and holomorphic.*

Proof. By lemma (7.6) the restriction of c to W factors to a holomorphic curve $c|W : W \to E_B$ for a suitable bounded absolutely convex set $B \subseteq E$. Since B is contained and bounded in some E_α one has $c|W : W \to E_B = (E_\alpha)_B \to E_\alpha$ is holomorphic. □

7.8. Definition. Let E and F be convenient vector spaces and let $U \subseteq E$ be c^∞-open. A mapping $f : U \to F$ is called *holomorphic*, if it maps holomorphic curves in U to holomorphic curves in F.

It is remarkable that [Fantappié, 1930] already gave this definition. Connections to other concepts of holomorphy are discussed in [Kriegl, Nel, 1985, 2.19].

So by (7.4) f is holomorphic if and only if $\ell \circ f \circ c : \mathbb{D} \to \mathbb{C}$ is a holomorphic function for all $\ell \in F^*$ and holomorphic curve c.

Clearly, any composition of holomorphic mappings is again holomorphic.

For finite dimensions this coincides with the usual notion of holomorphic mappings, by the finite dimensional Hartogs' theorem.

7.9. Hartogs' Theorem. *Let E_1, E_2, and F be convenient vector spaces with U c^∞-open in $E_1 \times E_2$. Then a mapping $f : U \to F$ is holomorphic if and only if it is separately holomorphic, i.e. $f(\quad, y)$ and $f(x, \quad)$ are holomorphic.*

Proof. If f is holomorphic then $f(\quad, y)$ is holomorphic on the c^∞-open set $E_1 \times \{y\} \cap U = \operatorname{incl}_y^{-1}(U)$, likewise for $f(x, \quad)$.

If f is separately holomorphic, for any holomorphic curve $(c_1, c_2) : \mathbb{D} \to U \subseteq E_1 \times E_2$ we consider the holomorphic mapping $c_1 \times c_2 : \mathbb{D}^2 \to E_1 \times E_2$. Since the c_k are smooth by (7.4.8) also $c_1 \times c_2$ is smooth and thus $(c_1 \times c_2)^{-1}(U)$ is open in $\mathbb{C}^2$. For each $\lambda \in F^*$ the mapping $\lambda \circ f \circ (c_1 \times c_2) : (c_1 \times c_2)^{-1}(U) \to \mathbb{C}$ is separately holomorphic and so holomorphic by the usual Hartogs' theorem. By composing with the diagonal mapping we see that $\lambda \circ f \circ (c_1, c_2)$ is holomorphic, thus f is holomorphic. □

7.10. Lemma. *Let $f : E \supseteq U \to F$ be holomorphic from a c^∞-open subset in a convenient vector space to another convenient vector space. Then the derivative $(df)^\wedge : U \times E \to F$ is again holomorphic and complex linear in the second variable.*

Proof. $(z, v, w) \mapsto f(v + zw)$ is holomorphic. We test with all holomorphic curves and linear functionals and see that $(v, w) \mapsto \frac{\partial}{\partial z}|_{z=0} f(v + zw) =: df(v)w$ is again holomorphic, $\mathbb{C}$-homogeneous in w by (7.4).

Now $w \mapsto df(v)w$ is a holomorphic and $\mathbb{C}$-homogeneous mapping $E \to F$. But any such mapping is automatically $\mathbb{C}$-linear: Composed with a bounded linear functional on F and restricted to any two dimensional subspace of E this is a finite dimensional assertion. □

7.11. Remark. In the definition of holomorphy (7.8) one could also have admitted subsets U which are only open in the final topology with respect to holomorphic curves. But then there is a counterexample to (7.10), see [Kriegl, Nel, 1985, 2.5].

7.12. Theorem. *A multilinear mapping between convenient vector spaces is holomorphic if and only if it is bounded.*

This result is false for not c^∞-complete vector spaces, see [Kriegl, Nel, 1985, 1.4].

Proof. Since both conditions can be tested in each factor separately by Hartogs' theorem (7.9) and by (5.19), and by testing with linear functionals, we may restrict our attention to linear mappings $f : E \to \mathbb{C}$ only.

By theorem (7.4.2) a bounded linear mapping is holomorphic. Conversely, suppose that $f : E \to \mathbb{C}$ is a holomorphic but unbounded linear functional. So there exists a sequence (a_n) in E with $|f(a_n)| > 1$ and $\{2^n a_n\}$ bounded. Consider the power series $\sum_{n=0}^\infty (a_n - a_{n-1})(2z)^n$. This describes a holomorphic curve c in E, by (7.3) and (7.4.2). Then $f \circ c$ is holomorphic and thus has a power series expansion $f(c(z)) = \sum_{n=0}^\infty b_n z^n$. On the other hand

$$f(c(z)) = \sum_{n=0}^{N} (f(a_n) - f(a_{n-1}))(2z)^n + (2z)^N f\left(\sum_{n>N} (a_n - a_{n-1})(2z)^{n-N}\right).$$

So $b_n = 2^n(f(a_n) - f(a_{n-1}))$ and we get the contradiction

$$0 = f(0) = f(c(1/2)) = \sum_{n=0}^{\infty} (f(a_n) - f(a_{n-1})) = \lim_{n\to\infty} f(a_n). \qquad \square$$

Parts of the following results (7.13) to (10.2) can be found in [Bochnak, Siciak, 1971]. For x in any vector space E let x^k denote the element $(x, \dots, x) \in E^k$.

7.13. Lemma. Polarization formulas. *Let $f : E \times \dots \times E \to F$ be an k-linear symmetric mapping between vector spaces. Then we have:*

(1) $$f(x_1, \dots, x_k) = \tfrac{1}{k!} \sum_{\varepsilon_1,\dots,\varepsilon_k=0}^{1} (-1)^{k-\Sigma\varepsilon_j} f\left(\left(x_0 + \sum \varepsilon_j x_j\right)^k\right).$$

(2) $$f(x^k) = \tfrac{1}{k!} \sum_{j=0}^{k} (-1)^{k-j} \tbinom{k}{j} f((a + jx)^k).$$

(3) $$f(x^k) = \tfrac{k^k}{k!} \sum_{j=0}^{k} (-1)^{k-j} \tbinom{k}{j} f((a + \tfrac{j}{k}x)^k).$$

(4) $$f(x_1^0 + \lambda x_1^1, \dots, x_k^0 + \lambda x_k^1) = \sum_{\varepsilon_1,\dots,\varepsilon_k=0}^{1} \lambda^{\Sigma\varepsilon_j} f(x_1^{\varepsilon_1}, \dots, x_k^{\varepsilon_k}).$$

Formula (4) will mainly be used for $\lambda = \sqrt{-1}$ in the passage to the complexification.

Proof. (1). (see [Mazur, Orlicz, 1935]). By multilinearity and symmetry the right hand side expands to

$$\sum_{j_0+\dots+j_k=k} \frac{A_{j_0,\dots,j_k}}{j_0!\cdots j_k!} f(\underbrace{x_0,\dots,x_0}_{j_0},\dots,\underbrace{x_k,\dots,x_k}_{j_k}),$$

where the coefficients are given by

$$A_{j_0,\dots,j_k} = \sum_{\varepsilon_1,\dots,\varepsilon_k=0}^{1} (-1)^{k-\Sigma\varepsilon_j}\varepsilon_1^{j_1}\cdots\varepsilon_k^{j_k}.$$

The only nonzero coefficient is $A_{0,1,\dots,1} = 1$.

(2). In formula (1) we put $x_0 = a$ and all $x_j = x$.

(3). In formula (2) we replace a by ka and pull k out of the k-linear expression $f((ka + jx)^k)$.

(4) is obvious. □

7.14. Lemma. Power series. *Let E be a real or complex Fréchet space and let f_k be a k-linear symmetric scalar valued bounded functional on E, for each $k \in \mathbb{N}$. Then the following statements are equivalent:*

(1) *$\sum_k f_k(x^k)$ converges pointwise on an absorbing subset of E.*
(2) *$\sum_k f_k(x^k)$ converges uniformly and absolutely on some neighborhood of 0.*
(3) *$\{f_k(x^k) : k \in \mathbb{N}, x \in U\}$ is bounded for some neighborhood U of 0.*
(4) *$\{f_k(x_1,\dots,x_k) : k \in \mathbb{N}, x_j \in U\}$ is bounded for some neighborhood U of 0.*

If any of these statements are satisfied over the reals, then also for the complexification of the functionals f_k.

Proof. (1) ⇒ (3) The set $A_{K,r} := \{x \in E : |f_k(x^k)| \le Kr^k \text{ for all } k\}$ is closed in E since every bounded multilinear mapping is continuous. The countable union $\bigcup_{K,r} A_{K,r}$ is E, since the series converges pointwise on an absorbing subset. Since E is Baire there are $K > 0$ and $r > 0$ such that the interior U of $A_{K,r}$ is non void. Let $x_0 \in U$ and let V be an absolutely convex neighborhood of 0 contained in $U - x_0$

From (7.13) (3) we get for all $x \in V$ the following estimate:

$$\begin{aligned} |f(x^k)| &\le \tfrac{k^k}{k!}\sum_{j=0}^{k}\tbinom{k}{j}|f((x_0 + \tfrac{j}{k}x)^k)| \\ &\le \tfrac{k^k}{k!}2^k K r^k \le K(2re)^k. \end{aligned}$$

Now we replace V by $\frac{1}{2re}V$ and get the result.

(3) $\Rightarrow$ (4) From (7.13) (1) we get for all $x_j \in U$ the estimate:

$$\begin{aligned}
|f(x_1,\dots,x_k)| &\le \tfrac{1}{k!} \sum_{\varepsilon_1,\dots,\varepsilon_k=0}^{1} |f\left((\textstyle\sum \varepsilon_j x_j)^k\right)| \\
&= \tfrac{1}{k!} \sum_{\varepsilon_1,\dots,\varepsilon_k=0}^{1} (\textstyle\sum \varepsilon_j)^k \left| f\left(\left(\frac{\sum \varepsilon_j x_j}{\sum \varepsilon_j}\right)^k\right)\right| \\
&\le \tfrac{1}{k!} \sum_{\varepsilon_1,\dots,\varepsilon_k=0}^{1} (\textstyle\sum \varepsilon_j)^k C \\
&\le \tfrac{1}{k!} \sum_{j=0}^{k} \tbinom{k}{j} j^k C \le C(2e)^k.
\end{aligned}$$

Now we replace U by $\frac{1}{2e}U$ and get (4).

Proof of (4) $\Rightarrow$ (2) The series converges on rU uniformly and absolutely for any $0 < r < 1$.

(2) $\Rightarrow$ (1) is clear.

(4), real case, $\Rightarrow$ (4), complex case, by (7.13.4) for $\lambda = \sqrt{-1}$. □

7.15. Lemma. *Let E be a complex convenient vector space and let f_k be a k-linear symmetric scalar valued bounded functional on E, for each $k \in \mathbb{N}$. If $\sum_k f_k(x^k)$ converges pointwise on E and $x \mapsto f(x) := \sum_k f_k(x^k)$ is bounded on bounded sets, then the power series converges uniformly on bounded sets.*

Proof. Let B be an absolutely convex bounded set in E. For $x \in 2B$ we apply the vector valued Cauchy inequalities from (7.6) to the holomorphic curve $z \mapsto f(zx)$ at $z = 0$ for $r = 1$ and get that $f_k(x^k)$ is contained in the closed absolutely convex hull of $\{f(zx) : |z| = 1\}$. So $\{f_k(x^k) : x \in 2B, k \in \mathbb{N}\}$ is bounded and the series converges uniformly on B. □

7.16. Example. We consider the power series $\sum_k k(x_k)^k$ on the Hilbert space $\ell^2 = \{x = (x_k) : \sum_k |x_k|^2 < \infty\}$. This series converges pointwise everywhere, it yields a holomorphic function f on ℓ^2 by (7.19.5) which however is unbounded on the unit sphere, so convergence cannot be uniform on the unit sphere.

The function $g : \mathbb{C}^{(\mathbb{N})} \times \ell^2 \to \mathbb{C}$ given by $g(x, y) := \sum_k x_k f(kx_1 y)$ is holomorphic since it is a finite sum locally along each holomorphic curve by (7.7), but its Taylor series at 0 does not converge uniformly on any neighborhood of 0 in the locally convex topology: A typical neighborhood is of the form $\{(x, y) : |x_k| \le \varepsilon_k \text{ for all } k, \|y\|_2 \le \varepsilon\}$ and so it contains points (x, y) with $|x_k f(kx_1 y)| \ge 1$, for all large k. This shows that lemma (7.14) is not true for arbitrary convenient vector spaces.

7.17. Corollary. *Let E be a real or complex Fréchet space and let f_k be a k-linear symmetric scalar valued bounded functional on E, for each $k \in \mathbb{N}$ such that* *the power series* $\sum f_k(x^k)$ *converges to* $f(x)$ *for* x *near 0 in* E. *Let* $\sum_{k\geq 1} a_k z^k$ *be a power series in* E *which converges to* $a(z) \in E$ *for* z *near 0 in* $\mathbb{C}$.

Then the composite

$$\sum_{k\geq 0}\sum_{n\geq 0}\sum_{\substack{k_1,\dots,k_n\in\mathbb{N}\\ k_1+\dots+k_n=k}} f_n(a_{k_1},\dots,a_{k_n})\, z^k$$

of the power series converges to $f \circ a$ *near 0.*

Proof. By (7.14) there exists a 0-neighborhood U in E such that $\{f_k(x_1,\dots,x_k) : k\in\mathbb{N}, x_j\in U\}$ is bounded. Since the series for a converges there is $r>0$ such that $a_k r^k \in U$ for all k. For $|z| < \frac{r}{2}$ we have

$$\begin{aligned} f(a(z)) &= \sum_{n\geq 0} f_n\Big(\sum_{k_1\geq 1} a_{k_1} z^{k_1},\dots,\sum_{k_n\geq 1} a_{k_n} z^{k_n}\Big)\\ &= \sum_{n\geq 0}\sum_{k_1\geq 1}\cdots\sum_{k_n\geq 1} f_n\big(a_{k_1},\dots,a_{k_n}\big) z^{k_1+\dots+k_n}\\ &= \sum_{k\geq 0}\sum_{n\geq 0}\sum_{\substack{k_1,\dots,k_n\in\mathbb{N}\\ k_1+\dots+k_n=k}} f_n(a_{k_1},\dots,a_{k_n})\, z^k, \end{aligned}$$

since the last complex series converges absolutely: the coefficient of z^k is a sum of $2^k - 1$ terms which are bounded when multiplied by r^k. The second equality follows from boundedness of all f_k. □

7.18. Almost continuous functions. In the proof of the next theorem we will need the following notion: A (real valued) function on a topological space is called *almost continuous* if removal of a meager set yields a continuous function on the remainder.

Lemma. [Hahn, 1932, p. 221] *A pointwise limit of a sequence of almost continuous functions on a Baire space is almost continuous.*

Proof. Let (f_k) be a sequence of almost continuous real valued functions on a Baire space X which converges pointwise to f. Since the complement of a meager set in a Baire space is again Baire we may assume that each function f_k is continuous on X. We denote by X_n the set of all $x\in X$ such that there exists $N\in\mathbb{N}$ and a neighborhood U of x with $|f_k(y) - f(y)| < \frac{1}{n}$ for all $k\geq N$ and all $y\in U$. The set X_n is clearly open.

We claim that each X_n is dense: Let V be a nonempty open subset of X. For $N\in\mathbb{N}$ the set $V_N := \{x\in V : |f_k(x) - f_\ell(x)| \leq \frac{1}{2n}$ for all $k,\ell\geq N\}$ is closed in V and $V = \bigcup_N V_N$ since the sequence (f_k) converges pointwise. Since V is a Baire space, some V_N contains a nonempty open set W. For each $y\in W$ we have $|f_k(y) - f_\ell(y)| \leq \frac{1}{2n}$ for all k, $\ell\geq N$. We take the pointwise limit for $\ell\to\infty$ and see that $W\subseteq V\cap X_n$.

Since X is Baire, the set $\bigcap_n X_n$ has a meager complement and obviously the restriction of f on this set is continuous. □

7.19. Theorem. *Let $f : E \supseteq U \to F$ be a mapping from a c^∞-open subset in a convenient vector space to another convenient vector space. Then the following assertions are equivalent:*

(1) *f is holomorphic.*
(2) *For all $\ell \in F^*$ and absolutely convex closed bounded sets B the mapping $\ell \circ f : E_B \to \mathbb{C}$ is holomorphic.*
(3) *f is holomorphic along all affine (complex) lines and is c^∞-continuous.*
(4) *f is holomorphic along all affine (complex) lines and is bounded on bornologically compact sets (i.e. those compact in some E_B).*
(5) *f is holomorphic along all affine (complex) lines and at each point the first derivative is a bounded linear mapping.*
(6) *f is c^∞-locally a convergent series of bounded homogeneous complex polynomials.*
(7) *f is holomorphic along all affine (complex) lines and in every connected component for the c^∞-topology there is at least one point where all derivatives are bounded multilinear mappings.*
(8) *f is smooth and the derivative is complex linear at every point.*
(9) *f is $\mathcal{L}\text{ip}^1$ in the sense of* (12.1) *and the derivative is complex linear at every point.*

Proof. (1) $\Leftrightarrow$ (2) By (7.6) every holomorphic curve factors locally over some E_B and we test with linear functionals on F.

So for the rest of the proof we may assume that $F = \mathbb{C}$. We prove the rest of the theorem first for the case where E is a Banach space.

(1) $\Rightarrow$ (5) By lemma (7.10) the derivative of f is holomorphic and $\mathbb{C}$-linear in the second variable. By (7.12) $f'(z)$ is bounded.

(5) $\Rightarrow$ (6) Choose a fixed point $z \in U$. Since f is holomorphic along each complex line through z it is given there by a pointwise convergent power series. By the classical Hartogs' theorem f is holomorphic along each finite dimensional linear subspace. The mapping $f' : E \supseteq U \to E'$ is well defined by assumption and is also holomorphic along each affine line since we may test this by all point evaluations: using (5.18) we see that it is smooth and by (7.4.8) it is a holomorphic curve. So the mapping

$$\begin{aligned} v \mapsto f^{(n+1)}(z)(v, v_1, \dots, v_n) &= (f'(\quad)(v))^{(n)}(z)(v_1, \dots, v_n) \\ &= (f')^{(n)}(z)(v_1, \dots, v_n)(v). \end{aligned}$$

is bounded, and by symmetry of higher derivatives at z they are thus separately bounded in all variables. By (5.19) f is given by a power series of bounded homogeneous polynomials which converges pointwise on the open set $\{z + v : z + \lambda v \in U \text{ for all } |\lambda| \le 1\}$. Now (6) follows from lemma (7.14).

(6) $\Rightarrow$ (3) By lemma (7.14) the series converges uniformly and hence f is continuous.

(3) $\Rightarrow$ (4) is obvious.

(4) ⇒ (5) By the (1-dimensional) Cauchy integral formula we have

$$f'(z)v = \frac{1}{2\pi\sqrt{-1}} \int_{|\lambda|=1} \frac{f(z+\lambda v)}{\lambda^2} d\lambda.$$

So $f'(z)$ is a linear functional which is bounded on compact sets K for which $\{z + \lambda v : |\lambda| \leq 1, v \in K\} \subseteq U$, thus it is bounded, by lemma (5.4).

(6) ⇒ (1) follows by composing the two locally uniformly converging power series, see corollary (7.17).

Sublemma. *Let E be a Fréchet space and let $U \subseteq E$ be open. Let $f : U \to \mathbb{C}$ be holomorphic along affine lines which is also the pointwise limit on U of a power series with bounded homogeneous composants. Then f is holomorphic on U.*

Proof. By assumption, and the lemma in (7.18) the function f is almost continuous, since it is the pointwise limit of polynomials. For each z the derivative $f'(z) : E \to \mathbb{C}$ as pointwise limit of difference quotients is also almost continuous on $\{v : z + \lambda v \in U \text{ for } |\lambda| \leq 1\}$, thus continuous on E since it is linear and by the Baire property.

By (5) ⇒ (1) the function f is holomorphic on U. □

(6) ⇒ (7) is obvious.

(7) ⇒ (1) [Zorn, 1945] We treat each connected component of U separately and assume thus that U is connected. The set $U_0 := \{z \in U : f \text{ is holomorphic near } z\}$ is open. By (6) ⇒ (1) f is holomorphic near the point, where all derivatives are bounded, so U_0 is not empty. From the sublemma above we see that for any point z in U_0 the whole star $\{z + v : z + \lambda v \in U \text{ for all } |\lambda| \leq 1\}$ is contained in U. Since U is in particular polygonally connected, we have $U_0 = U$.

(8) ⇒ (9) is trivial.

(9) ⇒ (3) Clearly, f is holomorphic along affine lines and c^∞-continuous.

(1) ⇒ (8) All derivatives are again holomorphic by (7.10) and thus locally bounded. So f is smooth by (5.20).

Now we treat the case where E is a general convenient vector space. Restricting to suitable spaces E_B transforms each of the statements into the weaker corresponding one where E is a Banach space. These pairs of statements are equivalent: This is obvious except the following two cases.

For (6) we argue as follows. The function $f|(U \cap E_B)$ satisfies condition (6) (so all the others) for each bounded closed absolutely convex $B \subseteq E$. By (5.20) f is smooth and it remains to show that the Taylor series at z converges pointwise on a c^∞-open neighborhood of z. The star $\{z+v : z+\lambda v \in U \text{ for all } |\lambda| le 1\}$ with center z in U is again c^∞-open by (4.17) and on it the Taylor series of f at z converges pointwise.

For (7) replace on both sides the condition "at least one point" by the condition "for all points". □

7.20. Chain rule. *The composition of holomorphic mappings is holomorphic and the usual formula for the derivative of the composite holds.*

Proof. Use (7.19.1) ⇔ (7.19.8), and the real chain rule (3.18). □

7.21. Definition. For convenient vector spaces E and F and for a c^∞-open subset $U \subseteq E$ we denote by $\mathcal{H}(U, F)$ the *space of all holomorphic mappings* $U \to F$. It is a closed linear subspace of $C^\infty(U, F)$ by (7.19.8) and we give it the induced convenient vector space structure.

7.22. Theorem. Cartesian closedness. *For convenient vector spaces E_1, E_2, and F, and for c^∞-open subsets $U_j \subseteq E_j$ a mapping $f : U_1 \times U_2 \to F$ is holomorphic if and only if the canonically associated mapping $f^\vee : U_1 \to \mathcal{H}(U_2, F)$ is holomorphic.*

Proof. Obviously, $f^\vee$ has values in $\mathcal{H}(U_2, F)$ and is smooth by smooth cartesian closedness (3.12). Since its derivative is canonically associated to the first partial derivative of f, it is complex linear. So $f^\vee$ is holomorphic by (7.19.8).

If conversely $f^\vee$ is holomorphic, then it is smooth into $\mathcal{H}(U_2, F)$ by (7.19), thus also smooth into $C^\infty(U_2, F)$. Thus, $f : U_1 \times U_2 \to F$ is smooth by smooth cartesian closedness. The derivative $df(x,y)(u,v) = (df^\vee(x)v)(y) + (d \circ f^\vee)(x)(y)w$ is obviously complex linear, so f is holomorphic. □

7.23. Corollary. *Let E etc. be convenient vector spaces and let U etc. be c^∞-open subsets of such. Then the following canonical mappings are holomorphic.*

$$\begin{aligned}
&\mathrm{ev} : \mathcal{H}(U, F) \times U \to F, \quad \mathrm{ev}(f, x) = f(x)\\
&\mathrm{ins} : E \to \mathcal{H}(F, E \times F), \quad \mathrm{ins}(x)(y) = (x, y)\\
&(\quad)^\wedge : \mathcal{H}(U, \mathcal{H}(V, G)) \to \mathcal{H}(U \times V, G)\\
&(\quad)^\vee : \mathcal{H}(U \times V, G) \to \mathcal{H}(U, \mathcal{H}(V, G))\\
&\mathrm{comp} : \mathcal{H}(F, G) \times \mathcal{H}(U, F) \to \mathcal{H}(U, G)\\
&\mathcal{H}(\ ,\ ) : \mathcal{H}(F, F') \times \mathcal{H}(U', E) \to \mathcal{H}(\mathcal{H}(E, F), \mathcal{H}(U', F'))\\
&\qquad (f, g) \mapsto (h \mapsto f \circ h \circ g)\\
&\prod : \prod \mathcal{H}(E_i, F_i) \to \mathcal{H}(\prod E_i, \prod F_i)
\end{aligned}$$

Proof. Just consider the canonically associated holomorphic mappings on multiple products. □

7.24. Theorem (Holomorphic functions on Fréchet spaces).
Let $U \subseteq E$ be open in a complex Fréchet space E. The following statements on $f : U \to \mathbb{C}$ are equivalent:

(1) *f is holomorphic.*
(2) *f is smooth and is locally given by its uniformly and absolutely converging Taylor series.*
(3) *f is locally given by a uniformly and absolutely converging power series.*

Proof. (1) $\Rightarrow$ (2) follows from (7.14.1) $\Rightarrow$ (7.14.2) and (7.19.1) $\Rightarrow$ (7.19.6).

(2) $\Rightarrow$ (3) is obvious.

(3) $\Rightarrow$ (1) is the chain rule for converging power series (7.17). □

8. Spaces of Holomorphic Mappings and Germs

8.1. Spaces of holomorphic functions. For a complex manifold N (always assumed to be separable) let $\mathcal{H}(N,\mathbb{C})$ be the *space of all holomorphic functions* on N with the topology of uniform convergence on compact subsets of N.

Let $\mathcal{H}_b(N,\mathbb{C})$ denote the Banach space of bounded holomorphic functions on N equipped with the supremum norm.

For any open subset W of N let $\mathcal{H}_{bc}(N \supseteq W,\mathbb{C})$ be the closed subspace of $\mathcal{H}_b(W,\mathbb{C})$ of all holomorphic functions on W which extend to continuous functions on the closure $\overline{W}$.

For a poly-radius $r = (r_1,\dots,r_n)$ with $r_i > 0$ and for $1 \le p \le \infty$ let ℓ^p_r denote the real Banach space $\left\{ x \in \mathbb{R}^{\mathbb{N}^n} : \|(x_\alpha r^\alpha)_{\alpha\in\mathbb{N}^n}\|_p < \infty \right\}$.

8.2. Theorem (Structure of $\mathcal{H}(N,\mathbb{C})$ for complex manifolds N).
The space $\mathcal{H}(N,\mathbb{C})$ of all holomorphic functions on N with the topology of uniform convergence on compact subsets of N is a (strongly) nuclear Fréchet space and embeds bornologically as a closed subspace into $C^\infty(N,\mathbb{R})^2$.

Proof. By taking a countable covering of N with compact sets, one obtains a countable neighborhood basis of 0 in $\mathcal{H}(N,\mathbb{C})$. Hence, $\mathcal{H}(N,\mathbb{C})$ is metrizable.

That $\mathcal{H}(N,\mathbb{C})$ is complete, and hence a Fréchet space, follows since the limit of a sequence of holomorphic functions with respect to the topology of uniform convergence on compact sets is again holomorphic.

The vector space $\mathcal{H}(N,\mathbb{C})$ is a subspace of $C^\infty(N,\mathbb{R}^2) = C^\infty(N,\mathbb{R})^2$ since a function $N \to \mathbb{C}$ is holomorphic if and only if it is smooth and the derivative at every point is $\mathbb{C}$-linear. It is a closed subspace, since it is described by the continuous linear equations $df(x)(\sqrt{-1}\cdot v) = \sqrt{-1}\cdot df(x)(v)$. Obviously, the identity from $\mathcal{H}(N,\mathbb{C})$ with the subspace topology to $\mathcal{H}(N,\mathbb{C})$ is continuous, hence by the open mapping theorem (52.11) for Fréchet spaces it is an isomorphism.

That $\mathcal{H}(N,\mathbb{C})$ is nuclear and unlike $C^\infty(N,\mathbb{R})$ even strongly nuclear can be shown as follows. For N equal to the open polycylinder $\mathbb{D}^n \subseteq \mathbb{C}^n$ this result can be found in (52.36). For an arbitrary N the space $\mathcal{H}(N,\mathbb{C})$ carries the initial topology induced by the linear mappings $(u^{-1})^* : \mathcal{H}(N,\mathbb{C}) \to \mathcal{H}(u(U),\mathbb{C})$ for all charts (u,U) of N, for which we may assume $u(U) = \mathbb{D}^n$, and hence by the stability properties of strongly nuclear spaces, cf. (52.34), $\mathcal{H}(N,\mathbb{C})$ is strongly nuclear. □

8.3. Spaces of germs of holomorphic functions. For a subset $A \subseteq N$ let $\mathcal{H}(N \supseteq A, \mathbb{C})$ be the space of *germs along A of holomorphic functions* $W \to \mathbb{C}$ for open sets W in N containing A. We equip $\mathcal{H}(N \supseteq A, \mathbb{C})$ with the locally convex topology induced by the inductive cone $\mathcal{H}(W, \mathbb{C}) \to \mathcal{H}(N \supseteq A, \mathbb{C})$ for all W. This is Hausdorff, since iterated derivatives at points in A are continuous functionals and separate points. In particular, $\mathcal{H}(N \supseteq W, \mathbb{C}) = \mathcal{H}(W, \mathbb{C})$ for W open in N. For $A_1 \subseteq A_2 \subseteq N$ the "restriction" mappings $\mathcal{H}(N \supseteq A_2, \mathbb{C}) \to \mathcal{H}(N \supseteq A_1, \mathbb{C})$ are continuous.

The structure of $\mathcal{H}(S^2 \supseteq A, \mathbb{C})$, where $A \subseteq S^2$ is a subset of the Riemannian sphere, has been studied by [Toeplitz, 1949], [Sebastião e Silva, 1950b,] [Van Hove, 1952], [Köthe, 1953], and [Grothendieck, 1953].

8.4. Theorem (Structure of $\mathcal{H}(N \supseteq K, \mathbb{C})$ for compact subsets K of complex manifolds N). *The following inductive cones are cofinal to each other.*

$$\mathcal{H}(N \supseteq K, \mathbb{C}) \leftarrow \{\mathcal{H}(W, \mathbb{C}), N \supseteq W \supseteq K\}$$
$$\mathcal{H}(N \supseteq K, \mathbb{C}) \leftarrow \{\mathcal{H}_b(W, \mathbb{C}), N \supseteq W \supseteq K\}$$
$$\mathcal{H}(N \supseteq K, \mathbb{C}) \leftarrow \{\mathcal{H}_{bc}(N \supseteq W, \mathbb{C}), N \supseteq W \supseteq K\}$$

If $K = \{z\}$ these inductive cones and the following ones for $1 \leq p \leq \infty$ are cofinal to each other.

$$\mathcal{H}(N \supseteq \{z\}, \mathbb{C}) \leftarrow \{\ell^p_r \otimes \mathbb{C},\ r \in \mathbb{R}^n_+\}$$

So all inductive limit topologies coincide. Furthermore, the space $\mathcal{H}(N \supseteq K, \mathbb{C})$ is a Silva space, i.e. a countable inductive limit of Banach spaces, where the connecting mappings between the steps are compact, i.e. mapping bounded sets to relatively compact ones. The connecting mappings are even strongly nuclear. In particular, the limit is regular, i.e. every bounded subset is contained and bounded in some step, and $\mathcal{H}(N \supseteq K, \mathbb{C})$ is complete and (ultra-)bornological (hence a convenient vector space), webbed, strongly nuclear and thus reflexive, and its dual is a nuclear Fréchet space. The space $\mathcal{H}(N \supseteq K, \mathbb{C})$ is smoothly paracompact. It is however not a Baire space.

Proof. Let $K \subseteq V \subseteq \overline{V} \subseteq W \subseteq N$, where W and V are open and $\overline{V}$ is compact. Then the obvious mappings

$$\mathcal{H}_{bc}(N \supseteq W, \mathbb{C}) \to \mathcal{H}_b(W, \mathbb{C}) \to \mathcal{H}(W, \mathbb{C}) \to \mathcal{H}_{bc}(N \supseteq V, \mathbb{C})$$

are continuous. This implies the first cofinality assertion. For $q \leq p$ and multiradii $s < r$ the obvious maps $\ell^q_r \to \ell^p_r$, $\ell^\infty_r \to \ell^1_s$, and $\ell^1_r \otimes \mathbb{C} \to \mathcal{H}_b(\{w \in \mathbb{C}^n : |w_i - z_i| < r_i\}, \mathbb{C}) \to \ell^\infty_s \otimes \mathbb{C}$ are continuous, by the Cauchy inequalities from the proof of (7.7). So the remaining cofinality assertion follows.

Let us show next that the connecting mapping $\mathcal{H}_b(W, \mathbb{C}) \to \mathcal{H}_b(V, \mathbb{C})$ is strongly nuclear (hence nuclear and compact). Since the restriction mapping from $E := \mathcal{H}(W, \mathbb{C})$ to $\mathcal{H}_b(V, \mathbb{C})$ is continuous, it factors over $E \to \widetilde{E_{(U)}}$ for some zero neighborhood U in E, where $\widetilde{E_{(U)}}$ is the completed quotient of E with the Minkowski functional of U as norm, see (52.15). Since E is strongly nuclear by (8.2), there exists by definition some larger 0-neighborhood U' in E such that the natural mapping $\widetilde{E_{(U')}} \to \widetilde{E_{(U)}}$ is strongly nuclear. So the claimed connecting mapping is strongly nuclear, since it can be factorized as

$$\mathcal{H}_b(W,\mathbb{C}) \to \mathcal{H}(W,\mathbb{C}) = E \to \widetilde{E_{(U')}} \to \widetilde{E_{(U)}} \to \mathcal{H}_b(V,\mathbb{C}).$$

So $\mathcal{H}(N \supseteq K,\mathbb{C})$ is a Silva space. It is strongly nuclear by the permanence properties of strongly nuclear spaces (52.34). By (16.10) this also shows that $\mathcal{H}(N \supseteq K,\mathbb{C})$ is smoothly paracompact. The remaining properties follow from (52.37). □

Completeness of $\mathcal{H}(\mathbb{C}^n \supseteq K,\mathbb{C})$ was shown in [Van Hove, 1952, théorème II], and for regularity of the inductive limit $\mathcal{H}(\mathbb{C} \supseteq K,\mathbb{C})$ see e.g. [Köthe, 1953, Satz 12].

8.5. Lemma. *For a closed subset $A \subseteq \mathbb{C}$ the spaces $\mathcal{H}(A \subseteq S^2,\mathbb{C})$ and the space $\mathcal{H}_\infty(S^2 \supseteq S^2 \setminus A,\mathbb{C})$ of all germs vanishing at ∞ are strongly dual to each other.*

Proof. This is due to [Köthe, 1953, Satz 12] and has been generalized by [Grothendieck, 1953,] théorème 2 bis, to arbitrary subsets $A \subseteq S^2$. □

Compare also the modern theory of hyperfunctions, cf. [Kashiwara, Kawai, Kimura, 1986].

8.6. Theorem (Structure of $\mathcal{H}(N \supseteq A,\mathbb{C})$ for closed subsets A of complex manifolds N). *The inductive cone*

$$\mathcal{H}(N \supseteq A,\mathbb{C}) \leftarrow \{\, \mathcal{H}(W,\mathbb{C}) : A \subseteq W \underset{\text{open}}{\subseteq} N\}$$

is regular, i.e. every bounded set is contained and bounded in some step.

The projective cone

$$\mathcal{H}(N \supseteq A,\mathbb{C}) \to \{\, \mathcal{H}(N \supseteq K,\mathbb{C}) : K \text{ compact in } A\}$$

generates the bornology of $\mathcal{H}(N \supseteq A,\mathbb{C})$.

The space $\mathcal{H}(N \supseteq A,\mathbb{C})$ is Montel (hence quasi-complete and reflexive), and ultrabornological (hence a convenient vector space). Furthermore, it is webbed and conuclear.

Proof. Compare also with the proof of the more general theorem (30.6).

We choose a continuous function $f : N \to \mathbb{R}$ which is positive and proper. Then $(f^{-1}([n,n+1]))_{n\in\mathbb{N}_0}$ is an exhaustion of N by compact subsets and $(K_n := A \cap f^{-1}([n,n+1]))$ is a compact exhaustion of A.

Let $\mathcal{B} \subseteq \mathcal{H}(N \supseteq A,\mathbb{C})$ be bounded. Then $\mathcal{B}|K$ is also bounded in $\mathcal{H}(N \supseteq K,\mathbb{C})$ for each compact subset K of A. Since the cone

$$\{\mathcal{H}(W,\mathbb{C}) : K \subseteq W \underset{\text{open}}{\subseteq} N\} \to \mathcal{H}(N \supseteq K,\mathbb{C})$$

is regular by (8.4), there exist open subsets W_K of N containing K such that $\mathcal{B}|K$ is contained (so that the extension of each germ is unique) and bounded in $\mathcal{H}(W_K,\mathbb{C})$. In particular, we choose $W_{K_n\cap K_{n+1}} \subseteq W_{K_n} \cap W_{K_{n+1}} \cap f^{-1}((n,n+2))$. Then we let W be the union of those connected components of

$$W' := \bigcup_n (W_{K_n} \cap f^{-1}((n,n+1))) \cup \bigcup_n W_{K_n\cap K_{n+1}}$$

which meet A. Clearly, W is open and contains A. Each $f \in \mathcal{B}$ has an extension to W': Extend $f|K_n$ uniquely to f_n on W_{K_n}. The function $f|(K_n \cap K_{n+1})$ has also a unique extension $f_{n,n+1}$ on $W_{K_n\cap K_{n+1}}$, so we have $f_n|W_{K_n\cap K_{n+1}} = f_{n,n+1}$. This extension of $f \in \mathcal{B}$ has a unique restriction to W. $\mathcal{B}$ is bounded in $\mathcal{H}(W,\mathbb{C})$ if it is uniformly bounded on each compact subset K of W. Each K is covered by finitely many W_{K_n} and $\mathcal{B}|K_n$ is bounded in $\mathcal{H}(W_{K_n},\mathbb{C})$, so $\mathcal{B}$ is bounded as required.

The space $\mathcal{H}(N \supseteq A,\mathbb{C})$ is ultra-bornological, Montel and in particular quasi-complete, and conuclear, as regular inductive limit of the nuclear Fréchet spaces $\mathcal{H}(W,\mathbb{C})$.

And it is webbed because it is the (ultra-)bornologification of the countable projective limit of webbed spaces $\mathcal{H}(N \supseteq K,\mathbb{C})$, see (52.14) and (52.13). □

8.7. Lemma. *Let A be closed in $\mathbb{C}$. Then the dual generated by the projective cone*

$$\mathcal{H}(\mathbb{C} \supseteq A,\mathbb{C}) \to \{\, \mathcal{H}(\mathbb{C} \supseteq K,\mathbb{C}),\ K \text{ compact in } A \,\}$$

is just the topological dual of $\mathcal{H}(\mathbb{C} \supseteq A,\mathbb{C})$.

Proof. The induced topology is obviously coarser than the given one. So let λ be a continuous linear functional on $\mathcal{H}(\mathbb{C} \supseteq A,\mathbb{C})$. Then we have $\lambda \in \mathcal{H}_\infty(S^2 \supseteq S^2 \setminus A,\mathbb{C})$ by (8.5). Hence, $\lambda \in \mathcal{H}_\infty(U,\mathbb{C})$ for some open neighborhood U of $S^2 \setminus A$, so again by (8.5) λ is a continuous functional on $\mathcal{H}(S^2 \supseteq K,\mathbb{C})$, where $K = S^2 \setminus U$ is compact in A. So λ is continuous for the induced topology. □

Problem. *Does this cone generate even the topology of $\mathcal{H}(\mathbb{C} \supseteq A,\mathbb{C})$? This would imply that the bornological topology on $\mathcal{H}(\mathbb{C} \supseteq A,\mathbb{C})$ is complete and nuclear.*

8.8. Lemma (Structure of $\mathcal{H}(N \supseteq A,\mathbb{C})$ for smooth closed submanifolds A of complex manifolds N). *The projective cone*

$$\mathcal{H}(N \supseteq A,\mathbb{C}) \to \{\, \mathcal{H}(N \supseteq \{z\},\mathbb{C}) : z \in A\}$$

generates the bornology.

Proof. Let $\mathcal{B} \subseteq \mathcal{H}(N \supseteq A,\mathbb{C})$ be such that the set $\mathcal{B}$ is bounded in $\mathcal{H}(N \supseteq \{z\},\mathbb{C})$ for all $z \in A$. By the regularity of the inductive cone $\mathcal{H}(\mathbb{C}^n \supseteq \{0\},\mathbb{C}) \leftarrow \mathcal{H}(W,\mathbb{C})$ we find arbitrary small open neighborhoods W_z such that the set $\mathcal{B}_z$ of the germs at z of all germs in $\mathcal{B}$ is contained and bounded in $\mathcal{H}(W_z,\mathbb{C})$.

Now choose a tubular neighborhood $p : U \to A$ of A in N. We may assume that W_z is contained in U, has fibers which are star shaped with respect to the zero-section and the intersection with A is connected. The union W of all the W_z, is therefore an open subset of U containing A. And it remains to show that the germs in $\mathcal{B}$ extend to W. For this it is enough to show that the extensions of the germs at z_1 and z_2 agree on the intersection of W_{z_1} with W_{z_2}. So let w be a point in the intersection. It can be radially connected with the base point $p(w)$, which itself can be connected by curves in A with z_1 and z_2. Hence, the extensions of both germs to $p(w)$ coincide with the original germ, and hence their extensions to w are equal.

That $\mathcal{B}$ is bounded in $\mathcal{H}(W, \mathbb{C})$, follows immediately since every compact subset $K \subseteq W$ can be covered by finitely many W_z. □

8.9. The following example shows that (8.8) fails to be true for general closed subsets $A \subseteq N$.

Example. *Let $A := \{\frac{1}{n} : n \in \mathbb{N}\} \cup \{0\}$. Then A is compact in $\mathbb{C}$ but the projective cone $\mathcal{H}(\mathbb{C} \supseteq A, \mathbb{C}) \to \{\mathcal{H}(\mathbb{C} \supseteq \{z\}, \mathbb{C}) : z \in A\}$ does not generate the bornology.*

Proof. Let $\mathcal{B} \subseteq \mathcal{H}(\mathbb{C} \supseteq A, \mathbb{C})$ be the set of germs of the following locally constant functions $f_n : \{x + iy \in \mathbb{C} : x \neq r_n\} \to \mathbb{C}$, with $f_n(x + iy)$ equal to 0 for $x < r_n$ and equal to 1 for $x > r_n$, where $r_n := \frac{2}{2n+1}$, for $n \in \mathbb{N}$. Then $\mathcal{B} \subseteq \mathcal{H}(\mathbb{C} \supseteq A, \mathbb{C})$ is not bounded, otherwise there would exist a neighborhood W of A such that the germ of f_n extends to a holomorphic mapping on W for all n. Since every f_n is 0 on some neighborhood of 0, these extensions have to be zero on the component of W containing 0, which is not possible, since $f_n(\frac{1}{n}) = 1$.

But on the other hand the set $\mathcal{B}_z \subseteq \mathcal{H}(\mathbb{C} \supseteq \{z\}, \mathbb{C})$ of germs at z of all germs in $\mathcal{B}$ is bounded, since it contains only the germs of the constant functions 0 and 1. □

8.10. Theorem (Holomorphic uniform boundedness principle). *Let E and F be complex convenient vector spaces, and let $U \subseteq E$ be a c^∞-open subset. Then $\mathcal{H}(U, F)$ satisfies the uniform boundedness principle for the point evaluations ev_x, $x \in U$.*

For any closed subset $A \subseteq N$ of a complex manifold N the locally convex space $\mathcal{H}(N \supseteq A, \mathbb{C})$ satisfies the uniform $\mathcal{S}$-boundedness principle for every point separating set $\mathcal{S}$ of bounded linear functionals.

Proof. By definition (7.21) $\mathcal{H}(U, F)$ carries the structure induced from the embedding into $C^\infty(U, F)$ and hence satisfies the uniform boundedness principle (5.26) and (5.25).

The second part is an immediate consequence of (5.24) and (8.6), and (8.4). □

Direct proof of a particular case of the second part. We prove the theorem for a closed smooth submanifold $A \subseteq \mathbb{C}$ and the set $\mathcal{S}$ of all iterated derivatives at points in A.

Let us suppose first that A is the point 0. We will show that condition (5.22.3) is satisfied. Let (b_n) be an unbounded sequence in $\mathcal{H}(\{0\}, \mathbb{C})$ such that each Taylor coefficient $b_{n,k} = \frac{1}{k!} b_n^{(k)}(0)$ is bounded with respect to n:

$$\sup\{ |b_{n,k}| : n \in \mathbb{N} \} < \infty. \tag{1}$$

We have to find $(t_n) \in \ell^1$ such that $\sum_n t_n b_n$ is no longer the germ of a holomorphic function at 0.

Each b_n has positive radius of convergence, in particular there is an $r_n > 0$ such that

$$\sup\{ |b_{n,k} r_n^k| : k \in \mathbb{N} \} < \infty. \tag{2}$$

By theorem (8.4) the space $\mathcal{H}(\{0\}, \mathbb{C})$ is a regular inductive limit of spaces ℓ_r^∞. Hence, a subset $\mathcal{B}$ is bounded in $\mathcal{H}(\{0\}, \mathbb{C})$ if and only if there exists an $r > 0$ such that $\{ \frac{1}{k!} b^{(k)}(0)\, r^k : b \in \mathcal{B}, k \in \mathbb{N} \}$ is bounded. That the sequence (b_n) is unbounded thus means that for all $r > 0$ there are n and k such that $|b_{n,k}| > (\frac{1}{r})^k$. We can even choose $k > 0$ for otherwise the set $\{ b_{n,k} r^k : n, k \in \mathbb{N}, k > 0 \}$ is bounded, so only $\{ b_{n,0} : n \in \mathbb{N} \}$ can be unbounded. This contradicts (1).

Hence, for each m there are $k_m > 0$ such that $\mathcal{N}_m := \{ n \in \mathbb{N} : |b_{n,k_m}| > m^{k_m} \}$ is not empty. We can choose (k_m) strictly increasing, for if they were bounded, $|b_{n,k_m}| < C$ for some C and all n by (1), but $|b_{n_m,k_m}| > m^{k_m} \to \infty$ for some n_m.

Since by (1) the set $\{ b_{n,k_m} : n \in \mathbb{N} \}$ is bounded, we can choose $n_m \in \mathcal{N}_m$ such that

$$\begin{aligned} &|b_{n_m,k_m}| \geq \tfrac{1}{2} |b_{j,k_m}| \quad \text{for } j > n_m \\ &|b_{n_m,k_m}| > m^{k_m} \end{aligned} \tag{3}$$

We can choose also (n_m) strictly increasing, for if they were bounded we would get $|b_{n_m,k_m} r^{k_m}| < C$ for some $r > 0$ and C by (2). But $(\frac{1}{m})^{k_m} \to 0$.

We pass now to the subsequence (b_{n_m}) which we denote again by (b_m). We put

$$t_m := \operatorname{sign}\left(\frac{1}{b_{m,k_m}} \sum_{j<m} t_j\, b_{j,k_m} \right) \cdot \frac{1}{4^m}. \tag{4}$$

Assume now that $b_\infty = \sum_m t_m b_m$ converges weakly with respect to $\mathcal{S}$ to a holomorphic germ. Then its Taylor series is $b_\infty(z) = \sum_{k\geq 0} b_{\infty,k} z^k$, where the coefficients are given by $b_{\infty,k} = \sum_{m\geq 0} t_m b_{m,k}$. But we may compute as follows, using (3) and (4) :

$$|b_{\infty,k_m}| \geq \left| \sum_{j\leq m} t_j\, b_{j,k_m} \right| - \sum_{j>m} |t_j\, b_{j,k_m}|$$

$$= \left|\sum_{j<m} t_j\, b_{j,k_m}\right| + |t_m\, b_{m,k_m}| \qquad \text{(same sign)}$$

$$- \sum_{j>m} |t_j\, b_{j,k_m}| \geq$$

$$\geq 0 + |b_{m,k_m}| \cdot \left(|t_m| - 2\sum_{j>m} |t_j|\right)$$

$$= |b_{m,k_m}| \cdot \frac{1}{3\cdot 4^m} \geq \frac{m^{k_m}}{3\cdot 4^m}.$$

So $|b_{\infty,k_m}|^{1/k_m}$ goes to ∞, hence b_∞ cannot have a positive radius of convergence, a contradiction. So the theorem follows for the space $\mathcal{H}(\{t\},\mathbb{C})$.

Let us consider now an arbitrary closed smooth submanifold $A \subseteq \mathbb{C}$. By (8.8) the projective cone $\mathcal{H}(N \supseteq A, \mathbb{C}) \to \{\mathcal{H}(N \supseteq \{z\}, \mathbb{C}), z \in A\}$ generates the bornology. Hence, the result follows from the case where $A = \{0\}$ by (5.25). □

9. Real Analytic Curves

9.1. As for smoothness and holomorphy we would like to obtain cartesian closedness for real analytic mappings. Thus, one should have at least the following:

$f : \mathbb{R}^2 \to \mathbb{R}$ is real analytic in the classical sense if and only if $f^\vee : \mathbb{R} \to C^\omega(\mathbb{R},\mathbb{R})$ is real analytic in some appropriate sense.

The following example shows that there are some subtleties involved.

Example. *The mapping $f : \mathbb{R}^2 \ni (s,t) \mapsto \frac{1}{(st)^2+1} \in \mathbb{R}$ is real analytic, whereas there is no reasonable topology on $C^\omega(\mathbb{R},\mathbb{R})$, such that the mapping $f^\vee : \mathbb{R} \to C^\omega(\mathbb{R},\mathbb{R})$ is locally given by its convergent Taylor series.*

Proof. For a topology on $C^\omega(\mathbb{R},\mathbb{R})$ to be reasonable we require only that all evaluations $\mathrm{ev}_t : C^\omega(\mathbb{R},\mathbb{R}) \to \mathbb{R}$ are bounded linear functionals. Now suppose that $f^\vee(s) = \sum_{k=0}^\infty f_k s^k$ converges in $C^\omega(\mathbb{R},\mathbb{R})$ for small s, where $f_k \in C^\omega(\mathbb{R},\mathbb{R})$. Then the series converges even bornologically, see (9.5) below, so $f(s,t) = \mathrm{ev}_t(f^\vee(s)) = \sum f_k(t)\, s^k$ for all t and small s. On the other hand $f(s,t) = \sum_{k=0}^\infty (-1)^k (st)^{2k}$ for $|s| < 1/|t|$. So for all t we have $f_k(t) = (-1)^m t^k$ for $k = 2m$, and 0 otherwise, since for fixed t we have a real analytic function in one variable. Moreover, the series $\left(\sum f_k z^k\right)(t) = \sum (-1)^k t^{2k} z^{2k}$ has to converge in $C^\omega(\mathbb{R},\mathbb{R}) \otimes \mathbb{C}$ for $|z| \leq \delta$ and all t, see (9.5). This is not the case: use $z = \sqrt{-1}\,\delta$, $t = 1/\delta$. □

There is, however, another notion of real analytic curves.

Example. Let $f : \mathbb{R} \to \mathbb{R}$ be a real analytic function with finite radius of convergence at 0. Now consider the curve $c : \mathbb{R} \to \mathbb{R}^{\mathbb{N}}$ defined by $c(t) := (f(k \cdot t))_{k\in\mathbb{N}}$. Clearly, the composite of c with any continuous linear functional is real analytic, since these functionals depend only on finitely many coordinates. But the Taylor series of c at 0 does not converge on any neighborhood of 0, since the radii of convergence of the coordinate functions go to 0. For an even more natural example see (11.8).

9.2. Lemma. *For a formal power series $\sum_{k\ge 0} a_k t^k$ with real coefficients the following conditions are equivalent.*

(1) *The series has positive radius of convergence.*
(2) *$\sum a_k r_k$ converges absolutely for all sequences (r_k) with $r_k\, t^k \to 0$ for all $t > 0$.*
(3) *The sequence $(a_k r_k)$ is bounded for all (r_k) with $r_k\, t^k \to 0$ for all $t > 0$.*
(4) *For each sequence (r_k) satisfying $r_k > 0$, $r_k r_\ell \ge r_{k+\ell}$, and $r_k\, t^k \to 0$ for all $t > 0$ there exists an $\varepsilon > 0$ such that $(a_k\, r_k\, \varepsilon^k)$ is bounded.*

This bornological description of real analytic curves will be rather important for the theory presented here, since condition (3) and (4) are linear conditions on the coefficients of a formal power series enforcing local convergence.

Proof. (1) $\Rightarrow$ (2) The series $\sum a_k r_k = \sum (a_k t^k)(r_k t^{-k})$ converges absolutely for some small t.

(2) $\Rightarrow$ (3) $\Rightarrow$ (4) is clear.

(4) $\Rightarrow$ (1) If the series has radius of convergence 0, then $\sum_k |a_k|\,(\frac{1}{n^2})^k = \infty$ for all n. There are $k_n \nearrow \infty$ with

$$\sum_{k=k_{n-1}}^{k_n-1} |a_k|\,(\tfrac{1}{n^2})^k \ge 1.$$

We put $r_k := (\frac{1}{n})^k$ for $k_{n-1} \le k < k_n$, then $\sum_k |a_k|\, r_k (\frac{1}{n})^k = \infty$ for all n, so $(a_k r_k (\frac{1}{2n})^k)_k$ is not bounded for any n, but $r_k\, t^k$, which equals $(\frac{t}{n})^k$ for $k_{n-1} \le k < k_n$, converges to 0 for all $t > 0$, and the sequence (r_k) is subadditive as required. □

9.3. Theorem (Description of real analytic functions). *For a smooth function $c : \mathbb{R} \to \mathbb{R}$ the following statements are equivalent.*

(1) *The function f is real analytic.*
(2) *For each sequence (r_k) with $r_k\, t^k \to 0$ for all $t > 0$, and each compact set K in $\mathbb{R}$, the set $\{\frac{1}{k!}\, c^{(k)}(a)\, r_k : a \in K, k \in \mathbb{N}\}$ is bounded.*
(3) *For each sequence (r_k) satisfying $r_k > 0$, $r_k r_\ell \ge r_{k+\ell}$, and $r_k\, t^k \to 0$ for all $t > 0$, and each compact set K in $\mathbb{R}$, there exists an $\varepsilon > 0$ such that $\{\frac{1}{k!}\, c^{(k)}(a)\, r_k\, \varepsilon^k : a \in K, k \in \mathbb{N}\}$ is bounded.*
(4) *For each compact set $K \subset \mathbb{R}$ there exist constants $M, \rho > 0$ with the property that $|\frac{1}{k!} c^{(k)}(a)| < M\rho^k$ for all $k \in \mathbb{N}$ and $a \in K$.*

Proof. (1) $\Rightarrow$ (4) Clearly, c is smooth. Since the Taylor series of c converges at a there are constants M_a, ρ_a satisfying the claimed inequality for fixed a. For a' with

$|a-a'| \le \frac{1}{2\rho_a}$ we obtain by differentiating $c(a') = \sum_{\ell\ge 0} \frac{c^{(\ell)}(a)}{\ell!}(a'-a)^\ell$ with respect to a' the estimate

$$\left|\frac{c^{(k)}(a')}{k!}\right| \le M_a\rho_a^k \frac{1}{k!}\frac{\partial^k}{\partial t^k}\Big|_{t=\frac{1}{2}} \frac{1}{1-t},$$

hence the condition is satisfied locally with some new constants M_a', ρ_a' incorporating the estimates for the coefficients of $\frac{1}{1-t}$. Since K is compact the claim follows.

(4) $\Rightarrow$ (2) We have $|\frac{1}{k!}\, c^{(k)}(a)\, r_k| \le M r_k (\rho)^k$ which is bounded since $r_k\rho^k \to 0$, as required.

(2) $\Rightarrow$ (3) follows by choosing $\varepsilon = 1$.

(3) $\Rightarrow$ (1) Let $a_k := \sup_{a\in K} |\frac{1}{k!}\, c^{(k)}(a)|$. Using (9.2).(4$\Rightarrow$1) these are the coefficients of a power series with positive radius ρ of convergence. Hence, the remainder $\frac{1}{(k+1)!}c^{(k+1)}(a+\theta(a'-a))(a'-a)^{k+1}$ of the Taylor series goes locally to zero. $\square$

9.4. Corollary. Real analytic curves. *For a curve $c : \mathbb{R} \to E$ in a convenient vector space E are equivalent:*

(1) *$\ell \circ c : \mathbb{R} \to \mathbb{R}$ is real analytic for all ℓ in some family of bounded linear functionals, which generates the bornology of E.*
(2) *$\ell \circ c : \mathbb{R} \to \mathbb{R}$ is real analytic for all $\ell \in E'$*

A curve satisfying these equivalent conditions will be called real analytic.

Proof. The non-trivial implication is (1 $\Rightarrow$ 2). So assume (1). By (2.14.6) the curve c is smooth and hence $\ell \circ c$ is smooth for all bounded linear $\ell : E \to \mathbb{R}$ and satisfies $(\ell\circ c)^{(k)}(t) = \ell(c^{(k)}(t))$. In order to show that $\ell\circ c$ is real analytic, we have to prove boundedness of

$$\ell\Big(\Big\{\frac{1}{k!}c^{(k)}(a)r_k : a\in K, k\in\mathbb{N}\Big\}\Big) = \Big\{\frac{1}{k!}(\ell\circ c)^{(k)}(a)r_k : a\in K, k\in\mathbb{N}\Big\}$$

for all compact $K\subset\mathbb{R}$ and appropriate r_k, by (9.3). Since ℓ is bounded it suffices to show that $\{\frac{1}{k!}c^{(k)}(a)r_k : a\in K, k\in\mathbb{N}\}$ is bounded, we follows since its image under all ℓ mentioned in (1) is bounded, again by (9.3). $\square$

9.5. Lemma. *Let E be a convenient vector space and let $c : \mathbb{R} \to E$ be a curve. Then the following conditions are equivalent.*

(1) *The curve c is locally given by a power series converging with respect to the locally convex topology.*
(2) *The curve c factors locally over a topologically real analytic curve into E_B for some bounded absolutely convex set $B \subseteq E$.*
(3) *The curve c extends to a holomorphic curve from some open neighborhood U of $\mathbb{R}$ in $\mathbb{C}$ into the complexification $(E_\mathbb{C}, E_\mathbb{C}')$.*

Where a curve satisfying condition (1) will be called topologically real analytic. One that satisfies condition (2) will be called bornologically real analytic.

Proof. (1) $\Rightarrow$ (3) For every $t\in\mathbb{R}$ one has for some $\delta > 0$ and all $|s| < \delta$ a converging power series representation $c(t+s) = \sum_{k=1}^\infty x_k s^k$. For any complex number z with $|z| < \delta$ the series converges for $z = s$ in $E_{\mathbb{C}}$, hence c can be locally extended to a holomorphic curve into $E_{\mathbb{C}}$. By the 1-dimensional uniqueness theorem for holomorphic maps, these local extensions fit together to give a holomorphic extension as required.

(3) $\Rightarrow$ (2) A holomorphic curve factors locally over $(E_{\mathbb{C}})_B$ by (7.6), where B can be chosen of the form $B \times \sqrt{-1}B$. Hence, the restriction of this factorization to $\mathbb{R}$ is real analytic into E_B.

(2) $\Rightarrow$ (1) Let c be bornologically real analytic, i.e. c is locally real analytic into some E_B, which we may assume to be complete. Hence, c is locally even topologically real analytic in E_B by (9.6) and so also in E. □

Although topological real analyticity is a strictly stronger than real analyticity, cf. (9.4), sometimes the converse is true as the following slight generalization of [Bochnak, Siciak, 1971, Lemma 7.1] shows.

9.6. Theorem. *Let E be a convenient vector space and assume that a Baire vector space topology on E' exists for which the point evaluations ev_x for $x \in E$ are continuous. Then any real analytic curve $c : \mathbb{R} \to E$ is locally given by its Mackey convergent Taylor series, and hence is bornologically real analytic and topologically real analytic for every locally convex topology compatible with the bornology.*

Proof. Since c is real analytic, it is smooth and all derivatives exist in E, since E is convenient, by (2.14.6).

Let us fix $t_0 \in \mathbb{R}$, let $a_n := \frac{1}{n!}c^{(n)}(t_0)$. It suffices to find some $r > 0$ for which $\{r^n a_n : n \in \mathbb{N}_0\}$ is bounded; because then $\sum t^n a_n$ is Mackey-convergent for $|t| < r$, and its limit is $c(t_0 + t)$ since we can test this with functionals.

Consider the sets $A_r := \{\lambda \in E' : |\lambda(a_n)| \leq r^n \text{ for all } n \in \mathbb{N}\}$. These A_r are closed in the Baire topology, since the point evaluations at a_n are continuous. Since c is real analytic, $\bigcup_{r>0} A_r = E'$, and by the Baire property there is an $r > 0$ such that the interior U of A_r is not empty. Let $\lambda_0 \in U$, then for all λ in the open neighborhood $U - \lambda_0$ of 0 we have $|\lambda(a_n)| \leq |(\lambda + \lambda_0)(a_n)| + |\lambda_0(a_n)| \leq 2r^n$. The set $U - \lambda_0$ is absorbing, thus for every $\lambda \in E'$ some multiple $\varepsilon\lambda$ is in $U - \lambda_0$ and so $\lambda(a_n) \leq \frac{2}{\varepsilon} r^n$ as required. □

9.7. Theorem. Linear real analytic mappings. *Let E and F be convenient vector spaces. For any linear mapping $\lambda : E \to F$ the following assertions are equivalent.*

(1) *λ is bounded.*
(2) *$\lambda \circ c : \mathbb{R} \to F$ is real analytic for all real analytic $c : \mathbb{R} \to E$.*
(3) *$\lambda \circ c : \mathbb{R} \to F$ is bornologically real analytic for all bornologically real analytic curves $c : \mathbb{R} \to E$*
(4) *$\lambda \circ c : \mathbb{R} \to F$ is real analytic for all bornologically real analytic curves $c : \mathbb{R} \to E$*

This will be generalized in (10.4) to non-linear mappings.

Proof. (1) $\Rightarrow$ (3) $\Rightarrow$ (4), and (2) $\Rightarrow$ (4) are obvious.

(4) $\Rightarrow$ (1) Let λ satisfy (4) and suppose that λ is unbounded. By composing with an $\ell \in E'$ we may assume that $\lambda : E \to \mathbb{R}$ and there is a bounded sequence (x_k) such that $\lambda(x_k)$ is unbounded. By passing to a subsequence we may suppose that $|\lambda(x_k)| > k^{2k}$. Let $a_k := k^{-k}\, x_k$, then $(r^k\, a_k)$ is bounded and $(r^k\, \lambda(a_k))$ is unbounded for all $r > 0$. Hence, the curve $c(t) := \sum_{k=0}^{\infty} t^k\, a_k$ is given by a Mackey convergent power series. So $\lambda \circ c$ is real analytic and near 0 we have $\lambda(c(t)) = \sum_{k=0}^{\infty} b_k\, t^k$ for some $b_k \in \mathbb{R}$. But $\lambda(c(t)) = \sum_{k=0}^{N} \lambda(a_k)t^k + t^N \lambda(\sum_{k>N} a_k t^{k-N})$ and $t \mapsto \sum_{k>N} a_k t^{k-N}$ is still a Mackey converging power series in E. Comparing coefficients we see that $b_k = \lambda(a_k)$ and consequently $\lambda(a_k)r^k$ is bounded for some $r > 0$, a contradiction.

Proof of (1) $\Rightarrow$ (2) Let $c : \mathbb{R} \to E$ be real analytic. By theorem (9.3) the set $\{\frac{1}{k!}\, c^{(k)}(a)\, r_k : a \in K, k \in \mathbb{N}\}$ is bounded for all compact sets $K \subset \mathbb{R}$ and for all sequences (r_k) with $r_k\, t^k \to 0$ for all $t > 0$. Since c is smooth and bounded linear mappings are smooth by (2.11), the function $\lambda \circ c$ is smooth and $(\lambda \circ c)^{(k)}(a) = \lambda(c^{(k)}(a))$. By applying (9.3) we obtain that $\lambda \circ c$ is real analytic. □

9.8. Corollary. *For two convenient vector space structures on a vector space E the following statements are equivalent:*

(1) *They have the same bounded sets.*
(2) *They have the same smooth curves.*
(3) *They have the same real analytic curves.*

Proof. (1) $\Leftrightarrow$ (2) was shown in (2.11). The implication (1) $\Rightarrow$ (3) follows from (9.3), which shows that real analyticity is a bornological concept, whereas the implication (1) $\Leftarrow$ (3) follows from (9.7). □

9.9. Corollary. *If a cone of linear maps $T_\alpha : E \to E_\alpha$ between convenient vector spaces generates the bornology on E, then a curve $c : \mathbb{R} \to E$ is C^ω resp. C^∞ provided all the composites $T_\alpha \circ c : \mathbb{R} \to E_\alpha$ are.*

Proof. The statement on the smooth curves is shown in (3.8). That on the real analytic curves follows again from the bornological condition of (9.3). □

10. Real Analytic Mappings

10.1. Theorem (Real analytic functions on Fréchet spaces). *Let $U \subseteq E$ be open in a real Fréchet space E. The following statements on $f : U \to \mathbb{R}$ are equivalent:*

(1) *f is smooth and is real analytic along topologically real analytic curves.*
(2) *f is smooth and is real analytic along affine lines.*
(3) *f is smooth and is locally given by its pointwise converging Taylor series.*
(4) *f is smooth and is locally given by its uniformly and absolutely converging Taylor series.*

(5) *f is locally given by a uniformly and absolutely converging power series.*
(6) *f extends to a holomorphic mapping $\tilde{f} : \tilde{U} \to \mathbb{C}$ for an open subset $\tilde{U}$ in the complexification $E_{\mathbb{C}}$ with $\tilde{U} \cap E = U$.*

Proof. (1) $\Rightarrow$ (2) is obvious. The implication (2) $\Rightarrow$ (3) follows from (7.14), (1) $\Rightarrow$ (2), whereas (3) $\Rightarrow$ (4) follows from (7.14),(2) $\Rightarrow$ (3), and (4) $\Rightarrow$ (5) is obvious.

Proof of (5) $\Rightarrow$ (6) Locally we can extend converging power series into the complexification by (7.14). Then we take the union $\tilde{U}$ of their domains of definition and use uniqueness to glue $\tilde{f}$ which is holomorphic by (7.24).

Proof of (6) $\Rightarrow$ (1) Obviously, f is smooth. Any topologically real analytic curve c in E can locally be extended to a holomorphic curve in $E_{\mathbb{C}}$ by (9.5). So $f \circ c$ is real analytic. □

10.2. The assumptions 'f is smooth' cannot be dropped in (10.1.1) even in finite dimensions, as shown by the following example, due to [Boman, 1967].

Example. *The mapping $f : \mathbb{R}^2 \to \mathbb{R}$, defined by $f(x,y) := \frac{xy^{n+2}}{x^2+y^2}$ is real analytic along real analytic curves, is n-times continuous differentiable but is not smooth and hence not real analytic.*

Proof. Take a real analytic curve $t \mapsto (x(t), y(t))$ into $\mathbb{R}^2$. The components can be factored as $x(t) = t^k u(t)$, $y(t) = t^k v(t)$ for some k and real analytic curves u, v with $u(0)^2 + v(0)^2 \neq 0$. The composite $f \circ (x,y)$ is then the function $t \mapsto t^{k(n+1)} \frac{uv^{n+2}}{u^2+v^2}(t)$, which is obviously real analytic near 0. The mapping f is n-times continuous differentiable, since it is real analytic on $\mathbb{R}^2 \backslash \{0\}$ and the directional derivatives of order i are $(n+1-i)$-homogeneous, hence continuously extendable to $\mathbb{R}^2$. But f cannot be $(n+1)$-times continuous differentiable, otherwise the derivative of order $n+1$ would be constant, and hence f would be a polynomial. □

10.3. Definition (Real analytic mappings). Let E be a convenient vector space. Let us denote by $C^\omega(\mathbb{R}, E)$ the *space of all real analytic curves.*

Let $U \subseteq E$ be c^∞-open, and let F be a second convenient vector space. A mapping $f : U \to F$ will be called *real analytic* or C^ω for short, if f is real analytic along real analytic curves and is smooth (i.e. is smooth along smooth curves); so $f \circ c \in C^\omega(\mathbb{R}, F)$ for all $c \in C^\omega(\mathbb{R}, E)$ with $c(\mathbb{R}) \subseteq U$ and $f \circ c \in C^\infty(\mathbb{R}, F)$ for all $c \in C^\infty(\mathbb{R}, E)$ with $c(\mathbb{R}) \subseteq U$. Let us denote by $C^\omega(U, F)$ the *space of all real analytic mappings* from U to F.

10.4. Analogue of Hartogs' Theorem for real analytic mappings. *Let E and F be convenient vector spaces, let $U \subseteq E$ be c^∞-open, and let $f : U \to F$. Then f is real analytic if and only if f is smooth and $\lambda \circ f$ is real analytic along each affine line in E, for all $\lambda \in F'$.*

Proof. One direction is clear, and by definition (10.3) we may assume that $F = \mathbb{R}$.

Let $c : \mathbb{R} \to U$ be real analytic. We show that $f \circ c$ is real analytic by using theorem (9.3). So let (r_k) be a sequence such that $r_k r_\ell \geq r_{k+\ell}$ and $r_k t^k \to 0$ for all $t > 0$

and let $K \subset \mathbb{R}$ be compact. We have to show, that there is an $\varepsilon > 0$ such that the set $\{\frac{1}{\ell!}(f \circ c)^{(\ell)}(a)\, r_l\, (\frac{\varepsilon}{2})^\ell : a \in K, \ell \in \mathbb{N}\}$ is bounded.

By theorem (9.3) the set $\{\frac{1}{n!}c^{(n)}(a)\, r_n : n \geq 1, a \in K\}$ is contained in some bounded absolutely convex subset $B \subseteq E$, such that E_B is a Banach space. Clearly, for the inclusion $i_B : E_B \to E$ the function $f \circ i_B$ is smooth and real analytic along affine lines. Since E_B is a Banach space, by (10.1.2) $\Rightarrow$ (10.1.4) $f \circ i_B$ is locally given by its uniformly and absolutely converging Taylor series. Then for each $a \in K$ by (7.14.2) $\Rightarrow$ (7.14.4) there is an $\varepsilon > 0$ such that the set $\{\frac{1}{k!}d^k f(c(a))(x_1, \dots, x_k) : k \in \mathbb{N}, x_j \in \varepsilon B\}$ is bounded. For each $y \in \frac{1}{2}\varepsilon B$ termwise differentiation gives $d^p f(c(a) + y)(x_1, \dots, x_p) = \sum_{k \geq p} \frac{1}{(k-p)!} d^k f(c(a))(x_1, \dots, x_p, y, \dots, y)$, so we may assume that $\{d^k f(c(a))(x_1, \dots, x_k)/k! : k \in \mathbb{N}, x_j \in \varepsilon B, a \in K\}$ is contained in $[-C, C]$ for some $C > 0$ and some uniform $\varepsilon > 0$.

The Taylor series of $f \circ c$ at a is given by

$$(f \circ c)(a+t) = \sum_{\ell \geq 0} \sum_{k \geq 0} \tfrac{1}{k!} \sum_{\substack{(m_n) \in \mathbb{N}_0^{\mathbb{N}} \\ \sum_n m_n = k \\ \sum_n m_n n = \ell}} \frac{k!}{\prod_n m_n!} d^k f(c(a)) \Big(\prod_n (\tfrac{1}{n!} c^{(n)}(a))^{m_n} \Big) t^\ell,$$

$$\text{where } \prod_n x_n^{m_n} := (\underbrace{x_1, \dots, x_1}_{m_1}, \dots, \underbrace{x_n, \dots, x_n}_{m_n}, \dots).$$

This follows easily from composing the Taylor series of f and c and ordering by powers of t. Furthermore, we have

$$\sum_{\substack{(m_n) \in \mathbb{N}_0^{\mathbb{N}} \\ \sum_n m_n = k \\ \sum_n m_n n = \ell}} \frac{k!}{\prod_n m_n!} = \binom{\ell-1}{k-1}$$

by the following argument: It is the ℓ-th Taylor coefficient at 0 of the function $(\sum_{n \geq 0} t^n - 1)^k = (\frac{t}{1-t})^k = t^k \sum_{j=0}^{\infty} \binom{-k}{j}(-t)^j$, which turns out to be the binomial coefficient in question.

By the foregoing considerations we may estimate as follows.

$$\begin{aligned}
&\tfrac{1}{\ell!} |(f \circ c)^{(\ell)}(a)|\, r_l\, (\tfrac{\varepsilon}{2})^\ell \leq \\
&\leq \sum_{k \geq 0} \Big| \tfrac{1}{k!} \sum_{\substack{(m_n) \in \mathbb{N}_0^{\mathbb{N}} \\ \sum_n m_n = k \\ \sum_n m_n n = \ell}} \frac{k!}{\prod_n m_n!} d^k f(c(a)) \Big(\prod_n (\tfrac{1}{n!} c^{(n)}(a))^{m_n} \Big) \Big|\, r_\ell\, (\tfrac{\varepsilon}{2})^\ell \\
&\leq \sum_{k \geq 0} \Big| \tfrac{1}{k!} \sum_{\substack{(m_n) \in \mathbb{N}_0^{\mathbb{N}} \\ \sum_n m_n = k \\ \sum_n m_n n = \ell}} \frac{k!}{\prod_n m_n!} d^k f(c(a)) \Big(\prod_n (\tfrac{1}{n!} c^{(n)}(a)\, r_n\, \varepsilon^n)^{m_n} \Big) \Big| \tfrac{1}{2^\ell} \\
&\leq \sum_{k \geq 0} \binom{\ell-1}{k-1} C \tfrac{1}{2^\ell} = \tfrac{1}{2} C,
\end{aligned}$$

because

$$\sum_{\substack{(m_n)\in\mathbb{N}_0^{\mathbb{N}}\\ \sum_n m_n=k\\ \sum_n m_n n=\ell}} \frac{k!}{\prod_n m_n!}\prod_n \left(\tfrac{1}{n!}c^{(n)}(a)\,\varepsilon^n\, r_n\right)^{m_n} \in \tbinom{\ell-1}{k-1}\,(\varepsilon B)^k \subseteq (E_B)^k. \quad \square$$

10.5. Corollary. *Let E and F be convenient vector spaces, let $U \subseteq E$ be c^∞-open, and let $f : U \to F$. Then f is real analytic if and only if f is smooth and $\lambda \circ f \circ c$ is real analytic for every periodic (topologically) real analytic curve $c : \mathbb{R} \to U \subseteq E$ and all $\lambda \in F'$.*

Proof. By (10.4) f is real analytic if and only if f is smooth and $\lambda \circ f$ is real analytic along topologically real analytic curves $c : \mathbb{R} \to E$. Let $h : \mathbb{R} \to \mathbb{R}$ be defined by $h(t) = t_0 + \varepsilon \cdot \sin t$. Then $c \circ h : \mathbb{R} \to \mathbb{R} \to U$ is a (topologically) real analytic, periodic function with period 2π, provided c is (topologically) real analytic. If $c(t_0) \in U$ we can choose $\varepsilon > 0$ such that $h(\mathbb{R}) \subseteq c^{-1}(U)$. Since *sin* is locally around 0 invertible, real analyticity of $\lambda \circ f \circ c \circ h$ implies that $\lambda \circ f \circ c$ is real analytic near t_0. Hence, the proof is completed. $\square$

10.6. Corollary. Reduction to Banach spaces. *Let E be a convenient vector space, let $U \subseteq E$ be c^∞-open, and let $f : U \to \mathbb{R}$ be a mapping. Then f is real analytic if and only if the restriction $f : E_B \supseteq U \cap E_B \to \mathbb{R}$ is real analytic for all bounded absolutely convex subsets B of E.*

So any result valid on Banach spaces can be translated into a result valid on convenient vector spaces.

Proof. By theorem (10.4) it suffices to check f along bornologically real analytic curves. These factor by definition (9.4) locally to real analytic curves into some E_B. $\square$

10.7. Corollary. *Let U be a c^∞-open subset in a convenient vector space E and let $f : U \to \mathbb{R}$ be real analytic. Then for every bounded B there is some $r_B > 0$ such that the Taylor series $y \mapsto \sum \frac{1}{k!} d^k f(x)(y^k)$ converges to $f(x+y)$ uniformly and absolutely on $r_B B$.*

Proof. Use (10.6) and (10.1.4). $\square$

10.8. Scalar analytic functions on convenient vector spaces E are in general not germs of holomorphic functions from $E_{\mathbb{C}}$ to $\mathbb{C}$:

Example. *Let $f_k : \mathbb{R} \to \mathbb{R}$ be real analytic functions with radius of convergence at zero converging to 0 for $k \to \infty$. Let $f : \mathbb{R}^{(\mathbb{N})} \to \mathbb{R}$ be the mapping defined on the countable sum $\mathbb{R}^{(\mathbb{N})}$ of the reals by $f(x_0, x_1, \dots) := \sum_{k=1}^{\infty} x_k f_k(x_0)$. Then f is real* *analytic, but there is no complex valued holomorphic mapping $\tilde{f}$ on some neighborhood of 0 in $\mathbb{C}^{(\mathbb{N})}$ which extends f, and the Taylor series of f is not pointwise convergent on any c^∞-open neighborhood of 0.*

Proof. Claim. f is real analytic.
Since the limit $\mathbb{R}^{(\mathbb{N})} = \varinjlim_n \mathbb{R}^n$ is regular, every smooth curve (and hence every real analytic curve) in $\mathbb{R}^{(\mathbb{N})}$ is locally smooth (resp. real analytic) into $\mathbb{R}^n$ for some n. Hence, $f \circ c$ is locally just a finite sum of smooth (resp. real analytic) functions and is therefore smooth (resp. real analytic).

Claim. f has no holomorphic extension.
Suppose there exists some holomorphic extension $\tilde{f} : U \to \mathbb{C}$, where $U \subseteq \mathbb{C}^{(\mathbb{N})}$ is c^∞-open neighborhood of 0, and is therefore open in the locally convex Silva topology by (4.11.2). Then U is even open in the box-topology (52.7), i.e., there exist $\varepsilon_k > 0$ for all k, such that $\{(z_k) \in \mathbb{C}^{(\mathbb{N})} : |z_k| \leq \varepsilon_k \text{ for all } k\} \subseteq U$. Let U_0 be the open disk in $\mathbb{C}$ with radius ε_0 and let $\tilde{f}_k : U_0 \to \mathbb{C}$ be defined by $\tilde{f}_k(z) := \tilde{f}(z, 0, \dots, 0, \varepsilon_k, 0, \dots)\frac{1}{\varepsilon_k}$, where ε_k is inserted instead of the variable x_k. Obviously, $\tilde{f}_k$ is an extension of f_k, which is impossible, since the radius of convergence of f_k is less than ε_0 for k sufficiently large.

Claim. The Taylor series does not converge.
If the Taylor series would be pointwise convergent on some U, then the previous arguments for $\mathbb{R}^{(\mathbb{N})}$ instead of $\mathbb{C}^{(\mathbb{N})}$ would show that the radii of convergence of the f_k were bounded from below. □

11. The Real Analytic Exponential Law

11.1. Spaces of germs of real-analytic functions. Let M be a real analytic finite dimensional manifold. If $f : M \to M'$ is a mapping between two such manifolds, then f is real analytic if and only if f maps smooth curves into smooth ones and real analytic curves into real analytic ones, by (10.1).

For each real analytic manifold M of real dimension m there is a *complex manifold* $M_\mathbb{C}$ of complex dimension m containing M as a real analytic closed submanifold, whose germ along M is unique ([Whitney, Bruhat, 1959, Prop. 1]), and which can be chosen even to be a Stein manifold, see [Grauert, 1958, section 3]. The complex charts are just extensions of the real analytic charts of an atlas of M into the complexification of the modeling real vector space.

Real analytic mappings $f : M \to M'$ are the germs along M of holomorphic mappings $W \to M'_\mathbb{C}$ for open neighborhoods W of M in $M_\mathbb{C}$.

Let $C^\omega(M, F)$ be the *space of real analytic functions* $f : M \to F$, for any convenient vector space F, and let $\mathcal{H}(M_\mathbb{C} \supseteq M, \mathbb{C})$ be the space of germs along M of holomorphic functions as in (8.3). Furthermore, for a subset $A \subseteq M$ let $C^\omega(M \supseteq A, \mathbb{R})$ denotes the *space of germs of real analytic functions* along A, defined on some neighborhood of A.

11.2. Lemma. *For any subset A of M the complexification of the real vector space $C^\omega(M \supseteq A, \mathbb{R})$ is the complex vector space $\mathcal{H}(M_\mathbb{C} \supseteq A, \mathbb{C})$.*

Definition. For any $A \subseteq M$ of a real analytic manifold M we will topologize the space sections $C^\omega(M \supseteq A, \mathbb{R})$ as subspace of $\mathcal{H}(M_\mathbb{C} \supseteq A, \mathbb{C})$, in fact as the real part of it.

Proof. Let $f, g \in C^\omega(M \supseteq A, \mathbb{R})$. These are germs of real analytic mappings defined on some open neighborhood of A in M. Inserting complex numbers into the locally convergent Taylor series in local coordinates shows, that f and g can be considered as holomorphic mappings from some neighborhood W of A in $M_\mathbb{C}$, which have real values if restricted to $W \cap M$. The mapping $h := f + \sqrt{-1}g : W \to \mathbb{C}$ gives then an element of $\mathcal{H}(M_\mathbb{C} \supseteq A, \mathbb{C})$.

Conversely, let $h \in \mathcal{H}(M_\mathbb{C} \supseteq A, \mathbb{C})$. Then h is the germ of a holomorphic function $\widetilde{h} : W \to \mathbb{C}$ for some open neighborhood W of A in $M_\mathbb{C}$. The decomposition of h into real and imaginary part $f = \frac{1}{2}(h + \bar{h})$ and $g = \frac{1}{2\sqrt{-1}}(h - \bar{h})$, which are real analytic functions if restricted to $W \cap M$, gives elements of $C^\omega(M \supseteq A, \mathbb{R})$.

These correspondences are inverse to each other since a holomorphic germ is determined by its restriction to a germ of mappings $M \supseteq A \to \mathbb{C}$. □

11.3. Lemma. *For a finite dimensional real analytic manifold M the inclusion $C^\omega(M, \mathbb{R}) \to C^\infty(M, \mathbb{R})$ is continuous.*

Proof. Consider the following diagram, where W is an open neighborhood of M in its complexification $M_\mathbb{C}$.

$$\begin{array}{ccc}
C^\omega(M,\mathbb{R}) & \xhookrightarrow{\text{inclusion}} & C^\infty(M,\mathbb{R}) \\
{\scriptstyle\text{direct summand}}\big\downarrow{\scriptstyle(11.2)} & & \big\downarrow{\scriptstyle\text{direct summand}} \\
\mathcal{H}(M_\mathbb{C} \supseteq M,\mathbb{C}) & \xhookrightarrow{\text{inclusion}} & C^\infty(M,\mathbb{R}^2) \\
{\scriptstyle\text{restriction}}\big\uparrow{\scriptstyle(8.4)} & & \big\uparrow{\scriptstyle\text{restriction}} \\
\mathcal{H}(W,\mathbb{C}) & \xhookrightarrow[(8.2)]{\text{inclusion}} & C^\infty(W,\mathbb{R}^2)
\end{array}$$

11.4. Theorem (Structure of $C^\omega(M \supseteq A, \mathbb{R})$ for closed subsets A of real analytic manifolds M). *The inductive cone*

$$C^\omega(M \supseteq A, \mathbb{R}) \leftarrow \{\, C^\omega(W, \mathbb{R}) : A \subseteq W \underset{\text{open}}{\subseteq} M\}$$

is regular, i.e. every bounded set is contained and bounded in some step.

The projective cone

$$C^\omega(M \supseteq A, \mathbb{R}) \to \{\, C^\omega(M \supseteq K, \mathbb{R}) : K \text{ compact in } A\}$$

generates the bornology of $C^\omega(M \supseteq A, \mathbb{R})$.

If A is even a smooth submanifold, then the following projective cone also generates the bornology.

$$C^\omega(M \supseteq A, \mathbb{R}) \to \{\, C^\omega(M \supseteq \{x\}, \mathbb{R}) : x \in A\}$$

The space $C^\omega(\mathbb{R}^m \supseteq \{0\}, \mathbb{R})$ is also the regular inductive limit of the spaces $\ell^p_r(r \in \mathbb{R}^m_+)$ for all $1 \le p \le \infty$, see (8.1).

For general closed $A \subseteq N$ the space $C^\omega(M \supseteq A, \mathbb{R})$ is Montel (hence quasi-complete and reflexive), and ultra-bornological (hence a convenient vector space). It is also webbed and conuclear. If A is compact then it is even a strongly nuclear Silva space and its dual is a nuclear Fréchet space and it is smoothly paracompact. It is however not a Baire space.

Proof. This follows using (11.2) from (8.4), (8.6), and (8.8) by passing to the real parts and from the fact that all properties are inherited by complemented subspaces as $C^\omega(M \supseteq A, \mathbb{R})$ of $\mathcal{H}(M_\mathbb{C} \supseteq A, \mathbb{C})$. $\square$

11.5. Corollary. *A subset $\mathcal{B} \subseteq C^\omega(\mathbb{R}^m \supseteq \{0\}, \mathbb{R})$ is bounded if and only if there exists an $r > 0$ such that $\{\frac{f^{(\alpha)}(0)}{\alpha!} r^{|\alpha|} : f \in \mathbb{B}, \alpha \in \mathbb{N}_0^m\}$ is bounded in $\mathbb{R}$.*

Proof. The space $C^\omega(\mathbb{R}^m \supseteq \{0\}, \mathbb{R})$ is the regular inductive limit of the spaces ℓ^∞_r for $r \in \mathbb{R}^m_+$ by (11.4). Hence, $\mathcal{B}$ is bounded if and only if it is contained and bounded in ℓ^∞_r for some $r \in \mathbb{R}^m_+$, which is the looked for condition. $\square$

11.6. Theorem (Special real analytic uniform boundedness principle). *For any closed subset $A \subseteq M$ of a real analytic manifold M, the space $C^\omega(M \supseteq A, \mathbb{R})$ satisfies the uniform $\mathcal{S}$-boundedness principle for any point separating set $\mathcal{S}$ of bounded linear functionals.*

If A has no isolated points and M is 1-dimensional this applies to the set of all point evaluations ev_t, $t \in A$.

Proof. Again this follows from (5.24) using now (11.4). If A has no isolated points and M is 1-dimensional the point evaluations are separating, by the uniqueness theorem for holomorphic functions. $\square$

Direct proof of a particular case. We show that $C^\omega(\mathbb{R}, \mathbb{R})$ satisfies the uniform $\mathcal{S}$-boundedness principle for the set $\mathcal{S}$ of all point evaluations.

We check property (5.22.2). Let $\mathcal{B} \subseteq C^\omega(\mathbb{R}, \mathbb{R})$ be absolutely convex such that $\mathrm{ev}_t(\mathcal{B})$ is bounded for all t and such that $C^\omega(\mathbb{R}, \mathbb{R})_B$ is complete. We have to show that $\mathcal{B}$ is complete.

By lemma (11.3) the set $\mathcal{B}$ satisfies the conditions of (5.22.2) in the space $C^\infty(\mathbb{R}, \mathbb{R})$. Since $C^\infty(\mathbb{R}, \mathbb{R})$ satisfies the uniform $\mathcal{S}$-boundedness principle, cf. [Frölicher, Kriegl, 1988], the set $\mathcal{B}$ is bounded in $C^\infty(\mathbb{R}, \mathbb{R})$. Hence, all iterated derivatives at points are bounded on $\mathcal{B}$, and a fortiori the conditions of (5.22.2) are satisfied for $\mathcal{B}$ in $\mathcal{H}(\mathbb{R}, \mathbb{C})$. By the particular case of theorem (8.10) the set $\mathcal{B}$ is bounded in $\mathcal{H}(\mathbb{R}, \mathbb{C})$ and hence also in the direct summand $C^\omega(\mathbb{R}, \mathbb{R})$. $\square$

11.7. Theorem. *The real analytic curves in $C^\omega(\mathbb{R},\mathbb{R})$ correspond exactly to the real analytic functions $\mathbb{R}^2 \to \mathbb{R}$.*

Proof. ($\Rightarrow$) Let $f : \mathbb{R} \to C^\omega(\mathbb{R},\mathbb{R})$ be a real analytic curve. Then $f : \mathbb{R} \to C^\omega(\mathbb{R} \supseteq \{t\},\mathbb{R})$ is also real analytic. We use theorems (11.4) and (9.6) to conclude that f is even a topologically real analytic curve in $C^\omega(\mathbb{R} \supseteq \{t\},\mathbb{R})$. By lemma (9.5) for every $s \in \mathbb{R}$ the curve f can be extended to a holomorphic mapping from an open neighborhood of s in $\mathbb{C}$ to the complexification (11.2) $\mathcal{H}(\mathbb{C} \supseteq \{t\},\mathbb{C})$ of $C^\omega(\mathbb{R} \supseteq \{t\},\mathbb{R})$.

From (8.4) it follows that $\mathcal{H}(\mathbb{C} \supseteq \{t\},\mathbb{C})$ is the regular inductive limit of all spaces $\mathcal{H}(U,\mathbb{C})$, where U runs through some neighborhood basis of t in $\mathbb{C}$. Lemma (7.7) shows that f is a holomorphic mapping $V \to \mathcal{H}(U,\mathbb{C})$ for some open neighborhoods U of t and V of s in $\mathbb{C}$.

By the exponential law for holomorphic mappings (see (7.22)) the canonically associated mapping $f^\wedge : V \times U \to \mathbb{C}$ is holomorphic. So its restriction is a real analytic function $\mathbb{R} \times \mathbb{R} \to \mathbb{R}$ near (s,t) which coincides with $f^\wedge$ for the original f.

($\Leftarrow$) Let $f : \mathbb{R}^2 \to \mathbb{R}$ be a real analytic mapping. Then $f(t, \quad)$ is real analytic, so the associated mapping $f^\vee : \mathbb{R} \to C^\omega(\mathbb{R},\mathbb{R})$ makes sense. It remains to show that it is real analytic. Since the mappings $C^\omega(\mathbb{R},\mathbb{R}) \to C^\omega(\mathbb{R} \supseteq K,\mathbb{R})$ generate the bornology, by (11.4), it is by (9.9) enough to show that $f^\vee : \mathbb{R} \to C^\omega(\mathbb{R} \supseteq K,\mathbb{R})$ is real analytic for each compact $K \subseteq \mathbb{R}$, which may be checked locally near each $s \in \mathbb{R}$.

$f : \mathbb{R}^2 \to \mathbb{R}$ extends to a holomorphic function on an open neighborhood $V \times U$ of $\{s\} \times K$ in $\mathbb{C}^2$. By cartesian closedness for the holomorphic setting the associated mapping $f^\vee : V \to \mathcal{H}(U,\mathbb{C})$ is holomorphic, so its restriction $V \cap \mathbb{R} \to C^\omega(U \cap \mathbb{R},\mathbb{R}) \to C^\omega(K,\mathbb{R})$ is real analytic as required. □

11.8. Remark. From (11.7) it follows that the curve $c : \mathbb{R} \to C^\omega(\mathbb{R},\mathbb{R})$ defined in (9.1) is real analytic, but it is not topologically real analytic. In particular, it does not factor locally to a real analytic curve into some Banach space $C^\omega(\mathbb{R},\mathbb{R})_B$ for a bounded subset B and it has no holomorphic extension to a mapping defined on a neighborhood of $\mathbb{R}$ in $\mathbb{C}$ with values in the complexification $\mathcal{H}(\mathbb{R},\mathbb{C})$ of $C^\omega(\mathbb{R},\mathbb{R})$, cf. (9.5).

11.9. Lemma. *For a real analytic manifold M, the bornology on $C^\omega(M,\mathbb{R})$ is induced by the following cone:*
$C^\omega(M,\mathbb{R}) \xrightarrow{c^*} C^\alpha(\mathbb{R},\mathbb{R})$ *for all C^α-curves $c : \mathbb{R} \to M$, where α equals ∞ and ω.*

Proof. The maps c^* are bornological since $C^\omega(M,\mathbb{R})$ is convenient by (11.4), and by the uniform $\mathcal{S}$-boundedness principle (11.6) for $C^\omega(\mathbb{R},\mathbb{R})$ and by (5.26) for $C^\infty(\mathbb{R},\mathbb{R})$ it suffices to check that $\mathrm{ev}_t \circ c^* = \mathrm{ev}_{c(t)}$ is bornological, which is obvious.

Conversely, we consider the identity mapping i from the space E into $C^\omega(M,\mathbb{R})$, where E is the vector space $C^\omega(M,\mathbb{R})$, but with the locally convex structure induced by the cone.

Claim. The bornology of E is complete.

The spaces $C^\omega(\mathbb{R},\mathbb{R})$ and $C^\infty(\mathbb{R},\mathbb{R})$ are convenient by (11.4) and (2.15), respectively. So their product

$$\prod_{c\in C^\omega(\mathbb{R},M)} C^\omega(\mathbb{R},\mathbb{R}) \times \prod_{c\in C^\infty(\mathbb{R},M)} C^\infty(\mathbb{R},\mathbb{R})$$

is also convenient. By theorem (10.1.1) $\Leftrightarrow$ (10.1.5) the embedding of E into this product has closed image, hence the bornology of E is complete.

Now we may apply the uniform S-boundedness principle for $C^\omega(M,\mathbb{R})$ (11.6), since obviously $\mathrm{ev}_p \circ i = \mathrm{ev}_0 \circ c_p^*$ is bounded, where c_p is the constant curve with value p, for all $p \in M$. □

11.10. Structure on $C^\omega(U,F)$. Let E be a real convenient vector space and let U be c^∞-open in E. We equip the space $C^\omega(U,\mathbb{R})$ of all real analytic functions (cf. (10.3)) with the locally convex topology induced by the families of mappings

$$C^\omega(U,\mathbb{R}) \xrightarrow{c^*} C^\omega(\mathbb{R},\mathbb{R}), \text{ for all } c \in C^\omega(\mathbb{R},U)$$

$$C^\omega(U,\mathbb{R}) \xrightarrow{c^*} C^\infty(\mathbb{R},\mathbb{R}), \text{ for all } c \in C^\infty(\mathbb{R},U).$$

For a finite dimensional vector spaces E this definition gives the same bornology as the one defined in (11.1), by lemma (11.9).

If F is another convenient vector space, we equip the space $C^\omega(U,F)$ of all real analytic mappings (cf. (10.3)) with the locally convex topology induced by the family of mappings

$$C^\omega(U,F) \xrightarrow{\lambda_*} C^\omega(U,\mathbb{R}), \text{ for all } \lambda \in F'.$$

Obviously, the injection $C^\omega(U,F) \to C^\infty(U,F)$ is bounded and linear.

11.11. Lemma. *Let E and F be convenient vector spaces and let $U \subseteq E$ be c^∞-open. Then $C^\omega(U,F)$ is also convenient.*

Proof. This follows immediately from the fact that $C^\omega(U,F)$ can be considered as closed subspace of the product of factors $C^\omega(U,\mathbb{R})$ indexed by all $\lambda \in F'$. And $C^\omega(U,\mathbb{R})$ can be considered as closed subspace of the product of the factors $C^\omega(\mathbb{R},\mathbb{R})$ indexed by all $c \in C^\omega(\mathbb{R},U)$ and the factors $C^\infty(\mathbb{R},\mathbb{R})$ indexed by all $c \in C^\infty(\mathbb{R},U)$. Since all factors are convenient so are the closed subspaces. □

11.12. Theorem (General real analytic uniform boundedness principle). *Let E and F be convenient vector spaces and $U \subseteq E$ be c^∞-open. Then $C^\omega(U,F)$ satisfies the uniform $\mathcal{S}$-boundedness principle, where $\mathcal{S} := \{ev_x : x \in U\}$.*

Proof. The convenient structure of $C^\omega(U,F)$ is induced by the cone of mappings $c^* : C^\omega(U,F) \to C^\omega(\mathbb{R},F)$ $(c \in C^\omega(\mathbb{R},U))$ together with the maps $c^* : C^\omega(U,F) \to C^\infty(\mathbb{R},F)$ $(c \in C^\infty(\mathbb{R},U))$. Both spaces $C^\omega(\mathbb{R},F)$ and $C^\infty(\mathbb{R},F)$ satisfy the uniform $\mathcal{T}$-boundedness principle, where $\mathcal{T} := \{ev_t : t \in \mathbb{R}\}$, by (11.6) and (5.26), respectively. Hence, $C^\omega(U,F)$ satisfies the uniform $\mathcal{S}$-boundedness principle by lemma (5.25), since $ev_t \circ c^* = ev_{c(t)}$. □

11.13. Remark. Let E and F be convenient vector spaces. Then $L(E,F)$, the space of bounded linear mappings from E to F, are by (9.7) exactly the real analytic ones.

11.14. Theorem. *Let E_i for $i = 1, \dots n$ and F be convenient vector spaces. Then the bornology on $L(E, \dots, E_n; F)$ (described in (5.1), see also (5.6)) is induced by the embedding $L(E_1, \dots, E_n; F) \to C^\omega(E_1 \times \dots E_n, F)$.*

Thus, mapping f into $L(E_1, \dots, E_n; F)$ is real analytic if and only if the composites $ev_x \circ f$ are real analytic for all $x \in E_1 \times \dots E_n$, by (9.9).

Proof. Let $\mathcal{S} = \{\mathrm{ev}_x : x \in E_1 \times \dots \times E_n\}$. Since $C^\omega(E_1 \times \dots \times E_n, F)$ satisfies the uniform $\mathcal{S}$-boundedness principle (11.12), the inclusion is bounded. On the other hand $L(E_1, \dots, E_n; F)$ also satisfies the uniform $\mathcal{S}$-boundedness principle by (5.18), so the identity from $L(E_1, \dots, E_n; F)$ with the bornology induced from $C^\omega(E_1 \times \dots \times E_n, F)$ into $L(E_1, \dots, E_n; F)$ is bounded as well.

Since to be real analytic depends only on the bornology by (9.4) and since the convenient vector space $L(E_1, \dots, E_n; F)$ satisfies the uniform $\mathcal{S}$-boundedness principle, the second assertion follows also. □

The following two results will be generalized in (11.20). At the moment we will make use of the following lemma only in case where $E = C^\infty(\mathbb{R}, \mathbb{R})$.

11.15. Lemma. *For any convenient vector space E the flip of variables induces an isomorphism $L(E, C^\omega(\mathbb{R}, \mathbb{R})) \cong C^\omega(\mathbb{R}, E')$ as vector spaces.*

Proof. For $c \in C^\omega(\mathbb{R}, E')$ consider $\tilde{c}(x) := \mathrm{ev}_x \circ c \in C^\omega(\mathbb{R}, \mathbb{R})$ for $x \in E$. By the uniform $\mathcal{S}$-boundedness principle (11.6) for $\mathcal{S} = \{\mathrm{ev}_t : t \in \mathbb{R}\}$ the linear mapping $\tilde{c}$ is bounded, since $\mathrm{ev}_t \circ \tilde{c} = c(t) \in E'$.

If conversely $\ell \in L(E, C^\omega(\mathbb{R}, \mathbb{R}))$, we consider $\tilde{\ell}(t) = \mathrm{ev}_t \circ \ell \in E' = L(E, \mathbb{R})$ for $t \in \mathbb{R}$. Since the bornology of E' is generated by $\mathcal{S} := \{ev_x : x \in E\}$, $\tilde{\ell} : \mathbb{R} \to E'$ is real analytic, for $\mathrm{ev}_x \circ \tilde{\ell} = \ell(x) \in C^\omega(\mathbb{R}, \mathbb{R})$, by (11.14). □

11.16. Corollary. *We have $C^\infty(\mathbb{R}, C^\omega(\mathbb{R}, \mathbb{R})) \cong C^\omega(\mathbb{R}, C^\infty(\mathbb{R}, \mathbb{R}))$ as vector spaces.*

Proof. The dual $C^\infty(\mathbb{R}, \mathbb{R})'$ is the free convenient vector space over $\mathbb{R}$ by (23.11), and $C^\omega(\mathbb{R}, \mathbb{R})$ is convenient, so we have

$$\begin{aligned} C^\infty(\mathbb{R}, C^\omega(\mathbb{R}, \mathbb{R})) &\cong L(C^\infty(\mathbb{R}, \mathbb{R})', C^\omega(\mathbb{R}, \mathbb{R})) \\ &\cong C^\omega(\mathbb{R}, C^\infty(\mathbb{R}, \mathbb{R})'') \qquad \text{by lemma (11.15)} \\ &\cong C^\omega(\mathbb{R}, C^\infty(\mathbb{R}, \mathbb{R})), \end{aligned}$$

by reflexivity of $C^\infty(\mathbb{R}, \mathbb{R})$, see (6.5.7). □

11.17. Theorem. *Let E be a convenient vector space, let U be c^∞-open in E, let $f : \mathbb{R} \times U \to \mathbb{R}$ be a real analytic mapping and let $c \in C^\infty(\mathbb{R}, U)$. Then $c^* \circ f^\vee : \mathbb{R} \to C^\omega(U, \mathbb{R}) \to C^\infty(\mathbb{R}, \mathbb{R})$ is real analytic.*

This result on the mixing of C^∞ and C^ω will become quite essential in the proof of cartesian closedness. It will be generalized in (11.21), see also (42.15).

Proof. Let $I \subseteq \mathbb{R}$ be open and relatively compact, let $t \in \mathbb{R}$ and $k \in \mathbb{N}$. Now choose an open and relatively compact $J \subseteq \mathbb{R}$ containing the closure $\bar{I}$ of I. There is a bounded subset $B \subseteq E$ such that $c \mid J : J \to E_B$ is a $\mathcal{L}ip^k$-curve in the Banach space E_B generated by B, by (1.8). Let U_B denote the open subset $U \cap E_B$ of the Banach space E_B. Since the inclusion $E_B \to E$ is continuous, f is real analytic as a function $\mathbb{R} \times U_B \to \mathbb{R} \times U \to \mathbb{R}$. Thus, by (10.1) there is a holomorphic extension $f : V \times W \to \mathbb{C}$ of f to an open set $V \times W \subseteq \mathbb{C} \times (E_B)_\mathbb{C}$ containing the compact set $\{t\} \times c(\bar{I})$. By cartesian closedness of the category of holomorphic mappings $f^\vee : V \to \mathcal{H}(W, \mathbb{C})$ is holomorphic. Now recall that the bornological structure of $\mathcal{H}(W, \mathbb{C})$ is induced by that of $C^\infty(W, \mathbb{C}) := C^\infty(W, \mathbb{R}^2)$. And $c^* : C^\infty(W, \mathbb{C}) \to \mathcal{L}ip^k(I, \mathbb{C})$ is a bounded $\mathbb{C}$-linear map, by the chain rule (12.8) for $\mathcal{L}ip^k$-mappings and by the uniform boundedness principle for the point evaluations (12.9). Thus, $c^* \circ f^\vee : V \to \mathcal{L}ip^k(I, \mathbb{C})$ is holomorphic, and hence its restriction to $\mathbb{R} \cap V$, which has values in $\mathcal{L}ip^k(I, \mathbb{R})$, is (even topologically) real analytic by (9.5). Since $t \in \mathbb{R}$ was arbitrary we conclude that $c^* \circ f^\vee : \mathbb{R} \to \mathcal{L}ip^k(I, \mathbb{R})$ is real analytic. But the bornology of $C^\infty(\mathbb{R}, \mathbb{R})$ is generated by the inclusions into $\mathcal{L}ip^k(I, \mathbb{R})$, by the uniform boundedness principles (5.26) for $C^\infty(\mathbb{R}, \mathbb{R})$ and (12.9) for $\mathcal{L}ip^k(\mathbb{R}, \mathbb{R})$, and hence $c^* \circ f^\vee : \mathbb{R} \to C^\infty(\mathbb{R}, \mathbb{R})$ is real analytic. □

11.18. Theorem. Cartesian closedness. *The category of real analytic mappings between convenient vector spaces is cartesian closed. More precisely, for convenient vector spaces E, F and G and c^∞-open sets $U \subseteq E$ and $W \subseteq G$ a mapping $f : W \times U \to F$ is real analytic if and only if $f^\vee : W \to C^\omega(U, F)$ is real analytic.*

Proof. Step 1. The theorem is true for $W = G = F = \mathbb{R}$.

($\Leftarrow$) Let $f^\vee : \mathbb{R} \to C^\omega(U, \mathbb{R})$ be C^ω. We have to show that $f : \mathbb{R} \times U \to \mathbb{R}$ is C^ω. We consider a curve $c_1 : \mathbb{R} \to \mathbb{R}$ and a curve $c_2 : \mathbb{R} \to U$.

If the c_i are C^∞, then $c_2^* \circ f^\vee : \mathbb{R} \to C^\omega(U, \mathbb{R}) \to C^\infty(\mathbb{R}, \mathbb{R})$ is C^ω by assumption, hence is C^∞, so $c_2^* \circ f^\vee \circ c_1 : \mathbb{R} \to C^\infty(\mathbb{R}, \mathbb{R})$ is C^∞. By cartesian closedness of smooth mappings, $(c_2^* \circ f^\vee \circ c_1)^\wedge = f \circ (c_1 \times c_2) : \mathbb{R}^2 \to \mathbb{R}$ is C^∞. By composing with the diagonal mapping $\Delta : \mathbb{R} \to \mathbb{R}^2$ we obtain that $f \circ (c_1, c_2) : \mathbb{R} \to \mathbb{R}$ is C^∞.

If the c_i are C^ω, then $c_2^* \circ f^\vee : \mathbb{R} \to C^\omega(U, \mathbb{R}) \to C^\omega(\mathbb{R}, \mathbb{R})$ is C^ω by assumption, so $c_2^* \circ f^\vee \circ c_1 : \mathbb{R} \to C^\omega(\mathbb{R}, \mathbb{R})$ is C^ω. By theorem (11.7) the associated map $(c_2^* \circ f^\vee \circ c_1)^\wedge = f \circ (c_1 \times c_2) : \mathbb{R}^2 \to \mathbb{R}$ is C^ω. So $f \circ (c_1, c_2) : \mathbb{R} \to \mathbb{R}$ is C^ω.

($\Rightarrow$) Let $f : \mathbb{R} \times U \to \mathbb{R}$ be C^ω. We have to show that $f^\vee : \mathbb{R} \to C^\omega(U, \mathbb{R})$ is real analytic. Obviously, $f^\vee$ has values in this space. We consider a curve $c : \mathbb{R} \to U$.

If c is C^∞, then by theorem (11.17) the associated mapping $c^* \circ f^\vee : \mathbb{R} \to C^\infty(\mathbb{R}, \mathbb{R})$ is C^ω.

If c is C^ω, then $f \circ (\mathrm{Id} \times c) : \mathbb{R} \times \mathbb{R} \to \mathbb{R} \times U \to \mathbb{R}$ is C^ω. By theorem (11.7) the associated mapping $(f \circ (\mathrm{Id} \times c))^\vee = c^* \circ f^\vee : \mathbb{R} \to C^\omega(\mathbb{R}, \mathbb{R})$ is C^ω.

Step 2. The theorem is true for $F = \mathbb{R}$.

($\Leftarrow$) Let $f^\vee : W \to C^\omega(U, \mathbb{R})$ be C^ω. We have to show that $f : W \times U \to \mathbb{R}$ is C^ω. We consider a curve $c_1 : \mathbb{R} \to W$ and a curve $c_2 : \mathbb{R} \to U$.

If the c_i are C^∞, then $c_2^* \circ f^\vee : W \to C^\omega(U, \mathbb{R}) \to C^\infty(\mathbb{R}, \mathbb{R})$ is C^ω by assumption, hence is C^∞, so $c_2^* \circ f^\vee \circ c_1 : \mathbb{R} \to C^\infty(\mathbb{R}, \mathbb{R})$ is C^∞. By cartesian closedness of smooth mappings, the associated mapping $(c_2^* \circ f^\vee \circ c_1)^\wedge = f \circ (c_1 \times c_2) : \mathbb{R}^2 \to \mathbb{R}$ is C^∞. So $f \circ (c_1, c_2) : \mathbb{R} \to \mathbb{R}$ is C^∞.

If the c_i are C^ω, then $f^\vee \circ c_1 : \mathbb{R} \to W \to C^\omega(U, \mathbb{R})$ is C^ω by assumption, so by step 1 the mapping $(f^\vee \circ c_1)^\wedge = f \circ (c_1 \times \mathrm{Id}_U) : \mathbb{R} \times U \to \mathbb{R}$ is C^ω. Hence, $f \circ (c_1, c_2) = f \circ (c_1 \times \mathrm{Id}_U) \circ (\mathrm{Id}, c_2) : \mathbb{R} \to \mathbb{R}$ is C^ω.

($\Rightarrow$) Let $f : W \times U \to \mathbb{R}$ be C^ω. We have to show that $f^\vee : W \to C^\omega(U, \mathbb{R})$ is real analytic. Obviously, $f^\vee$ has values in this space. We consider a curve $c_1 : \mathbb{R} \to W$.

If c_1 is C^∞, we consider a second curve $c_2 : \mathbb{R} \to U$. If c_2 is C^∞, then $f \circ (c_1 \times c_2) : \mathbb{R} \times \mathbb{R} \to W \times U \to \mathbb{R}$ is C^∞. By cartesian closedness the associated mapping $(f \circ (c_1 \times c_2))^\vee = c_2^* \circ f^\vee \circ c_1 : \mathbb{R} \to C^\infty(\mathbb{R}, \mathbb{R})$ is C^∞. If c_2 is C^ω, the mapping $f \circ (\mathrm{Id}_W \times c_2) : W \times \mathbb{R} \to \mathbb{R}$ and also the flipped one $(f \circ (\mathrm{Id}_W \times c_2))^\sim : \mathbb{R} \times W \to \mathbb{R}$ are C^ω, hence by theorem (11.17) $c_1^* \circ ((f \circ (\mathrm{Id}_W \times c_2))^\sim)^\vee : \mathbb{R} \to C^\infty(\mathbb{R}, \mathbb{R})$ is C^ω. By corollary (11.16) the associated mapping $(c_1^* \circ ((f \circ (\mathrm{Id}_W \times c_2))^\sim)^\vee)^\sim = c_2^* \circ f^\vee \circ c_1 : \mathbb{R} \to C^\omega(\mathbb{R}, \mathbb{R})$ is C^∞. So for both families describing the structure of $C^\omega(U, \mathbb{R})$ we have shown that the composite with $\check{f} \circ c_1$ is C^∞, so $f^\vee \circ c_1$ is C^∞.

If c_1 is C^ω, then $f \circ (c_1 \times \mathrm{Id}_U) : \mathbb{R} \times U \to W \times U \to \mathbb{R}$ is C^ω. By step 1 the associated mapping $(f \circ (c_1 \times \mathrm{Id}_U))^\vee = f^\vee \circ c_1 : \mathbb{R} \to C^\omega(U, \mathbb{R})$ is C^ω.

Step 3. The general case.

$$\begin{aligned} & f : W \times U \to F \text{ is } C^\omega \\ \Leftrightarrow\quad & \lambda \circ f : W \times U \to \mathbb{R} \text{ is } C^\omega \text{ for all } \lambda \in F' \\ \Leftrightarrow\quad & (\lambda \circ f)^\vee = \lambda_* \circ f^\vee : W \to C^\omega(U, \mathbb{R}) \text{ is } C^\omega, \text{ by step 2 and (11.10)} \\ \Leftrightarrow\quad & f^\vee : W \to C^\omega(U, F) \text{ is } C^\omega. \quad \square \end{aligned}$$

11.19. Corollary. Canonical mappings are real analytic. *The following mappings are C^ω:*

(1) $\mathrm{ev} : C^\omega(U, F) \times U \to F$, $(f, x) \mapsto f(x)$,
(2) $\mathrm{ins} : E \to C^\omega(F, E \times F)$, $x \mapsto (y \mapsto (x, y))$,
(3) $(\quad)^\wedge : C^\omega(U, C^\omega(V, G)) \to C^\omega(U \times V, G)$,
(4) $(\quad)^\vee : C^\omega(U \times V, G) \to C^\omega(U, C^\omega(V, G))$,
(5) $\mathrm{comp} : C^\omega(F, G) \times C^\omega(U, F) \to C^\omega(U, G)$, $(f, g) \mapsto f \circ g$,
(6) $C^\omega(\quad,\quad) : C^\omega(E_2, E_1) \times C^\omega(F_1, F_2) \to$
$\to C^\omega(C^\omega(E_1, F_1), C^\omega(E_2, F_2))$, $(f, g) \mapsto (h \mapsto g \circ h \circ f)$.

Proof. Just consider the canonically associated smooth mappings on multiple products, as in (3.13). □

11.20. Lemma. Canonical isomorphisms. *One has the following natural isomorphisms:*

(1) $C^\omega(W_1, C^\omega(W_2, F)) \cong C^\omega(W_2, C^\omega(W_1, F))$,
(2) $C^\omega(W_1, C^\infty(W_2, F)) \cong C^\infty(W_2, C^\omega(W_1, F))$.
(3) $C^\omega(W_1, L(E, F)) \cong L(E, C^\omega(W_1, F))$.
(4) $C^\omega(W_1, \ell^\infty(X, F)) \cong \ell^\infty(X, C^\omega(W_1, F))$.
(5) $C^\omega(W_1, \mathcal{L}\text{ip}^k(X, F)) \cong \mathcal{L}\text{ip}^k(X, C^\omega(W_1, F))$.

In (4) *the space* X *is a* ℓ^∞*-space, i.e. a set together with a bornology induced by a family of real valued functions on* X, *cf.* [Frölicher, Kriegl, 1988, 1.2.4]. *In* (5) *the space* X *is a* $\mathcal{L}\text{ip}^k$*-space, cf.* [Frölicher, Kriegl, 1988, 1.4.1]. *The spaces* $\ell^\infty(X, F)$ *and* $\mathcal{L}\text{ip}^k(W, F)$ *are defined in* [Frölicher, Kriegl, 1988, 3.6.1 and 4.4.1].

Proof. All isomorphisms, as well as their inverse mappings, are given by the flip of coordinates: $f \mapsto \tilde{f}$, where $\tilde{f}(x)(y) := f(y)(x)$. Furthermore, all occurring function spaces are convenient and satisfy the uniform $\mathcal{S}$-boundedness theorem, where $\mathcal{S}$ is the set of point evaluations, by (11.11), (11.14), (11.12), and by [Frölicher, Kriegl, 1988, 3.6.1, 4.4.2, 3.6.6, and 4.4.7].

That $\tilde{f}$ has values in the corresponding spaces follows from the equation $\tilde{f}(x) = ev_x \circ f$. One only has to check that $\tilde{f}$ itself is of the corresponding class, since it follows that $f \mapsto \tilde{f}$ is bounded. This is a consequence of the uniform boundedness principle, since

$$(\text{ev}_x \circ (\tilde{\ \ }))(f) = \text{ev}_x(\tilde{f}) = \tilde{f}(x) = \text{ev}_x \circ f = (\text{ev}_x)_*(f).$$

That $\tilde{f}$ is of the appropriate class in (1) and (2) follows by composing with $c_1 \in C^{\beta_1}(\mathbb{R}, W_1)$ and $C^{\beta_2}(\lambda, c_2) : C^{\alpha_2}(W_2, F) \to C^{\beta_2}(\mathbb{R}, \mathbb{R})$ for all $\lambda \in F'$ and $c_2 \in C^{\beta_2}(\mathbb{R}, W_2)$, where β_k and α_k are in $\{\infty, \omega\}$ and $\beta_k \le \alpha_k$ for $k \in \{1, 2\}$. Then $C^{\beta_2}(\lambda, c_2) \circ \tilde{f} \circ c_1 = (C^{\beta_1}(\lambda, c_1) \circ f \circ c_2)^\sim : \mathbb{R} \to C^{\beta_2}(\mathbb{R}, \mathbb{R})$ is C^{β_1} by (11.7) and (11.16), since $C^{\beta_1}(\lambda, c_1) \circ f \circ c_2 : \mathbb{R} \to W_2 \to C^{\alpha_1}(W_1, F) \to C^{\beta_1}(\mathbb{R}, \mathbb{R})$ is C^{β_2}.

That $\tilde{f}$ is of the appropriate class in (3) follows, since $L(E, F)$ is the c^∞-closed subspace of $C^\omega(E, F)$ formed by the linear C^ω-mappings.

That $\tilde{f}$ is of the appropriate class in (4) or (5) follows from (3), using the free convenient vector spaces $\ell^1(X)$ or $\lambda^k(X)$ over the ℓ^∞-space X or the the $\mathcal{L}\text{ip}^k$-space X, see [Frölicher, Kriegl, 1988, 5.1.24 or 5.2.3], satisfying $\ell^\infty(X, F) \cong L(\ell^1(X), F)$ or satisfying $\mathcal{L}\text{ip}^k(X, F) \cong L(\lambda^k(X), F)$. Existence of these free convenient vector spaces can be proved in a similar way as (23.6). □

Definition. By a $C^{\infty,\omega}$-mapping $f : U \times V \to F$ we mean a mapping f for which $f^\vee \in C^\infty(U, C^\omega(V, F))$.

11.21. Theorem. Composition of $C^{\infty,\omega}$-mappings. *Let $f : U \times V \to F$ and $g : U_1 \times V_1 \to V$ be $C^{\infty,\omega}$, and $h : U_1 \to U$ be C^∞. Then $f\circ(h\circ pr_1, g) : U_1 \times V_1 \to F$, $(x,y) \mapsto f(h(x), g(x,y))$ is $C^{\infty,\omega}$.*

Proof. We have to show that the mapping $x \mapsto (y \mapsto f(h(x), g(x,y)))$, $U_1 \to C^\omega(V_1, F)$ is C^∞. It is well-defined, since f and g are C^ω in the second variable. In order to show that it is C^∞ we compose with $\lambda_* : C^\omega(V_1, F) \to C^\omega(V_1, \mathbb{R})$, where $\lambda \in F'$ is arbitrary. Thus, it is enough to consider the case $F = \mathbb{R}$. Furthermore, we compose with $c^* : C^\omega(V_1, \mathbb{R}) \to C^\alpha(\mathbb{R}, \mathbb{R})$, where $c \in C^\alpha(\mathbb{R}, V_1)$ is arbitrary for α equal to ω and ∞.

In case $\alpha = \infty$ the composite with c^* is C^∞, since the associated mapping $U_1 \times \mathbb{R} \to \mathbb{R}$ is $f \circ (h \circ pr_1, g \circ (id \times c))$ which is C^∞.

Now the case $\alpha = \omega$. Let $I \subseteq \mathbb{R}$ be an arbitrary open bounded interval. Then $c^* \circ g^\vee : U_1 \to C^\omega(\mathbb{R}, G)$ is C^∞, where G is the convenient vector space containing V as an c^∞-open subset, and has values in $\{\gamma : \gamma(\bar{I}) \subseteq V\} \subseteq C^\omega(\mathbb{R}, G)$. This set is c^∞-open, since it is open for the topology of uniform convergence on compact sets which is coarser than the bornological topology on $C^\infty(\mathbb{R}, E)$ and hence than the c^∞-topology on $C^\omega(\mathbb{R}, G)$, see (11.10).

Thus, the composite with c^*, $\operatorname{comp}\circ(f^\vee \circ h, c^* \circ g^\vee)$ is C^∞, since $f^\vee \circ h : U_1 \to U \to C^\omega(V, F)$ is C^∞, $c^* \circ g^\vee : U_1 \to C^\omega(\mathbb{R}, G)$ is C^∞ and $\operatorname{comp} : C^\omega(V, \mathbb{R}) \times \{\gamma \in C^\omega(\mathbb{R}, G) : \gamma(\bar{I}) \subseteq V\} \to C^\omega(I, \mathbb{R})$ is C^ω, because it is associated to $\operatorname{ev}\circ(id \times \operatorname{ev}) : C^\omega(V, F) \times \{\gamma \in C^\omega(\mathbb{R}, G) : \gamma(\bar{I}) \subseteq V\} \times I \to V$. That $\operatorname{ev} : \{\gamma \in C^\omega(\mathbb{R}, G) : \gamma(\bar{I}) \subseteq V\} \times I \to \mathbb{R}$ is C^ω follows, since the associated mapping is the restriction mapping $C^\omega(\mathbb{R}, G) \to C^\omega(I, G)$. □

11.22. Corollary. *Let $w : W_1 \to W$ be C^ω, let $u : U \to U_1$ be smooth, let $v : V \to V_1$ be C^ω, and let $f : U_1 \times V_1 \to W_1$ be $C^{\infty,\omega}$. Then $w \circ f \circ (u \times v) : U \times V \to W$ is again $C^{\infty,\omega}$.*

This is generalization of theorem (11.17).

Proof. Use (11.21) twice. □

11.23. Corollary. *Let $f : E \supseteq U \to F$ be C^ω, let $I \subseteq \mathbb{R}$ be open and bounded, and α be ω or ∞. Then $f_* : C^\alpha(\mathbb{R}, E) \supseteq \{c : c(\bar{I}) \subseteq U\} \to C^\alpha(I, F)$ is C^ω.*

Proof. Obviously, $f_*(c) := f \circ c \in C^\alpha(I, F)$ is well-defined for all $c \in C^\alpha(\mathbb{R}, E)$ satisfying $c(\bar{I}) \subseteq U$.

Furthermore, the composite of f_* with any C^β-curve $\gamma : \mathbb{R} \to \{c : c(\bar{I}) \subseteq U\} \subseteq C^\alpha(\mathbb{R}, E)$ is a C^β-curve in $C^\alpha(I, F)$ for β equal to ω or ∞. For $\beta = \alpha$ this follows from cartesian closedness of the C^α-maps. For $\alpha \neq \beta$ this follows from (11.22).

Finally, $\{c : c(\bar{I}) \subseteq U\} \subseteq C^\alpha(\mathbb{R}, E)$ is c^∞-open, since it is open for the topology of uniform convergence on compact sets which is coarser than the bornological and hence than the c^∞-topology on $C^\alpha(\mathbb{R}, E)$. Here is the only place where we make use of the boundedness of I. □

11.24. Lemma. Derivatives. *The derivative d, where $df(x)(v) := \frac{d}{dt}|_{t=0} f(x+tv)$, is bounded and linear $d: C^\omega(U,F) \to C^\omega(U, L(E,F))$.*

Proof. The differential $df(x)(v)$ makes sense and is linear in v, because every real analytic mapping f is smooth. So it remains to show that $(f,x,v) \mapsto df(x)(v)$ is real analytic. So let f, x, and v depend real analytically (resp. smoothly) on a real parameter s. Since $(t,s) \mapsto x(s) + tv(s)$ is real analytic (resp. smooth) into $U \subseteq E$, the mapping $r \mapsto ((t,s) \mapsto f(r)(x(s)+tv(s))$ is real analytic into $C^\omega(\mathbb{R}^2, F)$ (resp. smooth into $C^\infty(\mathbb{R}^2, F)$. Composing with $\frac{\partial}{\partial t}|_{t=0}$: $C^\omega(\mathbb{R}^2,F) \to C^\omega(\mathbb{R},F)$ (resp. : $C^\infty(\mathbb{R}^2,F) \to C^\infty(\mathbb{R},F)$) shows that $r \mapsto (s \mapsto d(f(r))(x(s))(v(s)))$, $\mathbb{R} \to C^\omega(\mathbb{R},F)$ is real analytic. Considering the associated mapping on $\mathbb{R}^2$ composed with the diagonal map shows that $(f,x,v) \mapsto df(x)(v)$ is real analytic. □

The following examples as well as several others can be found in [Frölicher, Kriegl, 1988, 5.3.6].

11.25. Example. *Let $T: C^\infty(\mathbb{R},\mathbb{R}) \to C^\infty(\mathbb{R},\mathbb{R})$ be given by $T(f) = f'$. Then the continuous linear differential equation $x'(t) = T(x(t))$ with initial value $x(0) = x_0$ has a unique smooth solution $x(t)(s) = x_0(t+s)$ which is however not real analytic.*

Note the curious form $x'(t) = x(t)'$ of this differential equation. Beware of careless notation!

Proof. A smooth curve $x: \mathbb{R} \to C^\infty(\mathbb{R},\mathbb{R})$ is a solution of the differential equation $x'(t) = T(x(t))$ if and only if $\frac{\partial}{\partial t}\hat{x}(t,s) = \frac{\partial}{\partial s}\hat{x}(t,s)$. Hence, we have $\frac{d}{dt}\hat{x}(t, r-t) = 0$, i.e. $\hat{x}(t, r-t)$ is constant and hence equal to $\hat{x}(0,r) = x_0(r)$. Thus, $\hat{x}(t,s) = x_0(t+s)$.

Suppose $x: \mathbb{R} \to C^\infty(\mathbb{R},\mathbb{R})$ were real analytic. Then the composite with $ev_0: C^\infty(\mathbb{R},\mathbb{R}) \to \mathbb{R}$ were a real analytic function. But this composite is just $x_0 = ev_0 \circ x$, which is not in general real analytic. □

11.26. Example. *Let E be either $C^\infty(\mathbb{R},\mathbb{R})$ or $C^\omega(\mathbb{R},\mathbb{R})$. Then the mapping $\exp_*: E \to E$ is C^ω, has invertible derivative at every point, but the image does not contain an open neighborhood of $\exp_*(0)$.*

Proof. The mapping $\exp_*$ is real analytic by (11.23). Its derivative is given by $(\exp_*)'(f)(g): t \mapsto g(t)e^{f(t)}$ and hence is invertible with $g \mapsto (t \mapsto g(t)e^{-f(t)})$ as inverse mapping. Now consider the real analytic curve $c: \mathbb{R} \to E$ given by $c(t)(s) = 1 - (ts)^2$. One has $c(0) = 1 = \exp_*(0)$, but $c(t)$ is not in the image of $\exp_*$ for any $t \neq 0$, since $c(t)(\frac{1}{t}) = 0$ but $\exp_*(g)(t) = e^{g(t)} > 0$ for all g and t. □

Historical Remarks on Holomorphic and Real Analytic Calculus

The notion of holomorphic mappings used in section (15) was first defined by the italian Luigi Fantappié in the papers [Fantappié, 1930] and [Fantappié, 1933]:

> S.1: "Wenn jeder Funktion $y(t)$ einer Funktionenmenge H eine bestimmte Zahl f entspricht, d.h. die Zahl f von der Funktion $y(t)$ (unabhängige Veränderliche in der Menge H) abhängt, werden wir sagen, daß ein Funktional von $y(t)$:
>
> $$f = F[y(t)]$$
>
> ist; H heißt das Definitionsfeld des Funktionals F.
> [...] gemischtes Funktional [...]
>
> $$f = F[y_1(t_1,\dots),\dots,y_n(t_1,\dots);z_1,\dots,z_m]\text{"}$$

He also considered the 'functional transform' and noticed the relation

$$f = F[y(t);z] \text{ corresponds to } y \mapsto f(z)$$

> S.4: "Sei jetzt $F(y(t))$ ein Funktional, das in einem Funktionenbereich H (von analytischen Funktionen) definiert ist, und $y_0(t)$ ein Funktion von H, die mit einer Umgebung (r) oder (r,σ) zu H angehört. Wenn für jede analytische Mannigfaltigkeit $y(t;\alpha_1,\dots,\alpha_m)$, die in diese Umgebung eindringt (d.h. eine solche, die für alle Wertesysteme $\alpha_1,\dots,\alpha_m$) eines Bereichs Γ eine Funktion von t der Umgebung liefert), der Wert des Funktionals
>
> $$F_t[y(t;\alpha_1,\dots,\alpha_m)] = f(\alpha_1,\dots,\alpha_m)$$
>
> immer eine Funktion der Parameter $\alpha_1,\dots,\alpha_m$ ist, die nicht nur in Γ definiert, sondern dort noch eine analytische Funktion ist, werden wir sagen, daß das Funktional F regulär ist in der betrachteten Umgebung $y_0(t)$. Wenn ein Funktional F regulär ist in einer Umgebung jeder Funktion seines Definitionsbereiches, so heißt F analytisch."

The development in the complex case was much faster than in the smooth case since one did not have to explain the concept of higher derivatives.

The portuguese José Sebastião e Silva showed that analyticity in the sense of Fantappié coincides with other concepts, in his dissertation [Sebastião e Silva, 1948], published as [Sebastião e Silva, 1950a], and in [Sebastião e Silva, 1953]. An overview over various notions of holomorphicity was given by the Brasilian Domingos Pisanelli in [Pisanelli, 1972a] and [Pisanelli, 1972b].

Chapter III
Partitions of Unity

The main aim of this chapter is to discuss the abundance or scarcity of smooth functions on a convenient vector space: E.g. existence of bump functions and partitions of unity. This question is intimately related to differentiability of seminorms and norms, and in many examples these are, if at all, only finitely often differentiable. So we start this chapter with a short (but complete) account of finite order differentiability, based on Lipschitz conditions on higher derivatives, since with this notion we can get as close as possible to exponential laws. A more comprehensive exposition of finite order Lipschitz differentiability can be found in the monograph [Frölicher, Kriegl, 1988].

Then we treat differentiability of seminorms and convex functions, and we have tried to collect all relevant information from the literature. We give full proofs of all what will be needed later on or is of central interest. We also collect related results, mainly on 'generic differentiability', i.e. differentiability on a dense G_δ-set.

If enough smooth bump functions exist on a convenient vector space, we call it 'smoothly regular'. Although the smooth (i.e. bounded) linear functionals separate points on any convenient vector space, stronger separation properties depend very much on the geometry. In particular, we show that ℓ^1 and $C[0,1]$ are not even C^1-regular. We also treat more general 'smooth spaces' here since most results do not depend on a linear structure, and since we will later apply them to manifolds.

In many problems like E. Borel's theorem (15.4) that any power series appears as Taylor series of a smooth function, or the existence of smooth functions with given carrier (15.3), one uses in finite dimensions the existence of smooth functions with globally bounded derivatives. These do not exist in infinite dimensions in general; even for bump functions this need not be true globally. Extreme cases are Hilbert spaces where there are smooth bump functions with globally bounded derivatives, and c_0 which does not even admit C^2-bump functions with globally bounded derivatives.

In the final section of this chapter a space which admits smooth partitions of unity subordinated to any open cover is called smoothly paracompact. Fortunately, a wide class of convenient vector spaces has this property, among them all spaces of smooth sections of finite dimensional vector bundles which we shall need later as modeling spaces for manifolds of mappings. The theorem (16.15) of [Toruńczyk, 1973] characterizes smoothly paracompact metrizable spaces, and we will give a full proof. It is the only tool for investigating whether non-separable spaces are smoothly paracompact and we give its main applications.

12. Differentiability of Finite Order

12.1. Definition. A mapping $f : E \supseteq U \to F$, where E and F are convenient vector spaces, and $U \subseteq E$ is c^∞-open, is called $\mathcal{L}\mathrm{ip}^k$ if $f \circ c$ is a $\mathcal{L}\mathrm{ip}^k$-curve (see (1.2)) for each $c \in C^\infty(\mathbb{R}, U)$.

This is equivalent to the property that $f\circ c$ is $\mathcal{L}\mathrm{ip}^k$ on $c^{-1}(U)$ for each $c \in C^\infty(\mathbb{R}, E)$. This can be seen by reparameterization.

12.2. General curve lemma. *Let E be a convenient vector space, and let $c_n \in C^\infty(\mathbb{R}, E)$ be a sequence of curves which converges fast to 0, i.e., for each $k \in \mathbb{N}$ the sequence $n^k c_n$ is bounded. Let $s_n \geq 0$ be reals with $\sum_n s_n < \infty$.*

Then there exists a smooth curve $c \in C^\infty(\mathbb{R}, E)$ and a converging sequence of reals t_n such that $c(t + t_n) = c_n(t)$ for $|t| \leq s_n$, for all n.

Proof. Let $r_n := \sum_{k<n}(\frac{2}{k^2} + 2s_k)$ and $t_n := \frac{r_n + r_{n+1}}{2}$. Let $h : \mathbb{R} \to [0,1]$ be smooth with $h(t) = 1$ for $t \geq 0$ and $h(t) = 0$ for $t \leq -1$, and put $h_n(t) := h(n^2(s_n + t)).h(n^2(s_n - t))$. Then we have $h_n(t) = 0$ for $|t| \geq \frac{1}{n^2} + s_n$ and $h_n(t) = 1$ for $|t| \leq s_n$, and for the derivatives we have $|h_n^{(j)}(t)| \leq n^{2j}.H_j$, where $H_j := \max\{|h^{(j)}| : t \in \mathbb{R}\}$. Thus, in the sum

$$c(t) := \sum_n h_n(t - t_n).c_n(t - t_n)$$

at most one summand is non-zero for each $t \in \mathbb{R}$, and c is a smooth curve since for each $\ell \in E'$ we have

$$(\ell \circ c)(t) = \sum_n f_n(t), \qquad \text{where } f_n(t + t_n) := h_n(t).\ell(c_n(t)),$$

$$n^2.\sup_t |f_n^{(k)}(t)| = n^2.\sup\left\{|f_n^{(k)}(s + t_n)| : |s| \leq \tfrac{1}{n^2} + s_n\right\}$$

$$\leq n^2 \sum_{j=0}^{k} \tbinom{k}{j} n^{2j} H_j.\sup\left\{|(\ell \circ c_n)^{(k-j)}(s)| : |s| \leq \tfrac{1}{n^2} + s_n\right\}$$

$$\leq \left(\sum_{j=0}^{k} \tbinom{k}{j} n^{2j+2} H_j\right).\sup\left\{|(\ell \circ c_n)^{(i)}(s)| : |s| \leq \max_n(\tfrac{1}{n^2} + s_n) \text{ and } i \leq k\right\},$$

which is uniformly bounded with respect to n, since c_n converges to 0 fast. □

12.3. Corollary. *Let $c_n : \mathbb{R} \to E$ be polynomials of bounded degree with values in a convenient vector space E. If for each $\ell \in E'$ the sequence $n \mapsto \sup\{|(\ell \circ c_n)(t) : |t| \le 1\}$ converges to 0 fast, then the sequence c_n converges to 0 fast in $C^\infty(\mathbb{R}, E)$, so the conclusion of* (12.2) *holds.*

Proof. The structure on $C^\infty(\mathbb{R}, E)$ is the initial one with respect to the cone $\ell_* : C^\infty(\mathbb{R}, E) \to C^\infty(\mathbb{R}, \mathbb{R})$ for all $\ell \in E'$, by (3.9). So we only have to show the result for $E = \mathbb{R}$. On the finite dimensional space of all polynomials of degree at most d the expression in the assumption is a norm, and the inclusion into $C^\infty(\mathbb{R}, \mathbb{R})$ is bounded. □

12.4. Difference quotients. For a curve $c : \mathbb{R} \to E$ with values in a vector space E the *difference quotient* $\delta^k c$ of order k is given recursively by

$$\delta^0 c := c,$$
$$\delta^k c(t_0, \dots, t_k) := k \, \frac{\delta^{k-1} c(t_0, \dots, t_{k-1}) - \delta^{k-1} c(t_1, \dots, t_k)}{t_0 - t_k},$$

for pairwise different t_i. The constant factor k in the definition of δ^k is chosen in such a way that δ^k approximates the k-th derivative. By induction, one can easily see that

$$\delta^k c(t_0, \dots, t_k) = k! \sum_{i=0}^{k} c(t_i) \prod_{\substack{0 \le j \le k \\ j \ne i}} \tfrac{1}{t_i - t_j}.$$

We shall mainly need the *equidistant difference quotient* $\delta^k_{\text{eq}} c$ of order k, which is given by

$$\delta^k_{\text{eq}} c(t; v) := \delta^k c(t, t+v, \dots, t+kv) = \frac{k!}{v^k} \sum_{i=0}^{k} c(t+iv) \prod_{\substack{0 \le j \le k \\ j \ne i}} \tfrac{1}{i-j}.$$

Lemma. *For a convenient vector space E and a curve $c : \mathbb{R} \to E$ the following conditions are equivalent:*

1. *c is $\mathcal{L}\text{ip}^{k-1}$.*
2. *The difference quotient $\delta^k c$ of order k is bounded on bounded sets.*
3. *$\ell \circ c$ is continuous for each $\ell \in E'$, and the equidistant difference quotient $\delta^k_{\text{eq}} c$ of order k is bounded on bounded sets in $\mathbb{R} \times (\mathbb{R} \setminus \{0\})$.*

Proof. All statements can be tested by composing with bounded linear functionals $\ell \in E'$, so we may assume that $E = \mathbb{R}$.

(3) ⇒ (2) Let $I \subset \mathbb{R}$ be a bounded interval. Then there is some $K > 0$ such that $|\delta^k_{\text{eq}} c(x; v)| \le K$ for all $x \in I$ and $kv \in I$. Let $t_i \in I$ be pairwise different points. We claim that $|\delta^k c(t_0, \dots, t_k)| \le K$. Since $\delta^k c$ is symmetric we may assume that $t_0 < t_1 < \dots < t_k$, and since it is continuous (c is continuous) we may assume that all $\frac{t_i - t_0}{t_k - t_0}$ are of the form $\frac{n_i}{N}$ for $n_i, N \in \mathbb{N}$. Put $v := \frac{t_k - t_0}{N}$, then $\delta^k c(t_0, \dots, t_k) =$

$\delta^k c(t_0, t_0+n_1 v, \dots, t_0+n_k v)$ is a convex combination of $\delta^k_{\text{eq}} c(t_0+rv; v)$ for $0 \le r \le \max_i n_i - k$. This follows by recursively inserting intermediate points of the form $t_0 + mv$, and using

$$\begin{aligned}\delta^k(t_0+m_0 v, \dots, \widehat{t_0+m_i v}, \dots, t_0+m_{k+1}v) &= \\ &= \frac{m_i - m_0}{m_{k+1}-m_0}\delta^k(t_0+m_0 v, \dots, t_0+m_k v) \\ &\quad + \frac{m_{k+1}-m_i}{m_{k+1}-m_0}\delta^k(t_1+m_1 v, \dots, t_0+m_{k+1}v)\end{aligned}$$

which itself may be proved by induction on k.

(2) $\Rightarrow$ (1) We have to show that c is k times differentiable and that $\delta^1 c^{(k)}$ is bounded on bounded sets. We use induction, $k=0$ is clear.

Let $T \neq S$ be two subsets of $\mathbb{R}$ of cardinality $j+1$. Then there exist enumerations $T = \{t_0, \dots, t_j\}$ and $S = \{s_0, \dots, s_j\}$ such that $t_i \neq s_j$ for $i \le j$; then we have

$$\delta^j c(t_0, \dots, t_j) - \delta^j c(s_0, \dots, s_j) = \tfrac{1}{j+1}\sum_{i=0}^{j}(t_i - s_i)\delta^{j+1}c(t_0, \dots, t_i, s_i, \dots, s_j).$$

For the enumerations we put the elements of $T \cap S$ at the end in T and at the beginning in S. Using the recursive definition of $\delta^{j+1}c$ and symmetry the right hand side becomes a telescoping sum.

Since $\delta^k c$ is bounded we see from the last equation that all $\delta^j c$ are also bounded, in particular this is true for $\delta^2 c$. Then

$$\frac{c(t+s)-c(t)}{s} - \frac{c(t+s')-c(t)}{s'} = \tfrac{s-s'}{2}\,\delta^2 c(t, t+s, t+s')$$

shows that the difference quotient of c forms a Mackey Cauchy net, and hence the limit $c'(t)$ exists.

Using the easily checked formula

$$c(t_j) = \sum_{i=0}^{j}\tfrac{1}{i!}\prod_{l=0}^{i-1}(t_j - t_l)\,\delta^i c(t_0, \dots, t_j),$$

induction on j and differentiability of c one shows that

$$\delta^j c'(t_0, \dots, t_j) = \tfrac{1}{j+1}\sum_{i=0}^{j}\delta^{j+1}c(t_0, \dots, t_j, t_i), \tag{4}$$

where $\delta^{j+1}c(t_0, \dots, t_j, t_i) := \lim_{t \to t_i}\delta^{j+1}c(t_0, \dots, t_j, t)$. The right hand side of (4) is bounded, so c' is $\mathcal{L}\text{ip}^{k-2}$ by induction on k.

(1) $\Rightarrow$ (2) For a differentiable function $f : \mathbb{R} \to \mathbb{R}$ and $t_0 < \dots < t_j$ there exist s_i with $t_i < s_i < t_{i+1}$ such that

$$\delta^j f(t_0, \dots, t_j) = \delta^{j-1}f'(s_0, \dots, s_{j-1}). \tag{5}$$

Let p be the interpolation polynomial

$$(6)\qquad p(t) := \sum_{i=0}^{j} \tfrac{1}{i!} \prod_{l=0}^{i-1}(t - t_l)\; \delta^j f(t_0, \dots, t_j).$$

Since f and p agree on all t_j, by Rolle's theorem the first derivatives of f and p agree on some intermediate points s_i. So p' is the interpolation polynomial for f' at these points s_i. Comparing the coefficient of highest order of p' and of the interpolation polynomial (6) for f' at the points s_i (5) follows.

Applying (5) recursively for $f = c^{(k-2)}, c^{(k-3)}, \dots, c$ shows that $\delta^k c$ is bounded on bounded sets, and (2) follows.

(2) $\Rightarrow$ (3) is obvious. □

12.5. Let $r_0, \dots, r_k$ be the unique rational solution of the linear equation

$$\sum_{i=0}^{k} i^j r_i = \begin{cases} 1 & \text{for } j = 1 \\ 0 & \text{for } j = 0, 2, 3, \dots, k. \end{cases}$$

Lemma. *If $f : \mathbb{R}^2 \to \mathbb{R}$ is $\mathcal{L}ip^k$ for $k \geq 1$ and I is a compact interval then there exists M such that for all $t, v \in I$ we have*

$$\left| \tfrac{\partial}{\partial s}|_0 f(t,s).v - \sum_{i=0}^{k} r_i f(t, iv) \right| \leq M|v|^{k+1}.$$

Proof. We consider first the case $0 \notin I$ so that v stays away from 0. For this it suffices to show that the derivative $\frac{\partial}{\partial s}|_0 f(t,s)$ is locally bounded. If it is unbounded near some point x_∞, there are x_n with $|x_n - x_\infty| \leq \frac{1}{2^n}$ such that $\frac{\partial}{\partial s}|_0 f(x_n, s) \geq n.2^n$. We apply the general curve lemma (12.2) to the curves $c_n : \mathbb{R} \to \mathbb{R}^2$ given by $c_n(t) := (x_n, \frac{t}{2^n})$ and to $s_n := \frac{1}{2^n}$ in order to obtain a smooth curve $c : \mathbb{R} \to \mathbb{R}^2$ and scalars $t_n \to 0$ with $c(t + t_n) = c_n(t)$ for $|t| \leq s_n$. Then $(f \circ c)'(t_n) = \frac{1}{2^n} \frac{\partial}{\partial s}|_0 f(x_n, s) \geq n$, which contradicts that f is $\mathcal{L}ip^1$.

Now we treat the case $0 \in I$. If the assertion does not hold there are $x_n, v_n \in I$, such that $\left|\frac{\partial}{\partial s}|_0 f(x_n, s).v_n - \sum_{i=0}^{k} r_i f(x_n, iv_n)\right| \geq n.2^{n(k+1)}|v_n|^{k+1}$. We may assume $x_n \to x_\infty$, and by the case $0 \notin I$ we may assume that $v_n \to 0$, even with $|x_n - x_\infty| \leq \frac{1}{2^n}$ and $|v_n| \leq \frac{1}{2^n}$. We apply the general curve lemma (12.2) to the curves $c_n : \mathbb{R} \to \mathbb{R}^2$ given by $c_n(t) := (x_n, \frac{t}{2^n})$ and to $s_n := \frac{1}{2^n}$ to obtain a smooth curve $c : \mathbb{R} \to \mathbb{R}^2$ and scalars $t_n \to 0$ with $c(t + t_n) = c_n(t)$ for $|t| \leq s_n$. Then we have

$$\left|(f \circ c)'(t_n) 2^n v_n - \sum_{i=0}^{k} r_i (f \circ c)(t_n + i 2^n v_n)\right| =$$

$$= \left|(f \circ c_n)'(0) 2^n v_n - \sum_{i=0}^{k} r_i (f \circ c_n)(i 2^n v_n)\right|$$

$$= \left|\tfrac{1}{2^n} \tfrac{\partial}{\partial s}|_0 f(x_n, s) 2^n v_n - \sum_{i=0}^{k} r_i f(x_n, iv_n)\right| \geq n(2^n |v_n|)^{k+1}.$$

This contradicts the next claim for $g = f \circ c$.

Claim. If $g : \mathbb{R} \to \mathbb{R}$ is $\mathcal{L}\mathrm{ip}^k$ for $k \geq 1$ and I is a compact interval then there is $M > 0$ such that for $t, v \in I$ we have $\left|g'(t).v - \sum_{i=0}^k r_i g(t+iv)\right| \leq M|v|^{k+1}$.

Consider $g_t(v) := g'(t).v - \sum_{i=0}^k r_i g(t+iv)$. Then the derivatives up to order k at $v = 0$ of g_t vanish by the choice of the r_i. Since $g^{(k)}$ is locally Lipschitzian there exists an M such that $|g_t^{(k)}(v)| \leq M|v|$ for all $t, v \in I$, which we may integrate in turn to obtain $|g_t(v)| \leq M\frac{|v|^{k+1}}{(k+1)!}$. $\square$

12.6. Lemma. *Let $f : \mathbb{R}^2 \to \mathbb{R}$ be $\mathcal{L}\mathrm{ip}^{k+1}$. Then $t \mapsto \frac{\partial}{\partial s}|_0 f(t,s)$ is $\mathcal{L}\mathrm{ip}^k$.*

Proof. Suppose that $g : t \mapsto \frac{\partial}{\partial s}|_0 f(t,s)$ is not $\mathcal{L}\mathrm{ip}^k$. Then by lemma (12.4) the equidistant difference quotient $\delta_{\mathrm{eq}}^{k+1} g$ is not locally bounded at some point which we may assume to be 0. Then there are x_n and v_n with $|x_n| \leq 1/4^n$ and $0 < v_n < 1/4^n$ such that

$$(1) \qquad |\delta_{\mathrm{eq}}^{k+1} g(x_n; v_n)| > n.2^{n(k+2)}.$$

We apply the general curve lemma (12.2) to the curves $c_n : \mathbb{R} \to \mathbb{R}^2$ given by $c_n(t) := e_n(\frac{t}{2^n} + x_n) := (\frac{t}{2^n} + x_n - v_n, \frac{t}{2^n})$ and to $s_n := \frac{k+2}{2^n}$ in order to obtain a smooth curve $c : \mathbb{R} \to \mathbb{R}^2$ and scalars $t_n \to 0$ with $c(t + t_n) = c_n(t)$ for $0 \leq t \leq s_n$.

Put $f_0(t,s) := \sum_{i=0}^k r_i\, f(t, is)$ for r_i as in (12.5), put $f_1(t,s) := g(t)s$, finally put $f_2 := f_1 - f_0$. Then f_0 in $\mathcal{L}\mathrm{ip}^{k+1}$, so $f_0 \circ c$ is $\mathcal{L}\mathrm{ip}^{k+1}$, hence the equidistant difference quotient $\delta_{\mathrm{eq}}^{k+2}(f_0 \circ c)(x_n; 2^n v_n)$ is bounded.

By lemma (12.5) there exists $M > 0$ such that $|f_2(t,s)| \leq M|s|^{k+2}$ for all $t, s \in [-(k+1), k+1]$, so we get

$$\begin{aligned}
|\delta_{\mathrm{eq}}^{k+2}(f_2 \circ c)(x_n; 2^n v_n)| &= |\delta_{\mathrm{eq}}^{k+2}(f_2 \circ c_n)(0; 2^n v_n)| \\
&= \tfrac{1}{2^{n(k+2)}} |\delta_{\mathrm{eq}}^{k+2}(f_2 \circ e_n)(x_n; v_n)| \\
&\leq \tfrac{(k+2)!}{2^{n(k+2)}} \sum_{i=1}^{k+2} \frac{|f_2((i-1)v_n + x_n, iv_n)|}{|iv_n|^{(k+2)}} \frac{i^{(k+2)}}{\prod_{j\neq i}|i-j|} \\
&\leq \tfrac{(k+2)!}{2^{n(k+2)}} \sum_{i=1}^{k+2} M \frac{i^{(k+2)}}{\prod_{j\neq i}|i-j|}.
\end{aligned}$$

This is bounded, and so for $f_1 = f_0 + f_2$ the expression $|\delta_{\mathrm{eq}}^{k+2}(f_1 \circ c)(x_n; 2^n v_n)|$ is also bounded, with respect to n. However, on the other hand we get

$$\begin{aligned}
\delta_{\mathrm{eq}}^{k+2}(f_1 \circ c)(x_n; 2^n v_n) &= \delta_{\mathrm{eq}}^{k+2}(f_1 \circ c_n)(0; 2^n v_n) \\
&= \tfrac{1}{2^{n(k+2)}} \delta_{\mathrm{eq}}^{k+2}(f_1 \circ e_n)(x_n; v_n) \\
&= \tfrac{(k+2)!}{2^{n(k+2)}} \sum_{i=0}^{k+2} \frac{f_1((i-1)v_n + x_n, iv_n)}{v_n^{(k+2)}} \prod_{\substack{0\leq j\leq k+2 \\ j\neq i}} \tfrac{1}{i-j}
\end{aligned}$$

$$= \frac{(k+2)!}{2^{n(k+2)}} \sum_{i=0}^{k+2} \frac{g((i-1)v_n + x_n)iv_n}{v_n^{(k+2)}} \prod_{\substack{0\le j\le k+2\\ j\ne i}} \frac{1}{i-j}$$

$$= \frac{(k+2)!}{2^{n(k+2)}} \sum_{l=0}^{k+1} \frac{g(lv_n + x_n)}{v_n^{(k+1)}} \prod_{\substack{0\le j\le k+1\\ j\ne l}} \frac{1}{l-j}$$

$$= \frac{k+2}{2^{n(k+2)}} \delta_{\text{eq}}^{k+1} g(x_n; v_n),$$

which in absolute value is larger than $(k+2)n$ by (1), a contradiction. □

12.7. Lemma. *Let E be a normed space and F be a convenient vector space, U open in E. Then, a mapping $f : U \to F$ is $\mathcal{L}ip^0$ if and only if f is locally Lipschitz, i.e., $\frac{f(x)-f(y)}{\|x-y\|}$ is locally bounded.*

Proof. ($\Rightarrow$) If f is $\mathcal{L}ip^0$ but not locally Lipschitz near $z \in U$, there are $\ell \in F'$ and points $x_n \neq y_n$ in U with $\|x_n - z\| \le 1/2^n$ and $\|y_n - z\| \le 1/2^n$, such that $\ell(f(y_n) - f(x_n)) \ge n.\|y_n - x_n\|$. Now we apply the general curve lemma (12.2) with $s_n := \|y_n - x_n\|$ and $c_n(t) := x_n - z + t(y_n - x_n)$ to get a smooth curve c with $c(t + t_n) = c_n(t)$ for $0 \le t \le s_n$. Then $\frac{1}{s_n}((\ell \circ f \circ c)(t_n + s_n) - (\ell \circ f \circ c)(t_n)) = \frac{1}{\|y_n - x_n\|}\ell(f(y_n) - f(x_n)) \ge n$.

($\Leftarrow$) This is obvious, since the composition of locally Lipschitzian mappings is again locally Lipschitzian. □

12.8. Theorem. *Let $f : E \supseteq U \to F$ be a mapping, where E and F are convenient vector spaces, and $U \subseteq E$ is c^∞-open. Then the following assertions are equivalent for each $k \ge 0$:*

1. *f is $\mathcal{L}ip^{k+1}$.*
2. *The directional derivative*

$$(d_v f)(x) := \tfrac{\partial}{\partial t}|_{t=0}(f(x + tv))$$

exists for $x \in U$ and $v \in E$ and defines a $\mathcal{L}ip^k$-mapping $U \times E \to F$.

Note that this result gives a different (more algebraic) proof of Boman's theorem (3.4) and (3.14).

Proof. (1) $\Rightarrow$ (2) Clearly, $t \mapsto f(x+tv)$ is $\mathcal{L}ip^{k+1}$, and so the directional derivative exists and is the Mackey-limit of the difference quotients, by lemma (1.7). In order to show that $df : (x,v) \mapsto d_v f(x)$ is $\mathcal{L}ip^k$ we take a smooth curve $(x,v) : \mathbb{R} \to U \times E$ and $\ell \in F'$, and we consider $g(t,s) := x(t) + s.v(t)$, $g : \mathbb{R}^2 \to E$. Then $\ell \circ f \circ g : \mathbb{R}^2 \to \mathbb{R}$ is $\mathcal{L}ip^{k+1}$, so by lemma (12.6) the curve

$$t \mapsto \ell(df(x(t), v(t))) = \ell\left(\tfrac{\partial}{\partial s}|_0 f(g(t,s))\right) = \tfrac{\partial}{\partial s}|_0 \ell(f(g(t,s)))$$

is of class $\mathcal{L}ip^k$.

(2) ⇒ (1) If $c \in C^\infty(\mathbb{R}, U)$ then

$$\frac{f(c(t)) - f(c(0))}{t} - df(c(0), c'(0)) =$$
$$= \int_0^1 \Big(df(c(0) + s(c(t) - c(0)), \tfrac{c(t)-c(0)}{t}) - df(c(0), c'(0)) \Big)\, ds$$

converges to 0 for $t \to 0$ since $g : (t, s) \mapsto df(c(0) + s(c(t) - c(0)), \frac{c(t)-c(0)}{t}) - df(c(0), c'(0))$ is $\mathcal{L}\mathrm{ip}^k$, thus by lemma (12.7) g is locally Lipschitz, so the set of all $\frac{g(t_1,s)-g(t_2,s)}{t_1-t_2}$ is locally bounded, and finally $t \mapsto \int_0^1 g(t, s)ds$ is locally Lipschitz. Thus, $f \circ c$ is differentiable with derivative $(f \circ c)'(0) = df(c(0), c'(0))$.

Since df is $\mathcal{L}\mathrm{ip}^k$ and (c, c') is smooth we get that $(f \circ c)'$ is $\mathcal{L}\mathrm{ip}^k$, hence $f \circ c$ is $\mathcal{L}\mathrm{ip}^{k+1}$. □

12.9. Corollary. Chain rule. *The composition of $\mathcal{L}\mathrm{ip}^k$-mappings is again $\mathcal{L}\mathrm{ip}^k$, and the usual formula for the derivative of the composite holds.*

Proof. We have to compose $f \circ g$ with a smooth curve c, but then $g \circ c$ is a $\mathcal{L}\mathrm{ip}^k$-curve, thus it is sufficient to show that the composition of a $\mathcal{L}\mathrm{ip}^k$ curve $c : \mathbb{R} \to U \subseteq E$ with a $\mathcal{L}\mathrm{ip}^k$-mapping $f : U \to F$ is again $\mathcal{L}\mathrm{ip}^k$, and that $(f \circ c)'(t) = df(c(t), c'(t))$.

This follows by induction on k for $k \geq 1$ in the same way as we proved theorem (12.8.2) ⇒ (12.8.1), using theorem (12.8) itself. □

12.10. Definition and Proposition. *Let F be a convenient vector space. The space $\mathcal{L}\mathrm{ip}^k(\mathbb{R}, F)$ of all $\mathcal{L}\mathrm{ip}^k$-curves in F is again a convenient vector space with the following equivalent structures:*

1. *The initial structure with respect to the $k + 2$ linear mappings (for $0 \leq j \leq k+1$) $c \mapsto \delta^j c$ from $\mathcal{L}\mathrm{ip}^k(\mathbb{R}, F)$ into the space of all F-valued maps in $j + 1$ pairwise different real variables $(t_0, \dots, t_j)$ which are bounded on bounded subsets, with the c^∞-complete locally convex topology of uniform convergence on bounded subsets. In fact, the mappings δ^0 and δ^{k+1} are sufficient.*
2. *The initial structure with respect to the $k + 2$ linear mappings (for $0 \leq j \leq k+1$) $c \mapsto \delta^j_{\mathrm{eq}} c$ from $\mathcal{L}\mathrm{ip}^k(\mathbb{R}, F)$ into the space of all maps from $\mathbb{R} \times (\mathbb{R} \setminus \{0\})$ into F which are bounded on bounded subsets, with the c^∞-complete locally convex topology of uniform convergence on bounded subsets. In fact, the mappings δ^0_{eq} and $\delta^{k+1}_{\mathrm{eq}}$ are sufficient.*
3. *The initial structure with respect to the derivatives of order $j \leq k$ considered as linear mappings into the space of $\mathcal{L}\mathrm{ip}^0$-curves, with the locally convex topology of uniform convergence of the curve on bounded subsets of $\mathbb{R}$ and of the difference quotient on bounded subsets of $\{(t, s) \in \mathbb{R}^2 : t \neq s\}$.*

The convenient vector space $\mathcal{L}\mathrm{ip}^k(\mathbb{R}, F)$ satisfies the uniform boundedness principle with respect to the point evaluations.

Proof. All three structures describe closed embeddings into finite products of spaces, which in (1) and (2) are obviously c^∞-complete. For (3) this follows, since by (1) the structure on $\mathcal{L}\mathrm{ip}^0(\mathbb{R}, E)$ is convenient.

All structures satisfy the uniform boundedness principle for the point evaluations by (5.25), and since spaces of all bounded mappings on some (bounded) set satisfy this principle. This can be seen by composing with ℓ_* for all $\ell \in E'$, since Banach spaces do this by (5.24).

By applying this uniform boundedness principle one sees that all these structures are indeed equivalent. □

12.11. Definition and Proposition. *Let E and F be convenient vector spaces and $U \subseteq E$ be c^∞-open. The space $\mathcal{L}\text{ip}^k(U,F)$ of all $\mathcal{L}\text{ip}^k$-mappings from U to F is again a convenient vector space with the following equivalent structures:*

(1) *The initial structure with respect to the linear mappings $c^* : \mathcal{L}\text{ip}^k(U,F) \to \mathcal{L}\text{ip}^k(\mathbb{R},F)$ for all $c \in C^\infty(\mathbb{R},F)$.*
(2) *The initial structure with respect to the linear mappings $c^* : \mathcal{L}\text{ip}^k(U,F) \to \mathcal{L}\text{ip}^k(\mathbb{R},F)$ for all $c \in \mathcal{L}\text{ip}^k(\mathbb{R},F)$.*

This space satisfies the uniform boundedness principle with respect to the evaluations $\mathrm{ev}_x : \mathcal{L}\text{ip}^k(U,F) \to F$ for all $x \in U$.

Proof. The structure (1) is convenient since by (12.1) it is a closed subspace of the product space which is convenient by (12.10). The structure in (2) is convenient since it is closed by (12.9). The uniform boundedness principle for the point evaluations now follows from (5.25) and (12.10), and this in turn gives us the equivalence of the two structures. □

12.12. Remark. We want to call the attention of the reader to the fact that there is no general exponential law for $\mathcal{L}\text{ip}^k$-mappings. In fact, if $f \in \mathcal{L}\text{ip}^k(\mathbb{R},\mathcal{L}\text{ip}^k(\mathbb{R},F))$ then $(\frac{\partial}{\partial t})^p(\frac{\partial}{\partial s})^q f^\wedge(t,s)$ exists if $\max(p,q) \le k$. This describes a smaller space than $\mathcal{L}\text{ip}^k(\mathbb{R}^2,F)$, which is not invariantly describable.

However, some partial results still hold, namely for convenient vector spaces E, F, and G, and for c^∞-open sets $U \subseteq E$, $V \subseteq F$ we have

$$\mathcal{L}\text{ip}^k(U, L(F,G)) \cong L(F, \mathcal{L}\text{ip}^k(U,G)),$$
$$\mathcal{L}\text{ip}^k(U, \mathcal{L}\text{ip}^l(V,G)) \cong \mathcal{L}\text{ip}^l(V, \mathcal{L}\text{ip}^k(U,G)),$$

see [Frölicher, Kriegl, 1988, 4.4.5, 4.5.1, 4.5.2]. For a mapping $f : U \times F \to G$ which is linear in F we have: $f \in \mathcal{L}\text{ip}^k(U \times F, G)$ if and only if $f^\vee \in \mathcal{L}\text{ip}^k(U, L(E,F))$, see [Frölicher, Kriegl, 1988, 4.3.5]. The last property fails if we weaken Lipschitz to continuous, see the following example.

12.13. Smolyanov's Example. Let $f : \ell^2 \to \mathbb{R}$ be defined by $f := \sum_{k\ge 1} \frac{1}{k^2} f_k$, where $f_k(x) := \varphi(k(kx_k - 1)) \cdot \prod_{j<k} \varphi(jx_j)$ and $\varphi : \mathbb{R} \to [0,1]$ is smooth with $\varphi(0) = 1$ and $\varphi(t) = 0$ for $|t| \ge \frac{1}{4}$. We shall show that

(1) $f : \ell^2 \to \mathbb{R}$ is Fréchet differentiable.
(2) $f' : \ell^2 \to (\ell^2)'$ is not continuous.
(3) $f' : \ell^2 \times \ell^2 \to \mathbb{R}$ is continuous.

Proof. Let $A := \{x \in \ell^2 : |kx_k| \le \frac{1}{4} \text{ for all } k\}$. This is a closed subset of ℓ^2.

(1) Remark that for $x \in \ell^2$ at most one $f_k(x)$ can be unequal to 0. In fact $f_k(x) \ne 0$ implies that $|kx_k - 1| \le \frac{1}{4k} \le \frac{1}{4}$, and hence $kx_k \ge \frac{3}{4}$ and thus $f_j(x) = 0$ for $j > k$. For $x \notin A$ there exists a $k > 0$ with $|kx_k| > \frac{1}{4}$ and the set of points satisfying this condition is open. It follows that $\varphi(kx_k) = 0$ and hence $f = \sum_{j<k} \frac{1}{j^2} f_j$ is smooth on this open set.

On the other hand let $x \in A$. Then $|kx_k - 1| \ge \frac{3}{4} > \frac{1}{4}$ and hence $\varphi(k(kx_k - 1)) = 0$ for all k and thus $f(x) = 0$. Let $v \in \ell^2$ be such that $f(x+v) \ne 0$. Then there exists a unique k such that $f_k(x+v) \ne 0$ and therefore $|j(x_j + v_j)| < \frac{1}{4}$ for $j < k$ and $|k(x_k + v_k) - 1| < \frac{1}{4k} \le \frac{1}{4}$. Since $|kx_k| \le \frac{1}{4}$ we conclude $|kv_k| \ge 1 - |k(x_k + v_k) - 1| - |kx_k| \ge 1 - \frac{1}{4} - \frac{1}{4} = \frac{1}{2}$. Hence $|f(x+v)| = \frac{1}{k^2}|f_k(x+v)| \le \frac{1}{k^2} \le (2|v_k|)^2 \le 4\|v\|^2$. Thus $\frac{\|f(x+v) - 0 - 0\|}{\|v\|} \le 4\|v\| \to 0$ for $\|v\| \to 0$, i.e. f is Fréchet differentiable at x with derivative 0.

(2) If fact take $a \in \mathbb{R}$ with $\varphi'(a) \ne 0$. Then $f'(t\,e^k)(e^k) = \frac{d}{dt}\frac{1}{k^2} f_k(t\,e^k) = \frac{d}{dt}\frac{1}{k^2}\varphi(k^2\,t - k) = \varphi'(k\,(k\,t - 1)) = \varphi'(a)$ if $t = t_k := \frac{1}{k}\left(\frac{a}{k} + 1\right)$, which goes to 0 for $k \to \infty$. However $f'(0)(e^k) = 0$ since $0 \in A$.

(3) We have to show that $f'(x^n)(v^n) \to f'(x)(v)$ for $(x^n, v^n) \to (x, v)$. For $x \notin A$ this is obviously satisfied, since then there exists a k with $|kx_k| > \frac{1}{4}$ and hence $f = \sum_{j \le k} \frac{1}{j^2} f_j$ locally around x.

If $x \in A$ then $f'(x) = 0$ and thus it remains to consider the case, where $x^n \notin A$. Let $\varepsilon > 0$ be given. Locally around x^n at most one summand f_k does not vanish: If $x^n \notin A$ then there is some k with $|kx^k| > 1/4$ which we may choose minimal. Thus $|jx^j| \le 1/4$ for all $j < k$, so $|j(jx^j - 1)| \ge 3j/4$ and hence $f_j = 0$ locally since the first factor vanishes. For $j > k$ we get $f_j = 0$ locally since the second factor vanishes. Thus we can evaluate the derivative:

$$|f'(x^n)(v^n)| = \left|\frac{1}{k^2} f_k'(x^n)(v^n)\right| \le \frac{\|\varphi'\|_\infty}{k^2}\Big(k^2|v_k^n| + \sum_{j<k} j|v_j^n|\Big).$$

Since $v \in \ell^2$ we find a K_1 such that $(\sum_{j \ge K_1} |v_j|^2)^{1/2} \le \frac{\varepsilon}{2\|\varphi'\|_\infty}$. Thus we conclude from $\|v^n - v\|_2 \to 0$ that $|v_j^n| \le \frac{\varepsilon}{\|\varphi'\|_\infty}$ for $j \ge K_1$ and large n. For the finitely many small n we can increase K_1 such that for these n and $j \ge K_1$ also $|v_j^n| \le \frac{\varepsilon}{\|\varphi'\|_\infty}$. Furthermore there is a constant $K_2 \ge 1$ such that $\|v^n\|_\infty \le \|v^n\|_2 \le K_2$ for all n. Now choose $N \ge K_1$ so large that $N^2 \ge \frac{1}{\varepsilon}\|\varphi'\|_\infty K_2\, K_1^2$. Obviously $\sum_{n<N} \frac{1}{n^2} f_n$ is smooth. So it remains to consider those n for which the non-vanishing term has index $k \ge N$. For those terms we have

$$\begin{aligned}
|f'(x^n)(v^n)| &= \left|\frac{1}{k^2} f_k'(x^n)(v^n)\right| \le \|\varphi'\|_\infty\Big(|v_k^n| + \frac{1}{k^2}\sum_{j<k} j|v_j^n|\Big) \\
&\le |v_k^n|\|\varphi'\|_\infty + \|\varphi'\|_\infty \frac{1}{k^2}\sum_{j<K_1} j|v_j^n| + \frac{1}{k^2}\sum_{K_1 \le j<k} j|v_j^n|\,\|\varphi'\|_\infty \\
&\le \varepsilon + \|\varphi'\|_\infty \frac{K_1^2}{N^2}\|v^n\|_\infty + \frac{1}{k^2}\sum_{K_1 \le j<k} j\,\varepsilon \le \varepsilon + \varepsilon + \varepsilon = 3\varepsilon
\end{aligned}$$

This shows the continuity. □

13. Differentiability of Seminorms

A desired separation property is that the smooth functions generate the topology. Since a locally convex topology is generated by the continuous seminorms it is natural to look for smooth seminorms. Note that every seminorm $p : E \to \mathbb{R}$ on a vector space E factors over $E_p := E/\ker p$ and gives a norm on this space. Hence, it can be extended to a norm $\tilde{p} : \tilde{E}_p \to \mathbb{R}$ on the completion $\tilde{E}_p$ of the space E_p which is normed by this factorization. If E is a locally convex space and p is continuous, then the canonical quotient mapping $E \to E_p$ is continuous. Thus, smoothness of $\tilde{p}$ off 0 implies smoothness of p on its carrier, and so the case where E is a Banach space is of central importance.

Obviously, every seminorm is a convex function, and hence we can generalize our treatment slightly by considering convex functions instead. The question of their differentiability properties is exactly the topic of this section.

Note that since the smooth functions depend only on the bornology and not on the locally convex topology the same is true for the initial topology induced by all smooth functions. Hence, it is appropriate to make the following

Convention. In this chapter the locally convex topology on all convenient vector spaces is assumed to be the bornological one.

13.1. Remark. It can be easily seen that for a function $f : E \to \mathbb{R}$ on a vector space E the following statements are equivalent (see for example [Frölicher, Kriegl, 1988, p. 199]):

(1) The function f is convex, i.e. $f(\sum_{i=1}^n \lambda_i x_i) \le \sum_{i=1}^n \lambda_i f(x_i)$ for $\lambda_i \ge 0$ with $\sum_{i=1}^n \lambda_i = 1$;
(2) The set $U_f := \{(x,t) \in E \times \mathbb{R} : f(x) \le t\}$ is convex;
(3) The set $A_f := \{(x,t) \in E \times \mathbb{R} : f(x) < t\}$ is convex.

Moreover, for any translation invariant topology on E (and hence in particular for the locally convex topology or the c^∞-topology on a convenient vector space) and any convex function $f : E \to \mathbb{R}$ the following statements are equivalent:

(1) The function f is continuous;
(2) The set A_f is open in $E \times \mathbb{R}$;
(3) The set $f_{<t} := \{x \in E : f(x) < t\}$ is open in E for all $t \in \mathbb{R}$.

13.2. Result. Convex Lipschitz functions. *Let $f : E \to \mathbb{R}$ be a convex function on a convenient vector space E. Then the following statements are equivalent:*

(1) *It is locally Lipschitzian;*
(2) *It is continuous for the locally convex topology;*
(3) *It is continuous for the c^∞-topology;*
(4) *It is bounded on Mackey converging sequences;*

If f is a seminorm, then these further are equivalent to

(5) *It is bounded on bounded sets.*

If E is normed this further is equivalent to

(6) *It is locally bounded.*

The proof is due to [Aronszajn, 1976] for Banach spaces and [Frölicher, Kriegl, 1988, p. 200], for convenient vector spaces.

13.3. Basic definitions. Let $f : E \supseteq U \to F$ be a mapping defined on a c^∞-open subset of a convenient vector space E with values in another one F. Let $x \in U$ and $v \in E$. Then the (one sided) *directional derivative* of f at x in direction v is defined as

$$f'(x)(v) = d_v f(x) := \lim_{t \searrow 0} \frac{f(x + t\,v) - f(x)}{t}.$$

Obviously, if $f'(x)(v)$ exists, then so does $f'(x)(s\,v)$ for $s > 0$ and equals $s\,f'(x)(v)$. Even if $f'(x)(v)$ exists for all $v \in E$ the mapping $v \mapsto f'(x)(v)$ may not be linear in general, and if it is linear it will not be bounded in general. Hence, f is called *Gâteaux-differentiable* at x, if the directional derivatives $f'(x)(v)$ exist for all $v \in E$ and $v \mapsto f'(x)(v)$ is a bounded linear mapping from $E \to F$.

Even for Gâteaux-differentiable mappings the difference quotient $\frac{f(x+t\,v)-f(x)}{t}$ need not converge uniformly for v in bounded sets (or even in compact sets). Hence, one defines f to be *Fréchet-differentiable* at x if f is Gâteaux-differentiable at x and $\frac{f(x+t\,v)-f(x)}{t} - f'(x)(v) \to 0$ uniformly for v in any bounded set. For a Banach space E this is equivalent to the existence of a bounded linear mapping denoted $f'(x) : E \to F$ such that

$$\lim_{v \to 0} \frac{f(x + v) - f(x) - f'(x)(v)}{\|v\|} = 0.$$

If $f : E \supseteq U \to F$ is Gâteaux-differentiable and the derivative $f' : E \supseteq U \to L(E, F)$ is continuous, then f is Fréchet-differentiable, and we will call such a function C^1. In fact, the fundamental theorem applied to $t \mapsto f(x + t\,v)$ gives us

$$f(x + v) - f(x) = \int_0^1 f'(x + t\,v)(v)\,dt,$$

and hence

$$\frac{f(x + s\,v) - f(x)}{s} - f'(x)(v) = \int_0^1 \Big(f'(x + t\,s\,v) - f'(x)\Big)(v)\,dt \to 0,$$

which converges to 0 for $s \to 0$ uniformly for v in any bounded set, since $f'(x + t\,s\,v) \to f'(x)$ uniformly on bounded sets for $s \to 0$ and uniformly for $t \in [0, 1]$ and v in any bounded set, since f' is assumed to be continuous.

Recall furthermore that a mapping $f : E \supseteq U \to F$ on a Banach space E is called *Lipschitz* if

$$\left\{ \frac{f(x_1) - f(x_2)}{\|x_1 - x_2\|} : x_1, x_2 \in U, x_1 \neq x_2 \right\} \text{ is bounded in } F.$$

It is called *Hölder* of order $0 < p \le 1$ if

$$\left\{ \frac{f(x_1) - f(x_2)}{\|x_1 - x_2\|^p} : x_1, x_2 \in U, x_1 \neq x_2 \right\} \text{ is bounded in } F.$$

13.4. Lemma. Gâteaux-differentiability of convex functions. *Every convex function $q : E \to \mathbb{R}$ has one sided directional derivatives. The derivative $q'(x)$ is sublinear and locally bounded (or continuous) if q is locally bounded (or continuous). In particular, such a function is Gâteaux-differentiable at x if and only if $q'(x)$ is an odd function, i.e. $q'(x)(-v) = -q'(x)(v)$.*

If E is not normed, then locally bounded-ness should mean bounded on bornologically compact sets.

Proof. For $0 < t < t'$ we have by convexity that

$$q(x + t\,v) = q\big((1 - \frac{t}{t'})x + \frac{t}{t'}\,(x + t'\,v)\big) \leq (1 - \frac{t}{t'})\,q(x) + \frac{t}{t'}\,q(x + t'v).$$

Hence $\frac{q(x+t\,v)-q(x)}{t} \leq \frac{q(x+t'\,v)-q(x)}{t'}$. Thus, the difference quotient is monotone falling for $t \to 0$. It is also bounded from below, since for $t' < 0 < t$ we have

$$\begin{aligned} q(x) &= q\big(\frac{t}{t-t'}\,(x + t'\,v) + (1 - \frac{t}{t-t'})\,(x + t\,v)\big) \\ &\leq \frac{t}{t-t'}\,q(x + t'\,v) + (1 - \frac{t}{t-t'})\,q(x + t\,v), \end{aligned}$$

and hence $\frac{q(x+t'\,v)-q(x)}{t'} \leq \frac{q(x+t\,v)-q(x)}{t}$. Thus, the one sided derivative

$$\lim_{t \searrow 0} \frac{q(x + t\,v) - q(x)}{t}$$

exists.

As a derivative $q'(x)$ automatically satisfies $q'(x)(t\,v) = t\,q'(x)(v)$ for all $t \geq 0$. The derivative $q'(x)$ is convex as limit of the convex functions $v \mapsto \frac{q(x+tv)-q(x)}{t}$. Hence it is sublinear.

The convexity of q implies that

$$q(x) - q(x - v) \leq q'(x)(v) \leq q(x + v) - q(x).$$

Therefore, the local boundedness of q at x implies that of $q'(x)$ at 0. Let $\ell := f'(x)$, then subadditivity and odd-ness implies $\ell(a) \leq \ell(a + b) + \ell(-b) = \ell(a + b) - \ell(b)$ and hence the converse triangle inequality. □

Remark. If q is a seminorm, then $\frac{q(x+tv)-q(x)}{t} \leq \frac{q(x)+t\,q(v)-q(x)}{t} = q(v)$, hence $q'(x)(v) \leq q(v)$, and furthermore $q'(x)(x) = \lim_{t\searrow 0} \frac{q(x+t\,x)-q(x)}{t} = \lim_{t\searrow 0} q(x) = q(x)$. Hence we have

$$\|q'(x)\| := \sup\{|q'(x)(v)| : q(v) \leq 1\} = 1.$$

Convention. Let $q \neq 0$ be a seminorm and let $q(x) = 0$. Then there exists a $v \in E$ with $q(v) \neq 0$, and we have $q(x + tv) = |t|\,q(v)$, hence $q'(x)(\pm v) = q(v)$. So q is not Gâteaux differentiable at x. Therefore, we call a seminorm *smooth* for some differentiability class, if and only if it is smooth on its carrier $\{x : q(x) > 0\}$.

13.5. Differentiability properties of convex functions f can be translated in geometric properties of A_f:

Lemma. Differentiability of convex functions. *Let $f : E \to \mathbb{R}$ be a continuous convex function on a Banach space E, and let $x_0 \in E$. Then the following statements are equivalent:*

(1) *The function f is Gâteaux (Fréchet) differentiable at x_0;*
(2) *There exists a unique $\ell \in E'$ with*

$$\ell(v) \leq f(x_0 + v) - f(x_0) \text{ for all } v \in E;$$

(3) *There exists a unique affine hyperplane through $(x_0, f(x_0))$ which is tangent to A_f.*
(4) *The Minkowski functional of A_f is Gâteaux differentiable at $(x_0, f(x_0))$.*

Moreover, for a sublinear function f the following statements are equivalent:

(5) *The function f is Gâteaux (Fréchet) differentiable at x;*
(6) *The point x_0 (strongly) exposes the closed unit ball $\{x : f(x) \leq 1\}$.*

In particular, the following statements are equivalent for a convex function f:

(7) *The function f is Gâteaux (Fréchet) differentiable at x_0;*
(8) *The Minkowski functional of A_f is Gâteaux (Fréchet) differentiable at the point $(x_0, f(x_0))$;*
(9) *The point $(x_0, f(x_0))$ (strongly) exposes the polar $(A_f)^o$.*

An element $x^* \in E^*$ is said to *expose a subset* $K \subseteq E$ if there exists a unique point $k_0 \in K$ with $x^*(k_0) = \sup\{x^*(k) : k \in K\}$. It is said to *strongly expose* K, if satisfies in addition that $x^*(x_n) \to x^*(k_0)$ implies $x_n \to k_0$.

By an affine hyperplane H *tangent* to a convex set K at a point $x \in K$ we mean that $x \in H$ and K lies on one side of H.

Proof. Let f be a convex function. By (13.4) and continuity we know that f is Gâteaux-differentiable if and only if the sub-linear mapping $f'(x_0)$ is linear. This is exactly the case if $f'(x_0)$ is minimal among all sub-linear mappings. From this follows (1) $\Rightarrow$ (2) by the following arguments: We have $f'(x_0)(v) \leq f(x_0+v)-f(x_0)$, and $\ell(v) \leq f(x_0 + v) - f(x_0)$ implies $\ell(v) \leq \frac{f(x_0+t\,v)-f(x_0)}{t}$, and hence $\ell(v) \leq f'(x)(v)$.

(2) $\Rightarrow$ (1) The uniqueness of ℓ implies $f'(x_0) = \ell$ since otherwise we had a linear functional $\mu \neq 0$ with $u \leq f'(x) - \ell$. Then $\mu + \ell$ contradicts uniqueness.

(2) $\Leftrightarrow$ (3) Any hyperplane tangent to A_f at $(x_0, f(x_0))$ is described by a functional $(\ell, s) \in E' \times \mathbb{R}$ such that $\ell(x) + s\,t \geq \ell(x_0) + s\,f(x_0)$ for all $t \geq f(x)$. Note that the scalar s cannot be 0, since this would imply that $\ell(x) \geq \ell(x_0)$ for all x. It has to be positive, since otherwise the left side would go to $-\infty$ for $f(x) \leq t \to +\infty$. Without loss of generality we may thus assume that $s = 1$, so the linear functional is uniquely determined by the hyperplane. Moreover, $\ell(x - x_0) \geq f(x_0) - f(x)$ or, by replacing ℓ by $-\ell$, we have $\ell(x_0 + v) \geq f(x_0) + \ell(v)$ for all $v \in E$.

(3) $\Leftrightarrow$ (4) Since the graph of a sublinear functional p is just the cone of $\{(y,1) : p(y) = 1\}$, the set A_p has exactly one tangent hyperplane at $(x,1)$ if and only if the set $\{y : p(y) \le 1\}$ has exactly one tangent hyperplane at x. Applying this to the Minkowski-functional p of A_f gives the desired result.

(5) $\Leftrightarrow$ (6) We show this first for Gâteaux-differentiability. We have to show that there is a unique tangent hyperplane to $x_0 \in K := \{x : f(x) \le 1\}$ if and only if x_0 exposes $K^o := \{\ell \in E^* : \ell(x) \le 1 \text{ for all } x \in K\}$. Let us assume $0 \in K$ and $0 \ne x_0 \in \partial K$. Then a tangent hyperplane to K at x_0 is uniquely determined by a linear functional $\ell \in E^*$ with $\ell(x_0) = 1$ and $\ell(x) \le 1$ for all $x \in K$. This is equivalent to $\ell \in K^o$ and $\ell(x_0) = 1$, since by Hahn-Banach there exists an $\ell \in K^o$ with $\ell(x_0) = 1$. From this the result follows.

This shows also (7) $\Leftrightarrow$ (8) $\Leftrightarrow$ (9) for Gâteaux-differentiability.

In order to show the statements for Fréchet-differentiability one has to show that $\ell = f'(x)$ is a Fréchet derivative if and only if x_0 is a strongly exposing point. This is left to the reader, see also (13.19) for a more general result. □

13.6. Lemma. Duality for convex functions. [Moreau, 1965].
Let $\langle\ ,\ \rangle : F \times G \to \mathbb{R}$ be a dual pairing.

(1) *For $f : F \to \mathbb{R} \cup \{+\infty\}$, $f \ne +\infty$ one defines the dual function*

$$f^* : G \to \mathbb{R} \cup \{+\infty\}, \quad f^*(z) := \sup\{\langle z, y\rangle - f(y) : y \in F\}.$$

(2) *The dual function f^* is convex and lower semi-continuous with respect to the weak topology. Since a convex function g is lower semi-continuous if and only if for all $a \in \mathbb{R}$ the set $\{x : g(x) > a\}$ is open, equivalently the convex set $\{x : g(x) \le a\}$ is closed, this is equivalent for every topology which is compatible with the duality.*
(3) $f_1 \le f_2 \Rightarrow f_1^* \ge f_2^*$.
(4) $f^* \le g \Leftrightarrow g^* \le f$.
(5) *$f^{**} = f$ if and only if f is lower semi-continuous and convex.*
(6) *Suppose $z \in G$ satisfies $f(x+v) \ge f(x) + \langle z, v\rangle$ for all v (in particular, this is true if $z = f'(x)$). Then $f(x) + f^*(z) = \langle z, x\rangle$.*
(7) *If $f_1(y) = f(y-a)$, then $f_1^*(z) = \langle z, a\rangle + f^*(z)$.*
(8) *If $f_1(y) = f(y) + a$, then $f_1^*(z) = f^*(z) - a$.*
(9) *If $f_1(y) = f(y) + \langle b, x\rangle$, then $f_1^*(z) = f^*(z-b)$.*
(10) *If $E = F = \mathbb{R}$ and $f \ge 0$ with $f(0) = 0$, then $f^*(t) = \sup\{ts - f(s) : t \ge 0\}$ for $t \ge 0$.*
(11) *If $\gamma \ge 0$ is convex and $\frac{\gamma(t)}{t} \to 0$, then $\gamma(t) > 0$ for $t > 0$.*
(12) *Let (F, G) be a Banach space and its dual. If $\gamma \ge 0$ is convex and $\gamma(0) = 0$, and $f(y) := \gamma(\|y\|)$, then $f^*(z) = \gamma^*(\|z\|)$.*
(13) *A convex function f on a Banach space is Fréchet differentiable at a with derivative $b := f'(a)$ if and only if there exists a convex non-negative function γ, with $\gamma(0) = 0$ and $\lim_{t\to 0} \frac{\gamma(t)}{t} = 0$, such that*

$$f(a+h) \le f(a) + \langle f'(a), h\rangle + \gamma(\|h\|).$$

Proof. (1) Since $f \neq +\infty$, there is some y for which $\langle z, y\rangle - f(y)$ is finite, hence $f^*(z) > -\infty$.

(2) The function $z \mapsto \langle z, y\rangle - f(y)$ is continuous and linear, and hence the supremum $f^*(z)$ is lower semi-continuous and convex. It remains to show that f^* is not constant $+\infty$: This is not true. In fact, take $f(t) = -t^2$ then $f^*(s) = \sup\{s\,t - f(t) : t \in \mathbb{R}\} = \sup\{s\,t + t^2 : t \in \mathbb{R}\} = +\infty$. More generally, $f^* \neq +\infty \Leftrightarrow f$ lies above some affine hyperplane, see (5).

(3) If $f_1 \le f_2$ then $\langle z, y\rangle - f_1(z) \ge \langle z, y\rangle - f_2(z)$, and hence $f_1^*(z) \ge f_2^*(z)$.

(4) One has

$$\begin{aligned} \forall z : f^*(z) \le g(z) &\Leftrightarrow \forall z, y : \langle z, y\rangle - f(y) \le g(z) \\ &\Leftrightarrow \forall z, y : \langle z, y\rangle - g(z) \le f(y) \\ &\Leftrightarrow \forall y : g^*(y) \le f(y). \end{aligned}$$

(5) Since $(f^*)^*$ is convex and lower semi-continuous, this is true for f provided $f = (f^*)^*$. Conversely, let $g(b) = -a$ and $g(z) = +\infty$ otherwise. Then $g^*(y) = \sup\{\langle z, y\rangle - g(z) : z \in G\} = \langle b, y\rangle + a$. Hence, $a + \langle b, \quad\rangle \le f \Leftrightarrow f^*(b) \le -a$. If f is convex and lower semi-continuous, then it is the supremum of all continuous linear functionals $a + \langle b, \quad\rangle$ below it, and this is exactly the case if $f^*(b) \le -a$. Hence, $f^{**}(y) = \sup\{\langle z, y\rangle - f^*(z) : z \in G\} \ge \langle b, y\rangle + a$ and thus $f = f^{**}$.

(6) Let $f(a+y) \ge f(a) + \langle b, y\rangle$. Then $f^*(b) = \sup\{\langle b, y\rangle - f(y) : y \in F\} = \sup\{\langle b, a + v\rangle - f(a+v) : v \in F\} \le \sup\{\langle b, a\rangle + \langle b, v\rangle - f(a) - \langle b, v\rangle : v \in F\} = \langle b, a\rangle - f(a)$.

(7) Let $f_1(y) = f(y-a)$. Then

$$\begin{aligned} f_1^*(z) &= \sup\{\langle z, y\rangle - f(y-a) : y \in F\} \\ &= \sup\{\langle z, y+a\rangle - f(y) : y \in F\} = \langle z, a\rangle + f^*(z). \end{aligned}$$

(8) Let $f_1(y) = f(y) + a$. Then

$$f_1^*(z) = \sup\{\langle z, y\rangle - f(y) - a : y \in F\} = f^*(z) - a.$$

(9) Let $f_1(y) = f(y) + \langle b, y\rangle$. Then

$$\begin{aligned} f_1^*(z) &= \sup\{\langle z, y\rangle - f(y) - \langle b, y\rangle : y \in F\} \\ &= \sup\{\langle z - b, y\rangle - f(y) : y \in F\} = f^*(z-b). \end{aligned}$$

(10) Let $E = F = \mathbb{R}$ and $f \ge 0$ with $f(0) = 0$, and let $s \ge 0$. Using that $s\,t - f(t) \le 0$ for $t \le 0$ and that $s\,0 - f(0) = 0$ we obtain

$$f^*(s) = \sup\{s\,t - f(t) : t \in \mathbb{R}\} = \sup\{s\,t - f(t) : t \ge 0\}.$$

(11) Let $\gamma \ge 0$ with $\lim_{t \searrow 0} \frac{\gamma(t)}{t} = 0$, and let $s > 0$. Then there are t with $s > \frac{\gamma(t)}{t}$, and hence

$$\gamma^*(s) = \sup\{st - \gamma(t) : t \ge 0\} = \sup\{t(s - \frac{\gamma(t)}{t}) : t \ge 0\} > 0.$$

(12) Let $f(y) = \gamma(\|y\|)$. Then

$$\begin{aligned} f^*(z) &= \sup\{\langle z, y\rangle - \gamma(\|y\|) : y \in F\} \\ &= \sup\{t\langle z, y\rangle - \gamma(t) : \|y\| = 1, t \geq 0\} \\ &= \sup\{\sup\{t\langle z, y\rangle - \gamma(t) : \|y\| = 1\}, t \geq 0\} \\ &= \sup\{t\|z\| - \gamma(t) : \|y\| = 1, t \geq 0\} \\ &= \gamma^*(\|z\|). \end{aligned}$$

(13) If $f(a+h) \leq f(a) + \langle b, h\rangle + \gamma(\|h\|)$, then we have

$$\frac{f(a+t\,h) - f(a)}{t} \leq \langle b, h\rangle + \frac{\gamma(t\,\|h\|)}{t},$$

hence $f'(a)(h) \leq \langle b, h\rangle$. Since $h \mapsto f'(a)(h)$ is sub-linear and the linear functionals are minimal among the sublinear ones, we have equality. By convexity we have

$$\frac{f(a+t\,h) - f(a)}{t} \geq \langle b, h\rangle = f'(a)(h).$$

So f is Fréchet-differentiable at a with derivative $f'(a)(h) = \langle b, h\rangle$, since the remainder is bounded by $\gamma(\|h\|)$ which satisfies $\frac{\gamma(\|h\|)}{\|h\|} \to 0$ for $\|h\| \to 0$.

Conversely, assume that f is Fréchet-differentiable at a with derivative b. Then

$$\frac{|f(a+h) - f(a) - \langle b, h\rangle|}{\|h\|} \to 0 \text{ for } h \to 0,$$

and by convexity

$$g(h) := f(a+h) - f(a) - \langle b, h\rangle \geq 0.$$

Let $\gamma(t) := \sup\{g(u) : \|u\| = |t|\}$. Since g is convex γ is convex, and obviously $\gamma(t) \in [0, +\infty]$, $\gamma(0) = 0$ and $\frac{\gamma(t)}{t} \to 0$ for $t \to 0$. This is the required function. □

13.7. Proposition. Continuity of the Fréchet derivative. [Asplund, 1968]. *The differential f' of any continuous convex function f on a Banach space is continuous on the set of all points where f is Fréchet differentiable. In general, it is however neither uniformly continuous nor bounded, see* (15.8).

Proof. Let $f'(x)(h)$ denote the one sided derivative. From convexity we conclude that $f(x+v) \geq f(x) + f'(x)(v)$. Suppose $x_n \to x$ are points where f is Fréchet differentiable. Then we obtain $f'(x_n)(v) \leq f(x_n + v) - f(x_n)$ which is bounded in n. Hence, the $f'(x_n)$ form a bounded sequence. We get

$$\begin{aligned} f(x) &\geq \langle f'(x_n), x\rangle - f^*(f'(x_n)) && \text{since } f(y) + f^*(z) \geq \langle z, y\rangle \\ &= \langle f'(x_n), x\rangle + f(x_n) - \langle f'(x_n), x_n\rangle && \text{since } f^*(f'(z)) + f(z) = f'(z)(z) \\ &\geq \langle f'(x_n), x - x_n\rangle + f(x) + \langle f'(x), x_n - x\rangle && \text{since } f(x+h) \geq f(x) + f'(x)(h) \\ &= \langle f'(x_n) - f'(x), x - x_n\rangle + f(x). \end{aligned}$$

Since $x_n \to x$ and $f'(x_n)$ is bounded, both sides converge to $f(x)$, hence

$$\lim_{n\to\infty} \langle f'(x_n), x\rangle - f^*(f'(x_n)) = f(x).$$

Since f is convex and Fréchet-differentiable at $a := x$ with derivative $b := f'(x)$, there exists by (13.6.13) a γ with

$$f(h) \le f(a) + \langle b, h-a\rangle + \gamma(\|h-a\|).$$

By duality we obtain using (13.6.3)

$$f^*(z) \ge \langle z, a\rangle - f(a) + \gamma^*(\|z-b\|).$$

If we apply this to $z := f'(x_n)$ we obtain

$$f^*(f'(x_n)) \ge \langle f'(x_n), x\rangle - f(x) + \gamma^*(\|f'(x_n) - f'(x)\|).$$

Hence

$$\gamma^*(\|f'(x_n) - f'(x)\|) \le f^*(f'(x_n)) - \langle f'(x_n), x\rangle + f(x),$$

and since the right side converges to 0, we have that $\gamma^*(\|f'(x_n) - f'(x)\|) \to 0$. Then $\|f'(x_n) - f'(x)\| \to 0$ where we use that γ is convex, $\gamma(0) = 0$, and $\gamma(t) > 0$ for $t > 0$, thus γ is strictly monotone increasing. □

13.8. Asplund spaces and generic Fréchet differentiability. From (13.4) it follows easily that a convex function $f : \mathbb{R} \to \mathbb{R}$ is differentiable at all except countably many points. This has been generalized by [Rademacher, 1919] to: Every Lipschitz mapping from an open subset of $\mathbb{R}^n$ to $\mathbb{R}$ is differentiable almost everywhere. Recall that a locally bounded convex function is locally Lipschitz, see (13.2).

Proposition. *For a Banach space E the following statements are equivalent:*

(1) *Every continuous convex function $f : E \to \mathbb{R}$ is Fréchet-differentiable on a dense G_δ-subset of E;*
(2) *Every continuous convex function $f : E \to \mathbb{R}$ is Fréchet-differentiable on a dense subset of E;*
(3) *Every locally Lipschitz function $f : E \to \mathbb{R}$ is Fréchet-differentiable on a dense subset of E;*
(4) *Every equivalent norm is Fréchet-differentiable at least at one point;*
(5) *E has no equivalent rough norm;*
(6) *Every (closed) separable subspace has a separable dual;*
(7) *The dual E^* has the Radon-Nikodym property;*
(8) *Every linear mapping $E \to L^1(X, \Omega, \mu)$ which is integral is nuclear;*
(9) *Every closed convex bounded subset of E^* is the closed convex hull of its extremal points;*
(10) *Every bounded subset of E^* is dentable.*

A Banach space satisfying these equivalent conditions is called Asplund space. *Every Banach space with a Fréchet differentiable bump function is Asplund,* [Ekeland, Lebourg, 1976, p. 203]. *It is an open question whether the converse is true.* *Every WCG-space (i.e. a Banach space for which a weakly compact subset K exists, whose linear hull is the whole space) is Asplund,* [John, Zizler, 1976].
The Asplund property is inherited by subspaces, quotients, and short exact sequences, [Stegall, 1981].

About the proof. (1) [Asplund, 1968]: If a convex function is Fréchet differentiable on a dense subset then it is so on a dense G_δ-subset, i.e. a dense countable intersection of open subsets.

(2) is in fact a local property, since in [Borwein, Fitzpatrick, Kenderov, 1991] it is mentioned that for a Lipschitz function $f : E \to \mathbb{R}$ with Lipschitz constant L defined on a convex open set U the function

$$\tilde{f}(x) := \inf\{f(y) + L\|x - y\| : y \in U\}$$

is a Lipschitz extension with constant L, and it is convex if f is.

(2) $\Rightarrow$ (3) is due to [Preiss, 1990], Every locally Lipschitz function on an Asplund space is Fréchet differentiable at points of a dense subset.

(3) $\Rightarrow$ (2) follows from the fact that continuous convex functions are locally Lipschitz, see (13.2).

(2) $\Leftrightarrow$ (4) is mentioned in [Preiss, 1990] without any proof or reference.

(2) $\Leftrightarrow$ (10) is due to [Stegall, 1975]. A subset D of a Banach space is called *dentable*, if and only if for every $x \in D$ there exists an $\varepsilon > 0$ such that x is not in the closed convex hull of $\{y \in D : \|y - x\| \geq \varepsilon\}$.

(2) $\Leftrightarrow$ (5) is due to [John, Zizler, 1978]. A norm p is called *rough*, see also (13.23), if and only if there exists an $\varepsilon > 0$ such that arbitrary close to each $x \in X$ there are points x_i and u with $\|u\| = 1$ such that $|p'(x_2)(u) - p'(x_1)(u)| \geq \varepsilon$. The usual norms on $C[0,1]$ and on ℓ^1 are rough by (13.12) and (13.13). A norm is not rough if and only if the dual ball is w^*-dentable. The unit ball is dentable if and only if the dual norm is not rough.

(2) $\Leftrightarrow$ (6) is due to [Stegall, 1975].

(2) $\Leftrightarrow$ (7) is due to [Stegall, 1978]. A closed bounded convex subset K of a Banach space E is said to have the *Radon-Nikodym property* if for any finite measure space (Ω, Σ, μ) every μ-continuous countably additive function $m : \Sigma \to E$ of finite variation with average range $\{\frac{m(A)}{\mu(S)} : S \in \Sigma, \mu(S) > 0\}$ contained in K is representable by a Bochner integrable function, i.e. there exists a Borel-measurable essentially separably valued function $f : \Omega \to E$ with $m(S) = \int_S f\, d\mu$. This function f is then called the Radon-Nikodym derivative of m. A Banach space is said to have the Radon-Nikodym property if every closed bounded convex subset has it. See also [Diestel, 1975]. A subset K is a Radon-Nikodym set if and only if every closed convex subset of K is the closed convex hull of its strongly exposed points.

(7) ⇔ (8) can be found in [Stegall, 1975] and is due to [Grothendieck, 1955]. A linear mapping $E \to F$ is called *integral* if and only if it has a factorization

$$\begin{array}{ccccc} E & \longrightarrow & F & \longrightarrow & F^{**} \\ \downarrow & & & & \uparrow \\ C(K) & & \longrightarrow & & L^1(K,\mu) \end{array}$$

for some Radon-measure μ on a compact space K.
A linear mapping $E \to F$ is called *nuclear* if and only if there are $x_n^* \in E^*$ and $y_n \in F$ such that $\sum_n \|x_n^*\| \, \|y_n\| < \infty$ and $T = \sum_n x_n^* \otimes y_n$.

(2) ⇔ (9) is due to [Stegall, 1981, p.516]. □

13.9. Results on generic Gâteaux differentiability of Lipschitz functions.

(1) [Mazur, 1933] *&* [Asplund, 1968] *A Banach space E with the property that every continuous convex function $f : E \to \mathbb{R}$ is Gâteaux-differentiable on a dense G_δ-subset is called weakly Asplund. Separable Banach spaces are weakly Asplund.*

(2) *In* [Živkov, 1983] *it is mentioned that there are Lipschitz functions on $\mathbb{R}$, which fail to be differentiable on a dense G_δ-subset.*

(3) *A Lipschitz function on a separable Banach space is "almost everywhere" Gâteaux-differentiable,* [Aronszajn, 1976].

(4) [Preiss, 1990] *If the norm on a Banach space is $\mathcal{B}$-differentiable then every Lipschitz function is $\mathcal{B}$-differentiable on a dense set. A function $f : E \supseteq U \to F$ is called $\mathcal{B}$-differentiable at $x \in U$ for some family $\mathcal{B}$ of bounded subsets, if there exists a continuous linear mapping (denoted $f'(x)$) in $L(E,F)$ such that for every $B \in \mathcal{B}$ one has $\frac{f(x+tv)-f(x)}{t} - f'(x)(v) \to 0$ for $t \to 0$ uniformly for $v \in B$.*

(5) [Kenderov, 1974], *see* [Živkov, 1983]. *Every locally Lipschitzian function on a separable Banach space which has one sided directional derivatives for each direction in a dense subset is Gâteaux differentiable on a non-meager subset.*

(6) [Živkov, 1983]. *For every space with Fréchet differentiable norm any locally Lipschitzian function which has directional derivatives for a dense set of directions is generically Gâteaux differentiable.*

(7) *There exists a Lipschitz Gâteaux differentiable function $f : L^1[0,1] \to \mathbb{R}$ which is nowhere Fréchet differentiable,* [Sova, 1966a], *see also* [Gieraltowska-Kedzierska, Van Vleck, 1991]. *Hence, this is an example of a weakly Asplund but not Asplund space.*

Further references on generic differentiability are: [Phelps, 1989], [Preiss, 1984], and [Zhivkov, 1987].

13.10. Lemma. Smoothness of $2n$-norm. *For $n \in \mathbb{N}$ the $2n$-norm is smooth on $L^{2n} \setminus \{0\}$.*

Proof. Since $t \mapsto t^{1/2n}$ is smooth on $\mathbb{R}^+$ it is enough to show that $x \mapsto (\|x\|_{2n})^{2n}$ is smooth. Let $p := 2n$. Since $(x_1, \dots, x_n) \mapsto x_1 \cdot \ldots \cdot x_n$ is a n-linear contraction from $L^p \times \ldots \times L^p \to L^1$ by the Hölder-inequality ($\sum_{i=1}^p \frac{1}{p} = 1$) and $\int : L^1 \to \mathbb{R}$ is a linear contraction the mapping $x \mapsto (x, \dots, x) \mapsto \int x^{2n}$ is smooth. Note that since we have a real Banach space and $p = 2n$ is even we can drop the absolute value in the formula of the norm. □

13.11. Derivative of the 1-norm. Let $x \in \ell^1$ and $j \in \mathbb{N}$ be such that $x_j = 0$. Let e_j be the characteristic function of $\{j\}$. Then $\|x + t\, e_j\|_1 = \|x\|_1 + |t|$ since the supports of x and e_j are disjoint. Hence, the directional derivative of the norm $p : v \mapsto \|v\|_1$ is given by $p'(x)(e_i) = 1$ and $p'(x)(-e_i) = 1$, and p is not differentiable at x. More generally we have:

Lemma. [Mazur, 1933, p.79]. *Let Γ be some set, and let p be the 1-norm given by $\|x\|_1 = p(x) := \sum_{\gamma\in\Gamma} |x_\gamma|$ for $x \in \ell^1(\Gamma)$. Then $p'(x)(h) = \sum_{x_\gamma=0} |h_\gamma| + \sum_{x_\gamma\neq 0} h_\gamma \operatorname{sign} x_\gamma$.*

The basic idea behind this result is, that the unit sphere of the 1-norm is a hyper-octahedra, and the points on the faces are those, for which no coordinate vanishes.

Proof. Without loss of generality we may assume that $p(x) = 1 = p(h)$, since for $r > 0$ and $s \geq 0$ we have $p'(r\,x)(s\,h) = \frac{d}{dt}|_{t=0} p(r\,x + t\,s\,h) = \frac{d}{dt}|_{t=0} r\,p(x + t\,(\frac{s}{r}h)) = r\,p'(x)(\frac{s}{r}h) = s\,p'(x)(h)$.

We have $|x_\gamma + h_\gamma| - |x_\gamma| = ||x_\gamma| + h_\gamma \operatorname{sign} x_\gamma| - |x_\gamma| \geq |x_\gamma| + h_\gamma \operatorname{sign} x_\gamma - |x_\gamma| = h_\gamma \operatorname{sign} x_\gamma$, and is equal to $|h_\gamma|$ if $x_\gamma = 0$. Summing up these (in)equalities we obtain

$$p(x+h) - p(x) - \sum_{x_\gamma=0} |h_\gamma| - \sum_{x_\gamma\neq 0} h_\gamma \operatorname{sign} x_\gamma \geq 0.$$

For $\varepsilon > 0$ choose a finite set $F \subset \Gamma$, such that $\sum_{\gamma\notin F} |h_\gamma| < \frac{\varepsilon}{2}$. Now choose t so small that

$$|x_\gamma| + t\,h_\gamma \operatorname{sign} x_\gamma \geq 0 \text{ for all } \gamma \in F \text{ with } x_\gamma \neq 0.$$

We claim that

$$\frac{q(x+t\,h) - q(x)}{t} - \sum_{x_\gamma=0} |h_\gamma| - \sum_{x_\gamma\neq 0} h_\gamma \operatorname{sign} x_\gamma \leq \varepsilon.$$

Let first γ be such that $x_\gamma = 0$. Then $\frac{|x_\gamma + t\,h_\gamma| - |x_\gamma|}{t} = |h_\gamma|$, hence these terms cancel with $-\sum_{x_\gamma=0} |h_\gamma|$.

Let now $x_\gamma \neq 0$. For $|x_\gamma| + t\,h_\gamma \operatorname{sign} x_\gamma \geq 0$ (hence in particular for $\gamma \in F$ with $x_\gamma \neq 0$) we have

$$\frac{|x_\gamma + t\,h_\gamma| - |x_\gamma|}{t} = \frac{|x_\gamma| + t\,h_\gamma \operatorname{sign} x_\gamma - |x_\gamma|}{t} = h_\gamma \operatorname{sign} x_\gamma.$$

Thus, these terms sum up to the corresponding sum $\sum_\gamma h_\gamma \operatorname{sign} x_\gamma$.

It remains to consider γ with $x_\gamma \neq 0$ and $|x_\gamma| + t\,h_\gamma \operatorname{sign} x_\gamma < 0$. Then $\gamma \notin F$ and

$$\frac{|x_\gamma + t\,h_\gamma| - |x_\gamma|}{t} - h_\gamma \operatorname{sign} x_\gamma = \frac{-|x_\gamma| - t\,h_\gamma \operatorname{sign} x_\gamma - |x_\gamma| - t\,h_\gamma \operatorname{sign} x_\gamma}{t}$$
$$\leq -2h_\gamma \operatorname{sign} x_\gamma,$$

and since $\sum_{\gamma\notin F} |h_\gamma| < \frac{\varepsilon}{2}$ these remaining terms sum up to something smaller than ε. □

Remark. The 1-norm is rough. This result shows that the 1-norm is Gâteaux-differentiable exactly at those points, where all coordinates are non-zero. Thus, if Γ is uncountable, the 1-norm is nowhere Gâteaux-differentiable.

In contrast to what is claimed in [Mazur, 1933, p.79], the 1-norm is nowhere Fréchet differentiable. In fact, take $0 \neq x \in \ell^1(\Gamma)$. For γ with $x_\gamma \neq 0$ and $t > 0$ we have that

$$p(x + t\,(-\operatorname{sign} x_\gamma\, e_\gamma)) - p(x) - t\,p'(x)(-\operatorname{sign} x_\gamma\, e_\gamma) =$$
$$= |x_\gamma - t \operatorname{sign} x_\gamma| - |x_\gamma| + t = \big||x_\gamma| - t\big| - |x_\gamma| + t \geq t \cdot 1,$$

provided $t \geq 2\,|x_\gamma|$, since then $\big||x_\gamma| - t\big| = t - |x_\gamma| \geq |x_\gamma|$. Obviously, for every $t > 0$ there are γ satisfying this required condition; either $x_\gamma = 0$ then we have a corner, or $x_\gamma \neq 0$ then it gets arbitrarily small. Thus, the directional difference quotient does not converge uniformly on the unit-sphere.

The set of points x in ℓ^1 where at least for one n the coordinate x_n vanishes is dense, and one has

$$p(x + t\,e^n) = p(x) + |t|, \text{ hence } p'(x + t\,e^n)(e^n) = \begin{cases} +1 & \text{for } t \geq 0 \\ -1 & \text{for } t < 0 \end{cases}.$$

Hence the derivative of p is uniformly discontinuous, i.e., in every non-empty open set there are points x_1, x_2 for which there exists an $h \in \ell^1$ with $\|h\| = 1$ and $|p'(x_1)(h) - p'(x_2)(h)| \geq 2$.

13.12. Derivative of the ∞-norm. On c_0 the norm is not differentiable at points x, where the norm is attained in at least two points. In fact let $|x(a)| = \|x\| = |x(b)|$ and let $h := \operatorname{sign} x(a)\, e_a$. Then $p(x + th) = |(x + th)(a)| = \|x\| + t$ for $t \geq 0$ and $p(x + th) = |(x + th)(b)| = \|x\|$ for $t \leq 0$. Thus, $t \mapsto p(x + th)$ is not differentiable at 0 and thus p not at x.

If the norm of x is attained at a single coordinate a, then p is differentiable at x. In fact $p(x + th) = |(x + th)(a)| = |\operatorname{sign}(x(a))\|x\| + th(a)\operatorname{sign}^2(x(a))| = |\|x\| + th(a)\operatorname{sign}(x(a))| = \|x\| + th(a)\operatorname{sign}(x(a))$ for $|t|\,\|h\| \leq \|x\| - \sup\{|x(t)| : t \neq a\}$. Hence the directional difference-quotient converges uniformly for h in the unit-ball.

Let $x \in C[0,1]$ be such that $\|x\|_\infty = |x(a)| = |x(b)|$ for $a \neq b$. Choose a y with $y(s)$ between 0 and $x(s)$ for all s and $y(a) = x(a)$ but $y(b) = 0$. For $t \geq 0$ we have $|(x + t\,y)(s)| \leq |x(a) + t\,y(a)| = (1+t)\,\|x\|_\infty$ and hence $\|x + t\,y\|_\infty = (1+t)\,\|x\|_\infty$. For $-1 \leq t \leq 0$ we have $|(x + t\,y)(s)| \leq |x(a)|$ and $\|(x + t\,y)(b)\| = \|x(a)\|$ and hence $\|x + t\,y\|_\infty = \|x\|_\infty$. Thus the directional derivative is given by $p'(x)(y) = \|x\|_\infty$ and $p'(x)(-y) = 0$. More precisely we have the following results.

Lemma. [Banach, 1932, p. 168]. *Let T be a compact metric space. Let $x \in C(T,\mathbb{R}) \setminus \{0\}$ and $h \in C(T,\mathbb{R})$. By p we denote the ∞-norm $\|x\|_\infty = p(x) := \sup\{|x(t)| : t \in T\}$. Then $p'(x)(h) = \sup\{h(t) \operatorname{sign} x(t) : |x(t)| = p(x)\}$.*

The idea here is, that the unit-ball is a hyper-cube, and the points on the faces are exactly those for which the supremum is attained only in one point.

Proof. Without loss of generality we may assume that $p(x) = 1 = p(h)$, since for $r > 0$ and $s \geq 0$ we have $p'(r\,x)(s\,h) = \frac{d}{dt}|_{t=0} p(r\,x + t\,s\,h) = \frac{d}{dt}|_{t=0} r\,p(x + t\,(\frac{s}{r}h)) = r\,p'(x)(\frac{s}{r}h) = s\,p'(x)(h)$.

Let $A := \{t \in T : |x(t)| = p(x)\}$. For given $\varepsilon > 0$ we find by the uniform continuity of x and h a δ_1 such that $|x(t) - x(t')| < \frac{1}{2}$ and $|h(t) - h(t')| < \varepsilon$ for $\operatorname{dist}(t,t') < \delta_1$. Then $\{t : \operatorname{dist}(t,A) \geq \delta_1\}$ is closed, hence compact. Therefore $\delta := \|x\|_\infty - \sup\{|x(t)| : \operatorname{dist}(t,A) \geq \delta_1\} > 0$.

Now we claim that for $0 < t < \min\{\delta, 1\}$ we have

$$0 \leq \frac{p(x + t\,h) - p(x)}{t} - \sup\{h(r) \operatorname{sign} x(r) : r \in A\} \leq \varepsilon.$$

For all $s \in A$ we have

$$\begin{aligned} p(x + t\,h) &\geq |(x + t\,h)(s)| = \big||x(s)| \operatorname{sign} x(s) + t\,h(s) \operatorname{sign} x(s)^2\big| \\ &= \big||x(s)| + t\,h(s) \operatorname{sign} x(s)\big| = p(x) + t\,h(s) \operatorname{sign} x(s) \end{aligned}$$

for $0 \leq t \leq 1$, since $|h(s)| \leq p(h) = p(x)$. Hence

$$\frac{p(x + t\,h) - p(x)}{t} \geq \sup\{h(t) \operatorname{sign} x(t) : t \in A\}.$$

This shows the left inequality.

Let s be a point where the supremum $p(x + t\,h)$ is attained. From the left inequality it follows that

$$\begin{aligned} p(x + t\,h) &\geq p(x) + t \sup\{h(r) \operatorname{sign} x(r) : r \in A\}, \qquad \text{and hence} \\ |x(s)| &\geq |(x + th)(s)| - t\,|h(s)| \geq p(x + t\,h) - t\,p(h) \\ &\geq p(x) - t\,\underbrace{\Big(p(h) - \sup\{h(r) \operatorname{sign} x(r) : r \in A\}\Big)}_{\leq 1} \\ &> p(x) - \delta = \sup\{|x(r)| : \operatorname{dist}(r,A) \geq \delta_1\}. \end{aligned}$$

Therefore $\operatorname{dist}(s,A) < \delta_1$, and thus there exists an $a \in A$ with $\operatorname{dist}(s,a) < \delta_1$ and consequently $|x(s) - x(a)| < \frac{1}{2}$ and $|h(s) - h(a)| < \varepsilon$. In particular, $\operatorname{sign} x(s) = \operatorname{sign} x(a) \neq 0$. So we get

$$\begin{aligned} \frac{p(x + t\,h) - p(x)}{t} &= \frac{|(x + t\,h)(s)| - p(x)}{t} = \frac{\big||x(s)| + t\,h(s) \operatorname{sign} x(s)\big| - p(x)}{t} \\ &= \frac{|x(s)| + t\,h(s) \operatorname{sign} x(s) - p(x)}{t} \leq h(s) \operatorname{sign} x(a) \\ &\leq |h(s) - h(a)| + h(a) \operatorname{sign} x(a) \\ &< \varepsilon + \sup\{h(r) \operatorname{sign} x(r) : r \in A\}. \end{aligned}$$

This proves the claim which finally implies

$$p'(x)(v) = \lim_{t\searrow 0} \frac{p(x+t\,h)-p(x)}{t} = \sup\{h(r)\,\operatorname{sign} x(r) : r \in A\}. \quad \square$$

Remark. The ∞-norm is rough. This result shows that the points where the ∞-norm is Gâteaux-differentiable are exactly those x where the supremum $p(x)$ is attained in a single point a. The Gâteaux-derivative is then given by $p'(x)(h) = h(a)\,\operatorname{sign} x(a)$. In general, this is however not the Fréchet derivative:
Let $x \neq 0$. Without loss we may assume (that $p(x) = 1$ and) that there is a unique point a, where $|x(a)| = p(x)$. Moreover, we may assume $x(a) > 0$. Let $a_n \to a$ be such that $0 < x(a_n) < x(a)$ and let $0 < \delta_n := x(a) - x(a_n) < x(a)$. Now choose $s_n := 2\delta_n \to 0$ and $h_n \in C[0,1]$ with $p(h_n) \leq 1$, $h_n(a) = 0$ and $h_n(a_n) := 1$ and $p(x + s_n h_n) = (x + s_n h_n)(a_n) = x(a_n) + 2(x(a) - x(a_n)) = 2x(a) - x(a_n)$. For this choose $(x + s_n h_n)(t) \leq (x + s_n h_n)(a_n)$ locally, i.e.. $h_n(t) \leq 1 + (x(a_n) - x(t))/s_n$ and 0 far away from x. Then $p'(x)(h_n) = 0$ by (13.12) and

$$\begin{aligned}\frac{p(x + s_n\,h_n) - p(x)}{s_n} - p'(x)(h_n) &= \frac{2x(a) - x(a_n) - x(a)}{s_n}\\ &= \frac{\delta_n}{2\delta_n} = \frac{1}{2} \not\to 0\end{aligned}$$

Thus the limit is not uniform and p is not Fréchet differentiable at x.

The set of vectors $x \in C[0,1]$ which attain their norm at least at two points a and b is dense, and one has for appropriately chosen h with $h(a) = -x(a)$, $h(b) = x(b)$ that

$$p(x + t\,h) = (1 + \max\{t, -t\})\,p(x), \text{ hence } p'(x + t\,h)(h) = \begin{cases} +1 & \text{for } t \geq 0 \\ -1 & \text{for } t < 0 \end{cases}.$$

Therefore, the derivative of the norm is uniformly discontinuous, i.e., in every non-empty open set there are points x_1, x_2 for which there exists an $h \in C[0,1]$ with $\|h\| = 1$ and $|p'(x_1)(h) - p'(x_2)(h)| \geq 2$.

13.13. Results on the differentiability of p-norms. [Bonic, Frampton, 1966, p.887].
For $1 < p < \infty$ not an even integer the function $t \mapsto |t|^p$ is differentiable of order n if $n < p$, and the highest derivative ($t \mapsto p\,(p-1)\dots(p-n+1)\,|t|^{p-n}$) satisfies a Hölder-condition with modulus $p - n$, one can show that the p-norm has exactly these differentiability properties, i.e.

1. *It is $(p-1)$-times differentiable with Lipschitzian highest derivative if p is an integer.*
2. *It is $[p]$-times differentiable with highest derivative being Hölderian of order $p - [p]$, otherwise.*
3. *The norm has no higher Hölder-differentiability properties.*

That the norm on L^p is C^1 for $1 < p < \infty$ was already shown by [Mazur, 1933].

13.14. Proposition. Smooth norms on a Banach space. *A norm on a Banach space is of class C^n on $E \setminus \{0\}$ if and only if its unit sphere is a C^n-submanifold of E.*

Proof. Let $p : E \to \mathbb{R}$ be a smooth norm. Since $p'(x)(x) = \frac{d}{dt}|_{t=0} p(x + tx) = \frac{d}{dt}|_{t=0}(1 + t)p(x) = p(x)$, we see that $p(x) = 1$ is a regular equation and hence the unit sphere $S := p^{-1}(1)$ is a smooth submanifold (of codimension 1), see (27.11).

Explicitly, this can be shown as follows: For $a \in S$ let $\Phi : \ker(p'(a)) \times \mathbb{R}^+ \to E^+ := \{x \in E : p'(a)(x) > 0\}$ be given by $(v, t) \mapsto t \frac{a+v}{p(a+v)}$. This is well-defined, since $p(a + v) \geq p(a) + p'(a)(v) = p(a) = 0$ for $v \in \ker(p'(a))$. Note that $\Phi(v, t) = y$ implies that $t = p(y)$ and $v \in \ker(p'(a))$ is such that $a + v = \mu y$ for some $\mu \neq 0$, i.e. $\mu p'(a)(y) = p'(a)(a + v) = p'(a)(a) = p(a) = 1$ and hence $v = \frac{1}{p'(a)(y)} y - a$. Thus Φ is a diffeomorphism that maps $\ker(p'(a)) \times \{1\}$ onto $S \cap E^+$.

Conversely, let $x_0 \in E \setminus \{0\}$ and $a := \frac{x_0}{p(x_0)}$. Then a is in the unit sphere, hence there exists locally around a a diffeomorphism $\Phi : E \supseteq U \to \Phi(U) \subseteq E$ which maps $S \cap U \to F \cap \Phi(U)$ for some closed linear subspace $F \subseteq U$. Let $\lambda : E \to \mathbb{R}$ be a continuous linear functional with $\lambda(a) = 1$ and $\lambda \leq p$. Note that $b := \Phi'(a)(a) \neq F$, since otherwise $t \mapsto \Phi^{-1}(tb)$ is in S, but then $\lambda(\Phi^{-1}(tb)) \leq 0$ and hence $0 = \frac{d}{dt}|_{t=0}\lambda(\Phi^{-1}(tb)) = \lambda(\Phi'(a)^{-1}b) = \lambda(a) = 1$ gives a contradiction. Choose $\mu \in E'$ with $\mu|_F = 0$ and $\mu(b) = 1$. We have to show that $x \mapsto p(x)$ is C^n locally around x_0, or equivalently that this is true for $g : x \mapsto \frac{1}{p(x)}$. Then $g(x)$ is solution of the implicit equation $\varphi(x, g(x)) = 0$, where $\varphi : E \times \mathbb{R} \to F$ is given by $(x, g) \mapsto f(g \cdot x)$ with $f := \mu \circ \Phi$. This solution is C^n by the implicit function theorem, since $\partial_2\varphi(x_0, g(x_0)) = f'(g(x_0)x_0)(x_0) = p(x_0) f'(a)(a) = p(x_0) \mu(b) = p(x) \neq 0$, because f is a regular equation at a. $\square$

Although this proof uses the implicit function theorem on Banach spaces we can do without as the following theorem shows:

13.15. Theorem. Characterization of smooth seminorms. *Let E be a convenient vector space.*

(1) Let $p : E \to \mathbb{R}$ be a convex function which is smooth on a neighborhood of $p^{-1}(1)$, and assume that $U := \{x \in E : p(x) < 1\}$ is not empty. Then U is open, and its boundary ∂U equals $\{x : p(x) = 1\}$, a smooth splitting submanifold of E.

(2) If U is a convex absorbing open subset of E whose boundary is a smooth submanifold of E then the Minkowski functional p_U is a smooth sublinear mapping, and $U = \{x \in E : p_U(x) < 1\}$.

Proof. (1) The set U is obviously convex and open by (4.5) and (13.1). Let $M := \{x : p(x) = 1\}$. We claim that $M = \partial U$. Let $x_0 \in U$ and $x_1 \in M$. Since $t \mapsto p(x_1 + t(x_0 - x_1))$ is convex it is strictly decreasing in a neighborhood of 0. Hence, there are points x close to x_1 with $p(x) < p(x_1)$ and such with $p(x) \geq 1$, i.e. x belongs to ∂U. Conversely, let $x \in \partial U$. Since U is open we have $p(x_1) \geq 1$. Suppose $p(x_1) > 1$, then $p(x) > 1$ locally around x_1, a contradiction to $x_1 \in \partial U$.

Now we show that M is a smooth splitting submanifold of E, i.e. every point has a neighborhood, in which M is up to a diffeomorphism a complemented subspace. Let $x_1 \in M = \partial U$. We consider the convex mapping $t \mapsto p(x_0 + t(x_1 - x_0))$. It is locally around 1 differentiable, and its value at 0 is strictly less than that at 1. Thus, $p'(x_1)(x_1 - x_0) \geq p(x_1) - p(x_0) > 0$, and hence we may replace x_0 by some point on the segment from x_0 to x_1 closer to x_1, such that $p'(x_0)(x_1 - x_0) > 0$. Without loss of generality we may assume that $x_0 = 0$. Let $U := \{x \in E : p'(0)x > 0 \text{ and } p'(x_1)x > 0\}$ and $V := (U - x_1) \cap \ker p'(x_1) \subseteq \ker p'(x_1)$. A smooth mapping from the open set $U \subseteq E$ to the open set $V \times \mathbb{R} \subseteq \ker p'(x_1) \times (p(0), +\infty)$ is given by $x \mapsto (tx - x_1, p(x))$, where $t := \frac{p'(x_1)(x_1)}{p'(x_1)(x)}$. This mapping is a diffeomorphism, since for $(y, r) \in \ker p'(x_1) \times \mathbb{R}$ the inverse image is given as $t(y + x_1)$ where t can be calculated from $r = p(t\,(y+x_1))$. Since $t \mapsto p(t\,(y+x_1))$ is a diffeomorphism between the intervals $(0, +\infty) \to (p(0), +\infty)$ this t is uniquely determined. Furthermore, t depends smoothly on (y, r): Let $s \mapsto (y(s), r(s))$ be a smooth curve, then $t(s)$ is given by the implicit equation $p(t\,(y(s) + x_1)) = r(s)$, and by the 2-dimensional implicit function theorem the solution $s \mapsto t(s)$ is smooth.

(2) By general principles p_U is a sublinear mapping, and $U = \{x : p_U(x) < 1\}$ since U is open. Thus it remains to show that p_U is smooth on its open carrier. So let c be a smooth curve in the carrier. By assumption, there is a diffeomorphism v, locally defined on E near an intersection point a of the ray through $c(0)$ with the boundary $\partial U = \{x : p(x) = 1\}$, such that ∂U corresponds to a closed linear subspace $F \subseteq E$. Since U is convex there is a bounded linear functional $\lambda \in E'$ with $\lambda(a) = 1$ and $U \subseteq \{x \in E : \lambda(x) \leq 1\}$ by the theorem of Hahn-Banach. Then $\lambda(T_a(\partial U)) = 0$ since any smooth curve in ∂U through a stays inside $\{x : \lambda(x) \leq 1\}$. Furthermore, $b : \frac{\partial}{\partial t}|_1 v(ta) \notin F$, since otherwise $t \mapsto v^{-1}(tb) \in \partial U$ but $\frac{\partial}{\partial t}|_1 \lambda(v^{-1}(tb)) = \lambda(a) = 1$. Put $f := 1/p_U \circ c : \mathbb{R} \to \mathbb{R}$. Then f is a solution of the implicit equation $(\lambda \circ dv^{-1}(0) \circ v)(f(t)c(t)) = 0$ which has a unique smooth solution by the implicit function theorem in dimension 2 since

$$\tfrac{\partial}{\partial s}|_{s=f(t)} (\lambda \circ dv^{-1}(0) \circ v)(sc(t)) = \lambda dv^{-1}(0) dv(f(t)c(t))c(t) \neq 0$$

for t near 0, since for $t = 0$ we get $\lambda(c(0)) = \frac{1}{f(0)}$. So p_U is smooth on its carrier. □

13.16. The space $c_0(\Gamma)$. For an arbitrary set Γ the space $c_0(\Gamma)$ is the closure of all functions on Γ with finite support in the Banach space $\ell^\infty(\Gamma)$ of globally bounded functions on Γ with the supremum norm. The supremum norm on $c_0(\Gamma)$ is not differentiable on its carrier, see (13.12). Nevertheless, it was shown in [Bonic, Frampton, 1965] that c_0 is C^∞-regular.

Proposition. Smooth norm on c_0. Due to Kuiper according to [Bonic, Frampton, 1966]. *There exists an equivalent norm on $c_0(\Gamma)$ which is smooth off 0.*

Proof. To prove this let $h : \mathbb{R} \to \mathbb{R}$ be an unbounded symmetric smooth convex function vanishing near 0. Let $f : c_0(\Gamma) \to \mathbb{R}$ be given by $f(x) := \sum_{\gamma \in \Gamma} h(x_\gamma)$. Locally on $c_0(\Gamma)$ the function f is just a finite sum, hence f is smooth. In fact let

$h(t) = 0$ for $|t| \le \delta$. For $x \in c_0(\Gamma)$ the set $F := \{\gamma : |x_\gamma| \ge \delta/2\}$ is finite, and for $\|y - x\| < \delta$ we have that $f(y) = \sum_{\gamma \in F} h(y_\gamma)$.

The set $U := \{x : f(x) < 1\}$ is open, and bounded: Let $h(t) \ge 1$ for $|t| \ge \Delta$ and $f(x) < 1$, then $h(x_\gamma) < 1$ and thus $|x_\gamma| \le \Delta$ for all γ. The set U is also absolutely convex: Since h is convex, so is f and hence U. Since h is symmetric, so is f and hence U.

The boundary $\partial U = f^{-1}(1)$ is a splitting submanifold of $c_0(\Gamma)$ by the implicit function theorem on Banach spaces, since $df(x)x \ne 0$ for $x \in \partial U$. In fact $df(x)(x) = \sum_\gamma h'(x_\gamma)x_\gamma \ge 0$ and at least for one γ we have $h(x_\gamma) > 0$ and thus $h'(x_\gamma) \ne 0$. So by (13.14) the Minkowski functional p_U is smooth off 0. Obviously, it is an equivalent norm. □

13.17. Proposition. Inheritance properties for differentiable norms.

(1) *The product of two spaces with C^n-norm has again a C^n-norm given by $\|(x_1, x_2)\| := \sqrt{\|x_1\|^2 + \|x_2\|^2}$. More generally, the ℓ^2-sum of C^n-normable Banach spaces is C^n-normable.*

(2) *A subspace of a space with a C^n-norm has a C^n-norm.*

(3) [Godefroy, Pelant, et. al., 1988]. *If $c_0(\Gamma) \to E \to F$ is a short exact sequence of Banach spaces, and F has a C^k-norm, then E has a C^k-norm. See also* (14.12.1) *and* (16.19).

(4) *For a compact space K let K' be the set of all accumulation points of K. The operation $K \mapsto K'$ has the following properties:*

$$(a) \quad A \subseteq B \Rightarrow A' \subseteq B'$$
$$(b) \quad (A \cup B)' = A' \cup B'$$
$$(c) \quad (A \times B)' = (A' \times B) \cup (A \times B')$$
$$(d) \quad \big(\{0\} \cup \{\tfrac{1}{n} : n \in \mathbb{N}\}\big)' = \{0\}$$
$$(e) \quad K' = \emptyset \Leftrightarrow K \text{ discrete.}$$

(5) *If K is compact and $K^{(\omega)} = \emptyset$ then $C(K)$ has an equivalent C^∞-norm, see also* (16.20).

Proof. (1) and (2) are obvious.

(4) (a) is obvious, since if $\{x\}$ is open in B and $x \in A$, then it is also open in A in the trace topology, hence $A \cap (B \setminus B') \subseteq A \setminus A'$ and hence $A' = A \setminus (A \setminus A') \subseteq (A \setminus A \cap (B \setminus B')) = A \cap B' \subseteq B'$.

(b) By monotonicity we have '$\supseteq$'. Conversely let $x \in A' \cup B'$, w.l.o.g. $x \in A'$, suppose $x \notin (A \cup B)'$, then $\{x\}$ is open in $A \cup B$ and hence $\{x\} = \{x\} \cap A$ would be open in A, i.e. $x \notin A'$, a contradiction.

(c) is obvious, since $\{(x, y)\}$ is open in $A \times B \Leftrightarrow \{x\}$ is open in A and $\{y\}$ is open in B.

(d) and (e) are trivial.

For (3) a construction is used similar to that of Kuiper's smooth norm for c_0. Let $\pi : E \to F$ be the quotient mapping and $\| \quad \|$ the quotient norm on F. The dual sequence $\ell^1(A) \leftarrow E^* \leftarrow F^*$ splits (just define $T : \ell^1(A) \to E^*$ by selection of $x_a^* := T(e_a) \in E^*$ with $\|x_a^*\| = 1$ and $x_a^*|_{c_0(A)} = \mathrm{ev_a}$ using Hahn Banach). Note that for every $x \in E$ and $\varepsilon > 0$ the set $\{\alpha : |x_\alpha^*(x)| \geq \|\pi(x)\| + \varepsilon\}$ is finite. In fact, by definition of the quotient norm $\|\pi(x)\| := \sup\{\|x+y\| : y \in c_0(\Gamma)\}$ there is a $y \in c_0(\Gamma)$ such that $\|x+y\| \leq \|\pi(x)\| + \varepsilon/2$. The set $\Gamma_0 := \{\alpha : |y_\alpha| \geq \varepsilon/2\}$ is finite. For all other α we have

$$\begin{aligned} |x_\alpha^*(x)| \leq |x_\alpha^*(x+y)| + |x_\alpha^*(y)| \leq \|x_\alpha^*\| \, \|x+y\| + |y_\alpha| < \\ < 1\,(\|\pi(x)\| + \varepsilon/2) + \varepsilon/2 = \|\pi(x)\| + \varepsilon. \end{aligned}$$

Furthermore, we have

$$\|x\| \leq 2\|\pi(x)\| + \sup\{|x_\alpha^*(x)| : \alpha\}.$$

In fact,

$$\begin{aligned} \|x\| &= \sup\{|\langle x^*, x\rangle| : \|x^*\| \leq 1\} \\ &\leq \sup\{|\langle T(\lambda) + y^* \circ \pi, x\rangle| : \|\lambda\|_1 \leq 1, \|y^*\| \leq 2\} \\ &= \sup\{|x_\alpha^*(x)| : \alpha\} + 2\|\pi(x)\|, \end{aligned}$$

since $x^* = T(\lambda) + x^* - T(\lambda)$, where $\lambda := x^*|_{c_0(\Gamma)}$ and hence $\|\lambda\|_1 \leq \|x^*\| \leq 1$, and $|T(\lambda)(x)| \leq \|\lambda\|_1 \sup\{|x_\alpha^*(x)| : \alpha\} \leq \|x\|$ hence $\|T(\lambda)\| \leq \|\lambda\|_1$, and $y^* \circ \pi = x^* - T(\lambda)$. Let $\| \quad \|$ denote a norm on F which is smooth and is larger than the quotient norm. Analogously to (13.16) we define

$$f(x) := h(4\|\pi(x)\|) \prod_{a \in A} h(x_a^*(x)),$$

where $h : \mathbb{R} \to [0,1]$ is smooth, even, 1 for $|t| \leq 1$, 0 for $|t| \geq 2$ and concave on $\{t : h(t) \geq 1/2\}$. Then f is smooth, since if $\pi(x) > 1/2$ then the first factor vanishes locally, and if $\|\pi(x)\| < 1$ we have that $\Gamma_0 := \{\alpha : |x_\alpha^*(x)| \geq 1 - \varepsilon\}$ is finite, where $\varepsilon := (1 - \|\pi(x)\|)/2$, for $\|y - x\| < \varepsilon$ also $|x_\alpha^*(y) - x_\alpha^*(x)| < \varepsilon$ and hence $|x_\alpha^*(y)| < 1 - \varepsilon + \varepsilon = 1$ for all $\alpha \notin \Gamma_0$. So the product is locally finite. The set $\{x : f(x) > \frac{1}{2}\}$ is open, bounded and absolutely convex and has a smooth boundary $\{x : f(x) = \frac{1}{2}\}$. It is symmetric since f is symmetric. It is bounded, since $f(x) > 1/2$ implies $h(4\|\pi(x)\|) \geq 1/2$ and $h(x_a^*(x)) \geq 1/2$ for all a. Thus $4\|\pi(x)\| \leq 2$ and $|x_a^*(x)| \leq 2$ and thus $\|x\| \leq 2 \cdot 1/2 + 2 = 3$. For the convexity note that $x_i \geq 0$, $y_i \geq 0$, $0 \leq t \leq 1$, $\prod_i x_i \geq 1/2$, $\prod_i y_i \geq 1/2$ imply $\prod_i (tx_i + (1-t)y_i) \geq 1/2$, since log is concave. Since all factors of f have to be $\geq 1/2$ and h is concave on this set, convexity follows. Since one factor of $f(x) = \prod_\alpha f_\alpha(x)$ has to be unequal to 1, the derivative $f'(x)(x) < 0$, since $f_\alpha'(x)(x) \leq 0$ for all α by concavity and $f_\alpha'(x)(x) < 0$ for all x with $f_\alpha(x) < 1$. So its Minkowski-functional is an equivalent smooth norm on E.

Statement (5) follows from (3). First recall that K' is the set of accumulation points of K, i.e. those points x for which every neighborhood meets $K \setminus \{x\}$, i.e. x is not open. Thus $K \setminus K'$ is discrete. For successor ordinals $\alpha = \beta + 1$ one defines $K^{(\alpha)} := (K^{(\beta)})'$ and for limit ordinals α as $\bigcap_{\beta<\alpha} K^{(\beta)}$. For a compact space K the equality $K^{(\omega)} = \emptyset$ implies $K^{(n)} = \emptyset$ for some $n \in \omega$, since $K^{(n)}$ is closed. Now one shows this by induction. Let $E := \{f \in C(K) : f|_{K'} = 0\}$. By the Tietze-Urysohn theorem one has a short exact sequence $c_0(K \setminus K') \cong E \to C(K) \to C(K')$. The equality $E = c_0(K \setminus K_0)$ can be seen as follows:

Let $f \in C(K)$ with $f|_{K'} = 0$. Suppose there is some $\varepsilon > 0$ such that $\{x : |f(x)| \ge \varepsilon\}$ is not finite. Then there is some accumulation point x_∞ of this set and hence $|f(x_\infty)| \ge \varepsilon$ but $x_\infty \in K'$ and so $f(x_\infty) = 0$. Conversely let $f \in c_0(K \setminus K')$ and define $\tilde f$ by $\tilde f|_{K'} := 0$ and $\tilde f|_{K\setminus K'} = f$. Then $\tilde f$ is continuous on $K \setminus K'$, since $K \setminus K'$ is discrete. For $x \in K'$ we have that $\tilde f(x) = 0$ and for each $\varepsilon > 0$ the set $\{y : |\tilde f(y)| \ge \varepsilon\}$ is finite, hence its complement is a neighborhood of x, and $\tilde f$ is continuous at x. So the result follows by induction. □

13.18. Results.

(1) *We do not know whether the quotient of a C^n-normable space is again C^n-normable. Compare however with* [Fitzpatrick, 1980].

(2) *The statement* (13.17.5) *is quite sharp, since by* [Haydon, 1990] *there is a compact space K with $K^{(\omega)} = \{\infty\}$ but without a Gâteaux-differentiable norm.*

(3) [Talagrand, 1986] *proved that for every ordinal number γ, the compact and scattered space $[0, \gamma]$ with the order topology is C^1-normable.*

(4) *It was shown by* [Toruńczyk, 1981] *that two Banach spaces are homeomorphic if and only if their density number is the same. Hence, one can view Banach spaces as exotic (differentiable or linear) structures on Hilbert spaces. If two Banach spaces are even C^1-diffeomorphic then the differential (at 0) gives a continuous linear homeomorphism. It was for some time unknown if also uniformly homeomorphic (or at least Lipschitz homeomorphic) Banach spaces are already linearly homeomorphic. By* [Enflo, 1970] *a Banach space which is uniformly homeomorphic to a Hilbert space is linearly homeomorphic to it. A counter-example to the general statement was given by* [Aharoni, Lindenstrauss, 1978], *and another one is due to* [Ciesielski, Pol, 1984]*: There exists a short exact sequence $c_0(\Gamma_1) \to C(K) \to c_0(\Gamma_2)$ where $C(K)$ cannot be continuously injected into some $c_0(\Gamma)$ but is Lipschitz equivalent to $c_0(\Gamma)$. For these and similar questions see* [Tzafriri, 1980].

(5) *A space all of whose closed subspaces are complemented is a Hilbert space,* [Lindenstrauss, Tzafriri, 1971].

(6) [Enflo, Lindenstrauss, Pisier, 1975] *There exists a Banach space E not isomorphic to a Hilbert space and a short exact sequence $\ell^2 \to E \to \ell^2$.*

(7) [Bonic, Reis, 1966]. *If the norm of a Banach space and its dual norm are C^2 then the space is a Hilbert space.*

(8) [Deville, Godefroy, Zizler, 1990]. *This yields also an example that existence*

of smooth norms is not a three-space property, cf. (14.12).

Notes. (2) Note that $K \setminus K'$ is discrete, open and dense in K. So we get for every $n \in \mathbb{N}$ by induction a space K_n with $K_n^{(n)} \neq \emptyset$ and $K_n^{(n+1)} = \emptyset$. In fact $(A \times B)^{(n)} = \bigcup_{i+j=n} A^{(i)} \times B^{(j)}$. Next consider the 1-point compactification K_∞ of the locally compact space $\bigsqcup_{n\in\mathbb{N}} K_n$. Then $K'_\infty = \{\infty\} \cup \bigsqcup_{n\in\mathbb{N}} K'_n$. In fact every neighborhood of $\{\infty\}$ contains all but finitely many of the K_n, thus we have $\supseteq$. The obvious relation is clear. Hence $K_\infty^{(n)} = \{\infty\} \cup \bigsqcup_{i\geq n} K_n^{(i)}$. And $K_\infty^{(\omega)} = \bigcap_{n<\omega} K_\infty^{(n)} = \{\infty\} \neq \emptyset$. The space of [Haydon, 1990] is the one-point compactification of a locally compact space L given as follows: $L := \bigsqcup_{\alpha<\omega_1} \omega_1^\alpha$, i.e. the space of functions $\omega_1 \to \omega_1$, which are defined on some countable ordinal. It is ordered by restriction, i.e. $s \preceq t :\Leftrightarrow \operatorname{dom} s \subseteq \operatorname{dom} t$ and $t|_{\operatorname{dom} s} = s$.

(3) The order topology on $X := [0, \gamma]$ has the sets $\{x : x < a\}$ and $\{x : x > a\}$ as basis. In particular open intervals $(a, b) := \{x : a < x < b\}$ are open. It is compact, since every subset has a greatest lower bound. In fact let $\mathcal{U}$ on X be a covering. Consider $S := \{x \in X : [\inf X, x)$ is covered by finitely many $U \in \mathcal{U}\}$. Let $s_\infty := \sup S$. Note that $x \in S$ implies that $[\inf X, x]$ is covered by finitely many sets in $\mathcal{U}$. We have that $s_\infty \in S$, since there is an $U \in \mathcal{U}$ with $s_\infty \in U$. Then there is an x with $s_\infty \in (x, s_\infty] \subseteq U$, hence $[\inf X, x]$ is covered by finitely many sets in $\mathcal{U}$ since there is an $s \in S$ with $x < s$, so $[\inf X, s_\infty] = [\inf X, x] \cup (x, s_\infty]$ is covered by finitely many sets, i.e. $s_\infty \in S$.

The space X is *scattered*, i.e. $X^{(\alpha)} = \emptyset$ for some ordinal α. For this we have to show that every closed non-empty subset $K \subseteq X$ has open points. For every subset K of X there is a minimum $\min K \in K$, hence $[\inf X, \min K + 1) \cap K = \{\min K\}$ is open in K.

For γ equal to the first infinite ordinal ω we have $[0, \gamma] = \mathbb{N}_\infty$, the one-point compactification of the discrete space $\mathbb{N}$. Thus $C([0, \gamma]) \cong c_0 \times \mathbb{R}$ and the result follows in this case from (13.16).

(5) For splitting short exact sequences the result analogous to (13.17.3) is by (13.17.1) obviously true. By (5) there are non-splitting exact sequences $0 \to F \to E \to E/F \to 0$ for every Banach space which is not Hilbertizable.

(8) By (6) there is a sort exact sequence with hilbertizable ends, but with middle term E not hilbertizable. So neither the sequence nor the dualized sequence splits. If E and E' would have a C^2-norm then E would be hilbertizable by (7).

13.19. Proposition. *Let E be a Banach space, $\|x\| = 1$. Then the following statements are equivalent:*

(1) *The norm is Fréchet differentiable at x;*
(2) *The following two equivalent conditions hold:*

$$\lim_{h\to 0} \frac{\|x+h\| + \|x-h\| - 2\|x\|}{\|h\|} = 0,$$

$$\lim_{t\to 0} \frac{\|x+th\| + \|x-th\| - 2\|x\|}{t} = 0 \textit{ uniformly in } \|h\| \leq 1;$$

(3) $\|y_n^*\| = 1$, $\|z_n^*\| = 1$, $y_n^*(x) \to 1$, $z_n^*(x) \to 1 \Rightarrow y_n^* - z_n^* \to 0$.

Proof. (1)⇒(2) This is obvious, since for the derivative ℓ of the norm at x we have $\lim_{h\to 0} \frac{\|x\pm h\|-\|x\|-l(\pm h)}{\|h\|} = 0$ and adding these equations gives (2).

(2) ⇒ (1) Since $\ell(h) := \lim_{t\searrow 0} \frac{\|x+th\|-\|x\|}{t}$ always exists, and since

$$\frac{\|x+th\| + \|x-th\| - 2\|x\|}{t} = \frac{\|x+th\| - \|x\|}{t} + \frac{\|x+t(-h)\| - \|x\|}{t}$$
$$\geq l(h) + l(-h) \geq 0$$

we have $\ell(-h) = \ell(h)$, thus ℓ is linear. Moreover $\frac{\|x\pm th\|-\|x\|}{t} - \ell(\pm h) \geq 0$, so the limit is uniform for $\|h\| \leq 1$.

(2) ⇒ (3) By (2) we have that for $\varepsilon > 0$ there exists a δ such that $\|x+h\|+\|x-h\| \leq 2+\varepsilon\|h\|$ for all $\|h\| < \delta$. For $\|y_n^*\| = 1$ and $\|z_n^*\| = 1$ we have

$$y_n^*(x+h) + z_n^*(x-h) \leq \|x+h\| + \|x-h\|.$$

Since $y_n^*(x) \to 1$ and $z_n^*(x) \to 1$ we get for large n that

$$(y_n^* - z_n^*)(h) \leq 2 - y_n^*(x) - z_n^*(x) + \varepsilon\|h\| \leq 2\varepsilon\delta,$$

hence $\|y_n^* - z_n^*\| \leq 2\varepsilon$, i.e. $z_n^* - y_n^* \to 0$.

(3) ⇒ (2) Otherwise, there exists an $\varepsilon > 0$ and $0 \neq h_n \to 0$, such that

$$\|x + h_n\| + \|x - h_n\| \geq 2 + \varepsilon\|h_n\|.$$

Now choose $\|y_n^*\| = 1$ and $\|z_n^*\| = 1$ with

$$y_n^*(x+h_n) \geq \|x+h_n\| - \frac{1}{n}\|h_n\| \text{ and } z_n^*(x-h_n) \geq \|x-h_n\| - \frac{1}{n}\|h_n\|.$$

Then $y_n^*(x) = y_n^*(x+h_n) - y_n^*(h_n) \to 1$ and similarly $z_n^*(x) \to 1$. Furthermore

$$y_n^*(x+h_n) + z_n^*(x-h_n) \geq 2 + (\varepsilon - \frac{2}{n})\,\|h_n\|,$$

hence

$$(y_n^* - z_n^*)(h_n) \geq 2 + (\varepsilon - \frac{2}{n})\,\|h_n\| - (y_n^* + z_n^*)(x) \geq (\varepsilon - \frac{2}{n})\,\|h_n\|,$$

thus $\|y_n^* - z_n^*\| \geq \varepsilon - \frac{2}{n}$, a contradiction. □

13.20. Proposition. Fréchet differentiable norms via locally uniformly rotund duals. [Lovaglia, 1955] *If the dual norm of a Banach space E is locally uniformly rotund on E' then the norm is Fréchet differentiable on E.*

A norm is called *locally uniformly rotund* if $\|x_n\| \to \|x\|$ and $\|x + x_n\| \to 2\|x\|$ implies $x_n \to x$. This is equivalent to $2(\|x\|^2 + \|x_n\|^2) - \|x + x_n\|^2 \to 0$ implies $x_n \to x$, since

$$2(\|x\|^2 + 2\|x_n\|^2) - \|x + x_n\|^2 \geq 2\|x\|^2 + 2\|x_n\|^2 - (\|x\| + \|x_n\|)^2 = (\|x\| - \|x_n\|)^2.$$

Proof. We use (13.19), so let $\|x\| = 1$, $\|y_n^*\| = 1$, $\|z_n^*\| = 1$, $y_n^*(x) \to 1$, $z_n^*(x) \to 1$. Let $\|x^*\| = 1$ with $x^*(x) = 1$. Then $2 \geq \|x^* + y_n^*\| \geq (x^* + y_n^*)(x) \to 2$. Since $\|\ \|_{E'}$ is locally uniformly rotund we get $y_n^* \to x$ and similarly $z_n^* \to z$, hence $y_n^* - z_n^* \to 0$. □

13.21. Remarks on locally uniformly rotund spaces. By [Kadec, 1959] and [Kadec, 1961] every separable Banach space is isomorphic to a locally uniformly rotund Banach space. By [Day, 1955] the space $\ell^\infty(\Gamma)$ is not isomorphic to a locally uniformly rotund Banach space. Every Banach space admitting a continuous linear injection into some $c_0(\Gamma)$ is locally uniformly rotund renormable, see [Troyanski, 1971]. By (53.21) every WCG-Banach space has such an injection, which is due to [Amir, Lindenstrauss, 1968]. By [Troyanski, 1968] every Banach space with unconditional basis (see [Jarchow, 1981, 14.7]) is isomorphic to a locally uniformly rotund Banach space.

In particular, it follows from these results that every reflexive Banach space has an equivalent Fréchet differentiable norm. In particular L^p has a Fréchet differentiable norm for $1 < p < \infty$ and in fact the p-norm is itself Fréchet differentiable, see (13.13).

13.22. Proposition. *If E' is separable then E admits an equivalent norm, whose dual norm is locally uniform rotund.*

Proof. Let E' be separable. Then there exists a bounded linear operator $T : E \to \ell^2$ such that $T^*((\ell^2)')$ is dense in E' (and obviously T^* is weak*-continuous): Take a dense subset $\{x_i^* : i \in \mathbb{N}\} \subseteq E'$ of $\{x^* \in E' : \|x^*\| \leq 1\}$ with $\|x_i^*\| \leq 1$. Define $T : E \to \ell^2$ by

$$T(x)_i := \frac{x_i^*(x)}{2^i}.$$

Then for the basic unit vector $e_i \in (\ell^2)'$ we have

$$T^*(e_i)(x) = e_i(T(x)) = T(x)_i = \frac{x_i^*(x)}{2^i},$$

i.e. $T^*(e_i) = 2^{-i}\, x_i^*$.

Note that the canonical norm on ℓ^2 is locally uniformly rotund. We now claim that E' has a dual locally uniform rotund norm. For $x^* \in E'$ and $n \in \mathbb{N}$ we define

$$\|x^*\|_n^2 := \inf\{\|x^* - T^*y^*\|^2 + \frac{1}{n}\|y^*\|^2 : y^* \in (\ell^2)'\} \text{ and}$$

$$\|x^*\|_\infty := \sum_{n=1}^{\infty} \frac{1}{2^n}\|x^*\|_n.$$

We claim that $\| \quad \|_\infty$ is the required norm.

So we show first, that it is an equivalent norm. For $\|x^*\| = 1$ we have $\|x^*\|_n \geq \min\{1/(2\sqrt{n}\|T^*\|), 1/2\}$. In fact if $\|y^*\| \geq 1/(2\|T^*\|)$ then $\|x^* - T^*y^*\|^2 + \frac{1}{n}\|y^*\|^2 \geq 1/(2n^2\|T^*\|^2)$ and if $\|y^*\| \leq 1/(2\|T^*\|)$ then $\|x^* - T^*y^*\| \geq \|x\| - \|T^*y^*\| \geq 1 - \frac{1}{2} = \frac{1}{2}$. Furthermore if we take $y := 0$ then we see that $\|x^*\|_n \leq \|x\|$. Thus $\| \quad \|_n$ and $\| \quad \|$ are equivalent norms, and hence also $\| \quad \|_\infty$.

Note first, that a dual norm is the supremum of the weak* (lower semi-)continuous functions $x^* \mapsto |x^*(x)|$ for $\|x\| \leq 1$. Conversely the unit ball B has to be weak* closed in E' since the norm is assumed to be weak* lower semi-continuous and B is convex. Let B_o be its polar in E. By the bipolar-theorem $(B_o)^o = B$, and thus the dual of the Minkowski functional of B_o is the given norm.

Next we show that the infimum defining $\| \quad \|_n$ is in fact a minimum, i.e. for each n and x^* there exists a y^* with $\|x^*\|_2^n = \|x^* - T^*y^*\|^2 + \frac{1}{n}\|y^*\|^2$. Since $f_x : y^* \mapsto \|x^* - T^*y^*\|^2 + \frac{1}{n}\|y^*\|^2$ is weak* lower semi-continuous and satisfies $\lim_{y^*\to\infty} f_x(y^*) = +\infty$, hence it attains its minimum on some large (weak*-compact) ball.

We have that $\|x\|_n \to 0$ for $n \to \infty$.
In fact since the image of T^* is dense in E', there is for every $\varepsilon > 0$ a y^* with $\|x^* - T^*y^*\| < \varepsilon$, and so for large n we have $\|x^*\|_n^2 \le \|x^* - T^*y^*\|^2 + \frac{1}{n}\|y\|^2 < \varepsilon^2$.

Let us next show that $\| \quad \|_\infty$ is a dual norm. For this it is enough to show that $\| \quad \|_n$ is a dual norm, i.e. is weak* lower semi-continuous. So let x_i^* be a net converging weak* to x^*. Then we may choose y_i^* with $\|x_i^*\|_n^2 = \|x_i^* - T^*y_i^*\|^2 + \frac{1}{n}\|y_i^*\|^2$. Then $\{x_i^* : i\}$ is bounded, and hence also $\|y_i^*\|^2$. Let thus y^* be a weak* cluster point of the (y_i^*). Without loss of generality we may assume that $y_i^* \to y^*$. Since the original norms are weak* lower semicontinuous we have

$$\|x^*\|_n^2 \le \|x^* - T^*y^*\|^2 + \frac{1}{n}\|y^*\|^2 \le \liminf_i (\|x_i^* - T^*y_i^*\|^2 + \frac{1}{n}\|y_i^*\|^2) = \liminf_i \|x_i^*\|_2^n.$$

So $\| \quad \|_n$ is weak* lower semicontinuous.
Here we use that a function $f : E \to \mathbb{R}$ is lower semicontinuous if and only if $x_\infty = \lim_i x_i \Rightarrow f(x_\infty) \le \liminf_i f(x_i)$.
$(\Rightarrow)$ otherwise for some subnet (which we again denote by x_i) we have $f(x_\infty) > \lim_i f(x_i)$ and this contradicts the fact that $f^{-1}((a, \infty))$ has to be a neighborhood of x_∞ for $2a := f(x_\infty) + \lim_i f(x_i)$.
$(\Rightarrow)$ otherwise there exists some x_∞ and an $a < f(x_\infty)$ such that in every neighborhood U of x_∞ there is some x_U with $f(x_U) \le a$. Hence $\lim_U x_U = x_\infty$ and $\liminf_U f(x_U) \le \limsup_U f(x_U) \le a < f(x_\infty)$.

Let us finally show that $\| \quad \|_\infty$ is locally uniform rotund.
So let $x^*, x_j^* \in E'$ with

$$2(\|x^*\|_\infty^2 + \|x_j^*\|_\infty^2) - \|x^* + x_j^*\|_\infty^2 \to 0,$$

or equivalently

$$\|x_j^*\|_\infty \to \|x^*\|_\infty \text{ and } \|x^* + x_j^*\|_\infty \to 2\|x^*\|_\infty.$$

Thus also

$$\|x_j^*\|_n \to \|x^*\|_n \text{ and } \|x^* + x_j^*\|_n \to 2\|x^*\|_n$$

and equivalently

$$2(\|x^*\|_n^2 + \|x_j^*\|_n^2) - \|x^* + x_j^*\|_n^2 \to 0.$$

Now we may choose y^* and y_j^* such that

$$\|x^*\|_2^n = \|x^* - T^*y^*\|^2 + \frac{1}{n}\|y^*\|^2 \text{ and } \|x_j^*\|_2^n = \|x_j^* - T^*y_j^*\|^2 + \frac{1}{n}\|y_j^*\|^2.$$

We calculate as follows:

$$\begin{aligned}
2(\|x^*\|_n^2 + \|x_j^*\|_n^2) - \|x^* + x_j^*\|^2 \geq \\
&\geq 2(\|x^* - T^*y^*\|^2 + \frac{1}{n}\|y^*\|^2 + \|x_j^* - T^*y_j^*\|^2 + \frac{1}{n}\|y_j^*\|^2 \\
&\quad - \|x^* + x_j^* - T^*(y^* + y_j^*)\|^2 - \frac{1}{n}\|y^* + y_j^*\|^2 \\
&\geq 2(\|x^* - T^*y^*\|^2 + \frac{1}{n}\|y^*\|^2 + \|x_j^* - T^*y_j^*\|^2 + \frac{1}{n}\|y_j^*\|^2 \\
&\quad - (\|x^* - T^*(y^*)\| + \|x_j^* - T^*(y_j^*)\|)^2 - \frac{1}{n}\|y^* + y_j^*\|^2 \\
&\geq (\|x^* - T^*y^*\| - \|x_j^* - T^*y_j^*\|)^2 + \\
&\quad + \frac{1}{n}(2\|y^*\|^2 + 2\|y_j^*\|^2 - \|y^* + y_j^*\|^2) \geq 0,
\end{aligned}$$

hence

$$\|x_j^* - T^*y_j^*\| \to \|x^* - T^*y^*\| \text{ and } 2(\|y^*\|^2 + \|y_j^*\|^2) - \|y^* + y_j^*\|^2 \to 0.$$

Since $\| \quad \|$ is locally uniformly rotund on $(\ell^2)^*$ we get that $y_j^* \to y^*$. Hence

$$\begin{aligned}
\limsup_j \|x^* - x_j^*\| &\leq \limsup_j (\|x^* - T^*y^*\| + \|T^*(y^* - y_j^*)\| + \|x_j^* - T^*y_j^*\|) \\
&= 2\,\|x^* - T^*y^*\| \leq 2\|x^*\|_n.
\end{aligned}$$

Since $\|x^*\|_n \to 0$ for $n \to \infty$ we get $x_j^* \to x^*$. □

13.23. Proposition. [Leach, Whitfield, 1972]. *For the norm $\| \quad \| = p$ on a Banach space E the following statements are equivalent:*

(1) *The norm is rough, i.e. p' is uniformly discontinuous, see* (13.8.5).

(2) *There exists an $\varepsilon > 0$ such that for all $x \in E$ with $\|x\| = 1$ and all y_n^*, $z_n^* \in E'$ with $\|y_n^*\| = 1 = \|z_n^*\|$ and $\lim_n y_n^*(x) = 1 = \lim_n z_n^*(x)$ we have:*

$$\limsup_n \|y_n^* - z_n^*\| \geq \varepsilon;$$

(3) *There exists an $\varepsilon > 0$ such that for all $x \in E$ with $\|x\| = 1$ we have that*

$$\limsup_{h \to 0} \frac{\|x + h\| + \|x - h\| - 2}{\|h\|} \geq \varepsilon;$$

(4) *There exists an $\varepsilon > 0$ such that for every $x \in E$ with $\|x\| = 1$ and $\delta > 0$ there is an $h \in E$ with $\|h\| \leq 1$ and $\|x + th\| \geq \|x\| + \varepsilon|t| - \delta$ for all $|t| \leq 1$.*

Note that we always have

$$0 \leq \frac{\|x + h\| + \|x - h\| - 2\|x\|}{\|x\|} \leq 2,$$

hence ε in (3) satisfies $\varepsilon \leq 2$. For ℓ^1 and $C[0,1]$ the best choice is $\varepsilon = 2$, see (13.11) and (13.12).

Proof. (3)⇒(2) is due to [Cudia, 1964]. Let $\varepsilon > 0$ such that for all $\|x\| = 1$ there are $0 \neq h_n \to 0$ with $\|x + h_n\| + \|x - h_n\| - 2 \geq \varepsilon\|h_n\|$. Now choose y_n^*, $z_n^* \in E'$ with $\|y_n^*\| = 1 = \|z_n\|^*$, $y_n^*(x + h_n) = \|x + h_n\|$ and $z_n^*(x - h_n) = \|x - h_n\|$. Then $\lim_n y_n^*(x) = \|x\| = 1$ and also $\lim_n z_n^*(x) = 1$. Moreover,

$$y_n^*(x + h_n) + z_n^*(x - h_n) \geq 2 + \varepsilon\|h_n\|$$

and hence

$$(y_n^* - z_n^*)(h_n) \geq 2 - y_n^*(x) - z_n^*(x) + \varepsilon\|h_n\| \geq \varepsilon\|h_n\|,$$

thus (2) is satisfied.

(2)⇒(1) By (2) we have an $\varepsilon > 0$ such that for all $\|x\| = 1$ there are y_n^* and z_n^* with $\|y_n^*\| = 1 = \|z_n^*\|$, $\lim_n y_n^*(x) = 1 = \lim_n z_n^*(x)$ and h_n with $\|h_n\| = 1$ and $(y_n^* - z_n^*)(h_n) \geq \varepsilon$. Let $0 < \delta < \varepsilon/2$ and $t > 0$. Then

$$y_n^*(x) > 1 - \frac{\delta^2}{4} \quad \text{and} \quad z_n^*(x) > 1 - \frac{\delta^2}{4} \text{ for large } n.$$

Thus

$$\|x + th_n\| \geq y_n^*(x + th_n) \geq 1 - \frac{\delta^2}{4} + ty_n^*(h_n)$$

and hence

$$\begin{aligned} t\,p'(x + th_n)(h_n) &\geq \|x + th_n\| - \|x\| \geq ty_n^*(h_n) - \frac{\delta^2}{4} \quad \Rightarrow \\ p'(x + th_n)(h_n) &\geq y_n^*(h_n) - \frac{\delta^2}{4t} \\ \text{and similarly } -p'(x - th_n)(h_n) &\geq -z_n^*(h_n) - \frac{\delta^2}{4t} \end{aligned}$$

If we choose $0 < t < \delta$ such that $\delta^2/(2t) < \delta$ we get

$$p'(x + th_n)(h_n) - p'(x - th_n)(h_n) \geq (y_n^* - z_n^*)(h_n) - \frac{\delta^2}{2t} > \varepsilon - \delta > \frac{\varepsilon}{2}.$$

(1)⇒(4) Using the uniform discontinuity assumption of p' we get $x_j \in E$ with $p(x_j - x) \leq \eta/4$ and $u \in E$ with $p(u) = 1$ such that $(p'(x_2) - p'(x_1))(u) \geq \varepsilon$. Let $\mu := (p'(x_1) + p'(x_2))(u)/(2p(x))$ and $v := u - \mu\, x$.

Since $p'(x_1)(u) \leq p'(x_2)(u) - \varepsilon$ we get $(p'(x_1) + p'(x_2))(u))/2 \leq p'(x_2)(u) - \varepsilon/2 \leq p(u) - \varepsilon/2 < 1$ and $(p'(x_1) + p'(x_2))(u)/2 \geq p'(x_1)(u) + \varepsilon \geq -p(u) + \varepsilon/2 > 1$, i.e. $|(p'(x_1) + p'(x_2))(u)/2| < 1$, so $0 < p(v) < 2$. For $0 \leq t \leq p(x)$ and $s := 1 - t\,\mu$ we get

$$x + tv = sx + tu = s(x + \frac{t}{s}u) = s\Big((x_2 + \frac{t}{s}u) + (x - x_2)\Big).$$

Thus $0 < s < 2$ and

$$\begin{aligned} p(x+tv) &\geq s(p(x_2+\frac{t}{s}u)-p(x-x_2)) \\ &> s\Big(p(x_2)+\frac{t}{s}p'(x_2)u-\eta/4\Big) \quad \text{since } p(y+w)\geq p(y)+p'(y)(w) \\ &> s\,p(x)+t\,p'(x_2)(u)-s\,\eta/2 \quad \text{since } p(x)\leq p(x_2)+p(x-x_2) \\ &= p(x)+(t/2)\,(p'(x_2)-p'(x_1))(u)-s\,\eta/2 \\ &> p(x)+t\varepsilon/2-\eta. \end{aligned}$$

If $-p(x)\leq t<0$ we proceed with the role of x_1 and x_2 exchanged and obtain

$$\begin{aligned} p(x+tv) &> s\,p(x)+t\,p'(x_1)(u)-s\,\eta/2 \\ &= p(x)+(-t/2)\,(p'(x_2)-p'(x_1))(u)-s\,\eta/2 \\ &> p(x)+|t|\,\varepsilon/2-\eta. \end{aligned}$$

Thus

$$p(x+tv)\geq p(x)+|t|\,\varepsilon/2-\eta.$$

(4)$\Rightarrow$(3) By (4) there exists an $\varepsilon>0$ such that for every $x\in E$ with $\|x\|=1$ and $\delta>0$ there is an $h\in E$ with $\|h\|\leq 1$ and $\|x+th\|\geq\|x\|+\varepsilon|t|-\delta$ for all $|t|\leq 1$. If we put $t:=1/n$ we have

$$n(\|x+h_n/n\|+\|x-h_n/n\|-2)\geq\varepsilon-1/n>\varepsilon/2 \text{ for large } n. \quad \square$$

13.24. Results on the non-existence of C^1-norms on certain spaces.

(1) [Restrepo, 1964] *and* [Restrepo, 1965]. *A separable Banach space has an equivalent C^1-norm if and only if E^* is separable. This will be proved in* (16.11).

(2) [Kadec, 1965]. *More generally, if for a Banach space* $\operatorname{dens} E < \operatorname{dens} E^*$ *then no C^1-norm exists. This will be proved by showing the existence of a rough norm in* (14.10) *and then using* (14.9). *The density number* $\operatorname{dens} X$ *of a topological space X is the minimum of the cardinalities of all dense subsets of X.*

(3) [Haydon, 1990]. *There exists a compact space K, such that $K^{(\omega_1)}=\{*\}$, in particular $K^{(\omega_1+1)}=\emptyset$, but $C(K)$ has no equivalent Gâteaux differentiable norm, see also* (13.18.2).

One can interpret these results by saying that in these spaces every convex body necessarily has corners.

14. Smooth Bump Functions

In this section we return to the original question whether the smooth functions generate the topology. Since we will use the results given here also for manifolds, and since the existence of charts is of no help here, we consider fairly general non-linear spaces. This allows us at the same time to treat all considered differentiability classes in a unified way.

14.1. Convention. We consider a Hausdorff topological space X with a subalgebra $\mathcal{S} \subseteq C(X,\mathbb{R})$, whose elements will be called the *smooth* or *$\mathcal{S}$-functions* on X. We assume that for functions $h \in C^\infty(\mathbb{R},\mathbb{R})$ (at least for those being constant off some compact set, in some cases) one has $h_*(\mathcal{S}) \subseteq \mathcal{S}$, and that $f \in \mathcal{S}$ provided it is locally in $\mathcal{S}$, i.e., there exists an open covering $\mathcal{U}$ such that for every $U \in \mathcal{U}$ there exists an $f_U \in \mathcal{S}$ with $f = f_U$ on U.

In particular, we will use for $\mathcal{S}$ the classes of C^∞- and of $\mathcal{L}\mathrm{ip}^k$-mappings on c^∞-open subsets X of convenient vector spaces with the c^∞-topology and the class of C^n-mappings on open subsets of Banach spaces, as well as subclasses formed by boundedness conditions on the derivatives or their difference quotients.

Under these assumptions on $\mathcal{S}$ one has that $\frac{1}{f} \in \mathcal{S}$ provided $f \in \mathcal{S}$ with $f(x) > 0$ for all $x \in X$: Just choose everywhere positive $h_n \in C^\infty(\mathbb{R},\mathbb{R})$ with $h_n(t) = \frac{1}{t}$ for $t \geq \frac{1}{n}$. Then $h_n \circ f \in \mathcal{S}$ and $\frac{1}{f} = h_n \circ f$ on the open set $\{x : f(x) > \frac{1}{n}\}$. Hence, $\frac{1}{f} \in \mathcal{S}$.

For a (convenient) vector space F the *carrier* $\operatorname{carr}(f)$ of a mapping $f : X \to F$ is the set $\{x \in X : f(x) \neq 0\}$. The *zero set* of f is the set where f vanishes, $\{x \in X : f(x) = 0\}$. The *support* of f $\operatorname{support}(f)$ is the closure of $\operatorname{carr}(f)$ in X.

We say that X is *smoothly regular* (with respect to $\mathcal{S}$) or *$\mathcal{S}$-regular* if for any neighborhood U of a point x there exists a smooth function $f \in \mathcal{S}$ such that $f(x) = 1$ and $\operatorname{carr}(f) \subseteq U$. Such a function f is called a *bump function.*

14.2. Proposition. Bump functions and regularity. [Bonic, Frampton, 1966]. *A Hausdorff space is $\mathcal{S}$-regular if and only if its topology is initial with respect to $\mathcal{S}$.*

Proof. The initial topology with respect to $\mathcal{S}$ has as a subbasis the sets $f^{-1}(I)$, where $f \in \mathcal{S}$ and I is an open interval in $\mathbb{R}$. Let $x \in U$, with U open for the initial topology. Then there exist finitely many open intervals $I_1, \dots, I_n$ and $f_1, \dots, f_n \in \mathcal{S}$ with $x \in \bigcap_{i=1}^n f_i^{-1}(I_i)$. Without loss of generality we may assume that $I_i = \{t : |f_i(x) - t| < \varepsilon_i\}$ for certain $\varepsilon_i > 0$. Let $h \in C^\infty(\mathbb{R},\mathbb{R})$ be chosen such that $h(0) = 1$ and $h(t) = 0$ for $|t| \geq 1$. Set $f(x) := \prod_{i=1}^n h(\frac{f_i(x)}{\varepsilon_i})$. Then f is the required bump function. □

14.3. Corollary. Smooth regularity is inherited by products and subspaces. *Let X_i be topological spaces and $\mathcal{S}_i \subseteq C(X_i,\mathbb{R})$. On a space X we consider the initial topology with respect to mappings $f_i : X \to X_i$, and we assume that $\mathcal{S} \subseteq C(X,\mathbb{R})$ is given such that $f_i^*(\mathcal{S}_i) \subseteq \mathcal{S}$ for all i. If each X_i is $\mathcal{S}_i$-regular, then X is $\mathcal{S}$-regular.* □

Note however that the c^∞-topology on a locally convex subspace is not the trace of the c^∞-topology in general, see (4.33) and (4.36.5). However, for c^∞-closed subspaces this is true, see (4.28).

14.4. Proposition. [Bonic, Frampton, 1966]. *Every Banach space with $\mathcal{S}$-norm is $\mathcal{S}$-regular.*

More general, a convenient vector space is smoothly regular if its c^∞-topology is generated by seminorms which are smooth on their respective carriers. For example, nuclear Fréchet spaces have this property.

Proof. Namely, $g \circ p$ is a smooth bump function with carrier contained in $\{x : p(x) < 1\}$ if g is a suitably chosen real function, i.e., $g(t) = 1$ for $t \leq 0$ and $g(t) = 0$ for $t \geq 1$.

Nuclear spaces have a basis of Hilbert-seminorms (52.34), and on Fréchet spaces the c^∞-topology coincides with the locally convex one (4.11.1), hence nuclear Fréchet spaces are c^∞-regular. $\square$

14.5. Open problem. Has every non-separable $\mathcal{S}$-regular Banach space an equivalent $\mathcal{S}$-norm? Compare with (16.11).

A partial answer is given in:

14.6. Proposition. *Let E be a C^∞-regular Banach space. Then there exists a smooth function $h : E \to \mathbb{R}_+$, which is positively homogeneous and smooth on $E \setminus \{0\}$.*

Proof. Let $f : E\setminus\{0\} \to \{t \in \mathbb{R} : t \geq 0\}$ be a smooth function, such that $\operatorname{carr}(f)$ is bounded in E and $f(x) \geq 1$ for x near 0. Let $U := \{x : f(tx) \neq 0 \text{ for some } t \geq 1\}$. Then there exists a smooth function $Mf : E \setminus \{0\} \to \mathbb{R}$ with $(Mf)'(x)(x) < 0$ for $x \in U$, $\lim_{x\to 0} f(x) = +\infty$ and $\operatorname{carr} Mf \subseteq U$.

The idea is to construct out of the smooth function $f \geq 0$ another smooth function Mf with $(Mf)'(x)(x) = -f(x) \leq 0$, i.e. $(Mf)'(tx)(tx) = -f(tx)$ and hence

$$\frac{d}{dt} Mf(tx) = (Mf)'(tx)(x) = -\frac{f(tx)}{t} \quad \text{for } t \neq 0.$$

Since we want bounded support for Mf, we get

$$Mf(x) = -\Big[Mf(tx)\Big]_{t=1}^{\infty} = -\int_1^\infty \frac{d}{dt} Mf(tx)\,dt = \int_1^\infty \frac{f(tx)}{t}\,dt,$$

and we take this as a definition of Mf. Since the support of f is bounded, we may replace the integral locally by $\int_1^N$ for some large N, hence Mf is smooth on $E\setminus\{0\}$ and $(Mf)'(x)(x) = -f(x)$.

Since $f(x) > \varepsilon$ for all $\|x\| < \delta$, we have that

$$Mf(x) \geq \int_1^N \frac{1}{t} f(tx)\,dt \geq \log(N)\varepsilon$$

for all $\|x\| < \frac{\delta}{N}$, i.e. $\lim_{x\to 0} Mf(x) = +\infty$.

Furthermore $\operatorname{carr}(Mf) \subseteq U$, since $f(tx) = 0$ for all $t \geq 1$ and $x \notin U$.

Now consider $M^2 f := M(Mf) : E \setminus \{0\} \to \mathbb{R}$. Since $(Mf)'(x)(x) \leq 0$, we have $(M^2 f)'(x)(x) = \int_1^\infty (Mf)'(tx)(x)\,dt \leq 0$ and it is < 0 if for some $t \geq 1$ we have $(Mf)'(tx)(x) < 0$, in particular this is the case if $M^2 f(x) > 0$.

Thus $U_\varepsilon := \{x : M^2 f(x) \geq \varepsilon\}$ is radial set with smooth boundary, and the Minkowski-functional is smooth on $E\setminus\{0\}$. Moreover $U_\varepsilon \cong E$ via $x \mapsto \frac{x}{M^2 f(x)}$. $\square$

14.7. Lemma. Existence of smooth bump functions.
For a class $\mathcal{S}$ on a Banach space E in the sense of (14.1) *the following statements are equivalent:*

(1) *E is not $\mathcal{S}$-regular;*
(2) *For every $f \in \mathcal{S}$, every $0 < r_1 < r_2$ and $\varepsilon > 0$ there exists an x with $r_1 \leq \|x\| \leq r_2$ and $|f(x) - f(0)| < \varepsilon$;*
(3) *For every $f \in \mathcal{S}$ with $f(0) = 0$ there exists an x with $1 \leq \|x\| \leq 2$ and $|f(x)| \leq \|x\|$*

Proof. (1) $\Rightarrow$ (2) Assume that there exists an f and $0 < r_1 < r_2$ and $\varepsilon > 0$ such that $|f(x) - f(0)| \geq \varepsilon$ for all $r_1 \leq \|x\| \leq r_2$. Let $h : \mathbb{R} \to \mathbb{R}$ be a smooth bump function on $\mathbb{R}$. Let $g(x) := h(\frac{1}{\varepsilon} f(r_1 x) - f(0))$. Then g is of the corresponding class, $g(0) = h(0) = 1$, and for all x with $1 \leq \|x\| \leq \frac{r_2}{r_1}$ we have $|f(r_1 x) - f(0)| \geq \varepsilon$, and hence $g(x) = 0$. By redefining g on $\{x : \|x\| \geq \frac{r_2}{r_1}\}$ as 0, we obtain the required bump function.

(2) $\Rightarrow$ (3) Take $r_1 = 1$ and $r_2 = 2$ and $\varepsilon = 1$.

(3) $\Rightarrow$ (1) Assume a bump function g exists, i.e., $g(0) = 1$ and $g(x) = 0$ for all $\|x\| \geq 1$. Take $f := 2 - g$. Then $f(0) = 0$ and $f(x) = 2$ for $\|x\| \geq 1$, a contradiction to (3). □

14.8. Proposition. Boundary values for smooth mappings. [Bonic, Frampton, 1966] *Let E and F be convenient vector spaces, let F be $\mathcal{S}$-regular but E not $\mathcal{S}$-regular. Let $U \subseteq E$ be c^∞-open and $f \in C(\overline{U}, F)$ with $f^*(\mathcal{S}) \subseteq \mathcal{S}$. Then $\overline{f(\partial U)} \supseteq f(\overline{U})$. Hence, $f = 0$ on ∂U implies $f = 0$ on U.*

Proof. Since $f(\overline{U}) \subseteq \overline{f(U)}$ it is enough to show that $f(U) \subseteq \overline{f(\partial U)}$. Suppose $f(x) \notin \overline{f(\partial U)}$ for some $x \in U$. Choose a smooth h on F such that $h(f(x)) = 1$ and $h = 0$ on a neighborhood of $f(\partial U)$. Let $g = h \circ f$ on U and 0 outside. Then g is a smooth bump function on E, a contradiction. □

14.9. Theorem. C^1-regular spaces admit no rough norm. [Leach, Whitfield, 1972]. *Let E be a Banach space whose norm $p = \|\ \ \|$ has uniformly discontinuous directional derivative. If f is Fréchet differentiable with $f(0) = 0$ then there exists an $x \in E$ with $1 \leq \|x\| < 2$ and $f(x) \leq \|x\|$.*

By (14.7) this result implies that on a Banach space with rough norm there exists no Fréchet differentiable bump function. In particular, $C([0,1])$ and ℓ^1 are not C^1-regular by (13.11) and (13.12), which is due to [Kurzweil, 1954].

Proof. We try to reach the exterior of the unit ball by a recursively defined sequence x_n in $\{x : f(x) \leq p(x)\}$ starting at 0 with large step-length ≤ 1 in directions, where p' is large. Given x_n we consider the set

$$\mathcal{M}_n := \left\{ y \in E : \begin{array}{l} (1)\ f(y) \leq p(y), \\ (2)\ p(y - x_n) \leq 1 \text{ and} \\ (3)\ p(y) - p(x_n) \geq (\varepsilon/8)\, p(y - x_n) \end{array} \right\}.$$

Since $x_n \in \mathcal{M}_n$, this set is not empty and hence $M_n := \sup\{p(y - x_n) : y \in \mathcal{M}_n\} \le 1$ is well-defined and it is possible to choose $x_{n+1} \in \mathcal{M}_n$ with

$$(4)\ p(x_{n+1} - x_n) \ge M_n/2.$$

We claim that $p(x_n) \ge 1$ for some n, since then $x := x_n$ for the minimal n satisfies the conclusion of the theorem:
Otherwise $p(x_n)$ is bounded by 1 and increasing by (3), hence a Cauchy-sequence. By (3) we then get that (x_n) is a Cauchy-sequence. So let z be its limit. If $z = 0$ then $\mathcal{M}_n = \{0\}$ and hence $f(y) > p(y)$ for all $|y| \le 1$. Thus f is not differentiable. Then $p(z) \le 1$ and $f(z) \le p(z)$. Since f is Fréchet-differentiable at z there exists a $\delta > 0$ such that

$$f(z+u) - f(z) - f'(z)(u) \le \varepsilon p(u)/8 \text{ for all } p(u) < \delta.$$

Without loss of generality let $\delta \le 1$ and $\delta \le 2p(z)$. By (13.23.4) there exists a v such that $p(v) < 2$ and $p(z + tv) > p(z) + \varepsilon|t|/2 - \varepsilon\delta/8$ for all $|t| \le p(z)$. Now let $t := -\operatorname{sign}(f'(z)(v))\, \delta/2$. Then

$$\begin{aligned} &(1) \quad p(z+tv) > p(z) + \varepsilon\delta/8 \ge f(z) + \varepsilon p(tv)/8 \ge f(z+tv), \\ &(2) \quad p(z+tv-z) = |t|p(v) < \delta \le 1, \\ &(3) \quad p(z+tv) - p(z) > \varepsilon\delta/8 > \varepsilon p(tv)/8. \end{aligned}$$

Since f and p are continuous the $z + tv$ satisfy (1)-(3) for large n and hence $M_n \ge p(z+tv-x_n)$. From $p(z+tv-z) > \varepsilon\delta/8$ we get $M_n > \varepsilon\delta/8$ and so $p(x_{n+1}-x_n) > \varepsilon\delta/16$ by (4) contradicts the convergence of x_n. □

14.10. Proposition. *Let E be a Banach-space with* $\operatorname{dens} E < \operatorname{dens} E'$. *Then there is an equivalent rough norm on E.*

Proof. The idea is to describe the unit ball of a rough norm as intersection of hyper planes $\{x \in E : x^*(x) \le 1\}$ for certain functionals $x^* \in E'$. The fewer functionals we use the more 'corners' the unit ball will have, but we have to use sufficiently many in order that this ball is bounded and hence that its Minkowski-functional is an equivalent norm. We call a set X large, if and only if $|X| > \operatorname{dens}(E)$ and small otherwise. For $x \in E$ and $\varepsilon > 0$ let $B_\varepsilon(x) := \{y \in E : \|x - y\| \le \varepsilon\}$. Now we choose using Zorn's lemma a subset $D \subseteq E'$ maximal with respect to the following conditions:

(1) $0 \in D$;
(2) $x^* \in D \Rightarrow -x^* \in D$;
(3) $x^*, y^* \in D,\ x^* \ne y^* \Rightarrow \|x^* - y^*\| > 1$.

Note that D is then also maximal with respect to (3) alone, since otherwise, we could add a point x^* with $\|x^* - y^*\| > 1$ for all $y^* \in D$ and also add the point $-x^*$, and obtain a larger set satisfying all three conditions.

Claim. $D_\infty := \bigcup_{n\in\mathbb{N}} \frac{1}{n} D$ is dense in E', and hence $|D_\infty| \ge \operatorname{dens}(E')$:
Assume indirectly, that there is some $x^* \in E'$ and $n \in \mathbb{N}$ with $B_{1/n}(x^*) \cap D_\infty =$

$\emptyset$. Then $B_1(nx^*) \cap D = \emptyset$ and hence we may add x^* to D, contradicting the maximality.

Without loss of generality we may assume that D is at least countable. Then $|D| = |\bigcup_{n\in\mathbb{N}} \frac{1}{n} D| \geq \operatorname{dens}(E') > \operatorname{dens}(E)$, i.e. D is large. Since $D = \bigcup_{n\in\mathbb{N}} D \cap B_n(0)$, we find some n such that $D \cap B_n(0)$ is large. Let $y^* \in E'$ be arbitrary and $w^* := \frac{1}{4n+2} y^*$. For every $x^* \in D$ there is a $z^* \in \frac{1}{2}D$ such that $\|x^* + w^* - z^*\| \leq \frac{1}{2}$ (otherwise we could add $2(x^* + w^*)$ to D). Thus we may define a mapping $D \to \frac{1}{2}D$ by $x^* \mapsto z^*$. This mapping is injective, since $\|x_j^* + w^* - z^*\| \leq \frac{1}{2}$ for $j \in \{1, 2\}$ implies $\|x_1^* - x_2^*\| \leq 1$ and hence $x_1^* = x_2^*$. If we restrict it to the large set $D \cap B_n(0)$ it has image in $\frac{1}{2}D \cap B_{n+1/2}(w^*)$, since $\|z^* - w^*\| \leq \|z^* + x^* - w^*\| + \|x^*\| \leq \frac{1}{2} + n$. Hence also $\frac{1}{2(4n+2)} D \cap B_{1/4}(y^*) = \frac{1}{4n+2}\frac{1}{2} D \cap B_{n+1/2}(w^*)$ is large.

In particular for $y^* := 0$ and $1/4$ replaced by 1 we get that $A := \frac{1}{4(2n+1)} D \cap B_1(0)$ is large. Now let

$$U := \Big\{x \in E : \exists A_0 \subseteq A \text{ small}, \forall x^* \in A \setminus A_0 : x^*(x) \leq 1\Big\}.$$

Since A is symmetric, the set U is absolutely convex (use that the union of two small exception sets is small). It is a 0-neighborhood, since $\{x : \|x\| \leq 1\} \subseteq U$ ($x^*(x) \leq \|x^*\| \cdot \|x\| = \|x\| \leq 1$ for $x^* \in A$). It is bounded, since for $x \in E$ we may find by Hahn-Banach an $x^* \in E'$ with $x^*(x) = \|x\|$ and $\|x^*\| = 1$. For all y^* in the large set $A \cap B_{1/4}(\frac{3}{4}x^*)$ we have $y^*(x) = (y^* - \frac{3}{4}x^*)(x) + \frac{3}{4}x^*(x) \geq \frac{3}{4}\|x\| - \frac{1}{4}\|x\| \geq \frac{1}{2}\|x\|$. For $\|x\| > 2$ we thus get $x \notin U$. Now let σ be the Minkowski-functional generated by U and σ^* the dual norm on E'. Let $\Delta \subseteq E$ be a small dense subset. Then $\{x^* \in A : \sigma^*(x^*) > 1\}$ is small, since $\sigma^*(x^*) > 1$ for $x^* \in A$ implies that there exists an $x \in \Delta$ with $x^*(x) > \sigma(x)$, but this is $\bigcup_{n\in\mathbb{N}}\{x^* \in A : x^*(x) > \sigma(x) + \frac{1}{n}\}$, and each of these sets is small by construction of $\sigma(x)$. Since Δ is small so is the union over all $x \in \Delta$. Thus $A_1 := \{x^* \in A : \sigma(x^*) \leq 1\}$ is large.

Now let $\varepsilon := \frac{1}{8(2n+1)}$, let $x \in E$, and let $0 < \eta < \varepsilon$. We may choose two different $x_i^* \in A_1$ for $i \in \{1, 2\}$ with $x_i^*(x) > \sigma(x) - \eta^2/2$. This is possible, since this is true for all but a small set of $x^* \in A$. Thus $\sigma^*(x_1^* - x_2^*) \geq \|x_1^* - x_2^*\| > 2\varepsilon$, and hence there is an $h \in E$ with $\sigma(h) = 1$ and $(x_1^* - x_2^*)(h) > 2\varepsilon$. Let now $t > 0$. Then

$$\sigma(x + th) \geq x_1^*(x + th) = x_1^*(x) + tx_1^*(h) > \sigma(x) - \frac{\eta^2}{2} + tx_1^*(h),$$

$$\sigma(x - th) \geq x_2^*(x - th) > \sigma(x) - \frac{\eta^2}{2} - tx_2^*(h).$$

Furthermore $\sigma(x) \geq \sigma(x + th) - t\sigma'(x + th)(h)$ implies

$$\sigma'(x + th)(h) \geq \frac{\sigma(x + th) - \sigma(x)}{t} > x_1^*(h) - \frac{\eta^2}{2t},$$

$$-\sigma'(x - th)(h) \geq -x_2^*(h) - \frac{\eta^2}{2t}.$$

Adding the last two inequalities gives

$$\sigma'(x + th)(h) - \sigma'(x - th)(h) \geq (x_2^* - x_1^*)(h) - \frac{\eta^2}{t} > \varepsilon,$$

since $(x_2^* - x_1^*)(h) > 2\varepsilon$ and we choose $t < \eta$ such that $\frac{\eta^2}{t} < \varepsilon$. □

14.11. Results. Spaces which are not smoothly regular. *For Banach spaces one has the following results:*

(1) [Bonic, Frampton, 1965]. *By* (14.9) *no Fréchet-differentiable bump function exists on* $C[0,1]$ *and on* ℓ^1. *Hence, most infinite dimensional* C^**-algebras are not regular for 1-times Fréchet-differentiable functions, in particular those for which a normal operator exists whose spectrum contains an open interval.*

(2) [Leduc, 1970]. *If* $\operatorname{dens} E < \operatorname{dens} E^*$ *then no* C^1*-bump function exists. This follows from* (14.10), (14.9), and (14.7). *See also* (13.24.2).

(3) [John, Zizler, 1978]. *A norm is called strongly rough if and only if there exists an* $\varepsilon > 0$ *such that for every* x *with* $\|x\| = 1$ *there exists a unit vector* y *with* $\limsup_{t\searrow 0} \frac{\|x+ty\|+\|x-ty\|-2}{t} \geq \varepsilon$. *The usual norm on* $\ell^1(\Gamma)$ *is strongly rough, if* Γ *is uncountable. There is however an equivalent non-rough norm on* $\ell^1(\Gamma)$ *with no point of Gâteaux-differentiability. If a Banach space has Gâteaux differentiable bump functions then it does not admit a strongly rough norm.*

(4) [Day, 1955]. *On* $\ell^1(\Gamma)$ *with uncountable* Γ *there is no Gâteaux differentiable continuous bump function.*

(5) [Bonic, Frampton, 1965]. $E < \ell^p$, $\dim E = \infty$: *If* $p = 2n+1$ *then* E *is not* D^p*-regular. If* $p \notin \mathbb{N}$ *then* E *is not* $\mathcal{S}$*-regular, where* $\mathcal{S}$ *denotes the* $C^{[p]}$*-functions whose highest derivative satisfies a Hölder like condition of order* $p - [p]$ *but with* $o(\ \ )$ *instead of* $O(\ \ )$.

14.12. Results.

(1) [Deville, Godefroy, Zizler, 1990]. *If* $c_0(\Gamma) \to E \to F$ *is a short exact sequence of Banach spaces and* F *has* C^k*-bump functions then also* E *has them. Compare with* (16.19).

(2) [Meshkov, 1978] *If a Banach space* E *and its dual* E^* *admit* C^2*-bump functions, then* E *is isomorphic to a Hilbert space. Compare with* (13.18.7).

(3) *Smooth bump functions are not inherited by short exact sequences.*

Notes. (1) As in (13.17.3) one chooses $x_a^* \in E^*$ with $x_a^*|_{c_0(\Gamma)} = \mathrm{ev}_a$. Let g be a smooth bump function on E/F and $h \in C^\infty(\mathbb{R}, [0,1])$ with compact support and equal to 1 near 0. Then $f(x) := g(x+F)\prod_{a\in\Gamma} h(x_a^*(x))$ is the required bump function.

(3) Use the example mentioned in (13.18.6), and apply (2).

Open problems. Is the product of C^∞-regular convenient vector spaces again C^∞-regular? Beware of the topology on the product!

Is every quotient of any $\mathcal{S}$-regular space again $\mathcal{S}$-regular?

15. Functions with Globally Bounded Derivatives

In many problems (like Borel's theorem (15.4), or the existence of smooth functions with given carrier (15.3)) one uses in finite dimensions the existence of smooth functions with bounded derivatives. In infinite dimensions C^k-functions have locally bounded k-th derivatives, but even for bump functions this need not be true globally.

15.1. Definitions. For normed spaces we use the following notation: $C^k_B := \{f \in C^k : \|f^{(k)}(x)\| \leq B \text{ for all } x \in E\}$ and $C^k_b := \bigcup_{B>0} C^k_B$. For general convenient vector spaces we may still define C^∞_b as those smooth functions $f : U \to F$ for which the image $d^k f(U)$ of each derivative is bounded in the space $L^k_{\text{sym}}(E, F)$ of bounded symmetric multilinear mappings.

Let $\mathcal{L}\text{ip}^k_K$ denote the space of C^k-functions with global Lipschitz-constant K for the k-th derivatives and $\mathcal{L}\text{ip}^k_{\text{global}} := \bigcup_{K>0} \mathcal{L}\text{ip}^k_K$. Note that $C^k_K = C^k \cap \mathcal{L}\text{ip}^{k-1}_K$.

15.2. Lemma. Completeness of C^n. *Let f_j be C^n-functions on some Banach space such that $f_j^{(k)}$ converges uniformly on bounded sets to some function f^k for each $k \leq n$. Then $f := f^0$ is C^n, and $f^{(k)} = f^k$ for all $k \leq n$.*

Proof. It is enough to show this for $n = 1$. Since $f'_n \to f^1$ uniformly, we have that f^1 is continuous, and hence $\int_0^1 f^1(x + t\,h)(h)\,dt$ makes sense and

$$f_n(x + h) - f_n(x) = \int_0^1 f'_n(x + t\,h)(h)\,dt \to \int_0^1 f^1(x + t\,h)(h)\,dt$$

for x and h fixed. Since $f_n \to f$ pointwise, this limit has to be $f(x + h) - f(x)$. Thus we have

$$\begin{aligned}\frac{\|f(x + h) - f(x) - f^1(x)(h)\|}{\|h\|} &= \frac{1}{\|h\|}\left\|\int_0^1 (f^1(x + t\,h) - f^1(x))(h)\,dt\right\| \\ &\leq \int_0^1 \|f^1(x + t\,h) - f^1(x))\|\,dt\end{aligned}$$

which goes to 0 for $h \to 0$ and fixed x, since f^1 is continuous. Thus, f is differentiable and $f' = f^1$. □

15.3. Proposition. When are closed sets zero-sets of smooth functions. [Wells, 1973]. *Let E be a separable Banach space and $n \in \mathbb{N}$. Then E has a C^n_b-bump function if and only if every closed subset of E is the zero-set of a C^n-function.*

For $n = \infty$ and E a convenient vector space we still have ($\Rightarrow$), provided all $L^k(E; \mathbb{R})$ satisfy the second countability condition of Mackey, i.e. for every countable family of bounded sets B_k there exist $t_k > 0$ such that $\bigcup_k t_k B_k$ is bounded.

Proof. ($\Rightarrow$) Suppose first that E has a C^n_b-bump function. Let $A \subseteq E$ be closed and $U := E \setminus A$ be the open complement. For every $x \in U$ there exists an $f_x \in$

$C_b^n(E)$ with $f_x(x) = 1$ and $\operatorname{carr}(f_x) \subseteq U$. The family of carriers of the f_x is an open covering of U. Since E is separable, those points in a countable dense subset that lie in U are dense in the metrizable space U. Thus, U is Lindelöf, and consequently we can find a sequence of points x_n such that for the corresponding functions $f_n := f_{x_n}$ the carriers still cover U. Now choose constants $t_n > 0$ such that $t_n \cdot \sup\{\|f_n^{(j)}(x)\| : x \in E\} \leq \frac{1}{2^{n-j}}$ for all $j < n$. Then $f := \sum_n t_n f_n$ converges uniformly in all derivatives, hence represents by (15.2) a C^n-function on E that vanishes on A. Since the carriers of the f_n cover U, it is strictly positive on U, and hence the required function has as 0-set exactly A.

($\Leftarrow$) Consider a vector $a \neq 0$, and let $A := E \setminus \bigcup_{n\in\mathbb{N}}\{x : \|x - \frac{1}{2^n}a\| < \frac{1}{2^{n+1}}\}$. Since A is closed there exists by assumption a C^n-function $f : E \to \mathbb{R}$ with $f^{-1}(0) = A$ (without loss of generality we may assume $f(E) \subseteq [0,1]$). By continuity of the derivatives we may assume that $f^{(n)}$ is bounded on some neighborhood U of 0. Choose n so large that $D := \{x : \|x - \frac{1}{2^n}a\| < \frac{1}{2^n}\} \subseteq U$, and let $g := f$ on $A \cup D$ and 0 on $E \setminus D$. Then $f \in C^n$ and $f^{(n)}$ is bounded. Up to affine transformations this is the required bump function. □

15.4. Borel's theorem. [Wells, 1973]. *Suppose a Banach space E has C_b^∞-bump functions. Then every formal power series with coefficients in $L^n_{\text{sym}}(E;F)$ for another Banach space F is the Taylor-series of a smooth mapping $E \to F$.*

Moreover, if G is a second Banach space, and if for some open set $U \subseteq G$ we are given $b_k \in C_b^\infty(U, L^k_{\text{sym}}(E,F))$, then there is a smooth $f \in C^\infty(E \times U, F)$ with $d^k(f(\ \ ,y))(0) = b_k(y)$ for all $y \in U$ and $k \in \mathbb{N}$. In particular, smooth curves can be lifted along the mapping $C^\infty(E,F) \to \prod_k L^k_{\text{sym}}(E;F)$.

Proof. Let $\rho \in C_b^\infty(E,\mathbb{R})$ be a C_b^∞-bump function, which equals 1 locally at 0. We shall use the notation $b_k(x,y) := b_k(y)(x^k)$. Define

$$f_k(x,y) := \frac{1}{k!} b_k(x,y)\,\rho(x)$$

and

$$f(x,y) := \sum_{k\geq 0} \frac{1}{t_k^k} f_k(t_k \cdot x, y)$$

with appropriately chosen $t_k > 0$. Then $f_k \in C^\infty(E \times U, F)$ and f_k has carrier inside of $\operatorname{carr}(\rho) \times U$, i.e. inside $\{x : \|x\| < 1\} \times U$. For the derivatives of b_k we have

$$\partial_1^j \partial_2^i b_k(x,y)(\xi,\eta) = k\,(k-1)\,\ldots\,(k-j)\,(d^i b_k(y)(\eta))(x^{k-j},\xi^j).$$

Hence, for $\|x\| \leq 1$ this derivative is bounded by

$$(k)_j \sup_{y\in U} \|d^i b_k(y)\|_{L(F,L^k_{\text{sym}}(E;G))},$$

where $(k)_j := k(k-1)\ldots(k-j)$. Using the product rule we see that for $j \geq k$ the derivative $\partial_1^j\partial_2^i f_k$ of f_k is globally bounded by

$$\sum_{l\leq k} \binom{j}{l} \sup\{\|\rho^{(j-l)}(x)\| : x \in E\}\,(k)_l \sup_{y\in U} \|d^i b_k(y)\| < \infty.$$

The partial derivatives of f would be

$$\partial_1^j \partial_2^i f_k(x,y) = \sum_k \frac{t_k^j}{t_k^k} \partial_1^j \partial_2^i f_k(t_k x, y).$$

We now choose the $t_k \geq 1$ such that these series converge uniformly. This is the case if,

$$\frac{1}{t_k^{k-j}} \sup\{\|\partial_1^j \partial_2^i f_k(x,y)\| : x \in E, y \in U\} \leq$$

$$\leq \frac{1}{t_k^{k-(j+i)}} \sup\{\|\partial_1^j \partial_2^i f_k(x,y)\| : x \in E, y \in U\} \leq \frac{1}{2^{k-(j+i)}},$$

and thus if

$$t_k \geq 2. \sup\{\|\partial_1^j \partial_2^i f_k(x,y)\|^{\frac{1}{k-(j+i)}} : x \in E, y \in U, j+i < k\}.$$

Since we have $\partial_1^j f_k(0,y)(\xi) = \frac{1}{k!}(k)_j b_k(y)(0^{k-j}, \xi^j)\, \rho(0) = \delta_k^j\, b_k(y)$, we conclude the desired result $\partial_1^j f(0,y) = b_k(y)$. $\square$

Remarks on Borel's theorem.

(1) [Colombeau, 1979]. Let E be a strict inductive limit of a non-trivial sequence of Fréchet spaces E_n. Then Borel's theorem is wrong for $f : \mathbb{R} \to E$. The idea is to choose $b_n = f^{(n)}(0) \in E_{n+1} \setminus E_n$ and to use that locally every smooth curve has to have values in some E_n.

(2) [Colombeau, 1979]. Let $E = \mathbb{R}^{\mathbb{N}}$. Then Borel's theorem is wrong for $f : E \to \mathbb{R}$. In fact, let $b_n : E \times \ldots \times E \to \mathbb{R}$ be given by $b_n := \mathrm{pr}_n \otimes \cdots \otimes \mathrm{pr}_n$. Assume $f \in C^\infty(E, \mathbb{R})$ exists with $f^{(n)}(0) = b_n$. Let f_n be the restriction of f to the n-th factor $\mathbb{R}$ in E. Then $f_n \in C^\infty(\mathbb{R}, \mathbb{R})$ and $f_n^{(n)}(0) = 1$. Since $f' : \mathbb{R}^n \to (\mathbb{R}^n)' = \mathbb{R}^{(\mathbb{N})}$ is continuous, the image of $B := \{x : |x_n| \leq 1 \text{ for all } n\}$ in $\mathbb{R}^{(\mathbb{N})}$ is bounded, hence contained in some $\mathbb{R}^{N-1}$. Since f_N is not constant on the interval $(-1,1)$ there exists some $|t_N| < 1$ with $f_N'(t_N) \neq 0$. For $x_N := (0, \ldots, 0, t_N, 0, \ldots)$ we obtain

$$f'(x_N)(y) = f_N'(t_N)(y_N) + \sum_{i \neq N} a_i\, y_i,$$

a contradiction to $f'(x_n) \in \mathbb{R}^{N-1}$.

(3) [Colombeau, 1979] showed that Borel's theorem is true for mappings $f : E \to F$, where E has a basis of Hilbert-seminorms and for any countable family of 0-neighborhoods U_n there exist $t_n > 0$ such that $\bigcap_{n=1}^\infty t_n\, U_n$ is a 0-neighborhood.

(4) If theorem (15.4) would be true for $G = \prod_k L^k_{\mathrm{sym}}(E;F)$ and $b_k = \mathrm{pr}_k$, then the quotient mapping $C^\infty(E,F) \to G = \prod_k L^k_{\mathrm{sym}}(E;F)$ would admit a smooth and hence a linear section. This is well know to be wrong even for $E = F = \mathbb{R}$, see (21.5).

15.5. Proposition. Hilbert spaces have C_b^∞-bump functions. [Wells, 1973] *If the norm is given by the n-th root of a homogeneous polynomial b of even degree n, then $x \mapsto \rho(b(x^n))$ is a C_b^∞-bump function, where $\rho : \mathbb{R} \to \mathbb{R}$ is smooth with $\rho(t) = 1$ for $t \leq 0$ and $\rho(t) = 0$ for $t \geq 1$.*

Proof. As before in the proof of (15.4) we see that the j-th derivative of $x \mapsto b(x^n)$ is bounded by $(n)_j$ on the closed unit ball. Hence, by the chain-rule and the global boundedness of all derivatives of ρ separately, the composite has bounded derivatives on the unit ball, and since it is zero outside, even everywhere. Obviously, $\rho(b(0)) = \rho(0) = 1$. □

In [Bonic, Frampton, 1966] it is shown that L^p is $\mathcal{L}\mathrm{ip}^n_{\mathrm{global}}$-smooth for all n if p is an even integer and is $\mathcal{L}\mathrm{ip}^{[p-1]}_{\mathrm{global}}$-smooth otherwise. This follows from the fact (see loc. cit., p. 140) that $d^{(p+1)}\|x\|^p = 0$ for even integers p and

$$\Big\| d^k\|x+h\|^p - d^k\|x\|^p \Big\| \leq \frac{p!}{k!}\|h\|^{p-k}$$

otherwise, cf. (13.13).

15.6. Estimates for the remainder in the Taylor-expansion. The Taylor formula of order k of a C^{k+1}-function is given by

$$f(x+h) = \sum_{j=0}^{k} \frac{1}{j!} f^{(j)}(x)(h^j) + \int_0^1 \frac{(1-t)^k}{k!}\, f^{(k+1)}(x+th)(h^{k+1})\, dt,$$

which can easily be seen by repeated partial integration of $\int_0^1 f'(x+th)(h)\,dt = f(x+h) - f(x)$.

For a C^2_B function we have

$$|f(x+h) - f(x) - f'(x)(h)| \leq \int_0^1 (1-t)\|f^{(2)}(x+th)\|\, \|h\|^2\, dt \leq B\frac{1}{2!}\,\|h\|^2.$$

If we take the Taylor formula of f up to order 0 instead, we obtain

$$f(x+h) = f(x) + \int_0^1 f'(x+th)(h)\, dt$$

and usage of $f'(x)(h) = \int_0^1 f'(x)(h)\,dt$ gives

$$|f(x+h) - f(x) - f'(x)(h)| \leq \int_0^1 \frac{\|f'(x+th) - f'(x)\|}{\|th\|}\, \|h\|^2\, dt \leq B\frac{1}{2!}\,\|h\|^2,$$

so it is in fact enough to assume $f \in C^1$ with f' satisfying a Lipschitz-condition with constant B.

For a C^3_B function we have

$$
\begin{aligned}
&|f(x+h) - f(x) - f'(x)(h) - \frac{1}{2}f''(x)(h^2)| \le \\
&\qquad\qquad \le \int_0^1 \frac{(1-t)^2}{2!} \|f^{(3)}(x+th)\| \, \|h\|^3 \, dt \le B\frac{1}{3!}\, \|h\|^3.
\end{aligned}
$$

If we take the Taylor formula of f up to order 1 instead, we obtain

$$
f(x+h) = f(x) + f'(x)(h) + \int_0^1 (1-t)\, f''(x+th)(h^2)\, dt,
$$

and using $\frac{1}{2}f''(x)(h^2) = \int_0^1 (1-t)\, f''(x)(h^2)\, dt$ we get

$$
\begin{aligned}
&|f(x+h) - f(x) - f'(x)(h) - \frac{1}{2}f''(x)(h^2)| \le \\
&\qquad\qquad \le \int_0^1 (1-t)t\, \frac{\|f''(x+th) - f''(x)\|}{\|th\|} \, \|h\|^3 \, dt \le B\frac{1}{3!}\, \|h\|^3.
\end{aligned}
$$

Hence, it is in fact enough to assume $f \in C^2$ with f'' satisfying a Lipschitz-condition with constant B.

Let $f \in C^k_B$ be flat of order k at 0. Applying $\|f(h) - f(0)\| = \| \int_0^1 f'(th)(h)\, dt\| \le \sup\{\|f'(th)\| : t \in [0,1]\}\, \|h\|$ to $f^{(j)}(\quad)(h_1, \dots, h_j)$ gives using $\|f^{(k)}(x)\| \le B$ inductively

$$
\begin{aligned}
\|f^{(k-1)}(x)\| &\le B \cdot \|x\| \\
\|f^{(k-2)}(x)\| &\le \int_0^1 \|f^{(k-1)}(tx)(x, \dots)\|\, dt \le B \int_0^1 t\, dt\, \|x\|^2 = \frac{B}{2}\, \|x\|^2 \\
&\vdots \\
\|f^{(j)}(x)\| &\le \frac{B}{(k-j)!}\, \|x\|^{k-j}.
\end{aligned}
$$

15.7. Lemma. $\mathcal{L}ip^1_{\text{global}}$-functions on $\mathbb{R}^n$. [Wells, 1973]. *Let $n := 2^N$ and $E = \mathbb{R}^n$ with the ∞-norm. Suppose $f \in \mathcal{L}ip^1_M(E, \mathbb{R})$ with $f(0) = 0$ and $f(x) \ge 1$ for $\|x\| \ge 1$. Then $M \ge 2N$.*

The idea behind the proof is to construct recursively a sequence of points $x_k := \sum_{j<k} \sigma_j h_j$ of norm $\frac{k-1}{N}$ starting at $x_0 = 0$, such that the increment along the segment is as small as possible. In order to evaluate this increment one uses the Taylor-formula and chooses the direction h_k such that the derivative at x_k vanishes.

Proof. Let A be the set of all edges of a hyper-cube, i.e.

$$
A := \{x : x_i = \pm 1 \text{ for all } i \text{ except one } i_0 \text{ and } |x_{i_0}| \le 1\}.
$$

Then A is symmetric. Let $x \in E$ be arbitrary. We want to find $h \in A$ with $f'(x)(h) = 0$. By permuting the coordinates we may assume that $i \mapsto |f'(x)(e^i)|$ is monotone decreasing. For $2 \le i \le n$ we choose recursively $h_i \in \{\pm 1\}$ such that $\sum_{j=2}^{i} h_j\, f'(x)(e_j)$ is an alternating sum. Then $|\sum_{j=2}^{i} f'(x)(e^j)h_j| \le |f'(x)(e^1)|$. Finally, we choose $\|h_1\| \le 1$ such that $f'(x)(h) = 0$.

Now we choose inductively $h_i \in \frac{1}{N}A$ and $\sigma_i \in \{\pm 1\}$ such that $f'(x_i)(h_i) = 0$ for $x := \sum_{j<i} \sigma_j h_j$ and x_i has at least 2^{N-i} coordinates equal to $\frac{i}{N}$. For the last statement we have that $x_{i+1} = x_i + \sigma_i\, h_i$ and at least 2^{N-i} coordinates of x_i are $\frac{i}{N}$. Among those coordinates all but at most 1 of the h_i are $\pm\frac{1}{N}$. Now let σ_i be the sign which occurs more often and hence at least $2^{N-i}/2$ times. Then those $2^{N-(i+1)}$ many coordinates of x_{i+1} are $\frac{i+1}{N}$.

Thus $\|x_i\| = \frac{i}{N}$ for $i \le N$, since at least one coordinate has this value. Furthermore we have

$$1 = |f(x_N) - f(x_0)| \le \sum_{k=0}^{N-1} |f(x_{k+1}) - f(x_k) - f'(x_k)(h_k)|$$
$$\le \sum_{k=1}^{N} \frac{M}{2}\, \|h_k\|^2 \le N\, \frac{M}{2}\, \frac{1}{N^2},$$

hence $M \ge 2N$. □

15.8. Corollary. c_0 is not $\mathcal{L}\mathrm{ip}^1_{\mathrm{global}}$-regular. [Wells, 1973]. *The space c_0 is not $\mathcal{L}\mathrm{ip}^1_{\mathrm{global}}$-smooth.*

Proof. Suppose there exists an $f \in \mathcal{L}\mathrm{ip}^1_{\mathrm{global}}$ with $f(0) = 1$ and $f(x) = 0$ for all $\|x\| \ge 1$. Then the previous lemma applied to $1 - f$ restricted to finite dimensional subspaces shows that the Lipschitz constant M of the derivative has to be greater or equal to N for all N, a contradiction. □

This shows even that there exist no differentiable bump functions on $c_0(A)$ which have uniformly continuous derivative. Since otherwise there would exist an $N \in \mathbb{N}$ such that

$$\|f(x+h) - f(x) - f'(x)h\| \le \int_0^1 \|f'(x + t\,h) - f'(x)\|\, \|h\|\, dt \le \frac{1}{2}\|h\|,$$

for $\|h\| \le \frac{1}{N}$. Hence, the estimation in the proof of (15.7) would give $1 \le N\frac{1}{2}\frac{1}{N} = \frac{1}{2}$, a contradiction.

15.9. Positive results on $\mathcal{L}\mathrm{ip}^1_{\mathrm{global}}$-functions. [Wells, 1973].

(1) *Every closed subset of a Hilbert space is the zero-set of a $\mathcal{L}\mathrm{ip}^1_{\mathrm{global}}$-function.*
(2) *For every two closed subsets of a Hilbert space which have distance $d > 0$ there exists a $\mathcal{L}\mathrm{ip}^1_{4/d^2}$-function which has value 0 on one set and 1 on the other.*
(3) *Whitney's extension theorem is true for $\mathcal{L}\mathrm{ip}^1_{\mathrm{global}}$-functions on closed subsets of Hilbert spaces.*

16. Smooth Partitions of Unity and Smooth Normality

16.1. Definitions. We say that a Hausdorff space X is *smoothly normal* with respect to a subalgebra $\mathcal{S} \subseteq C(X,\mathbb{R})$ or $\mathcal{S}$-*normal*, if for two disjoint closed subsets A_0 and A_1 of X there exists a function $f : X \to \mathbb{R}$ in $\mathcal{S}$ with $f|A_i = i$ for $i = 0, 1$. If an algebra $\mathcal{S}$ is specified, then by a smooth function we will mean an element of $\mathcal{S}$. Otherwise it is a C^∞-function.

A $\mathcal{S}$-*partition of unity* on a space X is a set $\mathcal{F}$ of smooth functions $f : X \to \mathbb{R}$ which satisfy the following conditions:

(1) For all $f \in \mathcal{F}$ and $x \in X$ one has $f(x) \geq 0$.
(2) The set $\{\operatorname{carr}(f) : f \in \mathcal{F}\}$ of all carriers is a locally finite covering of X.
(3) The sum $\sum_{f\in\mathcal{F}} f(x)$ equals 1 for all $x \in X$.

Since a family of open sets is locally finite if and only if the family of the closures is locally finite, the foregoing condition (2) is equivalent to:

(2') The set $\{\operatorname{supp}(f) : f \in \mathcal{F}\}$ of all supports is a locally finite covering of X.

The partition of unity is called *subordinated* to an open covering $\mathcal{U}$ of X, if for every $f \in \mathcal{F}$ there exists an $U \in \mathcal{U}$ with $\operatorname{carr}(f) \subseteq U$.

We say that X is *smoothly paracompact* with respect to $\mathcal{S}$ or $\mathcal{S}$-*paracompact* if every open cover $\mathcal{U}$ admits a $\mathcal{S}$-partition $\mathcal{F}$ of unity subordinated to it. This implies that X is $\mathcal{S}$-normal.

The partition of unity can then even be chosen in such a way that for every $f \in \mathcal{F}$ there exists a $U \in \mathcal{U}$ with $\operatorname{supp}(f) \subseteq U$. This is seen as follows. Since the family of carriers is a locally finite open refinement of $\mathcal{U}$, the topology of X is paracompact. So we may find a finer open cover $\{\tilde{U} : U \in \mathcal{U}\}$ such that the closure of $\tilde{U}$ is contained in U for all $U \in \mathcal{U}$, see [Bourbaki, 1966, IX.4.3]. The partition of unity subordinated to this finer cover has the support property for the original one.

Lemma. *Let $\mathcal{S}$ be an algebra which is closed under sums of locally finite families of functions. If $\mathcal{F}$ is an $\mathcal{S}$-partition of unity subordinated to an open covering $\mathcal{U}$, then we may find an $\mathcal{S}$-partition of unity $(f_U)_{U\in\mathcal{U}}$ with* $\operatorname{carr}(f_U) \subseteq U$.

Proof. For every $f \in \mathcal{F}$ we choose a $U_f \in \mathcal{U}$ with $\operatorname{carr}(f) \in U_f$. For $U \in \mathcal{U}$ put $\mathcal{F}_U := \{f : U_f = U\}$ and let $f_U := \sum_{f\in\mathcal{F}_U} f \in \mathcal{S}$. □

16.2. Proposition. Characterization of smooth normality. *Let X be a Hausdorff space with $\mathcal{S} \subseteq C(X,\mathbb{R})$ as in* (14.1) *Consider the following statements:*

(1) *X is $\mathcal{S}$-normal;*
(2) *For any two closed disjoint subsets $A_i \subseteq X$ there is a function $f \in \mathcal{S}$ with $f|A_0 = 0$ and $0 \notin f(A_1)$;*
(3) *Every locally finite open covering admits $\mathcal{S}$-partitions of unity subordinated to it.*
(4) *For any two disjoint zero-sets A_0 and A_1 of continuous functions there exists a function $g \in \mathcal{S}$ with $g|_{A_j} = j$ for $j = 0, 1$ and $g(X) \subseteq [0, 1]$;*

(5) *For any continuous function* $f : X \to \mathbb{R}$ *there exists a function* $g \in \mathcal{S}$ *with* $f^{-1}(0) \subseteq g^{-1}(0) \subseteq f^{-1}(\mathbb{R} \setminus \{1\})$.
(6) *The set* $\mathcal{S}$ *is dense in the algebra of continuous functions with respect to the topology of uniform convergence;*
(7) *The set of all bounded functions in* $\mathcal{S}$ *is dense in the algebra of continuous bounded functions on* X *with respect to the supremum norm;*
(8) *The bounded functions in* $\mathcal{S}$ *separate points in the Stone-Čech-compactification* βX *of* X.

The statements (1)-(3) are equivalent, and (4)-(8) are equivalent as well. If X *is metrizable all statements are equivalent.*
If every open set is the carrier set of a smooth function then X *is* $\mathcal{S}$*-normal. If* X *is* $\mathcal{S}$*-normal, then it is* $\mathcal{S}$*-regular.*
A space is $\mathcal{S}$*-paracompact if and only if it is paracompact and* $\mathcal{S}$*-normal.*

Proof. (2) $\Rightarrow$ (1). By assumption, there is a smooth function f_0 with $f_0|A_1 = 0$ and $0 \notin f_0(A_0)$, and again by assumption, there is a smooth function f_1 with $f_1|A_0 = 0$ and $0 \notin f_1(\{x : f_0(x) = 0\})$. The function $f = \frac{f_1}{f_0+f_1}$ has the required properties.

(1) $\Rightarrow$ (2) is obvious.

(3) $\Rightarrow$ (1) Let A_0 and A_1 be two disjoint closed subset. Then $\mathcal{U} := \{X \setminus A_1, X \setminus A_0\}$ admits a $\mathcal{S}$-partition of unity $\mathcal{F}$ subordinated to it, and

$$\sum \{f \in \mathcal{F} : \operatorname{carr} f \subseteq X \setminus A_0\}$$

is the required bump function.

(1) $\Rightarrow$ (3) Let $\mathcal{U}$ be a locally finite covering of X. The space X is $\mathcal{S}$-normal, so its topology is also normal, and therefore for every $U \in \mathcal{U}$ there exists an open set V_U such that $\overline{V_U} \subseteq U$ and $\{V_U : U \in \mathcal{U}\}$ is still an open cover. By assumption, there exist smooth functions $g_U \in \mathcal{S}$ such that $V_U \subseteq \operatorname{carr}(g_U) \subseteq U$, cf. (16.1). The function $g := \sum_U g_U$ is well defined, positive, and smooth since $\mathcal{U}$ is locally finite, and $\{f_U := g_U/g : U \in \mathcal{U}\}$ is the required partition of unity.

(5) $\Rightarrow$ (4) Let $A_j := f_j^{-1}(a_j)$ for $j = 0, 1$. By replacing f_j by $(f_j - a_j)^2$ we may assume that $f_j \geq 0$ and $A_j = f_j^{-1}(0)$. Then $(f_1 + f_2)(x) > 0$ for all $x \in X$, since $A_1 \cap A_2 = \emptyset$. Thus, $f := \frac{f_0}{f_0+f_1}$ is a continuous function in $C(X, [0, 1])$ with $f|_{A_j} = j$ for $j = 0, 1$.
Now we reason as in ((2) $\Rightarrow$ (1)): By (4) there exists a $g_0 \in \mathcal{S}$ with $A_0 \subseteq f^{-1}(0) \subseteq g_0^{-1}(0) \subseteq f^{-1}(\mathbb{R} \setminus \{1\}) = X \setminus f^{-1}(1) \subseteq X \setminus A_1$. By replacing g_0 by g_0^2 we may assume that $g_0 \geq 0$.
Applying the same argument to the zero-sets A_1 and $g_0^{-1}(0)$ we obtain a $g_1 \in \mathcal{S}$ with $A_1 \subseteq g_1^{-1}(0) \subseteq X \setminus g_0^{-1}(0)$. Thus, $(g_0 + g_1)(x) > 0$, and hence $g := \frac{g_0}{g_0+g_1} \in \mathcal{S}$ satisfies $g|_{A_j} = j$ for $j = 0, 1$ and $g(X) \subseteq [0, 1]$.

(4) $\Rightarrow$ (6) Let f be continuous. Without loss of generality we may assume $f \geq 0$ (decompose $f = f_+ - f_-$). Let $\varepsilon > 0$. Then choose $g_k \in \mathcal{S}$ with image in $[0, 1]$, and $g_k(x) = 0$ for all x with $f(x) \leq k\,\varepsilon$, and $g_k(x) = 1$ for all x with $f(x) \geq (k + 1)\,\varepsilon$.

Let k be the largest integer less or equal to $\frac{f(x)}{\varepsilon}$. Then $g_j(x) = 1$ for all $j < k$, and $g_j(x) = 0$ for all $j > k$. Hence, the sum $g := \varepsilon \sum_{k \in \mathbb{N}} g_k \in \mathcal{S}$ is locally finite, and $|f(x) - g(x)| < 2\varepsilon$.

(6) $\Rightarrow$ (7) This is obvious, since for any given bounded continuous f and for any $\varepsilon > 0$, by (6) there exists $g \in \mathcal{S}$ with $|f(x) - g(x)| < \varepsilon$ for all $x \in X$, hence $\|f - g\|_\infty \leq \varepsilon$ and $\|g\|_\infty \leq \|f\|_\infty + \|f - g\|_\infty < \infty$.

(7) $\Leftrightarrow$ (8) This follows from the Stone-Weierstraß theorem, since obviously the bounded functions in $\mathcal{S}$ form a subalgebra in $C_b(X) = C(\beta X)$. Hence, it is dense if and only if it separates points in the compact space βX.

(7) $\Rightarrow$ (4) By cutting off f at 0 and at 1, we may assume that f is bounded. By (7) there exists a bounded $g_0 \in \mathcal{S}$ with $\|f - g_0\|_\infty < \frac{1}{2}$. Let $h \in C^\infty(\mathbb{R}, \mathbb{R})$ be such that $h(t) = 0 \Leftrightarrow t \leq \frac{1}{2}$. Then $g := h \circ g_0 \in \mathcal{S}$, and $f(x) = 0 \Rightarrow g_0(x) \leq |g_0(x)| \leq |f(x)| + \|f - g_0\|_\infty \leq \frac{1}{2} \Rightarrow g(x) = h(g_0(x)) = 0$ and also $f(x) = 1 \Rightarrow g_0(x) \geq f(x) - \|f - g_0\|_\infty > 1 - \frac{1}{2} = \frac{1}{2} \Rightarrow g(x) \neq 0$.

If X is metrizable and $A \subseteq X$ is closed, then $\operatorname{dist}(\quad, A) : x \mapsto \sup\{\operatorname{dist}(x, a) : a \in A\}$ is a continuous function with $f^{-1}(0) = A$. Thus, (1) and (4) are equivalent.

Let every open subset be the carrier of a smooth mapping, and let A_0 and A_1 be closed disjoint subsets of X. By assumption, there is a smooth function f with $\operatorname{carr}(f) = X \setminus A_0$.

Obviously, every $\mathcal{S}$-normal space is $\mathcal{S}$-regular. Take as second closed set in (2) a single point. If we take instead the other closed set as single point, then we have what has been called small zero-sets in (19.8).

That a space is $\mathcal{S}$-paracompact if and only if it is paracompact and $\mathcal{S}$-normal can be shown as in the proof that a paracompact space admits continuous partitions of unity, see [Engelking, 1989, 5.1.9]. □

In [Kriegl, Michor, Schachermayer, 1989] it is remarked that in an uncountable product of real lines there are open subsets, which are not carrier sets of continuous functions.

Corollary. Denseness of smooth functions. *Let X be $\mathcal{S}$-paracompact, let F be a convenient vector space, and let $U \subseteq X \times F$ be open such that for all $x \in X$ the set $\iota_x^{-1}(U) \subseteq F$ is convex and non-empty, where $\iota_x : F \to X \times F$ is given by $y \mapsto (x, y)$. Then there exists an $f \in \mathcal{S}$ whose graph is contained in U.*

Under the following assumption this result is due to [Bonic, Frampton, 1966]: For $U := \{(x, y) : p(y - g(x)) < \varepsilon(x)\}$, where $g : X \to F$, $\varepsilon : X \to \mathbb{R}^+$ are continuous and p is a continuous seminorm on F.

Proof. For every $x \in X$ let y_x be chosen such that $(x, y_x) \in U$. Next choose open neighborhoods U_x of x such that $U_x \times \{y_x\} \subseteq U$. Since X is $\mathcal{S}$-paracompact there exists a $\mathcal{S}$-partition of unity $\mathcal{F}$ subordinated to the covering $\{U_x : x \in X\}$. In particular, for every $\varphi \in \mathcal{F}$ there exists an $x_\varphi \in X$ with $\operatorname{carr} \varphi \subseteq U_{x_\varphi}$. Now define

$f := \sum_{\varphi\in\mathcal{F}} y_{x_\varphi}\,\varphi$. Then $f \in \mathcal{S}$ and for every $x \in X$ we have

$$f(x) = \sum_{\varphi\in\mathcal{F}} y_{x_\varphi}\,\varphi(x) = \sum_{x\in\operatorname{carr}\varphi} y_{x_\varphi}\,\varphi(x) \in \iota_x^{-1}(U),$$

since $\iota_x^{-1}(U)$ is convex, contains y_{x_φ} for $x \in \operatorname{carr}(\varphi) \subseteq U_{x_\varphi}$, and $\varphi(x) \geq 0$ with $1 = \sum_\varphi \varphi(x) = \sum_{x\in\operatorname{carr}\varphi}\varphi(x)$. □

16.3. Lemma. $\mathcal{L}\mathrm{ip}^2$-functions on $\mathbb{R}^n$. [Wells, 1973]. *Let $B \in \mathbb{N}$ and $A := \{x \in \mathbb{R}^N : x_i \leq 0$ for all i and $\|x\| \leq 1\}$. Suppose that $f \in C^3_B(\mathbb{R}^N,\mathbb{R})$ with $f|_A = 0$ and $f(x) \geq 1$ for all x with* $\operatorname{dist}(x, A) \geq 1$. *Then $N < B^2 + 36\,B^4$.*

Proof. Suppose $N \geq B^2+36\,B^4$. We may assume that f is symmetric by replacing f with $x \mapsto \frac{1}{N!}\sum_\sigma f(\sigma^* x)$, where σ runs through all permutations, and σ^* just permutes the coordinates. Consider points $x^j \in \mathbb{R}^N$ for $j = 0,\dots,B^2$ of the form

$$x^j = \Big(\underbrace{\tfrac{1}{B},\dots,\tfrac{1}{B}}_{j},\underbrace{-\tfrac{1}{B},\dots,-\tfrac{1}{B}}_{B^2-j},\underbrace{0,\dots,0}_{>36\,B^4}\Big).$$

Then $\|x^j\| = 1$, $x^0 \in A$ and $d(x^{B^2}, A) \geq 1$. Since f is symmetric and $y^j := \frac{1}{2}(x^j + x^{j+1})$ has vanishing $j, B^2+1,\dots,N$ coordinates, we have for the partial derivatives $\partial_j f(y^j) = \partial_k f(y^j)$ for $k = B^2+1,\dots,N$. Thus

$$|\partial_j f(y^j)|^2 = \frac{1}{N-B^2}\sum_{k=B^2+1}^{N}|\partial_k f(y^j)|^2 \leq \frac{\|f'(y^j)\|_2^2}{36\,B^4} = \frac{\|f'(y^j)\|^2}{36\,B^4} \leq \frac{1}{36\,B^2},$$

since from $f|_A = 0$ we conclude that $f(0) = f'(0) = f''(0) = f'''(0)$ and hence $\|f^{(j)}(h)\| \leq B\,\|h\|^{3-j}$ for $j \leq 3$, see (15.6).

From $|f(x+h) - f(x) - f'(x)(h) - \frac{1}{2}f''(x)(h^2)| \leq B\frac{1}{3!}\,\|h\|^3$ we conclude that

$$\begin{aligned}|f(x+h) - f(x-h)| &\leq |f(x+h) - f(x) - f'(x)(h) - \tfrac{1}{2}f''(x)(h^2)|\\ &\quad + |f(x-h) - f(x) + f'(x)(h) - \tfrac{1}{2}f''(x)(h^2)|\\ &\quad + 2|f'(x)(h)|\\ &\leq \tfrac{2}{3!}\,B\|h\|^3 + 2|f'(x)(h)|.\end{aligned}$$

If we apply this to $x = y^j$ and $h = \frac{1}{B}e_j$, where e_j denotes the j-th unit vector, then we obtain

$$|f(x^{j+1}) - f(x^j)| \leq \frac{2}{3!}\,B\,\frac{1}{B^3} + 2|\partial_j f(y^j)|\,\frac{1}{B} \leq \frac{2}{3B^2}.$$

Summing up yields $1 \leq |f(x^{B^2})| = |f(x^{B^2}) - f(x^0)| \leq \frac{2}{3} < 1$, a contradiction. □

16.4. Corollary. ℓ^2 is not $\mathcal{L}\mathrm{ip}^2_{\mathrm{glob}}$-normal. [Wells, 1973]. *Let $A_0 := \{x \in \ell^2 : x_j \leq 0$ for all j and $\|x\| \leq 1\}$ and $A_1 := \{x \in \ell^2 : d(x, A) \geq 1\}$ and $f \in C^3(\ell^2,\mathbb{R})$ with $f|_{A_j} = j$ for $j = 0, 1$. Then $f^{(3)}$ is not bounded.*

Proof. By the preceding lemma a bound B of $f^{(3)}$ must satisfy for f restricted to $\mathbb{R}^N$, that $N < B^2 + 36B^4$. This is not for all N possible. □

16.5. Corollary. Whitney's extension theorem is false on ℓ^2. [Wells, 1973]. *Let $E := \mathbb{R} \times \ell^2 \cong \ell^2$ and $\pi : E \to \mathbb{R}$ be the projection onto the first factor. For subsets $A \subseteq \ell^2$ consider the cone $CA := \{(t, ta) : t \geq 0, a \in A\} \subseteq E$. Let $A := C(A_0 \cup A_1)$ with A_0 and A_1 as in (16.4). Let a jet (f^j) on A be defined by $f^j = 0$ on the cone CA_1 and $f^j(x)(v^1, \dots, v^j) = h^{(j)}(\pi(x))(\pi(v^1), \dots, \pi(v^j))$ for all x in the cone of CA_0, where $h \in C^\infty(\mathbb{R}, \mathbb{R})$ is infinite flat at 0 but with $h(t) \neq 0$ for all $t \neq 0$. This jet has no C^3-prolongation to E.*

Proof. Suppose that such a prolongation f exists. Then $f^{(3)}$ would be bounded locally around 0, hence $f_a(x) := 1 - \frac{1}{h(a)} f(a, ax)$ would be a C^3_B function on ℓ^2 for small a, which is 1 on A_1 and vanishes on A_0. This is a contradiction to (16.4).

So it remains to show that the following condition of Whitney (22.2) is satisfied:

$$\Big\| f^j(y) - \sum_{i=0}^{k-j} \frac{1}{i!} f^{j+i}(x)(y-x)^j \Big\| = o(\|x-y\|^{k-j}) \text{ for } A \ni x, y \to a.$$

Let $f_1^j := 0$ and $f_0^j(x) := h^{(j)}(\pi(x)) \circ (\pi \times \dots \times \pi)$. Then both are smooth on $\mathbb{R} \oplus \ell^2$, and thus Whitney's condition is satisfied on each cone separately. It remains to show this when x is in one cone and y in the other and both tend to 0. Thus, we have to replace f at some places by f_1 and at others by f_0. Since h is infinite flat at 0 we have $\|f_0^j(z)\| = o(\|z\|^n)$ for every n. Furthermore for $x_i \in CA_i$ for $i = 0, 1$ we have that $\|x_1 - x_0\| \geq \sin(\arctan 2 - \arctan 1) \max\{\|x_0\|, \|x_1\|\}$. Thus, we may replace $f_0^j(y)$ by $f_1^j(y)$ and vice versa. So the condition is reduced to the case, where y and z are in the same cone CA_i. □

16.6. Lemma. Smoothly regular strict inductive limits. *Let E be the strict inductive limit of a sequence of C^∞-normal convenient vector spaces E_n such that $E_n \hookrightarrow E_{n+1}$ is closed and has the extension property for smooth functions. Then E is C^∞-regular.*

Proof. Let U be open in E and $0 \in U$. Then $U_n := U \cap E_n$ is open in E_n. We choose inductively a sequence of functions $f_n \in C^\infty(E_n, \mathbb{R})$ such that $\operatorname{supp}(f_n) \subseteq U_n$, $f_n(0) = 1$, and $f_n|E_{n-1} = f_{n-1}$. If f_n is already constructed, we may choose by C^∞-normality a smooth $g : E_{n+1} \to \mathbb{R}$ with $\operatorname{supp}(g) \subseteq U_{n+1}$ and $g|_{\operatorname{supp}(f_n)} = 1$. By assumption, f_n extends to a function $\widetilde{f_n} \in C^\infty(E_{n+1}, \mathbb{R})$. The function $f_{n+1} := g \cdot \widetilde{f_n}$ has the required properties.

Now we define $f : E \to \mathbb{R}$ by $f|E_n := f_n$ for all n. It is smooth since any $c \in C^\infty(\mathbb{R}, E)$ locally factors to a smooth curve into some E_n by (1.8) since a strict inductive limit is regular by (52.8), so $f \circ c$ is smooth. Finally, $f(0) = 1$, and if $f(x) \neq 0$ then $x \in E_n$ for some n, and we have $f_n(x) = f(x) \neq 0$, thus $x \in U_n \subseteq U$. □

For counter-examples for the extension property see (21.7) and (21.11). However, for complemented subspaces the extension property obviously holds.

16.7. Proposition. C_c^∞ **is** C^∞**-regular.** *The space* $C_c^\infty(\mathbb{R}^m, \mathbb{R})$ *of smooth functions on* $\mathbb{R}^m$ *with compact support satisfies the assumptions of* (16.6).

Let $K_n := \{x \in \mathbb{R}^m : |x| \leq n\}$. Then $C_c^\infty(\mathbb{R}^m, \mathbb{R})$ is the strict inductive limit of the closed subspaces $C_{K_n}^\infty(\mathbb{R}^m, \mathbb{R}) := \{f : \operatorname{supp}(f) \subseteq K_n\}$, which carry the topology of uniform convergence in all partial derivatives separately. They are nuclear Fréchet spaces and hence separable, see (52.27). Thus they are C^∞-normal by (16.10) below.

In order to show the extension property for smooth functions we proof more general that for certain sets A the subspace $\{f \in C^\infty(E, \mathbb{R}) : f|_A = 0\}$ is a complemented subspace of $C^\infty(E, \mathbb{R})$. The first result in this direction is:

16.8. Lemma. [Seeley, 1964] *The subspace* $\{f \in C^\infty(\mathbb{R}, \mathbb{R}) : f(t) = 0 \text{ for } t \leq 0\}$ *of the Fréchet space* $C^\infty(\mathbb{R}, \mathbb{R})$ *is a direct summand.*

Proof. We claim that the following map is a bounded linear mapping being left inverse to the inclusion: $s(g)(t) := g(t) - \sum_{k\in\mathbb{N}} a_k h(-t2^k) g(-t2^k)$ for $t > 0$ and $s(g)(t) = 0$ for $t \leq 0$. Where $h : \mathbb{R} \to \mathbb{R}$ is a smooth function with compact support satisfying $h(t) = 1$ for $t \in [-1, 1]$ and (a_k) is a solution of the infinite system of linear equations $\sum_{k\in\mathbb{N}} a_k(-2^k)^n = 1$ $(n \in \mathbb{N})$ (the series is assumed to converge absolutely). The existence of such a solution is shown in [Seeley, 1964] by taking the limit of solutions of the finite subsystems. Let us first show that $s(g)$ is smooth. For $t > 0$ the series is locally around t finite, since $-t2^k$ lies outside the support of h for k sufficiently large. Its derivative $(sg)^{(n)}(t)$ is

$$g^{(n)}(t) - \sum_{k\in\mathbb{N}} a_k(-2^k)^n \sum_{j=0}^{n} h^{(j)}(-t2^k) g^{(n-j)}(-t2^k)$$

and this converges for $t \to 0$ towards $g^{(n)}(0) - \sum_{k\in\mathbb{N}} a_k(-2^k)^n g^{(n)}(0) = 0$. Thus $s(g)$ is infinitely flat at 0 and hence smooth on $\mathbb{R}$. It remains to show that $g \mapsto s(g)$ is a bounded linear mapping. By the uniform boundedness principle (5.26) it is enough to show that $g \mapsto (sg)(t)$ is bounded. For $t \leq 0$ this map is 0 and hence bounded. For $t > 0$ it is a finite linear combination of evaluations and thus bounded. □

Now the general result:

16.9. Proposition. *Let* E *be a convenient vector space, and let* p *be a smooth seminorm on* E. *Let* $A := \{x : p(x) \geq 1\}$. *Then the closed subspace* $\{f : f|_A = 0\}$ *in* $C^\infty(E, \mathbb{R})$ *is complemented.*

Proof. Let $g \in C^\infty(E, \mathbb{R})$ be a smooth reparameterization of p with support in $E \setminus A$ equal to 1 near $p^{-1}(0)$. By lemma (16.8), there is a bounded projection $P : C^\infty(\mathbb{R}, \mathbb{R}) \to C^\infty_{(-\infty,0]}(\mathbb{R}, \mathbb{R})$. The following mappings are smooth in turn by the properties of the cartesian closed smooth calculus, see (3.12):

$$\begin{aligned}
E \times \mathbb{R} \ni (x,t) &\mapsto f(e^t, x) \in \mathbb{R} \\
E \ni x &\mapsto f(e^{(\ )} x) \in C^\infty(\mathbb{R},\mathbb{R}) \\
E \ni x &\mapsto P(f(e^{(\ )} x)) \in C^\infty_{(-\infty,0]}(\mathbb{R},\mathbb{R}) \\
E \times \mathbb{R} \ni (x,r) &\mapsto P(f(e^{(\ )} x))(r) \in \mathbb{R} \\
\operatorname{carr} p \ni x &\mapsto \left(\frac{x}{p(x)}, \ln(p(x))\right) \mapsto P\left(f(e^{(\ )} \tfrac{x}{p(x)})\right)(\ln(p(x))) \in \mathbb{R}.
\end{aligned}$$

So we get the desired bounded linear projection

$$\begin{aligned}
\bar{P} &: C^\infty(E,\mathbb{R}) \to \{f \in C^\infty(E,\mathbb{R}) : f|_A = 0\}, \\
(\bar{P}(f))(x) &:= g(x)\, f(x) + (1 - g(x))\, P(f(e^{(\ )} \tfrac{x}{p(x)}))(\ln(p(x))). \quad \square
\end{aligned}$$

16.10. Theorem. Smoothly paracompact Lindelöf. [Wells, 1973]. *If X is Lindelöf and $\mathcal{S}$-regular, then X is $\mathcal{S}$-paracompact. In particular, all nuclear Fréchet spaces and strict inductive limits of sequences of such spaces are C^∞-paracompact. Furthermore, nuclear Silva spaces, see* (52.37), *are C^∞-paracompact.*

The first part was proved by [Bonic, Frampton, 1966] under stronger assumptions. The importance of the proof presented here lies in the fact that we need not assume that $\mathcal{S}$ is local and that $\frac{1}{f} \in \mathcal{S}$ for $f \in \mathcal{S}$. The only things used are that $\mathcal{S}$ is an algebra and for each $g \in \mathcal{S}$ there exists an $h : \mathbb{R} \to [0,1]$ with $h \circ g \in \mathcal{S}$ and $h(t) = 0$ for $t \le 0$ and $h(t) = 1$ for $t \ge 1$. In particular, this applies to $\mathcal{S} = \mathcal{L}\mathrm{ip}^p_{\mathrm{global}}$ and X a separable Banach space.

Proof. Let $\mathcal{U}$ be an open covering of X.

Claim. There exists a sequence of functions $g_n \in \mathcal{S}(X,[0,1])$ such that $\{\operatorname{carr} g_n : n \in \mathbb{N}\}$ is a locally finite family subordinated to $\mathcal{U}$ and $\{g_n^{-1}(1) : n \in \mathbb{N}\}$ is a covering of X.

For every $x \in X$ there exists a neighborhood $U \in \mathcal{U}$ (since $\mathcal{U}$ is a covering) and hence an $h_x \in \mathcal{S}(X,[0,2])$ with $h_x(x) = 2$ and $\operatorname{carr}(h_x) \subseteq U$ (since X is $\mathcal{S}$-regular). Since X is Lindelöf we find a sequence x_n such that $\{x : h_n(x) > 1 : n \in \mathbb{N}\}$ is a covering of X (we denote $h_n := h_{x_n}$). Now choose an $h \in C^\infty(\mathbb{R},[0,1])$ with $h(t) = 0$ for $t \le 0$ and $h(t) = 1$ for $t \ge 1$. Set

$$g_n(x) := h(n\,(h_n(x) - 1) + 1) \prod_{j<n} h(n\,(1 - h_j(x)) + 1).$$

Note that

$$\begin{aligned}
h(n\,(h_n(x) - 1) + 1) &= \begin{cases} 0 & \text{for } h_n(x) \le 1 - \frac{1}{n} \\ 1 & \text{for } h_n(x) \ge 1 \end{cases} \\
h(n\,(1 - h_j(x)) + 1) &= \begin{cases} 0 & \text{for } h_j(x) \ge 1 + \frac{1}{n} \\ 1 & \text{for } h_j(x) \le 1 \end{cases}
\end{aligned}$$

Then $g_n \in \mathcal{S}(X, [0,1])$ and $\operatorname{carr} g_n \subseteq \operatorname{carr} h_n$. Thus, the family $\{\operatorname{carr} g_n : n \in \mathbb{N}\}$ is subordinated to $\mathcal{U}$.

The family $\{g_n^{-1}(1) : n \in \mathbb{N}\}$ covers X since for each $x \in X$ there exists a minimal n with $h_n(x) \geq 1$, and thus $g_n(x) = 1$.

If we could divide in $\mathcal{S}$, then $f_n := g_n/\sum_j g_j$ would be the required partition of unity (and we do not need the last claim in this strong from).

Instead we proceed as follows. The family $\{\operatorname{carr} g_n : n \in \mathbb{N}\}$ is locally finite: Let n be such that $h_n(x) > 1$, and take $k > n$ so large that $1 + \frac{1}{k} < h_n(x)$, and let $U_x := \{y : h_n(y) > 1 + \frac{1}{k}\}$, which is a neighborhood of x. For $m \geq k$ and $y \in U_x$ we have that $h_n(y) > 1 + \frac{1}{k} \geq 1 + \frac{1}{m}$, hence the $(n+1)$-st factor of g_m vanishes at y, i.e. $\{j : \operatorname{carr} g_j \cap U_x \neq \emptyset\} \subseteq \{1, \dots, m-1\}$.

Now define $f_n := g_n \prod_{j<n}(1 - g_j) \in \mathcal{S}$. Then $\operatorname{carr} f_n \subseteq \operatorname{carr} g_n$, hence $\{\operatorname{carr} f_n : n \in \mathbb{N}\}$ is a locally finite family subordinated to $\mathcal{U}$. By induction, one shows that $\sum_{j \leq n} f_j = 1 - \prod_{j \leq n}(1 - g_j)$. In fact $\sum_{j \leq n} f_j = f_n + \sum_{j<n} f_j = g_n \prod_{j<n}(1 - g_j) + 1 - \prod_{j<n}(1 - g_j) = 1 + (g_n - 1)\prod_{j<n}(1 - g_j)$. For every $x \in U$ there exists an n with $g_n(x) = 1$, hence $f_k(x) = 0$ for $k > n$ and $\sum_{j=0}^{\infty} f_j(x) = \sum_{j \leq n} f_j(x) = 1 - \prod_{j \leq n}(1 - g_j(x)) = 1$.

Let us consider a nuclear Silva space. By (52.37) its dual is a nuclear Fréchet space. By (4.11.2) on the strong dual of a nuclear Fréchet space the c^∞-topology coincides with the locally convex one. Hence, it is C^∞-regular since it is nuclear, so it has a base of (smooth) Hilbert seminorms. A Silva space is an inductive limit of a sequence of Banach spaces with compact connecting mappings (see (52.37)), and we may assume that the Banach spaces are separable by replacing them by the closures of the images of the connecting mappings, so the topology of the inductive limit is Lindelöf. Therefore, by the first assertion we conclude that the space is C^∞-paracompact.

In order to obtain the statement on nuclear Fréchet spaces we note that these are separable, see (52.27), and thus Lindelöf. A strict inductive limit of a sequence of nuclear Fréchet spaces is C^∞-regular by (16.6), and it is also Lindelöf for its c^∞-topology, since this is the inductive limit of topological spaces (not locally convex spaces). □

Remark. In particular, every separable Hilbert space has $\mathcal{L}ip^2_{\text{global}}$-partitions of unity, thus there is such a $\mathcal{L}ip^2_{\text{global}}$-partition of functions φ subordinated to $\ell^2 \setminus A_0$ and $\ell^2 \setminus A_1$, with A_0 and A_1 mentioned in (16.4). Hence, $f := \sum_{\operatorname{carr} \varphi \cap A_0 = \emptyset} \varphi \in C^2$ satisfies $f|_{A_j} = j$ for $j = 0, 1$. However, $f \notin \mathcal{L}ip^2_{\text{global}}$. The reason behind this is that $\mathcal{L}ip^2_{\text{global}}$ is not a sheaf.

Open problem. *Classically, one proves the existence of continuous partitions of unity from the paracompactness of the space. So the question arises whether theorem* (16.10) *can be strengthened to: If the initial topology with respect to $\mathcal{S}$ is paracompact, do there exist $\mathcal{S}$-partitions of unity? Or equivalently: Is every paracompact $\mathcal{S}$-regular space $\mathcal{S}$-paracompact?*

16.11. Theorem. Smoothness of separable Banach spaces. *Let E be a separable Banach space. Then the following conditions are equivalent.*

(1) *E has a C^1-norm;*
(2) *E has C^1-bump functions, i.e., E is C^1-regular;*
(3) *The C^1-functions separate closed sets, i.e., E is C^1-normal;*
(4) *E has C^1-partitions of unity, i.e., E is C^1-paracompact;*
(5) *E has no rough norm, i.e. E is Asplund;*
(6) *E' is separable.*

Proof. The implications (1) ⇒ (2) and (4) ⇒ (3) ⇒ (2) are obviously true. The implication (2) ⇒ (4) is (16.10). (2) ⇒ (5) holds by (14.9). (5) ⇒ (6) follows from (14.10) since E is separable. (6) ⇒ (1) is (13.22) together with (13.20). □

A more general result is:

16.12. Result. [John, Zizler, 1976] *Let E be a WCG Banach space. Then the following statements are equivalent:*

(1) *E is C^1-normable;*
(2) *E is C^1-regular;*
(3) *E is C^1-paracompact;*
(4) *E has norm, whose dual norm is LUR;*
(5) *E has shrinking Markuševič basis, i.e. vectors $x_i \in E$ and $x_i^* \in E'$ with $x_i^*(x_j) = \delta_{i,j}$ and the span of the x_i is dense in E and the span of x_i^* is dense in E'.*

16.13. Results.

(1) [Godefroy, Pelant, et. al., 1983] *(*[Vanderwerff, 1992]*) Let E' is WCG Banach space (or even WCD, see* (53.8)*). Then E is C^1-regular.*
(2) [Vanderwerff, 1992] *Let K be compact with $K^{(\omega_1)} = \emptyset$. Then $C(K)$ is C^1-paracompact. Compare with* (13.18.2) *and* (13.17.5).
(3) [Godefroy, Troyanski, et. al., 1983] *Let E be a subspace of a WCG Banach space. If E is C^k-regular then it is C^k-paracompact. This will be proved in* (16.18).
(4) [MacLaughlin, 1992] *Let E' be a WCG Banach space. If E is C^k-regular then it is C^k-paracompact.*

16.14. Lemma. Smooth functions on $c_0(\Gamma)$. [Toruńczyk, 1973]. *The norm-topology of $c_0(\Gamma)$ has a basis which is a countable union of locally finite families of carriers of smooth functions, each of which depends locally only on finitely many coordinates.*

Proof. The open balls $B_r := \{x : \|x\|_\infty < r\}$ are carriers of such functions: In fact, similarly to (13.16) we choose a $h \in C^\infty(\mathbb{R}, \mathbb{R})$ with $h = 1$ locally around 0 and $\operatorname{carr} h = (-1, 1)$, and define $f(x) := \prod_{\gamma\in\Gamma} h(x_\gamma)$. Let

$$\mathcal{U}_{n,r,q} = \{B_r + q_1 e_{\gamma_1} + \cdots + q_n e_{\gamma_n} : \{\gamma_1, \ldots, \gamma_n\} \subseteq \Gamma\}$$

where $n \in \mathbb{N}$, $r \in \mathbb{Q}$, $q \in \mathbb{Q}^n$ with $|q_i| > 2r$ for $1 \le i \le n$. This is the required countable family.

Claim. The union $\bigcup_{n,r,q} \mathcal{U}_{n,r,q}$ is a basis for the topology.

Let $x \in c_0(\Gamma)$ and $\varepsilon > 0$. Choose $0 < r < \frac{\varepsilon}{2}$ such that $r \neq |x_\gamma|$ for all γ (note that $|x_\gamma| \ge \varepsilon/4$ only for finitely many γ). Let $\{\gamma_1, \dots, \gamma_n\} := \{\gamma : |x_\gamma| > r\}$. For q_i with $|q_i - x_{\gamma_i}| < r$ and $|q_i| > 2r$ we have

$$x - \sum_i q_i\, e_{\gamma_i} \in B_r,$$

and hence

$$x \in B_r + \sum_{i=1}^{n} q_i e_{\gamma_i} \subseteq x + B_{2r} \subseteq \{y : \|y - x\|_\infty \le \varepsilon\}.$$

Claim. Each family $\mathcal{U}_{n,r,q}$ is locally finite.

For given $x \in c_0(\Gamma)$, let $\{\gamma_1, \dots, \gamma_m\} := \{\gamma : |x_\gamma| > \frac{r}{2}\}$ and assume there exists a $y \in (x + B_{\frac{r}{2}}) \cap (B_r + \sum_{i=1}^n q_i e_{\beta_i}) \neq \emptyset$. For $y \in x + B_{\frac{r}{2}}$ we have $|y_a| < r$ for all $\gamma \notin \{\gamma_1, \dots, \gamma_m\}$ and for $y \in B_r + \sum_{i=1}^n q_i\, e_{\beta_i}$ we have $|y_\gamma| > r$ for all $\gamma \in \{\beta_1, \dots, \beta_n\}$. Hence, $\{\beta_1, \dots, \beta_n\} \subseteq \{\gamma_1, \dots, \gamma_m\}$ and $\mathcal{U}_{n,r,q}$ is locally finite. □

16.15. Theorem, Smoothly paracompact metrizable spaces. [Toruńczyk, 1973]. *Let X be a metrizable smooth space. Then the following are equivalent:*

(1) *X is $\mathcal{S}$-paracompact, i.e. admits $\mathcal{S}$-partitions of unity.*
(2) *X is $\mathcal{S}$-normal.*
(3) *The topology of X has a basis which is a countable union of locally finite families of carriers of smooth functions.*
(4) *There is a homeomorphic embedding $i : X \to c_0(A)$ for some A (with image in the unit ball) such that $ev_a \circ i$ is smooth for all $a \in A$.*

Proof. (1) ⇒ (3) Let $\mathcal{U}_n$ be the cover formed by all open balls of radius $1/n$. By (1) there exists a partition of unity subordinated to it. The carriers of these smooth functions form a locally finite refinement $\mathcal{V}_n$. The union of all $\mathcal{V}_n$ is clearly a base of the topology since that of all $\mathcal{U}_n$ is one.

(3) ⇒ (2) Let A_1 and A_2 be two disjoint closed subsets of X. Let furthermore $\mathcal{U}_n$ be a locally finite family of carriers of smooth functions such that $\bigcup_n \mathcal{U}_n$ is a basis. Let $W_n^i := \bigcup\{U \in \mathcal{U}_n : U \cap A_i = \emptyset\}$. This is the carrier of the smooth locally finite sum of the carrying functions of the U's. The family $\{W_n^i : i \in \{0,1\}, n \in \mathbb{N}\}$ forms a countable cover of X. By the argument used in the proof of (16.10) we may shrink the W_n^i to a locally finite cover of X. Then $W^1 = \bigcup_n W_n^1$ is a carrier containing A_2 and avoiding A_1. Now use (16.2.2).

(2) ⇒ (1) is lemma (16.2), since metrizable spaces are paracompact.

(3) ⇒ (4) Let $\mathcal{U}_n$ be a locally finite family of carriers of smooth functions such that $\mathcal{U} := \bigcup_n \mathcal{U}_n$ is a basis. For every $U \in \mathcal{U}_n$ let $f_U : X \to [0, \frac{1}{n}]$ be a smooth function with carrier U. We define a mapping $i : X \to c_0(\mathcal{U})$, by $i(x) = (f_U(x))_{U \in \mathcal{U}}$. It

is continuous at $x_0 \in X$, since for $n \in \mathbb{N}$ there exists a neighborhood V of x_0 which meets only finitely many sets $U \in \bigcup_{k\le 2n}\mathcal{U}_k$, and so $\|i(x)-i(x_0)\| \le \frac{1}{n}$ for those $x \in V$ with $|f_U(x)-f_U(x_0)| < \frac{1}{n}$ for all $U \in \bigcup_{k\le n}\mathcal{U}_k$ meeting V. The mapping i is even an embedding, since for $x_0 \in U \in \mathcal{U}$ and $x \notin U$ we have $\|i(x)-i(x_0)\| = f_U(x_0) > 0$.

(4) $\Rightarrow$ (3) By (16.14) the Banach space $c_0(A)$ has a basis which is a countable union of locally finite families of carriers of smooth functions, all of which depend locally only on finitely many coordinates. The pullbacks of all these functions via i are smooth on X, and their carriers furnish the required basis. □

16.16. Corollary. Hilbert spaces are C^∞-paracompact. [Toruńczyk, 1973]. *Every space $c_0(\Gamma)$ (for arbitrary index set Γ) and every Hilbert space (not necessarily separable) is C^∞-paracompact.*

Proof. The assertion for $c_0(\Gamma)$ is immediate from (16.15). For a Hilbert space $\ell^2(\Gamma)$ we use the embedding $i : \ell^2(\Gamma) \to c_0(\Gamma \cup \{*\})$ given by

$$i(x)_\gamma = \begin{cases} x_\gamma & \text{for } \gamma \in \Gamma \\ \|x\|^2 & \text{for } \gamma = * \end{cases}$$

This is an embedding: From $\|x^n - x\|_\infty \to 0$ we conclude by Hölder's inequality that $\langle y, x^n - x\rangle \to 0$ for all $y \in \ell^2$ and hence $\|x_n - x\|^2 = \|x_n\|^2 + \|x\|^2 - 2\langle x, x_n\rangle \to 2\|x\|^2 - 2\|x\|^2 = 0$. □

16.17. Corollary. *A countable product of $\mathcal{S}$-paracompact metrizable spaces is again $\mathcal{S}$-paracompact.*

Proof. By theorem (16.15) we have certain embeddings $i_n : X_n \to c_0(A_n)$ with images contained in the unit balls. We consider the embedding $i : \prod_n X_n \to c_0(\bigsqcup_n A_n)$ given by $i(x)_a = \frac{1}{n} i_n(x_n)$ for $a \in A_n$ which has the required properties for theorem (16.15). It is an embedding, since $i(x^n) \to i(x)$ if and only if $x_k^n \to x_k$ for all k (all but finitely many coordinates are small anyhow). □

16.18. Corollary. [Godefroy, Troyanski, et. al., 1983]
Let E be a Banach space with a separable projective resolution of identity, see (53.13). If E is C^k-regular, then it is C^k-paracompact.

Proof. By (53.20) there exists a linear, injective, norm 1 operator $T : E \to c_0(\Gamma_1)$ for some Γ_1 and by (53.13) projections P_α for $\omega \le \alpha \le \operatorname{dens} E$. Let $\Gamma_2 := \{\Delta : \Delta \subseteq [\omega, \operatorname{dens} E),\ \text{finite}\}$. For $\Delta \in \Gamma_2$ choose a dense sequence $(x_n^\Delta)_n$ in the unit sphere of $P_\omega(E) \oplus \bigoplus_{\alpha\in\Delta}(P_{\alpha+1} - P_\alpha)(E)$ and let $y_n^\Delta \in E'$ be such that $\|y_n^\Delta\| = 1$ and $y_n^\Delta(x_n^\Delta) = 1$. For $n \in \mathbb{N}$ let $\pi_n^\Delta : x \mapsto x - y_n^\Delta(x)x_n^\Delta$. Choose a smooth function $h \in C^\infty(E, [0,1])$ with $h(x) = 0$ for $\|x\| \le 1$ and $h(x) = 1$ for $\|x\| \ge 2$. Let $R_\alpha := (P_{\alpha+1} - P_\alpha)/\|P_{\alpha+1} - P_\alpha\|$.

Now define an embedding as follows: Let $\Gamma := \mathbb{N}^3 \times \Gamma_2 \sqcup \mathbb{N} \times [\omega, \operatorname{dens} E) \sqcup \mathbb{N} \sqcup \Gamma_1$ and let $u : E \to c_0(\Gamma)$ be given by

$$u(x)_\gamma := \begin{cases} \frac{1}{2^{n+m+l}} h(m\pi_n^\Delta x) \prod_{\alpha\in\Delta} h(lR_\alpha x) & \text{for } \gamma = (m,n,l,\Delta) \in \mathbb{N}^3 \times \Gamma_2, \\ \frac{1}{2^m} h(mR_\alpha x) & \text{for } \gamma = (m,\alpha) \in \mathbb{N} \times [\omega, \operatorname{dens} E), \\ \frac{1}{2} h(\frac{x}{m}) & \text{for } \gamma = m \in \mathbb{N}, \\ T(x)_\alpha & \text{for } \gamma = \alpha \in \Gamma_1. \end{cases}$$

Let us first show that u is well-defined and continuous. We do this only for the coordinates in the first row (for the others it is easier, the third has locally even finite support).
Let $x_0 \in E$ and $0 < \varepsilon < 1$. Choose n_0 with $1/2^{n_0} < \varepsilon$. Then $|u(x)_\gamma| < \varepsilon$ for all $x \in X$ and all $\alpha = (m,n,l,\Delta)$ with $m+n+l \geq n_0$.
For the remaining coordinates we proceed as follows: We first choose $\delta < 1/n_0$. By (53.13.8) there is a finite set $\Delta_0 \in \Gamma_2$ such that $\|R_\alpha x_0\| < \delta/2$ for all $\alpha \notin \Delta_0$. For those α and $\|x - x_0\| < \delta/2$ we get

$$\|R_\alpha(x)\| \leq \|R_\alpha(x_0)\| + \|R_\alpha(x - x_0)\| < \frac{\delta}{2} + \frac{\delta}{2} = \delta,$$

hence $u(x)_\gamma = 0$ for all $\gamma = (m,n,l,\Delta)$ with $m+n+l < n_0$ and $\Delta \cap ([\omega, \operatorname{dens} E \setminus \Delta_0) \neq \emptyset$.
For the remaining finitely many coordinates $\gamma = (m,n,l,\Delta)$ with $m+n+l < n_0$ and $\Delta \subseteq \Delta_0$ we may choose a $\delta_1 > 0$ such that $|u(x)_\gamma - u(x_0)_\gamma| < \varepsilon$ for all $\|x - x_0\| < \delta_1$. Thus for $\|x - x_0\| < \min\{\delta/2, \delta_1\}$ we have $|u(x)_\gamma - u(x_0)_\gamma| < 2\varepsilon$ for all $\gamma \in \mathbb{N}^3 \times \Gamma_2$ and $|u(x_0)_\gamma| \geq \varepsilon$ only for $\alpha = (m,n,l,\Delta)$ with $m+n+l < n_0$ and $\Delta \subseteq \Delta_0$.

Since T is injective, so is u. In order to show that u is an embedding let $x_\infty, x_p \in E$ with $u(x_p) \to u(x_\infty)$. Then x_p is bounded, since for $n_0 > \|x_\infty\|$ implies that $h(x_\infty/n_0) = 0$ and from $h(x_p/n_0) \to h(x_\infty/n_0)$ we conclude that $\|x_p/n_0\| \leq 2$ for large p.

Now we show that for any $\varepsilon > 0$ there is a finite ε-net for $\{x_p : p \in \mathbb{N}\}$: For this we choose $m_0 > 2/\varepsilon$. By (53.13.8) there is a finite set $\Delta_0 \subseteq \Lambda(x_\infty) := \bigcup_{\varepsilon>0}\{\alpha < \operatorname{dens} E : \|R_\alpha(x_\infty)\| \geq \varepsilon\}$ and an $n_0 := n \in \mathbb{N}$ such that $\|m_0\pi_n^{\Delta_0}(x_\infty)\| \leq 1$ and hence $h(m_0\pi_n^{\Delta_0}(x_\infty)) = 0$. In fact by (53.13.9) there is a finite linear combination of vectors $R_\alpha(x_\infty)$, which has distance less than ε from x_∞, let $\delta := \min\{\|R_\alpha(x)\| :$ for those $\alpha\} > 0$. Since the $y_n^{\Delta_0}$ are dense in the unit sphere of $P_\omega \oplus \bigoplus_{\alpha\in\Delta_0} R_\alpha E$ we may choose an n such that $\|x_\infty - \|x_\infty\| x_n^{\Delta_0}\| < \frac{1}{2m_0}$ and hence

$$\begin{aligned} \|\pi_n^{\Delta_0}(x_\infty)\| &= \|x_\infty - y_n^{\Delta_0}(x_\infty) x_n^{\Delta_0}\| \\ &\leq \left\| x_\infty - \|x_\infty\| x_n^{\Delta_0} \right\| + \|x_\infty\| \left\| x_n^{\Delta_0} - y_n^{\Delta_0}(x_n^{\Delta_0}) x_n^{\Delta_0} \right\| \\ &\qquad + \|y_n^{\Delta_0}\| \left\| \|x_\infty\| x_n^{\Delta_0} - x_\infty) \right\| \|x_n^{\Delta_0}\| \\ &\leq \frac{1}{2m_0} + 0 + \frac{1}{2m_0} = \frac{1}{m_0} \end{aligned}$$

Next choose $l_0 := l \in \mathbb{N}$ such that $l_0\delta_0 \geq 2$ and hence $\|l_0 R_\alpha x_\infty\| \geq 2$ for all $\alpha \in \Delta_0$. Then

$$h(m_0\pi_{n_0}^{\Delta_0}x_p) \prod_{\alpha\in\Delta_0} h(l_0\, R_\alpha x_p) \to h(m_0\pi_{n_0}^{\Delta_0}x_\infty) \prod_{\alpha\in\Delta_0} h(l_0\, R_\alpha x_\infty)$$

$$\text{and} \qquad h(l_0 R_\alpha x_p) \to h(l_0 R_\alpha x_\infty) = 1 \text{ for } \alpha \in \Delta_0$$

Hence

$$h(m_0\pi_{n_0}^{\Delta_0}x_p) \to h(m_0\pi_{n_0}^{\Delta_0}x_\infty) = 0,$$

and so $\|\pi_{n_0}^{\Delta_0}x_p\| \leq 2/m_0 < \varepsilon$ for all large p. Thus $d(x_p, \mathbb{R}\, x_{n_0}^{\Delta_0}) \leq \varepsilon$, hence $\{x_p : p \in \mathbb{N}\}$ has a finite ε-net, since its projection onto the one dimensional subspace $\mathbb{R}x_{n_0}^{\Delta_0}$ is bounded.

Thus $\{x_\infty, x_p : p \in \mathbb{N}\}$ is relatively compact, and hence u restricted to its closure is a homeomorphism onto the image. So $x_p \to x_\infty$.

Now the result follows from (16.15). □

16.19. Corollary. [Deville, Godefroy, Zizler, 1990]. *Let $c_0(\Gamma) \to E \to F$ be a short exact sequence of Banach spaces and assume F admits C^p-partitions of unity. Then E admits C^p-partitions of unity.*

Proof. Without loss of generality we may assume that the norm of E restricted to $c_0(\Gamma)$ is the supremum norm. Furthermore there is a linear continuous splitting $T : \ell^1(\Gamma) \to E'$ by (13.17.3) and a continuous splitting $S : F \to E$ by (53.22) with $S(0) = 0$. We put $T_\gamma := T(e_\gamma)$ for all $\gamma \in \Gamma$. For $n \in \mathbb{N}$ let $\mathcal{F}_n$ be a C^p-partition of unity on F with $\operatorname{diam}(\operatorname{carr}(f)) \leq 1/n$ for all $f \in \mathcal{F}_n$. Let $\mathcal{F} := \bigsqcup_n \mathcal{F}_n$ and let $\Gamma_2 := \{\Delta \subseteq \Gamma : \Delta \text{ is finite}\}$. For any $f \in \mathcal{F}$ choose $x_f \in S(\operatorname{carr}(f))$ and for any $\Delta \in \Gamma_2$ choose a dense sequence $\{y_{f,m}^\Delta : m \in \mathbb{N}\} \ni 0$ in the linear subspace generated by $\{x_f + e_\gamma : \gamma \in \Delta\}$. Let $\ell_{f,m}^\Delta \in E'$ be such that $\ell_{f,m}^\Delta(y_{f,m}^\Delta) = \|\ell_{f,m}^\Delta\| \cdot \|y_{f,m}^\Delta\| = 1$. Let $\pi_{f,m}^\Delta : E \to E$ be given by $\pi_{f,m}^\Delta(x) := x - \ell_{f,m}^\Delta(x)\, y_{f,m}^\Delta$. Let $h : E \to \mathbb{R}$ be C^p with $h(x) = 0$ for $\|x\| \leq 1$ and $h(x) = 1$ for $\|x\| \geq 2$. Let $g : \mathbb{R} \to [-1, 1]$ be C^p with $g(t) = 0$ for $|t| \leq 1$ and injective on $\{t : |t| > 1\}$. Now define a mapping $u : E \to c_0(\tilde{\Gamma})$, where

$$\tilde{\Gamma} := (\mathcal{F} \times \Gamma_2 \times \mathbb{N}^2) \sqcup (\mathcal{F} \times \Gamma) \sqcup (\mathcal{F} \times \mathbb{N}) \sqcup \mathcal{F} \sqcup \mathbb{N} \sqcup \mathbb{N}$$

by

$$u(x)_{\tilde{\gamma}} := \frac{1}{2^{n+m+j}}\, f(\hat{x})\, h(j\, \pi_{f,m}^\Delta(x)) \prod_{\gamma\in\Delta} g(n\, T_\gamma(x - x_f))$$

for $\tilde{\gamma} = (f, \Delta, j, m) \in \mathcal{F}_n \times \Gamma_2 \times \mathbb{N}^2$, and by

$$u(x)_{\tilde{\gamma}} := \begin{cases} \frac{1}{2^n}\, f(\hat{x})\, g(n\, T_\gamma(x - x_f)) & \text{for } \tilde{\gamma} = (f, \gamma) \in \mathcal{F}_n \times \Gamma \\ \frac{1}{2^{n+j}}\, f(\hat{x})\, h(j\,(x - x_f)) & \text{for } \tilde{\gamma} = (f, j) \in \mathcal{F}_n \times \mathbb{N} \\ \frac{1}{2^n}\, f(\hat{x}) & \text{for } \tilde{\gamma} = f \in \mathcal{F}_n \subseteq \mathcal{F} \\ \frac{1}{2^n}\, h(n\, x) & \text{for } \tilde{\gamma} = n \in \mathbb{N} \\ \frac{1}{2^n}\, h(x/n) & \text{for } \tilde{\gamma} = n \in \mathbb{N}. \end{cases}$$

We first claim that u is well-defined and continuous. Every coordinate $x \mapsto u(x)_\gamma$ is continuous, so it remains to show that for every $\varepsilon > 0$ locally in x the set of coordinates γ, where $|u(x)_\gamma| > \varepsilon$ is finite. We do this for the first type of coordinates. For this we may fix n, m and j (since the factors are bounded by 1). Since $\mathcal{F}_n$ is a partition of unity, locally $f(\hat{x}) \neq 0$ for only finitely many $f \in \mathcal{F}_n$, so we may also fix $f \in \mathcal{F}_n$. For such an f the set $\Delta_0 := \{\gamma : |T_\gamma(x - x_f)| \geq \pi(x - x_f) + \frac{1}{n}\}$ is finite by the proof of (13.17.3). Since $\|\hat{x} - x_f\| = \|\pi(x - x_f)\| \leq 1/n$ be have $g(n\, T_\gamma(x - x_f)) = 0$ for $\gamma \notin \Delta_0$.

Thus only for those Δ contained in the finite set Δ_0, we have that the corresponding coordinate does not vanish.

Next we show that u is injective. Let $x \neq y \in E$.
If $\hat{x} \neq \hat{y}$, then there is some n and a $f \in \mathcal{F}_n$ such that $f(\hat{x}) \neq 0 = f(\hat{y})$. Thus this is detected by the 4th row.
If $\hat{x} = \hat{y}$ then $S\hat{x} = S\hat{y}$ and since $x - S\hat{x}$, $y - S\hat{y} \in c_0(\Gamma)$ there is a $\gamma \in \Gamma$ with

$$T_\gamma(x - S\hat{x}) = (x - S\hat{x})_\gamma \neq (y - S\hat{y})_\gamma = T_\gamma(y - S\hat{y}).$$

We will make use of the following method repeatedly:
For every n there is a $f_n \in \mathcal{F}_n$ with $f_n(\hat{x}) \neq 0$ and hence $\|\hat{x} - \hat{x}_{f_n}\| \leq 1/n$. Since S is continuous we get $x_{f_n} = S(\hat{x}_{f_n}) \to S(\hat{x})$ and thus $\lim_n T_\gamma(x - x_{f_n}) = \lim_n T_\gamma(x - S(\hat{x}_{f_n})) = T_\gamma(x - S(\hat{x}))$.
So we get

$$\lim_n T_\gamma(x - x_{f_n}) = T_\gamma(x - S(\hat{x})) \neq T_\gamma(y - S(\hat{y})) = \lim_n T_\gamma(y - x_{f_n}).$$

If all coordinates for $u(x)$ and $u(y)$ in the second row would be equal, then

$$g(n\, T_\gamma(x - x_f)) = g(n\, T_\gamma(y - x_f))$$

since $f_\gamma(\hat{x}) \neq 0$, and hence $\|T_\gamma(x - x_f) - T_\gamma(y - x_f)\| \leq 2/n$, a contradiction.

Now let us show that u is a homeomorphism onto its image. We have to show $x_k \to x$ provided $u(x_k) \to u(x)$.
We consider first the case, where $x = S\hat{x}$. As before we choose $f_n \in \mathcal{F}_n$ with $f_n(\hat{x}) \neq 0$ and get $x_{f_n} = S(\hat{x}_{f_n}) \to S(\hat{x}) = x$. Let $\varepsilon > 0$ and $j > 3/\varepsilon$. Choose an n such that $\|x_{f_n} - x\| < 1/j$. Then $h(j\,(x_{f_n} - x)) = 0$. From the coordinates in the third and fourth row we conclude

$$f(\hat{x}_k)\, h(j\,(x_k - x_{f_n})) \to f(\hat{x})\, h(j\,(x - x_{f_n})) \quad \text{and} \quad f(\hat{x}_k) \to f(\hat{x}) \neq 0.$$

Hence

$$h(j\,(x_k - x_{f_n})) \to h(j\,(x - x_{f_n})) = 0.$$

Thus $\|x_k - x_{f_n}\| < 2/j$ for all large k. But then

$$\|x_k - x\| \leq \|x_k - x_{f_n}\| + \|x_{f_n} - x\| < \frac{3}{j} < \varepsilon,$$

i.e. $x_k \to x$.
Now the case, where $x \neq S\hat{x}$. We show first that $\{x_k : k \in \mathbb{N}\}$ is bounded. Pick $n > \|x\|$. From the coordinates in the last row we get that $\lim_k h(x_k/n) = 0$, i.e. $\|x_k\| \leq 2n$ for all large k.
We claim that for $j \in \mathbb{N}$ there is an $n \in \mathbb{N}$ and an $f \in \mathcal{F}_n$ with $f(\hat{x}) \neq 0$, a finite set $\Delta \subseteq \Gamma$ with $\prod_{\gamma\in\Delta} g(n\,T_\gamma(x - x_f)) \neq 0$ and an $m \in \mathbb{N}$ with $h(j\,\pi^\Delta_{f,m}(x)) = 0$.
From $0 \neq (x - S\hat{x}) \in c_0(\Gamma)$ we deduce that there is a finite set $\Delta \subseteq \Gamma$ with $T_\gamma(x - S\hat{x}) = (x - S\hat{x})_\gamma \neq 0$ for all $\gamma \in \Delta$ and $\operatorname{dist}(x - S\hat{x}, \langle e_\gamma : \gamma \in \Delta\rangle) < 1/(3j)$, i.e. $|(x - S\hat{x})_\gamma| \leq 1/(3j)$ for all $\gamma \notin \Delta$. As before we choose $f_n \in \mathcal{F}_n$ with $f_n(\hat{x}) \neq 0$ and get $x_{f_n} = S(\hat{x}_{f_n}) \to S(\hat{x})$ and

$$\lim_n T_\gamma(x - x_{f_n}) = (x - S\hat{x})_\gamma \neq 0 \text{ for } \gamma \in \Delta.$$

Thus $g(n\,(T_\gamma(x - x_{f_n}))) \neq 0$ for all large n and $\gamma \in \Delta$. Furthermore, $\operatorname{dist}(x, x_{f_n} + \langle e_\gamma : \gamma \in \Delta\rangle) = \operatorname{dist}(x - x_{f_n}, \langle e_\gamma : \gamma \in \Delta\rangle) < 1/(2j)$. Since $\{y^\Delta_{f_n,m} : m \in \mathbb{N}\}$ is dense in $\langle x_{f_n} + e_\gamma : \gamma \in \Delta\rangle$ there is an m such that $\|x - y^\Delta_{f_n,m}\| < 1/(2j)$. Since $\|\pi^\Delta_{f_n,m}\| \leq 2$ we get

$$\begin{aligned}\|\pi^\Delta_{f_n,m}(x)\| &\leq \|x - y^\Delta_{f_n,m}\| + |1 - \ell^\Delta_{f_n,m}(x)|\,\|y^\Delta_{f_n,m}\| \\ &\leq \frac{1}{2j} + \|\ell^\Delta_{f_n,m}\|\,\|x - y^\Delta_{f_n,m}\|\,\|y^\Delta_{f_n,m}\| \leq \frac{1}{2j} + \frac{1}{2j} = \frac{1}{j},\end{aligned}$$

hence $h(j\,\pi^\Delta_{f_n,m}(x)) = 0$.
We claim that for every $\varepsilon > 0$ there is a finite ε-net of $\{x_k : k \in \mathbb{N}\}$. Let $\varepsilon > 0$. We choose $j > 4/\varepsilon$ and we pick $n \in \mathbb{N}$, $f \in \mathcal{F}_n$, $\Delta \subseteq \Gamma$ finite, and $m \in \mathbb{N}$ satisfying the previous claim. From $u(x_k) \to u(x)$ we deduce from the coordinates in the first row, that

$$\begin{aligned}&f(\hat{x}_k)\,h(j\,\pi^\Delta_{f,m}(x_k)) \prod_{\gamma\in\Delta} g(n\,T_\gamma(x_k - x_f)) \to \\ &\qquad \to f(\hat{x})\,h(j\,\pi^\Delta_{f,m}(x)) \prod_{\gamma\in\Delta} g(n\,T_\gamma(x - x_f)) \text{ for } k \to \infty\end{aligned}$$

and since by the coordinates in the fourth row $f(\hat{x}_k) \to f(\hat{x}) \neq 0$ we obtain from the coordinates in the second row, that

$$g(n\,T_\gamma(x_k - x_f)) \to g(n\,T_\gamma(x - x_f)) \neq 0 \text{ for } \gamma \in \Delta.$$

Hence

$$h(j\,\pi^\Delta_{f,m}(x_k)) \to h(j\,\pi^\Delta_{f,m}(x)) = 0.$$

Therefore

$$\|x_k - \ell^\Delta_{f,m}(x_k)\,y^\Delta_{f,n}\| = \|\pi^\Delta_{f,m}(x_k)\| < \frac{1}{j} < \frac{\varepsilon}{4} \text{ for all large } k.$$

Thus there is a finite dimensional subspace in E spanned by $y^\Delta_{f,n}$ and finitely many x_k, such that all x_k have distance $\leq \varepsilon/4$ from it. Since $\{x_k : k \in \mathbb{N}\}$ are bounded, the compactness of the finite dimensional balls implies that $\{x_k : k \in \mathbb{N}\}$ has an ε-net, hence $\{x_k : k \in \mathbb{N}\}$ is relatively compact, and since u is injective we have $\lim_k x_k = x$.

Now the result follows from (16.15). □

Remark. In general, the existence of C^∞-partitions of unity is not inherited by the middle term of short exact sequences: Take a short exact sequence of Banach spaces with Hilbert ends and non-Hilbertizable E in the middle, as in (13.18.6). If both E and E^* admitted C^2-partitions of unity, then they would admit C^2-bump functions, hence E was isomorphic to a Hilbert space by [Meshkov, 1978], a contradiction.

16.20. Results on $C(K)$. *Let K be compact. Then for the Banach space $C(K)$ we have:*

1. [Deville, Godefroy, Zizler, 1990]. *If $K^{(\omega)} = \emptyset$ then $C(K)$ is C^∞-paracompact.*
2. [Vanderwerff, 1992] *If $K^{(\omega_1)} = \emptyset$ then $C(K)$ is C^1-paracompact.*
3. [Haydon, 1990] *In contrast to (2) there exists a compact space K with $K^{(\omega_1)} = \{*\}$, but such that $C(K)$ has no Gâteaux-differentiable norm. Nevertheless $C(K)$ is C^1-regular by* [Haydon, 1991]. *Compare with* (13.18.2).
4. [Namioka, Phelps, 1975]. *If there exists an ordinal number α with $K^{(\alpha)} = \emptyset$ then the Banach space $C(K)$ is Asplund (and conversely), hence it does not admit a rough norm, by* (13.8).
5. [Ciesielski, Pol, 1984] *There exists a compact K with $K^{(3)} = \emptyset$. Consequently, there is a short exact sequence $c_0(\Gamma_1) \to C(K) \to c_0(\Gamma_2)$, and the space $C(K)$ is Lipschitz homeomorphic to some $c_0(\Gamma)$. However, there is no continuous linear injection of $C(K)$ into some $c_0(\Gamma)$.*

Notes. (1) Applying theorem (16.19) recursively we get the result as in (13.17.5).

16.21. Some radial subsets are diffeomorphic to the whole space. We are now going to show that certain subsets of convenient vector spaces are diffeomorphic to the whole space. So if these subsets form a base of the c^∞-topology of the modeling space of a manifold, then we may choose charts defined on the whole modeling space. The basic idea is to 'blow up' subsets $U \subseteq E$ along all rays starting at a common center. Without loss of generality assume that the center is 0. In order for this technique to work, we need a positive function $\rho : U \to \mathbb{R}$, which should give a diffeomorphism $f : U \to E$, defined by $f(x) := \frac{1}{\rho(x)}x$. For this we need that ρ is smooth, and since the restriction of f to $U \cap \mathbb{R}^+x \to \mathbb{R}^+x$ has to be a diffeomorphism as well, and since the image set is connected, we need that the domain $U \cap \mathbb{R}^+x$ is connected as well, i.e., U has to be radial. Let $U_x := \{t > 0 : tx \in U\}$, and let $f_x : U_x \to \mathbb{R}$ be given by $f(tx) = \frac{t}{\rho(tx)}x =: f_x(t)x$. Since up to diffeomorphisms this is just the restriction of the diffeomorphism f, we need that $0 < f_x'(t) = \frac{\partial}{\partial t}\frac{t}{\rho(tx)} = \frac{\rho(tx)-t\rho'(tx)(x)}{\rho(tx)^2}$ for all $x \in U$ and $0 < t \leq 1$. This means that $\rho(y) > \rho'(y)(y)$ for all $y \in U$, which is quite a restrictive condition, and so we want to construct out of an arbitrary smooth function $\rho : U \to \mathbb{R}$, which tends to 0 towards the boundary, a new smooth function ρ satisfying the additional assumption.

Theorem. *Let $U \subseteq E$ be c^∞-open with $0 \in U$ and let $\rho : U \to \mathbb{R}^+$ be smooth, such that for all $x \notin U$ with $tx \in U$ for $0 \le t < 1$ we have $\rho(tx) \to 0$ for $t \nearrow 1$. Then* $\operatorname{star} U := \{x \in U : tx \in U$ *for all* $t \in [0,1]\}$ *is diffeomorphic to E.*

Proof. First remark that $\operatorname{star} U$ is c^∞-open. In fact, let $c : \mathbb{R} \to E$ be smooth with $c(0) \in \operatorname{star} U$. Then $\varphi : \mathbb{R}^2 \to E$, defined by $\varphi(t,s) := t\,c(s)$ is smooth and maps $[0,1] \times \{0\}$ into U. Since U is c^∞-open and $\mathbb{R}^2$ carries the c^∞-topology there exists a neighborhood of $[0,1] \times \{0\}$, which is mapped into U, and in particular there exists some $\varepsilon > 0$ such that $c(s) \in \operatorname{star} U$ for all $|s| < \varepsilon$. Thus $c^{-1}(\operatorname{star} U)$ is open, i.e., $\operatorname{star} U$ is c^∞-open. Note that ρ satisfies on $\operatorname{star} U$ the same boundary condition as on U. So we may assume without loss of generality that U is radial. Furthermore, we may assume that $\rho = 1$ locally around 0 and $0 < \rho \le 1$ everywhere, by composing with some function which is constantly 1 locally around $[\rho(0), +\infty)$.

Now we are going to replace ρ by a new function $\tilde\rho$, and we consider first the case, where $E = \mathbb{R}$. We want that $\tilde\rho$ satisfies $\tilde\rho'(t)t < \tilde\rho(t)$ (which says that the tangent to $\tilde\rho$ at t intersects the $\tilde\rho$-axis in the positive part) and that $\tilde\rho(t) \le \rho(t)$, i.e., $\log\circ\tilde\rho \le \log\circ\rho$, and since we will choose $\tilde\rho(0) = 1 = \rho(0)$ it is sufficient to have $\frac{\tilde\rho'}{\tilde\rho} = (\log\circ\tilde\rho)' \le (\log\circ\rho)' = \frac{\rho'}{\rho}$ or equivalently $\frac{\tilde\rho'(t)t}{\tilde\rho(t)} \le \frac{\rho'(t)t}{\rho(t)}$ for $t > 0$. In order to obtain this we choose a smooth function $h : \mathbb{R} \to \mathbb{R}$ which satisfies $h(t) < 1$, and $h(t) \le t$ for all t, and $h(t) = t$ for t near 0, and we take $\tilde\rho$ as solution of the following ordinary differential equation

$$\tilde\rho'(t) = \frac{\tilde\rho(t)}{t} \cdot h\left(\frac{\rho'(t)t}{\rho(t)}\right) \quad \text{with } \tilde\rho(0) = 1.$$

Note that for t near 0, we have $\frac{1}{t}h\left(\frac{\rho'(t)t}{\rho(t)}\right) = \frac{\rho'(t)}{\rho(t)}$, and hence locally a unique smooth solution $\tilde\rho$ exists. In fact, we can solve the equation explicitly, since $(\log\circ\tilde\rho)'(t) = \frac{\tilde\rho'(t)}{\tilde\rho(t)} = \frac{1}{t} \cdot h\left(\frac{\rho'(t)t}{\rho(t)}\right)$, and hence $\tilde\rho(s) = \exp(\int_0^s \frac{1}{t} \cdot h(\frac{\rho'(t)t}{\rho(t)})\,dt)$, which is smooth on the same interval as ρ is.

Note that if ρ is replaced by $\rho_s : t \mapsto \rho(ts)$, then the corresponding solution $\widetilde{\rho_s}$ satisfies $\widetilde{\rho_s} = \tilde\rho_s$. In fact,

$$(\log\circ\tilde\rho_s)'(t) = \frac{(\tilde\rho_s)'(t)}{\tilde\rho_s(t)} = \frac{s\tilde\rho'(st)}{\tilde\rho(st)} = \frac{1}{t} \cdot \frac{st\tilde\rho'(st)}{\tilde\rho(st)} = \frac{1}{t}h\Big(\frac{st\rho'(st)}{\rho(st)}\Big) = \frac{1}{t}h\Big(\frac{t(\rho_s)'(t)}{\rho_s(t)}\Big).$$

For arbitrary E and $x \in E$ let $\rho_x : U_x \to \mathbb{R}^+$ be given by $\rho_x(t) := \rho(tx)$, and let $\tilde\rho : U \to \mathbb{R}^+$ be given by $\tilde\rho(x) := \widetilde{\rho_x}(1)$, where $\widetilde{\rho_x}$ is the solution of the differential equation above with ρ_x in place of ρ.

Let us now show that $\tilde\rho$ is smooth. Since U is c^∞-open, it is enough to consider a smooth curve $x : \mathbb{R} \to U$ and show that $t \mapsto \tilde\rho(x(t)) = \tilde\rho_{(x(t))}(1)$ is smooth. This is the case, since $(t,s) \mapsto \frac{1}{s}h\left(\frac{\rho'_{x(t)}(s)s}{\rho_{x(t)}(s)}\right) = \frac{1}{s}h\left(\frac{\rho'(s\,x(t))(s\,x(t))}{\rho(s\,x(t))}\right)$ is smooth, since $\varphi(t,s) := \frac{\rho'(s\,x(t))(s\,x(t))}{\rho(s\,x(t))}$ satisfies $\varphi(t,0)=0$, and hence $\frac{1}{s}h(\varphi(t,s)) = \frac{\varphi(t,s)}{s} = \frac{\rho'(s\,x(t))(x(t))}{\rho(s\,x(t))}$ locally.

From $\rho_{sx}(t) = \rho(tsx) = \rho_x(ts)$ we conclude that $\widetilde{\rho_{sx}}(t) = \widetilde{\rho_x}(ts)$, and hence $\tilde\rho(sx) = \widetilde{\rho_x}(s)$. Thus, $\tilde\rho'(x)(x) = \frac{\partial}{\partial t}|_{t=1}\tilde\rho(tx) = \frac{\partial}{\partial t}|_{t=1}\tilde\rho_x(t) = \tilde\rho_x'(1) < \tilde\rho_x(1) = \tilde\rho(x)$. This shows that we may assume without loss of generality that $\rho : U \to (0,1]$ satisfies the additional assumption $\rho'(x)(x) < \rho(x)$.

Note that $f_x : t \mapsto \frac{t}{\rho(tx)}$ is bijective from $U_x := \{t > 0 : tx \in U\}$ to $\mathbb{R}^+$, since 0 is mapped to 0, the derivative is positive, and $\frac{t}{\rho(tx)} \to \infty$ if either $\rho(tx) \to 0$ or $t \to \infty$ since $\rho(tx) \le 1$.

It remains to show that the bijection $x \mapsto \frac{1}{\rho(x)}x$ is a diffeomorphism. Obviously, its inverse is of the form $y \mapsto \sigma(y)y$ for some $\sigma : E \to \mathbb{R}^+$. They are inverse to each other so $\frac{1}{\rho(\sigma(y)y)}\sigma(y)y = y$, i.e., $\sigma(y) = \rho(\sigma(y)y)$ for $y \neq 0$. This is an implicit equation for σ. Note that $\sigma(y) = 1$ for y near 0, since ρ has this property. In order to show smoothness, let $t \mapsto y(t)$ be a smooth curve in E. Then it suffices to show that the implicit equation $(\sigma \circ y)(t) = \rho((\sigma \circ y)(t) \cdot y(t))$ satisfies the assumptions of the 2-dimensional implicit function theorem, i.e., $0 \neq \frac{\partial}{\partial \sigma}(\sigma - \rho(\sigma \cdot y(t))) = 1 - \rho'(\sigma \cdot y(t))(y(t))$, which is true, since multiplied with $\sigma > 0$ it equals $\sigma - \rho'(\sigma \cdot y(t))(\sigma \cdot y(t)) < \sigma - \rho(\sigma \cdot y(t)) = 0$. □

Chapter IV
Smoothly Realcompact Spaces

As motivation for the developments in this chapter let us tell a mathematical short story which was posed as an exercise in [Milnor, Stasheff, 1974, p.11]. For a finite dimensional Hausdorff second countable manifold M, one can prove that the space of algebra homomorphisms $\operatorname{Hom}(C^\infty(M,\mathbb{R}),\mathbb{R})$ equals M as follows. The kernel of a homomorphism $\varphi : C^\infty(M,\mathbb{R}) \to \mathbb{R}$ is an ideal of codimension 1 in $C^\infty(M,\mathbb{R})$. The zero sets $Z_f := f^{-1}(0)$ for $f \in \ker\varphi$ form a filter of closed sets, since $Z_f \cap Z_g = Z_{f^2+g^2}$, which contains a compact set Z_f for a function f which is proper (i.e., compact sets have compact inverse images). Thus $\bigcap_{f\in\ker\varphi} Z_f$ is not empty, it contains at least one point $x_0 \in M$. But then for any $f \in C^\infty(M,\mathbb{R})$ the function $f - \varphi(f)1$ belongs to the kernel of φ, so vanishes on x_0 and we have $f(x_0) = \varphi(f)$.

This question has many rather complicated (partial) answers in any infinite dimensional setting which are given in this chapter. One is able to prove that the answer is positive surprisingly often, but the proofs are involved and tied intimately to the class of spaces under consideration. The existing counter-examples are based on rather trivial reasons. We start with setting up notation and listing some interesting algebras of functions on certain spaces.

First we recall the topological theory of realcompact spaces from the literature and discuss the connections to the concept of smooth realcompactness. For an algebra homomorphism $\varphi : \mathcal{A} \to \mathbb{R}$ on some algebra of functions on a space X we investigate when $\varphi(f) = f(x)$ for some $x \in X$ for one function f, later for countably many, and finally for all $f \in \mathcal{A}$. We study stability of smooth realcompactness under pullback along injective mappings, and also under (left) exact sequences. Finally we discuss the relation between smooth realcompactness and bounding sets, i.e. sets on which every function of the algebra is bounded. In this chapter, the ordering principle for sections and results is based on the amount of evaluating properties obtained and we do not aim for linearly ordered proofs. So we will often use results presented later in the text. We believe that this is here a more transparent presentation than the usual one. Most of the material in this chapter can also be found in the theses' [Biström, 1993] and [Adam, 1993].

17. Basic Concepts and Topological Realcompactness

17.1. The setting. In [Hewitt, 1948, p.85] those completely regular topological spaces were considered under the name Q-spaces, for which each real valued algebra homomorphism on the algebra of all continuous functions is the evaluation at some point of the space. Later on these spaces where called realcompact spaces. Accordingly, we call a 'space' *smoothly realcompact* if this is true for 'the' algebra of smooth functions. There are other algebras for which this question is interesting, like polynomials, real analytic functions, C^k-functions. So we will treat the question in the following setting. Let

X be a set;

$\mathcal{A} \subseteq \mathbb{R}^X$ a point-separating subalgebra with unit; If X is a topological space we also require that $\mathcal{A} \subseteq C(X, \mathbb{R})$; If $X = E$ is a locally convex vector space we also assume that $\mathcal{A}$ is invariant under all translations and contains the dual E^* of all continuous linear functionals;

$X_\mathcal{A}$ the set X equipped with the initial topology with respect to $\mathcal{A}$;

$\varphi : \mathcal{A} \to \mathbb{R}$ an algebra homomorphism preserving the unit;

$Z_f := \{x \in X : f(x) = \varphi(f)\}$ for $f \in \mathcal{A}$;

$\operatorname{Hom} \mathcal{A}$ be the set of all real valued algebra homomorphisms $\mathcal{A} \to \mathbb{R}$ preserving the unit.

Moreover,

φ is called $\mathcal{F}$-*evaluating* for some subset $\mathcal{F} \subseteq \mathcal{A}$ if there exists an $x \in X$ with $\varphi(f) = f(x)$ for all $f \in \mathcal{F}$; equivalently $\bigcap_{f \in \mathcal{F}} Z_f \neq \emptyset$;

φ is called $\mathfrak{m}$-*evaluating* for a cardinal number $\mathfrak{m}$ if φ is $\mathcal{F}$-evaluating for all $\mathcal{F} \subseteq \mathcal{A}$ with cardinality of $\mathcal{F}$ at most $\mathfrak{m}$; This is most important for $\mathfrak{m} = 1$ and for $\mathfrak{m} = \omega$, the first infinite cardinal number;

φ is said to be $\bar{1}$-evaluating if $\varphi(f) \in \overline{f(X)}$ for all $f \in \mathcal{A}$.

φ is said to be evaluating if φ is $\mathcal{A}$-evaluating, i.e., $\varphi = \operatorname{ev}_x$ for some $x \in X$;

$\operatorname{Hom}_\omega \mathcal{A}$ is the set of all ω-evaluating homomorphisms in $\operatorname{Hom} \mathcal{A}$;

$\mathcal{A}$ is called $\mathfrak{m}$-evaluating if φ is $\mathfrak{m}$-evaluating for each algebra homomorphism $\varphi \in \operatorname{Hom} \mathcal{A}$;

$\mathcal{A}$ is called evaluating if φ is evaluating for algebra homomorphism $\varphi \in \operatorname{Hom} \mathcal{A}$;

X is called $\mathcal{A}$-realcompact if $\mathcal{A}$ is evaluating; i.e., each algebra homomorphism $\varphi \in \operatorname{Hom} \mathcal{A}$ is the evaluation at some point in X.

The algebra $\mathcal{A}$ is called

inversion closed if $1/f \in \mathcal{A}$ for all $f \in \mathcal{A}$ with $f(x) > 0$ for every $x \in X$; equivalently, if $1/f \in \mathcal{A}$ for all $f \in \mathcal{A}$ with f nowhere 0 (use $f^2 > 0$).

bounded inversion closed if $1/f \in \mathcal{A}$ for $f \in \mathcal{A}$ with $f(x) > \varepsilon$ for some $\varepsilon > 0$ and all $x \in X$;

$C^{(\infty)}$-algebra if $h \circ f \in \mathcal{A}$ for all $f \in \mathcal{A}$ and $h \in C^\infty(\mathbb{R}, \mathbb{R})$;

C^∞-algebra if $h \circ (f_1, \dots, f_n) \in \mathcal{A}$ for all $f_j \in \mathcal{A}$ and $h \in C^\infty(\mathbb{R}^n, \mathbb{R})$;

C^∞_{lfs}-algebra if it is a C^∞-algebra which is closed under locally finite sums, with respect to a specified topology on X. This holds if $\mathcal{A}$ is local, i.e., it contains any function f such that for each $x \in X$ there is some $f_x \in \mathcal{A}$ with $f = f_x$ near x.

C^∞_{lfcs}-algebra if it is a C^∞-algebra which is closed under locally finite countable sums.

Interesting algebras are the following, where in this chapter in the notation we shall generally omit the range space $\mathbb{R}$.

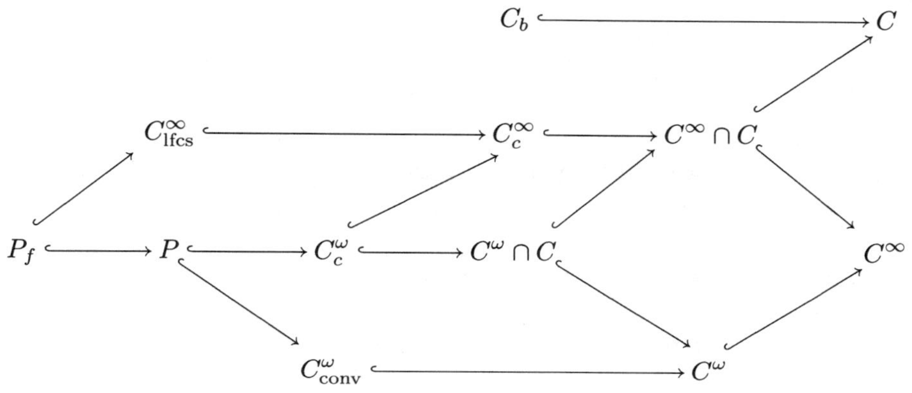

$C(X) = C(X, \mathbb{R})$, the algebra of continuous functions on a topological space X. It has all the properties from above.

$C_b(X) = C_b(X, \mathbb{R})$, the algebra of bounded continuous functions on a topological space X. It is only bounded inversion closed and a C^∞-algebra, in general.

$C^\infty(X) = C^\infty(X, \mathbb{R})$, the algebra of smooth functions on a Frölicher space X, see (23.1), or on a smooth manifold X, see section (27). It has all properties from above, where we may use the c^∞-topology.

$C^\infty(E) \cap C(E)$, the algebra of smooth and continuous functions on a locally convex space E. It has all properties from above, where we use the locally convex topology on E.

$C^\infty_c(E) = C^\infty_c(E, \mathbb{R})$, the algebra of smooth functions, all of whose derivatives are continuous on a locally convex space E. It has all properties from above, again for the locally convex topology on E.

$C^\omega(X) = C^\omega(X, \mathbb{R})$, the algebra of real analytic functions on a real analytic manifold X. It is only inversion closed.

$C^\omega(E) \cap C(E)$, the algebra of real analytic and continuous functions on a locally convex space E. It is only inversion closed.

$C^\omega_c(E) = C^\omega_c(E, \mathbb{R})$, the algebra of real analytic functions, all of whose derivatives are continuous on a locally convex space E. It is only inversion closed.

$C^\omega_{\text{conv}}(E) = C^\omega_{\text{conv}}(E, \mathbb{R})$, the algebra of globally convergent power series on a locally convex space E.

$P_f(E) = \text{Poly}_f(E, \mathbb{R})$, the algebra of finite type polynomials on a locally convex space E, i.e. the algebra $\langle E' \rangle_{\text{Alg}}$ generated by E'. This is the free commutative algebra generated by the vector space E', see (18.12). It has none of the properties from above.

$P(E) = \text{Poly}(E, \mathbb{R})$, the algebra of polynomials on a locally convex space E, see (5.15), (5.17), i.e. the homogeneous parts are given by bounded symmetric multilinear mappings. No property from above holds.

$C^\infty_{\text{lfcs}}(E) = C^\infty_{\text{lfcs}}(E, \mathbb{R})$, the C^∞_{lfcs}-algebra (see below) generated by E', and hence also called $(E')^\infty_{\text{lfcs}}$. Only the C^∞_{lfs}-property does not hold.

17.2. Results. *For completely regular topological spaces X and $\mathcal{A} = C(X)$ the following holds:*

(1) *Due to* [Hewitt, 1948, p.85 + p.60] *&* [Shirota, 1952, p.24], *see also* [Engelking, 1989, 3.12.22.g & 3.11.3]. *The space X is called realcompact if all algebra homomorphisms in* $\operatorname{Hom} C(X)$ *are evaluations at points of X, equivalently, if X is a closed subspace of a product of $\mathbb{R}$'s.*

(2) *Due to* [Hewitt, 1948, p.61] *&* [Katětov, 1951, p.82], *see also* [Engelking, 1989, 3.11.4 & 3.11.5]. *Hence every closed subspace of a product of realcompact spaces is realcompact.*

(3) *Due to* [Hewitt, 1948, p.85], *see also* [Engelking, 1989, 3.11.12]. *Each Lindelöf space is realcompact.*

(4) *Due to* [Katětov, 1951, p.82], *see also* [Engelking, 1989, 5.5.10]. *Paracompact spaces are realcompact if and only if all closed discrete subspaces are realcompact.*

(5) *Due to* [Hewitt, 1950, p.170, p.175] *&* [Mackey, 1944], *see also* [Engelking, 1989, 3.11.D.a]. *Discrete spaces are realcompact if and only if their cardinality is non-measurable.*

(6) *Hence Banach spaces are realcompact if and only if their density (i.e., the cardinality of a maximal discrete or of a minimal dense subset) or their cardinality is non-measurable.*

(7) [Shirota, 1952], *see also* [Engelking, 1989, 5.5.10 & 8.5.13.h]. *A space of non-measurable cardinality is realcompact if and only if it admits a complete uniformity.*

(8) *Due to* [Dieudonné, 1939,] *see also* [Engelking, 1989, 8.5.13.a]. *A space admits a complete uniformity, i.e. is Dieudonné complete, if and only if it is a closed subspace of a product of metrizable spaces*

Realcompact spaces where introduced by [Hewitt, 1948, p.85] under the name Q-compact spaces. The equivalence in (1) is due to [Shirota, 1952, p.24]. The results (1) and (2) are proved in [Engelking, 1989] for a different notion of realcompactness, which was shown to be equivalent to the original one by [Katětov, 1951], see also [Engelking, 1989, 3.12.22.g].

17.3. Lemma. [Kriegl, Michor, Schachermayer, 1989, 2.2, 2.3]. *Let $\mathcal{A}$ be $\bar{1}$-evaluating. Then we have a topological embedding*

$$\delta : X_{\mathcal{A}} \hookrightarrow \prod_{\mathcal{A}} \mathbb{R}, \quad \operatorname{pr}_f \circ \delta := f,$$

with dense image in the closed subset $\operatorname{Hom} \mathcal{A} \subseteq \prod_{\mathcal{A}} \mathbb{R}$. *Hence X is $\mathcal{A}$-realcompact if and only if δ has closed image.*

Proof. The topology of $X_{\mathcal{A}}$ is by definition initial with respect to all $f = \mathrm{pr}_f \circ \delta$, hence δ is an embedding. Obviously $\operatorname{Hom}\mathcal{A} \subseteq \prod_{\mathcal{A}} \mathbb{R}$ is closed. Let $\varphi : \mathcal{A} \to \mathbb{R}$ be an algebra-homomorphism. For $f \in \mathcal{A}$ consider Z_f. If $\mathcal{A}$ is 1-evaluating then by (18.8) for any finite subset $\mathcal{F} \subseteq \mathcal{A}$ there exists an $x_{\mathcal{F}} \in \bigcap_{f\in\mathcal{F}} Z_f$. Thus $\delta(x_{\mathcal{F}})_f = \varphi(f)$ for all $f \in \mathcal{F}$. If $\mathcal{A}$ is only $\bar{1}$-evaluating, then we get as in the proof of (18.3) for every $\varepsilon > 0$ a point $x_{\mathcal{F}} \in X$ such that $|f(x_{\mathcal{F}}) - \varphi(f)| < \varepsilon$ for all $f \in \mathcal{F}$. Thus $\delta(x_{\mathcal{F}})$ lies in the corresponding neighborhood of $(\varphi(f))_f$. Thus $\delta(X)$ is dense in $\operatorname{Hom}\mathcal{A}$.

Now X is $\mathcal{A}$-realcompact if and only if δ has $\operatorname{Hom}\mathcal{A}$ as image, and hence if and only if the image of δ is closed. □

17.4. Theorem. [Kriegl, Michor, Schachermayer, 1989, 2.4] *&* [Adam, Biström, Kriegl, 1995, 3.1]. *The topology of pointwise convergence on* $\operatorname{Hom}_\omega \mathcal{A}$ *is realcompact. If* $X_{\mathcal{A}}$ *is not realcompact then there exists an* ω*-evaluating homomorphism* φ *which is not evaluating.*

Proof. We first show the weaker statement, that: If $X_{\mathcal{A}}$ is not realcompact then there exists a non-evaluating φ, i.e., X is not $\mathcal{A}$-realcompact.
Assume that X is $\mathcal{A}$-realcompact, then $\mathcal{A}$ is 1-evaluating and hence by lemma (17.3) $\delta : X_{\mathcal{A}} \to \prod_{\mathcal{A}} \mathbb{R}$ is a closed embedding. Thus by (17.2.1) the space $X_{\mathcal{A}}$ is realcompact.

Now we give a proof of the stronger statement that $\operatorname{Hom}_\omega(\mathcal{A}, \mathbb{R})$ is realcompact:
Assume that all sets of homomorphisms are endowed with the pointwise topology. Let $\mathcal{M} \subseteq 2^{\mathcal{A}}$ be the family of all countable subsets of $\mathcal{A}$ containing the unit. For $M \in \mathcal{M}$, consider the topological space $\operatorname{Hom}_\omega\langle M\rangle$, where $\langle M\rangle$ denotes the subalgebra generated by M. Obviously the family $(\delta_f)_{f\in M}$, where $\delta_f(\varphi) = \varphi(f)$, is a countable subset of $C(\operatorname{Hom}_\omega\langle M\rangle)$ that separates the points in $\operatorname{Hom}_\omega\langle M\rangle$. Hence $\operatorname{Hom}_\omega\langle M\rangle = \operatorname{Hom} C(\operatorname{Hom}_\omega\langle M\rangle)$ by (18.25), since $C(\operatorname{Hom}_\omega\langle M\rangle)$ is ω-evaluating by (18.11), i.e. $\operatorname{Hom}_\omega\langle M\rangle$ is realcompact. Now $\operatorname{Hom}_\omega \mathcal{A}$ is an inverse limit of the spaces $\operatorname{Hom}_\omega\langle M\rangle$ for $M \in \mathcal{M}$. Since $\operatorname{Hom}_\omega\langle M\rangle$ is Hausdorff, we obtain that $\operatorname{Hom}_\omega \mathcal{A}$ as a closed subset of a product of realcompact spaces is realcompact by (17.2.2).

Since X is not realcompact in the topology $X_{\mathcal{A}}$, which is that induced from the embedding into $\operatorname{Hom}_\omega \mathcal{A}$, we have that $X \neq \operatorname{Hom}_\omega \mathcal{A}$ and the statement is proved. □

17.5. Counter-example. [Kriegl, Michor, 1993, 3.6.2]. *The locally convex space* $\mathbb{R}^\Gamma_{count}$ *of all points in the product with countable carrier is not* C^∞*-realcompact, if* Γ *is uncountable and none-measurable.*

Proof. By [Engelking, 1989, 3.10.17 & 3.11.2] the space $\mathbb{R}^\Gamma_{count}$ is not realcompact, in fact every c^∞-continuous function on it extends to a continuous function on $\mathbb{R}^\Gamma$, see the proof of (4.27). Since the projections are smooth, X_{C^∞} is the product topology. So the result follows from (17.4). □

17.6. Theorem. [Kriegl, Michor, Schachermayer, 1989, 3.2] *&* [Garrido, Gómez, Jaramillo, 1994, 1.8]. *Let* X *be a realcompact and completely regular topological space, let* $\mathcal{A}$ *be uniformly dense in* $C(X)$ *and* $\bar{1}$*-evaluating.*

Then X is $\mathcal{A}$-realcompact. Moreover, if X is $\mathcal{A}$-paracompact then $\mathcal{A}$ is uniformly dense in $C(X)$.

In [Kriegl, Michor, Schachermayer, 1989] it is shown that C^∞_{lfcs}-algebra $\mathcal{A}$ is uniformly dense in $C(X)$ if and only if $\mathcal{A} \cap C_b(X)$ is uniformly dense in $C_b(X)$. One may find also other equivalent conditions there.

Proof. Since $\mathcal{A} \subseteq C(X)$ we have that the identity $X_\mathcal{A} \to X$ is continuous, and hence $A \subseteq C(X_\mathcal{A}) \subseteq C(X)$. For each of these point-separating algebras we consider the natural inclusion δ of X into the product of factors $\mathbb{R}$ over the algebra, given by $\text{pr}_f \circ \delta = f$. It is a uniform embedding for the uniformity induced on X by this algebra and the complete product uniformity on $\prod \mathbb{R}$ with basis formed by the sets $U_{f,\varepsilon} := \{(u,v) : |\,\text{pr}_f(u) - \text{pr}_f(v)| < \varepsilon\}$ with $\varepsilon > 0$.

The condition that $\mathcal{A} \subseteq C$ is dense implies that the uniformities generated by $C(X)$, by $C(X_\mathcal{A})$ and by $\mathcal{A}$ coincide and hence we will consider X as a uniform space endowed with this uniform structure in the sequel. In fact for an arbitrarily given continuous map f and $\varepsilon > 0$ choose a $g \in \mathcal{A}$ such that $|g - f| < \varepsilon$. Then

$$\begin{aligned} \{(x,y) : |f(x) - f(y)| < \varepsilon\} &\subseteq \{(x,y) : |g(x) - g(y)| < 3\varepsilon\} \\ &\subseteq \{(x,y) : |f(x) - f(y)| < 5\varepsilon\}. \end{aligned}$$

Since $X_\mathcal{A}$ is realcompact, $\delta_C(X_\mathcal{A}) = \text{Hom}(C(X_\mathcal{A}))$ and hence is closed and so the uniform structure on X is complete. And similarly also if X is realcompact. Thus, the image $\delta_\mathcal{A}(X)$ is a complete uniform subspace of $\prod_\mathcal{A} \mathbb{R}$ and so it is closed with respect to the product topology, i.e. X is $\mathcal{A}$-realcompact by (17.3). □

17.7. In the case of a locally convex vector space the last result (17.6) can be slightly generalized to:

Result. [Biström, Lindström, 1993b, Thm.6]. *For E a realcompact locally convex vector space, let $E' \subseteq \mathcal{A} \subseteq C(E)$ be a ω-evaluating $C^{(\infty)}$-algebra which is invariant under translations and homotheties. Moreover, we assume that there exists a 0-neighborhood U in E such that for each $f \in C(E)$ there exists $g \in \mathcal{A}$ with* $\sup_{x \in U} |f(x) - g(x)| < \infty$.

Then E is $\mathcal{A}$-realcompact.

18. Evaluation Properties of Homomorphisms

In this section we consider first properties near the evaluation property at single functions, then evaluation properties for homomorphisms on countable many functions, and finally direct situations where all homomorphisms are point evaluations.

18.1. Remark. *If φ in* $\text{Hom}\,\mathcal{A}$ *is 1-evaluating (i.e., $\varphi(f) \in f(X)$ for all f in $\mathcal{A}$), then φ is $\bar{1}$-evaluating.* □

18.2. Lemma. [Biström, Bjon, Lindström, 1991, p.181]. *For a topological space X the following assertions are equivalent:*

(1) *φ is $\bar{1}$-evaluating;*
(2) *There exists $\tilde{x}$ in the Stone-Čech compactification βX with $\varphi(f) = \tilde{f}(\tilde{x})$ for all $f \in \mathcal{A}$.*

Here $\tilde{f}$ denotes the extension of $f : X \to \mathbb{R} \hookrightarrow \mathbb{R}_\infty$ to the Stone-Čech-compactification βX with values in the 1-point compactification $\mathbb{R}_\infty$ of $\mathbb{R}$.

In [Garrido, Gómez, Jaramillo, 1994, 1.3] it is shown for a subalgebra of $C_b(\mathbb{R})$ that $\tilde{x}$ need not be unique.

Proof. For $f \in \mathcal{A}$ and $\varepsilon > 0$ let $U(f, \varepsilon) := \{x \in X : |\varphi(f) - f(x)| < \varepsilon\}$. Then $\mathcal{U} := \{U(f, \varepsilon) : f \in \mathcal{A}, \varepsilon > 0\}$ is a filter basis on X. Consider X as embedded into βX and take an ultrafilter $\tilde{\mathcal{U}}$ on βX that is finer than $\mathcal{U}$. For $f := (f_1 - \varphi(f_1))^2 + (f_2 - \varphi(f_2))^2$ we have in fact

$$U(f_1, \varepsilon_1) \cap U(f_2, \varepsilon_2) \supseteq U(f, \min\{\varepsilon_1, \varepsilon_2\}^2).$$

Let $\tilde{x} \in \beta X$ be the point to which $\tilde{\mathcal{U}}$ converges. For an arbitrary function f in $\mathcal{A}$ the filter $f(\mathcal{U})$ converges to $\varphi(f)$ by construction. But $\tilde{f}(\tilde{\mathcal{U}}) \geq \tilde{f}(\mathcal{U}) = f(\mathcal{U})$, so $\varphi(f) = \tilde{f}(\tilde{x})$. The converse is obvious since $\varphi(f) = \tilde{f}(\tilde{x}) \in \tilde{f}(\beta X) \subseteq \overline{f(X)} \subseteq \mathbb{R}_\infty$, and $\varphi(f) \in \mathbb{R}$. $\square$

18.3. Lemma. [Adam, Biström, Kriegl, 1995, 4.1]. *An algebra homomorphism φ is $\bar{1}$-evaluating if and only if φ extends (uniquely) to an algebra homomorphism on $\mathcal{A}^\infty$, the C^∞-algebra generated by $\mathcal{A}$.*

Proof. For C^∞-algebras $\mathcal{A}$, we have that

$$\varphi(h \circ (f_1, \dots, f_n)) = h(\varphi(f_1), \dots, \varphi(f_n))$$

for all $h \in C^\infty(\mathbb{R}^n, \mathbb{R})$ and $f_1, \dots, f_n$ in $\mathcal{A}$.

In fact set $a := (\varphi(f_1), \dots, \varphi(f_n)) \in \mathbb{R}^n$. Then

$$h(x) - h(a) = \int_0^1 \sum_{j \leq n} \partial_j h(a + t(x - a))\, dt \cdot (x_j - a_j) = \sum_{j \leq n} h_j^a(x) \cdot (x_j - a_j),$$

where $h_j^a(x) := \int_0^1 \partial_j h(a + t(x - a))dt$. Applying φ to this equation composed with the f_i one obtains

$$\varphi(h \circ (f_1, \dots, f_n)) - h(\varphi(f_1), \dots, \varphi(f_n)) =$$
$$= \sum_{j \leq n} \varphi(h_j^a \circ (f_1, \dots, f_n)) \cdot (\varphi(f_j) - \varphi(f_j)) = 0.$$

($\Rightarrow$) We define $\tilde{\varphi}(h \circ (f_1, \dots, f_n)) := h(\varphi(f_1), \dots, \varphi(f_n))$. By what we have shown above (1-preserving) algebra homomorphisms are C^∞-algebra homomorphisms and hence this is the only candidate for an extension. This map is well defined. Indeed, let $h \circ (f_1, \dots, f_n) = k \circ (g_1, \dots, g_m)$. For each $\varepsilon > 0$ there is a point $x \in E$ such that $|\varphi(f_i) - f_i(x)| < \varepsilon$ for $i = 1, ..., n$, and $|\varphi(g_j) - g_j(x)| < \varepsilon$ for $j = 1, ..., m$. In fact by (18.2) there is a point $\tilde{x} \in \beta X$ with $\varphi(f) = \tilde{f}(\tilde{x})$ for

$$f := \sum_{i=1}^{n} (f_i - \varphi(f_i))^2 + \sum_{j=1}^{m} (g_j - \varphi(g_j))^2,$$

and hence $\varphi(f_i) = \tilde{f}_i(\tilde{x})$ and $\varphi(g_j) = \tilde{g}_j(\tilde{x})$. Now approximate $\tilde{x}$ by $x \in X$. By continuity of h and k we obtain that

$$h(\varphi(f_1), \dots, \varphi(f_n)) = k(\varphi(f_1), \dots, \varphi(f_m)),$$

and we therefore have a well defined extension of φ. This extension is a homomorphism, since for every polynomial θ on $\mathbb{R}^m$ (or even for $\theta \in C^\infty(\mathbb{R}^m)$) and $g_i := h_i \circ (f_1^i, \dots, f_{n_i}^i) \in \mathcal{A}^\infty$ we have

$$\begin{aligned}
\tilde{\varphi}(\theta \circ (g_1, \dots, g_m)) &= \tilde{\varphi}(\theta \circ (h_1 \times \dots \times h_m) \circ (f_1^1, \dots, f_{n_m}^m)) \\
&= (\theta \circ (h_1 \times \dots \times h_m))(\varphi(f_1^1), \dots, \varphi(f_{n_m}^m)) \\
&= \theta(h_1(\varphi(f_1^1), \dots, \varphi(f_{n_1}^1)), \dots, h_m(\varphi(f_1^m), \dots, \varphi(f_{n_m}^m)) \\
&= \theta(\tilde{\varphi}(g_1), \dots, \tilde{\varphi}(g_m)).
\end{aligned}$$

($\Leftarrow$) Suppose there is some $f \in \mathcal{A}$ with $\varphi(f) \notin \overline{f(X)}$. Then we may find an $h \in C^\infty(\mathbb{R})$ with $h(\varphi(f)) = 1$ and $\operatorname{carr} h \cap f(X) = \emptyset$. Since $\mathcal{A}^\infty$ is a C^∞-algebra, we conclude from what we said above that $\tilde{\varphi}(h \circ f) = h(\varphi(f)) = 1$. But since $h \circ f = 0$ we arrive at a contradiction. □

18.4. Proposition. [Garrido, Gómez, Jaramillo, 1994, 1.2]. *If $\mathcal{A}$ is bounded inversion closed and $\varphi \in \operatorname{Hom} \mathcal{A}$ then φ is $\bar{1}$-evaluating.*

Proof. We assume indirectly that there is a function $f \in \mathcal{A}$ with $\varphi(f) \notin \overline{f(X)}$. Let $\varepsilon := \inf_{x \in X} |\varphi(f) - f(x)|$ and $g(x) := \frac{1}{\varepsilon}(\varphi(f) - f(x))$. Then $g \in \mathcal{A}$, $\varphi(g) = 0$ and $|g(x)| = \frac{1}{\varepsilon}|\varphi(f) - f(x)| \geq 1$ for each $x \in X$. Thus $1/g \in \mathcal{A}$. But then $1 = \varphi(g \cdot 1/g) = \varphi(g)\varphi(1/g) = 0$ gives a contradiction. □

18.5. Lemma. *Any $C^{(\infty)}$-algebra is bounded inversion closed.*

Moreover, it is stable under composition with smooth locally defined functions, which contain the closure of the image in its domain of definition.

Proof. Let $\mathcal{A}$ be a C^∞-algebra (resp. $C^{(\infty)}$-algebra), n a natural number (resp. $n = 1$), $U \subseteq \mathbb{R}^n$ open, $h \in C^\infty(U, \mathbb{R})$, $f := (f_1, \dots, f_n)$, with $f_i \in \mathcal{A}$ such that $\overline{f(X)} \subseteq U$, then $h \circ f \in \mathcal{A}$. Indeed, choose $\rho \in C^\infty(\mathbb{R})$ with $\rho|_{\overline{f(X)}} = 1$ and $\operatorname{supp} \rho \subseteq U$. Then $k := \rho \cdot h$ is a globally smooth function and $h \circ f = k \circ f \in \mathcal{A}$. □

18.6. Lemma. *Any inverse closed algebra $\mathcal{A}$ is 1-evaluating.*

By (18.10) the converse is wrong.

Proof. Let $f \in \mathcal{A}$ and assume indirectly that $Z_f = \emptyset$. Let $g := f - \varphi(f)$. Then $g \in \mathcal{A}$ and $g(x) \neq 0$ for all $x \in X$, by which $1/g \in \mathcal{A}$ since $\mathcal{A}$ is inverse-closed. But then $1 = \varphi(g \cdot 1/g) = \varphi(g)\varphi(1/g) = 0$, which is a contradiction. □

18.7. Proposition. [Biström, Jaramillo, Lindström, 1995, Lem.14] *&* [Adam, Biström, Kriegl, 1995, 4.2]. *For φ in* Hom $\mathcal{A}$ *the following statements are equivalent:*

(1) *φ is 1-evaluating.*
(2) *φ extends to a unique (1-evaluating) homomorphism on the algebra $\mathcal{RA} := \{f/g :\ f, g \in \mathcal{A},\ 0 \notin g(X)\}$.*
(3) *φ extends to a unique (1-evaluating) homomorphism on the following C^∞-algebra $\mathcal{A}^{\langle\infty\rangle}$ constructed from $\mathcal{A}$:*

$$\mathcal{A}^{\langle\infty\rangle} := \{h \circ (f_1, \dots, f_n) : f_i \in \mathcal{A}, (f_1, \dots, f_n)(X) \subseteq U,\\ U \text{ open in some } \mathbb{R}^n, h \in C^\infty(U)\}.$$

Proof. (1) ⇒ (3) We define $\varphi(h \circ (f_1, \dots, f_n)) := h(\varphi(f_1), \dots, \varphi(f_n))$. Since there exists by (18.8) an x with $\varphi(f_i) = f_i(x)$, we have $(\varphi(f_1), \dots, \varphi(f_n)) \in U$, hence the right side makes sense. The rest follows in the same way as in the proof of (18.3).

(3) ⇒ (2) Existence is obvious, since $\mathcal{RA} \subseteq \mathcal{A}^{\langle\infty\rangle}$, and uniqueness follows from the definition of $\mathcal{RA}$.

(2) ⇒ (1) Since $\mathcal{RA}$ is inverse-closed, the extension of φ to this algebra is 1-evaluating by (18.6), hence the same is true for φ on $\mathcal{A}$. □

18.8. Lemma. *Every 1-evaluating homomorphism is finitely evaluating.*

Proof. Let $\mathcal{F}$ be a finite subset of $\mathcal{A}$. Define a function $f : X \to \mathbb{R}$ by

$$f := \sum_{g \in \mathcal{F}} (g - \varphi(g))^2.$$

Then $f \in \mathcal{A}$ and $\varphi(f) = 0$. By assumption there is a point $x \in X$ with $\varphi(f) = f(x)$. Hence $g(x) = \varphi(g)$ for all $g \in \mathcal{F}$. □

18.9. Theorem. Automatic boundedness. [Kriegl, Michor, 1993] *&* [Arias-de-Reyna, 1988] *Every 1-evaluating homomorphism $\varphi \in$* Hom $\mathcal{A}$ *is positive, i.e., $0 \leq \varphi(f)$ for all $0 \leq f \in \mathcal{A}$. Moreover we even have $\varphi(f) > 0$ for $f \in \mathcal{A}$ with $f(x) > 0$ for all $x \in X$.*

Every positive homomorphism $\varphi \in$ Hom $\mathcal{A}$ *is bounded for any convenient algebra structure on $\mathcal{A}$.*

A convenient algebra structure on A is a locally convex topology, which turns $\mathcal{A}$ into a convenient vector space and such that the multiplication $\mathcal{A} \times \mathcal{A} \to \mathcal{A}$ is bounded, compare (5.21).

Proof. Positivity: Let $f_1 \leq f_2$. By (17) and (18.8) there exists an $x \in X$ such that $\varphi(f_i) = f_i(x)$ for $i = 1, 2$. Thus $\varphi(f_1) = f_1(x) \leq f_2(x) = \varphi(f_2)$. Note that if $f(x) > 0$ for all x, then $\varphi(f) > 0$.

Boundedness: Suppose f_n is a bounded sequence, but $|\varphi(f_n)|$ is unbounded. By replacing f_n by f_n^2 we may assume that $f_n \geq 0$ and hence also $\varphi(f_n) \geq 0$. Choosing a subsequence we may even assume that $\varphi(f_n) \geq 2^n$. Now consider $\sum_n \frac{1}{2^n} f_n$. This series converges Mackey, and since the bornology on $\mathcal{A}$ is by assumption complete the limit is an element $f \in \mathcal{A}$. Applying φ yields

$$\varphi(f) = \varphi\left(\sum_{n=0}^{N} \frac{1}{2^n} f_n + \sum_{n>N} \frac{1}{2^n} f_n\right) = \sum_{n=0}^{N} \frac{1}{2^n} \varphi(f_n) + \varphi\left(\sum_{n>N} \frac{1}{2^n} f_n\right) \geq$$
$$\geq \sum_{n=0}^{N} \frac{1}{2^n} \varphi(f_n) + 0 = \sum_{n=0}^{N} \frac{1}{2^n} \varphi(f_n),$$

where we used the monotonicity of φ applied to $\sum_{n>N} \frac{1}{2^n} f_n \geq 0$. Thus the series $N \mapsto \sum_{n=0}^{N} \frac{1}{2^n} \varphi(f_n)$ is bounded and increasing, hence converges, but its summands are bounded by 1 from below. This is a contradiction. □

18.10. Lemma. *For a locally convex vector space E the algebra $P_f(E)$ is 1-evaluating.*

More on the algebra $P_f(E)$ can be found in (18.27), (18.28), and (18.12).

Proof. Every finite type polynomial p is a polynomial in a finite number of linearly independent functionals $\ell_1, \ldots, \ell_n$ in E'. So there is for each $i = 1, \ldots, n$ some point $a_i \in E$ such that $\ell_i(a_i) = \varphi(\ell_i)$ and $\ell_j(a_i) = 0$ for all $j \neq i$. Let $a = a_1 + \cdots + a_n \in E$. Then $\ell_i(a) = \ell_i(a_i) = \varphi(\ell_i)$ for $i = 1, \ldots, n$ hence $\varphi(p) = p(a)$. □

Countably Evaluating Homomorphisms

18.11. Theorem. *Idea of* [Arias-de-Reyna, 1988, proof of thm.8], [Adam, Biström, Kriegl, 1995, 2.5]. *For a topological space X any C^∞_{lfcs}-algebra $\mathcal{A} \subseteq C(X)$ is closed under composition with local smooth functions and is ω-evaluating.*

Note that this does not apply to C^ω.

Proof. We first show closedness under local smooth functions (and hence in particular under inversion), i.e. if $h \in C^\infty(U)$, where $U \subseteq \mathbb{R}^n$ is open and $f := (f_1, \ldots, f_n)$ with $f_i \in \mathcal{A}$ has values in U, then $h \circ f \in \mathcal{A}$.
Consider a smooth partition of unity $\{h_j : j \in \mathbb{N}\}$ of U, such that $\operatorname{supp} h_j \subseteq U$. Then $h_j \cdot h$ is a smooth function on $\mathbb{R}^n$ vanishing outside $\operatorname{supp} h_j$. Hence $(h_j \cdot h) \circ f \in \mathcal{A}$. Since we have

$$\operatorname{carr}\Big((h_j \cdot h) \circ f\Big) \subseteq f^{-1}(\operatorname{carr} h_j),$$

the family $\{\operatorname{carr}((h_j \cdot h) \circ f) : j \in \mathbb{N}\}$ is locally finite, f is continuous, and since $1 = \sum_{j\in\mathbb{N}} h_j$ on U we obtain that $h \circ f = \sum_{j\in\mathbb{N}}(h_j \cdot h) \circ f \in \mathcal{A}$.

By (18.6) we have that φ is 1-evaluating, hence finitely evaluating by (18.8). We now show that φ is countably evaluating:

For this take a sequence $(f_n)_n$ in $\mathcal{A}$. Then $h_n : x \mapsto (f_n(x) - \varphi(f_n))^2$ belongs to $\mathcal{A}$ and $\varphi(h_n) = 0$. We have to show that there exists an $x \in X$ with $h_n(x) = 0$ for all n. Assume that this were not true, i.e. for all $x \in X$ there exists an n with $h_n(x) > 0$. Take $h \in C^\infty(\mathbb{R}, [0,1])$ with $\operatorname{carr} h = \{t : t > 0\}$ and let $g_n : x \mapsto h(h_n(x)) \cdot h(\frac{1}{n} - h_1(x)) \cdot \dots \cdot h(\frac{1}{n} - h_{n-1}(x))$. Then $g_n \in \mathcal{A}$ and the sum $\sum_n \frac{1}{2^n} g_n$ is locally finite, hence defines a function $g \in \mathcal{A}$. Since φ is 1-evaluating there exists for any n an $x_n \in X$ with $h_n(x_n) = \varphi(h_n) = 0$ and $\varphi(g_n) = g_n(x_n)$. Hence

$$\varphi(g_n) = g_n(x_n) = h(h_n(x_n)) \cdot h(\tfrac{1}{n} - h_1(x_n)) \cdot \dots \cdot h(\tfrac{1}{n} - h_{n-1}(x_n)) = 0.$$

By assumption on the h_n and h we have that $g > 0$. Hence by (18.9) $\varphi(g) > 0$, since φ is 1-evaluating. Let N be so large that $1/2^N < \varphi(g)$. Again since $\mathcal{A}$ is 1-evaluating, there is some $a \in X$ such that $\varphi(g) = g(a)$ and $\varphi(g_j) = g_j(a)$ for $j \le N$. Then

$$\frac{1}{2^N} < \varphi(g) = g(a) = \sum_n \frac{1}{2^n} g_n(a) = \sum_{n\le N} \frac{1}{2^n}\varphi(g_n) + \sum_{n>N} \frac{1}{2^n} g_n(a) \le 0 + \frac{1}{2^N}$$

gives a contradiction. □

18.12. Counter-example. [Biström, Jaramillo, Lindström, 1995, Prop.17]. *For any non-reflexive weakly realcompact locally convex space (and any non-reflexive Banach space) E the algebra $P_f(E)$ of finite type polynomials is not ω-evaluating.*

Moreover, $E_{\mathcal{A}}$ is realcompact, but E is not $\mathcal{A}$-realcompact, for $\mathcal{A} = P_f(E)$, so that the converse of the assertion in (17.4) holds only under the additional assumptions of (17.6).

As example we may take $E = \ell^1$, which is non-reflexive, but by (18.27) weakly realcompact.

By (18.10) the algebra $P_f(E)$ is 1-evaluating and hence by (18.7) it has the same homomorphisms as $RP_f(E)$, $P_f(E)^\infty$ or even $P_f(E)^{\langle\infty\rangle}$. So these algebras are not ω-evaluating for spaces E as above.

Proof. By the universal property (5.10) of $P_f(E)$ we get $\operatorname{Hom} P_f(E) \cong (E')^\times$, the space of (not necessarily bounded) linear functionals on E'. For weakly realcompact E by (18.27) we have $\operatorname{Hom}_\omega P_f(E) = E$. So if $P_f(E)$ were ω-evaluating then even $E = \operatorname{Hom} P_f(E)$. Any bounded subset of E is obviously P_f-bounding and hence by (20.2) relatively compact in the weak topology, since $E_{P_f(E)} = (E, \sigma(E, E'))$. Since E is not semi-reflexive, this is a contradiction, see [Jarchow, 1981, 11.4.1].

If we have a (not necessarily weakly compact) Banach space, we can replace in the argument above (20.2) by the following version given in [Biström, 1993, 5.10]: If $\operatorname{Hom}_\omega P_f(E) = \operatorname{Hom} P_f(E)$ then every $\mathcal{A}$-bounding set with complete closed convex hull is relatively compact in the weak topology. □

18.13. Lemma. *The C^∞_{lfs}-algebra $\mathcal{A}^\infty_{\text{lfs}}$ generated by an algebra $\mathcal{A}$ can be obtained in two steps as $(\mathcal{A}^\infty)_{\text{lfs}}$. Also the C^∞_{lfcs}-algebra $\mathcal{A}^\infty_{\text{lfcs}}$ can be obtained in two steps as $(\mathcal{A}^\infty)_{\text{lfcs}}$.*

Proof. We prove the result only for countable sums, the general case is easier. We have to show that $(\mathcal{A}^\infty)_{\text{lfcs}}$ is closed under composition with smooth mappings. So take $h \in C^\infty(\mathbb{R}^n)$ and $\sum_{j\geq 1} f_{i,j} \in (\mathcal{A}^\infty)_{\text{lfcs}}$ for $i = 1, \dots, n$. We put $h_0 := 0$ and $h_k := h \circ (\sum_{j=1}^k f_{1,j}, \dots, \sum_{j=1}^k f_{n,j}) \in \mathcal{A}^\infty$ and obtain

$$h \circ (\sum_{j\geq 1} f_{1,j}, \dots, \sum_{j\geq 1} f_{n,j}) = \sum_{k\geq 1}(h_k - h_{k-1}),$$

where the right member is locally finite and hence an element of $(\mathcal{A}^\infty)_{\text{lfcs}}$. □

18.14. Theorem. [Adam, Biström, Kriegl, 1995, 4.3]. *A homomorphism φ in* Hom $\mathcal{A}$ *is ω-evaluating if and only if φ extends (uniquely) to an algebra homomorphism on the C^∞_{lfcs}-algebra $\mathcal{A}^\infty_{\text{lfcs}}$ generated by $\mathcal{A}$, which can be obtained in two steps as $(\mathcal{A}^\infty)_{\text{lfcs}}$ (and this extension is ω-evaluating by* (18.11)*).*

Proof. ($\Rightarrow$) The algebra $\mathcal{A}^\infty_{\text{lfcs}}$ is the union of the algebras obtained by a finite iteration of passing to $\mathcal{A}_{lfcs}$ and $\mathcal{A}^\infty$, where $\mathcal{A}_{\text{lfcs}} := \{f : f = \sum_n f_n, f_n \in \mathcal{A}, \text{ the sum is locally finite}\}$. To $\mathcal{A}^\infty$ it extends by (18.3). It is countably evaluating there, since in any $f \in \mathcal{A}^\infty$ only finitely many elements of $\mathcal{A}$ are involved. Remains to show that φ can be extended to $\mathcal{A}_{lfcs}$ and that this extension is also countably evaluating.
For a locally finite sum $f = \sum_k f_k$ we define $\varphi(f) := \sum_k \varphi(f_k)$. This makes sense, since there exists an $x \in X$ with $\varphi(f_n) = f_n(x)$, and since $\sum_n f_n$ is point finite, we have that the sum $\sum_n \varphi(f_n) = \sum_n f_n(x)$ is in fact finite. It is well defined, since for $\sum_n f_n = \sum_n g_n$ we can choose an $x \in X$ with $\varphi(f_n) = f_n(x)$ and $\varphi(g_n) = g_n(x)$ for all n, and hence $\sum_n \varphi(f_n) = \sum_n f_n(x) = \sum_n g_n(x) = \sum_n \varphi(g_n)$. The extension is a homomorphism, since for the product for example we have

$$\varphi\Big((\sum_n f_n)(\sum_k g_k)\Big) = \varphi\Big(\sum_{n,k} f_n\, g_k\Big) = \sum_{n,k} \varphi(f_n\, g_k) =$$
$$= \sum_{n,k} \varphi(f_n)\, \varphi(g_k) = \Big(\sum_n \varphi(f_n)\Big)\Big(\sum_k \varphi(g_k)\Big).$$

Remains to show that the extension is countably evaluating. So let $f^k = \sum_n f_n^k$ be given. By assumption there exists an x such that $\varphi(f_n^k) = f_n^k(x)$ for all n and all k. Thus $\varphi(f^k) = \sum_n \varphi(f_n^k) = \sum_n f_n^k(x) = f^k(x)$ for all k.

($\Leftarrow$) Since $\mathcal{A}^\infty_{lfcs}$ is a C^∞_{lfcs}-algebra we conclude from (18.11) that the extension of φ is countably evaluating. □

18.15. Proposition. [Garrido, Gómez, Jaramillo, 1994, 1.10]. *Let φ in* Hom $\mathcal{A}$ *be 1-evaluating, and let $f_n \in \mathcal{A}$ be such that $\sum_n \lambda_n f_n^j \in \mathcal{A}$ for all $\lambda \in \ell^1$ and $j \in \{1, 2\}$.*

Then φ is $\{f_n : n \in \mathbb{N}\}$-evaluating.

For a convenient algebra structure on $\mathcal{A}$ and $\{f_n : n \in \mathbb{N}\}$ bounded in $\mathcal{A}$ the second condition holds, as used in (18.26).

It would be interesting to know if the assumption for $j = 2$ can be removed, since then the application in (18.26) to finite type polynomials would work.

Proof. Choose a positive absolutely summable sequence $(\lambda_n)_{n\in\mathbb{N}}$ such that the sequences $(\lambda_n\,\varphi(f_n))_{n\in\mathbb{N}}$ and $(\lambda_n\,\varphi(f_n)^2)_{n\in\mathbb{N}}$ are summable. Then the sum

$$g := \sum_{j=1}^{\infty} \lambda_j (f_j - \varphi(f_j))^2 \in \mathcal{A}.$$

If there exists $x \in X$ with $g(x) = 0$, we are done. If not, then consider the (positive) function

$$h := \sum_{j=1}^{\infty} \frac{1}{2^j} \lambda_j (f_j - \varphi(f_j))^2 \in \mathcal{A}.$$

For every $n \in \mathbb{N}$ there exists $x_n \in X$ such that $\varphi(f_j) = f_j(x_n)$ for all $j \le n$, $\varphi(g) = g(x_n)$ and $\varphi(h) = h(x_n)$. But then for all $n \in \mathbb{N}$ we have by (18.9) that

$$0 < 2^n \varphi(h) = \sum_{j>n} 2^{n-j} \lambda_j\, \varphi(f_j - \varphi(f_j))^2 \le \sum_{j>n} \lambda_j\, \varphi(f_j - \varphi(f_j))^2 = \varphi(g),$$

a contradiction. □

18.16. Corollary. [Biström, Jaramillo, Lindström, 1995, Prop.9]. *Let E be a Banach space and $\mathcal{A}$ a 1-evaluating algebra containing $P(E)$. Then for each $\varphi \in$* Hom $\mathcal{A}$, *each $f \in \mathcal{A}$, and each sequence $(p_n)_{n\in\mathbb{N}}$ in $P(E)$ with uniformly bounded degree, there exists $a \in E$ with $\varphi(f) = f(a)$ and $\varphi(p_n) = p_n(a)$ for all $n \in \mathbb{N}$.*

Proof. Let $(\lambda_n)_{n\in\mathbb{N}}$ be a sequence of positive reals such that $\{\lambda_n p_n : n \in \mathbb{N}\}$ is bounded. Then by (18.15) the set $\{f, p_n\}$ is evaluated. □

18.17. Theorem. [Adam, Biström, Kriegl, 1995, 3.3]. *Let $(f_\gamma)_{\gamma\in\Gamma}$ be a family in $\mathcal{A}$ such that $\sum_{\gamma\in\Gamma} z_\gamma f_\gamma^j$ is a pointwise convergent sum in $\mathcal{A}$ for all $z = (z_\gamma) \in \ell^\infty(\Gamma)$ and $j = 1, 2$. Let $|\Gamma|$ be non-measurable, and let φ be ω-evaluating.*

Then φ is $\{f_\gamma : \gamma \in \Gamma\}$-evaluating.

We will apply this in particular if $\{f_\gamma : \gamma \in \Gamma\}$ is locally finite, and $\mathcal{A}$ stable under locally finite sums. Note that we can always add finitely many $f \in \mathcal{A}$ to $\{f_\gamma : \gamma \in \Gamma\}$.

Again it would be nice to get rid of the assumption for $j = 2$.

Proof. Let $x \in X$ and set $z_\gamma := \operatorname{sign}(f_\gamma(x))$ for all $\gamma \in \Gamma$. Then $z = (z_\gamma) \in \ell^\infty(\Gamma)$ and $\sum_{\gamma\in\Gamma} |f_\gamma(x)| = \sum_{\gamma\in\Gamma} z_\gamma f_\gamma(x) < \infty$, i.e. $(f_\gamma(x))_{\gamma\in\Gamma} \in \ell^1(\Gamma)$. Next observe that $(\varphi(f_\gamma))_{\gamma\in\Gamma} \in c_0(\Gamma)$, since otherwise there exists some $\varepsilon > 0$ and a countable set $\Lambda \subseteq \Gamma$ with $|\varphi(f_\gamma)| \geq \varepsilon$ for each $\gamma \in \Lambda$. By the countably evaluating property of φ there is a point $x \in X$ with $|f_\gamma(x)| = |\varphi(f_\lambda| \geq \varepsilon$ for each $\gamma \in \Lambda$, violating the condition $(f_\gamma(x))_{\gamma\in\Gamma} \in \ell^1(\Gamma)$. Since as a vector in $c_0(\Gamma)$ it has countable support and since φ is countably evaluating we get even $(\varphi(f_\gamma))_{\gamma\in\Gamma} \in \ell^1(\Gamma)$. Therefore we may consider g, defined by

$$X \ni x \mapsto g(x) := \Big((f_\gamma(x) - \varphi(f_\gamma))^2\Big)_{\gamma\in\Gamma} \in \ell^1(\Gamma).$$

This gives a map $g^* : \ell^\infty(\Gamma) = \ell^1(\Gamma)' \to \mathcal{A}$, by

$$g^*(z) : x \mapsto \langle z, g(x)\rangle = \sum_{\gamma\in\Gamma} z_\gamma \cdot (f_\gamma(x) - \varphi(f_\gamma))^2,$$

since $(\varphi(f_\gamma)_{\gamma\in\Gamma} \in \ell^1(\Gamma)$. Let $\Phi : \ell^\infty(\Gamma) \to \mathbb{R}$ be the linear map $\Phi := \varphi \circ g^* : \ell^\infty(\Gamma) \to \mathcal{A} \to \mathbb{R}$. By the countably evaluating property of φ, for any sequence (z_n) in $\ell^\infty(\Gamma)$ there exists an $x \in X$ such that $\Phi(z_n) = \varphi(g^*(z_n)) = g^*(z_n)(x) = \langle z_n, g(x)\rangle$ for all n. For non-measurable $|\Gamma|$ the weak topology on $\ell^1(\Gamma)$ is realcompact by [Edgar, 1979, p.575]. By (18.19) there exists a point $c \in \ell^1(\Gamma)$ such that $\Phi(z) = \langle z, c\rangle$ for all $z \in \ell^\infty(\Gamma)$. For each standard unit vector $e_\gamma \in \ell^\infty(\Gamma)$ we have $0 = \Phi(e_\gamma) = \langle e_\gamma, c\rangle = c_\gamma$. Hence $c = 0$ and therefore $\Phi = 0$. For the constant vector $\mathbf{1}$ in $\ell^\infty(\Gamma)$, we get $0 = \Phi(\mathbf{1}) = \varphi(g^*(\mathbf{1}))$. Since φ is 1-evaluating there exists an $a \in X$ with $\varphi(g^*(\mathbf{1})) = g^*(\mathbf{1})(a) = \langle \mathbf{1}, g(a)\rangle = \sum_{\gamma\in\Gamma}(f_\gamma(a) - \varphi(f_\gamma))^2$, hence $\varphi(f_\gamma) = f_\gamma(a)$ for each $\gamma \in \Gamma$. □

18.18. Valdivia gives in [Valdivia, 1982] a characterization of the locally convex spaces which are realcompact in their weak topologies. Let us mention some classes of spaces that are *weakly realcompact*:

Result.

(1) *All locally convex spaces E with $\sigma(E', E)$-separable E'.*

(2) *All weakly Lindelöf locally convex spaces, and hence in particular all weakly countably determined Banach spaces, see* [Vašák, 1981]. *In particular this applies to $c_0(X)$ for locally compact metrizable X by* [Corson, 1961, p.5].

(3) *The Banach spaces E with angelic weak* dual unit ball* [Edgar, 1979, p.564].

Note that $(E^*, weak^*)$ is angelic $:\Leftrightarrow$ for $B \subseteq E^*$ bounded the weak-closure is obtained by weak*-convergent sequences in B, i.e. sequentially for the weak*-topology.*

(4) *$\ell^1(\Gamma)$ for $|\Gamma|$ non-measurable. Furthermore the spaces $C[0,1]$, ℓ^∞, $L^\infty[0,1]$, the space JL of* [Johnson, Lindenstrauss, 1974] *(a short exact sequence $c_0 \to JL \to \ell^2(\Gamma)$ exists), the space $D[0,1]$ or right-continuous functions having left sided limits, by* [Edgar, 1979, p.575] *and* [Edgar, 1977]. *All these spaces are not weakly Lindelöf.*

(5) *All closed subspaces of products of the spaces listed above.*

(6) *Not weakly realcompact are $C[0,\omega_1]$ and $\ell^\infty_{\text{count}}[0,1]$, the space of bounded functions on $[0,1]$ with countable support, by* [Edgar, 1979].

18.19. Lemma. [Corson, 1961]. *If E is a weakly realcompact locally convex space, then every linear countably evaluating $\Phi : E' \to \mathbb{R}$ is given by a point-evaluation ev_x on E' with $x \in E$.*

Proof. Since $\Phi : E' \to \mathbb{R}$ is countably evaluating it is linear and $\mathcal{F} := \{Z_K : K \subseteq E' \text{ countable}\}$ does not contain the empty set and generates a filter. We claim that this filter is Cauchy with respect to the uniformity defined by the weakly continuous real functions on E:

To see this, let $f : E \to \mathbb{R}$ be weakly continuous. For each $r \in \mathbb{R}$, let $L_r := \{x \in E : f(x) < r\}$ and similarly $U_r := \{x \in E : f(x) > r\}$. By [Jarchow, 1981, 8.1.4] we have that E is $\sigma(E'^*, E')$-dense in E'^*. Thus there are open disjoint subsets $\tilde{L}_r$ and $\tilde{U}_r$ on E'^* having trace L_r and U_r on E (take the complements of the closures of the complements). Let $\mathcal{B} \subseteq E'$ be an algebraic basis of E'. Then the map $\chi : E'^* \to \mathbb{R}^{\mathcal{B}}$, $l \mapsto (l(x'))_{x' \in \mathcal{B}}$ is a topological isomorphism for $\sigma(E'^*, E')$. By [Bockstein, 1948] there exists a countable subset $K_r \subseteq \mathcal{B} \subseteq E'$, such that the images under $\mathrm{pr}_{K_r} : \mathbb{R}^{\mathcal{B}} \to \mathbb{R}^{K_r}$ of the open sets $\tilde{L}_r$ and $\tilde{U}_r$ are disjoint. Let $K = \bigcup_{r \in \mathbb{Q}} K_r$. For $\varepsilon > 0$ we have that $Z_K \times Z_K \subseteq \{(x_1, x_2) : f(x_1) = f(x_2)\} \subseteq \{(x_1, x_2) : |f(x_1) - f(x_2)| < \varepsilon\}$, i.e. the filter generated by $\mathcal{F}$ is Cauchy. In fact, let $x_1, x_2 \in Z_K$. Then $x'(x_1) = \varphi(x') = x'(x_2)$ for all $x' \in K$. Suppose $f(x_1) \neq f(x_2)$. Without loss of generality we find a $r \in \mathbb{Q}$ with $f(x_1) < r < f(x_2)$, i.e. $x_1 \in L_r$ and $x_2 \in U_r$. But then $x'(x_1) \neq x'(x_2)$ for all $x' \in K_r \subseteq K$ gives a contradiction.

By realcompactness of $(E, \sigma(E, E'))$ the uniform structure generated by the weakly continuous functions $E \to \mathbb{R}$ is complete (see [Gillman, Jerison, 1960, p.226]) and hence the filter $\mathcal{F}$ converges to a point $a \in E$. Thus $a \in Z_K$ for all countable $K \subseteq E'$, and in particular $\Phi(x') = x'(a)$ for all $x' \in E'$. □

18.20. Proposition. [Biström, Jaramillo, Lindström, 1995, Thm.10]. *Let E be a Banach space, let $\mathcal{A} \supseteq C^{\omega}_{\mathrm{conv}}(E)$ be 1-evaluating, let $f \in \mathcal{A}$, and let $\mathcal{F}$ be a countable subset of $C^{\omega}_{\mathrm{conv}}(E)$.*

Then $\{f\} \cup \mathcal{F}$ is evaluating. In particular, $\mathcal{R}C^{\omega}_{\mathrm{conv}}(E)$ (see (18.7.2)) is ω-evaluating for every Banach space E.

Proof. Let $(p_n)_{n \in \mathbb{N}}$ be a sequence in $P(E)$ and $(k_n)_{n \in \mathbb{N}}$ a sequence of odd natural numbers with $k_1 = 1$ and $k_{n+1} > 2k_n(1 + \deg p_n)$ for $n \in \mathbb{N}$. Then $|p_n^{k_n}(x)| \leq \|p_n\|^{k_n} \cdot \|x\|^{k_n \deg p_n}$ for every $x \in E$. Set

$$g := \sum_{n=1}^{\infty} \frac{1}{\lambda_n} \cdot \frac{1}{2^n} \cdot \frac{1}{n^{2k_n \deg p_n}} (p_n^{k_n} - \varphi(p_n^{k_n}))^2,$$

where $(\lambda_n)_{n \in \mathbb{N}}$ is a sequence of reals with

$$\lambda_n > \|p_n\|^{2k_n} + 2|\varphi(p_n^{k_n})| \cdot \|p_n\|^{k_n} + (\varphi(p_n^{2k_n}))^2 \text{ for all } n \in \mathbb{N}.$$

Then

$$\begin{aligned} g(x) &\leq \sum_{n=1}^{\infty} \frac{1}{2^n} \cdot \frac{1}{n^{2k_n \deg p_n}} \left(\|x\|^{k_n 2 \deg p_n} + \|x\|^{k_n \deg p_n} + 1\right) \\ &\leq \sum_{n=1}^{\infty} \frac{1}{2^n} \left(\|\frac{x}{n}\|^{2k_n \deg p_n} + \|\frac{x}{n}\|^{k_n \deg p_n} + 1\right) < \infty \text{ for all } x \in E \end{aligned}$$

Since g is pointwise convergent, it is a function in $C^\omega_{\text{conv}}(E)$. By the technique used in (18.15) we obtain that there exists $x \in E$ with $\varphi(f) = f(x)$ and $\varphi(p_n^{k_n}) = p_n^{k_n}(x)$ for all $n \in \mathbb{N}$. As for each $n \in \mathbb{N}$ the number k_n is odd, it follows that $\varphi(p_n) = p_n(x)$ for all $n \in \mathbb{N}$. Since each $g \in \mathcal{F}$ is a sum $\sum_{n\in\mathbb{N}} p_{n,g}$ of homogeneous polynomials $p_{n,g} \in P(E)$ of degree n for $n \in \mathbb{N}$, there exists $x \in E$ with $\varphi(g) = g(x)$ for all $g \in \mathcal{F}$, and $\varphi(p_{n,g}) = p_{n,g}(x)$ for all $n \in \mathbb{N}$, whence $\varphi(g) = \sum_{n\in\mathbb{N}} \varphi(p_{n,g})$ for all $g \in \mathcal{F}$. Let $a \in E$ with $\varphi(f) = f(a)$ and $\varphi(p_{n,g}) = p_{n,g}(a)$ for all $n \in \mathbb{N}$ and all $g \in \mathcal{F}$. Then

$$\varphi(g) = \sum_{n\in\mathbb{N}} \varphi(p_{n,g}) = \sum_{n\in\mathbb{N}} p_{n,g}(a) = g(a) \text{ for all } g \in \mathcal{F}. \quad \square$$

18.21. Result. [Adam, Biström, Kriegl, 1995, 2.1]. *Given two infinite cardinals* $\mathfrak{m} < \mathfrak{n}$, *let* $E := \{x \in \mathbb{R}^{\mathfrak{n}} : |\operatorname{supp} x| \le \mathfrak{m}\}$ *Then for any algebra* $\mathcal{A} \subseteq C(E)$, *containing the natural projections* $(\operatorname{pr}_\gamma)_{\gamma\in\mathfrak{n}}$, *there is a homomorphism* φ *on* $\mathcal{A}$ *that is* $\mathfrak{m}$*-evaluating but not* $\mathfrak{n}$*-evaluating.*

Evaluating Homomorphisms

18.22. Proposition. [Garrido, Gómez, Jaramillo, 1994, 1.7]. *Let* X *be a closed subspace of a product* $\mathbb{R}^\Gamma$. *Let* $\mathcal{A} \subseteq C(X)$ *be a subalgebra containing the projections* $\operatorname{pr}_\gamma|_X : X \subseteq \mathbb{R}^\Gamma \to \mathbb{R}$, *and let* $\varphi \in \operatorname{Hom}\mathcal{A}$ *be* $\bar{1}$*-evaluating.*

Then φ *is* $\mathcal{A}$*-evaluating.*

Proof. Set $a_\gamma = \varphi(\operatorname{pr}_\gamma|_X)$. Then the point $a = (a_\gamma)_{\gamma\in\Gamma}$ is an element in X. Otherwise, since X is closed there exists a finite set $J \subseteq \Gamma$ and $\varepsilon > 0$ such that no point y with $|y_\gamma - a_\gamma| < \varepsilon$ for all $\gamma \in J$ is contained in X. Set $p(x) := \sum_{\gamma\in J}(\operatorname{pr}_\gamma(x) - a_\gamma)^2$ for $x \in X$. Then $p \in \mathcal{A}$ and $\varphi(p) = 0$. By assumption there is an $x \in X$, such that $|\varphi(p) - p(x)| < \varepsilon^2$, but then $|\operatorname{pr}_\gamma(x) - a_\gamma| < \varepsilon$ for all $\gamma \in J$, a contradiction. Thus $a \in X$ and $\varphi(g) = g(a)$ for all g in the algebra $\mathcal{A}_0$ generated by all functions $\operatorname{pr}_\gamma|_X$.

By the assumption and by (18.2) there exists a point $\tilde{x}$ in the Stone-Čech compactification βX such that $\varphi(f) = \tilde{f}(\tilde{x})$ for all $f \in \mathcal{A}$, where $\tilde{f}$ is the unique continuous extension $\beta X \to \mathbb{R}_\infty$ of f. We claim that $\tilde{x} = a$. This holds if $\tilde{x} \in X$ since the pr_γ separate points on X. So let $\tilde{x} \in \beta X \setminus X$. Then $\tilde{x}$ is the limit of an ultrafilter $\mathcal{U}$ in X. Since $\mathcal{U}$ does not converge to a, there is a neighborhood of a in X, without loss of generality of the form $U = \{x \in X : f(x) > 0\}$ for some $f \in \mathcal{A}_0$. But then the complement of U is in the ultrafilter $\mathcal{U}$, thus $\tilde{f}(\tilde{x}) \le 0$. But this contradicts $\tilde{f}(\tilde{x}) = \varphi(f) = f(a)$ for all $f \in \mathcal{A}_0$. $\square$

18.23. Corollary. [Kriegl, Michor, 1993, 1]. *If* $\mathcal{A}$ *is finitely generated then each 1-evaluating* $\varphi \in \operatorname{Hom}\mathcal{A}$ *is evaluating.*

Finitely generated can even be meant in the sense of $C^{\langle\infty\rangle}$-algebra, see the proof. This applies to the algebras $\mathcal{R}P$, C^ω, C^ω_{conv} and C^∞ on $\mathbb{R}^n$ (or a closed submanifold of $\mathbb{R}^n$).

Proof. Let $\mathcal{F} \subseteq \mathcal{A}$ be a finite subset which generates $\mathcal{A}$ in the sense that $\mathcal{A} \subseteq \mathcal{F}^{\langle\infty\rangle} := (\langle\mathcal{F}\rangle_{\mathrm{Alg}})^{\langle\infty\rangle}$, compare (18.7.3). By (18.7) again we have that φ restricted to $\langle\mathcal{F}\rangle_{\mathrm{Alg}}$ extends to $\tilde{\varphi} \in \operatorname{Hom} \mathcal{F}^{\langle\infty\rangle}$ by $\varphi(h \circ (f_1, \dots, f_n)) = h(\varphi(f_1), \dots, \varphi(f_n))$ for $f_i \in \mathcal{F}$, $h \in C^\infty(U, \mathbb{R})$ where $(f_1, \dots, f_n)(X) \subseteq U$ and U is open in $\mathbb{R}^n$. For $f \in \mathcal{A}$ there exists $x \in X$ such that $\varphi = \mathrm{ev}_x$ on f and on $\mathcal{F}$, which implies that $\tilde{\varphi}(f) = f(x) = \varphi(f)$. Finally note that if $\varphi = \mathrm{ev}_x$ on $\mathcal{F}$ then $\tilde{\varphi} = \mathrm{ev}_x$ on $\mathcal{F}^{\langle\infty\rangle}$, thus $\varphi = \mathrm{ev}_x$ on $\mathcal{A}$. □

18.24. Proposition. [Biström, Bjon, Lindström, 1992, Prop.4]. *Let $\varphi \in \operatorname{Hom} \mathcal{A}$ be ω-evaluating and X be Lindelöf (for some topology finer than $X_{\mathcal{A}}$).*

Then φ is evaluating.

This applies to any ω-evaluating algebra on a separable Fréchet space, [Arias-de-Reyna, 1988, 8].
It applies also to $\mathcal{A} = C^\infty_{\mathrm{lfcs}}(E)$ for any weakly Lindelöf space by (18.27). In particular, for $1 < p \leq \infty$ the space $\ell^p(\Gamma)$ is weakly Lindelöf by (18.18.1) as weak*-dual of the normed space ℓ^q with $q := 1/(1 - \frac{1}{p})$ and the same holds for the spaces $(\ell^1(\Gamma), \sigma(\ell^1(\Gamma), c_0(\Gamma)))$. Furthermore it is true for $(\ell^1(\Gamma), \sigma(\ell^1(\Gamma), \ell^\infty(\Gamma)))$ by [Edgar, 1979], and for $(c_0(\Gamma), \sigma(c_0(\Gamma), \ell^1(\Gamma)))$ by [Corson, 1961, p.5].

Proof. By the sequentially evaluating property of $\mathcal{A}$ the family $(Z_f)_{f\in\mathcal{A}}$ of closed sets $Z_f = \{x \in X : f(x) = \varphi(f)\}$ has the countable intersection property. Since X is Lindelöf, the intersection of all sets in this collection is non-empty. Thus φ is a point evaluation with a point in this intersection. □

18.25. Proposition. *Let $\mathcal{A}$ be an algebra which contains a countable point-separating subset.*

Then every ω-evaluating φ in $\operatorname{Hom} \mathcal{A}$ is $\mathcal{A}$ is evaluating.

If a Banach space E has weak*-separable dual and $D \subseteq E'$ is countable and weak*-dense, then D is point-separating, since for $x \neq 0$ there is some $\ell \in E'$ with $\ell(x) = 1$ and since $\{x' \in E' : x'(x) > 0\}$ is open in the weak*-topology also an $\ell \in D$ with $\ell(x) > 0$. The converse is true as well, see [Biström, 1993, p.28].

Thus (18.25) applies to all Banach-spaces with weak*-separable dual and the algebras RP, C^ω, $RC^\omega_{\mathrm{conv}}$, C^∞.

Proof. Let $\{f_n\}_n$ be a countable subset of $\mathcal{A}$ separating the points of X. Let $f \in \mathcal{A}$. Since $\mathcal{A}$ is ω-evaluating there exists a point $x_f \in X$ with $f(x_f) = \varphi(f)$ and $f_n(x_f) = \varphi(f_n)$. Since the f_n are point-separating this point x_f is uniquely determined and hence independent on $f \in \mathcal{A}$. □

18.26. Proposition. [Arias-de-Reyna,1988, Thm.8] *for C^m on separable Banach spaces;* [Gómez, Llavona, 1988, Thm.1] *for ω-evaluating algebras on locally convex spaces with w^*-separable dual;* [Adam, 1993, 6.40]. *Let E be a convenient vector space, let $\mathcal{A} \supseteq P$ be an algebra containing a point separating bounded sequence of homogeneous polynomials of fixed degree.*

Then each 1-evaluating homomorphism is evaluating.

In particular this applies to c_0 and ℓ^p for $1 \le p \le \infty$. It also applies to a dual of a separable Fréchet space, since then any dense countable subset of E can be made equicontinuous on E' by [Biström, 1993, 4.13].

Proof. Let $\{p_n : n \in \mathbb{N}\}$ be a point-separating bounded sequence. By the polarization formulas given in (7.13) this is equivalent to boundedness of the associated multilinear symmetric mappings, hence $\{p_n : n \in \mathbb{N}\}$ satisfies the assumptions of (18.15) and thus $\{p_n : n \in \mathbb{N}\}$ is evaluated. Now the result follows as in (18.25). □

18.27. Theorem. [Adam, Biström, Kriegl, 1995, 5.1]. *A locally convex space E is weakly realcompact if and only if $E = \operatorname{Hom}_\omega P_f(E) (= \operatorname{Hom} C^\infty_{\text{lfcs}}(E))$.*

Proof. By (18.14) we have $\operatorname{Hom}_\omega P_f(E) = \operatorname{Hom} C^\infty_{\text{lfcs}}(E)$.
($\Rightarrow$) Let E be weakly realcompact. Since E is $\sigma(E'^*, E')$-dense in E'^* (see [Jarchow, 1981, 8.1.4]), it follows from (18.19) that any $\varphi \in \operatorname{Hom}_\omega P_f(E) = \operatorname{Hom} C^\infty_{\text{lfcs}}(E)$ is E'-evaluating and hence also evaluating on the algebra $P_f(E)$ generated by E'.
($\Leftarrow$) By (17.4) the space $\operatorname{Hom}_\omega(P_f(E))$ is realcompact in the topology of pointwise convergence. Since $E = \operatorname{Hom}_\omega P_f(E)$ and $\sigma(E, E')$ equals the topology of pointwise convergence on $\operatorname{Hom}_\omega(P_f(E))$, we have that $(E, \sigma(E, E'))$ is realcompact. □

18.28. Proposition. [Biström, Jaramillo, Lindström, 1995, Thm.13]. *Let E be a Banach space with the Dunford-Pettis property that does not contain a copy of ℓ^1. Then $P_f(E)$ is dense in $P(E)$ for the topology of uniform convergence on bounded sets.*

A Banach space E is said to have the *Dunford-Pettis property* [Diestel, 1984, p.113] if $x_n^* \to 0$ in $\sigma(E', E''))$ and $x_n \to 0$ in $\sigma(E, E')$ implies $x_n^*(x_n) \to 0$. Well known Banach spaces with the Dunford-Pettis property are $L^1(\mu)$, $C(K)$ for any compact K, and $\ell^\infty(\Gamma)$ for any Γ. Furthermore $c_0(\Gamma)$ and $\ell^1(\Gamma)$ belong to this class since if E' has the Dunford-Pettis property then also E has. According to [Aron, 1976, p.215], the space ℓ^1 is not contained in $C(K)$ if and only if K is dispersed, i.e. $K^{(\alpha)} = \emptyset$ for some α, or equivalently whenever its closed subsets admit isolated points.

Proof. According to [Carne, Cole, Gamelin, 1989, theorem 7.1], the restriction of any $p \in P(E)$ to a weakly compact set is weakly continuous if E has the Dunford-Pettis property and, consequently, sequentially weakly continuous. By [Llavona, 1986, theorems 4.4.7 and 4.5.9], such a polynomial p is weakly uniformly continuous on bounded sets if E, in addition, does not contain a copy of ℓ_1. The assertion therefore follows from [Llavona, 1986, theorem 4.3.7]. □

18.29. Theorem. [Garrido, Gómez, Jaramillo, 1994, 2.4] *&* [Adam, Biström, Kriegl, 1995, 3.4]. *Let E be $\ell^{2n}(\Gamma)$ for some n and some Γ of non-measurable cardinality. Let $P(E) \subseteq \mathcal{A} \subseteq C(E)$.*

Then every 1-evaluating homomorphism φ is evaluating.

Proof. For $f \in \mathcal{A}$ let $\mathcal{A}_f$ be the algebra generated by f and all i-homogeneous polynomials in $P(E)$ with degree $i \le 4n+2$. Take a sequence (p_n) of continuous polynomials with degree $i \le 2n+1$. Then there is a sequence (t_n) in $\mathbb{R}_+$ such that $\{t_n p_n : n \in \mathbb{N}\}$ is bounded, hence φ is by (18.15) evaluating on it, i.e. φ is ω-evaluating on $\mathcal{A}_f$.

Given $z = (z_\gamma) \in \ell^\infty(\Gamma)$ and $x \in E$, set

$$f_{z,j}(x) := f(x)^j + \sum_{\gamma \in \Gamma} z_\gamma \operatorname{pr}_\gamma(x)^{(2n+1)j},$$

where $j = 1, 2$. Then $f_{z,j} \in \mathcal{A}_f$ and we can apply (18.17). Thus there is a point $x_f \in E$ with $\varphi(f) = f(x_f)$ and $\varphi(\operatorname{pr}_\gamma)^{2n+1} = \operatorname{pr}_\gamma(x_f)^{2n+1}$ for all $\gamma \in \Gamma$. Hence $\varphi(\operatorname{pr}_\gamma) = \operatorname{pr}_\gamma(x_f)$, and since $(\operatorname{pr}_\gamma)_{\gamma\in\Gamma}$ is point separating, x_f is uniquely determined and thus not depending on f and we are finished. □

18.30. Proposition. *Let $E = c_0(\Gamma)$ with Γ non-measurable. If one of the following conditions is satisfied, then φ is evaluating:*

1. [Biström, 1993, 2.22] *&* [Adam, Biström, Kriegl, 1995, 5.4]. $C^\infty_{\text{lfs}}(E) \subseteq \mathcal{A}$ *and φ is ω-evaluating.*
2. [Garrido, Gómez, Jaramillo, 1994, 2.7]. *$P(E) \subseteq \mathcal{A}$, every $f \in \mathcal{A}$ depends only on countably many coordinates and φ is 1-evaluating.*

Proof. (1) Since φ is ω-eval, it follows that $(\varphi(\operatorname{pr}_\gamma))_{\gamma\in\Gamma} \in c_0(\Gamma)$, where $\operatorname{pr}_\gamma : c_0(\Gamma) \to \mathbb{R}$ are the natural coordinate projections (see the proof of (18.17)). Fix n and consider the function $f_n : c_0(\Gamma) \to \mathbb{R}$ defined by the locally finite product

$$f_n(x) := \prod_{\gamma\in\Gamma} h\Big(n \cdot (\operatorname{pr}_\gamma(x) - \varphi(\operatorname{pr}_\gamma))\Big),$$

where $h \in C^\infty(\mathbb{R}, [0,1])$ is chosen such that $h(t) = 1$ for $|t| \le 1/2$ and $h(t) = 0$ for $|t| \ge 1$. Note that a locally finite product $f := \prod_{i\in I} f_i$ (i.e. locally only finitely many factors f_i are unequal to 1) can be written as locally finite sum $f = \sum_J g_J$, where $g_i := f_i - 1$ and for finite subsets $J \subseteq I$ let $g_J := \prod_{j\in J} g_j \in \mathcal{A}$ and the index J runs through all finite subsets of I. Since I is at least countable, the set of these indices has the same cardinality as I has.

By means of (18.17) $\varphi(f_n) = \prod_{\gamma\in\Gamma} h(0) = 1$ for all n. Now let $f \in \mathcal{A}$. Then there exists a $x_f \in E$ with $\varphi(f) = f(x_f)$ and $1 = \varphi(f_n) = f_n(x_f)$. Hence $|n \cdot (\operatorname{pr}_\gamma(x_f) - \varphi(\operatorname{pr}_\gamma))| \le 1$ for all n, i.e. $\operatorname{pr}_\gamma(x_f) = \varphi(\operatorname{pr}_\gamma)$ for all $\gamma \in \Gamma$. Since $(\operatorname{pr}_\gamma)_{\gamma\in\Gamma}$ is point separating, the point $x_f \in E$ is unique and thus does not depend on f.

(2) By (18.15) or (18.16) the restriction of φ to the algebra generated by $\{\operatorname{pr}_\gamma : \gamma \in \Gamma\}$ is ω-evaluating. Since $c_0(K)$ is weakly-realcompact by [Corson, 1961] for locally compact metrizable K and hence in particular for discrete K, we have by (18.19) that φ is evaluating on this algebra, i.e. there exists $a = (a_\gamma)_{\gamma\in\Gamma} \in E$ with $a_\gamma = \operatorname{pr}_\gamma(a) = \varphi(\operatorname{pr}_\gamma)$ for all $\gamma \in \Gamma$.

Every $f \in \mathcal{A}(E)$ depends only on countably many coordinates, i.e. there exists a countable $\Gamma_f \subseteq \Gamma$ and a function $\tilde{f} : c_0(\Gamma_f) \to \mathbb{R}$ with $\tilde{f} \circ \mathrm{pr}_{\Gamma_f} = f$. Let

$$\mathcal{A}_f := \{g \in \mathbb{R}^{c_0(\Gamma_f)} : g \circ \mathrm{pr}_{\Gamma_f} \in \mathcal{A}\}$$

and let $\tilde{\varphi} : \mathcal{A}_f \to \mathbb{R}$ be given by $\tilde{\varphi} = \varphi \circ \mathrm{pr}_{\Gamma_f}$. Since Γ_f is countable there is by (18.15) an $x^f \in c_0(\Gamma_f)$ with $\tilde{\varphi}(\tilde{f}) = \tilde{f}(x^f)$ and $a_\gamma = \varphi(\mathrm{pr}_\gamma) = \tilde{\varphi}(\mathrm{pr}_\gamma) = \mathrm{pr}_\gamma(x^f) = x^f_\gamma$ for all $\gamma \in \Gamma_f$. Thus $\mathrm{pr}_{\Gamma_f}(a) = x$ and

$$\varphi(f) = \varphi(\tilde{f} \circ \mathrm{pr}_{\Gamma_f}) = \tilde{\varphi}(\tilde{f}) = \tilde{f}(\mathrm{pr}_{\Gamma_f}(a)) = f(a). \quad \square$$

18.31. Proposition. [Garrido, Gómez, Jaramillo, 1994, 2.7]. *Each $f \in C^\omega(c_0(\Gamma))$ depends only on countably many coordinates.*

Proof. Let $f : c_0(\Gamma) \to \mathbb{R}$ be real analytic. So there is a ball $B_\varepsilon(0) \subseteq c_0(\Gamma)$ such that $f(x) = \sum_{n=1}^\infty p_n(x)$ for all $x \in B_r(0)$, where $p_n \in L^n_{\mathrm{sym}}(c_0(\Gamma); \mathbb{R})$ for all $n \in \mathbb{N}$. By (18.28) the space $P_f(c_0(\Gamma))$ is dense in $P(c_0(\Gamma))$ for the topology of uniform convergence on bounded sets, since $c_0(\Gamma)$ has the Dunford-Pettis property and does not contain ℓ^1 as topological linear subspace. Thus we have that for any $n, k \in \mathbb{N}$ there is some $q_{nk} \in P_f(c_0(\Gamma))$ with

$$\sup\{|p_n(x) - q_{nk}(x)| : x \in B_\varepsilon(0)\} < \frac{1}{k}.$$

Since each $q \in P_f(c_0(\Gamma))$ is a polynomial form in elements of $\ell^1(\Gamma)$, there is a countable set $\Lambda_{nk} \subseteq \Gamma$ such that q_{nk} only depends on the coordinates with index in Λ_{nk}, whence p_n on $B_\varepsilon(0)$ only depends on coordinates with index in $\Lambda_n := \bigcup_{k\in\mathbb{N}} \Lambda_{nk}$. Set $\Lambda := \bigcup_{n\in\mathbb{N}} \Lambda_n$ and let $\iota_\Lambda : c_0(\Lambda) \to c_0(\Gamma)$ denote the natural injection given by $(\iota_\Lambda(x))_\gamma = x_\gamma$ if $\gamma \in \Lambda$ and $(\iota_\Lambda(x))_\gamma = 0$ otherwise. By construction $f = f \circ \iota_\Lambda \circ \mathrm{pr}_\Lambda$ on $B_\varepsilon(0)$. Since both functions are real analytic and agree on $B_\varepsilon(0)$, they also agree on $c_0(\Gamma)$. $\square$

18.32. Example. [Garrido, Gómez, Jaramillo, 1994, 2.6]. *For uncountable Γ the space $c_0(\Gamma) \setminus \{0\}$ is not C^ω-realcompact.*

But for non-measurable Γ the whole space $c_0(\Gamma)$ is C^ω-evaluating by (18.30) and (18.31).

Proof. Let $\Omega := c_0(\Gamma) \setminus \{0\}$, let $f : \Omega \to \mathbb{R}$ be real analytic and consider any sequence $(u^m)_{m\in\mathbb{N}}$ in Ω with $u^m \to 0$. For each $m \in \mathbb{N}$ there exists $\varepsilon_m > 0$ and homogeneous P^n_m in $P(c_0(\Gamma))$ of degree n for all n, such that, for $\|h\| < \varepsilon_m$

$$f(u^m + h) = f(u^m) + \sum_{n=1}^\infty P^n_m(h).$$

As carried out in (18.31), each P^n_m only depends on coordinates with index in some countable set $\Lambda^n_m \subseteq \Gamma$. The set $\Lambda := (\bigcup_{n,m\in\mathbb{N}} \Lambda^n_m) \cup (\bigcup_{m\in\mathbb{N}} \operatorname{supp} u^m)$ is countable.

Let $\gamma \in \Gamma \setminus \Lambda$. Then, since $P^n_m(e_\gamma) = 0$ and $u^m + te_\gamma \neq 0$ for all $m, n \in \mathbb{N}$ and $t \in \mathbb{R}$, we get $f(u^m + te_\gamma) = f(u^m)$ for all $|t| < \varepsilon_m$. Thus $f(u^m + te_\gamma) = f(u^m)$ for every $t \in \mathbb{R}$, since the function $t \mapsto f(u^m + te_\gamma)$ is real analytic on $\mathbb{R}$. In particular, $f(u^m + e_\gamma) = f(u^m)$ and, since $u^m + e_\gamma \to e_\gamma$, there exists

$$\varphi(f) := \lim_{m\in\mathbb{N}} f(u^m) = \lim_{m\in\mathbb{N}} f(u^m + e_\gamma) = f(e_\gamma).$$

Then φ is an algebra homomorphism, since a common γ can be found for finitely many f. And since $\ell_1(\Gamma) \subseteq C^\omega(\Omega)$ is point separating the homomorphism φ cannot be an evaluation at some point of Ω. □

18.33. Example. [Biström, Jaramillo, Lindström, 1995, Prop.16]. *The algebra $C^\omega_{\text{conv}}(\ell^\infty)$ is not 1-evaluating.*

Proof. Suppose that $C^\omega_{\text{conv}}(\ell^\infty)$ is 1-evaluating. By (20.3) the unit ball B_{c_0} of c_0 is C^ω_{conv}-bounding in ℓ^∞. By (18.20) the algebra $C^\omega_{\text{conv}}(\ell^\infty)$ is ω-evaluating and, since $(\ell^\infty)'$ admits a point separating sequence, we have $\ell^\infty = \operatorname{Hom}(C^\omega_{\text{conv}}(\ell^\infty))$ by (18.25). Hence by (20.2), every C^ω_{conv}-bounding set in ℓ^∞ is relatively compact in the initial topology induced by $C^\omega_{\text{conv}}(\ell^\infty)$ and in particular relatively $\sigma(\ell^\infty, (\ell^\infty)')$-compact. Therefore, since the topologies $\sigma(c_0, \ell^1)$ and $\sigma(\ell^\infty, (\ell^\infty)')$ coincide on c_0, we have that B_{c_0} is $\sigma(c_0, \ell^1)$-compact, which contradicts the non-reflexivity of c_0 by by [Jarchow, 1981, 11.4.4]. □

19. Stability of Smoothly Realcompact Spaces

In this section stability of evaluation properties along certain mappings are studied which furnish some large classes of smoothly realcompact spaces.

19.1. Proposition. *Let $\mathcal{A}_X$ and $\mathcal{A}_Y$ be algebras of functions on sets X and Y as in (17.1), let $T : X \to Y$ be injective with $T^*(\mathcal{A}_Y) \subseteq \mathcal{A}_X$, and suppose that Y is $\mathcal{A}_Y$-realcompact. Then we have:*

(1) [Jaramillo, 1992, 5]. *If $\mathcal{A}_X$ is 1-evaluating and $\mathcal{A}_Y$ is 1-isolating on Y, then X is $\mathcal{A}_X$-realcompact and $\mathcal{A}_X$ is 1-isolating on X.*

(2) [Biström, Lindström, 1993a, Thm.2]. *If $\mathcal{A}_X$ is ω-evaluating and $\mathcal{A}_Y$ is ω-isolating on Y, then X is $\mathcal{A}_X$-realcompact and $\mathcal{A}_X$ is ω-isolating on X.*

We say that $\mathcal{A}_X$ is *1-isolating* on X if for every $x \in X$ there is an $f \in \mathcal{A}_X$ with $\{x\} = f^{-1}(f(x))$.

Similarly $\mathcal{A}_X$ is called *ω-isolating* on X if for every $x \in X$ there exists a sequence $(f_n)_n$ in $\mathcal{A}_X$ such that $\{x\} = \bigcap_n f_n^{-1}(f_n(x))$. This was called $\mathcal{A}$-countably separated in [Biström, Lindström, 1993a].

Proof. There is a point $y \in Y$ with $\psi = \mathrm{ev}_y$. Let $\mathcal{G} \subseteq \mathcal{A}_Y$ be such that $\{y\} = \bigcap_{g\in\mathcal{G}} g^{-1}(g(y))$, where $\mathcal{G}$ is either a single function or countably many functions. Let $f \in \mathcal{A}_X$ be arbitrary. By assumption there exists $x_f \in X$ with

$\varphi(f) = f(x_f)$ and $\varphi(T^*(g)) = T^*(g)(x_f)$ for all $g \in \mathcal{G}$. Since $g(y) = \psi(g) = \varphi(T^*(g) = T^*(g)(x_f) = g(T(x_f))$ for all $g \in \mathcal{G}$, we obtain that $y = T(x_f)$. Since T is injective, we get that x_f does not depend on f, and hence φ is evaluating. □

19.2. Lemma. *If E is a convenient vector space which admits a bounded point-separating sequence in the dual E' then the algebra $P(E)$ of polynomials is 1-isolating on E.*

Proof. Let $\{x'_n : n \in \mathbb{N}\} \subseteq E'$ be such a sequence and let $a \in E$ be arbitrary. Then the series $x \mapsto \sum_{n=1}^{\infty} 2^{-n} x'_n(x-a)^2$ converges in $P(E)$, since $x'_n(\quad -a)^2$ is bounded and $\sum_{n=1}^{\infty} 2^{-n} < \infty$. It gives a polynomial which vanishes exactly at a. □

19.3. Examples. [Garrido, Gómez, Jaramillo, 1994, 2.4 and 2.5.2]. *Any super-reflexive Banach space X of non-measurable cardinality is $\mathcal{A}_X$-realcompact, for each 1-isolating and 1-evaluating algebra $\mathcal{A}_X$ as in* (17.1) *which contains the algebra of rational functions $\mathcal{R}P(X)$, see* (18.7.2).

A Banach-space E is called *super-reflexive*, if all Banach-spaces F which are finitely representable in E (i.e. for any finite dimensional subspace F_1 and $\varepsilon > 0$ there exists a isomorphism $T : F_1 \cong E_1 \subseteq E$ onto a subspace E_1 of E with $\|T\| \cdot \|T^{-1}\| \leq 1+\varepsilon$) are reflexive (see [Enflo, Lindenstrauss, Pisier, 1975]). This is by [Enflo, 1972] equivalent to the existence of an equivalent *uniformly convex norm*, i.e. $\inf\{2 - \|x+y\| : \|x\| = \|y\| = 1, \|x-y\| \geq \varepsilon\} > 0$ for all $0 < \varepsilon < 2$. In [Enflo, Lindenstrauss, Pisier, 1975] it is shown that superreflexivity has the 3-space property.

Proof. A super-reflexive Banach space injects continuously and linearly into $\ell^p(\Gamma)$ for some $p > 1$ and some Γ by [John, Torunczyk, Zizler, 1981, p.133] and hence into some $\ell^{2n}(\Gamma)$. We apply (19.1.1) to the situation $X := E \to \ell^{2n}(\Gamma) =: Y$, which is possible because the algebra $P(Y)$ is 1-isolating on Y, since the $2n$-th power of the norm is a polynomial and can be used as isolating function. By (18.6) the algebra $\mathcal{R}P(Y)$ is 1-evaluating, and by (18.29) it is thus evaluating on Y. □

19.4. Lemma.

(1) *Every 1-isolating algebra is ω-isolating.*
(2) *If X is $\mathcal{A}$-regular and $X_{\mathcal{A}}$ has first countable topology then $\mathcal{A}$ is ω-isolating.*
(3) *If for a convenient vector space the dual $(E', \sigma(E', E))$ is separable then the algebra $P_f(E)$ of finite type polynomials is ω-isolating on E.*

Proof. (1) is trivial.

(2) Let $x \in X$ be given and consider a countable neighborhood base $(U_n)_n$ of x. Since X is assumed to be $\mathcal{A}$-regular, there exist $f_n \in \mathcal{A}$ with $f_n(y) = 0$ for $y \notin U_n$ and $f_n(x) = 1$. Thus $\bigcap_n f_n^{-1}(f_n(x)) = \{x\}$.

(3) Let $\{x'_n : n \in \mathbb{N}\}$ be dense in $(E', \sigma(E', E))$ and $0 \neq x \in E$. Then there is some $x' \in E'$ with $x'(x) = 1$. By the denseness there is some n such that $|x'_n(x) - x'(x)| < 1$ and hence $x'_n(x) > 0$. So $\{0\} = \bigcap_n (x'_n)^{-1}(0)$. □

19.5. Example. *For Γ of non-measurable cardinality, the Banach space $E := c_0(\Gamma)$ is $C^\infty_{\text{lfs}}(E)$-paracompact by* (16.15), *and hence any ω-evaluating algebra $\mathcal{A} \supseteq C^\infty_{\text{lfs}}(E)$ is ω-isolating and evaluating.*

Proof. The Banach space E is $C^\infty_{\text{lfs}}(E)$-paracompact by (16.16). By (17.6) the space E is $\mathcal{A}$-realcompact for any $\mathcal{A} \supseteq C^\infty_{\text{lfs}}(E)$ and is ω-isolating by (19.4.2). □

19.6. Example. *Let K be a compact space of non-measurable cardinality with $K^{(\omega_0)} = \emptyset$.*

Then the Banach space $C(K)$ is C^∞-paracompact by (16.20.1), *hence $C^\infty(C(K))$ is ω-isolating and $C(K)$ is C^∞-realcompact.*

Proof. We use the exact sequence

$$c_0(K \setminus K') \cong \{f \in C(K) : F|_{K'} = 0\} \to C(K) \to C(K')$$

to obtain that $C(K)$ is C^∞-paracompact, see (16.19). By (17.6) the space E is C^∞-realcompact, is ω-isolating by (19.4.2). □

19.7. Example. [Biström, Lindström, 1993a, Corr.3bac]. *The following locally convex space are $\mathcal{A}$-realcompact for each ω-evaluating algebra $\mathcal{A} \supseteq C^\infty_{\text{lfs}}$, if their cardinality is non-measurable.*

(1) *Weakly compactly generated (WCG) Banach spaces, in particular separable Banach spaces and reflexive ones. More generally weakly compactly determined (WCD) Banach spaces.*
(2) *$C(K)$ for Valdivia-compact spaces K, i.e. compact subsets $K \subseteq \mathbb{R}^\Gamma$ with $K \cap \{x \in \mathbb{R}^\Gamma : \operatorname{supp} x$ countable$\}$ being dense in K.*
(3) *The dual of any realcompact Asplund Banach space.*

Proof. All three classes of spaces inject continuous and linearly into some $c_0(\Gamma)$ with non-measurable Γ by (53.21). Now we apply (19.5) for the algebra C^∞_{lfs} on $c_0(\Gamma)$ to see that the conditions of (19.1.2) for the range space $Y = c_0(\Gamma)$ are satisfied. So (19.1.2) implies the result. □

19.8. Proposition. *Let $T : X \to Y$ be a closed embedding between topological spaces equipped with algebras of continuous functions such that $T^*(\mathcal{A}_Y) \subseteq \mathcal{A}_X$. Let $\varphi \in \operatorname{Hom} \mathcal{A}_X$ such that $\psi := \varphi \circ T^*$ is $\mathcal{A}_Y$-evaluating.*

(1) [Kriegl, Michor, 1993, 8]. *If φ is 1-evaluating on $\mathcal{A}_X$ and $\mathcal{A}_Y$ has 1-small zerosets on Y then φ is $\mathcal{A}_X$-evaluating, and $\mathcal{A}_X$ has 1-small zerosets on X.*
(2) [Biström, Lindström, 1993b, p.178]. *If φ is ω-evaluating on $\mathcal{A}_X$ and $\mathcal{A}_Y$ has ω-small zerosets on Y then φ is $\mathcal{A}_X$-evaluating, and $\mathcal{A}_X$ has ω-small zerosets on X.*

Let $\mathfrak{m}$ be a cardinal number (often 1 or ω). We say that there are $\mathfrak{m}$-*small* $\mathcal{A}_Y$-zerosets on Y or $\mathcal{A}_Y$ has $\mathfrak{m}$-small zerosets on Y if for every $y \in Y$ and neighborhood U of y there exists a subset $\mathcal{G} \subseteq \mathcal{A}_Y$ with $\bigcap_{g \in \mathcal{G}} g^{-1}(g(y)) \subseteq U$ and $|\mathcal{G}| \leq \mathfrak{m}$.

For $\mathfrak{m} = 1$ this was called large $\mathcal{A}$-carriers in [Kriegl, Michor, 1993], and for $\mathfrak{m} = \omega$ it was called weakly $\mathcal{A}$-countably separated in [Biström, Lindström, 1993b, p.178].

Proof. Let $y \in Y$ be a point with $\psi = \text{ev}_y$. Since Y admits $\mathfrak{m}$-small $\mathcal{A}_Y$-zerosets there exists for every neighborhood U of y a set $\mathcal{G} \subseteq \mathcal{A}_Y$ of functions with $\bigcap_{g\in\mathcal{G}} g^{-1}(g(y)) \subseteq U$ with $|\mathcal{G}| \leq \mathfrak{m}$. Let now $f \in \mathcal{A}_X$ be arbitrary. Since $\mathcal{A}_X$ is assumed to be $\mathfrak{m}$-evaluating, there exists a point $x_{f,U}$ such that $f(x_{f,U}) = \varphi(f)$ and $g(T(x_{f,U})) = T^*(g)(x_{f,U}) = \varphi(T^*g) = \psi(g) = g(y)$ for all $g \in \mathcal{G}$, hence $T(x_{f,U}) \in U$. Thus the net $T(x_{f,U})$ converges to y for $U \to y$ and since T is a closed embedding there exists a unique x with $T(x) = y$ and $x = \lim_U x_{f,U}$. Consequently $f(x) = f(\lim_U x_{f,U}) = \lim_U f(x_{f,U}) = \lim_U \varphi(f) = \varphi(f)$ since f is continuous.

The additional assertions are obvious. □

19.9. Corollary. [Adam, Biström, Kriegl, 1995, 5.6]. *Let E be a locally convex space, $\mathcal{A} \supseteq E'$, and let $\varphi \in \operatorname{Hom} \mathcal{A}$ be ω-evaluating. Assume φ is E'-evaluating (this holds if $(E, \sigma(E, E'))$ is realcompact by* (18.27), *e.g.). Let E admit ω-small $((E')^\infty \cap \mathcal{A})_{\text{lfs}} \cap \mathcal{A}$-zerosets. Then φ is evaluating on $\mathcal{A}$.*

In particular, if E is realcompact in the weak topology and admits ω-small C^∞_{lfs}-zerosets then $E = \operatorname{Hom}_\omega C^\infty_{\text{lfs}}(E)$.

Proof. We may apply (19.8.2) to $X = Y := E$, $\mathcal{A}_X = \mathcal{A}$ and $\mathcal{A}_Y := ((E')^\infty \cap \mathcal{A})_{\text{lfs}} \cap \mathcal{A}$. Note that φ is evaluating on $\mathcal{A}_Y$ by (18.17) and that $C^\infty_{\text{lfs}}(E) = ((E')^\infty)_{\text{lfs}}$ by (18.13). □

19.10. Lemma. [Adam, Biström, Kriegl, 1995, 5.5].

(1) *If a space is $\mathcal{A}$-regular then it admits 1-small $\mathcal{A}$-zerosets (and in turn also ω-small $\mathcal{A}$-zerosets).*
(2) *For any cardinality $\mathfrak{m}$, any $\mathfrak{m}$-isolating algebra $\mathcal{A}$ has $\mathfrak{m}$-small $\mathcal{A}$-zerosets.*
(3) *If a topological space X is first countable and admits ω-small $\mathcal{A}$-zerosets then $\mathcal{A}$ is ω-isolating.*
(4) *Any Lindelöf locally convex space admits ω-small P_f-zerosets.*

The converse to (1) is false for $P_f(E)$, where E is an infinite dimensional separable Banach space E, see [Adam, Biström, Kriegl, 1995, 5.5].
The converse to (2) is false for $P_f(\mathbb{R}^\Gamma)$ with uncountable Γ, see [Adam, Biström, Kriegl, 1995, 5.5].

Proof. (1) and (2) are obvious.

(3) Let $x \in X$ and $\mathcal{U}$ a countable neighborhood basis of x. For every $U \in \mathcal{U}$ there is a countable set $\mathcal{G}_U \subseteq \mathcal{A}$ with $\bigcap_{g\in\mathcal{G}_U} g^{-1}(g(y)) \subseteq U$. Then $\mathcal{G} := \bigcup_{U\in\mathcal{U}} \mathcal{G}_U$ is countable and

$$\bigcap_{g\in\mathcal{G}} g^{-1}(g(y)) \subseteq \bigcap_{U\in\mathcal{U}} \bigcap_{g\in\mathcal{G}_U} g^{-1}(g(y)) \subseteq \bigcap_{U\in\mathcal{U}} U = \{y\}$$

(4) Take a point x and an open set U with $x \in U \subseteq E$. For each $y \in E \setminus U$ let $p_y \in E' \subseteq P_f(E)$ with $p_y(x) = 0$ and $p_y(y) = 1$. Set $V_y := \{z \in E : p_y(z) > 0\}$. By the Lindelöf property, there is a sequence (y_n) in $E \setminus U$ such that $\{U\} \cup \{V_{y_n}\}_{n\in\mathbb{N}}$ is a cover of E. Hence for each $y \in E \setminus U$ there is some $n \in \mathbb{N}$ such that $y \in V_{y_n}$, i.e. $p_{y_n}(y) > 0 = p_{y_n}(x)$. □

19.11. Theorem. [Kriegl, Michor, 1993] & [Biström, Lindström, 1993b, Prop.4]. *Let $\mathfrak{m}$ be 1 or an infinite cardinal and let X be a closed subspace of $\prod_{i\in I} X_i$, let $\mathcal{A}$ be an algebra of functions on X and let $\mathcal{A}_i$ be algebras on X_i, respectively, such that $\mathrm{pr}_i^*(\mathcal{A}_i) \subseteq \mathcal{A}$ for all i.*

If each X_i admits $\mathfrak{m}$-small $\mathcal{A}_i$-zerosets then X admits $\mathfrak{m}$-small $\mathcal{A}$-zerosets.

If in addition $\varphi \in \operatorname{Hom}\mathcal{A}$ is $\mathfrak{m}$-eval on $\mathcal{A}$ and $\varphi_i := \varphi \circ \mathrm{pr}_i^* \in \operatorname{Hom}\mathcal{A}_i$ is evaluating on $\mathcal{A}_i$ for all i, then φ is evaluating $\mathcal{A}$ on X.

Proof. We consider $Y := \prod_i X_i$ and the algebra $\mathcal{A}_Y$ generated by $\bigcup_i \{f_i \circ \mathrm{pr}_i : f_i \in \mathcal{A}_{X_i}\}$, where $\mathrm{pr}_j : \prod_i X_i \to X_j$ denotes the canonical projection.

Now we prove the first statement for $\mathcal{A}_Y$. Let $x \in Y$ and U a neighborhood of $x = (x_i)_i$ in Y. Thus there exists a neighborhood in $\prod_i X_i$ contained in U, which we may assume to be of the form $\prod_i U_i$ with $U_i = X_i$ for all but finitely many i. Let $\mathcal{F}$ be the finite set of those exceptional i. For each $i \in \mathcal{F}$ we choose a set $\mathcal{G}_i \subseteq \mathcal{A}$ with $\bigcap_{g\in\mathcal{G}_i} g^{-1}(g(x_i)) \subseteq U_i$. Without loss of generality we may assume $g(x_i) = 0$ and $g \geq 0$ (replace g by $x \mapsto (g(x) - g(x_i))^2$). For any $g \in \prod_{i\in\mathcal{F}} \mathcal{G}_i$ we define $\tilde{g} \in \mathcal{A}_Y$ by $\tilde{g} := \sum_{i\in\mathcal{F}} g_i \circ \mathrm{pr}_i \in \mathcal{A}_Y$. Then $\tilde{g}(x) = \sum_{i\in\mathcal{F}} g_i(x) = 0$

$$\bigcap_{g\in\prod_{i\in\mathcal{F}} \mathcal{G}_i} \tilde{g}^{-1}(0) \subseteq U,$$

since for $z \notin U$ we have $z_i \notin U_i$ for at least one $i \in \mathcal{F}$. Note that $|\prod_{i\in\mathcal{F}} \mathcal{G}_i| \leq \mathfrak{m}$, since $\mathfrak{m}$ is either 1 or infinite.

That $\mathcal{A}_Y$ is evaluating follows trivially since $\varphi_i := \varphi \circ {\mathrm{pr}_i}^* : \mathcal{A}_{X_i} \to \mathcal{A}_X \to \mathbb{R}$ is an algebra homomorphism and $\mathcal{A}_{X_i}$ is evaluating, so there exists a point $a_i \in X_i$ such that $\varphi(f_i \circ \mathrm{pr}_i) = (\varphi \circ {\mathrm{pr}_i}^*)(f_i) = f_i(a_i)$ for all $f_i \in \mathcal{A}_{X_i}$. Let $a := (a_i)_i$. Then obviously every $f \in \mathcal{A}_Y$ is evaluated at a.

If now X is a closed subspace of the product $Y := \prod_i X_i$ then we can apply (19.8.1) and (19.8.2). □

19.12. Theorem (19.11) is usually applied as follows. Let $\mathcal{U}$ be a zero-neighborhood basis of a locally convex space E. Then E embeds into $\prod_{U\in\mathcal{U}} \widehat{E_{(U)}}$, where $\widehat{E_{(U)}}$ denotes the completion of the Banach space $E_{(U)} := E/\ker p_U$, where p_U denotes the Minkowski functional of U. If E is complete, then this is a closed embedding, and in order to apply (19.11) we have to find an appropriate basis $\mathcal{U}$ and for each $U \in \mathcal{U}$ an algebra $\mathcal{A}_U$ on $\widehat{E_{(U)}}$, which pulls back to $\mathcal{A}$ along the canonical projections $\pi_U : E \to E_{(U)} \subseteq \widehat{E_{(U)}}$, such that the Banach space $\widehat{E_{(U)}}$ is $\mathcal{A}_U$-realcompact and has $\mathfrak{m}$-small $\mathcal{A}_U$-zerosets.

Examples.

(1) [Kriegl, Michor, 1993]. *A complete, trans-separable (i.e. contained in product of separable normed spaces) locally convex space is $\mathcal{A}$-realcompact for every 1-evaluating algebra $\mathcal{A} \supseteq \bigcup_U \pi_U^*(P_f)$.*

Note that for products of separable Banach spaces one has $C^\infty = C^\infty_c$, see [Adam, 1993, 9.18] *&* [Kriegl, Michor, 1993].

(2) [Biström, 1993, 4.5]. *A complete, Hilbertizable (i.e. there exists a basis of Hilbert seminorms, in particular nuclear spaces) locally convex space is $\mathcal{A}$-realcompact for every 1-evaluating $\mathcal{A} \supseteq \bigcup_U \pi_U^*(P)$.*

(3) [Biström, Lindström, 1993b, Cor.3]. *Every complete non-measurable WCG locally convex space is C^∞-realcompact.*

(4) [Biström, Lindström, 1993b, Cor.5]. *Any reflexive non-measurable Fréchet space is $C^\infty = C^\infty_c$-realcompact.*

(5) [Biström, Lindström, 1993b, Cor.4]. *Any complete non-measurable infra-Schwarz space is C^∞_c-realcompact.*

(6) [Biström, 1993, 4.16-4.18]. *Every countable coproduct of locally convex spaces, and every countable ℓ^p-sum or c_0-sum of Banach-spaces injects continuously into the corresponding product. Thus from $\mathcal{A}$ being ω-isolating and evaluating on each factor, we deduce the same for the total space by (19.1.2) if $\mathcal{A}$ is ω-evaluating on it.*

A locally convex space is usually called WCG if there exists a sequence of absolutely convex, weakly-compact subsets, whose union is dense.

Proof. (1) We have for $\widehat{E_{(U)}}$ that it is $\mathcal{A}$-realcompact for every 1-evaluating $\mathcal{A} \supseteq P$ by (18.26) and P_f is 1-isolating by (19.2) and hence has 1-small zero sets by (19.10.2).

For a product E of metrizable spaces the two algebras $C^\infty(E)$ and $C^\infty_c(E)$ coincide: For every countable subset $\mathcal{A}$ of the index set, the corresponding product is separable and metrizable, hence C^∞-realcompact. Thus there exists a point $x_{\mathcal{A}}$ in this countable product such that $\varphi(f) = f(x_{\mathcal{A}})$ for all f which factor over the projection to that countable subproduct. Since for $\mathcal{A}_1 \subseteq \mathcal{A}_2$ the projection of $x_{\mathcal{A}_2}$ to the product over $\mathcal{A}_1$ is just $x_{\mathcal{A}_1}$ (use the coordinate projections for f), there is a point x in the product, whose projection to the subproduct with index set $\mathcal{A}$ is just $x_{\mathcal{A}}$. Every Mackey continuous function, and in particular every C^∞-function, depends only on countable many coordinates, thus factors over the projection to some subproduct with countable index set $\mathcal{A}$, hence $\varphi(f) = f(x_{\mathcal{A}}) = f(x)$. This can be shown by the same proof as for a product of factors $\mathbb{R}$ in (4.27), since the result of [Mazur, 1952] is valid for a product of separable metrizable spaces.

(2) By (19.3) we have that $\ell^2(\Gamma)$ is $\mathcal{A}$-realcompact for every 1-evaluating $\mathcal{A} \supseteq P$ and P is 1-isolating.

(3) For every U the Banach space $\widehat{E_{(U)}}$ is then WCG, hence as in (19.7.1) there is a SPRI, and by (53.20) a continuous linear injection into some $c_0(\Gamma)$. By (19.5) any ω-evaluating algebra $\mathcal{A}$ on $c_0(\Gamma)$ which contains C^∞_{lfs} is evaluating and ω-isolating.

By (19.1.2) this is true for such stable algebras on $\widehat{E_{(U)}}$, and hence by (19.11) for E.

(4) Here $E_{(U)}$ embeds into $C(K)$, where $K := (U^o, \sigma(E', E''))$ is Talagrand compact [Cascales, Orihuela, 1987, theorem 12] and hence Corson compact [Negrepontis, 1984, 6.23]. Thus by (19.7.2) we have PRI. Now we proceed as in (3).

(5) Any complete infra-Schwarz space is a closed subspace of a product of reflexive and hence WCG Banach spaces, since weakly compact mappings factor over such spaces by [Jarchow, 1981, p.374]. Hence we may proceed as in (3). □

Short Exact Sequences

In the following we will consider exact sequences of locally convex spaces

$$0 \to H \xrightarrow{\iota} E \xrightarrow{\pi} F,$$

i.e. $\iota : H \to E$ is a embedding of a closed subspace and π has $\iota(H)$ as kernel. Let algebras $\mathcal{A}_H$, $\mathcal{A}_E$ and $\mathcal{A}_F$ on H, E and F be given, which satisfy $\pi^*(\mathcal{A}_F) \subseteq \mathcal{A}_E$ and $\iota^*(\mathcal{A}_E) \supseteq \mathcal{A}_H$, the latter one telling us that $\mathcal{A}_H$ functions on H can be extended to $\mathcal{A}_E$ functions on E. This is a very strong requirement, since by (21.11) not even polynomials of degree 2 on a closed subspace of a Banach space can be extended to a smooth function. The only algebra, where we have the extension property in general is that of finite type polynomials. So we will apply the following theorem in (19.14) and (19.15) to situations, where $\mathcal{A}_H$ is of quite different type then $\mathcal{A}_E$ and $\mathcal{A}_F$.

19.13. Theorem. [Adam, Biström, Kriegl, 1995, 6.1]. *Let $0 \to H \xrightarrow{\iota} E \xrightarrow{\pi} F$ be an exact sequence of locally convex spaces equipped with algebras satisfying*

(i) $\pi^*(\mathcal{A}_F) \subseteq \mathcal{A}_E$ *and* $\iota^*(\mathcal{A}_E) \supseteq \mathcal{A}_H$.
(ii) $\mathcal{A}_F$ *is ω-isolating on F.*
(iii) $\mathcal{A}_E$ *is translation invariant.*

Then we have:

(1) *If $\mathcal{A}_H$ is ω-isolating on H then $\mathcal{A}_E$ is ω-isolating on E.*
(2) *If H has ω-small $\mathcal{A}_H$-zerosets then E has ω-small $\mathcal{A}_E$-zerosets.*

If in addition

(iv) $\operatorname{Hom}_\omega \mathcal{A}_F = F$ *and* $\operatorname{Hom}_\omega \mathcal{A}_H = H$,

then we have:

(3) *If $\varphi \in \operatorname{Hom} \mathcal{A}_E$ is ω-evaluating on $\mathcal{A}_E$ then φ is evaluating on $\mathcal{A}_0 := \{f \in \mathcal{A}_E : \iota^*(f) \in \mathcal{A}_H\}$.*
(4) *If $\varphi \in \operatorname{Hom} \mathcal{A}_E$ is ω-evaluating on $\mathcal{A}_E$ and if $\mathcal{A}_H$ is ω-isolating on H then φ is evaluating on $\mathcal{A}_E$; i.e., $E = \operatorname{Hom}_\omega \mathcal{A}_E$.*

Proof. Let $x \in E$. Since $\mathcal{A}_E$ is translation invariant, we may assume $x = 0$. By (ii) there is a sequence (g_n) in $\mathcal{A}_F$ which isolates $\pi(x)$ in F, i.e. $g_n(\pi(x)) = 0$ and $\bigcap g_n^{-1}(0) = \{\pi(x)\}$.

(1) By the special assumption in (19.13.1) there exist countable many $h_n \in \mathcal{A}_H$ which isolate 0 in H. According to (i) $\pi^*(g_n) \in \mathcal{A}_E$ and there exist $\tilde{h}_n \in \mathcal{A}_E$ with $\tilde{h}_n \circ \iota = h_n$. By (iii) we have that $f_n := \tilde{h}_n(\ \ -x) \in \mathcal{A}_E$. Now the functions $\pi^*(g_n)$ together with the sequence (f_n) isolate x. Indeed, if $x' \in E$ is such that $(g_n \circ \pi)(x') = (g_n \circ \pi)(x)$ for all n, then $\pi(x') = \pi(x)$, i.e. $x' - x \in H$. From $f_n(x') = f_n(x)$ we conclude that $h_n(x'-x) = \tilde{h}_n(x'-x) = f_n(x') = f_n(x) = h_n(0)$, and hence $x' = x$.

(2) Let U be a 0-neighborhood in E. By the special assumption there are countably many $h_n \in \mathcal{A}_H$ with $0 \in \bigcap_n Z(h_n) \subseteq U \cap H$. As before consider the sequence of functions $f_n := \tilde{h}_n(\ \ -x)$. The common kernel of the functions in the sequences (f_n) and $(\pi^*(g_n))$ contains x and is contained in $\pi^{-1}(\pi(x)) = x + H$ and hence in $(x+U) \cap (x+H) \subseteq x + U$.

Now the remaining two statements:
Let $\varphi \in \operatorname{Hom}_\omega \mathcal{A}_E$. Then $\varphi \circ \pi^* : \mathcal{A}_F \to \mathbb{R}$ is a ω-evaluating homomorphism, and hence by (iv) given by the evaluation at a point $y \in F$. By (ii) there is a sequence (g_n) in $\mathcal{A}_F$ which isolates y. Since φ is ω-evaluating there exists a point $x \in E$, such that $g_n(y) = \varphi(\pi^*(g_n)) = \pi^*(g_n)(x) = g_n(\pi(x))$ for all n. Hence $y = \pi(x)$. Since φ obviously evaluates each countable set in $\mathcal{A}_E$ at a point in $\pi^{-1}(y) \cong K$, φ induces a ω-evaluating homomorphism $\varphi_H : \mathcal{A}_H \to \mathbb{R}$ by $\varphi_H(\iota^*(f)) := \varphi(f(\ \ -x))$ for $f \in \mathcal{A}_0$. In fact let $f, \bar{f} \in \mathcal{A}_0$ with $\iota^*(f) = \iota^*(\bar{f})$. Then φ evaluates $f(\ \ -x)$, $\bar{f}(\ \ -x)$ and all $\pi^*(g_n)$ at some common point $\bar{x}$. So $\pi(\bar{x}) = y = \pi(x)$, hence $\bar{x} - x \in H$ and $f(\bar{x} - x) = \bar{f}(\bar{x} - x)$.
By (iv), φ_H is given by the evaluation at a point $z \in H$.

(3) Here we have that $\mathcal{A}_H = \iota^*(\mathcal{A}_0)$, and hence

$$\varphi(f) = \varphi_H(\iota^*(f(\ \ +x)) = \iota^*(f(\ \ +x)(z)) = f(\iota(z) + x)$$

for all $f \in \mathcal{A}_0$. So φ is evaluating on $\mathcal{A}_0$.

(4) We show that $\varphi = \delta_{\iota(z)+x}$ on $\mathcal{A}_E$. Indeed, by the special assumption there is a sequence (h_n) in $\mathcal{A}_H$ which isolates z. By (i) and (iii), we may find $f_n \in \mathcal{A}_E$ such that $h_n = \iota^*(f_n(\ \ +x))$. The sequences $(\pi^*(g_n))$ and (f_n) isolate $z + x$. So let $f \in \mathcal{A}_E$ be arbitrary. Then there exists a point $z' \in E$, such that $\varphi = \delta_{z'}$ for all these functions, hence $z' = \iota(z) + x$. □

19.14. Corollary. [Adam, Biström, Kriegl, 1995, 6.3]. *Let $0 \to H \xrightarrow{\iota} E \xrightarrow{\pi} F$ be a left exact sequence of locally convex spaces and let $\mathcal{A}_F$ and $\mathcal{A}_E \supseteq E'$ be algebras on F and E, respectively, that satisfy all the assumptions (i-iv) of (19.13) not involving $\mathcal{A}_H$. Let furthermore $\varphi : \mathcal{A}_E \to \mathbb{R}$ be ω-evaluating and $\varphi \circ \pi^*$ be evaluating on $\mathcal{A}_F$. Then we have*

1. *The homomorphism φ is $\mathcal{A}_E$-evaluating if $(H, \sigma(H, H'))$ is realcompact and admits ω-small P_f-zerosets.*
2. *The homomorphism φ is $\mathcal{A}_0$-evaluating if $(H, \sigma(H, \iota^*(\mathcal{A}_0)))$ is Lindelöf and $\mathcal{A}_0 \subseteq \mathcal{A}_E$ is some subalgebra.*
3. *The homomorphism φ is E'-evaluating if $(H, \sigma(H, H'))$ is realcompact.*

Proof. We will apply (19.13.3). For this we choose appropriate subalgebras $\mathcal{A}_0 \subseteq \mathcal{A}_E$ and put $\mathcal{A}_H := \iota^*(\mathcal{A}_0)$. Then (i-iii) of (19.13) is satisfied. Remains to show for (iv) that $\operatorname{Hom}_\omega(\mathcal{A}_H) = H$ in the three cases:

(1) Let $\mathcal{A}_0 := \mathcal{A}_E$. Then we have $\operatorname{Hom}_\omega(\mathcal{A}_H) = H$ by (19.9) using (18.27).

(2) If $H_{\mathcal{A}_H} = (H, \sigma(H, \mathcal{A}_H))$ is Lindelöf, then $H = \operatorname{Hom}_\omega(\mathcal{A}_H)$, by (18.24).

(3) Let $\mathcal{A}_0 := P_f(E)$. Then $\mathcal{A}_H := \iota^*(\mathcal{A}_0) = P_f(H)$ by Hahn-Banach. If H is $\sigma(H, H')$-realcompact, then $H = \operatorname{Hom}_\omega(\mathcal{A}_H)$, by (18.27). □

19.15. Theorem. [Adam, Biström, Kriegl, 1995, 6.4 and 6.5]. *Let $c_0(\Gamma) \xrightarrow{\iota} E \xrightarrow{\pi} F$ be a short exact sequence of locally convex spaces where $\mathcal{A}_E$ is translation invariant and contains $(\pi^*(\mathcal{A}_F) \cup E')^\infty_{\text{lfs}}$, and where F is $\mathcal{A}_F$-regular.*

Then $\iota^*(\mathcal{A}_E)$ contains the algebra $\mathcal{A}_{c_0(\Gamma)}$ which is generated by all functions $x \mapsto \prod_{\gamma\in\Gamma} \eta(x_\gamma)$, where $\eta \in C^\infty(\mathbb{R}, \mathbb{R})$ is 1 near 0.

If $\mathcal{A}_F$ is ω-isolating on F then $\mathcal{A}_E$ is ω-isolating on E. If in addition $F = \operatorname{Hom}_\omega \mathcal{A}_F$ and Γ is non-measurable then $E = \operatorname{Hom}_\omega \mathcal{A}_E$.

Proof. Let us show that the function $x \mapsto \prod_{\gamma\in\Gamma} \eta(x_\gamma)$ can be extended to a function in $\mathcal{A}_E$.
Remark that this product is locally finite, since $x \in c_0(\Gamma)$ and $\eta = 1$ locally around 0. Let p be an extension of the supremum norm $\| \;\; \|_\infty$ on $c_0(\Gamma)$ to a continuous seminorm on E, and let $\tilde{p}$ be the corresponding quotient seminorm on F defined by $\tilde{p}(y) := \inf\{p(x) : \pi(x) = y\}$. Let furthermore ℓ_γ be a continuous linear extensions of $\operatorname{pr}_\gamma : c_0(\Gamma) \to \mathbb{R}$ which satisfy $|\ell_\gamma(x)| \le p(x)$ for all $x \in E$. Finally let $\varepsilon > 0$ be such that $\eta(t) = 1$ for $|t| \le \varepsilon$.

We show first, that for the open subset $\{x \in E : \tilde{p}(\pi(x)) < \varepsilon\}$ the product $\prod_{\gamma\in\Gamma} \eta(\ell_\gamma(x))$ is locally finite as well. So let $\tilde{p}(\pi(x)) < \varepsilon$ and $3\,\delta := \varepsilon - \tilde{p}(\pi(x))$. We claim that

$$\Gamma_x := \{\gamma : |\ell_\gamma(x)| \ge \tilde{p}(\pi(x)) + 2\delta\}$$

is finite. In fact by definition of the quotient seminorm $\tilde{p}(\pi(x)) := \inf\{p(x+y) : y \in c_0(\Gamma)\}$ there is a $y \in c_0(\Gamma)$ such that $p(x+y) \le \tilde{p}(\pi(x)) + \delta$. Since $y \in c_0(\Gamma)$ the set $\Gamma_0 := \{\gamma : |y_\gamma| \ge \delta\}$ is finite. For all $\gamma \notin \Gamma_0$ we have

$$|\ell_\gamma(x)| \le |\ell_\gamma(x+y)| + |\ell_\gamma(y)| \le p(x+y) + |y_\gamma| < \tilde{p}(\pi(x)) + 2\,\delta,$$

hence $\Gamma_x \subseteq \Gamma_0$ is finite.
Now take $z \in E$ with $p(z-x) \le \delta$. Then for $\gamma \notin \Gamma_x$ we have

$$|\ell_\gamma(z)| \le |\ell_\gamma(x)| + |\ell_\gamma(z-x)| < \tilde{p}(\pi(x)) + 2\,\delta + p(z-x) \le \tilde{p}(\pi(x)) + 3\,\delta = \varepsilon,$$

hence $\eta(\ell_\gamma(z)) = 1$ and the product is a locally finite.

In order to obtain the required extension to all of E, we choose $0 < \varepsilon' < \varepsilon$ and a function $g \in \mathcal{A}_F$ with carrier contained inside $\{z : \tilde{p}(z) \le \varepsilon'\}$ and with $g(0) = 1$. Then $f : E \to \mathbb{R}$ defined by

$$f(x) := g(\pi(x)) \prod_{\gamma\in\Gamma} \eta(\ell_\gamma(x))$$

is an extension belonging to $\langle \pi^*(\mathcal{A}_F) \cup (E')^\infty_{\text{lfs}} \rangle_{\text{Alg}} \subseteq (\pi^*(\mathcal{A}_F) \cup E')^\infty_{\text{lfs}} \subseteq \mathcal{A}_E$.

Let us now show that we can find such an extension with arbitrary small carrier, and hence that E is $\mathcal{A}_E$-regular.
So let an arbitrary seminorm p on E be given. Then there exists a $\delta > 0$ such that $\delta\, p|_{c_0(\Gamma)} \leq \|\ \ \|_\infty$. Let q be an extension of $\|\ \ \|_\infty$ to a continuous seminorm on E. By replacing p with $\max\{q, \delta\, p\}$ we may assume that $p|_{c_0(\Gamma)} = \|\ \ \|_\infty$ and the unit ball of the original p contains the δ-ball of the new p. Let again $\tilde{p}$ be the corresponding quotient norm on F.

Then the construction above with some $0 < \varepsilon' < \varepsilon < \varepsilon'' \leq \delta/3$, for $\eta \in C^\infty(\mathbb{R}, \mathbb{R})$ with $\eta(t) = 1$ for $|t| \leq \varepsilon$ and $\eta(t) = 0$ for $|t| > \varepsilon'' > \varepsilon$ and $g \in C^\infty(F, \mathbb{R})$ with $\operatorname{carr}(g) \subseteq \{y \in F : \tilde{p}(y) \leq \varepsilon' < \varepsilon\}$ gives us a function $f \in \mathcal{A}_E$ and it remains to show that the carrier of f is contained in the δ-ball of p. So let $x \in E$ be such that $f(x) \neq 0$. Then on one hand $g(\pi(x)) \neq 0$ and hence $\tilde{p}(\pi(x)) \leq \varepsilon'$ and on the other hand $\eta(\ell_\gamma(x)) \neq 0$ for all $\gamma \in \Gamma$ and hence $|\ell_\gamma(x)| \leq \varepsilon''$. We have a unique continuous linear mapping $T : \ell^1(\Gamma) \to E'$, which maps pr_γ to ℓ_γ, and satisfies $|T(y^*)(z)| \leq \|y^*\|\, p(z)$ for all $z \in E$ since the unit ball of $\ell^1(\Gamma)$ is the closed absolutely convex hull of $\{\operatorname{pr}_\gamma : \gamma \in \Gamma\}$. By Hahn-Banach there is some $\ell \in E'$ be such that $|\ell(z)| \leq p(z)$ for all z and $\ell(x) = p(x)$. Hence $\iota^*(\ell) = \ell|_{c_0(\Gamma)}$ is in the unit ball of $\ell^1(\Gamma)$, and hence $|T(\iota^*(\ell))(x)| \leq \varepsilon''$, since $|\ell_\gamma(x)| \leq \varepsilon''$. Moreover $|T(\iota^*(\ell))(z)| \leq p(z)$. Then $\ell_0 := (T \circ \iota^* - 1)(\ell) = T(\ell|_{c_0(\Gamma)}) - \ell \in E'$ vanishes on $c_0(\Gamma)$ and $|\ell_0(z)| \leq 2\, p(z)$ for all z. Hence $|\ell_0(x)| \leq 2\,\tilde{p}(\pi(x)) \leq 2\,\varepsilon'$. So $p(x) = |\ell(x)| \leq |T(\iota^*(\ell))(x)| + |\ell_0(x)| \leq \varepsilon'' + 2\,\varepsilon' < \delta$.

Because of the extension property $\mathcal{A}_{c_0(\Gamma)} \subseteq \iota^*(\mathcal{A}_E)$ and since $c_0(\Gamma)$ is $\mathcal{A}_{c_0(\Gamma)}$-regular and hence by (19.10.1) ω-isolated, we can apply (19.13.1) to obtain the statement on ω-isolatedness. The evaluating property now follows from (19.13.4) using that $\operatorname{Hom}_\omega \mathcal{A}_{c_0(\Gamma)} = c_0(\Gamma)$ by (18.30.1). □

19.16. The class c_0-ext. We shall show in (19.18) that in the short exact sequence of (19.15) we can in fact replace $c_0(\Gamma)$ by spaces from a huge class which we now define.

Definition. Let c_0-*ext* be the class of spaces H, for which there are short exact sequences $c_0(\Gamma_j) \to H_j \to H_{j+1}$ for $j = 1, ..., n$, with $|\Gamma_j|$ non-measurable, $H_{n+1} = \{0\}$ and $T : H \to H_1$ an operator whose kernel is weakly realcompact and has ω-small P_f-zerosets (By (18.18.1) and (19.2) these two conditions are satisfied, if it has for example a weak*-separable dual).

Of course all spaces which admit a continuous linear injection into some $c_0(\Gamma)$ with non-measurable Γ belong to c_0-ext. Besides these there are other natural spaces in c_0-ext. For example let K be a compact space with $|K|$ non-measurable and $K^{(\omega_0)} = \emptyset$, where ω_0 is the first infinite ordinal and $K^{(\omega_0)}$ the corresponding ω_0-th derived set. Then the Banach space $C(K)$ belongs to c_0-ext, but is in general not even injectable into some $c_0(\Gamma)$, see [Godefroy, Pelant, et. al., 1988]. In fact, from $K^{(\omega)} = \emptyset$ and the compactness of K, we conclude that $K^{(n)} = \emptyset$ for some integer

n. We have the short exact sequence

$$c_0(K \setminus K^{(1)}) \cong E \xrightarrow{\iota} C(K) \xrightarrow{\pi} C(K)/E \cong C(K^{(1)}),$$

where $E := \{f \in C(K) : f|_{K^{(1)}} = 0\}$. By using (19.15) inductively the space $C(K)$ is C^∞_{lfs}-regular. Also it is again an example of a Banach space E with $E = \operatorname{Hom} C^\infty(E)$ that we are able to obtain without using the quite complicated result (16.20.1) that it admits C^∞-partition of unity.

19.17. Lemma. Pushout. [Adam, Biström, Kriegl, 1995, 6.6]. *Let a closed subspace $\iota : H \hookrightarrow E$ and a continuous linear mapping $T : H \to H_1$ of locally convex spaces be given.*

Then the pushout of ι and T is the locally convex space $E_1 := H_1 \times E/\{(Tz, -z) : z \in H\}$. The natural mapping $\iota_1 : H \to E_1$, given by $u \mapsto [(u,0)]$ is a closed embedding and the natural mapping $T_1 : E \to E_1$ given by $T_1(x) := [(0,x)]$ is continuous and linear. Moreover, if T is a quotient mapping then so is T_1.

Given a short exact sequence $H \xrightarrow{\iota} E \xrightarrow{\pi} F$ of locally convex vector spaces and a continuous linear map $T : H \to H_1$ then we obtain by this construction a short exact sequence $H_1 \xrightarrow{\iota_1} E_1 \xrightarrow{\pi_1} F$ and a (unique) extension $T_1 : E \to E_1$ of T, with $\ker T = \ker T_1$, *such that the following diagram commutes*

$$\begin{array}{ccccc}
\ker T & = & \ker T_1 & \twoheadrightarrow & 0 \\
\downarrow & & \downarrow & & \downarrow \\
H & \xrightarrow{\iota} & E & \xrightarrow{\pi} & F \\
\downarrow T & & \downarrow T_1 & & \| \\
H_1 & \xrightarrow{\iota_1} & E_1 & \xrightarrow{\pi_1} & F
\end{array}$$

Proof. Since H is closed in E the space E_1 is a Hausdorff locally convex space. The mappings ι_1 and T_1 are clearly continuous and linear. And ι_1 is injective, since $(u,0) \in \{(T(z), -z) : z \in H\}$ implies $0 = z$ and $u = T(z) = T(0) = 0$. In order to see that ι_1 is a topological embedding let U be an absolutely convex 0-neighborhood in H_1. Since ι is a topological embedding there is a 0-neighborhood W in E with $W \cap H = T^{-1}(U)$. Now consider the image of $U \times W \subseteq H_1 \times E$ under the quotient map $H_1 \times E \to E_1$. This is a 0-neighborhood in E_1 and its inverse image under ι_1 is contained in $2U$. Indeed, if $[(u,0)] = [(x,z)]$ with $u \in H_1$, $x \in U$ and $z \in W$, then $x - u = T(z)$ and $z \in H \cap W$, by which $u = x - T(z) \in U - U = 2U$. Hence ι_1 embeds H_1 topologically into E_1.

We have the universal property of a pushout, since for any two continuous linear mappings $\alpha : E \to G$ and $\beta : H_1 \to G$ with $\beta \circ T = \alpha \circ \iota$, there exists a unique linear mapping $\gamma : E_1 \to G$, given by $[(u,x)] \mapsto \alpha(x) - \beta(u)$ with $\gamma \circ T_1 = \alpha$ and $\gamma \circ \iota_1 = \beta$. Since $H_1 \oplus E \to E_1$ is a quotient mapping γ is continuous as well.

Let now $\pi : E \to F$ be a continuous linear mapping with kernel H, e.g. π the natural quotient mapping $E \to F := E/H$. Then by the universal property we get a unique continuous linear $\pi_1 : E_1 \to F$ with $\pi_1 \circ T_1 = \pi$ and $\pi_1 \circ \iota_1 = 0$. We have $\iota_1(H_1) = \ker(\pi_1)$, since $0 = \pi_1[(u,z)] = \pi(z)$ if and only if $z \in H$, i.e. if and only if $[(u,z)] = [(u + Tz, 0)]$ lies in the image of ι_1. If π is a quotient map then clearly so is π_1. In particular the image of ι_1 is closed.

Since $T(x) = 0$ if and only if $[(0,x)] = [(0,0)]$, we have that $\ker T = \ker T_1$. Assume now, in addition, that T is a quotient map. Given any $[(y,x)] \in E_1$, there is then some $z \in H$ with $T(z) = y$. Thus $T_1(x+z) = [(0, x+z)] = [(T(z), x)] = [(y,x)]$ and T_1 is onto. Remains to prove that T_1 is final, which follows by categorical reasoning. In fact let $g : E_1 \to G$ be a mapping with $g \circ T_1$ continuous and linear. Then $g \circ \iota_1 : H_1 \to G$ is a mapping with $(g \circ \iota_1) \circ T = g \circ T_1 \circ \iota$ continuous and linear and since T is final also $g \circ \iota_1$ is continuous. Thus g composed with the quotient mapping $H_1 \oplus E \to E_1$ is continuous and linear and thus also g itself. □

19.18. Theorem. [Adam, Biström, Kriegl, 1995, 6.7]. *Let $H \xrightarrow{\iota} E \xrightarrow{\pi} F$ be a short exact sequence of locally convex spaces, let F be C^∞_{lfs}-regular and let H be of class c_0-ext, see* (19.16).

If $C^\infty_{\text{lfs}}(F)$ is ω-isolating on F then $C^\infty_{\text{lfs}}(E)$ is ω-isolating on E. If, in addition, $F = \operatorname{Hom}_\omega C^\infty_{\text{lfs}}(F)$ then $E = \operatorname{Hom}_\omega C^\infty_{\text{lfs}}(E)$.

Proof. Since H is of class c_0-ext there are short exact sequences $c_0(\Gamma_j) \to H_j \to H_{j+1}$ for $j = 1, \ldots, n$ such that $|\Gamma_j|$ is non-measurable, $H_{n+1} = \{0\}$, and $T : H \to H_1$ is an operator whose kernel is weakly realcompact and has ω-small P_f-zerosets. We proceed by induction on the length of the resolution

$$H_0 := H \to H_1 \twoheadrightarrow \cdots \twoheadrightarrow H_{n+1} = \{0\}.$$

According to (19.17) we have for every continuous linear $T : H_j \to H_{j+1}$ the following diagram

$$\begin{array}{ccccc}
\ker T & = & \ker T_1 & \twoheadrightarrow & 0 \\
\downarrow & & \downarrow & & \downarrow \\
H_j & \xrightarrow{\iota_j} & E_j & \xrightarrow{p_j} & F \\
\downarrow T & & \downarrow T_1 & & \| \\
H_{j+1} & \xrightarrow{\iota_{j+1}} & E_{j+1} & \xrightarrow{\pi_{j+1}} & F
\end{array}$$

For $j > 0$ we have that $\ker T = c_0(\Gamma)$ for some none-measurable Γ, and T and T_1 are quotient mappings. So let as assume that we have already shown for the bottom row, that E_{j+1} has the required properties and is in addition C^∞_{lfs}-regular. Then by the exactness of the middle column we get the same properties for E_j using (19.15). If $j = 0$, then the kernel is by assumption weakly paracompact and admits ω-small P_f-zerosets. Thus applying (19.14.1) and (19.13.1) to the left exact middle column we get the required properties for $E = E_0$. □

A Class of C^∞_{lfs}-Realcompact Locally Convex Spaces

19.19. Definition. Following [Adam, Biström, Kriegl, 1995] let RZ denote the class of all locally convex spaces E which admit ω-small C^∞_{lfs}-zerosets and have the property that $E = \operatorname{Hom}_\omega \mathcal{A}$ for each translation invariant algebra $\mathcal{A}$ with $C^\infty_{\text{lfs}}(E) \subseteq \mathcal{A} \subseteq C(E)$. In particular this applies to the algebras C, C^∞_c and $C^\infty \cap C$.

Note that for every continuous linear $T : E \to F$ we have $T^* : C^\infty_{\text{lfs}}(F) \to C^\infty_{\text{lfs}}(E)$. In fact we have $T^*(F') \subseteq E'$, hence $T^* : (F')^\infty \to (E')^\infty$ and $T^*(\sum_i f_i)$ is again locally finite, if T is continuous and $\sum_i f_i$ is it.

A locally convex space E with ω-small C^∞_{lfs}-zerosets belongs to RZ if and only if $E = \operatorname{Hom}_\omega C^\infty_{\text{lfs}}(E) = \operatorname{Hom} C^\infty_{\text{lfs}}(E)$. In fact by (18.11) we have $\operatorname{Hom}_\omega C^\infty_{\text{lfs}}(E) = \operatorname{Hom} C^\infty_{\text{lfs}}(E)$. Now let $\mathcal{A} \supseteq C^\infty_{\text{lfs}}(E)$ and let $\varphi \in \operatorname{Hom}_\omega \mathcal{A}$ be countably evaluating. Then by (19.8.2) applied to $X = Y = E$, $\mathcal{A}_X := \mathcal{A}$ and $\mathcal{A}_Y := C^\infty_{\text{lfs}}(E)$ the homomorphism φ is evaluating on $\mathcal{A}$.

Note that by (19.10.3) for metrizable E the condition of having ω-small C^∞_{lfs}-zerosets can be replaced by C^∞_{lfs} being ω-isolating. Moreover, by (19.10.1) it is enough to assume that E is C^∞_{lfs}-regular in order that E belongs to RZ.

19.20. Proposition. *The class RZ is closed under formation of arbitrary products and closed subspaces.* □

Proof. This is a direct corollary of (19.11). □

19.21. Proposition. [Adam, Biström, Kriegl, 1995]. *Every locally convex space that admits a linear continuous injection into a metrizable space of class RZ is itself of class RZ.*

Proof. Use (19.1.2) and (19.10.3). □

19.22. Corollary. [Adam, Biström, Kriegl, 1995]. *The countable locally convex direct sum of a sequence of metrizable spaces in RZ belongs to RZ.*

The class of Banach spaces in RZ is closed under forming countable c_0-sums and ℓ_p-sums with $1 \le p \le \infty$.

Proof. By (19.20) the class RZ is stable under (countable) products. And (19.21) applies since a countable product of metrizable is again metrizable. □

19.23. Corollary. [Adam, Biström, Kriegl, 1995]. *Among the complete locally convex spaces the following belong to the class RZ:*

(1) *All trans-separable (i.e. subspaces of products of separable Banach spaces) locally convex spaces;*
(2) *All Hilbertizable locally convex spaces;*
(3) *All non-measurable WCG locally convex spaces;*
(4) *All non-measurable reflexive Fréchet spaces;*
(5) *All non-measurable infra-Schwarz locally convex spaces.*

Proof. By (19.20), (19.5), and (19.21) we see that every complete locally convex space E belongs to RZ, if it admits a zero-neighborhood basis $\mathcal{U}$ such that each Banach space $\widehat{E_{(U)}}$ for $U \in \mathcal{U}$ injects into some $c_0(\Gamma_U)$ with non-measurable Γ_U. Apply this to the examples (19.12.1)-(19.12.5). □

19.24. Proposition. [Adam, Biström, Kriegl, 1995]. *Let $0 \to H \hookrightarrow E \to F$ be an exact sequence. Let F be in RZ and let C^∞_{lfs} be ω-isolating on F.*

Then E is in RZ under any of the following assumptions.

(1) *The sequence $0 \to H \to E \to F \to 0$ is exact, H is in c_0-ext and F is C^∞_{lfs}-regular; Here it follows also that C^∞_{lfs} is ω-isolating on E.*
(2) *The sequence $0 \to H \to E \to F \to 0$ is exact, $H = c_0(\Gamma)$ for some none-measurable Γ and F is C^∞_{lfs}-regular; Here it follows also that E is C^∞_{lfs}-regular.*
(3) *The weak topology on H is realcompact and H admits ω-small P_f-zerosets.*
4 *The class c_0-ext is a subclass of RZ.*

Proof. (1) This is (19.18).

(2) follows directly from (19.15) applied to the algebra $\mathcal{A} = C^\infty_{\text{lfs}}$.

(3) By (19.13.2) the space E has ω-small C^∞_{lfs}-zerosets. By (19.14.1) we have assumption (iv) in (19.13), and then by (19.13.4) we have $E = \operatorname{Hom}_\omega(C^\infty_{\text{lfs}}(E))$. Thus E belongs to RZ.

(4) Since every space E in c_0-ext is obtained by applying finitely many constructions as in (2) and a last one as in (3) we get it for E. □

19.25. Remark. [Adam, Biström, Kriegl, 1995]. The class RZ is 'quite big'. By (19.24.4) we have that c_0-ext is a subclass of RZ. Also the following spaces are in RZ:

The space $C(K)$ where K is the one-point compactification of the topological disjoint union of a sequence of compact spaces K_n with $K_n^{(\omega)} = \emptyset$. In fact we have a continuous injection given by the countable product of the restriction maps $C(K) \to C(K_n)$. Hence the result follows from (19.24.4) using also the remark in (19.16) for the $C(K_n)$, followed by (19.20) for the product and by (19.21) for $C(K)$. Remark that in such a situation we might have $K^{(\omega)} = \{\infty\} \neq \emptyset$.

The space $D[0,1]$ of all functions $f : [0,1] \to \mathbb{R}$ which are right continuous and have left limits and endowed with the sup norm is in RZ. Indeed it contains $C[0,1]$ as a subspace and $D[0,1]/C[0,1] \cong c_0[0,1]$ according to [Corson, 1961]. By (18.27) we have that $C[0,1]$ is weakly Lindelöf and P_f is ω-isolating, since $\{\text{ev}_t : t \in \mathbb{Q} \cap [0,1]\}$ are point-separating. Now we use (19.24.3).

Open Problem. Is $\ell^\infty(\Gamma)$ in RZ for $|\Gamma|$ non-measurable, i.e. is $C^\infty_{\text{lfs}}(\ell^\infty(\Gamma))$ ω-isolating on $\ell^\infty(\Gamma)$ and is $\operatorname{Hom}_\omega C^\infty_{\text{lfs}}(\ell^\infty(\Gamma)) = \ell^\infty(\Gamma)$?

If this is true, then every complete locally convex space E of non-measurable cardinality would be in RZ, since every Banach space E is a closed subspace of $\ell^\infty(\Gamma)$, where Γ is the closed unit-ball of E'.

20. Sets on which all Functions are Bounded

In this last section the relationship of evaluation properties and bounding sets, i.e. sets on which every function of the algebra is bounded, are studied.

20.1. Proposition. [Kriegl, Nel, 1990, 2.2]. *Let $\mathcal{A}$ be a convenient algebra, and $B \subseteq X$ be $\mathcal{A}$-bounding. Then $p_B : f \mapsto \sup\{|f(x)| : x \in B\}$ is a bounded seminorm on $\mathcal{A}$.*

A subset $B \subseteq X$ is called *$\mathcal{A}$-bounding* if $f(B) \subseteq \mathbb{R}$ is bounded for all $f \in \mathcal{A}$.

Proof. Since B is bounding, we have that $p_B(f) < \infty$. Now assume there is some bounded set $\mathcal{F} \subseteq \mathcal{A}$, for which $p_B(\mathcal{F})$ is not bounded. Then we may choose $f_n \in \mathcal{F}$, such that $p_B(f_n) \geq \sqrt{n2^n}$. Note that $\{f^2 : f \in \mathcal{F}\}$ is bounded, since multiplication is assumed to be bounded. Furthermore $p_B(f^2) = \sup\{|f(x)|^2 : x \in B\} = \sup\{|f(x)| : x \in B\}^2 = p_B(f)^2$, since $t \mapsto t^2$ is a monotone bijection $\mathbb{R}_+ \to \mathbb{R}_+$, hence $p_B(f_n^2) \geq n2^n$. Now consider the series $\sum_{n=0}^\infty \frac{1}{2^n} f_n^2$. This series is Mackey-Cauchy, since $(2^{-n})_n \in \ell^1$ and $\{f_n^2 : n \in \mathbb{N}\}$ is bounded. Since $\mathcal{A}$ is assumed to be convenient, we have that this series is Mackey convergent. Let $f \in \mathcal{A}$ be its limit. Since all summands are non-negative we have

$$p_B(f) = p_B\Big(\sum_{n=0}^\infty \frac{1}{2^n} f_n^2\Big) \geq p_B(\frac{1}{2^n} f_n^2) = \frac{1}{2^n} p_B(f_n)^2 \geq n,$$

for all $n \in \mathbb{N}$, a contradiction. □

20.2. Proposition. [Kriegl, Nel, 1990, 2.3] *for $\mathcal{A}$-paracompact,* [Biström, Bjon, Lindström, 1993, Prop.2]. *If X is $\mathcal{A}$-realcompact then every $\mathcal{A}$-bounding subset of X is relatively compact in $X_\mathcal{A}$.*

Proof. Consider the diagram

$$X_\mathcal{A} \xhookrightarrow{\cong} \operatorname{Hom}(\mathcal{A}) \hookrightarrow \prod_\mathcal{A} \mathbb{R}$$

and let $B \subseteq X$ be $\mathcal{A}$-bounding. Then its image in $\prod_\mathcal{A} \mathbb{R}$ is relatively compact by Tychonoff's theorem. Since $\operatorname{Hom}(\mathcal{A}) \subseteq \prod_\mathcal{A} \mathbb{R}$ is closed, we have that B is relatively compact in $X_\mathcal{A}$. □

20.3. Proposition. [Biström, Jaramillo, Lindström, 1995, Prop.7]. *Every function $f = \sum_{n=0}^\infty p_n \in C^\omega_{\text{conv}}(\ell^\infty)$ converges uniformly on the bounded sets in c_0. In particular, each bounded set in c_0 is C^ω_{conv}-bounding in l^∞.*

Proof. Take $f = \sum_{n=0}^\infty p_n \in C^\omega_{\text{conv}}(\ell^\infty)$. According to (7.14), the function f may be extended to a holomorphic function $\tilde{f} \in H(\ell^\infty \otimes \mathbb{C})$ on the complexification. [Josefson, 1978] showed that each holomorphic function on $\ell^\infty \otimes \mathbb{C}$ is bounded on every bounded set in $c_0 \otimes \mathbb{C}$. Hence, the restriction $\tilde{f}|_{c_0 \otimes \mathbb{C}}$ is a holomorphic function on $c_0 \otimes \mathbb{C}$ which is bounded on bounded subsets. By (7.15) its Taylor series at zero $\sum_{n=0}^\infty \widetilde{p_n}$ converges uniformly on each bounded subset of $c_0 \otimes \mathbb{C}$. The statement then follows by restricting to the bounded subsets of the real space c_0. □

20.4. Result. [Biström, Jaramillo, Lindström, 1995, Corr.8]. *Every weakly compact set in c_0, in particular the set $\{e_n : n \in \mathbb{N}\} \cup \{0\}$ with e_n the unit vectors, is $RC^{\omega}_{\text{conv}}$-bounding in l^∞.*

20.5. Result. [Biström, Jaramillo, Lindström, 1995, Thm.5]. *Let $\mathcal{A}$ be a functorial algebra on the category of continuous linear maps between Banach spaces with $RP \subseteq \mathcal{A}$. Then, for every Banach space E, the $\mathcal{A}$-bounding sets are relatively compact in E if there is a function in $\mathcal{A}(\ell^\infty)$ that is unbounded on the set of unit vectors in ℓ^∞.*

20.6. Result.

(1) [Biström, Jaramillo, 1994, Thm.2] *&* [Biström, 1993, p.73, Thm.5.23]. *In all Banach spaces the C^∞_{lfcs}-bounding sets are relatively compact.*

(2) [Biström, Jaramillo, 1994, p.5] *&* [Biström, 1993, p.74,Cor.5.24]. *Any C^∞_{lfcs}-bounding set in a locally convex space E is precompact and therefore relatively compact if E, in addition, is quasi-complete.*

(3) [Biström, Jaramillo, 1994, Cor.4] *&* [Biström, 1993, p.74, 5.25]. *Let E be a quasi-complete locally convex space. Then E and $E_{C^\infty_{\text{lfcs}}}$ have the same compact sets. Furthermore $x_n \to x$ in E if and only if $f(x_n) \to f(x)$ for all $f \in C^\infty_{\text{lfcs}}(E)$.*

Chapter V
Extensions and Liftings of Mappings

In this chapter we will consider various extension and lifting problems. In the first section we state the problems and give several counter-examples: We consider the subspace F of all functions which vanish of infinite order at 0 in the nuclear Fréchet space $E := C^\infty(\mathbb{R}, \mathbb{R})$, and we construct a smooth function on F that has no smooth extension to E, and a smooth curve $\mathbb{R} \to F'$ that has not even locally a smooth lifting along $E' \to F'$. These results are based on E. Borel's theorem which tells us that $\mathbb{R}^{\mathbb{N}}$ is isomorphic to the quotient E/F and the fact that this quotient map $E \to \mathbb{R}^{\mathbb{N}}$ has no continuous right inverse. Also the result (16.8) of [Seeley, 1964] is used which says that, in contrast to F, the subspace $\{f \in C^\infty(\mathbb{R}, \mathbb{R}) : f(t) = 0 \text{ for } t \leq 0\}$ of E is complemented.

In section (22) we characterize in terms of a simple boundedness condition on the difference quotients those functions $f : A \to \mathbb{R}$ on an arbitrary subset $A \subseteq \mathbb{R}$ which admit a smooth extension $\tilde{f} : \mathbb{R} \to \mathbb{R}$ as well as those which admit an m-times differentiable extension $\tilde{f}$ having locally Lipschitzian derivatives. This results are due to [Frölicher, Kriegl, 1993] and are much stronger than Whitney's extension theorem, which holds for closed subsets only and needs the whole jet and conditions on it. There is, however, up to now no analog in higher dimensions, since difference quotients are defined only on lattices.

Section (23) gives an introduction to smooth spaces in the sense of Frölicher. These are sets together with curves and functions which compose into $C^\infty(\mathbb{R}, \mathbb{R})$ and determine each other by this. They are very useful for chasing smoothness of mappings which sometimes leave the realm of manifolds.

In section (23) it is shown that there exist free convenient vector spaces over Frölicher spaces, this means that to every such space X one can associate a convenient vector space λX together with a smooth map $\iota_X : X \to \lambda X$ such that for any convenient vector space E the map $(\iota_X)^* : L(\lambda X, E) \to C^\infty(X, E)$ is a bornological isomorphism. The space λX can be obtained as the c^∞-closure of the linear subspace spanned by the image of the canonical map $X \to C^\infty(X, \mathbb{R})'$. In the case where X is a finite dimensional smooth manifold we prove that the linear subspace generated by $\{\ell \circ \mathrm{ev}_x : x \in X, \ell \in E'\}$ is c^∞-dense in $C^\infty(X, E)'$. From this we conclude that the free convenient vector space over a manifold X is the space of distributions with compact support on X.

In the last 3 sections we discuss germs of smooth, holomorphic, and real analytic functions on convex sets with non-empty interior, following [Kriegl, 1997]. Let us recall the finite dimensional situation for smooth maps, so let first $E = F = \mathbb{R}$ and X be a non-trivial closed interval. Then a map $f : X \to \mathbb{R}$ is usually called smooth, if it is infinite often differentiable on the interior of X and the one-sided derivatives of all orders exist. The later condition is equivalent to the condition, that all derivatives extend continuously from the interior of X to X. Furthermore, by Whitney's extension theorem these maps turn out to be the restrictions to X of smooth functions on (some open neighborhood of X in) $\mathbb{R}$. In case where $X \subseteq \mathbb{R}$ is more general, these conditions fall apart. Now what happens if one changes to $X \subseteq \mathbb{R}^n$. For closed convex sets with non-empty interior the corresponding conditions to the one dimensional situation still agree. In case of holomorphic and real analytic maps the germ on such a subset is already defined by the values on the subset. Hence, we are actually speaking about germs in this situation. In infinite dimensions we will consider maps on just those convex subsets. So we do not claim greatest achievable generality, but rather restrict to a situation which is quite manageable. We will show that even in infinite dimensions the conditions above often coincide, and that real analytic and holomorphic maps on such sets are often germs of that class. Furthermore, we have exponential laws for all three classes, more precisely, the maps on a product correspond uniquely to maps from the first factor into the corresponding function space on the second.

21. Extension and Lifting Properties

21.1. Remark. The extension property. The general extension problem is to find an arrow $\tilde{f}$ making a diagram of the following form commutative:

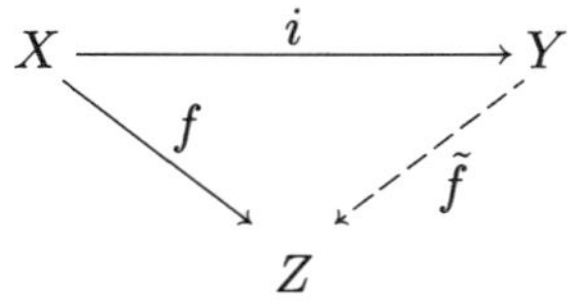

We will consider problems of this type for smooth, for real-analytic and for holomorphic mappings between appropriate spaces, e.g., Frölicher spaces as treated in section (23).

Let us first sketch a step by step approach to the general problem for the smooth mappings at hand.

If for a given mapping $i : X \to Y$ an extension $\tilde{f} : Y \to Z$ exists for all $f \in C^\infty(X, Z)$, then this says that the restriction operator $i^* : C^\infty(Y, Z) \to C^\infty(X, Z)$ is surjective.

Note that a mapping $i : X \to Y$ has the extension property for all $f : X \to Z$ with values in an arbitrary space Z if and only if i is a section, i.e. there exists a mapping $\widetilde{\mathrm{Id}_X} : Y \to X$ with $\widetilde{\mathrm{Id}_X} \circ i = \mathrm{Id}_X$. (Then $\tilde{f} := f \circ \widetilde{\mathrm{Id}_X}$ is the extension of a general mapping f).

A particularly interesting case is $Z = \mathbb{R}$. A mapping $i : X \to Y$ with the extension property for all $f : X \to \mathbb{R}$ is said to have the *scalar valued extension property.* Such a mapping is necessarily initial: In fact let $g : Z \to X$ be a mapping with $i \circ g : X \to Y$ being smooth. Then $f \circ g = \tilde{f} \circ i \circ g$ is smooth for all $f \in C^\infty(X, \mathbb{R})$ and hence g is smooth, since the functions $f \in C^\infty(X, \mathbb{R})$ generate the smooth structure on the Frölicher space X.

More generally, we consider the same question for any convenient vector space $Z = E$. Let us call this the *vector valued extension property.* Assume that we have already shown the scalar valued extension property for $i : X \to Y$, and thus we have an operator $S : C^\infty(X, \mathbb{R}) \to C^\infty(Y, \mathbb{R})$ between convenient vector spaces, which is a right inverse to $i^* : C^\infty(Y, \mathbb{R}) \to C^\infty(X, \mathbb{R})$. It is reasonable to hope that S will be linear (which can be easily checked). So the next thing would be to check, whether it is bounded. By the uniform boundedness theorem it is enough to show that $\mathrm{ev}_y \circ S : C^\infty(X, \mathbb{R}) \to C^\infty(Y, \mathbb{R}) \to Y$ given by $f \mapsto \tilde{f}(y)$ is smooth, and usually this is again easily checked. By dualization we get a bounded linear operator $S^* : C^\infty(Y, \mathbb{R})' \to C^\infty(X, \mathbb{R})'$ which is a left inverse to $i^{**} : C^\infty(X, \mathbb{R})' \to C^\infty(Y, \mathbb{R})'$. Now in order to solve the vector valued extension problem we use the free convenient vector space λX over a smooth space X given in (23.6). Thus any $f \in C^\infty(X, E)$ corresponds to a bounded linear $\tilde{f} : \lambda X \to E$. It is enough to extend $\tilde{f}$ to a bounded linear operator $\lambda Y \to E$ given by $\tilde{f} \circ S^*$. So we only need that $S^*|_{\lambda Y}$ has values in λX, or equivalently, that $S^* \circ \delta_Y : Y \to C^\infty(Y, \mathbb{R})' \to C^\infty(X, \mathbb{R})'$, given by $y \mapsto (f \mapsto \tilde{f}(y))$, has values in λX. In the important cases (e.g. finite dimensional manifolds X), where $\lambda X = C^\infty(X, \mathbb{R})'$, this is automatically satisfied. Otherwise it is by the uniform boundedness principle enough to find for given $y \in Y$ a bounding sequence $(x_k)_k$ in X (i.e. every $f \in C^\infty(X, \mathbb{R})$ is bounded on $\{x_k : k \in \mathbb{N}\}$) and an absolutely summable sequence $(a_k)_k \in \ell^1$ such that $\tilde{f}(y) = \sum_k a_k f(x_k)$ for all $f \in C^\infty(X, \mathbb{R})$. Again we can hope that this can be achieved in many cases.

21.2. Proposition. *Let $i : X \to Y$ be a smooth mapping, which satisfies the vector valued extension property. Then there exists a bounded linear extension operator $C^\infty(X, E) \to C^\infty(Y, E)$.*

Proof. Since i is smooth, the mapping $i^* : C^\infty(Y, E) \to C^\infty(X, E)$ is a bounded linear operator between convenient vector spaces. Its kernel is $\ker(i^*) = \{f \in C^\infty(Y, E) : f \circ i = 0\}$. And we have to show that the sequence

$$0 \longrightarrow \ker(i^*) \hookrightarrow C^\infty(Y, E) \xrightarrow{\ i^*\ } C^\infty(X, E) \longrightarrow 0$$

splits via a bounded linear operator $\sigma : C^\infty(X, E) \ni f \mapsto \tilde{f} \in C^\infty(Y, E)$, i.e. a bounded linear extension operator.

By the exponential law (3.13) a mapping $\sigma \in L(C^\infty(X, E), C^\infty(Y, E))$ would correspond to $\tilde{\sigma} \in C^\infty(Y, L(C^\infty(X, E), E))$ and $\sigma \circ i^* = \mathrm{Id}$ translates to $\tilde{\sigma} \circ i = \widetilde{\mathrm{Id}} =$

$\delta : X \to L(C^\infty(X,E),E)$, given by $x \mapsto (f \mapsto f(x))$, i.e. $\tilde{\sigma}$ must be a solution of the following vector valued extension problem:

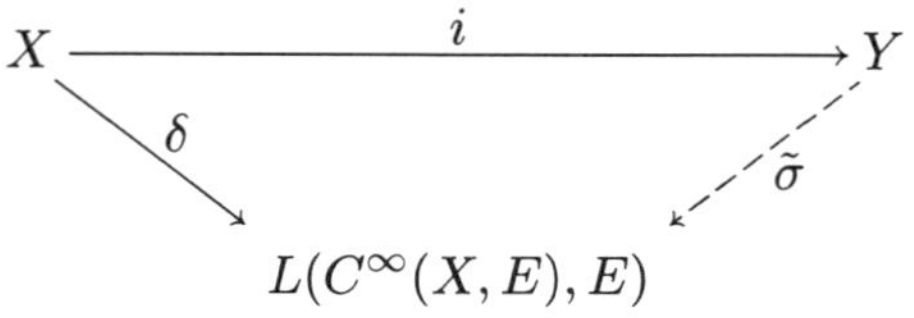

By the vector valued extension property such a $\tilde{\sigma}$ exists. □

21.3. The lifting property. Dual to the extension problem, we have the lifting problem, i.e. we want to find an arrow $\tilde{f}$ making a diagram of the following form commutative:

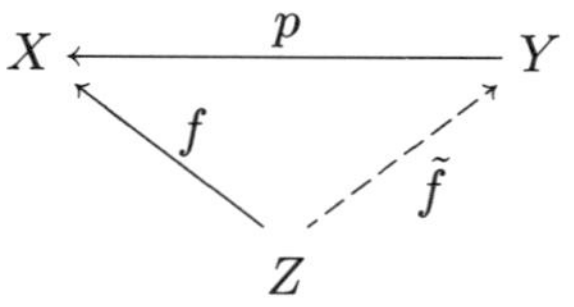

Note that in this situation it is too restrictive to search for a bounded linear or even just a smooth lifting operator $T : C^\infty(Z,X) \to C^\infty(Z,Y)$. If such an operator exists for some $Z \neq \emptyset$, then $p : Y \to X$ has a smooth right inverse namely the dashed arrow in the following diagram:

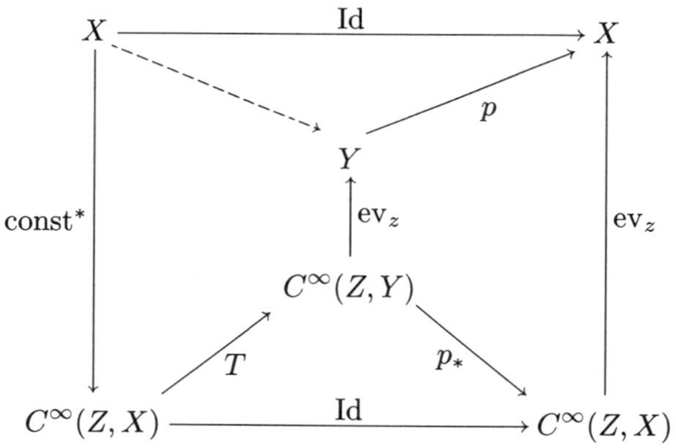

Again the first important case is, when $Z = \mathbb{R}$. If X and Y are even convenient vector spaces, then we know that the image of a convergent sequence $t_n \to t$ under a smooth curve $c : \mathbb{R} \to Y$ is Mackey convergent. And since one can find by the general curve lemma a smooth curve passing through sufficiently fast falling subsequences of a Mackey convergent sequence, the first step could be to check whether such sequences can be lifted. If bounded sets (or at least sequences) can be lifted, then the same is true for Mackey convergent sequences. However, this is not always true as we will show in (21.9).

21.4. Remarks. The scalar valued extension property for bounded linear mappings on a c^∞-dense linear subspace is true if and only if the embedding represents the c^∞-completion by (4.30). In this case it even has the vector valued extension property by (4.29).

That in general bounded linear functionals on a (dense or c^∞-closed subspace) may not be extended to bounded (equivalently, smooth) linear functionals on the whole space was shown in (4.36.6).

The scalar valued extension problem is true for the c^∞-closed subspace of an uncountable product formed by all points with countable support, see (4.27) (and (4.12)). As a consequence this subspace is not smoothly real compact, see (17.5).

Let E be not smoothly regular and U be a corresponding 0-neighborhood. Then the closed subset $X := \{0\} \cup (E \setminus U) \subseteq Y := E$ does not have the extension property for the smooth mapping $f = \chi_{\{0\}} : X \to \mathbb{R}$.

Let E be not smoothly normal and A_0, A_1 be the corresponding closed subsets. Then the closed subset $X := A_1 \cup A_2 \subseteq Y := E$ does not have the extension property for the smooth mapping $f = \chi_{A_1} : X \to \mathbb{R}$.

If $q : E \to F$ is a quotient map of convenient vector spaces one might expect that for every smooth curve $c : \mathbb{R} \to F$ there exists (at least locally) a smooth lifting, i.e. a smooth curve $c : \mathbb{R} \to E$ with $q \circ c = c$. And if $\iota : F \to E$ is an embedding of a convenient subspace one might expect that for every smooth function $f : F \to \mathbb{R}$ there exists a smooth extension to E. In this section we give examples showing that both properties fail. As convenient vector spaces we choose spaces of smooth real functions and their duals. We start with some lemmas.

21.5. Lemma. *Let $E := C^\infty(\mathbb{R}, \mathbb{R})$, let $q : E \to \mathbb{R}^\mathbb{N}$ be the infinite jet mapping at 0, given by $q(f) := (f^{(n)}(0))_{n \in \mathbb{N}}$, and let $F \xrightarrow{\iota} E$ be the kernel of q, consisting of all smooth functions which are flat of infinite order at 0.*

Then the following sequence is exact:

$$0 \to F \xrightarrow{\iota} E \xrightarrow{q} \mathbb{R}^\mathbb{N} \to 0.$$

Moreover, $\iota^* : E' \to F'$ is a quotient mapping between the strong duals. Every bounded linear mapping $s : \mathbb{R}^\mathbb{N} \to E$ the composite $q \circ s$ factors over $pr_N : \mathbb{R}^\mathbb{N} \to \mathbb{R}^N$ for some $N \in \mathbb{N}$, and so the sequence does not split.

Proof. The mapping $q : E \to \mathbb{R}^\mathbb{N}$ is a quotient mapping by the open mapping theorem (52.11) & (52.12), since both spaces are Fréchet and q is surjective by Borel's theorem (15.4). The inclusion ι is an embedding of Fréchet spaces, so the adjoint ι^* is a quotient mapping for the strong duals (52.28). Note that these duals are bornological by (52.29).

Let $s : \mathbb{R}^\mathbb{N} \to E$ be an arbitrary bounded linear mapping. Since $\mathbb{R}^\mathbb{N}$ is bornological s has to be continuous. The set $U := \{g \in E : |g(t)| \le 1 \text{ for } |t| \le 1\}$ is a 0-neighborhood in the locally convex topology of E. So there has to exist an $N \in \mathbb{N}$ such that $s(V) \subseteq U$ with $V := \{x \in \mathbb{R}^\mathbb{N} : |x_n| < \frac{1}{N} \text{ for all } n \le N\}$. We show that $q \circ s$ factors over $\mathbb{R}^N$. So let $x \in \mathbb{R}^\mathbb{N}$ with $x_n = 0$ for all $n \le N$. Then $k \cdot x \in V$

for all $k \in \mathbb{N}$, hence $k \cdot s(x) \in U$, i.e. $|s(x)(t)| \leq \frac{1}{k}$ for all $|t| \leq 1$ and $k \in \mathbb{N}$. Hence $s(x)(t) = 0$ for $|t| \leq 1$ and therefore $q(s(x)) = 0$.

Suppose now that there exists a bounded linear mapping $\rho : E \to F$ with $\rho \circ \iota = \mathrm{Id}_F$. Define $s(q(x)) := x - \iota\rho x$. This definition makes sense, since q is surjective and $q(x) = q(x')$ implies $x - x' \in F$ and thus $x - x' = \rho(x - x')$. Moreover s is a bounded linear mapping, since q is a quotient map, as surjective continuous map between Fréchet spaces; and $(q \circ s)(q(x)) = q(x) - q(\iota(\rho(x))) = q(x) - 0$. □

21.6. Proposition. [Frölicher, Kriegl, 1988], *7.1.5 Let $\iota^* : E' \to F'$ the quotient map of* (21.5). *The curve $c : \mathbb{R} \to F'$ defined by $c(t) := \mathrm{ev}_t$ for $t \geq 0$ and $c(t) = 0$ for $t < 0$ is smooth but has no smooth lifting locally around 0. In contrast, bounded sets and Mackey convergent sequences are liftable.*

Proof. By the uniform boundedness principle (5.18) c is smooth provided $\mathrm{ev}_f \circ c : \mathbb{R} \to \mathbb{R}$ is smooth for all $f \in F$. Since $(\mathrm{ev}_f \circ c)(t) = f(t)$ for $t \geq 0$ an $(\mathrm{ev}_f \circ c)(t) = 0$ for $t \leq 0$ this obviously holds.

Assume first that there exists a global smooth lifting of c, i.e. a smooth curve $e : \mathbb{R} \to E'$ with $\iota^* \circ e = c$. By exchanging the variables, c corresponds to a bounded linear mapping $\tilde{c} : F \to E$ and e corresponds to a bounded linear mapping $\tilde{e} : E \to E$ with $\tilde{e} \circ \iota = \tilde{c}$. The curve c was chosen in such a way that $\tilde{c}(f)(t) = f(t)$ for $t \geq 0$ and $\tilde{c}(f)(t) = 0$ for $t \leq 0$.

We show now that such an extension $\tilde{e}$ of $\tilde{c}$ cannot exist. In (16.8) we have shown the existence of a retraction s to the embedding of the subspace $F_+ := \{f \in F : f(t) = 0$ for $t \leq 0\}$ of E. For $f \in F$ one has $s(\tilde{e}(f)) = s(\tilde{c}(f)) = \tilde{c}(f)$ since $\tilde{c}(E) \subseteq F_+$. Now let $\Psi : E \to E$, $\Psi(f)(t) := f(-t)$ be the reflection at 0. Then $\Psi(F) \subseteq F$ and $f = \tilde{c}(f) + \Psi(\tilde{c}(\Psi(f)))$ for $f \in F$. We claim that $\rho := s \circ \tilde{e} + \Psi \circ s \circ \tilde{e} \circ \Psi : E \to F$ is a retraction to the inclusion, and this is a contradiction with (21.5). In fact

$$\rho(f) = (s \circ \tilde{e})(f) + (\Psi \circ s \circ \tilde{e} \circ \Psi)(f) = \tilde{c}(f) + \Psi(\tilde{c}(\Psi(f))) = f$$

for all $f \in F$. So we have proved that c has no global smooth lifting.

Assume now that $c|_I$ has a smooth lifting $e_0 : U \to E'$ for some open neighborhood I of 0. Trivially $c|_{\mathbb{R}\setminus\{0\}}$ has a smooth lifting e_1 defined by the same formula as c. Take now a smooth partition $\{f_0, f_1\}$ of the unity subordinated to the open covering $\{(-\varepsilon, \varepsilon), \mathbb{R} \setminus \{0\}\}$ of $\mathbb{R}$, i.e. $f_0 + f_1 = 1$ with $\mathrm{supp}(f_0) \subseteq (-\varepsilon, \varepsilon)$ and $0 \notin \mathrm{supp}(f_1)$. Then $f_0 e_0 + f_1 e_1$ gives a global smooth lifting of c, in contradiction with the case treated above.

Let now $B \subseteq F'$ be bounded. Without loss of generality we may assume that $B = U^o$ for some 0-neighborhood U in F. Since F is a subspace of the Fréchet space E, the set U can be written as $U = F \cap V$ for some 0-neighborhood V in E. Then the bounded set $V^o \subseteq E'$ is mapped onto $B = U^o$ by the Hahn-Banach theorem. □

21.7. Proposition. [Frölicher, Kriegl, 1988], *7.1.7 Let $\iota : F \to E$ be as in* (21.5). *The function $\varphi : F \to \mathbb{R}$ defined by $\varphi(f) := f(f(1))$ for $f(1) \geq 0$ and $\varphi(f) := 0$ for $f(1) < 0$ is smooth but has no smooth extension to E and not even to a neighborhood of F in E.*

Proof. We first show that φ is smooth. Using the bounded linear $\tilde{c} : F \to E$ associated to the smooth curve $c : \mathbb{R} \to F'$ of (21.6) we can write φ as the composite $\operatorname{ev} \circ (\tilde{c}, \operatorname{ev}_1)$ of smooth maps.

Assume now that a smooth global extension $\psi : E \to \mathbb{R}$ of φ exists. Using a fixed smooth function $h : \mathbb{R} \to [0,1]$ with $h(t) = 0$ for $t \leq 0$ and $h(t) = 1$ for $t \geq 1$, we then define a map $\sigma : E \to E$ as follows:

$$(\sigma g)(t) := \psi\big(g + (t - g(1))h\big) - \big(t - g(1)\big)h(t).$$

Obviously $\sigma g \in E$ for any $g \in E$, and using cartesian closedness (3.12) one easily verifies that σ is a smooth map. For $f \in F$ one has, using that $\big(f + (t - f(1))h\big)(1) = t$, the equations

$$(\sigma f)(t) = \big(f + \big(t - f(1)\big)h\big)(t) - \big(t - f(1)\big)h(t) = f(t)$$

for $t \geq 0$ and $(\sigma f)(t) = 0 - (t - f(1))h(t) = 0$ for $t \leq 0$. This means $\sigma f = \tilde{c} f$ for $f \in F$. So one has $\tilde{c} = \sigma \circ \iota$ with σ smooth. Differentiation gives $\tilde{c} = \tilde{c}'(0) = \sigma'(0) \circ \iota$, and $\sigma'(0)$ is a bounded linear mapping $E \to E$. But in the proof of (21.6) it was shown that such an extension of $\tilde{c}$ does not exist.

Let us now assume that a local extension to some neighborhood of F in E exists. This extension could then be multiplied with a smooth function $E \to \mathbb{R}$ being 1 on F and having support inside the neighborhood (E as nuclear Fréchet space has smooth partitions of unity see (16.10)) to obtain a global extension. □

21.8. Remark. As a corollary it is shown in [Frölicher, Kriegl, 1988, 7.1.6] that the category of smooth spaces is not locally cartesian closed, since pullbacks do not commute with coequalizers.

Furthermore, this examples shows that the structure curves of a quotient of a Frölicher space need not be liftable as structure curves and the structure functions on a subspace of a Frölicher space need not be extendable as structure functions.

In fact, since Mackey-convergent sequences are liftable in the example, one can show that every $f : F' \to \mathbb{R}$ is smooth, provided $f \circ \iota^*$ is smooth, see [Frölicher,Kriegl, 1988, 7.1.8].

21.9. Example. In [Jarchow, 1981, 11.6.4] a Fréchet Montel space is given, which has ℓ^1 as quotient. The standard basis in ℓ^1 cannot have a bounded lift, since in a Montel space every bounded set is by definition relatively compact, hence the standard basis would be relatively compact.

21.10. Result. [Jarchow, 1981, remark after 9.4.5]. *Let $q : E \to F$ be a quotient map between Fréchet spaces. Then (Mackey) convergent sequences lift along q.*

This is not true for general spaces. In [Frölicher, Kriegl, 1988, 7.2.10] it is shown that the quotient map $\coprod_{\text{dens}\, A=0} \mathbb{R}^A \to E := \{x \in \mathbb{R}^{\mathbb{N}} : \text{dens}(\text{carr}(x)) = 0\}$ does not lift Mackey-converging sequences. Note, however, that this space is not convenient. We do not know whether smooth curves can be lifted over quotient mappings, even in the case of Banach spaces.

21.11. Example. *There exists a short exact sequence $\ell^2 \xrightarrow{\iota} E \to \ell^2$, which does not split, see* (13.18.6). *The square of the norm on the subspace ℓ^2 does not extend to a smooth function on E.*

Proof. Assume indirectly that a smooth extension of the square of the norm exists. Let $2b$ be the second derivative of this extension at 0, then $b(x, y) = \langle x, y\rangle$ for all $x, y \in \ell^2$, and hence the following diagram commutes

$$\begin{array}{ccc} \ell^2 & \xhookrightarrow{\iota} & E \\ {\cong}\downarrow{\sharp} & & \downarrow{b^\vee} \\ (\ell^2)^* & \xleftarrow{\iota^*} & E^* \end{array}$$

giving a retraction to ι. $\square$

22. Whitney's Extension Theorem Revisited

Whitney's extension theorem [Whitney, 1934] concerns extensions of jets and not of functions. In particular it says, that a real-valued function f from a closed subset $A \subseteq \mathbb{R}$ has a smooth extension if and only if there exists a (not uniquely determined) sequence $f_n : A \to \mathbb{R}$, such that the formal Taylor series satisfies the appropriate remainder conditions, see (22.1). Following [Frölicher, Kriegl, 1993], we will characterize in terms of a simple boundedness condition on the difference quotients those functions $f : A \to \mathbb{R}$ on an arbitrary subset $A \subseteq \mathbb{R}$ which admit a smooth extension $\tilde{f} : \mathbb{R} \to \mathbb{R}$ as well as those which admit an m-times differentiable extension $\tilde{f}$ having locally Lipschitzian derivatives.

We shall use Whitney's extension theorem in the formulation given in [Stein, 1970]. In order to formulate it we recall some definitions.

22.1. Notation on jets. An *m-jet* on A is a family $F = (F^k)_{k\le m}$ of continuous functions on A. With $J^m(A, \mathbb{R})$ one denotes the vector space of all m-jets on A.

The canonical map $j^m : C^\infty(\mathbb{R}, \mathbb{R}) \to J^m(A, \mathbb{R})$ is given by $f \mapsto (f^{(k)}|_A)_{k\le m}$.

For $k \le m$ one has the *'differentiation operator'* $D^k : J^m(A, \mathbb{R}) \to J^{m-k}(A, \mathbb{R})$ given by $D^k : (F^i)_{i\le m} \mapsto (F^{i+k})_{i\le m-k}$.

For $a \in A$ the *Taylor-expansion operator* $T_a^m : J^m(A,\mathbb{R}) \to C^\infty(\mathbb{R},\mathbb{R})$ of order m at a is defined by $T_a^m((F^i)_{i\le m}) : x \mapsto \sum_{k\le m} \frac{(x-a)^k}{k!} F^k(a)$.

Finally the *remainder operator* $R_a^m : J^m(A,\mathbb{R}) \to J^m(A,\mathbb{R})$ at a of order m is given by $F \mapsto F - j^m(T_a^m F)$.

In [Stein, 1970, p.176] the space $\mathcal{L}\mathrm{ip}(m+1,A)$ denotes all m-jets on A for which there exists a constant $M > 0$ such that

$$|F^j(a)| \le M \text{ and } \left|(R_a^m F)^j(b)\right| \le M\,|a-b|^{m+1-j}$$

for all $a, b \in A$ and all $j \le m$.

The smallest constant M defines a norm on $\mathcal{L}\mathrm{ip}(m+1,A)$.

22.2. Whitney's Extension. The construction of Whitney for finite order m goes as follows, see [Malgrange, 1966], [Tougeron, 1972] or [Stein, 1970]:

First one picks a special partition of unity Φ for $\mathbb{R}^n \setminus A$ satisfying in particular $\mathrm{diam}(\mathrm{supp}\,\varphi) \le 2\,d(\mathrm{supp}\,\varphi, A)$ for $\varphi \in \Phi$. For every $\varphi \in \Phi$ one chooses a nearest point $a_\varphi \in A$, i.e. a point a_φ with $d(\mathrm{supp}\,\varphi, A) = d(\mathrm{supp}\,\varphi, a_\varphi)$. The extension $\tilde{F}$ of the jet F is then defined by

$$\tilde{F}(x) := \begin{cases} F^0(x) & \text{for } x \in A \\ \sum_{\varphi\in\Phi'} \varphi(x) T_{a_\varphi}^m F(x) & \text{otherwise,} \end{cases}$$

where the set Φ' consists of all $\varphi \in \Phi$ such that $d(\mathrm{supp}\,\varphi, A) \le 1$.

The version of [Stein, 1970, theorem 4, p. 177] of Whitney's extension theorem is:

Whitney's Extension Theorem. *Let m be an integer and A a compact subset of $\mathbb{R}$. Then the assignment $F \mapsto \tilde{F}$ defines a bounded linear mapping $\mathcal{E}^m : \mathcal{L}\mathrm{ip}(m+1,A) \to \mathcal{L}\mathrm{ip}(m+1,\mathbb{R})$ such that $\mathcal{E}^m(F)|_A = F^0$.*

In order that $\mathcal{E}^m$ makes sense, one has to identify $\mathcal{L}\mathrm{ip}(m+1,\mathbb{R})$ with a space of functions (and not jets), namely those functions on $\mathbb{R}$ which are m-times differentiable on $\mathbb{R}$ and the m-th derivative is Lipschitzian. In this way $\mathcal{L}\mathrm{ip}(m+1,\mathbb{R})$ is identified with the space $\mathcal{L}\mathrm{ip}^m(\mathbb{R},\mathbb{R})$ in (1.2) (see also (12.10)).

Remark. The original condition of [Whitney, 1934] which guarantees a C^m-extension is:

$$(R_a^m F)^k(b) = o(|a-b|^{m-k}) \text{ for } a, b \in A \text{ with } |a-b| \to 0 \text{ and } k \le m.$$

In the following A will be an arbitrary subset of $\mathbb{R}$.

22.3. Difference Quotients. The definition of difference quotients $\delta^k f$ given in (12.4) works also for functions $f : A \to \mathbb{R}$ defined on arbitrary subsets $A \subseteq \mathbb{R}$. The natural domain of definition of $\delta^k f$ is the subset $A^{<k>}$ of A^{k+1} of pairwise distinct points, i.e.

$$A^{<k>} := \left\{(t_0, \dots, t_k) \in A^{k+1} : t_i \ne t_j \text{ for all } i \ne j\right\}.$$

The following product rule can be found for example in [Verde-Star, 1988] or [Frölicher, Kriegl, 1993, 3.3].

22.4. The Leibniz product rule for difference quotients.

$$\delta^k(f\cdot g)(t_0,\dots,t_k)=\sum_{i=0}^{k}\binom{k}{i}\delta^i f(t_0,\dots,t_i)\cdot\delta^{k-i}g(t_i,\dots,t_k)$$

Proof. This is easily proved by induction on k. □

We will make strong use of interpolation polynomials as they have been already used in the proof of lemma (12.4). The following descriptions are valid for them:

22.5. Lemma. Interpolation polynomial. *Let $f:A\to E$ be a function with values in a vector space E and let $(t_0,\dots,t_m)\in A^{<m>}$. Then there exists a unique polynomial $P^m_{(t_0,\dots,t_m)}f$ of degree at most m which takes the values $f(t_j)$ on t_j for all $j=0,\dots,m$. It can be written in the following ways:*

$$P^m_{(t_0,\dots,t_m)}f: t\mapsto\sum_{k=0}^{m}\frac{1}{k!}\delta^k f(t_0,\dots,t_k)\prod_{j=0}^{k-1}(t-t_j)\qquad (\textit{Newton})$$

$$t\mapsto\sum_{k=0}^{m}f(t_k)\prod_{j\neq k}\frac{t-t_j}{t_k-t_j}\qquad (\textit{Lagrange}).$$

See, for example, [Frölicher, Kriegl, 1988, 1.3.7] for a proof of the first description. The second one is obvious.

22.6. Lemma. *For pairwise distinct points $a,b,t_1,\dots,t_m$ and $k\le m$ one has:*

$$\begin{aligned}&\Big(P^m_{(a,t_1,\dots,t_m)}f-P^m_{(b,t_1,\dots,t_m)}f\Big)^{(k)}(t)=\\&=(a-b)\tfrac{1}{(m+1)!}\,\delta^{m+1}f(a,b,t_1,\dots,t_m)\cdot\\&\quad\cdot k!\sum_{i_1<\dots<i_k}(t-t_1)\cdot\ldots\cdot\widehat{(t-t_{i_1})}\cdot\ldots\cdot\widehat{(t-t_{i_k})}\cdot\ldots\cdot(t-t_m).\end{aligned}$$

Proof. For the interpolation polynomial we have

$$\begin{aligned}&P^m_{(a,t_1,\dots,t_m)}f(t)=P^m_{(t_1,\dots,t_m,a)}f(t)=\\&=f(t_1)+\dots+(t-t_1)\cdot\ldots\cdot(t-t_{m-1})\tfrac{1}{(m-1)!}\,\delta^{m-1}f(t_1,\dots,t_m)\\&\qquad+(t-t_1)\cdot\ldots\cdot(t-t_m)\tfrac{1}{m!}\,\delta^m f(t_1,\dots,t_m,a).\end{aligned}$$

Thus we obtain

$$\begin{aligned}&P^m_{(a,t_1,\dots,t_m)}f(t)-P^m_{(b,t_1,\dots,t_m)}f(t)=\\&=0+\dots+0+(t-t_1)\cdot\ldots\cdot(t-t_m)\tfrac{1}{m!}\,\delta^m f(t_1,\dots,t_m,a)\\&\qquad-(t-t_1)\cdot\ldots\cdot(t-t_m)\tfrac{1}{m!}\,\delta^m f(t_1,\dots,t_m,b)\\&=(t-t_1)\cdot\ldots\cdot(t-t_m)\tfrac{1}{m!}\tfrac{a-b}{m+1}\,\delta^{m+1}f(t_1,\dots,t_m,a,b)\\&=(a-b)\cdot(t-t_1)\cdot\ldots\cdot(t-t_m)\tfrac{1}{(m+1)!}\,\delta^{m+1}f(a,b,t_1,\dots,t_m).\end{aligned}$$

Differentiation using the product rule (22.4) gives the result. □

22.7. Proposition. *Let $f : A \to \mathbb{R}$ be a function, whose difference quotient of order $m+1$ is bounded on $A^{<m+1>}$. Then the approximation polynomial $P^m_{\mathbf{a}} f$ converges to some polynomial denoted by $P^m_{\mathbf{x}} f$ of degree at most m if the point $\mathbf{a} \in A^{<m>}$ converges to $\mathbf{x} \in A^{m+1}$.*

Proof. We claim that $P^m_{\mathbf{a}} f$ is a Cauchy net for $A^{<m>} \ni \mathbf{a} \to \mathbf{x}$. Since $P^m_{\mathbf{a}} f$ is symmetric in the entries of $\mathbf{a}$ we may assume without loss of generality that the entries x_j of $\mathbf{x}$ satisfy $x_0 \le x_1 \le \dots \le x_m$. For a point $\mathbf{a} \in A^{<m>}$ which is close to $\mathbf{x}$ and any two coordinates i and j with $x_i < x_j$ we have $a_i < a_j$. Let $\mathbf{a}$ and $\mathbf{b}$ be two points close to $\mathbf{x}$. Let J be a set of indices on which $\mathbf{x}$ is constant. If the set $\{a_j : j \in J\}$ differs from the set $\{b_j : j \in J\}$, then we may order them as in the proof of lemma (12.4) in such a way that $a_i \ne b_j$ for $i \le j$ in J. If the two sets are equal we order both strictly increasing and thus have $a_i < a_j = b_j$ for $i < j$ in J. Since $\mathbf{x}$ is constant on J the distance $|a_i - b_j| \le |a_i - x_i| + |x_j - b_j|$ goes to zero as $\mathbf{a}$ and $\mathbf{b}$ approach $\mathbf{x}$. Altogether we obtained that $a_i \ne b_j$ for all $i < j$ and applying now (22.6) for $k = 0$ inductively one obtains:

$$
\begin{aligned}
&P^m_{(a_0,\dots,a_m)} f(t) - P^m_{(b_0,\dots,b_m)} f(t) = \\
&\quad = \sum_{j=0}^m \Big(P^m_{(a_0,\dots,a_{j-1},b_j,\dots,b_m)} f(t) - P^m_{(a_0,\dots,a_j,b_{j+1},\dots,b_m)} f(t) \Big) \\
&\quad = \sum_{j=0}^m (a_j - b_j)(t-a_0)\dots(t-a_{j-1})(t-b_{j+1})\dots(t-b_m)\cdot \\
&\qquad\qquad \cdot \tfrac{1}{(m+1)!}\delta^{m+1} f(a_0,\dots,a_j,b_j,\dots,b_m).
\end{aligned}
$$

Where those summands with $a_j = b_j$ have to be defined as 0. Since $a_j - b_j \to 0$ the claim is proved and thus also the convergence of $P^m_{\mathbf{a}} f$. □

22.8. Definition of $\mathcal{L}\text{ip}^k$ function spaces. Let E be a convenient vector space, let A be a subset of $\mathbb{R}$ and k a natural number or 0. Then we denote with $\mathcal{L}\text{ip}^k_{\text{ext}}(A, E)$ the vector space of all maps $f : A \to E$ for which the difference quotient of order $k+1$ is bounded on bounded subsets of $A^{<k>}$. As in (12.10) – but now for arbitrary subsets $A \subseteq \mathbb{R}$ – we put on this space the initial locally convex topology induced by $f \mapsto \delta^j f \in \ell^\infty(A^{\langle j \rangle}, E)$ for $0 \le j \le k+1$, where the spaces ℓ^∞ carry the topology of uniform convergence on bounded subsets of $A^{\langle j \rangle} \subseteq \mathbb{R}^{j+1}$.

In case where $A = \mathbb{R}$ the elements of $\mathcal{L}\text{ip}^k_{\text{ext}}(A, \mathbb{R})$ are exactly the k-times differentiable functions on $\mathbb{R}$ having a locally Lipschitzian derivative of order $k+1$ and the locally convex space $\mathcal{L}\text{ip}^k_{\text{ext}}(A, \mathbb{R})$ coincides with the convenient vector space $\mathcal{L}\text{ip}^k(\mathbb{R}, \mathbb{R})$ studied in section (12).

If k is infinite, then $\mathcal{L}\text{ip}^\infty_{\text{ext}}(A, E)$ or alternatively $C^\infty_{\text{ext}}(A, E)$ denotes the intersection of $\mathcal{L}\text{ip}^j_{\text{ext}}(A, E)$ for all finite j.

If $A = \mathbb{R}$ then the elements of $C^\infty_{\text{ext}}(\mathbb{R}, \mathbb{R})$ are exactly the smooth functions on $\mathbb{R}$ and the space $C^\infty_{\text{ext}}(\mathbb{R}, \mathbb{R})$ coincides with the usual Fréchet space $C^\infty(\mathbb{R}, \mathbb{R})$ of all smooth functions.

22.9. Proposition. Uniform boundedness principle for $\mathcal{L}\mathrm{ip}^k_{\mathrm{ext}}$. *For any finite or infinite k and any convenient vector space E the space $\mathcal{L}\mathrm{ip}^k_{\mathrm{ext}}(A,E)$ is also convenient. It carries the initial structure with respect to*

$$\ell_* : \mathcal{L}\mathrm{ip}^k_{\mathrm{ext}}(A,E) \to \mathcal{L}\mathrm{ip}^k_{\mathrm{ext}}(A,\mathbb{R}) \textit{ for } \ell \in E'.$$

Moreover, it satisfies the $\{\mathrm{ev}_x : x \in A\}$-uniform boundedness principle. If E is Fréchet then so is $\mathcal{L}\mathrm{ip}^k_{\mathrm{ext}}(A,E)$.

Proof. We consider the commutative diagram

$$\begin{array}{ccc}
\mathcal{L}\mathrm{ip}^m_{\mathrm{ext}}(A,E) & \xrightarrow{\ \ell_*\ } & \mathcal{L}\mathrm{ip}^m_{\mathrm{ext}}(A,\mathbb{R}) \\
{\scriptstyle \delta^j}\big\downarrow & & \big\downarrow{\scriptstyle \delta^j} \\
\ell^\infty(A^{<j>},E) & \xrightarrow{\ \ell_*\ } & \ell^\infty(A^{<j>},\mathbb{R})
\end{array}$$

Obviously the bornology is initial with respect to the bottom arrows for $\ell \in E'$ and by definition also with respect to the vertical arrows for $j \le k+1$. Thus also the top arrows form an initial source. By (2.15) the spaces in the bottom row are c^∞-complete and are metrizable if E is metrizable. Since the boundedness of the difference quotient of order $k+1$ implies that of order $j \le k+1$, also $\mathcal{L}\mathrm{ip}^m_{\mathrm{ext}}(A,E)$ is convenient, and it is Fréchet provided E is. The uniform boundedness principle follows also from this diagram, using the stability property (5.25) and that the Fréchet and hence webbed space $\ell^\infty(A^{<j>},\mathbb{R})$ has it by (5.24). □

22.10. Proposition. *For a convenient vector space E the following operators are well-defined bounded linear mappings:*

(1) *The restriction operator $\mathcal{L}\mathrm{ip}^m_{\mathrm{ext}}(A_1,E) \to \mathcal{L}\mathrm{ip}^m_{\mathrm{ext}}(A_2,E)$ defined by $f \mapsto f|_{A_2}$ for $A_2 \subseteq A_1$.*

(2) *For $g \in \mathcal{L}\mathrm{ip}^m_{\mathrm{ext}}(A,\mathbb{R})$ the multiplication operator*

$$\begin{aligned}\mathcal{L}\mathrm{ip}^m_{\mathrm{ext}}(A,E) &\to \mathcal{L}\mathrm{ip}^m_{\mathrm{ext}}(A,E)\\ f &\mapsto g\cdot f.\end{aligned}$$

(3) *The gluing operator*

$$\mathcal{L}\mathrm{ip}^m_{\mathrm{ext}}(A_1,E) \times_{A_1\cap A_2} \mathcal{L}\mathrm{ip}^m_{\mathrm{ext}}(A_2,E) \to \mathcal{L}\mathrm{ip}^m_{\mathrm{ext}}(A,E)$$

defined by $(f_1,f_2) \mapsto f_1 \cup f_2$ for any covering of A by relatively open subsets $A_1 \subseteq A$ and $A_2 \subseteq A$.

The fibered product (pull back) $\mathcal{L}\mathrm{ip}^m_{\mathrm{ext}}(A_1,E)\times_{A_1\cap A_2}\mathcal{L}\mathrm{ip}^m_{\mathrm{ext}}(A_2,E) \to \mathcal{L}\mathrm{ip}^m_{\mathrm{ext}}(A,E)$ is the subspace of $\mathcal{L}\mathrm{ip}^m_{\mathrm{ext}}(A_1,E) \times \mathcal{L}\mathrm{ip}^m_{\mathrm{ext}}(A_2,E)$ formed by all (f_1,f_2) with $f_1 = f_2$ on $A_0 := A_1 \cap A_2$.

Proof. It is enough to consider the particular case where $E = \mathbb{R}$. The general case follows easily by composing with ℓ_* for each $\ell \in E'$.

(1) is obvious.
(2) follows from the Leibniz formula (22.4).
(3) First we show that the gluing operator has values in $\mathcal{L}ip^m_{\text{ext}}(A,\mathbb{R})$. Suppose the difference quotient $\delta^j f$ is not bounded for some $j \le m+1$, which we assume to be minimal. So there exists a bounded sequence $\mathbf{x}^n \in A^{<j>}$ such that $(\delta^j f)(\mathbf{x}^n)$ converges towards infinity. Since A is compact we may assume that $\mathbf{x}^n$ converges to some point $\mathbf{x}^\infty \in A^{(j+1)}$. If $\mathbf{x}^\infty$ does not lie on the diagonal, there are two indices $i_1 \neq i_2$ and some $\delta > 0$, such that $|\mathbf{x}^n{}_{i_1} - \mathbf{x}^n{}_{i_2}| \ge \delta$. But then

$$\delta^j f(\mathbf{x}^n)(\mathbf{x}^n{}_{i_1} - \mathbf{x}^n{}_{i_2}) = \tfrac{1}{j}\left(\delta^{j-1} f(\dots,\widehat{\mathbf{x}^n{}_{i_2}},\dots) - \delta^{j-1} f(\dots,\widehat{\mathbf{x}^n{}_{i_1}},\dots)\right).$$

Which is a contradiction to the boundedness of $\delta^{j-1} f$ and hence the minimality of j. So $\mathbf{x}^\infty = (x^\infty,\dots,x^\infty)$ and since the covering $\{A_1, A_2\}$ of A is open x^∞ lies in A_i for $i=1$ or $i=2$. Thus we have that $\mathbf{x}^n \in A_i{}^{<j>}$ for almost all n, and hence $\delta^j f(\mathbf{x}^n) = \delta^j f_i(\mathbf{x}^n)$, which is bounded by assumption on f_i.

Because of the uniform boundedness principle (22.9) it only remains to show that $(f_1, f_2) \mapsto f(a)$ is bounded, which is obvious since $f(a) = f_i(a)$ for some i depending on the location of a. □

22.11. Remark. If A is finite, we define an extension $\tilde{f} : \mathbb{R} \to E$ of the given function $f : A \to E$ as the interpolation polynomial of f at all points in A. For infinite compact sets $A \subset \mathbb{R}$ we will use Whitney's extension theorem (22.2), where we will replace the Taylor polynomial in the definition (22.2) of the extension by the interpolation polynomial at appropriately chosen points near a_φ. For this we associate to each point $a \in A$ a sequence $\mathbf{a} = (a_0, a_1, \dots)$ of points in A starting from the given point $a_0 = a$.

22.12. Definition of $a \mapsto \mathbf{a}$. Let A be a closed infinite subset of $\mathbb{R}$, and let $a \in A$. Our aim is to define a sequence $\mathbf{a} = (a_0, a_1, a_2, \dots)$ in a certain sense close to a. The construction is by induction and goes as follows: $a_0 := a$. For the induction step we choose for every non-empty finite subset $F \subset A$ a point a_F in the closure of $A \setminus F$ having minimal distance to F. In case F does not contain an accumulation point the set $A \setminus F$ is closed and hence $a_F \notin F$, otherwise the distance of $A \setminus F$ to F is 0 and a_F is an accumulation point in F. In both cases we have for the distances $d(a_F, F) = d(A \setminus F, F)$. Now suppose $(a_0, \dots, a_{j-1})$ is already constructed. Then let $F := \{a_0, \dots, a_{j-1}\}$ and define $a_j := a_F$.

Lemma. *Let $\mathbf{a} = (a_0, \dots)$ and $\mathbf{b} = (b_0, \dots)$ be constructed as above. If $\{a_0, \dots, a_k\} \neq \{b_0, \dots, b_k\}$ then we have for all $i, j \le k$ the estimates*

$$|a_i - b_j| \le (i + j + 1)\,|a_0 - b_0|$$
$$|a_i - a_j| \le \max\{i, j\}\,|a_0 - b_0|$$
$$|b_i - b_j| \le \max\{i, j\}\,|a_0 - b_0|.$$

Proof. First remark that if $\{a_0, \dots, a_i\} = \{b_0, \dots, b_i\}$ for some i, then the same is true for all larger i, since the construction of a_{i+1} depends only on the set

$\{a_0, \ldots, a_i\}$. Furthermore the set $\{a_0, \ldots, a_i\}$ contains at most one accumulation point, since for an accumulation point a_j with minimal index j we have by construction that $a_j = a_{j+1} = \cdots = a_i$.

We now show by induction on $i \in \{1, \ldots, k\}$ that

$$d(a_{i+1}, \{a_0, \ldots, a_i\}) \leq |a_0 - b_0|,$$
$$d(b_{i+1}, \{b_0, \ldots, b_i\}) \leq |a_0 - b_0|.$$

We proof this statement for a_{i+1}, it then follows for b_{i+1} by symmetry.
In case where $\{a_0, \ldots, a_i\} \supseteq \{b_0, \ldots, b_i\}$ we have that $\{a_0, \ldots, a_i\} \supset \{b_0, \ldots, b_i\}$ by assumption. Thus some of the elements of $\{b_0, \ldots, b_i\}$ have to be equal and hence are accumulation points. So $\{a_0, \ldots, a_i\}$ contains an accumulation point, and hence $a_{i+1} \in \{a_0, \ldots, a_i\}$ and the claimed inequality is trivially satisfied.
In the other case there exist some $j \leq i$ such that $b_j \notin \{a_0, \ldots, a_i\}$. We choose the minimal j with this property and obtain

$$d(a_{i+1}, \{a_0, \ldots, a_i\}) := d(A \setminus \{a_0, \ldots, a_i\}, \{a_0, \ldots, a_i\}) \leq d(b_j, \{a_0, \ldots, a_i\}).$$

If $j = 0$, then this can be further estimated as follows

$$d(b_j, \{a_0, \ldots, a_i\}) \leq |a_0 - b_0|.$$

Otherwise $\{b_0, \ldots, b_{j-1}\} \subseteq \{a_0, \ldots, a_j\}$ and hence we have

$$d(b_j, \{a_0, \ldots, a_i\}) \leq d(b_j, \{b_0, \ldots, b_{j-1}\}) \leq |a_0 - b_0|$$

by induction hypothesis. Thus the induction is completed.

From the proven inequalities we deduce by induction on $k := \max\{i, j\}$ that

$$|a_i - a_j| \leq \max\{i, j\}\, |a_0 - b_0|$$

and similarly for $|b_j - b_i|$:
For $k = 0$ this is trivial. Now for $k > 0$. We may assume that $i > j$. Let $i' < i$ be such that $|a_i - a_{i'}| = d(a_i, \{a_0, \ldots, a_{i-1}\}) \leq |a_0 - b_0|$. Thus by induction hypothesis $|a_{i'} - a_j| \leq (k-1)\, |a_0 - b_0|$ and hence

$$|a_i - a_j| \leq |a_i - a_{i'}| + |a_{i'} - a_j| \leq k\, |a_0 - b_0|.$$

By the triangle inequality we finally obtain

$$|a_i - b_j| \leq |a_i - a_0| + |a_0 - b_0| + |b_0 - b_j| \leq (i + 1 + j)\, |a_0 - b_0|. \quad \square$$

22.13. Finite Order Extension Theorem. *Let E be a convenient vector space, A a subset of $\mathbb{R}$ and m be a natural number or 0. A function $f : A \to E$ admits an extension to $\mathbb{R}$ which is m-times differentiable with locally Lipschitzian m-th derivative if and only if its difference quotient of order $m+1$ is bounded on bounded sets.*

Proof. Without loss of generality we may assume that A is infinite. We consider first the case that A is compact and $E = \mathbb{R}$.

So let $f : A \to \mathbb{R}$ be in $\mathcal{L}\mathrm{ip}^m_{\mathrm{ext}}$. We want to apply Whitney's extension theorem (22.2). So we have to find an m-jet F on A. For this we define

$$F^k(a) := (P^m_{\mathbf{a}} f)^{(k)}(a),$$

where $\mathbf{a}$ denotes the sequence obtained by this construction starting with the point $a$and where $P^m_{\mathbf{a}} f$ denotes the interpolation polynomial of f at the first $m+1$ points of $\mathbf{a}$ if these are all different; if not, at least one of these $m+1$ points is an accumulation point of A and then $P^m_{\mathbf{a}} f$ is taken as limit of interpolation polynomials according to (22.7).

Let Φ be the partition of unity mentioned in (22.2) and Φ' the subset specified there. Then we define $\tilde{f}$ analogously to (22.2) where $\mathbf{a}_\varphi$ denotes the sequence obtained by construction (22.12) starting with the point a_φ chosen in (22.2):

$$\tilde{f}(x) := \begin{cases} f(x) & \text{for } x \in A \\ \sum_{\varphi \in \Phi'} \varphi(x) P^m_{\mathbf{a}_\varphi} f(x) & \text{otherwise.} \end{cases}$$

In order to verify that F belongs to $\mathcal{L}\mathrm{ip}(m+1, A)$ we need the Taylor polynomial

$$T^m_a F(x) := \sum_{k=0}^m \frac{(x-a)^k}{k!} F^k(a) = \sum_{k=0}^m \frac{(x-a)^k}{k!} (P^m_{\mathbf{a}} f)^{(k)}(a) = P^m_{\mathbf{a}} f(x),$$

where the last equation holds since $P^m_{\mathbf{a}} f$ is a polynomial of degree at most m. This shows that our extension $\tilde{f}$ coincides with the classical extension $\tilde{F}$ given in (22.2) of the m-jet F constructed from f.

The remainder term $R^m_a F := F - j^m(T^m_a F)$ is given by:

$$(R^m_a F)^k(b) = F^k(b) - (T^m_a F)^{(k)}(b) = (P^m_{\mathbf{b}} f)^{(k)}(b) - (P^m_{\mathbf{a}} f)^{(k)}(b)$$

We have to show that for some constant M one has $\left|(R^m_a F)^k(b)\right| \le M|a-b|^{m+1-k}$ for all $a, b \in A$ and all $k \le m$.

In order to estimate this difference we write it as a telescoping sum of terms which can written by (22.6) as

$$\begin{aligned} &\left(P^m_{(a_0,\dots,a_{i-1},b_i,b_{i+1},\dots,b_m)} f - P^m_{(a_0,\dots,a_{i-1},a_i,b_{i+1},\dots,b_m)} f\right)^{(k)}(t) = \\ &= \frac{k!}{(m+1)!} \delta^{m+1} f(a_0, \dots, a_i, b_i, \dots, b_m) \cdot \\ &\quad \cdot (b_i - a_i) \sum_{i_1 < \dots < i_k} (t-a_0) \dots \widehat{(t-a_{i_1})} \dots \widehat{(t-b_{i_k})} \dots (t-b_m). \end{aligned}$$

Note that this formula remains valid also in case where the points are not pairwise different. This follows immediately by passing to the limit with the help of (22.7).

We have estimates for the distance of points in $\{a_0, \dots, a_m; b_0, \dots, b_m\}$ by (22.12) and so we obtain the required constant M as follows

$$
\begin{aligned}
|(R_a^m F)^k(b)| \leq \frac{k!}{(m+1)!} &\sum_{i=0}^{m} (2i+1)\, |b-a|^{m+1-k} \\
&\sum_{i_1 < \dots < i_k} 1 \cdot 2 \cdot \ldots \cdot \widehat{(1+i_1)} \ldots \widehat{i_k} \ldots \cdot m \cdot \\
&\cdot \max\{|\delta^{m+1} f(\{a_0, \dots, a_m, b_0, \dots, b_m\}^{<m+1>})|\}.
\end{aligned}
$$

In case, where E is an arbitrary convenient vector space we define an extension $\tilde{f}$ for $f \in \mathcal{L}ip^m_{\text{ext}}(A, E)$ by the same formula as before. Since Φ' is locally finite, this defines a function $\tilde{f} : \mathbb{R} \to E$. In order to show that $\tilde{f} \in \mathcal{L}ip^m(\mathbb{R}, E)$ we compose with an arbitrary $\ell \in E'$. Then $\ell \circ \tilde{f}$ is just the extension of $\ell \circ f$ given above, thus belongs to $\mathcal{L}ip^m(\mathbb{R}, \mathbb{R})$.

Let now A be a closed subset of $\mathbb{R}$. Then let the compact subsets $A_n \subset \mathbb{R}$ be defined by $A_1 := A \cap [-2, 2]$ and $A_n := [-n+1, n-1] \cup (A \cap [-n-1, n+1])$ for $n > 1$. We define recursively functions $f_n \in \mathcal{L}ip^m_{\text{ext}}(A_n, E)$ as follows: Let f_1 be a $\mathcal{L}ip^m$-extension of $f|_{A_1}$. Let $f_n : A_n \to \mathbb{R}$ be a $\mathcal{L}ip^m$-extension of the function which equals f_{n-1} on $[-n+1, n-1]$ and which equals f on $A \cap [-n-1, n+1]$. This definition makes sense, since the two sets

$$
\begin{aligned}
A_n \setminus [-n+1, n-1] &= A \cap \big([-n-1, n+1] \setminus [-n+1, n-1]\big), \\
A_n \setminus \big([-n-1, -n] \cup [n, n+1]\big) &= [-n+1, n-1] \cup \big(A \cap [-n, n]\big)
\end{aligned}
$$

form an open cover of A_n, and their intersection is contained in the set $A \cap [-n, n]$ on which f and f_{n-1} coincide. Now we apply (22.10). The sequence f_n converges uniformly on bounded subsets of $\mathbb{R}$ to a function $\tilde{f} : \mathbb{R} \to E$, since $f_j = f_n$ on $[-n, n]$ for all $j > n$. Since each f_n is $\mathcal{L}ip^m$, so is $\tilde{f}$. Furthermore, $\tilde{f}$ is an extension of f, since $\tilde{f} = f_n$ on $[-n, n]$ and hence on $A \cap [-n+1, n-1]$ equal to f.

Finally the case, where $A \subseteq \mathbb{R}$ is completely arbitrary. Let $\bar{A}$ denote the closure of A in $\mathbb{R}$. Since the first difference quotient is bounded on bounded subsets of A one concludes that f is Lipschitzian and hence uniformly continuous on bounded subsets of A, moreover, the values $f(a)$ form a Mackey Cauchy net for $A \ni a \to \bar{a} \in \mathbb{R}$. Thus f has a unique continuous extension $\tilde{f}$ to $\bar{A}$, since the limit $\tilde{f}(\bar{a}) := \lim_{a \to \bar{a}} f(a)$ exists in E, because E is convenient. Boundedness of the difference quotients of order j of $\tilde{f}$ can be tested by composition with linear continuous functionals, so we may assume $E = \mathbb{R}$. Its value at $(\tilde{t}_0, \dots, \tilde{t}_j) \in \bar{A}^{<j>}$ is the limit of $\delta^j f(t_0, \dots, t_j)$, where $A^{<j>} \ni (t_0, \dots, t_j)$ converges to $(\tilde{t}_0, \dots, \tilde{t}_j)$, since in the explicit formula for δ^j the factors $f(t_i)$ converge to $\tilde{f}(\tilde{t}_i)$. Now we may apply the result for closed A to obtain the required extension. □

22.14. Extension Operator Theorem. *Let E be a convenient vector space and let m be finite. Then the space $\mathcal{L}\mathrm{ip}^m_{\mathrm{ext}}(A,E)$ of functions having an extension in the sense of (22.13) is a convenient vector space and there exists a bounded linear extension operator from $\mathcal{L}\mathrm{ip}^m_{\mathrm{ext}}(A,E)$ to $\mathcal{L}\mathrm{ip}^m(\mathbb{R},E)$.*

Proof. This follows from (21.2).

Explicitly the proof runs as follows: For any convenient vector space E we have to construct a bounded linear operator

$$T:\mathcal{L}\mathrm{ip}^m_{\mathrm{ext}}(A,E)\to\mathcal{L}\mathrm{ip}^m(\mathbb{R},E)$$

satisfying $T(f)|_A=f$ for all $f\in\mathcal{L}\mathrm{ip}^m_{\mathrm{ext}}(A,E)$. Since $\mathcal{L}\mathrm{ip}^m_{\mathrm{ext}}(A,E)$ is a convenient vector space, this is by (12.12) via a flip of variables equivalent to the existence of a $\mathcal{L}\mathrm{ip}^m$-curve

$$\tilde{T}:\mathbb{R}\to L(\mathcal{L}\mathrm{ip}^m_{\mathrm{ext}}(A,E),E)$$

satisfying $\tilde{T}(a)(f)=T(f)(a)=f(a)$. Thus $\tilde{T}$ should be a $\mathcal{L}\mathrm{ip}^m$-extension of the map $e:A\to L(\mathcal{L}\mathrm{ip}^m_{\mathrm{ext}}(A,E),E)$ defined by $e(a)(f):=f(a)=\mathrm{ev}_a(f)$.

By the vector valued finite order extension theorem (22.13) it suffices to show that this map e belongs to $\mathcal{L}\mathrm{ip}^m_{\mathrm{ext}}(A,L(\mathcal{L}\mathrm{ip}^m_{\mathrm{ext}}(A,E),E))$. So consider the difference quotient $\delta^{m+1}e$ of e. Since, by the linear uniform boundedness principle (5.18), boundedness in $L(F,E)$ can be tested pointwise, we consider

$$\begin{aligned}\delta^{m+1}e(a_0,\dots,a_{m+1})(f)&=\delta^{m+1}(\mathrm{ev}_f\circ e)(a_0,\dots,a_{m+1})\\&=\delta^{m+1}f(a_0,\dots,a_{m+1}).\end{aligned}$$

This expression is bounded for $(a_0,\dots,a_{m+1})$ varying in bounded sets, since $f\in\mathcal{L}\mathrm{ip}^m_{\mathrm{ext}}(A,E)$. □

In order to obtain a extension theorem for smooth mappings, we use a modification of the original construction of [Whitney, 1934]. In particular we need the following result.

22.15. Result. [Malgrange, 1966, lemma 4.2], also [Tougeron, 1972, lemme 3.3]. *There exist constants c_k, such that for any compact set $K\subset\mathbb{R}$ and any $\delta>0$ there exists a smooth function h_δ on $\mathbb{R}$ which satisfies*

(1) *$h_\delta=1$ locally around K and $h_\delta(x)=0$ for $d(x,K)\ge\delta$;*
(2) *for all $x\in\mathbb{R}$ and $k\ge0$ one has: $\left|h_\delta^{(k)}(x)\right|\le\frac{c_k}{\delta^k}$.*

22.16. Lemma. *Let A be compact and A_{acc} be the compact set of accumulation points of A. We denote by $C^\infty_A(\mathbb{R},\mathbb{R})$ the set of smooth functions on $\mathbb{R}$ which vanish on A. For finite m we denote by $C^m_A(\mathbb{R},\mathbb{R})$ the set of C^m-functions on $\mathbb{R}$, which vanish on A, are m-flat on A_{acc} and are smooth on the complement of A_{acc}. Then $C^\infty_A(\mathbb{R},\mathbb{R})$ is dense in $C^{m+1}_A(\mathbb{R},\mathbb{R})$ with respect to the structure of $C^m(\mathbb{R},\mathbb{R})$.*

Proof. Let $\varepsilon>0$ and let $g\in C^{m+1}_A(\mathbb{R},\mathbb{R})$ be the function which we want to approximate. By Taylor's theorem we have for $f\in C^{m+1}(\mathbb{R},\mathbb{R})$ the equation

$$f(x)-\sum_{i=0}^{k}\frac{f^{(i)}(a)}{i!}(x-a)^i=(x-a)^{k+1}\frac{f^{(k+1)}(\xi)}{(k+1)!}$$

for some ξ between a and x. If we apply this equation for $j \leq m$ and $k = m - j$ to $g^{(j)}$ for some point $a \in A_{acc}$ we obtain

$$\left|g^{(j)}(x) - 0\right| \leq |x-a|^{m+1-j} \left|\frac{g^{(m+1)}}{(m+1-j)!}(\xi)\right|$$

Taking the infimum over all $a \in A_{acc}$ we obtain a constant

$$K := \sup\left\{\left|\frac{g^{(m+1)}}{(m+1-j)!}(\xi)\right| : d(\xi, A_{acc}) \leq 1\right\}$$

$$\text{satisfying} \quad \left|g^{(j)}(x)\right| \leq K \cdot d(x, A_{acc})^{m+1-j}$$

for all x with $d(x, A) \leq 1$.

We choose $0 < \delta < 1$ depending on ε such that $\delta \cdot \max\{c_i : i \leq m\} \cdot K \cdot 2^m \leq \varepsilon$, and let h_δ be the function given in (22.15) for $K := A_{acc}$. The function $(1 - h_\delta) \cdot g$ is smooth, since on $\mathbb{R} \setminus A_{acc}$ both factors are smooth and on a neighborhood of A_{acc} one has $h_\delta = 1$. The function $(1 - h_\delta) \cdot g$ equals g on $\{x : d(x, A_{acc}) \geq \delta\}$, since h_δ vanishes on this set. So it remains to show that the derivatives of $h_\delta \cdot g$ up to order m are bounded by ε on $\{x : d(x, A_{acc}) \leq \delta\}$. By the Leibniz rule we have:

$$(h_\delta \cdot g)^{(j)} = \sum_{i=0}^{j} \binom{j}{i} h_\delta^{(i)} g^{(j-i)}.$$

The i-th summand can be estimated as follows:

$$\left|h_\delta^{(i)}(x) g^{(j-i)}(x)\right| \leq \frac{c_i}{\delta^i} K\, d(x, A_{acc})^{m+1+i-j} \leq c_i\, K\, \delta^{m+1-j}$$

An estimate for the derivative now is

$$\begin{aligned}\left|(h_\delta \cdot g)^{(j)}(x)\right| &\leq \sum_{i=0}^{j} \binom{j}{i} c_i\, K\, \delta^{m+1-j} \\ &\leq 2^j\, K\, \delta^{m+1-j} \max\{c_i : 0 \leq i \leq j\} \leq \varepsilon. \qquad \square\end{aligned}$$

22.17. Smooth Extension Theorem. *Let E be a Fréchet space (or, slightly more general, a convenient vector space satisfying Mackey's countability condition) A function $f : A \to E$ admits a smooth extension to $\mathbb{R}$ if and only if each of its difference quotients is bounded on bounded sets.*

A convenient vector space is said to satisfy *Mackey's countability condition* if for every sequence of bounded sets $B_n \subseteq E$ there exists a sequence $\lambda_n > 0$ such that $\bigcup_{n \in \mathbb{N}} \lambda_n B_n$ is bounded in E.

Proof. We consider first the case, where $E = \mathbb{R}$. For $k \geq 0$ let $\tilde{f}^k$ be a $\mathcal{L}\text{ip}^k$-extension of f according to (22.13). The difference $\tilde{f}^{k+1} - \tilde{f}^k$ is an element of $C^k_A(\mathbb{R}, \mathbb{R})$: It is by construction C^k and on $\mathbb{R} \setminus A$ smooth. At an accumulation point a

of A the Taylor expansion of $\tilde{f}^k$ of order $j \leq k$ is just the approximation polynomial $P^j_{(a,\dots,a)}f$ by (22.13). Thus the derivatives up to order k of $\tilde{f}^{k+1}$ and $\tilde{f}^k$ are equal in a, and hence the difference is k-flat at a. Locally around any isolated point of A, i.e. a point $a \in A \setminus A_{acc}$, the extension $\tilde{f}^k$ is just the approximation polynomial $P^k_{\mathbf{a}}$ and hence smooth. In order to see this, use that for x with $|x-a| < \frac{1}{4}d(a, A\setminus\{a\})$ the point $\mathbf{a}_\varphi$ has as first entry a for every φ with $x \in \operatorname{supp}\varphi$: Let $b \in A \setminus \{a\}$ and $y \in \operatorname{supp}\varphi$ be arbitrary, then

$$\begin{aligned}|b-x| &\geq |b-a| - |a-x| \geq d(a, A\setminus\{a\}) - |a-x| > (4-1)\,|a-x| \\ |b-y| &\geq |b-x| - |x-y| > 3\,|a-x| - \operatorname{diam}(\operatorname{supp}\varphi) \\ &\geq 3\,d(a, \operatorname{supp}\varphi) - 2\,d(a,\operatorname{supp}\varphi) = d(a,\operatorname{supp}\varphi) \\ &\Rightarrow\ d(b,\operatorname{supp}\varphi) > d(a,\operatorname{supp}\varphi) \ \Rightarrow\ a_\varphi = a.\end{aligned}$$

By lemma (22.16) there exists an $h_k \in C^\infty_A(\mathbb{R},\mathbb{R})$ such that

$$\left|(\tilde{f}^{k+1} - \tilde{f}^k - h_k)^{(j)}(x)\right| \leq \frac{1}{2^k} \text{ for all } j \leq k-1.$$

Now we consider the function $\tilde{f} := \tilde{f}^0 + \sum_{k\geq 0}(\tilde{f}^{k+1} - \tilde{f}^k - h_k)$. It is the required smooth extension of f, since the summands $\tilde{f}^{k+1} - \tilde{f}^k - h_k$ vanish on A, and since for any n it can be rewritten as $\tilde{f} = \tilde{f}^n + \sum_{k<n} h_k + \sum_{k\geq n}(\tilde{f}^{k+1} - \tilde{f}^k - h_k)$, where the first summand is C^n, the first sum is C^∞, and the derivatives up to order $n-1$ of the terms of the second sum are uniformly summable.

Now we prove the vector valued case, where E satisfies Mackey's countability condition. It is enough to show the result for compact subsets $A \subset \mathbb{R}$, since the generalization arguments given in the proof of (22.13) can be applied equally in the smooth case. First one has to give a vector valued version of (22.16): Let a function $g \in \mathcal{L}ip^m(\mathbb{R}, E)$ with compact support be given, which vanishes on A, is m-flat on A_{acc} and smooth on the complement of A_{acc}. Then for every $\varepsilon > 0$ there exists a $h \in C^\infty(\mathbb{R},\mathbb{R})$, which equals 1 on a neighborhood of A_{acc} and such that $\delta^m(h\cdot g)(\mathbb{R}^{m+1})$ is contained in ε times the absolutely convex hull of the image of $\delta^{m+1}g$.

The proof of this assertion is along the lines of that of (22.16). One only has to define K as the absolutely convex hull of the image of $\delta^{m+1}g$ and choose $0 < \delta < 1$ such that $\delta \cdot \max\{c_i : i \leq m\} \cdot 2^m \leq \varepsilon$.

Now one proceeds as in scalar valued part: Let $\tilde{f}^k$ be the $\mathcal{L}ip^k$-extension of f according to (22.13). Then $g_k := \tilde{f}^{k+1} - \tilde{f}^k$ satisfies the assumption of the vector valued version of (22.16). Let K_k be the absolutely convex hull of the bounded image of $\delta^{k+1}g_k$. By assumption on E there exist $\lambda_n > 0$ such that $K := \bigcup_{k\in\mathbb{N}} \lambda_k \cdot K_k$ is bounded. Hence we may choose an $h_k \in C^\infty_A(\mathbb{R},\mathbb{R})$ such that $\delta^k(h_k\cdot g_k)(\mathbb{R}^{\langle k+1\rangle}) \subseteq \frac{\lambda_k}{2^k}K_k$. Now the extension $\tilde{f}$ is given by

$$\tilde{f} = \tilde{f}^0 + \sum_{k\geq 0} h_k \cdot g_k = \tilde{f}^n + \sum_{k<n}(1-h_k)\cdot g_k + \sum_{k\geq n} h_k \cdot g_k$$

and the result follows as above using convergence in the Banach space E_K. □

22.18. Remark. The restriction operator $\mathcal{L}\mathrm{ip}^m(\mathbb{R}, E) \to \mathcal{L}\mathrm{ip}^m_{\mathrm{ext}}(A, E)$ is a quotient mapping. We constructed a section for it, which is bounded and linear in the finite order case. It is unclear, whether it is possible to obtain a bounded linear section also in the smooth case, even if $E = \mathbb{R}$.

If the smooth extension theorem were true for any arbitrary convenient vector space E, then it would also give the extension operator theorem for the smooth case. Thus in order to obtain a counter-example to the latter one, the first step might be to find a counter-example to the vector valued extension theorem. In the particular cases, where the values lie in a Fréchet space E the vector valued smooth extension theorem is however true.

22.19. Proposition. *Let A be the image of a strictly monotone bounded sequence $\{a_n : n \in \mathbb{N}\}$. Then a map $f : A \to \mathbb{R}$ has a $\mathcal{L}\mathrm{ip}^m$-extension to $\mathbb{R}$ if and only if the sequence $\delta^k f(a_n, a_{n+1}, \dots, a_{n+k})$ is bounded for $k = m + 1$ if m is finite, respectively for all k if $m = \infty$.*

Proof. By [Frölicher, Kriegl, 1988, 1.3.10], the difference quotient $\delta^k f(a_{i_0}, \dots, a_{i_k})$ is an element of the convex hull of the difference quotients $\delta^k f(a_n, \dots, a_{n+k})$ for all $\min\{i_0, \dots, i_k\} \le n \le n + k \le \max\{i_0, \dots, i_k\}$. So the result follows from the extension theorems (22.13) and (22.17). □

For explicit descriptions of the boundedness condition for $\mathcal{L}\mathrm{ip}^k$-mappings defined on certain sequences and low k see [Frölicher, Kriegl, 1993, Sect. 6].

23. Frölicher Spaces and Free Convenient Vector Spaces

The central theme of this book is 'infinite dimensional manifolds'. But many natural examples suggest that this is a quite restricted notion, and it will be very helpful to have at hand a much more general and also easily useable concept, namely smooth spaces as they were introduced by [Frölicher, 1980, 1981]. We follow his line of development, replacing technical arguments by simple use of cartesian closedness of smooth calculus on convenient vector spaces, and we call them Frölicher spaces.

23.1. The category of Frölicher spaces.
A *Frölicher space* or a space with *smooth structure* is a triple $(X, \mathcal{C}_X, \mathcal{F}_X)$ consisting of a set X, a subset $\mathcal{C}_X$ of the set of all mappings $\mathbb{R} \to X$, and a subset $\mathcal{F}_X$ of the set of all functions $X \to \mathbb{R}$, with the following two properties:

(1) A function $f : X \to \mathbb{R}$ belongs to $\mathcal{F}_X$ if and only if $f \circ c \in C^\infty(\mathbb{R}, \mathbb{R})$ for all $c \in \mathcal{C}_X$.
(2) A curve $c : \mathbb{R} \to X$ belongs to $\mathcal{C}_X$ if and only if $f \circ c \in C^\infty(\mathbb{R}, \mathbb{R})$ for all $f \in \mathcal{F}_X$.

Note that a set X together with any subset $\mathcal{F}$ of the set of functions $X \to \mathbb{R}$

generates a unique Frölicher space $(X, \mathcal{C}_X, \mathcal{F}_X)$, where we put in turn:

$$\mathcal{C}_X := \{c : \mathbb{R} \to X : f \circ c \in C^\infty(\mathbb{R}, \mathbb{R}) \text{ for all } f \in \mathcal{F}\},$$
$$\mathcal{F}_X := \{f : X \to \mathbb{R} : f \circ c \in C^\infty(\mathbb{R}, \mathbb{R}) \text{ for all } c \in \mathcal{C}_X\},$$

so that $\mathcal{F} \subseteq \mathcal{F}_X$. The set $\mathcal{F}$ will be called a *generating set of functions for the Frölicher space.* A locally convex space is convenient if and only if it is a Frölicher space with the smooth curves and smooth functions from section (1) by (2.14). Furthermore, c^∞-open subsets U of convenient vector spaces E are Frölicher spaces, where $\mathcal{C}_U = C^\infty(\mathbb{R}, U)$ and $\mathcal{F}_U = C^\infty(U, \mathbb{R})$. Here we can use as generating set $\mathcal{F}$ of functions the restrictions of any set of bounded linear functionals which generates the bornology of E, see (2.14.4).

A mapping $\varphi : X \to Y$ between two Frölicher spaces is called *smooth* if the following three equivalent conditions hold

(3) For each $c \in \mathcal{C}_X$ the composite $\varphi \circ c$ is in $\mathcal{C}_Y$.
(4) For each $f \in \mathcal{F}_Y$ the composite $f \circ \varphi$ is in $\mathcal{F}_X$.
(5) For each $c \in \mathcal{C}_X$ and for each $f \in \mathcal{F}_Y$ the composite $f \circ \varphi \circ c$ is in $C^\infty(\mathbb{R}, \mathbb{R})$.

Note that $\mathcal{F}_Y$ can be replaced by any generating set of functions. The set of all smooth mappings from X to Y will be denoted by $C^\infty(X, Y)$. Then we have $C^\infty(\mathbb{R}, X) = \mathcal{C}_X$ and $C^\infty(X, \mathbb{R}) = \mathcal{F}_X$. Frölicher spaces and smooth mappings form a category.

23.2. Theorem. *The category of Frölicher spaces and smooth mappings has the following properties:*

(1) *Complete, i.e., arbitrary limits exist. The underlying set is formed as in the category of sets as a certain subset of the cartesian product, and the smooth structure is generated by the smooth functions on the factors.*
(2) *Cocomplete, i.e., arbitrary colimits exist. The underlying set is formed as in the category of set as a certain quotient of the disjoint union, and the smooth functions are exactly those which induce smooth functions on the cofactors.*
(3) *Cartesian closed, which means: The set $C^\infty(X, Y)$ carries a canonical smooth structure described by*

$$C^\infty(X, Y) \xrightarrow{C^\infty(c,f)} C^\infty(\mathbb{R}, \mathbb{R}) \xrightarrow{\lambda} \mathbb{R}$$

where $c \in C^\infty(\mathbb{R}, X)$, where f is in $C^\infty(Y, \mathbb{R})$ or in a generating set of functions, and where $\lambda \in C^\infty(\mathbb{R}, \mathbb{R})'$. With this structure the exponential law holds:

$$C^\infty(X \times Y, Z) \cong C^\infty(X, C^\infty(Y, Z)).$$

Proof. Obviously, the limits and colimits described above have all required universal properties.

We have the following implications:

$\varphi^\vee : X \to C^\infty(Y, Z)$ is smooth.

$\Leftrightarrow \varphi^\vee \circ c_X : \mathbb{R} \to C^\infty(Y, Z)$ is smooth for all smooth curves $c_X \in C^\infty(\mathbb{R}, X)$, by definition.

$\Leftrightarrow C^\infty(c_Y, f_Z) \circ \varphi^\vee \circ c_X : \mathbb{R} \to C^\infty(\mathbb{R}, \mathbb{R})$ is smooth for all smooth curves $c_X \in C^\infty(\mathbb{R}, X)$, $c_Y \in C^\infty(\mathbb{R}, Y)$, and smooth functions $f_Z \in C^\infty(Z, \mathbb{R})$, by definition.

$\Leftrightarrow f_Z \circ \varphi \circ (c_X \times c_Y) = f_Z \circ (c_Y^* \circ \varphi^\vee \circ c_X)^\wedge : \mathbb{R}^2 \to \mathbb{R}$ is smooth for all smooth curves c_X, c_Y, and smooth functions f_Z, by the simplest case of cartesian closedness of smooth calculus (3.10).

$\Rightarrow \varphi : X \times Y \to Z$ is smooth, since each curve into $X \times Y$ is of the form $(c_X, c_Y) = (c_X \times c_Y) \circ \Delta$, where Δ is the diagonal mapping.

$\Rightarrow \varphi \circ (c_X \times c_Y) : \mathbb{R}^2 \to Z$ is smooth for all smooth curves c_X and c_Y, since the product and the composite of smooth mappings is smooth.

As in the proof of (3.13) it follows in a formal way that the exponential law is a diffeomorphism for the smooth structures on the mapping spaces. □

23.3. Remark. By [Frölicher, Kriegl, 1988, 2.4.4] the convenient vector spaces are exactly the linear Frölicher spaces for which the smooth linear functionals generate the smooth structure, and which are separated and 'complete'. On a locally convex space which is not convenient, one has to saturate to the scalarwise smooth curves and the associated functions in order to get a Frölicher space.

23.4 Proposition. *Let X be a Frölicher space and E a convenient vector space. Then $C^\infty(X, E)$ is a convenient vector space with the smooth structure described in* (23.2.3).

Proof. We consider the locally convex topology on $C^\infty(X, E)$ induced by $c^* : C^\infty(X, E) \to C^\infty(\mathbb{R}, E)$ for all $c \in C^\infty(\mathbb{R}, X)$. As in (3.11) one shows that this describes $C^\infty(X, E)$ as inverse limit of spaces $C^\infty(\mathbb{R}, E)$, which are convenient by (3.7). Thus also $C^\infty(X, E)$ is convenient by (2.15). By (2.14.4), (3.8), (3.9) and (3.7) its smooth curves are exactly those $\gamma : \mathbb{R} \to C^\infty(X, E)$, for which

$$\mathbb{R} \xrightarrow{\gamma} C^\infty(X, E) \xrightarrow{c^*} C^\infty(\mathbb{R}, E) \xrightarrow{f_*} C^\infty(\mathbb{R}, \mathbb{R}) \xrightarrow{\lambda} \mathbb{R}$$

is smooth for all $c \in C^\infty(\mathbb{R}, X)$, for all f in the generating set E' of functions, and all $\lambda \in C^\infty(\mathbb{R}, \mathbb{R})$. This is the smooth structure described in (23.2.3). □

23.5. Related concepts: Holomorphic Frölicher spaces. They can be defined in a way similar as smooth Frölicher spaces in (23.1), with the following changes: As curves one has to take mappings from the complex unit disk. Then the results analogous to (23.2) hold, where for the proof one has to use the holomorphic exponential law (7.22) instead of the smooth one (3.10), see [Siegl, 1995] and [Siegl, 1997].

The concept of holomorphic Frölicher spaces is not without problems: Namely finite dimensional complex manifolds are holomorphic Frölicher spaces if they are Stein, and compact complex manifolds are never holomorphic Frölicher spaces. But arbitrary subsets A of complex convenient vector spaces E are holomorphic Frölicher spaces with the initial structure, again generated by the restrictions of bounded complex linear functionals. Note that *analytic subsets* of complex convenient spaces, i.e., locally zero sets of holomorphic mappings, are holomorphic spaces. But usually, as analytic sets, holomorphic functions on them are restrictions of holomorphic functions defined on neighborhoods, whereas as holomorphic spaces they admit more holomorphic functions, as the following example shows:

Example. Neil's parabola $P := \{z_1^2 - z_2^3 = 0\} \subset \mathbb{C}^2$ has the holomorphic curves $a : \mathbb{D} \to P \subset \mathbb{C}^2$ of the form $a = (b^3, b^2)$ for holomorphic $b : \mathbb{D} \to \mathbb{C}$: If $a(z) = (z^k a_1(z), z^l a_2(z))$ with $a(0) = 0$ and $a_i(0) \neq 0$, then $k = 3n$ and $l = 2n$ for some $n > 0$ and (a_1, a_2) is still a holomorphic curve in $P \setminus 0$, so $(a_1, a_2) = (c^3, c^2)$ by the implicit function theorem, then $b(z) = z^n c(z)$ is the solution. Thus, $z \mapsto (z^3, z^2)$ is biholomorphic $\mathbb{C} \to P$. So z is a holomorphic function on P which cannot be extended to a holomorphic function on a neighborhood of 0 in $\mathbb{C}^2$, since this would have infinite differential at 0.

23.6. Theorem. Free Convenient Vector Space. [Frölicher, Kriegl, 1988], *5.1.1 For every Frölicher space X there exists a free convenient vector space λX, i.e. a convenient vector space λX together with a smooth mapping $\delta_X : X \to \lambda X$, such that for every smooth mapping $f : X \to G$ with values in a a convenient vector space G there exists a unique linear bounded mapping $\tilde{f} : \lambda X \to G$ with $\tilde{f} \circ \delta_X = f$. Moreover $\delta^* : L(\lambda X, G) \cong C^\infty(X, G)$ is an isomorphisms of convenient vector spaces and δ is an initial morphism.*

Proof. In order to obtain a candidate for λX, we put $G := \mathbb{R}$ and thus should have $(\lambda X)' = L(\lambda X, \mathbb{R}) \cong C^\infty(X, \mathbb{R})$ and hence λX should be describable as subspace of $(\lambda X)'' \cong C^\infty(X, \mathbb{R})'$. In fact every $f \in C^\infty(E, \mathbb{R})$ acts as bounded linear functional $\mathrm{ev}_f : C^\infty(X, \mathbb{R})' \to \mathbb{R}$ and if we define $\delta_X : X \to C^\infty(X, \mathbb{R})'$ to be $\delta_X : x \mapsto \mathrm{ev}_x$ then $\mathrm{ev}_f \circ \delta_X = f$ and δ_X is smooth, since by the uniform boundedness principle (5.18) it is sufficient to check that $\mathrm{ev}_f \circ \delta_X = f : X \to C^\infty(X, \mathbb{R})' \to \mathbb{R}$ is smooth for all $f \in C^\infty(X, \mathbb{R})$. In order to obtain uniqueness of the extension $\tilde{f} := \mathrm{ev}_f$, we have to restrict it to the c^∞-closure of the linear span of $\delta_X(X)$. So let λX be this closure and let $f : X \to G$ be an arbitrary smooth mapping with values in some convenient vector space. Since δ belongs to C^∞ we have that $\delta^* : L(\lambda X, G) \to C^\infty(X, G)$ is well defined and it is injective since the linear subspace generated by the image of δ is c^∞-dense in λX by construction. To show surjectivity consider the following diagram:

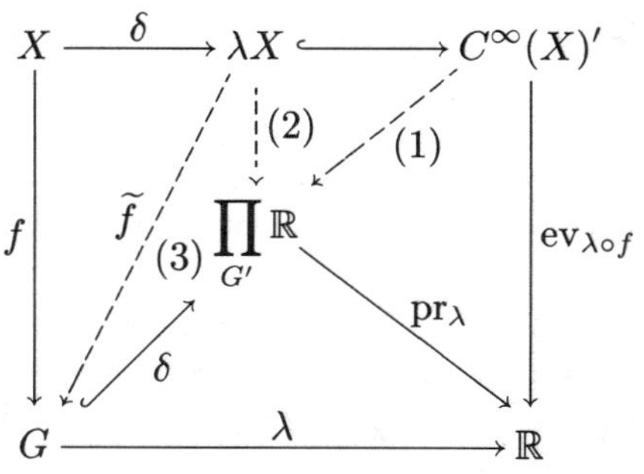

Note that (2) has values in $\delta(G)$, since this is true on the ev_x, which generate by definition a c^∞-dense subspace of λX.

Remains to show that this bijection is a bornological isomorphism. In order to show that the linear mapping $C^\infty(X,G)\to L(\lambda X,G)$ is bounded we can reformulate this equivalently using (3.12), the universal property of λX and the uniform boundedness principle (5.18) in turn:

$$\begin{aligned}
&C^\infty(X,G)\to L(\lambda X,G) \text{ is } L\\
\iff&\lambda X\to L(C^\infty(X,G),G) \text{ is } L\\
\iff&X\to L(C^\infty(X,G),G) \text{ is } C^\infty\\
\iff&X\to L(C^\infty(X,G),G)\xrightarrow{\mathrm{ev}_f} G \text{ is } C^\infty
\end{aligned}$$

and since the composition is just f we are done.

Conversely we have to show that $L(\lambda X,G)\to C^\infty(X,G)$ belongs to L. Composed with $\mathrm{ev}_x : C^\infty(X,G)\to G$ this yields the bounded linear map $\mathrm{ev}_{\delta(x)} : L(\lambda X,G)\to G$. Thus this follows from the uniform boundedness principle (5.26).

That δ_X is initial follows immediately from the fact that the structure of X is initial with respect to family $\{f=\mathrm{ev}_o\circ\delta_X : f\in C^\infty(X,\mathbb{R})\}$. $\square$

Remark. The corresponding result with the analogous proof is true for holomorphic Frölicher spaces, $\mathcal{L}\mathrm{ip}^k$-spaces, and ℓ^∞-spaces. For the first see [Siegl, 1997] for the last two see [Frölicher, Kriegl, 1988].

23.7. Corollary. *Let X be a Frölicher space such that the functions in $C^\infty(X,\mathbb{R})$ separate points on X. Then X is diffeomorphic as Frölicher space to a subspace of the convenient vector space $\lambda(X)\subseteq C^\infty(X,\mathbb{R})'$ with the initial smooth structure (generated by the restrictions of linear bounded functionals, among other possibilities).* $\square$

We have constructed the free convenient vector space λX as the c^∞-closure of the linear subspace generated by the point evaluations in $C^\infty(X,\mathbb{R})'$. This is not very constructive, in particular since adding Mackey-limits of sequences (or even nets) of a subspace does not always give its Mackey-closure. In important cases (like when X is a finite dimensional smooth manifold) one can show however that not only $\lambda X=C^\infty(X,\mathbb{R})'$, but even that every element of λX is the Mackey-limit of a sequence of linear combinations of point evaluations, and that $C^\infty(X,\mathbb{R})'$ is the space of distributions of compact support.

23.8. Proposition. *Let E be a convenient vector space and X a finite dimensional smooth separable manifold. Then for every $\ell\in C^\infty(X,E)'$ there exists a compact set $K\subseteq X$ such that $\ell(f)=0$ for all $f\in C^\infty(X,E)$ with $f|_K=0$.*

Proof. Since X is separable its compact bornology has a countable basis $\{K_n : n\in\mathbb{N}\}$ of compact sets. Assume now that no compact set has the claimed property. Then for every $n\in\mathbb{N}$ there has to exist a function $f_n\in C^\infty(X,E)$ with $f_n|_{K_n}=0$

but $\ell(f_n) \neq 0$. By multiplying f_n with $\frac{n}{\ell(f_n)}$ we may assume that $\ell(f_n) = n$. Since every compact subset of X is contained in some K_n one has that $\{f_n : n \in \mathbb{N}\}$ is bounded in $C^\infty(X, E)$, but $\ell(\{f_n : n \in \mathbb{N}\}$ is not; this contradicts the assumption that ℓ is bounded. $\square$

23.9. Remark. The proposition above remains true if X is a finite dimensional smooth paracompact manifold with non-measurably many components. In order to show this generalization one uses that for the partition $\{X_j : j \in J\}$ by the non-measurably many components one has $C^\infty(X, E) \cong \prod_{j\in J} C^\infty(X_j, E)$, and the fact that an ℓ belongs to the dual of such a product if it is a finite sum of elements of the duals of the factors. Now the result follows from (23.8) since the components of a paracompact manifold are paracompact and hence separable.

For such manifolds X the dual $C^\infty(X, \mathbb{R})'$ is the space of distributions with compact support. In fact, in case X is connected, $C^\infty(X, \mathbb{R})'$ is the space of all linear functionals which are continuous for the classically considered topology on $C^\infty(X, \mathbb{R})$ by (6.1); and in case of an arbitrary X this result follows using the isomorphism $C^\infty(X, \mathbb{R}) \cong \prod_j C^\infty(X_j, \mathbb{R})$ where the X_j denote the connected components of X.

23.10. Theorem. [Frölicher, Kriegl, 1988], *5.1.7 Let E be a convenient vector space and X a finite dimensional separable smooth manifold. Then the Mackey-adherence of the linear subspace generated by $\{\ell \circ \mathrm{ev}_x : x \in X,\, \ell \in E'\}$ is $C^\infty(X, E)'$.*

Proof. The proof is in several steps.

(Step 1) There exist $g_n \in C^\infty(\mathbb{R}, \mathbb{R})$ with $\operatorname{supp}(g_n) \subseteq [-\frac{2}{n}, \frac{2}{n}]$ such that for every $f \in C^\infty(\mathbb{R}, E)$ the set $\{n \cdot \big(f - \sum_{k\in\mathbb{Z}} f(r_{n,k}) g_{n,k}\big) : n \in \mathbb{N}\}$ is bounded in $C^\infty(\mathbb{R}, E)$, where $r_{n,k} := \frac{k}{2^n}$ and $g_{n,k}(t) := g_n(t - r_{n,k})$.

We choose a smooth $h : \mathbb{R} \to [0, 1]$ with $\operatorname{supp}(h) \subseteq [-1, 1]$ and $\sum_{k\in\mathbb{Z}} h(t - k) = 1$ for all $t \in \mathbb{R}$ and we define $Q^n : C^\infty(\mathbb{R}, E) \to C^\infty(\mathbb{R}, E)$ by setting

$$Q^n(f)(t) := \sum_k f(\tfrac{k}{n}) h(tn - k).$$

Let $K \subseteq \mathbb{R}$ be compact. Then

$$n(Q^n(f) - f)(t) = \sum_k (f(\tfrac{k}{n}) - f(t)) \cdot n \cdot h(tn - k) \in B_1(f, K + \tfrac{1}{n} \operatorname{supp}(h))$$

for $t \in K$, where $B_n(f, K_1)$ denotes the absolutely convex hull of the bounded set $\delta^n f(K_1^{\langle n\rangle})$.

To get similar estimates for the derivatives we use convolution. Let $h_1 : \mathbb{R} \to \mathbb{R}$ be a smooth function with support in $[-1, 1]$ and $\int_\mathbb{R} h_1(s) ds = 1$. Then for $t \in K$ one has

$$(f * h_1)(t) := \int_\mathbb{R} f(t - s) h_1(s) ds \;\in\; B_0(f, K + \operatorname{supp}(h_1)) \cdot \|h_1\|_1,$$

where $\|h_1\|_1 := \int_\mathbb{R} |h_1(s)| ds$. For smooth functions $f, h : \mathbb{R} \to \mathbb{R}$ one has $(f * h)^{(k)} = f * h^{(k)}$; one immediately deduces that the same holds for smooth functions $f : \mathbb{R} \to$

E and one obtains $(f*h_1)(t)-f(t)=\int_{\mathbb{R}}(f(t-s)-f(t))h_1(s)ds \in \operatorname{diam}(\operatorname{supp}(h_1))\cdot \|h_1\|_1\cdot B_1(f,K+\operatorname{supp}(h_1))$ for $t\in K$, where $\operatorname{diam}(S):=\sup\{|s|: s\in S\}$. Using now $h_n(t):=n\cdot h_1(nt)$ we obtain for $t\in K$:

$$\begin{aligned}(Q^m(f)*h_n-f)^{(k)}(t)&=(Q^m(f)*h_n^{(k)}-f*h_n^{(k)})(t)+(f^{(k)}*h_n-f^{(k)})(t)\\&=(Q^m(f)-f)*h_n^{(k)}(t)+(f^{(k)}*h_n-f^{(k)})(t)\\&\in B_0(Q^m(f)-f,K+\operatorname{supp}(h_n))\cdot\|h_n^{(k)}\|_1+\\&\quad+B_1(f^{(k)},K+\operatorname{supp}(h_n))\cdot\operatorname{diam}(\operatorname{supp}(h_n))\cdot\|h_n\|_1\\&\subseteq \tfrac{1}{m}n^k\cdot B_1(f,K+\operatorname{supp}(h_n)+\tfrac{1}{m}\operatorname{supp}(h))\cdot\|h_1^{(k)}\|_1\\&\quad+n\cdot B_1(f^{(k)},K+\operatorname{supp}(h_n))\cdot\|h_n\|_1.\end{aligned}$$

Let now $m:=2^n$ and $P^n(f):=Q^m(f)*h_n$. Then

$$\begin{aligned}n\cdot\big(P^n(f)-f\big)^{(k)}(t)&\in n^{k+1}2^{-n}\cdot B_1(f,K+(\tfrac{1}{n}+\frac{1}{2^n})[-1,1])\|h_1^{(k)}\|_1\\&\quad+B_1(f^{(k)},K+\tfrac{1}{n}[-1,1])\|h_1\|_1\end{aligned}$$

for $t\in K$ and the right hand side is uniformly bounded for $n\in\mathbb{N}$.

With $g_n(t):=\int_{\mathbb{R}}h(s2^n-k)h_n(t+k2^{-n}-s)ds=\int_{\mathbb{R}}h(s2^n)h_n(t-s)ds$ we obtain

$$\begin{aligned}P^n(f)(t)=(Q^{2^n}(f)*h_n)(t)&=\sum_k f(k2^{-n})h(t2^n-k)*h_n\\&=\sum_k f(k2^{-n})\int_{\mathbb{R}}h(s2^n-k)h_n(t-s)ds\\&=\sum_k f(k2^{-n})g_n(t-k2^{-n}).\end{aligned}$$

Thus $r_{n,k}:=k2^{-n}$ and the g_n have all the claimed properties.

(Step 2) For every $m\in\mathbb{N}$ and every $f\in C^\infty(\mathbb{R}^m,E)$ the set

$$\Big\{n\cdot\Big(f-\sum_{k_1\in\mathbb{Z},\dots,k_m\in\mathbb{Z}}f(r_{n;k_1,\dots,k_m})g_{n;k_1,\dots,k_m}\Big):n\in\mathbb{N}\Big\}$$

is bounded in $C^\infty(\mathbb{R}^m,E)$, where $r_{n;k_1,\dots,k_m}:=(r_{n,k_1},\dots,r_{n,k_m})$ and

$$g_{n;k_1,\dots,k_m}(x_1,\dots,x_m):=g_{n,k_1}(x_1)\cdot\ldots\cdot g_{n,k_m}(x_m).$$

We prove this statement by induction on m. For $m=1$ it was shown in step 1. Now assume that it holds for m and $C^\infty(\mathbb{R},E)$ instead of E. Then by induction hypothesis applied to $f^\vee: C^\infty(\mathbb{R}^m,C^\infty(\mathbb{R},E))$ we conclude that

$$\Big\{n\cdot\Big(f-\sum_{k_1\in\mathbb{Z},\dots,k_m\in\mathbb{Z}}f(r_{n;k_1,\dots,k_m},\quad)g_{n;k_1,\dots,k_m}\Big):n\in\mathbb{N}\Big\}$$

is bounded in $C^\infty(\mathbb{R}^{m+1}, E)$. Thus it remains to show that

$$\Big\{n \sum_{k_1,\dots,k_m} g_{n;k_1,\dots,k_m}\Big(f(r_{n;k_1,\dots,k_m}, \quad) - \sum_{k_{m+1}} f(r_{n;k_1,\dots,k_m}, r_{k_{m+1}}) g_{n,k_{m+1}}\Big) : n \in \mathbb{N}\Big\}$$

is bounded in $C^\infty(\mathbb{R}^{m+1}, E)$. Since the support of the $g_{n;k_1,\dots,k_m}$ is locally finite only finitely many summands of the outer sum are non-zero on a given compact set. Thus it is enough to consider each summand separately. By step (1) we know that the linear operators $h \mapsto n\big(h - \sum_k h(r_{n,k}) g_{n,k}\big)$, $n \in \mathbb{N}$, are pointwise bounded. So they are bounded on bounded sets, by the linear uniform boundedness principle (5.18). Hence

$$\Big\{n \cdot \Big(f(r_{n;k_1,\dots,k_m}, \quad) - \sum_{k_{m+1}} f(r_{n;k_1,\dots,k_m}, r_{k_{m+1}}) g_{n,k_{m+1}}\Big) : n \in \mathbb{N}\Big\}$$

is bounded in $C^\infty(\mathbb{R}^{m+1}, E)$. Using that the multiplication $\mathbb{R} \times E \to E$ is bounded one concludes immediately that also the multiplication with a map $g \in C^\infty(X, \mathbb{R})$ is bounded from $C^\infty(X, E) \to C^\infty(X, E)$ for any Frölicher space X. Thus the proof of step (2) is complete.

(Step 3) For every $\ell \in C^\infty(X, E)'$ there exist $x_{n,k} \in X$ and $\ell_{n,k} \in E'$ such that $\{n(\ell - \sum_k \ell_{n,k} \circ \mathrm{ev}_{x_{n,k}}) : n \in \mathbb{N}\}$ is bounded in $C^\infty(X, E)'$, where in the sum only finitely many terms are non-zero. In particular the subspace generated by $\ell_E \circ \mathrm{ev}_x$ for $\ell_E \in E'$ and $x \in X$ is c^∞-dense.

By (23.8) there exists a compact set K with $f|_K = 0$ implying $\ell(f) = 0$. One can cover K by finitely many relatively compact $U_j \cong \mathbb{R}^m$ ($j = 1 \dots N$). Let $\{h_j : j = 0 \dots N\}$ be a partition of unity subordinated to $\{X \setminus K, U_1, \dots, U_N\}$. Then $\ell(f) = \sum_{j=1}^N \ell(h_j \cdot f)$ for every f. By step (2) the set

$$\Big\{n(h_j f - \sum h_j f(r_{n,k_1,\dots,k_m}) g_{n,k_1,\dots,k_m} : n \in \mathbb{N}\Big\}$$

is bounded in $C^\infty(U_j, E)$. Since $\mathrm{supp}(h_j)$ is compact in U_j this is even bounded in $C^\infty(X, E)$ and for fixed n only finitely many $r_{n,k_1,\dots,k_m}$ belong to $\mathrm{supp}(h_j)$. Thus the above sum is actually finite and the supports of all functions in the bounded subset of $C^\infty(U_j, E)$ are included in a common compact subset. Applying ℓ to this subset yields that $\big\{n\big((\ell(h_j f) - \sum \ell_{n,k_1,\dots,k_m} \circ \mathrm{ev}(r_{n,k_1,\dots,k_m})\big) : n \in \mathbb{N}\big\}$ is bounded in $\mathbb{R}$, where $\ell_{n,k_1,\dots,k_m}(x) := \ell\big(h_j(r_{n,k_1,\dots,k_m}) g_{n;k_1,\dots,k_m} \cdot x\big)$.

To complete the proof one only has to take as $x_{n,k}$ all the $r_{n,k_1,\dots,k_m}$ for the finitely many charts $U_j \cong \mathbb{R}^m$ and as $\ell_{n,k}$ the corresponding functionals $\ell_{n,k_1,\dots,k_m} \in E'$. □

23.11. Corollary. [Frölicher, Kriegl, 1988], *5.1.8 Let X be a finite dimensional separable smooth manifold. Then the free convenient vector space λX over X is equal to $C^\infty(X, \mathbb{R})'$.* □

23.12. Remark. In [Kriegl, Nel, 1990] it was shown that the free convenient vector space over the long line L is not $C^\infty(L, \mathbb{R})'$ and the same for the space E of points with countable support in an uncountable product of $\mathbb{R}$.

In [Adam, 1995, 2.2.6] it is shown that the isomorphism $\delta^* : L(C^\infty(X,\mathbb{R})', G) \cong C^\infty(X,G)$ is even a topological isomorphism for (the) natural topologies on all spaces under consideration provided X is a finite dimensional separable smooth manifold. Furthermore, the corresponding statement holds for holomorphic mappings, provided X is a separable complex manifold modeled on polycylinders. For Riemannian surfaces X it is shown in [Siegl, 1997, 2.11] that the free convenient vector space for holomorphic mappings is the Mackey adherence of the linear subspace of $\mathcal{H}(X,\mathbb{C})'$ generated by the point evaluations ev_x for $x \in X$. In [Siegl, 1997, 2.52] the same is shown for pseudo-convex subsets of $X \subseteq \mathbb{C}^n$. Reflexivity of the space of scalar valued functions implies that the linear space generated by the point evaluations is dense in the dual of the function space with respect to its bornological topology by [Siegl, 1997, 3.3]. And conversely if $\Lambda(X)$ is this dual, then the function space is reflexive. Thus $\Lambda(E) \neq C^\infty(E,\mathbb{R})'$ for non-reflexive convenient vector spaces E. Partial positive results for infinite dimensional spaces have been obtained in [Siegl, 1997, section 3].

23.13. Remark. On can define *convenient co-algebras* dually to convenient algebras, as a convenient vector space E together with a compatible co-algebra structure, i.e. two bounded linear mappings

> $\mu : E \to E\tilde{\otimes}_\beta E$, called co-multiplication, into the c^∞-completion (4.29) of the bornological tensor product (5.9);
> and $\varepsilon : E \to \mathbb{R}$, called co-unit,

such that one has the following commutative diagrams:

$$\begin{array}{ccccc} E\tilde{\otimes}_\beta E & \xrightarrow{\mu\tilde{\otimes}_\beta \mathrm{Id}} & (E\tilde{\otimes}_\beta E)\tilde{\otimes}_\beta E & \xrightarrow{\cong} & E\tilde{\otimes}_\beta(E\tilde{\otimes}_\beta E) \\ {\scriptstyle\mu}\big\downarrow & & & & \big\downarrow{\scriptstyle \mathrm{Id}\tilde{\otimes}_\beta\mu} \\ E & & \xrightarrow{\qquad\qquad\mu\qquad\qquad} & & E\tilde{\otimes}_\beta E \end{array}$$

$$\begin{array}{ccccc} & & E\tilde{\otimes}_\beta E & & \\ & {\scriptstyle\mu}\nearrow & & \searrow{\scriptstyle\varepsilon\tilde{\otimes}\,\mathrm{Id}} & \\ E & & \xrightarrow{\qquad\cong\qquad} & & \mathbb{R}\tilde{\otimes}_\beta E \end{array}$$

In words, the co-multiplication has to be co-associative and ε has to be a co-unit with respect to μ.

If, in addition, the following diagram commutes

$$\begin{array}{ccccc} E\tilde{\otimes}_\beta E & & \xrightarrow{\qquad\cong\qquad} & & E\tilde{\otimes}_\beta E \\ & {\scriptstyle\mu}\nwarrow & & \nearrow{\scriptstyle\mu} & \\ & & E & & \end{array}$$

then the co-algebra is called *co-commutative.*

Morphisms $g : E \to F$ between convenient co-algebras E and F are bounded linear mappings for which the following diagrams commute:

$$\begin{array}{ccc} E\tilde{\otimes}_\beta E & \xrightarrow{g\tilde{\otimes}g} & F\tilde{\otimes}_\beta F \\ {\scriptstyle\mu_E}\uparrow & & \uparrow{\scriptstyle\mu_F} \\ E & \xrightarrow{g} & F \end{array} \qquad \begin{array}{ccc} \mathbb{R} & \xrightarrow{\mathrm{Id}} & \mathbb{R} \\ {\scriptstyle\varepsilon_E}\uparrow & & \uparrow{\scriptstyle\varepsilon_F} \\ E & \xrightarrow{g} & F \end{array}$$

A *co-idempotent* in a convenient co-algebra E, is an element $x \in E$ satisfying $\varepsilon(x) = 1$ and $\mu(x) = x \otimes x$. They correspond bijectively to convenient co-algebra morphisms $\mathbb{R} \to E$, see [Frölicher, Kriegl, 1988, 5.2.7].

In [Frölicher, Kriegl, 1988, 5.2.4] it was shown that $\lambda(X \times Y) \cong \lambda(X)\tilde{\otimes}\lambda(Y)$ using only the universal property of the free convenient vector space. Thus $\lambda(\Delta) : \lambda(X) \to \lambda(X \times X) \cong \lambda(X)\tilde{\otimes}\lambda(X)$ of the diagonal mapping $\Delta : X \to X \times X$ defines a co-multiplication on $\lambda(X)$ with co-unit $\lambda(\text{const}) : \lambda(X) \to \lambda(\{*\}) \cong \mathbb{R}$. In this way λ becomes a functor from the category of Frölicher spaces into that of convenient co-algebras, see [Frölicher, Kriegl, 1988, 5.2.5]. In fact this functor is left-adjoint to the functor I, which associates to each convenient co-algebra the Frölicher space of co-idempotents with the initial structure inherited from the co-algebra, see [Frölicher, Kriegl, 1988, 5.2.9].

Furthermore, it was shown in [Frölicher, Kriegl, 1988, 5.2.18] that any co-idempotent element e of $\lambda(X)$ defines an algebra-homomorphism $C^\infty(X, \mathbb{R}) \cong \lambda(X)' \xrightarrow{\mathrm{ev}_e} \mathbb{R}$. Thus the equality $I(\lambda(X)) = X$, i.e. every co-idempotent $e \in \lambda(X)$ is given by ev_x for some $x \in X$, is thus satisfied for smoothly realcompact spaces X, as they are treated in chapter IV.

24. Smooth Mappings on Non-Open Domains

In this section we will discuss smooth maps $f : E \supseteq X \to F$, where E and F are convenient vector spaces and X are certain not necessarily open subsets of E. We consider arbitrary subsets $X \subseteq E$ as Frölicher spaces with the initial smooth structure induced by the inclusion into E, i.e., a map $f : E \supseteq X \to F$ is smooth if and only if for all smooth curves $c : \mathbb{R} \to X \subseteq E$ the composite $f \circ c : \mathbb{R} \to F$ is a smooth curve.

24.1. Lemma. Convex sets with non-void interior.
Let $K \subseteq E$ be a convex set with non-void c^∞-interior K^o. Then the segment $(x, y] := \{x + t(y - x) : 0 < t \leq 1\}$ is contained in K^o for every $x \in K$ and $y \in K^0$. The interior K^o is convex and open even in the locally convex topology. And K is closed if and only if it is c^∞-closed.

Proof. Let $y_0 := x + t_0(y - x)$ be an arbitrary point on the segment $(x, y]$, i.e., $0 < t_0 \leq 1$. Then $x + t_0(K^o - x)$ is an c^∞-open neighborhood of y_0, since homotheties are c^∞-continuous. It is contained in K, since K is convex.

In particular, the c^∞-interior K^o is convex, hence it is not only c^∞-open but open in the locally convex topology (4.5).

Without loss of generality we now assume that $0 \in K^o$. We claim that the closure of K is the set $\{x : tx \in K^o \text{ for } 0 < t < 1\}$. This implies the statement on closedness. Let $U := K^o$ and consider the Minkowski-functional $p_U(x) := \inf\{t > 0 : x \in tU\}$. Since U is convex, the function p_U is convex, see (52.2). Using that U is c^∞-open it can easily be shown that $U = \{x : p_U(x) < 1\}$. From (13.2) we conclude that p_U is c^∞-continuous, and furthermore that it is even continuous for the locally convex topology. Hence, the set $\{x : tx \in K^o \text{ for } 0 < t < 1\} = \{x : p_U(x) \le 1\} = \{x : p_K(x) \le 1\}$ is the closure of K in the locally convex topology by (52.3). □

24.2. Theorem. Derivative of smooth maps.
Let $K \subseteq E$ be a convex subset with non-void interior K^o, and let $f : K \to \mathbb{R}$ be a smooth map. Then $f|_{K^o} : K^o \to F$ is smooth, and its derivative $(f|_{K^o})'$ extends (uniquely) to a smooth map $K \to L(E, F)$.

Proof. Only the extension property is to be shown. Let us first try to find a candidate for $f'(x)(v)$ for $x \in K$ and $v \in E$ with $x + v \in K^o$. By convexity the smooth curve $c_{x,v} : t \mapsto x + t^2 v$ has for $0 < |t| < 1$ values in K^o and $c_{x,v}(0) = x \in K$, hence $f \circ c_{x,v}$ is smooth. In the special case where $x \in K^o$ we have by the chain rule that $(f \circ c_{x,v})'(t) = f'(x)(c_{x,v}(t))(c'_{x,v}(t))$, hence $(f \circ c_{x,v})''(t) = f''(c_{x,v}(t))(c'_{x,v}(t), c'_{x,v}(t)) + f'(c_{x,v}(t))(c''_{x,v}(t))$, and for $t = 0$ in particular $(f \circ c_{x,v})''(0) = 2\, f'(x)(v)$. Thus we define

$$2\, f'(x)(v) := (f \circ c_{x,v})''(0) \text{ for } x \in K \text{ and } v \in K^o - x.$$

Note that for $0 < \varepsilon < 1$ we have $f'(x)(\varepsilon\, v) = \varepsilon\, f'(x)(v)$, since $c_{x,\varepsilon\, v}(t) = c_{x,v}(\sqrt{\varepsilon}\, t)$.

Let us show next that $f'(\quad)(v) : \{x \in K : x + v \in K^o\} \to \mathbb{R}$ is smooth. So let $s \mapsto x(s)$ be a smooth curve in K, and let $v \in K^0 - x(0)$. Then $x(s) + v \in K^o$ for all sufficiently small s. And thus the map $(s, t) \mapsto c_{x(s),v}(t)$ is smooth from some neighborhood of $(0, 0)$ into K. Hence $(s, t) \mapsto f(c_{x(s),v}(t))$ is smooth and also its second derivative $s \mapsto (f \circ c_{x(s),v})''(0) = 2\, f'(x(s))(v)$.

In particular, let $x_0 \in K$ and $v_0 \in K^o - x_0$ and $x(s) := x_0 + s^2 v_0$. Then

$$2f'(x_0)(v) := (f \circ c_{x_0,v})''(0) = \lim_{s\to 0}(f \circ c_{x(s),v})''(0) = \lim_{s\to 0} 2\, f'(x(s))(v),$$

with $x(s) \in K^o$ for $0 < |s| < 1$. Obviously this shows that the given definition of $f'(x_0)(v)$ is the only possible smooth extension of $f'(\quad)(v)$ to $\{x_0\} \cup K^o$.

Now let $v \in E$ be arbitrary. Choose a $v_0 \in K^o - x_0$. Since the set $K^o - x_0 - v_0$ is a c^∞-open neighborhood of 0, hence absorbing, there exists some $\varepsilon > 0$ such that $v_0 + \varepsilon v \in K^o - x_0$. Thus

$$f'(x)(v) = \tfrac{1}{\varepsilon} f'(x)(\varepsilon v) = \tfrac{1}{\varepsilon}\big(f'(x)(v_0 + \varepsilon v) - f'(x)(v_0)\big)$$

for all $x \in K^0$. By what we have shown above the right side extends smoothly to $\{x_0\} \cup K^o$, hence the same is true for the left side. I.e. we define $f'(x_0)(v) :=$

$\lim_{s\to 0} f'(x(s))(v)$ for some smooth curve $x : (-1,1) \to K$ with $x(s) \in K^o$ for $0 < |s| < 1$. Then $f'(x)$ is linear as pointwise limit of $f'(x(s)) \in L(E,\mathbb{R})$ and is bounded by the Banach-Steinhaus theorem (applied to E_B). This shows at the same time, that the definition does not depend on the smooth curve x, since for $v \in x_0 + K^o$ it is the unique extension.

In order to show that $f' : K \to L(E,F)$ is smooth it is by (5.18) enough to show that

$$s \mapsto f'(x(s))(v), \quad \mathbb{R} \xrightarrow{x} K \xrightarrow{f'} L(E,F) \xrightarrow{\mathrm{ev}_x} F$$

is smooth for all $v \in E$ and all smooth curves $x : \mathbb{R} \to K$. For $v \in x_0 + K^o$ this was shown above. For general $v \in E$, this follows since $f'(x(s))(v)$ is a linear combination of $f'(x(s))(v_0)$ for two $v_0 \in x_0 + K^o$ not depending on s locally. □

By (24.2) the following lemma applies in particular to smooth maps.

24.3. Lemma. Chain rule. *Let $K \subseteq E$ be a convex subset with non-void interior K^o, let $f : K \to \mathbb{R}$ be smooth on K^o and let $f' : K \to L(E,F)$ be an extension of $(f|_{K^o})'$, which is continuous for the c^∞-topology of K, and let $c : \mathbb{R} \to K \subseteq E$ be a smooth curve. Then $(f \circ c)'(t) = f'(c(t))(c'(t))$.*

Proof.

Claim Let $g : K \to L(E,F)$ be continuous along smooth curves in K, then $\hat{g} : K \times E \to F$ is also continuous along smooth curves in $K \times E$.
In order to show this let $t \mapsto (x(t),v(t))$ be a smooth curve in $K \times E$. Then $g \circ x : \mathbb{R} \to L(E,F)$ is by assumption continuous (for the bornological topology on $L(E,F)$) and $v^* : L(E,F) \to C^\infty(\mathbb{R},F)$ is bounded and linear (3.13) and (3.17). Hence, the composite $v^* \circ g \circ x : \mathbb{R} \to C^\infty(\mathbb{R},F) \to C(\mathbb{R},F)$ is continuous. Thus, $(v^* \circ g \circ x)^\wedge : \mathbb{R}^2 \to F$ is continuous, and in particular when restricted to the diagonal in $\mathbb{R}^2$. But this restriction is just $g \circ (x,v)$.

Now choose a $y \in K^o$. And let $c_s(t) := c(t) + s^2(y - c(t))$. Then $c_s(t) \in K^o$ for $0 < |s| \le 1$ and $c_0 = c$. Furthermore, $(s,t) \mapsto c_s(t)$ is smooth and $c_s'(t) = (1-s^2)c'(t)$. And for $s \neq 0$

$$\frac{f(c_s(t)) - f(c_s(0))}{t} = \int_0^1 (f \circ c_s)'(t\tau)d\tau = (1-s^2)\int_0^1 f'(c_s(t\tau))(c'(t\tau))d\tau\,.$$

Now consider the specific case where $c(t) := x + tv$ with $x,\ x+v \in K$. Since f is continuous along $(t,s) \mapsto c_s(t)$, the left side of the above equation converges to $\frac{f(c(t))-f(c(0))}{t}$ for $s \to 0$. And since $f'(\cdot)(v)$ is continuous along $(t,\tau,s) \mapsto c_s(t\tau)$ we have that $f'(c_s(t\tau))(v)$ converges to $f'(c(t\tau))(v)$ uniformly with respect to $0 \le \tau \le 1$ for $s \to 0$. Thus, the right side of the above equation converges to $\int_0^1 f'(c(t\tau))(v)d\tau$. Hence, we have

$$\frac{f(c(t)) - f(c(0))}{t} = \int_0^1 f'(c(t\tau))(v)d\tau \to \int_0^1 f'(c(0))(v)d\tau = f'(c(0))(c'(0))$$

for $t \to 0$.

Now let $c : \mathbb{R} \to K$ be an arbitrary smooth curve. Then $(s,t) \mapsto c(0)+s(c(t)-c(0))$ is smooth and has values in K for $0 \le s \le 1$. By the above consideration we have for $x = c(0)$ and $v = (c(t)-c(0))/t$ that

$$\frac{f(c(t)) - f(c(0))}{t} = \int_0^1 f'\Big(c(0) + \tau(c(t) - c(0))\Big)\Big(\frac{c(t) - c(0)}{t}\Big)$$

which converges to $f'(c(0))(c'(0))$ for $t \to 0$, since f' is continuous along smooth curves in K and thus $f'(c(0)+\tau(c(t)-c(0))) \to f'(c(0))$ uniformly on the bounded set $\{\frac{c(t)-c(0)}{t} : t \text{ near } 0\}$. Thus, $f \circ c$ is differentiable with derivative $(f \circ c)'(t) = f'(c(t))(c'(t))$. $\square$

Since f' can be considered as a map $df : E \times E \supseteq K \times E \to F$ it is important to study sets $A \times B \subseteq E \times F$. Clearly, $A \times B$ is convex provided $A \subseteq E$ and $B \subseteq F$ are. Remains to consider the openness condition. In the locally convex topology $(A \times B)^o = A^o \times B^o$, which would be enough to know in our situation. However, we are also interested in the corresponding statement for the c^∞-topology. This topology on $E \times F$ is in general not the product topology $c^\infty E \times c^\infty F$. Thus, we cannot conclude that $A \times B$ has non-void interior with respect to the c^∞-topology on $E \times F$, even if $A \subseteq E$ and $B \subseteq F$ have it. However, in case where $B = F$ everything is fine.

24.4. Lemma. Interior of a product.
Let $X \subseteq E$. Then the interior $(X \times F)^o$ of $X \times F$ with respect to the c^∞-topology on $E \times F$ is just $X^o \times F$, where X^o denotes the interior of X with respect to the c^∞-topology on E.

Proof. Let W be the saturated hull of $(X \times F)^o$ with respect to the projection $\mathrm{pr}_1 : E \times F \to E$, i.e. the c^∞-open set $(X \times F)^o + \{0\} \times F \subseteq X \times F$. Its projection to E is c^∞-open, since it agrees with the intersection with $E \times \{0\}$. Hence, it is contained in X^o, and $(X \times F)^o \subseteq X^o \times F$. The converse inclusion is obvious since pr_1 is continuous. $\square$

24.5. Theorem. Smooth maps on convex sets.
Let $K \subseteq E$ be a convex subset with non-void interior K^o, and let $f : K \to F$ be a map. Then f is smooth if and only if f is smooth on K^o and all derivatives $(f|_{K^o})^{(n)}$ extend continuously to K with respect to the c^∞-topology of K.

Proof. ($\Rightarrow$) It follows by induction using (24.2) that $f^{(n)}$ has a smooth extension $K \to L^n(E;F)$.

($\Leftarrow$) By (24.3) we conclude that for every $c : \mathbb{R} \to K$ the composite $f \circ c : \mathbb{R} \to F$ is differentiable with derivative $(f \circ c)'(t) = f'(c(t))(c'(t)) =: df(c(t), c'(t))$.

The map df is smooth on the interior $K^o \times E$, linear in the second variable, and its derivatives $(df)^{(p)}(x,w)(y_1,w_1;\dots,y_p,w_p)$ are universal linear combinations of

$$f^{(p+1)}(x)(y_1,\dots,y_p;w) \text{ and of } f^{(k+1)}(x)(y_{i_1},\dots,y_{i_k};w_{i_0}) \text{ for } k \le p.$$

These summands have unique extensions to $K \times E$. The first one is continuous along smooth curves in $K \times E$, because for such a curve $(t \mapsto (x(t), w(t))$ the extension $f^{(k+1)} : K \to L(E^k, L(E,F))$ is continuous along the smooth curve x, and $w^* : L(E,F) \to C^\infty(\mathbb{R}, F)$ is continuous and linear, so the mapping $t \mapsto (s \mapsto f^{(k+1)}(x(t))(y_{i_1}, \dots, y_{i_k}; w(s)))$ is continuous from $\mathbb{R} \to C^\infty(\mathbb{R}, F)$ and thus as map from $\mathbb{R}^2 \to F$ it is continuous, and in particular if restricted to the diagonal. And the other summands only depend on x, hence have a continuous extension by assumption.

So we can apply (24.3) inductively using (24.4), to conclude that $f \circ c : \mathbb{R} \to F$ is smooth. □

In view of the preceding theorem (24.5) it is important to know the c^∞-topology $c^\infty X$ of X, i.e. the final topology generated by all the smooth curves $c : \mathbb{R} \to X \subseteq E$. So the first question is whether this is the trace topology $c^\infty E|_X$ of the c^∞-topology of E.

24.6. Lemma. The c^∞-topology is the trace topology.
In the following cases of subsets $X \subseteq E$ the trace topology $c^\infty E|X$ equals the topology $c^\infty X$:

(1) *X is $c^\infty E$-open.*
(2) *X is convex and locally c^∞-closed.*
(3) *The topology $c^\infty E$ is sequential and $X \subseteq E$ is convex and has non-void interior.*

(3) applies in particular to the case where E is metrizable, see (4.11). A topology is called sequential if and only if the closure of any subset equals its adherence, i.e. the set of all accumulation points of sequences in it. By (2.13) and (2.8) the adherence of a set X with respect to the c^∞-topology, is formed by the limits of all Mackey-converging sequences in X.

Proof. Note that the inclusion $X \to E$ is by definition smooth, hence the identity $c^\infty X \to c^\infty E|_X$ is always continuous.

(1) Let $U \subseteq X$ be $c^\infty X$-open and let $c : \mathbb{R} \to E$ be a smooth curve with $c(0) \in U$. Since X is $c^\infty E$-open, $c(t) \in X$ for all small t. By composing with a smooth map $h : \mathbb{R} \to \mathbb{R}$ which satisfies $h(t) = t$ for all small t, we obtain a smooth curve $c \circ h : \mathbb{R} \to X$, which coincides with c locally around 0. Since U is $c^\infty X$-open we conclude that $c(t) = (c \circ h)(t) \in U$ for small t. Thus, U is $c^\infty E$-open.

(2) Let $A \subseteq X$ be $c^\infty X$-closed. And let $\bar{A}$ be the $c^\infty E$-closure of A. We have to show that $\bar{A} \cap X \subseteq A$. So let $x \in \bar{A} \cap X$. Since X is locally $c^\infty E$-closed, there exists a $c^\infty E$-neighborhood U of $x \in X$ with $U \cap X$ c^∞-closed in U. For every $c^\infty E$-neighborhood U of x we have that x is in the closure of $A \cap U$ in U with respect to the $c^\infty E$-topology (otherwise some open neighborhood of x in U does not meet $A \cap U$, hence also not A). Let $a_n \in A \cap U$ be Mackey converging to $a \in U$. Then $a_n \in X \cap U$ which is closed in U thus $a \in X$. Since X is convex the infinite polygon through the a_n lies in X and can be smoothly parameterized by the special

curve lemma (2.8). Using that A is $c^\infty X$-closed, we conclude that $a \in A$. Thus, $A \cap U$ is $c^\infty U$-closed and $x \in A$.

(3) Let $A \subseteq X$ be $c^\infty X$-closed. And let $\bar{A}$ denote the closure of A in $c^\infty E$. We have to show that $\bar{A} \cap X \subseteq A$. So let $x \in \bar{A} \cap X$. Since $c^\infty E$ is sequential there is a Mackey converging sequence $A \ni a_n \to x$. By the special curve lemma (2.8) the infinite polygon through the a_n can be smoothly parameterized. Since X is convex this curve gives a smooth curve $c : \mathbb{R} \to X$ and thus $c(0) = x \in A$, since A is $c^\infty X$-closed. □

24.7. Example. The c^∞-topology is not trace topology.
Let $A \subseteq E$ be such that the c^∞-adherence Adh(A) *of A is not the whole c^∞-closure $\bar{A}$ of A. So let $a \in \bar{A} \setminus \operatorname{Adh}(A)$. Then consider the convex subset $K \subseteq E \times \mathbb{R}$ defined by $K := \{(x,t) \in E \times \mathbb{R} : t \geq 0$ and $(t = 0 \Rightarrow x \in A \cup \{a\})\}$ which has non-empty interior $E \times \mathbb{R}^+$. However, the topology $c^\infty K$ is not the trace topology of $c^\infty(E \times \mathbb{R})$ which equals $c^\infty(E) \times \mathbb{R}$ by* (4.15).

Note that this situation occurs quite often, see (4.13) and (4.36) where A is even a linear subspace.

Proof. Consider $A = A \times \{0\} \subseteq K$. This set is closed in $c^\infty K$, since $E \cap K$ is closed in $c^\infty K$ and the only point in $(K \cap E) \setminus A$ is a, which cannot be reached by a Mackey converging sequence in A, since $a \notin \operatorname{Adh}(A)$.

It is however not the trace of a closed subset in $c^\infty(E) \times \mathbb{R}$. Since such a set has to contain A and hence $\bar{A} \ni a$. □

24.8. Theorem. Smooth maps on subsets with collar.
Let $M \subseteq E$ have a smooth collar, i.e., the boundary ∂M of M is a smooth submanifold of E and there exists a neighborhood U of ∂M and a diffeomorphism $\psi : \partial M \times \mathbb{R} \to U$ which is the identity on ∂M and such that $\psi(M \times \{t \in \mathbb{R} : t \geq 0\}) = M \cap U$. Then every smooth map $f : M \to F$ extends to a smooth map $\tilde{f} : M \cup U \to F$. Moreover, one can choose a bounded linear extension operator $C^\infty(M,F) \to C^\infty(M \cup U, F)$, $f \mapsto \tilde{f}$.

Proof. By (16.8) there is a continuous linear right inverse S to the restriction map $C^\infty(\mathbb{R},\mathbb{R}) \to C^\infty(I,\mathbb{R})$, where $I := \{t \in \mathbb{R} : t \geq 0\}$. Now let $x \in U$ and $(p_x, t_x) := \psi^{-1}(x)$. Then $f(\psi(p_x, \cdot)) : I \to F$ is smooth, since $\psi(p_x, t) \in M$ for $t \geq 0$. Thus, we have a smooth map $S(f(\psi(p_x, \cdot))) : \mathbb{R} \to F$ and we define $\tilde{f}(x) := S(f(\psi(p_x, \cdot)))(t_x)$. Then $\tilde{f}(x) = f(x)$ for all $x \in M \cap U$, since for such an x we have $t_x \geq 0$. Now we extend the definition by $\tilde{f}(x) = f(x)$ for $x \in M^o$. Remains to show that $\tilde{f}$ is smooth (on U). So let $s \mapsto x(s)$ be a smooth curve in U. Then $s \mapsto (p_s, t_s) := \psi^{-1}(x(s))$ is smooth. Hence, $s \mapsto (t \mapsto f(\psi(p_s, t))$ is a smooth curve $\mathbb{R} \to C^\infty(I, F)$. Since S is continuous and linear the composite $s \mapsto (t \mapsto S(f\psi(p_s, \cdot))(t))$ is a smooth curve $\mathbb{R} \to C^\infty(\mathbb{R}, F)$ and thus the associated map $\mathbb{R}^2 \to F$ is smooth, and also the composite $\tilde{f}(x_s)$ of it with $s \mapsto (s, t_s)$.

The existence of a bounded linear extension operator follows now from (21.2). □

In particular, the previous theorem applies to the following convex sets:

24.9. Proposition. Convex sets with smooth boundary have a collar. *Let $K \subseteq E$ be a closed convex subset with non-empty interior and smooth boundary ∂K. Then K has a smooth collar as defined in* (24.8).

Proof. Without loss of generality let $0 \in K^o$.

In order to show that the set $U := \{x \in E : tx \notin K \text{ for some } t > 0\}$ is c^∞-open let $s \mapsto x(s)$ be a smooth curve $\mathbb{R} \to E$ and assume that $t_0 x(0) \notin K$ for some $t_0 > 0$. Since K is closed we have that $t_0 x(s) \notin K$ for all small $|s|$.

For $x \in U$ let $r(x) := \sup\{t \geq 0 : tx \in K^o\} > 0$, i.e. $r = \frac{1}{p_{K^o}}$ as defined in the proof of (24.1) and $r(x)x$ is the unique intersection point of $\partial K \cap (0, +\infty)x$. We claim that $r : U \to \mathbb{R}^+$ is smooth. So let $s \mapsto x(s)$ be a smooth curve in U and $x_0 := r(x(0))x(0) \in \partial K$. Choose a local diffeomorphism $\psi : (E, x_0) \to (E, 0)$ which maps ∂K locally to some closed hyperplane $F \subseteq E$. Any such hyperplane is the kernel of a continuous linear functional $\ell : E \to \mathbb{R}$, hence $E \cong F \times \mathbb{R}$.

We claim that $v := \psi'(x_0)(x_0) \notin F$. If this were not the case, then we consider the smooth curve $c : \mathbb{R} \to \partial K$ defined by $c(t) = \psi^{-1}(-tv)$. Since $\psi'(x_0)$ is injective its derivative is $c'(0) = -x_0$ and $c(0) = x_0$. Since $0 \in K^o$, we have that $x_0 + \frac{c(t)-c(0)}{t} \in K^o$ for all small $|t|$. By convexity $c(t) = x_0 + t\frac{c(t)-c(0)}{t} \in K^o$ for small $t > 0$, a contradiction.

So we may assume that $\ell(\psi'(x)(x)) \neq 0$ for all x in a neighborhood of x_0.

For s small $r(x(s))$ is given by the implicit equation $\ell(\psi(r(x(s))x(s))) = 0$. So let $g : \mathbb{R}^2 \to \mathbb{R}$ be the locally defined smooth map $g(t,s) := \ell(\psi(tx(s)))$. For $t \neq 0$ its first partial derivative is $\partial_1 g(t,s) = \ell(\psi'(tx(s))(x(s))) \neq 0$. So by the classical implicit function theorem the solution $s \mapsto r(x(s))$ is smooth.

Now let $\Psi : U \times \mathbb{R} \to U$ be the smooth map defined by $(x,t) \mapsto e^{-t} r(x)x$. Restricted to $\partial K \times \mathbb{R} \to U$ is injective, since $tx = t'x'$ with $x, x' \in \partial K$ and $t, t' > 0$ implies $x = x'$ and hence $t = t'$. Furthermore, it is surjective, since the inverse mapping is given by $x \mapsto (r(x)x, \ln(r(x)))$. Use that $r(\lambda x) = \frac{1}{\lambda} r(x)$. Since this inverse is also smooth, we have the required diffeomorphism Ψ. In fact, $\Psi(x,t) \in K$ if and only if $e^{-t} r(x) \leq r(x)$, i.e. $t \leq 0$. □

That (24.8) is far from being best possible shows the

24.10. Proposition. *Let $K \subseteq \mathbb{R}^n$ be the quadrant $K := \{x = (x_1, \dots, x_n) \in \mathbb{R}^n : x_1 \geq 0, \dots, x_n \geq 0\}$. Then there exists a bounded linear extension operator $C^\infty(K, F) \to C^\infty(\mathbb{R}^n, F)$ for each convenient vector space F.*

This can be used to obtain the same result for submanifolds with convex corners sitting in smooth finite dimensional manifolds.

Proof. Since $K = (\mathbb{R}_+)^n \subseteq \mathbb{R}^n$ and the inclusion is the product of inclusions $\iota : \mathbb{R}_+ \hookrightarrow \mathbb{R}$ we can use the exponential law (23.2.3) to obtain $C^\infty(K, F) \cong C^\infty((\mathbb{R}_+)^{n-1}, C^\infty(\mathbb{R}_+, F))$. By Seeley's theorem (16.8) we have a bounded linear extension operator $S : C^\infty(\mathbb{R}_+, F) \to C^\infty(\mathbb{R}, F)$. We now proceed by induction on n. So we have an extension operator $S_{n-1} : C^\infty((\mathbb{R}_+)^{n-1}, G) \to C^\infty(\mathbb{R}^{n-1}, G)$ for the convenient vector space $G := C^\infty(\mathbb{R}, F)$ by induction hypothesis. The composite gives up to natural isomorphisms the required extension operator

$$C^\infty(K,F) \cong C^\infty((\mathbb{R}_+)^{n-1}, C^\infty(\mathbb{R}_+,F)) \xrightarrow{S_*} C^\infty((\mathbb{R}_+)^{n-1}, C^\infty(\mathbb{R},F)) \to$$
$$\xrightarrow{S_{n-1}} C^\infty(\mathbb{R}^{n-1}, C^\infty(\mathbb{R},F)) \cong C^\infty(\mathbb{R}^n, F). \quad \square$$

25. Real Analytic Mappings on Non-Open Domains

In this section we will consider real analytic mappings defined on the same type of convex subsets as in the previous section.

25.1. Theorem. Power series in Fréchet spaces. *Let E be a Fréchet space and (F, F') be a dual pair. Assume that a Baire vector space topology on E' exists for which the point evaluations are continuous. Let f_k be k-linear symmetric bounded functionals from E to F, for each $k \in \mathbb{N}$. Assume that for every $\ell \in F'$ and every x in some open subset $W \subseteq E$ the power series $\sum_{k=0}^\infty \ell(f_k(x^k))t^k$ has positive radius of convergence. Then there exists a 0-neighborhood U in E, such that $\{f_k(x_1, \dots, x_k) : k \in \mathbb{N}, x_j \in U\}$ is bounded and thus the power series $x \mapsto \sum_{k=0}^\infty f_k(x^k)$ converges Mackey on some 0-neighborhood in E.*

Proof. Choose a fixed but arbitrary $\ell \in F'$. Then $\ell \circ f_k$ satisfy the assumptions of (7.14) for an absorbing subset in a closed cone C with non-empty interior. Since this cone is also complete metrizable we can proceed with the proof as in (7.14) to obtain a set $A_{K,r} \subseteq C$ whose interior in C is non-void. But this interior has to contain a non-void open set of E and as in the proof of (7.14) there exists some $\rho_\ell > 0$ such that for the ball U_{ρ_ℓ} in E with radius ρ_ℓ and center 0 the set $\{\ell(f_k(x_1, \dots, x_k)) : k \in \mathbb{N}, x_j \in U_{\rho_\ell}\}$ is bounded.

Now let similarly to (9.6)

$$A_{K,r,\rho} := \bigcap_{k\in\mathbb{N}} \bigcap_{x_1,\dots x_n \in U_\rho} \{\ell \in F' : |\ell(f_k(x_1, \dots, x_k))| \le Kr^k\}$$

for $K, r, \rho > 0$. These sets $A_{K,r,\rho}$ are closed in the Baire topology, since evaluation at $f_k(x_1, \dots, x_k)$ is assumed to be continuous.

By the first part of the proof the union of these sets is F'. So by the Baire property, there exist $K, r, \rho > 0$ such that the interior U of $A_{K,r,\rho}$ is non-empty. As in the proof of (9.6) we choose an $\ell_0 \in U$. Then for every $\ell \in F'$ there exists some $\varepsilon > 0$ such that $\ell_\varepsilon := \varepsilon\ell \in U - \ell_0$. So $|\ell(y)| \le \frac{1}{\varepsilon}(|\ell_\varepsilon(y) + \ell_0(y)| + |\ell_0(y)|) \le \frac{2}{\varepsilon}Kr^n$ for every $y = f_k(x_1, \dots, x_k)$ with $x_i \in U_\rho$. Thus, $\{f_k(x_1, \dots, x_k) : k \in \mathbb{N}, x_i \in U_{\frac{\rho}{r}}\}$ is bounded.

On every smaller ball we have therefore that the power series with terms f_k converges Mackey. $\square$

Note that if the vector spaces are real and the assumption above hold, then the conclusion is even true for the complexified terms by (7.14).

25.2. Theorem. Real analytic maps $I \to \mathbb{R}$ are germs.
Let $f : I := \{t \in \mathbb{R} : t \geq 0\} \to \mathbb{R}$ be a map. Suppose $t \mapsto f(t^2)$ is real analytic $\mathbb{R} \to \mathbb{R}$. Then f extends to a real analytic map $\tilde{f} : \tilde{I} \to \mathbb{R}$, where $\tilde{I}$ is an open neighborhood of I in $\mathbb{R}$.

Proof. We show first that f is smooth. Consider $g(t) := f(t^2)$. Since $g : \mathbb{R} \to \mathbb{R}$ is assumed to be real analytic it is smooth and clearly even. We claim that there exists a smooth map $h : \mathbb{R} \to \mathbb{R}$ with $g(t) = h(t^2)$ (this is due to [Whitney, 1943]). In fact, by $h(t^2) := g(t)$ a continuosu map $h : \{t :\in \mathbb{R} : t \geq 0\} \to \mathbb{R}$ is uniquely determined. Obviously, $h|_{\{t \in \mathbb{R}: t>0\}}$ is smooth. Differentiating for $t \neq 0$ the defining equation gives $h'(t^2) = \frac{g'(t)}{2t} =: g_1(t)$. Since g is smooth and even, g' is smooth and odd, so $g'(0) = 0$. Thus

$$t \mapsto g_1(t) = \frac{g'(t) - g'(0)}{2t} = \frac{1}{2} \int_0^1 g''(ts)\, ds$$

is smooth. Hence, we may define h' on $\{t \in \mathbb{R} : t \geq 0\}$ by the equation $h'(t^2) = g_1(t)$ with even smooth g_1. By induction we obtain continuous extensions of $h^{(n)} : \{t \in \mathbb{R} : t > 0\} \to \mathbb{R}$ to $\{t \in \mathbb{R} : t \geq 0\}$, and hence h is smooth on $\{t \in \mathbb{R} : t \geq 0\}$ and so can be extended to a smooth map $h : \mathbb{R} \to \mathbb{R}$.

From this we get $f(t^2) = g(t) = h(t^2)$ for all t. Thus, $h : \mathbb{R} \to \mathbb{R}$ is a smooth extension of f.

Composing with the exponential map $\exp : \mathbb{R} \to \mathbb{R}^+$ shows that f is real analytic on $\{t : t > 0\}$, and has derivatives $f^{(n)}$ which extend by (24.5) continuously to maps $I \to \mathbb{R}$. It is enough to show that $a_n := \frac{1}{n!} f^{(n)}(0)$ are the coefficients of a power series p with positive radius of convergence and for $t \in I$ this map p coincides with f.

Claim. We show that a smooth map $f : I \to \mathbb{R}$, which has a real analytic composite with $t \mapsto t^2$, is the germ of a real analytic mapping.
Consider the real analytic curve $c : \mathbb{R} \to I$ defined by $c(t) = t^2$. Thus, $f \circ c$ is real analytic. By the chain rule the derivative $(f \circ c)^{(p)}(t)$ is for $t \neq 0$ a universal linear combination of terms $f^{(k)}(c(t))c^{(p_1)}(t) \cdots c^{(p_k)}(t)$, where $1 \leq k \leq p$ and $p_1 + \ldots + p_k = p$. Taking the limit for $t \to 0$ and using that $c^{(n)}(0) = 0$ for all $n \neq 2$ and $c''(0) = 2$ shows that there is a universal constant c_p satisfying $(f \circ c)^{(2p)}(0) = c_p \cdot f^{(p)}(0)$. Take as $f(x) = x^p$ to conclude that $(2p)! = c_p \cdot p!$. Now we use (9.2) to show that the power series $\sum_{k=0}^{\infty} \frac{1}{k!} f^{(k)}(0) t^k$ converges locally. So choose a sequence (r_k) with $r_k t^k \to 0$ for all $t > 0$. Define a sequence $(\bar{r}_k)$ by $\bar{r}_{2n} = \bar{r}_{2n+1} := r_n$ and let $\bar{t} > 0$. Then $\bar{r}_k \bar{t}^k = r_n t^n$ for $2n = k$ and $\bar{r}_k \bar{t}^k = r_n t^n \bar{t}$ for $2n+1 = k$, where $t := \bar{t}^2 > 0$, hence $(\bar{r}_k)$ satisfies the same assumptions as (r_k) and thus by (9.3) ($1 \Rightarrow 3$) the sequence $\frac{1}{k!}(f \circ c)^{(k)}(0)\bar{r}_k$ is bounded. In particular, this is true for the subsequence

$$\tfrac{1}{(2p)!}(f \circ c)^{(2p)}(0)\bar{r}_{2p} = \tfrac{c_p}{(2p)!} f^{(p)}(0) r_p = \tfrac{1}{p!} f^{(p)}(0) r_p.$$

Thus, by (9.3) ($1 \Leftarrow 3$) the power series with coefficients $\frac{1}{p!} f^{(p)}(0)$ converges locally to a real analytic function $\tilde{f}$.

Remains to show that $\tilde{f} = f$ on J. But since $\tilde{f} \circ c$ and $f \circ c$ are both real analytic near 0, and have the same Taylor series at 0, they have to coincide locally, i.e. $\tilde{f}(t^2) = f(t^2)$ for small t. □

Note however that the more straight forward attempt of a proof of the first step, namely to show that $f \circ c$ is smooth for all $c : \mathbb{R} \to \{t \in \mathbb{R} : t \geq 0\}$ by showing that for such c there is a smooth map $h : \mathbb{R} \to \mathbb{R}$, satisfying $c(t) = h(t)^2$, is doomed to fail as the following example shows.

25.3. Example. A smooth function without smooth square root. *Let $c : \mathbb{R} \to \{t \in \mathbb{R} : t \geq 0\}$ be defined by the general curve lemma* (12.2) *using pieces of parabolas $c_n : t \mapsto \frac{2n}{2^n} t^2 + \frac{1}{4^n}$. Then there is no smooth square root of c.*

Proof. The curve c constructed in (12.2) has the property that there exists a converging sequence t_n such that $c(t + t_n) = c_n(t)$ for small t. Assume there were a smooth map $h : \mathbb{R} \to \mathbb{R}$ satisfying $c(t) = h(t)^2$ for all t. At points where $c(t) \neq 0$ we have in turn:

$$\begin{aligned} c'(t) &= 2h(t)h'(t) \\ c''(t) &= 2h(t)h''(t) + 2h'(t)^2 \\ 2c(t)c''(t) &= 4h(t)^3 h''(t) + c'(t)^2. \end{aligned}$$

Choosing t_n for t in the last equation gives $h''(t_n) = 2n$, which is unbounded in n. Thus h cannot be C^2. □

25.4. Definition. (Real analytic maps $I \to F$)
Let $I \subseteq \mathbb{R}$ be a non-trivial interval. Then a map $f : I \to F$ is called real analytic if and only if the composites $\ell \circ f \circ c : \mathbb{R} \to \mathbb{R}$ are real analytic for all real analytic $c : \mathbb{R} \to I \subseteq \mathbb{R}$ and all $\ell \in F'$. If I is an open interval then this definition coincides with (10.3).

25.5. Lemma. Bornological description of real analyticity. *Let $I \subseteq \mathbb{R}$ be a compact interval. A curve $c : I \to E$ is real analytic if and only if c is smooth and the set $\{\frac{1}{k!} c^{(k)}(a) r_k : a \in I, k \in \mathbb{N}\}$ is bounded for all sequences (r_k) with $r_k t^k \to 0$ for all $t > 0$.*

Proof. We use (9.3). Since both sides can be tested with $\ell \in E'$ we may assume that $E = \mathbb{R}$.

($\Rightarrow$) By (25.2) we may assume that $c : \tilde{I} \to \mathbb{R}$ is real analytic for some open neighborhood $\tilde{I}$ of I. Thus, the required boundedness condition follows from (9.3).

($\Leftarrow$) By (25.2) we only have to show that $f : t \mapsto c(t^2)$ is real analytic. For this we use again (9.3). So let $K \subseteq \mathbb{R}$ be compact. Then the Taylor series of f is obtained by that of c composed with t^2. Thus, the composite f satisfies the required boundedness condition, and hence is real analytic. □

This characterization of real analyticity can not be weakened by assuming the boundedness conditions only for single pointed K as the map $c(t) := e^{-1/t^2}$ for

$t \neq 0$ and $c(0) = 0$ shows. It is real analytic on $\mathbb{R} \setminus \{0\}$ thus the condition is satisfied at all points there, and at 0 the power series has all coefficients equal to 0, hence the condition is satisfied there as well.

25.6. Corollary. Real analytic maps into inductive limits.
Let $T_\alpha : E \to E_\alpha$ be a family of bounded linear maps that generates the bornology on E. Then a map $c : I \to F$ is real analytic if and only if all the composites $T_\alpha \circ c : I \to F_\alpha$ are real analytic.

Proof. This follows either directly from (25.5) or from (25.2) by using the corresponding statement for maps $\mathbb{R} \to E$, see (9.9). □

25.7. Definition. (Real analytic maps $K \to F$)
For an arbitrary subset $K \subseteq E$ let us call a map $f : E \supseteq K \to F$ real analytic if and only if $\lambda \circ f \circ c : I \to \mathbb{R}$ is a real analytic (resp. smooth) for all $\lambda \in F'$ and all real analytic (resp. smooth) maps $c : I \to K$, where $I \subset \mathbb{R}$ is some compact non-trivial interval. Note however that it is enough to use all real analytic (resp. smooth) curves $c : \mathbb{R} \to K$ by (25.2).

With $C^\omega(K, F)$ we denote the vector space of all real analytic maps $K \to F$. And we topologize this space with the initial structure induced by the cone $c^* : C^\omega(K, F) \to C^\omega(\mathbb{R}, F)$ (for all real analytic $c : \mathbb{R} \to K$) and the cone $c^* : C^\omega(K, F) \to C^\infty(\mathbb{R}, F)$ (for all smooth $c : \mathbb{R} \to K$). The space $C^\omega(\mathbb{R}, F)$ should carry the structure of (11.2) and the space $C^\infty(\mathbb{R}, F)$ that of (3.6).

For an open $K \subseteq E$ the definition for $C^\omega(K, F)$ given here coincides with that of (10.3).

25.8. Proposition. $C^\omega(K, F)$ is convenient. *Let $K \subseteq E$ and F be arbitrary. Then the space $C^\omega(K, F)$ is a convenient vector space and satisfies the $\mathcal{S}$-uniform boundedness principle* (5.22), *where $\mathcal{S} := \{\mathrm{ev}_x : x \in K\}$.*

Proof. Since both spaces $C^\omega(\mathbb{R}, \mathbb{R})$ and $C^\infty(\mathbb{R}, \mathbb{R})$ are c^∞-complete and satisfy the uniform boundedness principle for the set of point evaluations the same is true for $C^\omega(K, F)$, by (5.25). □

25.9. Theorem. Real analytic maps $K \to F$ are often germs.
Let $K \subseteq E$ be a convex subset with non-empty interior of a Fréchet space and let (F, F') be a complete dual pair for which a Baire topology on F' exists, as required in (25.1). *Let $f : K \to F$ be a real analytic map. Then there exists an open neighborhood $U \subseteq E_\mathbb{C}$ of K and a holomorphic map $\tilde{f} : U \to F_\mathbb{C}$ such that $\tilde{f}|_K = f$.*

Proof. By (24.5) the map $f : K \to F$ is smooth, i.e. the derivatives $f^{(k)}$ exist on the interior K^0 and extend continuously (with respect to the c^∞-topology of K) to the whole of K. So let $x \in K$ be arbitrary and consider the power series with coefficients $f_k = \frac{1}{k!} f^{(k)}(x)$. This power series has the required properties of (25.1), since for every $\ell \in F'$ and $v \in K^o - x$ the series $\sum_k \ell(f_k(v^k))t^k$ has positive radius of convergence. In fact, $\ell(f(x + tv))$ is by assumption a real analytic germ $I \to \mathbb{R}$, by (24.8) hence locally around any point in I it is represented by its converging Taylor series at that point. Since $(x, v-x] \subseteq K^o$ and f is smooth on this set, $(\frac{d}{dt})^k(\ell(f(x+tv)) = \ell(f^{(k)}(x+tv)(v^k)$ for $t > 0$. Now take the limit for $t \to 0$ to conclude that the Taylor coefficients of $t \mapsto \ell(f(x+tv))$ at $t=0$ are exactly $k!\ell(f_k)$. Thus, by (25.1) the power series converges locally and hence represents a holomorphic map in a neighborhood of x. Let $y \in K^o$ be an arbitrary point in this neighborhood. Then $t \mapsto \ell(f(x+t(y-x)))$ is real analytic $I \to \mathbb{R}$ and hence the series converges at $y-x$ towards $f(y)$. So the restriction of the power series to the interior of K coincides with f.

We have to show that the extensions f_x of $f : K \cap \tilde{U}_x \to F_{\mathbb{C}}$ to star shaped neighborhoods $\tilde{U}_x$ of x in $E_{\mathbb{C}}$ fit together to give an extension $\tilde{f} : \tilde{U} \to F_{\mathbb{C}}$. So let $\tilde{U}_x$ be such a domain for the extension and let $U_x := \tilde{U}_x \cap E$.

For this we claim that we may assume that U_x has the following additional property: $y \in U_x \Rightarrow [0,1]y \subseteq K^o \cup U_x$. In fact, let $U_0 := \{y \in U_x : [0,1]y \subseteq K^o \cup U_x\}$. Then U_0 is open, since $f : (t,s) \mapsto ty(s)$ being smooth, and $f(t,0) \in K^o \cup U_x$ for $t \in [0,1]$, implies that a $\delta > 0$ exists such that $f(t,s) \in K^o \cup U_x$ for all $|s| < \delta$ and $-\delta < t < 1+\delta$. The set U_0 is star shaped, since $y \in U_0$ and $s \in [0,1]$ implies that $t(x+s(y-x)) \in [x, t'y]$ for some $t' \in [0,1]$, hence lies in $K^o \cup U_x$. The set U_0 contains x, since $[0,1]x = \{x\} \cup [0,1)x \subseteq \{x\} \cup K^o$. Finally, U_0 has the required property, since $z \in [0,1]y$ for $y \in U_0$ implies that $[0,1]z \subseteq [0,1]y \subseteq K^o \cup U_x$, i.e. $z \in U_0$.

Furthermore, we may assume that for $x+iy \in \tilde{U}_x$ and $t \in [0,1]$ also $x+ity \in \tilde{U}_x$ (replace $\tilde{U}_x$ by $\{x+iy : x+ity \in \tilde{U}_x$ for all $t \in [0,1]\}$).

Now let $\tilde{U}_1$ and $\tilde{U}_2$ be two such domains around x_1 and x_2, with corresponding extensions f_1 and f_2. Let $x+iy \in \tilde{U}_1 \cap \tilde{U}_2$. Then $x \in U_1 \cap U_2$ and $[0,1]x \subseteq K^o \cup U_i$ for $i=1,2$. If $x \in K^o$ we are done, so let $x \notin K^o$. Let $t_0 := \inf\{t > 0 : tx \notin K^o\}$. Then $t_0 x \in U_i$ for $i=1,2$ and by taking t_0 a little smaller we may assume that $x_0 := t_0 x \in K^o \cap U_1 \cap U_2$. Thus, $f_i = f$ on $[x_0, x_i]$ and the f_i are real analytic on $[x_0, x]$ for $i=1,2$. Hence, $f_1 = f_2$ on $[x_0, x]$ and thus $f_1 = f_2$ on $[x, x+iy]$ by the 1-dimensional uniqueness theorem. □

That the result corresponding to (24.8) is not true for manifolds with real analytic boundary shows the following

25.10. Example. No real analytic extension exists.
Let $I := \{t \in \mathbb{R} : t \geq 0\}$, $E := C^\omega(I, \mathbb{R})$, and let $ev : E \times \mathbb{R} \supseteq E \times I \to \mathbb{R}$ be the real analytic map $(f,t) \mapsto f(t)$. Then there is no real analytic extension of ev to a neighborhood of $E \times I$.

Proof. Suppose there is some open set $U \subseteq E \times \mathbb{R}$ containing $\{(0,t) : t \geq 0\}$ and a C^ω-extension $\varphi : U \to \mathbb{R}$. Then there exists a c^∞-open neighborhood V of 0 and some $\delta > 0$ such that U contains $V \times (-\delta, \delta)$. Since V is absorbing in E, we have for every $f \in E$ that there exists some $\varepsilon > 0$ such that $\varepsilon f \in V$ and hence $\frac{1}{\varepsilon}\varphi(\varepsilon f, \cdot) : (-\delta, \delta) \to \mathbb{R}$ is a real analytic extension of f. This cannot be true, since there are $f \in E$ having a singularity inside $(-\delta, \delta)$. □

The following theorem generalizes (11.17).

25.11. Theorem. Mixing of C^∞ and C^ω.
Let (E, E') be a complete dual pair, let $X \subseteq E$, let $f : \mathbb{R}\times X \to \mathbb{R}$ be a mapping that extends for every B locally around every point in $\mathbb{R}\times(X\cap E_B)$ to a holomorphic map $\mathbb{C}\times(E_B)_\mathbb{C} \to \mathbb{C}$, and let $c \in C^\infty(\mathbb{R}, X)$. Then $c^* \circ f^\vee : \mathbb{R} \to C^\omega(X,\mathbb{R}) \to C^\infty(\mathbb{R},\mathbb{R})$ is real analytic.

Proof. Let $I \subseteq \mathbb{R}$ be open and relatively compact, let $t \in \mathbb{R}$ and $k \in \mathbb{N}$. Now choose an open and relatively compact $J \subseteq \mathbb{R}$ containing the closure $\bar{I}$ of I. By (1.8) there is a bounded subset $B \subseteq E$ such that $c|_J : J \to E_B$ is a $\mathcal{L}\text{ip}^k$-curve in the Banach space E_B generated by B. Let X_B denote the subset $X \cap E_B$ of the Banach space E_B. By assumption on f there is a holomorphic extension $f : V \times W \to \mathbb{C}$ of f to an open set $V \times W \subseteq \mathbb{C} \times (E_B)_\mathbb{C}$ containing the compact set $\{t\} \times c(\bar{I})$. By cartesian closedness of the category of holomorphic mappings $f^\vee : V \to \mathcal{H}(W, \mathbb{C})$ is holomorphic. Now recall that the bornological structure of $\mathcal{H}(W, \mathbb{C})$ is induced by that of $C^\infty(W,\mathbb{C}) := C^\infty(W,\mathbb{R}^2)$. Furthermore, $c^* : C^\infty(W,\mathbb{C}) \to \mathcal{L}\text{ip}^k(I,\mathbb{C})$ is a bounded $\mathbb{C}$-linear map (see tyhe proof of (11.17)). Thus, $c^* \circ f^\vee : V \to \mathcal{L}\text{ip}^k(I,\mathbb{C})$ is holomorphic, and hence its restriction to $\mathbb{R} \cap V$, which has values in $\mathcal{L}\text{ip}^k(I,\mathbb{R})$, is (even topologically) real analytic by (9.5). Since $t \in \mathbb{R}$ was arbitrary we conclude that $c^* \circ f^\vee : \mathbb{R} \to \mathcal{L}\text{ip}^k(I,\mathbb{R})$ is real analytic. But the bornology of $C^\infty(\mathbb{R},\mathbb{R})$ is generated by the inclusions into $\mathcal{L}\text{ip}^k(I,\mathbb{R})$, by the uniform boundedness principles (5.26) for $C^\infty(\mathbb{R},\mathbb{R})$ and (12.9) for $\mathcal{L}\text{ip}^k(\mathbb{R},\mathbb{R})$, and hence $c^* \circ f^\vee : \mathbb{R} \to C^\infty(\mathbb{R},\mathbb{R})$ is real analytic. □

This can now be used to show cartesian closedness with the same proof as in (11.18) for certain non-open subsets of convenient vector spaces. In particular, the previous theorem applies to real analytic mappings $f : \mathbb{R} \times X \to \mathbb{R}$, where $X \subseteq E$ is convex with non-void interior. Since for such a set the intersection X_B with E_B has the same property and since E_B is a Banach space, the real analytic mapping is the germ of a holomorphic mapping.

25.12. Theorem. Exponential law for real analytic germs.
Let K and L be two convex subsets with non-empty interior in convenient vector spaces. A map $f : K \to C^\omega(L, F)$ is real analytic if and only if the associated mapping $\hat{f} : K \times L \to F$ is real analytic.

Proof. ($\Rightarrow$) Let $c = (c_1, c_2) : \mathbb{R} \to K \times L$ be C^α (for $\alpha \in \{\infty, \omega\}$) and let $\ell \in F'$. We have to show that $\ell \circ \hat{f} \circ c : \mathbb{R} \to \mathbb{R}$ is C^α. By cartesian closedness of C^α it is enough to show that the map $\ell \circ \hat{f} \circ (c_1 \times c_2) : \mathbb{R}^2 \to \mathbb{R}$ is C^α. This map however is associated to $\ell_* \circ (c_2)_* \circ f \circ c_1 : \mathbb{R} \to K \to C^\omega(L, F) \to C^\alpha(\mathbb{R},\mathbb{R})$, hence is C^α by assumption on f and the structure of $C^\omega(L, F)$.

($\Leftarrow$) Let conversely $f : K \times L \to F$ be real analytic. Then obviously $f(x, \cdot) : L \to F$ is real analytic, hence $f^\vee : K \to C^\omega(L, F)$ makes sense. Now take an arbitrary C^α-map $c_1 : \mathbb{R} \to K$. We have to show that $f^\vee \circ c_1 : \mathbb{R} \to C^\omega(L, F)$ is C^α. Since the structure of $C^\omega(L, F)$ is generated by $C^\beta(c_1, \ell)$ for C^β-curves

$c_2 : \mathbb{R} \to L$ (for $\beta \in \{\infty, \omega\}$) and $\ell \in F'$, it is by (9.3) enough to show that $C^\beta(c_2, \ell) \circ f^\vee \circ c_1 : \mathbb{R} \to C^\beta(\mathbb{R}, \mathbb{R})$ is C^α. For $\alpha = \beta$ it is by cartesian closedness of C^α maps enough to show that the associate map $\mathbb{R}^2 \to \mathbb{R}$ is C^α. Since this map is just $\ell \circ f \circ (c_1 \times c_2)$, this is clear. In fact, take for $\gamma \leq \alpha, \gamma \in \{\infty, \omega\}$ an arbitrary C^γ-curve $d = (d_1, d_2) : \mathbb{R} \to \mathbb{R}^2$. Then $(c_1 \times c_2) \circ (d_1, d_2) = (c_1 \circ d_1, c_2 \circ d_2)$ is C^γ, and so the composite with $\ell \circ f$ has the same property.

It remains to show the mixing case, where c_1 is real analytic and c_2 is smooth or conversely. First the case c_1 real analytic, c_2 smooth. Then $\ell \circ f \circ (c_1 \times \mathrm{Id}) : \mathbb{R} \times L \to \mathbb{R}$ is real analytic, hence extends to some holomorphic map by (25.9), and by (25.11) the map

$$C^\infty(c_2, \ell) \circ f^\vee \circ c_1 = c_2^* \circ (\ell \circ f \circ (c_1 \times \mathrm{Id}))^\vee : \mathbb{R} \to C^\infty(\mathbb{R}, \mathbb{R})$$

is real analytic. Now the case c_1 smooth and c_2 real analytic. Then $\ell \circ f \circ (\mathrm{Id} \times c_2) : K \times \mathbb{R} \to \mathbb{R}$ is real analytic, so by the same reasoning as just before applied to $\tilde{f}$ defined by $\tilde{f}(x, y) := f(y, x)$, the map

$$C^\infty(c_1, \ell) \circ (\tilde{f})^\vee \circ c_2 = c_1^* \circ (\ell \circ \tilde{f} \circ (\mathrm{Id} \times c_2))^\vee : \mathbb{R} \to C^\infty(\mathbb{R}, \mathbb{R})$$

is real analytic. By (11.16) the associated mapping

$$(c_1^* \circ (\ell \circ \tilde{f} \circ (\mathrm{Id} \times c_2))^\vee)^\sim = C^\omega(c_2, \ell) \circ \tilde{f} \circ c_1 : \mathbb{R} \to C^\omega(\mathbb{R}, \mathbb{R})$$

is smooth. □

The following example shows that theorem (25.12) does not extend to arbitrary domains.

25.13. Example. The exponential law for general domains is false. *Let $X \subseteq \mathbb{R}^2$ be the graph of the map $h : \mathbb{R} \to \mathbb{R}$ defined by $h(t) := e^{-t^{-2}}$ for $t \neq 0$ and $h(0) = 0$. Let, furthermore, $f : \mathbb{R} \times X \to \mathbb{R}$ be the mapping defined by $f(t, s, r) := \frac{r}{t^2+s^2}$ for $(t, s) \neq (0, 0)$ and $f(0, 0, r) := 0$. Then $f : \mathbb{R} \times X \to \mathbb{R}$ is real analytic, however the associated mapping $f^\vee : \mathbb{R} \to C^\omega(X, \mathbb{R})$ is not.*

Proof. Obviously, f is real analytic on $\mathbb{R}^3 \setminus \{(0,0)\} \times \mathbb{R}$. If $u \mapsto (t(u), s(u), r(u))$ is real analytic $\mathbb{R} \to \mathbb{R} \times X$, then $r(u) = h(s(u))$. Suppose s is not constant and $t(0) = s(0) = 0$, then we have that $r(u) = h(u^n s_0(u))$ cannot be real analytic, since it is not constant but the Taylor series at 0 is identical 0, a contradiction. Thus, $s = 0$ and $r = h \circ s = 0$, therefore $u \mapsto f(t(u), s(u), r(u)) = 0$ is real analytic.

Remains to show that $u \mapsto f(t(u), s(u), r(u))$ is smooth for all smooth curves $(t, s, r) : \mathbb{R} \to \mathbb{R} \times X$. Since $f(t(u), s(u), r(u)) = \frac{h(s(u))}{t(u)^2+s(u)^2}$ it is enough to show that $\varphi : \mathbb{R}^2 \to \mathbb{R}$ defined by $\varphi(t, s) = \frac{h(s)}{t^2+s^2}$ is smooth. This is obviously the case, since each of its partial derivatives is of the form $h(s)$ multiplied by some rational function of t and s, hence extends continuously to $\{(0, 0)\}$.

Now we show that $f^\vee : \mathbb{R} \to C^\omega(X, \mathbb{R})$ is not real analytic. Take the smooth curve $c : u \mapsto (u, h(u))$ into X and consider $c^* \circ f^\vee : \mathbb{R} \to C^\infty(\mathbb{R}, \mathbb{R})$, which is given by $t \mapsto$

$(s \mapsto f(t, c(s)) = \frac{h(s)}{t^2+s^2})$. Suppose it is real analytic into $C([-1,+1],\mathbb{R})$. Then it has to be locally representable by a converging power series $\sum a_n t^n \in C([-1,+1],\mathbb{R})$. So there has to exist a $\delta > 0$ such that $\sum a_n(s)z^n = \frac{h(s)}{s^2}\sum_{k=0}^{\infty}(-1)^k(\frac{z}{s})^{2k}$ converges for all $|z| < \delta$ and $|s| < 1$. This is impossible, since at $z = si$ there is a pole. □

26. Holomorphic Mappings on Non-Open Domains

In this section we will consider holomorphic maps defined on two types of convex subsets. First the case where the set is contained in some real part of the vector space and has non-empty interior there. Recall that for a subset $X \subseteq \mathbb{R} \subseteq \mathbb{C}$ the space of germs of holomorphic maps $X \to \mathbb{C}$ is the complexification of that of germs of real analytic maps $X \to \mathbb{R}$, (11.2). Thus, we give the following

26.1. Definition. (Holomorphic maps $K \to F$)
Let $K \subseteq E$ be a convex set with non-empty interior in a real convenient vector space. And let F be a complex convenient vector space. We call a map $f : E_{\mathbb{C}} \supseteq K \to F$ holomorphic if and only if $f : E \supseteq K \to F$ is real analytic.

26.2. Lemma. Holomorphic maps can be tested by functionals.
Let $K \subseteq E$ be a convex set with non-empty interior in a real convenient vector space. And let F be a complex convenient vector space. Then a map $f : K \to F$ is holomorphic if and only if the composites $\ell \circ f : K \to \mathbb{C}$ are holomorphic for all $\ell \in L_{\mathbb{C}}(E, \mathbb{C})$, where $L_{\mathbb{C}}(E, \mathbb{C})$ denotes the space of $\mathbb{C}$-linear maps.

Proof. ($\Rightarrow$) Let $\ell \in L_{\mathbb{C}}(F, \mathbb{C})$. Then the real and imaginary part $\operatorname{Re}\ell, \operatorname{Im}\ell \in L_{\mathbb{R}}(F, \mathbb{R})$ and since by assumption $f : K \to F$ is real analytic so are the composites $\operatorname{Re}\ell \circ f$ and $\operatorname{Im}\ell \circ f$, hence $\ell \circ f : K \to \mathbb{R}^2$ is real analytic, i.e. $\ell \circ f : K \to \mathbb{C}$ is holomorphic.

($\Leftarrow$) We have to show that $\ell \circ f : K \to \mathbb{R}$ is real analytic for every $\ell \in L_{\mathbb{R}}(F, \mathbb{R})$. So let $\tilde{\ell} : F \to \mathbb{C}$ be defined by $\tilde{\ell}(x) = i\ell(x) + \ell(ix)$. Then $\tilde{\ell} \in L_{\mathbb{C}}(F, \mathbb{C})$, since $i\tilde{\ell}(x) = -\ell(x) + i\ell(ix) = \tilde{\ell}(ix)$. Note that $\ell = \operatorname{Im} \circ \tilde{\ell}$. By assumption, $\tilde{\ell} \circ f : K \to \mathbb{C}$ is holomorphic, hence its imaginary part $\ell \circ f : K \to \mathbb{R}$ is real analytic. □

26.3. Theorem. Holomorphic maps $K \to F$ are often germs.
Let $K \subseteq E$ be a convex subset with non-empty interior in a real Fréchet space E and let F be a complex convenient vector space such that F' carries a Baire topology as required in (25.1). Then a map $f : E_{\mathbb{C}} \supseteq K \to F$ is holomorphic if and only if it extends to a holomorphic map $\tilde{f} : \tilde{K} \to F$ for some neighborhood $\tilde{K}$ of K in $E_{\mathbb{C}}$.

Proof. Using (25.9) we conclude that f extends to a holomorphic map $\tilde{f} : \tilde{K} \to F_{\mathbb{C}}$ for some neighborhood $\tilde{K}$ of K in $E_{\mathbb{C}}$. The map $\operatorname{pr} : F_{\mathbb{C}} \to F$, given by $\operatorname{pr}(x, y) = x + iy \in F$ for $(x, y) \in F^2 = F \otimes_{\mathbb{R}} \mathbb{C}$, is $\mathbb{C}$-linear and restricted to $F \times \{0\} = F$ it is the identity. Thus, $\operatorname{pr} \circ \tilde{f} : \tilde{K} \to F_{\mathbb{C}} \to F$ is a holomorphic extension of f.

Conversely, let $\tilde{f} : \tilde{K} \to F$ be a holomorphic extension to a neighborhood $\tilde{K}$ of K. So it is enough to show that the holomorphic map $\tilde{f}$ is real analytic. By (7.19) it is smooth. So it remains to show that it is real analytic. For this it is enough to consider a topological real analytic curve in $\tilde{K}$ by (10.4). Such a curve is extendable to a holomorphic curve $\tilde{c}$ by (9.5), hence the composite $\tilde{f} \circ \tilde{c}$ is holomorphic and its restriction $\tilde{f} \circ c$ to $\mathbb{R}$ is real analytic. □

26.4. Definition. (Holomorphic maps on complex vector spaces)
Let $K \subseteq E$ be a convex subset with non-empty interior in a complex convenient vector space. And map $f : E \supseteq K \to F$ is called holomorphic iff it is real analytic and the derivative $f'(x)$ is $\mathbb{C}$-linear for all $x \in K^o$.

26.5. Theorem. Holomorphic maps are germs.
Let $K \subseteq E$ be a convex subset with non-empty interior in a complex convenient vector space. Then a map $f : E \supseteq K \to F$ into a complex convenient vector space F is holomorphic if and only if it extends to a holomorphic map defined on some neighborhood of K in E.

Proof. Since $f : K \to F$ is real analytic, it extends by (25.9) to a real analytic map $\tilde{f} : E \supseteq U \to F$, where we may assume that U is connected with K by straight line segments. We claim that $\tilde{f}$ is in fact holomorphic. For this it is enough to show that $f'(x)$ is $\mathbb{C}$-linear for all $x \in U$. So consider the real analytic mapping $g : U \to F$ given by $g(x) := if'(x)(v) - f'(x)(iv)$. Since it is zero on K^o it has to be zero everywhere by the uniqueness theorem. □

26.6. Remark. (There is no definition for holomorphy analogous to (25.7))
In order for a map $K \to F$ to be holomorphic it is not enough to assume that all composites $f \circ c$ for holomorphic $c : \mathbb{D} \to K$ are holomorphic, where $\mathbb{D}$ is the open unit disk. Take as K the closed unit disk, then $c(\mathbb{D}) \cap \partial K = \phi$. In fact let $z_0 \in \mathbb{D}$ then $c(z) = (z - z_0)^n(c_n + (z - z_0)\sum_{k>n} c_k(z - z_0)^{k-n-1})$ for z close to z_0, which covers a neighborhood of $c(z_0)$. So the boundary values of such a map would be completely arbitrary.

26.7. Lemma. Holomorphy is a bornological concept.
Let $T_\alpha : E \to E_\alpha$ be a family of bounded linear maps that generates the bornology on E. Then a map $c : K \to F$ is holomorphic if and only if all the composites $T_\alpha \circ c : I \to F_\alpha$ are holomorphic.

Proof. It follows from (25.6) that f is real analytic. And the $\mathbb{C}$-linearity of $f'(x)$ can certainly be tested by point separating linear functionals. □

26.8. Theorem. Exponential law for holomorphic maps.
Let K and L be convex subsets with non-empty interior in complex convenient vector spaces. Then a map $f : K \times L \to F$ is holomorphic if and only if the associated map $f^\vee : K \to H(L, F)$ is holomorphic.

Proof. This follows immediately from the real analytic result (25.12), since the $\mathbb{C}$-linearity of the involved derivatives translates to each other, since we obviously have $f'(x_1, x_2)(v_1, v_2) = ev_{x_2}((f^\vee)'(x_1)(v_1)) + (f^\vee(x_1))'(x_2)(v_2)$ for $x_1 \in K$ and $x_2 \in L$. □

Chapter VI
Infinite Dimensional Manifolds

This chapter is devoted to the foundations of infinite dimensional manifolds. We treat here only manifolds described by charts onto c^∞-open subsets of convenient vector spaces.

Note that this limits cartesian closedness of the category of manifolds. For finite dimensional manifolds M, N, P we will show later that $C^\infty(N, P)$ is not locally contractible (not even locally pathwise connected) for the compact-open C^∞-topology if N is not compact, so one has to use a finer structure to make it a manifold $\mathfrak{C}^\infty(N, P)$, see (42.1). But then $C^\infty(M, \mathfrak{C}^\infty(N, P)) \cong C^\infty(M \times N, P)$ if and only if N is compact see (42.14). Unfortunately, $\mathfrak{C}^\infty(N, P)$ cannot be generalized to infinite dimensional N, since this structure becomes discrete. Let us mention, however, that there exists a theory of manifolds and vector bundles, where the structure of charts is replaced by the set of smooth curves supplemented by other requirements, where one gets a cartesian closed category of manifolds and has the compact-open C^∞-topology on $C^\infty(N, P)$ for finite dimensional N, P, see [Seip, 1981], [Kriegl, 1980], [Michor, 1984a].

We start by treating the basic concept of manifolds, existence of smooth bump functions and smooth partitions of unity. Then we investigate tangent vectors seen as derivations or kinematically (via curves): these concepts differ, and we show in (28.4) that even on Hilbert spaces there exist derivations which are not tangent to any smooth curve. In particular, we have different kinds of tangent bundles, the most important ones are the kinematic and the operational one. We treat smooth, real analytic, and holomorphic vector bundles and spaces of sections of vector bundles, we give them structures of convenient vector spaces; they will become important as modeling spaces for manifolds of mappings in chapter IX.

Finally, we discuss Weil functors (certain product preserving functors of manifolds) as generalized tangent bundles. This last section is due to [Kriegl, Michor, 1997].

27. Differentiable Manifolds

27.1. Manifolds. A chart (U, u) on a set M is a bijection $u : U \to u(U) \subseteq E_U$ from a subset $U \subseteq M$ onto a c^∞-open subset of a convenient vector space E_U.

For two charts (U_α, u_α) and (U_β, u_β) on M the mapping $u_{\alpha\beta} := u_\alpha \circ u_\beta^{-1} : u_\beta(U_{\alpha\beta}) \to u_\alpha(U_{\alpha\beta})$ for α, $\beta \in A$ is called the *chart changing*, where $U_{\alpha\beta} := U_\alpha \cap U_\beta$. A family $(U_\alpha, u_\alpha)_{\alpha\in A}$ of charts on M is called an *atlas* for M, if the U_α form a cover of M and all chart changings $u_{\alpha\beta}$ are defined on c^∞-open subsets.

An atlas $(U_\alpha, u_\alpha)_{\alpha\in A}$ for M is said to be a C^∞*-atlas*, if all chart changings $u_{\alpha\beta} : u_\beta(U_{\alpha\beta}) \to u_\alpha(U_{\alpha\beta})$ are smooth. Two C^∞-atlas are called C^∞*-equivalent*, if their union is again a C^∞-atlas for M. An equivalence class of C^∞-atlas is sometimes called a C^∞*-structure* on M. The union of all atlas in an equivalence class is again an atlas, the *maximal atlas* for this C^∞-structure. A C^∞*-manifold* M is a set together with a C^∞-structure on it.

Atlas, structures, and manifolds are called *real analytic* or *holomorphic*, if all chart changings are real analytic or holomorphic, respectively. They are called topological, if the chart changings are only continuous in the c^∞-topology.

A holomorphic manifold is real analytic, and a real analytic one is smooth. By a manifold we will henceforth mean a smooth one.

27.2. A mapping $f : M \to N$ between manifolds is called smooth if for each $x \in M$ and each chart (V, v) on N with $f(x) \in V$ there is a chart (U, u) on M with $x \in U$, $f(U) \subseteq V$, such that $v \circ f \circ u^{-1}$ is smooth. This is the case if and only if $f \circ c$ is smooth for each smooth curve $c : \mathbb{R} \to M$.

We will denote by $C^\infty(M, N)$ the space of all C^∞-mappings from M to N.

Likewise, we have the spaces $C^\omega(M, N)$ of real analytic mappings and $\mathcal{H}(M, N)$ of holomorphic mappings between manifolds of the corresponding type. This can be also tested by composing with the relevant types of curves.

A smooth mapping $f : M \to N$ is called a *diffeomorphism* if f is bijective and its inverse is also smooth. Two manifolds are called *diffeomorphic* if there exists a diffeomorphism between them. Likewise, we have *real analytic* and *holomorphic diffeomorphisms*. The latter ones are also called *biholomorphic mappings*.

27.3. Products. Let M and N be smooth manifolds described by smooth atlas $(U_\alpha, u_\alpha)_{\alpha\in A}$ and $(V_\beta, v_\beta)_{\beta\in B}$, respectively. Then the family $(U_\alpha \times V_\beta, u_\alpha \times v_\beta : U_\alpha \times V_\beta \to E_\alpha \times F_\beta)_{(\alpha,\beta)\in A\times B}$ is a smooth atlas for the cartesian product $M \times N$. Beware, however, the manifold topology (27.4) of $M \times N$ may be finer than the product topology, see (4.22). If M and N are metrizable, then it coincides with the product topology, by (4.19). Clearly, the projections

$$M \xleftarrow{\mathrm{pr}_1} M \times N \xrightarrow{\mathrm{pr}_2} N$$

are also smooth. The *product* $(M \times N, pr_1, pr_2)$ has the following universal property:

For any smooth manifold P and smooth mappings $f : P \to M$ and $g : P \to N$ the mapping $(f, g) : P \to M \times N$, $(f, g)(x) = (f(x), g(x))$, is the unique smooth mapping with $\mathrm{pr}_1 \circ (f, g) = f$, $\mathrm{pr}_2 \circ (f, g) = g$.

Clearly, we can form products of finitely many manifolds. The reader may now wonder why we do not consider infinite products of manifolds. These have charts which are open for the so called 'box topology'. But then we get 'box products' without the universal property of products. The 'box products', however, have the universal product property for families of mappings such that locally almost all members are constant.

27.4. The topology of a manifold. The *natural topology* on a manifold M is the identification topology with respect to some (smooth) atlas $(u_\alpha : M \supseteq U_\alpha \to u_\alpha(U_\alpha) \subseteq E_\alpha)$, where a subset $W \subseteq M$ is open if and only if $u_\alpha(U_\alpha \cap W)$ is c^∞-open in E_α for all α. This topology depends only on the structure, since diffeomorphisms are homeomorphisms for the c^∞-topologies. It is also the final topology with respect to all inverses of chart mappings in one atlas. It is also the final topology with respect to all smooth curves. For a (smooth) manifold we will require certain properties for the natural topology, which will be specified when needed, like smoothly regular (14.1), smoothly normal (16.1), or smoothly paracompact (16.1).

Let us now discuss the relevant notions of Hausdorff.

(1) M is (topologically) Hausdorff, equivalently the diagonal is closed in the product topology on $M \times M$.
(2) The diagonal is closed in the manifold $M \times M$.
(3) The smooth functions in $C^\infty(M, \mathbb{R})$ separate points in M. Let us call M *smoothly Hausdorff* if this property holds.

We have the obvious implications (3)$\Rightarrow$(1)$\Rightarrow$(2). We have no counterexamples for the converse implications.

The three separation conditions just discussed do not depend on properties of the modeling convenient vector spaces, whereas properties like smoothly regular, smoothly normal, or smoothly paracompact do. Smoothly Hausdorff is the strongest of the three. But it is not so clear which separation property should be required for a manifold. In order to make some decision, from now on **we require that manifolds are smoothly Hausdorff.** Each convenient vector space has this property. But we will have difficulties with permanence of the property 'smoothly Hausdorff', in particular with quotient manifolds, see for example the discussion (27.14) on covering spaces below. For important examples (manifolds of mappings, diffeomorphism groups, etc.) we will prove that they are even smoothly paracompact.

The isomorphism type of the *modeling convenient vector spaces* E_α is constant on the connected components of the manifold M, since the derivatives of the chart changings are linear isomorphisms. A manifold M is called *pure* if this isomorphism type is constant on the whole of M.

Corollary. *If a smooth manifold (which is smoothly Hausdorff) is Lindelöf, and if all modeling vector spaces are smoothly regular, then it is smoothly paracompact.*

If a smooth manifold is metrizable and smoothly normal then it is smoothly paracompact.

Proof. See (16.10) for the first statement and (16.15) for the second one. □

27.5. Lemma. *Let M be a smoothly regular manifold. Then for any manifold N a mapping $f : N \to M$ is smooth if and only if $g \circ f : N \to \mathbb{R}$ is smooth for all $g \in C^\infty(M, \mathbb{R})$. This means that $(M, C^\infty(\mathbb{R}, M), C^\infty(M, \mathbb{R}))$ is a Frölicher space, see* (23.1).

Proof. Let $x \in N$ and let $(U, u : U \to E)$ be a chart of M with $f(x) \in U$. Choose some smooth bump function $g : M \to \mathbb{R}$ with $\operatorname{supp}(g) \subset U$ and $g = 1$ in a neighborhood V of $f(x)$. Then $f^{-1}(\operatorname{carr}(g)) = \operatorname{carr}(g \circ f)$ is an open neighborhood of x in N. Thus f is continuous, so $f^{-1}(V)$ is open. Moreover, $(g.(\ell \circ u)) \circ f$ is smooth for all $\ell \in E'$ and on $f^{-1}(V)$ this equals $\ell \circ u \circ (f|f^{-1}(V))$. Thus $u \circ (f|f^{-1}(V))$ is smooth since E is convenient, by (2.14.4), so f is smooth near x. □

27.6. Non-regular manifold. [Margalef, Outerelo, 1982] Let $0 \neq \lambda \in (\ell^2)^*$, let X be $\{x \in \ell^2 : \lambda(x) \geq 0\}$ with the Moore topology, i.e. for $x \in X$ we take $\{y \in \ell^2 \setminus \ker \lambda : \|y - x\| < \varepsilon\} \cup \{x\}$ for $\varepsilon > 0$ as neighborhood-basis. We set $X^+ := \{x \in \ell^2 : \lambda(x) > 0\} \subseteq \ell^2$.

Then obviously X is Hausdorff (its topology is finer than that of ℓ^2) but not regular: In fact the closed subspace $\ker \lambda \setminus \{0\}$ cannot be separated by open sets from $\{0\}$.

It remains to show that X is a C^∞-manifold. We use the following diffeomorphisms

(1) $S := \{x \in \ell^2 : \|x\| = 1\} \cong_{C^\infty} \ker \lambda$.
(2) $\varphi : \ell^2 \setminus \{0\} \cong_{C^\infty} \ker \lambda \times \mathbb{R}^+$.
(3) $S \cap X^+ \cong_{C^\infty} \ker \lambda$.
(4) $\psi : X^+ \to \ker \lambda \times \mathbb{R}^+$.

(1) This is due to [Bessaga, 1966].

(2) Let $f : S \to \ker \lambda$ be the diffeomorphism of (1) and define the required diffeomorphism to be $\varphi(x) := (f(x/\|x\|), \|x\|)$ with inverse $\varphi^{-1}(y, t) := t\, f^{-1}(y)$.

(3) Take an $a \in (\ker \lambda)^\perp$ with $\lambda(a) = 1$. Then the orthogonal projection $\ell^2 \to \ker \lambda$ is given by $x \mapsto x - \lambda(x)a$. This is a diffeomorphism of $S \cap X^+ \to \{x \in \ker \lambda : \|x\| < 1\}$, which in turn is diffeomorphic to $\ker \lambda$.

(4) Let $g : S \cap X^+ \to \ker \lambda$ be the diffeomorphism of (3) then the desired diffeomorphism is $\psi : x \mapsto (g(x/\|x\|), \|x\|)$.

We now show that there is a homeomorphism of $h : X^+ \cup \{0\} \to \ell^2$, such that $h(0) = 0$ and $h|_{X^+} : X^+ \to \ell^2 \setminus \{0\}$ is a diffeomorphism. We take

$$h(x) := \begin{cases} (\varphi^{-1} \circ \psi)(x) & \text{for } x \in X^+ \\ 0 & \text{for } x = 0 \end{cases}.$$

$$
\begin{array}{ccc}
X^+ & \xleftarrow[\cong]{\psi^{-1}} \ker\lambda\times\mathbb{R}^+ \xleftarrow[\cong]{\varphi} & \ell^2\setminus\{0\} \\
\big\downarrow & & \big\downarrow \\
X^+\cup\{0\} & \xleftarrow{\quad h \quad} & E \\
\big\uparrow & & \big\uparrow \\
\{0\} & \longleftarrow\!\!\longrightarrow & \{0\}
\end{array}
$$

Now we use translates of h as charts $\ell^2 \to X^+ \cup \{x\}$. The chart changes are then diffeomorphisms of $\ell^2 \setminus \{0\}$ and we thus obtained a smooth atlas for $X := \bigcup_{x\in\ker\lambda}(X^+ \cup \{x\})$. The topology described by this atlas is obviously the Moore topology.

If we use instead of X the union $\bigcup_{x\in D}(X^+ \cup \{x\})$, where $D \subseteq \ker\lambda$ is dense and countable. Then the same results are valid, but X is now even second countable.

Note however that a regular space which is locally metrizable is completely regular.

27.7. Proposition. *Let M be a manifold modeled on smoothly regular convenient vector spaces. Then M admits an atlas of charts defined globally on convenient vector spaces.*

Proof. That a convenient vector space is smoothly regular means that the c^∞-topology has a base of carrier sets of smooth functions, see (14.1). These functions satisfy the assumptions of theorem (16.21), and hence the stars of these sets with respect to arbitrary points in the sets are diffeomorphic to the whole vector space and still form a base of the c^∞-topology. □

27.8. Lemma. *A manifold M is metrizable if and only if it is paracompact and modeled on Fréchet spaces.*

Proof. A topological space is metrizable if and only if it is paracompact and locally metrizable. c^∞-open subsets of the modeling vector spaces are metrizable if and only if the spaces are Fréchet, by (4.19). □

27.9. Lemma. *Let M and N be smoothly paracompact metrizable manifolds. Then $M \times N$ is smoothly paracompact.*

Proof. By (16.15) there are embeddings into $c_0(\Gamma)$ and $c_0(\Lambda)$ for some sets Γ and Λ which pull back the coordinate projections to smooth functions. Then $M \times N$ embeds into $c_0(\Gamma) \times c_0(\Lambda) \cong c_0(\Gamma \sqcup \Lambda)$ in the same way and hence again by (16.15) the manifold $M \times N$ is smoothly paracompact. □

27.10. Facts on finite dimensional manifolds. A manifold M is called finite dimensional if it has finite dimensional modeling vector spaces. By (4.19), this is the case if and only if M is locally compact. Then the dimensions of the modeling spaces give a locally constant function on M.

If the manifold M is finite dimensional, then Hausdorff implies smoothly regular. We require then that the natural topology is in addition to Hausdorff also paracompact. It is then smoothly paracompact by (27.7), since all connected components are Lindelöf if M is paracompact.

Let us finally add some remarks on finite dimensional separable topological manifolds M: From differential topology we know that if M has a C^1-structure, then it also has a C^1-equivalent C^∞-structure and even a C^1-equivalent C^ω-structure. But there are manifolds which do not admit differentiable structures. For example, every 4-dimensional manifold is smooth off some point, but there are some which are not smooth, see [Quinn, 1982], [Freedman, 1982]. Note, finally, that any such manifold M admits a finite atlas consisting of $\dim M + 1$ (not necessarily connected) charts. This is a consequence of topological dimension theory, a proof may be found in [Greub, Halperin, Vanstone, 1972].

If there is a C^1-diffeomorphism between M and N, then there is also a C^∞-diffeomorphism. There are manifolds which are homeomorphic but not diffeomorphic: on $\mathbb{R}^4$ there are uncountably many pairwise non-diffeomorphic differentiable structures; on every other $\mathbb{R}^n$ the differentiable structure is unique. There are finitely many different differentiable structures on the spheres S^n for $n \geq 7$. See [Kervaire, Milnor, 1963].

27.11. Submanifolds. A subset N of a manifold M is called a *submanifold*, if for each $x \in N$ there is a chart (U, u) of M such that $u(U \cap N) = u(U) \cap F_U$, where F_U is a closed linear subspace of the convenient model space E_U. Then clearly N is itself a manifold with $(U \cap N, u|U \cap N)$ as charts, where (U, u) runs through all these *submanifold charts* from above.

A submanifold N of M is called a *splitting submanifold* if there is a cover of N by submanifold charts (U, u) as above such that the $F_U \subset E_U$ are complemented (i.e. splitting) linear subspaces. Then every submanifold chart is splitting.

Note that a closed submanifold of a smoothly paracompact manifold is again smoothly paracompact. Namely, the trace topology is the intrinsic topology on the submanifold since this is true for closed linear subspaces of convenient vector spaces, (4.28).

A mapping $f : N \to M$ between manifolds is called *initial* if it has the following property:

> A mapping $g : P \to N$ from a manifold P ($\mathbb{R}$ suffices) into N is smooth if and only if $f \circ g : P \to M$ is smooth.

Clearly, an initial mapping is smooth and injective. The embedding of a submanifold is always initial. The notion of initial smooth mappings will play an important role in this book whereas that of immersions will be used in finite dimensions only.

In a similar way we shall use the (now obvious) notion of *initial real analytic mappings* between real analytic manifolds and also *initial holomorphic mappings* between complex manifolds.

If $h : \mathbb{R} \to \mathbb{R}$ is a function such that h^p and h^q are smooth for some p, q which are relatively prime in $\mathbb{N}$, then h itself turns out to be smooth, see [Joris, 1982.] So the mapping $f : t \mapsto (t^p, t^q)$, $\mathbb{R} \to \mathbb{R}^2$, is initial, but f is not an immersion at 0.

Smooth mappings $f : N \to M$ which admit local smooth retracts are initial. By this we mean that for each $x \in N$ there are an open neighborhood U of $f(x)$ in M and a smooth mapping $r_x : U \to N$ such that $r \circ f|(f^{-1}(U)) = \mathrm{Id}_{f^{-1}U}$. We shall meet this class of initial mappings in (43.19).

27.12. Example. We now give an example of a smooth mapping f with the following properties:

(1) f is a topological embedding and the derivative at each point is an embedding of a closed linear subspace.
(2) The image of f is not a submanifold.
(3) The image of f cannot be described locally by a regular smooth equation.

This shows that the notion of an *embedding* is quite subtle in infinite dimensions.

Proof. For this let $\ell^2 \xrightarrow{\iota} E \to \ell^2$ be a short exact sequence, which does not split, see (13.18.6) Then the square of the norm on ℓ^2 does not extend to a smooth function on E by (21.11).

Choose a $0 \neq \lambda \in E^*$ with $\lambda \circ \iota = 0$ and choose a v with $\lambda(v) = 1$. Now consider $f : \ell^2 \to E$ given by $x \mapsto \iota(x) + \|x\|^2 v$.

(1) Since f is polynomial it is smooth. We have $(\lambda \circ f)(x) = \|x\|^2$, hence $g \circ f = \iota$, where $g : E \to E$ is given by $g(y) := y - \lambda(y)\, v$. Note however that g is no diffeomorphism, hence we don't have automatically a submanifold. Thus f is a topological embedding and also the differential at every point. Moreover the image is closed, since $f(x_n) \to y$ implies $\iota(x_n) = g(f(x_n)) \to g(y)$, hence $x_n \to x_\infty$ for some x_∞ and thus $f(x_n) \to f(x_\infty) = y$. Finally f is initial. Namely, let $h : G \to \ell^2$ be such that $f \circ h$ is smooth, then $g \circ f \circ h = \iota \circ h$ is smooth. As a closed linear embedding ι is initial, so h is smooth. Note that λ is an extension of $\|\ \ \|^2$ along $f : \ell^2 \to E$.

(2) Suppose there were a local diffeomorphism Φ around $f(0) = 0$ and a closed subspace $F < E$ such that locally Φ maps F onto $f(\ell^2)$. Then Φ factors as follows

$$\begin{array}{ccc} \ell^2 & \xhookrightarrow{\ f\ } & E \\ {\scriptstyle\varphi}\uparrow & & \uparrow{\scriptstyle \cong\ \Phi} \\ F & \xhookrightarrow{\ \text{incl}\ } & E \end{array}$$

In fact since $\Phi(F) \subseteq f(\ell^2)$, and f is injective, we have φ as mapping, and since f is initial φ is smooth. By using that $\text{incl} : F \to E$ is initial, we could deduce that φ is a local diffeomorphism. However we only need that $\varphi'(0) : F \to \ell^2$ is a linear isomorphism. Since $f'(0) \circ \varphi'(0) = \Phi'(0)|_F$ is a closed embedding, we have that $\varphi'(0)$ is a closed embedding. In order to see that $\varphi'(0)$ is onto, pick $v \in \ell^2$ and consider the curve $t \mapsto tv$. Then $w : t \mapsto \Phi^{-1}(f(tv)) \in F$ is smooth, and

$$\begin{aligned} f'(0)(\varphi'(0)(w'(0))) &= \frac{d}{dt}|_{t=0}(f \circ \varphi)(w(t)) \\ &= \frac{d}{dt}|_{t=0}\Phi(w(t)) = \frac{d}{dt}|_{t=0}f(tv) = f'(0)(v) \end{aligned}$$

and since $f'(0) = \iota$ is injective, we have $\varphi'(0)(w'(0)) = v$.

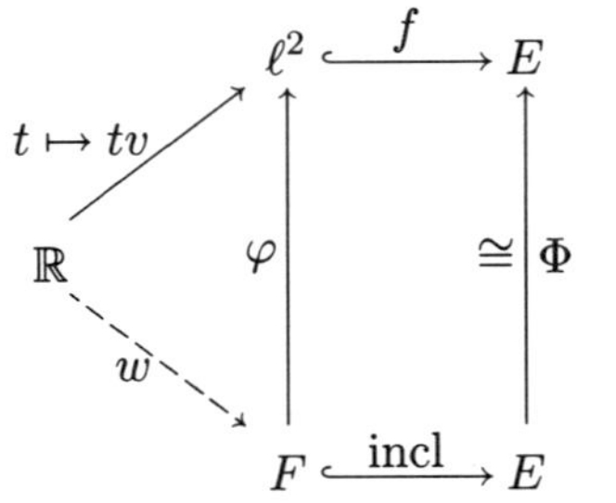

Now consider the diagram

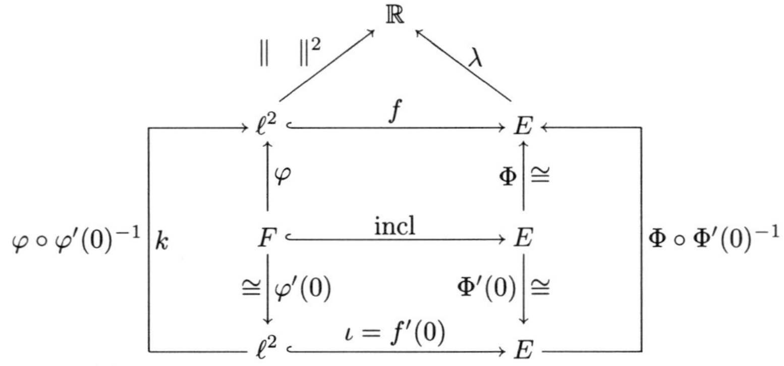

i.e.

$$\begin{aligned} (\lambda \circ \Phi \circ \Phi'(0)^{-1}) \circ \iota \circ \varphi'(0) &= \lambda \circ \Phi \circ \Phi'(0)^{-1} \circ f'(0) \circ \varphi'(0) \\ &= \lambda \circ \Phi \circ \Phi'(0)^{-1} \circ \Phi'(0) \circ \mathrm{incl} \\ &= \lambda \circ \Phi \circ \mathrm{incl} = \lambda \circ f \circ \varphi = \| \quad \|^2 \circ \varphi. \end{aligned}$$

By composing with $\varphi'(0)^{-1} : \ell^2 \to F$ we get an extension $\tilde{q}$ of $q := \| \quad \|^2 \circ k$ to E, where the locally defined mapping $k := \varphi \circ \varphi'(0)^{-1} : \ell^2 \to \ell^2$ is smooth and $k'(0) = \mathrm{Id}$. Now $\tilde{q}''(0) : E \times E \to \mathbb{R}$ is an extension of $q''(0) : \ell^2 \times \ell^2 \to \mathbb{R}$ given by $(v, w) \mapsto 2\langle k'(0)v, k'(0)w\rangle$. Hence the associated $\tilde{q}''(0)^\vee : E \to E^*$ fits into

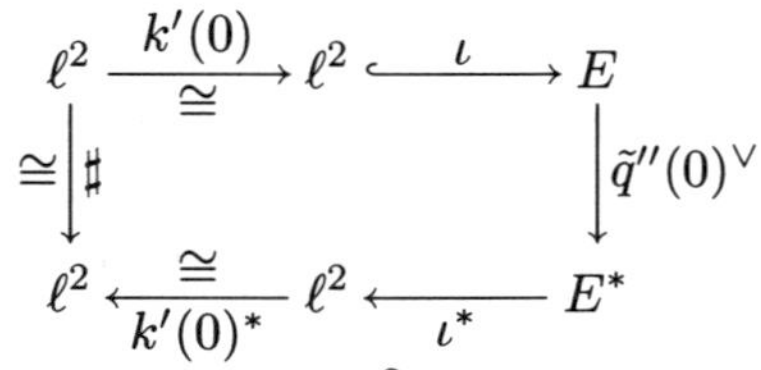

and in this way we get a retraction for $\iota : \ell^2 \to E$. This is a contradiction.

(3) Let us show now the even stronger statement that there is no local regular equation $\rho : E \rightsquigarrow G$ with $f(\ell^2) = \rho^{-1}(0)$ locally and $\ker \rho'(0) = \iota(\ell^2)$. Otherwise we have $\rho'(0)(v) \neq 0$ and hence there is a $\mu \in G'$ with $\mu(\rho'(0)(v)) = 1$. Thus $\mu \circ \rho : E \rightsquigarrow \mathbb{R}$ is smooth $\mu \circ \rho \circ f = 0$ and $(\mu \circ \rho)'(0)(v) = 1$. Moreover

$$\begin{aligned} 0 &= (\tfrac{d}{dt})^2|_{t=0}(\mu \circ \rho \circ f)(tx) \\ &= \frac{d}{dt}|_{t=0}(\mu \circ \rho)'(f(tx)) \cdot f'(tx) \cdot x \\ &= (\mu \circ \rho)''(0)(f'(0)x, f'(0)x) + (\mu \circ \rho)'(0) \cdot f''(0)(x,x) \\ &= (\mu \circ \rho)''(0)(\iota(x), \iota(x)) + 2\|x\|^2\, (\mu \circ \rho)'(0) \cdot v, \end{aligned}$$

hence $-(\mu \circ \rho)''(0)/2$ is an extension of $\|\ \ \|^2$ along ι, which is a contradiction. □

27.13. Theorem. Embedding of smooth manifolds. *If M is a smooth manifold modeled on a C^∞-regular convenient vector space E, which is Lindelöf. Then there exists a smooth embedding onto a splitting submanifold of $s \times E^{(\mathbb{N})}$ where s is the space of rapidly decreasing real sequences.*

Proof. We choose a countable atlas (U_n, u_n) and a subordinated smooth partition (h_n) of unity which exists by (16.10). Then the embedding is given by

$$x \mapsto ((h_n(x))_n, (h_n(x).u_n(x))_n) \in s \times E^{(\mathbb{N})}.$$

Local smooth retracts to this embedding are given by $((t_n),(x_n)) \mapsto u_k^{-1}(\frac{1}{t_k}x_k)$ defined for $t_k \neq 0$. □

27.14. Coverings. A surjective smooth mapping $p : N \to M$ between smooth manifolds is called a covering if it is the projection of a fiber bundle (see (37.1)) with discrete fiber. Note that on a product of a discrete space with a manifold the product topology equals the manifold topology. A product of two coverings is again a covering.

A smooth manifold M is locally contractible since we may choose charts with star-shaped images, and since the c^∞-topology on a product with $\mathbb{R}$ is the product of the c^∞-topologies. Hence the *universal covering space* $\tilde{M}$ of a connected smooth manifold M exists as a topological space. By pulling up charts it turns out to be a smooth manifold also, whose topology is the one of $\tilde{M}$. Since $\tilde{M} \times \tilde{M}$ is the universal covering of $M \times M$, the manifold $\tilde{M}$ is Hausdorff even in the sense of (27.4.2). If M is smoothly regular then $\tilde{M}$ is also smoothly regular, thus smoothly Hausdorff. As usual, the fundamental group $\pi(M, x_0)$ acts free and strictly discontinuously on $\tilde{M}$ in the sense that each $x \in \tilde{M}$ admits an open neighborhood U such that $g.U \cap U = \emptyset$ for all $g \neq e$ in $\pi(M, x_0)$.

Note that the universal covering space $\tilde{M}$ of a connected smooth manifold M can be viewed as the Frölicher space (see (23.1), (24.10)) $C^\infty((I,0),(M,x_0))$ of all smooth curves $c : [0,1] = I \to M$, such that $c(0) = x_0$ for a base point $x_0 \in M$ modulo smooth homotopies fixing endpoints. This can be shown by the usual topological proof, where one uses only smooth curves and homotopies, and smoothes by reparameterization those which are pieced together. Note that $\mathrm{ev}_1 : C^\infty((I,0),(M,x_0)) \to M$ is a final (27.15) smooth mapping since we may construct local smooth sections near any point in M: choose a chart $u : U \to u(U)$ on M with $u(U)$ a radial open set in the modeling space of M. Then let $\varphi(x)$ be the smooth curve which follows a smooth curve from x_0 to $u^{-1}(0)$ during the time from 0 to $\frac{1}{2}$ and stops infinitely flat at $\frac{1}{2}$, so the curve $t \mapsto u^{-1}(\psi(t).u(x))$ where $\psi : [\frac{1}{2}, 1] \to [0,1]$ is smooth, flat at $\frac{1}{2}$, $\psi(\frac{1}{2}) = 0$, and with $\psi(1) = 1$. These local smooth sections lift to smooth sections of $C^\infty((I,0),(M,x_0)) \to \tilde{M}$, thus the final smooth structure on $\tilde{M}$ coincides with that induced from the manifold structure.

If conversely a group G acts strictly discontinuously on a smooth manifold M, then the orbit space M/G turns out to be a smooth manifold (with $G.U$'s as above as charts), but it might be not Hausdorff, as the following example shows: $M = \mathbb{R}^2 \setminus 0$, $G = \mathbb{Z}$ acting by Fl^X_z where $X = x\partial_x - y\partial_y$.

The orbit space is Hausdorff if and only if $R := \{(g.x, x) : g \in G, x \in M\}$ is closed in $M \times M$ with the product topology: $M \to M/G$ is an open mapping, thus the product $M \times M \to M/G \times M/G$ is also open for the product topologies, and $(M \times M) \setminus R$ is mapped onto the complement of the diagonal in $M/G \times M/G$.

The orbit space has property (27.4.2) if and only if $R := \{(g.x, x) : g \in G, x \in M\}$ is closed in $M \times M$ with the manifold topology: the same proof as above works, where $M \times M \to M/G \times M/G$ is also open for the manifold topologies since we may lift smooth curves.

We were unable to find a condition on the action which would ensure that M/G is smoothly Hausdorff or inherits a stronger separation property from M. Classical results always use locally compact M.

27.15. Final smooth mappings. A mapping $f : M \to N$ between smooth manifolds is called *final* if:

> A mapping $g : N \to P$ into a manifold P is smooth if and only if $g \circ f : M \to P$ is smooth.

Clearly, a final mapping $f : M \to N$ is smooth, and surjective if N is connected. Coverings (27.14) are always final, as are projections of fiber bundles (37.1). Between finite dimensional separable manifolds without isolated points the final mappings are exactly the surjective submersions. We will use the notion submersion in finite dimensions only.

27.16. Foliations. Let F be a c^∞-closed linear subspace of a convenient vector space E. Let E_F be the smooth manifold modeled on F, which is the disjoint union of all affine subspaces of E which are translates of F. A diffeomorphism $f : U \to V$ between c^∞-open subsets of E is called F*-foliated* if it is also a homeomorphism (equivalently diffeomorphism) between the open subsets U and V of E_F.

Let M be a smooth manifold modeled on the convenient vector space E. A *foliation* on M is then given by a c^∞-closed linear subspace F in E and a smooth (maximal) atlas of M such that all chart changings are F-foliated. Each chart of this maximal atlas is called a *distinguished chart*. A connected component of the inverse image under a distinguished chart of an affine translate of F is called a *plaque*.

A foliation on M induces on the set M another structure of a smooth manifold, sometimes denoted by M_F, modeled on F, where we take as charts the restrictions of distinguished charts to plaques (with the image translated into F). The identity on M induces a smooth bijective mapping $M_F \to M$. Clearly, M_F is smoothly Hausdorff (if M is it). A *leaf of the foliation* is then a connected component of M_F.

The notion of foliation will be used in (39.2) below.

27.17. Lemma. *For a convenient vector space E and any smooth manifold M the set $C^\infty(M,E)$ of smooth E-valued functions on M is also a convenient vector space in any of the following isomorphic descriptions, and it satisfies the uniform boundedness principle for the point evaluations.*

(1) *The initial structure with respect to the cone*

$$C^\infty(M,E) \xrightarrow{c^*} C^\infty(\mathbb{R},E)$$

for all $c \in C^\infty(\mathbb{R},M)$.

(2) *The initial structure with respect to the cone*

$$C^\infty(M,E) \xrightarrow{(u_\alpha^{-1})^*} C^\infty(u_\alpha(U_\alpha),E),$$

where (U_α,u_α) is a smooth atlas with $u_\alpha(U_\alpha) \subset E_\alpha$.

Moreover, with this structure, for two manifolds M, N, the exponential law holds:

$$C^\infty(M,C^\infty(N,E)) \cong C^\infty(M\times N,E).$$

For a real analytic manifold M the set $C^\omega(M,E)$ of real analytic E-valued functions on M is also a convenient vector space in any of the following isomorphic descriptions, and it satisfies the uniform boundedness principle for the point evaluations.

(1) *The initial structure with respect to the cone*

$$C^\omega(M,E) \xrightarrow{c^*} C^\infty(\mathbb{R},E) \text{ for all } c \in C^\infty(\mathbb{R},M)$$
$$C^\omega(M,E) \xrightarrow{c^*} C^\omega(\mathbb{R},E) \text{ for all } c \in C^\omega(\mathbb{R},M).$$

(2) *The initial structure with respect to the cone*

$$C^\omega(M,E) \xrightarrow{(u_\alpha^{-1})^*} C^\omega(u_\alpha(U_\alpha),E),$$

where (U_α,u_α) is a real analytic atlas with $u_\alpha(U_\alpha) \subset E_\alpha$.

Moreover, with this structure, for two real analytic manifolds M, N, the exponential law holds:

$$C^\omega(M, C^\omega(N, E)) \cong C^\omega(M \times N, E).$$

For a complex convenient vector space E and any complex holomorphic manifold M the set $\mathcal{H}(M, E)$ of holomorphic E-valued functions on M is also a convenient vector space in any of the following isomorphic descriptions, and it satisfies the uniform boundedness principle for the point evaluations.

(1) *The initial structure with respect to the cone*

$$\mathcal{H}(M, E) \xrightarrow{c^*} \mathcal{H}(\mathbb{D}, E)$$

for all $c \in \mathcal{H}(\mathbb{D}, M)$.

(2) *The initial structure with respect to the cone*

$$\mathcal{H}(M, E) \xrightarrow{(u_\alpha^{-1})^*} \mathcal{H}(u_\alpha(U_\alpha), E),$$

where (U_α, u_α) is a holomorphic atlas with $u_\alpha(U_\alpha) \subset E_\alpha$.

Moreover, with this structure, for two manifolds M, N, the exponential law holds:

$$\mathcal{H}(M, \mathcal{H}(N, E)) \cong \mathcal{H}(M \times N, E).$$

Proof. For all descriptions the initial locally convex topology is convenient, since the spaces are closed linear subspaces in the relevant products of the right hand sides. Thus, the uniform boundedness principle for the point evaluations holds for all structures since it holds for all right hand sides. So the identity is bibounded between all respective structures.

The exponential laws now follow from the corresponding ones: use (3.12) for c^∞-open subsets of convenient vector spaces and description (2), for the real analytic case use (11.18), and for the holomorphic case use (7.22). □

27.18. Germs. Let M and N be manifolds, and let $A \subset M$ be a closed subset. We consider all smooth mappings $f : U_f \to N$, where U_f is some open neighborhood of A in M, and we put $f \underset{A}{\sim} g$ if there is some open neighborhood V of A with $f|V = g|V$. This is an equivalence relation on the set of functions considered. The equivalence class of a function f is called the *germ of f along A*, sometimes denoted by $\operatorname{germ}_A f$. As in (8.3) we will denote the space of all these germs by $C^\infty(M \supset A, N)$.

If we consider functions on M, i.e. if $N = \mathbb{R}$, we may add and multiply germs, so we get the real commutative algebra of germs of smooth functions. If $A = \{x\}$, this algebra $C^\infty(M \supset \{x\}, \mathbb{R})$ is sometimes also denoted by $C^\infty_x(M, \mathbb{R})$. We may consider the inductive locally convex vector space topology with respect to the cone

$$C^\infty(M \supseteq \{x\}, \mathbb{R}) \leftarrow C^\infty(U, \mathbb{R}),$$

where U runs through some neighborhood basis of x consisting of charts, so that each $C^\infty(U,\mathbb{R})$ carries a convenient vector space topology by (2.15).

This inductive topology is Hausdorff only if x is isolated in M, since the restriction to some one dimensional linear subspace of a modeling space is a projection on a direct summand which is not Hausdorff, by (27.19). Nevertheless, multiplication is a bounded bilinear operation on $C^\infty(M \supseteq \{x\},\mathbb{R})$, so the closure of 0 is an ideal. The quotient by this ideal is thus an algebra with bounded multiplication, denoted by $\operatorname{Tay}_x(M,\mathbb{R})$.

27.19. Lemma. *Let M be a smooth manifold modeled on Banach spaces which admit bump functions of class C_b^∞ (see (15.1)). Then the closure of 0 in $C^\infty(M \supseteq \{x\},\mathbb{R})$ is the ideal of all germs which are flat at x of infinite order.*

Proof. This is a local question, so let $x = 0$ in a modeling Banach space E. Let f be a representative in some open neighborhood U of 0 of a flat germ. This means that all iterated derivatives of f at 0 vanish. Let $\rho \in C_b^\infty(E,[0,1])$ be 0 on a neighborhood of 0 and $\rho(x) = 1$ for $\|x\| > 1$. For $f_n(x) := f(x)\rho(n.x)$ we have $\operatorname{germ}_0(f_n) = 0$, and it remains to show that $n(f - f_n)$ is bounded in $C^\infty(U,\mathbb{R})$. For this we fix a derivative d^k and choose N such that $\|d^{k+1}f(x)\| \le 1$ for $\|x\| \le \frac{1}{N}$. Then for $n \ge N$ we have the following estimate:

$$\begin{aligned}\|nd^k(f-f_n)(x)\| &\le \sum_{l=0}^{k}\binom{k}{l} n\|d^{k-l}f(x)\|n^l\|d^l(1-\rho)(nx)\| \\ &\le \sum_{l=0}^{k}\binom{k}{l} n\int_0^1 \frac{(1-t)^{l+1}}{(l+1)!}\|d^{k+1}f(tx)\|\,dt\,\|x\|^{l+1}n^l\|d^l(1-\rho)(nx)\| \\ &\le \begin{cases} 0 & \text{for } \|nx\|>1 \\ \sum_{l=0}^k \binom{k}{l}\frac{1}{l!}\|d^l(1-\rho)\|_\infty & \text{for } \|nx\| \le 1.\end{cases} \qquad \square\end{aligned}$$

27.20. Corollary. *For any C_b^∞-regular Banach space E and $a \in E$ the canonical mapping*

$$\operatorname{Tay}_a(E,\mathbb{R}) \to \prod_{k=0}^{\infty} L^k_{\text{sym}}(E,\mathbb{R})$$

is a bornological isomorphism.

Proof. For every open neighborhood U of a in E we have a continuous linear mapping $C^\infty(U,\mathbb{R}) \to \prod_{k=0}^\infty L^k_{\text{sym}}(E,\mathbb{R})$ into the space of formal power series, hence also $C^\infty(E \supseteq \{a\},\mathbb{R}) \to \prod_{k=0}^\infty L^k_{\text{sym}}(E,\mathbb{R})$, and finally from $\operatorname{Tay}_a(E,\mathbb{R}) \to \prod_{k=0}^\infty L^k_{\text{sym}}(E,\mathbb{R})$. Since E is Banach, the space of formal power series is a Fréchet space and since E is $C_b^\infty(E,\mathbb{R})$-regular the last mapping is injective by (27.19). By E. Borel's theorem (15.4) every bounded subset of the space of formal power series is the image of a bounded subset of $C^\infty(E,\mathbb{R})$. Hence this mapping is a bornological isomorphism and the inductive limit $C^\infty(E \supseteq \{a\},\mathbb{R})$ is regular. $\square$

27.21. Lemma. *If M is smoothly regular then each germ at a point of a smooth function has a representative which is defined on the whole of M.*

If M is smoothly paracompact then this is true for germs along closed subsets. □

For germs of real analytic or holomorphic functions this is not true.

If M is as in the lemma, $C^\infty(M \supseteq \{x\}, \mathbb{R})$ is the quotient of the algebra $C^\infty(M, \mathbb{R})$ by the ideal of all smooth functions $f : M \to \mathbb{R}$ which vanish on some neighborhood (depending on f) of x.

The assumption in the lemma is not necessary as is shown by the following example: By (14.9) the Banach space $E := C([0,1], \mathbb{R})$ is not C^∞-regular, in fact not even C^1-regular. For $h \in C^\infty(\mathbb{R}, \mathbb{R})$ the push forward $h_* : C^\infty(\mathbb{R}, \mathbb{R}) \to C^\infty(\mathbb{R}, \mathbb{R})$ is smooth, thus continuous, so $(h_*)_* : C([0,1], C^\infty(\mathbb{R}, \mathbb{R})) \to C([0,1], C^\infty(\mathbb{R}, \mathbb{R}))$ is continuous. The arguments in the proof of theorem (3.2) show that

$$C([0,1], C^\infty(\mathbb{R}, \mathbb{R})) \cong C^\infty(\mathbb{R}, C([0,1], \mathbb{R})),$$

thus $h_* : E \to E$ is smooth. Let $h(t) := t$ for $|t| \le \frac{1}{2}$ and $|h(t)| \le 1$ for all $t \in \mathbb{R}$. In particular h_* is the identity on $\{f \in E : \|f\| \le \frac{1}{2}\}$. Let U be a neighborhood of 0 in E. Choose $\varepsilon > 0$ such that the closed ball with radius $\varepsilon > 0$ is contained in U. Then $h_\varepsilon := \varepsilon\, h_* \frac{1}{\varepsilon} : E \to E$ has values in U and is the identity near 0. Thus $(h_\varepsilon)^* : C^\infty(U, \mathbb{R}) \to C^\infty(E, \mathbb{R})$ is a bounded algebra homomorphism, which respects the corresponding germs at 0.

28. Tangent Vectors

28.1. The tangent spaces of a convenient vector space E. Let $a \in E$. A *kinematic tangent vector* with foot point a is simply a pair (a, X) with $X \in E$. Let $T_aE = E$ be the space of all kinematic tangent vectors with foot point a. It consists of all derivatives $c'(0)$ at 0 of smooth curves $c : \mathbb{R} \to E$ with $c(0) = a$, which explains the choice of the name kinematic.

For each open neighborhood U of a in E (a, X) induces a linear mapping $X_a : C^\infty(U, \mathbb{R}) \to \mathbb{R}$ by $X_a(f) := df(a)(X)$, which is continuous for the convenient vector space topology on $C^\infty(U, \mathbb{R})$ and satisfies $X_a(f \cdot g) = X_a(f) \cdot g(a) + f(a) \cdot X_a(g)$, so it is a *continuous derivation over* ev_a. The value $X_a(f)$ depends only on the germ of f at a.

An *operational tangent vector* of E with foot point a is a bounded derivation $\partial : C^\infty(E \supseteq \{a\}, \mathbb{R}) \to \mathbb{R}$ over ev_a. Let D_aE be the vector space of all these derivations. Any $\partial \in D_aE$ induces a bounded derivation $C^\infty(U, \mathbb{R}) \to \mathbb{R}$ over ev_a for each open neighborhood U of a in E. Moreover any family of bounded derivations $\partial_U : C^\infty(U, \mathbb{R}) \to \mathbb{R}$ over ev_a, which is coherent with respect to the restriction maps, defines an $\partial \in D_aE$. So the vector space D_aE is a closed linear subspace of the convenient vector space $\prod_U L(C^\infty(U, \mathbb{R}), \mathbb{R})$. We equip D_aE with the induced convenient vector space structure. Note that the spaces D_aE are isomorphic for all $a \in E$.

Taylor expansion induces the dashed arrows in the following diagram.

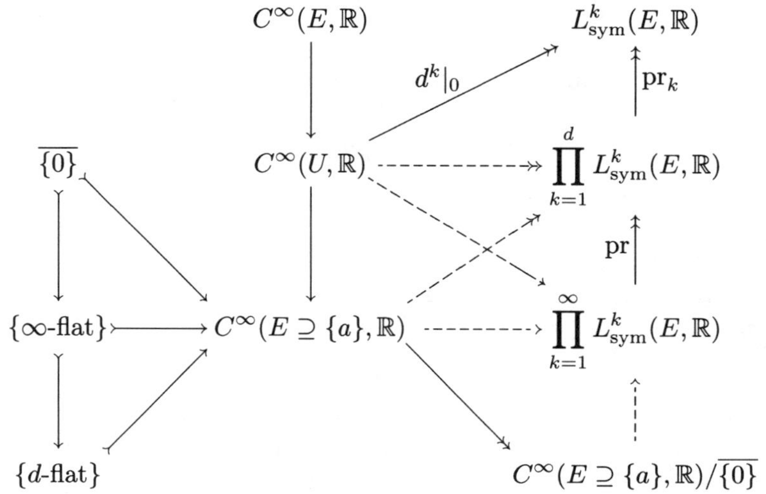

Note that all spaces in the right two columns except the top right corner are algebras, the finite product with truncated multiplication. The mappings are algebra-homomorphisms. And the spaces in the left column are the respective kernels. If E is a $C^\infty_b(E,\mathbb{R})$-regular Banach space, then by (27.20) the vertical dashed arrow is bibounded. Since $\mathbb{R}$ is Hausdorff every $\partial \in D_aE$ factors over $\text{Tay}_a(E,\mathbb{R}) := C^\infty(E \supseteq \{a\},\mathbb{R})/\overline{\{0\}}$, so in this case we can view ∂ as derivation on the algebra of formal power series. Any continuous linear functional on a countable product is a sum of continuous linear functionals on finitely many factors.

28.2. Degrees of operational tangent vectors. A derivation ∂ is said to have *order* at most d, if it factors over the space $\prod_{k=0}^d L^k_{\text{sym}}(E,\mathbb{R})$ of polynomials of degree at most d, i.e. it vanishes on all d-flat germs. If no such d exists, then it will be called of infinite order; this may happen only if ∂ does not factor over the space of formal power series, since if it factors to a bounded linear functional on the latter space it depends only on finitely many factors. If ∂ factors it must vanish first on the ideal of all flat germs, and secondly the resulting linear functional on $\text{Tay}_a(E,\mathbb{R})$ must extend to a bounded linear functional on the space of formal power series. For a results and examples in this direction see (28.3), (28.4), and (28.5). An open question is to find operational tangent vectors of infinite order.

An operational tangent vector is said to be *homogeneous of order d* if it factors over $L^d_{\text{sym}}(E,\mathbb{R})$, i.e. it corresponds to a continuous linear functional $\ell \in L^d_{\text{sym}}(E,\mathbb{R})'$ via $\partial(f) = \ell(\frac{f^{(d)}(0)}{d!})$. In order that such a functional defines a derivation, we need exactly that

$$\ell\Big(\text{Sym}\Big(\sum_{j=1}^{d-1} L^j_{\text{sym}}(E,\mathbb{R}) \otimes L^{d-j}_{\text{sym}}(E,\mathbb{R})\Big)\Big) = \{0\},$$

i.e. that ℓ vanishes on on the subspace

$$\sum_{i=1}^{j-1} L^i_{sym}(E;\mathbb{R}) \vee L^{j-i}_{sym}(E;\mathbb{R})$$

of decomposable elements of $L^j_{sym}(E;\mathbb{R})$. Here $L^i_{sym}(E;\mathbb{R}) \vee L^{j-i}_{sym}(E;\mathbb{R})$ denotes the linear subspace generated by all symmetric products $\Phi \vee \Psi$ of the corresponding elements. Any such ℓ defines an operational tangent vector $\partial^j_\ell|_a \in D_aE$ of order j by

$$\partial^j_\ell|_a(f) := \ell(\tfrac{1}{j!}d^j f(a)).$$

Since ℓ vanishes on decomposable elements we see from the Leibniz rule that ∂^j_ℓ is a derivation, and it is obviously of order j. The inverse bijection is given by $\partial \mapsto (\Phi \mapsto \partial((\Phi \circ \operatorname{diag})(\quad - a)))$, since the complete polarization of a homogeneous polynomial p of degree j is given by $\frac{1}{j!}d^j p(0)(v_1, \dots, v_j)$, and since the remainder of the Taylor expansion is flat of order $j-1$ at a.

Obviously every derivation of order at most d is a unique sum of homogeneous derivations of order j for $1 \le j \le d$. For $d > 0$ we denote by $D^{[d]}_a E$ the linear subspace of D_aE of operational tangent vectors of homogeneous order d and by $D^{(d)}_a E := \bigoplus_{j=1}^d D^{[j]}$ the subspace of (non homogeneous) operational tangent vectors of order at most d.

In more detail any operational tangent vector $\partial \in D_aE$ has a decomposition

$$\partial = \sum_{i=1}^{k-1} \partial^{[i]} + \partial^{[k,\infty]},$$

which we obtain by applying ∂ to the Taylor formula with remainder of order k, see (5.12),

$$f(a+y) = \sum_{i=0}^{k-1} \frac{1}{i!} d^i f(a) y^i + \int_0^1 \frac{(1-t)^{k-1}}{(k-1)!} d^k f(a+ty) y^k \, dt.$$

Thus, we have

$$\partial^{[i]}(f) := \partial\left(x \mapsto \frac{1}{i!} d^i f(a)(x-a)^i\right),$$

$$\partial^{[k,\infty]}(f) := \partial\left(x \mapsto \int_0^1 \frac{(1-t)^{(k-1)}}{k-1!} d^k f(a + t(x-a))(x-a)^k \, dt\right).$$

A simple computation shows that all $\partial^{[i]}$ are derivations. In fact

$$\begin{aligned}
\partial^{[k]}(f \cdot g) &= \partial\Big(\frac{1}{k!}\sum_{j=0}^{k}\binom{k}{j}(f^{(j)}(0)\circ\Delta)\cdot(g^{(k-j)}(0)\circ\Delta)\Big) \\
&= \sum_{j=0}^{k} \partial\left(\frac{f^{(j)}(0)\circ\Delta}{j!}\right)\cdot\frac{g^{(k-j)}(0)(0^{(k-j)})}{(k-j)!} \\
&\quad + \sum_{j=0}^{k} \frac{f^{(j)}(0)(0^j)}{j!}\cdot\partial\left(\frac{g^{(k-j)}(0)\circ\Delta}{(k-j)!}\right) \\
&= g(0)\cdot\partial^{[k]}(f) + 0 + \dots + 0 + f(0)\cdot\partial^{[k]}(g).
\end{aligned}$$

Hence also $\partial^{[k,\infty]}$ is a derivation. Obviously, $\partial^{[i]}$ is of order i, and hence we get a decomposition

$$D_aE = \bigoplus_{j=1}^{d} D_a^{[j]} \oplus D_a^{[d+1,\infty]},$$

where $D_a^{[d+1,\infty]}$ denotes the linear subspace of derivations which vanish on polynomials of degree at most d.

28.3. Examples. Queer operational tangent vectors. Let $Y \in E''$ be an element in the bidual of E. Then for each $a \in E$ we have an operational tangent vector $Y_a \in D_aE$, given by $Y_a(f) := Y(df(a))$. So we have a canonical injection $E'' \to D_aE$.

Let $\ell : L^2(E;\mathbb{R}) \to \mathbb{R}$ be a bounded linear functional which vanishes on the subset $E' \otimes E'$. Then for each $a \in E$ we have an operational tangent vector $\partial_\ell^{[2]}|_a \in D_aE$ given by $\partial_\ell^{[2]}|_a(f) := \ell(d^2f(a))$, since

$$\begin{aligned}\ell(d^2(fg)(a)) &= \ell(d^2f(a)g(a) + df(a)\otimes dg(a) + dg(a)\otimes df(a) + f(a)d^2g(a))\\ &= \ell(d^2f(a))g(a) + 0 + f(a)\ell(d^2g(a)).\end{aligned}$$

Let $E = (\ell^2)^{\mathbb{N}}$ be a countable product of copies of an infinite dimensional Hilbert space. A smooth function on E depends locally only on finitely many Hilbert space variables. Thus, $f \mapsto \sum_n \partial_{X_n}^{[k_n]}(f \circ \mathrm{inj}_n)$ is a well defined operational tangent vector in D_0E for arbitrary operational tangent vectors X_n of order k_n. If (k_n) is an unbounded sequence and if $X_n \neq 0$ for all n it is not of finite order. But only for $k = 1, 2, 3$ we know that nonzero tangent vectors of order k exist, see (28.4) below.

28.4. Lemma. *If E is an infinite dimensional Hilbert space, there exist nonzero operational tangent vectors of order* $2, 3$.

Proof. We may assume that $E = \ell^2$. For $k = 2$ one knows that the closure of $L(\ell^2,\mathbb{R}) \vee L(\ell^2,\mathbb{R})$ in $L^2_{\mathrm{sym}}(\ell^2,\mathbb{R})$ consists of all symmetric compact operators, and the identity is not compact.

For $k = 3$ we show that for any A in the closure of $L(\ell^2,\mathbb{R}) \vee L^2_{\mathrm{sym}}(\ell^2,\mathbb{R})$ the following condition holds:

(1) $$A(e_i, e_j, e_k) \to 0 \quad \text{for} \quad i, j, k \to \infty.$$

Since this condition is invariant under symmetrization it suffices to consider $A \in \ell^2 \otimes L(\ell^2, \ell^2)$, which we may view as a finite dimensional and thus compact operator $\ell^2 \to L(\ell^2, \ell^2)$. Then $\|A(e_i)\| \to 0$ for $i \to \infty$, since this holds for each continuous linear functional on ℓ^2. The trilinear form $A(x, y, z) := \sum_i x_iy_iz_i$ is in $L^3_{\mathrm{sym}}(\ell^2,\mathbb{R})$ and obviously does not satisfy (1). □

28.5. Proposition. *Let E be a convenient vector space with the following two properties:*

(1) *The closure of 0 in $C^\infty(E \supseteq \{0\}, \mathbb{R})$ consists of all flat germs.*
(2) *The quotient $\operatorname{Tay}_0(E, \mathbb{R}) = C^\infty(E \supseteq \{0\}, \mathbb{R})/\overline{\{0\}}$ with the bornological topology embeds as topological linear subspace into the space $\prod_k L^k_{\text{sym}}(E; \mathbb{R})$ of formal power series.*

Then each operational tangent vector on E is of finite order.

Any C^∞_b-regular Banach space, in particular any Hilbert space has these properties.

Proof. Let $\partial \in D_0E$ be an operational tangent vector. By property (1) it factors to a bounded linear mapping on $\operatorname{Tay}_0(E, \mathbb{R})$, it is continuous in the bornological topology, and by property (2) and the theorem of Hahn-Banach ∂ extends to a continuous linear functional on the space of all formal power series and thus depends only on finitely many factors.

A C^∞_b-regular Banach space E has property (1) by (27.19), and it has property (2) by E. Borel's theorem (15.4). Hilbert spaces are C^∞_b-regular by (15.5). □

28.6. Definition. A convenient vector space is said to have the *(bornological) approximation property* if $E' \otimes E$ is dense in $L(E, E)$ in the bornological locally convex topology.

For a list of spaces which have the bornological approximation property see (6.6)–(6.14).

28.7. Theorem. *Let E be a convenient vector space which has the approximation property. Then we have $D_aE \cong E''$. So if E is in addition reflexive, each operational tangent vector is a kinematic one.*

Proof. We may suppose that $a = 0$. Let $\partial : C^\infty(E \supseteq \{0\}, \mathbb{R}) \to \mathbb{R}$ be a derivation at 0, so it is bounded linear and satisfies $\partial(f \cdot g) = \partial(f) \cdot g(0) + f(0) \cdot \partial(g)$. Then we have $\partial(1) = \partial(1 \cdot 1) = 2\partial(1)$, so ∂ is zero on constant functions.

Since $E' = L(E, \mathbb{R})$ is continuously embedded into $C^\infty(E, \mathbb{R})$, $\partial|_{E'}$ is an element of the bidual E''. Obviously, $\partial - (\partial|_{E'})_0$ is a derivation which vanishes on affine functions. We have to show that it is zero. We call this difference again ∂. For $f \in C^\infty(U, \mathbb{R})$ where U is some radial open neighborhood of 0 we have

$$f(x) = f(0) + \textstyle\int_0^1 df(tx)(x)dt,$$

thus $\partial(f) = \partial(g)$, where $g(x) := \int_0^1 df(tx)(x)dt$. By assumption, there is a net $\ell_\alpha \in E' \otimes E \subset L(E, E)$ of bounded linear operators with finite dimensional image, which converges to Id_E in the bornological topology of $L(E, E)$. We consider $g_\alpha \in C^\infty(U, \mathbb{R})$, given by $g_\alpha(x) := \int_0^1 df(tx)(\ell_\alpha x)dt$.

Claim. $g_\alpha \to g$ in $C^\infty(U, \mathbb{R})$.

We have $g(x) = h(x, x)$ where $h \in C^\infty(U \times E, \mathbb{R})$ is just $h(x, y) = \int_0^1 df(tx)(y)dt$. By cartesian closedness, the associated mapping $h^\vee : U \to E' \subset C^\infty(E, \mathbb{R}))$ is smooth.

Since $' : L(E,E) \to L(E',E')$ is bounded linear, the net ℓ'_α converges to $\mathrm{Id}_{E'}$ in $L(E',E')$. The mapping $(h^\vee)^* : L(E',E') \subset C^\infty(E',E') \to C^\infty(U,E')$ is bounded linear, thus $(h^\vee)^*(\ell'_\alpha)$ converges to $h^\vee$ in $C^\infty(U,E')$. By cartesian closedness, the net $((h^\vee)^*(\ell'_\alpha))^\wedge$ converges to h in $C^\infty(U \times E, \mathbb{R})$. Since the diagonal mapping $\delta : U \to U \times E$ is smooth, the mapping $\delta^* : C^\infty(U \times E, \mathbb{R}) \to C^\infty(U,\mathbb{R})$ is continuous and linear, so finally $g_\alpha = \delta^*(((h^\vee)^*(\ell'_\alpha))^\wedge)$ converges to $\delta^*(h) = g$.

Claim. $\partial(g_\alpha) = 0$ for all α. This finishes the proof.

Let $\ell_\alpha = \sum_{i=1}^n \varphi_i \otimes x_i \in E' \otimes E \subset L(E,E)$. We have

$$\begin{aligned} g_\alpha(x) &= \int_0^1 df(tx)\Big(\sum_i \varphi_i(x)x_i\Big)dt \\ &= \sum_i \varphi_i(x) \int_0^1 df(tx)(x_i)dt =: \sum_i \varphi_i(x)h_i(x), \end{aligned}$$

$$\partial(g_\alpha) = \partial\Big(\sum_i \varphi_i \cdot h_i\Big) = \sum_i \Big(\partial(\varphi_i)h_i(0) + \varphi_i(0)\partial(h_i)\Big) = 0. \quad \square$$

28.8. Remark. There are no nonzero operational tangent vectors of order 2 on E if and only if $E' \vee E' \subset L^2_{sym}(E;\mathbb{R})$ is dense in the bornological topology. This seems to be rather near the bornological approximation property, and one may suspect that theorem (28.7) remains true under this weaker assumption.

28.9. Let $U \subseteq E$ be an open subset of a convenient vector space E. The operational tangent bundle DU of U is simply the disjoint union $\bigsqcup_{a\in U} D_a E$. Then DU is in bijection to the open subset $U \times D_0E$ of $E \times D_0(E)$ via $\partial_a \mapsto (a, \partial \circ (\ \ -a)^*)$. We use this bijection to put a smooth structure on DU. Let now $g : E \supset U \to V \subset F$ be a smooth mapping, then $g^* : C^\infty(W,\mathbb{R}) \to C^\infty(g^{-1}(W),\mathbb{R})$ is bounded and linear for all open $W \subset V$. The adjoints of these mappings uniquely define a mapping $Dg : DU \to DV$ by $(Dg.\partial)(f) := \partial(f \circ g)$.

Lemma. *$Dg : DU \to DV$ is smooth.*

Proof. Via the canonical bijections $DU \cong U \times D_0E$ and $DV \cong V \times D_0F$ the mapping Dg corresponds to

$$\begin{aligned} U \times D_0E &\to V \times D_0F \\ (a,\partial) &\mapsto \big(g(a), \partial \circ (\ \ +a)^* \circ g^* \circ (\ \ -g(a))^*\big) \\ &= \big(g(a), \partial \circ (g(\ \ +a) - g(a))^*\big). \end{aligned}$$

In order to show that this is smooth, its enough to consider the second component and we compose it with the embedding $D_0F \hookrightarrow \prod_{W \ni 0} C^\infty(W,\mathbb{R})'$. The associated mapping $U \times D_0E \times C^\infty(W,\mathbb{R}) \to \mathbb{R}$ is given by

(1) $$(a,\partial,f) \mapsto \partial\big(f \circ (g(\ \ +a) - g(a))\big),$$

where $f \circ (g(\quad +a) - g(a))$ is smooth on the open 0-neighborhood $W_a := \{y \in E : g(y+a) - g(a) \in W\} = g^{-1}(g(a)+W) - a$ in E. Now let $a : \mathbb{R} \to U$ be a smooth curve and I a bounded interval in $\mathbb{R}$. Then there exists an open neighborhood $U_{I,W}$ of 0 in E such that $U_{I,W} \subseteq W_{a(t)}$ for all $t \in I$. Then the mapping (1), composed with $a : I \to U$, factors as

$$I \times D_0E \times C^\infty(W, \mathbb{R}) \to C^\infty(U_{I,W}, \mathbb{R})' \times C^\infty(U_{I,W}, \mathbb{R}) \to \mathbb{R},$$

given by

$$(t, \partial, f) \mapsto \Big(\partial_{U_{I,W}}, f \circ \big(g(\quad +a(t)) - g(a(t))\big)\Big) \mapsto \partial_{U_{I,W}}(f \circ (g(\quad +a(t)) - g(a(t)))),$$

which is smooth by cartesian closedness. $\square$

28.10. Let E be a convenient vector space. Recall from (28.2) that $D_a^{(k)}E$ is the space of all operational tangent vectors of order $\leq k$. For an open subset U in a convenient vector space E and $k > 0$ we consider the disjoint union

$$D^{(k)}U := \bigsqcup_{a \in U} D_a^{(k)}E \cong U \times D_0^{(k)}E \subseteq E \times D_0^{(k)}E.$$

Lemma. *For a smooth mapping $f : E \supset U \to V \subset F$ the smooth mapping $Df : DU \to DV$ from (28.9) induces smooth mappings $D^{(k)}f : D^{(k)}U \to D^{(k)}V$.*

Proof. We only have to show that D_af maps $D_a^{(k)}E$ into $D_{f(a)}^{(k)}F$, because smoothness follows then by restriction.

The pullback $f^* : C^\infty(V, \mathbb{R}) \to C^\infty(U, \mathbb{R})$ maps functions which are flat of order k at $f(a)$ to functions which are flat of the same order at a. Thus, D_af maps the corresponding annihilator $D_a^{(k)}U$ into the annihilator $D_{f(a)}^{(k)}V$. $\square$

28.11. Lemma.

(1) *The chain rule holds in general: $D(f \circ g) = Df \circ Dg$ and $D^{(k)}(f \circ g) = D^{(k)}f \circ D^{(k)}g$.*

(2) *If $g : E \to F$ is a bounded affine mapping then D_xg commutes with the restriction and the projection to the subspaces of derivations which are homogeneous of degree $k > 1$.*

(3) *If $g : E \to F$ is a bounded affine mapping with linear part $\ell = g - g(0) : E \to F$ then $D_xg : D_x^{[k]}E \to D_{g(x)}^{[k]}F$ is induced by the linear mappings $(L_{sym}^k(\ell; \mathbb{R}))^* : L_{sym}^k(E, \mathbb{R})^* \to L_{sym}^k(F, \mathbb{R})^*$.*

(4) *If $g : E \to \mathbb{R}$ is bounded linear we have $Dg.X_x = D^{(1)}g.X_x^{[1]}$.*

Remark that if g is not affine then in general Dg does not respect the subspaces of derivations which are homogeneous of degree $k > 1$:

In fact let $g : E \to \mathbb{R}$ be a homogeneous polynomial of degree k on which $\partial \in D_0^{[k]}E$ does not vanish. Then by (4) we have that $0 \neq \partial(g) = Dg(\partial) \in \mathbb{R} \cong D_0^{[1]}\mathbb{R} = D_0\mathbb{R}$.

Proof. (1) is obvious.

For (2) let $X_x \in D_x E$ and $f \in C^\infty(F, \mathbb{R})$. Then we have

$$\begin{aligned}
(Dg.X_x)^{[k]}(f) &= (Dg.X_x)(\tfrac{1}{k!} d^k f(g(x))(\quad - g(x))^k) \\
&= \tfrac{1}{k!} X_x(d^k f(g(x))(g(\quad) - g(x))^k) \\
(Dg.X_x^{[k]})(f) &= X_x^{[k]}(f \circ g) \\
&= X_x(\tfrac{1}{k!} d^k (f \circ g)(x)(\quad - x)^k) \\
&= \tfrac{1}{k!} X_x(d^k f(g(x))(\ell(\quad - x))^k).
\end{aligned}$$

These expressions are equal.

$$\begin{array}{ccccc}
D_x E & \twoheadrightarrow & D_x^{[k]} E & \hookrightarrow & D_x E \\
\downarrow{\scriptstyle Dg} & & \downarrow{\scriptstyle Dg} & & \downarrow{\scriptstyle Dg} \\
D_{g(x)} F & \twoheadrightarrow & D_{g(x)}^{[k]} F & \hookrightarrow & D_x F
\end{array}$$

For (3) we take $\varphi \in L^k_{\text{sym}}(E; \mathbb{R})'$ which vanishes on all decomposable forms, and let $X_x = \partial^k_\varphi|_x \in D_x^{[k]} E$ be the corresponding homogeneous derivation. Then

$$\begin{aligned}
(Dg.\partial^k_\varphi|_x)(f) &= \partial^k_\varphi|_x(f \circ g) \\
&= \varphi(\tfrac{1}{k!} d^k (f \circ g)(x)) \\
&= \varphi(\tfrac{1}{k!} d^k f(g(x)) \circ \ell^k) \\
&= (L^k_{\text{sym}}(\ell; \mathbb{R})^* \varphi)(\tfrac{1}{k!} d^k f(g(x))) \\
&= \partial^{[k]}_{L^k_{\text{sym}}(\ell;\mathbb{R})^* \varphi}|_{g(x)}(f).
\end{aligned}$$

$$\begin{array}{ccc}
D_x^{[k]} E & \hookrightarrow & L^k_{\text{Sym}}(E, \mathbb{R})' \\
{\scriptstyle D^{[k]} g}\downarrow & & \downarrow{\scriptstyle L^k_{\text{Sym}}(\ell, \mathbb{R})^*} \\
D_{g(x)}^{[k]} F & \hookrightarrow & L^k_{\text{Sym}}(F, \mathbb{R})'
\end{array}$$

(4) is a special case of (2). □

28.12. The operational and the kinematic tangent bundles. Let M be a manifold with a smooth atlas $(M \supset U_\alpha \xrightarrow{u_\alpha} E_\alpha)_{\alpha \in A}$. We consider the following equivalence relation on the disjoint union

$$\begin{gathered}
\bigsqcup_{\alpha \in A} D(u_\alpha(U_\alpha)) := \bigcup_{\alpha \in A} D(u_\alpha(U_\alpha)) \times \{\alpha\}, \\
(\partial, \alpha) \sim (\partial', \beta) \iff D(u_{\alpha\beta})\partial' = \partial.
\end{gathered}$$

We denote the quotient set by DM and call it the *operational tangent bundle* of M. Let $\pi_M : DM \to M$ be the obvious foot point projection, let $DU_\alpha =$

$\pi_M^{-1}(U_\alpha) \subset DM$, and let $Du_\alpha : DU_\alpha \to D(u_\alpha(U_\alpha))$ be given by $Du_\alpha([\partial, \alpha]) = \partial$. So $Du_\alpha([\partial', \beta]) = D(u_{\alpha\beta})\partial'$.

The charts (DU_α, Du_α) form a smooth atlas for DM, since the chart changings are given by

$$Du_\alpha \circ (Du_\beta)^{-1} = D(u_{\alpha\beta}) : D(u_\beta(U_{\alpha\beta})) \to D(u_\alpha(U_{\alpha\beta})).$$

This chart changing formula also implies that the smooth structure on DM depends only on the equivalence class of the smooth atlas for M.

The mapping $\pi_M : DM \to M$ is obviously smooth. The natural topology is automatically Hausdorff: X, $Y \in DM$ can be separated by open sets of the form $\pi_M^{-1}(V)$ for $V \subset M$, if $\pi_M(X) \neq \pi_M(Y)$, since M is Hausdorff, and by open subsets of the form $(Tu_\alpha)^{-1}(E_\alpha \times W)$ for W open in E_α, if $\pi_M(X) = \pi_M(Y) \in U_\alpha$.

For $x \in M$ the set $D_xM := \pi_M^{-1}(x)$ is called the *operational tangent space* at x or the *fiber* over x of the operational tangent bundle. It carries a canonical convenient vector space structure induced by $D_x(u_\alpha) := Du_\alpha|_{D_xM} : D_{u_\alpha(x)}E_\alpha \cong D_0(E_\alpha)$ for some (equivalently any) α with $x \in U_\alpha$.

Let us construct now the *kinematic tangent bundle.* We consider the following equivalence relation on the disjoint union

$$\bigcup_{\alpha \in A} U_\alpha \times E_\alpha \times \{\alpha\},$$

$$(x, v, \alpha) \sim (y, w, \beta) \iff x = y \text{ and } d(u_{\alpha\beta})(u_\beta(x))w = v$$

and denote the quotient set by TM, the kinematic tangent bundle of M. Let $\pi_M : TM \to M$ be given by $\pi_M([x, v, \alpha]) = x$, let $TU_\alpha = \pi_M^{-1}(U_\alpha) \subset TM$, and let $Tu_\alpha : TU_\alpha \to u_\alpha(U_\alpha) \times E_\alpha$ be given by $Tu_\alpha([x, v, \alpha]) = (u_\alpha(x), v)$. So $Tu_\alpha([x, w, \beta]) = (u_\alpha(x), d(u_{\alpha\beta})(u_\beta(x))w)$.

The charts (TU_α, Tu_α) form a smooth atlas for TM, since the chart changings are given by

$$Tu_\alpha \circ (Tu_\beta)^{-1} : u_\beta(U_{\alpha\beta}) \times E_\beta \to u_\alpha(U_{\alpha\beta}) \times E_\alpha,$$
$$(x, v) \mapsto (u_{\alpha\beta}(x), d(u_{\alpha\beta})(x)v).$$

This chart changing formula also implies that the smooth structure on TM depends only on the equivalence class of the smooth atlas for M.

The mapping $\pi_M : TM \to M$ is obviously smooth. It is called the *(foot point) projection* of M. The natural topology is automatically Hausdorff; this follows from the bundle property and the proof is the same as for DM above.

For $x \in M$ the set $T_xM := \pi_M^{-1}(x)$ is called the *kinematic tangent space* at x or the *fiber* over x of the tangent bundle. It carries a canonical convenient vector space structure induced by $T_x(u_\alpha) := Tu_\alpha|_{T_xM} : T_xM \to \{x\} \times E_\alpha \cong E_\alpha$ for some (equivalently any) α with $x \in U_\alpha$.

Note that the kinematic tangent bundle TM embeds as a subbundle into DM; also for each $k \in \mathbb{N}$ the same construction as above gives us tangent bundles $D^{(k)}M$ which are subbundles of DM.

28.13. Let us now give an obvious description of TM as the space of all velocity vectors of curves, which explains the name 'kinematic tangent bundle': We put on $C^\infty(\mathbb{R}, M)$ the equivalence relation : $c \sim e$ if and only if $c(0) = e(0)$ and in one (equivalently each) chart (U, u) with $c(0) = e(0) \in U$ we have $\frac{d}{dt}|_0(u \circ c)(t) = \frac{d}{dt}|_0(u \circ e)(t)$. We have the following diagram

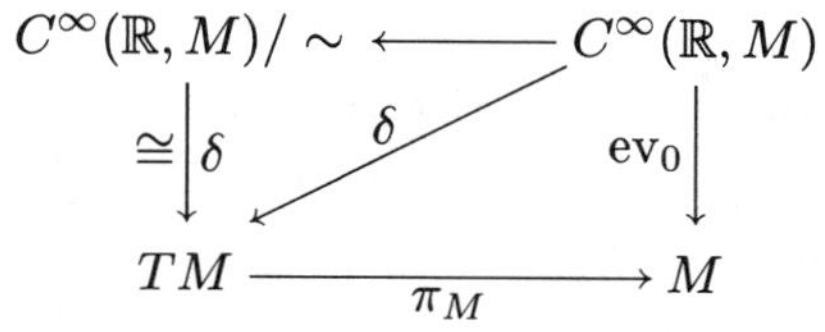

where to $c \in C^\infty(\mathbb{R}, M)$ we associate the tangent vector $\delta(c) := [c(0), \frac{\partial}{\partial t}|_0 (u_\alpha \circ c)(t), \alpha]$. It factors to a bijection $C^\infty(\mathbb{R}, M)/\sim \to TM$, whose inverse associates to $[x, v, \alpha]$ the equivalence class of $t \mapsto u_\alpha^{-1}(u_\alpha(x) + h(t)v)$ for h a small function with $h(t) = t$ near 0.

Since the c^∞-topology on $\mathbb{R} \times E_\alpha$ is the product topology by corollary (4.15), we can choose h uniformly for (x, v) in a piece of a smooth curve. Thus, a mapping $g : TM \to N$ into another manifold is smooth if and only if $g \circ \delta : C^\infty(\mathbb{R}, M) \to N$ maps 'smooth curves' to smooth curves, by which we mean $C^\infty(\mathbb{R}^2, M)$ to $C^\infty(\mathbb{R}, N)$.

28.14. Lemma. *If a smooth manifold M and the squares of its model spaces are smoothly paracompact, then also the kinematic tangent bundle TM is smoothly paracompact.*

If a smooth manifold M and $V \times D_0 V$ for any of its model spaces V are smoothly paracompact, then also the operational tangent bundle DM is smoothly paracompact.

Proof. This is a particular case of (29.7) below. □

28.15. Tangent mappings. Let $f : M \to N$ be a smooth mapping between manifolds. Then f induces a linear mapping $D_x f : D_x M \to D_{f(x)} N$ for each $x \in M$ by $(D_x f . \partial_x)(h) = \partial_x(h \circ f)$ for $h \in C^\infty(N \supseteq \{f(x)\}, \mathbb{R})$. These give a mapping $Df : DM \to DN$. If (U, u) is a chart around x and (V, v) is one around $f(x)$, then $Dv \circ Df \circ (Du)^{-1} = D(v \circ f \circ u^{-1})$ is smooth by lemma (28.9). So $Df : DM \to DN$ is smooth.

By lemma (28.10), Df restricts to smooth mappings $D^{(k)} f : D^{(k)} M \to D^{(k)} N$ and to $Tf : TM \to TN$. We check the last statement for open subsets M and N of convenient vector spaces. $(Df . X_a)(g) = X_a(g \circ f) = d(g \circ f)(a)(X) = dg(f(a)) df(a) X = (df(a)X)_{f(a)}(g)$.

If $f \in C^\infty(M, E)$ for a convenient vector space E, then $Df : DM \to DE = E \times D_0 E$. We then define the *differential* of f by $df := pr_2 \circ Df : DM \to D_0 E$. It restricts to smooth fiberwise linear mappings $D^{(k)} M \to D_0^{(k)} E$ and $df : TM \to E$.

If $f \in C^\infty(M, \mathbb{R})$, then $df : DM \to \mathbb{R}$. Let Id denote the identity function on $\mathbb{R}$, then $(Tf . \partial_x)(\mathrm{Id}) = \partial_x(\mathrm{Id} \circ f) = \partial_x(f)$, so we have $df(\partial_x) = \partial_x(f)$.

The mapping $f \mapsto df$ is bounded linear $C^\infty(M,\mathbb{R}) \to C^\infty(DM,\mathbb{R})$. That it is linear and has values in this space is obvious. So by the smooth uniform boundedness principle (5.26) it is enough to show that $f \mapsto df.X_x = X_x(f)$ is bounded for all $X_x \in DM$, which is true by definition of DM.

28.16. Remark. From the construction of the tangent bundle in (28.12) it is immediately clear that

$$TM \xleftarrow{T(\mathrm{pr}_1)} T(M \times N) \xrightarrow{T(\mathrm{pr}_2)} TN$$

is also a product, so that $T(M \times N) = TM \times TN$ in a canonical way.

We investigate $D_0(E \times F)$ for convenient vector spaces. Since D_0 is a functor for 0 preserving maps, we obtain linear sections $D_0(\mathrm{inj}_k) : D_0(E_k) \to D_0(E_1 \times E_2)$ and hence a section $D_0(\mathrm{inj}_1) + D_0(\mathrm{inj}_2) : D_0(E_1) \oplus D_0(E_2) \to D_0(E_1 \oplus E_2)$. The complement of the image is given by the kernel of the linear mapping $(D_0(\mathrm{pr}_1), D_0(\mathrm{pr}_2)) : D_0(E_1 \oplus E_2) \to D_0(E_1) \oplus D_0(E_2)$.

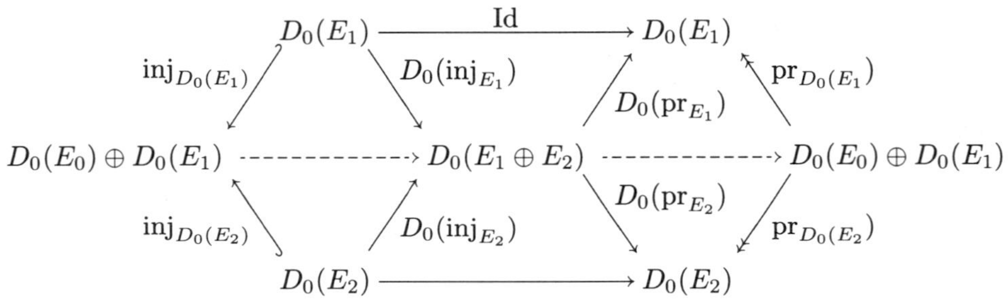

Lemma. *In the case $E_1 = \ell^2 = E_2$ this mapping is not injective.*

Proof. The space $L^2(E_1 \times E_2, E_1 \times E_2;\mathbb{R})$ can be viewed as $L^2(E_1, E_1;\mathbb{R}) \times L^2(E_1, E_2;\mathbb{R}) \times L^2(E_2, E_1;\mathbb{R}) \times L^2(E_2, E_2;\mathbb{R})$ and the subspace formed by those forms whose (2,1) and (1,2) components with respect to this decomposition are compact considered as operators in $L(\ell^2, \ell^2) \cong L^2(\ell^2, \ell^2;\mathbb{R})$ is a closed subspace. So, by Hahn-Banach, there is a non-trivial continuous linear functional $\ell : L^2(\ell^2 \times \ell^2, \ell^2 \times \ell^2;\mathbb{R}) \to \mathbb{R}$ vanishing on this subspace. We claim that the linear mapping $\partial : C^\infty(\ell^2 \times \ell^2, \mathbb{R}) \ni f \mapsto \ell(f''(0,0)) \in \mathbb{R}$ is an operational tangent vector of $\ell^2 \times \ell^2$ but not a direct sum of two operational tangent vectors on ℓ^2. In fact, the second derivative of a product h of two functions f and g is given by

$$\begin{aligned} d^2h(0,0)(w_1,w_2) &= d^2f(0,0)(w_1,w_2)\, g(0,0) \\ &\quad + df(0,0)(w_1)\, dg(0,0)(w_2) \\ &\quad + df(0,0)(w_2)\, dg(0,0)(w_1) \\ &\quad + f(0,0)\, d^2g(0,0)(w_1,w_2). \end{aligned}$$

Thus ∂ is a derivation since the middle terms give finite dimensional operators in $L^2(\ell^2, \ell^2;\mathbb{R})$. It is not a direct sum of two operational tangent vectors on ℓ^2 since functions f depending only on the j-th factor have as second derivative forms with nonzero (j,j) entry only. Hence $D_0(\mathrm{pr}_j)(\partial)(f) = \partial(f \circ \mathrm{pr}_j) = \ell((f \circ \mathrm{pr}_j)''(0)) = 0$, but $\partial \neq 0$. □

29. Vector Bundles

29.1. Vector bundles. Let $p : E \to M$ be a smooth mapping between manifolds. By a *vector bundle chart* on (E, p, M) we mean a pair (U, ψ), where U is an open subset in M, and where ψ is a fiber respecting diffeomorphism as in the following diagram:

$$\begin{array}{ccc} E|U := p^{-1}(U) & \xrightarrow{\quad\psi\quad} & U \times V \\ & {\scriptstyle p}\searrow \quad \swarrow{\scriptstyle \mathrm{pr}_1} & \\ & U. & \end{array}$$

Here V is a fixed convenient vector space, called the *standard fiber* or the *typical fiber*, real for the moment.

Two vector bundle charts (U_1, ψ_1) and (U_2, ψ_2) are called *compatible*, if $\psi_1 \circ \psi_2^{-1}$ is a fiber linear isomorphism, i.e., $(\psi_1 \circ \psi_2^{-1})(x, v) = (x, \psi_{1,2}(x)v)$ for some mapping $\psi_{1,2} : U_{1,2} := U_1 \cap U_2 \to GL(V)$. The mapping $\psi_{1,2}$ is then unique and smooth into $L(V, V)$, and it is called the *transition function* between the two vector bundle charts.

A *vector bundle atlas* $(U_\alpha, \psi_\alpha)_{\alpha \in A}$ for $p : E \to M$ is a set of pairwise compatible vector bundle charts (U_α, ψ_α) such that $(U_\alpha)_{\alpha \in A}$ is an open cover of M. Two vector bundle atlas are called *equivalent*, if their union is again a vector bundle atlas.

A (smooth) *vector bundle* $p : E \to M$ consists of manifolds E (the *total space*), M (the *base*), and a smooth mapping $p : E \to M$ (the *projection*) together with an equivalence class of vector bundle atlas: We must know at least one vector bundle atlas. The projection p turns out to be a surjective smooth mapping which has the 0-section as global smooth right inverse. Hence it is a final smooth mapping, see (27.15).

If all mappings mentioned above are real analytic we call $p : E \to M$ a *real analytic vector bundle*. If all mappings are holomorphic and V is a complex vector space we speak of a *holomorphic vector bundle*.

29.2. Remark. Let $p : E \to M$ be a finite dimensional real analytic vector bundle. If we extend the transition functions $\psi_{\alpha\beta}$ to $\widetilde{\psi}_{\alpha\beta} : \widetilde{U}_{\alpha\beta} \to GL(V_{\mathbb{C}}) = GL(V)_{\mathbb{C}}$, we see that there is a holomorphic vector bundle $(E_{\mathbb{C}}, p_{\mathbb{C}}, M_{\mathbb{C}})$ over a complex (even Stein) manifold $M_{\mathbb{C}}$ such that E is isomorphic to a real part of $E_{\mathbb{C}}|M$, compare (11.1). The germ of it along M is unique. Real analytic sections $s : M \to E$ coincide with certain germs along M of holomorphic sections $W \to E_{\mathbb{C}}$ for open neighborhoods W of M in $M_{\mathbb{C}}$.

Note that every smooth finite dimensional vector bundle admits a compatible real analytic structure, see [Hirsch, 1976, p. 101].

29.3. We will now give a formal description of the set vector bundles with fixed base M and fixed standard fiber V, up to equivalence. We only treat smooth vector bundles; similar descriptions are possible for real analytic and holomorphic vector bundles.

Let us first fix an open cover $(U_\alpha)_{\alpha\in A}$ of M. If $p : E \to M$ is a vector bundle which admits a vector bundle atlas (U_α, ψ_α) with the given open cover, then we have $\psi_\alpha \circ \psi_\beta^{-1}(x,v) = (x, \psi_{\alpha\beta}(x)v)$ for transition functions $\psi_{\alpha\beta} : U_{\alpha\beta} = U_\alpha \cap U_\beta \to GL(V) \subset L(V,V)$, which are smooth. This family of transition functions satisfies

$$(1) \qquad \begin{cases} \psi_{\alpha\beta}(x)\cdot\psi_{\beta\gamma}(x) = \psi_{\alpha\gamma}(x) & \text{for each } x \in U_{\alpha\beta\gamma} = U_\alpha \cap U_\beta \cap U_\gamma, \\ \psi_{\alpha\alpha}(x) = e & \text{for all } x \in U_\alpha. \end{cases}$$

Condition (1) is called a *cocycle condition*, and thus we call the family $(\psi_{\alpha\beta})$ the *cocycle of transition functions* for the vector bundle atlas (U_α, ψ_α).

Let us now suppose that the same vector bundle $p : E \to M$ is described by an equivalent vector bundle atlas $(U_\alpha, \varphi_\alpha)$ with the same open cover (U_α). Then the vector bundle charts (U_α, ψ_α) and $(U_\alpha, \varphi_\alpha)$ are compatible for each α, so $\varphi_\alpha \circ \psi_\alpha^{-1}(x,v) = (x, \tau_\alpha(x)v)$ for some $\tau_\alpha : U_\alpha \to GL(V)$. But then we have

$$\begin{aligned} (x, \tau_\alpha(x)\psi_{\alpha\beta}(x)v) &= (\varphi_\alpha \circ \psi_\alpha^{-1})(x, \psi_{\alpha\beta}(x)v) \\ &= (\varphi_\alpha \circ \psi_\alpha^{-1} \circ \psi_\alpha \circ \psi_\beta^{-1})(x,v) = (\varphi_\alpha \circ \psi_\beta^{-1})(x,v) \\ &= (\varphi_\alpha \circ \varphi_\beta^{-1} \circ \varphi_\beta \circ \psi_\beta^{-1})(x,v) = (x, \varphi_{\alpha\beta}(x)\tau_\beta(x)v). \end{aligned}$$

So we get

$$(2) \qquad \tau_\alpha(x)\psi_{\alpha\beta}(x) = \varphi_{\alpha\beta}(x)\tau_\beta(x) \quad \text{for all } x \in U_{\alpha\beta}.$$

We say that the two cocycles $(\psi_{\alpha\beta})$ and $(\varphi_{\alpha\beta})$ of transition functions over the cover (U_α) are *cohomologous*. The *cohomology classes* of cocycles $(\psi_{\alpha\beta})$ over the open cover (U_α) (where we identify cohomologous ones) form a set $\check{H}^1((U_\alpha), \underline{GL}(V))$ the first *Čech cohomology set* of the open cover (U_α) with values in the sheaf $C^\infty(\quad, GL(V)) =: \underline{GL}(V)$.

Now let $(W_i)_{i\in I}$ be an open cover of M refining (U_α) with $W_i \subset U_{\varepsilon(i)}$, where $\varepsilon : I \to A$ is some refinement mapping. Then for any cocycle $(\psi_{\alpha\beta})$ over (U_α) we define the cocycle $\varepsilon^*(\psi_{\alpha\beta}) =: (\varphi_{ij})$ by the prescription $\varphi_{ij} := \psi_{\varepsilon(i),\varepsilon(j)}|W_{ij}$. The mapping ε^* respects the cohomology relations and thus induces a mapping $\varepsilon^\sharp : \check{H}^1((U_\alpha), \underline{GL}(V)) \to \check{H}^1((W_i), \underline{GL}(V))$. One can show that the mapping ε^* depends on the choice of the refinement mapping ε only up to cohomology (use $\tau_i = \psi_{\varepsilon(i),\eta(i)}|W_i$ if ε and η are two refinement mappings), so we may form the inductive limit $\varinjlim \check{H}^1(\mathcal{U}, \underline{GL}(V)) =: \check{H}^1(M, \underline{GL}(V))$ over all open covers of M directed by refinement.

Theorem. *$\check{H}^1(M, \underline{GL}(V))$ is bijective to the set of all isomorphism classes of vector bundles over M with typical fiber V.*

Proof. Let $(\psi_{\alpha\beta})$ be a cocycle of transition functions $\psi_{\alpha\beta} : U_{\alpha\beta} \to GL(V)$ over some open cover (U_α) of M. We consider the disjoint union $\bigsqcup_{\alpha\in A}\{\alpha\} \times U_\alpha \times V$

and the following relation on it: $(\alpha, x, v) \sim (\beta, y, w)$ if and only if $x = y$ and $\psi_{\beta\alpha}(x)v = w$.

By the cocycle property (1) of $(\psi_{\alpha\beta})$, this is an equivalence relation. The space of all equivalence classes is denoted by $E = VB(\psi_{\alpha\beta})$, and it is equipped with the quotient topology. We put $p : E \to M$, $p[(\alpha, x, v)] = x$, and we define the vector bundle charts (U_α, ψ_α) by $\psi_\alpha[(\alpha, x, v)] = (x, v)$, $\psi_\alpha : p^{-1}(U_\alpha) =: E|U_\alpha \to U_\alpha \times V$. Then the mapping $\psi_\alpha \circ \psi_\beta^{-1}(x, v) = \psi_\alpha[(\beta, x, v)] = \psi_\alpha[(\alpha, x, \psi_{\alpha\beta}(x)v)] = (x, \psi_{\alpha\beta}(x)v)$ is smooth, so E becomes a smooth manifold. E is Hausdorff: let $u \neq v$ in E; if $p(u) \neq p(v)$ we can separate them in M and take the inverse image under p; if $p(u) = p(v)$, we can separate them in one chart. Hence $p : E \to M$ is a vector bundle.

Now suppose that we have two cocycles $(\psi_{\alpha\beta})$ over (U_α), and (φ_{ij}) over (V_i). Then there is a common refinement (W_γ) for the two covers (U_α) and (V_i). The construction described a moment ago gives isomorphic vector bundles if we restrict the cocycle to a finer open cover. So we may assume that $(\psi_{\alpha\beta})$ and $(\varphi_{\alpha\beta})$ are cocycles over the same open cover (U_α). If the two cocycles are cohomologous, i.e., $\tau_\alpha \cdot \psi_{\alpha\beta} = \varphi_{\alpha\beta} \cdot \tau_\beta$ on $U_{\alpha\beta}$, then a fiber linear diffeomorphism $\tau : VB(\psi_{\alpha\beta}) \to VB(\varphi_{\alpha\beta})$ is given by $\tau[(\alpha, x, v)] = [(\alpha, x, \tau_\alpha(x)v)]$. By relation (2), this is well defined, so the vector bundles $VB(\psi_{\alpha\beta})$ and $VB(\varphi_{\alpha\beta})$ are isomorphic.

Most of the converse direction has already been shown above, and the argument given can easily be refined to show that isomorphic vector bundles give cohomologous cocycles. □

Remark. If $GL(V)$ is an abelian group (if V is real or complex 1-dimensional), then $\check{H}^1(M, \underline{GL}(V))$ is a usual cohomology group with coefficients in the sheaf $\underline{GL}(V)$, and it can be computed with the methods of algebraic topology. If $GL(V)$ is not abelian, then the situation is rather mysterious: there is no accepted definition for $\check{H}^2(M, \underline{GL}(V))$ for example. So $\check{H}^1(M, \underline{GL}(V))$ is more a notation than a mathematical concept.

A coarser relation on vector bundles (stable equivalence) leads to the concept of topological K-theory, which can be handled much better, but is only a quotient of the true situation.

29.4. Let $p : E \to M$ and $q : F \to N$ be vector bundles. A *vector bundle homomorphism* $\varphi : E \to F$ is a fiber respecting, fiber linear smooth mapping

$$\begin{array}{ccc} E & \xrightarrow{\varphi} & F \\ {\scriptstyle p}\downarrow & & \downarrow{\scriptstyle q} \\ M & \xrightarrow{\underline{\varphi}} & N \end{array}$$

i.e., we require that $\varphi_x : E_x \to F_{\underline{\varphi}(x)}$ is linear. We say that φ covers $\underline{\varphi}$, which turns out to be smooth. If φ is invertible, it is called a *vector bundle isomorphism.*

29.5. Constructions with vector bundles. Let $\mathcal{F}$ be a covariant functor from the category of convenient vector spaces and bounded linear mappings into itself, such that $\mathcal{F} : L(V, W) \to L(\mathcal{F}(V), \mathcal{F}(W))$ is smooth. Then $\mathcal{F}$ will be called a *smooth functor* for shortness' sake. Well known examples of smooth functors are $F(V) = \tilde{\bigotimes}_\beta{}^k V$, the k-th iterated convenient tensor product, $\mathcal{F}(V) = \Lambda^k(V)$ (the k-th exterior product, the skew symmetric elements in $\tilde{\bigotimes}_\beta{}^k V$), or $\mathcal{F}(V) = L^k_{\text{sym}}(V'; \mathbb{R})$, in particular $\mathcal{F}(V) = V''$, also $\mathcal{F}(V) = D_0 V$ (see the proof of lemma (28.9)), and similar ones.

If $p : E \to M$ is a vector bundle, described by a vector bundle atlas with cocycle of transition functions $\varphi_{\alpha\beta} : U_{\alpha\beta} \to GL(V)$, where (U_α) is an open cover of M, then we may consider the functions $\mathcal{F}(\varphi_{\alpha\beta}) : x \mapsto \mathcal{F}(\varphi_{\alpha\beta}(x))$, $U_{\alpha\beta} \to GL(\mathcal{F}(V))$, which are smooth into $L(\mathcal{F}(V), \mathcal{F}(V))$. Since $\mathcal{F}$ is a covariant functor, $\mathcal{F}(\varphi_{\alpha\beta})$ satisfies again the cocycle condition (29.3.1), and cohomology of cocycles (29.3.2) is respected, so there exists a unique vector bundle $\mathcal{F}(E) := VB(\mathcal{F}(\varphi_{\alpha\beta}) \xrightarrow{p} M$, the value at the vector bundle $p : E \to M$ of the canonical extension of the functor $\mathcal{F}$ to the category of vector bundles and their homomorphisms.

If $\mathcal{F}$ is a contravariant smooth functor like the duality functor $\mathcal{F}(V) = V'$, then we have to consider the new cocycle $\mathcal{F}(\varphi^{-1}_{\alpha\beta}) = \mathcal{F}(\varphi_{\beta\alpha})$ instead.

If $\mathcal{F}$ is a contra-covariant smooth bifunctor like $L(V, W)$, then the rule

$$\mathcal{F}(VB(\psi_{\alpha\beta}), VB(\varphi_{\alpha\beta})) := VB(\mathcal{F}(\psi^{-1}_{\alpha\beta}, \varphi_{\alpha\beta}))$$

describes the induced canonical vector bundle construction.

So for vector bundles $p : E \to M$ and $q : F \to M$ we have the following vector bundles with base M: $\Lambda^k E$, $E \oplus F$, E^*, $\Lambda E := \bigoplus_{k \geq 0} \Lambda^k E$, $E \tilde{\otimes}_\beta F$, $L(E, F)$, and so on.

29.6. Pullback of vector bundles. Let $p : E \to M$ be a vector bundle, and let $f : N \to M$ be smooth. Then the *pullback vector bundle* $f^*p : f^*E \to N$ with the same typical fiber and a vector bundle homomorphism

$$\begin{array}{ccc} f^*E & \xrightarrow{p^*f} & E \\ {\scriptstyle f^*p}\downarrow & & \downarrow{\scriptstyle p} \\ N & \xrightarrow{f} & M \end{array}$$

is defined as follows. Let E be described by a cocycle $(\psi_{\alpha\beta})$ of transition functions over an open cover (U_α) of M, $E = VB(\psi_{\alpha\beta})$. Then $(\psi_{\alpha\beta} \circ f)$ is a cocycle of transition functions over the open cover $(f^{-1}(U_\alpha))$ of N, and the bundle is given by $f^*E := VB(\psi_{\alpha\beta} \circ f)$. As a manifold we have $f^*E = N \underset{(f,M,p)}{\times} E$.

The vector bundle f^*E has the following universal property: For any vector bundle $q : F \to P$, vector bundle homomorphism $\varphi : F \to E$, and smooth $g : P \to N$ such that $f \circ g = \underline{\varphi}$, there is a unique vector bundle homomorphism $\psi : F \to f^*E$ with $\underline{\psi} = g$ and $p^*f \circ \psi = \varphi$.

$$\begin{array}{ccccc} F & \xrightarrow{\psi} & f^*E & \xrightarrow{p^*f} & E \\ \downarrow{\scriptstyle q} & & \downarrow{\scriptstyle f^*p} & & \downarrow{\scriptstyle p} \\ P & \xrightarrow{g} & N & \xrightarrow{f} & M \end{array}$$

29.7. Proposition. *Let $p : E \to M$ be a smooth vector bundle with standard fiber V, and suppose that M and the product of the model space of M and V are smoothly paracompact. In particular this holds if M and V are metrizable and smoothly paracompact.*

Then the total space E is smoothly paracompact.

Proof. If M and V are metrizable and smoothly paracompact then by (27.9) the product $M \times V$ is smoothly paracompact. Let M be modeled on the convenient vector space F. Let (U_α) be an open cover of E. We choose $W_\beta \subset \overline{W_\beta} \subset W'_\beta$ in M such that the (W_β) are an open cover of M and the W'_β are open, trivializing for the vector bundle E, and domains of charts for M. We choose a partition of unity (f_β) on M which is subordinated to (W_β). Then $E|W'_\beta \cong W'_\beta \times V$ is diffeomorphic to an open subset of the smoothly paracompact convenient vector space $F \times V$. We consider the open cover of $F \times V$ consisting of $(U_\alpha \cap E|W_\beta)_\alpha$ and $(F \setminus \operatorname{supp}(f_\beta)) \times V$ and choose a subordinated partition of unity consisting of $(g_{\alpha\beta})_\alpha$ and one irrelevant function. Since the $g_{\alpha\beta}$ have support with respect to $E|W'_\beta$ in $U_\alpha \cap E|W_\beta$ they extend to smooth functions on the whole of E. Then $(\sum_\beta g_{\alpha\beta}(f_\beta \circ p))_\alpha$ is a partition of unity which is subordinated to U_α. □

29.8. Theorem. *For any vector bundle $p : E \to M$ with M smoothly regular there is a smooth vector bundle embedding into a trivial vector bundle over M with locally (over M) splitting image. If the fibers are Banach spaces, and M is smoothly paracompact then the fiber of the trivial bundle can be chosen as Banach space as well.*

A fiberwise short exact sequence of vector bundles over a smoothly paracompact manifold M which is locally splitting is even globally splitting.

Proof. We choose first a vector bundle atlas, then smooth bump functions with supports in the base sets of the atlas such that the carriers still cover M, then we refine the atlas such that in the end we have an atlas $(U_\alpha, \psi_\alpha : E|U_\alpha \to U_\alpha \times E_\alpha)$ and functions $f_\alpha \in C^\infty(M, \mathbb{R})$ with $U_\alpha \supset \operatorname{supp}(f_\alpha)$ such that $(\operatorname{carr}(f_\alpha))$ is an open cover of M.

Then we define a smooth vector bundle homomorphism

$$\Phi : E \to M \times \prod_\alpha E_\alpha$$
$$\Phi(u) = (p(u), (f_\alpha(p(u)) \cdot \psi_\alpha(u))_\alpha).$$

This gives a locally splitting embedding with the following inverse

$$(x,(v_\beta)_\beta)\mapsto \frac{1}{f_\alpha(x)}\psi_\alpha^{-1}(x,v_\alpha)$$

over $\operatorname{carr}(f_\alpha)$.

If the fibers are Banach spaces and M is smoothly paracompact, we may assume that the family $(\psi_\alpha)_\alpha$ is a smooth partition of unity. Then we may take as fiber of the trivial bundle the space $\{(x_\alpha)_\alpha\in\prod_\alpha E_\alpha : (\|x_\alpha\|)_\alpha\in c_0\}$ supplied with the supremum norm of the norms of the coordinates.

The second assertion follows since we may glue the local splittings with the help of a partition of unity. □

29.9. The kinematic tangent bundle of a vector bundle. Let $p: E\to M$ be a vector bundle with fiber addition $+_E : E\times_M E\to E$ and fiber scalar multiplication $m_t^E : E\to E$. Then $\pi_E : TE\to E$, the tangent bundle of the manifold E, is itself a vector bundle, with fiber addition $+_{TE}$ and scalar multiplication m_t^{TE}

If $(U_\alpha,\psi_\alpha : E|U_\alpha\to U_\alpha\times V)_{\alpha\in A}$ is a vector bundle atlas for E, and if $(u_\alpha : U_\alpha\to u_\alpha(U_\alpha)\subset F)$ is a manifold atlas for M, then $(E|U_\alpha,\psi'_\alpha)_{\alpha\in A}$ is an atlas for the manifold E, where

$$\psi'_\alpha := (u_\alpha\times \mathrm{Id}_V)\circ\psi_\alpha : E|U_\alpha\to U_\alpha\times V\to u_\alpha(U_\alpha)\times V\subset F\times V.$$

Hence, the family $(T(E|U_\alpha),T\psi'_\alpha : T(E|U_\alpha)\to T(u_\alpha(U_\alpha)\times V)=(u_\alpha(U_\alpha)\times V\times F\times V)_{\alpha\in A}$ is the atlas describing the canonical vector bundle structure of $\pi_E : TE\to E$. The transition functions are:

$$\begin{aligned}
(\psi_\alpha\circ\psi_\beta^{-1})(x,v) &= (x,\psi_{\alpha\beta}(x)v)\\
(u_\alpha\circ u_\beta^{-1})(x) &= u_{\alpha\beta}(x)\\
(\psi'_\alpha\circ(\psi'_\beta)^{-1})(x,v) &= (u_{\alpha\beta}(x),\psi_{\alpha\beta}(u_\beta^{-1}(x))v)\\
(T\psi'_\alpha\circ T(\psi'_\beta)^{-1})(x,v;\xi,w) &=\\
=\big(u_{\alpha\beta}(x),\psi_{\alpha\beta}(u_\beta^{-1}(x))v; d(u_{\alpha\beta})(x)\xi,&\,(d(\psi_{\alpha\beta}\circ u_\beta^{-1})(x)\xi)v+\psi_{\alpha\beta}(u_\beta^{-1}(x))w\big).
\end{aligned}$$

So we see that for fixed (x,v) the transition functions are linear in $(\xi,w)\in F\times V$. This describes the vector bundle structure of the tangent bundle $\pi_E : TE\to E$.

For fixed (x,ξ) the transition functions of TE are also linear in $(v,w)\in V\times V$. This gives a vector bundle structure on $Tp : TE\to TM$. Its fiber addition will be denoted by $T(+_E) : T(E\times_M E)=TE\times_{TM}TE\to TE$, since it is the tangent mapping of $+_E$. Likewise, its scalar multiplication will be denoted by $T(m_t^E)$. One might say that the vector bundle structure on $Tp : TE\to TM$ is the derivative of the original one on E.

The subbundle $\{\Xi\in TE : Tp.\Xi=0 \text{ in } TM\}=(Tp)^{-1}(0)\subseteq TE$ is denoted by VE and is called the *vertical bundle* over E. The local form of a vertical vector Ξ is $T\psi'_\alpha.\Xi=(x,v;0,w)$, so the transition functions look like

$$(T\psi'_\alpha\circ T(\psi'_\beta)^{-1})(x,v;0,w)=(u_{\alpha\beta}(x),\psi_{\alpha\beta}(u_\beta^{-1}(x)v;0,\psi_{\alpha\beta}(u_\beta^{-1}(x)w).$$

They are linear in $(v,w) \in V \times V$ for fixed x, so VE is a vector bundle over M. It coincides with $0_M^*(TE, Tp, TM)$, the pullback of the bundle $TE \to TM$ over the zero section. We have a canonical isomorphism $\mathrm{vl}_E : E \times_M E \to VE$, called the *vertical lift*, given by $\mathrm{vl}_E(u_x, v_x) := \frac{d}{dt}|_0(u_x + tv_x)$, which is fiber linear over M. The local representation of the vertical lift is $(T\psi_\alpha \circ \mathrm{vl}_E \circ (\psi_\alpha \times \psi_\alpha)^{-1})((x,u),(x,v)) = (x,u;0,v)$.

If (and only if) $\varphi : (E,p,M) \to (F,q,N)$ is a vector bundle homomorphism, then we have $\mathrm{vl}_F \circ (\varphi \times_M \varphi) = T\varphi \circ \mathrm{vl}_E : E \times_M E \to VF \subset TF$. So vl is a natural transformation between certain functors on the category of vector bundles and their homomorphisms.

The mapping $\mathrm{vpr}_E := pr_2 \circ \mathrm{vl}_E^{-1} : VE \to E$ is called the *vertical projection*. Note also the relation $\mathrm{pr}_1 \circ \mathrm{vl}_E^{-1} = \pi_E | VE$.

29.10. The second kinematic tangent bundle of a manifold. All of (29.9) is valid for the second tangent bundle $T^2M = TTM$ of a manifold, but here we have one more natural structure at our disposal. The *canonical flip* or *involution* $\kappa_M : T^2M \to T^2M$ is defined locally by

$$(T^2u \circ \kappa_M \circ T^2u^{-1})(x,\xi;\eta,\zeta) = (x,\eta;\xi,\zeta),$$

where (U,u) is a chart on M. Clearly, this definition is invariant under changes of charts.

The flip κ_M has the following properties:

(1) $\kappa_N \circ T^2 f = T^2 f \circ \kappa_M$ for each $f \in C^\infty(M,N)$.
(2) $T(\pi_M) \circ \kappa_M = \pi_{TM}$.
(3) $\pi_{TM} \circ \kappa_M = T(\pi_M)$.
(4) $\kappa_M^{-1} = \kappa_M$.
(5) κ_M is a linear isomorphism from $T(\pi_M) : TTM \to TM$ to $\pi_{TM} : TTM \to TM$, so it interchanges the two vector bundle structures on TTM.
(6) κ_M is the unique smooth mapping $TTM \to TTM$ satisfying $\frac{\partial}{\partial t}\frac{\partial}{\partial s}c(t,s) = \kappa_M \frac{\partial}{\partial s}\frac{\partial}{\partial t}c(t,s)$ for each $c : \mathbb{R}^2 \to M$.

All this follows from the local formula given above.

29.11. Remark. In (28.16) we saw that in general $D_0(E \times F) \neq D_0E \times D_0F$. So the constructions of (29.9) and (29.10) do not carry over to the operational tangent bundles.

30. Spaces of Sections of Vector Bundles

30.1. Let us fix a vector bundle $p : E \to M$ for the moment. On each fiber $E_x := p^{-1}(x)$ (for $x \in M$) there is a unique structure of a convenient vector space, induced by any vector bundle chart (U_α, ψ_α) with $x \in U_\alpha$. So $0_x \in E_x$ is a special element, and $0 : M \to E$, $0(x) = 0_x$, is a smooth mapping, the *zero section*.

A *section* u of $p : E \to M$ is a smooth mapping $u : M \to E$ with $p \circ u = \mathrm{Id}_M$. The *support* of the section u is the closure of the set $\{x \in M : u(x) \neq 0_x\}$ in M. The space of all smooth sections of the bundle $p : E \to M$ will be denoted by either $C^\infty(M \leftarrow E) = C^\infty(E, p, M) = C^\infty(E)$. Also the notation $\Gamma(E \to M) = \Gamma(p) = \Gamma(E)$ is used in the literature. Clearly, it is a vector space with fiber wise addition and scalar multiplication.

If $(U_\alpha, \psi_\alpha)_{\alpha \in A}$ is a vector bundle atlas for $p : E \to M$, then any smooth mapping $f_\alpha : U_\alpha \to V$ (the standard fiber) defines a local section $x \mapsto \psi_\alpha^{-1}(x, f_\alpha(x))$ on U_α. If $(g_\alpha)_{\alpha \in A}$ is a partition of unity subordinated to (U_α), then a global section can be formed by $x \mapsto \sum_\alpha g_\alpha(x) \cdot \psi_\alpha^{-1}(x, f_\alpha(x))$. So a smooth vector bundle has "many" smooth sections if M admits enough smooth partitions of unity.

We equip the space $C^\infty(M \leftarrow E)$ with the structure of a convenient vector space given by the closed embedding

$$\begin{aligned} C^\infty(M \leftarrow E) &\to \prod_\alpha C^\infty(U_\alpha, V) \\ s &\mapsto \mathrm{pr}_2 \circ \psi_\alpha \circ (s|U_\alpha), \end{aligned}$$

where $C^\infty(U_\alpha, V)$ carries the natural structure described in (27.17), see also (3.11). This structure is independent of the choice of the vector bundle atlas, because $C^\infty(U_\alpha, V) \to \prod_\beta C^\infty(U_{\alpha\beta}, V)$ is a closed linear embedding for any other atlas $(U_\beta)_\beta$.

Proposition. *The space $C^\infty(M \leftarrow E)$ of sections of the vector bundle (E, p, M) with this structure satisfies the uniform boundedness principle with respect to the point evaluations $\mathrm{ev}_x : C^\infty(M \leftarrow E) \to E_x$ for all $x \in M$.*

If M is a separable manifold modeled on duals of nuclear Fréchet spaces, and if each fiber E_x is a nuclear Fréchet space then $C^\infty(M \leftarrow E)$ is a nuclear Fréchet space and thus smoothly paracompact.

Proof. By definition of the structure on $C^\infty(M \leftarrow E)$ the uniform boundedness principle follows from (5.26) via (5.25).

For the statement about nuclearity note that by (6.1) the spaces $C^\infty(U_\alpha, V)$ are nuclear since we may assume that the U_α form a countable cover of M by charts which are diffeomorphic to c^∞-open subsets of duals of nuclear Fréchet spaces, and closed subspaces of countable products of nuclear Fréchet spaces are again nuclear Fréchet. By (16.10) nuclear Fréchet spaces are smoothly paracompact. □

30.2. Lemma. *Let M be a smooth manifold and let $f : M \to L(E, F)$ be smooth, where E and F are convenient vector spaces.*

Then $f_*(h)(x) := f(x)(h(x))$ is a linear bounded $C^\infty(M, E) \to C^\infty(M, F)$ with the natural structure of convenient vector spaces described in (27.17). *The corresponding statements in the real analytic and holomorphic cases are also true.*

Proof. This follows from the uniform boundedness principles and the exponential laws of (27.17). □

30.3. Lemma. *Under additional assumptions we have alternative descriptions of the convenient structure on the vector space of sections $C^\infty(M \leftarrow E)$:*

(1) *If M is smoothly regular, choose a smooth closed embedding $E \to M \times F$ into a trivial vector bundle with fiber a convenient vector space F by (29.8). Then $C^\infty(M \leftarrow E)$ can be considered as a closed linear subspace of $C^\infty(M, F)$, with the natural structure from (27.17).*

(2) *If there exists a smooth linear covariant derivative ∇ with unique parallel transport on $p : E \to M$, see , then we equip $C^\infty(M \leftarrow E)$ with the initial structure with respect to the cone:*

$$C^\infty(M \leftarrow E) \xrightarrow{\operatorname{Pt}(c,\ )^*} C^\infty(\mathbb{R}, E_{c(0)}),$$
$$s \mapsto (t \mapsto \operatorname{Pt}(c,t)^{-1} s(c(t))),$$

where $c \in C^\infty(\mathbb{R}, M)$ and Pt denotes the parallel transport.

The space $C^\infty(M \leftarrow E)$ of sections of the vector bundle $p : E \to M$ with this structure satisfies the uniform boundedness principle with respect to the point evaluations $\operatorname{ev}_x : C^\infty(M \leftarrow E) \to E_x$ for all $x \in M$.

If M is a separable manifold modeled on duals of nuclear Fréchet spaces, and if each fiber E_x is a nuclear Fréchet space then $C^\infty(M \leftarrow E)$ is a nuclear Fréchet space and thus smoothly paracompact.

If in (1) M is even smoothly paracompact we may choose a 'complementary' smooth vector bundle $p' : E' \to M$ such that the Whitney sum is trivial $E \oplus_M E' \cong M \times F$, see also (29.8).

For a linear covariant derivative $\nabla : \mathfrak{X}(M) \times C^\infty(M \leftarrow E) \to C^\infty(M \leftarrow E)$ with unique parallel transport we require that the parallel transport $\operatorname{Pt}(c,t)v \in E_{c(t)}$ along each smooth curve $c : \mathbb{R} \to M$ for all $v \in E_{c(0)}$ and $t \in \mathbb{R}$ is the unique solution of the differential equation $\nabla_{\partial_t} \operatorname{Pt}(c,t)v = 0$. See (32.12) till (32.16).

Proof. This structure is independent of the choice of the vector bundle atlas, because $C^\infty(U_\alpha, V) \to \prod_\beta C^\infty(U_{\alpha\beta}, V)$ is a closed linear embedding for any other atlas (U_β).

The structures from (30.1) and (1) give even the same locally convex topology if we equip $C^\infty(M, F)$ with the initial topology given by the following diagram.

$$\begin{array}{ccc} C^\infty(M \leftarrow E) & \xrightarrow{\operatorname{inj}_*} & C^\infty(M, F) \\ {\scriptstyle (\operatorname{pr}_2 \circ \psi_\alpha)_*} \downarrow & & \downarrow \\ C^\infty(U_\alpha, V) & \longrightarrow & C^\infty(U_\alpha, F) \end{array}$$

where the bottom arrow is a push forward with the vector bundle embedding $h : U_\alpha \to L(V, F)$ of trivial bundles, given by $h^\wedge := pr_2 \circ \operatorname{inj} \circ \psi_\alpha^{-1} : U_\alpha \times V \to F$, which is bounded by (30.2).

We now show that the identity from description (2) to description (30.1) is bounded. The restriction mapping $C^\infty(M \leftarrow E) \to C^\infty(U_\alpha \leftarrow E|U_\alpha)$ is obviously bounded for description (2) on both sides. Hence, it suffices to check for a trivial bundle $E = M \times V$, that the identity from description (2) to description (30.1) is bounded. For the constant parallel transport $\operatorname{Pt}^{\text{const}}$ the result follows from proposition (27.17).

The change to an arbitrary parallel transport is done as follows: For each C^∞-curve $c : \mathbb{R} \to M$ the diagram

$$\begin{array}{ccc} C^\infty(M,V) & \xrightarrow{\ \operatorname{Pt}^\nabla(c,\ \ )^*\ } & C^\infty(\mathbb{R},V) \\ & \searrow^{\operatorname{Pt}^{\text{const}}(c,\ \ )^* = c^*} & \Big\downarrow h_* \\ & & C^\infty(\mathbb{R},V) \end{array}$$

commutes, where $h : \mathbb{R} \to GL(V)$ is given by $h(t)(v) = \operatorname{Pt}(c,t)v$ with inverse $h^{-1}(t)(w) = \operatorname{Pt}(c,t)^{-1}w = \operatorname{Pt}(c(\ \ +t),-t)w$, and its push forward is bibounded by (30.2).

Finally, we show that the identity from description (30.1) to description (2) is bounded. The structure on $C^\infty(\mathbb{R}, E_{c(0)})$ is initial with respect to the restriction maps to a covering by intervals I which is subordinated to the cover $c^{-1}(U_\alpha)$ of $\mathbb{R}$. Thus, it suffices to show that the map $C^\infty(M \leftarrow E) \xrightarrow{\operatorname{Pt}(c,\ \ )^*} C^\infty(I, E_{c(0)})$ is bounded for the structure (30.1) on $C^\infty(M \leftarrow E)$. This map factors as

$$\begin{array}{ccc} C^\infty(M \leftarrow E) & \xrightarrow{\ (\operatorname{pr}_2 \circ \psi_\alpha)_*\ } & C^\infty(U_\alpha, V) \\ {\scriptstyle \operatorname{Pt}(c,\ \ )^*}\Big\downarrow & & \Big| \\ C^\infty(\mathbb{R}, E_{c(0)}) & & \Big| (c|I)^* \\ \Big\downarrow & & \Big\downarrow \\ C^\infty(I, E_{c(0)}) & \xleftarrow{\ \ h_*\ \ } & C^\infty(I, V) \end{array}$$

where

$$h(t)(v) := \operatorname{Pt}(c,t)^{-1}(\psi_\alpha^{-1}(c(t),v)) = \operatorname{Pt}(c(\ \ +t),-t)(\psi_\alpha^{-1}(c(t),v))$$

is again a smooth map $I \to L(V, E_{c(0)})$. $\square$

30.4. Spaces of smooth sections with compact supports. For a smooth vector bundle $p : E \to M$ with finite dimensional second countable base M and standard fiber V we denote by $C_c^\infty(M \leftarrow E)$ the vector space of all smooth sections with compact supports in M.

Lemma. *The following structures of a convenient vector space on $C_c^\infty(M \leftarrow E)$ are all equivalent:*

(1) *Let $C_K^\infty(M \leftarrow E)$ be the space of all smooth sections of $E \to M$ with supports contained in the fixed compact subset $K \subset M$, a closed linear subspace of*

$C^\infty(M \leftarrow E)$. Consider the final convenient vector space structure on $C_c^\infty(M \leftarrow E)$ induced by the cone

$$C_K^\infty(M \leftarrow E) \to C_c^\infty(M \leftarrow E)$$

where K runs through a basis for the compact subsets of M. Then $C_c^\infty(M \leftarrow E)$ is even the strict and regular inductive limit of spaces $C_K^\infty(M \leftarrow E)$ where K runs through a countable base of compact sets.

(2) *Choose a second smooth vector bundle $q : E' \to M$ such that the Whitney sum is trivial (29.8): $E \oplus E' \cong M \times F$. Then $C_c^\infty(M \leftarrow E)$ can be considered as a closed direct summand of $C_c^\infty(M, F)$.*

The space $C_c^\infty(M \leftarrow E)$ satisfies the uniform boundedness principle with respect to the point evaluations. Moreover, if the standard fiber V is a nuclear Fréchet space and the base M is in addition separable then $C_c^\infty(M \leftarrow E)$ is smoothly paracompact.

Proof. Since $C_K^\infty(M \leftarrow E)$ is closed in $C^\infty(M \leftarrow E)$ the inductive limit $C_K^\infty(M \leftarrow E) \to C_c^\infty(M \leftarrow E)$ is strict. So the limit is regular (52.8) and hence $C_c^\infty(M \leftarrow E)$ is convenient with the structure in (1). The direct sum property $C_K^\infty(M \leftarrow E) \subset C_K^\infty(M, F)$ from (30.3.1) passes through the direct limits, so the equivalence of statements (1) and (2) follows.

We now show that $C_c^\infty(M \leftarrow E)$ satisfies the uniform boundedness principle for the point evaluations. Using description (2) and (5.25) for a direct sum we may assume that the bundle is trivial, hence we only have to consider $C_c^\infty(M, V)$ for a convenient vector space V. Now let F be a Banach space, and let $f : F \to C_c^\infty(M, V)$ be a linear mapping, such that $\mathrm{ev}_x \circ f : F \to V$ is bounded for each $x \in M$. Then by the uniform boundedness principle (27.17) it is bounded into $C^\infty(M, V)$. We claim that f has values even in $C_K^\infty(M, V)$ for some K, so it is bounded therein, and hence in $C_c^\infty(M, V)$, as required.

If not we can recursively construct the following data: a discrete sequence (x_n) in M, a bounded sequence (y_n) in the Banach space F, and linear functionals $\ell_n \in V'$ such that

$$|\ell_k(f(y_n)(x_k))| \begin{cases} = 0 & \text{if } n < k, \\ = 1 & \text{if } n = k, \\ < 1 & \text{if } n > k. \end{cases}$$

Namely, we choose $y_n \in F$ and $x_n \in M$ such that $f(y_n)(x_n) \neq 0$ in V, and x_n has distance 1 to $\bigcup_{m<n} \mathrm{supp}(f(y_m))$ (in a complete Riemannian metric, where closed bounded subsets are compact). By shrinking y_n we may get $|\ell_m(f(y_n)(x_m))| < 1$ for $m < n$. Then we choose $\ell_n \in V'$ such that $\ell_n(f(y_n)(x_n)) = 1$.

Then $y := \sum_n \frac{1}{2^n} y_n \in F$, and $f(y)(x_k) \neq 0$ for all k since $|\ell_k(f(y)(x_k))| > 0$. So $f(y) \notin C_c^\infty(M, V)$.

For the last assertion, if the standard fiber V is a nuclear Fréchet space and the base M is separable then $C^\infty(M \leftarrow E)$ is a nuclear Fréchet space by the proposition in (30.1), so each closed linear subspace $C_K^\infty(M \leftarrow E)$ is a nuclear Fréchet space, and by (16.10) the countable strict inductive limit $C_c^\infty(M \leftarrow E)$ is smoothly paracompact. □

30.5. Spaces of holomorphic sections. Let $q : F \to N$ be a holomorphic vector bundle over a complex (i.e., holomorphic) manifold N with standard fiber V, a complex convenient vector space. We denote by $\mathcal{H}(N \leftarrow F)$ the vector space of all holomorphic sections $s : N \to F$, equipped with the topology which is initial with respect to the cone

$$\mathcal{H}(N \leftarrow F) \to \mathcal{H}(U_\alpha \leftarrow F|U_\alpha) \xrightarrow{(\mathrm{pr}_2 \circ \psi_\alpha)_*} \mathcal{H}(U_\alpha, V)$$

where the convenient structure on the right hand side is described in (27.17), see also (7.21).

By (5.25) and (8.10) the space $\mathcal{H}(N \leftarrow F)$ of sections satisfies the uniform boundedness principle for the point evaluations.

For a finite dimensional holomorphic vector bundle the topology on $\mathcal{H}(N \leftarrow F)$ turns out to be nuclear and Fréchet by (8.2), so by (16.10) $\mathcal{H}(N \leftarrow F)$ is smoothly paracompact.

30.6. Spaces of real analytic sections. Let $p : E \to M$ be a real analytic vector bundle with standard fiber V. We denote by $C^\omega(M \leftarrow E)$ the vector space of all real analytic sections. We will equip it with one of the equivalent structures of a convenient vector space described in the next lemma.

Lemma. *The following structures of a convenient vector space on the space of sections $C^\omega(M \leftarrow E)$ are all equivalent:*

(1) *Choose a vector bundle atlas (U_α, ψ_α), and consider the initial structure with respect to the cone*

$$C^\omega(M \leftarrow E) \to C^\omega(U_\alpha \leftarrow E|U_\alpha) \xrightarrow{(\mathrm{pr}_2 \circ \psi_\alpha)_*} C^\omega(U_\alpha, V),$$

where the spaces $C^\omega(U_\alpha, V)$ are equipped with the structure of (27.17).

(2) *If M is smoothly regular, choose a smooth closed embedding $E \to M \times F$ into a trivial vector bundle with fiber a convenient vector space F. Then $C^\omega(M \leftarrow E)$ can be considered as a closed linear subspace of $C^\omega(M, F)$.*

The space $C^\omega(M \leftarrow E)$ satisfies the uniform boundedness principle for the point evaluations $\mathrm{ev}_x : C^\omega(M \leftarrow E) \to E_x$.

If the base manifold is compact finite dimensional real analytic, and if the standard fiber is a finite dimensional vector space, then $C^\omega(M \leftarrow E)$ is smoothly paracompact.

Proof. We use the following diagram

$$\begin{array}{ccc} C^\omega(M \leftarrow E) & \xrightarrow{\mathrm{inj}_*} & C^\omega(M, F) \\ {\scriptstyle (\mathrm{pr}_2 \circ \psi_\alpha)_*} \downarrow & & \downarrow \\ C^\omega(U_\alpha, V) & \longrightarrow & C^\omega(U_\alpha, F), \end{array}$$

where the bottom arrow is a push forward with the vector bundle embedding $h : U_\alpha \to L(V, F)$ of trivial bundles, given by $h^\wedge := pr_2 \circ \text{inj} \circ \psi_\alpha^{-1} : U_\alpha \times V \to F$, which is bounded by (30.2). The uniform boundedness principle follows from (11.12).

For proving that $C^\omega(M \leftarrow E)$ is smoothly paracompact we use the second description. Then $C^\omega(M \leftarrow E)$ is a direct summand in a space $C^\omega(M, V)$, where M is a compact real analytic manifold and V is a finite dimensional real vector space. The function space $C^\omega(M, V)$ is smoothly paracompact by (11.4). □

30.7. $C^{\infty,\omega}$-mappings. Let M and N be real analytic manifolds. A mapping $f : \mathbb{R} \times M \to N$ is said to be of class $C^{\infty,\omega}$ if for each $(t, x) \in \mathbb{R} \times M$ and each real analytic chart (V, v) of N with $f(t, x) \in V$ there are a real analytic chart (U, u) of M with $x \in U$, an open interval $t \in I \subset \mathbb{R}$ such that $f(I \times U) \subset V$, and $v \circ f \circ (I \times u^{-1}) : I \times u(U) \to v(V)$ is of class $C^{\infty,\omega}$ in the sense of (11.20), i.e., the canonical associate is a smooth mapping $(v \circ f \circ (I \times u^{-1}))^\vee : I \to C^\omega(u(U), v(V))$.

The mapping is said to be $C^{\omega,\infty}$ if the canonical associate is a real analytic mapping $(v \circ f \circ (I \times u^{-1}))^\vee : I \to C^\infty(u(U), v(V))$, see (11.20.2).

These notions are well defined by the composition theorem for $C^{\infty,\omega}$-mappings (11.22), and the obvious generalization of (11.21) is true.

We choose one factor to be $\mathbb{R}$ because we need the c^∞-topology of the product to be the product of the c^∞-topologies, see (4.15) and (4.22).

30.8. Lemma. Curves in spaces of sections.

(1) *For a smooth vector bundle $p : E \to M$ a curve $c : \mathbb{R} \to C^\infty(M \leftarrow E)$ is smooth if and only if $c^\wedge : \mathbb{R} \times M \to E$ is smooth.*
(2) *For a holomorphic vector bundle $p : E \to M$ a curve $c : \mathbb{D} \to \mathcal{H}(M \leftarrow E)$ is holomorphic if and only if $c^\wedge : \mathbb{D} \times M \to E$ is holomorphic.*
(3) *For a real analytic vector bundle $p : E \to M$ a curve $c : \mathbb{R} \to C^\omega(M \leftarrow E)$ is real analytic if and only if the associated mapping $c^\wedge : \mathbb{R} \times M \to E$ is real analytic.*
(4) *For a real analytic vector bundle $p : E \to M$ a curve $c : \mathbb{R} \to C^\omega(M \leftarrow E)$ is smooth if and only if $c^\wedge : \mathbb{R} \times M \to E$ is $C^{\infty,\omega}$, see (30.7). A curve $c : \mathbb{R} \to C^\infty(M \leftarrow E)$ is real analytic if and only if $c^\wedge : \mathbb{R} \times M \to E$ is $C^{\omega,\infty}$, see (11.20).*

Proof. By the descriptions of the structures ((30.1) for the smooth case, (30.5) for the holomorphic case, and (30.6) for the real analytic case) we may assume that M is open in a convenient vector space F, and we may consider functions with values in the standard fiber instead of sections. The statements then follow from the respective exponential laws ((3.12) for the smooth case, (7.22) for the holomorphic case, (11.18) for the real analytic case, and the definition in (11.20) for the $C^{\infty,\omega}$ and $C^{\omega,\infty}$ cases). □

30.9. Lemma (Curves in spaces of sections with compact support).

(1) *For a smooth vector bundle $p : E \to M$ with finite dimensional base manifold M a curve $c : \mathbb{R} \to C_c^\infty(M \leftarrow E)$ is smooth if and only if $c^\wedge : \mathbb{R} \times M \to E$*

is smooth and satisfies the following condition:

> *For each compact interval* $[a,b] \subset \mathbb{R}$ *there is a compact subset* $K \subset M$ *such that* $c^\wedge(t,x)$ *is constant in* $t \in [a,b]$ *for all* $x \in M \setminus K$.

(2) *For a real analytic finite dimensional vector bundle* $p : E \to M$ *a curve* $c : \mathbb{R} \to C_c^\infty(M \leftarrow E)$ *is real analytic if and only if* $c^\wedge$ *satisfies the condition of (1) above and* $c^\wedge : \mathbb{R} \times M \to E$ *is* $C^{\omega,\infty}$, *see* (30.7).

Compare this with (42.5) and (42.12).

Proof. By lemma (30.4.1) a curve $c : \mathbb{R} \to C_c^\infty(M \leftarrow E)$ is smooth if it factors locally as a smooth curve into some step $C_K^\infty(M \leftarrow E)$ for some compact K in M, and this is by (30.8.1) equivalent to smoothness of $c^\wedge$ and to condition (1). An analogous proof applies to the real analytic case. □

30.10. Corollary. *Let* $p : E \to M$ *and* $p' : E' \to M$ *be smooth vector bundles with finite dimensional base manifold. Let* $W \subseteq E$ *be an open subset, and let* $f : W \to E'$ *be a fiber respecting smooth (nonlinear) mapping. Then* $C_c^\infty(M \leftarrow W) := \{s \in C_c^\infty(M \leftarrow E) : s(M) \subseteq W\}$ *is open in the convenient vector space* $C_c^\infty(M \leftarrow E)$. *The mapping* $f_* : C_c^\infty(M \leftarrow W) \to C_c^\infty(M \leftarrow E')$ *is smooth with derivative* $(d_v f)_* : C_c^\infty(M \leftarrow W) \times C_c^\infty(M \leftarrow E) \to C_c^\infty(M \leftarrow E')$, *where the vertical derivative* $d_v f : W \times_M E \to E'$ *is given by* $d_v f(u,w) := \frac{d}{dt}|_0 f(u+tw)$.

If the vector bundles and f *are real analytic then* $f_* : C_c^\infty(M \leftarrow W) \to C_c^\infty(M \leftarrow E')$ *is real analytic with derivative* $(d_v f)_*$.

If M *is compact and the vector bundles and* f *are real analytic then* $C^\omega(M \leftarrow W) := \{s \in C^\omega(M \leftarrow E) : s(M) \subseteq W\}$ *is open in the convenient vector space* $C^\omega(M \leftarrow E)$, *and the mapping* $f_* : C^\omega(M \leftarrow W) \to C^\omega(M \leftarrow E')$ *is real analytic with derivative* $(d_v f)_*$.

Proof. The set $C_c^\infty(M \leftarrow W)$ is open in $C_c^\infty(M \leftarrow E)$ since its intersection with each $C_K^\infty(M \leftarrow E)$ is open therein, see corollary (4.16), and the colimit is strict and regular by (30.4). Then f_* has all the stated properties, since it preserves (by (30.7) for $C^{\infty,\omega}$) the respective classes of curves which are described in (30.8) and (30.9). The derivative can be computed pointwise on M. □

30.11. Relation between spaces of real analytic and holomorphic sections in finite dimensions. Now let us assume that $p : F \to N$ is a finite dimensional holomorphic vector bundle over a finite dimensional complex manifold N. For a subset $A \subseteq N$ let $\mathcal{H}(M \supseteq A \leftarrow F|A)$ be the space of germs along A of holomorphic sections $W \to F|W$ for open sets W in N containing A. We equip $\mathcal{H}(M \supseteq A \leftarrow F|A)$ with the locally convex topology induced by the inductive cone $\mathcal{H}(M \supseteq W \leftarrow F|W) \to \mathcal{H}(M \supseteq A \leftarrow F|A)$ for all such W. This is Hausdorff since jet prolongations at points in A separate germs.

For a real analytic finite dimensional vector bundle $p : E \to M$ let $C^\omega(M \leftarrow E)$ be the space of real analytic sections $s : M \to E$. Furthermore, let $C^\omega(M \supseteq A \leftarrow E|A)$

denote the space of germs at a subset $A \subseteq M$ of real analytic sections defined near A. The complexification of this real vector space is the complex vector space $\mathcal{H}(M \supseteq A \leftarrow E_{\mathbb{C}}|A)$, because germs of real analytic sections $s : A \to E$ extend uniquely to germs along A of holomorphic sections $W \to E_{\mathbb{C}}$ for open sets W in $M_{\mathbb{C}}$ containing A, compare (11.2).

We topologize $C^\omega(M \supseteq A \leftarrow E|A)$ as subspace of $\mathcal{H}(M \supseteq A \leftarrow E_{\mathbb{C}}|A)$.

Theorem. Structure on spaces of germs of sections. *If $p : E \to M$ is a real analytic finite dimensional vector bundle and A a closed subset of M, then the space $C^\omega(M \supseteq A \leftarrow E|A)$ is convenient. Its bornology is generated by the cone*

$$C^\omega(M \supseteq A \leftarrow E|A) \xrightarrow{(\psi_\alpha)_*} C^\omega(U_\alpha \supseteq U_\alpha \cap A, \mathbb{R})^k,$$

where $(U_\alpha, \psi_\alpha)_\alpha$ is an arbitrary real analytic vector bundle atlas of E. If A is compact, the space $C^\omega(M \supseteq A \leftarrow E|A)$ is nuclear.

The uniform boundedness principle for all point-evaluations holds if these separate points. This follows from (11.6).

Proof. We show the corresponding result for holomorphic germs. By taking real parts the theorem then follows. So let $q : F \to N$ be a holomorphic finite dimensional vector bundle, and let A be a closed subset of N. Then $\mathcal{H}(M \supseteq A \leftarrow F|A)$ is a bornological locally convex space, since it is an inductive limit of the spaces $\mathcal{H}(W \leftarrow F|W)$ for open sets W containing A, which are nuclear and Fréchet by (30.5). If A is compact, $\mathcal{H}(M \supseteq A \leftarrow F|A)$ is nuclear as countable inductive limit.

Let $(U_\alpha, \psi_\alpha)_\alpha$ be a holomorphic vector bundle atlas for F. Then we consider the cone

$$\mathcal{H}(M \supseteq A \leftarrow F|A) \xrightarrow{(\psi_\alpha)_*} \mathcal{H}(U_\alpha \supseteq U_\alpha \cap A, \mathbb{C}^k) = \mathcal{H}(U_\alpha \supseteq U_\alpha \cap A, \mathbb{C})^k.$$

Obviously, each mapping is continuous, so the cone induces a bornology which is coarser than the given one, and which is complete by (11.4).

It remains to show that every subset $\mathcal{B} \subseteq \mathcal{H}(M \supseteq A \leftarrow F|A)$, such that $(ps_\alpha)_*(\mathcal{B})$ is bounded in every $\mathcal{H}(U_\alpha \supseteq U_\alpha \cap A, \mathbb{C})^k$, is bounded in $\mathcal{H}(F|W)$ for some open neighborhood W of A in N.

Since all restriction mappings to smaller subsets are continuous, it suffices to show the assertions of the theorem for some refinement of the atlas (U_α). Let us pass first to a relatively compact refinement. By topological dimension theory, there is a further refinement such that any $\dim_{\mathbb{R}} N + 2$ different sets have empty intersection. We call the resulting atlas again (U_α). Let (K_α) be a cover of N consisting of compact subsets $K_\alpha \subseteq U_\alpha$ for all α.

For any finite set $\mathcal{A}$ of indices let us now consider all non empty intersections $U_{\mathcal{A}} := \bigcap_{\alpha \in \mathcal{A}} U_\alpha$ and $K_{\mathcal{A}} := \bigcap_{\alpha \in \mathcal{A}} K_\alpha$. Since by (8.4) (or (8.6)) the space $\mathcal{H}(U_{\mathcal{A}} \supseteq A \cap K_{\mathcal{A}}, \mathbb{C})$ is a regular inductive limit, there are open sets $W_{\mathcal{A}} \subseteq U_{\mathcal{A}}$ containing

$A \cap K_{\mathcal{A}}$, such that $\mathcal{B}|(A \cap K_{\mathcal{A}})$ (more precisely $(\psi_{\mathcal{A}})_*(\mathcal{B}|(A \cap K_{\mathcal{A}}))$ for some suitable vector bundle chart mapping $\psi_{\mathcal{A}}$) is contained and bounded in $\mathcal{H}(W_{\mathcal{A}}, \mathbb{C})^k$. By passing to smaller open sets, we may assume that $W_{\mathcal{A}_1} \subseteq W_{\mathcal{A}_2}$ for $\mathcal{A}_1 \supseteq \mathcal{A}_2$. Now we define the subset

$$W := \bigcup_{\mathcal{A}} \widehat{W}_{\mathcal{A}}, \text{ where } \widehat{W}_{\mathcal{A}} := W_{\mathcal{A}} \setminus \bigcup_{\alpha \notin \mathcal{A}} K_\alpha.$$

The set W is open since (K_α) is a locally finite cover. For $x \in A$ let $\mathcal{A} := \{\alpha : x \in K_\alpha\}$, then $x \in \widehat{W}_{\mathcal{A}}$.

Now we show that every germ $s \in \mathcal{B}$ has a unique extension to W. For every $\mathcal{A}$ the germ of s along $A \cap K_{\mathcal{A}}$ has a unique extension $s_{\mathcal{A}}$ to a section over $W_{\mathcal{A}}$ and for $\mathcal{A}_1 \subseteq \mathcal{A}_2$ we have $s_{\mathcal{A}_1}|W_{\mathcal{A}_2} = s_{\mathcal{A}_2}$. We define the extension s_W by $s_W|\widehat{W}_{\mathcal{A}} = s_{\mathcal{A}}|\widehat{W}_{\mathcal{A}}$. This is well defined since one may check that $\widehat{W}_{\mathcal{A}_1} \cap \widehat{W}_{\mathcal{A}_2} \subseteq \widehat{W}_{\mathcal{A}_1 \cap \mathcal{A}_2}$.

$\mathcal{B}$ is bounded in $\mathcal{H}(M \supseteq W \leftarrow F|W)$ if it is uniformly bounded on each compact subset K of W. This is true since each K is covered by finitely many W_α and $\mathcal{B}|A \cap K_\alpha$ is bounded in $\mathcal{H}(W_\alpha, \mathbb{C})$. $\square$

30.12. Real analytic sections are dense. Let $p : E \to M$ be a real analytic finite dimensional vector bundle. Then there is another real analytic vector bundle $p' : E' \to M$ such that the Whitney sum $E \oplus E' \to M$ is real analytically isomorphic to a trivial bundle $M \times \mathbb{R}^k \to M$. This is seen as follows: By [Grauert, 1958, theorem 3] there is a closed real analytic embedding $i : E \to \mathbb{R}^k$ for some k. Now the fiber derivative along the zero section gives a fiberwise linear and injective real analytic mapping $E \to \mathbb{R}^k$, which induces a real analytic embedding j of the vector bundle $p : E \to M$ into the trivial bundle $M \times \mathbb{R}^k \to M$. The standard inner product on $\mathbb{R}^k$ gives rise to the real analytic orthogonal complementary vector bundle $E' := E^\perp$ and a real analytic Riemannian metric on the vector bundle E.

Now we can easily show that the space $C^\omega(M \leftarrow E)$ of real analytic sections of the vector bundle $E \to M$ is dense in the space of smooth sections, in the Whitney C^∞-topology: A smooth section corresponds to a smooth function $M \to \mathbb{R}^k$, which we may approximate by a real analytic function in the Whitney C^∞-topology, using [Grauert, 1958, Proposition 8]. The latter one can be projected to a real analytic approximating section of E.

Clearly, an embedding of the real analytic vector bundle into another one induces a linear embedding of the spaces of real analytic sections onto a direct summand. In this situation the orthogonal projection onto the vertical bundle VE within $T(M \times \mathbb{R}^k)$ gives rise to a real analytic linear connection (covariant derivative) $\nabla : C^\omega(M \leftarrow TM) \times C^\omega(M \leftarrow E) \to C^\omega(M \leftarrow E)$. If $c : \mathbb{R} \to M$ is a smooth or real analytic curve in M then the parallel transport $\operatorname{Pt}(c,t)v \in E_{c(t)}$ along c is smooth or real analytic, respectively, in $(t,v) \in \mathbb{R} \times E_{c(0)}$. It is given by the differential equation $\nabla_{\partial_t} \operatorname{Pt}(c,t)v = 0$.

More generally, for fiber bundles we get a similar result.

Lemma. *Let $p : E \to M$ be a locally trivial real analytic finite dimensional fiber bundle. Then the set $C^\omega(M \leftarrow E)$ of real analytic sections is dense in the space $C^\infty(M \leftarrow E)$ of smooth sections, in the Whitney C^∞-topology.*

The Whitney topology, even in infinite dimensions, will be explained in (41.10).

Proof. By the results of Grauert cited above, we choose a real analytic embedding $i : E \to \mathbb{R}^k$ onto a closed submanifold. Let $i_x : E_x \to \mathbb{R}^k$ be the restriction to the fiber over $x \in M$. Using the standard inner product on $\mathbb{R}^k$ and the affine structure, we consider the orthogonal tubular neighborhood $T(i_x(E_x))^\perp \supset V_x \cong U_x \subset \mathbb{R}^k$, with projection $q_x : U_x \to i_x(E_x)$, where we choose V_x so small that $U := \bigcup_{x \in M} \{x\} \times U_x$ is open in $M \times \mathbb{R}^k$. Then $q : U \to (p, i)(E) \subset M \times \mathbb{R}^k$ is real analytic.

Now a smooth section of E corresponds to a smooth function $f : M \to \mathbb{R}^k$ with $f(x) \in i_x(E_x)$. We may approximate f by a real analytic function $g : M \to \mathbb{R}^k$ such that $g(x) \in U_x$ for each x. Then $h(x) = q_x(g(x))$ corresponds to a real analytic approximating section. □

By looking at the trivial fiber bundle $\mathrm{pr}_1 : N \times N \to M$ this lemma says that for finite dimensional real analytic manifolds M and N the space $C^\omega(M, N)$ of real analytic mappings is dense in $C^\infty(M, N)$, in the Whitney C^∞-topology. Moreover, for a smooth finite dimensional vector bundle $p : E \to M$ there is a smoothly isomorphic structure of a real analytic vector bundle. Namely, as smooth vector bundle E is the pullback $f^*E(k, n)$ of the universal bundle $E(k, n) \to G(k, n)$ over the Grassmann manifold $G(k, n)$ for n high enough via a suitable smooth mapping $f : M \to G(k, n)$. Choose a smoothly compatible real analytic structure on M and choose a real analytic mapping $g : M \to G(k, n)$ which is near enough to f in the Whitney C^∞-topology to be smoothly homotopic to it. Then $g^*E(k, n)$ is a real analytic vector bundle and is smoothly isomorphic to $E = f^*E(k, n)$.

30.13. Corollary. *Let ∇ be a real analytic linear connection on a finite dimensional vector bundle $p : E \to M$, which exists by* (30.12). *Then the following cone generates the bornology on $C^\omega(M \leftarrow E)$.*

$$C^\omega(M \leftarrow E) \xrightarrow{\mathrm{Pt}(c,\ \ )^*} C^\alpha(\mathbb{R}, E_{c(0)}),$$
$$s \mapsto (t \mapsto \mathrm{Pt}(c, t)^{-1} s(c(t))),$$

for all $c \in C^\alpha(\mathbb{R}, M)$ and $\alpha = \omega, \infty$.

Proof. The bornology induced by the cone is coarser that the given one by (30.6). A still coarser bornology is induced by all curves subordinated to some vector bundle atlas. Hence, by theorem (30.6) it suffices to check for a trivial bundle that this bornology coincides with the given one. So we assume that E is trivial. For the constant parallel transport the result follows from lemma (11.9).

The change to an arbitrary real analytic parallel transport is done as follows: For each C^α-curve $c:\mathbb{R}\to M$ the diagram

$$\begin{array}{ccc} C^\omega(M\leftarrow E) & \xrightarrow{\ \operatorname{Pt}^\nabla(c,\ \ )^*\ } & C^\alpha(\mathbb{R},E_{c(0)}) \\ {\scriptstyle c^*}\big\downarrow & & \big\uparrow{\scriptstyle \cong} \\ C^\alpha(c^*E) & \xrightarrow[\ \operatorname{Pt}^{c^*\nabla}(\mathrm{Id},\ \ )^*\ ]{} & C^\alpha(\mathbb{R}\times E_{c(0)}) \end{array}$$

commutes and the the bottom arrow is an isomorphism by (30.10), so the structure induced by the cone does not depend on the choice of the connection. □

30.14. Lemma. Curves in spaces of sections.

(1) *For a real analytic finite dimensional vector bundle $p:E\to M$ a curve $c:\mathbb{R}\to C^\omega(M\leftarrow E)$ is smooth if and only if $c^\wedge:\mathbb{R}\times M\to E$ satisfies the following condition:*

> *For each n there is an open neighborhood U_n of $\mathbb{R}\times M$ in $\mathbb{R}\times M_\mathbb{C}$ and a (unique) C^n-extension $\tilde c:U_n\to E_\mathbb{C}$ (29.2) such that $\tilde c(t,\ \ )$ is holomorphic for all $t\in\mathbb{R}$.*

(2) *For a smooth finite dimensional vector bundle $p:E\to M$ a curve $c:\mathbb{R}\to C^\infty(M\leftarrow E)$ is real analytic if and only if $c^\wedge$ satisfies the following condition:*

> *For each n there is an open neighborhood U_n of $\mathbb{R}\times M$ in $\mathbb{C}\times M$ and a (unique) C^n-extension $\tilde c:U_n\to E\otimes\mathbb{C}$ such that $\tilde c(\ \ ,x):U_n\cap(\mathbb{C}\times\{x\})\to E_x\otimes\mathbb{C}$ is holomorphic for all $x\in M$.*

Proof. (1) By theorem (30.6) we may assume that M is open in $\mathbb{R}^n$, and we consider $C^\infty(M,\mathbb{R})$ instead of $C^\infty(M\leftarrow E)$. We note that $C^\omega(M,\mathbb{R})$ is the real part of $\mathcal{H}(\mathbb{C}^m\supseteq M,\mathbb{C})$ by (11.2), which is a regular inductive limit of spaces $\mathcal{H}(W,\mathbb{C})$ for open neighborhoods W of M in $\mathbb{C}^m$ by (8.6). By (1.8) the curve c is smooth if and only if for each n and each bounded interval $J\subset\mathbb{R}$ it factors to a C^n-curve $J\to\mathcal{H}(W,\mathbb{C})$, which sits continuously embedded in $C^\infty(W,\mathbb{R}^2)$. So the associated mapping $\mathbb{R}\times M_\mathbb{C}\supseteq J\times W\to\mathbb{C}$ is C^n and holomorphic in the second variable, and conversely.

(2) By (30.1) we may assume that M is open in $\mathbb{R}^m$, and again we consider $C^\infty(M,\mathbb{R})$ instead of $C^\infty(M\leftarrow E)$. We note that $C^\infty(M,\mathbb{R})$ is the projective limit of the Banach spaces $C^n(M_i,\mathbb{R})$, where M_i is a covering of M by compact cubes. By (9.9) the curve c is real analytic if and only if it is real analytic into each $C^n(M_i,\mathbb{R})$. By (9.6) and (9.5) it extends locally to a holomorphic curve $\mathbb{C}\to C^n(M_i,\mathbb{C})$. Its associated mappings fit together to the C^n-extension $\tilde c$ we were looking for. □

30.15. Lemma (Curves in spaces of sections with compact support). *For a smooth finite dimensional vector bundle $p : E \to M$ a curve $c : \mathbb{R} \to C_c^\infty(M \leftarrow E)$ is real analytic if and only if $c^\wedge$ satisfies the following two conditions:*

(1) *For each compact interval $[a,b] \subset \mathbb{R}$ there is a compact subset $K \subset M$ such that $c^\wedge(t,x)$ is constant in $t \in [a,b]$ for all $x \in M \setminus K$.*
(2) *For each n there is an open neighborhood U_n of $\mathbb{R} \times M$ in $\mathbb{C} \times M$ and a (unique) C^n-extension $\tilde{c} : U_n \to E \otimes \mathbb{C}$ such that $\tilde{c}(\quad, x) : U_n \cap (\mathbb{C} \times \{x\}) \to E_x \otimes \mathbb{C}$ is holomorphic for all $x \in M$.*

Proof. By lemma (30.4.1) a curve $c : \mathbb{R} \to C_c^\infty(M \leftarrow E)$ is real analytic if it factors locally as a real analytic curve into some step $C_K^\infty(M \leftarrow E)$ for some compact K in M (this is equivalent to (1)), and real analyticity is equivalent to (2), by lemma (30.14.2). □

31. Product Preserving Functors on Manifolds

In this section, we discuss Weil functors as generalized tangent bundles, namely those product preserving functors of manifolds which can be described in some detail. The name Weil functor derives from the fundamental paper [Weil, 1953] who gave the construction in (31.5) in finite dimensions for the first time.

31.1. A real commutative algebra A with unit $1 \neq 0$ is called *formally real* if for any $a_1, \dots, a_n \in A$ the element $1 + a_1^2 + \dots + a_n^2$ is invertible in A. Let $E = \{e \in A : e^2 = e, e \neq 0\} \subset A$ be the set of all nonzero idempotent elements in A. It is not empty since $1 \in E$. An idempotent $e \in E$ is said to be *minimal* if for any $e' \in E$ we have $ee' = e$ or $ee' = 0$.

Lemma. *Let A be a real commutative algebra with unit which is formally real and finite dimensional as a real vector space.*

Then there is a decomposition $1 = e_1 + \dots + e_k$ into all minimal idempotents. Furthermore, A decomposes as a sum of ideals $A = A_1 \oplus \dots \oplus A_k$ where $A_i = e_i A = \mathbb{R} \cdot e_i \oplus N_i$, as vector spaces, and N_i is a nilpotent ideal.

Proof. First we remark that every system of nonzero idempotents $e_1, \dots, e_r$ satisfying $e_i e_j = 0$ for $i \neq j$ is linearly independent over $\mathbb{R}$. Indeed, if we multiply a linear combination $k_1 e_1 + \dots + k_r e_r = 0$ by e_i we obtain $k_i = 0$. Consider a non minimal idempotent $e \neq 0$. Then there exists $e' \in E$ with $e \neq ee' =: \bar{e} \neq 0$. Then both $\bar{e}$ and $e - \bar{e}$ are nonzero idempotents, and $\bar{e}(e - \bar{e}) = 0$. To deduce the required decomposition of 1 we proceed by recurrence. Assume that we have a decomposition $1 = e_1 + \dots + e_r$ into nonzero idempotents satisfying $e_i e_j = 0$ for $i \neq j$. If e_i is not minimal, we decompose it as $e_i = \bar{e}_i + (e_i - \bar{e}_i)$ as above. The new decomposition of 1 into $r + 1$ idempotents is of the same type as the original one. Since A is finite dimensional this procedure stabilizes. This yields $1 = e_1 + \dots + e_k$

with minimal idempotents. Multiplying this relation by a minimal idempotent e, we find that e appears exactly once in the right hand side. Then we may decompose A as $A = A_1 \oplus \cdots \oplus A_k$, where $A_i := e_i A$.

Now each A_i has only one nonzero idempotent, namely e_i, and it suffices to investigate each A_i separately. To simplify the notation, we suppose that $A = A_i$, so that now 1 is the only nonzero idempotent of A. Let $N := \{n \in A : n^k = 0 \text{ for some } k\}$ be the ideal of all nilpotent elements in A.

We claim that any $x \in A \setminus N$ is invertible. Since A is finite dimensional the decreasing sequence

$$A \supset xA \supset x^2 A \supset \cdots$$

of ideals must become stationary. If $x^k A = 0$ then $x \in N$, thus there is a k such that $x^{k+\ell} A = x^k A \neq 0$ for all $\ell > 0$. Then $x^{2k} A = x^k A$, and there is some $y \in A$ with $x^k = x^{2k} y$. So we have $(x^k y)^2 = x^k y \neq 0$, and since 1 is the only nontrivial idempotent of A we have $x^k y = 1$. So $x^{k-1} y$ is an inverse of x as required.

Thus, the quotient algebra A/N is a finite dimensional field, so A/N equals $\mathbb{R}$ or $\mathbb{C}$. If $A/N = \mathbb{C}$, let $x \in A$ be such that $x + N = \sqrt{-1} \in \mathbb{C} = A/N$. Then $1 + x^2 + N = N = 0$ in $\mathbb{C}$, so $1 + x^2$ is nilpotent, and A cannot be formally real. Thus $A/N = \mathbb{R}$, and $A = \mathbb{R} \cdot 1 \oplus N$ as required. □

31.2. Definition. A *Weil algebra* A is a real commutative algebra with unit which is of the form $A = \mathbb{R} \cdot 1 \oplus N$, where N is a finite dimensional ideal of nilpotent elements.

So by lemma (31.1), a formally real and finite dimensional unital commutative algebra is the direct sum of finitely many Weil algebras.

31.3. Remark. The evaluation mapping $ev : M \to \operatorname{Hom}(C^\infty(M, \mathbb{R}), \mathbb{R})$, given by $ev(x)(f) := f(x)$, is bijective if and only if M is smoothly realcompact, see (17.1).

31.4. Corollary. *For two manifolds M_1 and M_2, with M_2 smoothly real compact and smoothly regular, the mapping*

$$\begin{aligned} C^\infty(M_1, M_2) &\to \operatorname{Hom}(C^\infty(M_2, \mathbb{R}), C^\infty(M_1, \mathbb{R})) \\ f &\mapsto (f^* : g \mapsto g \circ f) \end{aligned}$$

is bijective.

Proof. Let $x_1 \in M_1$ and $\varphi \in \operatorname{Hom}(C^\infty(M_2, \mathbb{R}), C^\infty(M_1, \mathbb{R}))$. Then $\mathrm{ev}_{x_1} \circ \varphi$ is in $\operatorname{Hom}(C^\infty(M_2, \mathbb{R}), \mathbb{R})$, so by (17.1) there is a unique $x_2 \in M_2$ such that $\mathrm{ev}_{x_1} \circ \varphi = \mathrm{ev}_{x_2}$. If we write $x_2 = f(x_1)$, then $f : M_1 \to M_2$ and $\varphi(g) = g \circ f$ for all $g \in C^\infty(M_2, \mathbb{R})$. This implies that f is smooth, since M_2 is smoothly regular, by (27.5). □

31.5. Chart description of Weil functors. Let $A = \mathbb{R}\cdot 1 \oplus N$ be a Weil algebra. We want to associate to it a functor $T_A : \mathcal{M}f \to \mathcal{M}f$ from the category $\mathcal{M}f$ of all smooth manifolds modeled on convenient vector spaces into itself.

Step 1. If $f \in C^\infty(\mathbb{R},\mathbb{R})$ and $\lambda 1 + n \in \mathbb{R}\cdot 1 \oplus N = A$, we consider the Taylor expansion $j^\infty f(\lambda)(t) = \sum_{j=0}^\infty \frac{f^{(j)}(\lambda)}{j!} t^j$ of f at λ, and we put

$$T_A(f)(\lambda 1 + n) := f(\lambda)1 + \sum_{j=1}^\infty \frac{f^{(j)}(\lambda)}{j!} n^j,$$

which is a finite sum, since n is nilpotent. Then $T_A(f) : A \to A$ is smooth, and we get $T_A(f\circ g) = T_A(f)\circ T_A(g)$ and $T_A(\mathrm{Id}_\mathbb{R}) = \mathrm{Id}_A$.

Step 2. If $f \in C^\infty(\mathbb{R}, F)$ for a convenient vector space F and $\lambda 1 + n \in \mathbb{R}\cdot 1 \oplus N = A$, we consider the Taylor expansion $j^\infty f(\lambda)(t) = \sum_{j=0}^\infty \frac{f^{(j)}(\lambda)}{j!} t^j$ of f at λ, and we put

$$T_A(f)(\lambda 1 + n) := 1\otimes f(\lambda) + \sum_{j=1}^\infty n^j \otimes \frac{f^{(j)}(\lambda)}{j!},$$

which is a finite sum, since n is nilpotent. Then $T_A(f) : A \to A\otimes F =: T_A F$ is smooth.

Step 3. For $f \in C^\infty(E,F)$, where E, F are convenient vector spaces, we want to define the value of $T_A(f)$ at an element of the convenient vector space $T_A E = A\otimes E$. Such an element may be uniquely written as $1\otimes x_1 + \sum_j n_j \otimes x_j$, where 1 and the $n_j \in N$ form a fixed finite linear basis of A, and where the $x_i \in E$. Let again $j^\infty f(x_1)(y) = \sum_{k\geq 0} \frac{1}{k!} d^k f(x_1)(y^k)$ be the Taylor expansion of f at $x_1 \in E$ for $y \in E$. Then we put

$$\begin{aligned} T_A(f)(1\otimes x_1 + \sum_j n_j \otimes x_j) &:= \\ &= 1\otimes f(x_1) + \sum_{k\geq 0} \frac{1}{k!} \sum_{j_1,\dots,j_k} n_{j_1}\dots n_{j_k} \otimes d^k f(x_1)(x_{j_1},\dots,x_{j_k}), \end{aligned}$$

which also is a finite sum. A change of basis in N induces the transposed change in the x_i, namely $\sum_i(\sum_j a_i^j n_j)\otimes \bar{x}_i = \sum_j n_j \otimes (\sum_i a_i^j \bar{x}_i)$, so the value of $T_A(f)$ is independent of the choice of the basis of N. Since the Taylor expansion of a composition is the composition of the Taylor expansions we have $T_A(f\circ g) = T_A(f)\circ T_A(g)$ and $T_A(\mathrm{Id}_E) = \mathrm{Id}_{T_A E}$.

If $\varphi : A \to B$ is a homomorphism between two Weil algebras we have $(\varphi\otimes F)\circ T_A f = T_B f\circ(\varphi\otimes E)$ for $f \in C^\infty(E,F)$.

Step 4. Let $\pi = \pi_A : A \to A/N = \mathbb{R}$ be the projection onto the quotient field of the Weil algebra A. This is a surjective algebra homomorphism, so by step 3 the following diagram commutes for $f \in C^\infty(E,F)$:

$$\begin{CD} A\otimes E @>{T_A f}>> A\otimes F \\ @V{\pi\otimes E}VV @VV{\pi\otimes F}V \\ E @>{f}>> F \end{CD}$$

If $U \subset E$ is a c^∞-open subset we put $T_A(U) := (\pi \otimes E)^{-1}(U) = (1 \otimes U) \times (N \otimes E)$, which is a c^∞-open subset in $T_A(E) := A \otimes E$. If $f : U \to V$ is a smooth mapping between c^∞-open subsets U and V of E and F, respectively, then the construction of step 3 applied to the Taylor expansion of f at points in U, produces a smooth mapping $T_A f : T_A U \to T_A V$, which fits into the following commutative diagram:

$$\begin{array}{ccccccc} U \times (N \otimes E) & = & T_A U & \xrightarrow{T_A f} & T_A V & = & V \times (N \otimes F) \\ & {\scriptstyle \mathrm{pr}_1}\searrow & \downarrow{\scriptstyle \pi\otimes E} & & {\scriptstyle \pi\otimes F}\downarrow & \swarrow{\scriptstyle \mathrm{pr}_1} & \\ & & U & \xrightarrow{f} & V & & \end{array}$$

We have $T_A(f \circ g) = T_A f \circ T_A g$ and $T_A(\mathrm{Id}_U) = \mathrm{Id}_{T_A U}$, so T_A is now a covariant functor on the category of c^∞-open subsets of convenient vector spaces and smooth mappings between them.

Step 5. Let M be a smooth manifold, let $(U_\alpha, u_\alpha : U_\alpha \to u_\alpha(U_\alpha) \subset E_\alpha)$ be a smooth atlas of M with chart changings $u_{\alpha\beta} := u_\alpha \circ u_\beta^{-1} : u_\beta(U_{\alpha\beta}) \to u_\alpha(U_{\alpha\beta})$. Then the smooth mappings

$$\begin{CD} T_A(u_\beta(U_{\alpha\beta})) @>{T_A(u_{\alpha\beta})}>> T_A(u_\alpha(U_{\alpha\beta})) \\ @V{\pi\otimes E_\beta}VV @VV{\pi\otimes E_\alpha}V \\ u_\beta(U_{\alpha\beta}) @>{u_{\alpha\beta}}>> u_\alpha(U_{\alpha\beta}) \end{CD}$$

form likewise a cocycle of chart changings, and we may use them to glue the c^∞-open sets $T_A(u_\alpha(U_\alpha)) = u_\alpha(U_\alpha) \times (N \otimes E_\alpha) \subset T_A E_\alpha$ together in order to obtain a smooth manifold which we denote by $T_A M$. By the diagram above, we see that $T_A M$ will be the total space of a fiber bundle $T(\pi_A, M) = \pi_{A,M} : T_A M \to M$, since the atlas $(T_A(U_\alpha), T_A(u_\alpha))$ constructed just now is already a fiber bundle atlas, see (37.1) below. So if M is Hausdorff then also $T_A M$ is Hausdorff, since two points x_i can be separated in one chart if they are in the same fiber, or they can be separated by inverse images under $\pi_{A,M}$ of open sets in M separating their projections. This construction does not depend on the choice of the atlas, because two atlas have a common refinement and one may pass to this.

If $f \in C^\infty(M, M')$ for two manifolds M, M', we apply the functor T_A to the local representatives of f with respect to suitable atlas. This gives local representatives which fit together to form a smooth mapping $T_A f : T_A M \to T_A M'$. Clearly, we again have $T_A(f \circ g) = T_A f \circ T_A g$ and $T_A(\mathrm{Id}_M) = \mathrm{Id}_{T_A M}$, so that $T_A : \mathcal{M}f \to \mathcal{M}f$ is a covariant functor.

31.6. Remark. If we apply the construction of (31.5), step 5 to the algebra $A = 0$, which we did not allow ($1 \neq 0 \in A$), then $T_0 M$ depends on the choice of the atlas. If each chart is connected, then $T_0 M = \pi_0(M)$, computing the connected components of M. If each chart meets each connected component of M, then $T_0 M$ is one point.

31.7. Theorem. Main properties of Weil functors. *Let $A = \mathbb{R} \cdot 1 \oplus N$ be a Weil algebra, where N is the maximal ideal of nilpotents. Then we have:*

(1) *The construction of (31.5) defines a covariant functor $T_A : \mathcal{M}f \to \mathcal{M}f$ such that $\pi_{A:T_A M \to M}, M$ is a smooth fiber bundle with standard fiber $N \otimes E$ if M is modeled on the convenient space E. For any $f \in C^\infty(M, M')$ we have a commutative diagram*

$$\begin{array}{ccc} T_A M & \xrightarrow{T_A f} & T_A M' \\ {\scriptstyle \pi_{A,M}} \downarrow & & \downarrow {\scriptstyle \pi_{A,M'}} \\ M & \xrightarrow{f} & M'. \end{array}$$

So (T_A, π_A) is a bundle functor on $\mathcal{M}f$, which gives a vector bundle functor on $\mathcal{M}f$ if and only if N is nilpotent of order 2.

(2) *The functor $T_A : \mathcal{M}f \to \mathcal{M}f$ is multiplicative: it respects products. It maps the following classes of mappings into itself: embeddings of (splitting) submanifolds, surjective smooth mappings admitting local smooth sections, fiber bundle projections. For fixed manifolds M and M' the mapping $T_A : C^\infty(M, M') \to C^\infty(T_A M, T_A M')$ is smooth, so it maps smoothly parameterized families to smoothly parameterized families.*

(3) *If (U_α) is an open cover of M then $T_A(U_\alpha)$ is an open cover of $T_A M$.*

(4) *Any algebra homomorphism $\varphi : A \to B$ between Weil algebras induces a natural transformation $T(\varphi, \quad) = T_\varphi : T_A \to T_B$. If φ is injective, then $T(\varphi, M) : T_A M \to T_B M$ is a closed embedding for each manifold M. If φ is surjective, then $T(\varphi, M)$ is a fiber bundle projection for each M. So we may view T as a co-covariant bifunctor from the category of Weil algebras times $\mathcal{M}f$ to $\mathcal{M}f$.*

Proof. (1) The main assertion is clear from (31.5). The fiber bundle $\pi_{A,M} : T_A M \to M$ is a vector bundle if and only if the transition functions $T_A(u_{\alpha\beta})$ are fiber linear $N \otimes E_\alpha \to N \otimes E_\beta$. So only the first derivatives of $u_{\alpha\beta}$ should act on N, hence any product of two elements in N must be 0, thus N has to be nilpotent of order 2.

(2) The functor T_A respects finite products in the category of c^∞-open subsets of convenient vector spaces by (31.5), step 3 and 5. All the other assertions follow by looking again at the chart structure of $T_A M$ and by taking into account that f is part of $T_A f$ (as the base mapping).

(3) This is obvious from the chart structure.

(4) We define $T(\varphi, E) := \varphi \otimes E : A \otimes E \to B \otimes E$. By (31.5), step 3, this restricts to a natural transformation $T_A \to T_B$ on the category of c^∞-open subsets of convenient

vector spaces, and – by gluing – also on the category $\mathcal{M}f$. Obviously, T is a co-covariant bifunctor on the indicated categories. Since $\pi_B \circ \varphi = \pi_A$ (φ respects the identity), we have $T(\pi_B, M) \circ T(\varphi, M) = T(\pi_A, M)$, so $T(\varphi, M) : T_A M \to T_B M$ is fiber respecting for each manifold M. In each fiber chart it is a linear mapping on the typical fiber $N_A \otimes E \to N_B \otimes E$.

So if φ is injective, $T(\varphi, M)$ is fiberwise injective and linear in each canonical fiber chart, so it is a closed embedding.

If φ is surjective, let $N_1 := \ker \varphi \subseteq N_A$, and let $V \subset N_A$ be a linear complement to N_1. Then if M is modeled on convenient vector spaces E_α and for the canonical charts we have the commutative diagram:

$$\begin{array}{ccc}
T_A M & \xrightarrow{\quad T(\varphi, M) \quad} & T_B M \\
\uparrow & & \uparrow \\
T_A(U_\alpha) & \xrightarrow{\quad T(\varphi, U_\alpha) \quad} & T_B(U_\alpha) \\
{\scriptstyle T_A(u_\alpha)}\downarrow & & \downarrow{\scriptstyle T_B(u_\alpha)} \\
u_\alpha(U_\alpha) \times (N_A \otimes E_\alpha) & \xrightarrow{\operatorname{Id} \times ((\varphi|N_A) \otimes E_\alpha)} & u_\alpha(U_\alpha) \times (N_B \otimes E_\alpha) \\
\| & & \| \\
u_\alpha(U_\alpha) \times (N_1 \otimes E_\alpha) \times (V \otimes E_\alpha) & \xrightarrow{\operatorname{Id} \times 0 \times \text{iso}} & u_\alpha(U_\alpha) \times 0 \times (N_B \otimes E_\alpha)
\end{array}$$

Hence $T(\varphi, M)$ is a fiber bundle projection with standard fiber $E_\alpha \otimes \ker \varphi$. □

31.8. Theorem. *Let A and B be Weil algebras. Then we have:*

(1) *We get the algebra A back from the Weil functor T_A by $T_A(\mathbb{R}) = A$ with addition $+_A = T_A(+_\mathbb{R})$, multiplication $m_A = T_A(m_\mathbb{R})$ and scalar multiplication $m_t = T_A(m_t) : A \to A$.*

(2) *The natural transformations $T_A \to T_B$ correspond exactly to the algebra homomorphisms $A \to B$.*

Proof. (1) is obvious. (2) For a natural transformation $\varphi : T_A \to T_B$ its value $\varphi_\mathbb{R} : T_A(\mathbb{R}) = A \to T_B(\mathbb{R}) = B$ is an algebra homomorphisms. The inverse of this mapping has already been described in theorem (31.7.4). □

31.9. Remark. If M is a smoothly real compact and smoothly regular manifold we consider the set $D_A(M) := \operatorname{Hom}(C^\infty(M, \mathbb{R}), A)$ of all bounded homomorphisms from the convenient algebra of smooth functions on M into a Weil algebra A. Obviously we have a natural mapping $T_A M \to D_A M$ which is given by $X \mapsto (f \mapsto T_A(f).X)$, using (3.5) and (3.6).

Let $\mathbb{D}$ be the algebra of Study numbers $\mathbb{R}.1 \oplus \mathbb{R}.\delta$ with $\delta^2 = 0$. Then $T_\mathbb{D} M = TM$, the tangent bundle, and $D_\mathbb{D}(M)$ is the smooth bundle of all operational tangent vectors, i.e. bounded derivations at a point x of the algebra of germs $C^\infty(M \supseteq \{x\}, \mathbb{R})$ see (28.12).

It would be nice if $D_A(M)$ were a smooth manifold not only for $A = \mathbb{D}$. We do not know whether this is true. The obvious method of proof hits severe obstacles, which we now explain.

Let $A = \mathbb{R}.1 \oplus N$ be a Weil algebra and let $\pi : A \to \mathbb{R}$ be the corresponding projection. Then for $\varphi \in D_A(M) = \operatorname{Hom}(C^\infty(M,\mathbb{R}), A)$ the character $\pi \circ \varphi$ equals ev_x for a unique $x \in M$, since M is smoothly real compact. Moreover, $X := \varphi - \operatorname{ev}_x .1 : C^\infty(M,\mathbb{R}) \to N$ satisfies the *expansion property* at x:

$$(1) \qquad X(fg) = X(f).g(x) + f(x).X(g) + X(f).X(g).$$

Conversely, a bounded linear mapping $X : C^\infty(M,\mathbb{R}) \to N$ with property (1) is called an *expansion at* x. Clearly each expansion at x defines a bounded homomorphism φ with $\pi \circ \varphi = \operatorname{ev}_x$. So we view $D_A(M)_x$ as the set of all expansions at x. Note first that for an expansion $X \in D_A(M)_x$ the value $X(f)$ depends only on the germ of f at x: If $f|U = 0$ for a neighborhood U of x, choose a smooth function h with $h = 1$ off U and $h(x) = 0$. Then $h^k f = f$ and $X(f) = X(h^k f) = 0 + 0 + X(h^k)X(f) = \dots = X(h)^k X(f)$, which is 0 for k larger than the nilpotence index of N.

Suppose now that $M = U$ is a c^∞-open subset of a convenient vector space E. We can ask whether $D_A(U)_x$ is a smooth manifold. Let us sketch the difficulty. A natural way to proceed would be to apply by induction on the nilpotence index of N. Let $N_0 := \{n \in N : n.N = 0\}$, which is an ideal in A. Consider the short exact sequence

$$0 \to N_0 \to N \xrightarrow{p} N/N_0 \to 0$$

and a linear section $s : N/N_0 \to N$. For $X : C^\infty(U,\mathbb{R}) \to N$ we consider $\bar X := p \circ X$ and $X_0 := X - s \circ \bar X$. Then X is an expansion at $x \in U$ if and only if

(2) $\bar X$ is an expansion at x with values in N/N_0, and X_0 satisfies

$$X_0(fg) = X_0(f)g(x) + f(x)X_0(g) + s(\bar X(f)).s(\bar X(g)) - s(\bar X(f).\bar X(g)).$$

Note that (2) is an affine equation in X_0 for fixed $\bar X$. By induction, the $\bar X \in D_{A/N_0}(U)_x$ form a smooth manifold, and the fiber over a fixed $\bar X$ consists of all $X = X_0 + s \circ \bar X$ with X_0 in the closed affine subspace described by (2), whose model vector space is the space of all derivations at x. If we were able to find a (local) section $D_{A/N_0}(U) \to D_A(U)$ and if these sections fitted together nicely we could then conclude that $D_A(U)$ was the total space of a smooth affine bundle over $D_{A/N_0}(U)$, so it would be smooth. But this translates to a lifting problem as follows: A homomorphism $C^\infty(U,\mathbb{R}) \to A/N_0$ has to be lifted in a 'natural way' to $C^\infty(U,\mathbb{R}) \to A$. But we know that in general $C^\infty(U,\mathbb{R})$ is not a free C^∞-algebra, see (31.16) for comparison.

31.10. The basic facts from the theory of Weil functors are completed by the following assertion.

Proposition. *Given two Weil algebras A and B, the composed functor $T_A \circ T_B$ is a Weil functor generated by the tensor product $A \otimes B$.*

Proof. For a convenient vector space E we have $T_A(T_B E) = A \otimes B \otimes E$, and this is compatible with the action of smooth mappings, by (31.5). □

Corollary. *There is a canonical natural equivalence $T_A \circ T_B \cong T_B \circ T_A$ generated by the exchange algebra isomorphism $A \otimes B \cong B \otimes A$.*

31.11. Examples. Let A be the algebra $\mathbb{R}.1+\mathbb{R}.\delta$ with $\delta^2 = 0$. Then $T_A M = TM$, the tangent bundle, and consequently we get $T_{A\otimes A} M = T^2 M$, the second tangent bundle.

31.12. Weil functors and Lie groups. We have (compare (38.10)) that the tangent bundle TG of a Lie group G is again a Lie group, the semidirect product $\mathfrak{g} \ltimes G$ of G with its Lie algebra $\mathfrak{g}$.

Now let A be a Weil algebra, and let T_A be its Weil functor. Then in the notation of (36.1) the space $T_A(G)$ is also a Lie group with multiplication $T_A(\mu)$ and inversion $T_A(\nu)$. By the properties (31.7), of the Weil functor T_A we have a surjective homomorphism $\pi_A : T_A G \to G$ of Lie groups. Following the analogy with the tangent bundle, for $a \in G$ we will denote its fiber over a by $(T_A)_a G \subset T_A G$, likewise for mappings. With this notation we have the following commutative diagram, where we assume that G is a regular Lie group (38.4):

$$\begin{array}{ccccccccc}
 & & \mathfrak{g}\otimes N & \longrightarrow & \mathfrak{g}\otimes A & & & & \\
 & & \| & & \| & & & & \\
0 & \longrightarrow & (T_A)_0\mathfrak{g} & \longrightarrow & T_A\mathfrak{g} & \longrightarrow & \mathfrak{g} & \longrightarrow & 0 \\
 & & {\scriptstyle (T_A)_0 \exp^G}\downarrow & & {\scriptstyle T_A \exp^G}\downarrow & & \downarrow{\scriptstyle \exp^G} & & \\
e & \longrightarrow & (T_A)_e G & \longrightarrow & T_A G & \xrightarrow{\pi_A} & G & \longrightarrow & e
\end{array}$$

The structural mappings (Lie bracket, exponential mapping, evolution operator, adjoint action) are determined by multiplication and inversion. Thus, their images under the Weil functor T_A are the same structural mappings. But note that the canonical flip mappings have to be inserted like follows. So for example

$$\mathfrak{g} \otimes A \cong T_A\mathfrak{g} = T_A(T_e G) \xrightarrow{\kappa} T_e(T_A G)$$

is the Lie algebra of $T_A G$, and the Lie bracket is just $T_A([\ \ ,\ \ ])$. Since the bracket is bilinear, the description of (31.5) implies that $[X \otimes a, Y \otimes b]_{T_A\mathfrak{g}} = [X, Y]_{\mathfrak{g}} \otimes ab$. Also $T_A \exp^G = \exp^{T_A G}$. If $\exp^G$ is a diffeomorphism near 0, $(T_A)_0(\exp^G) : (T_A)_0\mathfrak{g} \to (T_A)_e G$ is also a diffeomorphism near 0, since T_A is local. The natural transformation $0_G : G \to T_A G$ is a homomorphism which splits the bottom row of the diagram, so $T_A G$ is the semidirect product $(T_A)_0\mathfrak{g} \ltimes G$ via the mapping $T_A\rho : (u, g) \mapsto T_A(\rho_g)(u)$. So from (38.9) we may conclude that $T_A G$ is also a regular Lie group, if G is regular. If $\omega^G : TG \to T_eG$ is the Maurer Cartan form of G (i.e., the left logarithmic derivative of Id_G) then

$$\kappa_0 \circ T_A\omega^G \circ \kappa : TT_AG \cong T_ATG \to T_AT_eG \cong T_eT_AG$$

is the Maurer Cartan form of T_AG.

Product preserving functors from finite dimensional manifolds to infinite dimensional ones

31.13. Product preserving functors. Let $\mathcal{M}f_{\text{fin}}$ denote the category of all finite dimensional separable Hausdorff smooth manifolds, with smooth mappings as morphisms. Let $F : \mathcal{M}f_{\text{fin}} \to \mathcal{M}f$ be a functor which preserves products in the following sense: The diagram

$$F(M_1) \xleftarrow{F(\mathrm{pr}_1)} F(M_1 \times M_2) \xrightarrow{F(\mathrm{pr}_2)} F(M_2)$$

is always a product diagram.

Then $F(\text{point}) = \text{point}$, by the following argument:

$$\begin{array}{ccccc} F(\text{point}) & \underset{\cong}{\xleftarrow{F(\mathrm{pr}_1)}} & F(\text{point} \times \text{point}) & \underset{\cong}{\xrightarrow{F(\mathrm{pr}_2)}} & F(\text{point}) \\ & \nwarrow{\scriptstyle f_1} & \uparrow{\scriptstyle f} & \nearrow{\scriptstyle f_2} & \\ & & \text{point} & & \end{array}$$

Each of f_1, f, and f_2 determines each other uniquely, thus there is only one mapping $f_1 : \text{point} \to F(\text{point})$, so the space $F(\text{point})$ is a single point.

We also require that F has the following two properties:

(1) The map on morphisms $F : C^\infty(\mathbb{R}^n, \mathbb{R}) \to C^\infty(F(\mathbb{R}^n), F(\mathbb{R}))$ is smooth, where we regard $C^\infty(F(\mathbb{R}^n), F(\mathbb{R}))$ as Frölicher space, see section (23). Equivalently, the associated map $C^\infty(\mathbb{R}^n, \mathbb{R}) \times F(\mathbb{R}^n) \to F(\mathbb{R})$ is smooth.

(2) There is a natural transformation $\pi : F \to \mathrm{Id}$ such that for each M the mapping $\pi_M : F(M) \to M$ is a fiber bundle, and for an open submanifold $U \subset M$ the mapping $F(\mathrm{incl}) : F(U) \to F(M)$ is a pullback.

31.14. C^∞-algebras. An $\mathbb{R}$-algebra is a commutative ring A with unit together with a ring homomorphism $\mathbb{R} \to A$. Then every map $p : \mathbb{R}^n \to \mathbb{R}^m$ which is given by an m-tuple of real polynomials $(p_1, \dots, p_m)$ can be interpreted as a mapping $A(p) : A^n \to A^m$ in such a way that projections, composition, and identity are preserved, by just evaluating each polynomial p_i on an n-tuple $(a_1, \dots, a_n) \in A^n$. Compare with (17.1).

A C^∞-algebra A is a real algebra in which we can moreover interpret all smooth mappings $f : \mathbb{R}^n \to \mathbb{R}^m$. There is a corresponding map $A(f) : A^n \to A^m$, and again projections, composition, and the identity mapping are preserved.

More precisely, a C^∞-algebra A is a product preserving functor from the category C^∞ to the category of sets, where C^∞ has as objects all spaces $\mathbb{R}^n$, $n \geq 0$, and all smooth mappings between them as arrows. Morphisms between C^∞-algebras are then natural transformations: they correspond to those algebra homomorphisms which preserve the interpretation of smooth mappings.

Let us explain how one gets the algebra structure from this interpretation. Since A is product preserving, we have $A(\text{point}) = \text{point}$. All the laws for a commutative ring with unit can be formulated by commutative diagrams of mappings between products of the ring and the point. We do this for the ring $\mathbb{R}$ and apply the product preserving functor A to all these diagrams, so we get the laws for the commutative ring $A(\mathbb{R})$ with unit $A(1)$ with the exception of $A(0) \neq A(1)$ which we will check later for the case $A(\mathbb{R}) \neq \text{point}$. Addition is given by $A(+)$ and multiplication by $A(m)$. For $\lambda \in \mathbb{R}$ the mapping $A(m_\lambda) : A(\mathbb{R}) \to A(\mathbb{R})$ equals multiplication with the element $A(\lambda) \in A(\mathbb{R})$, since the following diagram commutes:

$$
\begin{array}{ccccc}
A(\mathbb{R}) & & \xrightarrow{\quad A(m_\lambda) \quad} & & \\
\cong\downarrow & & & & \\
A(\mathbb{R}) \times \text{point} & \xrightarrow{\text{Id} \times A(\lambda)} & A(\mathbb{R}) \times A(\mathbb{R}) & \longrightarrow & A(\mathbb{R}) \\
\cong\downarrow & & \| & \nearrow{\scriptstyle A(m)} & \\
A(\mathbb{R} \times \text{point}) & \xrightarrow{A(\text{Id} \times \lambda)} & A(\mathbb{R} \times \mathbb{R}) & &
\end{array}
$$

We may investigate now the difference between $A(\mathbb{R}) = \text{point}$ and $A(\mathbb{R}) \neq \text{point}$. In the latter case for $\lambda \neq 0$ we have $A(\lambda) \neq A(0)$ since multiplication by $A(\lambda)$ equals $A(m_\lambda)$ which is a diffeomorphism for $\lambda \neq 0$ and factors over a one pointed space for $\lambda = 0$. So for $A(\mathbb{R}) \neq \text{point}$ which we assume from now on, the group homomorphism $\lambda \mapsto A(\lambda)$ from $\mathbb{R}$ into $A(\mathbb{R})$ is actually injective.

This definition of C^∞-algebras is due to [Lawvere, 1967], for a thorough account see [Moerdijk, Reyes, 1991], for a discussion from the point of view of functional analysis see [Kainz, Kriegl, Michor, 1987]. In particular there on a C^∞-algebra A the natural topology is defined as the finest locally convex topology on A such that for all $a = (a_1, \dots, a_n) \in A^n$ the evaluation mappings $\varepsilon_a : C^\infty(\mathbb{R}^n, \mathbb{R}) \to A$ are continuous. In [Kainz, Kriegl, Michor, 1987, 6.6] one finds a method to recognize C^∞-algebras among locally-m-convex algebras. In [Michor, Vanžura, 1996] one finds a characterization of the algebras of smooth functions on finite dimensional algebras among all C^∞-algebras.

31.15. Theorem. *Let $F : \mathcal{M}f_{\text{fin}} \to \mathcal{M}f$ be a product preserving functor. Then either $F(\mathbb{R})$ is a point or $F(\mathbb{R})$ is a C^∞-algebra. If $\varphi : F_1 \to F_2$ is a natural transformation between two such functors, then $\varphi_{\mathbb{R}} : F_1(\mathbb{R}) \to F_2(\mathbb{R})$ is an algebra homomorphism.*

If F has property ((31.13.1)) then the natural topology on $F(\mathbb{R})$ is finer than the given manifold topology and thus is Hausdorff if the latter is it.

If F has property ((31.13.2)) then $F(\mathbb{R})$ is a local algebra with an algebra homomorphism $\pi = \pi_{\mathbb{R}} : F(\mathbb{R}) \to \mathbb{R}$ whose kernel is the maximal ideal.

Proof. By definition F restricts to a product preserving functor from the category of all $\mathbb{R}^n$'s and smooth mappings between them, thus it is a C^∞-algebra.

If F has property ((31.13.1)) then for all $a=(a_1,\dots,a_n)\in F(\mathbb{R})^n$ the evaluation mappings are given by

$$\varepsilon_a = \operatorname{ev}_a \circ F : C^\infty(\mathbb{R}^n,\mathbb{R}) \to C^\infty(F(\mathbb{R})^n, F(\mathbb{R})) \to F(\mathbb{R})$$

and thus are even smooth.

If F has property ((31.13.2)) then obviously $\pi_{\mathbb{R}} = \pi : F(\mathbb{R}) \to \mathbb{R}$ is an algebra homomorphism. It remains to show that the kernel of π is the largest ideal. So if $a\in A$ has $\pi(a)\neq 0\in\mathbb{R}$ then we have to show that a is invertible in A. Since the following diagram is a pullback,

$$\begin{array}{ccc} F(\mathbb{R}\setminus\{0\}) & \xrightarrow{F(i)} & F(\mathbb{R}) \\ {\scriptstyle\pi}\downarrow & & {\scriptstyle\pi}\downarrow \\ \mathbb{R}\setminus\{0\} & \xrightarrow{i} & \mathbb{R} \end{array}$$

we may assume that $a = F(i)(b)$ for a unique $b\in F(\mathbb{R}\setminus\{0\})$. But then $1/i : \mathbb{R}\setminus\{0\}\to\mathbb{R}$ is smooth, and $F(1/i)(b) = a^{-1}$, since $F(1/i)(b).a = F(1/i)(b).F(i)(b) = F(m)F(1/i,i)(b) = F(1)(b) = 1$, compare (31.14). □

31.16. Examples. Let A be an augmented local C^∞-algebra with maximal ideal N. Then A is quotient of a free C^∞-algebra $C^\infty_{\text{fin}}(\mathbb{R}^\Lambda)$ of smooth functions on some large product $\mathbb{R}^\Lambda$, which depend globally only on finitely many coordinates, see [Moerdijk, Reyes, 1991] or [Kainz, Kriegl, Michor, 1987]. So we have a short exact sequence

$$0\to I\to C^\infty_{\text{fin}}(\mathbb{R}^\Lambda)\xrightarrow{\varphi} A\to 0.$$

Then I is contained in the codimension 1 maximal ideal $\varphi^{-1}(N)$, which is easily seen to be $\{f\in C^\infty_{\text{fin}}(\mathbb{R}^\lambda) : f(x_0)=0\}$ for some $x_0\in\mathbb{R}^\Lambda$. Then clearly φ factors over the quotient of germs at x_0. If A has Hausdorff natural topology, then φ even factors over the Taylor expansion mapping, by the argument in [Kainz, Kriegl, Michor, 1987, 6.1], as follows. Let $f\in C^\infty_{\text{fin}}(\mathbb{R}^\Lambda)$ be infinitely flat at x_0. We shall show that f is in the closure of the set of all functions with germ 0 at x_0. Let $x_0=0$ without loss. Note first that f factors over some quotient $\mathbb{R}^\Lambda\to\mathbb{R}^N$, and we may replace $\mathbb{R}^\Lambda$ by $\mathbb{R}^N$ without loss. Define $g:\mathbb{R}^N\times\mathbb{R}^N\to\mathbb{R}^N$,

$$g(x,y)=\begin{cases} 0 & \text{if } |x|\le|y|,\\ (1-|y|/|x|)x & \text{if } |x|>|y|.\end{cases}$$

Since f is flat at 0, the mapping $y\mapsto (x\mapsto f_y(x) := f(g(x,y)))$ is a continuous mapping $\mathbb{R}^N\to C^\infty(\mathbb{R}^N,\mathbb{R})$ with the property that $f_0=f$ and f_y has germ 0 at 0 for all $y\neq 0$.

Thus the augmented local C^∞-algebras whose natural topology is Hausdorff are exactly the quotients of algebras of Taylor series at 0 of functions in $C^\infty_{\text{fin}}(\mathbb{R}^\Lambda)$.

It seems that local implies augmented: one has to show that a C^∞-algebra which is a field is 1-dimensional. Is this true?

31.17. Chart description of functors induced by C^∞-algebras. Let $A = \mathbb{R}\cdot 1 \oplus N$ be an augmented local C^∞-algebra which carries a compatible convenient structure, i.e. A is a convenient vector space and each mapping $A : C^\infty(\mathbb{R}^n, \mathbb{R}^m) \to C^\infty(A^n, A^m)$ is a well defined smooth mapping. As in the proof of (31.15) one sees that the natural topology on A is then finer than the given convenient one, thus is Hausdorff. Let us call this an *augmented local convenient C^∞-algebra.*

We want to associate to A a functor $T_A : \mathcal{M}f_{\text{fin}} \to \mathcal{M}f$ from the category $\mathcal{M}f_{\text{fin}}$ of all finite dimensional separable smooth manifolds to the category of smooth manifolds modeled on convenient vector spaces.

Step 1. Let $\pi = \pi_A : A \to A/N = \mathbb{R}$ be the augmentation mapping. This is a surjective homomorphism of C^∞-algebras, so the following diagram commutes for $f \in C^\infty(\mathbb{R}^n, \mathbb{R}^m)$:

$$\begin{array}{ccc} A^n & \xrightarrow{T_A f} & A^m \\ {\scriptstyle \pi^n}\downarrow & & \downarrow{\scriptstyle \pi^m} \\ \mathbb{R}^n & \xrightarrow{f} & \mathbb{R}^m \end{array}$$

If $U \subset \mathbb{R}^n$ is an open subset we put $T_A(U) := (\pi^n)^{-1}(U) = U \times N^n$, which is open subset in $T_A(\mathbb{R}^n) := A^n$.

Step 2. Now suppose that $f : \mathbb{R}^n \to \mathbb{R}^m$ vanishes on some open set $V \subset \mathbb{R}^n$. We claim that then $T_A f$ vanishes on the open set $T_A(V) = (\pi^n)^{-1}(V)$. To see this let $x \in V$, and choose a smooth function $g \in C^\infty(\mathbb{R}^n, \mathbb{R})$ with $g(x) = 1$ and support in V. Then $g.f = 0$ thus we have also $0 = A(g.f) = A(m) \circ A(g, f) = A(g).A(f)$, where the last multiplication is pointwise diagonal multiplication between A and A^m. For $a \in A^n$ with $(\pi^n)(a) = x$ we get $\pi(A(g)(a)) = g(\pi(a)) = g(x) = 1$, thus $A(g)(a)$ is invertible in the algebra A, and from $A(g)(a).A(f)(a) = 0$ we may conclude that $A(f)(a) = 0 \in A^m$.

Step 3. Now let $f : U \to W$ be a smooth mapping between open sets $U \subseteq \mathbb{R}^n$ and $W \subseteq \mathbb{R}^m$. Then we can define $T_A(f) : T_A(U) \to T_A(W)$ in the following way. For $x \in U$ let $f_x : \mathbb{R}^n \to \mathbb{R}^m$ be a smooth mapping which coincides with f in a neighborhood V of x in U. Then by step 2 the restriction of $A(f_x)$ to $T_A(V)$ does not depend on the choice of the extension f_x, and by a standard argument one can uniquely define a smooth mapping $T_A(f) : T_A(U) \to T_A(V)$. Clearly this gives us an extension of the functor A from the category of all $\mathbb{R}^n$'s and smooth mappings into convenient vector spaces to a functor from open subsets of $\mathbb{R}^n$'s and smooth mappings into the category of c^∞-open (indeed open) subsets of convenient vector spaces.

Step 4. Let M be a smooth finite dimensional manifold, let $(U_\alpha, u_\alpha : U_\alpha \to u_\alpha(U_\alpha) \subset \mathbb{R}^m)$ be a smooth atlas of M with chart changings $u_{\alpha\beta} := u_\alpha \circ u_\beta^{-1} : u_\beta(U_{\alpha\beta}) \to u_\alpha(U_{\alpha\beta})$. Then by step 3 we get smooth mappings between c^∞-open subsets of convenient vector spaces

$$\begin{array}{ccc} T_A(u_\beta(U_{\alpha\beta})) & \xrightarrow{T_A(u_{\alpha\beta})} & T_A(u_\alpha(U_{\alpha\beta})) \\ \pi\downarrow & & \downarrow\pi \\ u_\beta(U_{\alpha\beta}) & \xrightarrow{u_{\alpha\beta}} & u_\alpha(U_{\alpha\beta}) \end{array}$$

form again a cocycle of chart changings and we may use them to glue the c^∞-open sets $T_A(u_\alpha(U_\alpha)) = \pi_{\mathbb{R}^m}^{-1}(u_\alpha(U_\alpha)) \subset A^m$ in order to obtain a smooth manifold which we denote by $T_A M$. By the diagram above we see that $T_A M$ will be the total space of a fiber bundle $T(\pi_A, M) = \pi_{A,M} : T_A M \to M$, since the atlas $(T_A(U_\alpha), T_A(u_\alpha))$ constructed just now is already a fiber bundle atlas. So if M is Hausdorff then also $T_A M$ is Hausdorff, since two points x_i can be separated in one chart if they are in the same fiber, or they can be separated by inverse images under $\pi_{A,M}$ of open sets in M separating their projections.

This construction does not depend on the choice of the atlas. For two atlas have a common refinement and one may pass to this.

If $f \in C^\infty(M, M')$ for two manifolds M, M', we apply the functor T_A to the local representatives of f with respect to suitable atlas. This gives local representatives which fit together to form a smooth mapping $T_A f : T_A M \to T_A M'$. Clearly we again have $T_A(f \circ g) = T_A f \circ T_A g$ and $T_A(\mathrm{Id}_M) = \mathrm{Id}_{T_A M}$, so that $T_A : \mathcal{M}f_{\text{fin}} \to \mathcal{M}f$ is a covariant functor.

31.18. Theorem. Main properties. *Let $A = \mathbb{R} \cdot 1 \oplus N$ be a local augmented convenient C^∞-algebra. Then we have:*

(1) *The construction of (31.17) defines a covariant functor $T_A : \mathcal{M}f_{\text{fin}} \to \mathcal{M}f$ such that $\pi_A : T_A M \to M$ is a smooth fiber bundle with standard fiber N^m if $\dim M = m$. For any $f \in C^\infty(M, M')$ we have a commutative diagram*

$$\begin{array}{ccc} T_A M & \xrightarrow{T_A f} & T_A M' \\ \pi_{A,M}\downarrow & & \downarrow\pi_{A,M'} \\ M & \xrightarrow{f} & M'. \end{array}$$

Thus, (T_A, π_A) is a bundle functor on $\mathcal{M}f_{\text{fin}}$ whose fibers may be infinite dimensional. It gives a vector bundle functor on $\mathcal{M}f$ if and only if N is nilpotent of order 2.

(2) *The functor $T_A : \mathcal{M}f \to \mathcal{M}f$ is multiplicative: It respects products and preserves the same classes of smooth mappings as in (31.7.2): Embeddings of (splitting) submanifolds, surjective smooth mappings admitting local smooth sections, fiber bundle projections. For fixed manifolds M and M' the mapping $T_A : C^\infty(M, M') \to C^\infty(T_A M, T_A M')$ is smooth.*

(3) *Any bounded algebra homomorphism $\varphi : A \to B$ between augmented convenient C^∞-algebras induces a natural transformation $T(\varphi, \quad) = T_\varphi : T_A \to T_B$. If φ is split injective, then $T(\varphi, M) : T_A M \to T_B M$ is a split embedding*

for each manifold M. If φ is split surjective, then $T(\varphi, M)$ is a fiber bundle projection for each M. So we may view T as a co-covariant bifunctor from the category of augmented convenient C^∞-algebras algebras times $\mathcal{M}f_{\text{fin}}$ to $\mathcal{M}f$.

Proof. (1) is clear from (31.17). The fiber bundle $\pi_{A,M} : T_A M \to M$ is a vector bundle if and only if the transition functions $T_A(u_{\alpha\beta})$ are fiber linear $N \otimes E_\alpha \to N \otimes E_\beta$. So only the first derivatives of $u_{\alpha\beta}$ should act on N, so any product of two elements in N must be 0, thus N has to be nilpotent of order 2.

(2) The functor T_A respects finite products in the category of c^∞-open subsets of convenient vector spaces by (31.5), step 3 and 5. All the other assertions follow by looking again at the chart structure of $T_A M$ and by taking into account that f is part of $T_A f$ (as the base mapping).

(3) We define $T(\varphi, \mathbb{R}^n) := \varphi^n : A^n \to B^n$. By (31.17), step 3, this restricts to a natural transformation $T_A \to T_B$ on the category of open subsets of $\mathbb{R}^n$'s, and by gluing we may extend it to a functor on the category $\mathcal{M}f$. Obviously T is a co-covariant bifunctor on the indicated categories. Since $\pi_B \circ \varphi = \pi_A$ (φ respects the identity), we have $T(\pi_B, M) \circ T(\varphi, M) = T(\pi_A, M)$, so $T(\varphi, M) : T_A M \to T_B M$ is fiber respecting for each manifold M. In each fiber chart it is a linear mapping on the typical fiber $N_A^m \to N_B^m$.

So if φ is split injective, $T(\varphi, M)$ is fiberwise split injective and linear in each canonical fiber chart, so it is a splitting embedding.

If φ is split surjective, let $N_1 := \ker \varphi \subseteq N_A$, and let $V \subset N_A$ be a topological linear complement to N_1. Then for $m = \dim M$ and for the canonical charts we have the commutative diagram:

$$\begin{array}{ccc}
T_A M & \xrightarrow{\quad T(\varphi, M) \quad} & T_B M \\
\uparrow & & \uparrow \\
T_A(U_\alpha) & \xrightarrow{\quad T(\varphi, U_\alpha) \quad} & T_B(U_\alpha) \\
{\scriptstyle T_A(u_\alpha)}\downarrow & & \downarrow{\scriptstyle T_B(u_\alpha)} \\
u_\alpha(U_\alpha) \times N_A^m & \xrightarrow{\quad \mathrm{Id} \times \varphi | N_A^m \quad} & u_\alpha(U_\alpha) \times N_B^m \\
\| & & \| \\
u_\alpha(U_\alpha) \times N_1^m \times V^m & \xrightarrow{\quad \mathrm{Id} \times 0 \times \mathrm{iso} \quad} & u_\alpha(U_\alpha) \times 0 \times N_B^m
\end{array}$$

So $T(\varphi, M)$ is a fiber bundle projection with standard fiber $E_\alpha \otimes \ker \varphi$. □

31.19. Theorem. *Let A and B be augmented convenient C^∞-algebras. Then we have:*

1. *We get the convenient C^∞-algebra A back from the functor T_A by restricting to the subcategory of $\mathbb{R}^n$'s.*
2. *The natural transformations $T_A \to T_B$ correspond exactly to the bounded C^∞-algebra homomorphisms $A \to B$.*

Proof. (1) is obvious. (2) For a natural transformation $\varphi : T_A \to T_B$ (which is smooth) its value $\varphi_{\mathbb{R}} : T_A(\mathbb{R}) = A \to T_B(\mathbb{R}) = B$ is a C^∞-algebra homomorphism which is smooth and thus bounded. The inverse of this mapping is already described in theorem (31.18.3). □

31.20. Proposition. *Let $A = \mathbb{R} \cdot 1 \oplus N$ be a local augmented convenient C^∞-algebra and let M be a smooth finite dimensional manifold.*

Then there exists a bijection

$$\varepsilon : T_A(M) \to \operatorname{Hom}(C^\infty(M, \mathbb{R}), A)$$

onto the space of bounded algebra homomorphisms, which is natural in A and M. Via ε the expression $\operatorname{Hom}(C^\infty(\quad, \mathbb{R}), A)$ *describes the functor T_A in a coordinate free manner.*

Proof. Step 1. Let $M = \mathbb{R}^n$, so $T_A(\mathbb{R}^n) = A^n$. Then for $a = (a_1, \dots, a_n) \in A^n$ we have $\varepsilon(a)(f) = A(f)(a_1, \dots, a_n)$, which gives a bounded algebra homomorphism $C^\infty(\mathbb{R}^n, \mathbb{R}) \to A$. Conversely, for $\varphi \in \operatorname{Hom}(C^\infty(\mathbb{R}^n, \mathbb{R}), A)$ consider $a = (\varphi(\mathrm{pr}_1), \dots, \varphi(\mathrm{pr}_n)) \in A^n$. Since polynomials are dense in $C^\infty(\mathbb{R}^n, \mathbb{R})$, φ is bounded, and A is Hausdorff, φ is uniquely determined by its values on the coordinate functions pr_i (compare [Kainz, Kriegl, Michor, 1987, 2.4.(3)], thus $\varphi(f) = A(f)(a)$ and ε is bijective. Obviously ε is natural in A and $\mathbb{R}^n$.

Step 2. Now let $i : U \subset \mathbb{R}^n$ be an embedding of an open subset. Then the image of the mapping

$$\operatorname{Hom}(C^\infty(U, \mathbb{R}), A) \xrightarrow{(i^*)^*} \operatorname{Hom}(C^\infty(\mathbb{R}^n, \mathbb{R}), A) \xrightarrow{\varepsilon^{-1}_{\mathbb{R}^n, A}} A^n$$

is the set $\pi^{-1}_{A,\mathbb{R}^n}(U) = T_A(U) \subset A^n$, and $(i^*)^*$ is injective.

To see this let $\varphi \in \operatorname{Hom}(C^\infty(U, \mathbb{R}), A)$. Then $\varphi^{-1}(N)$ is the maximal ideal in $C^\infty(U, \mathbb{R})$ consisting of all smooth functions vanishing at a point $x \in U$, and $x = \pi(\varepsilon^{-1}(\varphi \circ i^*)) = \pi(\varphi(\mathrm{pr}_1 \circ i), \dots, \varphi(\mathrm{pr}_n \circ i))$, so that $\varepsilon^{-1}((i^*)^*(\varphi)) \in T_A(U) = \pi^{-1}(U) \subset A^n$.

Conversely for $a \in T_A(U)$ the homomorphism $\varepsilon_a : C^\infty(\mathbb{R}^n, \mathbb{R}) \to A$ factors over $i^* : C^\infty(\mathbb{R}^n, \mathbb{R}) \to C^\infty(U, \mathbb{R})$, by steps 2 and 3 of (31.17).

Step 3. The two functors $\operatorname{Hom}(C^\infty(\quad, \mathbb{R}), A)$ and $T_A : \mathcal{M}f \to \mathcal{S}et$ coincide on all open subsets of $\mathbb{R}^n$'s, so they have to coincide on all manifolds, since smooth manifolds are exactly the retracts of open subsets of $\mathbb{R}^n$'s see e.g. [Federer, 1969] or [Kolář, Michor, Slovák, 1993, 1.14.1]. Alternatively one may check that the gluing process described in (31.17), step 4, works also for the functor $\operatorname{Hom}(C^\infty(\quad, \mathbb{R}), A)$ and gives a unique manifold structure on it, which is compatible to $T_A M$. □

Chapter VII
Calculus on Infinite Dimensional Manifolds

In chapter VI we have found that some of the classically equivalent definitions of tangent vectors differ in infinite dimensions, and accordingly we have different kinds of tangent bundles and vector fields. Since this is the central topic of any treatment of calculus on manifolds we investigate in detail Lie brackets for all these notions of vector fields. Only kinematic vector fields can have local flows, and we show that the latter are unique if they exist (32.16). Note also theorem (32.18) that any bracket expression of length k of kinematic vector fields is given as the k-th derivative of the corresponding commutator expression of the flows, which is not well known even in finite dimensions.

We also have different kinds of differential forms, which we treat in a systematic way, and we investigate how far the usual natural operations of differential forms generalize. In the end (33.21) the most common type of kinematic differential forms turns out to be the right ones for calculus on manifolds; for them the theorem of De Rham is proved.

We also include a version of the Frölicher-Nijenhuis bracket in infinite dimensions. The Frölicher-Nijenhuis bracket is a natural extension of the Lie bracket for vector fields to a natural graded Lie bracket for tangent bundle valued differential forms (later called vector valued). Every treatment of curvature later in (37.3) and (37.20) is initially based on the Frölicher-Nijenhuis bracket.

32. Vector Fields

32.1. Vector fields. Let M be a smooth manifold. A *kinematic vector field* X on M is just a smooth section of the kinematic tangent bundle $TM \to M$. The space of all kinematic vector fields will be denoted by $\mathfrak{X}(M) = C^\infty(M \leftarrow TM)$.

By an *operational vector field* X on M we mean a bounded derivation of the sheaf $C^\infty(\quad, \mathbb{R})$, i.e. for the open $U \subset M$ we are given bounded derivations $X_U : C^\infty(U, \mathbb{R}) \to C^\infty(U, \mathbb{R})$ commuting with the restriction mappings.

We shall denote by $\operatorname{Der}(C^\infty(M,\mathbb{R}))$ the space of all operational vector fields on M. We shall equip $\operatorname{Der}(C^\infty(M,\mathbb{R}))$ with the convenient vector space structure induced by the closed linear embedding

$$\operatorname{Der}(C^\infty(M,\mathbb{R})) \hookrightarrow \prod_U L(C^\infty(U,\mathbb{R}), C^\infty(U,\mathbb{R})).$$

Convention. In (32.4) below we will show that for a smoothly regular manifold the space of derivations on the algebra $C^\infty(M,\mathbb{R})$ of globally defined smooth functions coincides with the derivations of the sheaf. Thus we shall follow the convention, that either the manifolds in question are smoothly regular, or that (as defined above) Der means the space of derivations of the corresponding sheaf also denoted by $C^\infty(M,\mathbb{R})$.

32.2. Lemma. *On any manifold M the operational vector fields correspond exactly to the smooth sections of the operational tangent bundle. Moreover we have an isomorphism of convenient vector spaces* $\operatorname{Der}(C^\infty(M,\mathbb{R})) \cong C^\infty(M \leftarrow DM)$.

Proof. Every smooth section $X \in C^\infty(M \leftarrow DM)$ defines an operational vector field by $\partial_U(f)(x) := X(x)(\operatorname{germ}_x f) = pr_2(Df(X(x)))$ for $f \in C^\infty(U,\mathbb{R})$ and $x \in U$. We have that $\partial_U(f) = \operatorname{pr}_2 \circ Df \circ X = df \circ X \in C^\infty(U,\mathbb{R})$ by (28.15). Then ∂_U is obviously a derivation, since $df(X_x) = X_x(f)$ by (28.15). The linear mapping $\partial_U : C^\infty(U,\mathbb{R}) \to C^\infty(U,\mathbb{R})$ is bounded if and only if $\operatorname{ev}_x \circ \partial_U : C^\infty(U,\mathbb{R}) \to \mathbb{R}$ is bounded, by the smooth uniform boundedness principle (5.26), and this is true by (28.15), since $(\operatorname{ev}_x \circ X)(f) = df(X_x)$.

Moreover, the mapping

$$C^\infty(M \leftarrow DM) \to \operatorname{Der}(C^\infty(M,\mathbb{R})) \hookrightarrow \prod_U L(C^\infty(U,\mathbb{R}), C^\infty(U,\mathbb{R}))$$

given by $X \mapsto (\partial_U)_U$ is linear and bounded, since by the uniform boundedness principle (5.26) this is equivalent to the boundedness of $X \mapsto \partial_U(f)(x) = df(X_x)$ for all open $U \subseteq M$, $f \in C^\infty(U,\mathbb{R})$ and $x \in X$.

Now let conversely ∂ be an operational vector field on M. Then the family $\operatorname{ev}_x \circ \partial_U : C^\infty(U,\mathbb{R}) \to \mathbb{R}$, where U runs through all open neighborhoods of x, defines a unique bounded derivation $X_x : C^\infty(M \supseteq \{x\},\mathbb{R}) \to \mathbb{R}$, i.e. an element of D_xM. We have to show that $x \mapsto X_x$ is smooth, which is a local question, so we assume that M is a c^∞-open subset of a convenient vector space E. The mapping

$$M \xrightarrow{X} DM \cong M \times D_0E \subseteq M \times \prod_U L(C^\infty(U,\mathbb{R}),\mathbb{R})$$

is smooth if and only if for every neighborhood U of 0 in E the component $M \to L(C^\infty(U,\mathbb{R}),\mathbb{R})$, given by $\partial \mapsto X_x(f(\quad - x)) = \partial_{U_x}(f(\quad - x))(x)$ is smooth, where $U_x := U + x$. By the smooth uniform boundedness principle (5.18) this is the case if and only if its composition with ev_f is smooth for all $f \in C^\infty(U,\mathbb{R})$. If $t \mapsto x(t)$

is a smooth curve in $M \subseteq E$, then there is a $\delta > 0$ and an open neighborhood W of $x(0)$ in M such that $W \subseteq U + x(t)$ for all $|t| < \delta$ and hence $X_{x(t)}(f(\quad - x(t))) = \partial_W(f(\quad - x(t)))(x(t))$, which is by the exponential law smooth in t.

Moreover, the mapping $\mathrm{Der}(C^\infty(M,\mathbb{R})) \to C^\infty(M \leftarrow DM)$ given by $\partial \mapsto X$ is linear and bounded, since by the uniform boundedness principle in proposition (30.1) this is equivalent to the boundedness of $\partial \mapsto X_x \in D_xM \hookrightarrow \prod_U C^\infty(U,\mathbb{R})'$ for all $x \in M$, i.e. to that of $\partial \mapsto X_x(f) = \partial_U(f)(x)$ for all open neighborhoods U of x and $f \in C^\infty(U,\mathbb{R})$, which is obviously true. □

32.3. Lemma. *There is a natural embedding of convenient vector spaces*

$$\mathfrak{X}(M) = C^\infty(M \leftarrow TM) \to C^\infty(M \leftarrow DM) \cong \mathrm{Der}(C^\infty(M,\mathbb{R})).$$

Proof. Since TM is a closed subbundle of DM this is obviously true. □

32.4. Lemma. *Let M be a smoothly regular manifold.*

Then each bounded derivation $X : C^\infty(M,\mathbb{R}) \to C^\infty(M,\mathbb{R})$ is already an operational vector field. Moreover, we have an isomorphism

$$C^\infty(M \leftarrow DM) \cong \mathrm{Der}(C^\infty(M,\mathbb{R}), C^\infty(M,\mathbb{R}))$$

of convenient vector spaces.

Proof. Let ∂ be a bounded derivation of the algebra $C^\infty(M,\mathbb{R})$. If $f \in C^\infty(M,\mathbb{R})$ vanishes on an open subset $U \subset M$ then also $\partial(f)$: For $x \in U$ we take a bump function $g_{x,U} \in C^\infty(M,\mathbb{R})$ at x, i.e. $g_{x,U} = 1$ near x and $\mathrm{supp}(g_{x,U}) \subset U$. Then $\partial(f) = \partial((1 - g_{x,U})f) = \partial(1 - g_{x,U})f + (1 - g_{x,U})\partial(f)$, and both summands are zero near x. So $\partial(f)|U = 0$.

Now let $f \in C^\infty(U,\mathbb{R})$ for a c^∞-open subset U of M. We have to show that we can define $\partial_U(f) \in C^\infty(U,\mathbb{R})$ in a unique manner. For $x \in U$ let $g_{x,U} \in C^\infty(M,\mathbb{R})$ be a bump function as before. Then $g_{x,U}f \in C^\infty(M,\mathbb{R})$, and $\partial(g_{x,U}f)$ makes sense. By the argument above, $\partial(gf)$ near x is independent of the choice of g. So let $\partial_U(f)(x) := \partial(g_{x,U}f)(x)$. It has all the required properties since the topology on $C^\infty(U,\mathbb{R})$ is initial with respect to all mappings $f \mapsto g_{x,U}f$ for $x \in U$.

This mapping $\partial \mapsto \partial_U$ is bounded, since by the uniform boundedness principles (5.18) and (5.26) this is equivalent with the boundedness of $\partial \mapsto \partial_U(f)(x) := \partial(g_{x,U}f)(x)$ for all $f \in C^\infty(U,\mathbb{R})$ and all $x \in U$ □

32.5. The operational Lie bracket. Recall that operational vector fields are the bounded derivations of the sheaf $C^\infty(\quad,\mathbb{R})$, see (32.1). This is a convenient vector space by (32.2) and (30.1).

If X, Y are two operational vector fields on M, then the mapping $f \mapsto X(Y(f)) - Y(X(f))$ is also a bounded derivation of the sheaf $C^\infty(\quad,\mathbb{R})$, as a simple computation shows. We denote it by $[X,Y] \in \mathrm{Der}(C^\infty(\quad,\mathbb{R})) \cong C^\infty(M \leftarrow DM)$.

The $\mathbb{R}$-bilinear mapping

$$[\ ,\ ]: C^\infty(M \leftarrow DM) \times C^\infty(M \leftarrow DM) \to C^\infty(M \leftarrow DM)$$

is called the *Lie bracket.* Note also that $C^\infty(M \leftarrow DM)$ is a module over the algebra $C^\infty(M,\mathbb{R})$ by pointwise multiplication $(f,X) \mapsto fX$, which is bounded.

Theorem. *The Lie bracket* $[\ ,\ ] : C^\infty(M \leftarrow DM) \times C^\infty(M \leftarrow DM) \to C^\infty(M \leftarrow DM)$ *has the following properties:*

$$\begin{aligned}
&[X,Y] = -[Y,X],\\
&[X,[Y,Z]] = [[X,Y],Z] + [Y,[X,Z]], \quad \textit{the Jacobi identity,}\\
&[fX,Y] = f[X,Y] - (Yf)X,\\
&[X,fY] = f[X,Y] + (Xf)Y.
\end{aligned}$$

The form of the Jacobi identity we have chosen says that $ad(X) = [X,\ \ ]$ is a derivation for the Lie algebra $(C^\infty(M \leftarrow DM), [\ ,\ ])$.

Proof. All these properties can be checked easily for the commutator $[X,Y] = X \circ Y - Y \circ X$ in the space of bounded derivations of the algebra $C^\infty(U,\mathbb{R})$. □

32.6. Lemma. *Let* $b : E_1 \times \ldots \times E_k \to \mathbb{R}$ *be a bounded multilinear mapping on a product of convenient vector spaces. Let* $f \in C^\infty(E,\mathbb{R})$, *let* $f_i : E \to E_i$ *be smooth mappings, and let* $X_x \in E'' = D^{(1)}_x E$.

Then we have

$$\begin{aligned}
X_x(f) &= \langle X_x, df(x)\rangle_{E'} = df(x)^{**}.X_x\\
X_x(b\circ(f_1,\ldots,f_k)) &= d(b\circ(f_1,\ldots,f_k))(x)^{**}.X_x\\
&= \sum_{1\le i\le k} b(f_1(x),\ldots,f_{i-1}(x),\quad,f_{i+1}(x),\ldots,f_k(x))^{**}.df_i(x)^{**}.X_x\\
&= \sum_{1\le i\le k} \langle df_i(x)^{**}.X_x, b(f_1(x),\ldots,f_{i-1}(x),\quad,f_{i+1}(x),\ldots,f_k(x))\rangle_{E_i'}.
\end{aligned}$$

If $B : E_1 \times \ldots \times E_k \to F$ *is a vector valued bounded multilinear mapping, and if* $g : E \to F$ *is a smooth mapping, then we have*

$$\begin{aligned}
&D^{(1)}_x g.X_x = dg(x)^{**}.X_x \in F''\\
&D^{(1)}_x (B\circ(f_1,\ldots,f_k)).X_x =\\
&\quad = \sum_{1\le i\le k} B(\ldots,f_{i-1}(x),\quad,f_{i+1}(x),\ldots)^{**}.df_i(x)^{**}.X_x \in D^{(1)}_{B(f_1(x),\ldots,f_k(x))}F.
\end{aligned}$$

Here $\langle\ ,\ \rangle_H : H' \times H \to \mathbb{R}$ is the duality pairing for any convenient vector space H. We will further denote by $\iota_H : H \to H''$ the canonical embedding into the bidual space.

Proof. The first equation is immediate.

We have

$$\begin{aligned}
d(b\circ(f_1,\dots,f_k))(x) &= \sum_{j=1}^{k} b(f_1(x),\dots,f_{j-1}(x),\quad,f_{j+1}(x),\dots,f_k(x))\circ df_j(x)\\
&= \sum_{j=1}^{k} df_j(x)^*\Big(b(f_1(x),\dots,f_{j-1}(x),\quad,f_{j+1}(x),\dots,f_k(x))\Big).
\end{aligned}$$

Thus for $X_x \in D_x^{(1)}E$ we have

$$\begin{aligned}
X_x(b\circ(f_1,\dots,f_k)) &= X_x\Big(d(b\circ(f_1,\dots,f_k))(x)\Big)\\
&= X_x\Big(\sum_{j=1}^{k} df_j(x)^*\Big(b(f_1(x),\dots,f_{j-1}(x),\quad,f_{j+1}(x),\dots,f_k(x))\Big)\Big)\\
&= \sum_{j=1}^{k} X_x\Big(df_j(x)^*\Big(b(f_1(x),\dots,f_{j-1}(x),\quad,f_{j+1}(x),\dots,f_k(x))\Big)\Big)\\
&= \sum_{j=1}^{k} df_j(x)^{**}(X_x)\Big(b(f_1(x),\dots,f_{j-1}(x),\quad,f_{j+1}(x),\dots,f_k(x))\Big).
\end{aligned}$$

For the second assertion we choose a test germ

$$h \in C^\infty(F \supseteq \{B(f_1(x),\dots,f_k(x))\},\mathbb{R})$$

and proceed as follows:

$$\begin{aligned}
(D_x^{(1)}g.X_x)(h) &= X_x(h\circ g) = \langle X_x, d(h\circ g)(x)\rangle_{E'}\\
&= \langle X_x, dh(g(x))\circ dg(x)\rangle_{E'} = \langle X_x, dg(x)^*.dh(g(x))\rangle_{E'}\\
&= \langle dg(x)^{**}.X_x, dh(g(x))\rangle_{E'} = (dg(x)^{**}.X_x)(h).\\
(D_x^{(1)}(B\circ(f_1,\dots,f_k))X_x)(h) &= X_x(h\circ B\circ(f_1,\dots,f_k))\\
&= d(h\circ B\circ(f_1,\dots,f_k))(x)^{**}.X_x\\
&= \Big(dh(B(f_1(x),\dots))\circ\sum_{i=1}^{k} B(\dots,f_{i-1}(x),\quad,f_{i+1}(x),\dots)\circ df_i(x)\Big)^{**}.X_x\\
&= dh(B(f_1(x),\dots))^{**}.\sum_{i=1}^{k} B(\dots,f_{i-1}(x),\quad,f_{i+1}(x),\dots)^{**}.df_i(x)^{**}.X_x\\
&= \Big(\sum_{i=1}^{k} B(\dots,f_{i-1}(x),\quad,f_{i+1}(x),\dots)^{**}.df_i(x)^{**}.X_x\Big)\Big|_{B(f_1(x),\dots)}(h). \qquad\square
\end{aligned}$$

32.7. The Lie bracket of operational vector fields of order 1. One could hope that the Lie bracket restricts to a Lie bracket on $C^\infty(D^{(1)}M)$. But this is not the case. We will see that for a c^∞-open set U in a convenient vector space E and for $X, Y \in C^\infty(U, E'')$ the bracket $[X, Y]$ has also components of order 2, in general.

For a bounded linear mapping $\ell : F \to G'$ the *transposed mapping* $\ell^t : G \to F'$ is given by $\ell^t := \ell^* \circ \iota_G$, where $\iota_G : G \to G''$ is the canonical embedding into the bidual. If $\langle\ \ ,\ \ \rangle_H : H' \times H \to \mathbb{R}$ is the duality pairing, then this may also be described by $\langle \ell(x), y\rangle_G = \langle \ell^t(y), x\rangle_F$.

For $X, Y \in C^\infty(U, E'')$, for $f \in C^\infty(U, E)$ and for $x \in U$ we get:

$$\begin{aligned}
X(f)(x) &= X_x(f) = X_x(df(x))\\
X(f) &= \operatorname{ev} \circ (X, df)\\
Y(X(f))(x) &= Y_x(X(f)) = Y_x(\operatorname{ev} \circ (X, df))\\
&= Y_x\Big(dX(x)^*\Big(\operatorname{ev}(\ \ , df(x))\Big) + d(df)(x)^*\Big(\operatorname{ev}(X(x),\ \ )\Big)\Big)\\
&= Y_x\Big(dX(x)^*\Big(\iota(df(x))\Big) + d(df)(x)^*(X_x)\Big)\\
&= Y_x\Big(\iota(df(x)) \circ dX(x) + X_x \circ d(df)(x)\Big)\\
&= Y_x\Big(dX(x)^t\Big(df(x)\Big) + X_x \circ d(df)(x)\Big)\\
&= \Big(Y_x \circ dX(x)^t\Big)\Big(df(x)\Big) + Y_x\Big(X_x \circ d(df)(x)\Big).
\end{aligned}$$

Here we used the equation:

$$\iota(y) \circ T = T^t(y) \text{ for } y \in F,\ T \in L(E, F'),$$

which is true since

$$\Big(\iota(y) \circ T\Big)(x) = \iota(y)(T(x)) = T(x)(y) = T^t(y)(x).$$

Note that for the symmetric bilinear form $b := d(df)(x)^\wedge : E \times E \to \mathbb{R}$ a canonical extension to a bilinear form $\tilde{b}$ on E'' is given by

$$\tilde{b}(X_x, Y_x) := X_x(Y_x \circ b^\vee)$$

However, this extension is not symmetric as the following remark shows: Let $b := \operatorname{ev} : E' \times E \to \mathbb{R}$. Then $\tilde{b} : E''' \times E'' \to \mathbb{R}$ is given by

$$\tilde{b}(X, Y) := X(Y \circ b^\vee) = X(Y \circ \operatorname{Id}) = X(Y) = \iota_{E''}(Y)(X)$$

For $b := \operatorname{ev} \circ \operatorname{flip} : E \times E' \to \mathbb{R}$ we have that $\tilde{b} : E'' \times E''' \to \mathbb{R}$ is given by

$$\tilde{b}(Y, X) := Y(X \circ b^\vee) = Y(X \circ \iota_E) = Y(\iota_E^*(X)) = (Y \circ \iota_E^*)(X) = (\iota_E)^{**}(Y)(X).$$

Thus, $\tilde{b}$ is not symmetric in general, since $\ker(\iota_E^{**} - \iota_{E''}) = \iota_E(E)$, at least for Banach spaces, see [Cigler, Losert, Michor, 1979, 1.15], applied to ι_E.

Lemma. *For $X \in C^\infty(TM)$ and $Y \in C^\infty(D^{(1)}M)$ we have $[X,Y] \in C^\infty(D^{(1)}M)$, and the bracket is given by the following local formula for $M = U$, a c^∞-open subset in a convenient vector space E:*

$$[X,Y](x) = Y(x) \circ dX(x)^* - dY(x).X(x) \in E''.$$

Proof. From the computation above we get:

$$\begin{aligned}
Y(X(f))(x) &= \langle (d(\iota_E \circ X)(x)^t)^*.Y(x), df(x)\rangle_{E'} + \langle d(df)(x)^{**}.Y(x), \iota_E.X(x)\rangle_{E''}\\
&= \langle Y(x), (\iota_E \circ dX(x))^t.df(x)\rangle_{E'} + \langle Y(x), d(df)(x)^*.\iota_E.X(x)\rangle_{E'}\\
&= \langle Y(x), dX(x)^*.df(x)\rangle_{E'} + \langle Y(x), d(df)(x)^t.X(x)\rangle_{E'}\\
&= \langle Y(x) \circ dX(x)^*, df(x)\rangle_{E'} + \langle Y(x), d(df)(x)^t.X(x)\rangle_{E'}\\
X(Y(f))(x) &= \langle (dY(x)^t)^*.\iota_E.X(x), df(x)\rangle_{E'} + \langle d(df)(x)^{**}.\iota_E.X(x), Y(x)\rangle_{E''}\\
&= \langle \iota_E.X(x), dY(x)^t.df(x)\rangle_{E'} + \langle \iota_E.X(x), d(df)(x)^*.Y(x)\rangle_{E'}\\
&= \langle dY(x)^t.df(x), X(x)\rangle_E + \langle d(df)(x)^*.Y(x), X(x)\rangle_E\\
&= \langle dY(x).X(x), df(x)\rangle_{E'} + \langle Y(x), d(df)(x).X(x)\rangle_{E'}
\end{aligned}$$

Since $d(df)(x) : E \to E'$ is symmetric in the sense that $d(df)(x)^t = d(df)(x)$, the result follows. □

32.8. Theorem. *The Lie bracket restricts to the following mappings between splitting subspaces*

$$[\quad,\quad] : C^\infty(M \leftarrow D^{(k)}M) \times C^\infty(M \leftarrow D^{(\ell)}M) \to C^\infty(M \leftarrow D^{(k+\ell)}M).$$

The spaces $\mathfrak{X}(M) = C^\infty(M \leftarrow TM)$ and $C^\infty(D^{[1,\infty)}M) := \bigcup_{1\le i<\infty} C^\infty(M \leftarrow D^{(i)}M)$ are sub Lie algebras of $C^\infty(M \leftarrow DM)$.

If $X \in \mathfrak{X}(M)$ is a kinematic vector field, then $[X,\quad]$ maps $C^\infty(M \leftarrow D^{(\ell)}M)$ into itself.

This suggests to introduce the notation $D^{(0)} := T$, but here it does not indicate the order of differentiation present in the tangent vector.

Proof. All assertions can be checked locally, so we may assume that $M = U$ is open in a convenient vector space E.

We prove first that the kinematic vector fields form a Lie subalgebra. For X, $Y \in C^\infty(U, E)$ we have then for the vector field $\partial_X|_x(f) = df(x)(X(x))$, compare the notation set up in (28.2)

$$\begin{aligned}
[\partial_X, \partial_Y](f) &= \partial_X(\partial_Y(f)) - \partial_Y(\partial_X(f))\\
&= d(df.Y).X - d(df.X).Y\\
&= d^2f.(X,Y) + df.(dY.X) - d^2f.(Y,X) - df.(dX.Y)\\
&= \partial_{dY.X - dX.Y} f.
\end{aligned}$$

Let $\partial_X \in C^\infty(U \leftarrow D^{(k)}U)$ for $X = \sum_{i=1}^k X^{[i]}$, where $X^{[i]} \in C^\infty(U, L^i_{\text{sym}}(E;\mathbb{R})')$ vanishes on decomposable forms. Similarly, let $\partial_Y \in C^\infty(U \leftarrow D^{(\ell)}U)$, and suppose that $f : (U, x) \to \mathbb{R}$ is a $(k+\ell)$-flat germ at x. Since $\partial_Y(f)(y) = \sum_{i=1}^\ell Y^{[i]}(y)(\frac{1}{i!}d^i f(y))$ the germ $\partial_Y(f)$ is still k-flat at x, so $\partial_X(\partial_Y(f))(x) = 0$. Thus, $[\partial_X, \partial_Y](f)(x) = \partial_X(\partial_Y(f))(x) - \partial_Y(\partial_X(f))(x) = 0$, and we conclude that $[\partial_X, \partial_Y] \in C^\infty(U \leftarrow D^{(k+\ell)}U)$.

Now we suppose that $X \in C^\infty(U, E)$ and $Y \in C^\infty(U, L^\ell_{\text{sym}}(E;\mathbb{R})')$. Let $f : (U, x) \to \mathbb{R}$ be an ℓ-flat germ at x. Then we have

$$\begin{aligned}
\partial_Y(\partial_X(f))(x) &= Y(x)\left(\tfrac{1}{\ell!}d^\ell\langle df, X\rangle_E(x)\right)\\
&= Y(x)\left(\tfrac{1}{\ell!}\sum_{k=0}^{\ell}\binom{\ell}{k}\operatorname{sym}\langle d^k(df)(x), d^{\ell-k}X(x)\rangle_E\right)\\
&= Y(x)\left(\tfrac{1}{\ell!}\langle d^\ell(df)(x), X(x)\rangle_E\right) + 0\\
&= \langle Y(x), \tfrac{1}{\ell!}d^{1+\ell}f(x)(\quad, X(x))\rangle_{L^\ell_{\text{sym}}(E;\mathbb{R})}\\
\partial_X(\partial_Y(f))(x) &= \partial_{X(x)}\langle Y, \tfrac{1}{\ell!}d^\ell f\rangle_{L^\ell_{\text{sym}}(E;\mathbb{R})}\\
&= d\langle Y, \tfrac{1}{\ell!}d^\ell f\rangle_{L^\ell_{\text{sym}}(E;\mathbb{R})}(x).X(x)\\
&= \langle dY(x).X(x), \tfrac{1}{\ell!}d^\ell f(x)\rangle_{L^\ell_{\text{sym}}(E;\mathbb{R})} + \langle Y(x), \tfrac{1}{\ell!}d(d^\ell f)(x).X(x)\rangle_{L^\ell_{\text{sym}}(E;\mathbb{R})}\\
&= 0 + \langle Y(x), \tfrac{1}{\ell!}d^{\ell+1}f(x)(X(x), \quad)\rangle_{L^\ell_{\text{sym}}(E;\mathbb{R})}
\end{aligned}$$

So $[\partial_X, \partial_Y](f)(x) = 0$. $\square$

Remark. In the notation of (28.2) we have shown that on a convenient vector space we have

$$[\quad,\quad] : C^\infty(E \leftarrow D^{[k]}E) \times C^\infty(E \leftarrow D^{[\ell]}E) \to \sum_{i=\min(k,\ell)}^{k+\ell} C^\infty(E \leftarrow D^{[i]}E).$$

Thus, the space $C^\infty(E \leftarrow D^{[k,\infty)}E) := \sum_{k\le i<\infty} C^\infty(E \leftarrow D^{[i]}E)$ for $k \ge 1$ is a sub Lie algebra. The (possibly larger) space $C^\infty(D^{[k,\infty]}E)$ of all operational tangent fields which vanish on all polynomials of degree less than k is obviously a sub Lie algebra. But beware, none of these spaces of vector fields is invariant under the action of diffeomorphisms.

32.9. f-related vector fields. Let D^α be one of the following functors D, $D^{(k)}$, T. If $f : M \to M$ is a diffeomorphism, then for any vector field $X \in C^\infty(M \leftarrow D^\alpha M)$ the mapping $D^\alpha f^{-1} \circ X \circ f$ is also a vector field, which we will denote by f^*X. Analogously, we put $f_*X := D^\alpha f \circ X \circ f^{-1} = (f^{-1})^*X$.

But if $f : M \to N$ is a smooth mapping and $Y \in C^\infty(N \leftarrow D^\alpha N)$ is a vector field there may or may not exist a vector field $X \in C^\infty(M \leftarrow D^\alpha M)$ such that the following diagram commutes:

(1)
$$\begin{array}{ccc}
D^\alpha M & \xrightarrow{D^\alpha f} & D^\alpha N\\
{\scriptstyle X}\uparrow & & \uparrow{\scriptstyle Y}\\
M & \xrightarrow{f} & N.
\end{array}$$

Definition. Let $f : M \to N$ be a smooth mapping. Two vector fields $X \in C^\infty(M \leftarrow D^\alpha M)$ and $Y \in C^\infty(N \leftarrow D^\alpha N)$ are called *f-related*, if $D^\alpha f \circ X = Y \circ f$ holds, i.e. if diagram (1) commutes.

32.10. Lemma. *Let $X_i \in C^\infty(M \leftarrow DM)$ and $Y_i \in C^\infty(N \leftarrow DN)$ be vector fields for $i = 1, 2$, and let $f : M \to N$ be smooth. If X_i and Y_i are f-related for $i = 1, 2$, then also $\lambda_1 X_1 + \lambda_2 X_2$ and $\lambda_1 Y_1 + \lambda_2 Y_2$ are f-related, and also $[X_1, X_2]$ and $[Y_1, Y_2]$ are f-related.*

Proof. The first assertion is immediate. To prove the second we choose $h \in C^\infty(N, \mathbb{R})$, and we view each vector field as operational. Then by assumption we have $Df \circ X_i = Y_i \circ f$, thus:

$$\begin{aligned}(X_i(h \circ f))(x) &= X_i(x)(h \circ f) = (D_x f.X_i(x))(h) = \\ &= (Df \circ X_i)(x)(h) = (Y_i \circ f)(x)(h) = Y_i(f(x))(h) = (Y_i(h))(f(x)),\end{aligned}$$

so $X_i(h \circ f) = (Y_i(h)) \circ f$, and we may continue:

$$\begin{aligned}[X_1, X_2](h \circ f) &= X_1(X_2(h \circ f)) - X_2(X_1(h \circ f)) = \\ &= X_1(Y_2(h) \circ f) - X_2(Y_1(h) \circ f) = \\ &= Y_1(Y_2(h)) \circ f - Y_2(Y_1(h)) \circ f = [Y_1, Y_2](h) \circ f.\end{aligned}$$

But this means $Df \circ [X_1, X_2] = [Y_1, Y_2] \circ f$. □

32.11. Corollary. *Let D^α be one of the following functors D, $D^{(k)}$, T. Let $f : M \to N$ be a local diffeomorphism so that $(T_x f)^{-1}$ makes sense for each $x \in M$.*

Then for $Y \in C^\infty(N \leftarrow D^\alpha N)$ a vector field $f^*Y \in C^\infty(M \leftarrow D^\alpha M)$ is defined by $(f^*Y)(x) = (T_x f)^{-1}.Y(f(x))$, and the linear mapping $f^* : C^\infty(N \leftarrow D^\beta N) \to C^\infty(M \leftarrow D^\beta M)$ is a Lie algebra homomorphism, i.e. $f^*[Y_1, Y_2] = [f^*Y_1, f^*Y_2]$, where D^β is one of D, T, $D^{[1,\infty)}$.

32.12. Integral curves. Let $c : J \to M$ be a smooth curve in a manifold M defined on an interval J. It will be called an *integral curve* or *flow line* of a kinematic vector field $X \in \mathfrak{X}(M)$ if $c'(t) = X(c(t))$ holds for all $t \in J$.

For a given kinematic vector field integral curves need not exist locally, and if they exist they need not be unique for a given initial value. This is due to the fact that the classical results on existence and uniqueness of solutions of equations like the inverse function theorem, the implicit function theorem, and the Picard-Lindelöf theorem on ordinary differential equations can be deduced essentially from one another, and all depend on Banach's fixed point theorem. Beyond Banach spaces these proofs do not work any more, since the reduction does no longer lead to a contraction on a metrizable space. We are now going to give examples, which show that almost everything that might fail indeed fails.

Example 1. Let $E := s$ be the Fréchet space of rapidly decreasing sequences. Note that $s = C^\infty(S^1, \mathbb{R})$ by the theory of Fourier series. Consider the continuous linear operator $T : E \to E$ given by $T(x_0, x_1, x_2, \dots) := (0, 1^2 x_1, 2^2 x_2, 3^2 x_3, \dots)$. The ordinary linear differential equation $x'(t) = T(x(t))$ with constant coefficients and initial value $x(0) := (1, 0, 0, \dots)$ has no solution, since the coordinates would have to satisfy the recursive relation $x_n'(t) = n^2 x_{n-1}(t)$ with $x_1'(t) = 0$, and hence we must have $x_n(t) = n! t^n$. But the so defined curve $t \mapsto x(t)$ has only for $t = 0$ values in E. Thus, no local solution exists. By recursion one sees that the solution for an arbitrary initial value $x(0)$ should be given by

$$x_n(t) = \sum_{i=0}^{n} \left(\tfrac{n!}{i!}\right)^2 x_i(0) \frac{t^{n-i}}{(n-i)!}.$$

If the initial value is a finite sequence, say $x_n(0) = 0$ for $n > N$ and $x_N(0) \neq 0$, then

$$\begin{aligned}
x_n(t) &= \sum_{i=0}^{N} \left(\tfrac{n!}{i!}\right)^2 x_i(0) \frac{t^{n-i}}{(n-i)!} \\
&= \frac{(n!)^2}{(n-N)!}\, t^{n-N} \sum_{i=0}^{N} \left(\tfrac{1}{i!}\right)^2 x_i(0) \tfrac{(n-N)!}{(n-i)!}\, t^{N-i} \\
|x_n(t)| &\geq \frac{(n!)^2}{(n-N)!}\, |t|^{n-N} \left(|x_N(0)| \left(\tfrac{1}{N!}\right)^2 - \sum_{i=0}^{N-1} \left(\tfrac{1}{i!}\right)^2 |x_i(0)| \tfrac{(n-N)!}{(n-i)!} |t|^{N-i} \right) \\
&\geq \frac{(n!)^2}{(n-N)!}\, |t|^{n-N} \left(|x_N(0)| \left(\tfrac{1}{N!}\right)^2 - \sum_{i=0}^{N-1} \left(\tfrac{1}{i!}\right)^2 |x_i(0)| |t|^{N-i} \right),
\end{aligned}$$

where the first factor does not lie in the space s of rapidly decreasing sequences, and where the second factor is larger than $\varepsilon > 0$ for t small enough. So at least for a dense set of initial values this differential equation has no local solution.

This also shows that the theorem of Frobenius is wrong in the following sense: The vector field $x \mapsto T(x)$ generates a 1-dimensional subbundle E of the tangent bundle on the open subset $s \setminus \{0\}$. It is involutive since it is 1-dimensional. But through points representing finite sequences there exist no local integral submanifolds (M with $TM = E|M$). Namely, if c were a smooth non-constant curve with $c'(t) = f(t).T(c(t))$ for some smooth function f, then $x(t) := c(h(t))$ would satisfy $x'(t) = T(x(t))$, where h is a solution of $h'(t) = 1/f(h(t))$.

Example 2. Next consider $E := \mathbb{R}^{\mathbb{N}}$ and the continuous linear operator $T : E \to E$ given by $T(x_0, x_1, \dots) := (x_1, x_2, \dots)$. The corresponding differential equation has solutions for every initial value $x(0)$, since the coordinates must satisfy the recursive relation $x_{k+1}(t) = x_k'(t)$, and hence any smooth function $x_0 : \mathbb{R} \to \mathbb{R}$ gives rise to a solution $x(t) := (x_0^{(k)}(t))_k$ with initial value $x(0) = (x_0^{(k)}(0))_k$. So by Borel's theorem there exist solutions to this equation for all initial values and the difference of any two functions with same initial value is an arbitrary infinite flat function. Thus, the solutions are far from being unique. Note that $\mathbb{R}^{\mathbb{N}}$ is a topological direct summand in $C^\infty(\mathbb{R},\mathbb{R})$ via the projection $f \mapsto (f(n))_n$, and hence the same situation occurs in $C^\infty(\mathbb{R},\mathbb{R})$.

Note that it is not possible to choose the solution depending smoothly on the initial value: suppose that x is a local smooth mapping $\mathbb{R}\times E \supset I\times U \to E$ with $x(0,y) = y$ and $\partial_t x(t,y) = T(x(t,y))$, where I is an open interval containing 0 and U is open in E. Then $x_0 : I \times U \to \mathbb{R}$ induces a smooth local mapping $x_0{}^\vee : U \to C^\infty(I,\mathbb{R})$, which is a right inverse to the linear infinite jet mapping $j_0^\infty : C^\infty(I,\mathbb{R}) \to \mathbb{R}^{\mathbb{N}} = E$. Then the derivative of $x_0{}^\vee$ at any point in U would be a continuous linear right inverse to j_0^∞, which does not exist (since $\mathbb{R}^{\mathbb{N}}$ does not admit a continuous norm, whereas $C^\infty(I,\mathbb{R})$ does for compact I, see also [Tougeron, 1972, IV.3.9]).

Also in this example the theorem of Frobenius is wrong, now in the following sense: On the complement of $T^{-1}(0) = \mathbb{R}\times 0$ we consider again the 1-dimensional subbundle generated by the vector field T. For every smooth function $f \in C^\infty(\mathbb{R},\mathbb{R})$ the infinite jet $t \mapsto j_t^\infty(f)$ is an integral curve of T. We show that integral curves through a fixed point sweep out arbitrarily high dimensional submanifolds of $\mathbb{R}^{\mathbb{N}}$: Let $\varphi : \mathbb{R} \to [0,1]$ be smooth, $\varphi(t) = 0$ near $t = 0$, and $\varphi(t) = 1$ near $t = 1$. For each $(s_2,\dots,s_N)$ we get an integral curve

$$t \mapsto j_t\Big(t + \frac{s_2}{2!}\varphi(t)(t-1)^2 + \frac{s_3}{3!}\varphi(t)(t-1)^3 + \dots + \frac{s_N}{N!}\varphi(t)(t-1)^N\Big)$$

connecting $(0,1,0,\dots)$ with $(1,1,s_2,s_3,\dots,s_N,0,\dots)$, and for small s this integral curve lies in $\mathbb{R}^{\mathbb{N}}\setminus 0$.

Problem: Can any two points be joined by an integral curve in $\mathbb{R}^{\mathbb{N}}\setminus 0$: One has to find a smooth function on $[0,1]$ with prescribed jets at 0 and 1 which is nowhere flat in between.

Example 3. Let now $E := C^\infty(\mathbb{R},\mathbb{R})$, and consider the continuous linear operator $T : E \to E$ given by $T(x) := x'$. Let $x : \mathbb{R} \to C^\infty(\mathbb{R},\mathbb{R})$ be a solution of the equation $x'(t) = T(x(t))$. In terms of $\hat{x} : \mathbb{R}^2 \to \mathbb{R}$ this says $\frac{\partial}{\partial t}\hat{x}(t,s) = \frac{\partial}{\partial s}\hat{x}(t,s)$. Hence, $r \mapsto \hat{x}(t-r, s+r)$ has vanishing derivative everywhere, and so this function is constant, and in particular $x(t)(s) = \hat{x}(t,s) = \hat{x}(0,s+t) = x(0)(s+t)$. Thus, we have a smooth solution x uniquely determined by the initial value $x(0) \in C^\infty(\mathbb{R},\mathbb{R})$, which even describes a flow for the vector field T in the sense of (32.13) below. In general however, this solution is not real-analytic, since for any $x(0) \in C^\infty(\mathbb{R},\mathbb{R})$ which is not real-analytic in a neighborhood of a point s the composite $\mathrm{ev}_s \circ x = x(s+\ \ )$ is not real-analytic around 0.

32.13. The flow of a vector field. Let $X \in \mathfrak{X}(M)$ be a kinematic vector field. A *local flow* Fl^X for X is a smooth mapping $M \times \mathbb{R} \supset U \xrightarrow{\mathrm{Fl}^X} M$ defined on a c^∞-open neighborhood U of $M \times 0$ such that

(1) $U \cap (\{x\}\times\mathbb{R})$ is a connected open interval.
(2) If $\mathrm{Fl}_s^X(x)$ exists then $\mathrm{Fl}_{t+s}^X(x)$ exists if and only if $\mathrm{Fl}_t^X(\mathrm{Fl}_s^X(x))$ exists, and we have equality.

(3) $\operatorname{Fl}^X_0(x) = x$ for all $x \in M$.
(4) $\frac{d}{dt} \operatorname{Fl}^X_t(x) = X(\operatorname{Fl}^X_t(x))$.

In formulas similar to (4) we will often omit the point x for sake of brevity, without signalizing some differentiation in a space of mappings. The latter will be done whenever possible in section (42).

32.14. Lemma. *Let $X \in \mathfrak{X}(M)$ be a kinematic vector field which admits a local flow Fl^X_t. Then each for each integral curve c of X we have $c(t) = \operatorname{Fl}^X_t(c(0))$, thus there exists a unique maximal flow. Furthermore, X is Fl^X_t-related to itself, i.e., $T(\operatorname{Fl}^X_t) \circ X = X \circ \operatorname{Fl}^X_t$.*

Proof. We compute

$$\begin{aligned}\tfrac{d}{dt} \operatorname{Fl}^X(-t, c(t)) &= -\tfrac{d}{ds}|_{s=-t} \operatorname{Fl}^X(s, c(t)) + \tfrac{d}{ds}|_{s=t} \operatorname{Fl}^X(-t, c(s)) \\ &= -\tfrac{d}{ds}|_{s=0} \operatorname{Fl}^X_{-t} \operatorname{Fl}^X(s, c(t)) + T(\operatorname{Fl}^X_{-t}).c'(t) \\ &= -T(\operatorname{Fl}^X_{-t}).X(c(t)) + T(\operatorname{Fl}^X_{-t}).X(c(t)) = 0.\end{aligned}$$

Thus, $\operatorname{Fl}^X_{-t}(c(t)) = c(0)$ is constant, so $c(t) = \operatorname{Fl}^X_t(c(0))$. For the second assertion we have $X \circ \operatorname{Fl}^X_t = \frac{d}{dt} \operatorname{Fl}^X_t = \frac{d}{ds}|_0 \operatorname{Fl}^X_{t+s} = \frac{d}{ds}|_0 (\operatorname{Fl}^X_t \circ \operatorname{Fl}^X_s) = T(\operatorname{Fl}^X_t) \circ \frac{d}{ds}|_0 \operatorname{Fl}^X_s = T(\operatorname{Fl}^X_t) \circ X$, where we omit the point $x \in M$ for the sake of brevity. □

32.15. The Lie derivative. For a vector field $X \in \mathfrak{X}(M)$ which has a local flow Fl^X_t and $f \in C^\infty(M, \mathbb{R})$ we have $\frac{d}{dt}(\operatorname{Fl}^X_t)^* f = \frac{d}{dt} f \circ \operatorname{Fl}^X_t = df \circ X \circ \operatorname{Fl}^X_t = X(f) \circ \operatorname{Fl}^X_t = (\operatorname{Fl}^X_t)^* X(f)$.

We will meet situations (in (37.19), e.g.) where we do not know that the flow of X exists but where we will be able to produce the following assumption: Suppose that $\varphi : \mathbb{R} \times M \supset U \to M$ is a smooth mapping such that $(t, x) \mapsto (t, \varphi(t, x) = \varphi_t(x))$ is a diffeomorphism $U \to V$, where U and V are open neighborhoods of $\{0\} \times M$ in $\mathbb{R} \times M$, and such that $\varphi_0 = \mathrm{Id}_M$ and $\frac{\partial}{\partial t}\big|_0 \varphi_t = X \in \mathfrak{X}(M)$. Then we have $\frac{\partial}{\partial t}\big|_0 \varphi_t^{-1} = -X$, and still we get $\frac{d}{dt}|_0 (\varphi_t)^* f = \frac{d}{dt}|_0 (f \circ \varphi_t) = df \circ X = X(f)$ and similarly $\frac{\partial}{\partial t}\big|_0 (\varphi_t^{-1})^* f = -X(f)$.

Lemma. *In this situation we have for $Y \in C^\infty(M \leftarrow DM)$:*

$$\begin{aligned}\tfrac{d}{dt}|_0 (\varphi_t)^* Y &= [X, Y], \\ \tfrac{d}{dt}|_0 (\operatorname{Fl}^X_t)^* Y &= [X, Y], \\ \tfrac{d}{dt} (\operatorname{Fl}^X_t)^* Y &= (\operatorname{Fl}^X_t)^* [X, Y].\end{aligned}$$

Proof. Let $f \in C^\infty(M, \mathbb{R})$ be a function, and let $\alpha(t, s) := Y(\varphi(t, x))(f \circ \varphi_s^{-1})$, which is locally defined near 0. It satisfies

$$\begin{aligned}&\alpha(t, 0) = Y(\varphi(t, x))(f), \\ &\alpha(0, s) = Y(x)(f \circ \varphi_s^{-1}), \\ &\tfrac{\partial}{\partial t}\alpha(0, 0) = \tfrac{\partial}{\partial t}\big|_0 Y(\varphi(t, x))(f) = \tfrac{\partial}{\partial t}\big|_0 (Yf)(\varphi(t, x)) = X(x)(Yf), \\ &\tfrac{\partial}{\partial s}\alpha(0, 0) = \tfrac{\partial}{\partial s}|_0 Y(x)(f \circ \varphi_s^{-1}) = Y(x)\tfrac{\partial}{\partial s}|_0 (f \circ \varphi_s^{-1}) = -Y(x)(Xf).\end{aligned}$$

Hence, $\frac{\partial}{\partial u}|_0 \alpha(u,u) = [X,Y]_x(f)$. But on the other hand we have

$$\begin{aligned}\tfrac{\partial}{\partial u}|_0 \alpha(u,u) &= \tfrac{\partial}{\partial u}|_0 Y(\varphi(u,x))(f\circ\varphi_u^{-1}) = \\ &= \tfrac{\partial}{\partial u}|_0 \left(D(\varphi_u^{-1})\circ Y\circ\varphi_u\right)_x(f) \\ &= (\tfrac{\partial}{\partial u}|_0 (\varphi_u)^*Y)_x(f),\end{aligned}$$

so the first two assertions follow. For the third claim we compute as follows:

$$\begin{aligned}\tfrac{\partial}{\partial t}(\mathrm{Fl}^X_t)^*Y &= \tfrac{\partial}{\partial s}|_0 \left(D(\mathrm{Fl}^X_{-t})\circ D(\mathrm{Fl}^X_{-s})\circ Y\circ \mathrm{Fl}^X_s\circ \mathrm{Fl}^X_t\right) \\ &= D(\mathrm{Fl}^X_{-t})\circ \tfrac{\partial}{\partial s}|_0 \left(D(\mathrm{Fl}^X_{-s})\circ Y\circ \mathrm{Fl}^X_s\right)\circ \mathrm{Fl}^X_t \\ &= D(\mathrm{Fl}^X_{-t})\circ [X,Y]\circ \mathrm{Fl}^X_t = (\mathrm{Fl}^X_t)^*[X,Y]. \quad \square\end{aligned}$$

32.16. Lemma. *Let $X \in \mathfrak{X}(M)$ and $Y \in \mathfrak{X}(N)$ be f-related vector fields for a smooth mapping $f : M \to N$ which have local flows Fl^X and Fl^Y. Then we have $f\circ \mathrm{Fl}^X_t = \mathrm{Fl}^Y_t \circ f$, whenever both sides are defined.*

Moreover, if f is a diffeomorphism we have $\mathrm{Fl}^{f^*Y}_t = f^{-1}\circ \mathrm{Fl}^Y_t \circ f$ in the following sense: If one side exists then also the other and they are equal.

For $f = \mathrm{Id}_M$ this again implies that if there exists a flow then there exists a unique maximal flow Fl^X_t.

Proof. We have $Y\circ f = Tf\circ X$, and thus for small t we get, using (32.13.1),

$$\begin{aligned}\tfrac{d}{dt}(\mathrm{Fl}^Y_t \circ f\circ \mathrm{Fl}^X_{-t}) &= Y\circ \mathrm{Fl}^Y_t \circ f\circ \mathrm{Fl}^X_{-t} - T(\mathrm{Fl}^Y_t)\circ Tf\circ X\circ \mathrm{Fl}^X_{-t} \\ &= T(\mathrm{Fl}^Y_t)\circ Y\circ f\circ \mathrm{Fl}^X_{-t} - T(\mathrm{Fl}^Y_t)\circ Tf\circ X\circ \mathrm{Fl}^X_{-t} = 0.\end{aligned}$$

So $(\mathrm{Fl}^Y_t \circ f \circ \mathrm{Fl}^X_{-t})(x) = f(x)$ or $f(\mathrm{Fl}^X_t(x)) = \mathrm{Fl}^Y_t(f(x))$ for small t. By the flow properties (32.13.2), we get the result by a connectedness argument as follows: In the common interval of definition we consider the closed subset $J_x := \{t : f(\mathrm{Fl}^X_t(x)) = \mathrm{Fl}^Y_t(f(x))\}$. This set is open since for $t \in J_x$ and small $|s|$ we have $f(\mathrm{Fl}^X_{t+s}(x)) = f(\mathrm{Fl}^X_s(\mathrm{Fl}^X_t(x))) = \mathrm{Fl}^Y_s(f(\mathrm{Fl}^X_t(x))) = \mathrm{Fl}^Y_s(\mathrm{Fl}^Y_t(f(x))) = \mathrm{Fl}^Y_{t+s}(f(x))$.

The existence of the unique maximal flow now follows since two local flows have to agree on their common domain of definition. $\square$

32.17. Corollary. *Take $X \in \mathfrak{X}(M)$ be a vector field with local flow, and let $Y \in C^\infty(M \leftarrow DM)$. Then the following assertions are equivalent*

(1) $[X,Y] = 0$.

(2) $(\mathrm{Fl}^X_t)^*Y = Y$, *wherever defined.*

If also Y is kinematic and has a local flow then these are also equivalent to

(3) $\mathrm{Fl}^X_t \circ \mathrm{Fl}^Y_s = \mathrm{Fl}^Y_s \circ \mathrm{Fl}^X_t$, *wherever defined.*

Proof. (1) $\Leftrightarrow$ (2) is immediate from lemma (32.15). To see (2) $\Leftrightarrow$ (3) we note that $\mathrm{Fl}^X_t \circ \mathrm{Fl}^Y_s = \mathrm{Fl}^Y_s \circ \mathrm{Fl}^X_t$ if and only if $\mathrm{Fl}^Y_s = \mathrm{Fl}^X_{-t}\circ \mathrm{Fl}^Y_s \circ \mathrm{Fl}^X_t = \mathrm{Fl}^{(\mathrm{Fl}^X_t)^*Y}_s$ by lemma (32.16), and this in turn is equivalent to $Y = (\mathrm{Fl}^X_t)^*Y$, by the uniqueness of flows. $\square$

32.18. Theorem. [Mauhart, Michor, 1992] *Let M be a manifold, and let $\varphi^i : \mathbb{R}\times M \supset U_{\varphi^i} \to M$ be smooth mappings for $i=1,\dots,k$ such that $(t,x)\mapsto (t,\varphi^i(t,x)=\varphi^i_t(x))$ is a diffeomorphism $U_{\varphi_i}\to V_{\varphi_i}$. Here the U_{φ^i} and V_{φ_i} are open neighborhoods of $\{0\}\times M$ in $\mathbb{R}\times M$ such that $\varphi^i_0=\operatorname{Id}_M$ and $\frac{\partial}{\partial t}\big|_0 \varphi^i_t = X_i\in\mathfrak{X}(M)$. We put $[\varphi^i,\varphi^j]_t = [\varphi^i_t,\varphi^j_t] := (\varphi^j_t)^{-1}\circ(\varphi^i_t)^{-1}\circ\varphi^j_t\circ\varphi^i_t$. Then for each formal bracket expression P of length k we have*

$$0 = \tfrac{\partial^\ell}{\partial t^\ell}|_0 P(\varphi^1_t,\dots,\varphi^k_t) \quad \text{for } 1\le \ell<k,$$
$$P(X_1,\dots,X_k) = \tfrac{1}{k!}\tfrac{\partial^k}{\partial t^k}|_0 P(\varphi^1_t,\dots,\varphi^k_t)\in\mathfrak{X}(M)$$

as first non-vanishing derivative in the sense explained in step (—2 of the proof. In particular, we have for vector fields $X,Y\in\mathfrak{X}(M)$ admitting local flows

$$0 = \tfrac{\partial}{\partial t}\big|_0 (\mathrm{Fl}^Y_{-t}\circ \mathrm{Fl}^X_{-t}\circ \mathrm{Fl}^Y_t\circ \mathrm{Fl}^X_t),$$
$$[X,Y] = \tfrac12 \tfrac{\partial^2}{\partial t^2}|_0 (\mathrm{Fl}^Y_{-t}\circ \mathrm{Fl}^X_{-t}\circ \mathrm{Fl}^Y_t\circ \mathrm{Fl}^X_t).$$

Proof. Step 1. Let $c:\mathbb{R}\to M$ be a smooth curve. If $c(0)=x\in M$, $c'(0)=0,\dots,c^{(k-1)}(0)=0$, then $c^{(k)}(0)$ is a well defined tangent vector in T_xM, which is given by the derivation $f\mapsto (f\circ c)^{(k)}(0)$ at x. Namely, we have

$$\begin{aligned}((f.g)\circ c)^{(k)}(0) &= ((f\circ c).(g\circ c))^{(k)}(0) = \sum_{j=0}^k \tbinom{k}{j}(f\circ c)^{(j)}(0)(g\circ c)^{(k-j)}(0)\\ &= (f\circ c)^{(k)}(0)g(x) + f(x)(g\circ c)^{(k)}(0),\end{aligned}$$

since all other summands vanish: $(f\circ c)^{(j)}(0)=0$ for $1\le j<k$. That $c^{(k)}(0)$ is a kinematic tangent vector follows from the chain rule in a local chart.

Step 2. Let $(\mathrm{pr}_1,\varphi):\mathbb{R}\times M\supset U_\varphi\to V_\varphi\subset\mathbb{R}\times M$ be a diffeomorphism between open neighborhoods of $\{0\}\times M$ in $\mathbb{R}\times M$, such that $\varphi_0=\operatorname{Id}_M$. We say that φ_t is a *curve of local diffeomorphisms* though Id_M. Note that a local flow of a kinematic vector field is always such a curve of local diffeomorphisms.

From step 1 we see that if $\frac{\partial^j}{\partial t^j}|_0\varphi_t = 0$ for all $1\le j<k$, then $X := \frac{1}{k!}\frac{\partial^k}{\partial t^k}|_0\varphi_t$ is a well defined vector field on M. We say that X is the *first non-vanishing derivative* at 0 of the curve φ_t of local diffeomorphisms. We may paraphrase this as $(\partial^k_t|_0\varphi^*_t)f = k!\mathcal{L}_X f$.

Claim 3. Let φ_t, ψ_t be curves of local diffeomorphisms through Id_M, and let $f\in C^\infty(M,\mathbb{R})$. Then we have

$$\partial^k_t|_0(\varphi_t\circ\psi_t)^*f = \partial^k_t|_0(\psi^*_t\circ\varphi^*_t)f = \sum_{j=0}^k \tbinom{k}{j}(\partial^j_t|_0\psi^*_t)(\partial^{k-j}_t|_0\varphi^*_t)f.$$

The multinomial version of this formula holds also:

$$\partial^k_t|_0(\varphi^1_t\circ\dots\circ\varphi^\ell_t)^*f = \sum_{j_1+\dots+j_\ell=k}\frac{k!}{j_1!\dots j_\ell!}(\partial^{j_\ell}_t|_0(\varphi^\ell_t)^*)\dots(\partial^{j_1}_t|_0(\varphi^1_t)^*)f.$$

We only show the binomial version. For a function $h(t,s)$ of two variables we have

$$\partial_t^k h(t,t) = \sum_{j=0}^{k} \binom{k}{j} \partial_t^j \partial_s^{k-j} h(t,s)|_{s=t},$$

since for $h(t,s) = f(t)g(s)$ this is just a consequence of the Leibniz rule, and linear combinations of such decomposable tensors are dense in the space of all functions of two variables in the compact C^∞-topology (41.9), so that by continuity the formula holds for all functions. In the following form it implies the claim:

$$\partial_t^k|_0 f(\varphi(t,\psi(t,x))) = \sum_{j=0}^{k} \binom{k}{j} \partial_t^j \partial_s^{k-j} f(\varphi(t,\psi(s,x)))|_{t=s=0}.$$

Claim 4. Let φ_t be a curve of local diffeomorphisms through Id_M with first non-vanishing derivative $k!X = \partial_t^k|_0 \varphi_t$. Then the inverse curve of local diffeomorphisms φ_t^{-1} has first non-vanishing derivative $-k!X = \partial_t^k|_0 \varphi_t^{-1}$.

Since we have $\varphi_t^{-1} \circ \varphi_t = \mathrm{Id}$, by claim 3 we get for $1 \le j \le k$

$$\begin{aligned} 0 = \partial_t^j|_0 (\varphi_t^{-1} \circ \varphi_t)^* f &= \sum_{i=0}^{j} \binom{j}{i} (\partial_t^i|_0 \varphi_t^*)(\partial_t^{j-i} (\varphi_t^{-1})^*) f = \\ &= \partial_t^j|_0 \varphi_t^* (\varphi_0^{-1})^* f + \varphi_0^* \partial_t^j|_0 (\varphi_t^{-1})^* f, \qquad \text{which says} \\ \partial_t^j|_0 \varphi_t^* f &= -\partial_t^j|_0 (\varphi_t^{-1})^* f, \qquad \text{as required.} \end{aligned}$$

Claim 5. Let φ_t be a curve of local diffeomorphisms through Id_M with first non-vanishing derivative $m!X = \partial_t^m|_0 \varphi_t$, and let ψ_t be a curve of local diffeomorphisms through Id_M with first non-vanishing derivative $n!Y = \partial_t^n|_0 \psi_t$. Then the curve of local diffeomorphisms $[\varphi_t, \psi_t] = \psi_t^{-1} \circ \varphi_t^{-1} \circ \psi_t \circ \varphi_t$ has first non-vanishing derivative

$$(m+n)![X,Y] = \partial_t^{m+n}|_0 [\varphi_t, \psi_t].$$

From this claim the theorem follows.

By the multinomial version of claim 3, we have

$$\begin{aligned} A_N f &:= \partial_t^N|_0 (\psi_t^{-1} \circ \varphi_t^{-1} \circ \psi_t \circ \varphi_t)^* f \\ &= \sum_{i+j+k+\ell=N} \frac{N!}{i!j!k!\ell!} (\partial_t^i|_0 \varphi_t^*)(\partial_t^j|_0 \psi_t^*)(\partial_t^k|_0 (\varphi_t^{-1})^*)(\partial_t^\ell|_0 (\psi_t^{-1})^*) f. \end{aligned}$$

Let us suppose that $1 \le n \le m$; the case $m \le n$ is similar. If $N < n$ all summands are 0. If $N = n$ we have by claim 4

$$A_N f = (\partial_t^n|_0 \varphi_t^*) f + (\partial_t^n|_0 \psi_t^*) f + (\partial_t^n|_0 (\varphi_t^{-1})^*) f + (\partial_t^n|_0 (\psi_t^{-1})^*) f = 0.$$

If $n < N \le m$ we have, using again claim 4:

$$\begin{aligned} A_N f &= \sum_{j+\ell=N} \frac{N!}{j!\ell!} (\partial_t^j|_0 \psi_t^*)(\partial_t^\ell|_0 (\psi_t^{-1})^*) f + \delta_N^m \left((\partial_t^m|_0 \varphi_t^*) f + (\partial_t^m|_0 (\varphi_t^{-1})^*) f \right) \\ &= (\partial_t^N|_0 (\psi_t^{-1} \circ \psi_t)^*) f + 0 = 0. \end{aligned}$$

Now we come to the difficult case $m, n < N \leq m+n$.

$$
\begin{aligned}
A_N f &= \partial_t^N|_0(\psi_t^{-1}\circ\varphi_t^{-1}\circ\psi_t)^* f + \tbinom{N}{m}(\partial_t^m|_0\varphi_t^*)(\partial_t^{N-m}|_0(\psi_t^{-1}\circ\varphi_t^{-1}\circ\psi_t)^*)f \\
&\quad + (\partial_t^N|_0\varphi_t^*)f, \qquad (6)
\end{aligned}
$$

by claim 3, since all other terms vanish, see (8) below. Again by claim 3 we get:

$$
\begin{aligned}
\partial_t^N|_0(\psi_t^{-1}\circ\varphi_t^{-1}\circ\psi_t)^* f &= \sum_{j+k+\ell=N}\frac{N!}{j!k!\ell!}(\partial_t^j|_0\psi_t^*)(\partial_t^k|_0(\varphi_t^{-1})^*)(\partial_t^\ell|_0(\psi_t^{-1})^*)f \\
&= \sum_{j+\ell=N}\tbinom{N}{j}(\partial_t^j|_0\psi_t^*)(\partial_t^\ell|_0(\psi_t^{-1})^*)f + \tbinom{N}{m}(\partial_t^{N-m}|_0\psi_t^*)(\partial_t^m|_0(\varphi_t^{-1})^*)f \qquad (7)\\
&\quad + \tbinom{N}{m}(\partial_t^m|_0(\varphi_t^{-1})^*)(\partial_t^{N-m}|_0(\psi_t^{-1})^*)f + (\partial_t^N|_0(\varphi_t^{-1})^*)f \\
&= 0 + \tbinom{N}{m}(\partial_t^{N-m}|_0\psi_t^*)m!\mathcal{L}_{-X}f + \tbinom{N}{m}m!\mathcal{L}_{-X}(\partial_t^{N-m}|_0(\psi_t^{-1})^*)f \\
&\quad + (\partial_t^N|_0(\varphi_t^{-1})^*)f \\
&= \delta_{m+n}^N(m+n)!(\mathcal{L}_X\mathcal{L}_Y - \mathcal{L}_Y\mathcal{L}_X)f + (\partial_t^N|_0(\varphi_t^{-1})^*)f \\
&= \delta_{m+n}^N(m+n)!\mathcal{L}_{[X,Y]}f + (\partial_t^N|_0(\varphi_t^{-1})^*)f
\end{aligned}
$$

From the second expression in (7) one can also read off that

$$
\partial_t^{N-m}|_0(\psi_t^{-1}\circ\varphi_t^{-1}\circ\psi_t)^* f = \partial_t^{N-m}|_0(\varphi_t^{-1})^* f. \qquad (8)
$$

If we put (7) and (8) into (6) we get, using claims 3 and 4 again, the final result which proves claim 5 and the theorem:

$$
\begin{aligned}
A_N f &= \delta_{m+n}^N(m+n)!\mathcal{L}_{[X,Y]}f + (\partial_t^N|_0(\varphi_t^{-1})^*)f \\
&\quad + \tbinom{N}{m}(\partial_t^m|_0\varphi_t^*)(\partial_t^{N-m}|_0(\varphi_t^{-1})^*)f + (\partial_t^N|_0\varphi_t^*)f \\
&= \delta_{m+n}^N(m+n)!\mathcal{L}_{[X,Y]}f + \partial_t^N|_0(\varphi_t^{-1}\circ\varphi_t)^* f \\
&= \delta_{m+n}^N(m+n)!\mathcal{L}_{[X,Y]}f + 0. \quad \square
\end{aligned}
$$

33. Differential Forms

This section is devoted to the search for the right notion of differential forms which are stable under Lie derivatives $\mathcal{L}_X$, exterior derivative d, and pullback f^*. Here chaos breaks out (as one referee has put it) since the classically equivalent descriptions of differential forms give rise to many different classes; in the table (33.21) we shall have 12 classes. But fortunately it will turn out in (33.22) that there is only one suitable class satisfying all requirements, namely

$$
\Omega^k(M) := C^\infty(L_{\text{alt}}^k(TM, M\times\mathbb{R})).
$$

33.1. Cotangent bundles. We consider the contravariant smooth functor which associates to each convenient vector space E its dual E' of bounded linear functionals, and we apply it to the kinematic tangent bundle TM described in (28.12) of a smooth manifold M (see (29.5)) to get the *kinematic cotangent bundle* $T'M$. A smooth atlas $(U_\alpha, u_\alpha : U_\alpha \to E_\alpha)$ of M gives the cocycle of transition functions

$$U_{\alpha\beta} \ni x \mapsto d(u_\beta \circ u_\alpha^{-1})(u_\alpha(x))^* \in GL(E'_\beta, E'_\alpha).$$

If we apply the same duality functor to the operational tangent bundle DM described in (28.12) we get the *operational cotangent bundle* $D'M$. A smooth atlas $(U_\alpha, u_\alpha : U_\alpha \to E_\alpha)$ of M now gives rise to the following cocycle of transition functions

$$U_{\alpha\beta} \ni x \mapsto D(u_\beta \circ u_\alpha^{-1})(u_\alpha(x))^* \in GL((D_0 E_\beta)', (D_0 E_\alpha)'),$$

see (28.9) and (28.12).

For each $k \in \mathbb{N}$ we get the *operational cotangent bundle* $(D^{(k)})'M$ *of order* $\leq k$, which is described by the same cocycle of transition functions but now restricted to have values in $GL((D_0^{(k)} E_\beta)', (D_0^{(k)} E_\alpha)')$, see (28.10).

33.2. 1-forms. Let M be a smooth manifold. A *kinematic 1-form* is just a smooth section of the kinematic cotangent bundle $T'M$. So $C^\infty(M \leftarrow T'M)$ denotes the convenient vector space (with the structure from (30.1)) of all kinematic 1-forms on M.

An *operational 1-form* is just a smooth section of the operational cotangent bundle $D'M$. So $C^\infty(M \leftarrow D'M)$ denotes the convenient vector space (with the structure from (30.1)) of all operational 1-forms on M.

For each $k \in \mathbb{N}$ we get the convenient vector space $C^\infty(M \leftarrow (D^{(k)})'(M))$ of all *operational 1-forms of order* $\leq k$, a closed linear subspace of $C^\infty(M \leftarrow D'M)$.

A *modular 1-form* is a bounded linear sheaf homomorphism $\omega : \operatorname{Der}(C^\infty(\quad, \mathbb{R})) \to C^\infty(\quad, \mathbb{R})$ which satisfies $\omega_U(f.X) = f.\omega_U(X)$ for $X \in \operatorname{Der}(C^\infty(U, \mathbb{R})) = C^\infty(U \leftarrow DU)$ and $f \in C^\infty(U, \mathbb{R})$ for each open $U \subset M$. We denote the space of all modular 1-forms by

$$\operatorname{Hom}_{C^\infty(M,\mathbb{R})}(C^\infty(M \leftarrow DM), C^\infty(M, \mathbb{R}))$$

and we equip it with the initial structure of a convenient vector space induced by the closed linear embedding

$$\operatorname{Hom}_{C^\infty(M,\mathbb{R})}(C^\infty(M \leftarrow DM), C^\infty(M, \mathbb{R})) \hookrightarrow \prod_U L(C^\infty(U \leftarrow DU), C^\infty(U, \mathbb{R})).$$

Convention. Similarly as in (32.1), we shall follow the convention that either the manifolds in question are smoothly regular or that Hom means the space of sheaf homomorphisms (as defined above) between the sheafs of sections like $C^\infty(M \leftarrow DM)$ of the respective vector bundles. This is justified by (33.3) below.

33.3. Lemma. *If M is smoothly regular, the bounded $C^\infty(M,\mathbb{R})$-module homomorphisms $\omega : C^\infty(M \leftarrow DM) \to C^\infty(M,\mathbb{R})$ are exactly the modular 1-forms and this identification is an isomorphism of the convenient vector spaces.*

Proof. If $X \in C^\infty(M \leftarrow DM)$ vanishes on an open subset $U \subset M$ then also $\omega(X)$: For $x \in U$ we take a bump function $g \in C^\infty(M,\mathbb{R})$ at x, i.e. $g = 1$ near x and $\operatorname{supp}(g) \subset U$. Then $\omega(X) = \omega((1-g)X) = (1-g)\omega(X)$ which is zero near x. So $\omega(X)|U = 0$.

Now let $X \in C^\infty(U \leftarrow DU)$ for a c^∞-open subset U of M. We have to show that we can define $\omega_U(X) \in C^\infty(U,\mathbb{R})$ in a unique manner. For $x \in U$ let $g \in C^\infty(M,\mathbb{R})$ be a bump function at x, i.e. $g = 1$ near x and $\operatorname{supp}(g) \subset U$. Then $gX \in C^\infty(M \leftarrow DM)$, and $\omega(gX)$ makes sense. By the argument above, $\omega(gX)(x)$ is independent of the choice of g. So let $\omega_U(X)(x) := \omega(gX)(x)$. It has all required properties since the topology on $C^\infty(U \leftarrow DU)$ is initial with respect to all mappings $X \mapsto gX$, where g runs through all bump functions as above.

That this identification furnishes an isomorphism of convenient vector spaces can be seen as in (32.4). □

33.4. Lemma. *On any manifold M the space $C^\infty(M \leftarrow D'M)$ of operational 1-forms is a closed linear subspace of modular 1-forms $\operatorname{Hom}_{C^\infty(M,\mathbb{R})}(C^\infty(M \leftarrow DM), C^\infty(M,\mathbb{R}))$.*

The closed vector bundle embedding $TM \to DM$ induces a bounded linear mapping $C^\infty(M \leftarrow D'M) \to C^\infty(M \leftarrow T'M)$.

We do not know whether $C^\infty(M \leftarrow D'M) \to C^\infty(M \leftarrow T'M)$ is surjective or even final.

Proof. A smooth section $\omega \in C^\infty(M \leftarrow D'M)$ defines a modular 1-form which assigns $\omega_U(X)(x) := \omega(x)(X(x))$ to $X \in C^\infty(U \leftarrow DU)$ and $x \in U$, by (32.2), since this gives a bounded sheaf homomorphism which is $C^\infty(\quad,\mathbb{R})$-linear.

To show that this gives an embedding onto a c^∞-closed linear subspace we consider the following diagram, where (U_α) runs through an open cover of charts of M. Then the vertical mappings are closed linear embeddings by (30.1), (33.1), and (32.2).

$$\begin{array}{ccc}
C^\infty(M \leftarrow D'M) & \longrightarrow & \operatorname{Hom}_{C^\infty(M,\mathbb{R})}(C^\infty(M \leftarrow DM), C^\infty(M,\mathbb{R})) \\
\downarrow & & \downarrow \\
\prod_\alpha C^\infty(U_\alpha, (D_0E_\alpha)') & \longrightarrow & \prod_\alpha L(C^\infty(U_\alpha \leftarrow DU_\alpha), C^\infty(U_\alpha,\mathbb{R})) \\
\downarrow & & \downarrow \\
\prod_\alpha C^\infty(U_\alpha \times D_0E_\alpha, \mathbb{R}) & \longrightarrow & \prod_\alpha C^\infty(C^\infty(U_\alpha, D_0E_\alpha) \times U_\alpha, \mathbb{R})
\end{array}$$

The horizontal bottom arrow is the mapping $f \mapsto ((X,x) \mapsto f(x, X(x)))$, which is an embedding since $(X,x) \mapsto (x, X(x))$ has $(x,Y) \mapsto (\text{const}(Y), x)$ as smooth right inverse. □

33.5. Lemma. *Let M be a smooth manifold such that for all model spaces E the convenient vector space D_0E has the bornological approximation property (28.6). Then*

$$C^\infty(M \leftarrow D'M) \cong \operatorname{Hom}_{C^\infty(M,\mathbb{R})}(C^\infty(M \leftarrow DM), C^\infty(M,\mathbb{R})).$$

If all model spaces E have the bornological approximation property then $D_0E = E''$, and the space E'' also has the bornological approximation property. So in this case,

$$\operatorname{Hom}_{C^\infty(M,\mathbb{R})}(C^\infty(M \leftarrow DM), C^\infty(M,\mathbb{R})) \cong C^\infty(M \leftarrow T'''M).$$

If, moreover, all E are reflexive, we have

$$\operatorname{Hom}_{C^\infty(M,\mathbb{R})}(C^\infty(M \leftarrow DM), C^\infty(M,\mathbb{R})) \cong C^\infty(M \leftarrow T'M),$$

as in finite dimensions.

Proof. By lemma (33.4) the space $C^\infty(M \leftarrow D'M)$ is a closed linear subspace of the convenient vector space $\operatorname{Hom}_{C^\infty(M,\mathbb{R})}(C^\infty(M \leftarrow DM), C^\infty(M,\mathbb{R}))$. We have to show that any sheaf homomorphism $\omega \in \operatorname{Hom}_{C^\infty(M,\mathbb{R})}(C^\infty(M \leftarrow DM), C^\infty(M,\mathbb{R}))$ lies in $C^\infty(M \leftarrow D'M)$. This is a local question, hence we may assume that M is a c^∞-open subset of E.

We have to show that for each $X \in C^\infty(U, D_0E)$ the value $\omega_U(X)(x)$ depends only on $X(x) \in D_0E$. So let $X(x) = 0$, and we have to show that $\omega_U(X)(x) = 0$.

By assumption, there is a net $\ell_\alpha \in (D_0E)' \otimes D_0E \subset L(D_0E, D_0E)$ of bounded linear operators with finite dimensional images, which converges to Id_{D_0E} in the bornological topology of $L(D_0E, D_0E)$. Then $X_\alpha := \ell_\alpha \circ X$ converges to X in $C^\infty(U, D_0E)$ since $X^* : L(D_0E, D_0E) \to C^\infty(U, D_0E)$ is continuous linear. It remains to show that $\omega_U(X_\alpha)(x) = 0$ for each α.

We have $\ell_\alpha = \sum_{i=1}^n \varphi_i \otimes \partial_i \in (D_0E)' \otimes D_0E$, hence $X_\alpha = \sum(\varphi_i \circ X).\partial_i$ and $\omega_U(X_\alpha)(x) = \sum \varphi_i(X(x)).\omega_U(\partial_i)(x) = 0$ since $X(x) = 0$.

So we get a fiber linear mapping $\omega : DM \to M \times \mathbb{R}$ which is given by $\omega(X_x) = (x, \omega_U(X)(x))$ for any $X \in C^\infty(U \leftarrow DU)$ with $X(x) = X_x$. Obviously, $\omega : DM \to M \times \mathbb{R}$ is smooth and gives rise to a smooth section of $D'M$.

If E has the bornological approximation property, then by (28.7) we have $D_0E = E''$. If ℓ_α is a net of finite dimensional bounded operators which converges to Id_E in $L(E,E)$, then the finite dimensional operators ℓ_α^{**} converge to $\operatorname{Id}_E'' = \operatorname{Id}_{E''}$ in $L(E'', E'')$, in the bornological topology. The rest follows from theorem (28.7) □

33.6. Queer 1-forms. Let E be a convenient vector space without the bornological approximation property, for example an infinite dimensional Hilbert space. Then there exists a bounded linear functional $\alpha \in L(E,E)'$ which vanishes on

$E' \otimes E$ such that $\alpha(\mathrm{Id}_E) = 1$. Then $\omega_U : C^\infty(U, E) \to C^\infty(U, \mathbb{R})$, given by $\omega_U(X)(x) := \alpha(dX(x))$, is a bounded sheaf homomorphism which is a module homomorphism, since $\omega_U(f.X)(x) = \alpha(df(x) \otimes X(x) + f(x).dX(x)) = f(x)\omega_U(X)(x)$. Note that $\omega_U(X)(x)$ does not depend only on $X(x)$. So there are many 'kinematic modular 1-forms' which are not kinematic 1-forms.

This process can be iterated to involve higher derivatives like for derivations, see (28.2), but we resist the temptation to pursue this task. It would be more interesting to produce queer modular 1-forms which are not operational 1-forms.

33.7. k-forms. For a smooth manifold M there are at least eight interesting spaces of k-forms, see the diagram below where $A := C^\infty(M, \mathbb{R})$, and where $C^\infty(E)$ denotes the space of smooth sections of the vector bundle $E \to M$:

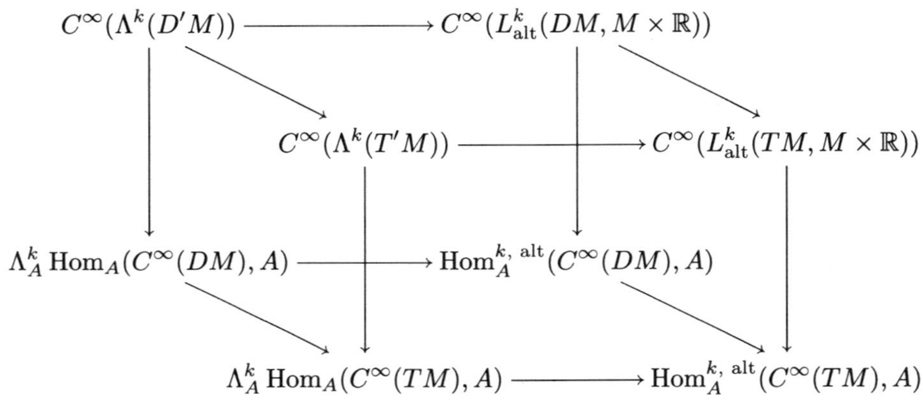

Here Λ^k is the bornological exterior product which was treated in (5.9). One could also start from other tensor products. By $\Lambda^k_A = \Lambda^k_{C^\infty(M,\mathbb{R})}$ we mean the convenient module exterior product, the subspace of all skew symmetric elements in the k-fold bornological tensor product over A, see (5.21). By $\operatorname{Hom}^k_{C^\infty(M,\mathbb{R}),\mathrm{alt}} = \operatorname{Hom}^{k,\,\mathrm{alt}}_{C^\infty(M,\mathbb{R})}$ we mean the convenient space of all bounded homomorphism between the respective sheaves of convenient modules over the sheaf of smooth functions.

33.8. Wedge product. For differential forms φ of degree k and ψ of degree ℓ and for (local) vector fields X_i (or tangent vectors) we put

$$(\varphi \wedge \psi)(X_1, \dots, X_{k+\ell}) =$$
$$= \tfrac{1}{k!\,\ell!} \sum_{\sigma \in \mathcal{S}_{k+\ell}} \operatorname{sign}\sigma \cdot \varphi(X_{\sigma 1}, \dots, X_{\sigma k}).\psi(X_{\sigma(k+1)}, \dots, X_{\sigma(k+\ell)}).$$

This is well defined for differential forms in each of the spaces in (33.7) and others (see (33.12) below) and gives a differential form of the same type of degree $k+\ell$. The wedge product is associative, i.e $(\varphi \wedge \psi) \wedge \tau = \varphi \wedge (\psi \wedge \tau)$, and graded commutative, i. e. $\varphi \wedge \psi = (-1)^{k\ell}\psi \wedge \varphi$. These properties are proved in multilinear algebra. There arise several kinds of algebras of differential forms.

33.9. Pullback of differential forms. Let $f : N \to M$ be a smooth mapping between smooth manifolds, and let φ be a differential form on M of degree k in any of the following spaces: $C^\infty(L^k_{\text{alt}}(D^\alpha M, M \times \mathbb{R}))$ for $D^\alpha = D, D^{(k)}, D^{[1,\infty)}, T$. In this situation the pullback $f^*\varphi$ is defined for tangent vectors $X_i \in D^\alpha_x N$ by

$$(1) \qquad (f^*\varphi)_x(X_1, \dots, X_k) := \varphi_{f(x)}(D^\alpha_x f.X_1, \dots, D^\alpha_x f.X_k).$$

Then we have $f^*(\varphi \wedge \psi) = f^*\varphi \wedge f^*\psi$, so the linear mapping f^* is an algebra homomorphism. Moreover, we have $(g\circ f)^* = f^*\circ g^*$ if $g : M \to P$, and $(\text{Id}_M)^* = \text{Id}$, and $(f, \varphi) \mapsto f^*\varphi$ is smooth in all these cases.

If $f : N \to M$ is a local diffeomorphism, then we may define the pullback $f^*\varphi$ also for a modular differential form $\varphi \in \text{Hom}^{k,\text{ alt}}_{C^\infty(M,\mathbb{R})}(C^\infty(M \leftarrow D^\alpha M), C^\infty(M, \mathbb{R}))$, by

$$(2) \quad (f^*\varphi)|_U(X_1, \dots, X_k) := \varphi|_{f(U)}(D^\alpha f\circ X_1\circ(f|U)^{-1}, \dots, D^\alpha f\circ X_k\circ(f|U)^{-1})\circ f.$$

These two definitions are intertwined by the canonical mappings between different spaces of differential forms.

33.10. Insertion operator. For a vector field $X \in C^\infty(M \leftarrow D^\alpha M)$ where $D^\alpha = D, D^{(k)}, D^{[1,\infty)}, T$ we define the *insertion operator*

$$\begin{aligned} i_X = i(X) : \text{Hom}^{k,\text{ alt}}_{C^\infty(M,\mathbb{R})}(C^\infty(M \leftarrow D^\alpha M), C^\infty(M, \mathbb{R})) \to \\ \to \text{Hom}^{k-1,\text{ alt}}_{C^\infty(M,\mathbb{R})}(C^\infty(M \leftarrow D^\alpha M), C^\infty(M, \mathbb{R})) \\ (i_X\varphi)(Y_1, \dots, Y_{k-1}) := \varphi(X, Y_1, \dots, Y_{k-1}). \end{aligned}$$

It restricts to operators

$$i_X = i(X) : C^\infty(L^k_{\text{alt}}(D^\alpha M, M \times \mathbb{R})) \to C^\infty(L^{k-1}_{\text{alt}}(D^\alpha M, M \times \mathbb{R})).$$

33.11. Lemma. *i_X is a graded derivation of degree -1, so we have $i_X(\varphi \wedge \psi) = i_X\varphi \wedge \psi + (-1)^{\deg \varphi}\varphi \wedge i_X\psi$.*

Proof. We have

$$\begin{aligned} (i_{X_1}(\varphi \wedge \psi))(X_2, \dots, X_{k+\ell}) &= (\varphi \wedge \psi)(X_1, \dots, X_{k+\ell}) \\ &= \tfrac{1}{k!\,\ell!} \sum_\sigma \text{sign}(\sigma)\, \varphi(X_{\sigma 1}, \dots, X_{\sigma k})\psi(X_{\sigma(k+1)}, \dots, X_{\sigma(k+\ell)}). \end{aligned}$$

$$\begin{aligned} (i_{X_1}\varphi \wedge \psi &+ (-1)^k\varphi \wedge i_{X_1}\psi)(X_2, \dots, X_{k+\ell}) \\ &= \tfrac{1}{(k-1)!\,\ell!} \sum_\sigma \text{sign}(\sigma)\, \varphi(X_1, X_{\sigma 2}, \dots, X_{\sigma k})\psi(X_{\sigma(k+1)}, \dots, X_{\sigma(k+\ell)}) \\ &+ \frac{(-1)^k}{k!\,(\ell-1)!} \sum_\sigma \text{sign}(\sigma)\, \varphi(X_{\sigma 2}, \dots, X_{\sigma(k+1)})\psi(X_1, X_{\sigma(k+2)}, \dots). \end{aligned}$$

Using the skew symmetry of φ and ψ we may distribute X_1 to each position by adding an appropriate sign. These are $k + \ell$ summands. Since $\frac{1}{(k-1)!\,\ell!} + \frac{1}{k!\,(\ell-1)!} = \frac{k+\ell}{k!\,\ell!}$, and since we can generate each permutation in $\mathcal{S}_{k+\ell}$ in this way, the result follows. □

33.12. Exterior derivative. Let $U \subset E$ be c^∞-open in a convenient vector space E, and let $\omega \in C^\infty(U, L^k_{\text{alt}}(E;\mathbb{R}))$ be a kinematic k-form on U. We define the exterior derivative $d\omega \in C^\infty(U, L^{k+1}_{\text{alt}}(E;\mathbb{R}))$ as the skew symmetrization of the derivative $d\omega(x) : E \to L^k_{\text{alt}}(E;\mathbb{R})$ (sorry for the two notions of d, it's only local); i.e.

$$\begin{aligned}(1)\qquad (d\omega)(x)(X_0,\dots,X_k) &= \sum_{i=0}^k (-1)^i d\omega(x)(X_i)(X_0,\dots,\widehat{X_i},\dots,X_k)\\ &= \sum_{i=0}^k (-1)^i d(\omega(\quad)(X_0,\dots,\widehat{X_i},\dots,X_k))(x)(X_i)\end{aligned}$$

where $X_i \in E$. Next we view the X_i as 'constant vector fields' on U and try to replace them by kinematic vector fields. Let us compute first for $X_j \in C^\infty(U,E)$, where we suppress obvious evaluations at $x \in U$:

$$\begin{aligned}&\sum_i (-1)^i X_i(\omega\circ(X_0,\dots,\widehat{X_i},\dots,X_k))(x) =\\
&\quad= \sum_i (-1)^i (d\omega(x).X_i)(X_0,\dots,\widehat{X_i},\dots,X_k)+\\
&\qquad+\sum_{j<i} (-1)^i \omega\circ(X_0,\dots,dX_j(x).X_i,\dots,\widehat{X_i},\dots,X_k)+\\
(2)\qquad&\qquad+\sum_{i<j} (-1)^i \omega\circ(X_0,\dots,\widehat{X_i},\dots,dX_j(x).X_i,\dots,X_k) =\\
&\quad= \sum_i (-1)^i (d\omega(x).X_i)(X_0,\dots,\widehat{X_i},\dots,X_k)+\\
&\qquad+\sum_{j<i} (-1)^{i+j} \omega\circ(dX_j(x).X_i - dX_i(x).X_j, X_0,\dots,\widehat{X_j},\dots,\widehat{X_i},\dots,X_k)\\
&\quad= \sum_i (-1)^i (d\omega(x).X_i)(X_0,\dots,\widehat{X_i},\dots,X_k)+\\
&\qquad+\sum_{j<i} (-1)^{i+j} \omega\circ([X_i,X_j], X_0,\dots,\widehat{X_j},\dots,\widehat{X_i},\dots,X_k).\end{aligned}$$

Combining (2) and (1) gives the *global formula for the exterior derivative*

$$\begin{aligned}(3)\qquad (d\omega)(x)(X_0,\dots,X_k) &= \sum_{i=0}^k (-1)^i X_i(\omega\circ(X_0,\dots,\widehat{X_i},\dots,X_k))+\\ &\quad+\sum_{i<j} (-1)^{i+j} \omega\circ([X_i,X_j], X_0,\dots,\widehat{X_i},\dots,\widehat{X_j},\dots,X_k).\end{aligned}$$

Formula (3) defines the exterior derivative for modular forms on $\mathfrak{X}(M)$, $C^\infty(M \leftarrow DM)$, and $C^\infty(M \leftarrow D^{[1,\infty)}M)$, since it gives multilinear module homomorphisms by the Lie module properties of the Lie bracket, see (32.5) and (32.8).

The local formula (1) gives the exterior derivative on $C^\infty(L^k_{\text{alt}}(TM, M\times\mathbb{R}))$: Local expressions (1) for two different charts describe the same differential form since both can be written in the global form (3), and the canonical mapping $C^\infty(L^k_{\text{alt}}(TM, M\times\mathbb{R})) \to \operatorname{Hom}^{k,\text{ alt}}_{C^\infty(M,\mathbb{R})}(\mathfrak{X}(M), C^\infty(M,\mathbb{R}))$ is injective, since we use sheaves on the right hand side.

The first line of the local formula (1) gives an exterior derivative d^{loc} also on the space $C^\infty(L^k_{\text{alt}}(DU,\mathbb{R}))$, where U is an open subset in a convenient vector space E, if we replace $d\omega(x)$ by $D_x\omega : D_0E \to D_0(L^k_{\text{alt}}(D_0E,\mathbb{R}))$ composed with the canonical mapping

$$\begin{aligned} D_0(L^k_{\text{alt}}(D_0E,\mathbb{R})) \xrightarrow{(\ )^{[1]}} D_0(L^k_{\text{alt}}(D_0E,\mathbb{R})) \xrightarrow{(\partial^{[1]})^{-1}} L^k_{\text{alt}}(D_0E,\mathbb{R})'' = \\ = (\Lambda^k(D_0E))''' \xrightarrow{\iota^*} (\Lambda^k(D_0E))' = L^k_{\text{alt}}(D_0E,\mathbb{R}). \end{aligned}$$

Here $\iota : \Lambda^k D_0E \to (\Lambda^k D_0E)''$ is the canonical embedding into the bidual. If we replace d by D in the second expression of the local formula (1) we get the same expression. For $\omega \in C^\infty(U, L^k_{\text{alt}}(D_0E,\mathbb{R}))$ we have

$$\begin{aligned} (d^{\text{loc}}\omega)(x)(X_0,\dots,X_k) &= \sum_{i=0}^k (-1)^i D_x(\omega(\ )(X_0,\dots,\widehat{X_i},\dots,X_k))(X_i) \\ &= \sum_{i=0}^k (-1)^i D_x(\operatorname{ev}_{(X_0,\dots,\widehat{X_i},\dots,X_k)} \circ \omega)(X_i) \\ &= \sum_{i=0}^k (-1)^i D_{\omega(x)}(\operatorname{ev}_{(X_0,\dots,\widehat{X_i},\dots,X_k)}).D_x\omega.X_i \\ &= \sum_{i=0}^k (-1)^i (D^{(1)}_{\omega(x)}\operatorname{ev}_{(X_0,\dots,\widehat{X_i},\dots,X_k)}).(D_x\omega.X_i)^{[1]} \qquad \text{by (28.11.4)} \\ &= \sum_{i=0}^k (-1)^i (\operatorname{ev}_{(X_0\wedge\dots\widehat{X_i}\dots\wedge X_k)})^{**}.(\partial^{[1]})^{-1}.(D_x\omega.X_i)^{[1]} \qquad \text{by (28.11.3)} \\ &= \sum_{i=0}^k (-1)^i \operatorname{ev}_{(X_0\wedge\dots\widehat{X_i}\dots\wedge X_k)}.\iota^*.(\partial^{[1]})^{-1}.(D_x\omega.X_i)^{[1]} \\ &= \sum_{i=0}^k (-1)^i \left(\iota^*\circ(\partial^{[1]})^{-1}\circ(\ )^{[1]}\circ D_x\omega\right)(X_i)(X_0,\dots,\widehat{X_i},\dots,X_k), \end{aligned}$$

since the following diagram commutes:

$$\begin{array}{ccc} (\Lambda^k D_0E)' & \xrightarrow{\operatorname{ev}_{(X_0\wedge\dots\widehat{X_i}\dots\wedge X_k)}} & \mathbb{R} \\ \uparrow{\scriptstyle \iota^*} & & \| \\ (\Lambda^k D_0E)''' & \xrightarrow{(\operatorname{ev}_{(X_0\wedge\dots\widehat{X_i}\dots\wedge X_k)})^{**}} & \mathbb{R} \end{array}$$

The local formula (1) describes by a similar procedure the local exterior derivative d^{loc} also on $C^\infty(L^k_{\text{alt}}(D^{[1,\infty)}M,\mathbb{R}))$.

For the forms of tensorial type (involving Λ^k) there is no exterior derivative in general, since the derivative is not tensorial in general.

For a manifold M let us now consider the following diagram of certain spaces of differential forms.

$$\begin{array}{ccc}
C^\infty(L^k_{\text{alt}}(DM, M\times\mathbb{R})) & \dashrightarrow & \operatorname{Hom}^{k,\text{alt}}_{C^\infty(M,\mathbb{R})}(C^\infty(M\leftarrow DM), C^\infty(M,\mathbb{R})) \\
\downarrow & & \downarrow \\
C^\infty(L^k_{\text{alt}}(D^{[1,\infty)}M, M\times\mathbb{R})) & \dashrightarrow & \operatorname{Hom}^{k,\text{alt}}_{C^\infty(M,\mathbb{R})}(C^\infty(D^{[1,\infty)}M), C^\infty(M,\mathbb{R})) \\
\downarrow & & \downarrow \\
C^\infty(L^k_{\text{alt}}(TM, M\times\mathbb{R}) & \longrightarrow & \operatorname{Hom}^{k,\text{alt}}_{C^\infty(M,\mathbb{R})}(\mathfrak{X}(M), C^\infty(M,\mathbb{R}))
\end{array}$$

If M is a c^∞-open subset in a convenient vector space E, on the two upper left spaces there exists only the local (from formula (1)) exterior derivative d^{loc}. On all other spaces the global (from formula (3)) exterior derivative d makes sense. All canonical mappings in this diagram commute with the exterior derivatives except the dashed ones. The following example (33.13) shows that

(1) The dashed arrows do not commute with the respective exterior derivatives.
(2) The (global) exterior derivative does not respect the spaces on the left hand side of the diagram except the bottom one.
(3) The dashed arrows are not surjective.

The example (33.14) shows that the local exterior derivative on the two upper left spaces does not commute with pullbacks of smooth mappings, not even of diffeomorphisms, in general. So it does not even exist on manifolds. Furthermore, $d^{\text{loc}}\circ d^{\text{loc}}$ is more interesting than 0, see example (33.16).

33.13. Example. *Let U be c^∞-open in a convenient vector space E. If $\omega\in C^\infty(U,E''') = C^\infty(U, L(D^{(1)}_0 E,\mathbb{R}))$ then in general the exterior derivative*

$$d\omega \in \operatorname{Hom}^{2,\text{ alt}}_{C^\infty(U,\mathbb{R})}(C^\infty(U\leftarrow DU), C^\infty(U,\mathbb{R}))$$

is not contained in $C^\infty(U\leftarrow L^2_{\text{alt}}(DU, U\times\mathbb{R}))$.

Proof. Let $X, Y\in C^\infty(U,E'')$. The Lie bracket $[X,Y]$ is given in (32.7), and ω depends only on the $D^{(1)}$-part of the bracket. Thus, we have

$$\begin{aligned}
d\omega(X,Y)(x) &= X(\omega(Y))(x) - Y(\omega(X))(x) - \omega([X,Y])(x)\\
&= \langle X(x), d\langle\omega, Y\rangle_{E''}(x)\rangle_{E'} - \langle Y(x), d\langle\omega, X\rangle_{E''}(x)\rangle_{E'}\\
&\quad - \langle\omega(x), (dY(x)^t)^*.X(x) - (dX(x)^t)^*.Y(x)\rangle_{E''}\\
&= \langle X(x), \langle d\omega(x), Y(x)\rangle_{E''}\rangle_{E'} + \langle X(x), \langle\omega(x), dY(x)\rangle_{E''}\rangle_{E'} -\\
&\quad - \langle Y(x), \langle d\omega(x), X(x)\rangle_{E''}\rangle_{E'} - \langle Y(x), \langle\omega(x), dX(x)\rangle_{E''}\rangle_{E'}\\
&\quad - \langle\omega(x), (dY(x)^*\circ\iota_{E'})^*.X(x)\rangle_{E''} + \langle\omega(x), (dX(x)^*\circ\iota_{E'})^*.Y(x)\rangle_{E''}.
\end{aligned}$$

Let us treat the terms separately which contain derivatives of X or Y. Choosing X constant (but arbitrary) we have to consider only the following expression:

$$\begin{aligned}
&\langle X(x), \langle\omega(x), dY(x)\rangle_{E''}\rangle_{E'} - \langle\omega(x), (dY(x)^* \circ \iota_{E'})^*.X(x)\rangle_{E''} = \\
&\quad = \langle X(x), \omega(x) \circ dY(x)\rangle_{E'} - \langle\omega(x), \iota_{E'}^*.dY(x)^{**}.X(x)\rangle_{E''} \\
&\quad = \langle X(x), dY(x)^*.\omega(x)\rangle_{E'} - \langle\iota_{E'}^{**}.\omega(x), dY(x)^{**}.X(x)\rangle_{E''''} \\
&\quad = \langle\iota_{E'''}.\omega(x), dY(x)^{**}.X(x)\rangle_{E''''} - \langle\iota_{E'}^{**}.\omega(x), dY(x)^{**}.X(x)\rangle_{E''''} \\
&\quad = \langle(\iota_{E'''} - \iota_{E'}^{**}).\omega(x), dY(x)^{**}.X(x)\rangle_{E''''},
\end{aligned}$$

which is not 0 in general since $\ker(\iota_{E'''} - \iota_{E'}^{**}) = \iota_{E'}(E')$ at least for Banach spaces, see [Cigler, Losert, Michor, 1979, 1.15], applied to $\iota_{E'}$. So we may assume that $(\iota_{E'''} - \iota_{E'}^{**}).\omega(x) \neq 0 \in E''''$. We choose a non-reflexive Banach space which is isomorphic to its bidual ([James, 1951]) and we choose as $dY(x)$ this isomorphism, then $dY(x)^{**}$ is also an isomorphism, and a suitable $X(x)$ makes the expression nonzero. □

Note that this also shows that for general convenient vector spaces E the exterior derivative $d\omega$ is in $C^\infty(U, L^2_{\text{alt}}(D_0^{(1)}E, \mathbb{R}))$ only if $\omega \in C^\infty(M \leftarrow T'M)$. Note that even for $\omega : U \to E'''$ a constant 1-form of order 1 we need not have $d\omega = 0$.

33.14. Example. *There exist c^∞-open subsets U and V in a Banach space E, a diffeomorphism $f : U \to V$, and a 1-form $\omega \in C^\infty(U, L(E'', \mathbb{R}))$ such that $d^{\text{loc}} f^*\omega \neq f^* d^{\text{loc}}\omega$.*

Proof. We start in a more general situation. Let $f : U \to V \subset F$ be a smooth mapping, and let $X_x, Y_x \in D_x^{(1)}U = E''$. Then we have

$$\begin{aligned}
d^{\text{loc}}(f^*\omega)_x(X_x, Y_x) &= D_x(f^*\omega(\ \ ).Y_x).X_x - D_x(f^*\omega(\ \ ).X_x).Y_x \\
&= D_x(\omega(f(\ \ )).D_{(\ \ )}f.Y_x).X_x - \dots \\
&= X_x\langle\omega \circ f, D_{(\ \ )}f.Y_x\rangle_{F''} - \dots \\
&= d\langle\omega \circ f, df(\ \ )^{**}.Y_x\rangle_{F''}(x)^{**}.X_x - \dots \text{ by (32.6)} \\
&= d\langle\omega(f(\ \ )), df(x)^{**}.Y_x\rangle_{F''}(x)^{**}.X_x + \\
&\qquad + d\langle\omega(f(x)), df(\ \ )^{**}.Y_x\rangle_{F''}(x)^{**}.X_x - \dots \quad \text{by (32.6)} \\
f^*(d^{\text{loc}}\omega)_x(X_x, Y_x) &= (d^{\text{loc}}\omega)_{f(x)}(D_x f.X_x, D_x f.Y_x) \\
&= D_{f(x)}(\omega_{(\ \ )}.D_x f.Y_x).D_x f.X_x - D_{f(x)}(\omega_{(\ \ )}.D_x f.X_x).D_x f.Y_x \\
&= d\langle\omega(\ \ ), df(x)^{**}.Y_x\rangle_{F''}(f(x))^{**}.df(x)^{**}.X_x - \dots
\end{aligned}$$

Recall that for $\ell \in H' = L(H, \mathbb{R})$ the bidual mapping satisfies $L(H'', \mathbb{R}) \ni \ell^{**} = \iota_{H'}(\ell) \in H'''$. Then for the difference we get

$$\begin{aligned}
&d^{\text{loc}}(f^*\omega)_x(X_x, Y_x) - f^*(d^{\text{loc}}\omega)_x(X_x, Y_x) \\
&\quad = d\langle\omega(f(x)), df(\ \ )^{**}.Y_x\rangle_{F''}(x)^{**}.X_x - d\langle\omega(f(x)), df(\ \ )^{**}.X_x\rangle_{F''}(x)^{**}.Y_x \\
&\quad = \langle i_{F'''}\omega(f(x)), d(df(\ \ )^{**}.Y_x)(x)^{**}.X_x - d(df(\ \ )^{**}.X_x)(x)^{**}.Y_x\rangle_{F''''}.
\end{aligned}$$

This expression does not vanish in general, e.g., when the following choices are made: We put $\omega(f(x)) = \iota_{F'}.\ell = \ell^{**}$ for $\ell \in F'$, and we have

$$\begin{aligned} d(d(\ell\circ f)(\quad)^{**}Y_x)(x)^{**}.X_x &= d(d\langle \ell, f\rangle_F(\quad)^{**}Y_x)(x)^{**}.X_x \\ &= d\langle \iota_{F'}\ell, df(\quad)^{**}Y_x\rangle_{F''}(x)^{**}.X_x \\ &= \langle \iota_{F'''}\ell^{**}, d(df(\quad)^{**}Y_x)(x)^{**}.X_x\rangle_{F''''}, \end{aligned}$$

which is not symmetric in general for $\ell\circ f = \mathrm{ev} : G'\times G\to\mathbb{R}$ (for a non reflexive Banach space G) by the argument in (32.7). It remains to show that such a factorization of ev over a diffeomorphism f and $\ell\in(G'\times G)'$ is possible. Choose $(\alpha, x)\in G'\times G$ such that $\langle\alpha, x\rangle = 1$, and consider

$$\begin{aligned} G'\times G = G'\times\ker\alpha\times\mathbb{R}.x &\xrightarrow{f} G'\times\ker\alpha\times\mathbb{R}.x \xrightarrow{\ell}\mathbb{R} \\ (\beta, y, tx) &\mapsto (\beta, y, \langle\beta, y+tx\rangle_G.x)\mapsto\langle\beta, y+tx\rangle_G \\ (\beta, y, \tfrac{t-\langle\beta,y\rangle}{\langle\beta,x\rangle}.x) &\leftarrow (\beta, y, tx). \quad\square \end{aligned}$$

33.15. Proposition. *Let $f : M\to N$ be a smooth mapping between smooth manifolds. Then we have*

$$f^*\circ d = d\circ f^* : C^\infty(L^k_{\text{alt}}(TN, N\times\mathbb{R}))\to C^\infty(L^{k+1}_{\text{alt}}(TM, M\times\mathbb{R})).$$

Proof. Since by (33.12) the local and global formula for the exterior derivative coincide on spaces $C^\infty(L^k_{\text{alt}}(D^\alpha M, M\times\mathbb{R}))$ we shall prove the result with help of the local formula. So we may assume that $f : U\to V$ is smooth between c^∞-open sets in convenient vector spaces E and F, respectively. Note that we may use the global formula only if f is a local diffeomorphism, see (33.9).

For $\omega\in C^\infty(V, L^k_{\text{alt}}(F,\mathbb{R}))$, $x\in U$, and $X_i\in E$ we have

$$(f^*\omega)(x)(X_1,\dots,X_k) = \omega(f(x))(df(x).X_1,\dots,df(x).X_k),$$

so by (33.12.1) we may compute

$$(df^*\omega)(x)(X_0,\dots,X_k) = \sum_{i=0}^k(-1)^i d(f^*\omega)(x)(X_i)(X_0,\dots,\widehat{X_i},\dots,X_k)$$

$$\begin{aligned} &= \sum_{i=0}^k(-1)^i\,(d\omega(f(x)).df(x).X_i)(df(x).X_0,\dots,\widehat{i},\dots,df(x).X_k) \\ &+\sum_{i=0}^k(-1)^i\sum_{j<i}\omega(f(x))(df(x).X_0,\dots,d^2f(x).(X_i,X_j),\dots,\widehat{i},\dots,df(x).X_k) \\ &+\sum_{i=0}^k(-1)^i\sum_{j>i}\omega(f(x))(df(x).X_0,\dots,\widehat{i},\dots,d^2f(x).(X_i,X_j),\dots,df(x).X_k) \end{aligned}$$

$$= \sum_{i=0}^{k}(-1)^i\, d\omega(f(x))(df(x).X_0,\dots,df(x).X_k)$$
$$+ \sum_{j<i}(-1)^{i+j}\omega(f(x))(d^2f(x).(X_i,X_j) - d^2f(x).(X_j,X_i),$$
$$df(x).X_0,\dots,\widehat{j},\dots,\widehat{i},\dots,df(x).X_k)$$
$$= (f^*d\omega)(x)(X_0,\dots,X_k) + 0. \quad \square$$

33.16. Example. *There exists a smooth function*

$$f \in C^\infty(E,\mathbb{R}) = C^\infty(E, L^0_{\text{alt}}(D^{(1)}E,\mathbb{R}))$$

such that

$$0 \neq d^{\text{loc}}d^{\text{loc}}f \in C^\infty(E, L^2_{\text{alt}}(D^{(1)}E,\mathbb{R})).$$

Proof. Let $f \in C^\infty(E,\mathbb{R})$, $X_x, Y_x \in D^{(1)}_x E = E''$. Then we have

$$\begin{aligned}
(d^{\text{loc}}f)_x(X_x) &= df(x)^{**}.X_x = \langle \iota_{F'}.df(x), X_x\rangle_{E''}\\
&= \langle X_x, df(x)\rangle_{E'}\\
(d^{\text{loc}}d^{\text{loc}}f)_x(X_x,Y_x) &=\\
&= d\langle Y_x, df(\quad)\rangle_{E'}(x)^{**}.X_x - d\langle X_x, df(\quad)\rangle_{E'}(x)^{**}.Y_x\\
&= \langle \iota_{E''}.Y_x, d(df)(x)^{**}.X_x\rangle_{E'''} - \langle \iota_{E''}.X_x, d(df)(x)^{**}.Y_x\rangle_{E'''}\\
&= \langle d(df)(x)^{**}.X_x, Y_x\rangle_{E''} - \langle d(df)(x)^{**}.Y_x, X_x\rangle_{E''},
\end{aligned}$$

which does not vanish in general by the argument in (32.7). $\square$

33.17. Lie derivatives. Let D^α denote one of T, D, or $D^{[1,\infty)}$. For a vector field $X \in C^\infty(M \leftarrow D^\alpha M)$ and $\omega \in \operatorname{Hom}^{k,\,\text{alt}}_{C^\infty(M,\mathbb{R})}(C^\infty(M \leftarrow D^\alpha M), C^\infty(M,\mathbb{R}))$ we define the *Lie derivative* $\mathcal{L}_X\omega$ of ω along X by

$$(\mathcal{L}_X\omega)|_U(Y_1,\dots,Y_k) = X(\omega(Y_1,\dots,Y_k)) - \sum_{i=1}^{k}\omega|_U(Y_1,\dots,[X,Y_i],\dots,Y_k),$$

for $Y_1,\dots,Y_k \in C^\infty(U \leftarrow D^\alpha U)$. From (32.5) it follows that

$$\mathcal{L}_X\omega \in \operatorname{Hom}^{k,\,\text{alt}}_{C^\infty(M,\mathbb{R})}(C^\infty(M \leftarrow D^\alpha M), C^\infty(M,\mathbb{R})).$$

33.18. Theorem. *The following formulas hold for $C^\infty(L^k_{\text{alt}}(TM, M\times\mathbb{R}))$ and for the spaces $\operatorname{Hom}^{k,\,\text{alt}}_{C^\infty(M,\mathbb{R})}(C^\infty(M \leftarrow D^\alpha M), C^\infty(M,\mathbb{R}))$ where D^α is any of D, $D^{[1,\infty)}$, or T.*

(1) $i_X(\varphi\wedge\psi) = i_X\varphi\wedge\psi + (-1)^{\deg\varphi}\varphi\wedge i_X\psi$.
(2) $\mathcal{L}_X(\varphi\wedge\psi) = \mathcal{L}_X\varphi\wedge\psi + \varphi\wedge\mathcal{L}_X\psi$.
(3) $d(\varphi\wedge\psi) = d\varphi\wedge\psi + (-1)^{\deg\varphi}\varphi\wedge d\psi$.

(4) $d^2 = d \circ d = \frac{1}{2}[d, d] = 0$.
(5) $[\mathcal{L}_X, d] = \mathcal{L}_X \circ d - d \circ \mathcal{L}_X = 0$.
(6) $[i_X, d] = i_X \circ d + d \circ i_X = \mathcal{L}_X$.
(7) $[\mathcal{L}_X, \mathcal{L}_Y] = \mathcal{L}_X \circ \mathcal{L}_Y - \mathcal{L}_Y \circ \mathcal{L}_X = \mathcal{L}_{[X,Y]}$.
(8) $[\mathcal{L}_X, i_Y] = \mathcal{L}_X i_Y - i_Y \mathcal{L}_X = i_{[X,Y]}$.
(9) $[i_X, i_Y] = i_X i_Y + i_Y i_X = 0$.
(10) $\mathcal{L}_{f.X}\varphi = f.\mathcal{L}_X\varphi + df \wedge i_X\varphi$.

Remark. In this theorem we used the graded commutator for graded derivations $[D_1, D_2] := D_1 \circ D_2 - (-1)^{\deg(D_1)\deg(D_2)} D_2 \circ D_1$. We will elaborate this notion in (35.1) below.

The left hand side of (6) maps the subspace $C^\infty(L^k_{\text{alt}}(TM, M \times \mathbb{R}))$ of the space of modular differential forms $\operatorname{Hom}^{k,\text{ alt}}_{C^\infty(M,\mathbb{R})}(\mathfrak{X}(M), C^\infty(M, \mathbb{R}))$ into itself, thus the Lie derivative $\mathcal{L}_X$ also does. We do not know whether this is true for the other spaces on the left hand side of the diagram in (33.12).

Proof. All results will be proved in $\operatorname{Hom}^{k,\text{ alt}}_{C^\infty(M,\mathbb{R})}(C^\infty(M \leftarrow D^\alpha M), C^\infty(M, \mathbb{R}))$, so they also hold in the subspace $C^\infty(M \leftarrow L^k_{\text{alt}}(TM, M \times \mathbb{R}))$.

(9) is obvious and (1) was shown in (33.11).

(8) Take the difference of the following two expressions:

$$(\mathcal{L}_X i_Y \omega)(Z_1, \dots, Z_k) = X((i_Y\omega)(Z_1, \dots, Z_k)) - \sum_{i=1}^k (i_Y\omega)(Z_1, \dots, [X, Z_i], \dots, Z_k)$$
$$= X(\omega(Y, Z_1, \dots, Z_k)) - \sum_{i=1}^k \omega(Y, Z_1, \dots, [X, Z_i], \dots, Z_k)$$
$$(i_Y \mathcal{L}_X \omega)(Z_1, \dots, Z_k) = \mathcal{L}_X \omega(Y, Z_1, \dots, Z_k)$$
$$= X(\omega(Y, Z_1, \dots, Z_k)) - \omega([X, Y], Z_1, \dots, Z_k) -$$
$$- \sum_{i=1}^k \omega(Y, Z_1, \dots, [X, Z_i], \dots, Z_k).$$

(2) Let φ be of degree p and ψ of degree q. We prove the result by induction on $p+q$. Suppose that (2) is true for $p+q < k$. Then for X we have by part (8), by 1, and by induction

$$(i_Y \mathcal{L}_X)(\varphi \wedge \psi) = (\mathcal{L}_X i_Y)(\varphi \wedge \psi) - i_{[X,Y]}(\varphi \wedge \psi)$$
$$= \mathcal{L}_X(i_Y\varphi \wedge \psi + (-1)^p \varphi \wedge i_Y\psi) - i_{[X,Y]}\varphi \wedge \psi - (-1)^p \varphi \wedge i_{[X,Y]}\psi$$
$$= \mathcal{L}_X i_Y \varphi \wedge \psi + i_Y\varphi \wedge \mathcal{L}_X\psi + (-1)^p \mathcal{L}_X\varphi \wedge i_Y\psi +$$
$$+ (-1)^p \varphi \wedge \mathcal{L}_X i_Y \psi - i_{[X,Y]}\varphi \wedge \psi - (-1)^p \varphi \wedge i_{[X,Y]}\psi$$
$$i_Y(\mathcal{L}_X\varphi \wedge \psi + \varphi \wedge \mathcal{L}_X\psi) = i_Y \mathcal{L}_X \varphi \wedge \psi + (-1)^p \mathcal{L}_X\varphi \wedge i_Y\psi +$$
$$+ i_Y\varphi \wedge \mathcal{L}_X\psi + (-1)^p \varphi \wedge i_Y \mathcal{L}_X \psi.$$

Using again (8), we get the result since the i_Y for all local vector fields Y together act point separating on each space of differential forms, in both cases of the convention (33.2).

(6) follows by summing up the following parts.

$$
\begin{aligned}
(\mathcal{L}_{X_0}\varphi)(X_1,\dots,X_k) &= X_0(\varphi(X_1,\dots,X_k))+\\
&\quad+\sum_{j=1}^{k}(-1)^{0+j}\varphi([X_0,X_j],X_1,\dots,\widehat{X_j},\dots,X_k)\\
(i_{X_0}d\varphi)(X_1,\dots,X_k) &= d\varphi(X_0,\dots,X_k)\\
&=\sum_{i=0}^{k}(-1)^i X_i(\varphi(X_0,\dots,\widehat{X_i},\dots,X_k))+\\
&\quad+\sum_{0\le i<j}(-1)^{i+j}\varphi([X_i,X_j],X_0,\dots,\widehat{X_i},\dots,\widehat{X_j},\dots,X_k).
\end{aligned}
$$

$$
\begin{aligned}
(di_{X_0}\varphi)(X_1,\dots,X_k) &= \sum_{i=1}^{k}(-1)^{i-1}X_i((i_{X_0}\varphi)(X_1,\dots,\widehat{X_i},\dots,X_k))+\\
&\quad+\sum_{1\le i<j}(-1)^{i+j-2}(i_{X_0}\varphi)([X_i,X_j],X_1,\dots,\widehat{X_i},\dots,\widehat{X_j},\dots,X_k)\\
&=-\sum_{i=1}^{k}(-1)^i X_i(\varphi(X_0,X_1,\dots,\widehat{X_i},\dots,X_k))-\\
&\quad-\sum_{1\le i<j}(-1)^{i+j}\varphi([X_i,X_j],X_0,X_1,\dots,\widehat{X_i},\dots,\widehat{X_j},\dots,X_k).
\end{aligned}
$$

(3) We prove the result again by induction on $p+q$. Suppose that (3) is true for $p+q<k$. Then for each local vector field X we have by (6), (2), 1, and by induction

$$
\begin{aligned}
i_X\, d(\varphi\wedge\psi) &= \mathcal{L}_X(\varphi\wedge\psi) - d\, i_X(\varphi\wedge\psi)\\
&= \mathcal{L}_X\varphi\wedge\psi + \varphi\wedge\mathcal{L}_X\psi - d(i_X\varphi\wedge\psi + (-1)^p\varphi\wedge i_X\psi)\\
&= i_Xd\varphi\wedge\psi + di_X\varphi\wedge\psi + \varphi\wedge i_Xd\psi + \varphi\wedge di_X\psi - di_X\varphi\wedge\psi\\
&\qquad - (-1)^{p-1}i_X\varphi\wedge d\psi - (-1)^p d\varphi\wedge i_X\psi - \varphi\wedge di_X\psi\\
&= i_X(d\varphi\wedge\psi + (-1)^p\varphi\wedge d\psi).
\end{aligned}
$$

Since X is arbitrary, the result follows.

(4) This follows by a long but straightforward computation directly from the the global formula (33.12.3), using only the definition of the Lie bracket as a commutator, the Jacobi identity, and cancellation.

(5) $d\mathcal{L}_X = d\,i_X\,d + dd i_X = di_Xd + i_Xdd = \mathcal{L}_Xd$.

(7) By the (graded) Jacobi identity, by (5), by (6), and by (8) we have $[\mathcal{L}_X,\mathcal{L}_Y] = [\mathcal{L}_X,[i_Y,d]] = [[\mathcal{L}_X,i_Y],d] + [i_Y,[\mathcal{L}_X,d]] = [i_{[X,Y]},d] + 0 = \mathcal{L}_{[X,Y]}$.

(10) $\mathcal{L}_{f.X}\varphi = [i_{f.X},d]\varphi = [f.i_X,d]\varphi = fi_Xd\varphi + d(fi_X\varphi) = fi_Xd\varphi + df\wedge i_X\varphi + fdi_X\varphi = f\mathcal{L}_X\varphi + df\wedge i_X\varphi$. □

33.19. Lemma. *Let $X \in \mathfrak{X}(M)$ be a kinematic vector field which has a local flow Fl^X_t. Or more generally, let us suppose that $\varphi : \mathbb{R} \times M \supset U \to M$ is a smooth mapping such that $(t,x) \mapsto (t, \varphi(t,x) = \varphi_t(x))$ is a diffeomorphism $U \to V$, where U and V are open neighborhoods of $\{0\} \times M$ in $\mathbb{R} \times M$, and such that $\varphi_0 = \operatorname{Id}_M$ and $\frac{\partial}{\partial t}\big|_0 \varphi_t = X \in \mathfrak{X}(M)$.*

Then for ω any k-form in $\operatorname{Hom}^{k,\text{ alt}}_{C^\infty(M,\mathbb{R})}(C^\infty(M \leftarrow DM), C^\infty(M,\mathbb{R}))$ we have

$$\begin{aligned}
\tfrac{\partial}{\partial t}|_0 (\varphi_t)^*\omega &= \mathcal{L}_X\omega,\\
\tfrac{\partial}{\partial t}|_0 (\operatorname{Fl}^X_t)^*\omega &= \mathcal{L}_X\omega,\\
\tfrac{\partial}{\partial t} (\operatorname{Fl}^X_t)^*\omega &= (\operatorname{Fl}^X_t)^*\mathcal{L}_X\omega = \mathcal{L}_X(\operatorname{Fl}^X_t)^*\omega.
\end{aligned}$$

In particular, for a vector field X with a local flow the Lie derivative $\mathcal{L}_X$ maps the spaces $C^\infty(L^k_{\text{alt}}(D^\alpha M, M \times \mathbb{R}))$ into themselves, for $D^\alpha = T$, D, and $D^{(p)}$.

Proof. For $Y_i \in C^\infty(M \leftarrow D^\alpha M)$ we have

$$\begin{aligned}
(\tfrac{\partial}{\partial t}|_0 (\varphi_t)^*\omega)(Y_1,\dots,Y_k) &= \tfrac{\partial}{\partial t}|_0 (\omega((\varphi_t^{-1})^*Y_1,\dots,(\varphi_t^{-1})^*Y_k)\circ\varphi_t)\\
&= \sum_{i=1}^k \omega(Y_1,\dots,\tfrac{\partial}{\partial t}|_0 (\varphi_t^{-1})^*Y_i,\dots,Y_k) + \tfrac{\partial}{\partial t}|_0 (\varphi_t)^*(\omega(Y_1,\dots,Y_p))\\
&= X(\omega(Y_1,\dots,Y_k)) - \sum_{i=1}^k \omega|_U(Y_1,\dots,[X,Y_i],\dots,Y_k),
\end{aligned}$$

where at the end we used (32.15). This proves the first two assertions.

For the third assertion we proceed as follows

$$\begin{aligned}
\tfrac{d}{dt}(\operatorname{Fl}^X_t)^*\omega &= \tfrac{d}{ds}|_0 (\operatorname{Fl}^X_t)^*(\operatorname{Fl}^X_s)^*\omega\\
&= (\operatorname{Fl}^X_t)^*\tfrac{d}{ds}|_0 (\operatorname{Fl}^X_s)^*\omega\\
&= (\operatorname{Fl}^X_t)^*\mathcal{L}_X\omega\\
\tfrac{d}{dt}(\operatorname{Fl}^X_t)^*\omega &= \tfrac{d}{ds}|_0 (\operatorname{Fl}^X_s)^*(\operatorname{Fl}^X_t)^*\omega\\
&= \mathcal{L}_X(\operatorname{Fl}^X_t)^*\omega.
\end{aligned}$$

We may commute $\frac{d}{ds}|_0$ with the bounded linear mapping $(\operatorname{Fl}^X_t)^*$ from the space of differential forms on U to that of forms on V, where V is open in U such that $\operatorname{Fl}^X_r(V) \subset U$ for all $r \in [0,t]$. We may find such open U and V because the c^∞-topology on $\mathbb{R} \times M$ is the product of the c^∞-topologies, by corollary (4.15). □

33.20. Lemma of Poincaré. *Let $\omega \in C^\infty(U, L^{k+1}_{\text{alt}}(D^\alpha_0 E; F))$ be a closed form i.e., $d\omega = 0$, where U is a star-shaped c^∞-open subset of a convenient vector space E, with values in a convenient vector space F. Here D^α may be any of T, D, $D^{(k)}$, etc.*

Then ω is exact, i.e. $\omega = d\varphi$ where

$$\varphi(x)(v_1,\dots,v_k) = \int_0^1 t^k\omega(tx)(x,v_1,\dots,v_k)dt$$

is a differential form $\varphi \in C^\infty(U, L^k_{\text{alt}}(D^\alpha_0 E, F))$.

Proof. We consider $\mu : \mathbb{R} \times E \to E$, given by $\mu(t,x) = \mu_t(x) = tx$. Let $I \in \mathfrak{X}(E)$ be the vector field $I(x) = x$, then $\mu(e^t, x) = \text{Fl}^I_t(x)$. So for (x,t) in a neighborhood of $U \times (0,1]$, in $\text{Hom}^{k,\text{ alt}}_{C^\infty(U,\mathbb{R})}(C^\infty(U \leftarrow D^\alpha U), C^\infty(U,\mathbb{R}))$ we have

$$\begin{aligned}\tfrac{d}{dt}\mu_t^*\omega &= \tfrac{d}{dt}(\text{Fl}^I_{\log t})^*\omega = \tfrac{1}{t}(\text{Fl}^I_{\log t})^*\mathcal{L}_I\omega \text{ by (33.19)}\\ &= \tfrac{1}{t}\mu_t^*(i_I d\omega + d i_I\omega) = \tfrac{1}{t} d\mu_t^* i_I\omega.\end{aligned}$$

For $X_1, \dots, X_k \in D_0 E$ we may compute

$$\begin{aligned}(\tfrac{1}{t}\mu_t^* i_I\omega)_x(X_1,\dots,X_k) &= \tfrac{1}{t}(i_I\omega)_{tx}(D_x\mu_t.X_1,\dots,D_x\mu_t.X_k)\\ &= \tfrac{1}{t}\omega_{tx}(tx, D_x\mu_t.X_1,\dots,D_x\mu_t.X_k) = \omega_{tx}(x, D_x\mu_t.X_1,\dots,D_x\mu_t.X_k).\end{aligned}$$

Since $T_x(\mu_t) = t.\text{Id}_E$ and $D^{(1)}\mu_t = \mu_t^{**} = t.\text{Id}_{E''}$ we can make the last computation more explicit if all $X_i \in E$ or E''. So if $k \geq 0$, the k-form $\frac{1}{t}\mu_t^* i_I\omega$ is defined and smooth in (t,x) for all $t \in [0,1]$ and describes a smooth curve in $C^\infty(U, L^k_{\text{alt}}(D^\alpha_0 E, F))$. Clearly, $\mu_1^*\omega = \omega$ and $\mu_0^*\omega = 0$, thus

$$\begin{aligned}\omega = \mu_1^*\omega - \mu_0^*\omega &= \int_0^1 \tfrac{d}{dt}\mu_t^*\omega dt\\ &= \int_0^1 d(\tfrac{1}{t}\mu_t^* i_I\omega)dt = d\left(\int_0^1 \tfrac{1}{t}\mu_t^* i_I\omega dt\right) = d\varphi. \quad \square\end{aligned}$$

Remark. We were unable to prove the Lemma of Poincaré for modular forms which are given by module homomorphisms, because $\mu_t^*\omega$ does not make sense in a differentiable way for $t = 0$. One may ask whether a closed modular differential form $\omega \in \text{Hom}^{k,\text{ alt}}_{C^\infty(M,\mathbb{R})}(C^\infty(M \leftarrow D^\alpha M), C^\infty(M,\mathbb{R}))$ already has to be in $C^\infty(L^k_{\text{alt}}(D^\alpha M, M \times \mathbb{R}))$.

33.21. Review of operations on differential forms.

Space	$\mathcal{L}_X$	d	f^*
$C^\infty(M \leftarrow \Lambda^*(D'M))$	–	–	+
$C^\infty(M \leftarrow \Lambda^*(T'M))$	–	–	+
$\Lambda^*_{C^\infty(M,\mathbb{R})} \text{Hom}_{C^\infty(M,\mathbb{R})}(C^\infty(M \leftarrow DM), C^\infty(M,\mathbb{R}))$	–	–	diff
$\Lambda^*_{C^\infty(M,\mathbb{R})} C^\infty(M \leftarrow T'M)$	–	–	+
$C^\infty(L^*_{\text{alt}}(DM, M \times \mathbb{R}))$	flow	–	+
$\text{Hom}^{*,\text{ alt}}_{C^\infty(M,\mathbb{R})}(C^\infty(M \leftarrow DM), C^\infty(M,\mathbb{R}))$	+	+	diff
$C^\infty(L^*_{\text{alt}}(D^{[1,\infty)}M, M \times \mathbb{R}))$	flow	–	+
$\text{Hom}^{*,\text{ alt}}_{C^\infty(M,\mathbb{R})}(C^\infty(D^{[1,\infty)}M), C^\infty(M,\mathbb{R}))$	+	+	diff
$C^\infty(M \leftarrow L^*_{\text{alt}}(D^{(1)}M, M \times \mathbb{R}))$	flow	–	+
$\text{Hom}^{*,\text{ alt}}_{C^\infty(M,\mathbb{R})}(C^\infty(M \leftarrow D^{(1)}M), C^\infty(M,\mathbb{R}))$	flow	?	diff
$C^\infty(M \leftarrow L^*_{\text{alt}}(TM, M \times \mathbb{R}))$	+	+	+
$\text{Hom}^{*,\text{ alt}}_{C^\infty(M,\mathbb{R})}(\mathfrak{X}(M), C^\infty(M,\mathbb{R}))$	+	+	diff

In this table a '–' means that the space is not invariant under the operation on top of the column, a '+' means that it is invariant, 'diff' means that it is invariant under f^* only for diffeomorphisms f, and 'flow' means that it is invariant under $\mathcal{L}_X$ for all kinematic vector fields X which admit local flows.

33.22. Remark. From the table (33.21) we see that for many purposes only one space of differential forms is fully suited. We will denote from now on by

$$\Omega^k(M) := C^\infty(M \leftarrow L^k_{\text{alt}}(TM, M \times \mathbb{R}))$$

the *space of differential forms*, for a smooth manifold M. By (30.1) it carries the structure of a convenient vector space induced by the closed embedding

$$\begin{aligned} \Omega^k(M) &\to \prod_\alpha C^\infty(U_\alpha, L^k_{\text{alt}}(E, \mathbb{R})) \\ s &\mapsto pr_2 \circ \psi_\alpha \circ (s|U_\alpha), \end{aligned}$$

where $(U_\alpha, u_\alpha : U_\alpha \to E)$ is a smooth atlas for the manifold M, and where $\psi_\alpha := L^k_{\text{alt}}(Tu_\alpha^{-1}, \mathbb{R}))$ is the induced vector bundle chart.

Similarly, we denote by

$$\Omega^k(M, V) := C^\infty(M \leftarrow L^k_{\text{alt}}(TM, M \times V))$$

the *space of differential forms with values in a convenient vector space* V, and by

$$\Omega^k(M; E) := C^\infty(M \leftarrow L^k_{\text{alt}}(TM, E))$$

the *space of differential forms with values in a vector bundle* $p : E \to M$.

Lemma. *The space $\Omega^k(M)$ is isomorphic as convenient vector space to the closed linear subspace of $C^\infty(TM \times_M \ldots \times_M TM, \mathbb{R})$ consisting of all fiberwise k-linear alternating smooth functions in the vector bundle structure $TM \oplus \cdots \oplus TM$ from* (29.5).

Proof. By (27.17), the space $C^\infty(TM \times_M \ldots \times_M TM, \mathbb{R})$ carries the initial structure with respect to the closed linear embedding

$$C^\infty(TM \times_M \ldots \times_M TM, \mathbb{R}) \to \prod_\alpha C^\infty(u_\alpha(U_\alpha) \times E \times \ldots \times E, \mathbb{R}),$$

and $C^\infty(u_\alpha(U_\alpha) \times E \times \ldots \times E, \mathbb{R})$ contains an isomorphic copy of $C^\infty(U_\alpha, L^k_{\text{alt}}(E, \mathbb{R}))$ as closed linear subspace by cartesian closedness. $\square$

Corollary. *All the important mappings are smooth:*

$$\begin{aligned} d &: \Omega^k(M) \to \Omega^{k+1}(M) \\ i &: \mathfrak{X}(M) \times \Omega^k(M) \to \Omega^{k-1}(M) \\ \mathcal{L} &: \mathfrak{X}(M) \times \Omega^k(M) \to \Omega^k(M) \\ f^* &: \Omega^k(M) \to \Omega^k(N) \end{aligned}$$

where $f : N \to M$ is a smooth mapping. The last mappings is even smooth considered as mapping $(f, \omega) \mapsto f^*\omega$, $C^\infty(N, M) \times \Omega^k(M) \to \Omega^k(N)$.

Recall once more the formulas for $\omega \in \Omega^k(M)$ and $X_i \in \mathfrak{X}(M)$, from (33.12.3), (33.10), (33.17) *:*

$$(d\omega)(x)(X_0, \dots, X_k) = \sum_{i=0}^{k}(-1)^i X_i(\omega(X_0, \dots, \widehat{X_i}, \dots, X_k)) +$$
$$+ \sum_{i<j}(-1)^{i+j}\omega([X_i, X_j], X_0, \dots, \widehat{X_i}, \dots, \widehat{X_j}, \dots, X_k),$$
$$(i_X\varphi)(X_1, \dots, X_{k-1}) = \varphi(X, X_1, \dots, X_{k-1}),$$
$$(\mathcal{L}_X\omega)(X_1, \dots, X_k) = X(\omega(X_1, \dots, X_k)) - \sum_{i=1}^{k}\omega(X_1, \dots, [X, X_i], \dots, X_k).$$

Proof. For d we use the local formula (33.12.1), smoothness of i is obvious, and for the Lie derivative we may use formula (33.18.6). The pullback mapping f^* is induced from $Tf \times \dots \times Tf$. □

34. De Rham Cohomology

Section (33) provides us with several graded commutative differential algebras consisting of various kinds of differential forms for which we can define De Rham cohomology, namely all those from the list (33.21) which have + in the d-column. But among these only $C^\infty(L^*_{\text{alt}}(TM, M \times \mathbb{R}))$ behaves functorially for all smooth mappings; the others are only functors over categories of manifolds where the morphisms are just the local diffeomorphisms. So we treat here cohomology only for these differential forms.

34.1. De Rham cohomology. Recall that for a smooth manifold M we have denoted

$$\Omega^k(M) := C^\infty(L^k_{\text{alt}}(TM, M \times \mathbb{R})).$$

We now consider the graded algebra $\Omega(M) = \bigoplus_{k\geq 0}\Omega^k(M)$ of all differential forms on M. Then the space $Z(M) := M) := \{\omega \in \Omega(M) : d\omega = 0\}$ of *closed forms* is a graded subalgebra of Ω (i. e. it is a subalgebra, and $\Omega^k(M) \cap Z(M) = Z^k(M)$), and the space $B(M) := \{d\varphi : \varphi \in \Omega(M)\}$ of *exact forms* is a graded ideal in $Z(M)$. This follows directly from $d^2 = 0$ and the derivation property $d(\varphi \wedge \psi) = d\varphi \wedge \psi + (-1)^{\deg\varphi}\varphi \wedge d\psi$ of the exterior derivative.

Definition. The algebra

$$H^*(M) := \frac{Z(M)}{B(M)} = \frac{\{\omega \in \Omega(M) : d\omega = 0\}}{\{d\varphi : \varphi \in \Omega(M)\}}$$

is called the *De Rham cohomology algebra* of the manifold M. It is graded by

$$H^*(M) = \bigoplus_{k\geq 0} H^k(M) = \bigoplus_{k\geq 0} \frac{\ker(d:\Omega^k(M)\to\Omega^{k+1}(M))}{\operatorname{im}(d:\Omega^{k-1}(M)\to\Omega^k(M))}.$$

If $f: M\to N$ is a smooth mapping between manifolds then $f^*:\Omega(N)\to\Omega(M)$ is a homomorphism of graded algebras by (33.9), which satisfies $d\circ f^* = f^*\circ d$ by (33.15). Thus, f^* induces an algebra homomorphism which we also call $f^*: H^*(N)\to H^*(M)$. Obviously, each H^k is a contravariant functor from the category of smooth manifolds and smooth mappings into the category of real vector spaces.

34.2. Lemma. *Let f, $g: M\to N$ be smooth mappings between manifolds which are C^∞-homotopic, i.e., there exists $h\in C^\infty(\mathbb{R}\times M, N)$ with $h(0,x)=f(x)$ and $h(1,x)=g(x)$. Then f and g induce the same mapping in cohomology $f^*=g^*: H^*(N)\to H^*(M)$.*

Remark. f, $g\in C^\infty(M,N)$ are called homotopic if there exists a continuous mapping $h:[0,1]\times M\to N$ with $h(0,x)=f(x)$ and $h(1,x)=g(x)$. For finite dimensional manifolds this apparently looser relation in fact coincides with the relation of C^∞-homotopy. We sketch a proof of this statement: let $\varphi:\mathbb{R}\to[0,1]$ be a smooth function with $\varphi(t)=0$ for $t\leq 1/4$, $\varphi(t)=1$ for $t\geq 3/4$, and φ monotone in between. Then consider $\bar h:\mathbb{R}\times M\to N$, given by $\bar h(t,x)=h(\varphi(t),x)$. Now we may approximate $\bar h$ by smooth functions $\tilde h:\mathbb{R}\times M\to N$ without changing it on $(-\infty,1/8)\times M$ where it equals f, and on $(7/8,\infty)\times M$, where it equals g. This is done chartwise by convolution with a smooth function with small support on $\mathbb{R}^m$. See [Bröcker, Jänich, 1973] for a careful presentation of the approximation. It is an open problem to extend this to some infinite dimensional manifolds.

The lemma of Poincaré (33.20) is an immediate consequence of this result.

Proof. For $\omega\in\Omega^k(M)$ we have $h^*\omega\in\Omega^k(\mathbb{R}\times M)$. We consider the insertion operator $\operatorname{ins}_t: M\to\mathbb{R}\times M$, given by $\operatorname{ins}_t(x)=(t,x)$. For $\varphi\in\Omega^k(\mathbb{R}\times M)$ we then have a smooth curve $t\mapsto\operatorname{ins}_t^*\varphi$ in $\Omega^k(M)$.

Consider the integral operator $I_0^1:\Omega^k(\mathbb{R}\times M)\to\Omega^k(M)$ given by $I_0^1(\varphi):=\int_0^1\operatorname{ins}_t^*\varphi\,dt$. Let $T:=\frac{\partial}{\partial t}\in C^\infty(\mathbb{R}\times M\leftarrow T(\mathbb{R}\times M))$ be the unit vector field in direction $\mathbb{R}$.

We have $\operatorname{ins}_{t+s}=\operatorname{Fl}_t^T\circ\operatorname{ins}_s$ for s, $t\in\mathbb{R}$, so

$$\begin{aligned}\tfrac{\partial}{\partial s}\operatorname{ins}_s^*\varphi &= \tfrac{\partial}{\partial t}\big|_0(\operatorname{Fl}_t^T\circ\operatorname{ins}_s)^*\varphi = \tfrac{\partial}{\partial t}\big|_0\operatorname{ins}_s^*(\operatorname{Fl}_t^T)^*\varphi\\ &= \operatorname{ins}_s^*\tfrac{\partial}{\partial t}\big|_0(\operatorname{Fl}_t^T)^*\varphi = (\operatorname{ins}_s)^*\mathcal{L}_T\varphi \qquad \text{by (33.19).}\end{aligned}$$

We have used that $(\operatorname{ins}_s)^*:\Omega^k(\mathbb{R}\times M)\to\Omega^k(M)$ is linear and continuous, and so one may differentiate through it by the chain rule. Then we have in turn

$$\begin{aligned} d\, I_0^1\, \varphi &= d \int_0^1 \operatorname{ins}_t^* \varphi\, dt = \int_0^1 d \operatorname{ins}_t^* \varphi\, dt \\ &= \int_0^1 \operatorname{ins}_t^* d\varphi\, dt = I_0^1\, d\,\varphi \qquad \text{by (33.15).} \\ (\operatorname{ins}_1^* - \operatorname{ins}_0^*)\varphi &= \int_0^1 \tfrac{\partial}{\partial t} \operatorname{ins}_t^* \varphi\, dt = \int_0^1 \operatorname{ins}_t^* \mathcal{L}_T \varphi\, dt \\ &= I_0^1\, \mathcal{L}_T \varphi = I_0^1 (d\, i_T + i_T\, d)\varphi \qquad \text{by (33.18.6).} \end{aligned}$$

Now we define the *homotopy operator* $\bar{h} := I_0^1 \circ i_T \circ h^* : \Omega^k(M) \to \Omega^{k-1}(M)$. Then we get

$$\begin{aligned} g^* - f^* &= (h \circ \operatorname{ins}_1)^* - (h \circ \operatorname{ins}_0)^* = (\operatorname{ins}_1^* - \operatorname{ins}_0^*) \circ h^* \\ &= (d \circ I_0^1 \circ i_T + I_0^1 \circ i_T \circ d) \circ h^* = d \circ \bar{h} - \bar{h} \circ d, \end{aligned}$$

which implies the desired result since for $\omega \in \Omega^k(M)$ with $d\omega = 0$ we have $g^*\omega - f^*\omega = d\bar{h}\omega - \bar{h}d\omega = d\bar{h}\omega$. □

34.3. Lemma. *If a manifold is decomposed into a disjoint union $M = \bigsqcup_\alpha M_\alpha$ of open submanifolds, then $H^k(M) = \prod_\alpha H^k(M_\alpha)$ for all k.*

Proof. $\Omega^k(M)$ is isomorphic to $\prod_\alpha \Omega^k(M_\alpha)$ via $\varphi \mapsto (\varphi|M_\alpha)_\alpha$. This isomorphism commutes with the exterior derivative d and induces the result. □

34.4. The setting for the Mayer-Vietoris Sequence. Let M be a smooth manifold, let U, $V \subset M$ be open subsets which cover M and admit a subordinated smooth partition of unity $\{f_U, f_V\}$ with $\operatorname{supp}(f_U) \subset U$ and $\operatorname{supp}(f_V) \subset V$. We consider the following embeddings:

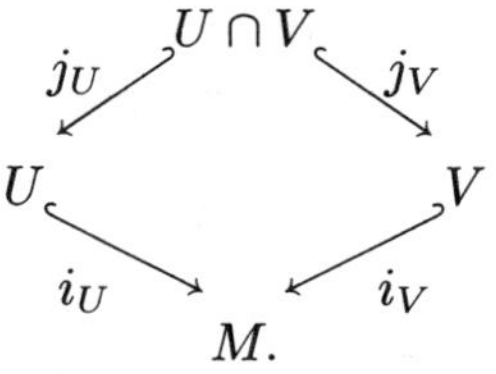

Lemma. *In this situation, the sequence*

$$0 \to \Omega(M) \xrightarrow{\alpha} \Omega(U) \oplus \Omega(V) \xrightarrow{\beta} \Omega(U \cap V) \to 0$$

is exact, where $\alpha(\omega) := (i_U^*\omega, i_V^*\omega)$ and $\beta(\varphi, \psi) = j_U^*\varphi - j_V^*\psi$. We also have $(d \oplus d) \circ \alpha = \alpha \circ d$ and $d \circ \beta = \beta \circ (d \oplus d)$.

Proof. We have to show that α is injective, $\ker \beta = \operatorname{im} \alpha$, and that β is surjective. The first two assertions are obvious. For $\varphi \in \Omega(U \cap V)$ we consider $f_V \varphi \in \Omega(U \cap V)$. Note that $\operatorname{supp}(f_V \varphi)$ is closed in the closed subset $\operatorname{supp}(f_V) \cap U$ of U and contained in the open subset $U \cap V$ of U, so we may extend $f_V \varphi$ by 0 to a smooth form $\varphi_U \in \Omega(U)$. Likewise, we extend $-f_U \varphi$ by 0 to $\varphi_V \in \Omega(V)$. Then we have $\beta(\varphi_U, \varphi_V) = (f_U + f_V)\varphi = \varphi$. □

34.5. Theorem. Mayer-Vietoris sequence. *Let U and V be open subsets in a manifold M which cover M and admit a subordinated smooth partition of unity. Then there is an exact sequence*

$$\dots \to H^k(M) \xrightarrow{\alpha_*} H^k(U) \oplus H^k(V) \xrightarrow{\beta_*} H^k(U \cap V) \xrightarrow{\delta} H^{k+1}(M) \to \cdots$$

It is natural in the triple (M, U, V). The homomorphisms α_* and β_* are algebra homomorphisms, but δ is not.

Proof. This follows from (34.4) and standard homological algebra. □

Since we shall need it later we will now give a detailed description of the connecting homomorphism δ. Let $\{f_U, f_V\}$ be a partition of unity with $\operatorname{supp} f_U \subset U$ and $\operatorname{supp} f_V \subset V$. Let $\omega \in \Omega^k(U \cap V)$ with $d\omega = 0$ so that $[\omega] \in H^k(U \cap V)$. Then $(f_V.\omega, -f_U.\omega) \in \Omega^k(U) \oplus \Omega^k(V)$ is mapped to ω by β, and we have

$$\begin{aligned}\delta[\omega] &= [\alpha^{-1}(d \oplus d)(f_V.\omega, -f_U.\omega)] = [\alpha^{-1}(df_V \wedge \omega, -df_U \wedge \omega)] \\ &= [df_V \wedge \omega] = -[df_U \wedge \omega)],\end{aligned}$$

where we have used the following fact: $f_U + f_V = 1$ implies that on $U \cap V$ we have $df_V = -df_U$, thus $df_V \wedge \omega = -df_U \wedge \omega$, and off $U \cap V$ both are 0.

34.6. Theorem. *Let M be a smooth manifold which is smoothly paracompact. Then the De Rham cohomology of M coincides with the sheaf cohomology of M with coefficients in the constant sheaf $\mathbb{R}$ on M.*

Proof. Since M is smoothly paracompact it is also paracompact, and thus the usual theory of sheaf cohomology using the notion of fine sheafs is applicable. For each k we consider the sheaf Ω^k_M on M which associates to each c^∞-open set $U \subset M$ the convenient vector space $\Omega^k(U)$. Then the following sequence of sheaves

$$\mathbb{R} \to \Omega^0_M \xrightarrow{d} \Omega^1_M \xrightarrow{d} \dots$$

is a resolution of the constant sheaf $\mathbb{R}$ by the lemma of Poincaré (33.20). Since we have smooth partitions of unity on M, each sheaf Ω^k_M is a fine sheaf, so this resolution is acyclic [Godement, 1958], [Hirzebruch, 1962, 2.11.1], and the sequence of global sections may be used to compute the sheaf cohomology of the constant sheaf $\mathbb{R}$. But this is the De Rham cohomology. □

34.7. Theorem. *Let M be a smooth manifold which is smoothly paracompact. Then the De Rham cohomology of M coincides with the singular cohomology with coefficients in $\mathbb{R}$ via a canonical isomorphism which is induced by integration of p-forms over smooth singular simplices.*

Proof. Denote by $\mathcal{S}^k_\infty$ the sheaf which is generated by the presheaf of singular smooth cochains with real coefficients. In more detail: let us put $S^k_\infty(U, \mathbb{R}) = \prod_\sigma \mathbb{R} = \mathbb{R}^{C^\infty(\Delta_k, U)}$, where $\sigma : \Delta_k \to U$ is any mapping which extends to a smooth mapping from a neighborhood of the standard k-simplex $\Delta_k \subset \mathbb{R}^{k+1}$ into U, where U is c^∞-open in M. This defines a presheaf. The associated sheaf is denoted by $\mathcal{S}^k_\infty$. The sequence

$$\mathbb{R} \to \mathcal{S}^0_\infty \xrightarrow{\delta^*} \mathcal{S}^1_\infty \xrightarrow{\delta^*} \mathcal{S}^2_\infty \to \dots$$

of sheafs is a resolution, because if U is a small open set, say diffeomorphic to a radial neighborhood of 0 in the modeling convenient vector space, then U is smoothly contractible to a point. Smooth mappings induce mappings in the $\mathcal{S}^*_\infty$-cohomology, thus $H^k(\mathcal{S}^*_\infty(U,\mathbb{R}),\partial) = 0$ for $k > 0$. This implies that the associated sequence of stalks is exact, so the sequence above is a resolution. A standard argument of sheaf theory shows that each sheaf $\mathcal{S}^k_\infty$ is a fine sheaf, so they form an acyclic resolution, and $H^k(\mathcal{S}^*_\infty(M,\mathbb{R}),\partial)$ coincides with the sheaf cohomology with coefficients in the constant sheaf $\mathbb{R}$.

Furthermore, integration of p-forms over smooth singular p-simplices defines a mapping of resolutions

$$\begin{array}{ccccccccc} & \nearrow & \Omega^0 & \longrightarrow & \Omega^1 & \longrightarrow & \Omega^2 & \longrightarrow & \dots \\ \mathbb{R} & & \downarrow & & \downarrow & & \downarrow & & \\ & \searrow & \mathcal{S}^0_\infty & \longrightarrow & \mathcal{S}^1_\infty & \longrightarrow & \mathcal{S}^2_\infty & \longrightarrow & \dots \end{array}$$

which induces an isomorphism from the De Rham cohomology of M to the cohomology $H^*(\mathcal{S}^*_\infty(M,\mathbb{R}),\partial)$.

Now we consider the resolution

$$\mathbb{R} \to \mathcal{S}^0 \xrightarrow{\delta^*} \mathcal{S}^1 \xrightarrow{\delta^*} \mathcal{S}^2 \to \dots$$

of the constant sheaf $\mathbb{R}$, where $\mathcal{S}^k$ is the usual sheaf induced by the singular continuous cochains. Since M is (even smoothly) paracompact and locally contractible, this is an acyclic resolution, and the embedding of smooth singular chains into continuous singular chains defines a mapping of resolutions

$$\begin{array}{ccccccccc} & \nearrow & \mathcal{S}^0 & \longrightarrow & \mathcal{S}^1 & \longrightarrow & \mathcal{S}^2 & \longrightarrow & \dots \\ \mathbb{R} & & \downarrow & & \downarrow & & \downarrow & & \\ & \searrow & \mathcal{S}^0_\infty & \longrightarrow & \mathcal{S}^1_\infty & \longrightarrow & \mathcal{S}^2_\infty & \longrightarrow & \dots \end{array}$$

which induces an isomorphism from the singular cohomology of M to the cohomology $H^*(\mathcal{S}^*_\infty(M,\mathbb{R}),\partial)$. $\square$

35. Derivations on Differential Forms and the Frölicher-Nijenhuis Bracket

35.1. In this section let M be a smooth manifold. We consider the graded commutative algebra

$$\Omega(M) = \bigoplus_{k\geq 0} \Omega^k(M) = \bigoplus_{k=0}^{\infty} C^\infty(M \leftarrow L^k_{\text{alt}}(TM, M\times\mathbb{R})) = \bigoplus_{k=-\infty}^{\infty} \Omega^k(M)$$

of differential forms on M, see (33.22), where $\Omega^0(M) = C^\infty(M,\mathbb{R})$, and where we put $\Omega^k(M) = 0$ for $k < 0$. We denote by $\operatorname{Der}_k \Omega(M)$ the space of all *(graded) derivations* of degree k, i.e., all bounded linear mappings $D : \Omega(M) \to \Omega(M)$ with $D(\Omega^l(M)) \subset \Omega^{k+l}(M)$ and $D(\varphi\wedge\psi) = D(\varphi)\wedge\psi + (-1)^{kl}\varphi\wedge D(\psi)$ for $\varphi\in\Omega^l(M)$.

Convention. In general, derivations need not be of local nature. Thus, we consider each derivation and homomorphism to be a sheaf morphism (compare (32.1) and the definition of modular 1-forms in (33.2)), or we assume that all manifolds in question are again smoothly regular. This is justified by the obvious extension of (32.4) and (33.3).

Lemma. *Then the space* $\operatorname{Der}\Omega(M) = \bigoplus_k \operatorname{Der}_k \Omega(M)$ *is a graded Lie algebra with the graded commutator* $[D_1, D_2] := D_1\circ D_2 - (-1)^{k_1k_2} D_2\circ D_1$ *as bracket. This means that the bracket is graded anticommutative and satisfies the graded Jacobi identity:*

$$[D_1, D_2] = -(-1)^{k_1k_2}[D_2, D_1],$$
$$[D_1,[D_2,D_3]] = [[D_1,D_2],D_3] + (-1)^{k_1k_2}[D_2,[D_1,D_3]]$$

(so that $\operatorname{ad}(D_1) = [D_1, \quad]$ *is itself a derivation of degree* k_1*).*

Proof. Plug in the definition of the graded commutator and compute. □

In section (33) we have already met some graded derivations: for a vector field X on M the derivation i_X is of degree -1, $\mathcal{L}_X$ is of degree 0, and d is of degree 1. In (33.18) we already met some some graded commutators like $\mathcal{L}_X = d\,i_X + i_X\,d = [i_X, d]$.

35.2. A derivation $D\in\operatorname{Der}_k\Omega(M)$ is called *algebraic* if $D \mid \Omega^0(M) = 0$. Then $D(f.\omega) = f.D(\omega)$ for $f\in C^\infty(M,\mathbb{R})$ and $\omega\in\Omega(M)$.

If the spaces $L^k_{\text{alt}}(T_xM;\mathbb{R})$ are all reflexive and have the bornological approximation property, then an algebraic derivation D induces for each $x\in M$ a derivation $D_x \in \operatorname{Der}_k(L^*_{\text{alt}}(T_xM;\mathbb{R}))$, by a method used in (33.5). It is not clear whether it suffices to assume that just the model spaces of M are all reflexive and have the bornological approximation property.

In the sequel, we will consider the space of all *vector valued kinematic differential forms*, which we will define by

$$\Omega(M;TM) = \bigoplus_{k\geq 0} \Omega^k(M;TM) = \bigoplus_{k\geq 0} C^\infty(M \leftarrow L^k_{\text{alt}}(TM;TM))$$

Note that $\Omega^0(M;TM) = \mathfrak{X}(M) = C^\infty(M \leftarrow TM)$. For simplicity's sake, we will not treat other kinds of vector valued differential forms.

Theorem. *(1) For $K \in \Omega^{k+1}(M;TM)$ the formula*

$$\begin{aligned}
(i_K\omega)(X_1,\dots,X_{k+l}) &= \\
&= \tfrac{1}{(k+1)!\,(l-1)!} \sum_{\sigma\in\mathcal{S}_{k+l}} \operatorname{sign}\sigma\,.\omega(K(X_{\sigma 1},\dots,X_{\sigma(k+1)}),X_{\sigma(k+2)},\dots)\\
&= \sum_{i_1<\dots<i_{k+1}} (-1)^{i_1+\dots+i_{k+1}-\frac{(k+1)(k+2)}{2}}\omega(K(X_{i_1},\dots,X_{i_{k+1}}),X_1,\dots,\widehat{X_{i_1}},\dots),\\
i_K f = 0 \quad & \textit{for } f\in C^\infty(M,\mathbb{R}) = \Omega^0(M),
\end{aligned}$$

for $\omega \in \Omega^l(M)$, $X_i \in \mathfrak{X}(M)$ defines an algebraic graded derivation $i_K = i(K) \in \operatorname{Der}_k \Omega(M)$.

(2) We define a bracket $[\ \ ,\ \ ]^\wedge$ on $\Omega^{*+1}(M;TM)$ for $K \in \Omega^{k+1}(M;TM)$, $L \in \Omega^{l+1}(M;TM)$ by

$$[K,L]^\wedge := i_K L - (-1)^{kl} i_L K,$$

where $i_K(L)$ is given by the same formula as in (1). This defines a graded Lie algebra structure with the grading as indicated, and we have $i([K,L]^\wedge) = [i_K, i_L] \in \operatorname{Der}\Omega(M)$. Thus, $i : \Omega^{*+1}(M;TM) \to \operatorname{Der}_*\Omega(M)$ is a homomorphism of graded Lie algebras, which is injective under the assumptions of (35.1).

The concomitant $[\ \ ,\ \ ]^\wedge$ is called the *algebraic bracket* or the *Nijenhuis-Richardson bracket*, compare [Nijenhuis, Richardson, 1967].

Proof. (1) We know that i_X is a derivation of degree -1 for a vector field $X \in \mathfrak{X}(M) = \Omega^0(M;TM)$ by (33.11). By direct evaluation, one gets

$$[i_X, i_K] = i(i_X K). \tag{3}$$

Using this and induction on the sum of the degrees of $K \in \Omega^k(M;TM)$, $\varphi \in \Omega(M)$, and $\psi \in \Omega(M)$, one can then show that

$$i_X i_K(\varphi\wedge\psi) = i_X(i_K\varphi\wedge\psi + (-1)^{k\deg\varphi}\varphi\wedge i_K\psi)$$

holds, which implies that i_K is a derivation of degree k.

(2) By induction on the sum of $k = \deg K - 1$, $l = \deg L - 1$, and $p = \deg \varphi$, and by (3) we have

$$\begin{aligned}
[i_X, [i_K, i_L]]\varphi &= [[i_X, i_K], i_L]\varphi + (-1)^k [i_K, [i_X, i_L]]\varphi \\
&= [i(i_X K), i_L]\varphi + (-1)^k [i_K, i(i_X L)]\varphi \\
&= i\left(i(i_X K)L - (-1)^{(k-1)l} i_L i_X K\right)\varphi \\
&\quad + (-1)^k i\left(i_K i_X L - (-1)^{k(l-1)} i(i_X L)K\right)\varphi \\
&= i\left(i_X i_K L - (-1)^{kl} i_X i_L K\right)\varphi = i\left(i_X [K, L]^\wedge\right)\varphi, \\
i_X [i_K, i_L]\varphi &= [i_X, [i_K, i_L]]\varphi + (-1)^{k+l} [i_K, i_L] i_X \varphi \\
&= i\left(i_X [K, L]^\wedge\right)\varphi + (-1)^{k+l} i([K, L]^\wedge) i_X \varphi \\
&= i\left(i_X [K, L]^\wedge\right)\varphi - i(i_X [K, L]^\wedge)\varphi + i_X i([K, L]^\wedge)\varphi \\
&= i_X i([K, L]^\wedge)\varphi.
\end{aligned}$$

This implies $i([K, L]^\wedge) = [i_K, i_L]$ since the i_X for $X \in TM$ separate points, in both cases of the convention (35.1). From $i_K df = df \circ K$ it follows that the mapping $i : \Omega(M; TM) \to \operatorname{Der}(\Omega(M))$ is injective, so $(\Omega^{*+1}(M; TM), [\quad,\quad]^\wedge)$ is a graded Lie algebra. □

35.3. The exterior derivative d is an element of $\operatorname{Der}_1 \Omega(M)$. In view of the formula $\mathcal{L}_X = [i_X, d] = i_X\, d + d\, i_X$ for vector fields X (see (33.18.6)), we define for $K \in \Omega^k(M; TM)$ the *Lie derivative* $\mathcal{L}_K = \mathcal{L}(K) \in \operatorname{Der}_k \Omega(M)$ by

$$\mathcal{L}_K := [i_K, d] = i_K\, d - (-1)^{k-1} d\, i_K.$$

Since the 1-forms df for all local functions on M separate points on each $T_x M$, the mapping $\mathcal{L} : \Omega(M; TM) \to \operatorname{Der} \Omega(M)$ is injective, because $\mathcal{L}_K f = i_K df = df \circ K$ for $f \in C^\infty(M, \mathbb{R})$.

From (35.2.1) it follows that $i(\mathrm{Id}_{TM})\omega = k\omega$ for $\omega \in \Omega^k(M)$. Hence, $\mathcal{L}(\mathrm{Id}_{TM})\omega = i(\mathrm{Id}_{TM})d\omega - d\, i(\mathrm{Id}_{TM})\omega = (k+1)d\omega - kd\omega = d\omega$, and thus $\mathcal{L}(\mathrm{Id}_{TM}) = d$.

35.4. Proposition. *For $K \in \Omega^k(M; TM)$ and $\omega \in \Omega^l(M)$ the Lie derivative of ω along K is given by the following formula, where the X_i are (local) vector fields on M.*

$$\begin{aligned}
&(\mathcal{L}_K \omega)(X_1, \dots, X_{k+l}) = \\
&= \tfrac{1}{k!\, l!} \sum_\sigma \operatorname{sign}\sigma\ \mathcal{L}_{K(X_{\sigma 1}, \dots, X_{\sigma k})}(\omega(X_{\sigma(k+1)}, \dots, X_{\sigma(k+l)})) \\
&+ (-1)^k \Bigg(\tfrac{1}{k!\,(l-1)!} \sum_\sigma \operatorname{sign}\sigma\ \omega([X_{\sigma 1}, K(X_{\sigma 2}, \dots, X_{\sigma(k+1)})], X_{\sigma(k+2)}, \dots) \\
&\qquad - \tfrac{1}{(k-1)!\,(l-1)!\,2!} \sum_\sigma \operatorname{sign}\sigma\ \omega(K([X_{\sigma 1}, X_{\sigma 2}], X_{\sigma 3}, \dots), X_{\sigma(k+2)}, \dots) \Bigg)
\end{aligned}$$

Proof. Consider $\mathcal{L}_K \omega = [i_K, d]\omega = i_K d\omega - (-1)^{k-1} d i_K \omega$, and plug into this the definitions (35.2.1), second version, and (33.12.3). After computing some signs the expression above follows. □

35.5. Definition and theorem. *For $K \in \Omega^k(M;TM)$ and $L \in \Omega^l(M;TM)$ we define the Frölicher-Nijenhuis bracket $[K,L]$ by the following formula, where the X_i are vector fields on M.*

$$
\begin{aligned}
(1)\quad & [K,L](X_1,\dots,X_{k+l}) = \\
&= \tfrac{1}{k!\,l!}\sum_\sigma \operatorname{sign}\sigma\ [K(X_{\sigma1},\dots,X_{\sigma k}),L(X_{\sigma(k+1)},\dots,X_{\sigma(k+l)})] \\
&+(-1)^k\Big(\tfrac{1}{k!\,(l-1)!}\sum_\sigma \operatorname{sign}\sigma\ L([X_{\sigma1},K(X_{\sigma2},\dots,X_{\sigma(k+1)})],X_{\sigma(k+2)},\dots) \\
&\qquad -\tfrac{1}{(k-1)!\,(l-1)!\,2!}\sum_\sigma \operatorname{sign}\sigma\ L(K([X_{\sigma1},X_{\sigma2}],X_{\sigma3},\dots),X_{\sigma(k+2)},\dots)\Big) \\
&-(-1)^{kl+l}\Big(\tfrac{1}{(k-1)!\,l!}\sum_\sigma \operatorname{sign}\sigma\ K([X_{\sigma1},L(X_{\sigma2},\dots,X_{\sigma(l+1)})],X_{\sigma(l+2)},\dots) \\
&\qquad -\tfrac{1}{(k-1)!\,(l-1)!\,2!}\sum_\sigma \operatorname{sign}\sigma\ K(L([X_{\sigma1},X_{\sigma2}],X_{\sigma3},\dots),X_{\sigma(l+2)},\dots)\Big).
\end{aligned}
$$

Then $[K,L]\in\Omega^{k+l}(M;TM)$, and we have

$$[\mathcal{L}(K),\mathcal{L}(L)] = \mathcal{L}([K,L]) \in \operatorname{Der}\Omega(M).$$

Therefore, the space $\Omega(M;TM)=\bigoplus_{k=0}^{\dim M}\Omega^k(M;TM)$ with its usual grading is a graded Lie algebra for the Frölicher-Nijenhuis bracket. So we have

$$
\begin{aligned}
[K,L] &= -(-1)^{kl}[L,K] \\
[K_1,[K_2,K_3]] &= [[K_1,K_2],K_3] + (-1)^{k_1k_2}[K_2,[K_1,K_3]]
\end{aligned}
$$

$\mathrm{Id}_{TM}\in\Omega^1(M;TM)$ *is in the center, i.e., $[K,\mathrm{Id}_{TM}]=0$ for all K.*

For vector fields the Frölicher-Nijenhuis bracket coincides with the Lie bracket. The mapping $\mathcal{L}:\Omega^*(M;TM)\to \operatorname{Der}_*\Omega(M)$ is an injective homomorphism of graded Lie algebras.

Proof. We first show that $[K,L]\in\Omega^{k+l}(M;TM)$. By convention (35.1), this is a local question in M, thus we may assume that M is a c^∞-open subset of a convenient vector space E, that $X_i : M\to E$, that $K: M\to L^k_{\text{alt}}(E;E)$, and that $L: M\to L^l_{\text{alt}}(E;E)$. Then each expression in the formula is a kinematic vector field, and for such fields Y_1,Y_2 the Lie bracket is given by $[Y_1,Y_2]=dY_2.Y_1-dY_1.Y_2$, as shown in the beginning of the proof of (32.8). If we rewrite the formula in this way, all terms containing the derivative of one X_i cancel, and the following local expression for $[K,L]$ remains, which is obviously an element of $\Omega^{k+l}(M;TM)$.

$$
\begin{aligned}
&[K,L](X_1,\dots,X_{k+l}) = \\
&\quad= \tfrac{1}{k!\,l!}\sum_\sigma \operatorname{sign}\sigma\ \Big((dL.K(X_{\sigma1},\dots,X_{\sigma k}))(X_{\sigma(k+1)},\dots) \\
&\qquad\qquad -(dK.L(X_{\sigma(k+1)},\dots))(X_{\sigma1},\dots,X_{\sigma k}) \\
&\qquad\qquad +l\,L((dK.X_{\sigma(k+1)})(X_{\sigma1},\dots,X_{\sigma k}),X_{\sigma(k+2)},\dots) \\
&\qquad\qquad -k\,K((dL.X_{\sigma1})(X_{\sigma(k+1)},\dots,X_{\sigma(k+l)}),X_{\sigma2},\dots,X_{\sigma k})\Big).
\end{aligned}
$$

Next we show that $\mathcal{L}([K,L]) = [\mathcal{L}_K, \mathcal{L}_L]$ holds, by the following purely algebraic method, which is adapted from [Dubois-Violette, Michor, 1997]. The Chevalley coboundary operator for the adjoint representation of the Lie algebra $\mathfrak{X}(M)$ is given by [Koszul, 1950], see also [Cartan, Eilenberg, 1956]

$$\begin{aligned}
\partial K(X_1,\dots,X_{k+1}) &= \tfrac{1}{k!}\sum_\sigma \operatorname{sign}\sigma\,[X_{\sigma 1}, K(X_{\sigma 2},\dots,X_{\sigma(k+1)})] \\
&\quad - \tfrac{1}{(k-1)!\,2!}\sum_\sigma \operatorname{sign}\sigma\, K([X_{\sigma 1},X_{\sigma 2}],X_{\sigma 3},\dots,X_{\sigma(k+1)}), \\
\partial K(X_0,\dots,X_k) &= \sum_{0\le i\le k}(-1)^i [X_i, K(X_0,\dots,\widehat{X_i},\dots,X_k)] \\
&\quad + \sum_{0\le i<j\le k}(-1)^{i+j}K([X_i,X_j],X_0,\dots,\widehat{X_i},\dots,\widehat{X_j},\dots,X_k),
\end{aligned}$$

and it is well known that $\partial\partial = 0$. The following computation and close relatives will appear several times in the remainder of this proof, so we include it once.

$$\begin{aligned}
&(i_{\partial K}\omega)(X_1,\dots,X_{k+l}) = \\
&= \tfrac{1}{(k+1)!\,(l-1)!}\sum_{\sigma\in\mathcal{S}_{k+l}} \operatorname{sign}(\sigma)\omega(\partial K(X_{\sigma 1},\dots),X_{\sigma(k+2)},\dots) \\
&= \tfrac{1}{(k+1)!\,(l-1)!}\sum_\sigma \operatorname{sign}(\sigma)\Big(\sum_{i=1}^{k+1}(-1)^{i-1}\omega([X_{\sigma i},K(X_{\sigma 1},\dots,\widehat{X_{\sigma i}},\dots)],X_{\sigma(k+2)},\dots) \\
&\quad + \sum_{1\le i<j\le k+1}(-1)^{i+j}\omega(K([X_{\sigma i},X_{\sigma j}],X_{\sigma 1},\dots,\widehat{X_{\sigma i}},\dots,\widehat{X_{\sigma j}},\dots),X_{\sigma(k+2)},\dots)\Big) \\
&= \tfrac{1}{(k+1)!\,(l-1)!}\sum_\tau \operatorname{sign}(\tau)\Big((k+1)\omega([X_{\tau 1},K(X_{\tau 2},\dots)],X_{\tau(k+2)},\dots) \\
&\quad - \tfrac{k(k+1)}{2}\omega(K([X_{\tau 1},X_{\tau 2}],X_{\tau 3},\dots),X_{\tau(k+2)},\dots)\Big) \\
&= \tfrac{1}{k!\,(l-1)!}\sum_\sigma \operatorname{sign}(\sigma)\omega([X_{\sigma 1},K(X_{\sigma 2},\dots,X_{\sigma(k+1)})],X_{\sigma(k+2)},\dots) \\
&\quad - \tfrac{1}{(k-1)!\,(l-1)!\,2!}\sum_\sigma \operatorname{sign}(\sigma)\omega(K([X_{\sigma 1},X_{\sigma 2}],X_{\sigma 3},\dots),X_{\sigma(k+2)},\dots).
\end{aligned}$$

Then the Frölicher-Nijenhuis bracket (1) is given by

$$(2)\qquad [K,L] = [K,L]_\wedge + (-1)^k i(\partial K)L - (-1)^{kl+l} i(\partial L)K,$$

where we have put

$$\begin{aligned}
(3)\quad &[K,L]_\wedge(X_1,\dots,X_{k+l}) := \\
&\qquad = \tfrac{1}{k!\,l!}\sum_\sigma \operatorname{sign}(\sigma)[K(X_{\sigma 1},\dots,X_{\sigma k}),L(X_{\sigma(k+1)},\dots,X_{\sigma(k+l)})].
\end{aligned}$$

Formula (2) is the same as in [Nijenhuis, 1969, p. 100], where it is also stated that from this formula 'one can show (with a good deal of effort) that this bracket defines a graded Lie algebra structure'. Similarly, we can write the Lie derivative (35.4) as

$$\mathcal{L}_K = \mathcal{L}_\wedge(K) + (-1)^k i(\partial K), \tag{4}$$

where the action $\mathcal{L}$ of $\mathfrak{X}(M)$ on $C^\infty(M,\mathbb{R})$ is extended to $\mathcal{L}_\wedge : \Omega(M;TM)\times\Omega(M) \to \Omega(M)$ by

$$(5)\quad (\mathcal{L}_\wedge(K)\omega)(X_1,\dots,X_{q+k}) = \\ = \tfrac{1}{k!\,q!}\sum_\sigma \operatorname{sign}(\sigma)\mathcal{L}(K(X_{\sigma 1},\dots,X_{\sigma k}))(\omega(X_{\sigma(k+1)},\dots,X_{\sigma(k+q)})).$$

Using (4), we see that

$$\begin{aligned}(6)\qquad [\mathcal{L}_K,\mathcal{L}_L] &= \mathcal{L}_\wedge(K)\mathcal{L}_\wedge(L) - (-1)^{kl}\mathcal{L}_\wedge(L)\mathcal{L}_\wedge(K)\\ &\quad + (-1)^k i(\partial K)\mathcal{L}_\wedge(L) - (-1)^{kl+k}\mathcal{L}_\wedge(L)i(\partial K)\\ &\quad - (-1)^{kl+l} i(\partial L)\mathcal{L}_\wedge(K) + (-1)^l\mathcal{L}_\wedge(K)i(\partial L)\\ &\quad + (-1)^{k+l} i(\partial K)i(\partial L) - (-1)^{kl+k+l} i(\partial L)i(\partial K),\end{aligned}$$

and from (2) and (4) we get

$$\begin{aligned}(7)\qquad \mathcal{L}_{[K,L]} &= \mathcal{L}_{[K,L]_\wedge} + (-1)^k\mathcal{L}_{i(\partial K)L} - (-1)^{kl+l}\mathcal{L}_{i(\partial L)K}\\ &= \mathcal{L}_\wedge([K,L]_\wedge) + (-1)^{k+l} i(\partial[K,L]_\wedge)\\ &\quad + (-1)^k\mathcal{L}_\wedge(i(\partial K)L) + (-1)^k i(\partial i(\partial K)L)\\ &\quad - (-1)^{kl+l}\mathcal{L}_\wedge(i(\partial L)K) - (-1)^{kl+k} i(\partial i(\partial L)K).\end{aligned}$$

By a straightforward direct computation, one checks that

$$\mathcal{L}_\wedge(K)\mathcal{L}_\wedge(L) - (-1)^{kl}\mathcal{L}_\wedge(L)\mathcal{L}_\wedge(K) = \mathcal{L}_\wedge([K,L]_\wedge). \tag{8}$$

The derivation i_K of degree k is seeing the expression $\mathcal{L}_\wedge(L)\omega$ as a 'wedge product' $L\wedge_{\mathcal{L}}\omega$, as in (33.8). So we may apply theorem (35.2.1) and get

$$i_K\mathcal{L}_\wedge(L)\omega = \mathcal{L}_\wedge(i_K L)\omega + (-1)^{(k-1)l}\mathcal{L}_\wedge(L)i_K\omega. \tag{9}$$

By a long but straightforward combinatorial computation, one can check directly from the definitions that the following formula holds:

$$\partial(i_K L) = i_{\partial K}L + (-1)^{k-1} i_K\partial L + (-1)^k[K,L]_\wedge. \tag{10}$$

Moreover, it is well-known (and easy to check) that

$$\partial[K,L]_\wedge = [\partial K,L]_\wedge + (-1)^k[K,\partial L]_\wedge. \tag{11}$$

We have to show that (6) equals (7). This follows by using (8), twice (9), then the first three lines in (6) correspond to the first terms in the first three lines in (7). For the remaining terms use twice (10), (11), and $\partial\partial = 0$.

That the Frölicher-Nijenhuis bracket defines a graded Lie bracket now follows from the fact that $\mathcal{L} : \Omega(M;TM) \to \operatorname{Der}(\Omega(M))$ is injective, by convention (35.1).

Since we have $[d,d] = 2d^2 = 0$, by the graded Jacobi identity we obtain $0 = [i_K,[d,d]] = [[i_K,d],d] + (-1)^{k-1}[d,[i_K,d]] = 2[\mathcal{L}_K,d] = 2\mathcal{L}([K,\mathrm{Id}_{TM}])$. □

35.6. Lemma. *Moreover, the Chevalley coboundary operator is a homomorphism from the Frölicher-Nijenhuis bracket to the Nijenhuis-Richardson bracket:*

$$\partial[K,L] = [\partial K, \partial L]^\wedge.$$

Proof. This follows directly from (35.5.2), (35.5.11), and twice (35.5.10), and from (35.2.2):

$$\begin{aligned}\partial[K,L] &= \partial[K,L]_\wedge + (-1)^k \partial i(\partial K)L - (-1)^{kl+l}\partial i(\partial L)K \\ &= [\partial K, L]_\wedge + (-1)^k[K,\partial L]_\wedge + 0 + i(\partial K)\partial L - [\partial K, L]_\wedge \\ &\quad - 0 - (-1)^{kl} i(\partial L)\partial K + (-1)^{kl}[\partial L, K]_\wedge \\ &= [\partial K, \partial L]^\wedge. \quad \square\end{aligned}$$

35.7. Lemma. *For $K \in \Omega^k(M;TM)$ and $L \in \Omega^{l+1}(M;TM)$ we have*

$$\begin{aligned}&[\mathcal{L}_K, i_L] = i([K,L]) - (-1)^{kl}\mathcal{L}(i_L K), \text{ or} \\ &[i_L, \mathcal{L}_K] = \mathcal{L}(i_L K) + (-1)^k\, i([L,K]).\end{aligned}$$

Proof. The two equations are obviously equivalent by graded skew symmetry, and the second one follows by expanding the left hand side using (35.5.4), (35.5.9), and (35.2.2), and by expanding the right hand side using (35.5.4), (35.5.2), and then (35.5.10):

$$\begin{aligned}[i_L, \mathcal{L}_K] &= [i_L, \mathcal{L}_\wedge(K)] + (-1)^k [i_L, i_{\partial K}] \\ &= \mathcal{L}_\wedge(i_L K) + (-1)^k\, i(i_L \partial K - (-1)^{(l-1)k} i_{\partial K} L), \\ \mathcal{L}(i_L K) + (-1)^k\, i([L,K]) &= \mathcal{L}_\wedge(i_L K) - (-1)^{k+l} i(\partial i_L K) \\ &\quad + (-1)^k\, i([L,K]_\wedge + (-1)^l i_{\partial L} K - (-1)^{kl+k} i_{\partial K} L). \quad \square\end{aligned}$$

35.8. The space $\operatorname{Der}\Omega(M)$ is a graded module over the graded algebra $\Omega(M)$ with the action $(\omega\wedge D)\varphi = \omega\wedge D(\varphi)$, because $\Omega(M)$ is graded commutative.

Theorem. *Let the degrees of ω be q, of φ be k, and of ψ be l. Let the other degrees be given by the corresponding lower case letters. Then we have:*

(1) $[\omega\wedge D_1, D_2] = \omega\wedge[D_1,D_2] - (-1)^{(q+k_1)k_2} D_2(\omega)\wedge D_1.$

(2) $i(\omega\wedge L) = \omega\wedge i(L)$

(3) $\omega\wedge\mathcal{L}_K = \mathcal{L}(\omega\wedge K) + (-1)^{q+k-1} i(d\omega\wedge K).$

(4) $[\omega\wedge L_1, L_2]^\wedge = \omega\wedge[L_1,L_2]^\wedge -$
$\quad - (-1)^{(q+l_1-1)(l_2-1)} i(L_2)\omega\wedge L_1.$

(5)
$$[\omega\wedge K_1,K_2]=\omega\wedge[K_1,K_2]-(-1)^{(q+k_1)k_2}\mathcal{L}(K_2)\omega\wedge K_1 + (-1)^{q+k_1}d\omega\wedge i(K_1)K_2.$$

(6)
$$\begin{aligned}[\varphi\otimes X,\psi\otimes Y]&=\varphi\wedge\psi\otimes[X,Y]\\ &\quad-\big(i_Yd\varphi\wedge\psi\otimes X-(-1)^{kl}i_Xd\psi\wedge\varphi\otimes Y\big)\\ &\quad-\big(d(i_Y\varphi\wedge\psi)\otimes X-(-1)^{kl}d(i_X\psi\wedge\varphi)\otimes Y\big)\\ &=\varphi\wedge\psi\otimes[X,Y]+\varphi\wedge\mathcal{L}_X\psi\otimes Y-\mathcal{L}_Y\varphi\wedge\psi\otimes X\\ &\quad+(-1)^k\left(d\varphi\wedge i_X\psi\otimes Y+i_Y\varphi\wedge d\psi\otimes X\right).\end{aligned}$$

Proof. For (1), (2), (3) write out the definitions. For (4) compute $i([\omega\wedge L_1,L_2]^\wedge)$. For (5) compute $\mathcal{L}([\omega\wedge K_1,K_2])$. For (6) use (5). □

35.9. Theorem. *For $K_i\in\Omega^{k_i}(M;TM)$ and $L_i\in\Omega^{k_i+1}(M;TM)$ we have*

(1)
$$\begin{aligned}&[\mathcal{L}_{K_1}+i_{L_1},\mathcal{L}_{K_2}+i_{L_2}]=\\ &\quad=\mathcal{L}\left([K_1,K_2]+i_{L_1}K_2-(-1)^{k_1k_2}i_{L_2}K_1\right)\\ &\qquad+i\left([L_1,L_2]^\wedge+[K_1,L_2]-(-1)^{k_1k_2}[K_2,L_1]\right).\end{aligned}$$

Each summand of this formula looks like a semidirect product of graded Lie algebras, but the mappings

$$\begin{aligned}i&:\Omega(M;TM)\to\operatorname{End}(\Omega(M;TM),[\ \ ,\ \ ])\\ \operatorname{ad}&:\Omega(M;TM)\to\operatorname{End}(\Omega(M;TM),[\ \ ,\ \ ]^\wedge)\end{aligned}$$

do not take values in the subspaces of graded derivations. Instead we have for $K\in\Omega^k(M;TM)$ and $L\in\Omega^{l+1}(M;TM)$ the following relations:

(2)
$$i_L[K_1,K_2]=[i_LK_1,K_2]+(-1)^{k_1l}[K_1,i_LK_2] -\Big((-1)^{k_1l}i([K_1,L])K_2-(-1)^{(k_1+l)k_2}i([K_2,L])K_1\Big).$$

(3)
$$[K,[L_1,L_2]^\wedge]=[[K,L_1],L_2]^\wedge+(-1)^{kk_1}[L_1,[K,L_2]]^\wedge- -\Big((-1)^{kk_1}[i(L_1)K,L_2]-(-1)^{(k+k_1)k_2}[i(L_2)K,L_1]\Big).$$

The algebraic meaning of these relations and its consequences in group theory have been investigated in [Michor, 1989a]. The corresponding product of groups is well known to algebraists under the name 'Zappa-Szep'-product.

Proof. Equation (1) is an immediate consequence of (35.7). Equations (2) and (3) follow from (1) by writing out the graded Jacobi identity. □

35.10. Corollary of (28.6). *For $K,\ L\in\Omega^1(M;TM)$ we have*

$$\begin{aligned}[K,L](X,Y)&=[KX,LY]-[KY,LX]-L([KX,Y]-[KY,X])\\ &\quad-K([LX,Y]-[LY,X])+(LK+KL)([X,Y]).\end{aligned}$$

35.11. Curvature. Let $P \in \Omega^1(M;TM)$ satisfy $P \circ P = P$, i.e., P is a projection in each fiber of TM. This is the most general case of a (first order) *connection.* We call $\ker P$ the *horizontal space* and $\operatorname{im} P$ the *vertical space* of the connection. If $\operatorname{im} P$ is some fixed sub vector bundle or (tangent bundle of) a foliation, P can be called a connection for it. The following result is immediate from (35.10).

Lemma. *We have*

$$[P,P] = 2\mathcal{R} + 2\bar{\mathcal{R}},$$

where $\mathcal{R}$, $\bar{\mathcal{R}} \in \Omega^2(M;TM)$ *are given by* $\mathcal{R}(X,Y) = P[(\mathrm{Id} - P)X, (\mathrm{Id} - P)Y]$ *and* $\bar{\mathcal{R}}(X,Y) = (\mathrm{Id} - P)[PX, PY]$.

If $\operatorname{im}(P)$ is a sub vector bundle, then $\mathcal{R}$ is an obstruction against integrability of the horizontal bundle $\ker P$, and $\bar{\mathcal{R}}$ is an obstruction against integrability of the vertical bundle $\operatorname{im} P$. Thus, we call $\mathcal{R}$ the *curvature* and $\bar{\mathcal{R}}$ the *cocurvature* of the connection P.

35.12. Lemma. Bianchi identity. *If* $P \in \Omega^1(M;TM)$ *is a connection (fiber projection) with curvature* $\mathcal{R}$ *and cocurvature* $\bar{\mathcal{R}}$, *then we have*

$$[P, \mathcal{R} + \bar{\mathcal{R}}] = 0$$
$$[\mathcal{R}, P] = i_{\mathcal{R}}\bar{\mathcal{R}} + i_{\bar{\mathcal{R}}}\mathcal{R}.$$

Proof. We have $[P,P] = 2\mathcal{R} + 2\bar{\mathcal{R}}$ by (35.11), and $[P,[P,P]] = 0$ by the graded Jacobi identity. So the first formula follows. We have $2\mathcal{R} = P \circ [P,P] = i_{[P,P]}P$. By (35.9.2) we get $i_{[P,P]}[P,P] = 2[i_{[P,P]}P, P] - 0 = 4[\mathcal{R}, P]$. Therefore, $[\mathcal{R}, P] = \frac{1}{4}i_{[P,P]}[P,P] = i(\mathcal{R} + \bar{\mathcal{R}})(\mathcal{R} + \bar{\mathcal{R}}) = i_{\mathcal{R}}\bar{\mathcal{R}} + i_{\bar{\mathcal{R}}}\mathcal{R}$ since $\mathcal{R}$ has vertical values and kills vertical vectors, so $i_{\mathcal{R}}\mathcal{R} = 0$; likewise for $\bar{\mathcal{R}}$. □

35.13. f-relatedness of the Frölicher-Nijenhuis bracket. Let $f : M \to N$ be a smooth mapping between manifolds. Two vector valued forms $K \in \Omega^k(M;TM)$ and $K' \in \Omega^k(N;TN)$ are called *f-related* or *f-dependent*, if for all $X_i \in T_xM$ we have

(1) $$K'_{f(x)}(T_xf \cdot X_1, \dots, T_xf \cdot X_k) = T_xf \cdot K_x(X_1, \dots, X_k).$$

Theorem.

(2) *If* K *and* K' *as above are f-related then* $i_K \circ f^* = f^* \circ i_{K'} : \Omega(N) \to \Omega(M)$.
(3) *If* $i_K \circ f^* \mid B^1(N) = f^* \circ i_{K'} \mid B^1(N)$, *then* K *and* K' *are f-related, where* B^1 *denotes the space of exact* 1*-forms.*
(4) *If* K_j *and* K'_j *are f-related for* $j = 1, 2$, *then* $i_{K_1}K_2$ *and* $i_{K'_1}K'_2$ *are f-related, and also* $[K_1, K_2]^\wedge$ *and* $[K'_1, K'_2]^\wedge$ *are f-related.*
(5) *If* K *and* K' *are f-related then* $\mathcal{L}_K \circ f^* = f^* \circ \mathcal{L}_{K'} : \Omega(N) \to \Omega(M)$.
(6) *If* $\mathcal{L}_K \circ f^* \mid \Omega^0(N) = f^* \circ \mathcal{L}_{K'} \mid \Omega^0(N)$, *then* K *and* K' *are f-related.*
(7) *If* K_j *and* K'_j *are f-related for* $j = 1, 2$, *then their Frölicher-Nijenhuis brackets* $[K_1, K_2]$ *and* $[K'_1, K'_2]$ *are also f-related.*

Proof. (2) By (35.2), we have for $\omega \in \Omega^q(N)$ and $X_i \in T_xM$:

$$\begin{aligned}
&(i_K f^*\omega)_x(X_1,\dots,X_{q+k-1}) = \\
&= \tfrac{1}{k!\,(q-1)!}\sum_\sigma \operatorname{sign}\sigma\,(f^*\omega)_x(K_x(X_{\sigma 1},\dots,X_{\sigma k}),X_{\sigma(k+1)},\dots) \\
&= \tfrac{1}{k!\,(q-1)!}\sum_\sigma \operatorname{sign}\sigma\,\omega_{f(x)}(T_xf\cdot K_x(X_{\sigma 1},\dots),T_xf\cdot X_{\sigma(k+1)},\dots) \\
&= \tfrac{1}{k!\,(q-1)!}\sum_\sigma \operatorname{sign}\sigma\,\omega_{f(x)}(K'_{f(x)}(T_xf\cdot X_{\sigma 1},\dots),T_xf\cdot X_{\sigma(k+1)},\dots) \\
&= (f^* i_{K'}\omega)_x(X_1,\dots,X_{q+k-1}).
\end{aligned}$$

(3) follows from this computation, since the dg, $g \in C^\infty(M,\mathbb{R})$ separate points, by convention (35.1).

(4) follows from the same computation for K_2 instead of ω, the result for the bracket then follows by (35.2.2).

(5) The algebra homomorphism f^* intertwines the operators i_K and $i_{K'}$ by (2), and f^* commutes with the exterior derivative d. Thus, f^* intertwines the commutators $[i_K, d] = \mathcal{L}_K$ and $[i_{K'}, d] = \mathcal{L}_{K'}$.

(6) For $g \in \Omega^0(N)$ we have $\mathcal{L}_K\, f^*\, g = i_K\, d\, f^*\, g = i_K\, f^*\, dg$, and on the other hand $f^*\, \mathcal{L}_{K'}\, g = f^*\, i_{K'}\, dg$. By (3) the result follows.

(7) The algebra homomorphism f^* intertwines $\mathcal{L}_{K_j}$ and $\mathcal{L}_{K'_j}$, thus also their graded commutators, which are equal to $\mathcal{L}([K_1,K_2])$ and $\mathcal{L}([K'_1,K'_2])$, respectively. Then use (6). □

35.14. Let $f : M \to N$ be a local diffeomorphism. Then we can consider the pullback operator $f^* : \Omega(N;TN) \to \Omega(M;TM)$, given by

(1) $$(f^*K)_x(X_1,\dots,X_k) = (T_xf)^{-1}K_{f(x)}(T_xf\cdot X_1,\dots,T_xf\cdot X_k).$$

This is a special case of the pullback operator for sections of natural vector bundles. Clearly, K and f^*K are then f-related.

Theorem. *In this situation we have:*

(2) $f^*[K,L] = [f^*K, f^*L]$.
(3) $f^*\, i_K L = i_{f^*K} f^*L$.
(4) $f^*[K,L]^\wedge = [f^*K, f^*L]^\wedge$.
(5) *For a vector field $X \in \mathfrak{X}(M)$ admitting a local flow Fl^X_t and $K \in \Omega(M;TM)$ the Lie derivative $\mathcal{L}_X K = \frac{\partial}{\partial t}\big|_0 (\mathrm{Fl}^X_t)^*K$ is defined. Then we have $\mathcal{L}_X K = [X,K]$, the Frölicher-Nijenhuis-bracket.*

This is sometimes expressed by saying that the Frölicher-Nijenhuis bracket, the bracket $[\ \ ,\ \ ]^\wedge$, etc., are *natural bilinear concomitants.*

Proof. (2) – (4) are obvious from (35.13). They also follow directly from the geometrical constructions of the operators in question.

(5) By inserting $Y_i \in \mathfrak{X}(M)$ we get from (1) the following expression which we can differentiate using (32.15) repeatedly.

$$\begin{aligned}
(\mathrm{Fl}^X_{-t})^*(K(Y_1,\dots,Y_k)) &= ((\mathrm{Fl}^X_t)^* K)((\mathrm{Fl}^X_{-t})^*Y_1,\dots,(\mathrm{Fl}^X_{-t})^*Y_k)\\
[X,K(Y_1,\dots,Y_k)] &= \tfrac{\partial}{\partial t}\big|_0 (\mathrm{Fl}^X_{-t})^*(K(Y_1,\dots,Y_k))\\
&= \tfrac{\partial}{\partial t}\big|_0 ((\mathrm{Fl}^X_t)^* K)((\mathrm{Fl}^X_{-t})^*Y_1,\dots,(\mathrm{Fl}^X_{-t})^*Y_k)\\
&= (\tfrac{\partial}{\partial t}\big|_0 (\mathrm{Fl}^X_t)^* K)(Y_1,\dots,Y_k) - \sum_{1\le i\le k} K(Y_1,\dots,\tfrac{\partial}{\partial t}\big|_0 (\mathrm{Fl}^X_{-t})^*Y_i,\dots,Y_k)\\
&= (\tfrac{\partial}{\partial t}\big|_0 (\mathrm{Fl}^X_t)^* K)(Y_1,\dots,Y_k) - \sum_{1\le i\le k} K(Y_1,\dots,[X,Y_i],\dots,Y_k).
\end{aligned}$$

This leads to

$$\begin{aligned}
(\tfrac{\partial}{\partial t}\big|_0 (\mathrm{Fl}^X_t)^* K)(Y_1,\dots,Y_k) &= [X,K(Y_1,\dots,Y_k)]\\
&\quad + \sum_{1\le i\le k} K([X,Y_i],Y_1,\dots\widehat{Y_i}\dots,Y_k)\\
&= [X,K](Y_1,\dots,Y_k), \qquad \text{by (35.5.1).} \quad \square
\end{aligned}$$

35.15. Remark. Finally, we mention the best known application of the Frölicher-Nijenhuis bracket, which also led to its discovery. A vector valued 1-form $J \in \Omega^1(M;TM)$ with $J\circ J = -\,\mathrm{Id}$ is called an *almost complex structure.* If it exists, J can be viewed as a fiber multiplication with $\sqrt{-1}$ on TM. By (35.10) we have

$$[J,J](X,Y) = 2([JX,JY] - [X,Y] - J[X,JY] - J[JX,Y]).$$

The vector valued form $\frac{1}{2}[J,J]$ is also called the *Nijenhuis tensor* of J. In finite dimensions an almost complex structure J comes from a complex structure on the manifold if and only if the Nijenhuis tensor vanishes.

Chapter VIII
Infinite Dimensional Differential Geometry

The theory of infinite dimensional Lie groups can be pushed surprisingly far: Exponential mappings are unique if they exist. In general, they are neither locally surjective nor locally injective. A stronger requirement (leading to regular Lie groups) is that to assume that smooth curves in the Lie algebra integrate to smooth curves in the group in a smooth way (an 'evolution operator' exists). This is due to [Milnor, 1984] who weakened the concept of [Omori et al., 1982]. It turns out that regular Lie groups have strong permanence properties. In fact, up to now all known Lie groups are regular. Connections on smooth principal bundles with a regular Lie group as structure group have parallel transport (39.1), and for flat connections the horizontal distribution is integrable (39.2). So some (equivariant) partial differential equations in infinite dimensions are very well behaved, although in general there are counter-examples in every possible direction (some can be found in (32.12)).

The actual development is quite involved. We start with general infinite dimensional Lie groups in section (36), but for a detailed study of the evolution operator of regular Lie groups (38.4) we need in (38.10) the Maurer-Cartan equation for right (or left) logarithmic derivatives of mappings with values in the Lie group (38.1), and this we can only get by looking at principal connections. Thus, in the second section (37) bundles, connections, principal bundles, curvature, associated bundles, and all results of principal bundle geometry which do not involve parallel transport are developed. Finally, we then prove the strong existence results mentioned above and treat regular Lie groups in section (38), and principal bundles with regular structure groups in section (39). The material in this chapter is an extended version of [Kriegl, Michor, 1997].

36. Lie Groups

36.1. Definition. A *Lie group* G is a smooth manifold and a group such that the multiplication $\mu : G \times G \to G$ and the inversion $\nu : G \to G$ are smooth. If not stated otherwise, G may be infinite dimensional. If an implicit function theorem is available, then smoothness of ν follows from smoothness of μ.

We shall use the following notation:
$\mu : G \times G \to G$, multiplication, $\mu(x, y) = x.y$.
$\mu_a : G \to G$, left translation, $\mu_a(x) = a.x$.
$\mu^a : G \to G$, right translation, $\mu^a(x) = x.a$.
$\nu : G \to G$, inversion, $\nu(x) = x^{-1}$.
$e \in G$, the unit element.

36.2. Lemma. *The kinematic tangent mapping $T_{(a,b)}\mu : T_aG \times T_bG \to T_{ab}G$ is given by*

$$T_{(a,b)}\mu.(X_a, Y_b) = T_a(\mu^b).X_a + T_b(\mu_a).Y_b,$$

and $T_a\nu : T_aG \to T_{a^{-1}}G$ is given by

$$T_a\nu = -T_e(\mu^{a^{-1}}).T_a(\mu_{a^{-1}}) = -T_e(\mu_{a^{-1}}).T_a(\mu^{a^{-1}}).$$

Proof. Let $\mathrm{ins}_a : G \to G \times G$, $\mathrm{ins}_a(x) = (a, x)$ be the right insertion, and let $\mathrm{ins}^b : G \to G \times G$, $\mathrm{ins}^b(x) = (x, b)$ be the left insertion. Then we have

$$\begin{aligned}T_{(a,b)}\mu.(X_a, Y_b) &= T_{(a,b)}\mu.(T_a(\mathrm{ins}^b).X_a + T_b(\mathrm{ins}_a).Y_b) = \\ &= T_a(\mu \circ \mathrm{ins}^b).X_a + T_b(\mu \circ \mathrm{ins}_a).Y_b = T_a(\mu^b).X_a + T_b(\mu_a).Y_b.\end{aligned}$$

Now we differentiate the equation $\mu(a, \nu(a)) = e$; this gives in turn

$$0_e = T_{(a,a^{-1})}\mu.(X_a, T_a\nu.X_a) = T_a(\mu^{a^{-1}}).X_a + T_{a^{-1}}(\mu_a).T_a\nu.X_a,$$
$$T_a\nu.X_a = -T_e(\mu_a)^{-1}.T_a(\mu^{a^{-1}}).X_a = -T_e(\mu_{a^{-1}}).T_a(\mu^{a^{-1}}).X_a. \quad \square$$

36.3. Invariant vector fields and Lie algebras. Let G be a (real) Lie group. A (kinematic) vector field ξ on G is called *left invariant*, if $\mu_a^*\xi = \xi$ for all $a \in G$, where $\mu_a^*\xi = T(\mu_{a^{-1}})\circ\xi\circ\mu_a$ as in (32.9). Since by (32.11) we have $\mu_a^*[\xi, \eta] = [\mu_a^*\xi, \mu_a^*\eta]$, the space $\mathfrak{X}_L(G)$ of all left invariant vector fields on G is closed under the Lie bracket, so it is a sub Lie algebra of $\mathfrak{X}(G)$. Any left invariant vector field ξ is uniquely determined by $\xi(e) \in T_eG$, since $\xi(a) = T_e(\mu_a).\xi(e)$. Thus, the Lie algebra $\mathfrak{X}_L(G)$ of left invariant vector fields is linearly isomorphic to T_eG, and the Lie bracket on $\mathfrak{X}_L(G)$ induces a Lie algebra structure on T_eG, whose bracket is again denoted by $[\quad,\quad]$. This Lie algebra will be denoted as usual by $\mathfrak{g}$, sometimes by $\mathrm{Lie}(G)$.

We will also give a name to the isomorphism with the space of left invariant vector fields: $L : \mathfrak{g} \to \mathfrak{X}_L(G)$, $X \mapsto L_X$, where $L_X(a) = T_e\mu_a.X$. Thus, $[X,Y] = [L_X, L_Y](e)$.

A vector field η on G is called *right invariant*, if $(\mu^a)^*\eta = \eta$ for all $a \in G$. If ξ is left invariant, then $\nu^*\xi$ is right invariant, since $\nu \circ \mu^a = \mu_{a^{-1}} \circ \nu$ implies that $(\mu^a)^*\nu^*\xi = (\nu\circ\mu^a)^*\xi = (\mu_{a^{-1}}\circ\nu)^*\xi = \nu^*(\mu_{a^{-1}})^*\xi = \nu^*\xi$. The right invariant vector fields form a sub Lie algebra $\mathfrak{X}_R(G)$ of $\mathfrak{X}(G)$, which also is linearly isomorphic to T_eG and induces a Lie algebra structure on T_eG. Since $\nu^* : \mathfrak{X}_L(G) \to \mathfrak{X}_R(G)$ is an isomorphism of Lie algebras by (32.11), $T_e\nu = -\mathrm{Id} : T_eG \to T_eG$ is an isomorphism between the two Lie algebra structures. We will denote by $R : \mathfrak{g} = T_eG \to \mathfrak{X}_R(G)$ the isomorphism discussed, which is given by $R_X(a) = T_e(\mu^a).X$.

36.4. Remark. It would be tempting to apply also other kinds of tangent bundle functors like D and $D^{[1,\infty)}$, where one gets Lie algebras of smooth sections, see (32.8). Some results will stay true like (36.3), (36.5). In general, one gets strictly larger Lie algebras for Lie groups, see (28.4). But the functors D and $D^{[1,\infty)}$ do not respect products in general, see (28.16), so e.g. (36.2) is wrong for these functors.

36.5. Lemma. *If L_X is a left invariant vector field and R_Y is a right invariant one, then $[L_X, R_Y] = 0$. So if the flows of L_X and R_Y exist, they commute.*

Proof. We consider the vector field $0 \times L_X \in \mathfrak{X}(G \times G)$, given by $(0 \times L_X)(a, b) = (0_a, L_X(b))$. Then $T_{(a,b)}\mu.(0_a, L_X(b)) = T_a\mu^b.0_a + T_b\mu_a.L_X(b) = L_X(ab)$, so $0 \times L_X$ is μ-related to L_X. Likewise, $R_Y \times 0$ is μ-related to R_Y. But then $0 = [0 \times L_X, R_Y \times 0]$ is μ-related to $[L_X, R_Y]$ by (32.10). Since μ is surjective, $[L_X, R_Y] = 0$ follows. □

36.6. Lemma. *Let $\varphi : G \to H$ be a smooth homomorphism of Lie groups. Then $\varphi' := T_e\varphi : \mathfrak{g} = T_eG \to \mathfrak{h} = T_eH$ is a Lie algebra homomorphism.*

Proof. For $X \in \mathfrak{g}$ and $x \in G$ we have

$$\begin{aligned} T_x\varphi.L_X(x) &= T_x\varphi.T_e\mu_x.X = T_e(\varphi \circ \mu_x).X \\ &= T_e(\mu_{\varphi(x)} \circ \varphi).X = T_e(\mu_{\varphi(x)}).T_e\varphi.X = L_{\varphi'(X)}(\varphi(x)). \end{aligned}$$

So L_X is φ-related to $L_{\varphi'(X)}$. By (32.10), the field $[L_X, L_Y] = L_{[X,Y]}$ is φ-related to $[L_{\varphi'(X)}, L_{\varphi'(Y)}] = L_{[\varphi'(X),\varphi'(Y)]}$. So we have $T\varphi \circ L_{[X,Y]} = L_{[\varphi'(X),\varphi'(Y)]} \circ \varphi$. If we evaluate this at e the result follows. □

36.7. One parameter subgroups. Let G be a Lie group with Lie algebra $\mathfrak{g}$. A *one parameter subgroup* of G is a Lie group homomorphism $\alpha : (\mathbb{R}, +) \to G$, i.e. a smooth curve α in G with $\alpha(s + t) = \alpha(s).\alpha(t)$, and hence $\alpha(0) = e$.

Note that a smooth mapping $\beta : (-\varepsilon, \varepsilon) \to G$ satisfying $\beta(t)\beta(s) = \beta(t + s)$ for $|t|$, $|s|$, $|t + s| < \varepsilon$ is the restriction of a one parameter subgroup. Namely, choose $0 < t_0 < \varepsilon/2$. Any $t \in \mathbb{R}$ can be uniquely written as $t = N.t_0 + t'$ for $0 \le t' < t_0$ and $N \in \mathbb{Z}$. Put $\alpha(t) = \beta(t_0)^N\beta(t')$. The required properties are easy to check.

Lemma. *Let $\alpha : \mathbb{R} \to G$ be a smooth curve with $\alpha(0) = e$. Let $X \in \mathfrak{g}$. Then the following assertions are equivalent.*

(1) *α is a one parameter subgroup with $X = \frac{\partial}{\partial t}\big|_0 \alpha(t)$.*
(2) *$\alpha(t)$ is an integral curve of the left invariant vector field L_X and also an integral curve of the right invariant vector field R_X.*
(3) *$\operatorname{Fl}^{L_X}(t, x) := x.\alpha(t)$ (or $\operatorname{Fl}^{L_X}_t = \mu^{\alpha(t)}$) is the (unique by (32.16)) global flow of L_X in the sense of (32.13).*
(4) *$\operatorname{Fl}^{R_X}(t, x) := \alpha(t).x$ (or $\operatorname{Fl}^{R_X}_t = \mu_{\alpha(t)}$) is the (unique) global flow of R_X.*

Moreover, each of these properties determines α uniquely.

Proof. (1) ⇒ (3) We have

$$\begin{aligned} \tfrac{d}{dt}x.\alpha(t) &= \tfrac{d}{ds}|_0 x.\alpha(t + s) = \tfrac{d}{ds}|_0 x.\alpha(t).\alpha(s) \\ &= \tfrac{d}{ds}|_0 \mu_{x.\alpha(t)}\alpha(s) = T_e(\mu_{x.\alpha(t)}).\tfrac{d}{ds}|_0\alpha(s) = L_X(x.\alpha(t)). \end{aligned}$$

Since it is obviously a flow, we have (3).

(3) $\Leftrightarrow$ (4) We have $\operatorname{Fl}_t^{\nu^*\xi} = \nu^{-1} \circ \operatorname{Fl}_t^{\xi} \circ \nu$ by (32.16). Therefore, we have by (36.3)

$$\begin{aligned}(\operatorname{Fl}_t^{R_X}(x^{-1}))^{-1} &= (\nu \circ \operatorname{Fl}_t^{R_X} \circ \nu)(x) = \operatorname{Fl}_t^{\nu^* R_X}(x) \\ &= \operatorname{Fl}_t^{-L_X}(x) = \operatorname{Fl}_{-t}^{L_X}(x) = x.\alpha(-t).\end{aligned}$$

So $\operatorname{Fl}_t^{R_X}(x^{-1}) = \alpha(t).x^{-1}$, and $\operatorname{Fl}_t^{R_X}(y) = \alpha(t).y$.

(3) and (4) together clearly imply (2).

(2) $\Rightarrow$ (1) This is a consequence of the following result.

Claim. *Consider two smooth curves $\alpha, \beta : \mathbb{R} \to G$ with $\alpha(0) = e = \beta(0)$ which satisfy the two differential equations*

$$\begin{aligned}\tfrac{d}{dt}\alpha(t) &= L_X(\alpha(t)) \\ \tfrac{d}{dt}\beta(t) &= R_X(\beta(t)).\end{aligned}$$

Then $\alpha = \beta$, and it is a 1-parameter subgroup.

We have $\alpha = \beta$ since

$$\begin{aligned}\tfrac{d}{dt}(\alpha(t)\beta(-t)) &= T\mu^{\beta(-t)}.L_X(\alpha(t)) - T\mu_{\alpha(t)}.R_X(\beta(-t)) \\ &= T\mu^{\beta(-t)}.T\mu_{\alpha(t)}.X - T\mu_{\alpha(t)}.T\mu^{\beta(-t)}.X = 0.\end{aligned}$$

Next we calculate for fixed s

$$\tfrac{d}{dt}(\beta(t-s)\beta(s)) = T\mu^{\beta(s)}.R_X(\beta(t-s)) = R_X(\beta(t-s)\beta(s)).$$

Hence, by the first part of the proof $\beta(t-s)\beta(s) = \alpha(t) = \beta(t)$.

The statement about uniqueness follows from (32.16), or from the claim. □

36.8. Definition. Let G be a Lie group with Lie algebra $\mathfrak{g}$. We say that G admits an *exponential mapping* if there exists a smooth mapping $\exp : \mathfrak{g} \to G$ such that $t \mapsto \exp(tX)$ is the (unique by (36.7)) 1-parameter subgroup with tangent vector X at 0. Then we have by (36.7)

(1) $\operatorname{Fl}^{L_X}(t, x) = x.\exp(tX)$.

(2) $\operatorname{Fl}^{R_X}(t, x) = \exp(tX).x$.

(3) $\exp(0) = e$ and $T_0 \exp = \operatorname{Id} : T_0\mathfrak{g} = \mathfrak{g} \to T_eG = \mathfrak{g}$ since $T_0 \exp.X = \frac{d}{dt}|_0 \exp(0 + t.X) = \frac{d}{dt}|_0 \operatorname{Fl}^{L_X}(t, e) = X$.

(4) Let $\varphi : G \to H$ be a smooth homomorphism between Lie groups admitting exponential mappings. Then the diagram

$$\begin{array}{ccc} \mathfrak{g} & \xrightarrow{\varphi'} & \mathfrak{h} \\ {\scriptstyle \exp^G}\downarrow & & \downarrow{\scriptstyle \exp^H} \\ G & \xrightarrow{\varphi} & H \end{array}$$

commutes, since $t \mapsto \varphi(\exp^G(tX))$ is a one parameter subgroup of H, and $\frac{d}{dt}|_0\varphi(\exp^G tX) = \varphi'(X)$, so $\varphi(\exp^G tX) = \exp^H(t\varphi'(X))$.

36.9. Remarks. [Omori et al., 1982, 1983, etc.] gave conditions under which a smooth Lie group modeled on Fréchet spaces admits an exponential mapping. We shall elaborate on this notion in (38.4) below. They called such groups 'regular Fréchet Lie groups'. We do not know any smooth Fréchet Lie group which does not admit an exponential mapping.

If G admits an exponential mapping, it follows from (36.8.3) that exp is a diffeomorphism from a neighborhood of 0 in $\mathfrak{g}$ onto a neighborhood of e in G, if a suitable inverse function theorem is applicable. This is true, for example, for smooth Banach Lie groups, also for gauge groups, see (42.21) but it is wrong for diffeomorphism groups, see (43.3).

If E is a Banach space, then in the Banach Lie group $GL(E)$ of all bounded linear automorphisms of E the exponential mapping is given by the series $\exp(X) = \sum_{i=0}^{\infty} \frac{1}{i!} X^i$.

If G is connected with exponential mapping and $U \subset \mathfrak{g}$ is open with $0 \in U$, then one may ask whether the group generated by $\exp(U)$ equals G. Note that this is a normal subgroup. So if G is simple, the answer is yes. This is true for connected components of diffeomorphism groups and many of their important subgroups, see [Epstein, 1970], [Thurston, 1974], [Mather, 1974, 1975, 1984, 1985], [Banyaga, 1978].

Results on weakened versions of the Baker-Campbell-Hausdorff formula can be found in [Wojtyński, 1994].

36.10. The adjoint representation. Let G be a Lie group with Lie algebra $\mathfrak{g}$. For $a \in G$ we define $\operatorname{conj}_a : G \to G$ by $\operatorname{conj}_a(x) = axa^{-1}$. It is called the *conjugation* or the *inner automorphism* by $a \in G$. This defines a smooth action of G on itself by automorphisms.

The adjoint representation $\operatorname{Ad} : G \to GL(\mathfrak{g}) \subset L(\mathfrak{g}, \mathfrak{g})$ is given by $\operatorname{Ad}(a) = (\operatorname{conj}_a)' = T_e(\operatorname{conj}_a) : \mathfrak{g} \to \mathfrak{g}$ for $a \in G$. By (36.6), $\operatorname{Ad}(a)$ is a Lie algebra homomorphism, moreover

$$\operatorname{Ad}(a) = T_e(\operatorname{conj}_a) = T_a(\mu^{a^{-1}}).T_e(\mu_a) = T_{a^{-1}}(\mu_a).T_e(\mu^{a^{-1}}).$$

Finally, we define the (lower case) *adjoint representation* of the Lie algebra $\mathfrak{g}$, $\operatorname{ad} : \mathfrak{g} \to \mathfrak{gl}(\mathfrak{g}) := L(\mathfrak{g}, \mathfrak{g})$ by $\operatorname{ad} := \operatorname{Ad}' = T_e\operatorname{Ad}$.

We shall also use the *right Maurer-Cartan form* $\kappa^r \in \Omega^1(G, \mathfrak{g})$, given by $\kappa^r_g = T_g(\mu^{g^{-1}}) : T_gG \to \mathfrak{g}$; similarly the *left Maurer-Cartan form* $\kappa^l \in \Omega^1(G, \mathfrak{g})$ is given by $\kappa^l_g = T_g(\mu_{g^{-1}}) : T_gG \to \mathfrak{g}$.

Lemma.

(1) $L_X(a) = R_{\operatorname{Ad}(a)X}(a)$ *for* $X \in \mathfrak{g}$ *and* $a \in G$.
(2) $\operatorname{ad}(X)Y = [X, Y]$ *for* $X, Y \in \mathfrak{g}$.
(3) $d\operatorname{Ad} = (\operatorname{ad} \circ \kappa^r).\operatorname{Ad} = \operatorname{Ad}.(\operatorname{ad} \circ \kappa^l) : TG \to L(\mathfrak{g}, \mathfrak{g})$.

Proof. (1) $L_X(a) = T_e(\mu_a).X = T_e(\mu^a).T_e(\mu^{a^{-1}} \circ \mu_a).X = R_{\operatorname{Ad}(a)X}(a)$.

Proof of (2) We need some preparation. Let V be a convenient vector space. For $f \in C^\infty(G,V)$ we define the *left trivialized derivative* $D_l f \in C^\infty(G, L(\mathfrak{g}, V))$ by

$$D_l f(x).X := df(x).T_e\mu_x.X = (L_X f)(x). \tag{4}$$

For $f \in C^\infty(G,\mathbb{R})$ and $g \in C^\infty(G,V)$ we have

$$\begin{aligned} D_l(f.g)(x).X &= d(f.g)(T_e\mu_x.X) \\ &= df(T_e\mu_x.X).g(x) + f(x).dg(T_e\mu_x.X) \\ &= (f.D_l g + D_l f \otimes g)(x).X. \end{aligned} \tag{5}$$

From the formula

$$\begin{aligned} D_l D_l f(x)(X)(Y) &= D_l(D_l f(\quad).Y)(x).X \\ &= D_l(L_Y f)(x).X = L_X L_Y f(x) \end{aligned}$$

follows

$$D_l D_l f(x)(X)(Y) - D_l D_l f(x)(Y)(X) = L_{[X,Y]} f(x) = D_l f(x).[X,Y]. \tag{6}$$

We consider now the linear isomorphism $L : C^\infty(G,\mathfrak{g}) \to \mathfrak{X}(G)$ given by $L_f(x) = T_e\mu_x.f(x) = L_{f(x)}(x)$ for $f \in C^\infty(G,\mathfrak{g})$. If $h \in C^\infty(G,V)$ we get $(L_f h)(x) = D_l h(x).f(x)$. For $f, g \in C^\infty(G,\mathfrak{g})$ and $h \in C^\infty(G,\mathbb{R})$ we get in turn, using (6) and (5), generalized to the bilinear pairing $L(\mathfrak{g},\mathbb{R}) \times \mathfrak{g} \to \mathbb{R}$,

$$\begin{aligned} (L_f L_g h)(x) &= D_l(D_l h(\quad).g(\quad))(x).f(x) \\ &= D_l D_l h(x)(f(x))(g(x)) + D_l h(x).D_l g(x).f(x) \\ ([L_f, L_g]h)(x) &= D_l^2 h(x).(f(x), g(x)) + D_l h(x).D_l g(x).f(x) - \\ &\quad - D_l^2 h(x).(g(x), f(x)) - D_l h(x).D_l f(x).g(x) \\ &= D_l h(x).\Big([f(x), g(x)]_{\mathfrak{g}} + D_l g(x).f(x) - D_l f(x).g(x)\Big) \end{aligned}$$

$$[L_f, L_g] = L\Big([f,g]_{\mathfrak{g}} + D_l g.f - D_l f.g\Big). \tag{7}$$

Now we are able to prove the second assertion of the lemma. For $X, Y \in \mathfrak{g}$ we will apply (7) to $f(x) = X$ and $g(x) = \mathrm{Ad}(x^{-1}).Y$. We have $L_g = R_Y$ by (1), and $[L_f, L_g] = [L_X, R_Y] = 0$ by (36.5). So

$$\begin{aligned} 0 = [L_X, R_Y](x) &= [L_f, L_g](x) \\ &= L([X, (\mathrm{Ad}\circ\nu)Y]_{\mathfrak{g}} + D_l((\mathrm{Ad}\circ\nu)(\quad).X).Y - 0)(x) \\ [X,Y] &= [X, \mathrm{Ad}(e)Y] = -D_l((\mathrm{Ad}\circ\nu)(\quad).X)(e).Y \\ &= d(\mathrm{Ad}(\quad).X)(e).Y = \mathrm{ad}(X)Y. \end{aligned}$$

Proof of (3) Let $X, Y \in \mathfrak{g}$ and $g \in G$, and let $c : \mathbb{R} \to G$ be a smooth curve with $c(0) = e$ and $c'(0) = X$. Then we have

$$\begin{aligned} (d\mathrm{Ad}(R_X(g))).Y &= \tfrac{\partial}{\partial t}|_0 \mathrm{Ad}(c(t).g).Y = \tfrac{\partial}{\partial t}|_0 \mathrm{Ad}(c(t)).\mathrm{Ad}(g).Y \\ &= \mathrm{ad}(X)\mathrm{Ad}(g)Y = (\mathrm{ad}\circ\kappa^r)(R_X(g)).\mathrm{Ad}(g).Y, \end{aligned}$$

and similarly for the second formula. □

36.11. Let $\ell : G \times M \to M$ be a smooth left action of a Lie group G, so $\ell^\vee : G \to \operatorname{Diff}(M)$ is a group homomorphism. Then we have partial mappings $\ell_a : M \to M$ and $\ell^x : G \to M$, given by $\ell_a(x) = \ell^x(a) = \ell(a,x) = a.x$.

For any $X \in \mathfrak{g}$ we define the *fundamental vector field* $\zeta_X = \zeta_X^M \in \mathfrak{X}(M)$ by $\zeta_X(x) = T_e(\ell^x).X = T_{(e,x)}\ell.(X, 0_x)$.

Lemma. *In this situation, the following assertions hold:*

(1) $\zeta : \mathfrak{g} \to \mathfrak{X}(M)$ *is a linear mapping.*
(2) $T_x(\ell_a).\zeta_X(x) = \zeta_{Ad(a)X}(a.x)$.
(3) $R_X \times 0_M \in \mathfrak{X}(G \times M)$ *is* ℓ*-related to* $\zeta_X \in \mathfrak{X}(M)$.
(4) $[\zeta_X, \zeta_Y] = -\zeta_{[X,Y]}$.

Proof. (1) is clear.

(2) We have $\ell_a \ell^x(b) = abx = aba^{-1}ax = \ell^{ax}\operatorname{conj}_a(b)$, so

$$\begin{aligned} T_x(\ell_a).\zeta_X(x) &= T_x(\ell_a).T_e(\ell^x).X = T_e(\ell_a \circ \ell^x).X \\ &= T_e(\ell^{ax}).\operatorname{Ad}(a).X = \zeta_{\operatorname{Ad}(a)X}(ax). \end{aligned}$$

(3) We have $\ell \circ (\operatorname{Id} \times \ell_a) = \ell \circ (\mu^a \times \operatorname{Id}) : G \times M \to M$, so

$$\begin{aligned} \zeta_X(\ell(a,x)) &= T_{(e,ax)}\ell.(X, 0_{ax}) = T\ell.(\operatorname{Id} \times T(\ell_a)).(X, 0_x) \\ &= T\ell.(T(\mu^a) \times \operatorname{Id}).(X, 0_x) = T\ell.(R_X \times 0_M)(a,x). \end{aligned}$$

(4) $[R_X \times 0_M, R_Y \times 0_M] = [R_X, R_Y] \times 0_M = -R_{[X,Y]} \times 0_M$ is ℓ-related to $[\zeta_X, \zeta_Y]$ by (3) and by (32.10). On the other hand, $-R_{[X,Y]} \times 0_M$ is ℓ-related to $-\zeta_{[X,Y]}$ by (3) again. Since ℓ is surjective we get $[\zeta_X, \zeta_Y] = -\zeta_{[X,Y]}$. □

36.12. Let $r : M \times G \to M$ be a right action, so $r^\vee : G \to \operatorname{Diff}(M)$ is a group anti homomorphism. We will use the following notation: $r^a : M \to M$ and $r_x : G \to M$, given by $r_x(a) = r^a(x) = r(x,a) = x.a$.

For any $X \in \mathfrak{g}$ we define the *fundamental vector field* $\zeta_X = \zeta_X^M \in \mathfrak{X}(M)$ by $\zeta_X(x) = T_e(r_x).X = T_{(x,e)}r.(0_x, X)$.

Lemma. *In this situation the following assertions hold:*

(1) $\zeta : \mathfrak{g} \to \mathfrak{X}(M)$ *is a linear mapping.*
(2) $T_x(r^a).\zeta_X(x) = \zeta_{Ad(a^{-1})X}(x.a)$.
(3) $0_M \times L_X \in \mathfrak{X}(M \times G)$ *is* r*-related to* $\zeta_X \in \mathfrak{X}(M)$.
(4) $[\zeta_X, \zeta_Y] = \zeta_{[X,Y]}$. □

37. Bundles and Connections

37.1. Definition. A *(fiber) bundle* $(p : E \to M, S) = (E, p, M, S)$ consists of smooth manifolds E, M, S, and a smooth mapping $p : E \to M$. Furthermore, each

$x \in M$ has an open neighborhood U such that $E \mid U := p^{-1}(U)$ is diffeomorphic to $U \times S$ via a fiber respecting diffeomorphism:

$$\begin{array}{ccc} E \mid U & \xrightarrow{\psi} & U \times S \\ & {\scriptstyle p}\searrow \quad \swarrow{\scriptstyle \mathrm{pr}_1} & \\ & U. & \end{array}$$

E is called *total space*, M is called *base space* or *basis*, p is a final surjective smooth mapping, called *projection*, and S is called *standard fiber*. (U, ψ) as above is called a *fiber chart*.

A collection of fiber charts (U_α, ψ_α) such that (U_α) is an open cover of M, is called a *fiber bundle atlas*. If we fix such an atlas, then $\psi_\alpha \circ {\psi_\beta}^{-1}(x, s) = (x, \psi_{\alpha\beta}(x, s))$, where $\psi_{\alpha\beta} : (U_\alpha \cap U_\beta) \times S \to S$ is smooth, and where $\psi_{\alpha\beta}(x, \ )$ is a diffeomorphism of S for each $x \in U_{\alpha\beta} := U_\alpha \cap U_\beta$. These mappings $\psi_{\alpha\beta}$ are called *transition functions* of the bundle. They satisfy the *cocycle condition*: $\psi_{\alpha\beta}(x) \circ \psi_{\beta\gamma}(x) = \psi_{\alpha\gamma}(x)$ for $x \in U_{\alpha\beta\gamma}$ and $\psi_{\alpha\alpha}(x) = \mathrm{Id}_S$ for $x \in U_\alpha$. Therefore, the collection $(\psi_{\alpha\beta})$ is called a *cocycle of transition functions*.

Given an open cover (U_α) of a manifold M and a cocycle of transition functions $(\psi_{\alpha\beta})$ we may construct a fiber bundle $(p : E \to M, S)$, as in finite dimensions.

37.2. Let $(p : E \to M, S)$ be a fiber bundle. We consider the fiber linear tangent mapping $Tp : TE \to TM$ and its kernel $\ker Tp =: VE$, which is called the *vertical bundle* of E. It is a locally splitting vector subbundle of the tangent bundle TE, by the following argument: E looks locally like $U_M \times U_S$, where U_M is c^∞-open in a modeling space W_M of M and U_S in a modeling space W_S of S. Then TE looks locally like $U_M \times W_M \times U_S \times W_S$, and the mapping Tp corresponds to $(x, v, y, w) \mapsto (x, v)$, so that VE looks locally like $U_M \times 0 \times U_S \times W_S$.

Definition. A *connection* on the fiber bundle $(p : E \to M, S)$ is a vector valued 1-form $\Phi \in \Omega^1(E; VE)$ with values in the vertical bundle VE such that $\Phi \circ \Phi = \Phi$ and $\mathrm{im}\Phi = VE$; so Φ is just a projection $TE \to VE$.

The kernel $\ker \Phi$ is a sub vector bundle of TE, it is called the space of *horizontal vectors* or the *horizontal bundle*, and it is denoted by HE. Clearly, $TE = HE \oplus VE$ and $T_uE = H_uE \oplus V_uE$ for $u \in E$.

Now we consider the mapping $(Tp, \pi_E) : TE \to TM \times_M E$. Then by definition $(Tp, \pi_E)^{-1}(0_{p(u)}, u) = V_uE$, so $(Tp, \pi_E) \mid HE : HE \to TM \times_M E$ is a fiber linear isomorphism, which may be checked in a chart. Its inverse is denoted by

$$C := ((Tp, \pi_E) \mid HE)^{-1} : TM \times_M E \to HE \hookrightarrow TE.$$

So $C : TM \times_M E \to TE$ is fiber linear over E and a right inverse for (Tp, π_E). C is called the *horizontal lift* associated to the connection Φ.

Note the formula $\Phi(\xi_u) = \xi_u - C(Tp.\xi_u, u)$ for $\xi_u \in T_uE$. So we can equally well describe a connection Φ by specifying C. Then we call Φ *vertical projection* and $\chi := \mathrm{id}_{TE} - \Phi = C \circ (Tp, \pi_E)$ will be called *horizontal projection*.

37.3. Curvature. If $\Phi : TE \to VE$ is a connection on the bundle $(p : E \to M, S)$, then as in (35.11) the curvature $\mathcal{R}$ of Φ is given by

$$2\mathcal{R} = [\Phi, \Phi] = [\mathrm{Id} - \Phi, \mathrm{Id} - \Phi] \in \Omega^2(E; VE).$$

The cocurvature $\bar{\mathcal{R}}$ vanishes since the vertical bundle VE is integrable. We have $\mathcal{R}(X, Y) = \frac{1}{2}[\Phi, \Phi](X, Y) = \Phi[\chi X, \chi Y]$ by (35.11), so $\mathcal{R}$ is an obstruction against involutivity of the horizontal subbundle in the following sense: If the curvature $\mathcal{R}$ vanishes, then horizontal kinematic vector fields on E also have a horizontal Lie bracket. Note that for vector fields $\xi, \eta \in \mathfrak{X}(M)$ and their horizontal lifts $C\xi, C\eta \in \mathfrak{X}(E)$ we have $\mathcal{R}(C\xi, C\eta) = [C\xi, C\eta] - C([\xi, \eta])$. Since the vertical bundle VE is even integrable, by (35.12) we have the *Bianchi identity* $[\Phi, \mathcal{R}] = 0$.

37.4. Pullback. Let $(p : E \to M, S)$ be a fiber bundle, and consider a smooth mapping $f : N \to M$. Let us consider the *pullback* $N \times_{(f,M,p)} E := \{(n, e) \in N \times E : f(n) = p(e)\}$; we will denote it by f^*E. The following diagram sets up some further notation for it:

$$\begin{array}{ccc} f^*E & \xrightarrow{p^*f} & E \\ {\scriptstyle f^*p}\downarrow & & \downarrow{\scriptstyle p} \\ N & \xrightarrow{f} & M. \end{array}$$

Proposition. *In the situation above we have:*

(1) *(f^*E, f^*p, N, S) is a fiber bundle, and p^*f is a fiberwise diffeomorphism.*

(2) *If $\Phi \in \Omega^1(E; TE)$ is a connection on the bundle E, then the vector valued form $f^*\Phi$, given by $(f^*\Phi)_u(X) := T_u(p^*f)^{-1}.\Phi.T_u(p^*f).X$ for $X \in T_uE$, is a connection on the bundle f^*E. The forms $f^*\Phi$ and Φ are p^*f-related in the sense of* (35.13).

(3) *The curvatures of $f^*\Phi$ and Φ are also p^*f-related.*

Proof. (1) If (U_α, ψ_α) is a fiber bundle atlas of $(p : E \to M, S)$ in the sense of (37.1), then $(f^{-1}(U_\alpha), (f^*p, \mathrm{pr}_2 \circ \psi_\alpha \circ p^*f))$ is visibly a fiber bundle atlas for the pullback bundle (f^*E, f^*p, N, S). (2) is obvious. (3) follows from (2) and (35.13.7). □

37.5. Local description. Let Φ be a connection on $(p : E \to M, S)$. Let us fix a fiber bundle atlas (U_α) with transition functions $(\psi_{\alpha\beta})$, and let us consider the connection $((\psi_\alpha)^{-1})^*\Phi \in \Omega^1(U_\alpha \times S; U_\alpha \times TS)$, which may be written in the form

$$(((\psi_\alpha)^{-1})^*\Phi)(\xi_x, \eta_y) =: -\Gamma^\alpha(\xi_x, y) + \eta_y \text{ for } \xi_x \in T_xU_\alpha \text{ and } \eta_y \in T_yS,$$

since it reproduces vertical vectors. The Γ^α are given by

$$(0_x, \Gamma^\alpha(\xi_x, y)) := -T(\psi_\alpha).\Phi.T(\psi_\alpha)^{-1}.(\xi_x, 0_y).$$

We consider Γ^α as an element of the space $\Omega^1(U_\alpha; \mathfrak{X}(S))$, a 1-form on U^α with values in the Lie algebra $\mathfrak{X}(S)$ of all kinematic vector fields on the standard fiber. The Γ^α are called the *Christoffel forms* of the connection Φ with respect to the bundle atlas (U_α, ψ_α).

Lemma. *The transformation law for the Christoffel forms is*

$$T_y(\psi_{\alpha\beta}(x,\quad)).\Gamma^\beta(\xi_x,y)=\Gamma^\alpha(\xi_x,\psi_{\alpha\beta}(x,y))-T_x(\psi_{\alpha\beta}(\quad,y)).\xi_x.$$

The curvature $\mathcal{R}$ of Φ satisfies

$$(\psi_\alpha^{-1})^*\mathcal{R}=d\Gamma^\alpha+\tfrac{1}{2}[\Gamma^\alpha,\Gamma^\alpha]^\wedge_{\mathfrak{X}(S)}.$$

Here $d\Gamma^\alpha$ is the exterior derivative of the 1-form $\Gamma^\alpha\in\Omega^1(U_\alpha,\mathfrak{X}(S))$ with values in the convenient vector space $\mathfrak{X}(S)$.

The formula for the curvature is the *Maurer-Cartan formula* which in this general setting appears only on the level of local description.

Proof. From $(\psi_\alpha\circ(\psi_\beta)^{-1})(x,y)=(x,\psi_{\alpha\beta}(x,y))$ we get that $T(\psi_\alpha\circ(\psi_\beta)^{-1}).(\xi_x,\eta_y)=(\xi_x,T_{(x,y)}(\psi_{\alpha\beta}).(\xi_x,\eta_y))$, and thus

$$\begin{aligned}
&T(\psi_\beta^{-1}).(0_x,\Gamma^\beta(\xi_x,y))=-\Phi(T(\psi_\beta^{-1})(\xi_x,0_y))=\\
&=-\Phi(T(\psi_\alpha^{-1}).T(\psi_\alpha\circ\psi_\beta^{-1}).(\xi_x,0_y))=\\
&=-\Phi(T(\psi_\alpha^{-1})(\xi_x,T_{(x,y)}(\psi_{\alpha\beta})(\xi_x,0_y)))=\\
&=-\Phi(T(\psi_\alpha^{-1})(\xi_x,0_{\psi_{\alpha\beta}(x,y)}))-\Phi(T(\psi_\alpha^{-1})(0_x,T_{(x,y)}\psi_{\alpha\beta}(\xi_x,0_y)))=\\
&=T(\psi_\alpha^{-1}).(0_x,\Gamma^\alpha(\xi_x,\psi_{\alpha\beta}(x,y)))-T(\psi_\alpha^{-1})(0_x,T_x(\psi_{\alpha\beta}(\quad,y)).\xi_x).
\end{aligned}$$

This implies the transformation law.

For the curvature $\mathcal{R}$ of Φ we have by (37.3) and (37.4.3)

$$\begin{aligned}
&(\psi_\alpha^{-1})^*\mathcal{R}\,((\xi^1,\eta^1),(\xi^2,\eta^2))=\\
&\quad=(\psi_\alpha^{-1})^*\Phi\,[(\mathrm{Id}-(\psi_\alpha^{-1})^*\Phi)(\xi^1,\eta^1),(\mathrm{Id}-(\psi_\alpha^{-1})^*\Phi)(\xi^2,\eta^2)]=\\
&\quad=(\psi_\alpha^{-1})^*\Phi[(\xi^1,\Gamma^\alpha(\xi^1)),(\xi^2,\Gamma^\alpha(\xi^2))]=\\
&\quad=(\psi_\alpha^{-1})^*\Phi\,([\xi^1,\xi^2],\xi^1\Gamma^\alpha(\xi^2)-\xi^2\Gamma^\alpha(\xi^1)+[\Gamma^\alpha(\xi^1),\Gamma^\alpha(\xi^2)])=\\
&\quad=-\Gamma^\alpha([\xi^1,\xi^2])+\xi^1\Gamma^\alpha(\xi^2)-\xi^2\Gamma^\alpha(\xi^1)+[\Gamma^\alpha(\xi^1),\Gamma^\alpha(\xi^2)]=\\
&\quad=d\Gamma^\alpha(\xi^1,\xi^2)+[\Gamma^\alpha(\xi^1),\Gamma^\alpha(\xi^2)]_{\mathfrak{X}(S)}.\quad\square
\end{aligned}$$

37.6. Parallel transport. Let Φ be a connection on a bundle $(p:E\to M,S)$, and let $c:(a,b)\to M$ be a smooth curve with $0\in(a,b)$, $c(0)=x$. The *parallel transport* along c is a smooth mapping $\mathrm{Pt}_c:U\to E$, where U is a neighborhood of $\mathrm{Diag}(a,b)\times_{(\mathrm{copr}_2,M,p)}E$ in $(a,b)\times(a,b)\times_{(\mathrm{copr}_2,M,p)}E$, such that the following properties hold:

(1) $U\cap((a,b)\times\{s\}\times\{u_{c(s)}\})$ is connected for each $s\in(a,b)$ and each $u_{c(s)}\in E_{c(s)}$.
(2) $p(\mathrm{Pt}(c,t,s,u_{c(s)}))=c(t)$ if defined, and $\mathrm{Pt}(c,s,s,u_{c(s)})=u_{c(s)}$.
(3) $\Phi(\frac{d}{dt}\mathrm{Pt}(c,t,s,u_{c(s)}))=0$ if defined.

(4) If $\operatorname{Pt}(c,t,s,u_{c(s)})$ exists then $\operatorname{Pt}(c,r,s,u_{c(s)}) = \operatorname{Pt}(c,r,t,\operatorname{Pt}(c,t,s,u_{c(s)}))$ in the sense that existence of both sides is equivalent and we have equality.

(5) U is maximal for properties (1) to (4).

(6) Pt also depends smoothly on c in the Frölicher space $C^\infty((a,b),M)$, see (23.1), in the following sense: For any smooth mapping $c : \mathbb{R}\times(a,b)\to M$ we have: For each $s\in(a,b)$, each $r\in\mathbb{R}$, and each $u\in E_{c(r,s)}$ there are a neighborhood $U_{c,r,s,u}$ of (r,s,s,u) in $\mathbb{R}\times(a,b)\times(a,b)\times_{(\operatorname{copr}_{1,3},M,p)} E\subset \mathbb{R}\times(a,b)\times(a,b)\times E$ such that $U_{c,r,s,u}\ni(r',t,s',u')\mapsto \operatorname{Pt}(c(r',\quad),t,s',u')$ is defined and smooth.

(7) Reparameterization invariance: If $f : (a',b')\to(a,b)$ is smooth, then we have $\operatorname{Pt}(c,f(t),f(s),u_{c(f(s))}) = \operatorname{Pt}(c\circ f,t,s,u_{c(f(s))})$

Requirements (1) – (5) are essential. (6) is a further requirement which is not necessary for the uniqueness result below, and (7) is a consequence of this uniqueness result.

Proposition. *The parallel transport along c is unique if it exists.*

Proof. Consider the pullback bundle $(c^*E, c^*p, (a,b), S)$ and the pullback connection $c^*\Phi$ on it. We shall need the horizontal lift $C : T(a,b)\times_{(a,b)} c^*E\to T(c^*E) = (Tc)^*(TE)$ associated to $c^*\Phi$, from (37.2). Consider the constant vector field $\partial\in\mathfrak{X}((a,b))$ and its horizontal lift $C(\partial)\in\mathfrak{X}(c^*E)$ which is given by $C(\partial)(u_s) = C(\partial|_s,u_s)\in T_{u_s}(c^*E)$. Now from the properties of the parallel transport we see that $t\mapsto\operatorname{Pt}(c(s+\quad),t,s,u_s)$ is a flow line of the horizontal vector field $C(\partial)$ with initial value $u_s = (s,u_{c(s)})\in(c^*E)_s\cong\{s\}\times E_{c(s)}$. (3) says that it has the flow property, so that by uniqueness of the flow (32.16) we see that $\operatorname{Pt}(c,t) = \operatorname{Fl}_t^{C(\partial)}$ is unique if it exists. $\square$

At this place one could consider complete connections (those whose parallel transport exists globally), which then give rise to holonomy groups, even for fiber bundles without structure groups. In finite dimensions some deep results are available, see [Kolář, Michor, Slovák, 1993, pp81].

37.7. Definition. Let G be a Lie group, and let $(p : E\to M, S)$ be a fiber bundle as in (37.1). A *G-bundle structure* on the fiber bundle consists of the following data:

(1) A left action $\ell : G\times S\to S$ of the Lie group on the standard fiber.

(2) A fiber bundle atlas (U_α,ψ_α) whose transition functions $(\psi_{\alpha\beta})$ act on S via the G-action: There is a family of smooth mappings $(\varphi_{\alpha\beta} : U_{\alpha\beta}\to G)$ which satisfies the cocycle condition $\varphi_{\alpha\beta}(x)\varphi_{\beta\gamma}(x) = \varphi_{\alpha\gamma}(x)$ for $x\in U_{\alpha\beta\gamma}$ and $\varphi_{\alpha\alpha}(x) = e$, the unit in the group, such that $\psi_{\alpha\beta}(x,s) = \ell(\varphi_{\alpha\beta}(x),s) = \varphi_{\alpha\beta}(x).s$.

A fiber bundle with a G-bundle structure is called a *G-bundle*. A fiber bundle atlas as in (2) is called a *G-atlas*, and the family $(\varphi_{\alpha\beta})$ is also called a cocycle of transition functions, but now for the G-bundle.

To be more precise, two G-atlas are said to be equivalent (to describe the same G-bundle), if their union is also a G-atlas. This translates to the two cocycles of transition functions as follows, where we assume that the two coverings of M are the same (by passing to the common refinement, if necessary): $(\varphi_{\alpha\beta})$ and $(\varphi'_{\alpha\beta})$ are called *cohomologous* if there is a family $(\tau_\alpha : U_\alpha \to G)$ such that $\varphi_{\alpha\beta}(x) = \tau_\alpha(x)^{-1}.\varphi'_{\alpha\beta}(x).\tau_\beta(x)$ holds for all $x \in U_{\alpha\beta}$.

In (2) one should specify only an equivalence class of G-bundle structures or only a cohomology class of cocycles of G-valued transition functions. From any open cover (U_α) of M, some cocycle of transition functions $(\varphi_{\alpha\beta} : U_{\alpha\beta} \to G)$ for it, and a left G-action on a manifold S, we may construct a G-bundle, which depends only on the cohomology class of the cocycle. By some abuse of notation, we write $(p : E \to M, S, G)$ for a fiber bundle with specified G-bundle structure.

37.8. Definition. A *principal (fiber) bundle* $(p : P \to M, G)$ is a G-bundle with typical fiber a Lie group G, where the left action of G on G is just the left translation.

So by (37.7) we are given a bundle atlas $(U_\alpha, \varphi_\alpha : P|U_\alpha \to U_\alpha \times G)$ such that we have $(\varphi_\alpha \circ \varphi_\beta^{-1})(x, a) = (x, \varphi_{\alpha\beta}(x).a)$ for the cocycle of transition functions $(\varphi_{\alpha\beta} : U_{\alpha\beta} \to G)$. This is now called a *principal bundle atlas*. Clearly, the principal bundle is uniquely determined by the cohomology class of its cocycle of transition functions.

Each principal bundle admits a unique right action $r : P \times G \to P$, called the *principal right action*, given by $\varphi_\alpha(r(\varphi_\alpha^{-1}(x, a), g)) = (x, ag)$. Since left and right translation on G commute, this is well defined. We write $r(u, g) = u.g$ when the meaning is clear. The principal right action is obviously free and for any $u_x \in P_x$ the partial mapping $r_{u_x} = r(u_x, \quad) : G \to P_x$ is a diffeomorphism onto the fiber through u_x, whose inverse is denoted by $\tau_{u_x} : P_x \to G$. These inverses together give a smooth mapping $\tau : P \times_M P \to G$, whose local expression is $\tau(\varphi_\alpha^{-1}(x, a), \varphi_\alpha^{-1}(x, b)) = a^{-1}.b$. This mapping is uniquely determined by the implicit equation $r(u_x, \tau(u_x, v_x)) = v_x$, thus we also have $\tau(u_x.g, u'_x.g') = g^{-1}.\tau(u_x, u'_x).g'$ and $\tau(u_x, u_x) = e$.

37.9. Lemma. *Let $p : P \to M$ be a surjective smooth mapping admitting local smooth sections near each point in M, and let G be a Lie group which acts freely on P such that the orbits of the action are exactly the fibers $p^{-1}(x)$ of p. If the unique mapping $\tau : P \times_M P \to G$ satisfying $u_x.\tau(u_x, v_x) = v_x$ is smooth, then $(p : P \to M, G)$ is a principal fiber bundle.*

Proof. Let the action be a right action by using the group inversion if necessary. Let $s_\alpha : U_\alpha \to P$ be local sections (right inverses) for $p : P \to M$ such that (U_α) is an open cover of M. Let $\varphi_\alpha^{-1} : U_\alpha \times G \to P|U_\alpha$ be given by $\varphi_\alpha^{-1}(x, a) = s_\alpha(x).a$, with smooth inverse $\varphi_\alpha(u_x) = (x, \tau(s_\alpha(x), u_x))$, a fiber respecting diffeomorphism $\varphi_\alpha : P|U_\alpha \to U_\alpha \times G$. So $(U_\alpha, \varphi_\alpha)$ is already a fiber bundle atlas. Clearly, we have $\tau(u_x, u'_x.g) = \tau(u_x, u'_x).g$ and $\varphi_\alpha(u_x) = (x, \tau(s_\alpha(x), u_x))$, so $\varphi_\alpha\varphi_\beta^{-1}(x, g) = \varphi_\alpha(s_\beta(x).g) = (x, \tau(s_\alpha(x), s_\beta(x).g)) = (x, \tau(s_\alpha(x), s_\beta(x)).g)$, and $(U_\alpha, \varphi_\alpha)$ is a principal bundle atlas. □

37.10. Remarks. In the proof of Lemma (37.9) we have seen that a principal bundle atlas of a principal fiber bundle $(p : P \to M, G)$ is already determined if we specify a family of smooth sections of P whose domains of definition cover the base M. Lemma (37.9) could serve as an equivalent definition for a principal bundle. From the lemma follows, that the pullback f^*P over a smooth mapping $f : M' \to M$ is also a principal fiber bundle.

37.11. Homomorphisms. Let $\chi : (p : P \to M, G) \to (p' : P' \to M', G)$ be a *principal fiber bundle homomorphism*, i.e., a smooth G-equivariant mapping $\chi : P \to P'$. Then, obviously, the diagram

(a)
$$\begin{array}{ccc} P & \xrightarrow{\ \chi\ } & P' \\ {\scriptstyle p}\downarrow & & \downarrow{\scriptstyle p'} \\ M & \xrightarrow[\bar\chi]{} & M' \end{array}$$

commutes for a uniquely determined smooth mapping $\bar\chi : M \to M'$. For each $x \in M$ the mapping $\chi_x := \chi|P_x : P_x \to P'_{\bar\chi(x)}$ is G-equivariant and therefore a diffeomorphism, so diagram (a) is a pullback diagram.

But the most general notion of a homomorphism of principal bundles is the following. Let $\Phi : G \to G'$ be a homomorphism of Lie groups. $\chi : (p : P \to M, G) \to (p' : P' \to M', G')$ is called a *homomorphism over Φ of principal bundles*, if $\chi : P \to P'$ is smooth and $\chi(u.g) = \chi(u).\Phi(g)$ holds. Then χ is fiber respecting, so diagram (a) again makes sense, but it is not a pullback diagram in general.

If χ covers the identity on the base, it is called a *reduction of the structure group G' to G for the principal bundle $(p' : P' \to M', G')$* — the name comes from the case, when Φ is the embedding of a subgroup.

By the universal property of the pullback any general homomorphism χ of principal fiber bundles over a group homomorphism can be written as the composition of a reduction of structure groups and a pullback homomorphism as follows, where we also indicate the structure groups:

(b)
$$\begin{array}{ccccc} (P,G) & \longrightarrow & (\bar\chi^* P', G') & \longrightarrow & (P', G') \\ & {\scriptstyle p}\searrow & \downarrow & & \downarrow{\scriptstyle p'} \\ & & M & \xrightarrow{\ \bar\chi\ } & M'. \end{array}$$

37.12. Associated bundles. Let $(p : P \to M, G)$ be a principal bundle, and let $\ell : G \times S \to S$ be a left action of the structure group G on a manifold S. We consider the right action $R : (P \times S) \times G \to P \times S$, given by $R((u,s),g) = (u.g, g^{-1}.s)$.

Theorem. *In this situation we have:*

(1) *The space $P \times_G S$ of orbits of the action R carries a unique smooth manifold structure such that the quotient map $q : P \times S \to P \times_G S$ is a final smooth mapping.*

(2) *$(P\times_G S, \bar p, M, S, G)$ is a G-bundle in a canonical way, where $\bar p : P\times_G S \to M$ is given by*

(a)
$$\begin{array}{ccc} P\times S & \xrightarrow{\;q\;} & P\times_G S \\ \Big\downarrow{\scriptstyle \mathrm{pr}_1} & & \Big\downarrow{\scriptstyle \bar p} \\ P & \xrightarrow{\;p\;} & M. \end{array}$$

In this diagram $q_u : \{u\}\times S \to (P\times_G S)_{p(u)}$ is a diffeomorphism for each $u\in P$.

(3) *$(P\times S, q, P\times_G S, G)$ is a principal fiber bundle with principal action R.*

(4) *If $(U_\alpha, \varphi_\alpha : P|U_\alpha \to U_\alpha\times G)$ is a principal bundle atlas with cocycle of transition functions $(\varphi_{\alpha\beta} : U_{\alpha\beta}\to G)$, then together with the left action $\ell : G\times S\to S$ this is also a cocycle for the G-bundle $(P\times_G S, \bar p, M, S, G)$.*

Notation. $(P\times_G S, \bar p, M, S, G)$ is called the *associated bundle* for the action $\ell : G\times S\to S$. We will also denote it by $P[S,\ell]$ or simply $P[S]$, and we will write p for $\bar p$ if no confusion is possible. We also define the smooth mapping $\tau = \tau^S : P\times_M P[S,\ell]\to S$ by $\tau(u_x, v_x) := q_{u_x}^{-1}(v_x)$. It satisfies $\tau(u, q(u,s)) = s$, $q(u_x, \tau(u_x, v_x)) = v_x$, and $\tau(u_x.g, v_x) = g^{-1}.\tau(u_x, v_x)$. In the special situation, where $S = G$ and the action is left translation, so that $P[G] = P$, this mapping coincides with τ considered in (37.8).

Proof. In the setting of diagram (a) the mapping $p\circ pr_1$ is constant on the R-orbits, so $\bar p$ exists as a mapping. Let $(U_\alpha, \varphi_\alpha : P|U_\alpha \to U_\alpha\times G)$ be a principal bundle atlas with transition functions $(\varphi_{\alpha\beta} : U_{\alpha\beta}\to G)$. We define $\psi_\alpha^{-1} : U_\alpha\times S \to \bar p^{-1}(U_\alpha)\subset P\times_G S$ by $\psi_\alpha^{-1}(x,s) = q(\varphi_\alpha^{-1}(x,e), s)$, which is fiber respecting. For each orbit in $\bar p^{-1}(x)\subset P\times_G S$ there is exactly one $s\in S$ such that this orbit passes through $(\varphi_\alpha^{-1}(x,e), s)$, namely $s = \tau^G(u_x, \varphi_\alpha^{-1}(x,e))^{-1}.s'$ if (u_x, s') is the orbit, since the principal right action is free. Thus, $\psi_\alpha^{-1}(x, \quad) : S\to \bar p^{-1}(x)$ is bijective. Furthermore,

$$\begin{aligned} \psi_\beta^{-1}(x,s) &= q(\varphi_\beta^{-1}(x,e), s) \\ &= q(\varphi_\alpha^{-1}(x, \varphi_{\alpha\beta}(x).e), s) = q(\varphi_\alpha^{-1}(x,e).\varphi_{\alpha\beta}(x), s) \\ &= q(\varphi_\alpha^{-1}(x,e), \varphi_{\alpha\beta}(x).s) = \psi_\alpha^{-1}(x, \varphi_{\alpha\beta}(x).s), \end{aligned}$$

so $\psi_\alpha\psi_\beta^{-1}(x,s) = (x, \varphi_{\alpha\beta}(x).s)$. Therefore, (U_α, ψ_α) is a G-atlas for $P\times_G S$ and makes it a smooth manifold and a G-bundle. The defining equation for ψ_α shows that q is smooth and admits local smooth sections, so it is final, consequently the smooth structure on $P\times_G S$ is uniquely defined, and $\bar p$ is smooth. By the definition of ψ_α, the diagram

(b)
$$\begin{array}{ccc} p^{-1}(U_\alpha)\times S & \xrightarrow{\;\varphi_\alpha\times \mathrm{Id}\;} & U_\alpha\times G\times S \\ {\scriptstyle q}\Big\downarrow & & \Big\downarrow{\scriptstyle \mathrm{Id}\times\ell} \\ \bar p^{-1}(U_\alpha) & \xrightarrow{\;\psi_\alpha\;} & U_\alpha\times S \end{array}$$

commutes; since its horizontal arrows are diffeomorphisms we conclude that $q_u : \{u\} \times S \to \bar{p}^{-1}(p(u))$ is a diffeomorphism. So (1), (2), and (4) are checked.

(3) follows directly from lemma (37.9). We give below an explicit chart construction. We rewrite diagram (b) in the following form:

$$\text{(c)}\qquad \begin{array}{ccccc} p^{-1}(U_\alpha) \times S & \xrightarrow{=} & q^{-1}(V_\alpha) & \xrightarrow{\lambda_\alpha} & V_\alpha \times G \\ & & \Big\downarrow q & & \Big\downarrow \mathrm{pr}_1 \\ & & \bar{p}^{-1}(U_\alpha) & \xrightarrow{=} & V_\alpha \end{array}$$

Here $V_\alpha := \bar{p}^{-1}(U_\alpha) \subset P \times_G S$, and the diffeomorphism λ_α is defined by stipulating $\lambda_\alpha^{-1}(\psi_\alpha^{-1}(x,s),g) := (\varphi_\alpha^{-1}(x,g), g^{-1}.s)$. Then we have

$$\begin{aligned} \lambda_\beta^{-1}(\psi_\alpha^{-1}(x,s),g) &= \lambda_\beta^{-1}(\psi_\beta^{-1}(x,\varphi_{\beta\alpha}(x).s),g) \\ &= (\varphi_\beta^{-1}(x,g), g^{-1}.\varphi_{\beta\alpha}(x).s) \\ &= (\varphi_\alpha^{-1}(x,\varphi_{\alpha\beta}(x).g), g^{-1}.\varphi_{\alpha\beta}(x)^{-1}.s) \\ &= \lambda_\alpha^{-1}(\psi_\alpha^{-1}(x,s),\varphi_{\alpha\beta}(x).g), \end{aligned}$$

so $\lambda_\alpha\lambda_\beta^{-1}(\psi_\alpha^{-1}(x,s),g) = (\psi_\alpha^{-1}(x,s),\varphi_{\alpha\beta}(x).g)$, and $(q : P \times S \to P \times_G S, G)$ is a principal bundle with structure group G and the same cocycle $(\varphi_{\alpha\beta})$ we started with. □

37.13. Corollary. *Let $(p : E \to M, S, G)$ be a G-bundle, specified by a cocycle of transition functions $(\varphi_{\alpha\beta})$ with values in G and a left action ℓ of G on S. Then from the cocycle of transition functions we may glue a unique principal bundle $(p : P \to M, G)$ such that $E = P[S,\ell]$.* □

37.14. Equivariant mappings and associated bundles.

(1) Let $(p : P \to M, G)$ be a principal fiber bundle, and consider two left actions of G, $\ell : G \times S \to S$ and $\ell' : G \times S' \to S'$. Let furthermore $f : S \to S'$ be a G-equivariant smooth mapping, so $f(g.s) = g.f(s)$ or $f \circ \ell_g = \ell'_g \circ f$. Then $\mathrm{Id}_P \times f : P \times S \to P \times S'$ is equivariant for the actions $R : (P \times S) \times G \to P \times S$ and $R' : (P \times S') \times G \to P \times S'$ and is thus a homomorphism of principal bundles, so there is an induced mapping

$$\text{(a)}\qquad \begin{array}{ccc} P \times S & \xrightarrow{\mathrm{Id} \times f} & P \times S' \\ \Big\downarrow q & & \Big\downarrow q' \\ P \times_G S & \xrightarrow{\mathrm{Id} \times_G f} & P \times_G S', \end{array}$$

which is fiber respecting over M and a homomorphism of G-bundles in the sense of the definition (37.15) below.

(2) Let $\chi : (p : P \to M, G) \to (p' : P' \to M', G)$ be a principal fiber bundle homomorphism as in (37.11). Furthermore, we consider a smooth left action ℓ :

$G \times S \to S$. Then $\chi \times \mathrm{Id}_S : P \times S \to P' \times S$ is G-equivariant and induces a mapping $\chi \times_G \mathrm{Id}_S : P \times_G S \to P' \times_G S$, which is fiber respecting over M, fiberwise a diffeomorphism, and a homomorphism of G-bundles in the sense of definition (37.15) below.

(3) Now we consider the situations of (1) and (2) at the same time. We have two associated bundles $P[S,\ell]$ and $P'[S',\ell']$. Let $\chi : (p : P \to M, G) \to (p' : P' \to M', G)$ be a principal fiber bundle homomorphism, and let $f : S \to S'$ be a G-equivariant mapping. Then $\chi \times f : P \times S \to P' \times S'$ is clearly G-equivariant and therefore induces a mapping $\chi \times_G f : P[S,\ell] \to P'[S',\ell']$ which also is a homomorphism of G-bundles.

(4) Let S be a point. Then $P[S] = P \times_G S = M$. Furthermore, let $y \in S'$ be a fixpoint of the action $\ell' : G \times S' \to S'$, then the inclusion $i : \{y\} \hookrightarrow S'$ is G-equivariant, thus $\mathrm{Id}_P \times i$ induces $\mathrm{Id}_P \times_G i : M = P[\{y\}] \to P[S']$, which is a global section of the associated bundle $P[S']$.

If the action of G on S is trivial, i.e., $g.s = s$ for all $s \in S$, then the associated bundle is trivial: $P[S] = M \times S$. For a trivial principal fiber bundle any associated bundle is trivial.

37.15. Definition. In the situation of (37.14), a smooth fiber respecting mapping $\gamma : P[S,\ell] \to P'[S',\ell']$ covering a smooth mapping $\bar\gamma : M \to M'$ of the bases is called a *homomorphism of G-bundles*, if the following conditions are satisfied: P is isomorphic to the pullback $\bar\gamma^* P'$, and the local representations of γ in pullback-related fiber bundle atlas belonging to the two G-bundles are fiberwise G-equivariant.

Let us describe this in more detail now. Let $(U'_\alpha, \psi'_\alpha)$ be a G-atlas for $P'[S',\ell']$ with cocycle of transition functions $(\varphi'_{\alpha\beta})$, belonging to the principal fiber bundle atlas $(U'_\alpha, \varphi'_\alpha)$ of $(p' : P' \to M', G)$. Then the pullback-related principal fiber bundle atlas $(U_\alpha = \bar\gamma^{-1}(U'_\alpha), \varphi_\alpha)$ for $P = \bar\gamma^* P'$, as described in the proof of (37.4), has the cocycle of transition functions $(\varphi_{\alpha\beta} = \varphi'_{\alpha\beta} \circ \bar\gamma)$. It induces the G-atlas (U_α, ψ_α) for $P[S,\ell]$. Then $(\psi'_\alpha \circ \gamma \circ \psi_\alpha^{-1})(x,s) = (\bar\gamma(x), \gamma_\alpha(x,s))$, and $\gamma_\alpha(x, \quad) : S \to S'$ is required to be G-equivariant for all α and all $x \in U_\alpha$.

Lemma. *Let $\gamma : P[S,\ell] \to P'[S',\ell']$ be a homomorphism of G-bundles as defined above. Then there is a homomorphism $\chi : (p : P \to M, G) \to (p' : P' \to M', G)$ of principal bundles and a G-equivariant mapping $f : S \to S'$ such that $\gamma = \chi \times_G f : P[S,\ell] \to P'[S',\ell']$.*

Proof. The homomorphism $\chi : (p : P \to M, G) \to (p' : P' \to M', G)$ of principal fiber bundles is already determined by the requirement that $P = \bar\gamma^* P'$, and we have $\bar\gamma = \bar\chi$. The G-equivariant mapping $f : S \to S'$ can be read off the following diagram

$$\text{(a)} \qquad \begin{array}{ccc} P \times_M P[S] & \xrightarrow{\tau^S} & S \\ {\scriptstyle \chi \times_M \gamma}\big\downarrow & & \big\downarrow{\scriptstyle f} \\ P' \times_{M'} P'[S'] & \xrightarrow{\tau^{S'}} & S', \end{array}$$

which by the assumptions is well defined in the right column. □

So a homomorphism of G-bundles is described by the whole triple $(\chi : P \to P', f : S \to S'$ (G-equivariant)$, \gamma : P[S] \to P'[S'])$, such that diagram (a) commutes.

37.16. Sections of associated bundles. Let $(p : P \to M, G)$ be a principal fiber bundle and $\ell : G \times S \to S$ a left action. Let $C^\infty(P, S)^G$ denote the space of all smooth mappings $f : P \to S$ which are G-equivariant in the sense that $f(u.g) = g^{-1}.f(u)$ holds for $g \in G$ and $u \in P$.

Theorem. *The sections of the associated bundle $P[S, \ell]$ correspond exactly to the G-equivariant mappings $P \to S$; we have a bijection $C^\infty(P, S)^G \cong C^\infty(M \leftarrow P[S])$.*

This result follows from (37.14) and (37.15). Since it is very important we include a direct proof. That this is in general not an isomorphism of smooth structures will become clear in the proof of (42.21) below.

Proof. If $f \in C^\infty(P, S)^G$ we construct $s_f \in C^\infty(M \leftarrow P[S])$ in the following way. $\operatorname{graph}(f) = (\mathrm{Id}, f) : P \to P \times S$ is G-equivariant, since we have $(\mathrm{Id}, f)(u.g) = (u.g, f(u.g)) = (u.g, g^{-1}.f(u)) = ((\mathrm{Id}, f)(u)).g$. So it induces a smooth section $s_f \in C^\infty(M \leftarrow P[S])$ as seen from (37.14) and the diagram:

(a)
$$\begin{array}{ccc} P \times \{\text{point}\} \cong P & \xrightarrow{(\mathrm{Id}, f)} & P \times S \\ \Big\downarrow{\scriptstyle p} & & \Big\downarrow{\scriptstyle q} \\ M & \xrightarrow{s_f} & P[S]. \end{array}$$

If, conversely, $s \in C^\infty(M \leftarrow P[S])$ we define $f_s \in C^\infty(P, S)^G$ by $f_s := \tau^S \circ (\mathrm{Id}_P \times_M s) : P = P \times_M M \to P \times_M P[S] \to S$. This is G-equivariant since $f_s(u_x.g) = \tau^S(u_x.g, s(x)) = g^{-1}.\tau^S(u_x, s(x)) = g^{-1}.f_s(u_x)$ by (37.12). The two constructions are inverse to each other since we have

$$f_{s(f)}(u) = \tau^S(u, s_f(p(u))) = \tau^S(u, q(u, f(u))) = f(u),$$
$$s_{f(s)}(p(u)) = q(u, f_s(u)) = q(u, \tau^S(u, s(p(u)))) = s(p(u)). \quad \square$$

37.17. The bundle of gauges. If $(p : P \to M, G)$ is a principal fiber bundle we denote by $\operatorname{Aut}(P)$ the group of all G-equivariant diffeomorphisms $\chi : P \to P$. Then $p \circ \chi = \bar\chi \circ p$ for a unique diffeomorphism $\bar\chi$ of M, so there is a group homomorphism from $\operatorname{Aut}(P)$ into the group $\operatorname{Diff}(M)$ of all diffeomorphisms of M. The kernel of this homomorphism is called $\operatorname{Gau}(P)$, the group of *gauge transformations.* So $\operatorname{Gau}(P)$ is the space of all diffeomorphisms $\chi : P \to P$ which satisfy $p \circ \chi = p$ and $\chi(u.g) = \chi(u).g$.

Theorem. *The group* $\operatorname{Gau}(P)$ *of gauge transformations is equal to the space of sections* $C^\infty(P,(G,\operatorname{conj}))^G \cong C^\infty(M \leftarrow P[G,\operatorname{conj}])$.

If $(p : P \to M, G)$ is a finite dimensional principal bundle then there exists a structure of a Lie group on $\operatorname{Gau}(P) = C^\infty(M \leftarrow P[G,\operatorname{conj}])$, modeled on $C^\infty_c(M \leftarrow P[\mathfrak{g},\operatorname{Ad}])$. This will be proved in (42.21) below.

Proof. We again use the mapping $\tau : P \times_M P \to G$ from (37.8). For $\chi \in \operatorname{Gau}(P)$ we define $f_\chi \in C^\infty(P,(G,\operatorname{conj}))^G$ by $f_\chi := \tau \circ (\operatorname{Id},\chi)$. Then $f_\chi(u.g) = \tau(u.g,\chi(u.g)) = g^{-1}.\tau(u,\chi(u)).g = \operatorname{conj}_{g^{-1}} f_\chi(u)$, so f_χ is indeed G-equivariant.

If conversely $f \in C^\infty(P,(G,\operatorname{conj}))^G$ is given, we define $\chi_f : P \to P$ by $\chi_f(u) := u.f(u)$. It is easy to check that χ_f is indeed in $\operatorname{Gau}(P)$ and that the two constructions are inverse to each other, namely

$$\begin{aligned}
&\chi_f(ug) = ugf(ug) = ugg^{-1}f(u)g = \chi_f(u)g,\\
&f_{\chi_f}(u) = \tau(u,\chi_f(u)) = \tau(u,u.f(u)) = \tau(u,u)f(u) = f(u),\\
&\chi_{f_\chi}(u) = uf_\chi(u) = u\tau(u,\chi(u)) = \chi(u). \quad \square
\end{aligned}$$

37.18. Tangent bundles and vertical bundles. Let $(p : E \to M, S)$ be a fiber bundle. Recall the vertical subbundle $\pi_E : VE = \{\xi \in TE : Tp.\xi = 0\} \to E$ of TE from (37.2).

Theorem. *Let* $(p : P \to M, G)$ *be a principal fiber bundle with principal right action* $r : P \times G \to P$. *Let* $\ell : G \times S \to S$ *be a left action. Then the following assertions hold:*

(1) $(Tp : TP \to TM, TG)$ *is a principal fiber bundle with principal right action* $Tr : TP \times TG \to TP$, *where the structure group* TG *is the tangent group of* G, *see* (38.10).
(2) *The vertical bundle* $(\pi : VP \to P, \mathfrak{g})$ *of the principal bundle is trivial as a vector bundle over* P: $VP \cong P \times \mathfrak{g}$.
(3) *The vertical bundle of the principal bundle as bundle over* M *is a principal bundle:* $(p \circ \pi : VP \to M, TG)$.
(4) *The tangent bundle of the associated bundle* $P[S,\ell]$ *is given by* $T(P[S,\ell]) = TP[TS,T\ell]$.
(5) *The vertical bundle of the associated bundle* $P[S,\ell]$ *is given by* $V(P[S,\ell]) = P[TS,T_2\ell] = P \times_G TS$.

Proof. Let $(U_\alpha, \varphi_\alpha : P|U_\alpha \to U_\alpha \times G)$ be a principal fiber bundle atlas with cocycle of transition functions $(\varphi_{\alpha\beta} : U_{\alpha\beta} \to G)$. Since T is a functor which respects products, $(TU_\alpha, T\varphi_\alpha : TP|TU_\alpha \to TU_\alpha \times TG)$ is a principal fiber bundle atlas with cocycle of transition functions $(T\varphi_{\alpha\beta} : TU_{\alpha\beta} \to TG)$, describing the principal fiber bundle $(Tp : TP \to TM, TG)$. The assertion about the principal action is obvious. So (1) follows. For completeness' sake, we include here the transition formula for this atlas in the right trivialization of TG:

$$T(\varphi_\alpha \circ \varphi_\beta^{-1})(\xi_x, T_e(\mu^g).X) = (\xi_x, T_e(\mu^{\varphi_{\alpha\beta}(x).g}).(\delta^r\varphi_{\alpha\beta}(\xi_x) + \operatorname{Ad}(\varphi_{\alpha\beta}(x))X)),$$

where $\delta^r \varphi_{\alpha\beta} \in \Omega^1(U_{\alpha\beta}, \mathfrak{g})$ is the right logarithmic derivative of $\varphi_{\alpha\beta}$, see (38.1) below.

(2) The mapping $(u, X) \mapsto T_e(r_u).X = T_{(u,e)}r.(0_u, X)$ is a vector bundle isomorphism $P \times \mathfrak{g} \to VP$ over P.

(3) Obviously, $Tr : TP \times TG \to TP$ is a free right action which acts transitively on the fibers of $Tp : TP \to TM$. Since $VP = (Tp)^{-1}(0_M)$, the bundle $VP \to M$ is isomorphic to $TP|0_M$ and Tr restricts to a free right action, which is transitive on the fibers, so by lemma (37.9) the result follows.

(4) The transition functions of the fiber bundle $P[S, \ell]$ are given by the expression $\ell \circ (\varphi_{\alpha\beta} \times \mathrm{Id}_S) : U_{\alpha\beta} \times S \to G \times S \to S$. Then the transition functions of $T(P[S, \ell])$ are $T(\ell \circ (\varphi_{\alpha\beta} \times \mathrm{Id}_S)) = T\ell \circ (T\varphi_{\alpha\beta} \times \mathrm{Id}_{TS}) : TU_{\alpha\beta} \times TS \to TG \times TS \to TS$, from which the result follows.

(5) Vertical vectors in $T(P[S, \ell])$ have local representations $(0_x, \eta_s) \in TU_{\alpha\beta} \times TS$. Under the transition functions of $T(P[S, \ell])$ they transform as $T(\ell \circ (\varphi_{\alpha\beta} \times \mathrm{Id}_S)).(0_x, \eta_s) = T\ell.(0_{\varphi_{\alpha\beta}(x)}, \eta_s) = T(\ell_{\varphi_{\alpha\beta}(x)}).\eta_s = T_2\ell.(\varphi_{\alpha\beta}(x), \eta_s)$, and this implies the result $\square$

37.19. Principal connections. Let $(p : P \to M, G)$ be a principal fiber bundle. Recall from (37.2) that a (general) connection on P is a fiber projection $\Phi : TP \to VP$, viewed as a 1-form in $\Omega^1(P; VP) \subset \Omega^1(P; TP)$. Such a connection Φ is called a *principal connection* if it is G-equivariant for the principal right action $r : P \times G \to P$, so that $T(r^g).\Phi = \Phi.T(r^g)$ and Φ is r^g-related to itself, or $(r^g)^*\Phi = \Phi$ in the sense of (35.13), for all $g \in G$. By theorem (35.13.7), the curvature $\mathcal{R} = \frac{1}{2}.[\Phi, \Phi]$ is then also r^g-related to itself for all $g \in G$.

Recall from (37.18.2) that the vertical bundle of P is trivialized as a vector bundle over P by the principal action. So $\omega(X_u) := T_e(r_u)^{-1}.\Phi(X_u) \in \mathfrak{g}$, and in this way we get a $\mathfrak{g}$-valued 1-form $\omega \in \Omega^1(P, \mathfrak{g})$, which is called the *(Lie algebra valued) connection form* of the connection Φ. Recall from (36.12) the fundamental vector field mapping $\zeta : \mathfrak{g} \to \mathfrak{X}(P)$ for the principal right action. The defining equation for ω can be written also as $\Phi(X_u) = \zeta_{\omega(X_u)}(u)$.

Lemma. *If $\Phi \in \Omega^1(P; VP)$ is a principal connection on the principal fiber bundle $(p : P \to M, G)$ then the connection form has the following three properties:*

(1) *ω reproduces the generators of fundamental vector fields, so that we have $\omega(\zeta_X(u)) = X$ for all $X \in \mathfrak{g}$.*
(2) *ω is G-equivariant, $((r^g)^*\omega)(X_u) = \omega(T_u(r^g).X_u) = \mathrm{Ad}(g^{-1}).\omega(X_u)$ for all $g \in G$ and $X_u \in T_uP$.*
(3) *We have for the Lie derivative $\mathcal{L}_{\zeta_X}\omega = -\mathrm{ad}(X).\omega$.*

Conversely, a 1-form $\omega \in \Omega^1(P, \mathfrak{g})$ satisfying (1) defines a connection Φ on P by $\Phi(X_u) = T_e(r_u).\omega(X_u)$, which is a principal connection if and only if (2) is satisfied.

Proof. (1) $T_e(r_u).\omega(\zeta_X(u)) = \Phi(\zeta_X(u)) = \zeta_X(u) = T_e(r_u).X$. Since $T_e(r_u) : \mathfrak{g} \to V_uP$ is an isomorphism, the result follows.

(2) Both directions follow from

$$\begin{aligned} T_e(r_{ug}).\omega(T_u(r^g).X_u) &= \zeta_{\omega(T_u(r^g).X_u)}(ug) = \Phi(T_u(r^g).X_u) \\ T_e(r_{ug}).\mathrm{Ad}(g^{-1}).\omega(X_u) &= \zeta_{\mathrm{Ad}(g^{-1}).\omega(X_u)}(ug) = T_u(r^g).\zeta_{\omega(X_u)}(u) \\ &= T_u(r^g).\Phi(X_u). \end{aligned}$$

(3) Let $g(t)$ be a smooth curve in G with $g(0) = e$ and $\frac{\partial}{\partial t}|_0 g(t) = X$. Then $\varphi_t(u) = r(u, g(t))$ is a smooth curve of diffeomorphisms on P with $\frac{\partial}{\partial t}|_0 \varphi_t = \zeta_X$, and by the first claim of lemma (33.19) we have

$$\mathcal{L}_{\zeta_X}\omega = \tfrac{\partial}{\partial t}|_0 (r^{g(t)})^*\omega = \tfrac{\partial}{\partial t}|_0 Ad(g(t)^{-1})\omega = -ad(X)\omega. \quad \square$$

37.20. Curvature. Let Φ be a principal connection on the principal fiber bundle $(p : P \to M, G)$ with connection form $\omega \in \Omega^1(P, \mathfrak{g})$. We have already noted in (37.19) that also the curvature $\mathcal{R} = \frac{1}{2}[\Phi, \Phi]$ is then G-equivariant, $(r^g)^*\mathcal{R} = \mathcal{R}$ for all $g \in G$. Since $\mathcal{R}$ has vertical values we may define a $\mathfrak{g}$-valued 2-form $\Omega \in \Omega^2(P, \mathfrak{g})$ by $\Omega(X_u, Y_u) := -T_e(r_u)^{-1}.\mathcal{R}(X_u, Y_u)$, which is called the *(Lie algebra-valued) curvature form* of the connection. We also have $\mathcal{R}(X_u, Y_u) = -\zeta_{\Omega(X_u,Y_u)}(u)$. We take the negative sign here to get in finite dimensions the usual curvature form.

We equip the space $\Omega(P, \mathfrak{g})$ of all $\mathfrak{g}$-valued forms on P in a canonical way with the structure of a graded Lie algebra by

$$\begin{aligned} (1)\quad & [\Psi, \Theta]^{\mathfrak{g}}_{\wedge}(X_1, \dots, X_{p+q}) = \\ &= \tfrac{1}{p!\,q!} \sum_{\sigma} \operatorname{sign}\sigma\, [\Psi(X_{\sigma 1}, \dots, X_{\sigma p}), \Theta(X_{\sigma(p+1)}, \dots, X_{\sigma(p+q)})]_{\mathfrak{g}} \end{aligned}$$

or equivalently by $[\psi\otimes X, \theta\otimes Y]_\wedge := \psi\wedge\theta\otimes[X, Y]_{\mathfrak{g}}$. From the latter description it is clear that $d[\Psi, \Theta]_\wedge = [d\Psi, \Theta]_\wedge + (-1)^{\deg\Psi}[\Psi, d\Theta]_\wedge$. In particular, for $\omega \in \Omega^1(P, \mathfrak{g})$ we have $[\omega, \omega]_\wedge(X, Y) = 2[\omega(X), \omega(Y)]_{\mathfrak{g}}$.

Theorem. *The curvature form Ω of a principal connection with connection form ω has the following properties:*

(2) *Ω is horizontal, i.e., it kills vertical vectors.*
(3) *Ω is G-equivariant in the following sense: $(r^g)^*\Omega = \mathrm{Ad}(g^{-1}).\Omega$. Consequently, $\mathcal{L}_{\zeta_X}\Omega = -\mathrm{ad}(X).\Omega$.*
(4) *The Maurer-Cartan formula holds: $\Omega = d\omega + \frac{1}{2}[\omega, \omega]_\wedge$.*

Proof. (2) is true for $\mathcal{R}$ by (37.3). For (3) we compute as follows:

$$\begin{aligned} T_e(r_{ug}).((r^g)^*\Omega)(X_u, Y_u) &= T_e(r_{ug}).\Omega(T_u(r^g).X_u, T_u(r^g).Y_u) = \\ &= -\mathcal{R}_{ug}(T_u(r^g).X_u, T_u(r^g).Y_u) = -T_u(r^g).((r^g)^*\mathcal{R})(X_u, Y_u) = \\ &= -T_u(r^g).\mathcal{R}(X_u, Y_u) = T_u(r^g).\zeta_{\Omega(X_u,Y_u)}(u) = \\ &= \zeta_{\mathrm{Ad}(g^{-1}).\Omega(X_u,Y_u)}(ug) = T_e(r_{ug}).\mathrm{Ad}(g^{-1}).\Omega(X_u, Y_u), \quad \text{by (36.12.2)}. \end{aligned}$$

Proof of (4) For $X \in \mathfrak{g}$ we have $i_{\zeta_X}\mathcal{R} = 0$ by (2), and using (37.19.3) we get

$$\begin{aligned} i_{\zeta_X}(d\omega + \frac{1}{2}[\omega,\omega]_\wedge) &= i_{\zeta_X}d\omega + \frac{1}{2}[i_{\zeta_X}\omega,\omega]_\wedge - \frac{1}{2}[\omega, i_{\zeta_X}\omega]_\wedge = \\ &= \mathcal{L}_{\zeta_X}\omega + [X,\omega]_\wedge = -\operatorname{ad}(X)\omega + \operatorname{ad}(X)\omega = 0. \end{aligned}$$

So the formula holds if one vector is vertical, and for horizontal vector fields $X, Y \in C^\infty(P \leftarrow H(P))$ we have

$$\mathcal{R}(X,Y) = \Phi[X - \Phi X, Y - \Phi Y] = \Phi[X,Y] = \zeta_{\omega([X,Y])},$$

$$(d\omega + \frac{1}{2}[\omega,\omega]_\wedge)(X,Y) = X\omega(Y) - Y\omega(X) - \omega([X,Y]) + 0 = -\omega([X,Y]). \quad \square$$

37.21. Lemma. *Any principal fiber bundle $(p : P \to M, G)$ with smoothly paracompact basis M admits principal connections.*

Proof. Let $(U_\alpha, \varphi_\alpha : P|U_\alpha \to U_\alpha \times G)_\alpha$ be a principal fiber bundle atlas. Let us define $\gamma_\alpha(T\varphi_\alpha^{-1}(\xi_x, T_e\mu_g.X)) := X$ for $\xi_x \in T_xU_\alpha$ and $X \in \mathfrak{g}$. An easy computation involving lemma (36.12) shows that $\gamma_\alpha \in \Omega^1(P|U_\alpha, \mathfrak{g})$ satisfies the requirements of lemma (37.19) and thus is a principal connection on $P|U_\alpha$. Now let (f_α) be a smooth partition of unity on M which is subordinated to the open cover (U_α), and let $\omega := \sum_\alpha (f_\alpha \circ p)\gamma_\alpha$. Since both requirements of lemma (37.19) are invariant under convex linear combinations, ω is a principal connection on P. $\square$

37.22. Local descriptions of principal connections. We consider a principal fiber bundle $(p : P \to M, G)$ with some principal fiber bundle atlas $(U_\alpha, \varphi_\alpha : P|U_\alpha \to U_\alpha \times G)$ and corresponding cocycle $(\varphi_{\alpha\beta} : U_{\alpha\beta} \to G)$ of transition functions. We consider the sections $s_\alpha \in C^\infty(U_\alpha \leftarrow P|U_\alpha)$ which are given by $\varphi_\alpha(s_\alpha(x)) = (x, e)$ and satisfy $s_\alpha.\varphi_{\alpha\beta} = s_\beta$, since we have in turn:

$$\begin{aligned} \varphi_\alpha(s_\beta(x)) &= \varphi_\alpha\varphi_\beta^{-1}(x,e) = (x, \varphi_{\alpha\beta}(x)) \\ s_\beta(x) &= \varphi_\alpha^{-1}(x, e\varphi_{\alpha\beta}(x)) = \varphi_\alpha^{-1}(x,e)\varphi_{\alpha\beta}(x) = s_\alpha(x)\varphi_{\alpha\beta}(x). \end{aligned}$$

Let $\Phi = \zeta \circ \omega \in \Omega^1(P; VP)$ be a principal connection with connection form $\omega \in \Omega^1(P, \mathfrak{g})$. We may associate the following local data to the connection:

(1) $\omega_\alpha := s_\alpha{}^*\omega \in \Omega^1(U_\alpha, \mathfrak{g})$, the physicists' version of the connection.
(2) The Christoffel forms $\Gamma^\alpha \in \Omega^1(U_\alpha; \mathfrak{X}(G))$ from (37.5), which are given by $(0_x, \Gamma^\alpha(\xi_x, g)) = -T(\varphi_\alpha).\Phi.T(\varphi_\alpha)^{-1}(\xi_x, 0_g)$.

Lemma. *These local data have the following properties and are related by the following formulas.*

(3) *The forms $\omega_\alpha \in \Omega^1(U_\alpha, \mathfrak{g})$ satisfy the transition formulas*

$$\omega_\alpha = \operatorname{Ad}(\varphi_{\beta\alpha}^{-1})\omega_\beta + (\varphi_{\beta\alpha})^*\kappa^l,$$

where $\kappa^l \in \Omega^1(G, \mathfrak{g})$ is the left Maurer-Cartan form from (36.10). Any set of such forms with this transition behavior determines a unique principal connection.

(4) *The local expression of ω is given by*

$$(\varphi_\alpha^{-1})^*\omega(\xi_x, T\mu_g.X) = (\varphi_\alpha^{-1})^*\omega(\xi_x, 0_g) + X = \mathrm{Ad}(g^{-1})\omega_\alpha(\xi_x) + X.$$

(5) *The Christoffel form Γ^α and ω_α are related by*

$$\Gamma^\alpha(\xi_x, g) = -T_e(\mu_g).\mathrm{Ad}(g^{-1})\omega_\alpha(\xi_x) = -T_e(\mu^g)\omega_\alpha(\xi_x),$$

thus the Christoffel form is right invariant: $\Gamma^\alpha(\xi_x) = -R_{\omega_\alpha(\xi_x)} \in \mathfrak{X}(G)$.

(6) *The local expression of Φ is given by*

$$\begin{aligned}(((\varphi_\alpha)^{-1})^*\Phi)(\xi_x, \eta_g) &= -\Gamma_\alpha(\xi_x, g) + \eta_g = T_e(\mu^g).\omega_\alpha(\xi_x) + \eta_g \\ &= R_{\omega_\alpha(\xi_x)}(g) + \eta_g\end{aligned}$$

for $\xi_x \in T_xU_\alpha$ and $\eta_g \in T_gG$.

(7) *The local expression of the curvature $\mathcal{R}$ is given by*

$$((\varphi_\alpha)^{-1})^*\mathcal{R} = -R_{d\omega_\alpha + \frac{1}{2}[\omega_\alpha, \omega_\alpha]^\wedge_{\mathfrak{g}}}$$

so that $\mathcal{R}$ and Ω are indeed 'tensorial' 2-forms.

Proof. We start with (4).

$$\begin{aligned}(\varphi_\alpha^{-1})^*\omega(\xi_x, 0_g) &= (\varphi_\alpha^{-1})^*\omega(\xi_x, T_e(\mu^g)0_e) = (\omega \circ T(\varphi_\alpha)^{-1} \circ T(\mathrm{Id}_{U_\alpha} \times \mu^g))(\xi_x, 0_e) = \\ &= (\omega \circ T(r^g \circ \varphi_\alpha^{-1}))(\xi_x, 0_e) = \mathrm{Ad}(g^{-1})\omega(T(\varphi_\alpha^{-1})(\xi_x, 0_e)) \\ &= \mathrm{Ad}(g^{-1})(s_\alpha{}^*\omega)(\xi_x) = \mathrm{Ad}(g^{-1})\omega_\alpha(\xi_x).\end{aligned}$$

From this we get

$$\begin{aligned}(\varphi_\alpha^{-1})^*\omega(\xi_x, T\mu_g.X) &= (\varphi_\alpha^{-1})^*\omega(\xi_x, 0_g) + (\varphi_\alpha^{-1})^*\omega(0_x, T\mu_g.X) \\ &= \mathrm{Ad}(g^{-1})\omega_\alpha(\xi_x) + \omega(T(\varphi_\alpha)^{-1}(0_x, T\mu_g.X)) \\ &= \mathrm{Ad}(g^{-1})\omega_\alpha(\xi_x) + \omega(\zeta_X(\varphi_\alpha^{-1}(x, g))) \\ &= \mathrm{Ad}(g^{-1})\omega_\alpha(\xi_x) + X.\end{aligned}$$

(5) From the definition of the Christoffel forms we have

$$\begin{aligned}(0_x, \Gamma^\alpha(\xi_x, g)) &= -T(\varphi_\alpha).\Phi.T(\varphi_\alpha)^{-1}(\xi_x, 0_g) \\ &= -T(\varphi_\alpha).T_e(r_{\varphi_\alpha^{-1}(x,g)})\omega.T(\varphi_\alpha)^{-1}(\xi_x, 0_g) \\ &= -T_e(\varphi_\alpha \circ r_{\varphi_\alpha^{-1}(x,g)})\omega.T(\varphi_\alpha)^{-1}(\xi_x, 0_g) \\ &= -(0_x, T_e(\mu_g)\omega.T(\varphi_\alpha)^{-1}(\xi_x, 0_g)) \\ &= -(0_x, T_e(\mu_g)\mathrm{Ad}(g^{-1})\omega_\alpha(\xi_x)) = -(0_x, T_e(\mu^g)\omega_\alpha(\xi_x)).\end{aligned}$$

(3) Via (5) the transition formulas for the ω_α are easily seen to be equivalent to the transition formulas for the Christoffel forms in lemma (37.5). A direct proof goes as follows: We have $s_\alpha(x) = s_\beta(x)\varphi_{\beta\alpha}(x) = r(s_\beta(x), \varphi_{\beta\alpha}(x))$ and thus

$$\begin{aligned}
\omega_\alpha(\xi_x) &= \omega(T_x(s_\alpha).\xi_x) \\
&= (\omega \circ T_{(s_\beta(x),\varphi_{\beta\alpha}(x))}r)((T_x s_\beta.\xi_x, 0_{\varphi_{\beta\alpha}(x)}) - (0_{s_\beta(x)}, T_x\varphi_{\beta\alpha}.\xi_x)) \\
&= \omega(T_{s_\beta(x)}(r^{\varphi_{\beta\alpha}(x)}).T_x(s_\beta).\xi_x) + \omega(T_{\varphi_{\beta\alpha}(x)}(r_{s_\beta(x)}).T_x(\varphi_{\beta\alpha}).\xi_x) \\
&= \operatorname{Ad}(\varphi_{\beta\alpha}(x)^{-1})\omega(T_x(s_\beta).\xi_x) \\
&\qquad + \omega(T_{\varphi_{\beta\alpha}(x)}(r_{s_\beta(x)}).T(\mu_{\varphi_{\beta\alpha}(x)} \circ \mu_{\varphi_{\beta\alpha}(x)^{-1}})T_x(\varphi_{\beta\alpha}).\xi_x) \\
&= \operatorname{Ad}(\varphi_{\beta\alpha}(x)^{-1})\omega_\beta(\xi_x) + \omega(T_e(r_{s_\beta(x)\varphi_{\beta\alpha}(x)}).(\varphi_{\beta\alpha})^*\kappa^l.\xi_x) \\
&= \operatorname{Ad}(\varphi_{\beta\alpha}(x)^{-1})\omega_\beta(\xi_x) + (\varphi_{\beta\alpha})^*\kappa^l(\xi_x).
\end{aligned}$$

(6) This is clear from the definition of the Christoffel forms and from (5).

Second proof of (7) First note that the right trivialization or framing $(\kappa^r, \pi_G) : TG \to \mathfrak{g} \times G$ induces the isomorphism $R : C^\infty(G, \mathfrak{g}) \to \mathfrak{X}(G)$, given by $R_X(x) = T_e(\mu^x).X(x)$. For the Lie bracket we then have

$$\begin{aligned}
[R_X, R_Y] &= R_{-[X,Y]_\mathfrak{g} + dY.R_X - dX.R_Y}, \\
R^{-1}[R_X, R_Y] &= -[X,Y]_\mathfrak{g} + R_X(Y) - R_Y(X).
\end{aligned}$$

We write a vector field on $U_\alpha \times G$ as (ξ, R_X) where $\xi : G \to \mathfrak{X}(U_\alpha)$ and $X \in C^\infty(U_\alpha \times G, \mathfrak{g})$. Then the local expression of the curvature is given by

$$\begin{aligned}
&({\varphi_\alpha}^{-1})^*\mathcal{R}((\xi, R_X), (\eta, R_Y)) = (\varphi_\alpha^{-1})^*(\mathcal{R}((\varphi_\alpha)^*(\xi, R_X), (\varphi_\alpha)^*(\eta, R_Y))) \\
&\quad = (\varphi_\alpha^{-1})^*(\Phi[(\varphi_\alpha)^*(\xi, R_X) - \Phi(\varphi_\alpha)^*(\xi, R_X), \dots]) \\
&\quad = (\varphi_\alpha^{-1})^*(\Phi[(\varphi_\alpha)^*(\xi, R_X) - (\varphi_\alpha)^*(R_{\omega_\alpha(\xi)} + R_X), \dots]) \\
&\quad = (\varphi_\alpha^{-1})^*(\Phi(\varphi_\alpha)^*[(\xi, -R_{\omega_\alpha(\xi)}), (\eta, -R_{\omega_\alpha(\eta)})]) \\
&\quad = ((\varphi_\alpha^{-1})^*\Phi)([\xi,\eta]_{\mathfrak{X}(U_\alpha)} - R_{\omega_\alpha(\xi)}(\eta) + R_{\omega_\alpha(\eta)}(\xi), \\
&\qquad - \xi(R_{\omega_\alpha(\eta)}) + \eta(R_{\omega_\alpha(\xi)}) + R_{-[\omega_\alpha(\xi), \omega_\alpha(\eta)] + R_{\omega_\alpha(\xi)}(\omega_\alpha(\eta)) - R_{\omega_\alpha(\xi)}(\omega_\alpha(\eta))}) \\
&\quad = R_{\omega_\alpha([\xi,\eta]_{\mathfrak{X}(U_\alpha)} - R_{\omega_\alpha(\xi)}(\eta) + R_{\omega_\alpha(\eta)}(\xi))} - R_{\xi(\omega_\alpha(\eta))} + R_{\eta(\omega_\alpha(\xi))} \\
&\qquad + R_{-[\omega_\alpha(\xi), \omega_\alpha(\eta)] + R_{\omega_\alpha(\xi)}(\omega_\alpha(\eta)) - R_{\omega_\alpha(\xi)}(\omega_\alpha(\eta))} \\
&\quad = -R_{(d\omega_\alpha + \frac{1}{2}[\omega_\alpha, \omega_\alpha]^\wedge_\mathfrak{g})(\xi,\eta)}. \qquad \square
\end{aligned}$$

37.23. The covariant derivative. Let $(p : P \to M, G)$ be a principal fiber bundle with principal connection $\Phi = \zeta \circ \omega$. We consider the horizontal projection $\chi = \operatorname{Id}_{TP} - \Phi : TP \to HP$, cf. (37.2), which satisfies $\chi \circ \chi = \chi$, $\operatorname{im} \chi = HP$, $\ker \chi = VP$, and $\chi \circ T(r^g) = T(r^g) \circ \chi$ for all $g \in G$.

If W is a convenient vector space, we consider the mapping $\chi^* : \Omega(P, W) \to \Omega(P, W)$ which is given by

$$(\chi^*\varphi)_u(X_1, \dots, X_k) = \varphi_u(\chi(X_1), \dots, \chi(X_k)).$$

The mapping χ^* is a projection onto the subspace of *horizontal differential forms*, i.e. the space $\Omega_{hor}(P,W) := \{\psi \in \Omega(P,W) : i_X\psi = 0 \text{ for } X \in VP\}$. The notion of horizontal form is independent of the choice of a connection.

The projection χ^* has the following properties: $\chi^*(\varphi \wedge \psi) = \chi^*\varphi \wedge \chi^*\psi$ if one of the two forms has real values, $\chi^* \circ \chi^* = \chi^*$, $\chi^* \circ (r^g)^* = (r^g)^* \circ \chi^*$ for all $g \in G$, $\chi^*\omega = 0$, and $\chi^* \circ \mathcal{L}(\zeta_X) = \mathcal{L}(\zeta_X) \circ \chi^*$. All but the last easily follow from the corresponding properties of χ. The last property uses that for a smooth curve $g(t)$ in G with $g(0) = e$ and $\frac{\partial}{\partial t}\big|_0 g(t) = X$ by (33.19) we have $\mathcal{L}_{\zeta_X} = \frac{\partial}{\partial t}\big|_0 r^{g(t)}$.

We define the *covariant exterior derivative* $d_\omega : \Omega^k(P,W) \to \Omega^{k+1}(P,W)$ by the prescription $d_\omega := \chi^* \circ d$.

Theorem. *The covariant exterior derivative d_ω has the following properties.*

(1) $d_\omega(\varphi \wedge \psi) = d_\omega(\varphi) \wedge \chi^*\psi + (-1)^{\deg\varphi}\chi^*\varphi \wedge d_\omega(\psi)$ *if φ or ψ is real valued.*
(2) $\mathcal{L}(\zeta_X) \circ d_\omega = d_\omega \circ \mathcal{L}(\zeta_X)$ *for each $X \in \mathfrak{g}$.*
(3) $(r^g)^* \circ d_\omega = d_\omega \circ (r^g)^*$ *for each $g \in G$.*
(4) $d_\omega \circ p^* = d \circ p^* = p^* \circ d : \Omega(M,W) \to \Omega_{hor}(P,W)$.
(5) $d_\omega\omega = \Omega$, *the curvature form.*
(6) $d_\omega\Omega = 0$, *the Bianchi identity.*
(7) $d_\omega \circ \chi^* - d_\omega = \chi^* \circ i(\mathcal{R})$, *where $\mathcal{R}$ is the curvature.*
(8) $d_\omega \circ d_\omega = \chi^* \circ i(\mathcal{R}) \circ d$.
(9) *Let $\Omega_{\text{hor}}(P,\mathfrak{g})^G$ be the algebra of all horizontal G-equivariant $\mathfrak{g}$-valued forms, i.e., $(r^g)^*\psi = Ad(g^{-1})\psi$. Then for any $\psi \in \Omega_{\text{hor}}(P,\mathfrak{g})^G$ we have $d_\omega\psi = d\psi + [\omega,\psi]_\wedge$.*
(10) *The mapping $\psi \mapsto \zeta_\psi$, where $\zeta_\psi(X_1,\dots,X_k)(u) = \zeta_{\psi(X_1,\dots,X_k)(u)}(u)$, is an isomorphism between $\Omega_{\text{hor}}(P,\mathfrak{g})^G$ and the algebra $\Omega_{\text{hor}}(P,VP)^G$ of all horizontal G-equivariant forms with values in the vertical bundle VP. Then we have $\zeta_{d_\omega\psi} = -[\Phi,\zeta_\psi]$.*

Proof. (1) through (4) follow from the properties of χ^*.

(5) We have

$$\begin{aligned}(d_\omega\omega)(\xi,\eta) &= (\chi^*d\omega)(\xi,\eta) = d\omega(\chi\xi,\chi\eta)\\ &= (\chi\xi)\omega(\chi\eta) - (\chi\eta)\omega(\chi\xi) - \omega([\chi\xi,\chi\eta])\\ &= -\omega([\chi\xi,\chi\eta]) \text{ and}\\ -\zeta(\Omega(\xi,\eta)) &= \mathcal{R}(\xi,\eta) = \Phi[\chi\xi,\chi\eta] = \zeta_{\omega([\chi\xi,\chi\eta])}.\end{aligned}$$

(6) Using (37.20) we have

$$\begin{aligned}d_\omega\Omega &= d_\omega(d\omega + \tfrac{1}{2}[\omega,\omega]_\wedge)\\ &= \chi^*dd\omega + \tfrac{1}{2}\chi^*d[\omega,\omega]_\wedge\\ &= \tfrac{1}{2}\chi^*([d\omega,\omega]_\wedge - [\omega,d\omega]_\wedge) = \chi^*[d\omega,\omega]_\wedge\\ &= [\chi^*d\omega,\chi^*\omega]_\wedge = 0, \text{ since } \chi^*\omega = 0.\end{aligned}$$

(7) For $\varphi \in \Omega(P,W)$ we have

$$\begin{aligned}
(d_\omega \chi^*\varphi)(X_0,\dots,X_k) &= (d\chi^*\varphi)(\chi(X_0),\dots,\chi(X_k))\\
&= \sum_{0\le i\le k}(-1)^i\chi(X_i)((\chi^*\varphi)(\chi(X_0),\dots,\widehat{\chi(X_i)},\dots,\chi(X_k)))\\
&\quad+\sum_{i<j}(-1)^{i+j}(\chi^*\varphi)([\chi(X_i),\chi(X_j)],\\
&\qquad\qquad \chi(X_0),\dots,\widehat{\chi(X_i)},\dots,\widehat{\chi(X_j)},\dots,\chi(X_k))\\
&= \sum_{0\le i\le k}(-1)^i\chi(X_i)(\varphi(\chi(X_0),\dots,\widehat{\chi(X_i)},\dots,\chi(X_k)))\\
&\quad+\sum_{i<j}(-1)^{i+j}\varphi([\chi(X_i),\chi(X_j)]-\Phi[\chi(X_i),\chi(X_j)],\\
&\qquad\qquad \chi(X_0),\dots,\widehat{\chi(X_i)},\dots,\widehat{\chi(X_j)},\dots,\chi(X_k))\\
&= (d\varphi)(\chi(X_0),\dots,\chi(X_k)) + (i_{\mathcal R}\varphi)(\chi(X_0),\dots,\chi(X_k))\\
&= (d_\omega + \chi^* i_{\mathcal R})(\varphi)(X_0,\dots,X_k).
\end{aligned}$$

(8) $d_\omega d_\omega = d_\omega\chi^* d = (\chi^* i_{\mathcal R} + \chi^* d)d = \chi^* i_{\mathcal R} d$ holds by (7).

(9) If we insert one vertical vector field, say ζ_X for $X \in \mathfrak g$, into $d_\omega\psi$, we get 0 by definition. For the right hand side we use $i_{\zeta_X}\psi = 0$, and that by (33.19) for a smooth curve $g(t)$ in G with $g(0) = e$ and $\frac{\partial}{\partial t}\big|_0 g(t) = X$ we have $\mathcal L_{\zeta_X}\psi = \frac{\partial}{\partial t}\big|_0 (r^{g(t)})^*\psi = \frac{\partial}{\partial t}\big|_0 \operatorname{Ad}(g(t)^{-1})\psi = -ad(X)\psi$ in the computation

$$\begin{aligned}
i_{\zeta_X}(d\psi + [\omega,\psi]_\wedge) &= i_{\zeta_X}d\psi + di_{\zeta_X}\psi + [i_{\zeta_X}\omega,\psi] - [\omega, i_{\zeta_X}\psi]\\
&= \mathcal L_{\zeta_X}\psi + [X,\psi] = -ad(X)\psi + [X,\psi] = 0.
\end{aligned}$$

Let now all vector fields ξ_i be horizontal. Then we get

$$\begin{aligned}
(d_\omega\psi)(\xi_0,\dots,\xi_k) = (\chi^* d\psi)(\xi_0,\dots,\xi_k) = d\psi(\xi_0,\dots,\xi_k),\\
(d\psi + [\omega,\psi]_\wedge)(\xi_0,\dots,\xi_k) = d\psi(\xi_0,\dots,\xi_k).
\end{aligned}$$

(10) We proceed in a similar manner. Let Ψ be in the space $\Omega^\ell_{\text{hor}}(P,VP)^G$ of all horizontal G-equivariant forms with vertical values. Then for each $X \in \mathfrak g$ we have $i_{\zeta_X}\Psi = 0$. Furthermore, the G-equivariance $(r^g)^*\Psi = \Psi$ implies that $\mathcal L_{\zeta_X}\Psi = [\zeta_X,\Psi] = 0$ by (35.14.5). Using formula (35.9.2) we have

$$\begin{aligned}
i_{\zeta_X}[\Phi,\Psi] &= [i_{\zeta_X}\Phi,\Psi] - [\Phi, i_{\zeta_X}\Psi] + i([\Phi,\zeta_X])\Psi + i([\Psi,\zeta_X])\Phi\\
&= [\zeta_X,\Psi] - 0 + 0 + 0 = 0.
\end{aligned}$$

Let now all vector fields ξ_i again be horizontal, then from the huge formula (35.5.1) for the Frölicher-Nijenhuis bracket only the following terms in the fourth and fifth line survive:

$$\begin{aligned}
&[\Phi,\Psi](\xi_1,\dots,\xi_{\ell+1}) =\\
&\quad= \tfrac{(-1)^\ell}{\ell!}\sum_\sigma \operatorname{sign}\sigma\ \Phi([\Psi(\xi_{\sigma 1},\dots,\xi_{\sigma\ell}),\xi_{\sigma(\ell+1)}])\\
&\qquad+ \tfrac{1}{(\ell-1)!\,2!}\sum_\sigma \operatorname{sign}\sigma\ \Phi(\Psi([\xi_{\sigma 1},\xi_{\sigma 2}],\xi_{\sigma 3},\dots,\xi_{\sigma(\ell+1)})).
\end{aligned}$$

For $f : P \to \mathfrak{g}$ and horizontal ξ we have $\Phi[\xi, \zeta_f] = \zeta_{\xi(f)} = \zeta_{df(\xi)}$: It is $C^\infty(P,\mathbb{R})$-linear in ξ; or imagine it in local coordinates. So the last expression becomes

$$-\zeta_{d\psi(\xi_0,\dots,\xi_k)} = -\zeta_{(d\psi+[\omega,\psi]_\wedge)(\xi_0,\dots,\xi_k)},$$

as required. $\square$

37.24. Inducing principal connections on associated bundles. Let $(p : P \to M, G)$ be a principal bundle with principal right action $r : P\times G \to P$, and let $\ell : G \times S \to S$ be a left action of the structure group G on some manifold S. Then we consider the associated bundle $P[S] = P[S,\ell] = P \times_G S$, constructed in (37.12). Recall from (37.18) that its tangent and vertical bundles are given by $T(P[S,\ell]) = TP[TS,T\ell] = TP \times_{TG} TS$ and $V(P[S,\ell]) = P[TS,T_2\ell] = P \times_G TS$.

Let $\Phi = \zeta \circ \omega \in \Omega^1(P;TP)$ be a principal connection on the principal bundle P. We construct the *induced connection* $\bar\Phi \in \Omega^1(P[S];T(P[S]))$ by factorizing as in the following diagram:

$$\begin{array}{ccccc}
TP \times TS & \xrightarrow{\Phi\times \mathrm{Id}} & TP\times TS & \xrightarrow{=} & T(P\times S) \\
\Big\downarrow{\scriptstyle Tq = q'} & & \Big\downarrow{\scriptstyle q'} & & \Big\downarrow{\scriptstyle Tq} \\
TP\times_{TG} TS & \xrightarrow{\bar\Phi} & TP\times_{TG} TS & \xrightarrow{=} & T(P\times_G S).
\end{array}$$

Let us first check that the top mapping $\Phi \times \mathrm{Id}$ is TG-equivariant. For $g \in G$ and $X \in \mathfrak{g}$ the inverse of $T_e(\mu_g)X$ in the Lie group TG from (38.10) is denoted by $(T_e(\mu_g)X)^{-1}$. Furthermore, by (36.12) we have

$$Tr(\xi_u, T_e(\mu_g)X) = T_u(r^g)\xi_u + T_g(r_u)(T_e(\mu_g)X) = T_u(r^g)\xi_u + \zeta_X(ug).$$

We compute

$$\begin{aligned}
&(\Phi\times \mathrm{Id})(Tr(\xi_u, T_e(\mu_g)X), T\ell((T_e(\mu_g)X)^{-1}, \eta_s)) \\
&\quad = (\Phi(T_u(r^g)\xi_u + \zeta_X(ug)), T\ell((T_e(\mu_g)X)^{-1}, \eta_s)) \\
&\quad = (\Phi(T_u(r^g)\xi_u) + \Phi(\zeta_X(ug)), T\ell((T_e(\mu_g)X)^{-1}, \eta_s)) \\
&\quad = (T_u(r^g)\Phi\xi_u + \zeta_X(ug), T\ell((T_e(\mu_g)X)^{-1}, \eta_s)) \\
&\quad = (Tr(\Phi(\xi_u), T_e(\mu_g)X), T\ell((T_e(\mu_g)X)^{-1}, \eta_s)).
\end{aligned}$$

So the mapping $\Phi \times \mathrm{Id}$ factors to $\bar\Phi$ as indicated in the diagram, and we have $\bar\Phi \circ \bar\Phi = \bar\Phi$ from $(\Phi\times \mathrm{Id})\circ(\Phi\times\mathrm{Id}) = \Phi\times \mathrm{Id}$. The mapping $\bar\Phi$ is fiberwise linear, since $\Phi\times\mathrm{Id}$ and $q' = Tq$ are. The image of $\bar\Phi$ is

$$\begin{aligned}
q'(VP\times TS) &= q'(\ker(Tp : TP\times TS \to TM)) \\
&= \ker(Tp : TP\times_{TG} TS \to TM) = V(P[S,\ell]).
\end{aligned}$$

Thus, $\bar\Phi$ is a connection on the associated bundle $P[S]$. We call it the *induced connection.*

From the diagram also follows that the vector valued forms $\Phi\times\mathrm{Id}\in\Omega^1(P\times S;TP\times TS)$ and $\bar\Phi\in\Omega^1(P[S];T(P[S]))$ are $(q:P\times S\to P[S])$-related. So by (35.13) we have for the curvatures

$$\begin{aligned}\mathcal{R}_{\Phi\times\mathrm{Id}}&=\tfrac12[\Phi\times\mathrm{Id},\Phi\times\mathrm{Id}]=\tfrac12[\Phi,\Phi]\times 0=\mathcal{R}_\Phi\times 0,\\ \mathcal{R}_{\bar\Phi}&=\tfrac12[\bar\Phi,\bar\Phi]\end{aligned}$$

that they are also q-related, i.e., $Tq\circ(\mathcal{R}_\Phi\times 0)=R_{\bar\Phi}\circ(Tq\times_M Tq)$.

37.25. Recognizing induced connections. Let $(p:P\to M,G)$ be a principal fiber bundle, and let $\ell:G\times S\to S$ be a left action. We consider a connection $\Psi\in\Omega^1(P[S];T(P[S]))$ on the associated bundle $P[S]=P[S,\ell]$. Then the following question arises: When is the connection Ψ induced by a principal connection on P? If this is the case, we say that Ψ is compatible with the G-structure on $P[S]$. The answer is given in the following

Theorem. *Let Ψ be a (general) connection on the associated bundle $P[S]$. Let us suppose that the action ℓ is infinitesimally effective, i.e. the fundamental vector field mapping $\zeta:\mathfrak{g}\to\mathfrak{X}(S)$ is injective.*

Then the connection Ψ is induced from a principal connection ω on P if and only if the following condition is satisfied:

> *In some (equivalently any) fiber bundle atlas (U_α,ψ_α) of $P[S]$ belonging to the G-structure of the associated bundle the Christoffel forms $\Gamma^\alpha\in\Omega^1(U_\alpha;\mathfrak{X}(S))$ have values in the sub Lie algebra $\mathfrak{X}_{fund}(S)$ of fundamental vector fields for the action ℓ.*

Proof. Let $(U_\alpha,\varphi_\alpha:P|U_\alpha\to U_\alpha\times G)$ be a principal fiber bundle atlas for P. Then by the proof of theorem (37.12) the induced fiber bundle atlas $(U_\alpha,\psi_\alpha:P[S]|U_\alpha\to U_\alpha\times S)$ is given by

$$\psi_\alpha^{-1}(x,s)=q(\varphi_\alpha^{-1}(x,e),s),\tag{1}$$

$$(\psi_\alpha\circ q)(\varphi_\alpha^{-1}(x,g),s)=(x,g.s).\tag{2}$$

Let $\Phi=\zeta\circ\omega$ be a principal connection on P, and let $\bar\Phi$ be the induced connection on the associated bundle $P[S]$. By (37.5), its Christoffel symbols are given by

$$\begin{aligned}(0_x,\Gamma^\alpha_{\bar\Phi}(\xi_x,s))&=-(T(\psi_\alpha)\circ\bar\Phi\circ T(\psi_\alpha^{-1}))(\xi_x,0_s)\\ &=-(T(\psi_\alpha)\circ\bar\Phi\circ Tq\circ(T(\varphi_\alpha^{-1})\times\mathrm{Id}))(\xi_x,0_e,0_s) &&\text{by (1)}\\ &=-(T(\psi_\alpha)\circ Tq\circ(\Phi\times\mathrm{Id}))(T(\varphi_\alpha^{-1})(\xi_x,0_e),0_s) &&\text{by (37.24)}\\ &=-(T(\psi_\alpha)\circ Tq)(\Phi(T(\varphi_\alpha^{-1})(\xi_x,0_e)),0_s)\\ &=(T(\psi_\alpha)\circ Tq)(T(\varphi_\alpha^{-1})(0_x,\Gamma^\alpha_\Phi(\xi_x,e)),0_s) &&\text{by (37.22.2)}\\ &=-T(\psi_\alpha\circ q\circ(\varphi_\alpha^{-1}\times\mathrm{Id}))(0_x,\omega_\alpha(\xi_x),0_s) &&\text{by (37.22.5)}\\ &=-T_e(\ell^s)\omega_\alpha(\xi_x) &&\text{by (2)}\\ &=-\zeta_{\omega_\alpha(\xi_x)}(s).\end{aligned}$$

So the condition is necessary.

For the converse let us suppose that a connection Ψ on $P[S]$ is given such that the Christoffel forms Γ^α_Ψ with respect to a fiber bundle atlas of the G-structure have values in $\mathfrak{X}_{\text{fund}}(S)$. Then unique $\mathfrak{g}$-valued forms $\omega_\alpha \in \Omega^1(U_\alpha, \mathfrak{g})$ are given by the equation

$$\Gamma^\alpha_\Psi(\xi_x) = \zeta(\omega_\alpha(\xi_x)),$$

since the action is infinitesimally effective. From the transition formulas (37.5) for the Γ^α_Ψ follow the transition formulas (37.22.3) for the ω^α, so that they they combine to a unique principal connection on P, which by the first part of the proof induces the given connection Ψ on $P[S]$. □

37.26. Inducing principal connections on associated vector bundles. Let $(p : P \to M, G)$ be a principal fiber bundle, and let $\rho : G \to GL(W)$ be a representation of the structure group G on a convenient vector space W. See the beginning of section (49) for a discussion of such representations. We consider the associated vector bundle $(p : E := P[W, \rho] \to M, W)$, see (37.12).

Recall from (29.9) that the tangent bundle $TE = TP \times_{TG} TW$ has two vector bundle structures, with the projections

$$\pi_E : TE = TP \times_{TG} TW \to P \times_G W = E,$$
$$Tp \circ pr_1 : TE = TP \times_{TG} TW \to TM,$$

respectively. Recall the vertical bundle $VE = \ker(Tp)$ which is a vector subbundle of $\pi_E : TE \to E$, and recall the vertical lift mapping $\mathrm{vl}_E : E \times_M E \to VE$, which is an isomorphism, $(\mathrm{pr}_1\text{–}\pi_E)$–fiberwise linear and also $(p\text{–}Tp)$–fiberwise linear.

Now let $\Phi = \zeta \circ \omega \in \Omega^1(P; TP)$ be a principal connection on P. We consider the induced connection $\bar{\Phi} \in \Omega^1(E; TE)$ on the associated bundle E from (37.24). A glance at the following diagram shows that the induced connection is linear in both vector bundle structures. This property is expressed by calling it a *linear connection*, see (37.27), on the associated vector bundle.

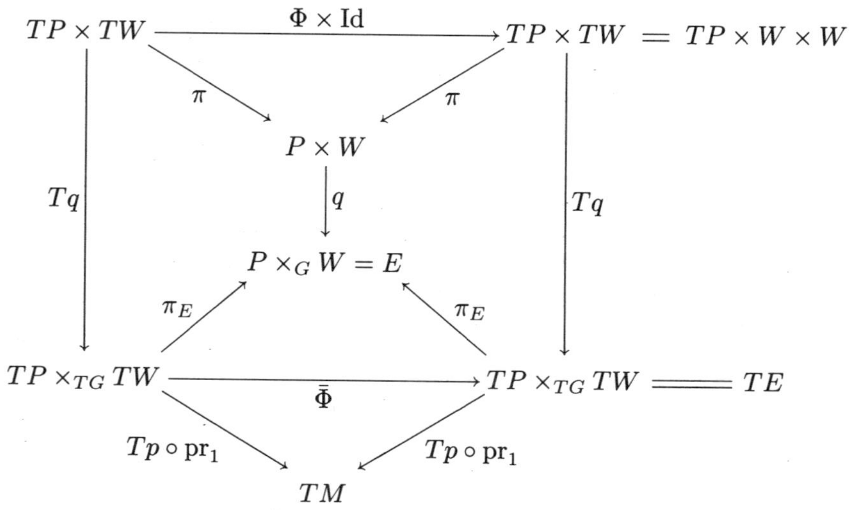

Now we define the *connector* K of the linear connection $\bar{\Phi}$ by

$$K := pr_2 \circ (\mathrm{vl}_E)^{-1} \circ \bar{\Phi} : TE \to VE \to E \times_M E \to E.$$

Lemma. *The connector $K : TE \to E$ is a vector bundle homomorphism for both vector bundle structures on TE and satisfies $K \circ \mathrm{vl}_E = \mathrm{pr}_2 : E \times_M E \to TE \to E$.*

So K is π_E–p–fiberwise linear and Tp–p–fiberwise linear.

Proof. This follows from the fiberwise linearity of the parts of K and from its definition. $\square$

37.27. Linear connections. If $p : E \to M$ is a vector bundle, a connection $\Psi \in \Omega^1(E; TE)$ such that $\Psi : TE \to VE \to TE$ is additionally Tp–Tp–fiberwise linear is called a *linear connection.*

Equivalently, a linear connection may be specified by a connector $K : TE \to E$ with the three properties of lemma (37.26). For then $HE := \{\xi_u : K(\xi_u) = 0_{p(u)}\}$ is a complement to VE in TE which is Tp–fiberwise linearly chosen.

37.28. Covariant derivative on vector bundles. Let $p : E \to M$ be a vector bundle with a linear connection, given by a connector $K : TE \to E$ with the properties in lemma (37.26).

For any manifold N, smooth mapping $s : N \to E$, and kinematic vector field $X \in \mathfrak{X}(N)$ we define the *covariant derivative* of s along X by

(1) $$\nabla_X s := K \circ Ts \circ X : N \to TN \to TE \to E.$$

If $f : N \to M$ is a fixed smooth mapping, let us denote by $C^\infty_f(N, E)$ the vector space of all smooth mappings $s : N \to E$ with $p \circ s = f$ — they are called sections of E along f. It follows from the universal property of the pullback that the vector space $C^\infty_f(N, E)$ is canonically linearly isomorphic to the space $C^\infty(N \leftarrow f^*E)$ of sections of the pullback bundle. Then the covariant derivative may be viewed as a bilinear mapping

(2) $$\nabla : \mathfrak{X}(N) \times C^\infty_f(N, E) \to C^\infty_f(N, E).$$

In particular, for $f = \mathrm{Id}_M$ we have

$$\nabla : \mathfrak{X}(M) \times C^\infty(M \leftarrow E) \to C^\infty(M \leftarrow E).$$

Lemma. *This covariant derivative has the following properties:*

(3) *$\nabla_X s$ is $C^\infty(N, \mathbb{R})$-linear in $X \in \mathfrak{X}(N)$. Moreover, for a tangent vector $X_x \in T_x N$ the mapping $\nabla_{X_x} : C^\infty_f(N, E) \to E_{f(x)}$ makes sense, and we have $(\nabla_X s)(x) = \nabla_{X(x)} s$. Thus, $\nabla s \in \Omega^1(N; f^*E)$.*

(4) *$\nabla_X s$ is $\mathbb{R}$-linear in $s \in C^\infty_f(N, E)$.*

(5) *$\nabla_X(h.s) = dh(X).s + h.\nabla_X s$ for $h \in C^\infty(N, \mathbb{R})$, the derivation property of ∇_X.*

(6) *For any manifold Q, smooth mapping $g : Q \to N$, and $Y_y \in T_y Q$ we have $\nabla_{Tg.Y_y} s = \nabla_{Y_y}(s \circ g)$. If $Y \in \mathfrak{X}(Q)$ and $X \in \mathfrak{X}(N)$ are g-related, then we have $\nabla_Y(s \circ g) = (\nabla_X s) \circ g$.*

Proof. All these properties follow easily from definition (1). $\square$

For vector fields $X, Y \in \mathfrak{X}(M)$ and a section $s \in C^\infty(M \leftarrow E)$ an easy computation shows that

$$\begin{aligned}\mathcal{R}^E(X,Y)s &:= \nabla_X\nabla_Y s - \nabla_Y\nabla_X s - \nabla_{[X,Y]}s\\ &= ([\nabla_X,\nabla_Y] - \nabla_{[X,Y]})s\end{aligned}$$

is $C^\infty(M,\mathbb{R})$-linear in X, Y, and s. By the method of (14.3), it follows that $\mathcal{R}^E$ is a 2-form on M with values in the vector bundle $L(E,E)$, i.e. $\mathcal{R}^E \in \Omega^2(M;L(E,E))$. It is called the *curvature* of the covariant derivative.

For $f : N \to M$, vector fields X, $Y \in \mathfrak{X}(N)$, and a section $s \in C^\infty_f(N,E)$ along f one can prove that

$$\nabla_X\nabla_Y s - \nabla_Y\nabla_X s - \nabla_{[X,Y]}s = (f^*\mathcal{R}^E)(X,Y)s := R^E(Tf.X,Tf.Y)s.$$

37.29. Covariant exterior derivative. Let $p : E \to M$ be a vector bundle with a linear connection, given by a connector $K : TE \to E$.

For a smooth mapping $f : N \to M$ let $\Omega(N;f^*E)$ be the vector space of all forms on N with values in the vector bundle f^*E. We can also view them as forms on N with values along f in E, but we do not introduce an extra notation for this.

As in (32.1), (33.2), and (35.1) we have to assume the
Convention. We consider each derivation and homomorphism to be a sheaf morphism (compare (32.1) and the definition of modular 1-forms in (33.2)), or we assume that all manifolds in question are smoothly regular.

The graded space $\Omega(N;f^*E)$ is a graded $\Omega(N)$-module via

$$\begin{aligned}&(\varphi\wedge\Phi)(X_1,\dots,X_{p+q}) =\\ &\qquad = \tfrac{1}{p!\,q!}\sum_\sigma \operatorname{sign}(\sigma)\;\varphi(X_{\sigma1},\dots,X_{\sigma p})\Phi(X_{\sigma(p+1)},\dots,X_{\sigma(p+q)}).\end{aligned}$$

Any $A \in \Omega^p(N;f^*L(E,E))$ defines a graded module homomorphism

$$\begin{aligned}(1)\qquad &\mu(A) : \Omega(N;f^*E) \to \Omega(N;f^*E),\\ &(\mu(A)\Phi)(X_1,\dots,X_{p+q}) =\\ &\qquad = \tfrac{1}{p!\,q!}\sum_\sigma \operatorname{sign}(\sigma)\;A(X_{\sigma1},\dots,X_{\sigma p})(\Phi(X_{\sigma(p+1)},\dots,X_{\sigma(p+q)})),\\ &\mu(A)(\varphi\wedge\Phi) = (-1)^{\deg A.\deg\varphi}\varphi\wedge\mu(A)(\Phi).\end{aligned}$$

But in general not all graded module homomorphisms are of this form, recall the distinction between modular differential forms and differential forms in (33.2). This is only true if the modeling spaces of N have the bornological approximation property; the proof is as in (33.5).

The *covariant exterior derivative* is given by

$$\begin{aligned}(2)\qquad & d_\nabla : \Omega^p(N;f^*E) \to \Omega^{p+1}(N;f^*E)\\ (d_\nabla\Phi)(X_0,\dots,X_p) &= \sum_{i=0}^{p}(-1)^i\nabla_{X_i}\Phi(X_0,\dots,\widehat{X_i},\dots,X_p)+\\ &+\sum_{0\le i<j\le p}(-1)^{i+j}\Phi([X_i,X_j],X_0,\dots,\widehat{X_i},\dots,\widehat{X_j},\dots,X_p),\end{aligned}$$

where the X_i are vector fields on N. It will be shown below that it is indeed well defined, i.e. that $d_\nabla\Phi \in \Omega^{p+1}(N;f^*E)$. Now we only see that it is a modular differential form.

The *covariant Lie derivative* along a vector field $X \in \mathfrak{X}(N)$ is given by

$$\begin{aligned}(3)\qquad & \mathcal{L}^\nabla_X : \Omega^p(N;f^*E) \to \Omega^p(N;f^*E)\\ (\mathcal{L}^\nabla_X\Phi)(X_1,\dots,X_p) &= \nabla_X(\Phi(X_1,\dots,X_p))-\\ &-\sum_i \Phi(X_1,\dots,[X,X_i],\dots,X_p).\end{aligned}$$

Again we will show below that it is well defined. Finally we recall the *insertion operator*

$$\begin{aligned}(4)\qquad & i_X : \Omega^p(N;f^*E) \to \Omega^{p-1}(N;f^*E)\\ (\mathcal{L}^\nabla_X\Phi)(X_1,\dots,X_{p-1}) &= \Phi(X,X_1,\dots,X_{p-1})\end{aligned}$$

Theorem. *The covariant exterior derivative d_∇, and the covariant Lie derivative are well defined and have the following properties.*

(5) *For $s \in C^\infty(N \leftarrow f^*E) = \Omega^0(N;f^*E)$ we have $(d_\nabla s)(X) = \nabla_X s$.*
(6) $d_\nabla(\varphi\wedge\Phi) = d\varphi\wedge\Phi + (-1)^{\deg\varphi}\varphi\wedge d_\nabla\Phi$.
(7) *For smooth $g : Q \to N$ and $\Phi \in \Omega(N;f^*E)$ we have $d_\nabla(g^*\Phi) = g^*(d_\nabla\Phi)$.*
(8) $d_\nabla d_\nabla\Phi = \mu(f^*\mathcal{R}^E)\Phi$.
(9) $i_X(\varphi\wedge\Phi) = i_X\varphi\wedge\Phi + (-1)^{\deg\varphi}\varphi\wedge i_X\Phi$.
(10) $\mathcal{L}^\nabla_X(\varphi\wedge\Phi) = \mathcal{L}_X\varphi\wedge\Phi + \varphi\wedge L^\nabla_X\Phi$.
(11) $[\mathcal{L}^\nabla_X, i_Y] = \mathcal{L}^\nabla_X\circ i_Y - i_Y\circ\mathcal{L}^\nabla_X = i_{[X,Y]}$.
(12) $[i_X, d_\nabla] = i_X\circ d_\nabla + d_\nabla\circ i_X = \mathcal{L}^\nabla_X$.

Proof. By the formula above $d_\nabla\Phi$ is a priori defined as a modular differential form and we have to show that it really lies in $\Omega(M;f^*E)$. For that let $s^* \in C^\infty(N \leftarrow f^*E')$ be a local smooth section on $U \subseteq N$ along $f|U : U \to M$ of the dual vector bundle $E' \to M$. Then $\langle\Phi,s^*\rangle \in \Omega^k(N)$, and for the canonical covariant derivative on the dual bundle (write down its connector!) we have

$$d\langle\Phi,s^*\rangle = \langle d_\nabla\Phi,s^*\rangle + (-1)^k\langle\Phi,\nabla^{E'}s^*\rangle_\wedge,$$

which shows that $d_\nabla\Phi \in \Omega^k(N,f^*E)$ since d respects $\Omega^*(N)$ by (33.12).

(5) is just (33.11). (7) follows from (37.28.6).

(11) Take the difference of the following two expressions:

$$\begin{aligned}(\mathcal{L}^\nabla_X i_Y\Phi)(Z_1,\dots,Z_k) &= \nabla_X((i_Y\Phi)(Z_1,\dots,Z_k))-\\ &\quad -\sum_i (i_Y\Phi)(Z_1,\dots,[X,Z_i],\dots,Z_k)\\ &= \nabla_X(\Phi(Y,Z_1,\dots,Z_k)) - \sum_i \Phi(Y,Z_1,\dots,[X,Z_i],\dots,Z_k)\\ (i_Y\mathcal{L}^\nabla_X\Phi)(Z_1,\dots,Z_k) &= \mathcal{L}^\nabla_X\Phi(Y,Z_1,\dots,Z_k)\\ &= \nabla_X(\Phi(Y,Z_1,\dots,Z_k)) - \Phi([X,Y],Z_1,\dots,Z_k)-\\ &\quad -\sum_i \Phi(Y,Z_1,\dots,[X,Z_i],\dots,Z_k).\end{aligned}$$

(10) Let φ be of degree p and Φ of degree q. We prove the result by induction on $p+q$. Suppose that (5) is true for $p+q<k$. Then for X we have by (9), by (11), and by induction

$$\begin{aligned}(i_Y\mathcal{L}^\nabla_X)(\varphi\wedge\Phi) &= (\mathcal{L}^\nabla_X i_Y)(\varphi\wedge\Phi) - i_{[X,Y]}(\varphi\wedge\Phi)\\ &= \mathcal{L}^\nabla_X(i_Y\varphi\wedge\Phi + (-1)^p\varphi\wedge i_Y\Phi) - i_{[X,Y]}\varphi\wedge\Phi - (-1)^p\varphi\wedge i_{[X,Y]}\Phi\\ &= \mathcal{L}_X i_Y\varphi\wedge\Phi + i_Y\varphi\wedge\mathcal{L}^\nabla_X\Phi + (-1)^p\mathcal{L}_X\varphi\wedge i_Y\Phi+\\ &\quad + (-1)^p\varphi\wedge\mathcal{L}^\nabla_X i_Y\Phi - i_{[X,Y]}\varphi\wedge\Phi - (-1)^p\varphi\wedge i_{[X,Y]}\Phi\\ i_Y(\mathcal{L}_X\varphi\wedge\Phi + \varphi\wedge\mathcal{L}^\nabla_X\Phi) &= i_Y\mathcal{L}_X\varphi\wedge\Phi + (-1)^p\mathcal{L}_X\varphi\wedge i_Y\Phi+\\ &\quad + i_Y\varphi\wedge\mathcal{L}^\nabla_X\Phi + (-1)^p\varphi\wedge i_Y\mathcal{L}^\nabla_X\Phi.\end{aligned}$$

Using again (11), we get the result since the i_Y for all local vector fields Y together act point separating on each space of differential forms, in both cases of the convention.

(12) We write out all relevant expressions.

$$\begin{aligned}(\mathcal{L}^\nabla_{X_0}\Phi)(X_1,\dots,X_k) &= X_0(\Phi(X_1,\dots,X_k))+\\ &\quad + \sum_{j=1}^k (-1)^{0+j}\Phi([X_0,X_j],X_1,\dots,\widehat{X_j},\dots,X_k)\\ (i_{X_0}d_\nabla\Phi)(X_1,\dots,X_k) &= d_\nabla\Phi(X_0,\dots,X_k)\\ &= \sum_{i=0}^k (-1)^i\nabla_{X_i}(\Phi(X_0,\dots,\widehat{X_i},\dots,X_k)) +\\ &\quad + \sum_{0\le i<j} (-1)^{i+j}\Phi([X_i,X_j],X_0,\dots,\widehat{X_i},\dots,\widehat{X_j},\dots,X_k).\end{aligned}$$

$$\begin{aligned}(d_\nabla i_{X_0}\Phi)(X_1,\dots,X_k) &= \sum_{i=1}^k (-1)^{i-1}\nabla_{X_i}((i_{X_0}\Phi)(X_1,\dots,\widehat{X_i},\dots,X_k)) +\\ &\quad + \sum_{1\le i<j} (-1)^{i+j-2}(i_{X_0}\Phi)([X_i,X_j],X_1,\dots,\widehat{X_i},\dots,\widehat{X_j},\dots,X_k)\end{aligned}$$

$$= -\sum_{i=1}^{k}(-1)^i\nabla_{X_i}(\Phi(X_0, X_1, \dots, \widehat{X_i}, \dots, X_k)) -$$
$$- \sum_{1\le i<j}(-1)^{i+j}\Phi([X_i, X_j], X_0, X_1, \dots, \widehat{X_i}, \dots, \widehat{X_j}, \dots, X_k).$$

By summing up, the result follows.

(6) We prove the result again by induction on $p+q$. Suppose that (6) is true for $p+q<k$. Then for each local vector field X we have by (10), (9), (12), and by induction

$$\begin{aligned} i_X\, d_\nabla(\varphi\wedge\Phi) &= \mathcal{L}_X^\nabla(\varphi\wedge\Phi) - d_\nabla i_X(\varphi\wedge\Phi) \\ &= \mathcal{L}_X^\nabla\varphi\wedge\Phi + \varphi\wedge\mathcal{L}_X^\nabla\Phi - d_\nabla(i_X\varphi\wedge\Phi + (-1)^p\varphi\wedge i_X\Phi) \\ &= i_X d\varphi\wedge\Phi + d_\nabla i_X\varphi\wedge\Phi + \varphi\wedge i_X d\Phi + \varphi\wedge d_\nabla i_X\Phi - d_\nabla i_X\varphi\wedge\Phi \\ &\qquad - (-1)^{p-1}i_X\varphi\wedge d_\nabla\Phi - (-1)^p d_\nabla\varphi\wedge i_X\Phi - \varphi\wedge d_\nabla i_X\Phi \\ &= i_X(d_\nabla\varphi\wedge\Phi + (-1)^p\varphi\wedge d_\nabla\Phi). \end{aligned}$$

Since X is arbitrary, the result follows.

(8) follows from a direct computation. The usual fast proofs are not conclusive in infinite dimensions. The computation is similar to the one for the proof of (33.18.4), and only the definitions (2) of d_∇ and (37.28) of $\mathcal{R}^E$, and the Jacobi identity enter. □

37.30. Let $(p : P \to M, G)$ be a principal fiber bundle, and let $\rho : G \to GL(W)$ be a representation of the structure group G on a convenient vector space W, as in (49.1).

Theorem. *There is a canonical isomorphism from the space of $P[W,\rho]$-valued differential forms on M onto the space of horizontal G-equivariant W-valued differential forms on P:*

$$\begin{aligned} q^\sharp : \Omega(M; P[W,\rho]) \to \Omega_{hor}(P,W)^G := \{\varphi\in\Omega(P,W) : i_X\varphi = 0 \\ \text{for all } X\in VP,\ (r^g)^*\varphi = \rho(g^{-1})\circ\varphi \text{ for all } g\in G\}, \end{aligned}$$

In particular, for $W=\mathbb{R}$ with trivial representation we see that

$$p^* : \Omega(M) \to \Omega_{hor}(P)^G = \{\varphi\in\Omega_{hor}(P) : (r^g)^*\varphi = \varphi\}$$

also is an isomorphism. We have $q^\sharp(\varphi\wedge\Phi) = p^*\varphi\wedge q^\sharp\Phi$ for $\varphi\in\Omega(M)$ and $\Phi\in\Omega(M;P[W])$.

The isomorphism

$$q^\sharp : \Omega^0(M;P[W]) = C^\infty(M\leftarrow P[W]) \to \Omega^0_{hor}(P,W)^G = C^\infty(P,W)^G$$

is a special case of the one from (37.16).

Proof. Let $\varphi \in \Omega^k_{hor}(P,W)^G$, $X_1,\dots,X_k \in T_uP$, and $X'_1,\dots,X'_k \in T_{u'}P$ such that $T_up.X_i = T_{u'}p.X'_i$ for each i. Then there is a $g \in G$ such that $ug = u'$:

$$\begin{aligned}q(u,\varphi_u(X_1,\dots,X_k)) &= q(ug,\rho(g^{-1})\varphi_u(X_1,\dots,X_k))\\ &= q(u',((r^g)^*\varphi)_u(X_1,\dots,X_k))\\ &= q(u',\varphi_{ug}(T_u(r^g).X_1,\dots,T_u(r^g).X_k))\\ &= q(u',\varphi_{u'}(X'_1,\dots,X'_k)), \text{ since } T_u(r^g)X_i - X'_i \in V_{u'}P.\end{aligned}$$

Thus, a vector bundle valued form $\Phi \in \Omega^k(M;P[W])$ is uniquely determined by

$$\Phi_{p(u)}(T_up.X_1,\dots,T_up.X_k) := q(u,\varphi_u(X_1,\dots,X_k)).$$

For the converse recall the smooth mapping $\tau^W : P\times_M P[W,\rho] \to W$ from (37.12), which satisfies $\tau^W(u,q(u,w)) = w$, $q(u_x,\tau^W(u_x,v_x)) = v_x$, and $\tau^W(u_xg,v_x) = \rho(g^{-1})\tau^W(u_x,v_x)$.
For $\Phi \in \Omega^k(M;P[W])$ we define $q^\sharp\Phi \in \Omega^k(P,W)$ as follows. For $X_i \in T_uP$ we put

$$(q^\sharp\Phi)_u(X_1,\dots,X_k) := \tau^W(u,\Phi_{p(u)}(T_up.X_1,\dots,T_up.X_k)).$$

Then $q^\sharp\Phi$ is smooth and horizontal. For $g \in G$ we have

$$\begin{aligned}((r^g)^*(q^\sharp\Phi))_u(X_1,\dots,X_k) &= (q^\sharp\Phi)_{ug}(T_u(r^g).X_1,\dots,T_u(r^g).X_k)\\ &= \tau^W(ug,\Phi_{p(ug)}(T_{ug}p.T_u(r^g).X_1,\dots,T_{ug}p.T_u(r^g).X_k))\\ &= \rho(g^{-1})\tau^W(u,\Phi_{p(u)}(T_up.X_1,\dots,T_up.X_k))\\ &= \rho(g^{-1})(q^\sharp\Phi)_u(X_1,\dots,X_k).\end{aligned}$$

Clearly, the two constructions are inverse to each other. □

37.31. Let $(p : P \to M, G)$ be a principal fiber bundle with a principal connection $\Phi = \zeta \circ \omega$, and let $\rho : G \to GL(W)$ be a representation of the structure group G on a convenient vector space W, as in (49.1). We consider the associated vector bundle $(p : E := P[W,\rho] \to M, W)$, the induced connection $\bar{\Phi}$ on it, and the corresponding covariant derivative.

Theorem. *The covariant exterior derivative d_ω from* (37.23) *on P and the covariant exterior derivative for $P[W]$-valued forms on M are connected by the mapping $q^\sharp$ from* (37.30)*, as follows:*

$$q^\sharp \circ d_\nabla = d_\omega \circ q^\sharp : \Omega(M;P[W]) \to \Omega_{hor}(P,W)^G.$$

Proof. Let us consider first $f \in \Omega^0_{hor}(P,W)^G = C^\infty(P,W)^G$, then $f = q^\sharp s$ for $s \in C^\infty(M \leftarrow P[W])$, and we have $f(u) = \tau^W(u,s(p(u)))$ and $s(p(u)) = q(u,f(u))$ by (37.30) and (37.16). Therefore, we have $Ts.Tp.X_u = Tq(X_u,Tf.X_u)$, where $Tf.X_u = (f(u),df(X_u)) \in TW = W\times W$. If $\chi : TP \to HP$ is the horizontal projection as in (37.23), we have $Ts.Tp.X_u = Ts.Tp.\chi.X_u = Tq(\chi.X_u, Tf.\chi.X_u)$. So we get

$$\begin{aligned}
(q^\sharp d_\nabla s)(X_u) &= \tau^W(u, (d_\nabla s)(Tp.X_u)) \\
&= \tau^W(u, \nabla_{Tp.X_u} s) && \text{by (37.29.5)} \\
&= \tau^W(u, K.Ts.Tp.X_u) && \text{by (37.28.1)} \\
&= \tau^W(u, K.Tq(\chi.X_u, Tf.\chi.X_u)) && \text{from above} \\
&= \tau^W(u, \operatorname{pr}_2.\operatorname{vl}^{-1}_{P[W]}.\bar{\Phi}.Tq(\chi.X_u, Tf.\chi.X_u)) && \text{by (37.26)} \\
&= \tau^W(u, \operatorname{pr}_2.\operatorname{vl}^{-1}_{P[W]}.Tq.(\Phi \times \operatorname{Id})(\chi.X_u, Tf.\chi.X_u)) && \text{by (37.24)} \\
&= \tau^W(u, \operatorname{pr}_2.\operatorname{vl}^{-1}_{P[W]}.Tq(0_u, Tf.\chi.X_u)) && \text{since } \Phi.\chi = 0 \\
&= \tau^W(u, q.\operatorname{pr}_2.\operatorname{vl}^{-1}_{P\times W}.(0_u, Tf.\chi.X_u)) && \text{since } q \text{ is fiber linear} \\
&= \tau^W(u, q(u, df.\chi.X_u)) = (\chi^* df)(X_u) \\
&= (d_\omega q^\sharp s)(X_u).
\end{aligned}$$

Now we turn to the general case. Let Y_i for $i = 0, \dots, k$ be local vector fields on M, and let CY_i be their horizontal lifts to P. Then $Tp.CY_i = y_i \circ p$, so Y_i and CY_i are p-related. Since both $q^\sharp d_\nabla \Phi$ and $d_\omega q^\sharp$ are horizontal, it suffices to check that they coincide on all local vector fields of the form CY_i. Since $C[Y_i, Y_j] = \chi[CY_i, CY_j]$, we get from the special case above and the definition of $q^\sharp$:

$$\begin{aligned}
(d_\omega q^\sharp \Phi)(CY_0, \dots, CY_k) &= \sum_{0\le i\le k} (-1)^i (CY_i)(q^\sharp\Phi)(CY_0, \dots, \widehat{CY_i}, \dots, CY_k) \\
&\quad + \sum_{i<j} (-1)^{i+j} (q^\sharp\Phi)([CY_i, CY_j], CY_0, \dots \widehat{CY_i} \dots \widehat{CY_j} \dots, CY_k) \\
&= \sum_{0\le i\le k} (-1)^i (CY_i)(q^\sharp(\Phi(Y_0, \dots, \widehat{Y_i}, \dots, Y_k)) \\
&\quad + \sum_{i<j} (-1)^{i+j} (q^\sharp\Phi)(C[Y_i, Y_j], CY_0, \dots \widehat{CY_i} \dots \widehat{CY_j} \dots, CY_k) \\
&= \sum_{0\le i\le k} (-1)^i q^\sharp(\nabla_{Y_i}(\Phi(Y_0, \dots, \widehat{Y_i}, \dots, Y_k))) \\
&\quad + \sum_{i<j} (-1)^{i+j} q^\sharp(\Phi([Y_i, Y_j], Y_0, \dots \widehat{Y_i} \dots \widehat{Y_j} \dots, Y_k)) \\
&= q^\sharp(d_\nabla \Phi(Y_0, \dots, Y_k)) = (q^\sharp d_\nabla \Phi)(CY_0, \dots, CY_k). \quad \square
\end{aligned}$$

37.32. Corollary. *In the situation of theorem* (37.31), *for the Lie algebra valued curvature form* $\Omega \in \Omega^2_{hor}(P, \mathfrak{g})$ *and the curvature* $\mathcal{R}^{P[W]} \in \Omega^2(M; L(P[W], P[W]))$ *we have the relation*

$$q^\sharp_{L(P[W],P[W])} \mathcal{R}^{P[W]} = \rho' \circ \Omega,$$

where $\rho' = T_e\rho : \mathfrak{g} \to L(W, W)$ *is the derivative of the representation* ρ.

Proof. We use the notation of the proof of theorem (37.31), by which we have for $X, Y \in T_uP$

$$\begin{aligned}(d_\omega d_\omega q^\sharp_{P[W]}s)_u(X,Y) &= (q^\sharp d_\nabla d_\nabla s)_u(X,Y)\\ &= (q^\sharp \mathcal{R}^{P[W]}s)_u(X,Y)\\ &= \tau^W(u, \mathcal{R}^{P[W]}(T_up.X, T_up.Y)s(p(u)))\\ &= (q^\sharp_{L(P[W],P[W])}\mathcal{R}^{P[W]})_u(X,Y)(q^\sharp_{P[W]}s)(u).\end{aligned}$$

On the other hand, let $g(t)$ be a smooth curve in G with $g(0)=e$ and $\frac{\partial}{\partial t}|_0 g(t) = \Omega_u(X,Y) \in \mathfrak{g}$. Then we have by theorem (37.23.8)

$$\begin{aligned}(d_\omega d_\omega q^\sharp s)_u(X,Y) &= (\chi^* i_{\mathcal{R}} dq^\sharp s)_u(X,Y)\\ &= (dq^\sharp s)_u(\mathcal{R}(X,Y)) \quad \text{since } \mathcal{R} \text{ is horizontal}\\ &= (dq^\sharp s)(-\zeta_{\Omega(X,Y)}(u)) \quad \text{by (37.20)}\\ &= \tfrac{\partial}{\partial t}|_0 (q^\sharp s)(r^{g(t)^{-1}}(u))\\ &= \tfrac{\partial}{\partial t}|_0 \tau^W(u.g(t)^{-1}, s(p(u.g(t)^{-1})))\\ &= \tfrac{\partial}{\partial t}|_0 \tau^W(u.g(t)^{-1}, s(p(u)))\\ &= \tfrac{\partial}{\partial t}|_0 \rho(g(t))\tau^W(u, s(p(u))) \quad \text{by (37.12)}\\ &= \rho'(\Omega_u(X,Y))(q^\sharp s)(u). \quad \square\end{aligned}$$

38. Regular Lie Groups

38.1. The right and left logarithmic derivatives. Let M be a manifold, and let $f : M \to G$ be a smooth mapping into a Lie group G with Lie algebra $\mathfrak{g}$. We define the mapping $\delta^r f : TM \to \mathfrak{g}$ by the formula

$$\delta^r f(\xi_x) := T_{f(x)}(\mu^{f(x)^{-1}}).T_xf.\xi_x = (f^*\kappa^r)(\xi_x) \text{ for } \xi_x \in T_xM,$$

where κ^r is the right Maurer-Cartan form from (36.10). Then $\delta^r f$ is a $\mathfrak{g}$-valued 1-form on M, $\delta^r f \in \Omega^1(M,\mathfrak{g})$. We call $\delta^r f$ the *right logarithmic derivative* of f, since for $f : \mathbb{R} \to (\mathbb{R}^+, \cdot)$ we have $\delta^r f(x).1 = \frac{f'(x)}{f(x)} = (\log \circ f)'(x)$.

Similarly, the *left logarithmic derivative* $\delta^l f \in \Omega^1(M,\mathfrak{g})$ of a smooth mapping $f : M \to G$ is given by

$$\delta^l f.\xi_x := T_{f(x)}(\mu_{f(x)^{-1}}).T_xf.\xi_x = (F^*\kappa^l)(\xi_x).$$

Lemma. *Let $f, g : M \to G$ be smooth. Then we have*

$$\delta^r(f.g)(x) = \delta^r f(x) + \mathrm{Ad}(f(x)).\delta^r g(x).$$

Moreover, the differential form $\delta^r f \in \Omega^1(M,\mathfrak{g})$ satisfies the 'left Maurer-Cartan equation' (left because it stems from the left action of G on itself)

$$\begin{aligned} &d\delta^r f(\xi,\eta) - [\delta^r f(\xi), \delta^r f(\eta)]^{\mathfrak{g}} = 0 \\ \text{or}\quad &d\delta^r f - \frac{1}{2}[\delta^r f, \delta^r f]^{\mathfrak{g}}_{\wedge} = 0, \end{aligned}$$

where $\xi,\eta \in T_xM$, and where the graded Lie bracket $[\ \ ,\ \ ]^{\mathfrak{g}}_{\wedge}$ was defined in (37.20.1).

The left logarithmic derivative also satisfies a 'Leibniz rule' and the 'right Maurer Cartan equation':

$$\begin{aligned} \delta^l(fg)(x) &= \delta^l g(x) + Ad(g(x)^{-1}).\delta^l f(x), \\ d\delta^l f &+ \frac{1}{2}[\delta^l f, \delta^l f]^{\mathfrak{g}}_{\wedge} = 0. \end{aligned}$$

For 'regular Lie groups' we will prove a converse to this statement later in (40.2).

Proof. We treat only the right logarithmic derivative, the proof for the left one is similar.

$$\begin{aligned} \delta^r(f.g)(x) &= T(\mu^{g(x)^{-1}.f(x)^{-1}}).T_x(f.g) \\ &= T(\mu^{f(x)^{-1}}).T(\mu^{g(x)^{-1}}).T_{(f(x),g(x))}\mu.(T_xf, T_xg) \\ &= T(\mu^{f(x)^{-1}}).T(\mu^{g(x)^{-1}}).\Big(T(\mu^{g(x)}).T_xf + T(\mu_{f(x)}).T_xg\Big) \\ &= \delta^r f(x) + \mathrm{Ad}(f(x)).\delta^r g(x). \end{aligned}$$

We shall now use principal bundle geometry from section (37). We consider the trivial principal bundle $\mathrm{pr}_1 : M \times G \to M$ with right principal action. Then the submanifolds $\{(x, f(x).g) : x \in M\}$ for $g \in G$ form a foliation of $M \times G$, whose tangent distribution is complementary to the vertical bundle $M \times TG \subset T(M\times G)$ and is invariant under the principal right G-action. So it is the horizontal distribution of a principal connection on $M \times G \to M$. For a tangent vector $(\xi_x, Y_g) \in T_xM \times T_gG$ the horizontal part is the right translate to the foot point (x,g) of $(\xi_x, T_xf.\xi_x)$. The decomposition in horizontal and vertical parts according to this distribution is

$$(\xi_x, Y_g) = (\xi_x, T(\mu^g).T(\mu^{f(x)^{-1}}).T_xf.\xi_x) + (0_x, Y_g - T(\mu^g).T(\mu^{f(x)^{-1}}).T_xf.\xi_x).$$

Since the fundamental vector fields for the right action on G are the left invariant vector fields, the corresponding connection form is given by

$$\begin{aligned} \omega^r(\xi_x, Y_g) &= T(\mu_{g^{-1}}).(Y_g - T(\mu^g).T(\mu^{f(x)^{-1}}).T_xf.\xi_x), \\ \omega^r_{(x,g)} &= T(\mu_{g^{-1}}) - \mathrm{Ad}(g^{-1}).\delta^r f_x, \\ \omega^r &= \kappa^l - (\mathrm{Ad}\circ\nu).\delta^r f, \qquad (1) \end{aligned}$$

where $\kappa^l : TG \to \mathfrak{g}$ is the *left Maurer-Cartan form* on G (the left trivialization), given by $\kappa^l_g = T(\mu_{g^{-1}})$. Note that κ^l is the principal connection form for the (unique) principal connection $p : G \to \{\text{point}\}$ with right principal action, which is flat so that the right (from right action) Maurer-Cartan equation holds in the form

$$d\kappa^l + \tfrac{1}{2}[\kappa^l, \kappa^l]_\wedge = 0. \tag{2}$$

The principal connection ω^r is flat since we got it via the horizontal leaves, so the principal curvature form vanishes:

$$\begin{aligned}
0 &= d\omega^r + \tfrac{1}{2}[\omega^r, \omega^r]_\wedge \\
&= d\kappa^l + \tfrac{1}{2}[\kappa^l, \kappa^l]_\wedge - d(\operatorname{Ad}\circ\nu)\wedge\delta^r f - (\operatorname{Ad}\circ\nu).d\delta^r f \\
&\quad - [\kappa^l, (\operatorname{Ad}\circ\nu).\delta^r f]_\wedge + \tfrac{1}{2}[(\operatorname{Ad}\circ\nu).\delta^r f, (\operatorname{Ad}\circ\nu).\delta^r f]_\wedge \\
&= -(\operatorname{Ad}\circ\nu).(d\delta^r f - \tfrac{1}{2}[\delta^r f, \delta^r f]_\wedge),
\end{aligned} \tag{3}$$

where we used (2) and the fact that for $\xi \in \mathfrak{g}$ and a smooth curve $c : \mathbb{R} \to G$ with $c(0) = e$ and $c'(0) = \xi$ we have

$$\begin{aligned}
d(\operatorname{Ad}\circ\nu)(T(\mu_g)\xi) &= \tfrac{\partial}{\partial t}\big|_0 \operatorname{Ad}(c(t)^{-1}.g^{-1}) = -\operatorname{ad}(\xi)\operatorname{Ad}(g^{-1}) \\
&= -\operatorname{ad}\Big(\kappa^l(T(\mu_g)\xi)\Big)(\operatorname{Ad}\circ\nu)(g),
\end{aligned}$$

$$d(\operatorname{Ad}\circ\nu) = -(\operatorname{ad}\circ\kappa^l).(\operatorname{Ad}\circ\nu). \tag{4}$$

So we have $d\delta^r f - \frac{1}{2}[\delta^r f, \delta^r f]_\wedge$ as asserted.

For the left logarithmic derivative $\delta^l f$ the proof is similar, and we discuss only the essential deviations. First note that on the trivial principal bundle $\operatorname{pr}_1 : M\times G \to M$ with left principal action of G the fundamental vector fields are the right invariant vector fields on G, and that for a principal connection form ω^l the curvature form is given by $d\omega^l - \frac{1}{2}[\omega^l, \omega^l]_\wedge$. Look at the proof of theorem (37.20) to see this. The connection form is then given by

$$\omega^l = \kappa^r - \operatorname{Ad}.\delta^l f, \tag{1'}$$

where the right Maurer-Cartan form $(\kappa^r)_g = T(\mu^{g^{-1}}) : T_g G \to \mathfrak{g}$ satisfies the left Maurer-Cartan equation

$$d\kappa^r - \frac{1}{2}[\kappa^r, \kappa^r]_\wedge = 0. \tag{2'}$$

Flatness of ω^l now leads to the computation

$$\begin{aligned}
0 &= d\omega^l - \tfrac{1}{2}[\omega^l, \omega^l]_\wedge \\
&= d\kappa^r - \tfrac{1}{2}[\kappa^r, \kappa^r]_\wedge - d\operatorname{Ad}\wedge\delta^l f - \operatorname{Ad}.d\delta^l f \\
&\quad + [\kappa^r, \operatorname{Ad}.\delta^l f]_\wedge - \tfrac{1}{2}[\operatorname{Ad}.\delta^l f, \operatorname{Ad}.\delta^l f]_\wedge \\
&= -\operatorname{Ad}.(d\delta^l f + \tfrac{1}{2}[\delta^l f, \delta^l f]_\wedge),
\end{aligned} \tag{3'}$$

where we have used $d\operatorname{Ad} = (\operatorname{ad}\circ\kappa^r)\operatorname{Ad}$ from (36.10.3) directly. □

Remark. The second half of the proof of lemma (38.1) can be shortened considerably. Namely, as soon as we know that κ^r satisfies the Maurer-Cartan equation $d\kappa^r - \frac{1}{2}[\kappa^r, \kappa^r]_\wedge$ we get it also for the right logarithmic derivative $\delta^r f = f^*\kappa^r$. But the computations in this proof will be used again in the proof of the converse, theorem (40.2) below.

38.2. Theorem. [Grabowski, 1993] *Let G be a Lie group with exponential mapping $\exp : \mathfrak{g} \to G$. Then for all $X, Y \in \mathfrak{g}$ we have*

$$\begin{aligned} T_X \exp .Y &= T_e\mu_{\exp X}. \int_0^1 \operatorname{Ad}(\exp(-tX))Y\,dt \\ &= T_e\mu^{\exp X}. \int_0^1 \operatorname{Ad}(\exp(tX))Y\,dt. \end{aligned}$$

Remark. If G is a Banach Lie group then we have from (36.8.4) and (36.9) the series $\operatorname{Ad}(\exp(tX)) = \sum_{i=0}^\infty \frac{t^i}{i!}\operatorname{ad}(X)^i$, so that we get the usual formula

$$T_X \exp = T_e\mu^{\exp X}. \sum_{i=0}^\infty \tfrac{1}{(i+1)!}\operatorname{ad}(X)^i.$$

Proof. We consider the smooth mapping

$$f : \mathbb{R}^2 \to G, \qquad f(s,t) := \exp(s(X+tY)).\exp(-sX).$$

Then $f(s,0) = e$ and

$$\begin{aligned} &\partial_0 f(s,0) = sT_{\exp(sX)}\mu^{\exp(-sX)}.T_{sX}\exp .Y, \\ (1)\qquad &T_X \exp .Y = T_e\mu^{\exp(X)}.\partial_t f(1,0). \end{aligned}$$

Moreover we get

$$\begin{aligned} \delta^r f(s,t).\partial_s &= T_{f(s,t)}\mu^{f(s,t)^{-1}}\Big(T\mu^{\exp(-sX)}.\partial_s \exp(s(X+tY)) \\ &\qquad + T\mu_{\exp(s(X+tY))}.\partial_s \exp(-sX)\Big) \\ &= T_{f(s,t)}\mu^{f(s,t)^{-1}}\Big(T\mu^{\exp(-sX)}.R_{X+tY}(\exp(s(X+tY))) \\ &\qquad - T\mu_{\exp(s(X+tY))}.L_X \exp(-sX)\Big) \\ &= X + tY - \operatorname{Ad}(f(s,t))X. \end{aligned}$$

$$(2)\qquad \partial_t|_0 \delta^r f(s,t).\partial_s = Y - \operatorname{ad}(\partial_t f(s,0)).X$$

Now we use (38.1) to get

$$\begin{aligned} 0 &= d(\delta^r f)(\partial_s, \partial_t) - [\delta^r f(\partial_s), \delta^r f(\partial_t)] \\ &= \partial_s(\delta^r f)(\partial_t) - \partial_t(\delta^r f)(\partial_s) - (\delta^r f)([\partial_s, \partial_t]) - [\delta^r f(\partial_s), \delta^r f(\partial_t)]. \end{aligned}$$

Since $(\delta^r f)(\partial_s)|_{t=0} = 0$ we get $\partial_s(\delta^r f)(s,0)(\partial_t) = \partial_t(\delta^r f)(s,0)(\partial_s)$, and from (2) we then conclude that the curve

$$c(s) = (\delta^r f)(s,0)(\partial_t) = \partial_t|_0 f(s,0) = s\delta^r \exp(sX).Y \in \mathfrak{g} \tag{3}$$

is a solution of the ordinary differential equation

$$c'(s) = Y + [X, c(s)] = Y + \operatorname{ad}(X).c(s), \quad c(0) = 0. \tag{4}$$

The unique solution for the homogeneous equation with $c(0) = c_0$ is

$$\begin{aligned} c(s) &= \operatorname{Ad}(\exp(sX)).c_0, \qquad \text{since} \\ c'(s) &= \partial_t|_{t=0}\operatorname{Ad}(\exp(tX))\operatorname{Ad}(\exp(sX))c_0 = [X, c(s)], \\ \partial_s(\operatorname{Ad}(\exp(-sX))C(s)) &= -\operatorname{Ad}(\exp(-sX)).\operatorname{ad}(X).C(s) + \\ &\quad + \operatorname{Ad}(\exp(-sX))[X, C(s)] = 0 \end{aligned}$$

for every other solution $C(t)$. Using the variation of constant ansatz we get the solution

$$c(s) = \operatorname{Ad}(\exp(sX)) \int_0^s \operatorname{Ad}(\exp(-tX))Y\,dt$$

of the inhomogeneous equation (4), which is unique for $c(0) = 0$ since 0 is the unique solution of the homogeneous equation with initial value 0. Finally, we have from (1)

$$\begin{aligned} T_X \exp .Y &= T_e\mu^{\exp(X)}.c(1) \\ &= T_e\mu^{\exp(X)}.\operatorname{Ad}(\exp(X)) \int_0^1 \operatorname{Ad}(\exp(-tX))Y\,dt \\ &= T_e\mu_{\exp(X)}. \int_0^1 \operatorname{Ad}(\exp(-tX))Y\,dt \\ T_X \exp .Y &= T_e\mu^{\exp(X)}.\operatorname{Ad}(\exp(X)) \int_0^1 \operatorname{Ad}(\exp(-tX))Y\,dt \\ &= T_e\mu^{\exp(X)} \int_0^1 \operatorname{Ad}(\exp((1-t)X))Y\,dt \\ &= T_e\mu^{\exp(X)} \int_0^1 \operatorname{Ad}(\exp(rX))Y\,dr. \quad \square \end{aligned}$$

38.3. Let G be a Lie group with Lie algebra $\mathfrak{g}$. For a closed interval $I \subset \mathbb{R}$ and for $X \in C^\infty(I, \mathfrak{g})$ we consider the ordinary differential equation

$$\begin{cases} g(t_0) = e \\ \frac{\partial}{\partial t} g(t) = T_e(\mu^{g(t)})X(t) = R_{X(t)}(g(t)), \quad \text{or } \kappa^r(\frac{\partial}{\partial t} g(t)) = X(t), \end{cases} \tag{1}$$

for local smooth curves g in G, where $t_0 \in I$.

Lemma.

(2) *Local solutions g of the differential equation (1) are uniquely determined.*

(3) *If for fixed X the differential equation (1) has a local solution near each $t_0 \in I$, then it has also a global solution $g \in C^\infty(I, G)$.*

(4) *If for all $X \in C^\infty(I, \mathfrak{g})$ the differential equation (1) has a local solution near one fixed $t_0 \in I$, then it has also a global solution $g \in C^\infty(I, G)$ for each X. Moreover, if the local solutions near t_0 depend smoothly on the vector fields X (see the proof for the exact formulation), then so does the global solution.*

(5) *The curve $t \mapsto g(t)^{-1}$ is the unique local smooth curve h in G which satisfies*

$$\begin{cases} h(t_0) = e \\ \frac{\partial}{\partial t} h(t) = T_e(\mu_{h(t)})(-X(t)) = L_{-X(t)}(h(t)), \qquad \textit{or } \kappa^l(\frac{\partial}{\partial t} h(t)) = -X(t). \end{cases}$$

Proof. (2) Suppose that $g(t)$ and $g_1(t)$ both satisfy (1). Then on the intersection of their intervals of definition we have

$$\begin{aligned} \tfrac{\partial}{\partial t}(g(t)^{-1} g_1(t)) &= -T(\mu^{g_1(t)}).T(\mu_{g(t)^{-1}}).T(\mu^{g(t)^{-1}}).T(\mu^{g(t)}).X(t) \\ &\quad + T(\mu_{g(t)^{-1}}).T(\mu^{g_1(t)}).X(t) = 0, \end{aligned}$$

so that $g = g_1$.

Proof of (3) It suffices to prove the claim for every compact subinterval of I, so let I be compact. If g is a local solution of (1) then $t \mapsto g(t).x$ is a local solution of the same differential equation with initial value x. By assumption, for each $s \in I$ there is a unique solution g_s of the differential equation with $g_s(s) = e$; so there exists $\delta_s > 0$ such that $g_s(s+t)$ is defined for $|t| < \delta_s$. Since I is compact there exist $s_0 < s_1 < \dots < s_k$ such that $I = [s_0, s_k]$ and $s_{i+1} - s_i < \delta_{s_i}$. Then we put

$$g(t) := \begin{cases} g_{s_0}(t) & \text{for } s_0 \le t \le s_1 \\ g_{s_1}(t).g_{s_0}(s_1) & \text{for } s_1 \le t \le s_2 \\ \dots & \\ g_{s_i}(t).g_{s_{i-1}}(s_i)\dots g_{s_0}(s_1) & \text{for } s_i \le t \le s_{i+1} \\ \dots & \end{cases}$$

which is smooth by the first case and solves the problem.

Proof of (4) Given $X : I \to \mathfrak{g}$ we first extend X to a smooth curve $\mathbb{R} \to \mathfrak{g}$, using (24.10). For $t_1 \in I$, by assumption, there exists a local solution g near t_0 of the translated vector field $t \mapsto X(t_1 - t_0 + t)$, thus $t \mapsto g(t_0 - t_1 + t)$ is a solution near t_1 of X. So by (3) the differential equation has a global solution for X on I.

Now we assume that the local solutions near t_0 depend smoothly on the vector field. So for any smooth curve $X : \mathbb{R} \to C^\infty(I, \mathfrak{g})$ we have:

For every compact interval $K \subset \mathbb{R}$ there is a neighborhood $U_{X,K}$ of t_0 in I and a smooth mapping $g : K \times U_{X,K} \to G$ with

$$\begin{cases} g(k, t_0) = e \\ \frac{\partial}{\partial t} g(k,t) = T_e(\mu^{g(k,t)}).X(k)(t) \quad \text{for all } k \in K, t \in U_{X,K}. \end{cases}$$

Given a smooth curve $X : \mathbb{R} \to C^\infty(I, \mathfrak{g})$ we extend (or lift) it smoothly to $X : \mathbb{R} \to C^\infty(\mathbb{R}, \mathfrak{g})$ by (24.10). Then the smooth parameter k from the compact interval K passes smoothly through the proofs given above to give a smooth global solution $g : K \times I \to G$. So the 'solving operation' respects smooth curves and thus is 'smooth'.

Proof of (5) One can show in a similar way that h is the unique solution of (5) by differentiating $h_1(t).h(t)^{-1}$. Moreover, the curve $t \mapsto g(t)^{-1} = h(t)$ satisfies (5), since

$$\tfrac{\partial}{\partial t}(g(t)^{-1}) = -T(\mu_{g(t)^{-1}}).T(\mu^{g(t)^{-1}}).T(\mu^{g(t)}).X(t) = T(\mu_{g(t)^{-1}}).(-X(t)). \quad \square$$

38.4. Definition. Regular Lie groups. If for each $X \in C^\infty(\mathbb{R}, \mathfrak{g})$ there exists $g \in C^\infty(\mathbb{R}, G)$ satisfying

$$(1) \qquad \begin{cases} g(0) = e \\ \tfrac{\partial}{\partial t} g(t) = T_e(\mu^{g(t)})X(t) = R_{X(t)}(g(t)), \\ \quad \text{or } \kappa^r(\tfrac{\partial}{\partial t} g(t)) = \delta^r g(\partial_t) = X(t) \end{cases}$$

then we write

$$\begin{aligned} \operatorname{evol}^r_G(X) = \operatorname{evol}_G(X) &:= g(1), \\ \operatorname{Evol}^r_G(X)(t) &:= \operatorname{evol}_G(s \mapsto tX(ts)) = g(t), \end{aligned}$$

and call them the *right evolution* of the curve X in G. By lemma (38.3), the solution of the differential equation (1) is unique, and for global existence it is sufficient that it has a local solution. Then

$$\operatorname{Evol}^r_G : C^\infty(\mathbb{R}, \mathfrak{g}) \to \{g \in C^\infty(\mathbb{R}, G) : g(0) = e\}$$

is bijective with inverse δ^r. The Lie group G is called a *regular Lie group* if $\operatorname{evol}^r : C^\infty(\mathbb{R}, \mathfrak{g}) \to G$ exists and is smooth. We also write

$$\begin{aligned} \operatorname{evol}^l_G(X) = \operatorname{evol}_G(X) &:= h(1), \\ \operatorname{Evol}^l_G(X)(t) &:= \operatorname{evol}^l_G(s \mapsto tX(ts)) = h(t), \end{aligned}$$

if h is the (unique) solution of

$$(2) \qquad \begin{cases} h(0) = e \\ \tfrac{\partial}{\partial t} h(t) = T_e(\mu_{h(t)})(X(t)) = L_{X(t)}(h(t)), \\ \quad \text{or } \kappa^l(\tfrac{\partial}{\partial t} h(t)) = \delta^l h(\partial_t) = X(t). \end{cases}$$

Clearly, $\operatorname{evol}^l : C^\infty(\mathbb{R}, \mathfrak{g}) \to G$ exists and is also smooth if evol^r does, since we have $\operatorname{evol}^l(X) = \operatorname{evol}^r(-X)^{-1}$ by lemma (38.3).

Let us collect some easily seen properties of the evolution mappings. If $f \in C^\infty(\mathbb{R},\mathbb{R})$ then we have

$$\begin{aligned}\operatorname{Evol}^r(X)(f(t)) &= \operatorname{Evol}^r(f'.(X\circ f))(t).\operatorname{Evol}^r(X)(f(0)),\\ \operatorname{Evol}^l(X)(f(t)) &= \operatorname{Evol}^l(X)(f(0)).\operatorname{Evol}^l(f'.(X\circ f))(t).\end{aligned}$$

If $\varphi : G \to H$ is a smooth homomorphism between regular Lie groups then the diagrams

$$(3)\qquad \begin{array}{ccc} C^\infty(\mathbb{R},\mathfrak{g}) & \xrightarrow{\varphi'_*} & C^\infty(\mathbb{R},\mathfrak{h}) \\ \Big\downarrow{\scriptstyle \operatorname{evol}_G} & & \Big\downarrow{\scriptstyle \operatorname{evol}_H} \\ G & \xrightarrow{\varphi} & H \end{array} \qquad\qquad \begin{array}{ccc} C^\infty(\mathbb{R},\mathfrak{g}) & \xrightarrow{\varphi'_*} & C^\infty(\mathbb{R},\mathfrak{h}) \\ \Big\downarrow{\scriptstyle \operatorname{Evol}_G} & & \Big\downarrow{\scriptstyle \operatorname{Evol}_H} \\ C^\infty(\mathbb{R},G) & \xrightarrow{\varphi_*} & C^\infty(\mathbb{R},H) \end{array}$$

commutes, since $\frac{\partial}{\partial t}\varphi(g(t)) = T\varphi.T(\mu^{g(t)}).X(t) = T(\mu^{\varphi(g(t))}).\varphi'.X(t)$. Note that each regular Lie group admits an exponential mapping, namely the restriction of evol^r to the constant curves $\mathbb{R}\to\mathfrak{g}$. A Lie group is regular if and only if its universal covering group is regular.

This notion of regularity is a weakening of the same notion of [Omori et al., 1982, 1983, etc.], who considered a sort of product integration property on a smooth Lie group modeled on Fréchet spaces. Our notion here is due to [Milnor, 1984]. Up to now the following statement holds:

All known Lie groups are regular.

Any Banach Lie group is regular since we may consider the time dependent right invariant vector field $R_{X(t)}$ on G and its integral curve $g(t)$ starting at e, which exists and depends smoothly on (a further parameter in) X. In particular, finite dimensional Lie groups are regular. For diffeomorphism groups the evolution operator is just integration of time dependent vector fields with compact support, see section (43) below.

38.5. Some abelian regular Lie groups. For $(E,+)$, where E is a convenient vector space, we have $\operatorname{evol}(X) = \int_0^1 X(t)dt$, so convenient vector spaces are regular abelian Lie groups. We shall need 'discrete' subgroups, which is not an obvious notion since $(E,+)$ is not a topological group: the addition is continuous only as a mapping $c^\infty(E\times E)\to c^\infty E$ and not for the cartesian product of the c^∞-topologies.

Next let Z be a 'discrete' subgroup of a convenient vector space E in the sense that there exists a c^∞-open neighborhood U of zero in E such that $U\cap(z+U)=\emptyset$ for all $0\neq z\in Z$ (equivalently $(U-U)\cap(Z\setminus 0)=\emptyset$). For that it suffices e.g., that Z is discrete in the bornological topology on E. Then E/Z is an abelian but possibly non Hausdorff Lie group. It does not suffice to take Z discrete in the c^∞-topology: Take as Z the subgroup generated by A in $\mathbb{R}^{\mathbb{N}\times c_0}$ in the proof of (4.26).(iv).

Let us assume that Z fulfills the stronger condition: there exists a symmetric c^∞-open neighborhood W of 0 such that $(W+W)\cap(z+W+W)=\emptyset$ for all $0\neq z\in Z$ (equivalently $(W+W+W+W)\cap(Z\setminus 0)=\emptyset$). Then E/Z is Hausdorff and thus an abelian regular Lie group, since its universal cover E is regular. Namely, for $x \notin Z$, we have to find neighborhoods U and V of 0 such that $(Z+U) \cap (x+Z+V) = \emptyset$. There are two cases. If $x \in Z+W+W$ then there is a unique $z \in Z$ with $x \in z+W+W$, and we may choose $U, V \subset W$ such that $(z+U) \cap (x+V) = \emptyset$; then $(Z+U) \cap (x+Z+V) = \emptyset$. In the other case, if $x \notin Z+W+W$, then we have $(Z+W) \cap (x+Z+W) = \emptyset$.

Notice that the two conditions above and their consequences also hold for general (non-abelian) (regular) Lie groups instead of E and their 'discrete' normal subgroups (which turn out to be central if G is connected).

It would be nice if any regular abelian Lie group would be of the form E/Z described above. A first result in this direction is that for an abelian Lie group G with Lie algebra $\mathfrak{g}$ which admits a smooth exponential mapping $\exp : \mathfrak{g} \to G$ one can easily check by using (38.2) that $\frac{\partial}{\partial t}(\exp(-tX).\exp(tX+Y)) = 0$, so that exp is a smooth homomorphism of Lie groups.

Let us consider some examples. More examples can be found in [Banaszczyk, 1984, 1986, 1991]. For the first one we consider a discrete subgroup $Z \subset \mathbb{R}^{\mathbb{N}}$. There exists a neighborhood of 0, without loss of generality of the form $U \times \mathbb{R}^{\mathbb{N}\setminus n}$ for $U \subset \mathbb{R}^n$, with $U \cap (Z \setminus 0) = \emptyset$. Then we consider the following diagram of Lie group homomorphisms

$$\begin{array}{ccccccc}
0 & \longrightarrow & \mathbb{R}^{\mathbb{N}\setminus n} & \longrightarrow & \mathbb{R}^{\mathbb{N}\setminus n} & & \\
\big\downarrow & & \big\downarrow & & \big\downarrow & & \\
Z & \longrightarrow & \mathbb{R}^{\mathbb{N}} & \longrightarrow & \mathbb{R}^{\mathbb{N}}/Z & = & (S^1)^k \times \mathbb{R}^{\mathbb{N}\setminus(n-k)} \\
\cong\big\downarrow & & \pi\big\downarrow & & \big\downarrow & & \big\downarrow \\
\pi(Z) & \longrightarrow & \mathbb{R}^n & \longrightarrow & \mathbb{R}^n/\pi(Z) & = & (S^1)^k \times \mathbb{R}^{n-k}
\end{array}$$

which has exact lines and columns. For the right hand column we use a diagram chase to see this. Choose a global linear section of π inverting $\pi|Z$. This factors to a global homomorphism of the right hand side column.

As next example we consider $\mathbb{Z}^{(\mathbb{N})} \subset \mathbb{R}^{(\mathbb{N})}$. Then, obviously, $\mathbb{R}^{(\mathbb{N})}/\mathbb{Z}^{(\mathbb{N})} = (S^1)^{(\mathbb{N})}$, which is a real analytic manifold modeled on $\mathbb{R}^{(\mathbb{N})}$, similar to the ones which are treated in section (47). The reader may convince himself that any Lie group covered by $\mathbb{R}^{(\mathbb{N})}$ is isomorphic to $(S^1)^{(A)} \times \mathbb{R}^{(\mathbb{N}\setminus A)}$ for $A \subseteq \mathbb{N}$.

As another example, one may check easily that $\ell^\infty/(\mathbb{Z}^{\mathbb{N}} \cap \ell^\infty) = (S^1)^{\mathbb{N}}$, equipped with the 'uniform box topology'; compare with the remark at the end of (27.3).

38.6. Extensions of Lie groups. Let H and K be Lie groups. A Lie group G is called a smooth *extension* of H with kernel K if we have a short exact sequence of groups

(1) $$\{e\} \to K \xrightarrow{i} G \xrightarrow{p} H \to \{e\},$$

such that i and p are smooth, and one of the following two equivalent conditions is satisfied:

(2) p admits a local smooth section s near e (equivalently near any point), and i is initial (27.11).

(3) i admits a local smooth retraction r near e (equivalently near any point), and p is final (27.15).

Of course, by $s(p(x))i(r(x)) = x$ the two conditions are equivalent, and then G is locally diffeomorphic to $K \times H$ via (r, p) with local inverse $(i \circ \mathrm{pr}_1).(s \circ \mathrm{pr}_2)$.

Not every smooth exact sequence of Lie groups admits local sections as required in (2). Let, for example, K be a closed linear subspace in a convenient vector space G which is not a direct summand, and let H be G/K. Then the tangent mapping at 0 of a local smooth splitting would make K a direct summand.

Theorem. *Let $\{e\} \to K \xrightarrow{i} G \xrightarrow{p} H \to \{e\}$ be a smooth extension of Lie groups. Then G is regular if and only if both K and H are regular.*

Proof. Clearly, the induced sequence of Lie algebras also is exact,

$$0 \to \mathfrak{k} \xrightarrow{i'} \mathfrak{g} \xrightarrow{p'} \mathfrak{h} \to 0,$$

with a bounded linear section $T_e s$ of p'. Thus, $\mathfrak{g}$ is isomorphic to $\mathfrak{k} \times \mathfrak{h}$ as convenient vector space.

Let us suppose that K and H are regular. Given $X \in C^\infty(\mathbb{R}, \mathfrak{g})$, we consider $Y(t) := p'(X(t)) \in \mathfrak{h}$ with evolution curve h satisfying $\frac{\partial}{\partial t} h(t) = T(\mu^{h(t)}).Y(t)$ and $h(0) = e$. By lemma (38.3) it suffices to find smooth local solutions g near 0 of $\frac{\partial}{\partial t} g(t) = T(\mu^{g(t)}).X(t)$ with $g(0) = e$, depending smoothly on X. We look for solutions of the form $g(t) = s(h(t)).i(k(t))$, where k is a local evolution curve in K of a suitable curve $t \mapsto Z(t)$ in $\mathfrak{k}$, i.e., $\frac{\partial}{\partial t} k(t) = T(\mu^{k(t)}).Z(t)$, and $k(0) = e$. For this ansatz we have

$$\begin{aligned}\tfrac{\partial}{\partial t} g(t) &= \tfrac{\partial}{\partial t}\Big(s(h(t)).i(k(t))\Big) = T(\mu_{s(h(t))}).Ti.\tfrac{\partial}{\partial t} k(t) + T(\mu^{i(k(t))}).Ts.\tfrac{\partial}{\partial t} h(t) \\ &= T(\mu_{s(h(t))}).Ti.T(\mu^{k(t)}).Z(t) + T(\mu^{i(k(t))}).Ts.T(\mu^{h(t)}).Y(t),\end{aligned}$$

and we want this to be

$$T(\mu^{g(t)}).X(t) = T(\mu^{s(h(t)).i(k(t))}).X(t) = T(\mu^{i(k(t))}).T(\mu^{s(h(t))}).X(t).$$

Using $i \circ \mu^k = \mu^{i(k)} \circ i$, one quickly sees that

$$i'.Z(t) := \mathrm{Ad}\Big(s(h(t))^{-1}\Big).\Big(X(t) - T(\mu^{s(h(t))^{-1}}).Ts.T(\mu^{h(t)}).Y(t)\Big) \in \ker p'$$

solves the problem, so G is regular.

Let now G be regular. If $Y \in C^\infty(\mathbb{R}, \mathfrak{h})$, then $p \circ \mathrm{Evol}^r_G(s' \circ Y) = \mathrm{Evol}_H(Y)$, by the diagram in (38.4.3). If $U \in C^\infty(\mathbb{R}, \mathfrak{k})$ then $p \circ \mathrm{Evol}_G(i' \circ U) = \mathrm{Evol}_H(0) = e$, so that $\mathrm{Evol}_G(i' \circ U)(t) \in i(K)$ for all t and thus equals $i(\mathrm{Evol}_K(U)(t))$. □

38.7. Subgroups of regular Lie groups. Let G and K be Lie groups, let G be regular, and let $i : K \to G$ be a smooth homomorphism which is initial (27.11) with $T_e i = i' : \mathfrak{k} \to \mathfrak{g}$ injective. We suspect that K is then regular, but we are only able to prove this under the following assumption.

> There are an open neighborhood $U \subset G$ of e and a smooth mapping $p : U \to E$ into a convenient vector space E such that $p^{-1}(0) = K \cap U$ and p is constant on left cosets $Kg \cap U$.

Proof. For $Z \in C^\infty(\mathbb{R}, \mathfrak{k})$ we consider $g(t) = \operatorname{Evol}_G(i' \circ Z)(t) \in G$. Then we have $\frac{\partial}{\partial t}(p(g(t))) = Tp.T(\mu^{g(t)}).i'(Z(t)) = 0$ by the assumption, so $p(g(t))$ is constant $p(e) = 0$, thus $g(t) = i(h(t))$ for a smooth curve h in H, since i is initial. Then $h = \operatorname{Evol}_H(Y)$ since $T_e i$ is injective, and h depends smoothly on Z since i is initial. □

38.8. Abelian and central extensions. From theorem (38.6), it is clear that any smooth extension G of a regular Lie group H with an abelian regular Lie group $(K, +)$ is regular. We shall describe Evol_G in terms of Evol_G, Evol_K, and in terms of the action of H on K and the cocycle $c : H \times H \to K$ if the latter exists.

Let us first recall these notions. If we have a smooth extension with abelian normal subgroup K,

$$\{e\} \to K \xrightarrow{i} G \xrightarrow{p} H \to \{e\},$$

then a unique smooth action $\alpha : H \times K \to K$ by automorphisms is given by $i(\alpha_h(k)) = s(h)i(k)s(h)^{-1}$, where s is any smooth local section of p defined near h. If moreover p admits a global smooth section $s : H \to G$, which we assume without loss of generality to satisfy $s(e) = e$, then we consider the smooth mapping $c : H \times H \to K$ given by $ic(h_1, h_2) := s(h_1).s(h_2).s(h_1.h_2)^{-1}$. Via the diffeomorphism $K \times H \to G$ given by $(k, h) \mapsto i(k).s(h)$ the identity corresponds to $(0, e)$, the multiplication and the inverse in G look as follows:

$$(1) \qquad \begin{aligned} (k_1, h_1).(k_2, h_2) &= (k_1 + \alpha_{h_1} k_2 + c(h_1, h_2), h_1 h_2), \\ (k, h)^{-1} &= (-\alpha_{h^{-1}}(k) - c(h^{-1}, h), h^{-1}). \end{aligned}$$

Associativity and $(0, e)^2 = (0, e)$ correspond to the fact that c satisfies the following *cocycle condition* and normalization

$$(2) \qquad \begin{gathered} \alpha_{h_1}(c(h_2, h_3)) - c(h_1 h_2, h_3) + c(h_1, h_2 h_3) - c(h_1, h_2) = 0, \\ c(e, e) = 0. \end{gathered}$$

These imply that $c(e, h) = 0 = c(h, e)$ and $\alpha_h(c(h^{-1}, h)) = c(h, h^{-1})$. For a central extension the action is trivial, $\alpha_h = \operatorname{Id}_K$ for all $h \in H$.

If conversely H acts smoothly by automorphisms on an abelian Lie group K and if $c : H \times H \to K$ satisfies (2), then (1) describes a smooth Lie group structure on $K \times H$, which is a smooth extension of H over K with a global smooth section.

For later purposes, let us compute

$$(0,h_1).(0,h_2)^{-1} = (-\alpha_{h_1}(c(h_2^{-1},h_2)) + c(h_1,h_2^{-1}), h_1h_2^{-1}),$$

$$T_{(0,h_1)}(\mu^{(0,h_2)^{-1}}).(0,Y_{h_1}) = (-T(\alpha^{c(h_2^{-1},h_2)}).Y_{h_1} + T(c(\quad,h_2^{-1})).Y_{h_1}, T(\mu^{h_2^{-1}}).Y_{h_1}).$$

Let us now assume that K and H are moreover regular Lie groups. We consider a curve $t \mapsto X(t) = (U(t), Y(t))$ in the Lie algebra $\mathfrak{g}$ which as convenient vector space equals $\mathfrak{k} \times \mathfrak{h}$. From the proof of (38.6) we get that

$$\begin{aligned}
g(t) &:= \operatorname{Evol}_G(U,Y)(t) = (0,h(t)).(k(t),e) = (\alpha_{h(t)}(k(t)), h(t)), \text{ where}\\
h(t) &:= \operatorname{Evol}_H(Y)(t) \in H,\\
(Z(t),0) &:= \operatorname{Ad}_G(0,h(t))^{-1}\Big((U(t),Y(t)) - T\mu^{(0,h(t))^{-1}}.(0,\tfrac{\partial}{\partial t}h(t))\Big),\\
Z(t) &= T_0(\alpha_{h(t)^{-1}}).\Big(U(t) + \big(T(\alpha^{c(h(t)^{-1},h(t))}) - T(c(\quad,h(t)^{-1}))\big).\tfrac{\partial}{\partial t}h(t)\Big),\\
k(t) &:= \operatorname{Evol}_K(Z)(t) \in K.
\end{aligned}$$

38.9. Semidirect products. From theorem (38.6) we see immediately that the semidirect product of regular Lie groups is regular. Since we shall need explicit formulas later we specialize the proof of (38.6) to this case.

Let H and K be regular Lie groups with Lie algebras $\mathfrak{h}$ and $\mathfrak{k}$, respectively. Let $\alpha : H \times K \to K$ be smooth such that $\alpha^\vee : H \to \operatorname{Aut}(K)$ is a group homomorphism. Then the semidirect product $K \rtimes H$ is the Lie group $K \times H$ with multiplication $(k,h).(k',h') = (k.\alpha_h(k'), h.h')$ and inverse $(k,h)^{-1} = (\alpha_{h^{-1}}(k)^{-1}, h^{-1})$. We have then $T_{(e,e)}(\mu^{(k',h')}).(U,Y) = (T(\mu^{k'}).U + T(\alpha^{k'}).Y, T(\mu^{h'}).Y)$.

Now we consider a curve $t \mapsto X(t) = (U(t), Y(t))$ in the Lie algebra $\mathfrak{k} \rtimes \mathfrak{h}$. Since $s : h \mapsto (e,h)$ is a smooth homomorphism of Lie groups, from the proof of (38.6) we get that

$$\begin{aligned}
g(t) &:= \operatorname{Evol}_{K\rtimes H}(U,Y)(t) = (e,h(t)).(k(t),e) = (\alpha_{h(t)}(k(t)), h(t)), \text{ where}\\
h(t) &:= \operatorname{Evol}_H(Y)(t) \in H,\\
(Z(t),0) &:= \operatorname{Ad}_{K\rtimes H}(e,h(t)^{-1})(U(t),0) = (T_e(\alpha_{h(t)^{-1}}).U(t),0),\\
k(t) &:= \operatorname{Evol}_K(Z)(t) \in K.
\end{aligned}$$

38.10. Corollary. *Let G be a Lie group. Then via right trivialization $(\kappa^r, \pi_G) : TG \to \mathfrak{g} \times G$ the tangent group TG is isomorphic to the semidirect product $\mathfrak{g} \rtimes G$, where G acts by $\operatorname{Ad} : G \to \operatorname{Aut}(\mathfrak{g})$.*

Therefore, if G is a regular Lie group, then $TG \cong \mathfrak{g} \rtimes G$ also is regular, and $T\operatorname{evol}^r_G$ corresponds to $\operatorname{evol}^r_{TG}$. In particular, for $(Y,X) \in C^\infty(\mathbb{R}, \mathfrak{g} \times \mathfrak{g}) = TC^\infty(\mathbb{R},\mathfrak{g})$, where X is the footpoint, we have

$$\operatorname{evol}^r_{\mathfrak{g}\rtimes G}(Y,X) = \Big(\operatorname{Ad}(\operatorname{evol}^r_G(X)) \int_0^1 \operatorname{Ad}(\operatorname{Evol}^r_G(X)(s)^{-1}).Y(s)\,ds,\ \operatorname{evol}^r_G(X)\Big)$$

$$T_X\operatorname{evol}^r_G.Y = T(\mu_{\operatorname{evol}^r_G(X)}).\int_0^1 \operatorname{Ad}(\operatorname{Evol}^r_G(X)(s)^{-1}).Y(s)\,ds,$$

$$T_X(\operatorname{Evol}^r_G(\quad)(t)).Y = T(\mu_{\operatorname{Evol}^r_G(X)(t)}).\int_0^t \operatorname{Ad}(\operatorname{Evol}^r_G(X)(s)^{-1}).Y(s)\,ds.$$

The expression in (38.2) for the derivative of the exponential mapping is a special case of the expression for $T\mathrm{evol}_G$, for constant curves in $\mathfrak{g}$. Note that in the semidirect product representation $TG \cong \mathfrak{g} \rtimes G$ the footpoint appears in the right factor G, contrary to our usual convention. We followed this also in $T\mathfrak{g} = \mathfrak{g} \rtimes \mathfrak{g}$.

Proof. Via right trivialization the tangent group TG *is the semidirect product* $\mathfrak{g} \rtimes G$, where G acts on the Lie algebra $\mathfrak{g}$ by $\mathrm{Ad} : G \to \mathrm{Aut}(\mathfrak{g})$, because by (36.2) we have for $g, h \in G$ and $X, Y \in \mathfrak{g}$, where $\mu = \mu_G$ is the multiplication on G:

$$\begin{aligned}
T_{(g,h)}\mu.(R_X(g), R_Y(h)) &= T(\mu^h).R_X(g) + T(\mu_g).R_Y(h)\\
&= T(\mu^h).T(\mu^g).X + T(\mu_g).T(\mu^h).Y\\
&= R_X(gh) + R_{\mathrm{Ad}(g)Y}(gh),\\
T_g\nu.R_X(g) &= -T(\mu^{g^{-1}}).T(\mu_{g^{-1}}).T(\mu^g).X\\
&= -R_{\mathrm{Ad}(g^{-1})X}(g^{-1}),
\end{aligned}$$

so that μ_{TG} and ν_{TG} are given by

(1)
$$\begin{aligned}
\mu_{\mathfrak{g}\rtimes G}((X, g), (Y, h)) &= (X + \mathrm{Ad}(g)Y, gh)\\
\nu_{\mathfrak{g}\rtimes G}(X, g) &= (-\mathrm{Ad}(g^{-1})X, g^{-1}).
\end{aligned}$$

Now we shall prove that the following diagram commutes and that the equations of the corollary follow. The lower triangle commutes by definition.

$$\begin{array}{ccc}
TC^\infty(\mathbb{R}, \mathfrak{g}) & \xrightarrow{\ \cong\ } & C^\infty(\mathbb{R}, \mathfrak{g} \rtimes \mathfrak{g})\\
{\scriptstyle T\mathrm{evol}_G}\big\downarrow & \swarrow{\scriptstyle \mathrm{evol}_{TG}} & \big\downarrow{\scriptstyle \mathrm{evol}_{\mathfrak{g}\rtimes G}}\\
TG & \xrightarrow[\ \cong\ ]{} & \mathfrak{g} \rtimes G
\end{array}$$

For that we choose $X, Y \in C^\infty(\mathbb{R}, \mathfrak{g})$. Let us first consider the evolution operator of the tangent group TG in the picture $\mathfrak{g} \rtimes G$. On $(\mathfrak{g}, +)$ the evolution mapping is the definite integral, so going through the prescription (38.9) for $\mathrm{evol}_{\mathfrak{g}\rtimes G}$ we have the following data:

(2)
$$\begin{aligned}
\mathrm{evol}_{\mathfrak{g}\rtimes G}(Y, X) &= (h(1), g(1)), \qquad \text{where}\\
g(t) &:= \mathrm{Evol}_G(X)(t) \in G,\\
Z(t) &:= \mathrm{Ad}(g(t)^{-1}).Y(t) \in \mathfrak{g},\\
h_0(t) &:= \mathrm{Evol}_{(\mathfrak{g},+)}(Z)(t) = \int_0^t \mathrm{Ad}(g(u)^{-1}).Y(u)\,du \in \mathfrak{g},\\
h(t) &:= \mathrm{Ad}(g(t))h_0(t) = \mathrm{Ad}(g(t)) \int_0^t \mathrm{Ad}(g(u)^{-1}).Y(u)\,du \in \mathfrak{g}.
\end{aligned}$$

This shows the first equation in the corollary. The differential equation for the curve $(h(t), g(t))$, which by lemma (38.3) has a unique solution starting at $(0, e)$, looks as follows, using (1):

$$\begin{aligned}\Big((h'(t),h(t)),g'(t)\Big) &= T_{(0,e)}(\mu_{\mathfrak{g}\rtimes G}^{(h(t),g(t))}).\Big((Y(t),0),X(t)\Big)\\ &= \Big(\Big(Y(t)+d\mathrm{Ad}(X(t)).h(t),0+\mathrm{Ad}(e).h(t)\Big),T(\mu_G^{g(t)}).X(t)\Big),\end{aligned}$$

$$\begin{aligned}&h'(t) = Y(t)+\mathrm{ad}(X(t))h(t),\\ &g'(t) = T(\mu_G^{g(t)}).X(t).\end{aligned}\tag{3}$$

For the computation of $T\mathrm{evol}_G$ we let

$$\begin{aligned}g(t,s) := \mathrm{evol}_G\Big(u\mapsto t(X(tu)+sY(tu))\Big) &= \mathrm{Evol}_G(X+sY)(t),\\ \text{satisfying}\quad \delta^r g(\partial_t(t,s)) &= X(t)+sY(t).\end{aligned}$$

Then $T\mathrm{evol}_G(Y,X) = \partial_s|_0 g(1,s)$, and the derivative $\partial_s|_0 g(t,s)$ in TG corresponds to the element

$$(T(\mu^{g(t,0)^{-1}}).\partial_s|_0 g(t,s), g(t,0)) = (\delta^r g(\partial_s(t,0)), g(t,0)) \in \mathfrak{g}\rtimes G$$

via right trivialization. For the right hand side we have $g(t,0)=g(t)$, so it remains to show that $\delta^r g(\partial_s(t,0)) = h(t)$. We will show that $\delta^r g(\partial_s(t,0))$ is the unique solution of the differential equation (3) for $h(t)$. Using the Maurer Cartan equation $d\delta^r g - \frac{1}{2}[\delta^r g,\delta^r g]_\wedge = 0$ from lemma (38.1) we get

$$\begin{aligned}\partial_t\delta^r g(\partial_s) &= \partial_s\delta^r g(\partial_t) + d(\delta^r g)(\partial_t,\partial_s) + \delta^r g([\partial_t,\partial_s])\\ &= \partial_s\delta^r g(\partial_t) + [\delta^r g(\partial_t),\delta^r g(\partial_s)]_{\mathfrak{g}} + 0\\ &= \partial_s(X(t)+sY(t)) + [X(t)+sY(t),\delta^r g(\partial_s)]_{\mathfrak{g}},\end{aligned}$$

so that for $s=0$ we get

$$\begin{aligned}\partial_t\delta^r g(\partial_s(t,0)) &= Y(t) + [X(t),\delta^r g(\partial_s(t,0))]_{\mathfrak{g}}\\ &= Y(t) + \mathrm{ad}(X(t))\delta^r g(\partial_s(t,0)).\end{aligned}$$

Thus, $\delta^r g(\partial_s(t,0))$ is a solution of the inhomogeneous linear ordinary differential equation (3), as required.

It remains to check the last formula. Note that $X \mapsto tX(t\ \ )$ is a bounded linear operator. So we have

$$\begin{aligned}&\mathrm{Evol}_G^r(X)(t) = \mathrm{evol}_G^r(s\mapsto tX(ts)),\\ &T_X(\mathrm{Evol}_G^r(\ \ )(t)).Y = T_{tX(t\ \ )}\mathrm{evol}_G^r.(tY(t\ \ ))\\ &\quad = T(\mu_{\mathrm{evol}_G^r(tX(t\ \ ))}).\int_0^1 \mathrm{Ad}_G\Big(\mathrm{Evol}_G^r(tX(t\ \ ))(s)^{-1}\Big).tY(ts)\,ds\\ &\quad = T(\mu_{\mathrm{Evol}_G^r(X)(t)}).\int_0^1 \mathrm{Ad}_G\Big(\mathrm{evol}_G^r(stX(st\ \ ))^{-1}\Big).tY(ts)\,ds\\ &\quad = T(\mu_{\mathrm{Evol}_G^r(X)(t)}).\int_0^t \mathrm{Ad}_G\Big(\mathrm{Evol}_G^r(X)(s)^{-1}\Big).Y(s)\,ds.\quad\square\end{aligned}$$

38.11. Current groups. We have another stability result: If G is regular and M is a finite dimensional manifold then also the space of all smooth mappings $M \to G$ is a a regular Lie group, denoted by $\mathfrak{C}^\infty(M, G)$, with $\operatorname{evol}_{\mathfrak{C}^\infty(M,G)} = \mathfrak{C}^\infty(M, \operatorname{evol}_G)$, see (42.21) below.

38.12. Theorem. *For a regular Lie group G we have*

$$\begin{aligned}\operatorname{evol}^r(X).\operatorname{evol}^r(Y) &= \operatorname{evol}^r\Big(t \mapsto X(t) + \operatorname{Ad}_G(\operatorname{Evol}^r(X)(t)).Y(t)\Big),\\ \operatorname{evol}^r(X)^{-1} &= \operatorname{evol}^r\Big(t \mapsto -\operatorname{Ad}_G(\operatorname{Evol}^r(X)(t)^{-1}).X(t)\Big),\end{aligned}$$

so that $\operatorname{evol}^r : C^\infty(\mathbb{R}, \mathfrak{g}) \to G$ *is a surjective smooth homomorphism of Lie groups, where on* $C^\infty(\mathbb{R}, \mathfrak{g})$ *we consider the operations*

$$\begin{aligned}(X * Y)(t) &= X(t) + \operatorname{Ad}_G(\operatorname{Evol}^r(X)(t)).Y(t),\\ X^{-1}(t) &= -\operatorname{Ad}_G(\operatorname{Evol}^r(X)(t)^{-1}).X(t).\end{aligned}$$

With this operations and with 0 as unit element $(C^\infty(\mathbb{R}, \mathfrak{g}), *)$ *becomes a regular Lie group. Its Lie algebra is* $C^\infty(\mathbb{R}, \mathfrak{g})$ *with bracket*

$$\begin{aligned}[X, Y]_{C^\infty(\mathbb{R},\mathfrak{g})}(t) &= \Big[\int_0^t X(s)\,ds, Y(t)\Big]_{\mathfrak{g}} + \Big[X(t), \int_0^t Y(s)\,ds\Big]_{\mathfrak{g}}\\ &= \tfrac{\partial}{\partial t}\Big[\int_0^t X(s)\,ds, \int_0^t Y(s)\,ds\Big]_{\mathfrak{g}}.\end{aligned}$$

Its evolution operator is given by

$$\begin{aligned}\operatorname{evol}_{(C^\infty(\mathbb{R},\mathfrak{g}),*)}(X) &:= \operatorname{Ad}_G(\operatorname{evol}_G(Y^s)).\int_0^1 \operatorname{Ad}_G(\operatorname{Evol}_G(Y^s)(v)^{-1}).X(v)(s)\,dv,\\ Y^s(t) &:= \int_0^s X(t)(u)du.\end{aligned}$$

Proof. For $X, Y \in C^\infty(\mathbb{R}, \mathfrak{g})$ we compute

$$\begin{aligned}&\tfrac{\partial}{\partial t}\Big(\operatorname{Evol}^r(X)(t).\operatorname{Evol}^r(Y)(t)\Big) =\\ &= T(\mu^{\operatorname{Evol}^r(Y)(t)}).T(\mu^{\operatorname{Evol}^r(X)(t)}).X(t) + T(\mu_{\operatorname{Evol}^r(X)(t)}).T(\mu^{\operatorname{Evol}^r(Y)(t)}).Y(t)\\ &= T(\mu^{\operatorname{Evol}^r(X)(t).\operatorname{Evol}^r(Y)(t)}).(X(t) + \operatorname{Ad}_G(\operatorname{Evol}^r(X)(t))Y(t)),\end{aligned}$$

which implies also

$$\operatorname{Evol}^r(X).\operatorname{Evol}^r(Y) = \operatorname{Evol}^r(X * Y), \quad \operatorname{Evol}^r(X)^{-1} = \operatorname{Evol}^r(X^{-1}).$$

Thus, $\operatorname{Evol}^r : C^\infty(\mathbb{R}, \mathfrak{g}) \to C^\infty(\mathbb{R}, G)$ is a group isomorphism onto the subgroup $\{c \in C^\infty(\mathbb{R}, G) : c(0) = e\}$ of $C^\infty(\mathbb{R}, G)$ with the pointwise product, which, however, is only a Frölicher space, see (23.1) Nevertheless, it follows that the product on $C^\infty(\mathbb{R},\mathfrak{g})$ is associative. It is clear that these operations are smooth, hence the convenient vector space $C^\infty(\mathbb{R},\mathfrak{g})$ becomes a Lie group and $C^\infty(\mathbb{R},G)$ becomes a manifold.

Now we aim for the Lie bracket. We have

$$\begin{aligned}(X*Y*X^{-1})(t) &= \Big(\big(X+\operatorname{Ad}(\operatorname{Evol}^r(X)).Y\big)*\big(-\operatorname{Ad}(\operatorname{Evol}^r(X)^{-1}).X\big)\Big)(t)\\ &= X(t)+\operatorname{Ad}(\operatorname{Evol}^r(X)(t)).Y(t)-\\ &\quad -\operatorname{Ad}\big(\operatorname{Evol}^r(X*Y)(t)\big).\operatorname{Ad}\big(\operatorname{Evol}^r(X)(t)^{-1}\big).X(t)\\ &= X(t)+\operatorname{Ad}\big(\operatorname{Evol}^r(X)(t)\big).Y(t)-\\ &\quad -\operatorname{Ad}\big(\operatorname{Evol}^r(X)(t)\big).\operatorname{Ad}\big(\operatorname{Evol}^r(Y)(t)\big).\operatorname{Ad}\big(\operatorname{Evol}^r(X)(t)^{-1}\big).X(t).\end{aligned}$$

We shall need

$$\begin{aligned}T_0\Big(\operatorname{Ad}_G(\operatorname{Evol}^r(\quad)(t))\Big).Y &= T_e\operatorname{Ad}_G.T_0(\operatorname{Evol}^r(\quad)(t)).Y\\ &= \operatorname{ad}_{\mathfrak{g}}\Big(\int_0^t Y(s)\,ds\Big),\qquad \text{by (38.10)}.\end{aligned}$$

Using this, we can differentiate the conjugation,

$$\begin{aligned}(\operatorname{Ad}_{C^\infty(\mathbb{R},\mathfrak{g})}(X).Y)(t) &= (T_0(X*(\quad)*X^{-1}).Y)(t)\\ &= 0+\operatorname{Ad}(\operatorname{Evol}^r(X)(t)).Y(t)-\\ &\quad -\operatorname{Ad}(\operatorname{Evol}^r(X)(t)).\Big(T_0(\operatorname{Ad}(\operatorname{Evol}^r(\quad)(t))).Y\Big).\operatorname{Ad}(\operatorname{Evol}^r(X)(t)^{-1}).X(t)\\ &= \operatorname{Ad}(\operatorname{Evol}^r(X)(t)).Y(t)-\\ &\quad -\operatorname{Ad}(\operatorname{Evol}^r(X)(t)).\operatorname{ad}_{\mathfrak{g}}\Big(\int_0^t Y(s)\,ds\Big).\operatorname{Ad}(\operatorname{Evol}^r(X)(t)^{-1}).X(t)\\ &= \operatorname{Ad}(\operatorname{Evol}^r(X)(t)).Y(t)-\operatorname{ad}_{\mathfrak{g}}.\Big(\operatorname{Ad}(\operatorname{Evol}^r(X)(t)).\int_0^t Y(s)\,ds\Big).X(t).\end{aligned}$$

Now we can compute the Lie bracket

$$\begin{aligned}[X,Y]_{C^\infty(\mathbb{R},\mathfrak{g})}(t) &= \Big(T_0(\operatorname{Ad}_{C^\infty(\mathbb{R},\mathfrak{g})}(\quad).Y).X\Big)(t)\\ &= \Big(T_0\Big(\operatorname{Ad}(\operatorname{Evol}^r(\quad)(t))\Big).X\Big).Y(t)-0-\Big[\operatorname{Ad}(\operatorname{Evol}^r(0)(t)).\int_0^t Y(s)\,ds, X(t)\Big]_{\mathfrak{g}}\\ &= \Big[\int_0^t X(s)\,ds, Y(t)\Big]_{\mathfrak{g}}-\Big[\int_0^t Y(s)\,ds, X(t)\Big]_{\mathfrak{g}}\\ &= \Big[\int_0^t X(s)\,ds, Y(t)\Big]_{\mathfrak{g}}+\Big[X(t),\int_0^t Y(s)\,ds\Big]_{\mathfrak{g}}\\ &= \tfrac{\partial}{\partial t}\Big[\int_0^t X(s)\,ds,\int_0^t Y(s)\,ds\Big]_{\mathfrak{g}}.\end{aligned}$$

We show that the Lie group $(C^\infty(\mathbb{R},\mathfrak{g}),*)$ is regular. Let $X^\vee \in C^\infty(\mathbb{R}, C^\infty(\mathbb{R},\mathfrak{g}))$ correspond to $X \in C^\infty(\mathbb{R}^2,\mathfrak{g})$. We look for $g \in C^\infty(\mathbb{R}^2,\mathfrak{g})$ which satisfies the equation (38.4.1):

$$\begin{aligned}
\mu^{g(t,\quad)}(Y)(s) &= (Y * g(t,\quad))(s) = Y(s) + \operatorname{Ad}_G(\operatorname{Evol}_G(Y)(s)).g(t,s)\\
\tfrac{\partial}{\partial t} g(t,s) &= \Big(T_0(\mu^{g(t,\quad)}).X(t,\quad)\Big)(s)\\
&= X(t,s) + \Big(T_0\Big(\operatorname{Ad}_G(\operatorname{Evol}_G(\quad)(s))\Big).X(t,\quad)\Big).g(t,s)\\
&= X(t,s) + \operatorname{ad}_{\mathfrak{g}}\Big(\int_0^s X(t,u)du\Big).g(t,s)\\
&= X(t,s) + \Big[\int_0^s X(t,u)du,\ g(t,s)\Big]_{\mathfrak{g}}.
\end{aligned}$$

This is the differential equation (38.10.3) for $h(t)$, depending smoothly on a further parameter s, which has the following unique solution given by (38.10.2)

$$\begin{aligned}
g(t,s) &:= \operatorname{Ad}_G(\operatorname{Evol}_G(Y^s)(t)).\int_0^t \operatorname{Ad}_G(\operatorname{Evol}_G(Y^s)(v)^{-1}).X(v,s)\,dv\\
Y^s(t) &:= \int_0^s X(t,u)du.
\end{aligned}$$

Since this solution is obviously smooth in X, the Lie group $C^\infty(\mathbb{R},\mathfrak{g})$ is regular. For convenience (yours, not ours) we show now (once more) that this, in fact, is a solution. Putting $Y^s(t) := \int_0^s X(t,u)du$ we have by (36.10.3)

$$\begin{aligned}
&\tfrac{\partial}{\partial t} g(t,s) =\\
&\quad= d\operatorname{Ad}(\tfrac{\partial}{\partial t}\operatorname{Evol}(Y^s)(t)).\int_0^t \operatorname{Ad}(\operatorname{Evol}(Y^s)(v)^{-1}).X(v,s)\,dv\\
&\qquad+ \operatorname{Ad}(\operatorname{Evol}(Y^s)(t)).\operatorname{Ad}(\operatorname{Evol}(Y^s)(t)^{-1}).X(t,s)\\
&\quad= ((\operatorname{ad}\circ\kappa^r).\operatorname{Ad})\Big(T(\mu^{\operatorname{Evol}(Y^s)(t)}).Y^s(t)\Big).\int_0^t \operatorname{Ad}(\operatorname{Evol}(Y^s)(v)^{-1}).X(v,s)\,dv\\
&\qquad+ X(t,s)\\
&\quad= \operatorname{ad}(Y^s(t)).\operatorname{Ad}(\operatorname{Evol}(Y^s)(t)).\int_0^t \operatorname{Ad}(\operatorname{Evol}(Y^s)(v)^{-1}).X(v,s)\,dv + X(t,s)\\
&\quad= \Big[\int_0^s X(t,u)du,\ g(t,s)\Big]_{\mathfrak{g}} + X(t,s). \quad\square
\end{aligned}$$

38.13. Corollary. *Let G be a regular Lie group. Then as Frölicher spaces and groups we have the following isomorphisms*

$$(C^\infty(\mathbb{R},\mathfrak{g}),*) \rtimes G \cong \{f \in C^\infty(\mathbb{R},G) : f(0)=e\} \rtimes G \cong C^\infty(\mathbb{R},G),$$

where $g \in G$ acts on f by $(\alpha_g(f))(t) = g.f(t).g^{-1}$, and on $X \in C^\infty(\mathbb{R},\mathfrak{g})$ by $\alpha_g(X)(t) = \operatorname{Ad}_G(g)(X(t))$. The leftmost space is a smooth manifold, thus all spaces are regular Lie groups.

For the Lie algebras we have an isomorphism

$$\begin{aligned} C^\infty(\mathbb{R},\mathfrak{g}) \rtimes \mathfrak{g} &\cong C^\infty(\mathbb{R},\mathfrak{g}),\\ (X,\eta) &\mapsto \Big(t\mapsto \eta + \int_0^t X(s)ds\Big)\\ (Y', Y(0)) &\leftarrow Y, \end{aligned}$$

where on the left hand side the Lie bracket is given by

$$\begin{aligned} &[(X_1,\eta_1),(X_2,\eta_2)] =\\ &= \Big(t\mapsto [\textstyle\int_0^t X_1(s)\,ds, X_2(t)]_{\mathfrak{g}} + [X_1(t), \int_0^t X_2(s)\,ds]_{\mathfrak{g}} + [\eta_1, X_2(t)]_{\mathfrak{g}} - [\eta_2, X_1(t)]_{\mathfrak{g}},\\ &\qquad\qquad [\eta_1,\eta_2]_{\mathfrak{g}}\Big), \end{aligned}$$

and where on the right hand side the bracket is given by

$$[X,Y](t) = [X(t),Y(t)]_{\mathfrak{g}}.$$

On the right hand sides the evolution operator is

$$\operatorname{Evol}^r_{C^\infty(\mathbb{R},G)} = C^\infty(\mathbb{R}, \operatorname{Evol}^r_G).$$

38.14. Remarks. Let G be a connected regular Lie group. The smooth homomorphism $\operatorname{evol}^r_G : C^\infty(\mathbb{R},\mathfrak{g}) \to G$ admits local smooth sections. Namely, using a smooth chart near e of G we can choose a smooth curve $c_g : \mathbb{R}\to G$ with $c_g(0) = e$ and $c_g(1) = g$, depending smoothly on g for g near e. Then $s(g) := \delta^r c_g$ is a local smooth section. We have an extension of groups

$$0 \to K \to C^\infty(\mathbb{R},\mathfrak{g}) \xrightarrow{\operatorname{evol}^r_G} G \to \{e\}$$

where $K = \ker(\operatorname{evol}^r_G)$ is isomorphic to the smooth group $\{f\in C^\infty(\mathbb{R},G) : f(0) = e, f(1) = e\}$ via the mapping Evol^r_G. We do not know whether K is a submanifold.

Next we consider the smooth group $C^\infty((S^1,1),(G,e))$ of all smooth mappings $f : S^1 \to G$ with $f(1) = e$. With pointwise multiplication this is a splitting closed normal subgroup of the regular Lie group $C^\infty(S^1,G)$ with the manifold structure given in (42.21). Moreover, $C^\infty(S^1,G)$ is the semidirect product $C^\infty((S^1,1),(G,e)) \rtimes G$, where G acts by conjugation on $C^\infty((S^1,1),(G,e))$. So by theorem (38.6) the subgroup $C^\infty((S^1,1),(G,e))$ is also regular.

The right logarithmic derivative for smooth loops $\delta^r : C^\infty(S^1,G) \to C^\infty(S^1,\mathfrak{g})$ restricts to a diffeomorphism $C^\infty((S^1,1),(G,e)) \to \ker(\operatorname{evol}_G) \subset C^\infty(S^1,\mathfrak{g})$, thus the kernel $\ker(\operatorname{evol}_G : C^\infty(S^1,\mathfrak{g}) \to G)$ is a regular Lie group which is isomorphic to $C^\infty((S^1,1),(G,e))$. It is also a subgroup (via pullback by the covering mapping $e^{2\pi i t} : \mathbb{R}\to S^1$) of the regular Lie group $(C^\infty(\mathbb{R},\mathfrak{g}),*)$. Note that $C^\infty(S^1,\mathfrak{g})$ is not a subgroup, since it is not closed under the product $*$ if G is not abelian.

39. Bundles with Regular Structure Groups

39.1. Theorem. *Let $(p : P \to M, G)$ be a smooth (locally trivial) principal bundle with a regular Lie group as structure group. Let $\omega \in \Omega^1(P, \mathfrak{g})$ be a principal connection form.*

Then the parallel transport for the principal connection exists, is globally defined, and is G-equivariant. In detail: For each smooth curve $c : \mathbb{R} \to M$ there is a unique smooth mapping $\operatorname{Pt}_c : \mathbb{R} \times P_{c(0)} \to P$ such that the following holds:

(1) $\operatorname{Pt}(c, t, u) \in P_{c(t)}$, $\operatorname{Pt}(c, 0) = \operatorname{Id}_{P_{c(0)}}$, *and* $\omega(\frac{d}{dt} \operatorname{Pt}(c, t, u)) = 0$.

It has the following further properties:

(2) $\operatorname{Pt}(c, t) : P_{c(0)} \to P_{c(t)}$ *is G-equivariant, i.e.* $\operatorname{Pt}(c, t, u.g) = \operatorname{Pt}(c, t, u).g$ *holds for all $g \in G$ and $u \in P$. Moreover, we have* $\operatorname{Pt}(c, t)^*(\zeta_X | P_{c(t)}) = \zeta_X | P_{c(0)}$ *for all $X \in \mathfrak{g}$.*

(3) *For any smooth function $f : \mathbb{R} \to \mathbb{R}$ we have*
$\operatorname{Pt}(c, f(t), u) = \operatorname{Pt}(c \circ f, t, \operatorname{Pt}(c, f(0), u))$.

(4) *The parallel transport is smooth as a mapping*

$$\operatorname{Pt} : C^\infty(\mathbb{R}, M) \times_{(\operatorname{ev}_0, M, p \circ \operatorname{pr}_2)} (\mathbb{R} \times P) \to P,$$

where $C^\infty(\mathbb{R}, M)$ is considered as a smooth space, see (23.1). If M is a smooth manifold with a local addition (see (42.4) below), then this holds for $C^\infty(\mathbb{R}, M)$ replaced by the smooth manifold $\mathfrak{C}^\infty(\mathbb{R}, M)$.

Proof. For a principal bundle chart $(U_\alpha, \varphi_\alpha)$ we have the data from (37.22)

$$\begin{aligned} s_\alpha(x) &:= \varphi_\alpha^{-1}(x, e), \\ \omega_\alpha &:= s_\alpha^* \omega, \\ \omega \circ T(\varphi_\alpha^{-1}) &= (\varphi_\alpha^{-1})^* \omega \in \Omega^1(U_\alpha \times G, \mathfrak{g}), \\ (\varphi_\alpha^{-1})^* \omega(\xi_x, T\mu_g.X) &= (\varphi_\alpha^{-1})^* \omega(\xi_x, 0_g) + X = \operatorname{Ad}(g^{-1}) \omega_\alpha(\xi_x) + X. \end{aligned}$$

For a smooth curve $c : \mathbb{R} \to M$ the horizontal lift $\operatorname{Pt}(c, \quad, u)$ through $u \in P_{c(0)}$ is given among all smooth lifts of c by the ordinary differential equation $\omega(\frac{d}{dt} \operatorname{Pt}(c, t, u)) = 0$ with initial condition $\operatorname{Pt}(c, 0, u) = u$. Locally, we have

$$\varphi_\alpha(\operatorname{Pt}(c, t, u)) = (c(t), \gamma(t)),$$

so that

$$\begin{aligned} 0 &= \operatorname{Ad}(\gamma(t)) \omega(\tfrac{d}{dt} \operatorname{Pt}(c, t, u)) = \operatorname{Ad}(\gamma(t)) (\omega \circ T(\varphi_\alpha^{-1}))(c'(t), \gamma'(t)) \\ &= \operatorname{Ad}(\gamma(t)) ((\varphi_\alpha^{-1})^* \omega)(c'(t), \gamma'(t)) = \omega_\alpha(c'(t)) + T(\mu^{\gamma(t)^{-1}}) \gamma'(t), \end{aligned}$$

i.e., $\gamma'(t) = -T(\mu^{\gamma(t)}).\omega_\alpha(c'(t))$. Thus, $\gamma(t)$ is given by

$$\gamma(t) = \operatorname{Evol}_G(-\omega_\alpha(c'))(t).\gamma(0) = \operatorname{evol}_G(s \mapsto -t\omega_\alpha(c'(ts))).\gamma(0).$$

By lemma (38.3), we may glue the local solutions over different bundle charts U_α, so Pt exists globally.

Properties (1) and (3) are now clear, and (2) can be checked as follows: The condition $\omega(\frac{d}{dt}\operatorname{Pt}(c,t,u).g) = \operatorname{Ad}(g^{-1})\omega(\frac{d}{dt}\operatorname{Pt}(c,t,u)) = 0$ implies $\operatorname{Pt}(c,t,u).g = \operatorname{Pt}(c,t,u.g)$. For the second assertion we compute for $u \in P_{c(0)}$:

$$\begin{aligned}\operatorname{Pt}(c,t)^*(\zeta_X|P_{c(t)})(u) &= T\operatorname{Pt}(c,t)^{-1}\zeta_X(\operatorname{Pt}(c,t,u))\\ &= T\operatorname{Pt}(c,t)^{-1}\tfrac{d}{ds}|_0\operatorname{Pt}(c,t,u).\exp(sX)\\ &= T\operatorname{Pt}(c,t)^{-1}\tfrac{d}{ds}|_0\operatorname{Pt}(c,t,u.\exp(sX))\\ &= \tfrac{d}{ds}|_0\operatorname{Pt}(c,t)^{-1}\operatorname{Pt}(c,t,u.\exp(sX))\\ &= \tfrac{d}{ds}|_0 u.\exp(sX) = \zeta_X(u).\end{aligned}$$

Proof of (4) It suffices to check that Pt respects smooth curves. So let $(f,g) : \mathbb{R} \to C^\infty(\mathbb{R},M) \times_M P \subset C^\infty(\mathbb{R},M) \times P$ be a smooth curve. By cartesian closedness (23.2.3), the smooth curve $f : \mathbb{R} \to C^\infty(\mathbb{R},M)$ corresponds to a smooth mapping $f^\wedge \in C^\infty(\mathbb{R}^2,M)$. For a principal bundle chart $(U_\alpha,\varphi_\alpha)$ as above we have $\varphi_\alpha(\operatorname{Pt}(f(s),t,g(s))) = (f(s)(t),\gamma(s,t))$, where γ is the evolution curve

$$\gamma(s,t) = \operatorname{Evol}_G\Big(-\omega_\alpha(\tfrac{\partial}{\partial t}f^\wedge(s,\quad))\Big)(t).\varphi_\alpha(g(s)),$$

which is clearly smooth in (s,t).

If M admits a local addition then $C^\infty(\mathbb{R},M)$ also carries the structure of a smooth manifold by (42.4), which is denoted by $\mathfrak{C}^\infty(\mathbb{R},M)$ there. Since the identity is smooth $\mathfrak{C}^\infty(\mathbb{R},M) \to C^\infty(\mathbb{R},M)$ by lemma (42.5), the result follows. □

39.2. Theorem. *Let $(p : P \to M, G)$ be a smooth principal bundle with a regular Lie group as structure group. Let $\omega \in \Omega^1(P,\mathfrak{g})$ be a principal connection form. If the connection is flat, then the horizontal subbundle $H^\omega(P) = \ker(\omega) \subset TP$ is integrable and defines a foliation in the sense of* (27.16).

If M is connected then each leaf of this horizontal foliation is a covering of M. All leaves are isomorphic via right translations. The principal bundle P is associated to the universal covering of M, which is viewed as principal fiber bundle with structure group the (discrete) fundamental group $\pi_1(M)$.

Proof. Let $(U_\alpha, u_\alpha : U_\alpha \to u_\alpha(U_\alpha) \subset E_\alpha)$ be a smooth chart of the manifold M and let $x_\alpha \in U_\alpha$ be such that $u_\alpha(x_\alpha) = 0$ and the c^∞-open subset $u_\alpha(U_\alpha)$ is star-shaped in E_α. Let us also suppose that we have a principal fiber bundle chart $(U_\alpha, \varphi_\alpha : P|U_\alpha \to U_\alpha \times G)$. We may cover M by such U_α.

We shall now construct for each $w_\alpha \in P_{x_\alpha}$ a smooth section $\psi_\alpha : U_\alpha \to P$ whose image is an integral submanifold for the horizontal subbundle $\ker(\omega)$. Namely, for $x \in U_\alpha$ let $c_x(t) := u_\alpha^{-1}(tu_\alpha(x))$ for $t \in [0,1]$. Then we put

$$\psi_\alpha(x) := \operatorname{Pt}(c_x, 1, w_\alpha).$$

We have to show that the image of $T\psi_\alpha$ is contained in the horizontal bundle $\ker(\omega)$. Then we get $T_x\psi_\alpha = Tp|H^\omega(p)^{-1}_{\psi_\alpha(x)}$. This is a consequence of the following notationally more suitable claim.

Let $h : \mathbb{R}^2 \to U_\alpha$ be smooth with $h(0,s) = x_\alpha$ for all s.

Claim: $\frac{\partial}{\partial s}\operatorname{Pt}(h(\quad, s), 1, w_\alpha)$ is horizontal.

Let $\varphi_\alpha(w_\alpha) = (x_\alpha, g_\alpha) \in U_\alpha \times G$. Then from the proof of theorem (39.1) we know that

$$\begin{aligned}
\varphi_\alpha \operatorname{Pt}(h(\quad, s), 1, w_\alpha) &= (h(1,s), \gamma(1,s)), \qquad \text{where}\\
\gamma(t,s) &= \tilde\gamma(t,s).g_\alpha\\
\tilde\gamma(t,s) &= \operatorname{evol}_G\Big(u \mapsto -t\omega_\alpha(\tfrac{\partial}{\partial t}h(tu,s))\Big)\\
&= \operatorname{Evol}_G\Big(-(h^*\omega_\alpha)(\partial_t(\quad, s))\Big)(t),\\
\omega_\alpha = s_\alpha^*\omega, &\qquad s_\alpha(x) = \varphi_\alpha^{-1}(x,e).
\end{aligned}$$

Since the curvature $\Omega = d\omega + \frac{1}{2}[\omega,\omega]_\wedge = 0$ we have

$$\begin{aligned}
\partial_s(h^*\omega_\alpha)(\partial_t) &= \partial_t(h^*\omega_\alpha)(\partial_s) - d(h^*\omega_\alpha)(\partial_t,\partial_s) - (h^*\omega_\alpha)([\partial_t,\partial_s])\\
&= \partial_t(h^*\omega_\alpha)(\partial_s) + [(h^*\omega_\alpha)(\partial_t), (h^*\omega_\alpha)(\partial_s)]_{\mathfrak g} - 0.
\end{aligned}$$

Using this and the expression for $T\operatorname{evol}_G$ from (38.10), we have

$$\begin{aligned}
\tfrac{\partial}{\partial s}\tilde\gamma(1,s) &= T_{-(h^*\omega_\alpha)(\partial_t)(\quad,s)}\operatorname{evol}_G.\Big(-\partial_s(h^*\omega_\alpha)(\partial_t)(\quad,s)\Big)\\
&= -T(\mu_{\tilde\gamma(1,s)}).\int_0^1 \operatorname{Ad}(\tilde\gamma(t,s)^{-1})\partial_s(h^*\omega_\alpha)(\partial_t)\,dt\\
&= -T(\mu_{\tilde\gamma(1,s)}).\Big(\int_0^1 \operatorname{Ad}(\tilde\gamma(t,s)^{-1})\partial_t(h^*\omega_\alpha)(\partial_s)\,dt +\\
&\qquad + \int_0^1 \operatorname{Ad}(\tilde\gamma(t,s)^{-1}).\operatorname{ad}((h^*\omega_\alpha)(\partial_t)).(h^*\omega_\alpha)(\partial_s)\,dt\Big).
\end{aligned}$$

Next we integrate by parts, use (36.10.3), and use $\kappa^l(\partial_t\tilde\gamma(t,s)^{-1}) = (h^*\omega_\alpha)(\partial_t)(t,s)$ from (38.3):

$$\begin{aligned}
&\int_0^1 \operatorname{Ad}(\tilde\gamma(t,s)^{-1})\partial_t(h^*\omega_\alpha)(\partial_s)\,dt =\\
&= -\int_0^1 \Big(\partial_t\operatorname{Ad}(\tilde\gamma(t,s)^{-1})\Big)(h^*\omega_\alpha)(\partial_s)\,dt + \operatorname{Ad}(\tilde\gamma(t,s)^{-1})(h^*\omega_\alpha)(\partial_s)\Big|_{t=0}^{t=1}\\
&= -\int_0^1 \operatorname{Ad}(\tilde\gamma(t,s)^{-1}).\operatorname{ad}\Big(\kappa^l\partial_t(\tilde\gamma(t,s)^{-1})\Big).(h^*\omega_\alpha)(\partial_s)\,dt\\
&\quad + \operatorname{Ad}(\tilde\gamma(1,s)^{-1})(h^*\omega_\alpha)(\partial_s)(1,s) - 0\\
&= -\int_0^1 \operatorname{Ad}(\tilde\gamma(t,s)^{-1}).\operatorname{ad}\Big((h^*\omega_\alpha)(\partial_t)\Big).(h^*\omega_\alpha)(\partial_s)\,dt\\
&\quad + \operatorname{Ad}(\tilde\gamma(1,s)^{-1})(h^*\omega_\alpha)(\partial_s)(1,s),
\end{aligned}$$

so that finally

$$\begin{aligned}
\tfrac{\partial}{\partial s}\tilde\gamma(1,s) &= -T(\mu_{\tilde\gamma(1,s)}).\operatorname{Ad}(\tilde\gamma(1,s)^{-1})(h^*\omega_\alpha)(\partial_s)(1,s)\\
&= -T(\mu^{\tilde\gamma(1,s)}).(h^*\omega_\alpha)(\partial_s)(1,s),\\
\tfrac{\partial}{\partial s}\gamma(1,s) &= T(\mu^{g_\alpha}).\tfrac{\partial}{\partial s}\tilde\gamma(1,s)\\
&= -T(\mu_{\gamma(1,s)}).\operatorname{Ad}(\gamma(1,s)^{-1})(h^*\omega_\alpha)(\partial_s)(1,s)\\
\omega\Big(\tfrac{\partial}{\partial s}\operatorname{Pt}(h(\ \ ,s),1,w_\alpha)\Big) &= ((\varphi_\alpha^{-1})^*\omega)\Big(\tfrac{\partial}{\partial s}h(1,s),\tfrac{\partial}{\partial s}\gamma(1,s)\Big)\\
&= \operatorname{Ad}(\gamma(1,s)^{-1})\omega_\alpha(\tfrac{\partial}{\partial s}h(1,s)) - \operatorname{Ad}(\gamma(1,s)^{-1})(h^*\omega_\alpha)(\partial_s)(1,s) = 0,
\end{aligned}$$

where at the end we used (37.22.4). Thus, the claim follows.

By the claim and by uniqueness of the parallel transport (39.1.1) for any smooth curve c in U_α the horizontal curve $\psi_\alpha(c(t))$ coincides with $\operatorname{Pt}(c,t,\psi_\alpha(c(0)))$.

To finish the proof, we may now glue overlapping right translations of $\psi_\alpha(U_\alpha)$ to maximal integral manifolds of the horizontal subbundle. As subset such an integral manifold consists of all endpoints of parallel transports of a fixed point. These are diffeomorphic covering spaces of M. Let us fix base points $x_0 \in M$ and $u_0 \in P_{x_0}$. The parallel transport $\operatorname{Pt}(c,1,u_0)$ depends only on the homotopy class relative to the ends of the curve c, by the claim above, so that a group homomorphism $\rho : \pi_1(M) \to G$ is given by $\operatorname{Pt}(\gamma,1,u_0) = u_0.\rho([\gamma])$. Now let $\tilde M \to M$ be the universal cover of M, a principal bundle with discrete structure group $\pi_1(M)$, viewed as the space of homotopy classes relative to the ends of smooth curves starting from x_0. Then the mapping

$$\begin{gathered}
\tilde M \times G \to P,\\
([c],g) \mapsto \operatorname{Pt}(c,1,u_0).g
\end{gathered}$$

factors to a smooth mapping from the associated bundle $\tilde M[G] = \tilde M \times_{\pi_1(M)} G$ to P which is a diffeomorphism, since we can find local smooth sections $P \to \tilde M \times G$ in the following way: For $u \in P$ choose a smooth curve c_u from x_0 to $p(u)$, and consider $([c_u],\tau(\operatorname{Pt}(c_u,1,u_0),u)) \in \tilde M \times G$. □

It is not clear, however, whether the integral submanifolds of the theorem are initial submanifolds of P, or whether they intersect each fiber in a totally disconnected subset, since M might have uncountable fundamental group.

39.3. Holonomy groups. Let $(p : P \to M, G)$ be a principal fiber bundle with regular structure group G so that the parallel transport exists along all curves by theorem (39.1). Let $\Phi = \zeta \circ \omega$ be a principal connection. We assume that M is connected, and we fix $x_0 \in M$.

Now let us fix $u_0 \in P_{x_0}$. Consider the subgroup $\operatorname{Hol}(\omega,u_0)$ of the structure group G which consists of all elements $\tau(u_0,\operatorname{Pt}(c,t,u_0)) \in G$ for c any piecewise smooth closed loop through x_0. Reparameterizing c by a function which is flat at each corner of c we may assume that any such c is smooth. We call $\operatorname{Hol}(\omega,u_0)$ the *holonomy group* of the connection. If we consider only those curves c which are null-homotopic, we obtain the *restricted holonomy group* $\operatorname{Hol}_0(\omega, u_0)$, a normal subgroup in $\operatorname{Hol}(\omega, u_0)$.

Theorem. (1) *We have* $\operatorname{Hol}(\omega, u_0.g) = \operatorname{conj}(g^{-1})\operatorname{Hol}(\omega, u_0)$ *and* $\operatorname{Hol}_0(\omega, u_0.g) = \operatorname{conj}(g^{-1})\operatorname{Hol}_0(\omega, u_0)$.

(2) *For every curve c in M with $c(0) = x_0$ we have* $\operatorname{Hol}(\omega, \operatorname{Pt}(c, t, u_0)) = \operatorname{Hol}(\omega, u_0)$ *and* $\operatorname{Hol}_0(\omega, \operatorname{Pt}(c, t, u_0)) = \operatorname{Hol}_0(\omega, u_0)$.

Proof. (1) This follows from the properties of the mapping τ from (37.8) and from the G-equivariance of the parallel transport:

$$\tau(u_0.g, \operatorname{Pt}(c, 1, u_0.g)) = \tau(u_0.g, \operatorname{Pt}(c, 1, u_0).g) = g^{-1}.\tau(u_0, \operatorname{Pt}(c, 1, u_0)).g.$$

(2) By reparameterizing the curve c we may assume that $t = 1$, and we put $\operatorname{Pt}(c, 1, u_0) =: u_1$. Then by definition for an element $g \in G$ we have $g \in \operatorname{Hol}(\omega, u_1)$ if and only if $g = \tau(u_1, \operatorname{Pt}(e, 1, u_1))$ for some closed smooth loop e through $x_1 := c(1) = p(u_1)$, that is,

$$\begin{aligned}\operatorname{Pt}(c,1)(r^g(u_0)) = r^g(\operatorname{Pt}(c,1)(u_0)) &= u_1 g = \operatorname{Pt}(e,1)(\operatorname{Pt}(c,1)(u_0))\\ u_0 g = \operatorname{Pt}(c,1)^{-1}\operatorname{Pt}(e,1)\operatorname{Pt}(c,1)(u_0) &= \operatorname{Pt}(c.e.c^{-1}, 3)(u_0),\end{aligned}$$

where $c.e.c^{-1}$ is the curve traveling along $c(t)$ for $0 \le t \le 1$, along $e(t-1)$ for $1 \le t \le 2$, and along $c(3-t)$ for $2 \le t \le 3$. This is equivalent to $g \in \operatorname{Hol}(\omega, u_0)$. Furthermore, e is null-homotopic if and only if $c.e.c^{-1}$ is null-homotopic, so we also have $\operatorname{Hol}_0(\omega, u_1) = \operatorname{Hol}_0(\omega, u_0)$. □

40. Rudiments of Lie Theory for Regular Lie Groups

40.1. From Lie algebras to Lie groups. It is not true in general that every convenient Lie algebra is the Lie algebra of a convenient Lie group. This is also wrong for Banach Lie algebras and Banach Lie groups; one of the first examples is from [Van Est, Korthagen, 1964], see also [de la Harpe, 1972].

To Lie subalgebras in the Lie algebra of a Lie group, in general, do not correspond Lie subgroups. We shall give easy examples in (43.6).

In principle, one should be able to tell whether a given convenient Lie algebra is the Lie algebra of a regular Lie group, but we have no idea how to do that.

40.2. The Cartan developing. Let G be a connected Lie group with Lie algebra $\mathfrak{g}$. For a smooth mapping $f : M \to G$ we have considered in (38.1) the right logarithmic derivative $\delta^r f \in \Omega^1(M, \mathfrak{g})$ which is given by $\delta^r f_x := T(\mu^{f(x)^{-1}}) \circ T_x f : T_x M \to T_{f(x)}G \to \mathfrak{g}$ and which satisfies the left (from the left action) Maurer-Cartan equation

$$d\delta^r f - \frac{1}{2}[\delta^r f, \delta^r f]^{\mathfrak{g}}_{\wedge} = 0.$$

Similarly, the left logarithmic derivative $\delta^l f \in \Omega^1(M, \mathfrak{g})$ of $f \in C^\infty(M, G)$ is given by $\delta^l f_x := T(\mu_{f(x)^{-1}}) \circ T_x f : T_x M \to T_{f(x)} G \to \mathfrak{g}$ and satisfies the right Maurer Cartan equation

$$d\delta^l f + \frac{1}{2}[\delta^l f, \delta^l f]^{\mathfrak{g}}_\wedge = 0.$$

For regular Lie groups we have the following converse, which for finite dimensional Lie groups can be found in [Onishchik, 1961, 1964, 1967], or in [Griffith, 1974] (proved with moving frames); see also [Alekseevsky, Michor, 1995b, 5.2].

Theorem. *Let G be a connected regular Lie group with Lie algebra $\mathfrak{g}$.*

If a 1-form $\varphi \in \Omega^1(M, \mathfrak{g})$ satisfies $d\varphi + \frac{1}{2}[\varphi, \varphi]^{\mathfrak{g}}_\wedge = 0$ then for each simply connected subset $U \subset M$ there exists a smooth mapping $f : U \to G$ with $\delta^l f = \varphi|U$, and f is uniquely determined up to a left translation in G.

If a 1-form $\psi \in \Omega^1(M, \mathfrak{g})$ satisfies $d\psi - \frac{1}{2}[\psi, \psi]^{\mathfrak{g}}_\wedge = 0$ then for each simply connected subset $U \subset M$ there exists a smooth mapping $f : U \to G$ with $\delta^r f = \psi|U$, and f is uniquely determined up to a right translation in G.

The mapping f is called the *left Cartan developing* of φ, or the *right Cartan developing* of ψ, respectively.

Proof. Let us treat the right logarithmic derivative since it leads to a principal connection for a bundle with right principal action. For the left logarithmic derivative the proof is similar, with the changes described in the second part of the proof of (38.1).

We put ourselves into the situation of the proof of (38.1). If we are given a 1-form $\varphi \in \Omega^1(M, \mathfrak{g})$ with $d\varphi - \frac{1}{2}[\varphi, \varphi]_\wedge = 0$ then we consider the 1-form $\omega^r \in \Omega^1(M \times G, \mathfrak{g})$, given by the analogue of (38.1.1) (where $\nu : G \to G$ is the inversion),

$$\omega^r = \kappa^l - (\mathrm{Ad} \circ \nu).\varphi \tag{1}$$

Then ω^r is a principal connection form on $M \times G$, since it reproduces the generators in $\mathfrak{g}$ of the fundamental vector fields for the principal right action, i.e., the left invariant vector fields, and ω^r is G-equivariant:

$$\begin{aligned}((\mu^g)^* \omega^r)_h &= \omega^r_{hg} \circ (\mathrm{Id} \times T(\mu^g)) = T(\mu_{g^{-1}.h^{-1}}).T(\mu^g) - \mathrm{Ad}(g^{-1}.h^{-1}).\varphi \\ &= \mathrm{Ad}(g^{-1}).\omega^r_h.\end{aligned}$$

The computation in (38.1.3) for φ instead of $\delta^r f$ shows that this connection is flat. Since the structure group G is regular, by theorem (39.2) the horizontal bundle is integrable, and $\mathrm{pr}_1 : M \times G \to M$, restricted to each horizontal leaf, is a covering. Thus, it may be inverted over each simply connected subset $U \subset M$, and the inverse $(\mathrm{Id}, f) : U \to M \times G$ is unique up to the choice of the branch of the covering and the choice of the leaf, i.e., f is unique up to a right translation by an element of G. The beginning of the proof of (38.1) then shows that $\delta^r f = \varphi|U$. □

40.3. Theorem. *Let G and H be Lie groups with Lie algebras $\mathfrak{g}$ and $\mathfrak{h}$, respectively. Let $f : \mathfrak{g} \to \mathfrak{h}$ be a bounded homomorphism of Lie algebras. If H is regular and if G is simply connected then there exists a unique homomorphism $F : G \to H$ of Lie groups with $T_eF = f$.*

Proof. We consider the 1-form

$$\psi \in \Omega^1(G;\mathfrak{h}), \quad \psi := f \circ \kappa^r, \quad \psi_g(\xi_g) = f(T(\mu^{g^{-1}}).\xi_g),$$

where κ^r is the right Maurer-Cartan form from (38.1). It satisfies the left Maurer-Cartan equation

$$\begin{aligned} d\psi - \tfrac{1}{2}[\psi,\psi]^{\mathfrak{h}}_{\wedge} &= d(f \circ \kappa^r) - \tfrac{1}{2}[f \circ \kappa^r, f \circ \kappa^r]^{\mathfrak{h}}_{\wedge} \\ &= f \circ (d\kappa^r - \tfrac{1}{2}[\kappa^r,\kappa^r]^{\mathfrak{g}}_{\wedge}) = 0, \end{aligned}$$

by (38.1).(2'). But then we can use theorem (40.2) to conclude that there exists a unique smooth mapping $F : G \to H$ with $F(e) = e$, whose right logarithmic derivative satisfies $\delta^r F = \psi$. For $g \in G$ we have $(\mu^g)^*\psi = \psi$, and thus

$$\delta^r(F \circ \mu^g) = \delta^r F \circ T(\mu^g) = (\mu^g)^*\psi = \psi.$$

By uniqueness in theorem (40.2), again, the mappings $F \circ \mu^g$ and $F : G \to H$ differ only by right translation in H by the element $(F \circ \mu^g)(e) = F(g)$, so that $F \circ \mu^g = \mu^{F(g)} \circ F$, or $F(g.g_1) = F(g).F(g_1)$. This also implies $F(g).F(g^{-1}) = F(g.g^{-1}) = F(e) = e$, hence that F is the unique homomorphism of Lie groups we have been looking for. □

Chapter IX
Manifolds of Mappings

Manifolds of smooth mappings between finite dimensional manifolds are the foremost examples of infinite dimensional manifolds, and in particular diffeomorphism groups can only be treated in a satisfactory manner at the level of generality developed in this book: One knows from [Omori, 1978b] that a Banach Lie group acting effectively on a finite dimensional manifold is necessarily finite dimensional. So there is **no way** to model the diffeomorphism group on Banach spaces as a manifold.

The space of smooth mappings $C^\infty(M,N)$ carries a natural atlas with charts induced by any exponential mapping on N (42.1), which permits us also to consider certain infinite dimensional manifolds N in (42.4). Unfortunately, for noncompact M, the space $C^\infty(M,N)$ is not locally contractible in the compact-open C^∞-topology, and the natural chart domains are quite small: Namely, the natural model spaces turn out to be convenient vector spaces of sections with compact support of vector bundles f^*TN, which have been treated in detail in section (30). Thus, the manifold topology on $C^\infty(M,N)$ is finer than the Whitney C^∞-topology, and we denote by $\mathfrak{C}^\infty(M,N)$ the resulting smooth manifold (otherwise, e.g. $C^\infty(\mathbb{R},\mathbb{R})$ would have two meanings).

With a careful description of the space of smooth curves (42.5) we can later often avoid the explicit use of the atlas, for example when we show that the composition mapping is smooth in (42.13). Since we insist on charts the exponential law for manifolds of mappings holds only for a compact source manifold M, (42.14).

If we insist that the exponential law should hold for manifolds of mappings between all (even only finite dimensional) manifolds, then one is quickly lead to a more general notion of a manifold, where an atlas of charts is replaced by the system of all smooth curves. One is lead to further requirements: tangent spaces should be convenient vector spaces, the tangent bundle should be trivial along smooth curves via a kind of parallel transport, and a local addition as in (42.4) should exist. In this way one obtains a cartesian closed category of smooth manifolds and smooth mappings between them, where those manifolds with Banach tangent spaces are exactly the classical smooth manifolds with charts. Theories along these lines can be found in [Kriegl, 1980], [Michor, 1984a], and [Kriegl, 1984]. Unfortunately they found no applications, and even the authors were not courageous enough to pursue them further and to include them in this book. But we still think that it is a valuable theory, since for instance the diffeomorphism group $\operatorname{Diff}(M)$ of a non-compact finite dimensional smooth manifold M with the compact-open C^∞-topology is a Lie group in this sense with the space of all vector fields on M as Lie algebra. Also, in section (45) results will appear which indicate that ultimately this is a more natural setting.

Let us return (after discussing non-contents) to describing the contents of this chapter. For the tangent space we have a natural diffeomorphism $T\mathfrak{C}^\infty(M,N) \cong \mathfrak{C}^\infty_c(M,TN) \subset \mathfrak{C}^\infty(M,TN)$, see (42.17). In the same manner we also treat manifolds of real analytic mappings from a compact manifold M into N.

In section (43) on diffeomorphism groups we first show that the group $\operatorname{Diff}(M)$ is a regular smooth Lie group (43.1). The proof clearly shows the power of our calculus: It is quite obvious that the inversion is smooth, whereas more traditional treatments as in [Leslie, 1967], [Michor, 1980a], and [Michor, 1980c] needed specially tailored inverse function theorems in infinite dimensions. The Lie algebra of the diffeomorphism group is the space $\mathfrak{X}_c(M)$ of all vector fields with compact support on M, with the negative of the usual Lie bracket. The exponential mapping exp is the flow mapping to time 1, but it is not surjective on any neighborhood of the identity (43.2), and $\operatorname{Ad}\circ\exp : \mathfrak{X}_c(M) \to L(\mathfrak{X}_c(M), \mathfrak{X}_c(M))$ is not real analytic, (43.3). Real analytic diffeomorphisms on a real analytic compact manifold form a regular real analytic Lie group (43.4). Also regular Lie groups are the subgroups of volume preserving (43.7), symplectic (43.12), exact symplectic (43.13), or contact diffeomorphisms (43.19).

In section (44) we treat principal bundles with a diffeomorphism group as structure group. The first example is the space of all embeddings between two manifolds (44.1), a sort of nonlinear Grassmann manifold, in particular if the image space is an infinite dimensional convenient vector space which leads to a smooth manifold which is a classifying space for the diffeomorphism group of a compact manifold (44.24). Another example is the nonlinear frame bundle of a fiber bundle with compact fiber (44.5), for which we investigate the action of the gauge group on the space of generalized connections (44.14) and show that in the smooth case there never exist slices (44.19), (44.20).

In section (45) we compute explicitly all geodesics for some natural (pseudo) Riemannian metrics on the space of all Riemannian metrics. Section (46) is devoted to the Korteweg–De Vrieß equation which is shown to be the geodesic equation of a certain right invariant Riemannian metric on the Virasoro group. Here we also compute the curvature (46.13) and the Jacobi equation (46.14).

41. Jets and Whitney Topologies

Jet spaces or jet bundles consist of the invariant expressions of Taylor developments up to a certain order of smooth mappings between manifolds. Their invention goes back to Ehresmann [Ehresmann, 1951.]

41.1. Jets between convenient vector spaces. Let E and F be convenient vector spaces, and let $U \subseteq E$ and $V \subseteq F$ be c^∞-open subsets. For $0 \le k \le \infty$ the space of *k-jets* from U to V is defined by

$$J^k(U,V) := U \times V \times \operatorname{Poly}^k(E,F), \text{ where } \operatorname{Poly}^k(E,F) = \prod_{j=1}^{k} L^j_{\text{sym}}(E;F).$$

We shall use the source and image projections $\alpha : J^k(U,V) \to U$ and $\beta : J^k(U,V) \to V$, and we shall consider $J^k(U,V) \to U \times V$ as a trivial bundle, with fibers $J^k_x(U,V)_y$ for $(x,y) \in U \times V$. Moreover, we have obvious projections $\pi^k_l : J^k(U,V) \to J^l(U,V)$ for $k > l$, given by truncation at order l. For a smooth mapping $f : U \to V$ the *k-jet extension* is defined by

$$j^k f(x) = j^k_x f := (x, f(x), df(x), \frac{1}{2!} d^2 f(x), \dots, \frac{1}{j!} d^j f(x), \dots),$$

the Taylor expansion of f at x of order k. If $k < \infty$ then $j^k : C^\infty(U,F) \to J^k(U,F)$ is smooth with a smooth right inverse (the polynomial), see (5.17). If $k = \infty$ then j^k need not be surjective for infinite dimensional E, see (15.4). For later use, we consider now the *truncated composition*

$$\bullet : \operatorname{Poly}^k(F,G) \times \operatorname{Poly}^k(E,F) \to \operatorname{Poly}^k(E,G),$$

where $p \bullet q$ is the composition $p \circ q$ of the polynomials p, q (formal power series in case $k = \infty$) without constant terms, and without all terms of order $> k$. Obviously, $\bullet$ is polynomial for finite k and is real analytic for $k = \infty$ since then each component is polynomial. Now let $U \subset E$, $V \subset F$, and $W \subset G$ be open subsets, and consider the fibered product

$$\begin{aligned} J^k(U,V) \times_U J^k(W,U) &= \{ (\sigma,\tau) \in J^k(U,V) \times J^k(W,U) : \alpha(\sigma) = \beta(\tau) \} \\ &= U \times V \times W \times \operatorname{Poly}^k(E,F) \times \operatorname{Poly}^k(G,E). \end{aligned}$$

Then the mapping

$$\begin{gathered} \bullet : J^k(U,V) \times_U J^k(W,U) \to J^k(W,V), \\ \sigma \bullet \tau = (\alpha(\sigma), \beta(\sigma), \bar\sigma) \bullet (\alpha(\tau), \beta(\tau), \bar\tau) := (\alpha(\tau), \beta(\sigma), \bar\sigma \bullet \bar\tau), \end{gathered}$$

is a real analytic mapping, called the *fibered composition of jets*.

Let U, $U' \subset E$ and $V \subset F$ be open subsets, and let $g : U' \to U$ be a smooth diffeomorphism. We define a mapping $J^k(g,V) : J^k(U,V) \to J^k(U',V)$ by $J^k(g,V)(\sigma) =$

$\sigma \bullet j^k g(g^{-1}(x))$, which also satisfies $J^k(g,V)(j^k f(x)) = j^k(f\circ g)(g^{-1}(\alpha(\sigma)))$. If $g' : U'' \to U'$ is another diffeomorphism, then clearly $J^k(g',V)\circ J^k(g,V) = J^k(g\circ g',V)$, and $J^k(\quad,V)$ is a contravariant functor acting on diffeomorphisms between open subsets of E. Since the truncated composition $\bar\sigma \mapsto \bar\sigma \bullet j^k_{g^{-1}(x)}g$ is linear, the mapping $J^k_x(g,F) := J^k(g,F)|J^k_x(U,F) : J^k_x(U,F) \to J^k_{g^{-1}(x)}(U',F)$ is also linear.

Now let $U\subset E$, $V\subset F$, and $W\subset G$ be c^∞-open subsets, and let $h : V\to W$ be a smooth mapping. Then we define $J^k(U,h) : J^k(U,V)\to J^k(U,W)$ by $J^k(U,h)\sigma = j^k h(\beta(\sigma))\bullet\sigma$, which satisfies $J^k(U,h)(j^k f(x)) = j^k(h\circ f)(x)$. Clearly, $J^k(U,\quad)$ is a covariant functor acting on smooth mappings between c^∞-open subsets of convenient vector spaces. The mapping $J^k_x(U,h)_y : J^k_x(U,V)_y \to J^k_x(U,W)_{h(y)}$ is linear if and only if h is affine or $k=1$ or $U=\emptyset$.

41.2. The differential group $GL^k(E)$. For a convenient vector space E, the k-jets at 0 of germs at 0 of diffeomorphisms of E which map 0 to 0 form a group under truncated composition, which will be denoted by $GL^k(E)$ and will be called the *differential group of order k*. Clearly, an arbitrary 0-respecting k-jet $\sigma\in \operatorname{Poly}^k(E,E)$ is in $GL^k(E)$ if and only if its linear part is invertible. Thus

$$GL^k(E) = GL(E)\times\prod_{j=2}^k L^j_{\text{sym}}(E;E) =: GL(E)\times P^k_2(E),$$

where we put $P^k_2(E) := \prod_{j=2}^k L^j_{\text{sym}}(E;E)$ for the space of all polynomial mappings of degree $\le k$ (formal power series for $k=\infty$) without constant and linear terms.

If the set $GL(E)$ of all bibounded linear isomorphisms of E is a Lie group contained in $L(E,E)$ (e.g., for E a Banach space), then since the truncated composition is real analytic, $GL^k(E)$ is also a Lie group. In general, $GL(E)$ may be viewed as a Frölicher space in the sense of (23.1) with the initial smooth structure with respect to $(\mathrm{Id},(\quad)^{-1}) : GL(E)\to L(E,E)\times L(E,E)$, where multiplication and inversion are now smooth: we call this a *smooth group*. Then $GL^k(E)$ is again a smooth group.

In both cases, clearly, for $k\ge l$ the mapping $\pi^k_l : GL^k(E)\to GL^l(E)$ is a homomorphism of smooth groups, thus its kernel $\ker(\pi^k_l) = \operatorname{Poly}^k_l(E,E) := \{\mathrm{Id}_E\}\times\{0\}\times\prod_{j=l+1}^k L^j_{\text{sym}}(E;E)$ is a closed normal subgroup for all l, which is a Lie group for $l\ge 1$. The exact sequence of groups

$$\{e\}\to\prod_{j=l+1}^k L^j_{\text{sym}}(E;E)\to GL^k(E)\to GL^l(E)\to\{e\}$$

splits if and only if $l=1$ for $\dim E>1$ or $l\le 2$ for $E=\mathbb{R}$, see [Kolář, Michor, Slovák, 1993, 13.8] for $E=\mathbb{R}^m$; only in this case this sequence describes a semidirect product.

41.3. Jets between manifolds. Now let M and N be smooth manifolds with smooth atlas (U_α,u_α) and (V_β,v_β), modeled on convenient vector spaces E and F, respectively. Then we may glue the open subsets $J^k(u_\alpha(U_\alpha), v_\beta(V_\beta))$ of convenient vector spaces via the chart change mappings

$$J^k(u_{\alpha'} \circ u_\alpha^{-1}, v_\beta \circ v_{\beta'}^{-1}) : J^k(u_{\alpha'}(U_\alpha \cap U_{\alpha'}), v_{\beta'}(V_\beta \cap V_{\beta'})) \to \\ \to J^k(u_\alpha(U_\alpha \cap U_{\alpha'}), v_\beta(V_\beta \cap V_{\beta'})),$$

and we obtain a smooth fiber bundle $J^k(M, N) \to M \times N$ with standard fiber $\operatorname{Poly}^k(E, F)$. With the identification topology $J^k(M, N)$ is Hausdorff, since it is a fiber bundle and the usual argument for gluing fiber bundles applies which was given, e.g., in (28.12).

Theorem. *If M and N are smooth manifolds, modeled on convenient vector spaces E and F, respectively. Let $0 \le k \le \infty$. Then the following results hold.*

(1) *$(J^k(M, N), (\alpha, \beta), M \times N, \operatorname{Poly}^k(E, F))$ is a fiber bundle with standard fiber $\operatorname{Poly}^k(E, F)$, with the smooth group $GL^k(E) \times GL^k(F)$ as structure group, where $(\gamma, \chi) \in GL^k(E) \times GL^k(F)$ acts on $\sigma \in \operatorname{Poly}^k(E, F)$ by $(\gamma, \chi).\sigma = \chi \bullet \sigma \bullet \gamma^{-1}$.*

(2) *If $f : M \to N$ is a smooth mapping then $j^k f : M \to J^k(M, N)$ is also smooth, called the k-jet extension of f. We have $\alpha \circ j^k f = \mathrm{Id}_M$ and $\beta \circ j^k f = f$.*

(3) *If $g : M' \to M$ is a diffeomorphism then also the induced mapping $J^k(g, N) : J^k(M, N) \to J^k(M', N)$ is a diffeomorphism.*

(4) *If $h : N \to N'$ is a smooth mapping then $J^k(M, h) : J^k(M, N) \to J^k(M, N')$ is also smooth. Thus, $J^k(M, \quad)$ is a covariant functor from the category of smooth manifolds and smooth mappings into itself which respects each of the following classes of mappings: initial mappings, embeddings, closed embeddings, splitting embeddings, fiber bundle projections. Furthermore, $J^k(\quad, \quad)$ is a contra-covariant bifunctor, where we have to restrict in the first variable to the category of diffeomorphisms.*

(5) *For $k \ge l$, the projections $\pi_l^k : J^k(M, N) \to J^l(M, N)$ are smooth and natural, i.e., they commute with the mappings from (3) and (4).*

(6) *$(J^k(M, N), \pi_l^k, J^l(M, N), \prod_{i=l+1}^k L_{\mathrm{sym}}^i(E; F))$ are fiber bundles for all $l \le k$. For finite k the bundle $(J^k(M, N), \pi_{k-1}^k, J^{k-1}(M, N), L_{\mathrm{sym}}^k(E, F))$ is an affine bundle. The first jet space $J^1(M, N) \to M \times N$ is a vector bundle. It is isomorphic to the bundle $(L(TM, TN), (\pi_M, \pi_N), M \times N)$, see (29.4) and (29.5). Moreover, we have $J_0^1(\mathbb{R}, N) = TN$ and $J^1(M, \mathbb{R})_0 = T^*M$.*

(7) *Truncated composition is a smooth mapping*

$$\bullet : J^k(N, P) \times_N J^k(M, N) \to J^k(M, P).$$

Proof. (1) is already proved. (2), (3), (5), and (7) are obvious from (41.1), mainly by the functorial properties of $J^k(\quad, \quad)$.

(4) It is clear from (41.1) that $J^k(M, h)$ is a smooth mapping. The rest follows by looking at special chart representations of h and the induced chart representations for $J^k(M, h)$.

It remains to show (6), and here we concentrate on the affine bundle. Let $a_1 + a \in GL(E) \times \prod_{i=2}^k L^i_{\text{sym}}(F;F)$, $\sigma + \sigma_k \in \text{Poly}^{k-1}(E,F) \times L^k_{\text{sym}}(E;F)$, and $b_1 + b \in GL(E) \times \prod_{i=2}^k L^i_{\text{sym}}(E;E)$, then the only term of degree k containing σ_k in $(a_1+a) \bullet (\sigma+\sigma_k) \bullet (b_1+b)$ is $a_1 \circ \sigma_k \circ b_1^k$, which depends linearly on σ_k. To this the degree k-components of compositions of the lower order terms of σ with the higher order terms of a and b are added, and these may be quite arbitrary. So an affine bundle results.

We have $J^1(M,N) = L(TM,TN)$ since both bundles have the same transition functions. Finally,

$$J^1_0(\mathbb{R},N) = L(T_0\mathbb{R},TN) = TN \quad \text{and} \quad J^1(M,\mathbb{R})_0 = L(TM,T_0\mathbb{R}) = T^*M. \quad \square$$

41.4. Jets of sections of fiber bundles. If $(p : E \to M, S)$ is a fiber bundle, let (U_α, u_α) be a smooth atlas of M such that $(U_\alpha, \psi_\alpha : E|U_\alpha \to U_\alpha \times S)$ is a fiber bundle atlas. If we glue the smooth manifolds $J^k(U_\alpha, S)$ via $(\sigma \mapsto j^k(\psi_{\alpha\beta}(\alpha(\sigma), \quad))) \bullet \sigma : J^k(U_\alpha \cap U_\beta, S) \to J^k(U_\alpha \cap U_\beta, S)$, we obtain the smooth manifold $J^k(E)$, which for finite k is the space of all k-jets of local sections of E.

Theorem. *In this situation we have:*

(1) *$J^k(E)$ is a splitting closed submanifold of $J^k(M,E)$, namely the set of all $\sigma \in J^k_x(M,E)$ with $J^k(M,p)(\sigma) = j^k(\text{Id}_M)(x)$.*

(2) *$J^1(E)$ of sections is an affine subbundle of the vector bundle $J^1(M,E) = L(TM,TE)$. In fact, we have*

$$J^1(E) = \{\, \sigma \in L(TM,TE) : Tp \circ \sigma = \text{Id}_{TM} \,\}.$$

(3) *For k finite $(J^k(E), \pi^k_{k-1}, J^{k-1}(E))$ is an affine bundle.*

(4) *If $p : E \to M$ is a vector bundle, then $(J^k(E), \alpha, M)$ is also a vector bundle. If $\phi : E \to E'$ is a homomorphism of vector bundles covering the identity, then $J^k(\varphi)$ is of the same kind.*

Proof. Locally $J^k(E)$ in $J^k(M,E)$ looks like $u_\alpha(U_\alpha) \times \text{Poly}^k(F_M, F_S)$ in $u_\alpha(U_\alpha) \times (u_\alpha(U_\alpha) \times v_\beta(V_\beta)) \times \text{Poly}^k(F_M, F_M \times F_S)$, where F_M and F_S are modeling spaces of M and S, respectively, and where (V_β, v_β) is a smooth atlas for S. The rest is clear. $\square$

41.5. The compact-open topology on spaces of continuous mappings. Let M and N be Hausdorff topological spaces. The best known topology on the space $C(M,N)$ of all continuous mappings is the *compact-open topology* or *CO-topology*. A subbasis for this topology consists of all sets of the form $\{f \in C(M,N) : f(K) \subseteq U\}$, where K runs through all compact subsets in M and U through all open subsets of N. This is a Hausdorff topology, since it is finer than the topology of pointwise convergence.

It is easy to see that if M has a countable basis of the compact sets and is compactly generated ((4.7).(i), i.e., M carries the final topology with respect to the inclusions of its compact subsets), and if N is a complete metric space, then there exists a complete metric on $(C(M,N), CO)$, so it is a Baire space.

41.6. The graph topology. For $f \in C(M,N)$ let $\operatorname{graph}_f : M \to M \times N$ be given by $\operatorname{graph}_f(x) = (x, f(x))$, the graph mapping of f.

The *WO-topology* or *wholly open topology* on $C(M,N)$ is given by the basis $\{f \in C(M,N) : f(M) \subset U\}$, where U runs through all open sets in N. It is not Hausdorff, since mappings with the same image cannot be separated.

The *graph topology* or *WO^0-topology* on $C(M,N)$ is induced by the mapping

$$\operatorname{graph} : C(M,N) \to (C(M, M \times N), \text{ WO-topology}).$$

A basis for it is given by all sets of the form $\{f \in C(M,N) : \operatorname{graph}_f(M) \subseteq U\}$, where U runs through all open sets in $M \times N$. This topology is Hausdorff since it is finer than the compact-open topology. Note that a continuous mapping $g : N \to P$ induces a continuous mapping $g_* : C(M,N) \to C(M,P)$ for the WO^0-topology, since $\operatorname{graph}_{g\circ f} = (\operatorname{Id} \times g) \circ \operatorname{graph}_f$.

If M is paracompact and (N,d) is a metric space, then for $f \in C(M,N)$ the sets $\{g \in C(M,N) : d(g(x), f(x)) < \varepsilon(x) \text{ for all } x \in M\}$ form a basis of neighborhoods, where ε runs through all positive continuous functions on M. This is easily seen.

41.7. Lemma. *Let N be metrizable, and let M satisfy one of the following conditions:*

(1) *M is locally compact with a countable basis of open sets.*
(2) $M = \mathbb{R}^{(\mathbb{N})}$.

Then for any sequence (f_n) in $C(M,N)$ the following holds: (f_n) converges to f in the WO^0-topology if and only if there exists a compact set $K \subseteq M$ such that f_n equals f off K for all but finitely many n, and $f_n|K$ converges to $f|K$ uniformly.

Note that in case (2) we get $f_n = f$ for all but finitely many n, since f differs from f_n on a c^∞-open subset.

Proof. Clearly, the condition above implies convergence. Conversely, let (f_n) and f in $C(M,N)$ be such that the condition does not hold. In case (1) let $K_n \subset K^o_{n+1}$ be a basis of the compact sets in M, and in case (2) let $K_n := \{x \in \mathbb{R}^n \subset \mathbb{R}^{(\mathbb{N})} : |x^i| \le n \text{ for } i \le n\}$. Then either f_n does not converge to f in the compact-open topology, or there exists $x_n \notin K_n$ with $d(f_n(x_n), f(x_n)) =: \varepsilon_n > 0$. Then (x_n) is without cluster point in M: This is obvious in case (1), and in case (2) this can be seen by the following argument: Assume that there exists a cluster point y. Let N be so large that $\operatorname{supp}(y) \subset \{0, \dots, N\}$ and $|y^i| \le N-1$ for all i. Then we define $k_n \in \mathbb{N}$ and $\delta_n > 0$ by

$$\begin{cases} k_n := n, \delta_n := 1 & \text{for } n \le N \text{ or } \operatorname{supp}(x_n) \subseteq \{1, \dots, n\} \\ k_n := \min\{i > n : x_n^i \ne 0\}, \delta_n := |x_n^{k_n}| & \text{otherwise} \end{cases}$$

Then $x_n - y \notin U := \{z : |z^{k_i}| < \delta_i \text{ for all } i\}$ for $n > N$, so y cannot be a cluster point.

Then by a paracompactness argument and the second description of the WO^0-topology the set $\{(x,y) \in M \times N : \text{ if } x = x_n \text{ then } d(f(x_n), y) < \varepsilon_n\}$ is an open neighborhood of $\operatorname{graph}_f(M)$ not containing any $\operatorname{graph}_{f_n}(M)$. So f_n cannot converge to f in the WO^0-topology. □

41.8. Lemma. *Let E be a convenient vector space, and suppose that M satisfies the following condition:*

(1) *Each neighborhood of each point contains a sequence without cluster point in M.*

Then for $f \in C(M, E)$ we have $tf \to 0$ in the WO^0-topology for $t \to 0$ in $\mathbb{R}$ if and only if $f = 0$.

Moreover, each open subset in an infinite dimensional locally convex space has property (1).

Proof. The mapping $f \mapsto g \circ f$ is continuous in the WO^0-topologies, so by composing with bounded linear functionals on E we may suppose that $E = \mathbb{R}$.

Suppose that $f \neq 0$, say $f(x) = 2$ for some x. Then $f(y) > 1$ for y in some neighborhood U of x, which contains a sequence x_n without cluster point in M. Then $\{(x, y) \in M \times \mathbb{R} : \text{ if } x = x_n \text{ then } y < 1/n\}$ is an open neighborhood of $\mathrm{graph}_0(M)$ not containing any $\mathrm{graph}_{tf}(M)$ for $t \neq 0$. So tf cannot converge to 0 in the WO^0-topology.

For the last assertion we have to show that the unit ball of each seminorm p in an infinite dimensional locally convex vector space M contains a sequence without cluster point. If the seminorm has non-trivial kernel $p^{-1}(0)$ then $(n.x)_n$ for $0 \neq x \in p^{-1}(0)$ has this property. If p has trivial kernel, it is a norm, and the unit ball in the normed space (M, p) contains a sequence without cluster point, since otherwise the unit ball would be compact, and (M, p) would be finite dimensional. This sequence has also no cluster point in M, since M has a finer topology. □

41.9. The CO^k-topology on spaces of smooth mappings. Let M and N be smooth manifolds, possibly infinite dimensional. For $0 \leq k \leq \infty$ the *compact-open C^k-topology* or *CO^k-topology* on the space $C^\infty(M, N)$ of all smooth mappings $M \to N$ is induced by the k-jet extension (41.3) from the CO-topology

$$j^k : C^\infty(M, N) \to (C(M, J^k(M, N)), \mathrm{CO}).$$

We conclude with some remarks. If M is infinite dimensional it would be more natural to replace the system of compact sets in M by the system of all subsets on which each smooth real valued function is bounded. Since these topologies will play only minor roles in this book we do not develop them here.

41.10. Whitney C^k-topology. Let M and N be smooth manifolds, possibly infinite dimensional. For $0 \leq k \leq \infty$ the *Whitney C^k-topology* or *WO^k-topology* on the space $C^\infty(M, N)$ of all smooth mappings $M \to N$ is induced by the k-jet extension (41.3) from the WO-topology

$$j^k : C^\infty(M, N) \to (C(M, J^k(M, N)), \mathrm{WO}).$$

A basis for the open sets is given by all sets of the form $\{f \in C^\infty(M, N) : j^k f(M) \subset U\}$, where U runs through all open sets in the smooth manifold $J^k(M, N)$. A smooth mapping $g : N \to P$ induces a smooth mapping $J^k(M, g) : J^k(M, N) \to J^k(M, P)$ by (41.3.4), and thus in turn a continuous mapping $g_* : C^\infty(M, N) \to C^\infty(M, P)$ for the WO^k-topologies for each k.

For a convenient vector space E and for a manifold M modeled on infinite dimensional Fréchet spaces (so that there the c^∞-topology coincides with the locally convex one) we see from (41.8) that for $f \in C^\infty(M, E)$ we have $t.f \to 0$ for $t \to 0$ in the WO^k-topology if and only if $f = 0$. So $(C^\infty(M, E), \mathrm{WO}^k)$ does not contain a non-trivial topological vector space if M is infinite dimensional.

If M is compact, then the WO^k-topology and the CO^k-topology coincide on the space $C^\infty(M, N)$ for all k.

41.11. Lemma. *Let M, N be smooth manifolds, where M is finite dimensional and second countable, and where N is metrizable. Then $J^\infty(M, N)$ is also a metrizable manifold. If, moreover, N is second countable then also $J^\infty(M, N)$ is also second countable.*

Let $K_n \subset K_{n+1}^o \subset K_{n+1}$ be a compact exhaustion of M. Then the following is a basis of open sets for the Whitney C^∞-topology:

$$M(U, m) := \{f \in C^\infty(M, N) : j^{m_n} f(M \setminus K_n^o) \subset U_n\},$$

where (m_n) is any sequence in $\mathbb{N}$ and where $U_n \subset J^{m_n}(M, N)$ is an open subset.

Proof. Looking at (41.3) we see that $J^\infty(M, N)$ is a bundle over $M \times N$ with Fréchet spaces as fibers, so it is metrizable. We can also write

$$M(U, m) := \{f \in C^\infty(M, N) : j^\infty f(M \setminus K_n^o) \subset (\pi^\infty_{m_n})^{-1} U_n\}.$$

By pulling up to higher jet bundles, we may assume that m_n is strictly increasing. If we put $V_n = (\pi^\infty_{m_n})^{-1} U_n$, we may then replace V_n by $V_0 \cap \dots \cap V_n$ without changing $M(U, m)$. But then we may replace $M \setminus K_n^o$ by $K_{n+1} \setminus K_n^o$ without changing the set. Using compactness of $j^\infty f(K_{n+1} \setminus K_n^o)$ and that $J^\infty(M, N)$ carries the initial topology with respect to all projections $\pi_l^\infty : J^\infty(M, N) \to J^l(M, N)$ by (41.3.6), we get an equivalent basis of open sets given by

$$M(U) := \{f \in C^\infty(M, N) : j^\infty f(K_{n+1} \setminus K_n^o) \subset U_n\},$$

where now $U_n \subset J^\infty(M, N)$ is a sequence of open sets. It is obvious that this basis generates a topology which is finer than the WO^∞-topology. To show the converse let $f \in M(U)$. Let d be a compatible metric on the metrizable manifold $J^\infty(M, N)$, and let $0 < \varepsilon_n$ be smaller than the distance between the compact set $j^\infty f(K_{n+1} \setminus K_n^o)$ and the complement of its open neighborhood U_n. Let ε be a positive continuous function on M such that $0 < \varepsilon(x) < \varepsilon_n$ for $x \in K_{n+1} \setminus K_n^o$, and consider the open set $W := \{\sigma \in J^\infty(M, N) : d(\sigma, j^\infty f(\alpha(\sigma))) < \varepsilon(\alpha(\sigma))\}$ in $J^\infty(M, N)$. Then $f \in \{g \in C^\infty(M, N) : j^\infty g(M) \subset W\} \subseteq M(U)$. $\square$

41.12. Corollary. *Let M, N be smooth manifolds, where M is finite dimensional and second countable, and where N is metrizable. Then the CO^k-topology is metrizable. If N is also second countable then so is the CO^k-topology.*

Proof. Use (41.11) and [Bourbaki, 1966, X, 3.3]. □

41.13. Comparison of topologies on $C^\infty(M,E)$. Let $p : E \to M$ be a smooth finite dimensional vector bundle over a finite dimensional second countable base manifold M. We consider the space $C^\infty_c(M \leftarrow E)$ of all smooth sections of E with compact support, equipped with the bornological locally convex topology from (30.4),

$$C^\infty_c(M \leftarrow E) = \varinjlim_K C^\infty_K(M \leftarrow E),$$

where K runs through all compact sets in M and each of the spaces $C^\infty_K(M \leftarrow f^*TN)$ is equipped with the topology of uniform convergence (on K) in all derivatives separately, as in (30.4), reformulated for the bornological topologies. Consider also the space $C^\infty(M,E)$ of all smooth mappings $M \to E$, equipped with the Whitney C^∞-topology, and the subspace $C^\infty(M \leftarrow E)$ of all smooth sections, with the induced topology.

Lemma. *Then the canonical injection*

$$C^\infty_c(M \leftarrow E) \to C^\infty(M,E)$$

is a topological embedding. The subspace $C^\infty(M \leftarrow E)$ is a vector space, but scalar multiplication is jointly continuous in the induced topology on it if and only if M is compact or the fiber is 0. The maximal topological vector space contained in $C^\infty(M \leftarrow E)$ is just $C^\infty_c(M \leftarrow E)$.

Proof. That the injection is an embedding is clear by contemplating the description of the Whitney C^∞-topology given in lemma (41.11), which obviously is the inductive limit topology $\varinjlim C^\infty_{K_n}(E)$. The rest follows from (41.7) since $t.f \to 0$ for $t \to 0$ in in $C^\infty(M,E)$ for WO^∞ if and only if $t.j^\infty f \to 0$ in $C^\infty(M, J^\infty(E))$ for the WO^0-topology. □

41.14. Tubular neighborhoods. Let M be an (embedded) submanifold of a smooth finite dimensional manifold N. Then the *normal bundle* of M in N is the vector bundle $\mathcal{N}(M) := (TN|M)/TM \xrightarrow{\pi} M$ with fiber T_xN/T_xM over a point $x \in M$. A *tubular neighborhood* of M in N consists of:

(1) A fiberwise radial open neighborhood $\tilde{U} \subset \mathcal{N}(M)$ of the 0-section in the normal bundle
(2) A diffeomorphism $\varphi : \tilde{U} \to U \subset N$ onto an open neighborhood U of M in N, which on the 0-section coincides with the projection of the normal bundle.

It is well known that tubular neighborhoods exist.

42. Manifolds of Mappings

42.1. Theorem. Manifold structure of $\mathfrak{C}^\infty(M,N)$. *Let M and N be smooth finite dimensional manifolds. Then the space $\mathfrak{C}^\infty(M,N)$ of all smooth mappings from M to N is a smooth manifold, modeled on spaces $C_c^\infty(M \leftarrow f^*TN)$ of smooth sections with compact support of pullback bundles along $f: M \to N$ over M.*

Proof. Choose a smooth Riemannian metric on N. Let $\exp : TN \supseteq U \to N$ be the smooth exponential mapping of this Riemannian metric, defined on a suitable open neighborhood of the zero section. We may assume that U is chosen in such a way that $(\pi_N, \exp) : U \to N \times N$ is a smooth diffeomorphism onto an open neighborhood V of the diagonal.

For $f \in C^\infty(M,N)$ we consider the pullback vector bundle

$$\begin{array}{ccccc} M \times_N TN & = & f^*TN & \xrightarrow{\pi_N^* f} & TN \\ & & {\scriptstyle f^*\pi_N}\downarrow & & \downarrow{\scriptstyle \pi_N} \\ & & M & \xrightarrow{f} & N. \end{array}$$

For f, $g \in C^\infty(M,N)$ we write $f \sim g$ if f and g agree off some compact subset in M. Then $C_c^\infty(M \leftarrow f^*TN)$ is canonically isomorphic to the space

$$C_c^\infty(M,TN)_f := \{h \in C^\infty(M,TN) : \pi_N \circ h = f,\ h \sim 0 \circ f\}$$

via $s \mapsto (\pi_N^* f) \circ s$ and $(\mathrm{Id}_M, h) \leftarrow\!\shortmid h$. We consider the space $C_c^\infty(M \leftarrow f^*TN)$ of all smooth sections with compact support and equip it with the inductive limit topology

$$C_c^\infty(M \leftarrow f^*TN) = \operatorname*{inj\,lim}_K C_K^\infty(M \leftarrow f^*TN),$$

where K runs through all compact sets in M and each of the spaces $C_K^\infty(M \leftarrow f^*TN)$ is equipped with the topology of uniform convergence (on K) in all derivatives separately, as in (30.4), reformulated for the bornological topology; see also (6.1). Now let

$$\begin{aligned} U_f &:= \{g \in C^\infty(M,N) : (f(x),\ g(x)) \in V \text{ for all } x \in M, g \sim f\}, \\ u_f &: U_f \to C_c^\infty(M \leftarrow f^*TN), \\ u_f(g)(x) &= (x, \exp_{f(x)}^{-1}(g(x))) = (x, ((\pi_N, \exp)^{-1} \circ (f,g))(x)). \end{aligned}$$

Then u_f is a bijective mapping from U_f onto the set $\{s \in C_c^\infty(M \leftarrow f^*TN) : s(M) \subseteq f^*U = (\pi_N^* f)^{-1}(U)\}$, whose inverse is given by $u_f^{-1}(s) = \exp \circ (\pi_N^* f) \circ s$, where we view $U \to N$ as fiber bundle. The set $u_f(U_f)$ is open in $C_c^\infty(M \leftarrow f^*TN)$ for the topology described above, see (30.10).

Now we consider the atlas $(U_f, u_f)_{f \in \mathfrak{C}^\infty(M,N)}$ for $\mathfrak{C}^\infty(M,N)$. Its chart change mappings are given for $s \in u_g(U_f \cap U_g) \subseteq C_c^\infty(M \leftarrow g^*TN)$ by

$$\begin{aligned} (u_f \circ u_g^{-1})(s) &= (\mathrm{Id}_M, (\pi_N, \exp)^{-1} \circ (f, \exp \circ (\pi_N^* g) \circ s)) \\ &= (\tau_f^{-1} \circ \tau_g)_*(s), \end{aligned}$$

where $\tau_g(x, Y_{g(x)}) := (x, \exp_{g(x)}(Y_{g(x)}))$ is a smooth diffeomorphism $\tau_g : g^*TN \supseteq g^*U \to (g \times \mathrm{Id}_N)^{-1}(V) \subseteq M \times N$ which is fiber respecting over M.

Smooth curves in $C_c^\infty(M \leftarrow f^*TN)$ are smooth sections of the bundle $\mathrm{pr}_2^* f^*TN \to \mathbb{R} \times M$, which have compact support in M locally in $\mathbb{R}$. The chart change $u_f \circ u_g^{-1} = (\tau_f^{-1} \circ \tau_g)_*$ is defined on an open subset and it is also smooth by (30.10).

Finally, following (27.1), the natural topology on $\mathfrak{C}^\infty(M, N)$ is the identification topology from this atlas (with the c^∞-topologies on the modeling spaces), which is obviously finer than the compact-open topology and thus Hausdorff.

The equation $u_f \circ u_g^{-1} = (\tau_f^{-1} \circ \tau_g)_*$ shows that the smooth structure does not depend on the choice of the smooth Riemannian metric on N. □

42.2. Remarks. We denote the manifold of all smooth mappings from M to N by $\mathfrak{C}^\infty(M, N)$ because otherwise the set $C^\infty(M, \mathbb{R}^n)$ would appear with two different convenient structures, see (6.1) or (30.1), where the other one was treated. From the last sentence of the proof above it follows that for a compact smooth M the manifold $\mathfrak{C}^\infty(M, \mathbb{R}^n)$ is diffeomorphic to the convenient vector space $C^\infty(M, \mathbb{R})^n$.

We describe now another topology on $\mathfrak{C}^\infty(M, N)$: Consider first the WO^∞-topology on $C^\infty(M, N)$ from (41.10) and refine it such that each equivalence class (of smooth mappings differing only on compact subsets) from the beginning of the proof above becomes open. For this topology all chart mappings are homeomorphisms into open subsets of $C_c^\infty(M \leftarrow f^*TN)$ with the bornological topology, and the chart changes are also homeomorphisms, by (41.10) and (41.13). With this topology $C^\infty(M, N)$ is also a topological manifold, modeled on locally convex spaces $C_c^\infty(M \leftarrow f^*TN)$, which, however, do not carry the c^∞-topologies. It is even a smooth manifold in a stronger sense (all derivatives of chart changes are continuous), and this is the structure used in [Michor, 1980c]. This smooth structure and the natural one described above in (42.1) have the same smooth curves (use (30.9) and (42.5) below). The natural topology is the final topology with respect to all these smooth curves. It is strictly finer if M is not compact.

42.3. Proposition. *For finite dimensional second countable manifolds M, N the smooth manifold $\mathfrak{C}^\infty(M, N)$ has separable connected components and is smoothly paracompact and Lindelöf. If M is compact, it is metrizable.*

Proof. Each connected component of a mapping f is contained in the open equivalence class $\{g : g \sim f\}$ of f consisting of those smooth mappings which differ from f only on compact subsets. This equivalence class is the countable inductive limit in the category of topological spaces of the sets $\{g : g = f \text{ off } K\}$ of all mappings which differ from f only on members K_n of a countable exhaustion of M with compact sets, see (30.9), since a smooth curve locally has values in these steps $\{g : g = f \text{ off } K_n\}$. By (41.12) the steps are metrizable and second countable. Thus, $\{g : g \sim f\}$ is Lindelöf and separable. Since its model spaces $C_c^\infty(M \leftarrow h^*TN)$ are smoothly paracompact by (30.4), by (16.10) the space $\{g : g \sim f\}$ is smoothly paracompact, and since $\mathfrak{C}^\infty(M, N)$ is the disjoint union of such open sets, it is smoothly paracompact, too. □

42.4. Manifolds of mappings with an infinite dimensional range space. The method of proof of theorem (42.1) carries over to spaces $C^\infty(M,\mathcal{N})$, where M is a finite dimensional smooth manifold, and where $\mathcal{N}$ is a possibly infinite dimensional manifold which is required to admit an analogue of the exponential mapping used above, i.e., a smooth mapping $\alpha : T\mathcal{N} \supset U \to \mathcal{N}$, defined on an open neighborhood of the zero section in $T\mathcal{N}$, which satisfies

(1) $(\pi_\mathcal{N}, \alpha) : T\mathcal{N} \supset U \to \mathcal{N} \times \mathcal{N}$ is a diffeomorphism onto a c^∞-open neighborhood of the diagonal.
(2) $\alpha(0_x) = x$ for all $x \in \mathcal{N}$.

A smooth mapping α with these properties is called a *local addition* on $\mathcal{N}$.

Each finite dimensional manifold M admits globally defined local additions. To see this, let $\exp : TM \supset U \to M$ be the exponential mapping with respect to a Riemannian metric g, where U is an open neighborhood of the 0-section, such that $(\pi_M, \exp) : U \to M \times M$ is a diffeomorphism onto an open neighborhood of the diagonal. Thus, exp is a local addition. One can do better. We construct a fiber respecting diffeomorphism $h : TM \to U$ with $h|0_M = \mathrm{Id}_M$ as follows. Let $\varepsilon : M \to (0,\infty)$ be a smooth positive function such that $U' := \{X \in TM : g(X,X) < \varepsilon(\pi_M(X))\} \subset U$. Let $h : TM \to U'$ be given by

$$h(X) := \frac{\varepsilon(\pi_M(X))}{\sqrt{1+g(X,X)}}X, \qquad h^{-1}(Y) := \frac{1}{\sqrt{\varepsilon(\pi_M(Y))^2 - g(Y,Y)}}Y.$$

Then $\alpha = \exp \circ h : TM \to M$ is a local addition.

If M is a real analytic finite dimensional manifold, then there exists a real analytic globally defined local addition $TM \to M$ constructed as above with a real analytic Riemannian metric g and real analytic ε; these exist by [Grauert, 1958, Prop. 8], see also (42.7) below.

The affine structure on each convenient vector space is a local addition, too.

Let G be a possibly infinite dimensional Lie group (36.1). Then G admits a local addition. Namely, let $v : V \to W \subseteq \mathfrak{g}$ be a chart defined on an open neighborhood V of e with $v(e) = 0 \in W$ where W is open in the Lie algebra $\mathfrak{g}$. Then put $TG \supseteq U := \bigcup_{g\in G} T(\mu_g)V \cong G \times V$ and let $\alpha : U \to G$ be given by $\alpha(\xi) := \pi_G(\xi).v^{-1}(T(\mu_{\pi(\xi)^{-1}}).\xi)$ be the local addition.

If a manifold $\mathcal{N}$ admits a local addition α, then it admits a 'spray', thus a torsionfree covariant derivative on $T\mathcal{N}$. Recall from [Ambrose, Palais, Singer, 1960] or [Lang, 1972] that a spray is a vector field S on TM such that $\pi_{TM} \circ S = \mathrm{Id}_{TM}$, $T(\pi_M)\circ S = \mathrm{Id}_{TM}$, so that in induced local charts as in (29.9) and (29.10) we have $S(x,y) = (x,y;y,\Gamma_x(y))$, where finally it is also required that $y \mapsto \Gamma_x(y)$ is quadratic. In order to see this, let $\varphi(X) := \frac{\partial}{\partial t}|_0 \alpha(tX)$. Then $\varphi : TM \to TM$ is a vector bundle automorphism with inverse (in local charts) $\varphi^{-1}(x,y) = \frac{\partial}{\partial t}|_0 (\mathrm{pr}_1, \alpha)^{-1}(x, x+ty)$. Then one checks easily that $S(X) := \frac{d^2}{dt^2}|_{t=0}\alpha(t\varphi^{-1}(X))$ is a spray.

Theorem. *Let M be a smooth finite dimensional manifold, and let $\mathcal{N}$ be a smooth manifold, possibly infinite dimensional, which admits a smooth local addition α.*

Then the space $\mathfrak{C}^\infty(M,\mathcal{N})$ of all smooth mappings from M to $\mathcal{N}$ is a smooth manifold, modeled on spaces $C_c^\infty(M \leftarrow f^*T\mathcal{N})$ of smooth sections with compact support of pullback bundles along $f : M \to \mathcal{N}$ over M. □

Let us remark again that for a compact smooth manifold M and a convenient vector space E the smooth manifold $\mathfrak{C}^\infty(M,E)$ is diffeomorphic to the convenient vector space $C^\infty(M,E)$, which is a special case of (30.1) for a trivial bundle with finite dimensional base.

42.5. Lemma. Smooth curves in $\mathfrak{C}^\infty(M,\mathcal{N})$. *Let M and $\mathcal{N}$ be smooth manifolds with M finite dimensional and $\mathcal{N}$ admitting a smooth local addition. Then the smooth curves c in $\mathfrak{C}^\infty(M,\mathcal{N})$ correspond exactly to the smooth mappings $c^\wedge \in C^\infty(\mathbb{R}\times M,\mathcal{N})$ which satisfy the following property:*

(1) *For each compact interval $[a,b] \subset \mathbb{R}$ there is a compact subset $K \subset M$ such that $c^\wedge(t,x)$ is constant in $t \in [a,b]$ for all $x \in M \setminus K$.*

In particular, the identity induces a smooth mapping $\mathfrak{C}^\infty(M,\mathcal{N}) \to C^\infty(M,\mathcal{N})$ into the Frölicher space $C^\infty(M,\mathcal{N})$ discussed in (23.2.3)*, which is a diffeomorphism if and only if M is compact or $\mathcal{N}$ is discrete.*

Proof. Since $\mathbb{R}$ is locally compact, property (1) is equivalent to

(2) For each $t \in \mathbb{R}$ there is an open neighborhood U of t in $\mathbb{R}$ and a compact $K \subset M$ such that the restriction has the property that $c^\wedge(t,x)$ is constant in $t \in U$ for all $x \in M \setminus K$.

Since this is a local condition on $\mathbb{R}$, and since smooth curves in $\mathfrak{C}^\infty(M,\mathcal{N})$ locally take values in charts as in the proof of theorem (42.1), it suffices to describe the smooth curves in the space $C_c^\infty(M \leftarrow E)$ of sections with compact support of a vector bundle $(p : E \to M, V)$ with finite dimensional base manifold M, with the structure described in (30.4). This was done in (30.9). □

42.6. Theorem. C^ω-manifold structure of $C^\omega(M,\mathcal{N})$. *Let M and $\mathcal{N}$ be real analytic manifolds, let M be compact, and let $\mathcal{N}$ be either finite dimensional, or let us assume that $\mathcal{N}$ admits a real analytic local addition in the sense of* (42.4)*.*

Then the space $C^\omega(M,\mathcal{N})$ of all real analytic mappings from M to $\mathcal{N}$ is a real analytic manifold, modeled on spaces $C^\omega(M \leftarrow f^*T\mathcal{N})$ of real analytic sections of pullback bundles along $f : M \to \mathcal{N}$ over M.

Proof. The proof is a variant of the proof of (42.4), using a real analytic Riemannian metric if $\mathcal{N}$ is finite dimensional, and the given real analytic local addition otherwise. For finite dimensional $\mathcal{N}$ a detailed proof can be found in [Kriegl, Michor, 1990]. □

42.7. Lemma. *Let M, N be real analytic finite dimensional manifolds. Then the space $C^\omega(M,N)$ of all real analytic mappings is dense in $C^\infty(M,N)$, in the Whitney C^∞-topology.*

This is not true in the manifold topology of $\mathfrak{C}^\infty(M,N)$ used in (42.1), if M is not compact, because of the compact support condition used there.

Proof. By [Grauert, 1958, theorem 3], there is a real analytic embedding $i : N \to \mathbb{R}^k$ on a closed submanifold, for some k. We use the constant standard inner product on $\mathbb{R}^k$ to obtain a real analytic tubular neighborhood U of $i(N)$ with projection $p : U \to i(N)$. By [Grauert, 1958, proposition 8] applied to each coordinate of $\mathbb{R}^k$, the space $C^\omega(M, \mathbb{R}^k)$ of real analytic $\mathbb{R}^k$-valued functions is dense in the space $C^\infty(M, \mathbb{R}^k)$ of smooth functions, in the Whitney C^∞-topology. If $f : M \to N$ is smooth we may approximate $i \circ f$ by real analytic mappings g in $C^\omega(M, U)$, then $p \circ g$ is real analytic $M \to i(N)$ and approximates $i \circ f$. $\square$

42.8. Theorem. **C^ω-manifold structure on $\mathfrak{C}^\infty(M, N)$.** *Let M and N be real analytic finite dimensional manifolds, with M compact. Then the smooth manifold $\mathfrak{C}^\infty(M, N)$ with the structure from* (42.1) *is even a real analytic manifold.*

Proof. For a fixed real analytic exponential mapping on N the charts (U_f, u_f) from (42.1) for $f \in C^\omega(M, N)$ form a smooth atlas for $\mathfrak{C}^\infty(M, N)$, since $C^\omega(M, N)$ is dense in $\mathfrak{C}^\infty(M, N)$ by (42.7)

The chart changings $u_f \circ u_g^{-1} = (\tau_f^{-1} \circ \tau_g)_*$ are real analytic by (30.10). $\square$

42.9. Corollary. *Let M_i and $\mathcal{N}_i$ be smooth manifolds with M_i finite dimensional for $i = 1, 2$ and $\mathcal{N}_i$ admitting smooth local additions. Then we have:*

(1) *If $f : \mathcal{N}_1 \to \mathcal{N}_2$ is initial* (27.11) *then the mapping*

$$\mathfrak{C}^\infty(M, f) : \mathfrak{C}^\infty(M, \mathcal{N}_1) \to \mathfrak{C}^\infty(M, \mathcal{N}_2)$$

is initial, too.

(2) *If $f : M_2 \to M_1$ is final* (27.15) *and proper then the mapping $\mathfrak{C}^\infty(f, \mathcal{N}) : \mathfrak{C}^\infty(M_1, \mathcal{N}) \to \mathfrak{C}^\infty(M_2, \mathcal{N})$ is initial.*

Proof. (1) Let $c : \mathbb{R} \to \mathfrak{C}^\infty(M, \mathcal{N}_1)$ be such that $f_* \circ c : \mathbb{R} \to \mathfrak{C}^\infty(M, \mathcal{N}_2)$ is smooth. By (42.5), the associated mapping $(f_* \circ c)^\wedge = f \circ c^\wedge : \mathbb{R} \times M \to \mathcal{N}_2$ is smooth and satisfies (42.5.1). Since f is initial, $c^\wedge$ is smooth, and since f is injective, $c^\wedge$ satisfies (42.5.1), hence c is smooth.

Proof of (2) Since f is final between finite dimensional manifolds, it is a surjective submersion, so $\mathbb{R} \times f$ is also a surjective submersion and thus final.

Let $c : \mathbb{R} \to \mathfrak{C}^\infty(M_1, \mathcal{N})$ be such that $f^* \circ c : \mathbb{R} \to \mathfrak{C}^\infty(M_2, \mathcal{N})$ is smooth. By (42.5), the associated mapping $(f^* \circ c)^\wedge = c^\wedge \circ (\mathbb{R} \times f) : \mathbb{R} \times M_2 \to \mathcal{N}$ is smooth and satisfies (42.5.1). Since $\mathbb{R} \times f$ is also final, $c^\wedge$ is smooth. Since f and thus also $\mathbb{R} \times f$ is proper, $c^\wedge$ satisfies (42.5.1), and thus c is smooth. $\square$

42.10. Lemma. *Let M and N be real analytic finite dimensional manifolds with M compact. Let (U_α, u_α) be a real analytic atlas for M, and let $i : N \to \mathbb{R}^n$ be a closed real analytic embedding into some $\mathbb{R}^n$. Let $\mathcal{M}$ be a possibly infinite dimensional real analytic manifold.*

Then $f : \mathcal{M} \to C^\omega(M, N)$ is real analytic or smooth if and only if $C^\omega(u_\alpha^{-1}, i) \circ f : \mathcal{M} \to C^\omega(u_\alpha(U_\alpha), \mathbb{R}^n)$ is real analytic or smooth, respectively.

Furthermore, $f : \mathcal{M} \to \mathfrak{C}^\infty(M,N)$ is real analytic or smooth if and only if the mapping $C^\infty(u_\alpha^{-1}, i) \circ f : \mathcal{M} \to C^\infty(u_\alpha(U_\alpha), \mathbb{R}^n)$ is real analytic or smooth, respectively.

Proof. The statement that i_* is initial is obvious. So we just have to treat $C^\infty(u_\alpha^{-1}, N)$. The corresponding statement for spaces of sections of vector bundles are (30.6) for the real analytic case and (30.1) for the smooth case. So if f takes values in a chart domain U_g of $C^\infty(M,N)$ for a real analytic $g : M \to N$, the result follows. Recall from the proof of (42.1) that $U_g = \{h \in C^\beta(M,N) : (g(x), h(x)) \in V\}$ where V is a fixed open neighborhood of the diagonal in $N \times N$, and where $\beta = \infty$ or ω. Let $f(z_0) \in U_g$ for $z_0 \in \mathcal{M}$. Since M is covered by finitely many of its charts U_α, and since by assumption $f(z)|U_\alpha$ is near $f(z_0)|U_\alpha$ for z near z_0, so $f(z) \in U_g$ for z near z_0 in $\mathcal{M}$. So f takes values locally in charts, and the result follows. □

42.11. Corollary. *Let M and N be finite dimensional real analytic manifolds with M compact. Then the inclusion $C^\omega(M,N) \to \mathfrak{C}^\infty(M,N)$ is real analytic.*

Proof. Use the chart description and lemma (11.3). □

42.12. Lemma. Curves in spaces of mappings. *Let M and N be finite dimensional real analytic manifolds with M compact.*

(1) *A curve $c : \mathbb{R} \to C^\omega(M,N)$ is real analytic if and only if the associated mapping $c^\wedge : \mathbb{R} \times M \to N$ is real analytic.*

   The curve $c : \mathbb{R} \to C^\omega(M,N)$ is smooth if and only if $c^\wedge : \mathbb{R} \times M \to N$ satisfies the following condition:

   > *For each n there is an open neighborhood U_n of $\mathbb{R} \times M$ in $\mathbb{R} \times M_\mathbb{C}$ and a (unique) C^n-extension $\tilde{c} : U_n \to N_\mathbb{C}$ such that $\tilde{c}(t, \quad)$ is holomorphic for all $t \in \mathbb{R}$.*

(2) *The curve $c : \mathbb{R} \to \mathfrak{C}^\infty(M,N)$ is real analytic if and only if $c^\wedge$ satisfies the following condition:*

   > *For each n there is an open neighborhood U_n of $\mathbb{R} \times M$ in $\mathbb{C} \times M$ and a (unique) C^n-extension $\tilde{c} : U_n \to N_\mathbb{C}$ such that $\tilde{c}(\quad, x)$ is holomorphic for all $x \in M$.*

Note that the two conditions are in fact local in $\mathbb{R}$. We need N finite dimensional since we need an extension $N_\mathbb{C}$ of N to a complex manifold.

Proof. This follows from the corresponding statement (30.8) for spaces of sections of vector bundles, and from the chart structure on $C^\omega(M,N)$ and $\mathfrak{C}^\infty(M,N)$. □

42.13. Theorem. Smoothness of composition. *If M, $\mathcal{N}$ are smooth manifolds with M finite dimensional and $\mathcal{N}$ admitting a smooth local addition, then the evaluation mapping $\mathrm{ev} : \mathfrak{C}^\infty(M,\mathcal{N}) \times M \to \mathcal{N}$ is smooth.*

If P is another smooth finite dimensional manifold, then the composition mapping

$$\operatorname{comp} : \mathfrak{C}^\infty(M,N) \times \mathfrak{C}^\infty_{\text{prop}}(P,M) \to \mathfrak{C}^\infty(P,N)$$

is smooth, where $\mathfrak{C}^\infty_{\text{prop}}(P,M)$ denotes the space of all proper smooth mappings $P \to M$ (i.e. compact sets have compact inverse images). This space is open in $\mathfrak{C}^\infty(P,M)$.

In particular, $f_* : \mathfrak{C}^\infty(M,\mathcal{N}) \to \mathfrak{C}^\infty(M,\mathcal{N}')$ and $g^* : \mathfrak{C}^\infty(M,\mathcal{N}) \to \mathfrak{C}^\infty(P,\mathcal{N})$ are smooth for $f \in C^\infty(\mathcal{N},\mathcal{N}')$ and $g \in \mathfrak{C}^\infty_{\text{prop}}(P,M)$.

The corresponding statement for real analytic mappings can be shown along similar lines, using (42.12). But we will give another proof in (42.15) below.

Proof. Using the description of smooth curves in $\mathfrak{C}^\infty(M,\mathcal{N})$ given in (42.5), we immediately see that $(\operatorname{ev}\circ(c_1,c_2))(t) = c_1^\wedge(t,c_2(t))$ is smooth for each smooth $(c_1,c_2) : \mathbb{R} \to \mathfrak{C}^\infty(M,\mathcal{N}) \times M$, so ev is smooth as claimed.

The space of proper mappings $\mathfrak{C}^\infty_{\text{prop}}(P,M)$ is open in the manifold $\mathfrak{C}^\infty(P,M)$ since changing a mapping only on a compact set does not change its property of being proper. Let $(c_1,c_2) : \mathbb{R} \to \mathfrak{C}^\infty(M,\mathcal{N}) \times \mathfrak{C}^\infty_{\text{prop}}(P,M)$ be a smooth curve. Then we have $(\operatorname{comp}\circ(c_1,c_2))(t)(p) = c_1^\wedge(t,c_2^\wedge(t,p))$, and one may check that this has again property (44.5.1), so it is a smooth curve in $\mathfrak{C}^\infty(P,\mathcal{N})$. Thus, comp is smooth. □

42.14. Theorem. Exponential law. *Let $\mathcal{M}$, M, and $\mathcal{N}$ be smooth manifolds with M finite dimensional and $\mathcal{N}$ admitting a smooth local addition.*

Then we have a canonical injection

$$C^\infty(\mathcal{M},\mathfrak{C}^\infty(M,\mathcal{N})) \subseteq C^\infty(\mathcal{M}\times M,\mathcal{N}),$$

where the image in the right hand side consists of all smooth mappings $f : \mathcal{M}\times M \to \mathcal{N}$ which satisfy the following property

(1) *If $\mathcal{M}$ is locally metrizable then for each point $x \in \mathcal{M}$ there is an open neighborhood $\mathcal{U}$ and a compact set $K \subset M$ such that $f(x,y)$ is constant in $x \in \mathcal{U}$ for all $y \in M \setminus K$.*

(2) *For general $\mathcal{M}$: For each $c \in C^\infty(\mathbb{R},\mathcal{M})$ and each $t \in \mathbb{R}$ there exists a neighborhood U of t and a compact set $K \subset M$ such that $f(c(s),y)$ is constant in $s \in U$ for each $y \in M \setminus K$.*

Under the assumption that $\mathcal{N}$ admits smooth functions which separate points, we have equality if and only if M is compact, or $\mathcal{N}$ is discrete, or each $f \in C^\infty(\mathcal{M},\mathbb{R})$ is constant along all smooth curves into $\mathcal{M}$.

If M and $\mathcal{N}$ are real analytic manifolds with M compact we have

$$C^\omega(\mathcal{M},C^\omega(M,\mathcal{N})) = C^\omega(\mathcal{M}\times M,\mathcal{N})$$

for each real analytic (possibly infinite dimensional) manifold $\mathcal{M}$.

Proof. The smooth case is simple: The equivalence for general $\mathcal{M}$ follows directly from the description of all smooth curves in $\mathfrak{C}^\infty(M,\mathcal{N})$ given in the proof of (42.5).

It remains to show that for locally metrizable $\mathcal{M}$ a smooth mapping $f : \mathcal{M} \to \mathfrak{C}^\infty(M, \mathcal{N})$ satisfies condition (1). Since f is smooth, locally it has values in a chart, so we may assume that $\mathcal{M}$ is open in a Fréchet space by (4.19), and that f has values in $C^\infty_c(M \leftarrow E)$ for some vector bundle $p : E \to M$.

We claim that f locally factors into some $C^\infty_{K_n}(E)$ where (K_n) is an exhaustion of M by compact subsets such that K_n is contained in the interior of K_{n+1}. If not there exist a (fast) converging sequence (y_n) in $\mathcal{M}$ and $x_n \notin K_n$ such that $f(y_n)(x_n) \neq 0$. One may find a proper smooth curve $e : \mathbb{R} \to M$ with $e(n) = x_n$ and a smooth curve $g : \mathbb{R} \to \mathcal{M}$ with $g(1/n) = y_n$. Then by (30.4), $\operatorname{Pt}(e, \quad)^* \circ f \circ g$ is a smooth curve into $C^\infty_c(\mathbb{R}, E_{e(0)})$. Since the latter space is a strict inductive limit of spaces $C^\infty_I(\mathbb{R}, E_{e(0)})$ for compact intervals I, the curve $\operatorname{Pt}(e, \quad)^* \circ f \circ g$ locally factors into some $C^\infty_I(\mathbb{R}, E_{e(0)})$, but $(e^* \circ f \circ g)(1/n)(n) = f(y_n)(x_n) \neq 0$, a contradiction.

We check now the statement on equality: if M is compact, or if $\mathcal{N}$ is discrete then (2) is automatically satisfied. If each $f \in C^\infty(\mathcal{M}, \mathbb{R})$ is constant along all smooth curves into $\mathcal{M}$, we may check global constancy in (2) by composing with smooth functions on $\mathcal{N}$ which separate points there.

For the converse, we may assume that there are a function $f \in C^\infty(\mathcal{M}, \mathbb{R})$, a curve $c \in C^\infty(\mathbb{R}, \mathcal{M})$ such that $f \circ c$ is not constant, and an injective smooth curve $e : \mathbb{R} \to \mathcal{N}$. Then $\mathcal{M} \times M \ni (x, y) \mapsto e(f(x))$ is in $C^\infty(\mathcal{M} \times M, \mathcal{N}) \setminus C^\infty(\mathcal{M}, \mathfrak{C}^\infty(M, \mathcal{N}))$ since condition (2) is violated for the curve c.

Now we treat the real analytic case. Let $f^\wedge \in C^\omega(\mathcal{M} \times M, \mathcal{N}) \subset C^\infty(\mathcal{M} \times M, \mathcal{N}) = C^\infty(M, \mathfrak{C}^\infty(M, \mathcal{N}))$. So we may restrict f to a neighborhood U in $\mathcal{M}$, where it takes values in a chart U_g of $C^\infty(M, \mathcal{N})$ for $g \in C^\omega(M, \mathcal{N})$. Then $f(U) \subset U_g \cap C^\omega(M, \mathcal{N})$, one of the canonical charts of $C^\omega(M, \mathcal{N})$. So we may assume that $f : U \to C^\omega(M \leftarrow g^*T\mathcal{N})$. For a real analytic vector bundle atlas (U_α, ψ_α) of $g^*T\mathcal{N}$ the composites $U \to C^\omega(M \leftarrow g^*T\mathcal{N}) \to C^\omega(U_\alpha, \mathbb{R}^n)$ are real analytic by applying cartesian closedness (11.18) to the mapping $(x, y) \mapsto \psi_\alpha(\pi_\mathcal{N}, \exp)^{-1}(g(y), f^\wedge(x, y))$. By the description (30.6) of the structure on $C^\omega(M \leftarrow g^*T\mathcal{N})$, the chart representation of f is real analytic, so f is it also.

Let conversely $f : \mathcal{M} \to C^\omega(M, \mathcal{N})$ be real analytic. Then its chart representation is real analytic and we may use càrtesian closedness in the other direction to conclude that $f^\wedge$ is real analytic. □

42.15. Corollary. *If M and $\mathcal{N}$ are real analytic manifolds with M compact and $\mathcal{N}$ admitting a real analytic local addition, then the evaluation mapping* $\operatorname{ev} : C^\omega(M, \mathcal{N}) \times M \to \mathcal{N}$ *is real analytic.*

If P is another compact real analytic manifold, then the composition mapping $\operatorname{comp} : C^\omega(M, \mathcal{N}) \times C^\omega(P, M) \to C^\omega(P, \mathcal{N})$ *is real analytic.*

In particular, $f_* : C^\omega(M, \mathcal{N}) \to C^\omega(M, \mathcal{N}')$ and $g^* : C^\omega(M, \mathcal{N}) \to C^\omega(P, \mathcal{N})$ are real analytic for real analytic $f : \mathcal{N} \to \mathcal{N}'$ and $g \in C^\omega(P, M)$.

Proof. The mapping $\operatorname{ev}^\vee = \operatorname{Id}_{C^\omega(M, \mathcal{N})}$ is real analytic, so ev too, by (42.14).

The mapping $\mathrm{comp}^\wedge = \mathrm{ev} \circ (\mathrm{Id}_{C^\omega(M,\mathcal{N})} \times \mathrm{ev}) : C^\omega(M,\mathcal{N}) \times C^\omega(P,M) \times P \to C^\omega(M,\mathcal{N}) \times M \to \mathcal{N}$ is real analytic, thus comp too. $\square$

42.16. Lemma. *Let M_i and N_i be finite dimensional real analytic manifolds with M_i compact. Then for $f \in C^\infty(N_1, N_2)$ the push forward $f_* : \mathfrak{C}^\infty(M, N_1) \to \mathfrak{C}^\infty(M, N_2)$ is real analytic if and only if f is real analytic. For $f \in C^\infty(M_2, M_1)$ the pullback $f^* : \mathfrak{C}^\infty(M_1, N) \to \mathfrak{C}^\infty(M_2, N)$ is, however, always real analytic.*

Proof. If f is real analytic and if $g \in C^\omega(M, N_1)$, then the mapping $u_{f\circ g} \circ f_* \circ u_g^{-1} : C^\infty(M \leftarrow g^*TN_1) \to C^\infty(M \leftarrow (f\circ g)^*TN_2)$ is a push forward by the real analytic mapping $(\mathrm{pr}_1, (\pi, \exp^{N_2})^{-1} \circ (f \circ g \circ \mathrm{pr}_1, f \circ \exp^{N_1} \circ \mathrm{pr}_2)) : g^*TN_1 \to (f \circ g)^*TN_2$, which is real analytic by (30.10).

The canonical mapping $\mathrm{ev}_x : \mathfrak{C}^\infty(M, N_2) \to N_2$ is real analytic since $\mathrm{ev}_x |U_g = \exp^{N_2} \circ \mathrm{ev}_x \circ u_g : U_g \to C^\infty(M \leftarrow g^*TN_2) \to T_{g(x)}N_2 \to N_2$, where the second ev_x is linear and bounded. Furthermore, $\mathrm{const} : N_1 \to \mathfrak{C}^\infty(M, N_1)$ is real analytic since the mapping $u_g \circ \mathrm{const} : y \mapsto (x \mapsto (\pi_{N_1}, \exp^{N_1})^{-1}(g(x), y))$ is locally real analytic into $C^\omega(M \leftarrow g^*TN_1)$ and hence into $C^\infty(M \leftarrow g^*TN_1)$.

If f_* is real analytic, also $f = \mathrm{ev}_x \circ f_* \circ \mathrm{const}$ is.

For the second statement choose real analytic atlas (U^i_α, u^i_α) of M_i such that $f(U^2_\alpha) \subseteq U^1_\alpha$ and a closed real analytic embedding $j : N \to \mathbb{R}^n$. Then the diagram

$$\begin{array}{ccc}
\mathfrak{C}^\infty(M_1, N) & \xrightarrow{\quad f^* \quad} & \mathfrak{C}^\infty(M_2, N) \\
{\scriptstyle \mathfrak{C}^\infty((u^1_\alpha)^{-1}, j)}\downarrow & & \downarrow{\scriptstyle \mathfrak{C}^\infty((u^2_\alpha)^{-1}, j)} \\
\mathfrak{C}^\infty(u^1_\alpha(U^1_\alpha), \mathbb{R}^n) & \xrightarrow{(u^2_\alpha \circ f \circ (u^1_\alpha)^{-1})^*} & \mathfrak{C}^\infty(u^2_\alpha(U^2_\alpha), \mathbb{R}^n)
\end{array}$$

commutes, the bottom arrow is a continuous and linear mapping, so it is real analytic. Thus, by (42.10), the mapping f^* is real analytic. $\square$

42.17. Theorem. *Let M and $\mathcal{N}$ be smooth manifolds with M compact and $\mathcal{N}$ admitting a local addition. Then the infinite dimensional smooth vector bundles $T\mathfrak{C}^\infty(M,\mathcal{N})$ and $\mathfrak{C}^\infty_c(M, T\mathcal{N}) \subset \mathfrak{C}^\infty(M, T\mathcal{N})$ over $\mathfrak{C}^\infty(M,\mathcal{N})$ are canonically isomorphic. The same assertion is true for $C^\omega(M,\mathcal{N})$ if M is compact.*

Here by $\mathfrak{C}^\infty_c(M, T\mathcal{N})$ we denote the space of all smooth mappings $f : M \to T\mathcal{N}$ such that $f(x) = 0_{\pi_M f(x)}$ for $x \notin K_f$, a suitable compact subset of M (equivalently, such that the associated section of the pull back bundle $(\pi_M \circ f)^*T\mathcal{N} \to M$ has compact support).

One can check directly that the atlas from (42.1) for $\mathfrak{C}^\infty(M,\mathcal{N})$ induces an atlas for $T\mathfrak{C}^\infty(M,\mathcal{N})$, which is equivalent to that for $\mathfrak{C}^\infty(M, T\mathcal{N})$ via some natural identifications in $TT\mathcal{N}$. This is carried out in great detail in [Michor, 1980c, 10.13]. We shall give here a simpler proof, using (42.5).

Proof. Recall from (28.13) the diagram

$$\begin{array}{ccc} C^\infty(\mathbb{R},\mathfrak{C}^\infty(M,\mathcal{N}))/\sim & \longleftarrow & C^\infty(\mathbb{R},\mathfrak{C}^\infty(M,\mathcal{N})) \\ \cong\Big\downarrow\delta & \swarrow\delta & \Big\downarrow \mathrm{ev}_0 \\ T\mathfrak{C}^\infty(M,\mathcal{N}) & \xrightarrow[\pi_{\mathfrak{C}^\infty(M,\mathcal{N})}]{} & \mathfrak{C}^\infty(M,\mathcal{N}), \end{array}$$

From (42.5) we see that $C^\infty(\mathbb{R},\mathfrak{C}^\infty(M,\mathcal{N}))$ corresponds to the space $C^\infty_{lc}(\mathbb{R}\times M,\mathcal{N})$ of all mappings $g^\wedge\in C^\infty(\mathbb{R}\times M,\mathcal{N})$ satisfying

(1) For each compact interval $[a,b]\subset\mathbb{R}$ there is a compact subset $K\subset M$ such that $g^\wedge(t,x)$ is constant in $t\in[a,b]$ for all $x\in M\setminus K$.

Now we consider the diagram

$$\begin{array}{ccccc} C^\infty(\mathbb{R},\mathfrak{C}^\infty(M,\mathcal{N})) & \xrightarrow[\varphi]{\cong} & C^\infty_{lc}(\mathbb{R}\times M,\mathcal{N}) & \hookrightarrow & C^\infty(\mathbb{R}\times M,\mathcal{N}) \\ \delta\Big\downarrow & & & & \\ C^\infty(\mathbb{R},\mathfrak{C}^\infty(M,\mathcal{N}))/\sim & & \Big\downarrow \frac{\partial}{\partial t}\Big|_0 & & \Big\downarrow \frac{\partial}{\partial t}\Big|_0 \\ \cong\Big\downarrow\delta & & & & \\ T\mathfrak{C}^\infty(M,\mathcal{N}) & \overset{\Phi}{\underset{\cong}{\dashrightarrow}} & \mathfrak{C}^\infty_c(M,T\mathcal{N}) & \hookrightarrow & \mathfrak{C}^\infty(M,T\mathcal{N}). \end{array}$$

The vertical mappings on the right hand side are $\frac{\partial}{\partial t}\big|_0 f = Tf\circ(\partial_t\times 0_M)|(0\times M)$. The middle one is surjective since $f(x)=\frac{\partial}{\partial t}\big|_0\exp(h(t).f(x))$ for suitable h, and h can be chosen uniformly for f in a piece of a smooth curve into $\mathfrak{C}^\infty(M,T\mathcal{N})$. By construction the top isomorphism factors to a bijection Φ.

The mapping Φ is smooth by (28.13) since $\Phi\circ\delta$ factors over φ, which maps the space $C^\infty(\mathbb{R}^2,\mathfrak{C}^\infty(M,\mathcal{N}))$ to $C^\infty_{lc,c}(\mathbb{R}\times M,T\mathcal{N})\cong C^\infty(\mathbb{R},\mathfrak{C}^\infty_c(M,T\mathcal{N}))$. The inverse of Φ is smooth by a similar argument, using again (28.13). $\square$

42.18. Corollary. Some tangent mappings. *For $f\in C^\infty(M_1,M_2)$ and $g\in C^\infty(\mathcal{N}_1,\mathcal{N}_2)$ we have*

$$\begin{array}{ccc} T\mathfrak{C}^\infty(M_2,\mathcal{N}) & \cong & \mathfrak{C}^\infty_c(M_2,T\mathcal{N}) \\ T\mathfrak{C}^\infty(f,\mathcal{N})\Big\downarrow & & \Big\downarrow\mathfrak{C}^\infty(f,T\mathcal{N}) \\ T\mathfrak{C}^\infty(M_1,\mathcal{N}) & \cong & \mathfrak{C}^\infty_c(M_1,T\mathcal{N}) \end{array}\qquad \begin{array}{ccc} T\mathfrak{C}^\infty(M,\mathcal{N}_1) & \cong & \mathfrak{C}^\infty_c(M,T\mathcal{N}_1) \\ T\mathfrak{C}^\infty(M,g)\Big\downarrow & & \Big\downarrow\mathfrak{C}^\infty(M,Tg) \\ T\mathfrak{C}^\infty(M,\mathcal{N}_2) & \cong & \mathfrak{C}^\infty_c(M,T\mathcal{N}_2). \end{array}$$

The tangent mapping of the composition

$$\mathrm{comp}:\mathfrak{C}^\infty(M,\mathcal{N})\times\mathfrak{C}^\infty_{\mathrm{prop}}(P,M)\to\mathfrak{C}^\infty(P,\mathcal{N})$$

at (f,g) in direction of $(X,Y)\in C^\infty_c(M\leftarrow f^*T\mathcal{N})\times C^\infty_c(P\leftarrow g^*TM)$ is given by

$$T_{(f,g)}\,\mathrm{comp}.(X,Y)=Tf\circ Y+X\circ g\in C^\infty_c(P\leftarrow(f\circ g)^*T\mathcal{N}).$$

The tangent mapping of the evaluation $\mathrm{ev} : \mathfrak{C}^\infty(M,\mathcal{N}) \times M \to \mathcal{N}$ *at* (f,x) *in direction of* $(X,\xi) \in C^\infty_c(M \leftarrow f^*T\mathcal{N}) \times T_xM$ *is given by* $T_{(f,x)}\mathrm{ev}.(X,\xi) = T_xf.\xi + X(x) \in T_{f(x)}\mathcal{N}$.

Proof. By (42.17), we may take a tangent vector $X \in T_{f(0,\ \ )}\mathfrak{C}^\infty(M,\mathcal{N}_1)$ of the form $X = \frac{\partial}{\partial t}\big|_0 f(t,\ \ ) \in C^\infty_c(M \leftarrow f^*T\mathcal{N}_1)$, where $f \in C^\infty_{lc}(\mathbb{R} \times M, \mathcal{N}_1)$. Then we have $(T_f(g_*).X)(x) = \frac{\partial}{\partial t}\big|_0 g(f(t,x)) = Tg.\frac{\partial}{\partial t}\big|_0 f(t,x) = Tg.X(x)$.

$T(g^*) = g^*$ is similar but easier, and the tangent mappings of the composition and the evaluation can be computed either from the partial derivatives, or directly by a variational computation as above. $\square$

42.19. The tangent mapping $T : \mathfrak{C}^\infty(M,\mathcal{N}) \to \mathfrak{C}^\infty(TM,T\mathcal{N})$ is not smooth, since the condition (42.5.1) is not preserved. But it is smooth as a mapping $T : \mathfrak{C}^\infty(M,\mathcal{N}) \to \mathfrak{C}^\infty(M, L(TM,T\mathcal{N}))$, and its tangent mapping is given by

$$\begin{array}{ccc} T\mathfrak{C}^\infty(M,\mathcal{N}) & \xrightarrow{\ \cong\ } & \mathfrak{C}^\infty_c(M,T\mathcal{N}) \\ {\scriptstyle T(T)}\downarrow & & \downarrow{\scriptstyle (\kappa_{\mathcal{N}})_*\circ T} \\ T\mathfrak{C}^\infty(M,L(TM,T\mathcal{N})) & \subseteq & C^\infty(TM,T^2\mathcal{N}), \end{array}$$

where $\kappa_{\mathcal{N}} : T^2\mathcal{N} \to T^2\mathcal{N}$ is the canonical flip mapping, compare with (29.10).

For the tangent mapping of the tangent mapping we consider $\xi_x = \frac{\partial}{\partial s}|_0 c(s) \in T_xM$, and $X \in T_{f(0,\ \ )}\mathfrak{C}^\infty(M,\mathcal{N})$ of the form $X = \frac{\partial}{\partial t}\big|_0 f(t,\ \ ) \in C^\infty_c(M \leftarrow f^*T\mathcal{N})$ as in the beginning of the proof. Then we have

$$\begin{aligned} T(f(t,\ \ )).\xi_x &= \tfrac{\partial}{\partial s}|_0 f(t,c(s)) \\ (T_{f(0,\ \ )}(T).X)(\xi_x) &= \big(\tfrac{\partial}{\partial t}\big|_0 T(f(t,\ \ ))\big)(\xi_x) = \tfrac{\partial}{\partial t}\big|_0 \big(T(f(t,\ \ )).\xi_x\big) \\ &= \tfrac{\partial}{\partial t}\big|_0 \tfrac{\partial}{\partial s}|_0 f(t,c(s)) = \kappa_{\mathcal{N}} \tfrac{\partial}{\partial s}|_0 \tfrac{\partial}{\partial t}\big|_0 f(t,c(s)) \\ &= \kappa_{\mathcal{N}} \tfrac{\partial}{\partial s}|_0 X(c(s)) = \kappa_{\mathcal{N}}.TX.\xi_x. \end{aligned}$$

42.20. Theorem. *Let* M *and* N *be smooth finite dimensional manifolds, and let* $q : N \to M$ *be smooth. Then the set* $\mathfrak{C}^\infty(q)$ *of all smooth sections of* q *is a splitting smooth submanifold of* $\mathfrak{C}^\infty(M,N)$, *whose tangent space is given by* $T\mathfrak{C}^\infty(q) = \mathfrak{C}^\infty_c(M, \ker(Tq)) \subset \mathfrak{C}^\infty_c(M,TN)$. *If* $q : E \to M$ *is a finite dimensional vector bundle, the convenient vector space* $C^\infty_c(M \leftarrow E)$ *is a splitting smooth submanifold of* $\mathfrak{C}^\infty(M,E)$.

Let now M *and* N *be real analytic finite dimensional manifolds with* M *compact, and let* $q : N \to M$ *be real analytic. Then the set* $C^\omega(q)$ *of all real analytic sections of* q *is a splitting real analytic submanifold of* $C^\omega(M,N)$, *and also* $\mathfrak{C}^\infty(q)$ *is a is a splitting real analytic submanifold of* $\mathfrak{C}^\infty(M,N)$. *If* $q : E \to M$ *is a real analytic finite dimensional vector bundle with* M *compact, the convenient vector space* $C^\omega(M \leftarrow E)$ *is a splitting real analytic submanifold of* $C^\omega(M,E)$, *and* $C^\infty(M \leftarrow E)$ *is a splitting real analytic submanifold of* $\mathfrak{C}^\infty(M,E)$.

It is possible to extend this result at least to the case of a fiber bundle $p : E \to M$ with infinite dimensional standard fiber by requiring certain properties. We do not present it here since in the only possible application (42.21) we have a simpler direct proof.

Proof. If a smooth section $s : M \to N$ of q exists, then q, restricted to an open neighborhood of $s(M)$, is a surjective submersion. Thus, there exists an open neighborhood W_s of $s(M)$ in N such that $p_s := s \circ q|W_s : W_s \to s(M)$ is a surjective submersion, and we may assume that W_s is a tubular neighborhood, so that $p_s : W_s \to s(M)$ is a vector bundle. Since $\mathfrak{C}^\infty(M, W_s)$ is open in $\mathfrak{C}^\infty(M, N)$, we may replace N by W_s or assume that $q : N \to M$ is a vector bundle, and that s is the zero section.

Claim. There exists a local addition $\alpha : TN \to N$ such that

(1) α restricts to a local addition $T0(M) \to 0(M)$ on the zero section.
(2) On each fiber N_x the local addition α restricts to the addition $TN_x \cong N_x \times N_x \to N_x$.

In fact, choose a second vector bundle $E \to M$ such that $N \oplus E = M \times \mathbb{R}^k$ is trivial, choose a local addition α_M on M, and let α_k be the addition on $\mathbb{R}^k$. Then $\alpha_M \times \alpha_k$ restricts to a local addition on N with the required properties.

Now we consider the atlas for $\mathfrak{C}^\infty(M, N)$ induced by α, as in (42.4), i.e., we use the formulas of (42.1) with exp replaced by α. In particular, for the zero section $s = 0$ and for $g \in U_0 \subset \mathfrak{C}^\infty(M, N)$ we have

$$\begin{aligned} u_0(g) = (\mathrm{Id}_M, (\pi_N, \alpha)^{-1} \circ (0, g)) \in C_c^\infty(M \leftarrow 0^*TN) \cong \\ \cong C_c^\infty(M \leftarrow TM \oplus N) \cong C_c^\infty(M \leftarrow TM) \times C_c^\infty(M \leftarrow N), \end{aligned}$$

so that $u_0(g) \in 0 \times C_c^\infty(M \leftarrow N)$ if and only if g is a section of the vector bundle. Moreover, $C_c^\infty(M \leftarrow N) \subset U_0$, so the second statement follows.

The statement about $T\mathfrak{C}^\infty(q)$ follows from (42.17) by noting that the derivative of smooth curves in $\mathfrak{C}^\infty(q)$ are precisely sections $s : M \to \ker(Tq)$ such that $s = 0 \circ \pi_E \circ s$ off some compact set in M.

This proof also works in the real analytic cases. □

42.21. Theorem. *Let $(p : P \to M, G)$ be a principal fiber bundle with finite dimensional base manifold M and a possibly infinite dimensional Lie group G as structure group.*

Then the gauge group $\mathrm{Gau}(P) = \mathfrak{C}^\infty(M \leftarrow P[G, \mathrm{conj}])$ *from* (37.17) *carries the structure of a smooth Lie group modeled on* $C_c^\infty(P[\mathfrak{g}, \mathrm{Ad}])$.

If G is a regular Lie group then $\mathrm{Gau}(P)$ *is regular, too. If G admits an exponential mapping then* $\mathrm{Gau}(P)$ *also admits an exponential mapping. If G is compact then* $\mathrm{Gau}(P)$ *is diffeomorphic to the splitting submanifold* $\mathfrak{C}^\infty(P, G)^G$ *of all G-equivariant smooth mappings in* $\mathfrak{C}^\infty(P, G)$.

If, moreover, M is compact and $(p : P \to M, G)$ is a real analytic principal bundle with real analytic Lie group G, possibly infinite dimensional, then $\mathrm{Gau}^\omega(P) :=$

$C^\omega(M \leftarrow P[G,\text{conj}])$ *is a real analytic Lie group with the corresponding properties as above.*

Proof. The associated bundle $P[G,\text{conj}] = P \times_{(G,\text{conj})} G$ is a group bundle over M with typical fiber G. It admits transition functions with values in $\text{Aut}(G)$. Therefore, the multiplication in G induces a smooth fiberwise group multiplication $\mu : P[G,\text{conj}] \times_M P[G,\text{conj}] \to P[G,\text{conj}]$, also the fiberwise inversion $\nu : P[G,\text{conj}] \to P[G,\text{conj}]$ is smooth.

The associated bundle $P[\mathfrak{g},\text{Ad}] = P \times_{(G,\text{Ad})} \mathfrak{g}$ is a bundle of Lie algebras with the same cocycle of transition functions. Thus, the bracket in $\mathfrak{g}$ induces a smooth fiberwise bilinear mapping $[\quad,\quad] : P[\mathfrak{g},\text{Ad}] \times_M P[\mathfrak{g},\text{Ad}] \to P[\mathfrak{g},\text{Ad}]$ and $C_c^\infty(M \leftarrow P[\mathfrak{g}\text{Ad}])$ is a convenient Lie algebra.

We shall use the canonical mappings $q : P \times \mathfrak{g} \to P \times_G \mathfrak{g} = p[\mathfrak{g},\text{Ad}]$ from (37.12.1), $\tau^G : P \times_M P \to G$ from (37.8), and $\tau^{\mathfrak{g}} : P \times_M P[\mathfrak{g},\text{Ad}] \to \mathfrak{g}$ from (37.12). We also recall the the bijection $C^\infty(P,\mathfrak{g})^G \cong C^\infty(M \leftarrow P[\mathfrak{g}])$ from (37.16), denoted by $f \mapsto s_f$ and given by $s_f(p(u)) = q(u, f(u))$, with inverse $s \mapsto f_s = \tau^{\mathfrak{g}} \circ (\text{Id}_P, s \circ p)$.

Let $u : U \to V \subseteq \mathfrak{g}$ be a chart of G centered at e. Note that any model space of a Lie group is isomorphic to the Lie algebra.

$$\begin{aligned}
U_\alpha &:= \{\chi \in \text{Gau}(P) : \tau(\alpha(z),\chi(z)) \in U \text{ for all } z \in P \\
&\qquad\qquad \text{and } p(\{x : \alpha(x) \neq \chi(x)\}) \text{ has compact closure in } M\},\\
\bar u_\alpha &: U_\alpha \to C^\infty(P,\mathfrak{g})^G,\\
\bar u_\alpha(\chi) &= u \circ \tau \circ (\alpha,\chi),\\
u_\alpha &: U_\alpha \to \tilde V := \{s \in C_c^\infty(M \leftarrow P[\mathfrak{g},\text{Ad}]) : \tau^{\mathfrak{g}}(z, s(p(z))) \in V \text{ for all } z \in P\}\\
&\qquad \subseteq C_c^\infty(M \leftarrow P[\mathfrak{g},\text{Ad}]),\\
u_\alpha(\chi) &= s_{\bar u_\alpha(\chi)} = s_{u\circ\tau\circ(\alpha,\chi)},\\
u_\alpha^{-1}(s)(z) &= \alpha(z).u^{-1}(\tau^{\mathfrak{g}}(z, s(p(z)))),\\
u_\alpha^{-1}(s) &= \alpha.(u^{-1}.\tau^{\mathfrak{g}}(\text{Id}_P, s\circ p)).
\end{aligned}$$

For the chart change we see that for $s \in u_\beta(U_\alpha \cap U_\beta)$ we have

$$(u_\alpha \circ u_\beta^{-1})(s)(p(z)) = q\big(z, u(\tau^P(\alpha(z), \beta(z).u^{-1}(\tau^{\mathfrak{g}}(z, s(p(z)))))))\big).$$

By (30.9.1), the space of smooth curves $C^\infty(\mathbb{R}, C_c^\infty(M \leftarrow P[\mathfrak{g},\text{Ad}]))$ consists of all sections c such that $c^\wedge : \mathbb{R} \times M \to P[\mathfrak{g},\text{Ad}]$ is smooth and the following condition holds:

(1) For each compact interval $[a,b] \subset \mathbb{R}$ there is a compact subset $K \subset M$ such that $c^\wedge(t,x)$ is constant in $t \in [a,b]$ for all $x \in M \setminus K$.

Obviously, the chart change respects the set of smooth curves and is smooth. Thus, the atlas (U_α, u_α) describes the structure of a smooth manifold which we denote by $\text{Gau}(P) \cong \mathfrak{C}^\infty(M \leftarrow P[G,\text{conj}])$, and we also see that the space of smooth curves $C^\infty(\mathbb{R}, \mathfrak{C}^\infty(M \leftarrow P[G,\text{conj}]))$ consists of all sections c such that the associated mapping $c^\wedge : \mathbb{R} \times M \to P[G, \mathrm{conj}]$ is smooth and condition (1) holds. Composition and inversion are smooth on $\mathrm{Gau}(P)$ since these correspond just to push forwards of sections via the smooth fiberwise group multiplication and inversion described at the beginning of the proof. The Lie algebra of $\mathfrak{C}^\infty(M \leftarrow P[G, \mathrm{conj}])$ is $C_c^\infty(M \leftarrow P[\mathfrak{g}, \mathrm{Ad}])$.

Let us now suppose that the Lie group G is regular with evolution operator $\mathrm{evol}_G : C^\infty(\mathbb{R}, \mathfrak{g}) \to G$. Since the smooth group bundle $P[G, \mathrm{conj}]$ is described by a cocycle of transition functions with values in the group of (inner) automorphisms of G and since by (38.4) we have $\mathrm{evol}_G \circ \varphi'_* = \varphi \circ \mathrm{evol}_G$ for any automorphism φ of G, there is an induced fiberwise evolution operator

$$\mathrm{evol} : P[C^\infty(\mathbb{R}, \mathfrak{g}), \mathrm{Ad}_*] \cong C^\infty(\mathbb{R}, p[\mathfrak{g}, \mathrm{Ad}]) \to P[G, \mathrm{conj}],$$

which, by push forward on sections, induces

$$\mathrm{evol}_{\mathrm{Gau}(P)} : C^\infty(\mathbb{R}, C_c^\infty(M \leftarrow P[\mathfrak{g}, \mathrm{Ad}])) \to C_c^\infty(M \leftarrow P[G, \mathrm{conj}]).$$

This maps smooth curves to smooth curves and is the evolution operator of $\mathrm{Gau}(P)$. The remaining assertions are easy to check. □

42.22. Manifolds of holomorphic mappings. It is a natural question whether the methods of this section carry over to spaces of holomorphic mappings between complex manifolds. The situation is described in the following result.

Lemma.

(1) *Each finite dimensional Stein manifold M admits holomorphic local additions $TM \supset U \to M$ in the sense of* (42.4).

(2) *Complex projective spaces do not admit holomorphic local additions.*

Proof. (1) A Stein manifold M is biholomorphically embedded as a closed complex submanifold of some $\mathbb{C}^n$ (where $n = 2\dim_{\mathbb{C}} M + 2$ suffices), see [Gunning, Rossi, 1965, p. 224], and there exists a holomorphic tubular neighborhood $p : V \to M$ in $\mathbb{C}^n$, see [Gunning, Rossi, 1965, p. 257], by an application of Cartan's theorem B that a coherent sheaf on a Stein manifold is acyclic. The affine addition $\varphi : T\mathbb{C}^n \to \mathbb{C}^n$, given by $\varphi(z, Z) := z + Z$ then gives a local addition $p \circ \varphi | TM : TM \supset U \to M$ for a suitable open neighborhood U of 0 in TM.

(2) First we show that $\mathbb{CP}^1$ does not admit a holomorphic local addition. The usual affine charts $u_0[z_0 : z_1] = \frac{z_1}{z_0}$ and $u_1[z_0 : z_1] = \frac{z_0}{z_1}$ have as chart change mapping $z \mapsto 1/z$ on $\mathbb{C} \setminus \{0\}$. Its tangent mapping is $(z, w) \mapsto (\frac{1}{z}, -\frac{1}{z^2}w)$. A local addition would be given by two holomorphic mappings $\alpha_i : T\mathbb{C} \supset U \to \mathbb{C}$ on an open neighborhood U of the zero section $\{(z, 0) : z \in \mathbb{C}\}$ with

$$\alpha_0(\tfrac{1}{z}, -\tfrac{1}{z^2}w) = \frac{1}{\alpha_1(z, w)} \text{ for } z \neq 0,$$
$$\alpha_i(z, 0) = z, \quad \tfrac{\partial}{\partial w}|_{w=0}\alpha_i(z, w) \neq 0 \text{ for all } z.$$

The derivatives $\frac{\partial}{\partial w}|_{w=0}$ and $\frac{\partial^2}{\partial w^2}|_{w=0}$ of $1=\alpha_0(\frac{1}{z},-\frac{1}{z^2}w)\alpha_1(z,w)$ are in turn

$$\begin{aligned}0&=\tfrac{1}{z}(\partial_2\alpha_1(z,0)-\partial_2\alpha_0(\tfrac{1}{z},0)),\\ 0&=\tfrac{1}{z}\partial_2^2\alpha_1(z,0)-\tfrac{2}{z^2}\partial_2\alpha_0(\tfrac{1}{z},0).\partial_2\alpha_1(z,0)+\tfrac{1}{z^3}\partial_2^2\alpha_0(\tfrac{1}{z},0)\\ &=\tfrac{1}{z^3}(z^2\partial_2^2\alpha_1(z,0)-z(\partial_2\alpha_1(z,0))^2+\partial_2^2\alpha_0(\tfrac{1}{z},0)).\end{aligned}$$

Hence, $\lim_{z\to 0}\partial_2^2\alpha_0(\frac{1}{z},0)=0$, and consequently $\partial_2^2\alpha_0(z,0)=0$ for all z since it is an entire function on $\mathbb{C}$ which vanishes at infinity. But then we get that $z\mapsto\partial_2^2\alpha_1(z,0)=\frac{1}{z}(\partial_2\alpha_1(z,0))^2-0$ has a pole at 0, a contradiction.

Now we treat $\mathbb{CP}^n$. Suppose that a holomorphic local addition $\alpha:T(\mathbb{CP}^n)\supset U\to\mathbb{CP}^n$ exists. Let us consider $\mathbb{CP}^1\subset\mathbb{CP}^n$, given by $[z_0:z_1]\mapsto[z_0:z_1:0:\dots:0]$. Then we have a holomorphic retraction $r:V\to\mathbb{CP}^1$ given by $r[z_0:\dots:z_n]=[z_0:z_1]$ for $V=\{[z_0:\dots:z_n]:(z_0,z_1)\neq(0,0)\}$. But then $r\circ\alpha|(U\cap T(\mathbb{CP}^1))$ is a holomorphic local addition on $\mathbb{CP}^1$, a contradiction. □

Results. From the argument given in (42.4) follows that a complex manifold admitting a holomorphic local addition also admits a holomorphic spray and thus a holomorphic linear connection on TM. Existence of the latter has been investigated in [Atiyah, 1957]. Let us sketch the relevant results. For a complex manifold M let TM be the complex tangent bundle, let $GL(\mathbb{C}^m,TM)$ be the linear frame bundle. Then using the local description from (29.9) and (29.10) we get in turn:

$$\begin{aligned}&GL(\mathbb{C}^m,TM) && (x,s)\in U\times GL(m,\mathbb{C}),\\ &T(GL(\mathbb{C}^m,TM)) && (x,s,\xi,\sigma)\in U\times GL(m,\mathbb{C})\times\mathbb{C}^m\times\mathfrak{gl}(m,\mathbb{C}),\\ &T(GL(\mathbb{C}^m,TM))= && \bigcup_{\xi\in TM}\{f\in L_{\mathbb{C}}(\mathbb{C}^m,T(\pi_M)^{-1}(\xi)):\\ &&& \qquad\qquad \pi_{TM}\circ f\in GL(\mathbb{C}^m,TM)\},\\ &T(GL(\mathbb{C}^m,TM))/GL(m,\mathbb{C}) && (x,\mathrm{Id},\xi,A)=(x,s\circ s^{-1},\xi,\sigma\circ s^{-1}),\\ &\frac{T(GL(\mathbb{C}^m,TM))}{GL(m,\mathbb{C})}= && \bigcup_{x\in M}\bigcup_{\xi\in T_xM}\{f\in L_{\mathbb{C}}(T_xM,T(\pi_M)^{-1}(\xi)):\\ &&& \qquad\qquad \pi_{TM}\circ f=\mathrm{Id}_{T_xM}\},\end{aligned}$$

which turns out to be a holomorphic vector bundle over M. Then we have the following exact sequence of holomorphic vector bundles over M:

$$0\to L(TM,TM)\xrightarrow{\mathrm{vl}_{TM}\circ(\mathrm{Id},\ \ )}T(GL(\mathbb{C}^m,TM))/GL(m,\mathbb{C})\xrightarrow[T]{T(\pi_M)}M\to 0.$$

A holomorphic splitting of this sequence is exactly a holomorphic linear connection on TM. This sequence defines an extension of the bundle TM by $L(TM,TM)$, i.e., an element $\mathfrak{b}(TM)$ in the sheaf cohomology $H^1(M;T^*M\otimes L(TM,TM))$. Thus:

[Atiyah, 1957, from theorems 2 and 5]. *A complex manifold M admits a holomorphic linear connection if and only if $\mathfrak{b}(TM)$ vanishes.*

Note that via Cartan's theorem B this again implies that Stein manifolds admit holomorphic local additions. Moreover, Atiyah proved the following results:

Result. [Atiyah, 1957, theorem 6]. *If M is a compact Kähler manifold, then the k-th Chern class of TM is given by*

$$c_k(TM) = (-2\pi\sqrt{-1})^{-k} S_k[\mathfrak{b}(TM)],$$

where S_k is the k characteristic coefficient $\mathfrak{gl}(m,\mathbb{C}) \to \mathbb{C}$.

Note that this also implies that $\mathbb{CP}^n$ does not admit local additions.

[Atiyah, 1957, proposition 22]. *Even if all characteristic classes of M vanish, M need not admit a holomorphic connection.*

43. Diffeomorphism Groups

43.1. Theorem. Diffeomorphism group. *For a smooth manifold M the group* $\operatorname{Diff}(M)$ *of all smooth diffeomorphisms of M is an open submanifold of $\mathfrak{C}^\infty(M,M)$, composition and inversion are smooth. It is a regular Lie group in the sense of* (38.4).

The Lie algebra of the smooth infinite dimensional Lie group $\operatorname{Diff}(M)$ *is the convenient vector space $C_c^\infty(M \leftarrow TM)$ of all smooth vector fields on M with compact support, equipped with the negative of the usual Lie bracket. The exponential mapping* $\exp : C_c^\infty(M \leftarrow TM) \to \operatorname{Diff}^\infty(M)$ *is the flow mapping to time 1, and it is smooth.*

Proof. We first show that $\operatorname{Diff}(M)$ is open in $\mathfrak{C}^\infty(M,M)$. Let $c : \mathbb{R} \to \mathfrak{C}^\infty(M,M)$ be a smooth curve such that $c(0)$ is a diffeomorphism. We have to show that then $c(t)$ also is a diffeomorphism for small t. The mapping $c(t)$ stays in the WO^1-open (and thus open by (42.1)) subset of immersions for t near 0, see (41.10).

The mapping $c(t)$ stays injective for t near 0: For $|t| \le 1$ we have $c(t)|(M \setminus K_1) = c(0)|(M \setminus K_1)$ for a compact subset $K_1 \subseteq M$, by (42.5). Let $K_2 := c(0)^{-1}(c^\wedge([-1,1] \times K_1)) \supset K_1$. If $c(t)$ does not stay injective for t near 0 then there are $t_n \to 0$ and $x_n \ne y_n$ in M with $c(t_n)(x_n) = c(t_n)(y_n)$. We claim that $x_n, y_n \in K_2$: If $x_n \notin K_2$ then $c(t_n)(x_n) = c(0)(x_n)$, so $y_n \in K_1$, since otherwise $c(t_n)(y_n) = c(0)(y_n) \ne c(0)(x_n)$; but then $c(t_n)(y_n) \in c^\wedge([-1,1] \times K_1) = c(0)(K_2) \not\ni c(0)(x_n)$. Passing to subsequences we may assume that $x_n \to x$ and $y_n \to y$ in K_2. By continuity of $c^\wedge$, we get $c(0)(x) = c(0)(y)$, so $x = y$. The mapping $(t,z) \mapsto (t, c(t)(z))$ is a diffeomorphism near $(0,x)$, since it is an immersion. But then $c(t_n)(x_n) \ne c(t_n)(y_n)$ for large n.

The mapping $c(t)$ stays surjective for t near 0: In the situation of the last paragraph $c(t)(M) = c(t)(K_2) \cup c(0)(M \setminus K_1^{\text{interior}})$ is closed in M for $|t| \le 1$ and also open for t near 0, since $c(t)$ is a local diffeomorphism. It meets each connected component of M since $c(t)$ is homotopic to $c(0)$. Thus, $c(t)(M) = M$.

Therefore, $\operatorname{Diff}(M)$ is an open submanifold of $\mathfrak{C}^\infty_{\text{prop}}(M,M)$, so composition is smooth by (42.13). To show that the inversion inv is smooth, we consider a smooth curve $c : \mathbb{R} \to \operatorname{Diff}(M) \subset \mathfrak{C}^\infty(M, M)$. Then the mapping $c^\wedge : \mathbb{R} \times M \to M$ satisfies (42.5.1), and $(\operatorname{inv} \circ c)^\wedge$ fulfills the finite dimensional implicit equation $c^\wedge(t, (\operatorname{inv} \circ c)^\wedge(t, m)) = m$ for all $t \in \mathbb{R}$ and $m \in M$. By the finite dimensional implicit function theorem, $(\operatorname{inv} \circ c)^\wedge$ is smooth in (t, m). Property (42.5.1) is obvious. Hence, inv maps smooth curves to smooth curves and is thus smooth. (This proof is by far simpler than the original one, see [Michor, 1980c], and shows the power of the Frölicher-Kriegl calculus.)

By the chart structure from (42.1), or directly from theorem (42.17), we see that the tangent space $T_e \operatorname{Diff}(M)$ equals the space $C_c^\infty(M \leftarrow TM)$ of all vector fields with compact support. Likewise $T_f \operatorname{Diff}(M) = C_c^\infty(M \leftarrow f^*TM)$, which we identify with the space of all vector fields with compact support along the diffeomorphism f. Right translation μ^f is given by $\mu^f(g) = f^*(g) = g \circ f$, thus $T(\mu^f).X = X \circ f$, and for the flow Fl_t^X of the vector field with compact support X we have $\frac{d}{dt} \operatorname{Fl}_t^X = X \circ \operatorname{Fl}_t^X = T(\mu^{\operatorname{Fl}_t^X}).X$. So the one parameter group $t \mapsto \operatorname{Fl}_t^X \in \operatorname{Diff}(M)$ is the integral curve of the right invariant vector field $R_X : f \mapsto T(\mu^f).X = X \circ f$ on $\operatorname{Diff}(M)$. Thus, the exponential mapping of the diffeomorphism group is given by $\exp = \operatorname{Fl}_1 : C_c^\infty(M \leftarrow TM) \to \operatorname{Diff}(M)$. To show that is smooth we consider a smooth curve in $C_c^\infty(M \leftarrow TM)$, i.e., a time dependent vector field with compact support X_t. We may view it as a complete vector field $(0_t, X_t)$ on $\mathbb{R} \times M$ whose smooth flow respects the level surfaces $\{t\} \times M$ and is smooth. Thus, $\exp \circ X = (\operatorname{pr}_2 \circ \operatorname{Fl}_1^{(0,X)})^\vee$ maps smooth curves to smooth curves and is smooth itself. Again one may compare this simple proof with the original one [Michor, 1983, section 4].

To see that $\operatorname{Diff}(M)$ is a regular Lie group note that the evolution is given by integrating time dependent vector fields with compact support,

$$\begin{aligned} \operatorname{evol}(t \mapsto X_t) &= \varphi(1, \quad) \\ \tfrac{\partial}{\partial t}\varphi(t, x) &= X(t, \varphi(t, x)), \quad \varphi(0, x) = x. \end{aligned}$$

Let us finally compute the Lie bracket on $C_c^\infty(M \leftarrow TM)$ viewed as the Lie algebra of $\operatorname{Diff}(M)$. For $X \in C_c^\infty(M \leftarrow TM)$ let L_X denote the left invariant vector field on $\operatorname{Diff}(M)$. Its flow is given by $\operatorname{Fl}_t^{L_X}(f) = f \circ \exp(tX) = f \circ \operatorname{Fl}_t^X = (\operatorname{Fl}_t^X)^*(f)$. From (32.15) we get $[L_X, L_Y] = \frac{d}{dt}|_0 (\operatorname{Fl}_t^{L_X})^* L_Y$, so for $e = \operatorname{Id}_M$ we have

$$\begin{aligned} [L_X, L_Y](e) &= (\tfrac{d}{dt}|_0 (\operatorname{Fl}_t^{L_X})^* L_Y)(e) \\ &= \tfrac{d}{dt}|_0 (T(\operatorname{Fl}_{-t}^{L_X}) \circ L_Y \circ \operatorname{Fl}_t^{L_X})(e) \\ &= \tfrac{d}{dt}|_0 T(\operatorname{Fl}_{-t}^{L_X})(L_Y(e \circ \operatorname{Fl}_t^X)) \\ &= \tfrac{d}{dt}|_0 T((\operatorname{Fl}_{-t}^X)^*)(T(\operatorname{Fl}_t^X) \circ Y) \\ &= \tfrac{d}{dt}|_0 (T(\operatorname{Fl}_t^X) \circ Y \circ \operatorname{Fl}_{-t}^X), \quad \text{by (42.18)} \\ &= \tfrac{d}{dt}|_0 (\operatorname{Fl}_{-t}^X)^* Y = -[X, Y]. \end{aligned}$$

Another proof using (36.10) is as follows:

$$\begin{aligned} \operatorname{Ad}(\exp(sX))Y &= \tfrac{\partial}{\partial t}\Big|_0 \exp(sX) \circ \exp(tY) \circ \exp(-sX) \\ &= T(\operatorname{Fl}_s^X) \circ Y \circ \operatorname{Fl}_{-s}^X = (\operatorname{Fl}_{-s}^X)^* Y, \end{aligned}$$

thus

$$\tfrac{\partial}{\partial t}\big|_0 \operatorname{Ad}(\exp(tX))Y = \tfrac{\partial}{\partial t}\big|_0 (\operatorname{Fl}^X_{-t})^* Y = -[X, Y]$$

is the negative of the usual Lie bracket on $C^\infty_c(M \leftarrow TM)$. □

It is well known that the space $\operatorname{Diff}(M)$ of all diffeomorphisms of M is open in $C^\infty(M, M)$ even for the Whitney C^∞-topology, see (41.10); proofs can be found in [Hirsch, 1976, p. 38] or [Michor, 1980c, section 5].

43.2. Example. The exponential mapping $\exp : C^\infty_c(M \leftarrow TM) \to \operatorname{Diff}(M)$ satisfies $T_0 \exp = \operatorname{Id}$, but it is not locally surjective near Id_M: This is due to [Freifeld, 1967] and [Koppell, 1970]. The strongest result in this direction is [Grabowski, 1988], where it is shown, that $\operatorname{Diff}(M)$ contains a smooth curve through Id_M contains an arcwise connected free subgroup on $2^{\aleph_0}$ generators which meets the image of exp only at the identity.

We shall prove only a weak version of this for $M = S^1$. For large $n \in \mathbb{N}$ we consider the diffeomorphism

$$f_n(\theta) = \theta + \tfrac{2\pi}{n} + \tfrac{1}{2^n}\sin^2(\frac{n\theta}{2}) \quad \mod 2\pi;$$

(the subgroup generated by) f_n has just one periodic orbit of period n, namely $\{\frac{2\pi k}{n} : k = 0, \dots, n-1\}$. For even n the diffeomorphism f_n cannot be written as $g \circ g$ for a diffeomorphism g (so f_n is not contained in a flow), by the following argument: If g has a periodic orbit of odd period, then this is also a periodic orbit of the same period of $g \circ g$, whereas a periodic orbit of g of period $2n$ splits into two disjoint orbits of period n each, of $g \circ g$. Clearly, a periodic orbit of $g \circ g$ is a subset of a periodic orbit of g. So if $g \circ g$ has only finitely many periodic orbits of some even order, there must be an even number of them.

Claim. Let $f \in \operatorname{Diff}(S^1)$ be fixed point free and in the image of exp. Then f is conjugate to some translation R_θ.

We have to construct a diffeomorphism $g : S^1 \to S^1$ such that $f = g^{-1} \circ R_\theta \circ g$. Since $p : \mathbb{R} \to \mathbb{R}/2\pi\mathbb{Z} = S^1$ is a covering map it induces an isomorphism $T_t p : \mathbb{R} \to T_{p(t)} S^1$. In the picture $S^1 \subseteq \mathbb{C}$ this isomorphism is given by $s \mapsto s\, p(t)^\perp$, where $p(t)^\perp$ is the normal vector obtained from $p(t) \in S^1$ via rotation by $\pi/2$. Thus, the vector fields on S^1 can be identified with the smooth functions $S^1 \to \mathbb{R}$ or, by composing with $p : \mathbb{R} \to S^1$, with the 2π-periodic functions $X : \mathbb{R} \to \mathbb{R}$. Let us first remark that the constant vector field $X^\theta \in \mathfrak{X}(S^1)$, $s \mapsto \theta$ has the flow $\operatorname{Fl}^{X^\theta} : (t, \varphi) \mapsto \varphi + t \cdot \theta$. Hence, $\exp(X^\theta) = \operatorname{Fl}^{X^\theta}(1, \quad) = R_\theta$.

Let $f = \exp(X)$ and suppose $g \circ f = R_\theta \circ g$. Then $g \circ \operatorname{Fl}^X(t, \quad) = \operatorname{Fl}^{X^\theta}(t, \quad) \circ g$ for $t = 1$. Let us assume that this is true for all t. Then differentiating at $t = 0$ yields $Tg(X_x) = X^\theta_{g(x)}$ for all $x \in S^1$. If we consider g as diffeomorphism $\mathbb{R} \to \mathbb{R}$ this means that $g'(t) \cdot X(t) = \theta$ for all $t \in \mathbb{R}$. Since f was assumed to be fixed point free the vector field X is nowhere vanishing, otherwise there would be a stationary point $x \in S^1$. So the condition on g is equivalent to $g(t) = g(0) + \int_0^t \frac{\theta}{X(s)}\, ds$. We take this as definition of g, where $g(0) := 0$, and where θ will be chosen such that g factors to an (orientation preserving) diffeomorphism on S^1, i.e. $\theta \int_t^{t+2\pi} \frac{ds}{X(s)} = g(t+2\pi) - g(t) = 1$. Since X is 2π-periodic this is true for $\theta = 1/\int_0^{2\pi} \frac{ds}{X(s)}$. Since the flow of a transformed vector field is nothing else but the transformed flow we obtain that $g(\mathrm{Fl}^X(t,x)) = \mathrm{Fl}^{X^\theta}(t, g(x))$, and hence $g \circ f = R_\theta \circ g$. □

Note that the formula from (38.2) for the tangent mapping of the exponential of a Lie group in the case $G = \mathrm{Diff}(M)$ looks as follows:

$$(1) \qquad T_X \exp .Y = \int_0^1 (\mathrm{Fl}^X_{-t})^* Y \, dt \circ \mathrm{Fl}^X_1,$$

by the formula for $\mathrm{Ad} \circ \exp$ in the proof of (43.1), and by (42.17).

The break-down of the inverse function theorem in this situation is explained by the following

Claim. [Grabowski, 1993] For each finite dimensional manifold M of dimension $m > 1$ and for $M = S^1$ the mapping $T_X \exp$ is not injective for some X arbitrarily near to 0. So $GL(\mathfrak{X}_c(M))$ is not open in $L(\mathfrak{X}_c(M), \mathfrak{X}_c(M))$.

For $M = \mathbb{R}$ this seems to be wrong for vector fields with compact support.

Proof. Let us start with $M = S^1$ and the vector fields $X_n(\theta) := \frac{1}{n}\frac{\partial}{\partial\theta}$ and $Y_n := \sin(n\theta)\frac{\partial}{\partial\theta}$ for $\theta \mod 2\pi$ in $\mathfrak{X}(S^1)$. Then $\mathrm{Fl}^{X_n}_t(\theta) = \theta + \frac{1}{n}t \mod 2\pi$, and hence we get $\int_0^1 (\mathrm{Fl}^{X_n}_{-t})^* Y_n \, dt = 0$.

For a manifold M of dimension $m > 1$ we now take an embedding $S^1 \times U \to M$ for an open ball $U \subset \mathbb{R}^{m-1}$, functions $g, h \in C^\infty_c(U, \mathbb{R})$ with $g.h = g$. Then the vector fields $\tilde{X}_n(\theta, x) = h(x) X_n(\theta)$ and $\tilde{Y}_n(\theta, x) = g(x) Y_n(\theta)$ in $\mathfrak{X}_c(S^1 \times U) \subset \mathfrak{X}_c(M)$ satisfy $\int_0^1 (\mathrm{Fl}^{\tilde{X}_n}_{-t})^* \tilde{Y}_n \, dt = 0$, since $h(x) = 1$ if $g(x) \neq 0$, too. □

43.3. Remarks. The mapping

$$Ad \circ \exp : C^\infty_c(M \leftarrow TM) \to \mathrm{Diff}(M) \to L(C^\infty_c(M \leftarrow TM), C^\infty_c(M \leftarrow TM))$$

is not real analytic since $\mathrm{Ad}(\exp(sX))Y(x) = (\mathrm{Fl}^X_{-s})^* Y(x) = T_x(\mathrm{Fl}^X_s)(Y(\mathrm{Fl}^X_{-s}(x)))$ is not real analytic in s in general: choose Y constant in a chart and X not real analytic.

For a real analytic compact manifold M the group $\mathrm{Diff}(M)$ is an open submanifold of the real analytic (see (42.8)) manifold $C^\infty(M, M)$. The composition mapping is, however, not real analytic by (42.16).

For $x \in M$ the mapping $\mathrm{ev}_x \circ \exp : C^\infty_c(M \leftarrow TM) \to \mathrm{Diff}(M) \to M$ is not real analytic since $(ev_x \circ \exp)(tX) = \mathrm{Fl}^X_t(x)$, which is not real analytic in t for general smooth X.

In contrast to this, one knows from [Omori, 1978b] that a Banach Lie group acting effectively on a finite dimensional manifold is necessarily finite dimensional. So there is no way to model the diffeomorphism group on Banach spaces as a manifold. There is, however, the possibility to view $\mathrm{Diff}(M)$ as an ILH-group (i.e. inverse limit of Hilbert manifolds), which sometimes permits to use an implicit function theorem. See [Omori, 1974] for this.

43.4. Theorem (Real analytic diffeomorphism group). *For a compact real analytic manifold M the group* $\operatorname{Diff}^\omega(M)$ *of all real analytic diffeomorphisms of M is an open submanifold of $C^\omega(M,M)$, composition and inversion are real analytic.*

Its Lie algebra is the space $C^\omega(M \leftarrow TM)$ of all real analytic vector fields on M, equipped with the negative of the usual Lie bracket. The associated exponential mapping $\exp : C^\omega(M \leftarrow TM) \to \operatorname{Diff}^\omega(M)$ *is the flow mapping to time 1, and it is real analytic.*

The real analytic Lie group $\operatorname{Diff}^\omega(M)$ *is regular in the sense of* (38.4), evol *is even real analytic.*

Proof. $\operatorname{Diff}^\omega(M)$ is open in $C^\omega(M,M)$ in the compact-open topology, thus also in the finer manifold topology. The composition is real analytic by (42.15), so it remains to show that the inversion ν is real analytic.

Let $c : \mathbb{R} \to \operatorname{Diff}^\omega(M)$ be a C^ω-curve. Then the associated mapping $c^\wedge : \mathbb{R} \times M \to M$ is C^ω by (42.14), and $(\nu \circ c)^\wedge$ is the solution of the implicit equation $c^\wedge(t, (\nu\circ c)^\wedge(t,x)) = x$ and therefore real analytic by the finite dimensional implicit function theorem. Hence, $\nu \circ c : \mathbb{R} \to \operatorname{Diff}^\omega(M)$ is real analytic, again by (42.14).

Let $c : \mathbb{R} \to \operatorname{Diff}^\omega(M)$ be a C^∞-curve. Then by lemma (42.12) the associated mapping $c^\wedge : \mathbb{R} \times M \to M$ has a unique extension to a C^n-mapping $\mathbb{R} \times M_{\mathbb{C}} \supseteq J \times W \to M_{\mathbb{C}}$ which is holomorphic on W (has $\mathbb{C}$-linear derivatives), for each $n \geq 1$. The same assertion holds for the curve $\nu \circ c$ by the finite dimensional implicit function theorem for C^n-mappings.

The tangent space at Id_M of $\operatorname{Diff}^\omega(M)$ is the space $C^\omega(M \leftarrow TM)$ of real analytic vector fields on M. The one parameter subgroup of a tangent vector is the flow $t \mapsto \mathrm{Fl}^X_t$ of the corresponding vector field $X \in C^\omega(M \leftarrow TM)$, so $\exp(X) = \mathrm{Fl}^X_1$ which exists since M is compact.

In order to show that $\exp : C^\omega(M \leftarrow TM) \to \operatorname{Diff}^\omega(M) \subseteq C^\omega(M,M)$ is real analytic, by the exponential law (42.14) it suffices to show that the associated mapping $\exp^\wedge = \mathrm{Fl}_1 : C^\omega(M \leftarrow TM) \times M \to M$ is real analytic. This follows from the finite dimensional theory of ordinary real analytic and smooth differential equations. The same is true for the evolution operator. □

43.5. Remark. *The exponential mapping* $\exp : C^\omega(S^1 \leftarrow TS^1) \to \operatorname{Diff}^\omega(S^1)$ *is not surjective on any neighborhood of the identity.*

Proof. The proof of (43.2) for the group of smooth diffeomorphisms of S^1 can be adapted to the real analytic case: $\varphi_n(\theta) = \theta + \frac{2\pi}{n} + \frac{1}{2^n}\sin^2(\frac{n\theta}{2}) \mod 2\pi$ is Mackey convergent (in U_{Id}) to Id_{S^1} in $\operatorname{Diff}^\omega(S^1)$, but φ_n is not in the image of the exponential mapping. □

43.6. Example 1. Let $\mathfrak{g} \subset \mathfrak{X}_c(\mathbb{R}^2)$ be the closed Lie subalgebra of all vector fields with compact support on $\mathbb{R}^2$ of the form $X(x,y) = f(x,y)\frac{\partial}{\partial x} + g(x,y)\frac{\partial}{\partial y}$ where g vanishes on the strip $0 \leq x \leq 1$.

Claim. There is no Lie subgroup G of $\operatorname{Diff}(\mathbb{R}^2)$ corresponding to $\mathfrak{g}$.

If G exists there is a smooth curve $t \mapsto f_t \in G \subset \operatorname{Diff}_c(\mathbb{R}^2)$. Then $X_t := (\frac{\partial}{\partial t} f_t) \circ f_t^{-1}$ is a smooth curve in $\mathfrak{g}$, and we may assume that $X_0 = f \frac{\partial}{\partial x}$ where $f = 1$ on a large ball. But then $\operatorname{Ad}^G(f_t) = f_t^* : \mathfrak{g} \not\to \mathfrak{g}$, a contradiction.

So we see that on any manifold of dimension greater than 2 there are closed Lie subalgebras of the Lie algebra of vector fields with compact support which do not admit Lie subgroups.

Example 2. The space $\mathfrak{X}_K(M)$ of all vector fields with support in some open set U is an ideal in $\mathfrak{X}_c(M)$, the corresponding Lie group is the connected component $\operatorname{Diff}_U(M)_0$ of the group of all diffeomorphisms which equal Id off some compact in U, but this is not a normal subgroup in the connected component $\operatorname{Diff}_c(M)_0$, since we may conjugate the support out of U.

Note that this examples do not work for the Lie group of real analytic diffeomorphisms on a compact manifold.

43.7. Theorem. [Ebin, Marsden, 1970] *Let M be a compact orientable manifold, let μ_0 be a positive volume form on M with total mass 1. Then the regular Lie group* $\operatorname{Diff}_+(M)$ *of all orientation preserving diffeomorphisms splits smoothly as* $\operatorname{Diff}_+(M) = \operatorname{Diff}(M, \mu_0) \times \operatorname{Vol}(M)$, *where* $\operatorname{Diff}(M, \mu_0)$ *is the regular Lie group of all μ_0-preserving diffeomorphisms, and* $\operatorname{Vol}(M)$ *is the space of all volume forms of total mass* 1.

If (M, μ_0) is real analytic, then $\operatorname{Diff}^\omega_+(M)$ splits real analytically as $\operatorname{Diff}^\omega_+(M) = \operatorname{Diff}^\omega(M, \mu_0) \times \operatorname{Vol}^\omega(M)$, where $\operatorname{Diff}^\omega(M, \mu_0)$ is the Lie group of all μ_0-preserving real analytic diffeomorphisms, and $\operatorname{Vol}^\omega(M)$ is the space of all real analytic volume forms of total mass 1.

Proof. We show first that there exists a smooth mapping $\tau : \operatorname{Vol}(M) \to \operatorname{Diff}_+(M)$ such that $\tau(\mu)^*\mu_0 = \mu$.

We put $\mu_t := \mu_0 + t(\mu - \mu_0)$. We want a smooth curve $t \mapsto f_t \in \operatorname{Diff}_+(M)$ with $f_t^*\mu_t = \mu_0$. We have $\frac{\partial}{\partial t} f_t = X_t \circ f_t$ for a time dependent vector field X_t on M. Then $0 = \frac{\partial}{\partial t} f_t^*\mu_t = f_t^*\mathcal{L}_{X_t}\mu_t + f_t^*\frac{\partial}{\partial t}\mu_t = f_t^*(\mathcal{L}_{X_t}\mu_t + (\mu - \mu_0))$, so $\mathcal{L}_{X_t}\mu_t = \mu_0 - \mu$ and $\mathcal{L}_{X_t}\mu_t = di_{X_t}\mu_t + i_{X_t}0 = d\omega$ for some $\omega \in \Omega^{\dim M - 1}(M)$. Now we choose ω such that $d\omega = \mu_0 - \mu$, and we choose it smoothly and in the real analytic case even real analytically depending on μ by the theorem of Hodge, as follows: For any $\alpha \in \Omega(M)$ we have $\alpha = \mathcal{H}\alpha + d\delta G\alpha + \delta G d\alpha$, where $\mathcal{H}$ is the projection on the space of harmonic forms, $\delta = *d*$ is the codifferential, $*$ is the Hodge-star operator, and G is the Green operator, see [Warner, 1971]. All these are bounded linear operators, G is even compact. So we may choose $\omega = \delta G(\mu_0 - \mu)$. Then the time dependent vector field X_t is uniquely determined by $i_{X_t}\mu_t = \omega$ since μ_t is nowhere 0. Let f_t be the evolution operator of X_t, and put $\tau(\mu) = f_1^{-1}$.

Now we may prove the theorem itself. We define a mapping $\Psi : \operatorname{Diff}_+(M) \to \operatorname{Diff}(M, \mu_0) \times \operatorname{Vol}(M)$ by $\Psi(f) := (f \circ \tau(f^*\mu_0)^{-1}, f^*\mu_0)$, which is smooth or real analytic by (42.15) and (43.4). An easy computation shows that the inverse is given by the smooth (or real analytic) mapping $\Psi^{-1}(g, \mu) = g \circ \tau(\mu)$.

That $\operatorname{Diff}(M, \mu_0)$ is regular follows from (38.7), where we use the mapping $p : \operatorname{Diff}_+(M) \to \Omega^{\max}(M)$, given by $p(f) := f^*\mu_0 - \mu_0$. □

We next treat the Lie group of symplectic diffeomorphisms.

43.8. Symplectic manifolds. Let M be a smooth manifold of dimension $2n \geq 2$. A *symplectic form* on M is a closed 2-form σ such that $\sigma^n = \sigma \wedge \cdots \wedge \sigma \in \Omega^{2n}(M)$ is nowhere 0. The pair (M, σ) is called a *symplectic manifold*. See section (48) for a treatment of infinite dimensional symplectic manifolds.

A symplectic form can be put into the following (Darboux) normal form: For each $x \in M$ there is a chart $M \supset U \xrightarrow{u} u(U) \subset \mathbb{R}^{2n}$ centered at x such that on U the form σ is given by $\sigma|U = u^1\, du^{n+1} + u^2\, du^{n+2} + \cdots + u^n\, du^{2n}$. This follows from proposition (43.11) below for $N = \{x\}$.

A diffeomorphism $f \in \operatorname{Diff}(M)$ with $f^*\sigma = \sigma$ is called a *symplectic diffeomorphism*; some authors also write *symplectomorphism*. The group of all symplectic diffeomorphisms will be denoted by $\operatorname{Diff}(M, \sigma)$.

A vector field $X \in \mathfrak{X}(M)$ will be called a *symplectic vector field* if $\mathcal{L}_X \sigma = 0$; some authors also write *locally Hamiltonian vector field*. The Lie algebra of all symplectic vector fields will be denoted by $\mathfrak{X}(M, \sigma)$.

For a finite dimensional symplectic manifold (M, σ) we have the following exact sequence of Lie algebras:

$$0 \to H^0(M) \to C^\infty(M, \mathbb{R}) \xrightarrow{\operatorname{grad}^\sigma} \mathfrak{X}(M, \sigma) \xrightarrow{\gamma} H^1(M) \to 0$$

Here $H^*(M)$ is the real De Rham cohomology of M, $\operatorname{grad}^\sigma f$ is the Hamiltonian vector field for $f \in C^\infty(M, \mathbb{R})$ given by $i(\operatorname{grad}^\sigma f)\sigma = df$, and $\gamma(\xi) = [i_\xi \sigma]$. The spaces $H^0(M)$ and $H^1(M)$ are equipped with the zero bracket, and the space $C^\infty(M, \mathbb{R})$ is equipped with the Poisson bracket

$$\begin{aligned}\{f, g\} &:= i(\operatorname{grad}^\sigma f) i(\operatorname{grad}^\sigma g)\sigma = \sigma(\operatorname{grad}^\sigma g, \operatorname{grad}^\sigma f) = \\ &= (\operatorname{grad}^\sigma f)(g) = dg(\operatorname{grad}^\sigma f).\end{aligned}$$

The image of $\operatorname{grad}^\sigma$ is the Lie subalgebra of *globally Hamiltonian vector fields*. We shall prove this for infinite dimensional manifolds in section (48) below.

A submanifold L of a symplectic manifold (M^{2n}, σ) is called a *Lagrange submanifold* if it is of dimension n and $\operatorname{incl}^* \sigma = 0$ on L.

43.9. Canonical example. Let Q be an n-dimensional manifold. Let us consider the natural 1-form θ_Q on the cotangent bundle T^*Q which is given by $\theta_Q(\Xi) := \langle \pi_{T^*Q}(\Xi), T(\pi_Q^*).\Xi \rangle_{TQ}$, where we used the projections $\pi_Q^* : T^*Q \to Q$ and

$$T^*Q \xleftarrow{\pi_{T^*Q}} T(T^*Q) \xrightarrow{T(\pi_Q^*)} TQ.$$

The canonical symplectic structure on T^*Q is then given by $\sigma_Q = -d\theta_Q$. If $q : U \to \mathbb{R}^n$ is a smooth chart on Q with induced chart $T^*q = (q, p) : T^*U = (\pi_Q^*)^{-1}(U) \to$

$\mathbb{R}^n \times \mathbb{R}^n$, we have $\theta_Q|T^*U = \sum p_i\, dq^i$ and $\sigma_Q|T^*U = \sum dq^i \wedge dp_i$. The canonical forms θ_Q and σ_Q on T^*Q have the following universal property and are determined by it: For any 1-form $\alpha \in \Omega^1(Q)$, viewed as a mapping $Q \to T^*Q$, we have $\alpha^*\theta = \alpha$ and $\alpha^*\sigma = -d\alpha$. Thus, the image $\alpha(Q) \subset T^*Q$ is a Lagrange submanifold if and only if $d\alpha = 0$. Moreover, each fiber T_x^*Q is a Lagrange submanifold.

43.10. Relative Poincaré Lemma. *Let M be a smooth finite dimensional manifold, let $N \subset M$ be a closed submanifold, and let $k \geq 0$. Let ω be a closed $(k+1)$-form on M which vanishes when pulled back to N. Then there exists a k-form φ on an open neighborhood U of N in M such that $d\varphi = \omega|U$ and $\varphi = 0$ along N. If moreover $\omega = 0$ along N, then we may choose φ such that the first derivatives of φ vanish on N.*

If all given data are real analytic then φ can be chosen real analytic, too.

Proof. By restricting to a tubular neighborhood of N in M, we may assume that $p : M =: E \to N$ is a smooth vector bundle and that $i : N \to E$ is the zero section of the bundle. We consider $\mu : \mathbb{R} \times E \to E$, given by $\mu(t,x) = \mu_t(x) = tx$, then $\mu_1 = \mathrm{Id}_E$ and $\mu_0 = i \circ p : E \to N \to E$. Let $V \in \mathfrak{X}(E)$ be the vertical vector field $V(x) = \mathrm{vl}(x,x) = \frac{\partial}{\partial t}\big|_0 x + tx$, then $\mathrm{Fl}_t^V = \mu_{e^t}$. So locally for t near $(0,1]$ we have

$$\begin{aligned}\tfrac{d}{dt}\mu_t^*\omega &= \tfrac{d}{dt}(\mathrm{Fl}^V_{\log t})^*\omega = \tfrac{1}{t}(\mathrm{Fl}^V_{\log t})^*\mathcal{L}_V\omega \text{ by (33.19)}\\ &= \tfrac{1}{t}\mu_t^*(i_V d\omega + d i_V\omega) = \tfrac{1}{t}d\mu_t^* i_V\omega.\end{aligned}$$

For $x \in E$ and $X_1, \dots, X_k \in T_xE$ we may compute

$$\begin{aligned}(\tfrac{1}{t}\mu_t^* i_V\omega)_x(X_1,\dots,X_k) &= \tfrac{1}{t}(i_V\omega)_{tx}(T_x\mu_t.X_1,\dots,T_x\mu_t.X_k)\\ &= \tfrac{1}{t}\omega_{tx}(V(tx), T_x\mu_t.X_1,\dots,T_x\mu_t.X_k)\\ &= \omega_{tx}(\mathrm{vl}(tx,x), T_x\mu_t.X_1,\dots,T_x\mu_t.X_k).\end{aligned}$$

So if $k \geq 0$, the k-form $\frac{1}{t}\mu_t^* i_V\omega$ is defined and smooth in (t,x) for all $t \in [0,1]$ and describes a smooth curve in $\Omega^k(E)$. Note that for $x \in N = 0_E$ we have $(\frac{1}{t}\mu_t^* i_V\omega)_x = 0$, and if $\omega = 0$ along N, then also all first derivatives of $\frac{1}{t}\mu_t^* i_V\omega$ vanish along N. Since $\mu_0^*\omega = p^*i^*\omega = 0$ and $\mu_1^*\omega = \omega$, we have

$$\begin{aligned}\omega = \mu_1^*\omega - \mu_0^*\omega &= \int_0^1 \tfrac{d}{dt}\mu_t^*\omega dt\\ &= \int_0^1 d(\tfrac{1}{t}\mu_t^* i_V\omega)dt = d\left(\int_0^1 \tfrac{1}{t}\mu_t^* i_V\omega dt\right) =: d\varphi.\end{aligned}$$

If $x \in N$, we have $\varphi_x = 0$, and also the last claim is obvious from the explicit form of φ. Finally, it is clear that this construction can be done in a real analytic way. □

43.11. Lemma. *Let M be a smooth finite dimensional manifold, let $N \subset M$ be a closed submanifold, and let σ_0 and σ_1 be symplectic forms on M which are equal along N.*

Then there exist: A diffeomorphism $f : U \to V$ between two open neighborhoods U and V of N in M which satisfies $f|N = \mathrm{Id}_N$, $Tf|(TM|N) = \mathrm{Id}_{TM|N}$, and $f^*\sigma_1 = \sigma_0$.

If all data are real analytic then the diffeomorphism can be chosen real analytic, too.

Proof. Let $\sigma_t = \sigma_0 + t(\sigma_1 - \sigma_0)$ for $t \in [0,1]$. Since the restrictions of σ_0 and σ_1 to $\Lambda^2 TM|N$ are equal, there is an open neighborhood U_1 of N in M such that σ_t is a symplectic form on U_1, for all $t \in [0,1]$. If $i : N \to M$ is the inclusion, we also have $i^*(\sigma_1 - \sigma_0) = 0$, so by lemma (43.10) there is a smaller open neighborhood U_2 of N such that $\sigma_1|U_2 - \sigma_0|U_2 = d\varphi$ for some $\varphi \in \Omega^1(U_2)$ with $\varphi_x = 0$ for $x \in N$, such that also all first derivatives of φ vanish along N.

Let us now consider the time dependent vector field $X_t := -(\sigma_t{}^\vee)^{-1} \circ \varphi$, which vanishes together with all first derivatives along N. Let f_t be the curve of local diffeomorphisms with $\frac{\partial}{\partial t} f_t = X_t \circ f_t$, then $f_t|N = \mathrm{Id}_N$ and $Tf_t|(TM|N) = \mathrm{Id}$. There is a smaller open neighborhood U of N such that f_t is defined on U for all $t \in [0,1]$. Then we have

$$\begin{aligned} \tfrac{\partial}{\partial t}(f_t^*\sigma_t) &= f_t^* \mathcal{L}_{X_t}\sigma_t + f_t^* \tfrac{\partial}{\partial t}\sigma_t = f_t^*(d i_{X_t}\sigma_t + \sigma_1 - \sigma_0) \\ &= f_t^*(-d\varphi + \sigma_1 - \sigma_0) = 0, \end{aligned}$$

so $f_t^*\sigma_t$ is constant in t, equals $f_0^*\sigma_0 = \sigma_0$, and finally $f_1^*\sigma_1 = \sigma_0$ as required. □

43.12. Theorem. *Let (M, σ) be a finite dimensional symplectic manifold. Then the group* $\mathrm{Diff}(M, \sigma)$ *of symplectic diffeomorphisms is a smooth regular Lie group and a closed submanifold of* $\mathrm{Diff}(M)$. *The Lie algebra of* $\mathrm{Diff}(M, \sigma)$ *agrees with* $\mathfrak{X}_c(M, \sigma)$.

If moreover (M, σ) is a compact real analytic symplectic manifold, then the group $\mathrm{Diff}^\omega(M, \sigma)$ *of real analytic symplectic diffeomorphisms is a real analytic regular Lie group and a closed submanifold of* $\mathrm{Diff}^\omega(M)$.

Proof. The smooth and the real analytic cases will be proved simultaneously; only once we will need an extra argument for the latter.

Consider a local addition $\alpha : TM \to M$ in the sense of (42.4), so that $(\pi_M, \alpha) : TM \to M \times M$ is a diffeomorphism onto an open neighborhood of the diagonal, and $\alpha(0_x) = x$. Let us compose α from the right with a fiber respecting diffeomorphism $TM^* \to TM$ (coming from the symplectic structure or from a Riemannian metric) and call the result again $\alpha : T^*M \to M$. Then $(\pi_M, \alpha) : T^*M \to M \times M$ also is a diffeomorphism onto an open neighborhood of the diagonal, and $\alpha(0_x) = x$.

We consider now two symplectic structures on T^*M, namely the canonical symplectic structure $\sigma_0 = \sigma_M$, and $\sigma_1 := (\pi_M, \alpha)^*(\mathrm{pr}_1^*\,\sigma - \mathrm{pr}_2^*\,\sigma)$. Both have vanishing pullbacks on the zero section $0_M \subset T^*M$.

Claim. In this situation, there exists a diffeomorphism $\varphi : V_0 \to V_1$ between two open neighborhoods V_0 and V_1 of the zero section in T^*M which is the identity on the zero section and satisfies $\varphi^*\sigma_1 = \sigma_0$.

First we solve the problem along the zero section, i.e., in $T(T^*M)|0_M$. There is a vector bundle isomorphism $\gamma : T(T^*M)|0_M \to T(T^*M)|0_M$ over the identity on 0_M, which is the identity on $T(0_M)$ and maps the symplectic structure σ_0 on each fiber to σ_1. In the smooth case, by using a partition of unity it suffices to construct γ locally. But locally σ_i can be described by choosing a Lagrange subbundle $L_i \subset T(T^*M)|0_M$ which is a complement to $T0_M$. Then σ_i is completely determined by the duality between $T0_M$ and W_i induced by it, and a smooth γ is then given by the resulting isomorphism $W_0 \to W_1$.

In the real analytic case, in order to get a real analytic γ, we consider the principal fiber bundle $P \to 0_M$ consisting of all $\gamma_x \in GL(T_{0_x}(T^*M))$ with $\gamma_x|T_{0_x}(0_M) = \mathrm{Id}$ and $\gamma_x^*\sigma_1 = (\sigma_0)_{0_x}$. The proof above shows that we may find a smooth section of P. By lemma (30.12), there also exist real analytic sections.

Next we choose a diffeomorphism $h : V_0 \to V_1$ between open neighborhoods of 0_M in T^*M such that $Th|0_M = \gamma$, which can be constructed as follows: Let $u : \mathcal{N}(0_M) \to V_0$ be a tubular neighborhood of the zero section, where $\mathcal{N}(0_M) = (T(T^*M)|0_M)/T(0_M)$ is the normal bundle of the zero section. Clearly, γ induces a vector bundle automorphism of this normal bundle, and $h = u \circ \gamma \circ u^{-1}$ satisfies all requirements.

Now σ_0 and $h^*\sigma_1$ agree along the zero section 0_M, so we may apply lemma (43.11), which implies the claim with possibly smaller V_i.

We consider the diffeomorphism $\rho := (\pi_M, \alpha) \circ \varphi : T^*M \supset V_0 \to V_2 \subset M \times M$ from an open neighborhood of the zero section to an open neighborhood of the diagonal, and we let $U \subseteq \mathrm{Diff}(M)$ be the open neighborhood of Id_M consisting of all $f \in \mathrm{Diff}(M)$ with compact support such that $(\mathrm{Id}_M, f)(M) \subset V_2$, i.e. the graph $\{(x, f(x)) : x \in M\}$ of f is contained in V_2, and $\pi_M : \rho^{-1}(\{(x, f(x)) : x \in M\}) \to M$ is still a diffeomorphism.

For $f \in U$ the mapping $(\mathrm{Id}_M, f) : M \to \mathrm{graph}(f) \subset M \times M$ is the natural diffeomorphism onto the graph of f, and the latter is a Lagrangian submanifold if and only if

$$0 = (\mathrm{Id}_M, f)^*(\mathrm{pr}_1^*\,\sigma - \mathrm{pr}_2^*\,\sigma) = \mathrm{Id}_M^*\,\sigma - f^*\sigma.$$

Therefore, $f \in \mathrm{Diff}(M, \sigma)$ if and only if the graph of f is a Lagrangian submanifold of $(M \times M, \mathrm{pr}_1^*\,\sigma - \mathrm{pr}_2^*\,\sigma)$. Since $\rho^*(\mathrm{pr}_1^*\,\sigma - \mathrm{pr}_2^*\,\sigma) = \sigma_0$ this is the case if and only if $\{\rho^{-1}(x, f(x)) : x \in M\}$ is a Lagrange submanifold of (T^*M, σ_0).

We consider now the following smooth chart of $\mathrm{Diff}(M)$ which is centered at the identity:

$$\mathrm{Diff}(M) \supset U \xrightarrow{u} u(U) \subset C_c^\infty(M \leftarrow T^*M) = \Omega_c^1(M),$$
$$u(f) := \rho^{-1} \circ (\mathrm{Id}_M, f) \circ (\pi_M \circ \rho^{-1} \circ (\mathrm{Id}_M, f))^{-1} : M \to T^*M.$$

Then $f \in U \cap \operatorname{Diff}(M,\sigma)$ if and only if $u(f)$ is a closed form, since $u(f)(M) = \{\rho^{-1}(x,f(x)) : x \in M\}$ is a Lagrange submanifold if and only if f is symplectic. Thus, (U,u) is a smooth chart of $\operatorname{Diff}(M)$ which is a submanifold chart for $\operatorname{Diff}(M,\sigma)$. For arbitrary $g \in \operatorname{Diff}(M,\sigma)$ we consider the smooth submanifold chart

$$\begin{aligned}\operatorname{Diff}(M) \supset U_g := \{f : f\circ g^{-1} \in U\} \xrightarrow{u_g} u_g(U_g) \subset C_c^\infty(M \leftarrow T^*M) = \Omega_c^1(M),\\ u_g(f) := u(f\circ g^{-1}).\end{aligned}$$

Hence, $\operatorname{Diff}(M,\sigma)$ is a closed smooth submanifold of $\operatorname{Diff}(M)$ and a smooth Lie group, since composition and inversion are smooth by restriction. If M is compact then the space of closed 1-forms is a direct summand in $\Omega^1(M)$ by Hodge theory, as in the proof of (43.7), so in this case $\operatorname{Diff}(M,\sigma)$ is even a splitting submanifold of $\operatorname{Diff}(M)$. The embedding $\operatorname{Diff}(M,\sigma) \to \operatorname{Diff}(M)$ is smooth, thus it induces a bounded injective homomorphism of Lie algebras which is an embedding onto a closed Lie subalgebra, which we shall soon identify with $\mathfrak{X}_c(M,\sigma)$.

Suppose that $X : \mathbb{R} \to \mathfrak{X}_c(M,\sigma)$ is a smooth curve, and consider the evolution curve $f(t) = \operatorname{Evol}^r_{\operatorname{Diff}(M)}(X)(t)$, which is the solution of the differential equation $\frac{\partial}{\partial t} f(t) = X(t)\circ f(t)$ on M. Then $f : \mathbb{R} \to \operatorname{Diff}(M)$ actually has values in $\operatorname{Diff}(M,\sigma)$, since $\frac{\partial}{\partial t} f_t^*\sigma = f_t^*\mathcal{L}_{X_t}\sigma = 0$. So the restriction of $\operatorname{evol}^r_{\operatorname{Diff}(M)}$ to $\mathfrak{X}_c(M,\sigma)$ is smooth into $\operatorname{Diff}(M,\sigma)$ and thus gives $\operatorname{evol}^r_{\operatorname{Diff}(M,\sigma)}$. We take now the right logarithmic derivative of $f(t)$ in $\operatorname{Diff}(M,\sigma)$ and get a smooth curve in the Lie algebra of $\operatorname{Diff}(M,\sigma)$ which maps to $X(t)$. Thus, the Lie algebra of $\operatorname{Diff}(M,\sigma)$ is canonically identified with $\mathfrak{X}_c(M,\sigma)$.

Note that this proof of regularity is an application of the method from (38.7), where $p(f) := f^*\sigma - \sigma$, $p : \operatorname{Diff}(M) \to \Omega^2(M)$. □

43.13. The regular Lie group of exact symplectic diffeomorphisms. Let us assume that (M,σ) is a connected finite dimensional separable symplectic manifold such that the space of exact 1-forms with compact support $B_c^1(M)$ on M is a convenient direct summand in the space $Z_c^1(M)$ of all closed forms. This is true if M is compact, by Hodge theory, as in (43.7).

In the setting of the last theorem (43.12) we consider the universal covering group $\widetilde{\operatorname{Diff}(M,\sigma)} \to \operatorname{Diff}(M,\sigma)$, which we view as the space of all smooth curves $c : [0,1] = I \to \operatorname{Diff}(M,\sigma)$ such that $c(0) = \operatorname{Id}_M$ modulo smooth homotopies fixing endpoints. For each such curve the right logarithmic derivative (38.1) $\delta^r c(\partial_t) : I \to \mathfrak{X}_c(M,\sigma)$ is given by $\delta^r c(\partial_t|_t) = \frac{\partial}{\partial t}|_0 c(t)\circ c(t)^{-1}$. Then $i(\delta^r c(\partial_t))\sigma$ is a curve of closed 1-forms with compact supports since $di(\delta^r c(\partial_t|_t))\sigma = \mathcal{L}_{\delta^r c(\partial_t|_t)}\sigma = 0$. For a smooth homotopy $(s,t) \mapsto h(s,t)$ with $h(0,t) = c(t)$ we have by the left Maurer Cartan equation $d\delta^r h - \frac{1}{2}[\delta^r h, \delta^r h] = 0$ in lemma (38.1)

$$\begin{aligned}\partial_s\delta^r h(\partial_t) &= \partial_t\delta^r h(\partial_s) + d(\delta^r h)(\partial_s,\partial_t) + \delta^r h([\partial_s,\partial_t])\\ &= \partial_t\delta^r h(\partial_s) + [\delta^r h(\partial_s), \delta^r h(\partial_t)]_{\mathfrak{X}_c(M,\sigma)} + 0.\end{aligned}$$

Then we get

$$\begin{aligned}
&\int_0^1 i_{\delta^r h(\partial_t|_{(1,t)})}\sigma\, dt - \int_0^1 i_{\delta^r h(\partial_t|_{(0,t)})}\sigma\, dt = \int_0^1\int_0^1 i_{\partial_s \delta^r h(\partial_t)}\sigma\, ds\, dt\\
&= \int_0^1\int_0^1 i_{\partial_t \delta^r h(\partial_s)+[\delta^r h(\partial_s),\delta^r h(\partial_t)]}\sigma\, ds\, dt\\
&= \int_0^1 \partial_t \int_0^1 i_{\delta^r h(\partial_s)}\sigma\, ds\, dt + \int_0^1\int_0^1 i_{[\delta^r h(\partial_s),\delta^r h(\partial_t)]}\sigma\, ds\, dt\\
&= \int_0^1 i_{\delta^r h(\partial_s|_{(s,1)})}\sigma\, ds - \int_0^1 i_{\delta^r h(\partial_s|_{(s,0)})}\sigma\, ds + \int_0^1\int_0^1 \left[\mathcal{L}_{\delta^r h(\partial_s)}, i_{\delta^r h(\partial_t)}\right]\sigma\, ds\, dt\\
&= 0 - 0 + \int_0^1\int_0^1 \mathcal{L}_{\delta^r h(\partial_s)} i_{\delta^r h(\partial_t)}\sigma\, ds\, dt - 0\\
&= \int_0^1\int_0^1 d\, i_{\delta^r h(\partial_s)} i_{\delta^r h(\partial_t)}\sigma\, ds\, dt + \int_0^1\int_0^1 i_{\delta^r h(\partial_s)}\mathcal{L}_{\delta^r h(\partial_t)}\sigma\, ds\, dt\\
&= d\int_0^1\int_0^1 \sigma\left(\delta^r h(\partial_t), \delta^r h(\partial_s)\right)\, ds\, dt.
\end{aligned}$$

Thus, we get a well defined smooth mapping into the de Rham cohomology with compact supports

$$\Gamma : \widetilde{\mathrm{Diff}(M,\sigma)} \to H^1_c(M),$$
$$\Gamma([c]) := \left[\int_0^1 i(\delta^r c(\partial_t))\sigma dt\right],$$

which is a homomorphism of regular Lie groups: the multiplication in $\widetilde{\mathrm{Diff}(M,\sigma)}$ is induced by pointwise multiplication of curves. But note that $t \mapsto c_1(t) \circ c_2(t)$ is homotopic to the curve which follows first c_2 and then $c_1(\quad) \circ c_2(1)$. The right logarithmic derivative does not feel the right translation, thus the integral $\Gamma([c_1].[c_2])$ equals $\Gamma([c_1]) + \Gamma([c_2])$.

Note that, under the assumption on M made above, Γ admits a global smooth section s as follows:

$$\begin{array}{ccc}
Z^1_c(M) & \xrightarrow{\quad\Psi\quad} & C^\infty(I, \mathrm{Diff}(M,\sigma))\\
\downarrow & & \downarrow\\
H^1_c(M) & \overset{s}{\dashrightarrow} & \widetilde{\mathrm{Diff}(M,\sigma)},
\end{array}$$

where $\Psi(\omega) = (\mathrm{Fl}_t^{\sigma^{-1}\omega})_{0\le t\le 1}$ is smooth. Since the canonical quotient mapping $Z^1_c(M) \to H^1_c(M)$ admits a section Ψ induces a section of Γ.

Claim. *The closed subgroup* $\ker\Gamma \subset \widetilde{\mathrm{Diff}(M,\sigma)}$ *is simply connected.*

First note that $\mathrm{Diff}(M,\sigma)$ is also a topological group in the the topology described in (42.2), thus a fortiori its universal covering $\widetilde{\mathrm{Diff}(M,\sigma)}$ is also a topological group. $H^1_c(M)$ is a direct summand in $Z^1_c(M)$, which is smoothly paracompact as a closed linear subspace of $\Omega^1_c(M)$ by (30.4). Since it admits a continuous section, $\Gamma : \widetilde{\mathrm{Diff}(M,\sigma)} \to H^1_c(M)$ is a fibration with contractible basis. The long exact homotopy sequence then implies that $\ker(\Gamma)$ is simply connected, too.

Theorem. *The subgroup* $\ker\Gamma \subset \widetilde{\mathrm{Diff}(M,\sigma)}$ *is a splitting regular Lie subgroup with Lie algebra* $C^\infty_c(M,\mathbb{R})/H^0_c(M)$.

Proof. Recall from the proof of (43.12) the chart (U,u) of $\mathrm{Diff}(M,\sigma)$ near the identity, which we consider also as a chart on the universal covering $\widetilde{\mathrm{Diff}(M,\sigma)}$. It is induced by a diffeomorphism $T^*M \supset V_0 \xrightarrow{\rho} V_2 \subset M\times M$ satisfying $\rho^*(\mathrm{pr}_1^*\sigma - \mathrm{pr}_2^*\sigma) = \sigma_M$ in the following way. To a symplectic diffeomorphism f near Id_M we first associate its graph, a Lagrange submanifold in $(V_2, \mathrm{pr}_1^*\sigma - \mathrm{pr}_2^*\sigma)$, then its inverse image L under ρ, a Lagrange submanifold in V_0, and finally a closed 1-form $u(f) = \omega \in \Omega^1_c(M)$ with $\omega(M) = L$. The form ω is exact if and only if the pullback of the natural 1-form $\theta_M \in \Omega^1(T^*M)$ (see (43.9)) on L is exact. Equivalently, the form $\theta_1 = (\rho^{-1})^*\theta_M$ on V_2 pulls back to an exact form on the graph of f, or $(\mathrm{Id}, f)^*\theta_1$ is exact on M. Let $f_t \in U$ for $t\in[0,1]$ with $f_0 = \mathrm{Id}_M$, and let $X_t = \frac{d}{dt}f_t\circ f_t^{-1}$. Then

$$\begin{aligned}
\Gamma(f) = \Gamma([f_t]) &= \left[\int_0^1 i(X_t)\sigma\, dt\right],\\
\tfrac{d}{dt}(\mathrm{Id}_M, f_t)^*\theta_1 &= (\mathrm{Id}_M, f_t)^*\mathcal{L}_{0\times X_t}\theta_1 \qquad \text{compare (33.19)}\\
&= (\mathrm{Id}_M, f_t)^* d i_{0\times X_t}\theta_1 + (\mathrm{Id}_M, f_t)^* i_{0\times X_t} d\theta_1\\
&= d(\mathrm{Id}_M, f_t)^* i_{0\times X_t}\theta_1 + f_t^* i_{X_t}\sigma,
\end{aligned}$$

since $-d\theta_1 = \mathrm{pr}_1^*\sigma - \mathrm{pr}_2^*\sigma$. Thus, $i_{X_t}\sigma$ is exact for all t if and only if $(\mathrm{Id}_M, f_t)^*\theta_1$ is exact for all t. If ω is exact let $f_t := u^{-1}(t\omega)$, and it follows that $\Gamma(f) = 0$. If conversely $f \in U\cap\ker\Gamma \subset \widetilde{\mathrm{Diff}(M,\sigma)}$, there exists a smooth curve $t\mapsto h_t$ in $U \subset \widetilde{\mathrm{Diff}(M,\sigma)}$ from Id_M to f. Then $\Gamma(h_t)$ is a closed smooth curve in $H^1_c(M)$, which we may lift smoothly to $g_t \in \widetilde{\mathrm{Diff}(M,\sigma)}$. Then $g_t^{-1}\circ h_t$ lies in $\ker(\Gamma)$ for all t. Thus, for f near Id_M in $\ker(\Gamma)$ we may find a smooth curve $t\mapsto f_t\in U$ which lies in $\ker(\Gamma)$. Then $\int_0^1 i(X_{ts})\sigma\,ds = t\int_0^t i(X_t)\sigma\,dt$ is exact, so $i(X_t)\sigma$ is exact, and finally $u(f_t)$ is exact in $Z^1_c(M)$. Hence, for some smaller U we have $f\in U\cap\ker\Gamma\subset\widetilde{\mathrm{Diff}(M,\sigma)}$ if and only if $\omega = u(f)\in u(U)\cap B^1_c(M)$, and $\ker(\Gamma)$ is a smooth splitting submanifold.

The Lie algebra of $\ker(\Gamma)$ consists of all globally Hamiltonian vector fields: for a smooth curve f_t in $\ker(\Gamma)$ we consider $X_t = \frac{d}{dt}f_t\circ f_t^{-1}$; from above we see that $i(X_t)\sigma = dh_t$ for some $h_t\in C^\infty_c(M,\mathbb{R})$ and then $\mathrm{grad}^\sigma(h_t) = X_t$. Conversely,

$$\Gamma([\mathrm{Fl}_t^{\mathrm{grad}^\sigma(h)}]) = \left[\int_0^1 i(\mathrm{grad}^\sigma(h))\sigma\,dt\right] = [dh] = 0 \in H^1_c(M).$$

That $\ker(\Gamma)$ is regular follows from (38.7), using $p = \Gamma$. $\square$

43.14. Remark. In the situation of (43.13) above, the fundamental group $\pi_1 = \pi_1(\text{Diff}(M,\sigma))$ is a discrete subgroup of the universal covering $\widetilde{\text{Diff}(M,\sigma)}$. Then $\Gamma(\pi_1) \subseteq H_c^1(M)$ is a subgroup, and we have an induced homomorphism $\underline{\Gamma}$ of groups:

$$\begin{array}{ccccc} & & \widetilde{\text{Diff}(M,\sigma)} & \xrightarrow{\Gamma} & H_c^1(M) \\ & & \downarrow & & \downarrow \\ \frac{\widetilde{\text{Diff}(M,\sigma)}}{\pi_1} & = & \text{Diff}(M,\sigma) & \xrightarrow{\underline{\Gamma}} & \frac{H_c^1(M)}{\Gamma(\pi_1)} \end{array}$$

Note that $H_c^1(M)/\Gamma(\pi_1)$ is Hausdorff if $\Gamma(\pi_1)$ is a 'discrete' subgroup of $H_c^1(M)$ in the sense of (38.5), and then $\underline{\Gamma}$ is a smooth homomorphism of regular Lie groups.

In any case, the group $\pi_1 \cap \ker(\Gamma)$ is a 'discrete' (in a sense analogous to (38.5)) central subgroup of $\ker(\Gamma)$, thus $\ker(\underline{\Gamma})$ is a regular Lie subgroup of $\text{Diff}(M,\sigma)$ with universal cover $\ker(\Gamma)$ and with Lie algebra the space of globally Hamiltonian vector fields.

It is known that $\Gamma(\pi_1)$ is 'discrete' in $H_c^1(M)$ if M is compact and either $\dim(M) = 2$ or M is a Kähler manifold or σ has integral periods on M. There seems to be no known example where $\Gamma(\pi)$ is not discrete, see [Banyaga, 1978] and [Banyaga, 1980].

The next topic is the Lie group of contact diffeomorphisms.

43.15. Contact manifolds. Let M be a smooth manifold of dimension $2n+1 \geq 3$. A *contact form* on M is a 1-form $\alpha \in \Omega^1(M)$ such that $\alpha \wedge (d\alpha)^n \in \Omega^{2n+1}(M)$ is nowhere zero. This is sometimes called an *exact contact structure.* The pair (M, α) is called a *contact manifold.*

A contact form can be put into the following normal form: For each $x \in M$ there is a chart $M \supset U \xrightarrow{u} u(U) \subset \mathbb{R}^{2n+1}$ centered at x such that $\alpha|U = u^1\, du^{n+1} + u^2\, du^{n+2} + \dots + u^n\, du^{2n} + du^{2n+1}$. This follows from proposition (43.18) below, for a simple direct proof see [Libermann, Marle, 1987].

The vector subbundle $\ker(\alpha) \subset TM$ is called the *contact distribution.* It is as non-involutive as possible, since $d\alpha$ is even non-degenerate on each fiber $\ker(\alpha)_x = \ker(\alpha_x) \subset T_xM$. The *characteristic vector field* $X_\alpha \in \mathfrak{X}(M)$ is the unique vector field satisfying $i_{X_\alpha}\alpha = 1$ and $i_{X_\alpha} d\alpha = 0$.

Note that $X \mapsto (i_X d\alpha, i_X \alpha)$ is isomorphic $TM \to \{\varphi \in T^*M : i_{X_\alpha}\varphi = 0\} \times \mathbb{R}$, but we shall use the isomorphism of vector bundles

$$(1) \qquad TM \to T^*M, \quad X \mapsto i_X d\alpha + \alpha(X).\alpha,$$

A diffeomorphism $f \in \text{Diff}(M)$ with $f^*\alpha = \lambda_f.\alpha$ for a nowhere vanishing function $\lambda_f \in C^\infty(M, \mathbb{R} \setminus 0)$ is called a *contact diffeomorphism.* The group of all contact diffeomorphisms will be denoted by $\text{Diff}(M, \alpha)$.

A vector field $X \in \mathfrak{X}(M)$ is called a contact vector field if $\mathcal{L}_X\alpha = \mu_X.\alpha$ for a smooth function $\mu_X \in C^\infty(M, \mathbb{R})$. The linear space of all contact vector fields will be denoted by $\mathfrak{X}(M,\alpha)$; it is clearly a Lie algebra. Contraction with α is a linear mapping also denoted by $\alpha : \mathfrak{X}(M,\alpha) \to C^\infty(M,\mathbb{R})$. It is bijective since we may apply i_{X_α} to $\mu_X.\alpha = \mathcal{L}_X\alpha = i_X\,d\alpha + d(\alpha(X))$ to get $\mu_X = 0 + X_\alpha(\alpha(X))$, and since by using (1) we may reconstruct X from $\alpha(X)$ as

$$\begin{aligned} i_X\,d\alpha + \alpha(X).\alpha &= \mu_X.\alpha - d(\alpha(X)) + \alpha(X).\alpha \\ &= X_\alpha(\alpha(X)).\alpha - d(\alpha(X)) + \alpha(X).\alpha. \end{aligned}$$

Note that the inverse $f \mapsto \operatorname{grad}^\alpha(f)$ of $\alpha : \mathfrak{X}(M,\alpha) \to C^\infty(M,\mathbb{R})$ is a linear differential operator of order 1.

A smooth mapping $f : L \to M$ is called a *Legendre mapping* if $f^*\alpha = 0$. If f is also an embedding and $\dim M = 2\dim L + 1$, then the image $f(L)$ is called a *Legendre submanifold* of M.

43.16. Lemma. *Let X_t be a time dependent vector field on M, and let f_t be the local curve of local diffeomorphisms with $\frac{\partial}{\partial t} f_t \circ f_t^{-1} = X_t$ and $f_0 = \mathrm{Id}$. Then $\mathcal{L}_{X_t}\alpha = \mu_t\alpha$ if and only if $f_t^*\alpha = \lambda_t.\alpha$, where λ_t and μ_t are related by $\frac{\partial_t\lambda_t}{\lambda_t} = f_t^*\mu_t$.*

Proof. The two following equations are equivalent:

$$\begin{aligned} \alpha &= \frac{1}{\lambda_t} f_t^*\alpha, \\ 0 &= \tfrac{\partial}{\partial t}\left(\frac{1}{\lambda_t} f_t^*\alpha\right) = -\frac{\frac{\partial}{\partial t}\lambda_t}{\lambda_t^2} f_t^*\alpha + \frac{1}{\lambda_t} f_t^*\mathcal{L}_{X_t}\alpha = \frac{1}{\lambda_t} f_t^*(-\mu_t.\alpha + \mathcal{L}_{X_t}\alpha). \quad \square \end{aligned}$$

43.17. Canonical example. Let N be an n-dimensional manifold, let $\theta \in \Omega^1(T^*N)$ be the canonical 1-form, which is given by $\theta(\xi) = \langle \pi_{T^*N}(\xi), T(\pi^*).\xi\rangle_{TN}$, and which has the following universal property: For any 1-form $\omega \in \Omega^1(N)$, viewed as a section of $T^*N \to N$, we have $\omega^*\theta = \theta$.

Then the 1-form $\theta - dt = \mathrm{pr}_1^*\,\theta - \mathrm{pr}_2^*\,dt \in \Omega^1(T^*N \times \mathbb{R})$ is a contact form. Note that $T^*N\times\mathbb{R} = J^1(N,\mathbb{R})$, the space of 1-jets of functions on N. A section s of $T^*N\times\mathbb{R} = J^1(N,\mathbb{R}) \to N$ is of the form $s = (\omega, f)$ for $\omega \in \Omega^1(N)$ and $f \in C^\infty(N,\mathbb{R})$. Thus, s is a Legendre mapping if and only if $0 = s^*(\theta - dt) = \omega^*\theta - f^*dt = \theta - df$ or $s = j^1 f$.

43.18. Proposition. ([Lychagin, 1977]) *If L is a Legendre submanifold of a (finite dimensional) contact manifold (M,α), then there exist:*

(1) *an open neighborhood U of L in M,*
(2) *an open neighborhood V of the zero section 0_L in $T^*L \times \mathbb{R}$,*
(3) *a diffeomorphism $\varphi : U \to V$ with $\varphi|L = \mathrm{Id}_L$ and $\varphi^*(\theta_L - dt) = \alpha$.*

If all data is real analytic then φ may be chosen real analytic, too.

Proof. By (41.14), there exists a tubular neighborhood $\mathcal{N}(L) = (TM|L)/TL \supset \tilde{U} \xrightarrow{\varphi} U \subset M$ of L in M.

Note that $\ker(d\alpha)$ is a trivial line bundle, framed by the characteristic vector field X_α, and that $TM = \ker(\alpha) \oplus \ker(d\alpha)$. Thus, for the normal bundle we have the following chain of natural isomorphisms of vector bundles:

$$\mathcal{N}(L) = (TM|L)/T(L) = \frac{\ker(\alpha)}{T(L)} \oplus \ker(d\alpha) \xrightarrow[\cong]{d\alpha \oplus \mathrm{Id}} T^*(L) \times \mathbb{R}.$$

Therefore, we may assume that the tubular neighborhood is given by $T^*L \times \mathbb{R} \supset V \xrightarrow{\varphi} U \subset M$, where now V is an open neighborhood of the 0-section 0_L (which we identify with L) in $T^*L \times \mathbb{R}$.

Next we consider the contact structure $\tilde{\alpha} := \varphi^*\alpha \in \Omega^1(V)$. Note that on the subbundle $TL = T(0_L) \subset TV$ both contact structures $\tilde{\alpha}$ and $\alpha_0 := \theta - dt$ vanish. We will first arrange that both contact structures agree on $TV|0_L$: We claim that there exists a vector bundle isomorphism $\gamma : TV|0_L \to TV|0_L$ which satisfies $\gamma^* d\tilde{\alpha} = d\alpha_0$, $\gamma^*\tilde{\alpha} = \alpha_0$, and such that $\gamma|T(0_L) = \mathrm{Id}$. Note that we have two symplectic vector subbundles $(\ker\tilde{\alpha}, d\tilde{\alpha})$ and $(\ker\alpha_0, d\alpha_0)$. We first choose a vector bundle isomorphism $\tilde{\gamma} : \ker\alpha_0 \to \ker\tilde{\alpha}$ with $\tilde{\gamma}^* d\tilde{\alpha} = d\alpha_0$ and $\tilde{\gamma}|T(0_L) = \mathrm{Id}$, as in the proof of the claim in (43.12), and then we complete $\tilde{\gamma}$ to γ in such a way that for the characteristic vector fields we have $\gamma(X_{\alpha_0}) = X_{\tilde{\alpha}}$.

There exists a diffeomorphism $\psi : V' \to V''$ between two open neighborhoods V' and V'' of 0_L in V such that $T\psi|0_L = \gamma$, and since $\gamma|T(0_L) = \mathrm{Id}$ we even have $\psi|0_L = \mathrm{Id}$. We put $\alpha_1 := \psi^*\tilde{\alpha} = \psi^*\varphi^*\alpha$. Then $\alpha_0|0_L = \alpha_1|0_L$, and we put

$$\alpha_t := (1-t)\alpha_0 + t\alpha_1,$$

and since $\alpha_t|T(0_L) = \alpha_0|T(0_L) = \alpha_1|T(0_L)$ the 1-form α_t is a contact structure on an (again smaller) neighborhood V of 0_L in $T^*M \times \mathbb{R}$.

Let us now suppose that f_t is a curve of diffeomorphisms near 0_L which satisfies $T(f_t)|T(0_L) = \mathrm{Id}_{TV|0_L}$, with time dependent vector field $X_t = (\frac{\partial}{\partial t} f_t) \circ f_t^{-1}$. Then we have

$$\begin{aligned} \tfrac{\partial}{\partial t} f_t^*\alpha_t &= \left(\tfrac{\partial}{\partial t} f_t^*\alpha_s\right)\big|_{s=t} + \left(f_s^*\tfrac{\partial}{\partial t}\alpha_t\right)\big|_{s=t} \\ &= f_t^*\mathcal{L}_{X_t}\alpha_t + f_t^*(\alpha_1 - \alpha_0) \\ &= f_t^*\left(i_{X_t} d\alpha_t + d\,i_{X_t}\alpha_t + \alpha_1 - \alpha_0\right). \end{aligned} \tag{4}$$

We want a time dependent vector field X_t with $i_{X_t} d\alpha_t + d\,i_{X_t}\alpha_t + \alpha_1 - \alpha_0 = 0$ near 0_L and we first look for a time dependent function h_t defined near 0_L such that $dh_t(X_{\alpha_t}) = i_{X_{\alpha_t}}(\alpha_1 - \alpha_0)$. Since $\alpha_1 = \alpha_0$ along 0_L and vanishes on $T(0_L)$, the vector field X_{α_t} equals X_{α_0} along 0_L and is not tangent to 0_L. So its flow lines leave 0_L and there is a submanifold S of codimension 1 in TM containing 0_L which is transversal to the flow of X_{α_t} for all $t \in [0,1]$, and we may take h_t as

$$h_t(\mathrm{Fl}_s^{X_{\alpha_t}}(z)) = \int_0^s (\alpha_1 - \alpha_0)(X_{\alpha_t})(\mathrm{Fl}_r^{X_{\alpha_t}}(z))\,dr \quad \text{for } z \in S.$$

Now we use (43.15.1) and choose the unique time dependent vector field X_t which satisfies

$$iX_t\,d\alpha_t + \alpha_t(X_t).\alpha_t = \alpha_0 - \alpha_1 - d\,h_t + h_t.\alpha_t.$$

Then for the curve of diffeomorphisms f_t which is determined by the ordinary differential equation $\frac{\partial}{\partial t} f_t = X_t \circ f_t^{-1}$ with initial condition $f_0 = \mathrm{Id}$ we have $\frac{\partial}{\partial t} f_t^* \alpha_t = 0$ by (4), so $f_t^*\alpha_t = f_0^*\alpha_0 = \alpha_0$ is constant in t. Since $(\alpha_1 - \alpha_0)|T(0_L) = 0$, also $h_t|0_L = 0$, and $dh_t|T(0_L) = 0$, the vector field X_t vanishes along 0_L, and thus the curve of diffeomorphisms f_t exists for all t near $[0,1]$ in a neighborhood of 0_L in $T^*L \times \mathbb{R}$. Then $f_1^*\alpha_1 = \alpha_0$ and $f_1 \circ \psi \circ \varphi$ is the looked for diffeomorphism. □

43.19. Theorem. *Let (M,α) be a finite dimensional contact manifold. Then the group $\mathrm{Diff}(M,\alpha)$ of contact diffeomorphisms is a smooth regular Lie group. The injection $i : \mathrm{Diff}(M,\alpha) \to \mathrm{Diff}(M)$ is smooth, $T_{\mathrm{Id}} i$ maps the Lie algebra of $\mathrm{Diff}(M,\alpha)$ isomorphically onto $\mathfrak{X}_c(M,\alpha)$ with the negative of the usual Lie bracket, and locally there exist smooth retractions to i, so i is an initial mapping, see* (27.11).

If (M,α) is in addition a real analytic and compact contact manifold then all assertions hold in the real analytic sense.

Proof. For a contact manifold (M,α) let $\widehat{M} = M \times M \times (\mathbb{R} \setminus 0)$, with the contact structure $\hat\alpha = t.\mathrm{pr}_1^*\alpha - \mathrm{pr}_2^*\alpha$, where $t = \mathrm{pr}_3 : M \times M \times (\mathbb{R}\setminus 0) \to \mathbb{R}$. Let $f \in \mathrm{Diff}(M,\alpha)$ be a contact diffeomorphism with $f^*\alpha = \lambda_f.\alpha$. Inserting the characteristic vector field X_α into this last equation we get

$$\lambda_f = i_{X_\alpha}\lambda_f\alpha = i_{X_\alpha}(f^*\alpha) = f^*(i_{f_* X_\alpha}\alpha). \tag{1}$$

Thus, f determines λ_f, and for an arbitrary diffeomorphism $f \in \mathrm{Diff}(M)$ we may define a smooth function λ_f by (1). Then $\lambda_f \in C^\infty(M, \mathbb{R}\setminus 0)$ if f is near a contact diffeomorphism in the Whitney C^0-topology. We consider its *contact graph* $\Gamma_f : M \to \widehat{M}$, given by $\Gamma_f(x) := (x, f(x), \lambda_f(x))$, a section of the surjective submersion $\mathrm{pr}_1 : \widehat{M} \to M$. Note that Γ_f is a Legendre mapping if and only if f is a contact diffeomorphism, $f \in \mathrm{Diff}(M,\alpha)$, since $\Gamma_f^*\hat\alpha = \lambda_f.\alpha - f^*\alpha$.

Let us now fix a contact diffeomorphism $f \in \mathrm{Diff}(M,\alpha)$ with $f^*\alpha = \lambda_f.\alpha$. By proposition (43.18), and also using the diffeomorphism $\Gamma_f : M \to \Gamma_f(M)$ there are: an open neighborhood U_f' of $\Gamma_f(M) \subset \widehat{M}$, an open neighborhood V_f' of the zero section 0_M in $T^*M \times \mathbb{R}$, and a diffeomorphism $\widehat{M} \supset U_f' \xrightarrow{\varphi_f} V_f' \subset T^*M \times \mathbb{R}$, such that the restriction $\varphi_f|\Gamma_f(M)$ equals the inverse of $\Gamma_f : 0_M \cong M \to \Gamma_f(M)$, and $\varphi_f^*(\theta_M - dt) = \hat\alpha$.

Now let $\tilde U_f$ be the open set of all diffeomorphisms $g \in \mathrm{Diff}(M)$ such that g equals f off some compact subset of M, $\Gamma_g(M) \subset U_f' \subset \widehat{M}$, and $\pi \circ \varphi_f \circ \Gamma_g : M \to M$ is a diffeomorphism, where $\pi : T^*M \times \mathbb{R} \to M$ is the vector bundle projection. For $g \in \tilde U_f$ and

$$\begin{aligned} s_f(g) &:= (\varphi_f \circ \Gamma_g) \circ (\pi \circ \varphi_f \circ \Gamma_g)^{-1} \in C_c^\infty(M \leftarrow T^*M \times \mathbb{R}) \\ &=: (\sigma_f(g), u_f(g)) \in \Omega_c^1(M) \times C_c^\infty(M, \mathbb{R}) \end{aligned}$$

the following conditions are equivalent:

(2) g is a contact diffeomorphism.

(3) $\Gamma_g(M)$ is a Legendre submanifold of $(\widehat{M}, \hat{\alpha})$.

(4) $\varphi_f(\Gamma_g(M))$ is a Legendre submanifold of $(T^*M \times \mathbb{R}, \theta_M - dt)$.

(5) The section $s_f(g)$ satisfies $s_f(g)^*(\theta_M - dt) = 0$, equivalently (by (43.17)) $\sigma_f(g) = d(u_f(g))$.

Let us now consider the following diagram:

$$\begin{array}{ccccccc}
\operatorname{Diff}(M) & \hookleftarrow & \tilde{U}_f & \xrightarrow{s_f} & \tilde{V}_f & \hookrightarrow & C_c^\infty(M \leftarrow T^*M \times \mathbb{R}) \\
\uparrow & & \uparrow & & \uparrow{\scriptstyle j} & & {\scriptstyle j}\uparrow \text{ linear, splitting} \\
\operatorname{Diff}(M, \alpha) & \hookleftarrow & U_f & \xrightarrow[\cong]{u_f} & V_f & \hookrightarrow & C_c^\infty(M, \mathbb{R}).
\end{array}$$

In this diagram we put $j(h) := (dh, h)$, a bounded linear splitting embedding. We let $\tilde{V}_f \subset C_c^\infty(M \leftarrow T^*M \times \mathbb{R})$ be the open set of all $(\omega, h) \in \Omega_c^1(M) \times C_c^\infty(M, \mathbb{R})$ with $(\omega, h)(M) \subset V_f'$ and such that $\operatorname{pr}_1 \circ \varphi_f^{-1} \circ (\omega, h) : M \to M$ is a diffeomorphism. We also consider the smooth mapping

$$w_f : \tilde{V}_f \to \operatorname{Diff}(M)$$
$$w_f(\omega, h) := \operatorname{pr}_2 \circ \varphi_f^{-1} \circ (\omega, h) \circ (\operatorname{pr}_1 \circ \varphi_f^{-1} \circ (\omega, h))^{-1} : M \to M,$$

and let $V_f = (w_f \circ j)^{-1} \tilde{U}_f$. Then $w_f \circ s_f = \operatorname{Id}$, and so we may use as chart mappings for $\operatorname{Diff}(M, \alpha)$:

$$u_f : U_f := \tilde{U}_f \cap \operatorname{Diff}(M, \alpha) \to V_f := (w_f \circ j)^{-1}(\tilde{U}_f) \subset C_c^\infty(M, \mathbb{R}),$$
$$u_f(g) := \operatorname{pr}_2 \circ (\varphi_f \circ \Gamma_g) \circ (\pi \circ \varphi_f \circ \Gamma_g)^{-1} \in C^\infty(M, \mathbb{R}),$$
$$u_f^{-1}(h) = (w_f \circ j)(h) = w_f(dh, h).$$

The chart change mapping $u_k \circ u_f^{-1}$ is defined on an open subset and is smooth, because $u_k \circ u_f^{-1} = \operatorname{pr}_2 \circ s_k \circ w_f \circ j$, and s_k and w_f are smooth by (42.13), (43.1), and by (42.20). Thus, the resulting atlas $(U_f, u_f)_{f \in \operatorname{Diff}(M, \alpha)}$ is smooth, and $\operatorname{Diff}(M, \alpha)$ is a smooth manifold in such a way that the injection $i : \operatorname{Diff}(M, \alpha) \to \operatorname{Diff}(M)$ is smooth.

Note that $s_f \circ w_f \neq \operatorname{Id}$, so we cannot construct (splitting) submanifold charts in this way.

But there exist local smooth retracts $u_f^{-1} \circ \operatorname{pr}_2 \circ s_f : (\operatorname{pr}_2 \circ s_f)^{-1}(V_f) \to U_f$. Therefore, the injection i has the property that a mapping into $\operatorname{Diff}(M, \alpha)$ is smooth if and only if its prolongation via i into $\operatorname{Diff}(M)$ is smooth. Thus, $\operatorname{Diff}(M, \alpha)$ is a Lie group, and from (38.7) we may conclude that it is a regular Lie group.

A direct proof of regularity goes as follows: From lemma (43.16) and (36.6) we see that $T_{\operatorname{Id}} i$ maps the Lie algebra of $\operatorname{Diff}(M, \alpha)$ isomorphically onto the Lie algebra $\mathfrak{X}_c(M, \alpha)$ of all contact vector fields with compact support. It also follows from lemma (43.16) that we have for the evolution operator

$$\operatorname{Evol}^r_{\operatorname{Diff}(M)} | C^\infty(\mathbb{R}, \mathfrak{X}_c(M, \alpha)) = \operatorname{Evol}^r_{\operatorname{Diff}(M, \alpha)}$$

so that $\operatorname{Diff}(M, \alpha)$ is a regular Lie group. $\square$

43.20. n-Transitivity. Let M be a connected smooth manifold with $\dim M \geq 2$. We say that a subgroup G of the group $\mathrm{Diff}(M)$ of all smooth diffeomorphisms acts *n-transitively on M*, if for any two ordered sets of n different points $(x_1, \dots, x_n)$ and $(y_1, \dots, y_n)$ in M there is a smooth diffeomorphism $f \in G$ such that $f(x_i) = y_i$ for each i.

Theorem. *Let M be a connected smooth (or real analytic) manifold of dimension $\dim M \geq 2$. Then the following subgroups of the group $\mathrm{Diff}(M)$ of all smooth diffeomorphisms act n-transitively on M, for every finite n:*

(1) *The group $\mathrm{Diff}_c(M)$ of all smooth diffeomorphisms with compact support.*
(2) *The group $\mathrm{Diff}^\omega(M)$ of all real analytic diffeomorphisms.*
(3) *If (M, σ) is a symplectic manifold, the group $\mathrm{Diff}_c(M, \sigma)$ of all symplectic diffeomorphisms with compact support, and even the subgroup of all globally Hamiltonian symplectic diffeomorphisms.*
(4) *If (M, σ) is a real analytic symplectic manifold, the group $\mathrm{Diff}^\omega(M, \sigma)$ of all real analytic symplectic diffeomorphisms, and even the subgroup of all globally Hamiltonian real analytic symplectic diffeomorphisms.*
(5) *If (M, μ) is a manifold with a smooth volume density, the group $\mathrm{Diff}_c(M, \mu)$ of all volume preserving diffeomorphisms with compact support.*
(6) *If (M, μ) is a manifold with a real analytic volume density, then the group $\mathrm{Diff}^\omega(M, \mu)$ of all real analytic volume preserving diffeomorphisms.*
(7) *If (M, α) is a contact manifold, the group $\mathrm{Diff}_c(M, \alpha)$ of all contact diffeomorphisms with compact support.*
(8) *If (M, α) is a real analytic contact manifold, the group $\mathrm{Diff}^\omega(M, \alpha)$ of all real analytic contact diffeomorphisms.*

Result (1) is folklore, the first trace is in [Milnor, 1965]. The results (3), (5), and (7) are due to [Hatakeyama, 1966] for 1-transitivity, and to [Boothby, 1969] in the general case. The results about real analytic diffeomorphisms and the proof given here is from [Michor, Vizman, 1994].

Proof. Let us fix a finite $n \in \mathbb{N}$. Let $M^{(n)}$ denote the open submanifold of all n-tuples $(x_1, \dots, x_n) \in M^n$ of pairwise distinct points. Since M is connected and of dimension ≥ 2, each $M^{(n)}$ is connected. The group $\mathrm{Diff}(M)$ acts on $M^{(n)}$ by the diagonal action, and we have to show that any of the subgroups G described above acts transitively. We shall show below that for each G the G-orbit through any n-tuple $(x_1, \dots, x_n) \in M^{(n)}$ contains an open neighborhood of $(x_1, \dots, x_n)$ in $M^{(n)}$, thus any orbit is open. Since $M^{(n)}$ is connected, there can be only one orbit.

The cases (2) and (1). We choose a complete Riemannian metric g on M, and we let $(Y_{ij})_{j=1}^m$ be an orthonormal basis of $T_{x_i}M$ with respect to g, for all i. Then we choose real analytic vector fields X_k for $1 \leq k \leq N = nm$ which satisfy:

$$(9) \qquad \begin{aligned} &|X_k(x_i) - Y_{ij}|_g < \varepsilon && \text{for } k = (i-1)m + j, \\ &|X_k(x_i)|_g < \varepsilon && \text{for all } k \notin [(i-1)m + 1, im], \\ &|X_k(x)|_g < 2 && \text{for all } x \in M \text{ and all } k. \end{aligned}$$

Since these conditions describe a Whitney C^0 open set, such vector fields exist by (30.12). The fields are bounded with respect to a complete Riemannian metric, so they have complete real analytic flows Fl^{X_k}, see e.g. [Hirsch, 1976.] We consider the real analytic mapping

$$f : \mathbb{R}^N \to M^{(n)},$$

$$f(t_1, \dots, t_N) := \begin{pmatrix} (\operatorname{Fl}^{X_1}_{t_1} \circ \dots \circ \operatorname{Fl}^{X_N}_{t_N})(x_1) \\ \dots \\ (\operatorname{Fl}^{X_1}_{t_1} \circ \dots \circ \operatorname{Fl}^{X_N}_{t_N})(x_n), \end{pmatrix}$$

which has values in the $\operatorname{Diff}^\omega(M)$-orbit through $(x_1, \dots, x_n)$. To get the tangent mapping at 0 of f we consider the partial derivatives

$$\tfrac{\partial}{\partial t_k}|_0 f(0, \dots, 0, t_k, 0, \dots, 0) = (X_k(x_1), \dots, X_k(x_n)).$$

If $\varepsilon > 0$ is small enough, this is near an orthonormal basis of $T_{(x_1,\dots,x_n)}M^{(n)}$ with respect to the product metric $g \times \dots \times g$. So $T_0 f$ is invertible, and the image of f thus contains an open subset.

In case (1), we can choose smooth vector fields X_k with compact support which satisfy conditions (9).

For the remaining cases we just indicate the changes which are necessary in this proof.

The cases (4) and (3) Let (M, σ) be a connected real analytic symplectic smooth manifold of dimension $m \geq 2$. We choose real analytic functions f_k for $1 \leq k \leq N = nm$ whose Hamiltonian vector fields $X_k = \operatorname{grad}^\sigma(f_k)$ satisfy conditions (9). Since these conditions describe Whitney C^1 open subsets, such functions exist by [Grauert, 1958, Proposition 8]. Now we may finish the proof as above.

The cases (8) and (7) Let (M, α) be a connected real analytic contact manifold of dimension $m \geq 3$. We choose real analytic functions f_k for $1 \leq k \leq N = nm$ such that their contact vector fields $X_k = \operatorname{grad}^\alpha(f_k)$ satisfy conditions (9). Since these conditions describe Whitney C^1 open subsets, such functions exist. Now we may finish the proof as above.

The cases (6) and (5) Let (M, μ) be a connected real analytic manifold of dimension $m \geq 2$ with a real analytic positive volume density. We can find a real analytic Riemannian metric γ on M whose volume density is μ. We also choose a complete Riemannian metric g.

First we assume that M is orientable. Then the divergence of a vector field $X \in \mathfrak{X}(M)$ is $\operatorname{div} X = *d{*}X^\flat$, where $X^\flat = \gamma(X) \in \Omega^1(M)$ (here we view $\gamma : TM \to T^*M$ and $*$ is the Hodge star operator of γ). We choose real analytic $(m-2)$-forms β_k for $1 \leq k \leq N = nm$ such that the vector fields $X_k = (-1)^{m+1}\gamma^{-1} * d\beta_k$ satisfy conditions (9). Since these conditions describe Whitney C^1 open subsets, such $(m-2)$-forms exist by (30.12). The real analytic vector fields X_k are then divergence free since $\operatorname{div} X_k = *d * \gamma X_k = *dd\beta_k = 0$. Now we may finish the proof as usual.

For non-orientable M, we let $\pi : \tilde{M} \to M$ be the real analytic connected oriented double cover of M, and let $\varphi : \tilde{M} \to \tilde{M}$ be the real analytic involutive covering map. We let $\pi^{-1}(x_i) = \{x_i^1, x_i^2\}$, and we pull back both metrics to $\tilde{M}$, so $\tilde{\gamma} := \pi^*\gamma$ and $\tilde{g} := \pi^* g$. We choose real analytic $(m-2)$-forms $\beta_k \in \Omega^{m-2}(\tilde{M})$ for $1 \le k \le N = nm$ whose vector fields $X_{\beta_k} = (-1)^{m+1}\tilde{\gamma}^{-1} * d\beta_k$ satisfy the following conditions, where we put $Y_{ij}^p := T_{x_{ij}^p}\pi^{-1}.Y_{ij}$ for $p = 1, 2$:

$$(10) \qquad \begin{aligned} &|X_{\beta_k}(x_i^p) - Y_{ij}^p|_{\tilde{g}} < \varepsilon && \text{for } k = (i-1)m + j, \quad p = 1, 2, \\ &|X_{\beta_k}(x_i^p)|_{\tilde{g}} < \varepsilon && \text{for all } k \notin [(i-1)m+1, im], \quad p = 1, 2, \\ &|X_{\beta_k}|_{\tilde{g}} < 2 && \text{for all } x \in \tilde{M} \text{ and all } k. \end{aligned}$$

Since these conditions describe Whitney C^1 open subsets, such $(m-2)$-forms exist by (30.12). Then the vector fields $\frac{1}{2}(X_{\beta_k} + \varphi_* X_{\beta_k})$ still satisfy the conditions (10), are still divergence free and induce divergence free vector fields $Z_{\beta_k} \in \mathfrak{X}(M)$, so that $\mathcal{L}_{Z_{\beta_k}}\mu$ is the zero density, which satisfy the conditions (9) on M as in the oriented case, and we may finish the proof as above. □

44. Principal Bundles with Structure Group a Diffeomorphism Group

44.1. Theorem. Principal bundle of embeddings. *Let M and N be smooth finite dimensional manifolds, connected and second countable without boundary such that* $\dim M \le \dim N$.

Then the set $\operatorname{Emb}(M, N)$ *of all smooth embeddings $M \to N$ is an open submanifold of $\mathfrak{C}^\infty(M, N)$. It is the total space of a smooth principal fiber bundle with structure group* $\operatorname{Diff}(M)$, *whose smooth base manifold is the space $B(M, N)$ of all submanifolds of N of type M.*

The open subset $\operatorname{Emb}_{\text{prop}}(M, N)$ *of proper (equivalently closed) embeddings is saturated under the* $\operatorname{Diff}(M)$*-action, and is thus the total space of the restriction of the principal bundle to the open submanifold $B_{\text{closed}}(M, N)$ of $B(M, N)$ consisting of all closed submanifolds of N of type M.*

This result is based on an idea implicitly contained in [Weinstein, 1971], it was fully proved by [Binz, Fischer, 1981] for compact M and for general M by [Michor, 1980b]. The clearest presentation was in [Michor, 1980c, section 13].

Proof. Let us fix an embedding $i \in \operatorname{Emb}(M, N)$. Let g be a fixed Riemannian metric on N, and let $\exp^N$ be its exponential mapping. Then let $p : \mathcal{N}(i) \to M$ be the *normal bundle* of i, defined in the following way: For $x \in M$ let $\mathcal{N}(i)_x := (T_x i(T_x M))^\perp \subset T_{i(x)}N$ be the g-orthogonal complement in $T_{i(x)}N$. Then we have an injective vector bundle homomorphism over i:

$$\begin{array}{ccc} \mathcal{N}(i) & \xrightarrow{\bar{\imath}} & TN \\ {\scriptstyle p=p^i}\downarrow & & \downarrow{\scriptstyle \pi_N} \\ M & \xrightarrow{i} & N. \end{array}$$

Now let $U^i = U$ be an open neighborhood of the zero section of $\mathcal{N}(i)$ which is so small that $(\exp^N \circ \bar{\imath})|U : U \to N$ is a diffeomorphism onto its image which describes a tubular neighborhood of the submanifold $i(M)$. Let us consider the mapping

$$\tau = \tau^i := (\exp^N \circ \bar{\imath})|U : \mathcal{N}(i) \supset U \to N,$$

a diffeomorphism onto its image, and the open set in $\operatorname{Emb}(M, N)$ which will serve us as a saturated chart,

$$\mathcal{U}(i) := \{j \in \operatorname{Emb}(M, N) : j(M) \subseteq \tau^i(U^i), j \sim i\},$$

where $j \sim i$ means that $j = i$ off some compact set in M. Then by (41.10) the set $\mathcal{U}(i)$ is an open neighborhood of i in $\operatorname{Emb}(M, N)$. For each $j \in \mathcal{U}(i)$ we define

$$\begin{aligned} &\varphi_i(j) : M \to U^i \subseteq \mathcal{N}(i), \\ &\varphi_i(j)(x) := (\tau^i)^{-1}(j(x)). \end{aligned}$$

Then $\varphi_i = ((\tau^i)^{-1})_* : \mathcal{U}(i) \to C^\infty(M, \mathcal{N}(i))$ is a smooth mapping which is bijective onto the open set

$$\mathcal{V}(i) := \{h \in \mathfrak{C}^\infty(M, \mathcal{N}(i)) : h(M) \subseteq U^i, h \sim 0\}$$

in $C^\infty(M, \mathcal{N}(i))$. Its inverse is given by the smooth mapping $\tau^i_* : h \mapsto \tau^i \circ h$.

We have $\tau^i_*(h \circ f) = \tau^i_*(h) \circ f$ for those $f \in \operatorname{Diff}(M)$ which are so near to the identity that $h \circ f \in \mathcal{V}(i)$. We consider now the open set

$$\{h \circ f : h \in \mathcal{V}(i), f \in \operatorname{Diff}(M)\} \subseteq \mathfrak{C}^\infty(M, U^i).$$

Obviously, we have a smooth mapping from this set into $C^\infty_c(M \leftarrow U^i) \times \operatorname{Diff}(M)$ given by $h \mapsto (h \circ (p \circ h)^{-1}, p \circ h)$, where $C^\infty_c(M \leftarrow U^i)$ is the space of sections with compact support of $U^i \to M$. So if we let $\mathcal{Q}(i) := \tau^i_*(C^\infty_c(M \leftarrow U^i) \cap \mathcal{V}(i)) \subset \operatorname{Emb}(M, N)$ we have

$$\mathcal{W}(i) := \mathcal{U}(i) \circ \operatorname{Diff}(M) \cong \mathcal{Q}(i) \times \operatorname{Diff}(M) \cong (C^\infty_c(M \leftarrow U^i) \cap \mathcal{V}(i)) \times \operatorname{Diff}(M),$$

since the action of $\operatorname{Diff}(M)$ on i is free. Furthermore, the restriction $\pi|\mathcal{Q}(i) : \mathcal{Q}(i) \to \operatorname{Emb}(M, N)/\operatorname{Diff}(M)$ is bijective onto an open set in the quotient.

We now consider $\varphi_i \circ (\pi|\mathcal{Q}(i))^{-1} : \pi(\mathcal{Q}(i)) \to C^\infty(M \leftarrow U^i)$ as a chart for the quotient space. In order to investigate the chart change, let $j \in \operatorname{Emb}(M, N)$ be such that $\pi(\mathcal{Q}(i)) \cap \pi(\mathcal{Q}(j)) \neq \emptyset$. Then there is an immersion $h \in \mathcal{W}(i) \cap \mathcal{Q}(j)$ and hence there exists a unique $f_0 \in \operatorname{Diff}(M)$ (given by $f_0 = p \circ \varphi_i(h)$) such that $h \circ f_0^{-1} \in \mathcal{Q}(i)$. If we consider $j \circ f_0^{-1}$ instead of j and call it again j, we have $\mathcal{Q}(i) \cap \mathcal{Q}(j) \neq \emptyset$, and consequently $\mathcal{U}(i) \cap \mathcal{U}(j) \neq \emptyset$. Then the chart change is given as follows:

$$\varphi_i \circ (\pi|\mathcal{Q}(i))^{-1} \circ \pi \circ (\tau^j)_* : C_c^\infty(M \leftarrow U^j) \to C_c^\infty(M \leftarrow U^i)$$
$$s \mapsto \tau^j \circ s \mapsto \varphi_i(\tau^j \circ s) \circ (p^i \circ \varphi_i(\tau^j \circ s))^{-1}.$$

This is of the form $s \mapsto \beta \circ s$ for a locally defined diffeomorphism $\beta : \mathcal{N}(j) \to \mathcal{N}(i)$ which is not fiber respecting, followed by $h \mapsto h \circ (p^i \circ h)^{-1}$. Both composants are smooth by the general properties of manifolds of mappings. Therefore, the chart change is smooth.

We show that the quotient space $B(M, N) = \operatorname{Emb}(M, N)/\operatorname{Diff}(M)$ is Hausdorff. Let i, $j \in \operatorname{Emb}(M, N)$ with $\pi(i) \neq \pi(j)$. Then $i(M) \neq j(M)$ in N for otherwise put $i(M) = j(M) =: L$, a submanifold of N; the mapping $i^{-1} \circ j : M \to L \to M$ is then a diffeomorphism of M and $j = i \circ (i^{-1} \circ j) \in i \circ \operatorname{Diff}(M)$, so $\pi(i) = \pi(j)$, contrary to the assumption.

Now we distinguish two cases.

Case 1. We may find a point $y_0 \in i(M) \setminus j(M)$, say, which is not a cluster point of $j(M)$. We choose an open neighborhood V of y_0 in N and an open neighborhood W of $j(M)$ in N such that $V \cap W = \emptyset$. Let $\mathcal{V} := \{k \in \operatorname{Emb}(M, N) : k(M) \subset V\}$ $\mathcal{W} := \{k \in \operatorname{Emb}(M, N) : k(M) \subset W\}$. Then $\mathcal{V}$ is obviously open in $\operatorname{Emb}(M, N)$, and $\mathcal{V}$ is even open in the coarser compact-open topology. Both $\mathcal{V}$ and $\mathcal{W}$ are $\operatorname{Diff}(M)$ saturated, $i \in \mathcal{W}$, $j \in \mathcal{V}$, and $\mathcal{V} \cap \mathcal{W} = \emptyset$. So $\pi(\mathcal{V})$ and $\pi(\mathcal{W})$ separate $\pi(i)$ and $\pi(j)$ in $B(M, N)$.

Case 2. Let $i(M) \subset \overline{j(M)}$ and $j(M) \subset \overline{i(M)}$. Let $y \in i(X)$, say. Let (V, v) be a chart of N centered at y which maps $i(M) \cap V$ into a linear subspace, $v(i(M) \cap V) \subseteq \mathbb{R}^m \cap v(V) \subset \mathbb{R}^n$, where $m = \dim M$, $n = \dim N$. Since $j(M) \subseteq \overline{i(M)}$ we conclude that we also have $v((i(M) \cup j(M)) \cap V) \subseteq \mathbb{R}^m \cap v(V)$. So we see that $L := i(M) \cup j(M)$ is a submanifold of N of the same dimension as N. Let (W_L, p_L, L) be a tubular neighborhood of L. Then $W_L|i(M)$ is a tubular neighborhood of $i(M)$ and $W_L|j(M)$ is one of $j(M)$. □

44.2. Result. [Cervera, Mascaro, Michor, 1991]. *Let M and N be smooth manifolds. Then the diffeomorphism group* $\operatorname{Diff}(M)$ *acts smoothly from the right on the manifold* $\operatorname{Imm}_{\text{prop}}(M, N)$ *of all smooth proper immersions $M \to N$, which is an open subset of* $\mathfrak{C}^\infty(M, N)$.

Then the space of orbits $\operatorname{Imm}_{prop}(M, N)/\operatorname{Diff}(M)$ *is Hausdorff in the quotient topology.*

Let $\operatorname{Imm}_{\text{free, prop}}(M, N)$ *be set of all proper immersions, on which* $\operatorname{Diff}(M)$ *acts freely. Then this is open in* $\mathfrak{C}^\infty(M, N)$ *and it is the total space of a smooth principal fiber bundle*

$$\operatorname{Imm}_{\text{free,prop}}(M, N) \to \operatorname{Imm}_{\text{free,prop}}(M, N)/\operatorname{Diff}(M).$$

44.3. Theorem (Principal bundle of real analytic embeddings). [Kriegl, Michor, 1990, section 6]. *Let M and N be real analytic finite dimensional manifolds, connected and second countable without boundary such that* $\dim M \le \dim N$, *with M compact. Then the set* $\mathrm{Emb}^\omega(M,N)$ *of all real analytic embeddings $M \to N$ is an open submanifold of* $C^\omega(M,N)$. *It is the total space of a real analytic principal fiber bundle with structure group* $\mathrm{Diff}^\omega(M)$, *whose real analytic base manifold $B^\omega(M,N)$ is the space of all real analytic submanifolds of N of type M.*

Proof. The proof of (44.1) is valid with the obvious changes. One starts with a real analytic Riemannian metric and uses its exponential mapping. The space of embeddings is open, since embeddings are open in $C^\infty(M,N)$, which induces a coarser topology. □

44.4. The nonlinear frame bundle of a fiber bundle. [Michor, 1988], [Michor, 1991]. Let now $(p: E \to M, S)$ be a fiber bundle, and let us fix a fiber bundle atlas (U_α) with transition functions $\psi_{\alpha\beta}: U_{\alpha\beta} \times S \to S$. By (42.14) we have

$$C^\infty(U_{\alpha\beta}, \mathfrak{C}^\infty(S,S)) \subseteq C^\infty(U_{\alpha\beta} \times S, S)$$

with equality if and only if S is compact. Let us therefore assume from now on that S **is compact**. Then we assume that the transition functions $\psi_{\alpha\beta}: U_{\alpha\beta} \to \mathrm{Diff}(S,S)$.

Now we define the nonlinear frame bundle of $(p: E \to M, S)$ as follows. We consider the set $\mathrm{Diff}\{S,E\} := \bigcup_{x\in M} \mathrm{Diff}(S, E_x)$ and equip it with the infinite dimensional differentiable structure which one gets by applying the functor $\mathrm{Diff}(S, \quad)$ to the cocycle of transition functions $(\psi_{\alpha\beta})$. Then the resulting cocycle of transition functions for $\mathrm{Diff}\{S,E\}$ induces the structure of a smooth principal bundle over M with structure group $\mathrm{Diff}(M)$. The principal action is just composition from the right.

We can now consider the smooth action $\mathrm{ev}: \mathrm{Diff}(S) \times S \to S$ and the associated bundle $\mathrm{Diff}\{S,E\}[S,\mathrm{ev}] = \frac{\mathrm{Diff}\{S,E\}\times S}{\mathrm{Diff}(S)}$. The mapping $\mathrm{ev}: \mathrm{Diff}\{S,E\} \times S \to E$ is invariant under the $\mathrm{Diff}(S)$-action and factors therefore to a smooth mapping $\mathrm{Diff}\{S,E\}[S,\mathrm{ev}] \to E$ as in the following diagram:

$$\begin{array}{ccc}
\mathrm{Diff}\{S,E\} \times S & \xrightarrow{\ \mathrm{pr}\ } & \dfrac{\mathrm{Diff}\{S,E\}\times S}{\mathrm{Diff}(S)} \\
{\scriptstyle \mathrm{ev}}\big\downarrow & & \Big\| \\
E & \longleftarrow & \mathrm{Diff}\{S,E\}[S,\mathrm{ev}].
\end{array}$$

The bottom mapping is easily seen to be a diffeomorphism. Thus, the bundle $\mathrm{Diff}\{S,E\}$ may in full right be called the (nonlinear) *frame bundle* of E.

44.5. Let now $\Phi \in \Omega^1(E;TE)$ be a connection on E, see (37.2). We want to lift Φ to a principal connection on $\mathrm{Diff}\{S,E\}$, and for this we need a good description of the tangent space $T\operatorname{Diff}\{S,E\}$. With the method of (42.17) one can easily show that

$$T\operatorname{Diff}\{S,E\} = \bigcup_{x\in M}\{f\in C^\infty(S,TE|E_x) : Tp\circ f = \text{ one point}$$
$$\text{in } T_xM \text{ and } \pi_E\circ f\in \operatorname{Diff}(S,E_x)\}.$$

Starting from the connection Φ we can then consider $\omega(f) := T(\pi_E\circ f)^{-1}\circ\Phi\circ f : S\to TE\to VE\to TS$ for $f\in T\operatorname{Diff}\{S,E\}$. Then $\omega(f)$ is a vector field on S, and we have:

Lemma. *$\omega\in\Omega^1(\operatorname{Diff}\{S,E\};\mathfrak{X}(S))$ is a principal connection, and the induced connection on $E = \operatorname{Diff}\{S,E\}[S,\operatorname{ev}]$ coincides with Φ.*

Proof. The fundamental vector field ζ_X on $\operatorname{Diff}\{S,E\}$ for $X\in\mathfrak{X}(S)$ is given by $\zeta_X(g) = Tg\circ X$. Then $\omega(\zeta_X(g)) = Tg^{-1}\circ\Phi\circ Tg\circ X = X$ since $Tg\circ X$ has vertical values. Hence, ω reproduces fundamental vector fields.

Now let $h\in\operatorname{Diff}(S)$, and denote by r^h the principal right action. Then we have

$$\begin{aligned}((r^h)^*\omega)(f) &= \omega(T(r^h)f) = \omega(f\circ h) = T(\pi_E\circ f\circ h)^{-1}\circ\Phi\circ f\circ h\\ &= Th^{-1}\circ\omega(f)\circ h = \operatorname{Ad}_{\operatorname{Diff}(S)}(h^{-1})\omega(f). \quad\square\end{aligned}$$

44.6. Theorem. *Let $(p : E\to M, S)$ be a fiber bundle with compact standard fiber S. Then connections on E and principal connections on $\operatorname{Diff}\{S,E\}$ correspond to each other bijectively, and their curvatures are related as in (37.24). Each principal connection on $\operatorname{Diff}\{S,E\}$ admits a global parallel transport. The holonomy groups and the restricted holonomy groups are equal as subgroups of $\operatorname{Diff}(S)$.*

Proof. This follows directly from (37.24) and (37.25). Each connection on E is complete since S is compact, and the lift to $\operatorname{Diff}\{S,E\}$ of its parallel transport is the global parallel transport of the lift of the connection, so the two last assertions follow. $\square$

44.7. Remark on the holonomy Lie algebra. Let M be connected, let $\rho = -d\omega - \frac{1}{2}[\omega,\omega]_{\mathfrak{X}(S)}$ be the usual $\mathfrak{X}(S)$-valued curvature of the lifted connection ω on $\operatorname{Diff}\{S,E\}$. Then we consider the $\mathbb{R}$-linear span of all elements $\rho(\xi_f,\eta_f)$ in $\mathfrak{X}(S)$, where ξ_f, $\eta_f\in T_f\operatorname{Diff}\{S,E\}$ are arbitrary (horizontal) tangent vectors, and we call this span $\operatorname{hol}(\omega)$. Then by the $\operatorname{Diff}(S)$-equivariance of ρ the vector space $\operatorname{hol}(\omega)$ is an ideal in the Lie algebra $\mathfrak{X}(S)$.

44.8. Lemma. *Let $f : S\to E_{x_0}$ be a diffeomorphism in $\operatorname{Diff}\{S,E\}_{x_0}$. Then $f_* : \mathfrak{X}(S)\to\mathfrak{X}(E_{x_0})$ induces an isomorphism between $\operatorname{hol}(\omega)$ and the $\mathbb{R}$-linear span of all $g^*R(CX,CY)$, X, $Y\in T_xM$, and $g : E_{x_0}\to E_x$ any diffeomorphism.*

The proof is obvious.

44.9. Gauge theory for fiber bundles. We consider the bundle $\operatorname{Diff}\{E, E\} := \bigcup_{x \in M} \operatorname{Diff}(E_x, E_x)$ which bears the smooth structure described by the cocycle of transition functions $\operatorname{Diff}(\psi_{\alpha\beta}^{-1}, \psi_{\alpha\beta}) = (\psi_{\alpha\beta})_*(\psi_{\beta\alpha})^*$, where $(\psi_{\alpha\beta})$ is a cocycle of transition functions for the fiber bundle $(p : E \to M, S)$.

44.10. Lemma. *The associated bundle* $\operatorname{Diff}\{S, E\}[\operatorname{Diff}(S), \operatorname{conj}]$ *is isomorphic to the fiber bundle* $\operatorname{Diff}\{E, E\}$.

Proof. The mapping $A : \operatorname{Diff}\{S, E\} \times \operatorname{Diff}(S) \to \operatorname{Diff}\{E, E\}$, given by $A(f, g) := f \circ g \circ f^{-1} : E_x \to S \to S \to E_x$ for $f \in \operatorname{Diff}(S, E_x)$, is $\operatorname{Diff}(S)$-invariant, so it factors to a smooth mapping $\operatorname{Diff}\{S, E\}[\operatorname{Diff}(S)] \to \operatorname{Diff}\{E, E\}$. It is bijective and admits locally over M smooth inverses, so it is a fiber respecting diffeomorphism. □

44.11. The *gauge group* $\operatorname{Gau}(E)$ of the finite dimensional fiber bundle $(p : E \to M, S)$ with compact standard fiber S is, by definition, the group of all principal bundle automorphisms of the $\operatorname{Diff}(S)$-bundle ($\operatorname{Diff}\{S, E\}$ which cover the identity of M. The usual reasoning (37.17) gives that $\operatorname{Gau}(E)$ equals the space of all smooth sections of the associated bundle $\operatorname{Diff}\{S, E\}[\operatorname{Diff}(S), \operatorname{conj}]$ which by (44.10) equals the space of sections of the bundle $\operatorname{Diff}\{E, E\} \to M$. We equip it with the topology and differentiable structure described in (42.21).

44.12. Theorem. *The gauge group* $\operatorname{Gau}(E) = \mathfrak{C}^\infty(M \leftarrow \operatorname{Diff}\{E, E\})$ *is a regular Lie group. Its exponential mapping is not surjective on any neighborhood of the identity. Its Lie algebra consists of all vertical vector fields with compact support on E (or M) with the negative of the usual Lie bracket. The obvious embedding* $\operatorname{Gau}(E) \to \operatorname{Diff}(E)$ *is a smooth homomorphism of regular Lie groups.*

Proof. The first statement has already been shown before the theorem. A curve through the identity of principal bundle automorphisms of $\operatorname{Diff}\{S, E\} \to M$ is a smooth curve through the identity in $\operatorname{Diff}(E)$ consisting of fiber respecting mappings. The derivative of such a curve is thus an arbitrary vertical vector field with compact support. The space of all these is therefore the Lie algebra of the gauge group, with the negative of the usual Lie bracket.

The exponential mapping is given by the flow operator of such vector fields. Since on each fiber it is just conjugate to the exponential mapping of $\operatorname{Diff}(S)$, it has all the properties of the latter. $\operatorname{Gau}(E) \to \operatorname{Diff}(E)$ is a smooth homomorphism since by (40.3) its prolongation to the universal cover of Gau(E) is smooth. □

44.13. Remark. If S is not compact we may circumvent the nonlinear frame bundle, and we may define the gauge group $\operatorname{Gau}(E)$ directly as the splitting closed subgroup of $\operatorname{Diff}(E)$ which consists of all fiber respecting diffeomorphisms which cover the identity of M. The Lie algebra of $\operatorname{Gau}(E)$ consists then of all vertical vector fields on E with compact support on E. We do not work out the details of this approach.

44.14. The space of connections. Let $J^1(E) \to E$ be the affine bundle of 1-jets of sections of $E \to M$. We have $J^1(E) = \{\ell \in L(T_xM, T_uE) : Tp \circ \ell =$

$\mathrm{Id}_{T_xM}, u \in E, p(u) = x\}$. Then a section of $J^1(E) \to E$ is just a horizontal lift mapping $TM \times_M E \to TE$ which is fiber linear over E, so it describes a connection as treated in (37.2), and we may view the space of sections $C^\infty(E \leftarrow J^1(E))$ as the space of all connections.

44.15. Theorem. *The action of the gauge group* $\mathrm{Gau}(E)$ *on the space of connections* $C^\infty(E \leftarrow J^1(E))$ *is smooth.*

Proof. This follows from (42.13) □

44.16. We will now give a different description of the action. We view a connection Φ again as a linear fiber wise projection $TE \to VE$, so the space of connections is now $\mathrm{Conn}(E) := \{\Phi \in \Omega^1(E;TE) : \Phi \circ \Phi = \Phi,\ \Phi(TE) = VE\}$. Since S is compact the canonical isomorphism $\mathrm{Conn}(E) \to C^\infty(E \leftarrow J^1(E))$ is even a diffeomorphism. Then the action of $f \in \mathrm{Gau}(E) \subset \mathrm{Diff}(E)$ on $\Phi \in \mathrm{Conn}(E)$ is given by $f_*\Phi = (f^{-1})^*\Phi = Tf \circ \Phi \circ Tf^{-1}$. Now it is very easy to describe the infinitesimal action. Let X be a vertical vector field with compact support on E and consider its global flow Fl^X_t.

Then we have $\frac{d}{dt}|_0(\mathrm{Fl}^X_t)^*\Phi = \mathcal{L}_X\Phi = [X,\Phi]$, the Frölicher Nijenhuis bracket, by (35.14.5). The tangent space of $\mathrm{Conn}(E)$ at Φ is the space $T_\Phi \mathrm{Conn}(E) = \{\Psi \in \Omega^1(E;TE) : \Psi|VE = 0\}$. The "infinitesimal orbit" at Φ in $T_\Phi \mathrm{Conn}(E)$ is $\{[X,\Phi] : X \in C^\infty_c(E \leftarrow VE)\}$.

The isotropy subgroup of a connection Φ is $\{f \in \mathrm{Gau}(E) : f^*\Phi = \Phi\}$. Clearly, this is just the group of all those f which respect the horizontal bundle $HE = \ker\Phi$. The most interesting object is of course the orbit space $\mathrm{Conn}(E)/\mathrm{Gau}(E)$.

44.17. Slices. [Palais, Terng, 1988] Let $\mathcal{M}$ be a smooth manifold, G a Lie group, $G \times \mathcal{M} \to \mathcal{M}$ a smooth action, $x \in \mathcal{M}$, and let $G_x = \{g \in G : g.x = x\}$ denote the isotropy group at x. A contractible subset $S \subseteq \mathcal{M}$ is called a *slice* at x, if it contains x and satisfies

(1) If $g \in G_x$ then $g.S = S$.
(2) If $g \in G$ with $g.S \cap S \neq \emptyset$ then $g \in G_x$.
(3) There exists a local continuous section $\chi : G/G_x \to G$ defined on a neighborhood V of the identity coset such that the mapping $F : V \times S \to \mathcal{M}$, defined by $F(v,s) := \chi(v).s$ is a homeomorphism onto a neighborhood of x.

This is a local version of the usual definition in finite dimensions, which is too narrow for the infinite dimensional situation. However, in finite dimensions the definition above is equivalent to the usual one where a subset $S \subseteq M$ is called a *slice* at x, if there is a G-invariant open neighborhood U of the orbit $G.x$ and a smooth equivariant retraction $r : U \to G.x$ such that $S = r^{-1}(x)$. In the general case we have the following properties:

(4) For $y \in F(V \times S) \cap S$ we get $G_y \subset G_x$, by (2).
(5) For $y \in F(V \times S)$ the isotropy group G_y is conjugate to a subgroup of G_x, by (3) and (4).

44.18. Counter-example. [Cerf, 1970], [Michor, Schichl, 1997]. *The right action of* $\operatorname{Diff}(S^1)$ *on* $C^\infty(S^1,\mathbb{R})$ *does not admit slices.*

Let $h(t) : S^1 = (\mathbb{R} \mod 1) \to \mathbb{R}$ be a smooth bump function with $h(t) = 0$ for $t \notin [0,\frac{1}{4}]$ and $h(t) > 0$ for $t \in (0,\frac{1}{4})$. Then put $h_n(t) = \frac{1}{4^n}h(4^n(t-(1-\frac{1}{4^n})/3))$ which is is nonzero in the interval $\big((1-\frac{1}{4^n})/3,(1-\frac{1}{4^{n+1}})/3\big)$, and consider

$$f_N(t) = \sum_{n=0}^{N} h_n(t)e^{-\frac{1}{(t-\frac{1}{3})^2}}, \qquad f(t) = \sum_{n=0}^{\infty} h_n(t)e^{-\frac{1}{(t-\frac{1}{3})^2}}.$$

Then $f \geq 0$ is a smooth function which in $(0,\frac{1}{3})$ has zeros exactly at $t = \frac{1-\frac{1}{4^n}}{3}$ and which is 0 for $t \notin (0,\frac{1}{3})$. In every neighborhood of f lies a function f_N which has only finitely many of the zeros of f and is identically zero in the interval $[(1-\frac{1}{4^{N+1}})/3, 1/3]$. All diffeomorphisms in the isotropy subgroup of f are also contained in the isotropy subgroup of f_N, but the latter group contains additionally all diffeomorphisms of S^1 which have support only on $[(1-\frac{1}{4^{N+1}})/3, 1/3]$. This contradicts (44.17.5).

44.19. Counter-example. [Michor, Schichl, 1997]. *The action of the gauge group* $\operatorname{Gau}(E)$ *on* $\operatorname{Conn}(E)$ *does not admit slices, for* $\dim M \geq 2$.

We will construct locally a connection, which satisfies that in any neighborhood there exist connections which have a bigger isotropy subgroup. Let $n = \dim S$, and let $h : \mathbb{R}^n \to \mathbb{R}$ be a smooth nonnegative bump function, which satisfies $\operatorname{carr} h = \{s \in \mathbb{R}^n |\ \|s-s_0\| < 1\}$. Put $h_r(s) := rh(s_0 + \frac{1}{r}(s-s_0))$, then $\operatorname{carr} h_r = \{s \in \mathbb{R}^n |\ \|s-s_0\| < r\}$. Then put $h_r^{s_1}(s) := h(s-(s_1-s_0))$ which implies $\operatorname{carr} h_r^{s_1} = \{s \in \mathbb{R}^n |\ \|s-s_1\| < r\}$. Using these functions, we can define new functions f_k for $k \in \mathbb{N}$ as

$$f_k(s) = \frac{1}{4^k}h^{s_k}_{\|z\|/2^k}(s),$$

where $z := \frac{s_\infty - s_0}{3}$ for some $s_\infty \in \mathbb{R}^n$ and $s_k := s_0 + z(2\sum_{l=0}^{k}\frac{1}{2^l} - 1 - \frac{1}{2^k})$. Further set

$$f^N(s) := e^{-\frac{1}{\|s-s_\infty\|^2}}\sum_{k=0}^{N} f_k(s), \quad f(s) := \lim_{N\to\infty} f^N(s).$$

The functions f^N and f are smooth, respectively, since all the functions f_k are smooth, on every point s at most one summand is nonzero, and the series is in each derivative uniformly convergent on a neighborhood of s_∞. The carriers are given by $\operatorname{carr} f^N = \bigcup_{k=0}^{N}\{s \in \mathbb{R}^n |\ \|s-s_k\| < \frac{1}{2^k}\|z\|\}$ and $\operatorname{carr} f = \bigcup_{k=0}^{\infty}\{s \in \mathbb{R}^n |\ \|s-s_k\| < \frac{1}{2^k}\|z\|\}$. The functions f^N and f vanish in all derivatives in all x_k, and f vanishes in all derivatives in s_∞.

Let $\psi : E|U \to U \times S$ be a fiber bundle chart of E with a chart $u : U \xrightarrow{\cong} \mathbb{R}^m$ on M, and let $v : V \xrightarrow{\cong} \mathbb{R}^n$ be a chart on S. Choose $g \in C_c^\infty(M,\mathbb{R})$ with $\emptyset \neq \operatorname{supp}(g) \subset U$ and $dg \wedge du^1 \neq 0$ on an open dense subset of $\operatorname{supp}(g)$. Then we can define a Christoffel form as in (37.5) by

$$\Gamma := g\,du^1 \otimes f(v)\partial_{v^1} \in \Omega^1(U, \mathfrak{X}(S)).$$

This defines a connection Φ on $E|U$ which can be extended to a connection Φ on E by the following method. Take a smooth functions $k_1, k_2 \geq 0$ on M satisfying $k_1 + k_2 = 1$ and $k_1 = 1$ on $\operatorname{supp}(g)$ and $\operatorname{supp}(k_1) \subset U$ and any connection Φ' on E, and set $\Phi = k_1\Phi^\Gamma + k_2\Phi'$, where Φ^Γ denotes the connection which is induced locally by Γ. In any neighborhood of Φ there exists a connection Φ^N defined by

$$\Gamma^N := g\, du^1 \otimes f^N(s)\partial_{v_1} \in \Omega^1(U, \mathfrak{X}(S)),$$

and extended like Φ.

Claim: There is no slice at Φ.

Proof: We have to consider the isotropy subgroups of Φ and Φ^N. Since the connections Φ and Φ^N coincide outside of U, we may investigate them locally on $W = \{u : k_1(u) = 1\} \subset U$. The curvature of Φ is given locally on W by (37.5) as

$$R_U := d\Gamma - \tfrac{1}{2}[\Gamma,\Gamma]^{\mathfrak{X}(S)}_\wedge = dg \wedge du^1 \otimes f(v)\partial_{v^1} - 0. \tag{1}$$

For every element of the gauge group $\operatorname{Gau}(E)$ which is in the isotropy group $\operatorname{Gau}(E)_\Phi$ the local representative over W which looks like $\tilde\gamma : (u,v) \mapsto (u, \gamma(u,v))$ by (37.5) satisfies

$$\begin{gathered} T_v(\gamma(u,\quad)).\Gamma(\xi_u, v) = \Gamma(\xi_u, \gamma(u,v)) - T_u(\gamma(\quad, v)).\xi_u, \\ g(u)du^1 \otimes f(v)\sum_i \frac{\partial\gamma^1}{\partial v^i}\partial_{v^i} = g(u)du^1 \otimes f(\gamma(u,v))\partial_{v^1} - \sum_{i,j}\frac{\partial\gamma^i}{\partial u^j}du^j \otimes \partial_{v^i}. \end{gathered} \tag{2}$$

Comparing the coefficients of $du^j \otimes \partial_{v^i}$ we get for γ over W the equations

$$\begin{gathered} \frac{\partial\gamma^i}{\partial u^j} = 0 \quad \text{for } (i,j) \neq (1,1), \\ g(u)f(v)\frac{\partial\gamma^1}{\partial v^1} = g(u)f(\gamma(u,v)) - \frac{\partial\gamma^1}{\partial u^1}. \end{gathered} \tag{3}$$

Considering next the transformation $\tilde\gamma^* R_U = R_U$ of the curvature (37.4.3), we get

$$\begin{gathered} T_v(\gamma(u,\quad)).R_U(\xi_u, \eta_u, v) = R_U(\xi_u, \eta_u, \gamma(u,v)), \\ dg \wedge du^1 \otimes f(v)\sum_i \frac{\partial\gamma^1}{\partial v^i}\partial_{v^i} = dg \wedge du^1 \otimes f(\gamma(u,v))\partial_{v^1}. \end{gathered} \tag{4}$$

Another comparison of coefficients yields the equations

$$\begin{gathered} f(v)\frac{\partial\gamma^1}{\partial v^i} = 0 \quad \text{for } i \neq 1, \\ f(v)\frac{\partial\gamma^1}{\partial v^1} = f(\gamma(u,v)), \end{gathered} \tag{5}$$

whenever $dg \wedge du^1 \neq 0$, but this is true on an open dense subset of $\operatorname{supp}(g)$. Finally, putting (5) into (3) shows

$$\frac{\partial\gamma^i}{\partial u^j} = 0 \quad \text{for all } i,\ j.$$

Collecting the results on $\operatorname{supp}(g)$, we see that γ has to be constant in all directions of u. Furthermore, wherever f is nonzero, γ^1 is a function of v^1 only and γ has to map zero sets of f to zero sets of f.

Replacing Γ by Γ^N we get the same results with f replaced by f^N. Since $f = f^N$ wherever f^N is nonzero or f vanishes, γ in the isotropy group of Φ obeys all these equations not only for f but also for f^N on $\operatorname{supp} f^N \cup f^{-1}(0)$. On $\operatorname{carr} f \setminus \operatorname{carr} f^N$ the gauge transformation γ is a function of v^1 only, hence it cannot leave the zero set of f^N by construction of f and f^N. Therefore, γ obeys all equations for f^N whenever it obeys all equations for f, thus every gauge transformation in the isotropy subgroup of Φ is in the isotropy subgroup of Φ^N.

On the other hand, any γ with support in $\operatorname{carr} f \setminus \operatorname{carr} f^N$ which changes only in the v^1 direction and does not keep the zero sets of f invariant, defines a gauge transformation in the isotropy subgroup of Φ^N which is not in the isotropy subgroup of Φ.

Therefore, there exists in every neighborhood of Φ a connection Φ^N whose isotropy subgroup is bigger than the isotropy subgroup of Φ. Thus, by property (44.17.5) no slice exists at Φ.

44.20. Counter-example. [Michor, Schichl, 1997]. *The action of the gauge group* $\operatorname{Gau}(E)$ *on* $\operatorname{Conn}(E)$ *also admits no slices for* $\dim M = 1$, *i.e. for* $M = S^1$.

The method of (44.19) is not applicable in this situation, since $dg \wedge du^1 \neq 0$ is not possible, any connection Φ on E is flat. Hence, the horizontal bundle is integrable, the horizontal foliation induced by Φ exists and determines Φ. Any gauge transformation leaving Φ invariant also has to map leaves of the horizontal foliation to other leaves of the horizontal foliation.

We shall construct connections $\Phi^{\lambda'}$ near Φ^λ such that the isotropy groups in $\operatorname{Gau}(E)$ look radically different near the identity, contradicting (44.17.5).

Let us assume without loss of generality that E is connected, and then, by replacing S^1 by a finite covering if necessary, that the fiber is connected. Then there exists a smooth global section $\chi : S^1 \to E$. By an argument given in the proof of (42.20) there exists a tubular neighborhood $\pi : U \subset E \to \operatorname{im}\chi$ such that $\pi = \chi \circ p|U$ (i.e. a tubular neighborhood with vertical fibers). This tubular neighborhood then contains an open thickened sphere bundle with fiber $S^1 \times \mathbb{R}^{n-1}$, and since we are only interested in gauge transformations near Id_E, which e.g. keep a smaller thickened sphere bundle inside the larger one, we may replace E by an S^1-bundle. By replacing the Klein bottle by a 2-fold covering we may finally assume that the bundle is $\operatorname{pr}_1 : S^1 \times S^1 \to S^1$.

Consider now connections where the horizontal foliation is a 1-parameter subgroup with slope λ we see that the isotropy group equals S^1 if λ is irrational, and equals S^1 times the diffeomorphism group of a closed interval if λ is rational.

44.21. A classifying space for the diffeomorphism group. Let ℓ^2 be the Hilbert space of square summable sequences, and let S be a compact manifold.

By a slight generalization of theorem (44.1) (we use a Hilbert space instead of a Riemannian manifold N), the space $\operatorname{Emb}(S,\ell^2)$ of all smooth embeddings is an open submanifold of $C^\infty(S,\ell^2)$, and it is also the total space of a smooth principal bundle with structure group $\operatorname{Diff}(S)$ acting from the right by composition. The base space $B(S,\ell^2) := \operatorname{Emb}(S,\ell^2)/\operatorname{Diff}(S)$ is a smooth manifold modeled on Fréchet spaces which are projective limits of Hilbert spaces. $B(S,\ell^2)$ is a Lindelöf space in the quotient topology, and the model spaces admit bump functions, thus $B(S,\ell^2)$ admits smooth partitions of unity, by (16.10). We may view $B(S,\ell^2)$ as the space of all submanifolds of ℓ^2 which are diffeomorphic to S, a nonlinear analog of the infinite dimensional Grassmannian.

44.22. Lemma. *The total space* $\operatorname{Emb}(S,\ell^2)$ *is contractible.*

Therefore, by the general theory of classifying spaces the base space $B(S,\ell^2)$ is a classifying space of $\operatorname{Diff}(S)$. We will give a detailed description of the classifying process in (44.24).

Proof. We consider the continuous homotopy $A:\ell^2\times[0,1]\to\ell^2$ through isometries which is given by $A_0=\operatorname{Id}$ and by

$$\begin{aligned}A_t(a_0,a_1,a_2,\dots) = (&a_0,\dots,a_{n-2},a_{n-1}\cos\theta_n(t),a_{n-1}\sin\theta_n(t),\\ &a_n\cos\theta_n(t),a_n\sin\theta_n(t),a_{n+1}\cos\theta_n(t),a_{n+1}\sin\theta_n(t),\dots)\end{aligned}$$

for $\frac{1}{n+1}\le t\le\frac{1}{n}$, where $\theta_n(t)=\varphi(n((n+1)t-1))\frac{\pi}{2}$ for a fixed smooth function $\varphi:\mathbb{R}\to\mathbb{R}$ which is 0 on $(-\infty,0]$, grows monotonely to 1 in $[0,1]$, and equals 1 on $[1,\infty)$.

Then $A_{1/2}(a_0,a_1,a_2,\dots)=(a_0,0,a_1,0,a_2,0,\dots)$ is in ℓ^2_{even} and on the other hand $A_1(a_0,a_1,a_2,\dots)=(0,a_0,0,a_1,0,a_2,0,\dots)$ is in ℓ^2_{odd}. The same homotopy makes sense as a mapping $A:\mathbb{R}^\infty\times\mathbb{R}\to\mathbb{R}^{(\mathbb{N})}$, and here it is easily seen to be smooth: a smooth curve in $\mathbb{R}^{(\mathbb{N})}$ is locally bounded and thus locally takes values in a finite dimensional subspace $\mathbb{R}^N\subset\mathbb{R}^{(\mathbb{N})}$. The image under A then has values in $\mathbb{R}^{2N}\subset\mathbb{R}^{(\mathbb{N})}$, and the expression is clearly smooth as a mapping into $\mathbb{R}^{2N}$. This is a variant of a homotopy constructed by [Ramadas, 1982].

Given two embeddings e_1 and $e_2\in\operatorname{Emb}(S,\ell^2)$ we first deform e_1 through embeddings to $e_1'\in\operatorname{Emb}(S,\ell^2_{\text{even}})$, and e_2 to $e_2'\in\operatorname{Emb}(S,\ell^2_{\text{odd}})$. Then we connect them by $te_1'+(1-t)e_2'$ which is a smooth embedding for all t since the values are always orthogonal. □

44.23. We consider the smooth action $\operatorname{ev}:\operatorname{Diff}(S)\times S\to S$ and the associated bundle $\operatorname{Emb}(S,\ell^2)[S,\operatorname{ev}]=\operatorname{Emb}(S,\ell^2)\times_{\operatorname{Diff}(S)}S$ which we call $E(S,\ell^2)$, a smooth fiber bundle over $B(S,\ell^2)$ with standard fiber S. In view of the interpretation of $B(S,\ell^2)$ as the nonlinear Grassmannian, we may visualize $E(S,\ell^2)$ as the "universal S-bundle" as follows: $E(S,\ell^2)=\{(N,x)\in B(S,\ell^2)\times\ell^2 : x\in N\}$ with the differentiable structure from the embedding into $B(S,\ell^2)\times\ell^2$.

The tangent bundle $TE(S,\ell^2)$ is then the space of all (N,x,ξ,v) where $N\in B(S,\ell^2)$, $x\in N$, ξ is a vector field along and normal to N in ℓ^2, and $v\in T_x\ell^2$ such that the part of v normal to T_xN equals $\xi(x)$. This follows from the description of the principal fiber bundle $\operatorname{Emb}(S,\ell^2)\to B(S,\ell^2)$ given in (44.1) combined with (42.17). Obviously, the vertical bundle $VE(S,\ell^2)$ consists of all (N,x,v) with $x\in N$ and $v\in T_xN$. The orthonormal projection $p_{(N,x)}:\ell^2\to T_xN$ defines a connection $\Phi^{\text{class}}:TE(S,\ell^2)\to VE(S,\ell^2)$ which is given by $\Phi^{\text{class}}(N,x,\xi,v)=(N,x,p_{(N,x)}v)$. It will be called the *classifying connection* for reasons to be explained in the next theorem.

44.24. Theorem. Classifying space for $\operatorname{Diff}(S)$**.**
The fiber bundle $(E(S,\ell^2)\to B(S,\ell^2),S)$ is classifying for S-bundles and Φ^{class} is a classifying connection:

For each finite dimensional bundle $(p:E\to M,S)$ and each connection Φ on E there is a smooth (classifying) mapping $f:M\to B(S,\ell^2)$ such that (E,Φ) is isomorphic to $(f^*E(S,\ell^2),f^*\Phi^{\text{class}})$. Homotopic maps pull back isomorphic S-bundles and conversely (the homotopy can be chosen smooth). The pulled back connection is invariant under a homotopy H if and only if $i(C^{\text{class}}T_{(x,t)}H.(0_x,\frac{d}{dt}))\mathcal{R}^{\text{class}}=0$ where C^{class} is the horizontal lift of Φ^{class}, and $\mathcal{R}^{\text{class}}$ is its curvature .

Since S is compact the classifying connection Φ^{class} is complete, and its parallel transport $\operatorname{Pt}^{\text{class}}$ has the following classifying property:

$$\tilde f\circ \operatorname{Pt}^{f^*\Phi^{\text{class}}}(c,t)=\operatorname{Pt}^{\text{class}}(f\circ c,t)\circ\tilde f,$$

where $\tilde f:E\cong f^*E(S,\ell^2)\to E(S,\ell^2)$ is the fiberwise diffeomorphic which covers the classifying mapping $f:M\to B(S,\ell^2)$.

Proof. We choose a Riemannian metric g_1 on the vector bundle $VE\to E$ and a Riemannian metric g_2 on the manifold M. We can combine these two into the Riemannian metric $g:=(Tp|\ker\Phi)^*g_2\oplus g_1$ on the manifold E, for which the horizontal and vertical spaces are orthogonal. By the theorem of [Nash, 1956], see also [Günther, 1989] for an easy proof, there is an isometric embedding $h:E\to\mathbb{R}^N$ for N large enough. We then embed $\mathbb{R}^N$ into the Hilbert space ℓ^2 and consider $f:M\to B(S,\ell^2)$, given by $f(x)=h(E_x)$. Then

$$\begin{array}{ccc} E & \xrightarrow{\tilde f=(f,h)} & E(S,\ell^2)\\ {\scriptstyle p}\downarrow & & \downarrow\\ M & \xrightarrow{\ \ f\ \ } & B(S,\ell^2)\end{array}$$

is fiberwise a diffeomorphism, so the diagram is a pullback and $f^*E(S,\ell^2)=E$. Since $T(f,h)$ maps horizontal and vertical vectors to orthogonal ones we have $(f,h)^*\Phi^{\text{class}}=\Phi$. If Pt denotes the parallel transport of the connection Φ and $c:[0,1]\to M$ is a (piecewise) smooth curve we have for $u\in E_{c(0)}$

$$\begin{aligned}\Phi^{\text{class}}\tfrac{\partial}{\partial t}\big|_0\tilde f(\operatorname{Pt}(c,t,u))&=\Phi^{\text{class}}.T\tilde f.\tfrac{\partial}{\partial t}\big|_0\operatorname{Pt}(c,t,u)\\ &=T\tilde f.\Phi.\tfrac{\partial}{\partial t}\big|_0\operatorname{Pt}(c,t,u)=0,\quad\text{so}\\ \tilde f(\operatorname{Pt}(c,t,u))&=\operatorname{Pt}^{\text{class}}(f\circ c,t,\tilde f(u)).\end{aligned}$$

Now let H be a continuous homotopy $M \times I \to B(S, \ell^2)$. Then we may approximate H by smooth mappings with the same H_0 and H_1, if they are smooth, see [Bröcker, Jänich, 1973], where the infinite dimensionality of $B(S, \ell^2)$ does not disturb. Then we consider the bundle $H^*E(S, \ell^2) \to M \times I$, equipped with the connection $H^*\Phi^{\text{class}}$, whose curvature is $H^*\mathcal{R}^{\text{class}}$. Let ∂_t be the vector field tangential to all $\{x\} \times I$ on $M \times I$. Parallel transport along the lines $t \mapsto (x, t)$ with respect $H^*\Phi^{\text{class}}$ is given by the flow of the horizontal lift $(H^*C^{\text{class}})(\partial_t)$ of ∂_t. Let us compute its action on the connection $H^*\Phi^{\text{class}}$ whose curvature is $H^*\mathcal{R}^{\text{class}}$ by (37.4.3). By lemma (44.25) below we have

$$\begin{aligned} \tfrac{\partial}{\partial t}\left(\mathrm{Fl}_t^{(H^*C^{\text{class}})(\partial_t)}\right)^* H^*\Phi^{\text{class}} &= -\tfrac{1}{2} i_{(H^*C^{\text{class}})(\partial_t)}(H^*\mathcal{R}^{\text{class}}) \\ &= -\frac{1}{2} H^*\left(i(C^{\text{class}} T_{(x,t)}H.(0_x, \tfrac{d}{dt}))\mathcal{R}^{\text{class}}\right), \end{aligned}$$

which implies the result. □

44.25. Lemma. *Let Φ be a connection on a finite dimensional fiber bundle $(p : E \to M, S)$ with curvature $\mathcal{R}$ and horizontal lift C. Let $X \in \mathfrak{X}(M)$ be a vector field on the base.*

Then for the horizontal lift $CX \in \mathfrak{X}(E)$ we have

$$\mathcal{L}_{CX}\Phi = \tfrac{\partial}{\partial t}\big|_0 (\mathrm{Fl}_t^{CX})^*\Phi = [CX, \Phi] = -\tfrac{1}{2} i_{CX}\mathcal{R}.$$

Proof. From (35.14.5) we get $\mathcal{L}_{CX}\Phi = \frac{\partial}{\partial t}\big|_0 (\mathrm{Fl}_t^{CX})^*\Phi = [CX, \Phi]$. From (35.9.2) we have

$$\begin{aligned} i_{CX}\mathcal{R} &= i_{CX}[\Phi, \Phi] \\ &= [i_{CX}\Phi, \Phi] - [\Phi, i_{CX}\Phi] + 2i_{[\Phi, CX]}\Phi \\ &= -2\Phi[CX, \Phi]. \end{aligned}$$

The vector field CX is p-related to X, and $\Phi \in \Omega^1(E; TE)$ is p-related to $0 \in \Omega^1(M; TM)$, so by (35.13.7) the form $[CX, \Phi] \in \Omega^1(E; TE)$ is also p-related to $0 = [X, 0] \in \Omega^1(M; TM)$. So $Tp.[CX, \Phi] = 0$, $[CX, \Phi]$ has vertical values, and $[CX, \Phi] = \Phi[CX, \Phi]$. □

44.26. A consequence of theorem (43.7) is that the classifying spaces of $\mathrm{Diff}(S)$ and $\mathrm{Diff}(S, \mu_0)$ are homotopy equivalent. So their classifying spaces are homotopy equivalent, too.

We now sketch a smooth classifying space for Diff_{μ_0}. Consider the space $B_1(S, \ell^2)$ of all submanifolds of ℓ^2 of type S and total volume 1 in the volume form induced from the inner product on ℓ^2. It is a closed splitting submanifold of codimension 1 of $B(S, \ell^2)$ by the Nash-Moser inverse function theorem (51.17). This theorem is applicable if we use ℓ^2 as image space, because the modeling spaces are then tame Fréchet spaces in the sense of (51.9). It is not applicable directly for $\mathbb{R}^{(\mathbb{N})}$ as image space.

44.27. Theorem. Classifying space for $\operatorname{Diff}^\omega(S)$. *Let S be a compact real analytic manifold. Then the space $\operatorname{Emb}^\omega(S,\ell^2)$ of real analytic embeddings of S into the Hilbert space ℓ^2 is the total space of a real analytic principal fiber bundle with structure group $\operatorname{Diff}^\omega(S)$ and real analytic base manifold $B^\omega(S,\ell^2)$, which is a classifying space for the Lie group $\operatorname{Diff}^\omega(S)$. It carries a universal $\operatorname{Diff}^\omega(S)$-connection.*

In other words:

$$\operatorname{Emb}^\omega(S,N)\times_{\operatorname{Diff}^\omega(S)} S\to B^\omega(S,\ell^2)$$

classifies real analytic fiber bundles with typical fiber S and carries a universal (generalized) connection.

The proof is similar to that of (44.24) with the appropriate changes to C^ω.

45. Manifolds of Riemannian Metrics

The usual metric on the space of all Riemannian metrics was considered by [Ebin, 1970], who used it to construct a slice for the action of the group of diffeomorphism on the space of all metrics. It was then reconsidered by [Freed, Groisser, 1989], and by [Gil-Medrano, Michor, 1991] for noncompact M. The results in this section are largely taken from the last paper and from [Gil-Medrano, Michor, Neuwirther, 1992].

45.1. Bilinear structures. Throughout this section let M be a smooth second countable finite dimensional manifold. Let $\otimes^2T^*M$ denote the vector bundle of all $\binom{0}{2}$-tensors on M, which we canonically identify with the bundle $L(TM,T^*M)$. Let $GL(TM,T^*M)$ denote the non degenerate ones. For any $b:T_xM\to T_x^*M$ we let the *transposed* be given by $b^t:T_xM\to T_x^{**}M\xrightarrow{b^*}T_x^*M$. As a bilinear structure b is skew symmetric if and only if $b^t=-b$, and b is symmetric if and only if $b^t=b$. In the latter case a frame (e_j) of T_xM can be chosen in such a way that in the dual frame (e^j) of T_x^*M we have

$$b=e^1\otimes e^1+\dots+e^p\otimes e^p-e^{p+1}\otimes e^{p+1}-e^{p+q}\otimes e^{p+q};$$

b has signature (p,q) and is non degenerate if and only if $p+q=n$, the dimension of M. In this case, q alone will be called the signature.

A section $b\in C^\infty(GL(TM,T^*M))$ will be called a non degenerate bilinear structure on M, and we will denote the space of all such structures by $\mathcal{B}(M)=\mathcal{B}:=C^\infty(GL(TM,T^*M))$. It is open in the space of sections $C^\infty(L(TM,T^*M))$ for the Whitney C^∞-topology, in which the latter space is, however, not a topological vector space, as explained in detail in (41.13). The space $\mathcal{B}_c:=C_c^\infty(L(TM,T^*M))$ of sections with compact support is the largest topological vector space contained in the topological group $(C^\infty(L(TM,T^*M)),+)$, and the trace of the Whitney C^∞-topology on it induces the convenient vector space structure described in (30.4).

So we declare the path components of $\mathcal{B} = C^\infty(GL(TM, T^*M))$ for the Whitney C^∞-topology also to be open; these are open in affine subspaces of the form $b + \mathcal{B}_c$ for some $b \in \mathcal{B}$ and we equip them with the translates of the c^∞-topology on $\mathcal{B}_c$. The resulting topology is finer than the Whitney topology and will be called the *natural topology*, similar as in (42.1).

45.2. The metrics. The tangent bundle of the space $\mathcal{B} = C^\infty(GL(TM, T^*M))$ of bilinear structures is $T\mathcal{B} = \mathcal{B} \times \mathcal{B}_c = C^\infty(GL(TM, T^*M)) \times C_c^\infty(L(TM, T^*M))$. Then $b \in \mathcal{B}$ induces two fiberwise bilinear forms on $L(TM, T^*M)$ which are given by $(h,k) \mapsto \operatorname{tr}(b^{-1}hb^{-1}k)$ and $(h,k) \mapsto \operatorname{tr}(b^{-1}h)\operatorname{tr}(b^{-1}k)$. We split each endomorphism $H = b^{-1}h : TM \to TM$ into its trace free part $H_0 := H - \frac{\operatorname{tr}(H)}{\dim M}\operatorname{Id}$ and its trace part which simplifies some formulas later on. Thus, we have $\operatorname{tr}(b^{-1}hb^{-1}k) = \operatorname{tr}((b^{-1}h)_0(b^{-1}k)_0) + \frac{1}{\dim M}\operatorname{tr}(b^{-1}h)\operatorname{tr}(b^{-1}k)$. The structure b also induces a volume density on the base manifold M by the local formula

$$\operatorname{vol}(b) = \sqrt{|\det(b_{ij})|}\,|dx_1 \wedge \dots \wedge dx_n|.$$

For each real α we have a smooth symmetric bilinear form on $\mathcal{B}$, given by

$$G_b^\alpha(h,k) = \int_M (\operatorname{tr}((b^{-1}h)_0(b^{-1}k)_0) + \alpha \operatorname{tr}(b^{-1}h)\operatorname{tr}(b^{-1}k))\operatorname{vol}(b).$$

It is invariant under the action of the diffeomorphism group $\operatorname{Diff}(M)$ on the space $\mathcal{B}$ of bilinear structures. The integral is defined since h and k have compact supports. For $n = \dim M$ we have

$$G_b(h,k) := G_b^{1/n}(h,k) = \int_M \operatorname{tr}(b^{-1}hb^{-1}k)\operatorname{vol}(b),$$

which for positive definite b is the usual metric on the space of all Riemannian metrics. We will see below in (45.3) that for $\alpha \neq 0$ it is weakly non degenerate, i.e. G_b^α defines a linear injective mapping from the tangent space $T_b\mathcal{B} = \mathcal{B}_c = C_c^\infty(L(TM, T^*M))$ into its dual $C_c^\infty(L(TM, T^*M))'$, the space of distributional densities with values in the dual bundle. This linear mapping is, however, never surjective. So we have a one parameter family of pseudo Riemannian metrics on the infinite dimensional space $\mathcal{B}$ which obviously is smooth in all appearing variables.

45.3. Lemma. *For $h, k \in T_b\mathcal{B}$ we have*

$$G_b^\alpha(h,k) = G_b(h + \tfrac{\alpha n - 1}{n}\operatorname{tr}(b^{-1}h)b, k),$$
$$G_b(h,k) = G_b^\alpha(h - \tfrac{\alpha n - 1}{\alpha n^2}\operatorname{tr}(b^{-1}h)b, k), \quad \text{if } \alpha \neq 0,$$

where $n = \dim M$. The pseudo Riemannian metric G^α is weakly non degenerate for all $\alpha \neq 0$.

Proof. The first equation is an obvious reformulation of the definition, the second follows since $h \mapsto h - \frac{\alpha n - 1}{\alpha n^2}\operatorname{tr}(b^{-1}h)b$ is the inverse of the transform $h \mapsto$

$h + \frac{\alpha n-1}{n}\operatorname{tr}(b^{-1}h)b$. Since $\operatorname{tr}(b_x^{-1}h_x(b_x^{-1}h_x)^{t,g}) > 0$ if $h_x \neq 0$, where $\ell^{t,g}$ is the transposed of a linear mapping with respect to an arbitrary fixed Riemannian metric g, we have

$$G_b(h, b(b^{-1}h)^{t,g}) = \int_M \operatorname{tr}(b^{-1}h(b^{-1}h)^{t,g})\operatorname{vol}(b) > 0$$

if $h \neq 0$. So G is weakly non degenerate, and by the second equation G^α is weakly non degenerate for $\alpha \neq 0$. □

45.4. Remark. Since G^α is only a weak pseudo Riemannian metric, all objects which are only implicitly given a priori lie in the Sobolev completions of the relevant spaces. In particular, this applies to the formula

$$\begin{aligned}2G^\alpha(\xi, \nabla^\alpha_\eta\zeta) =& \xi G^\alpha(\eta,\zeta) + \eta G^\alpha(\zeta,\xi) - \zeta G^\alpha(\xi,\eta)\\ &+ G^\alpha([\xi,\eta],\zeta) + G^\alpha([\eta,\zeta],\xi) - G^\alpha([\zeta,\xi],\eta),\end{aligned}$$

which a priori gives only uniqueness but not existence of the Levi Civita covariant derivative. We will show that it exists and we use it in the form explained in (37.28).

45.5. Lemma. *For $x \in M$ the pseudo metric on $GL(T_xM, T_x^*M)$ given by*

$$\gamma^\alpha_{b_x}(h_x, k_x) := \operatorname{tr}((b_x^{-1}h_x)_0(b_x^{-1}k_x)_0) + \alpha\operatorname{tr}(b_x^{-1}h_x)\operatorname{tr}(b_x^{-1}k_x)$$

has signature (the number of negative eigenvalues) $\frac{n(n-1)}{2}$ for $\alpha > 0$ and has signature $(\frac{n(n-1)}{2} + 1)$ for $\alpha < 0$.

Proof. In the framing $H = b_x^{-1}h_x$ and $K = b_x^{-1}k_x$ we have to determine the signature of the symmetric bilinear form $(H,K) \mapsto \operatorname{tr}(H_0K_0) + \alpha\operatorname{tr}(H)\operatorname{tr}(K)$. Since the signature is constant on connected components we have to determine it only for $\alpha = \frac{1}{n}$ and $\alpha = \frac{1}{n} - 1$.

For $\alpha = \frac{1}{n}$ we note first that on the space of matrices $(H,K) \mapsto \operatorname{tr}(HK^t)$ is positive definite, and since the linear isomorphism $K \mapsto K^t$ has the space of symmetric matrices as eigenspace for the eigenvalue 1 and the space of skew symmetric matrices as eigenspace for the eigenvalue -1, we conclude that the signature is $\frac{n(n-1)}{2}$ in this case.

For $\alpha = \frac{1}{n} - 1$ we proceed as follows: On the space of matrices with zeros on the main diagonal the signature of $(H,K) \mapsto \operatorname{tr}(HK)$ is $\frac{n(n-1)}{2}$ by the argument above and the form $(H,K) \mapsto -\operatorname{tr}(H)\operatorname{tr}(K)$ vanishes. On the space of diagonal matrices which we identify with $\mathbb{R}^n$ the whole bilinear form is given by $\langle x, y\rangle = \sum_i x^iy^i - (\sum_i x^i)(\sum_i y^i)$. Let (e_i) denote the standard basis of $\mathbb{R}^n$, and put $a_1 := \frac{1}{n}(e_1 + \dots + e_n)$ and

$$a_i := \frac{1}{\sqrt{i-1+(i-1)^2}}(e_1 + \dots + e_{i-1} - (i-1)e_i)$$

for $i > 1$. Then $\langle a_1, a_1\rangle = -1 + \frac{1}{n}$, and for $i > 1$ we get $\langle a_i, a_j\rangle = \delta_{i,j}$. So the signature is 1 in this case. □

45.6. Let $t \mapsto b(t)$ be a smooth curve in $\mathcal{B}$. So $b : \mathbb{R} \times M \to GL(TM, T^*M)$ is smooth, and by the choice of the topology on $\mathcal{B}$ made in (45.1) the curve $b(t)$ varies only in a compact subset of M, locally in t, by (30.9). Then its energy is given by

$$\begin{aligned} E_{a_1}^{a_2}(b) &:= \tfrac{1}{2} \int_{a_1}^{a_2} G_b^\alpha(b_t, b_t)dt \\ &= \tfrac{1}{2} \int_{a_1}^{a_2} \int_M \left(\operatorname{tr}((b^{-1}b_t)_0(b^{-1}b_t)_0) + \alpha \operatorname{tr}(b^{-1}b_t)^2\right) \operatorname{vol}(b)\, dt, \end{aligned}$$

where $b_t = \frac{\partial}{\partial t} b(t)$.

Now we consider a variation of this curve, so we assume that $(t, s) \mapsto b(t, s)$ is smooth in all variables and locally in (t, s) it only varies within a compact subset in M — this is again the effect of the topology chosen in (45.1). Note that $b(t, 0)$ is the original $b(t)$ above.

45.7. Lemma. *In the setting of* (45.6), *we have the first variation formula*

$$\begin{aligned} \tfrac{\partial}{\partial s}|_0 E(G^\alpha)_{a_0}^{a_1}(b(\quad, s)) &= G_b^\alpha(b_t, b_s)|_{t=a_0}^{t=a_1} + \\ &+ \int_{a_0}^{a_1} G(-b_{tt} + b_t b^{-1} b_t + \frac{1}{4}\operatorname{tr}(b^{-1}b_t b^{-1} b_t) b - \frac{1}{2}\operatorname{tr}(b^{-1}b_t) b_t + \\ &\qquad + \alpha\,(-\operatorname{tr}(b^{-1}b_{tt}) - \frac{1}{4}\operatorname{tr}(b^{-1}b_t)^2 + \operatorname{tr}(b^{-1}b_t b^{-1} b_t))b, b_s)\, dt = \\ &= G_b^\alpha(b_t, b_s)|_{t=a_0}^{t=a_1} + \\ &+ \int_{a_0}^{a_1} G^\alpha(-b_{tt} + b_t b^{-1} b_t - \frac{1}{2}\operatorname{tr}(b^{-1}b_t) b_t + \frac{1}{4\alpha n}\operatorname{tr}(b_t^{-1} b^{-1} b_t) b + \\ &\qquad + \frac{\alpha n - 1}{4\alpha n^2}\operatorname{tr}(b^{-1}b_t)^2 b, b_s)\, dt. \end{aligned}$$

Proof. We may interchange $\frac{\partial}{\partial s}|_0$ with the first integral describing the energy in (45.6) since this is finite dimensional analysis, and we may interchange it with the second one, since $\int_M$ is a continuous linear functional on the space of all smooth densities with compact support on M, by the chain rule. Then we use that tr_* is linear and continuous, $d(\operatorname{vol})(b)h = \frac{1}{2}\operatorname{tr}(b^{-1}h)\operatorname{vol}(b)$, and that $d((\quad)^{-1})_*(b)h = -b^{-1}hb^{-1}$, and partial integration. □

45.8. The geodesic equation. By lemma (45.7), the curve $t \mapsto b(t)$ is a geodesic if and only if we have

$$\begin{aligned} b_{tt} &= b_t b^{-1} b_t - \frac{1}{2}\operatorname{tr}(b^{-1}b_t) b_t + \frac{1}{4\alpha n}\operatorname{tr}(b^{-1}b_t b^{-1} b_t) b + \frac{\alpha n - 1}{4\alpha n^2}\operatorname{tr}(b^{-1}b_t)^2 b. \\ &= \Gamma_b(b_t, b_t), \end{aligned}$$

where the G^α*-Christoffel symbol* $\Gamma^\alpha : \mathcal{B} \times \mathcal{B}_c \times \mathcal{B}_c \to \mathcal{B}_c$ is given by symmetrization

$$\begin{aligned} \Gamma_b^\alpha(h, k) &= \frac{1}{2} h b^{-1} k + \frac{1}{2} k b^{-1} h - \frac{1}{4}\operatorname{tr}(b^{-1}h) k - \frac{1}{4}\operatorname{tr}(b^{-1}k) h + \\ &\quad + \frac{1}{4\alpha n}\operatorname{tr}(b^{-1}h b^{-1} k) b + \frac{\alpha n - 1}{4\alpha n^2}\operatorname{tr}(b^{-1}h)\operatorname{tr}(b^{-1}k) b. \end{aligned}$$

The sign of Γ^α is chosen in such a way that the horizontal subspace of $T^2\mathcal{B}$ is parameterized by $(x, y; z, \Gamma_x(y, z))$. If instead of the obvious framing we use $T\mathcal{B} = \mathcal{B} \times \mathcal{B}_c \ni (b, h) \mapsto (b, b^{-1}h) =: (b, H) \in \{b\} \times C_c^\infty(L(TM, TM))$, the Christoffel symbol looks like

$$\begin{aligned}\overline{\Gamma}_b^\alpha(H, K) &= \frac{1}{2}(HK + KH) - \frac{1}{4}\operatorname{tr}(H)K - \frac{1}{4}\operatorname{tr}(K)H \\ &\quad + \frac{1}{4\alpha n}\operatorname{tr}(HK)\operatorname{Id} + \frac{\alpha n - 1}{4\alpha n^2}\operatorname{tr}(H)\operatorname{tr}(K),\end{aligned}$$

and the G^α-geodesic equation for $B(t) := b^{-1}b_t$ becomes

$$B_t = \tfrac{\partial}{\partial t}(b^{-1}b_t) = \frac{1}{4\alpha n}\operatorname{tr}(BB)\operatorname{Id} - \frac{1}{2}\operatorname{tr}(B)B + \frac{\alpha n - 1}{4\alpha n^2}\operatorname{tr}(B)^2\operatorname{Id}.$$

45.9. The curvature. For another manifold N, for vector fields $X, Y \in \mathfrak{X}(N)$ and a vector field $s : N \to T\mathcal{M}$ along $f : N \to \mathcal{M}$ we have

$$\mathcal{R}(X, Y)s = (\nabla_{[X,Y]} - [\nabla_X, \nabla_Y])s = (K \circ TK - K \circ TK \circ \kappa_{TM}) \circ T^2 s \circ TX \circ Y,$$

where $K : TTM \to M$ is the connector (37.28), κ_{TM} is the canonical flip $TTTM \to TTTM$ (29.10), and where the second formula in local coordinates reduces to the usual formula

$$\text{(1)} \qquad \mathcal{R}(h, k)\ell = d\Gamma(h)(k, \ell) - d\Gamma(k)(h, \ell) - \Gamma(h, \Gamma(k, \ell)) + \Gamma(k, \Gamma(h, \ell)),$$

see [Kainz, Michor, 1987] or [Kolář, Michor, Slovak, 1993, 37.15].

45.10. Theorem. *The curvature for the pseudo Riemannian metric G^α on the manifold $\mathcal{B}$ of all non degenerate bilinear structures is given by*

$$\begin{aligned}b^{-1}\mathcal{R}_b^\alpha(h, k)l &= \frac{1}{4}[[H, K], L] + \frac{1}{16\alpha}(-\operatorname{tr}(HL)K + \operatorname{tr}(KL)H) + \\ &\quad + \frac{4\alpha n - 3\alpha n^2 + 4n - 4}{16\alpha n^2}(\operatorname{tr}(H)\operatorname{tr}(L)K - \operatorname{tr}(K)\operatorname{tr}(L)H) + \\ &\quad + \frac{4\alpha^2 n^2 - 4\alpha n + \alpha n^2 + 3}{16\alpha n^2}(\operatorname{tr}(HL)\operatorname{tr}(K)\operatorname{Id} - \operatorname{tr}(KL)\operatorname{tr}(H)\operatorname{Id}),\end{aligned}$$

where $H = b^{-1}h, K = b^{-1}k$ and $L = b^{-1}l$.

Proof. This is a long but direct computation starting from (45.9.1). □

The geodesic equation can be solved explicitly, and we have

45.11. Theorem. *Let $b^0 \in \mathcal{B}$ and $h \in T_{b^0}\mathcal{B} = \mathcal{B}_c$. Then the geodesic for the metric G^α in $\mathcal{B}$ starting at b^0 in the direction of h is the curve*

$$\exp^\alpha_{b^0}(th) = b^0 e^{(a(t)\operatorname{Id} + b(t)H_0)},$$

where H_0 is the traceless part of $H := (b^0)^{-1}h$ (i.e. $H_0 = H - \frac{\operatorname{tr}(H)}{n}\operatorname{Id}$), and where $a(t) = a_{\alpha,H}(t)$ and $b(t) = b_{\alpha,H}(t)$ in $C^\infty(M,\mathbb{R})$ are defined as follows:

$$a_{\alpha,H}(t) = \frac{2}{n}\log\left((1+\frac{t}{4}\operatorname{tr}(H))^2 + t^2\frac{\alpha^{-1}}{16}\operatorname{tr}(H_0^2)\right),$$

$$b_{\alpha,H}(t) = \begin{cases} \dfrac{4}{\sqrt{\alpha^{-1}\operatorname{tr}(H_0^2)}}\arctan\left(\dfrac{t\sqrt{\alpha^{-1}\operatorname{tr}(H_0^2)}}{4+t\operatorname{tr}(H)}\right) & \text{for } \alpha^{-1}\operatorname{tr}(H_0^2) > 0 \\ \dfrac{4}{\sqrt{-\alpha^{-1}\operatorname{tr}(H_0^2)}}\operatorname{artanh}\left(\dfrac{t\sqrt{-\alpha^{-1}\operatorname{tr}(H_0^2)}}{4+t\operatorname{tr}(H)}\right) & \text{for } \alpha^{-1}\operatorname{tr}(H_0^2) < 0 \\ \dfrac{t}{1+\frac{t}{4}\operatorname{tr}(H)} & \text{for } \operatorname{tr}(H_0^2) = 0. \end{cases}$$

Here arctan *is taken to have values in $(-\frac{\pi}{2},\frac{\pi}{2})$ for the points of the base manifold, where $\operatorname{tr}(H)\ge 0$, and on a point where $\operatorname{tr}(H)<0$ we define*

$$\arctan\left(\frac{t\sqrt{\alpha^{-1}\operatorname{tr}(H_0^2)}}{4+t\operatorname{tr}(H)}\right) = \begin{cases} \arctan \text{ in } [0,\frac{\pi}{2}) & \text{for } t\in[0,-\frac{4}{\operatorname{tr}(H)}) \\ \frac{\pi}{2} & \text{for } t = -\frac{4}{\operatorname{tr}(H)} \\ \arctan \text{ in } (\frac{\pi}{2},\pi) & \text{for } t\in(-\frac{4}{\operatorname{tr}(H)},\infty). \end{cases}$$

To describe the domain of definition of the exponential mapping we consider the sets

$$\begin{aligned} Z^h &:= \{x\in M : \tfrac{1}{\alpha}\operatorname{tr}_x(H_0^2) = 0 \text{ and } \operatorname{tr}_x(H)<0\}, \\ G^h &:= \{x\in M : 0 > \tfrac{1}{\alpha}\operatorname{tr}_x(H_0^2) > -\operatorname{tr}_x(H)^2 \text{ and } \operatorname{tr}_x(H)<0\} \\ &= \{x\in M : \alpha\gamma(h,h) \lesseqgtr \gamma^\alpha(h,h) \lesseqgtr 0 \text{ for } \alpha \lesseqgtr 0, \operatorname{tr}_x(H)<0\}, \\ E^h &:= \{x\in M : -\operatorname{tr}_x(H)^2 = \tfrac{1}{\alpha}\operatorname{tr}_x(H_0^2) \text{ and } \operatorname{tr}_x(H)<0\} \\ &= \{x\in M : \gamma^\alpha(h,h) = 0 \text{ and } \operatorname{tr}_x(H)<0\}, \\ L^h &:= \{x\in M : -\operatorname{tr}_x(H)^2 > \tfrac{1}{\alpha}\operatorname{tr}_x(H_0^2)\} \\ &= \{x\in M : \gamma^\alpha(h,h) \gtreqless 0 \text{ for } \alpha \lesseqgtr 0\}, \end{aligned}$$

where $\gamma(h,h) = \operatorname{tr}_x(H^2)$, and $\gamma^\alpha(h,h) = \operatorname{tr}_x(H_0^2) + \alpha\operatorname{tr}_x(H)^2$, see (45.5), are the integrands of $G_{b^0}(h,h)$ and $G^\alpha_{b^0}(h,h)$, respectively. Then we consider the numbers

$$\begin{aligned} z^h &:= \inf\left\{-\frac{4}{\operatorname{tr}_x(H)} : x\in Z^h\right\}, \\ g^h &:= \inf\left\{4\frac{-\alpha\operatorname{tr}_x(H) - \sqrt{-\alpha\operatorname{tr}_x(H_0^2)}}{\operatorname{tr}_x(H_0^2)+\alpha\operatorname{tr}(H)^2} : x\in G^h\right\}, \\ e^h &:= \inf\left\{-\frac{2}{\operatorname{tr}_x(H)} : x\in E^h\right\}, \\ l^h &:= \inf\left\{4\frac{-\alpha\operatorname{tr}_x(H) - \sqrt{-\alpha\operatorname{tr}_x(H_0^2)}}{\operatorname{tr}_x(H_0^2)+\alpha\operatorname{tr}(H)^2} : x\in L^h\right\}, \end{aligned}$$

if the corresponding set is not empty, with value ∞ if the set is empty. Denote $m^h := \inf\{z^h, g^h, e^h, l^h\}$. *Then* $\exp^{\alpha}_{b^0}(th)$ *is maximally defined for* $t \in [0, m^h)$.

The second representations of the sets G^h, L^h, and E^h clarifies how to take care of timelike, spacelike, and lightlike vectors, respectively.

Proof. Using $X(t) := g^{-1}g_t$ the geodesic equation reads as

$$X' = -\frac{1}{2}\operatorname{tr}(X)X + \frac{1}{4\alpha n}\operatorname{tr}(X^2)\operatorname{Id} + \frac{\alpha n - 1}{4\alpha n^2}\operatorname{tr}(X)^2\operatorname{Id},$$

and it is easy to see that a solution X satisfies

$$X_0' = -\frac{1}{2}\operatorname{tr}(X)X_0.$$

Then $X(t)$ is in the plane generated by H_0 and Id for all t and the solution has the form $g(t) = b^0\exp(a(t)\operatorname{Id} + b(t)H_0)$. Since $g_t = g(t)(a'(t)\operatorname{Id} + b'(t)H_0)$ we have

$$\begin{aligned} X(t) &= a'(t)\operatorname{Id} + b'(t)H_0 \text{ and} \\ X'(t) &= a''(t)\operatorname{Id} + b''(t)H_0, \end{aligned}$$

and the geodesic equation becomes

$$\begin{aligned} a''(t)\operatorname{Id} + b''(t)H_0 = &-\frac{1}{2}na'(t)(a'(t)\operatorname{Id} + b'(t)H_0) + \\ &+ \frac{1}{4\alpha n}(na'(t)^2 + b'(t)^2\operatorname{tr}(H_0^2))\operatorname{Id} + \\ &+ \frac{\alpha n - 1}{4\alpha n^2}(n^2 a'(t)^2)\operatorname{Id}. \end{aligned}$$

We may assume that Id and H_0 are linearly independent; if not $H_0 = 0$ and $b(t) = 0$. Hence, the geodesic equation reduces to the differential equation

$$\begin{cases} a'' = -\dfrac{n}{4}(a')^2 + \dfrac{\operatorname{tr}(H_0^2)}{4\alpha n}(b')^2 \\ b'' = -\dfrac{n}{2}a'\,b' \end{cases}$$

with initial conditions $a(0) = b(0) = 0$, $a'(0) = \frac{\operatorname{tr}(H)}{n}$, and $b'(0) = 1$.
If we take $p(t) = \exp(\frac{n}{2}a)$ it is easy to see that then p should be a solution of $p''' = 0$ and from the initial conditions

$$p(t) = 1 + \frac{t}{2}\operatorname{tr}(H) + \frac{t^2}{16}(\operatorname{tr}(H)^2 + \alpha^{-1}\operatorname{tr}(H_0^2)).$$

Using that the second equation becomes $b' = p^{-1}$, and then b is obtained just by computing the integral. The solutions are defined in $[0, m^h)$ where m^h is the infimum over the support of h of the first positive root of the polynomial p, if it exists, and ∞ otherwise. The description of m^h is now a technical fact. □

45.12. The exponential mapping. For $b^0 \in GL(T_xM, T_x^*M)$ and $H = (b^0)^{-1}h$ let C_{b^0} be the subset of $L(T_xM, T_x^*M)$ given by the union of the sets (compare with Z^h, G^h, E^h, L^h from (45.11))

$$\{h : \operatorname{tr}(H_0^2) = 0, \operatorname{tr}(H) \le -4\},$$
$$\left\{h : 0 > \tfrac{1}{\alpha}\operatorname{tr}(H_0^2) > -\operatorname{tr}(H)^2, 4\frac{-\alpha\operatorname{tr}(H) - \sqrt{-\alpha\operatorname{tr}(H_0^2)}}{\operatorname{tr}(H_0^2) + \alpha\operatorname{tr}(H)^2} \le 1, \operatorname{tr}(H) < 0\right\},$$
$$\left\{h : -\operatorname{tr}(H)^2 = \tfrac{1}{\alpha}\operatorname{tr}(H_0^2), \operatorname{tr}(H) < -2\right\},$$
$$\left\{h : -\operatorname{tr}(H)^2 > \tfrac{1}{\alpha}\operatorname{tr}(H_0^2), 4\frac{-\alpha\operatorname{tr}(H) - \sqrt{-\alpha\operatorname{tr}(H_0^2)}}{\operatorname{tr}(H_0^2) + \alpha\operatorname{tr}(H)^2} \le 1\right\}^{\text{closure}},$$

which by some limit considerations coincides with the union of the following two sets:

$$\left\{h : 0 > \tfrac{1}{\alpha}\operatorname{tr}(H_0^2) > -\operatorname{tr}(H)^2, 4\frac{-\alpha\operatorname{tr}(H) - \sqrt{-\alpha\operatorname{tr}(H_0^2)}}{\operatorname{tr}(H_0^2) + \alpha\operatorname{tr}(H)^2} \le 1, \operatorname{tr}(H) < 0\right\}^{\text{closure}},$$
$$\left\{h : -\operatorname{tr}(H)^2 > \tfrac{1}{\alpha}\operatorname{tr}(H_0^2), 4\frac{-\alpha\operatorname{tr}(H) - \sqrt{-\alpha\operatorname{tr}(H_0^2)}}{\operatorname{tr}(H_0^2) + \alpha\operatorname{tr}(H)^2} \le 1\right\}^{\text{closure}}$$

So C_{b^0} is closed. We consider the open sets $U_{b^0} := L(T_xM, T_x^*M) \setminus C_{b^0}$, $U'_{b^0} := \{(b^0)^{-1}h : h \in U_{b^0}\} \subset L(T_xM, T_xM)$, and finally the open sub fiber bundles over $GL(TM, T^*M)$

$$U := \bigcup \{\{b^0\} \times U_{b^0} : b^0 \in GL(TM, T^*M)\} \subset GL(TM, T^*M) \times_M L(TM, T^*M),$$
$$U' := \bigcup \{\{b^0\} \times U'_{b^0} : b^0 \in GL(TM, T^*M)\} \subset GL(TM, T^*M) \times_M L(TM, TM).$$

Then we consider the mapping $\Phi : U \to GL(TM, T^*M)$ which is given by the following composition

$$U \xrightarrow{\sharp} U' \xrightarrow{\varphi} GL(TM, T^*M) \times_M L(TM, TM) \xrightarrow{\operatorname{Id}\times_M \exp}$$
$$\xrightarrow{\operatorname{Id}\times_M \exp} GL(TM, T^*M) \times_M GL(TM, TM) \xrightarrow{\flat} GL(TM, T^*M),$$

where $\sharp(b^0, h) := (b^0, (b^0)^{-1}h)$ is a fiber respecting diffeomorphism, $\flat(b^0, H) := b^0 H$ is a diffeomorphism for fixed b^0, and where the other two mappings will be discussed below.

The usual fiberwise exponential mapping

$$\exp : L(TM, TM) \to GL(TM, TM)$$

is a diffeomorphism near the zero section on the ball of radius π centered at zero in a norm on the Lie algebra for which the Lie bracket is sub multiplicative, for example. If we fix a symmetric positive definite inner product g, then exp restricts to a global diffeomorphism from the linear subspace of g-symmetric endomorphisms onto the open subset of matrices which are positive definite with respect to g. If g has signature this is no longer true since then g-symmetric matrices may have non real eigenvalues.

On the open set of all matrices whose eigenvalues λ satisfy $|\operatorname{Im}\lambda| < \pi$, the exponential mapping is a diffeomorphism, see [Varadarajan, 1977].

The smooth mapping $\varphi : U' \to GL(TM, T^*M) \times_M L(TM, TM)$ is given by $\varphi(b^0, H) := (b^0, a_{\alpha,H}(1)\operatorname{Id} + b_{\alpha,H}(1)H_0)$ (see theorem (45.11)). It is a diffeomorphism onto its image with the following inverse:

$$\psi(H) := \begin{cases} \frac{4}{n}\left(e^{\frac{\operatorname{tr}(H)}{4}}\cos\left(\frac{\sqrt{\alpha^{-1}\operatorname{tr}(H_0^2)}}{4}\right) - 1\right)\operatorname{Id} + \\ \qquad + \frac{4}{\sqrt{\alpha^{-1}\operatorname{tr}(H_0^2)}} e^{\frac{\operatorname{tr}(H)}{4}} \sin\left(\frac{\sqrt{\alpha^{-1}\operatorname{tr}(H_0^2)}}{4}\right) H_0 & \text{if } \operatorname{tr}(H_0^2) \neq 0 \\ \frac{4}{n}\left(e^{\frac{\operatorname{tr}(H)}{4}} - 1\right)\operatorname{Id} & \text{otherwise,} \end{cases}$$

where cos is considered as a complex function, $\cos(iz) = i\cosh(z)$.

The mapping $(\mathrm{pr}_1, \Phi) : U \to GL(TM, T^*M) \times_M GL(TM, T^*M)$ is a diffeomorphism on an open neighborhood of the zero section in U.

45.13. Theorem. *In the setting of* (45.12) *the exponential mapping* $\exp^{\alpha}_{b^0}$ *for the metric* G^{α} *is a real analytic mapping defined on the open subset*

$$\mathcal{U}_{b^0} := \{h \in C_c^{\infty}(L(TM, T^*M)) : (b^0, h)(M) \subset U\},$$

and it is given by

$$\exp_{b^0}(h) = \Phi \circ (b^0, h).$$

The mapping $(\pi_{\mathcal{B}}, \exp) : T\mathcal{B} \to \mathcal{B} \times \mathcal{B}$ *is a real analytic diffeomorphism from an open neighborhood of the zero section in* $T\mathcal{B}$ *onto an open neighborhood of the diagonal in* $\mathcal{B} \times \mathcal{B}$. $\mathcal{U}_{b_0}$ *is the maximal domain of definition for the exponential mapping.*

Proof. Since $\mathcal{B}$ is a disjoint union of chart neighborhoods, it is trivially a real analytic manifold, even if M is not supposed to carry a real analytic structure.

From the consideration in (45.12) it follows that $\exp = \Phi_*$ and $(\pi_{\mathcal{M}}, \exp)$ are just push forwards by real analytic fiber respecting mappings of sections of bundles. So by (30.10) they are smooth, and this applies also to their inverses.

To show that these mappings are real analytic, by (10.3) it remains to check that they map real analytic curves into real analytic curves. These are described in (30.15). It is clear that Φ has a fiberwise extension to a holomorphic germ since Φ is fiber respecting from an open subset in a vector bundle and is fiberwise a real analytic mapping. So the push forward Φ_* preserves the description (30.15) and maps real analytic curves to real analytic curves. □

45.14. Submanifolds of pseudo Riemannian metrics. We denote by $\mathcal{M}^q$ the space of all pseudo Riemannian metrics on the manifold M of signature (the dimension of a maximal negative definite subspace) q. It is an open set in a closed locally affine subspace of $\mathcal{B}$ and thus a splitting submanifold of it with tangent bundle $T\mathcal{M}^q = \mathcal{M}^q \times C_c^\infty(M \leftarrow S^2T^*M)$.

We consider a geodesic $c(t) = c^0 e^{(a(t)\operatorname{Id} + b(t)H_0)}$ for the metric G^α in $\mathcal{B}$ starting at c^0 in the direction of h as in (45.11). If $c^0 \in \mathcal{M}^q$ then $h \in T_{c^0}\mathcal{M}^q$ if and only if $H = (c^0)^{-1}h \in L_{\text{sym},c^0}(TM, TM)$ is symmetric with respect to the pseudo Riemannian metric c^0. But then $e^{(a(t)\operatorname{Id} + b(t)H_0)} \in L_{\text{sym},c^0}(TM, TM)$ for all t in the domain of definition of the geodesic, so $c(t)$ is a curve of pseudo Riemannian metrics and thus of the same signature q as c^0. Thus, we have

45.15. Theorem. *For each $q \le n = \dim M$ the submanifold $\mathcal{M}^q$ of pseudo Riemannian metrics of signature q on M is a geodesically closed submanifold of $(\mathcal{B}, G^\alpha)$ for each $\alpha \neq 0$.*

Remark. The geodesics of $(\mathcal{M}^0, G^\alpha)$ have been studied for $\alpha = \frac{1}{n}$, in [Freed, Groisser, 1989], [Gil-Medrano, Michor, 1991] and from (45.15) and (45.11) we see that they are completely analogous for every positive α.

For fixed $x \in M$ there exists a family of homothetic pseudo metrics on the finite dimensional manifold $S^2_+T^*_xM$ whose geodesics are given by the evaluation of the geodesics of $(\mathcal{M}^0, G^\alpha)$ (see [Gil-Medrano, Michor, 1991] for more details). When α is negative, it is not difficult to see, from (45.15) and (45.11) again, that a geodesic of $(\mathcal{M}^0, G^\alpha)$ is defined for all t if and only if the initial velocity h satisfies $\gamma^\alpha(h, h) \le 0$ and $\operatorname{tr} H > 0$ at each point of M, and then the same is true for all the pseudo metrics on $S^2_+T^*_xM$. These results appear already in [DeWitt, 1967] for $n = 3$.

45.16. The local signature of G^α. Since G^α operates in infinite dimensional spaces, the usual definition of signature is not applicable. But for fixed $g \in \mathcal{M}^q$ the signature of

$$\gamma^\alpha_{g_x}(h_x, k_x) = \operatorname{tr}((g_x^{-1}h_x)_0(g_x^{-1}k_x)_0) + \alpha \operatorname{tr}(g_x^{-1}k_x)\operatorname{tr}(g_x^{-1}k_x)$$

on $T_g(S^2_qT^*_xM) = S^2T^*_xM$ is independent of $x \in M$ and the special choice of $g \in \mathcal{M}^q$. We will call it the *local signature* of G^α.

45.17. Lemma. *The signature of the quadratic form of* (45.16) *is*

$$Q(\alpha, q) = q(q-n) + \begin{cases} 0 & \text{for } \alpha > 0 \\ 1 & \text{for } \alpha < 0. \end{cases}$$

This result is due to [Schmidt, 1989].

Proof. Since the signature is constant on connected components we have to determine it only for $\alpha = \frac{1}{n}$ and $\alpha = \frac{1}{n} - 1$. In a basis for TM and its dual basis for T^*M the bilinear form $h \in S^2T^*_xM$ has a symmetric matrix. If the basis is orthonormal for g we have (for $A^t = A$ and $C^t = C$)

$$H = g^{-1}h = \begin{pmatrix} -\operatorname{Id}_q & 0 \\ 0 & \operatorname{Id}_{n-q} \end{pmatrix}\begin{pmatrix} A & B \\ B^t & C \end{pmatrix} = \begin{pmatrix} -A & -B \\ B^t & C \end{pmatrix},$$

which describes a typical matrix in the space $L_{\mathrm{sym},g}(T_xM, T_xM)$ of all matrices $H \in L(T_xM, T_xM)$ which are symmetric with respect to g_x.

Now we treat the case $\alpha = \frac{1}{n}$. The standard inner product $\operatorname{tr}(HK^t)$ is positive definite on $L_{\mathrm{sym},g}(T_xM, T_xM)$, and the linear mapping $K \mapsto K^t$ has an eigenspace of dimension $q(n-q)$ for the eigenvalue -1 in it and a complementary eigenspace for the eigenvalue 1. So $\operatorname{tr}(HK)$ has signature $q(n-q)$.

For the case $\alpha = \frac{1}{n} - 1$ we again split the space $L_{\mathrm{sym},g}(T_xM, T_xM)$ into the subspace with 0 on the main diagonal, where $\gamma^\alpha_g(h,k) = \operatorname{tr}(HK)$ and where $K \mapsto K^t$ has again an eigenspace of dimension $q(n-q)$ for the eigenvalue -1, and the space of diagonal matrices. There γ^α_g has signature 1 as determined in the proof of (45.5). □

45.18. The submanifold of almost symplectic structures.
A 2-form $\omega \in \Omega^2(M) = C^\infty(M \leftarrow \Lambda^2T^*M)$ can be non degenerate only if M is of even dimension $\dim M = n = 2m$. Then ω is non degenerate if and only if $\omega \wedge \dots \wedge \omega = \omega^m$ is nowhere vanishing. Usually this latter $2m$-form is regarded as the volume form associated with ω, but a short computation shows that we have

$$\operatorname{vol}(\omega) = \tfrac{1}{m!}|\omega^m|.$$

This implies $m\varphi \wedge \omega^{m-1} = \frac{1}{2}\operatorname{tr}(\omega^{-1}\varphi)\omega^m$ for all $\varphi \in \Omega^2(M)$.

45.19. Theorem. *The space $\Omega^2_{\mathrm{nd}}(M)$ of non degenerate 2-forms is a splitting geodesically closed submanifold of $(\mathcal{B}, G^\alpha)$ for each $\alpha \neq 0$.*

Proof. We consider a geodesic $c(t) = c^0 e^{(a(t)\operatorname{Id} + b(t)H_0)}$ for the metric G^α in $\mathcal{B}$ starting at c^0 in the direction of h as in (45.11). If $c^0 = \omega \in \Omega^2_{\mathrm{nd}}(M)$ then $h \in \Omega^2_c(M)$ if and only if $H = \omega^{-1}h$ is symmetric with respect to ω, since we have $\omega(HX,Y) = \langle \omega\omega^{-1}hX, Y\rangle = \langle hX, Y\rangle = h(X,Y) = -h(Y,X) = -\omega(HY,X) = \omega(X,HY)$. At a point $x \in M$ we may choose a Darboux frame (e_i) such that $\omega(X,Y) = Y^tJX$ where

$$J = \begin{pmatrix} 0 & \operatorname{Id} \\ -\operatorname{Id} & 0 \end{pmatrix}.$$

Then h is skew if and only if JH is a skew symmetric matrix in the Darboux frame, or $JH = H^tJ$. Since $(e^A)^t = e^{A^t}$ the matrix $e^{a(t)\operatorname{Id} + b(t)H_0}$ then has the same property, $c(t)$ is skew for all t. Thus, $\Omega^2_{\mathrm{nd}}(M)$ is a geodesically closed submanifold. □

45.20. Lemma. *For a non degenerate 2-form ω the signature of the quadratic form $\varphi \mapsto \operatorname{tr}(\omega^{-1}\varphi\omega^{-1}\varphi)$ on $\Lambda^2T^*_xM$ is $m^2 - m$ for $\alpha > 0$ and $m^2 - m + 1$ for $\alpha < 0$.*

Proof. Use the method of (45.5) and (45.17). The description of the space of matrices can be read off the proof of (45.19). □

45.21. Symplectic structures. The space $\operatorname{Symp}(M)$ of all symplectic structures is a closed submanifold of $(\mathcal{B}, G^\alpha)$. For a compact manifold M it is splitting by the Hodge decomposition theorem. For $\dim M = 2$ we have $\operatorname{Symp}(M) = \Omega^2_{\text{nd}}(M)$, so it is geodesically closed. But for $\dim M \geq 4$ the submanifold $\operatorname{Symp}(M)$ is not geodesically closed. For $\omega \in \operatorname{Symp}(M)$ and φ, $\psi \in T_\omega \operatorname{Symp}(M)$ the Christoffel form $\Gamma^\alpha_\omega(\varphi, \psi)$ is not closed in general.

46. The Korteweg – De Vrieß Equation as a Geodesic Equation

This section is based on [Michor, Ratiu, 1997], an overview of related ideas can be found in [Segal, 1991]. That the Korteweg – de Vrieß equation is a geodesic equation is attributed to [Gelfand, Dorfman, 1979], [Kirillov, 1981] or [Ovsienko, Khesin, 1987]. The curvature of a right invariant metric on an (infinite dimensional) Lie group was computed by [Arnold, 1966a, 1966b], see also [Arnold, 1978].

46.1. Recall from (44.1) the principal bundle of embeddings $\operatorname{Emb}(M, N)$, where M and N are smooth finite dimensional manifolds, connected and second countable without boundary such that $\dim M \leq \dim N$. The space $\operatorname{Emb}(M, N)$ of all embeddings from M into N is an open submanifold of $C^\infty(M, N)$, which is stable under the right action of the diffeomorphism group. Then $\operatorname{Emb}(M, N)$ is the total space of a smooth principal fiber bundle with structure group the diffeomorphism group. The base is called $B(M, N)$, it is a Hausdorff smooth manifold modeled on nuclear (LF)-spaces. It can be thought of as the "nonlinear Grassmannian" of all submanifolds of N which are of type M.

Recall from (44.24) that if we take a Hilbert space H instead of N, then $B(M, H)$ is the classifying space for $\operatorname{Diff}(M)$ if M is compact, and the classifying bundle $\operatorname{Emb}(M, H)$ carries also a universal connection.

46.2. If (N, g) is a Riemannian manifold then on the manifold $\operatorname{Emb}(M, N)$ we have an induced weak Riemannian metric given by

$$G_e(s_1, s_2) = \int_M g(s_1, s_2) \operatorname{vol}(e^* g).$$

Its covariant derivative and curvature were investigated in [Binz, 1980] for the case that $N = \mathbb{R}^{\dim M + 1}$ with the standard inner product, and in [Kainz, 1984] in the general case. We shall not reproduce the general formulas here. This weak Riemannian metric is invariant under the action of the diffeomorphism group $\operatorname{Diff}(M)$ by composition from the right, thus it induces a Riemannian metric on the base manifold $B(M, N)$, which can be viewed as an infinite dimensional non-linear analogue of the Fubini-Study metric on projective spaces and Grassmannians.

46.3. Example. Let us consider the metric on the space $\operatorname{Emb}(\mathbb{R}, \mathbb{R})$ of all embeddings of the real line into itself, which contains the diffeomorphism group $\operatorname{Diff}(\mathbb{R})$ as an open subset. We could also treat $\operatorname{Emb}(S^1, S^1)$, where the results are the same.

$$G_f(h,k) = \int_{\mathbb{R}} h(x)k(x)|f'(x)|\,dx, \qquad f \in \operatorname{Emb}(\mathbb{R},\mathbb{R}), h,k \in C_c^\infty(\mathbb{R},\mathbb{R}).$$

We shall compute the geodesic equation for this metric by variational calculus. The energy of a curve f of embeddings (without loss of generality orientation preserving) is the expression

$$E(f) = \tfrac{1}{2}\int_a^b G_f(f_t,f_t)dt = \tfrac{1}{2}\int_a^b \int_{\mathbb{R}} f_t^2 f_x\,dxdt.$$

If we assume that $f(x,t,s)$ depends smoothly on one variable more, so that we have a variation with fixed endpoints, then the derivative with respect to s of the energy is given by

$$\begin{aligned}
\tfrac{\partial}{\partial s}|_0 E(f(\quad,s)) &= \tfrac{\partial}{\partial s}|_0 \tfrac{1}{2}\int_a^b\int_{\mathbb{R}} f_t^2 f_x\,dxdt \\
&= \tfrac{1}{2}\int_a^b\int_{\mathbb{R}} (2f_t f_{ts} f_x + f_t^2 f_{xs})dxdt \\
&= -\tfrac{1}{2}\int_a^b\int_{\mathbb{R}} (2f_{tt} f_s f_x + 2 f_t f_s f_{tx} + 2 f_t f_{tx} f_s)dxdt \\
&= -\int_a^b\int_{\mathbb{R}} \left(f_{tt} + 2\frac{f_t f_{tx}}{f_x}\right) f_s f_x dxdt,
\end{aligned}$$

so that the geodesic equation with its initial data is

$$\begin{aligned}
(1)\qquad f_{tt} &= -2\frac{f_t f_{tx}}{f_x}, \qquad f(\quad,0)\in \operatorname{Emb}^+(\mathbb{R},\mathbb{R}), f_t(\quad,0) \in C_c^\infty(\mathbb{R},\mathbb{R}) \\
&= \Gamma_f(f_t,f_t),
\end{aligned}$$

where the Christoffel symbol $\Gamma : \operatorname{Emb}(\mathbb{R},\mathbb{R}) \times C_c^\infty(\mathbb{R},\mathbb{R}) \times C_c^\infty(\mathbb{R},\mathbb{R}) \to C_c^\infty(\mathbb{R},\mathbb{R})$ is given by symmetrization

$$(2)\qquad \Gamma_f(h,k) = -\frac{hk_x + h_x k}{f_x} = -\frac{(hk)_x}{f_x}.$$

For vector fields X, Y on $\operatorname{Emb}(\mathbb{R},\mathbb{R})$ the covariant derivative is given by the expression $\nabla_X^{\operatorname{Emb}} Y = dY(X) - \Gamma(X,Y)$. The Riemannian curvature $R(X,Y)Z = (\nabla_X\nabla_Y - \nabla_Y\nabla_X - \nabla_{[X,Y]})Z$ is then expressed in terms of the Christoffel symbol by the usual formula

$$\begin{aligned}
R_f(h,k)\ell &= -d\Gamma(f)(h)(k,\ell) + d\Gamma(f)(k)(h,\ell) + \Gamma_f(h,\Gamma_f(k,\ell)) - \Gamma_f(k,\Gamma_f(h,\ell)) \\
&= -\frac{h_x(k\ell)_x}{f_x^2} + \frac{1}{k_x(h\ell)_x} f_x^2 + \frac{\left(h\frac{(k\ell)_x}{f_x}\right)_x}{f_x} - \frac{\left(k\frac{(h\ell)_x}{f_x}\right)_x}{f_x} \\
(3)\qquad &= \frac{1}{f_x^3}\Big(f_{xx}h_x k\ell - f_{xx}hk_x\ell + f_x hk_{xx}\ell - f_x h_{xx}k\ell + 2f_x hk_x\ell - 2f_x h_x k\ell\Big)
\end{aligned}$$

The geodesic equation can be solved in the following way: If instead of the obvious framing we choose $T\operatorname{Emb} = \operatorname{Emb} \times C^\infty_c \ni (f,h) \mapsto (f, hf_x^2) =: (f,H)$ then the geodesic equation becomes $F_t = \frac{\partial}{\partial t}(f_t f_x^2) = f_x^2(f_{tt} + 2\frac{f_t f_{tx}}{f_x}) = 0$, so that $F = f_t f_x^2$ is constant in t, or $f_t(x,t)f_x(x,t)^2 = f_t(x,0)f_x(x,0)^2$. Using that one can then use separation of variables to solve the geodesic equation. The solution blows up in finite time in general.

Now let us consider the trivialization of $T\operatorname{Emb}(\mathbb{R},\mathbb{R})$ by right translation (this is clearest for $\operatorname{Diff}(\mathbb{R})$), then we have

$$\begin{aligned}
u &:= f_t \circ f^{-1}, \qquad \text{in more detail } u(y,t) = f_t(f(\quad,t)^{-1}(y),t)\\
u_x &= (f_{tx}\circ f^{-1})\frac{1}{f_x\circ f^{-1}},\\
u_t &= f_{tt}\circ f^{-1} - (f_{tx}\circ f^{-1})\frac{1}{f_x\circ f^{-1}}(f_t\circ f^{-1}) = -3\left(\frac{f_{tx}f_t}{f_x}\right)\circ f^{-1}\\
\text{(4)}\qquad u_t &= -3u_x u.
\end{aligned}$$

where we used $T_f(\operatorname{Inv})h = -T(f^{-1})\circ h \circ f^{-1}$.

46.4. Geodesics of a right invariant metric on a Lie group. Let G be a Lie group which may be infinite dimensional, with Lie algebra $\mathfrak{g}$. Recall (36.1) that $\mu : G\times G \to G$ denotes the multiplication with μ_x left translation and μ^x right translation by x, and (36.10) that $\kappa = \kappa^r \in \Omega^1(G,\mathfrak{g})$ denotes the right Maurer-Cartan form, $\kappa_x(\xi) = T_x(\mu^{x^{-1}}).\xi$. It satisfies (38.1) the right Maurer-Cartan equation $d\kappa - \frac12[\kappa,\kappa]_\wedge = 0$. Let $\langle\quad,\quad\rangle : \mathfrak{g}\times\mathfrak{g} \to \mathbb{R}$ be a positive definite bounded inner product. Then

$$\text{(1)}\qquad G_x(\xi,\eta) = \langle T(\mu^{x^{-1}}).\xi, T(\mu^{x^{-1}}).\eta\rangle = \langle\kappa(\xi),\kappa(\eta)\rangle$$

is a right invariant Riemannian metric on G, and any right invariant bounded Riemannian metric is of this form, for suitable $\langle\quad,\quad\rangle$.

Let $g : [a,b] \to G$ be a smooth curve. The velocity field of g, viewed in the right trivialization, is right logarithmic derivative $\delta^r g(\partial_t) = T(\mu^{g^{-1}})\partial_t g = \kappa(\partial_t g) = (g^*\kappa)(\partial_t)$. The energy of the curve g is given by

$$E(g) = \frac12\int_a^b G_g(g',g')dt = \frac12\int_a^b \langle g^*\kappa(\partial_t), g^*\kappa(\partial_t)\rangle\, dt.$$

For a variation $g(t,s)$ with fixed endpoints we have then, using the right Maurer-Cartan equation and integration by parts

$$\begin{aligned}
\partial_s E(g) &= \frac12\int_a^b 2\langle\partial_s(g^*\kappa)(\partial_t), g^*\kappa(\partial_t)\rangle\, dt\\
&= \int_a^b \langle\partial_t(g^*\kappa)(\partial_s) - d(g^*\kappa)(\partial_t,\partial_s), g^*\kappa(\partial_t)\rangle\, dt\\
&= \int_a^b (-\langle(g^*\kappa)(\partial_s), \partial_t(g^*\kappa)(\partial_t)\rangle - \langle[g^*\kappa(\partial_t), g^*\kappa(\partial_s)], g^*\kappa(\partial_t)\rangle)\, dt\\
&= -\int_a^b \langle(g^*\kappa)(\partial_s), \partial_t(g^*\kappa)(\partial_t) + \operatorname{ad}(g^*\kappa(\partial_t))^\top(g^*\kappa(\partial_t))\rangle\, dt
\end{aligned}$$

where $\mathrm{ad}(g^*\kappa(\partial_t))^\top : \mathfrak{g} \to \mathfrak{g}$ is the adjoint of $\mathrm{ad}(g^*\kappa(\partial_t))$ with respect to the inner product $\langle \quad , \quad \rangle$. In infinite dimensions one also has to check the existence of this adjoint. In terms of the right logarithmic derivative $u : [a,b] \to \mathfrak{g}$ of $g : [a,b] \to G$, given by $u(t) := g^*\kappa(\partial_t) = T_{g(t)}(\mu^{g(t)^{-1}})g'(t)$, the geodesic equation looks like

(2) $$u_t = -\mathrm{ad}(u)^\top u.$$

46.5. The covariant derivative. Our next aim is to derive the Riemannian curvature, and for that we develop the basis-free version of Cartan's method of moving frames in this setting, which also works in infinite dimensions. The right trivialization or framing $(\kappa, \pi_G) : TG \to \mathfrak{g} \times G$ induces the isomorphism $R : C^\infty(G, \mathfrak{g}) \to \mathfrak{X}(G)$, given by $R_X(x) = T_e(\mu^x).X(x)$. For the Lie bracket and the Riemannian metric we have

(1) $$\begin{aligned} [R_X, R_Y] &= R(-[X,Y]_{\mathfrak{g}} + dY.R_X - dX.R_Y), \\ R^{-1}[R_X, R_Y] &= -[X,Y]_{\mathfrak{g}} + R_X(Y) - R_Y(X), \\ G_x(R_X(x), R_Y(x)) &= \langle X(x), Y(x) \rangle. \end{aligned}$$

Lemma. *Assume that for all $X \in \mathfrak{g}$ the adjoint $\mathrm{ad}(X)^\top$ with respect to the inner product $\langle \quad , \quad \rangle$ exists and that $X \mapsto \mathrm{ad}(X)^\top$ is bounded. Then the Levi-Civita covariant derivative of the metric (1) exists and is given in terms of the isomorphism R by*

(2) $$\nabla_X Y = dY.R_X + \tfrac{1}{2}\mathrm{ad}(X)^\top Y + \tfrac{1}{2}\mathrm{ad}(Y)^\top X - \tfrac{1}{2}\mathrm{ad}(X)Y.$$

Proof. Easy computations shows that this covariant derivative respects the Riemannian metric,

$$R_X\langle Y, Z \rangle = \langle dY.R_X, Z \rangle + \langle Y, dZ.R_X \rangle = \langle \nabla_X Y, Z \rangle + \langle Y, \nabla_X Z \rangle,$$

and is torsionfree,

$$\nabla_X Y - \nabla_Y X + [X,Y]_{\mathfrak{g}} - dY.R_X + dX.R_Y = 0. \quad \square$$

Let us write $\alpha(X) : \mathfrak{g} \to \mathfrak{g}$, where $\alpha(X)Y = \mathrm{ad}(Y)^\top X$, then we have

(3) $$\nabla_X = R_X + \tfrac{1}{2}\mathrm{ad}(X)^\top + \tfrac{1}{2}\alpha(X) - \tfrac{1}{2}\mathrm{ad}(X)$$

46.6. The curvature. First note that we have the following relations:

(1) $$\begin{aligned} [R_X, \mathrm{ad}(Y)] &= \mathrm{ad}(R_X(Y)), & [R_X, \alpha(Y)] &= \alpha(R_X(Y)), \\ [R_X, \mathrm{ad}(Y)^\top] &= \mathrm{ad}(R_X(Y))^\top, & [\mathrm{ad}(X)^\top, \mathrm{ad}(Y)^\top] &= -\mathrm{ad}([X,Y])^\top. \end{aligned}$$

The Riemannian curvature is then computed by

(2)

$$
\begin{aligned}
\mathcal{R}(X,Y) &= [\nabla_X, \nabla_Y] - \nabla_{-[X,Y]_{\mathfrak g}+R_X(Y)-R_Y(X)}\\
&= [R_X + \tfrac12\operatorname{ad}(X)^\top + \tfrac12\alpha(X) - \tfrac12\operatorname{ad}(X), R_Y + \tfrac12\operatorname{ad}(Y)^\top + \tfrac12\alpha(Y) - \tfrac12\operatorname{ad}(Y)]\\
&\quad - R_{-[X,Y]_{\mathfrak g}+R_X(Y)-R_Y(X)} - \tfrac12\operatorname{ad}(-[X,Y]_{\mathfrak g} + R_X(Y) - R_Y(X))^\top\\
&\quad - \tfrac12\alpha(-[X,Y]_{\mathfrak g} + R_X(Y) - R_Y(X)) + \tfrac12\operatorname{ad}(-[X,Y]_{\mathfrak g} + R_X(Y) - R_Y(X))\\
&= -\tfrac14[\operatorname{ad}(X)^\top + \operatorname{ad}(X), \operatorname{ad}(Y)^\top + \operatorname{ad}(Y)]\\
&\quad + \tfrac14[\operatorname{ad}(X)^\top - \operatorname{ad}(X), \alpha(Y)] + \tfrac14[\alpha(X), \operatorname{ad}(Y)^\top - \operatorname{ad}(Y)]\\
&\quad + \tfrac14[\alpha(X), \alpha(Y)] + \tfrac12\alpha([X,Y]_{\mathfrak g}).
\end{aligned}
$$

If we plug in all definitions and use 4 times the Jacobi identity we get the following expression

$$
\begin{aligned}
\langle 4\mathcal{R}(X,Y)Z,U\rangle &= 2\langle [X,Y],[Z,U]\rangle - \langle [Y,Z],[X,U]\rangle + \langle [X,Z],[Y,U]\rangle\\
&- \langle Z,[U,[X,Y]]\rangle + \langle U,[Z,[X,Y]]\rangle - \langle Y,[X,[U,Z]]\rangle - \langle X,[Y,[Z,U]]\rangle\\
&+ \langle \operatorname{ad}(X)^\top Z, \operatorname{ad}(Y)^\top U\rangle + \langle \operatorname{ad}(X)^\top Z, \operatorname{ad}(U)^\top Y\rangle + \langle \operatorname{ad}(Z)^\top X, \operatorname{ad}(Y)^\top U\rangle\\
&- \langle \operatorname{ad}(U)^\top X, \operatorname{ad}(Y)^\top Z\rangle - \langle \operatorname{ad}(Y)^\top Z, \operatorname{ad}(X)^\top U\rangle - \langle \operatorname{ad}(Z)^\top Y, \operatorname{ad}(X)^\top U\rangle\\
&- \langle \operatorname{ad}(U)^\top X, \operatorname{ad}(Z)^\top Y\rangle + \langle \operatorname{ad}(U)^\top Y, \operatorname{ad}(Z)^\top X\rangle.
\end{aligned}
$$

46.7. Jacobi fields, I. We compute first the Jacobi equation via variations of geodesics. So let $g : \mathbb{R}^2 \to G$ be smooth, $t \mapsto g(t,s)$ a geodesic for each s. Let again $u = \kappa(\partial_t g) = (g^*\kappa)(\partial_t)$ be the velocity field along the geodesic in right trivialization which satisfies the geodesic equation $u_t = -\operatorname{ad}(u)^\top u$. Then $y := \kappa(\partial_s g) = (g^*\kappa)(\partial_s)$ is the Jacobi field corresponding to this variation, written in the right trivialization. From the right Maurer-Cartan equation we then have:

$$
\begin{aligned}
y_t = \partial_t(g^*\kappa)(\partial_s) &= d(g^*\kappa)(\partial_t,\partial_s) + \partial_s(g^*\kappa)(\partial_t) + 0\\
&= [(g^*\kappa)(\partial_t),(g^*\kappa)(\partial_s)]_{\mathfrak g} + u_s\\
&= [u,y] + u_s.
\end{aligned}
$$

From this, using the geodesic equation and (46.6.1) we get

$$
\begin{aligned}
u_{st} = u_{ts} = \partial_s u_t &= -\partial_s(\operatorname{ad}(u)^\top u) = -\operatorname{ad}(u_s)^\top u - \operatorname{ad}(u)^\top u_s\\
&= -\operatorname{ad}(y_t + [y,u])^\top u - \operatorname{ad}(u)^\top(y_t + [y,u])\\
&= -\alpha(u)y_t - \operatorname{ad}([y,u])^\top u - \operatorname{ad}(u)^\top y_t - \operatorname{ad}(u)^\top([y,u])\\
&= -\operatorname{ad}(u)^\top y_t - \alpha(u)y_t + [\operatorname{ad}(y)^\top, \operatorname{ad}(u)^\top]u - \operatorname{ad}(u)^\top\operatorname{ad}(y)u.
\end{aligned}
$$

Finally, we get the Jacobi equation as

$$
\begin{aligned}
y_{tt} &= [u_t,y] + [u,y_t] + u_{st}\\
&= \operatorname{ad}(y)\operatorname{ad}(u)^\top u + \operatorname{ad}(u)y_t - \operatorname{ad}(u)^\top y_t\\
&\qquad - \alpha(u)y_t + [\operatorname{ad}(y)^\top, \operatorname{ad}(u)^\top]u - \operatorname{ad}(u)^\top\operatorname{ad}(y)u
\end{aligned}
$$

(1)
$$
y_{tt} = [\operatorname{ad}(y)^\top + \operatorname{ad}(y), \operatorname{ad}(u)^\top]u - \operatorname{ad}(u)^\top y_t - \alpha(u)y_t + \operatorname{ad}(u)y_t.
$$

46.8. Jacobi fields, II. Let y be a Jacobi field along a geodesic g with right trivialized velocity field u. Then y satisfies the Jacobi equation

$$\nabla_{\partial_t}\nabla_{\partial_t}y + \mathcal{R}(y,u)u = 0$$

We want to show that this leads to same equation as (46.7). First note that from (46.5.2) we have

$$\nabla_{\partial_t}y = y_t + \tfrac12 \operatorname{ad}(u)^\top y + \tfrac12\alpha(u)y - \tfrac12\operatorname{ad}(u)y$$

so that we get, using $u_t = -\operatorname{ad}(u)^\top u$ heavily:

$$\begin{aligned}
\nabla_{\partial_t}\nabla_{\partial_t}y &= \nabla_{\partial_t}\Big(y_t + \tfrac12 \operatorname{ad}(u)^\top y + \tfrac12\alpha(u)y - \tfrac12\operatorname{ad}(u)y\Big)\\
&= y_{tt} + \tfrac12\operatorname{ad}(u_t)^\top y + \tfrac12\operatorname{ad}(u)^\top y_t + \tfrac12\alpha(u_t)y\\
&\quad + \tfrac12\alpha(u)y_t - \tfrac12\operatorname{ad}(u_t)y - \tfrac12\operatorname{ad}(u)y_t\\
&\quad + \tfrac12\operatorname{ad}(u)^\top\Big(y_t + \tfrac12 \operatorname{ad}(u)^\top y + \tfrac12\alpha(u)y - \tfrac12\operatorname{ad}(u)y\Big)\\
&\quad + \tfrac12\alpha(u)\Big(y_t + \tfrac12 \operatorname{ad}(u)^\top y + \tfrac12\alpha(u)y - \tfrac12\operatorname{ad}(u)y\Big)\\
&\quad - \tfrac12\operatorname{ad}(u)\Big(y_t + \tfrac12 \operatorname{ad}(u)^\top y + \tfrac12\alpha(u)y - \tfrac12\operatorname{ad}(u)y\Big)\\
&= y_{tt} + \operatorname{ad}(u)^\top y_t + \alpha(u)y_t - \operatorname{ad}(u)y_t\\
&\quad - \tfrac12\alpha(y)\operatorname{ad}(u)^\top u - \tfrac12\operatorname{ad}(y)^\top\operatorname{ad}(u)^\top u - \tfrac12\operatorname{ad}(y)\operatorname{ad}(u)^\top u\\
&\quad + \tfrac12\operatorname{ad}(u)^\top\Big(\tfrac12\alpha(y)u + \tfrac12\operatorname{ad}(y)^\top u + \tfrac12\operatorname{ad}(y)u\Big)\\
&\quad + \tfrac12\alpha(u)\Big(\tfrac12\alpha(y)u + \tfrac12\operatorname{ad}(y)^\top u + \tfrac12\operatorname{ad}(y)u\Big)\\
&\quad - \tfrac12\operatorname{ad}(u)\Big(\tfrac12\alpha(y)u + \tfrac12\operatorname{ad}(y)^\top u + \tfrac12\operatorname{ad}(y)u\Big)
\end{aligned}$$

In the second line of the last expression we use

$$-\tfrac12\alpha(y)\operatorname{ad}(u)^\top u = -\tfrac14\alpha(y)\operatorname{ad}(u)^\top u - \tfrac14\alpha(y)\alpha(u)u$$

and similar forms for the other two terms to get:

$$\begin{aligned}
\nabla_{\partial_t}\nabla_{\partial_t}y &= y_{tt} + \operatorname{ad}(u)^\top y_t + \alpha(u)y_t - \operatorname{ad}(u)y_t\\
&\quad + \tfrac14[\operatorname{ad}(u)^\top,\alpha(y)]u + \tfrac14[\operatorname{ad}(u)^\top,\operatorname{ad}(y)^\top]u + \tfrac14[\operatorname{ad}(u)^\top,\operatorname{ad}(y)]u\\
&\quad + \tfrac14[\alpha(u),\alpha(y)]u + \tfrac14[\alpha(u),\operatorname{ad}(y)^\top]u + \tfrac14[\alpha(u),\operatorname{ad}(y)]u\\
&\quad - \tfrac14[\operatorname{ad}(u),\alpha(y)]u - \tfrac14[\operatorname{ad}(u),\operatorname{ad}(y)^\top + \operatorname{ad}(y)]u,
\end{aligned}$$

where in the last line we also used $\operatorname{ad}(u)u = 0$. We now compute the curvature term:

$$\begin{aligned}
\mathcal{R}(y,u)u &= -\tfrac14[\operatorname{ad}(y)^\top + \operatorname{ad}(y),\operatorname{ad}(u)^\top + \operatorname{ad}(u)]u\\
&\quad + \tfrac14[\operatorname{ad}(y)^\top - \operatorname{ad}(y),\alpha(u)]u + \tfrac14[\alpha(y),\operatorname{ad}(u)^\top - \operatorname{ad}(u)]u\\
&\quad + \tfrac14[\alpha(y),\alpha(u)] + \tfrac12\alpha([y,u])u\\
&= -\tfrac14[\operatorname{ad}(y)^\top + \operatorname{ad}(y),\operatorname{ad}(u)^\top]u - \tfrac14[\operatorname{ad}(y)^\top + \operatorname{ad}(y),\operatorname{ad}(u)]u\\
&\quad + \tfrac14[\operatorname{ad}(y)^\top,\alpha(u)]u - \tfrac14[\operatorname{ad}(y),\alpha(u)]u + \tfrac14[\alpha(y),\operatorname{ad}(u)^\top - \operatorname{ad}(u)]u\\
&\quad + \tfrac14[\alpha(y),\alpha(u)] + \tfrac12\operatorname{ad}(u)^\top\operatorname{ad}(y)u
\end{aligned}$$

Summing up we get

$$\begin{aligned}\nabla_{\partial_t}\nabla_{\partial_t}y + \mathcal{R}(y,u)u &= y_{tt} + \operatorname{ad}(u)^\top y_t + \alpha(u)y_t - \operatorname{ad}(u)y_t\\ &\quad - \tfrac12[\operatorname{ad}(y)^\top + \operatorname{ad}(y), \operatorname{ad}(u)^\top]u\\ &\quad + \tfrac12[\alpha(u), \operatorname{ad}(y)]u + \tfrac12\operatorname{ad}(u)^\top\operatorname{ad}(y)u\end{aligned}$$

Finally, we need the following computation using (46.6.1):

$$\begin{aligned}\tfrac12[\alpha(u), \operatorname{ad}(y)]u &= \tfrac12\alpha(u)[y,u] - \tfrac12\operatorname{ad}(y)\alpha(u)u\\ &= \tfrac12\operatorname{ad}([y,u])^\top u - \tfrac12\operatorname{ad}(y)\operatorname{ad}(u)^\top u\\ &= -\tfrac12[\operatorname{ad}(y)^\top, \operatorname{ad}(u)^\top]u - \tfrac12\operatorname{ad}(y)\operatorname{ad}(u)^\top u.\end{aligned}$$

Inserting we get the desired result:

$$\begin{aligned}\nabla_{\partial_t}\nabla_{\partial_t}y + \mathcal{R}(y,u)u &= y_{tt} + \operatorname{ad}(u)^\top y_t + \alpha(u)y_t - \operatorname{ad}(u)y_t\\ &\quad - [\operatorname{ad}(y)^\top + \operatorname{ad}(y), \operatorname{ad}(u)^\top]u.\end{aligned}$$

46.9. The weak symplectic structure on the space of Jacobi fields. Let us assume now that the geodesic equation in $\mathfrak{g}$

$$u_t = -\operatorname{ad}(u)^\top u$$

admits a unique solution for some time interval, depending smoothly on the choice of the initial value $u(0)$. Furthermore, we assume that G is a regular Lie group (see (38.4)) so that each smooth curve u in $\mathfrak{g}$ is the right logarithmic derivative (see (38.1)) of a curve $\operatorname{evol}^G(u) = g$ in G, depending smoothly on u. Let us also assume that Jacobi fields exist on the same time interval on which u exists, depending uniquely on the initial values $y(0)$ and $y_t(0)$. So the space of Jacobi fields is isomorphic to $\mathfrak{g}\times\mathfrak{g}$.

There is the well known symplectic structure on the space $\mathcal{J}_u$ of all Jacobi fields along a fixed geodesic with velocity field u. It is given by the following expression which is constant in time t:

$$\begin{aligned}\sigma(y,z) &:= \langle y, \nabla_{\partial_t}z\rangle - \langle\nabla_{\partial_t}y, z\rangle\\ &= \langle y, z_t + \tfrac12\operatorname{ad}(u)^\top z + \tfrac12\alpha(u)z - \tfrac12\operatorname{ad}(u)z\rangle\\ &\quad - \langle y_t + \tfrac12\operatorname{ad}(u)^\top y + \tfrac12\alpha(u)y - \tfrac12\operatorname{ad}(u)y, z\rangle\\ &= \langle y, z_t\rangle - \langle y_t, z\rangle + \langle[u,y],z\rangle - \langle y,[u,z]\rangle - \langle[y,z],u\rangle\\ &= \langle y, z_t - \operatorname{ad}(u)z + \tfrac12\alpha(u)z\rangle - \langle y_t - \operatorname{ad}(u)y + \tfrac12\alpha(u)y, z\rangle.\end{aligned}$$

It is a nice exercise to derive directly from the equation of Jacobi fields (46.7.1) that $\sigma(y,z)$ is indeed constant in t: plug in all definitions and use the Jacobi equation (for the Lie bracket).

46.10. Geodesics and curvature on $\mathrm{Diff}(S^1)$ **revisited.** We consider again the Lie groups $\mathrm{Diff}(\mathbb{R})$ and $\mathrm{Diff}(S^1)$ with Lie algebras $\mathfrak{X}_c(\mathbb{R})$ and $\mathfrak{X}(S^1)$ where the Lie bracket $[X,Y]=X'Y-XY'$ is the negative of the usual one. For the inner product $\langle X,Y\rangle=\int X(x)Y(x)\,dx$ integration by parts gives

$$\langle [X,Y],Z\rangle=\int_{\mathbb{R}}(X'YZ-XY'Z)dx=\int_{\mathbb{R}}(2X'YZ+XYZ')dx=\langle Y,\mathrm{ad}(X)^\top Z\rangle,$$

which in turn gives rise to

$$\begin{aligned}\mathrm{ad}(X)^\top Z&=2X'Z+XZ',\\ \alpha(X)Z&=2Z'X+ZX',\\ (\mathrm{ad}(X)^\top+\mathrm{ad}(X))Z&=3X'Z,\\ (\mathrm{ad}(X)^\top-\mathrm{ad}(X))Z&=X'Z+2XZ'=\alpha(X)Z.\end{aligned}$$

The last equation means that $-\frac12\alpha(X)$ is the skew-symmetrization of $\mathrm{ad}(X)$ with respect to to the inner product $\langle\ ,\ \rangle$. From the theory of symmetric spaces one then expects that $-\frac12\alpha$ is a Lie algebra homomorphism and indeed one can check that

$$-\tfrac12\alpha([X,Y])=[-\tfrac12\alpha(X),-\tfrac12\alpha(Y)]$$

holds. From (46.4.2) we get the same geodesic equation as in (46.3.4):

$$u_t=-\mathrm{ad}(u)^\top u=-3u_xu.$$

Using the above relations and the curvature formula (46.6.2) the curvature becomes

$$\begin{aligned}\mathcal{R}(X,Y)Z&=-X''YZ+XY''Z-2X'YZ'+2XY'Z'=-2[X,Y]Z'-[X,Y]'Z.\\ &=-\alpha([X,Y])Z\end{aligned}$$

If we change the framing of the tangent bundle by

$$X=h\circ f^{-1},\quad X'=\left(\frac{h_x}{f_x}\right)\circ f^{-1},\quad X''=\left(\frac{h_{xx}f_x-h_xf_{xx}}{f_x^3}\right)\circ f^{-1},$$

and similarly for $Y=k\circ f^{-1}$ and $Z=\ell\circ f^{-1}$, then $(\mathcal{R}(X,Y)Z)\circ f$ coincides with formula (46.3.3) for the curvature.

46.11. Jacobi fields on $\mathrm{Diff}(S^1)$**.** A Jacobi field y on $\mathrm{Diff}(S^1)$ along a geodesic g with velocity field u is a solution of the partial differential equation (46.7.1), which in our case becomes

$$\text{(1)}\qquad\begin{aligned}y_{tt}&=[\mathrm{ad}(y)^\top+\mathrm{ad}(y),\mathrm{ad}(u)^\top]u-\mathrm{ad}(u)^\top y_t-\alpha(u)y_t+\mathrm{ad}(u)y_t\\ &=-3u^2y_{xx}-4uy_{tx}-2u_xy_t,\\ u_t&=-3u_xu.\end{aligned}$$

Since the geodesic equation has solutions, locally in time, see the hint in (46.3), and since $\mathrm{Diff}(S^1)$ and $\mathrm{Diff}(\mathbb{R})$ is a regular Lie group (see (43.1)), the space of all Jacoby fields exists and is isomorphic to $C^\infty(S^1,\mathbb{R})^2$ or $C^\infty_c(\mathbb{R},\mathbb{R})^2$, respectively. The weak symplectic structure on it is given by (46.9):

$$\begin{aligned}\sigma(y,z)&=\langle y,z_t-\tfrac12u_xz+2uz_x\rangle-\langle y_t-\tfrac12u_xy+2uy_x,z\rangle\\ \text{(2)}\qquad&=\int_{S^1\text{ or }\mathbb{R}}(yz_t-y_tz+2u(yz_x-y_xz))\,dx.\end{aligned}$$

46.12. Geodesics on the Virasoro-Bott group. For $\varphi \in \operatorname{Diff}^+(S^1)$ let $\varphi' : S^1 \to \mathbb{R}^+$ be the mapping given by $T_x\varphi.\partial_x = \varphi'(x)\partial_x$. Then

$$c : \operatorname{Diff}^+(S^1) \times \operatorname{Diff}^+(S^1) \to \mathbb{R}$$
$$c(\varphi,\psi) := \int_{S^1} \log(\varphi\circ\psi)' d\log\psi' = \int_{S^1} \log(\varphi'\circ\psi) d\log\psi'$$

is a smooth group cocycle with $c(\varphi,\varphi^{-1}) = 0$, and $S^1 \times \operatorname{Diff}^+(S^1)$ becomes a Lie group $S^1 \times_c \operatorname{Diff}(S^1)$ with the operations

$$\begin{pmatrix}\varphi\\ a\end{pmatrix}.\begin{pmatrix}\psi\\ b\end{pmatrix} = \begin{pmatrix}\varphi\circ\psi\\ ab\, e^{2\pi i c(\varphi,\psi)}\end{pmatrix}, \quad \begin{pmatrix}\varphi\\ a\end{pmatrix}^{-1} = \begin{pmatrix}\varphi^{-1}\\ a^{-1}\end{pmatrix}.$$

The Lie algebra of this Lie group turns out to be $\mathbb{R} \times_\omega \mathfrak{X}(S^1)$ with the bracket

$$\left[\begin{pmatrix}h\\ a\end{pmatrix},\begin{pmatrix}k\\ b\end{pmatrix}\right] = \begin{pmatrix}h'k - hk'\\ \omega(h,k)\end{pmatrix},$$

where $\omega : \mathfrak{X}(S^1) \times \mathfrak{X}(S^1) \to \mathbb{R}$ is the Lie algebra cocycle

$$\omega(h,k) = \omega(h)k = \int_{S^1} h'dk' = \int_{S^1} h'k''dx = \tfrac{1}{2}\int_{S^1} \det\begin{pmatrix}h' & k'\\ h'' & k''\end{pmatrix} dx,$$

a generator of the bounded Chevalley cohomology $H^2(\mathfrak{X}(S^1),\mathbb{R})$. Note that the Lie algebra cocycle makes sense on the Lie algebra $\mathfrak{X}_c(\mathbb{R})$ of all vector fields with compact support on $\mathbb{R}$, but it does not integrate to a group cocycle on $\operatorname{Diff}(\mathbb{R})$. The following considerations also make sense on $\mathfrak{X}_c(\mathbb{R})$. Note also that $H^2(\mathfrak{X}_c(M),\mathbb{R}) = 0$ for each finite dimensional manifold of dimension ≥ 2 (see [Fuks, 1984]), which blocks the way to find a higher dimensional analogue of the Korteweg – de Vrieß equation in a way similar to that sketched below. We shall use the following inner product on $\mathfrak{X}(S^1)$:

$$\left\langle\begin{pmatrix}h\\ a\end{pmatrix},\begin{pmatrix}k\\ b\end{pmatrix}\right\rangle := \int_{S^1} hk\,dx + a.b.$$

Integrating by parts we get

$$\begin{aligned}\left\langle \operatorname{ad}\begin{pmatrix}h\\ a\end{pmatrix}\begin{pmatrix}k\\ b\end{pmatrix},\begin{pmatrix}\ell\\ c\end{pmatrix}\right\rangle &= \left\langle\begin{pmatrix}h'k - hk'\\ \omega(h,k)\end{pmatrix},\begin{pmatrix}\ell\\ c\end{pmatrix}\right\rangle\\ &= \int_{S^1}(h'k\ell - hk'\ell + ch'k'')\,dx = \int_{S^1}(2h'\ell + h\ell' + ch''')k\,dx\\ &= \left\langle\begin{pmatrix}k\\ b\end{pmatrix}, \operatorname{ad}\begin{pmatrix}h\\ a\end{pmatrix}^\top\begin{pmatrix}\ell\\ c\end{pmatrix}\right\rangle,\end{aligned}$$

$$\operatorname{ad}\begin{pmatrix}h\\ a\end{pmatrix}^\top\begin{pmatrix}\ell\\ c\end{pmatrix} = \begin{pmatrix}2h'\ell + h\ell' + ch'''\\ 0\end{pmatrix},$$

so that in matrix notation we have (where $\partial := \partial_x$)

$$\begin{aligned}
\operatorname{ad}\begin{pmatrix} h \\ a \end{pmatrix} &= \begin{pmatrix} h' - h\partial & 0 \\ \omega(h) & 0 \end{pmatrix}, \\
\operatorname{ad}\begin{pmatrix} h \\ a \end{pmatrix}^\top &= \begin{pmatrix} 2h' + h\partial & h''' \\ 0 & 0 \end{pmatrix}, \\
\alpha\begin{pmatrix} h \\ a \end{pmatrix} = \operatorname{ad}\begin{pmatrix} \ \end{pmatrix}^\top \begin{pmatrix} h \\ a \end{pmatrix} &= \begin{pmatrix} h' + 2h\partial + a\partial^3 & 0 \\ 0 & 0 \end{pmatrix}, \\
\operatorname{ad}\begin{pmatrix} h \\ a \end{pmatrix}^\top + \operatorname{ad}\begin{pmatrix} h \\ a \end{pmatrix} &= \begin{pmatrix} 3h' & h''' \\ \omega(h) & 0 \end{pmatrix}, \\
\operatorname{ad}\begin{pmatrix} h \\ a \end{pmatrix}^\top - \operatorname{ad}\begin{pmatrix} h \\ a \end{pmatrix} &= \begin{pmatrix} h' + 2h\partial & h''' \\ -\omega(h) & 0 \end{pmatrix}.
\end{aligned}$$

From (46.4.2) we see that the geodesic equation on the Virasoro-Bott group is

$$\begin{pmatrix} u_t \\ a_t \end{pmatrix} = -\operatorname{ad}\begin{pmatrix} u \\ a \end{pmatrix}^\top \begin{pmatrix} u \\ a \end{pmatrix} = \begin{pmatrix} -3u'u - au''' \\ 0 \end{pmatrix},$$

so that c is a constant in time, and finally the geodesic equation is the periodic Korteweg-De Vrieß equation

$$u_t + 3u_x u + au_{xxx} = 0.$$

If we use $\mathfrak{X}_c(\mathbb{R})$ we get the usual Korteweg-De Vrieß equation.

46.13. The curvature. Now we compute the curvature. Recall from (46.12) the matrices $\operatorname{ad}\binom{h}{a}^\top$, $\alpha\binom{h}{a}$, and $\operatorname{ad}\binom{h}{a}$ whose entries are integro-differential operators, and insert them into formula (46.6.2). For the computation recall that the matrix is applied to vectors of the form $\binom{\ell}{c}$ where c a constant. Then we see that $4\mathcal{R}\big(\binom{h}{a}, \binom{k}{b}\big)$ is the following 2×2-matrix whose entries are integro-differential operators:

$$\begin{pmatrix}
\begin{matrix} 4(h_1h_2'' - h_1''h_2) + 2(a_1h_2^{(4)} - a_2h_1^{(4)}) \\ +(8(h_1h_2' - h_1'h_2) + 10(a_1h_2''' - a_2h_1'''))\partial \\ +18(a_1h_2'' - a_2h_1'')\partial^2 \\ +(12(a_1h_2' - a_2h_1') + 2\omega(h_1,h_2))\partial^3 \\ -h_1'''\omega(h_2) + h_2'''\omega(h_1) \end{matrix} & \begin{matrix} 2(h_1'''h_2' - h_1'h_2''') \\ +2(h_1h_2^{(4)} - h_1^{(4)}h_2) \\ +(a_1h_2^{(6)} - a_2h_1^{(6)}) \end{matrix} \\
\begin{matrix} \omega(h_2)(4h_1' + 2h_1\partial + a_1\partial^3) \\ -\omega(h_1)(4h_2' + 2h_2\partial + a_2\partial^3) \end{matrix} & 0
\end{pmatrix}.$$

This leads to the following expression for the sectional curvature:

$$\left\langle 4\mathcal{R}\left(\binom{h_1}{a_1}, \binom{h_2}{a_2}\right)\binom{h_1}{a_1}, \binom{h_2}{a_2}\right\rangle =$$

$$
\begin{aligned}
&= \int_{S^1} \Big(4(h_1h_2'' - h_1''h_2)h_1h_2 + 8(h_1h_2' - h_1'h_2)h_1'h_2 \\
&\qquad + 2(a_1h_2^{(4)} - a_2h_1^{(4)})h_1h_2 + 10(a_1h_2''' - a_2h_1''')h_1'h_2 \\
&\qquad + 18(a_1h_2'' - a_2h_1'')h_1''h_2 \\
&\qquad + 12(a_1h_2' - a_2h_1')h_1'''h_2 + 2\omega(h_1,h_2)h_1'''h_2 \\
&\qquad - h_1'''\omega(h_2,h_1)h_2 + h_2'''\omega(h_1,h_1)h_2 \\
&\qquad + 2(h_1'''h_2' - h_1'h_2''')a_1h_2 \\
&\qquad + 2(h_1h_2^{(4)} - h_1^{(4)}h_2)a_1h_2 \\
&\qquad + (a_1h_2^{(6)} - a_2h_1^{(6)})a_1h_2 \\
&\qquad + (4h_1'h_1h_2''' + 2h_1h_1'h_2''' + a_1h_1'''h_2''' \\
&\qquad\qquad - 4h_2'h_1h_1''' - 2h_2h_1'h_1''' - a_2h_1'''h_1''')a_2 \Big)\, dx \\
&= \int_{S^1} \Big(-4[h_1,h_2]^2 + 4(a_1h_2 - a_2h_1)(h_1h_2^{(4)} - h_1'h_2''' + h_1'''h_2' - h_1^{(4)}h_2) \\
&\qquad - (h_2''')^2a_1^2 + 2h_1'''h_2'''a_1a_2 - (h_1''')^2a_2^2 \Big)\, dx \\
&\quad +3\omega(h_1,h_2)^2.
\end{aligned}
$$

46.14. Jacobi fields. A Jacobi field $y = \binom{y}{b}$ along a geodesic with velocity field $\binom{u}{a}$ is a solution of the partial differential equation (46.7.1) which in our case looks as follows.

$$
\begin{aligned}
\begin{pmatrix} y_{tt} \\ b_{tt} \end{pmatrix} &= \left[\operatorname{ad}\begin{pmatrix} y \\ b \end{pmatrix}^\top + \operatorname{ad}\begin{pmatrix} y \\ b \end{pmatrix}, \operatorname{ad}\begin{pmatrix} u \\ a \end{pmatrix}^\top \right] \begin{pmatrix} u \\ a \end{pmatrix} \\
&\quad - \operatorname{ad}\begin{pmatrix} u \\ a \end{pmatrix}^\top \begin{pmatrix} y_t \\ b_t \end{pmatrix} - \alpha\begin{pmatrix} u \\ a \end{pmatrix}\begin{pmatrix} y_t \\ b_t \end{pmatrix} + \operatorname{ad}\begin{pmatrix} u \\ a \end{pmatrix}\begin{pmatrix} y_t \\ b_t \end{pmatrix} \\
&= \left[\begin{pmatrix} 3y_x & y_{xxx} \\ \omega(y) & 0 \end{pmatrix}, \begin{pmatrix} 2u_x + u\partial_x & u_{xxx} \\ 0 & 0 \end{pmatrix} \right] \begin{pmatrix} u \\ a \end{pmatrix} \\
&\quad + \begin{pmatrix} -2u_x - 4u\partial_x - a\partial_x^3 & -u_{xxx} \\ \omega(u) & 0 \end{pmatrix} \begin{pmatrix} y_t \\ b_t \end{pmatrix},
\end{aligned}
$$

which leads to

$$
\begin{aligned}
(1) \qquad y_{tt} &= -u(4y_{tx} + 3uy_{xx} + ay_{xxxx}) - u_x(2y_t + 2ay_{xxx}) \\
&\qquad - u_{xxx}(b_t + \omega(y,u) - 3ay_x) - ay_{txxx}, \\
(2) \qquad b_{tt} &= \omega(u, y_t) + \omega(y, 3u_xu) + \omega(y, au_{xxx}).
\end{aligned}
$$

Let us consider first equation (2):

$$
\begin{aligned}
(2') \qquad b_{tt} &= \omega(u, y_t) + \omega(y, 3u_xu) + \omega(y, au_{xxx}) \\
&= \int_{S^1} (-y_{txxx}u + y_{xxx}(3u_xu + au_{xxx}))dx
\end{aligned}
$$

Next we consider the disturbing integral term in equation (1), and using the geodesic equation for u we check that its derivative with respect to t equals equation (2'), so it is a constant:

$$(3) \qquad b_t + \omega(y,u) = b_t + \int_{S^1} y_{xxx} u\, dx =: B_1 \quad \text{since}$$

$$b_{tt} + \int_{S^1} (y_{txxx}u + y_{xxx}u_t)\, dx = b_{tt} + \int_{S^1} (y_{txxx}u + y_{xxx}(-3u_x u - a u_{xxx}))\, dx = 0.$$

Note that $b(t)$ can be explicitly solved as

$$(4) \qquad b(t) = B_0 + B_1 t - \int_a^t \int_{S^1} y_{xxx} u\, dx\, dt.$$

The first line of the Jacobi equation on the Virasoro-Bott group is a genuine partial differential equation and we get the following system of equations:

$$(5) \qquad \begin{aligned} y_{tt} &= -u(4y_{tx} + 3uy_{xx} + ay_{xxxx}) - u_x(2y_t + 2ay_{xxx}) \\ &\quad + (3ay_x - B_1)u_{xxx} - ay_{txxx}, \\ u_t &= -3u_x u - au_{xxx}, \\ a &= \text{ constant}, \end{aligned}$$

where $u(t,x), y(t,x)$ are either smooth functions in $(t,x) \in I \times S^1$ or in $(t,x) \in I \times \mathbb{R}$, where I is an interval or $\mathbb{R}$, and where in the latter case u, y, y_t have compact support with respect to x.

46.15. The weak symplectic structure on the space of Jacobi fields on the Virasoro Lie algebra. Since the Korteweg - de Vrieß equation has local solutions depending smoothly on the initial conditions (and global solutions if $a \neq 0$), the space of all Jacobi fields exists and is isomorphic to $(\mathbb{R} \times_\omega \mathfrak{X}(S^1)) \times (\mathbb{R} \times_\omega \mathfrak{X}(S^1))$. The weak symplectic structure is given by (46.9):

$$\begin{aligned} \sigma\left(\begin{pmatrix} y \\ b \end{pmatrix}, \begin{pmatrix} z \\ c \end{pmatrix}\right) &= \left\langle \begin{pmatrix} y \\ b \end{pmatrix}, \begin{pmatrix} z_t \\ c_t \end{pmatrix}\right\rangle - \left\langle \begin{pmatrix} y_t \\ b_t \end{pmatrix}, \begin{pmatrix} z \\ c \end{pmatrix}\right\rangle + \left\langle \left[\begin{pmatrix} u \\ a \end{pmatrix}, \begin{pmatrix} y \\ b \end{pmatrix}\right], \begin{pmatrix} z \\ c \end{pmatrix}\right\rangle \\ &\quad - \left\langle \begin{pmatrix} y \\ b \end{pmatrix}, \left[\begin{pmatrix} u \\ a \end{pmatrix}, \begin{pmatrix} z \\ b \end{pmatrix}\right]\right\rangle - \left\langle \left[\begin{pmatrix} y \\ b \end{pmatrix}, \begin{pmatrix} z \\ c \end{pmatrix}\right], \begin{pmatrix} u \\ a \end{pmatrix}\right\rangle \\ &= \int_{S^1 \text{ or } \mathbb{R}} (yz_t - y_t z + 2u(yz_x - y_x z))\, dx \\ &\quad + b(c_t + \omega(z,u)) - (b_t + \omega(y,u)) - a\omega(y,z) \end{aligned}$$

$$(1) \qquad \begin{aligned} &= \int_{S^1 \text{ or } \mathbb{R}} (yz_t - y_t z + 2u(yz_x - y_x z))\, dx \\ &\quad + bC_1 - B_1 c - \int_{S^1 \text{ or } \mathbb{R}} ay'z''a\, dx. \end{aligned}$$

Complements to Manifolds of Mappings

For a compact smooth finite dimensional manifold M, some results on the topological type of the diffeomorphism group $\operatorname{Diff}(M)$ are available in the literature. In [Smale, 1959] it is shown that $\operatorname{Diff}(S^2)$ is homotopy equivalent to $O(3,\mathbb{R})$. This result has been extended in [Hatcher, 1983] where it is shown that $\operatorname{Diff}(S^3)$ is homotopy equivalent to $O(4,\mathbb{R})$. The component group $\pi_0(\operatorname{Diff}_+(S^n))$ of the group of orientation preserving diffeomorphisms on the sphere S^n is isomorphic to the group of homotopy spheres of dimension $n+1$ for $n > 4$, see [Kervaire, Milnor, 1963]. If M is a product of spheres then $\pi_0(\operatorname{Diff}(M))$ has been computed by [Browder, 1967] and [Turner, 1969]. For a simply connected orientable compact manifold M in [Sullivan, 1978] it is shown that $\pi_0(\operatorname{Diff}(M))$ is commensurable to an arithmetic group, where two groups are said to be *commensurable* if there is a finite sequence of homomorphisms of groups between them with finite kernel and cokernel. By [Borel, Harish-Chandra, 1962], any arithmetic group is finitely presented, and by [Borel, Serre, 1973] it is even of *finite type*, which means that its classifying space is homotopy equivalent to a CW-complex with finitely many cells in each dimension.

Two diffeomorphisms f, g of M are called *pseudo-isotopic* if there is a diffeomorphism $F : M \times I \to M \times I$ restricting to f and g at the two ends of $M \times I$, respectively. Let $\mathcal{D}(M)$ be the group of pseudo-isotopy classes of diffeomorphisms of M, a quotient of $\operatorname{Diff}(M)/\operatorname{Diff}_0(M)$, where $\operatorname{Diff}_0(M)$ is the connected component. By [Cerf, 1970], if M is simply connected then $\pi_0(\operatorname{Diff}(M)) = \mathcal{D}(M)$, whereas for non simply connected M there is in general an abelian kernel A in an exact sequence

$$0 \to A \to \pi_0(\operatorname{Diff}(M)) \to \mathcal{D}(M) \to 0,$$

and A has been computed by [Hatcher, Wagoner, 1973] and [Igusa, 1984]. In particular, A is finitely generated if $\pi_0(M)$ is finite.

In [Triantafillou, 1994] the following result is announced: If M is a smooth compact orientable manifold of dimension ≥ 5 with finite fundamental group, then $\mathcal{D}(M)$ is commensurable to an arithmetic group. Moreover, $\pi_0(M)$ is of finite type.

Chapter X
Further Applications

In section (47) we show how to treat direct limit manifolds like $S^\infty = \varinjlim S^n$ or $SO(\infty, \mathbb{R}) = \varinjlim SO(n, \mathbb{R})$ as real analytic manifolds modeled on $\mathbb{R}^\infty = \mathbb{R}^{(\mathbb{N})} = \bigoplus_{\mathbb{N}} \mathbb{R}$. As topological spaces these are often used in algebraic topology, in particular the Grassmannians as classifying spaces of the groups $SO(k)$. The differential calculus is very well applicable, and the groups $GL(\infty, \mathbb{R})$, $SO(\infty, \mathbb{R})$ turn out to be regular Lie groups, where the exponential mapping is even locally a diffeomorphism onto a neighborhood of the identity, since it factors over finite dimensional exponential mappings.

In section (48) we consider a manifold with a closed 2-form σ inducing an injective (but in general not surjective) mapping $\sigma : TM \to T^*M$. This is called a weak symplectic manifold, and there are difficulties in defining the Poisson bracket for general smooth functions. We describe a natural subspace of functions for which the Poisson bracket makes sense, and which admit Hamiltonian vector fields.

For a (unitary) representation of a (finite dimensional) Lie group G in a Hilbert space H one wishes to have the infinitesimal representation of the Lie algebra at disposal. Classically this is given by unbounded operators and offers analytical difficulties. We show in (49.5) and (49.10) that the dense subspaces H_∞ and H_ω of smooth and real analytic vectors are invariant convenient vector spaces on which the action $G \times H_\infty \to H_\infty$ is smooth (resp. real analytic). These are well known results. Our proofs are transparent and surprisingly simple; they use, however, the uniform boundedness principles (5.18) and (11.12). Using this and the results from section (48), we construct the moment mapping of any unitary representation.

Section (50) on perturbation theory of operators is devoted to the background and proof of theorem (50.16) which says that a smooth curve of unbounded selfadjoint operators on Hilbert space with compact resolvent admits smooth parameterizations of its eigenvalues and eigenvectors, under some condition. The real analytic version of this theorem is due to [Rellich, 1940], see also [Kato, 1976, VII, 3.9], with formally stronger notions of real analyticity which are quite difficult to handle.

Again the power of convenient calculus shows in the ease with which this result is derived.

In section (51) we present a version of one of the hard implicit function theorems, which is applicable to some non-linear partial differential equations. Its origins are the result of John Nash about the existence of isometric embeddings of Riemannian manifolds into $\mathbb{R}^n$'s, see [Nash, 1956]. It was then identified by [Moser, 1961], [Moser, 1966] as an abstract implicit function theorem, and found the most elaborate presentations in [Hamilton, 1982], and [Gromov, 1986]. But the original application about the existence of isometric embeddings was finally reproved in a very simple way by [Günther, 1989, 1990], who composed the nonlinear perturbation problem with the inverse of a Laplace operator and then applied the Banach fixed point theorem. This is characteristic for applications of hard implicit function theorems: Each serious application is incredibly complicated, and finally a simple ad hoc method solves the problem. To our knowledge of the original applications only two have not yet found direct simpler proofs: The result by Hamilton [Hamilton, 1982], that a compact Riemannian 3-manifold with positive Ricci curvature also admits a metric with constant scalar curvature; and the application on the small divisor problem in celestial mechanics. We include here the hard implicit function theorem of Nash and Moser in the form of [Hamilton, 1982], in full generality and without any loss, in condensed form but with all details.

47. Manifolds for Algebraic Topology

47.1. Convention. In this section the space $\mathbb{R}^{(\mathbb{N})}$ of all finite sequences with the direct sum topology plays a an important role. It is also denoted by $\mathbb{R}^\infty$, mainly in in algebraic topology. It is a convenient vector space. We consider it equipped with the weak inner product $\langle x, y\rangle := \sum x_i y_i$, which is bilinear and bounded, therefore smooth. It is called weak, since it is non degenerate in the following sense: the associated linear mapping $\mathbb{R}^{(\mathbb{N})} \to (\mathbb{R}^{(\mathbb{N})})' = \mathbb{R}^{\mathbb{N}}$ is injective but far from being surjective. We will also use the weak Euclidean distance $|x| := \sqrt{\langle x, x\rangle}$, whose square is a smooth function.

47.2. Example: The sphere S^∞. This is the set $\{x \in R^{(\mathbb{N})} : \langle x, x\rangle = 1\}$, the usual infinite dimensional sphere used in algebraic topology, the topological inductive limit of $S^n \subset S^{n+1} \subset \dots$. The inductive limit topology coincides with the subspace topology since clearly $\varinjlim S^n \to S^\infty \subset \mathbb{R}^{(\mathbb{N})}$ is continuous, S^∞ as closed subset of $\mathbb{R}^{(\mathbb{N})}$ with the c^∞-topology is compactly generated, and since each compact set is contained in a step of the inductive limit.

We show that S^∞ is a smooth manifold by describing an explicit smooth atlas, the *stereographic atlas*. Choose $a \in S^\infty$ ("south pole"). Let

$$U_+ := S^\infty \setminus \{a\}, \qquad u_+ : U_+ \to \{a\}^\perp, \qquad u_+(x) = \tfrac{x-\langle x,a\rangle a}{1-\langle x,a\rangle},$$
$$U_- := S^\infty \setminus \{-a\}, \qquad u_- : U_- \to \{a\}^\perp, \qquad u_-(x) = \tfrac{x-\langle x,a\rangle a}{1+\langle x,a\rangle}.$$

From an obvious drawing in the 2-plane through 0, x, and a it is easily seen that u_+ is the usual stereographic projection. We also get

$$u_+^{-1}(y) = \tfrac{|y|^2-1}{|y|^2+1}a + \tfrac{2}{|y|^2+1}y \qquad \text{for } y \in \{a\}^\perp \setminus \{0\}$$

and $(u_- \circ u_+^{-1})(y) = \frac{y}{|y|^2}$. The latter equation can directly be seen from the drawing using the intersection theorem.

The two stereographic charts above can be extended to charts on open sets in $\mathbb{R}^{(\mathbb{N})}$ in such a way that S^∞ becomes a splitting submanifold of $\mathbb{R}^{(\mathbb{N})}$:

$$\begin{aligned}\tilde{u}_+ &: \mathbb{R}^{(\mathbb{N})} \setminus [0,+\infty)a \to a^\perp + (-1,+\infty)a \\ \tilde{u}_+(z) &:= u_+(\frac{z}{|z|}) + (|z|-1)a \\ &= (1 + \langle z, a\rangle)u_+^{-1}(z - \langle z,a\rangle a)\end{aligned}$$

Since the model space $\mathbb{R}^{(\mathbb{N})}$ of S^∞ has the bornological approximation property by (28.6), and is reflexive, by (28.7) the operational tangent bundle of S^∞ equals the kinematic one: $DS^\infty = TS^\infty$.

We claim that TS^∞ is diffeomorphic to $\{(x,v) \in S^\infty \times \mathbb{R}^{(\mathbb{N})} : \langle x, v\rangle = 0\}$.
The $X_x \in T_xS^\infty$ are exactly of the form $c'(0)$ for a smooth curve $c : \mathbb{R} \to S^\infty$ with $c(0) = x$ by (28.13). Then $0 = \frac{d}{dt}|_0 \langle c(t), c(t)\rangle = 2\langle x, X_x\rangle$. For $v \in x^\perp$ we use $c(t) = \cos(|v|t)x + \sin(|v|t)\frac{v}{|v|}$.

The construction of S^∞ works for any positive definite bounded bilinear form on any convenient vector space.

The sphere is smoothly contractible, by the following argument: We consider the homotopy $A : \mathbb{R}^{(\mathbb{N})} \times [0,1] \to \mathbb{R}^{(\mathbb{N})}$ through isometries which is given by $A_0 = \mathrm{Id}$ and by (44.22)

$$\begin{aligned}A_t(a_0, a_1, a_2, \dots) = (a_0, \dots, a_{n-2}, a_{n-1}\cos\theta_n(t), a_{n-1}\sin\theta_n(t),& \\ a_n\cos\theta_n(t), a_n\sin\theta_n(t), a_{n+1}\cos\theta_n(t), a_{n+1}\sin\theta_n(t), \dots)&\end{aligned}$$

for $\frac{1}{n+1} \le t \le \frac{1}{n}$, where $\theta_n(t) = \varphi(n((n+1)t-1))\frac{\pi}{2}$ for a fixed smooth function $\varphi : \mathbb{R} \to \mathbb{R}$ which is 0 on $(-\infty, 0]$, grows monotonely to 1 in $[0,1]$, and equals 1 on $[1,\infty)$. The mapping A is smooth since it maps smooth curves (which locally map into some $\mathbb{R}^N$) to smooth curves (which then locally have values in $\mathbb{R}^{2N}$). Then $A_{1/2}(a_0, a_1, a_2, \dots) = (a_0, 0, a_1, 0, a_2, 0, \dots)$ is in $\mathbb{R}^{(\mathbb{N})}_{\text{even}}$, and on the other hand $A_1(a_0, a_1, a_2, \dots) = (0, a_0, 0, a_1, 0, a_2, 0, \dots)$ is in $\mathbb{R}^{(\mathbb{N})}_{\text{odd}}$. This is a variant of a homotopy constructed by [Ramadas, 1982]. Now $A_t|S^\infty$ for $0 \le t \le 1/2$ is a smooth isotopy on S^∞ between the identity and $A_{1/2}(S^\infty) \subset \mathbb{R}^{(\mathbb{N})}_{\text{even}}$. The latter set is contractible in a chart.

One may prove in a simpler way that S^∞ is contractible with a real analytic homotopy with one corner: roll all coordinates one step to the right and then contract in the stereographic chart opposite to $(1, 0, \dots)$.

47.3. Example. The Grassmannians and the Stiefel manifolds. The *Grassmann manifold* $G(k,\infty;\mathbb{R}) = G(k,\infty)$ is the set of all k-dimensional linear subspaces of the space of all finite sequences $\mathbb{R}^{(\mathbb{N})}$. The *Stiefel manifold of orthonormal k-frames* $O(k,\infty;\mathbb{R}) = O(k,\infty)$ is the set of all linear isometries $\mathbb{R}^k \to \mathbb{R}^{(\mathbb{N})}$, where the latter space is again equipped with the standard weak inner product described at the beginning of (47.2). The *Stiefel manifold of all k-frames* $GL(k,\infty;\mathbb{R}) = GL(k,\infty;\mathbb{R})$ is the set of all injective linear mappings $\mathbb{R}^k \to \mathbb{R}^{(\mathbb{N})}$.

There is a canonical *transposition mapping* $(\quad)^t : L(\mathbb{R}^k, \mathbb{R}^{(\mathbb{N})}) \to L(\mathbb{R}^{(\mathbb{N})}, \mathbb{R}^k)$ which is given by

$$A^t : \mathbb{R}^{(\mathbb{N})} \xrightarrow{\text{incl}} \mathbb{R}^{\mathbb{N}} = \left(\mathbb{R}^{(\mathbb{N})}\right)' \xrightarrow{A'} (\mathbb{R}^k)' = \mathbb{R}^k$$

and satisfies $\langle A^t(x), y\rangle = \langle x, A(y)\rangle$. The transposition mapping is bounded and linear, so it is real analytic. Then we have

$$GL(k,\infty) = \{A \in L(\mathbb{R}^k, \mathbb{R}^{(\mathbb{N})}) : A^t \circ A \in GL(k)\},$$

since $A^t \circ A \in GL(k)$ if and only if $\langle Ax, Ay\rangle = \langle A^t Ax, y\rangle = 0$ for all y implies $x = 0$, which is equivalent to A injective. So in particular $GL(k,\infty)$ is open in $L(\mathbb{R}^k, \mathbb{R}^{(\mathbb{N})})$. The Lie group $GL(k)$ acts freely from the right on the space $GL(k,\infty)$. Two elements of $GL(k,\infty)$ lie in the same orbit if and only if they have the same image in $\mathbb{R}^{(\mathbb{N})}$. We have a surjective mapping $\pi : GL(k,\infty) \to G(k,\infty)$, given by $\pi(A) = A(\mathbb{R}^k)$, where the inverse images of points are exactly the $GL(k)$-orbits. Similarly, we have

$$O(k,\infty) = \{A \in L(\mathbb{R}^k, \mathbb{R}^{(\mathbb{N})}) : A^t \circ A = \mathrm{Id}_k\}.$$

The Lie group $O(k)$ of all isometries of $\mathbb{R}^k$ acts freely from the right on the space $O(k,\infty)$. Two elements of $O(k,\infty)$ lie in the same orbit if and only if they have the same image in $\mathbb{R}^{(\mathbb{N})}$. The projection $\pi : GL(k,\infty) \to G(k,\infty)$ restricts to a surjective mapping $\pi : O(k,\infty) \to G(k,\infty)$, and the inverse images of points are now exactly the $O(k)$-orbits.

47.4. Lemma. Iwasawa decomposition. *Let $T(k;\mathbb{R}) = T(k)$ be the group of all upper triangular $k \times k$-matrices with positive entries on the main diagonal. Then each $B \in GL(k,\infty)$ can be written in the form $B = p(B) \circ q(B)$, with unique $p(B) \in O(k,\infty)$ and $q(B) \in T(k)$. The mapping $q : GL(k,\infty) \to T(k)$ is real analytic, and $p : GL(k,\infty) \to O(k,\infty) \to GL(k,\infty)$ is real analytic, too.*

Proof. We apply the Gram Schmidt orthonormalization procedure to the vectors $B(e_1), \dots, B(e_k) \in \mathbb{R}^{(\mathbb{N})}$. The coefficients of this procedure form an upper triangular $k \times k$-matrix $q(B)$ whose entries are rational functions of the inner products $\langle B(e_i), B(e_j)\rangle$ and are positive on the main diagonal. So $(B \circ q(B)^{-1})(e_1), \dots, (B \circ q(B)^{-1})(e_k)$ is the orthonormalized frame $p(B)(e_1), \dots, p(B)(e_k)$. □

47.5. Theorem. *The following are real analytic principal fiber bundles:*

$$(\pi : O(k,\infty;\mathbb{R}) \to G(k,\infty;\mathbb{R}), O(k,\mathbb{R})),$$
$$(\pi : GL(k,\infty;\mathbb{R}) \to G(k,\infty;\mathbb{R}), GL(k,\mathbb{R})),$$
$$(p : GL(k,\infty;\mathbb{R}) \to O(k,\infty;\mathbb{R}), T(k;\mathbb{R})).$$

The last one is trivial. The embeddings $\mathbb{R}^n \to \mathbb{R}^{(\mathbb{N})}$ induce real analytic embeddings, which respect the principal right actions of all the structure groups

$$O(k,n) \to O(k,\infty),$$
$$GL(k,n) \to GL(k,\infty),$$
$$G(k,n) \to G(k,\infty).$$

All these cones are inductive limits in the category of real analytic (and smooth) manifolds. All manifolds are smoothly paracompact.

Proof. Step 1. $G(k,\infty)$ is a real analytic manifold.
For $A \in O(k,\infty)$ we consider the open subset $W_A := \{B \in GL(k,\infty) : A^t \circ B \in GL(k)\}$ of $L(\mathbb{R}^k, \mathbb{R}^{(\mathbb{N})})$, and we let $V_A := W_A \cap O(k,\infty) = \{B \in O(k,\infty) : A^t \circ B \in GL(k)\}$. Obviously, V_A is invariant under the action of $O(k)$ and $V_{AU} = V_A$ for $U \in O(k)$. So we may denote $U_{\pi(A)} := \pi(V_A)$. Let $P := \pi(A) = A(\mathbb{R}^k) \in G(k,\infty)$. We define the mapping

$$\begin{aligned} v_A &: V_A \to L(P, P^\perp), \\ v_A(B) &:= \left(B(A^tB)^{-1}A^t - AA^t\right)|P \\ &= \left((\mathrm{Id}_{\mathbb{R}^{(\mathbb{N})}} - AA^t)B(A^tB)^{-1}A^t\right)|P. \end{aligned}$$

In order to visualize this definition note that $A \circ A^t$ is the orthonormal projection $\mathbb{R}^{(\mathbb{N})} \to P$, and that the image of B in $\mathbb{R}^{(\mathbb{N})} = P \oplus P^\perp$ is the graph of $v_A(B)$. It is easily checked that $v_A(B) \in L(P, P^\perp)$ and that $v_A(BU) = v_A(B) = v_{AU}(B)$ for all $U \in O(k)$. So we may define

$$\begin{aligned} u_P &: U_P \to L(P, P^\perp), \\ u_P(\pi(B)) &:= v_A(B). \end{aligned}$$

For $C \in L(P, P^\perp)$ the mapping $A + C \circ A$ is a parameterization of the graph of C, it is in $GL(k,\infty)$, and we have (using p from lemma (47.3)) that $u_P^{-1}(C) = \pi(p(A+CA))$, since for $B \in V_A$ the image of B equals the graph of $C = u_P(\pi(B))$, which in turn is equal to $(A+CA)(\mathbb{R}^k) = (A+CA)\,q(A+CA)^{-1}(\mathbb{R}^k) = p(A+CA)(\mathbb{R}^k)$.

Now we check the chart changes: Let $P_1 = \pi(A_1)$, $P_2 = \pi(A_2)$, and $C \in L(P_1, P_1^\perp)$, then we have

$$\left(u_{P_2} \circ u_{P_1}^{-1}\right)(C) = (\mathrm{Id}_{\mathbb{R}^{(\mathbb{N})}} - A_2\,A_2^t)\,p(A_1 + C\,A_1)\,\left(A_2^t\,p(A_1 + C\,A_1)\right)^{-1}\, A_2^t|P_2,$$

which is defined on the open set of all $C \in L(P_1, P_1^\perp)$ for which $A_2^t\, p(A_1 + C\, A_1)$ is in $GL(k)$ and which is real analytic there.

Step 2. The principal bundles.
We fix $A \in O(k,\infty)$ and consider the section

$$\begin{aligned} s_A &: U_{\pi(A)} \to V_A, \\ s_A(Q) &:= p(A + u_{\pi(A)}(Q)\, A) \end{aligned}$$

and the principal fiber bundle chart

$$\begin{aligned} \psi_A &: V_A \to U_{\pi(A)} \times O(k), \\ \psi_A(B) &:= \left(\pi(B), s_A(\pi(B))^t\, B\right), \\ \psi_A^{-1}(Q,U) &= s_A(Q)\, U. \end{aligned}$$

Clearly, these charts give a principal fiber bundle atlas with cocycle of transition functions $Q \mapsto s_{A_2}(Q)^t\, s_{A_1}(Q) \in O(k)$.

The same formulas (for A still in $O(k,\infty)$) give fiber bundle charts $\psi_A : W_A \to U_{\pi(A)} \times GL(k)$ for $GL(k,\infty) \to G(k,\infty)$.

The injection $O(k,\infty) \to GL(k,\infty)$ is a real analytic section of the real analytic projection $p : GL(k,\infty) \to O(k,\infty)$, which by lemma (47.4) gives a trivial principal fiber bundle with structure group $T(k)$. This fact implies that $O(k,\infty)$ is a splitting real analytic submanifold of the convenient vector space $L(\mathbb{R}^k, \mathbb{R}^{(\mathbb{N})})$.

Since $\mathbb{R}^{(\mathbb{N})}$ is the inductive limit of the direct summands $\mathbb{R}^n$ in the category of convenient vector spaces and real analytic (smooth) mappings, and since the chart constructions above restrict to the usual ones on the finite dimensional Grassmannians and bundles, the assertion on the inductive limits follows.

All these manifolds are smoothly paracompact. For $\mathbb{R}^{(\mathbb{N})}$ this is in (16.10), so it holds for $L(\mathbb{R}^k, \mathbb{R}^{(\mathbb{N})})$ and for the closed subspace $O(k,\infty)$, see (47.3). Then it follows for $G(k,\infty)$ since $O(k,\infty) \to G(k,\infty)$ is a principal fiber bundle with compact structure group $O(k)$, by integrating the members of the partition over the fiber. Then we get the result for $GL(k,\infty)$ by bundle argumentation on $GL(k,\infty) \to G(k,\infty)$, since the fiber $GL(k)$ is finite dimensional, so the product is well behaved by (4.16). □

47.6. Theorem. *The principal bundle $(O(k,\infty), \pi, G(k,\infty))$ is classifying for finite dimensional principal $O(k)$-bundles and carries a universal real analytic $O(k)$-connection $\omega \in \Omega^1(O(k,\infty), \mathfrak{o}(k))$.*

This means: For each finite dimensional smooth or real analytic principal $O(k)$-bundle $P \to M$ with principal connection ω_P there is a smooth or real analytic mapping $f : M \to G(k,\infty)$ such that the pullback $O(k)$-bundle $f^*O(k,\infty)$ is isomorphic to P and the pullback connection $f^*\omega$ equals ω_P via this isomorphism.

For ∞ replaced by a large N and bundles where the dimension of the base is bounded this is due to [Schlafli, 1980].

Proof. Step 1. The tangent bundle of $O(k,\infty)$ is given by

$$TO(k,\infty) = \{(A,X) \in O(k,\infty) \times L(\mathbb{R}^k, \mathbb{R}^{(\mathbb{N})}) : X^t A \in \mathfrak{o}(k)\}.$$

We have $O(k,\infty) = \{A : A^t A = \mathrm{Id}_k\}$, thus $T_A O(k,\infty) \subseteq \{X : X^t A + A^t X = 0\}$. Since $A^t A = \mathrm{Id}_k$ is an equation of constant rank when restricted to $GL(k,n)$ for finite n, we have equality by the implicit function theorem.

Another argument which avoids the implicit function theorem is the following. By theorem (47.5) the vertical tangent space $\{AZ : Z \in \mathfrak{t}(k)\}$ at $A \in O(k,\infty)$ of the bundle $GL(k,\infty) \to O(k,\infty)$ is transversal to $T_A O(k,\infty)$, where $\mathfrak{t}(k)$ is the Lie algebra of $T(k)$. We have equality since an easy computation shows that $\{AZ : Z \in \mathfrak{t}(k)\} \cap \{X : X^t A + A^t X = 0\} = 0$.

Step 2. The inner product on $\mathbb{R}^k$ and the weak inner product on $\mathbb{R}^{(\mathbb{N})}$ induce a bounded weak inner product on the space $L(\mathbb{R}^k, \mathbb{R}^{(\mathbb{N})})$ by $\langle X, Y\rangle = \mathrm{Trace}(X^t Y) = \mathrm{Trace}(XY^t)$, where the second trace makes sense since XY^t has finite dimensional range. With respect to this inner product we consider the orthonormal projection $\Phi_A : T_A O(k,\infty) \to V_A O(k,\infty)$ onto the vertical tangent space $V_A O(k,\infty) = \{AY : Y \in \mathfrak{o}(k)\}$ of $O(k,\infty) \to G(k,\infty)$. Its kernel, the horizontal space, turns out to be

$$\begin{aligned}\{Z \in & T_A O(k,\infty) : \mathrm{Trace}(Z^t AY) = 0 \text{ for all } Y \in \mathfrak{o}(k)\} = \\ &= \{Z : Z^t A \text{ both skew and symmetric}\} \\ &= \{Z \in L(\mathbb{R}^k, \mathbb{R}^{(\mathbb{N})}) : Z^t A = 0\} \\ &= \{Z : Z(\mathbb{R}^k) \perp A(\mathbb{R}^k)\}.\end{aligned}$$

So $\Phi_A : T_A O(k,\infty) \to V_A O(k,\infty)$ turns out to be $\Phi_A(Z) = AA^t Z = A\omega_A(Z)$, where $\omega_A(Z) := A^t Z = -Z^t A$. Then $\omega \in \Omega^1(O(k,\infty), \mathfrak{o}(k))$ is an $O(k)$-equivariant form which reproduces the generators of fundamental vector fields of the principal right action, so it is a principal $O(k)$-connection (37.19):

$$\begin{aligned}((r^U)^*\omega)_A(Z) &= U^t A^t Z U = Ad(U^{-1})\omega_A(Z), \\ \omega_A(AY) &= Y \text{ for } Y \in \mathfrak{o}(k).\end{aligned}$$

Step 3. The classifying process.
Let $(p : P \to M, O(k))$ be a principal bundle with a principal connection form $\omega_P \in \Omega^1(P, \mathfrak{o}(k))$. We consider the obvious representation of $O(k)$ on $\mathbb{R}^k$ and the associated vector bundle $E = P[\mathbb{R}^k] = P \times_{O(k)} \mathbb{R}^k$ with its induced fiber Riemannian metric g_v and induced linear connection ∇.

Now we choose a Riemannian metric g_M on the base manifold M, which we pull back to the horizontal bundle $\mathrm{Hor}^\nabla E$ (with respect to ∇) in TE via the fiberwise isomorphism $Tp|\mathrm{Hor}^\nabla E : \mathrm{Hor}^\nabla E \to TM$. Then we use the vertical lift $\mathrm{vl}_E : E \times_M E \to VE \subset TE$ to heave the fiber metric g_v to the vertical bundle. Finally, we declare the horizontal and the vertical bundle to be orthogonal, and thus we get a Riemannian metric $g_E := (Tp|\mathrm{Hor}^\nabla E)^* g_M \oplus (\mathrm{vpr}_E)^* g_v$ on the total space

E. By the theorem of [Nash, 1956] (see also [Günther, 1989] and [Gromov, 1986]), there is an isometric embedding (which can be chosen real analytic, if all data are real analytic) $i : (E, g_E) \to \mathbb{R}^N$ into some high dimensional Euclidean space which in turn is contained in $\mathbb{R}^{(\mathbb{N})}$. Let $j := d_v i(0_E) : E \to \mathbb{R}^{(\mathbb{N})}$ be the vertical derivative along the zero section of E which is given by $d_v i(0_x)(u_x) = \frac{d}{dt}|_0 i(tu_x)$. Then $j : E \to \mathbb{R}^{(\mathbb{N})}$ is a fiber linear smooth mapping which is isometric on each fiber.

Let us now identify the principal bundle P with the orthonormal frame bundle $O(\mathbb{R}^k, (E, g_v))$ of its canonically associated Riemannian vector bundle. Then $j_* : P \ni u \mapsto j \circ u \in O(k, \infty)$ defines a smooth mapping which is $O(k)$-equivariant and therefore fits into the following pullback diagram

$$\begin{array}{ccc} P & \xrightarrow{j_*} & O(k,\infty) \\ {\scriptstyle p}\downarrow & & \downarrow{\scriptstyle \pi} \\ M & \xrightarrow{f} & G(k,\infty). \end{array}$$

The factored smooth mapping $f : M \to G(k, \infty)$ is therefore classifying for the bundle P, so that $f^*O(k, \infty) \cong P$.

In order to show that the canonical connection is pulled back to the given one we consider again the associated Riemannian vector bundle $E \to M$ from above. Note that $Ti(\operatorname{Hor}^\nabla E)$ is orthogonal to $Ti(VE)$ in $\mathbb{R}^{(\mathbb{N})}$, and we have to check that this is still true for Tj, see (37.26). This is a local question on M, so let $E = U \times \mathbb{R}^n$, then we have as in (29.9)

$$\begin{aligned} T(U \times \mathbb{R}^n) = U \times \mathbb{R}^n \times \mathbb{R}^m \times \mathbb{R}^n \ni (x, v; \xi, \omega) \mapsto& \\ \xrightarrow{Ti} (i(x,v), d_1 i(x,v).\xi + d_2 i(x,v)\omega) &\in \mathbb{R}^{(\mathbb{N})} \times \mathbb{R}^{(\mathbb{N})} \\ \xrightarrow{Tj} (d_2 i(x,0)v, d_1 d_2 i(x,0)(\xi, v) &+ d_2 i(x,0)\omega), \\ V(U \times \mathbb{R}^n) \ni (x, v; 0, \omega) \xrightarrow{Ti} (i(x,v), d_2 i(x,v)\omega),& \\ \xrightarrow{Tj} (d_2 i(x,0)v, d_2 i(x,0)\omega),& \\ \operatorname{Hor}^\nabla(U \times \mathbb{R}^n) \ni (x, v; \xi, \Gamma_x(\xi, v)) \mapsto (i(x,v), d_1 i(x,v).\xi &+ d_2 i(x,v)\Gamma_x(\xi, v)), \\ \xrightarrow{Tj} (d_2 i(x,0)v, d_1 d_2 i(x,0)(\xi, v) &+ d_2 i(x,0)\Gamma_x(\xi, v)), \\ 0 = \langle d_2 i(x,v)\omega,\ d_1 i(x,v).\xi + d_2 i(x,v)\Gamma_x(\xi, v)\rangle &\text{ for all } \xi, \omega. \end{aligned}$$

In the last equation we replace v and ω both by tv and apply $\partial_t|_0$ to get the required result

$$\begin{aligned} 0 &= \langle d_2 i(x,tv)tv,\ d_1 i(x,tv).\xi + d_2 i(x,tv)\Gamma_x(\xi, tv)\rangle \text{ for all } \xi, \omega. \\ 0 &= \langle d_2 i(x,0)v,\ d_2 d_1 i(x,0).(v,\xi) + d_2 i(x,0)\Gamma_x(\xi, v)\rangle \text{ for all } \xi, \omega. \quad \square \end{aligned}$$

47.7. The Lie group $GL(\infty;\mathbb{R})$. The canonical embeddings $\mathbb{R}^n \to \mathbb{R}^{n+1}$ onto the first n coordinates induce injections $GL(n) \to GL(n+1)$. The inductive limit is given by

$$GL(\infty;\mathbb{R}) = GL(\infty) := \varinjlim_{n\to\infty} GL(n)$$

in the category of sets or groups. Since each $GL(n)$ also injects into $L(\mathbb{R}^{(\mathbb{N})}, \mathbb{R}^{(\mathbb{N})})$ we can visualize $GL(\infty)$ as the set of all $\mathbb{N}\times\mathbb{N}$-matrices which are invertible and differ from the identity in finitely many entries only.

We also consider the Lie algebra $\mathfrak{gl}(\infty)$ of all $\mathbb{N}\times\mathbb{N}$-matrices with only finitely many nonzero entries, which is isomorphic to $\mathbb{R}^{(\mathbb{N}\times\mathbb{N})}$, and we equip it with this convenient vector space structure. Then

$$\mathfrak{gl}(\infty) = \varinjlim_{n\to\infty} \mathfrak{gl}(n)$$

in the category of real analytic mappings, since it is a regular inductive limit in the category of bounded linear mappings.

Claim. $\mathfrak{gl}(\infty) = L(\mathbb{R}^{\mathbb{N}}, \mathbb{R}^{(\mathbb{N})})$ *as convenient vector spaces. Composition is a bounded bilinear mapping on* $\mathfrak{gl}(\infty)$. *The transposition*

$$A \mapsto A^t = A' \circ i : \mathbb{R}^{\mathbb{N}} \to (\mathbb{R}^{\mathbb{N}})'' \to (\mathbb{R}^{\mathbb{N}})' = \mathbb{R}^{(\mathbb{N})}$$

on the space $L(\mathbb{R}^{\mathbb{N}}, \mathbb{R}^{(\mathbb{N})})$ *induces a bibounded linear isomorphism of* $\mathfrak{gl}(\infty)$ *which resembles the usual transposition of matrices.*

Proof. Let $T \in L(\mathbb{R}^{\mathbb{N}}, \mathbb{R}^{(\mathbb{N})})$. Then $T \in L(\mathbb{R}^{\mathbb{N}}, \mathbb{R}^{\mathbb{N}}) = \left(\mathbb{R}^{(\mathbb{N})}\right)^{\mathbb{N}}$ and hence is a matrix with finitely many non zero entries in every line. Since T has values in $\mathbb{R}^{(\mathbb{N})}$, there are also only finitely many non zero entries in each column, since $T(e_j) \in \mathbb{R}^{(\mathbb{N})}$. Suppose that T is not in $\mathfrak{gl}(\infty)$. Then the matrix of T has infinitely many nonzero entries, so there are $T^{i_k}_{j_k} \neq 0$ for $i_k \nearrow \infty$ and $j_k \nearrow \infty$ and such that j_k is the last index with nonzero entry in the line i_k. Now one can choose inductively an element $(x^i) \in \mathbb{R}^{\mathbb{N}}$ with $T(x) \notin \mathbb{R}^{(\mathbb{N})}$, a contradiction. For both spaces the evaluations $ev_{i,j}$ generate the convenient vector space structure by (5.18), so the convenient structures coincide.

Another argument leading to this conclusion is the following: Since both spaces are nuclear we have for the injective tensor product

$$L(\mathbb{R}^{\mathbb{N}}, \mathbb{R}^{(\mathbb{N})}) \cong \mathbb{R}^{(\mathbb{N})} \hat{\otimes}_\varepsilon \mathbb{R}^{(\mathbb{N})}.$$

By the same reason the injective and the projective tensor product coincide. Since both spaces are (DF), separately continuous bilinear functionals are jointly continuous, so the latter space coincides with the bornological tensor product $\mathbb{R}^{(\mathbb{N})} \tilde{\otimes}_\beta \mathbb{R}^{(\mathbb{N})}$, which commutes with direct sums, since it is a left adjoint functor, so finally we get $\mathbb{R}^{(\mathbb{N}\times\mathbb{N})}$.

Composition is bounded since it can be written as

$$L(\mathbb{R}^{\mathbb{N}},\mathbb{R}^{(\mathbb{N})})\times L(\mathbb{R}^{\mathbb{N}},\mathbb{R}^{(\mathbb{N})})\to L(\mathbb{R}^{\mathbb{N}},\mathbb{R}^{(\mathbb{N})})\times L(\mathbb{R}^{\mathbb{N}},\mathbb{R}^{\mathbb{N}})\xrightarrow{\text{comp}} L(\mathbb{R}^{\mathbb{N}},\mathbb{R}^{(\mathbb{N})}).$$

The assertion about transposition is obvious, using $L(\mathbb{R}^{\mathbb{N}},(\mathbb{R}^{\mathbb{N}})')\cong L^2(\mathbb{R}^{\mathbb{N}};\mathbb{R})$. □

Then the (convenient) affine space

$$\operatorname{Id}+\mathfrak{gl}(\infty)=\varinjlim_{n\to\infty}(\operatorname{Id}+\mathfrak{gl}(n))$$

is closed under composition, which is real analytic on it. The determinant is a real analytic function there, too.

Now obviously $GL(\infty)=\{A\in\operatorname{Id}+\mathfrak{gl}(\infty):\det(A)\neq 0\}$, so $GL(\infty)$ is an open subset in $\operatorname{Id}+\mathfrak{gl}(\infty)$ and is thus a real analytic manifold, in fact, it is the inductive limit of all the groups $GL(n)=\{A\in\operatorname{Id}_\infty+\mathfrak{gl}(n):\det(A)\neq 0\}$ in the category of real analytic manifolds.

We consider the *Killing form on* $\mathfrak{gl}(\infty)$, which is given by the trace

$$k(X,Y):=\operatorname{tr}(XY)\quad\text{for } X,Y\in\mathfrak{gl}(\infty).$$

This is the right concept, since for each n and $X,Y\in\mathfrak{gl}(n)\subset\mathfrak{gl}(\infty)$ we have

$$\operatorname{tr}_{\mathbb{R}^{(\mathbb{N})}}(XY)=\operatorname{tr}_{\mathbb{R}^n}(XY)=\tfrac{1}{2n}\left(\operatorname{tr}_{\mathfrak{gl}(n)}(\operatorname{ad}(X)\operatorname{ad}(Y))+2\operatorname{tr}_{\mathbb{R}^{(\mathbb{N})}}(X)\operatorname{tr}_{\mathbb{R}^{(\mathbb{N})}}(Y)\right),$$

but $\operatorname{ad}(X)\operatorname{ad}(Y)\in L(\mathbb{R}^{(\mathbb{N})},\mathbb{R}^{(\mathbb{N})})$ is not of trace class. We have the following short exact sequence of Lie algebras and Lie algebra homomorphisms

$$0\to\mathfrak{sl}(\infty)\to\mathfrak{gl}(\infty)\xrightarrow{\operatorname{tr}}\mathbb{R}\to 0.$$

It splits, using $t\mapsto\frac{t}{n}\cdot\operatorname{Id}_{\mathbb{R}^n}$ for an arbitrary n, but $\mathfrak{gl}(\infty)$ has no nontrivial abelian ideal $\mathfrak{a}$, since we would have $\mathfrak{a}\cap\mathfrak{gl}(n)\subset\mathbb{R}\cdot\operatorname{Id}_n$ for every n. So $\mathfrak{gl}(\infty)$ is only the semidirect product of $\mathbb{R}$ with the ideal $\mathfrak{sl}(\infty)$ and not the direct product.

47.8. Theorem. *$GL(\infty)$ is a real analytic regular Lie group modeled on $\mathbb{R}^{(\mathbb{N})}$ with Lie algebra $\mathfrak{gl}(\infty)$ and is the inductive limit of the Lie groups $GL(n)$ in the category of real analytic manifolds. The exponential mapping is well defined, real analytic, and a local real analytic diffeomorphism onto a neighborhood of the identity. The Campbell-Baker-Hausdorff formula gives a real analytic mapping near 0 and expresses the multiplication on $GL(\infty)$ via* exp. *The determinant* $\det: GL(\infty)\to\mathbb{R}\setminus 0$ *is a real analytic homomorphism. We have a real analytic left action $GL(\infty)\times\mathbb{R}^{(\mathbb{N})}\to\mathbb{R}^{(\mathbb{N})}$, such that $\mathbb{R}^{(\mathbb{N})}\setminus 0$ is one orbit, but the injection $GL(\infty)\hookrightarrow L(\mathbb{R}^{(\mathbb{N})},\mathbb{R}^{(\mathbb{N})})$ does not generate the topology.*

Proof. Since the exponential mappings are compatible with the inductive limits and are diffeomorphisms on open balls with radius π in norms in which the Lie brackets are submultiplicative, all these assertions follow from the inductive limit property. One may use the double of the operator norms.

Regularity is proved as follows: A smooth curve $X:\mathbb{R}\to\mathfrak{gl}(\infty)$ factors locally in $\mathbb{R}$ into some $\mathfrak{gl}(n)$, and we may integrate this piece of the resulting right invariant time dependent vector field on $GL(n)$. □

47.9. Theorem. *Let $\mathfrak{g}$ be a Lie subalgebra of $\mathfrak{gl}(\infty)$. Then there is a smoothly arcwise connected splitting regular Lie subgroup G of $GL(\infty)$ whose Lie algebra is $\mathfrak{g}$. The exponential mapping of $GL(\infty)$ restricts to that of G, which is a local real analytic diffeomorphism near zero. The Campbell-Baker-Hausdorff formula gives a real analytic mapping near 0 and has the usual properties, also on G.*

Proof. Let $\mathfrak{g}_n := \mathfrak{g} \cap \mathfrak{gl}(n)$, a finite dimensional Lie subalgebra of $\mathfrak{g}$. Then $\bigcup \mathfrak{g}_n = \mathfrak{g}$. The convenient structure $\mathfrak{g} = \varinjlim_n \mathfrak{g}_n$ coincides with the structure inherited as a complemented subspace, since $\mathfrak{gl}(\infty)$ carries the finest locally convex structure.

So for each n there is a connected Lie subgroup $G_n \subset GL(n)$ with Lie algebra $\mathfrak{g}_n$. Since $\mathfrak{g}_n \subset \mathfrak{g}_{n+1}$ we have $G_n \subset G_{n+1}$, and we may consider $G := \bigcup_n G_n \subset GL(\infty)$. Each $g \in G$ lies in some G_n and may be connected to Id via a smooth curve there, which is also smooth curve in G, so G is smoothly arcwise connected.

All mappings $\exp|\mathfrak{g}_n : \mathfrak{g}_n \to G_n$ are local real analytic diffeomorphisms near 0, so $\exp : \mathfrak{g} \to G$ is also a local real analytic diffeomorphism near zero onto an open neighborhood of the identity in G. A similar argument applies to evol so that G is regular. The rest is clear. □

47.10. Examples. In the following we list some of the well known examples of simple infinite dimensional Lie groups which fit into the picture treated in this section. The reader can easily continue this list, especially by complex versions.

The Lie group $SL(\infty)$ is the inductive limit

$$\begin{aligned} SL(\infty) &= \{A \in GL(\infty) : \det(A) = 1\} \\ &= \varinjlim_{n\to\infty} SL(n) \subset GL(\infty), \end{aligned}$$

the connected Lie subgroup with Lie algebra $\mathfrak{sl}(\infty) = \{X \in \mathfrak{gl}(\infty) : \operatorname{tr}(X) = 0\}$.

The Lie group $SO(\infty, \mathbb{R})$ is the inductive limit

$$\begin{aligned} SO(\infty) &= \{A \in GL(\infty) : \langle Ax, Ay\rangle = \langle x, y\rangle \text{ for all } x, y \in \mathbb{R}^{(\mathbb{N})} \text{ and } \det(A) = 1\} \\ &= \varinjlim_{n\to\infty} SO(n) \subset GL(\infty), \end{aligned}$$

the connected Lie subgroup of $GL(\infty)$ with the Lie algebra $\mathfrak{o}(\infty) = \{X \in \mathfrak{gl}(\infty) : X^t = -X\}$ of skew elements.

The Lie group $O(\infty)$ is the inductive limit

$$\begin{aligned} O(\infty) &= \{A \in GL(\infty) : \langle Ax, Ay\rangle = \langle x, y\rangle \text{ for all } x, y \in \mathbb{R}^{(\mathbb{N})}\} \\ &= \varinjlim_{n\to\infty} O(n) \subset GL(\infty). \end{aligned}$$

It has two connected components, that of the identity is $SO(\infty)$.

The Lie group $Sp(\infty,\mathbb{R})$ is the inductive limit

$$\begin{aligned} Sp(\infty,\mathbb{R}) &= \{A \in GL(\infty) : A^t J A = J\} \\ &= \varinjlim_{n\to\infty} Sp(2n,\mathbb{R}) \subset GL(\infty), \text{ where} \end{aligned}$$

$$J = \begin{pmatrix} 0 & 1 & & & \\ -1 & 0 & & & \\ & & 0 & 1 & \\ & & -1 & 0 & \\ & & & & \ddots \end{pmatrix} \in L(\mathbb{R}^{(\mathbb{N})}, \mathbb{R}^{(\mathbb{N})}).$$

It is the connected Lie subgroup of $GL(\infty)$ with the Lie algebra $\mathfrak{sp}(\infty,\mathbb{R}) = \{X \in \mathfrak{gl}(\infty) : X^t J + JX = 0\}$ of symplectically skew elements.

47.11. Theorem. *The following manifolds are real analytically diffeomorphic to the homogeneous spaces indicated:*

$$\begin{aligned} GL(k,\infty) &\cong GL(\infty) \Big/ \begin{pmatrix} \mathrm{Id}_k & L(\mathbb{R}^k, \mathbb{R}^{\infty-k}) \\ 0 & GL(\infty-k) \end{pmatrix}, \\ O(k,\infty) &\cong O(\infty)/(\mathrm{Id}_k \times O(\infty-k)), \\ G(k,\infty) &\cong O(\infty)/(O(k)\times O(\infty-k)). \end{aligned}$$

The universal vector bundle $(E(k,\infty),\pi,G(k,\infty),\mathbb{R}^k)$ *is defined as the associated bundle*

$$\begin{aligned} E(k,\infty) &= O(k,\infty)[\mathbb{R}^k] \\ &= \{(Q,x) : x \in Q\} \subset G(k,\infty)\times \mathbb{R}^{(\mathbb{N})}. \end{aligned}$$

The tangent bundle of the Grassmannian is then given by

$$TG(k,\infty) = L(E(k,\infty), E(k,\infty)^\perp).$$

Proof. This is a direct consequence of the chart construction of $G(k,\infty)$. □

48. Weak Symplectic Manifolds

48.1. Review. For a finite dimensional symplectic manifold (M,σ) we have the following exact sequence of Lie algebras, see also (45.7):

$$0 \to H^0(M) \to C^\infty(M,\mathbb{R}) \xrightarrow{\mathrm{grad}^\sigma} \mathfrak{X}(M,\sigma) \xrightarrow{\gamma} H^1(M) \to 0.$$

Here $H^*(M)$ is the real De Rham cohomology of M, the space $C^\infty(M,\mathbb{R})$ is equipped with the Poisson bracket $\{\ ,\ \}$, $\mathfrak{X}(M,\sigma)$ consists of all vector fields ξ with $\mathcal{L}_\xi\sigma = 0$ (the locally Hamiltonian vector fields), which is a Lie algebra for the

Lie bracket. Furthermore, $\operatorname{grad}^\sigma f$ is the Hamiltonian vector field for $f \in C^\infty(M,\mathbb{R})$ given by $i(\operatorname{grad}^\sigma f)\sigma = df$ and $\gamma(\xi) = [i_\xi \sigma]$. The spaces $H^0(M)$ and $H^1(M)$ are equipped with the zero bracket.

Given a symplectic left action $\ell : G \times M \to M$ of a connected Lie group G on M, the first partial derivative of ℓ gives a mapping $\ell' : \mathfrak{g} \to \mathfrak{X}(M,\sigma)$ which sends each element X of the Lie algebra $\mathfrak{g}$ of G to the fundamental vector field. This is a Lie algebra homomorphism.

$$\begin{array}{ccccccc} H^0(M) & \xrightarrow{\;i\;} & C^\infty(M,\mathbb{R}) & \xrightarrow{\;\operatorname{grad}^\sigma\;} & \mathfrak{X}(M,\sigma) & \xrightarrow{\;\gamma\;} & H^1(M) \\ & & \nwarrow{\scriptstyle\chi} & & \nearrow{\scriptstyle\ell'} & & \\ & & & \mathfrak{g} & & & \end{array}$$

A linear lift $\chi : \mathfrak{g} \to C^\infty(M,\mathbb{R})$ of ℓ' with $\operatorname{grad}^\sigma \circ \chi = \ell'$ exists if and only if $\gamma \circ \ell' = 0$ in $H^1(M)$. This lift χ may be changed to a Lie algebra homomorphism if and only if the 2-cocycle $\bar\chi : \mathfrak{g} \times \mathfrak{g} \to H^0(M)$, given by $(i \circ \bar\chi)(X,Y) = \{\chi(X), \chi(Y)\} - \chi([X,Y])$, vanishes in $H^2(\mathfrak{g}, H^0(M))$, for if $\bar\chi = \delta\alpha$ then $\chi - i \circ \alpha$ is a Lie algebra homomorphism.

If $\chi : \mathfrak{g} \to C^\infty(M,\mathbb{R})$ is a Lie algebra homomorphism, we may associate the *moment mapping* $\mu : M \to \mathfrak{g}' = L(\mathfrak{g},\mathbb{R})$ to it, which is given by $\mu(x)(X) = \chi(X)(x)$ for $x \in M$ and $X \in \mathfrak{g}$. It is G-equivariant for a suitably chosen (in general affine) action of G on $\mathfrak{g}'$. See [Weinstein, 1977] or [Libermann, Marle, 1987] for all this.

48.2. We now want to carry over to infinite dimensional manifolds the procedure of (48.1). First we need the appropriate notions in infinite dimensions. So let M be a manifold, which in general is infinite dimensional.

A 2-form $\sigma \in \Omega^2(M)$ is called a *weak symplectic structure* on M if it is closed ($d\sigma = 0$) and if its associated vector bundle homomorphism $\sigma^\vee : TM \to T^*M$ is injective.

A 2-form $\sigma \in \Omega^2(M)$ is called a *strong symplectic structure* on M if it is closed ($d\sigma = 0$) and if its associated vector bundle homomorphism $\sigma^\vee : TM \to T^*M$ is invertible with smooth inverse. In this case, the vector bundle TM has reflexive fibers T_xM: Let $i : T_xM \to (T_xM)''$ be the canonical mapping onto the bidual. Skew symmetry of σ is equivalent to the fact that the transposed $(\sigma^\vee)^t = (\sigma^\vee)^* \circ i : T_xM \to (T_xM)'$ satisfies $(\sigma^\vee)^t = -\sigma^\vee$. Thus, $i = -((\sigma^\vee)^{-1})^* \circ \sigma^\vee$ is an isomorphism.

48.3. Every cotangent bundle T^*M, viewed as a manifold, carries a canonical weak symplectic structure $\sigma_M \in \Omega^2(T^*M)$, which is defined as follows (see (43.9) for the finite dimensional version). Let $\pi_M^* : T^*M \to M$ be the projection. Then the *Liouville form* $\theta_M \in \Omega^1(T^*M)$ is given by $\theta_M(X) = \langle \pi_{T^*M}(X), T(\pi_M^*)(X)\rangle$ for $X \in T(T^*M)$, where $\langle\;\;,\;\;\rangle$ denotes the duality pairing $T^*M \times_M TM \to \mathbb{R}$. Then the symplectic structure on T^*M is given by $\sigma_M = -d\theta_M$, which of course in a local chart looks like $\sigma_E((v,v'),(w,w')) = \langle w', v\rangle_E - \langle v', w\rangle_E$. The associated mapping $\sigma^\vee : T_{(0,0)}(E \times E') = E \times E' \to E' \times E''$ is given by $(v,v') \mapsto (-v', i_E(v))$, where $i_E : E \to E''$ is the embedding into the bidual. So the canonical symplectic structure on T^*M is strong if and only if all model spaces of the manifold M are reflexive.

48.4. Let M be a weak symplectic manifold. The first thing to note is that the hamiltonian mapping $\operatorname{grad}^\sigma : C^\infty(M,\mathbb{R}) \to \mathfrak{X}(M,\sigma)$ does not make sense in general, since $\sigma^\vee : TM \to T^*M$ is not invertible. Namely, $\operatorname{grad}^\sigma f = (\sigma^\vee)^{-1} \circ df$ is defined only for those $f \in C^\infty(M,\mathbb{R})$ with $df(x)$ in the image of $\sigma^\vee$ for all $x \in M$. A similar difficulty arises for the definition of the Poisson bracket on $C^\infty(M,\mathbb{R})$.

Definition. For a weak symplectic manifold (M,σ) let $T^\sigma_x M$ denote the real linear subspace $T^\sigma_x M = \sigma^\vee_x(T_xM) \subset T^*_x M = L(T_xM,\mathbb{R})$, and let us call it the *smooth cotangent space* with respect to the symplectic structure σ of M at x in view of the embedding of test functions into distributions. These vector spaces fit together to form a subbundle of T^*M which is isomorphic to the tangent bundle TM via $\sigma^\vee : TM \to T^\sigma M \subseteq T^*M$. It is in general not a splitting subbundle.

48.5. Definition. For a weak symplectic vector space (E,σ) let

$$C^\infty_\sigma(E,\mathbb{R}) \subset C^\infty(E,\mathbb{R})$$

denote the linear subspace consisting of all smooth functions $f : E \to \mathbb{R}$ such that each iterated derivative $d^k f(x) \in L^k_{\text{sym}}(E;\mathbb{R})$ has the property that

$$d^k f(x)(\quad, y_2, \dots, y_k) \in E^\sigma$$

is actually in the smooth dual $E^\sigma \subset E'$ for all $x, y_2, \dots, y_k \in E$, and that the mapping

$$\begin{gathered} \prod^k E \to E \\ (x, y_2, \dots, y_k) \mapsto (\sigma^\vee)^{-1}(df(x)(\quad, y_2, \dots, y_k)) \end{gathered}$$

is smooth. By the symmetry of higher derivatives, this is then true for all entries of $d^k f(x)$, for all x.

48.6. Lemma. *For $f \in C^\infty(E,\mathbb{R})$ the following assertions are equivalent:*

(1) *$df : E \to E'$ factors to a smooth mapping $E \to E^\sigma$.*
(2) *f has a smooth σ-gradient $\operatorname{grad}^\sigma f \in \mathfrak{X}(E) = C^\infty(E,E)$ which satisfies $df(x)y = \sigma(\operatorname{grad}^\sigma f(x), y)$.*
(3) *$f \in C^\infty_\sigma(E,\mathbb{R})$.*

Proof. Clearly, (3) $\Rightarrow$ (2) $\Leftrightarrow$ (1). We have to show that (2) $\Rightarrow$ (3). Suppose that $f : E \to \mathbb{R}$ is smooth and $df(x)y = \sigma(\operatorname{grad}^\sigma f(x), y)$. Then

$$\begin{aligned} d^k f(x)(y_1, \dots, y_k) &= d^k f(x)(y_2, \dots, y_k, y_1) \\ &= (d^{k-1}(df))(x)(y_2, \dots, y_k)(y_1) \\ &= \sigma\big(d^{k-1}(\operatorname{grad}^\sigma f)(x)(y_2, \dots, y_k), y_1\big). \quad \square \end{aligned}$$

48.7. For a weak symplectic manifold (M,σ) let

$$C^\infty_\sigma(M,\mathbb{R})\subset C^\infty(M,\mathbb{R})$$

denote the linear subspace consisting of all smooth functions $f:M\to\mathbb{R}$ such that the differential $df:M\to T^*M$ factors to a smooth mapping $M\to T^\sigma M$. In view of lemma (48.6) these are exactly those smooth functions on M which admit a smooth σ-gradient $\operatorname{grad}^\sigma f\in\mathfrak{X}(M)$. Also the condition (48.6.1) translates to a local differential condition describing the functions in $C^\infty_\sigma(M,\mathbb{R})$.

48.8. Theorem. *The Hamiltonian mapping* $\operatorname{grad}^\sigma : C^\infty_\sigma(M,\mathbb{R})\to\mathfrak{X}(M,\sigma)$, *which is given by*

$$i_{\operatorname{grad}^\sigma f}\sigma = df \quad \text{or} \quad \operatorname{grad}^\sigma f := (\sigma^\vee)^{-1}\circ df$$

is well defined. Also the Poisson bracket

$$\begin{aligned}\{\ \ ,\ \ \} &: C^\infty_\sigma(M,\mathbb{R})\times C^\infty_\sigma(M,\mathbb{R})\to C^\infty_\sigma(M,\mathbb{R}),\\ \{f,g\} &:= i_{\operatorname{grad}^\sigma f} i_{\operatorname{grad}^\sigma g}\sigma = \sigma(\operatorname{grad}^\sigma g,\operatorname{grad}^\sigma f) =\\ &= dg(\operatorname{grad}^\sigma f) = (\operatorname{grad}^\sigma f)(g)\end{aligned}$$

is well defined and gives a Lie algebra structure to the space $C^\infty_\sigma(M,\mathbb{R})$, *which also fulfills*

$$\{f,gh\}=\{f,g\}h+g\{f,h\}.$$

We have the following long exact sequence of Lie algebras and Lie algebra homomorphisms:

$$0\to H^0(M)\to C^\infty_\sigma(M,\mathbb{R})\xrightarrow{\operatorname{grad}^\sigma}\mathfrak{X}(M,\sigma)\xrightarrow{\gamma} H^1_\sigma(M)\to 0,$$

where $H^0(M)$ *is the space of locally constant functions, and*

$$H^1_\sigma(M)=\frac{\{\varphi\in C^\infty(M\leftarrow T^\sigma M): d\varphi=0\}}{\{df: f\in C^\infty_\sigma(M,\mathbb{R})\}}$$

is the first symplectic cohomology space of (M,σ), *a linear subspace of the De Rham cohomology space* $H^1(M)$.

Proof. It is clear from lemma (48.6), that the Hamiltonian mapping $\operatorname{grad}^\sigma$ is well defined and has values in $\mathfrak{X}(M,\sigma)$, since by (33.18.6) we have

$$\mathcal{L}_{\operatorname{grad}^\sigma f}\sigma = i_{\operatorname{grad}^\sigma f}d\sigma + di_{\operatorname{grad}^\sigma f}\sigma = ddf = 0.$$

By (33.18.7), the space $\mathfrak{X}(M,\sigma)$ is a Lie subalgebra of $\mathfrak{X}(M)$. The Poisson bracket is well defined as a mapping $\{\ \ ,\ \ \}: C^\infty_\sigma(M,\mathbb{R})\times C^\infty_\sigma(M,\mathbb{R})\to C^\infty(M,\mathbb{R})$, and it only remains to check that it has values in the subspace $C^\infty_\sigma(M,\mathbb{R})$.

This is a local question, so we may assume that M is an open subset of a convenient vector space equipped with a (non-constant) weak symplectic structure. So let f, $g \in C^\infty_\sigma(M,\mathbb{R})$, then we have $\{f,g\}(x) = dg(x)(\operatorname{grad}^\sigma f(x))$, and we have

$$\begin{aligned} d(\{f,g\})(x)y &= d(dg(\quad)y)(x).\operatorname{grad}^\sigma f(x) + dg(x)(d(\operatorname{grad}^\sigma f)(x)y) \\ &= d(\sigma(\operatorname{grad}^\sigma g(\quad),y)(x).\operatorname{grad}^\sigma f(x) + \sigma\Big(\operatorname{grad}^\sigma g(x), d(\operatorname{grad}^\sigma f)(x)y\Big) \\ &= \sigma\Big(d(\operatorname{grad}^\sigma g)(x)(\operatorname{grad}^\sigma f(x)) - d(\operatorname{grad}^\sigma f)(x)(\operatorname{grad}^\sigma g(x)), y\Big), \end{aligned}$$

since $\operatorname{grad}^\sigma f \in \mathfrak{X}(M,\sigma)$ and for any $X \in \mathfrak{X}(M,\sigma)$ the condition $\mathcal{L}_X\sigma = 0$ implies $\sigma(dX(x)y_1,y_2) = -\sigma(y_1, dX(x)y_2)$. So (48.6.2) is satisfied, and thus $\{f,g\} \in C^\infty_\sigma(M,\mathbb{R})$.

If $X \in \mathfrak{X}(M,\sigma)$ then $di_X\sigma = \mathcal{L}_X\sigma = 0$, so $[i_X\sigma] \in H^1(M)$ is well defined, and by $i_X\sigma = \sigma^\vee\ oX$ we even have $\gamma(X) := [i_X\sigma] \in H^1_s i(M)$, so γ is well defined.

Now we show that the sequence is exact. Obviously, it is exact at $H^0(M)$ and at $C^\infty_\sigma(M,\mathbb{R})$, since the kernel of $\operatorname{grad}^\sigma$ consists of the locally constant functions. If $\gamma(X) = 0$ then $\sigma^\vee\ oX = i_X\sigma = df$ for $f \in C^\infty_\sigma(M,\mathbb{R})$, and clearly $X = \operatorname{grad}^\sigma f$. Now let us suppose that $\varphi \in C^\infty(M \leftarrow T^\sigma M) \subset \Omega^1(M)$ with $d\varphi = 0$. Then $X := (\sigma^\vee)^{-1} \circ \varphi \in \mathfrak{X}(M)$ is well defined and $\mathcal{L}_X\sigma = di_X\sigma = d\varphi = 0$, so $X \in \mathfrak{X}(M,\sigma)$ and $\gamma(X) = [\varphi]$.

Moreover, $H^1_\sigma(M)$ is a linear subspace of $H^1(M)$ since for $\varphi \in C^\infty(M \leftarrow T^\sigma M) \subset \Omega^1(M)$ with $\varphi = df$ for $f \in C^\infty(M,\mathbb{R})$ the vector field $X := (\sigma^\vee)^{-1} \circ \varphi \in \mathfrak{X}(M)$ is well defined, and since $\sigma^\vee\ oX = \varphi = df$ by (48.6.1) we have $f \in C^\infty_\sigma(M,\mathbb{R})$ with $X = \operatorname{grad}^\sigma f$.

The mapping $\operatorname{grad}^\sigma$ maps the Poisson bracket into the Lie bracket, since by (33.18) we have

$$\begin{aligned} i_{\operatorname{grad}^\sigma\{f,g\}}\sigma &= d\{f,g\} = d\mathcal{L}_{\operatorname{grad}^\sigma f}g = \mathcal{L}_{\operatorname{grad}^\sigma f}dg = \\ &= \mathcal{L}_{\operatorname{grad}^\sigma f}i_{\operatorname{grad}^\sigma g}\sigma - i_{\operatorname{grad}^\sigma g}\mathcal{L}_{\operatorname{grad}^\sigma f}\sigma \\ &= [\mathcal{L}_{\operatorname{grad}^\sigma f}, i_{\operatorname{grad}^\sigma g}]\sigma = i_{[\operatorname{grad}^\sigma f,\operatorname{grad}^\sigma g]}\sigma. \end{aligned}$$

Let us now check the properties of the Poisson bracket. By definition, it is skew symmetric, and we have

$$\begin{aligned} \{\{f,g\},h\} &= \mathcal{L}_{\operatorname{grad}^\sigma\{f,g\}}h = \mathcal{L}_{[\operatorname{grad}^\sigma f,\operatorname{grad}^\sigma g]}h = [\mathcal{L}_{\operatorname{grad}^\sigma f}, \mathcal{L}_{\operatorname{grad}^\sigma g}]h = \\ &= \mathcal{L}_{\operatorname{grad}^\sigma f}\mathcal{L}_{\operatorname{grad}^\sigma g}h - \mathcal{L}_{\operatorname{grad}^\sigma g}\mathcal{L}_{\operatorname{grad}^\sigma f}h = \{f,\{g,h\}\} - \{g,\{f,h\}\}, \\ \{f,gh\} &= \mathcal{L}_{\operatorname{grad}^\sigma f}(gh) = (\mathcal{L}_{\operatorname{grad}^\sigma f}g)h + g\mathcal{L}_{\operatorname{grad}^\sigma f}h = \\ &= \{f,g\}h + g\{f,h\}. \end{aligned}$$

Finally, it remains to show that all mappings in the sequence are Lie algebra homomorphisms, where we put the zero bracket on both cohomology spaces. For locally constant functions we have $\{c_1,c_2\} = \mathcal{L}_{\operatorname{grad}^\sigma c_1}c_2 = 0$. We have already checked that $\operatorname{grad}^\sigma$ is a Lie algebra homomorphism. For $X, Y \in \mathfrak{X}(M,\sigma)$

$$i_{[X,Y]}\sigma = [\mathcal{L}_X, i_Y]\sigma = \mathcal{L}_X i_Y\sigma + 0 = di_X i_Y\sigma + i_X\mathcal{L}_Y\sigma = di_X i_Y\sigma$$

is exact. □

48.9. Symplectic cohomology. The reader might be curious whether there exists a symplectic cohomology in all degrees extending $H^1_\sigma(M)$ which appeared in theorem (48.8). We should start by constructing a graded differential subalgebra of $\Omega(M)$ leading to this cohomology. Let (M,σ) be a weak symplectic manifold. The first space to be considered is $C^\infty(L^k_{\text{alt}}(TM,\mathbb{R})^\sigma)$, the space of smooth sections of the vector bundle with fiber $L^k_{\text{alt}}(T_xM,\mathbb{R})^{\sigma_x}$ consisting of all bounded skew symmetric forms ω with $\omega(\quad, X_2,\dots,X_k) \in T^\sigma_x M$ for all $X_i \in T_xM$. But these spaces of sections are not stable under the exterior derivative d, one should consider C^∞_σ-sections of vector bundles. For trivial bundles these could be defined as those sections which lie in $C^\infty_\sigma(M,\mathbb{R})$ after composition with a bounded linear functional. However, this definition is not invariant under arbitrary vector bundle isomorphisms, one should require that the transition functions are also in some sense C^∞_σ. So finally M should have, in some sense, C^∞_σ chart changings.

We try now a simpler approach. Let

$$\Omega^k_\sigma(M) := M) := \{\omega \in C^\infty(L^k_{\text{alt}}(TM,\mathbb{R})^\sigma) : d\omega \in C^\infty(L^{k+1}_{\text{alt}}(TM,\mathbb{R})^\sigma)\}.$$

Since $d^2 = 0$ and the wedge product of σ-dual forms is again a σ-dual form, we get a graded differential subalgebra $(\Omega_\sigma(M), d)$, whose cohomology will be denoted by $H^k_\sigma(M)$. Note that we have

$$er\{\omega \in \Omega^k_\sigma(M) : d\omega = 0\} == \{\omega \in C^\infty(L^k_{\text{alt}}(TM,\mathbb{R})^\sigma) : d\omega = 0\},$$
$$\Omega^0_\sigma(M) = C^\infty_\sigma(M,\mathbb{R}),$$

so that $H^1_\sigma(M)$ is the same space as in theorem (48.8).

Theorem. *If (M,σ) is a smooth weakly symplectic manifold which admits smooth partitions of unity in $C^\infty_\sigma(M,\mathbb{R})$, and which admits 'Darboux charts', then the symplectic cohomology equals the De Rham cohomology: $H^k_\sigma(M) = H^k(M)$.*

Proof. We use theorem (34.6) and its method of proof. We have to check that the sheaf Ω_σ satisfies the lemma of Poincaré and admits partitions of unity. The second requirement is immediate from the assumption. For the lemma of Poincaré let $\omega \in \Omega^{k+1}_\sigma(M)$ with $d\omega = 0$, and let $u : U \to u(U) \subset E$ be a smooth chart of M with $u(U)$ a disked c^∞-open set in a convenient vector space E. We may push σ to this subset of E and thus assume that U equals this subset. By the Lemma of Poincaré (33.20), we get $\omega = d\varphi$ where

$$\varphi(x)(v_1,\dots,v_k) = \int_0^1 t^k \omega(tx)(x,v_1,\dots,v_k)\,dt,$$

which is in $\Omega^k_\sigma(M)$ if σ is a constant weak symplectic form on $u(U)$. This is the case if we may choose a 'Darboux chart' on M. □

49. Applications to Representations of Lie Groups

This section is based on [Michor, 1990], see also [Michor, 1992]

49.1. Representations. Let G be any finite or infinite dimensional smooth real Lie group, and let E be a convenient vector space. Recall that $L(E,E)$, the space of all bounded linear mappings, is a convenient vector space, whose bornology is generated by the topology of pointwise convergence for any compatible locally convex topology on E, see for example (5.18). We shall need an explicit topology below in order to define representations, so we shall use on $L(E,E)$ the topology of pointwise convergence with respect to the bornological topology on E, that of bE. Let us call this topology the *strong operator topology* on $L(E,E)$, since this is the usual name if E is a Banach space.

A *representation* of G in E is a mapping ρ from G into the space of all linear mappings from E into E which satisfies $\rho(g.h) = \rho(g).\rho(h)$ for all $g, h \in G$ and $\rho(e) = \mathrm{Id}_E$, and which fulfills the following equivalent 'continuity requirements':

(1) ρ has values in $L(E,E)$ and is continuous from the c^∞-topology on G into the strong operator topology on $L(E,E)$.

(2) The associated mapping $\rho^\wedge : G \times bE \to bE$ is separately continuous.

The equivalence of (1) and (2) is due to the fact that $L(E,E)$ consists of all continuous linear mappings $bE \to bE$.

Lemma. *If G and bE are metrizable, and if ρ locally in G takes values in uniformly continuous subsets of $L(bE,bE)$, then the continuity requirements are equivalent to*

(3) $\rho^\wedge : G \times bE \to bE$ is (jointly) continuous.

A unitary representation of a metrizable Lie group on a Hilbert space H satisfies the requirements of the lemma.

Proof. We only have to show that (1) implies (3). Since on uniformly continuous subsets of $L(bE,bE)$ the strong operator topology coincides with the compact open topology, ρ is continuous $G \to L(bE,bE)_{co}$. By cartesian closedness of the category of compactly generated topological spaces (see [Brown, 1964], [Steenrod, 1967], or [Engelking, 1989]), $\rho^\wedge$ is continuous from the Kelley-fication $k(G \times bE)$ (compare (4.7)) of the topological product to bE. Since $G \times bE$ is metrizable it is compactly generated, so $\rho^\wedge$ is continuous on the topological product, which incidentally coincides with the manifold topology of the product manifold $G \times E$, see (27.3). □

49.2. The Space of Smooth Vectors. Let $\rho : G \to L(E,E)$ be a representation. A vector $x \in E$ is called *smooth* if the mapping $G \to E$ given by $g \mapsto \rho(g)x$ is smooth. Let us denote by E_∞ the linear subspace of all smooth vectors in E. Then we have an injection $j : E_\infty \to C^\infty(G,E)$, given by $x \mapsto (g \mapsto \rho(g)x)$. We equip $C^\infty(G,E)$ with the structure of a convenient vector space as described in (27.17), i.e., the initial structure with respect to the cone $C^\infty(M,E) \xrightarrow{c^*} C^\infty(\mathbb{R},E)$ for all $c \in C^\infty(\mathbb{R},G)$.

49.3. Lemma.

(1) *The image of the embedding* $j : E_\infty \to C^\infty(G, E)$ *is the closed subspace*

$$C^\infty(G,E)^G = \{f \in C^\infty(G,E) : f \circ \mu_g = \rho(g) \circ f \text{ for all } g \in G\}$$

of all G*-equivariant mappings. So with the induced structure* E_∞ *becomes a convenient vector space.*

(2) *The space of smooth vectors* E_∞ *is an invariant linear subspace of* E*, and we have* $j(\rho(g)x) = j(x) \circ \mu^g$*, or* $j \circ \rho(g) = (\mu^g)^* \circ j$*, where* μ^g *is the right translation on* G*.*

Proof. For $x \in E_\infty$ and $g, h \in G$ we have $j(x)\mu_g(h) = j(x)(gh) = \rho(gh)x = \rho(g)\rho(h)x = \rho(g)j(x)(h)$, so $j(x) \in C^\infty(G,E)^G$. If conversely $f \in C^\infty(G,E)^G$ then $f(g) = \rho(g)f(e) = j(f(e))(g)$. Moreover, for $x \in E_\infty$ the mapping $h \mapsto \rho(h)\rho(g)x = \rho(hg)x$ is smooth, so $\rho(g)x \in E_\infty$, and we have $j(\rho(g)x)(h) = \rho(h)\rho(g)x = \rho(hg)x = j(x)(hg) = j(x)(\mu^g(h))$. $\square$

49.4. Theorem. *If the Lie group* G *is finite dimensional and separable, and if the bornologification* bE *of the representation space* E *is sequentially complete, the space of smooth vectors* E_∞ *is dense in* bE*.*

Proof. Let $x \in E$, a continuous seminorm p on bE, and $\varepsilon > 0$ be given. Let $U = \{g \in G : p(\rho(g)x - x) < \varepsilon\}$, an open neighborhood of the identity in G. Let $f_U \in C^\infty(G, \mathbb{R})$ be a nonnegative smooth function with support in U with $\int_G f_U(g) d_L g = 1$, where d_L denotes left the Haar measure on G. We consider the element $\int_G f_U(g)\rho(g)x d_L g \in bE$. Note that this Riemann integral converges since bE is sequentially complete. We have

$$\begin{aligned} p\left(\int_G f_U(g)\rho(g)x d_L g - x\right) &\le \int_G f_U(g) p(\rho(g)x - x) d_L g \\ &\le \varepsilon \int_G f_U(g) d_L g = \varepsilon. \end{aligned}$$

So it remains to show that $\int_G f_U(g)\rho(g)x d_L g \in E_\infty$. We have

$$\begin{aligned} j\left(\int_G f_U(g)\rho(g)x d_L g\right)(h) &= \rho(h) \int_G f_U(g)\rho(g)x d_L g \\ &= \int_G f_U(g)\rho(h)\rho(g)x d_L g = \int_G f_U(g)\rho(hg)x d_L g \\ &= \int_G f_U(h^{-1}g)\rho(g)x d_L g, \end{aligned}$$

which is smooth as a function of h since we may view the last integral as having values in the vector space $C^\infty(G, bE)$ with a sequentially complete topology. The integral converges there, since $g \mapsto (h \mapsto f_U(h^{-1}g))$ is smooth, thus continuous $G \to C^\infty(G, \mathbb{R})$, and we multiply it by the continuous mapping $g \mapsto \rho(g)x$, $G \to bE$. It is easy to check that multiplication is continuous $C^\infty(G, \mathbb{R}) \times bE \to C^\infty(G, bE)$ for the topologies of compact convergence in all derivatives separately of composites with smooth curves, which is again sequentially complete. One may also use the compact C^∞-topology. $\square$

49.5. Theorem. *The mappings*

$$\rho^\wedge : G \times E_\infty \to E_\infty,$$
$$\rho : G \to L(E_\infty, E_\infty)$$

are smooth.

A proof analogous to that of (49.10) below would also work here.

Proof. We first show that $\rho^\wedge$ is smooth. By lemma (49.3), it suffices to show that

$$G\times C^\infty(G,E)^G \to C^\infty(G,E)^G \to C^\infty(G,E)$$
$$(g,f) \mapsto f \circ \mu^g$$

is smooth. This is the restriction of the mapping

$$G \times C^\infty(G,E) \to C^\infty(G,E)$$
$$(g,f) \mapsto f \circ \mu^g,$$

which by cartesian closedness (27.17) is smooth if and only if the canonically associated mapping

$$G \times C^\infty(G,E) \times G \to E$$
$$(g,f,h) \mapsto f(hg) = \operatorname{ev}(f, \mu(h,g))$$

is smooth. But this is holds by (3.13), extended to the manifold G. So $\rho^\wedge$ is smooth. By cartesian closedness (27.17) again $(\rho^\wedge)^\vee : G \to C^\infty(E_\infty, E_\infty)$ is smooth, and takes values in the closed linear subspace $L(E_\infty, E_\infty)$. So $\rho : G \to L(E_\infty, E_\infty)$ is smooth, too. □

49.6. Theorem. *Let $\rho : G \to L(E,E)$ be a representation of a Lie group G. Then the semidirect product*

$$E_\infty \rtimes_\rho G$$

from (38.9) *is a Lie group and is regular if G is regular. Its evolution operator is given by*

$$\operatorname{evol}^r_{E_\infty \rtimes G}(Y,X) = \Big(\rho(\operatorname{evol}^r_G(X)) \int_0^1 \rho(\operatorname{Evol}^r_G(X)(s)^{-1}).Y(s)\,ds,\ \operatorname{evol}^r_G(X)\Big)$$

for $(Y,X) \in C^\infty(\mathbb{R}, E_\infty \times \mathfrak{g})$.

Proof. This follows directly from (38.9) and (38.10). □

49.7. The space of analytic vectors. Let G now be a real analytic finite or infinite dimensional Lie group, let again $\rho : G \to L(E,E)$ be a representation as in in (49.1). A vector $x \in E$ is called *real analytic* if the mapping $G \to E$ given by $g \mapsto \rho(g)x$ is real analytic.

Let E_ω denote the vector space of all real analytic vectors in E. Then we have a linear embedding $j : E_\omega \to C^\omega(G,E)$ into the space of real analytic mappings, given by $x \mapsto (g \mapsto \rho(g)x)$. We equip $C^\omega(G,E)$ with the convenient vector space structure described in (27.17).

49.8. Lemma.

(1) *The image of the embedding* $j : E_\omega \to C^\omega(G,E)$ *is the space*

$$C^\omega(G,E)^G = \{f \in C^\omega(G,E) : f \circ \mu_g = \rho(g) \circ f \text{ for all } g \in G\}$$

of all G-equivariant mappings, and with the induced structure E_ω becomes a convenient vector space.

(2) *The space of analytic vectors E_ω is an invariant linear subspace of E, and we have $j(\rho(g)x) = j(x) \circ \mu^g$, or $j \circ \rho(g) = (\mu^g)^* \circ j$, where μ^g is the right translation on G.*

Proof. This is a transcription of the proof of lemma (49.3), replacing smooth by real analytic. □

49.9. Theorem. *If the Lie group G is finite dimensional and separable and if the bornologification bE of the representation space E is sequentially complete, the space of real analytic vectors E_ω is dense in bE.*

See [Warner, 1972, 4.4.5.7].

Proof. Let $x \in E$, a continuous seminorm p on bE, and $\varepsilon > 0$ be given. Let $U = \{g \in G : p(\rho(g)x - x) < \varepsilon\}$, an open neighborhood of the identity in G. Let $\varphi \in C(G,\mathbb{R})$ be a continuous positive function such that $\int_G \varphi(g) d_L(g) = 2$, where d_L denotes left Haar measure on G, and $\int_{G\setminus U} \varphi(g) p(\rho(g)x - x) d_L(g) < \varepsilon$.

Let $f \in C^\omega(G,\mathbb{R})$ be a real analytic function with $\frac{1}{2}\varphi(g) < f(g) < \varphi(g)$ for all $g \in G$, which exists by [Grauert, 1958]. Then $1 < \int_G f(g) d_L(g) < 2$, so if we replace f by $f/(\int_G f(g) d_L(g))$ we get $\int_G f(g) d_L(g) = 1$ and $\int_{G\setminus U} f(g) p(\rho(g)x - x) d_L(g) < \varepsilon$. We consider the element $\int_G f(g)\rho(g)x d_L g \in bE$. This Riemann integral converges since bE is sequentially complete. We have

$$\begin{aligned} p\Big(\int_G f(g)\rho(g)x d_L g - x\Big) &\le \int_U f(g)p(\rho(g)x - x) d_L g + \int_{G\setminus U} f(g)p(\rho(g)x - x) d_L g \\ &\le \varepsilon \int_G f(g) d_L g + \varepsilon < 2\varepsilon. \end{aligned}$$

So it remains to show that $\int_G f(g)\rho(g)x d_L g \in E_\omega$. We have

$$\begin{aligned} j\left(\int_G f(g)\rho(g)x d_L g\right)(h) &= \rho(h)\int_G f(g)\rho(g)x d_L g \\ &= \int_G f(g)\rho(h)\rho(g)x d_L g = \int_G f(g)\rho(hg)x d_L g \\ &= \int_G f(h^{-1}g)\rho(g)x d_L g, \end{aligned}$$

which is real analytic as a function of h, by the following argument: We have to check that the composition with any continuous linear functional on E maps this to a real analytic function on G, which is now a question of finite dimensional analysis.

We could also apply here the method of proof used at the end of (49.4), but describing a sequentially complete compatible topology on $C^\omega(G, bE)$ requires some effort. □

49.10. Theorem. *The mapping $\rho^\wedge : G \times E_\omega \to E_\omega$ is real analytic.*

We could also use a method analogous to that of (49.5), but we rather give a variant.

Proof. By cartesian closedness of the calculus (11.18) and (27.17), it suffices to show that the canonically associated mapping

$$\rho^{\wedge\vee} : G \to C^\omega(E_\omega, E_\omega)$$

is real analytic. It takes values in the closed linear subspace $L(E_\omega, E_\omega)$ of all bounded linear operators. So it suffices to check that the mapping $\rho : G \to L(E_\omega, E_\omega)$ is real analytic. Since E_ω is a convenient space, by the real analytic uniform boundedness principle (11.12), it suffices to show that

$$G \xrightarrow{\rho} L(E_\omega, E_\omega) \xrightarrow{\mathrm{ev}_x} E_\omega$$

is real analytic for each $x \in E_\omega$. Since the structure on E_ω is induced by the embedding into $C^\omega(G, E)$, we have to check, that

$$\begin{gathered} G \xrightarrow{\rho} L(E_\omega, E_\omega) \xrightarrow{\mathrm{ev}_x} E_\omega \xrightarrow{j} C^\omega(G, E), \\ g \mapsto \rho(g) \mapsto \rho(g)x \mapsto (h \mapsto \rho(h)\rho(g)x), \end{gathered}$$

is real analytic for each $x \in E_\omega$. Again by cartesian closedness (11.18), it suffices that the associated mapping

$$\begin{gathered} G \times G \to E \\ (g, h) \mapsto \rho(h)\rho(g)x = \rho(hg)x \end{gathered}$$

is real analytic, and this is the case since x is a real analytic vector. □

49.11. The model for the moment mapping. Let now $\rho : G \to U(H)$ be a unitary representation of a Lie group G on a Hilbert space H. We consider the space of smooth vectors H_∞ as a weak symplectic Fréchet manifold, equipped with the symplectic structure σ, the restriction of the imaginary part of the Hermitian inner product $\langle\ ,\ \rangle$ on H. See section (48) for the general notion of weak symplectic manifolds. So $\sigma \in \Omega^2(H_\infty)$ is a closed 2-form which is non degenerate in the sense that

$$\sigma^\vee : TH_\infty = H_\infty \times H_\infty \to T^*H_\infty = H_\infty \times {H_\infty}'$$

is injective (but not surjective), where ${H_\infty}' = L(H_\infty, \mathbb{R})$ denotes the real topological dual space. This is the meaning of 'weak' above.

49.12. Let $\langle x,y\rangle = \operatorname{Re}\langle x,y\rangle + \sqrt{-1}\sigma(x,y)$ be the decomposition of the Hermitian inner product into real and imaginary parts. Then $\operatorname{Re}\langle x,y\rangle = \sigma(\sqrt{-1}x, y)$, thus the real linear subspaces $\sigma^\vee(H_\infty) = \sigma(H_\infty,\ \ )$ and $\operatorname{Re}\langle H_\infty,\ \ \rangle$ of ${H_\infty}' = L(H_\infty,\mathbb{R})$ coincide.

Following (48.4), we let H_∞^σ denote the real linear subspace

$$H_\infty^\sigma = \sigma(H_\infty,\ \ ) = \operatorname{Re}\langle H_\infty,\ \ \rangle$$

of ${H_\infty}' = L(H_\infty,\mathbb{R})$, the *smooth dual* of H_∞ with respect to the weak symplectic structure σ. We have two canonical isomorphisms $H_\infty^\sigma \cong H_\infty$ induced by σ and $\operatorname{Re}\langle\ ,\ \rangle$, respectively. Both induce the same structure of a convenient vector space on H_∞^σ, which we fix from now on.

Following (48.7), we have the subspace $C^\infty_\sigma(H_\infty,\mathbb{R}) \subset C^\infty(H_\infty,\mathbb{R})$ consisting of all smooth functions $f : H_\infty \to \mathbb{R}$ admitting smooth σ-gradients $\operatorname{grad}^\sigma f$, see (48.6). Then by (48.8) the Poisson bracket

$$\begin{aligned}\{\ ,\ \} &: C^\infty_\sigma(H_\infty,\mathbb{R}) \times C^\infty_\sigma(H_\infty,\mathbb{R}) \to C^\infty_\sigma(H_\infty,\mathbb{R}),\\ \{f,g\} &:= i_{\operatorname{grad}^\sigma f} i_{\operatorname{grad}^\sigma g}\sigma = \sigma(\operatorname{grad}^\sigma g, \operatorname{grad}^\sigma f) =\\ &= (\operatorname{grad}^\sigma f)(g) = dg(\operatorname{grad}^\sigma f)\end{aligned}$$

is well defined and describes a Lie algebra structure on the space $C^\infty_\sigma(H_\infty,\mathbb{R})$. There is the long exact sequence of Lie algebras and Lie algebra homomorphisms:

$$0 \to H^0(H_\infty) \to C^\infty_\sigma(H_\infty,\mathbb{R}) \xrightarrow{\operatorname{grad}^\sigma} \mathfrak{X}(H_\infty,\sigma) \xrightarrow{\gamma} H^1(H_\infty) = 0.$$

49.13. We consider now like in (49.2) a unitary representation $\rho : G \to U(H)$. By theorem (49.5), the associated mapping $\rho^\wedge : G \times H_\infty \to H_\infty$ is smooth, so we have the infinitesimal mapping $\rho' : \mathfrak{g} \to \mathfrak{X}(H_\infty)$, given by $\rho'(X)(x) = T_e(\rho^\wedge(\ \ ,x))X$ for $X \in \mathfrak{g}$ and $x \in H_\infty$. Since ρ is a unitary representation, the mapping ρ' has values in the Lie subalgebra of all linear Hamiltonian vector fields $\xi \in \mathfrak{X}(H_\infty)$ which respect the symplectic form σ, i.e. $\xi : H_\infty \to H_\infty$ is linear and $\mathcal{L}_\xi\sigma = 0$.

49.14. Lemma. *The mapping $\chi : \mathfrak{g} \to C^\infty_\sigma(H_\infty, \mathbb{R})$ which is given by $\chi(X)(x) = \frac{1}{2}\sigma(\rho'(X)(x), x)$ for $X \in \mathfrak{g}$ and $x \in H_\infty$ is a Lie algebra homomorphism, and we have* $\operatorname{grad}^\sigma \circ \chi = \rho'$.

For $g \in G$ we have $\rho(g)^*\chi(X) = \chi(X) \circ \rho(g) = \chi(Ad(g^{-1})X)$, so χ is G-equivariant.

Proof. First we have to check that $\chi(X) \in C^\infty_\sigma(H_\infty, \mathbb{R})$. Since $\rho'(X) : H_\infty \to H_\infty$ is smooth and linear, i.e. bounded linear, this follows from (48.6.2). Furthermore,

$$\begin{aligned}\operatorname{grad}^\sigma(\chi(X))(x) &= (\sigma^\vee)^{-1}(d\chi(X)(x)) = \\ &= \tfrac{1}{2}(\sigma^\vee)^{-1}\left(\sigma(\rho'(X)(\quad), x) + \sigma(\rho'(X)(x), \quad)\right) = \\ &= (\sigma^\vee)^{-1}\left(\sigma(\rho'(X)(x), \quad)\right) = \rho'(X)(x),\end{aligned}$$

since $\sigma(\rho'(X)(x), y) = \sigma(\rho'(X)(y), x)$.

Clearly, $\chi([X,Y]) - \{\chi(X), \chi(Y)\}$ is a constant function by the long exact sequence. Since it also vanishes at $0 \in H_\infty$, the mapping $\chi : \mathfrak{g} \to C^\infty_\sigma(H_\infty)$ is a Lie algebra homomorphism.

For the last assertion we have

$$\begin{aligned}\chi(X)(\rho(g)x) &= \tfrac{1}{2}\sigma(\rho'(X)(\rho(g)x), \rho(g)x) \\ &= \tfrac{1}{2}(\rho(g)^*\sigma)(\rho(g^{-1})\rho'(X)(\rho(g)x), x) \\ &= \tfrac{1}{2}\sigma(\rho'(Ad(g^{-1})X)x, x) = \chi(Ad(g^{-1})X)(x). \quad \square\end{aligned}$$

49.15. The moment mapping. For a unitary representation $\rho : G \to U(H)$ we can now define the *moment mapping*

$$\begin{aligned}\mu : H_\infty &\to \mathfrak{g}' = L(\mathfrak{g}, \mathbb{R}), \\ \mu(x)(X) &:= \chi(X)(x) = \tfrac{1}{2}\sigma(\rho'(X)x, x),\end{aligned}$$

for $x \in H_\infty$ and $X \in \mathfrak{g}$.

49.16. Theorem. *The moment mapping $\mu : H_\infty \to \mathfrak{g}'$ has the following properties:*

1. *We have $(d\mu(x)y)(X) = \sigma(\rho'(X)x, y)$ for $x, y \in H_\infty$ and $X \in \mathfrak{g}$. Consequently, we have $\operatorname{ev}_X \circ \mu \in C^\infty_\sigma(H_\infty, \mathbb{R})$ for all $X \in \mathfrak{g}$.*
2. *If G is a finite dimensional Lie group, for $x \in H_\infty$ the image of $d\mu(x) : H_\infty \to \mathfrak{g}'$ is the annihilator $\mathfrak{g}_x^\circ$ of the Lie algebra $\mathfrak{g}_x = \{X \in \mathfrak{g} : \rho'(X)(x) = 0\}$ of the isotropy group $G_x = \{g \in G : \rho(g)x = x\}$ in $\mathfrak{g}'$. If G is infinite dimensional we can only assert that $d\mu(x)(H_\infty) \subseteq \mathfrak{g}_x^\circ$.*
3. *For $x \in H_\infty$ the kernel of the differential $d\mu(x)$ is $(T_x(\rho(G)x))^\sigma = \{y \in H_\infty : \sigma(y, T_x(\rho(G)x)) = 0\}$, the σ-annihilator of the 'tangent space' at x of the G-orbit through x.*
4. *The moment mapping is equivariant: $Ad^*(g) \circ \mu = \mu \circ \rho(g)$ for all $g \in G$, where $Ad^*(g) = Ad(g^{-1})^* : \mathfrak{g}' \to \mathfrak{g}'$ is the coadjoint action.*

(5) *If G is finite dimensional the pullback operator*

$$\mu^* : C^\infty(\mathfrak{g}', \mathbb{R}) \to C^\infty(H_\infty, \mathbb{R})$$

actually has values in the subspace $C^\infty_\sigma(H_\infty, \mathbb{R})$. It is also a Lie algebra homomorphism for the Poisson brackets involved.

Proof. (1) Differentiating the defining equation, we get

$$\text{(a)} \qquad (d\mu(x)y)(X) = \tfrac{1}{2}\sigma(\rho'(X)y, x) + \tfrac{1}{2}\sigma(\rho'(X)x, y) = \sigma(\rho'(X)x, y).$$

From lemma (48.6) we see that $\operatorname{ev}_X \circ \mu \in C^\infty_\sigma(H_\infty, \mathbb{R})$ for all $X \in \mathfrak{g}$.

(2) and (3) are immediate consequences of this formula.

(4) We have

$$\begin{aligned}\mu(\rho(g)x)(X) &= \chi(X)(\rho(g)x) = \chi(Ad(g^{-1})X)(x) \text{ by lemma (49.14)}\\ &= \mu(x)(Ad(g^{-1})X) = (Ad(g^{-1})'\mu(x))(X).\end{aligned}$$

(5) Take $f \in C^\infty(\mathfrak{g}', \mathbb{R})$, then we have

$$\text{(b)} \qquad \begin{aligned} d(\mu^* f)(x)y &= d(f \circ \mu)(x)y = df(\mu(x))d\mu(x)y \\ &= (d\mu(x)y)(df(\mu(x))) = \sigma(\rho'(df(\mu(x)))x, y)\end{aligned}$$

by (a), which is smooth in x as a mapping into $H_\infty \cong H^\sigma_\infty \subset H'_\infty$ since $\mathfrak{g}'$ is finite dimensional. From lemma (48.6) we have that $f \circ \mu \in C^\infty_\sigma(H_\infty, \mathbb{R})$.

$$\sigma(\operatorname{grad}^\sigma(\mu^* f)(x), y) = d(\mu^* f)(x)y = \sigma(\rho'(df(\mu(x)))x, y)$$

by (b), so $\operatorname{grad}^\sigma(\mu^* f)(x) = \rho'(df(\mu(x)))x$. The Poisson structure on $\mathfrak{g}'$ is given as follows: We view the Lie bracket on $\mathfrak{g}$ as a linear mapping $\Lambda^2\mathfrak{g} \to \mathfrak{g}$. Its adjoint $P : \mathfrak{g}' \to \Lambda^2\mathfrak{g}'$ is then a section of the bundle $\Lambda^2 T\mathfrak{g}' \to \mathfrak{g}'$, which is called the Poisson structure on $\mathfrak{g}'$. If for $\alpha \in \mathfrak{g}'$ we view $df(\alpha) \in L(\mathfrak{g}', \mathbb{R})$ as an element in $\mathfrak{g}$, the Poisson bracket for $f_i \in C^\infty(\mathfrak{g}', \mathbb{R})$ is given by $\{f_1, f_2\}_{\mathfrak{g}'}(\alpha) = (df_1 \wedge df_2)(P)|_\alpha = \alpha([df_1(\alpha), df_2(\alpha)])$. Then we may compute as follows.

$$\begin{aligned}(\mu^*\{f_1, f_2\}_{\mathfrak{g}'})(x) &= \{f_1, f_2\}_{\mathfrak{g}'}(\mu(x))\\ &= \mu(x)([df_1(\mu(x)), df_2(\mu(x))])\\ &= \chi([df_1(\mu(x)), df_2(\mu(x))])(x)\\ &= \{\chi(df_1(\mu(x))), \chi(df_2(\mu(x)))\}(x) && \text{by lemma (49.14)}\\ &= \sigma(\operatorname{grad}^\sigma \chi(df_2(\mu(x)))(x), \operatorname{grad}^\sigma \chi(df_1(\mu(x)))(x))\\ &= \sigma(\rho'(df_2(\mu(x)))x, \rho'(df_1(\mu(x)))x)\\ &= \sigma(\operatorname{grad}^\sigma(\mu^* f_2)(x), \operatorname{grad}^\sigma(\mu^* f_1)(x)) && \text{by (b)}\\ &= \{\mu^* f_1, \mu^* f_2\}_{H_\infty}(x). \quad \square\end{aligned}$$

Remark. Assertion (5) of the last theorem also remains true for infinite dimensional Lie groups G, in the following sense:

We define $C^\infty_\sigma(\mathfrak{g}',\mathbb{R})$ as the space of all $f\in C^\infty(\mathfrak{g}',\mathbb{R})$ such that the following condition is satisfied (compare with lemma (48.6)):

$df:\mathfrak{g}'\to\mathfrak{g}''$ factors to a smooth mapping $\mathfrak{g}'\to\mathfrak{g}\xrightarrow{\iota}\mathfrak{g}''$, where $\iota:\mathfrak{g}\to\mathfrak{g}''$ is the canonical injection into the bidual.

Then the Poisson bracket on $C^\infty_\sigma(\mathfrak{g}',\mathbb{R})$ is defined by $\{f,g\}(\alpha)=\alpha([df(\alpha),dg(\alpha)])$, and the pullback $\mu^*:C^\infty(\mathfrak{g}',\mathbb{R})\to C^\infty(H_\infty,\mathbb{R})$ induces a Lie algebra homomorphism $\mu^*:C^\infty_\sigma(\mathfrak{g}',\mathbb{R})\to C^\infty_\sigma(H_\infty,\mathbb{R})$ for the Poisson brackets involved. The proof is as above, with obvious changes.

49.17. Let now G be a real analytic Lie group, and let $\rho:G\to U(H)$ be a unitary representation on a Hilbert space H. Again we consider H_ω as a weak symplectic real analytic manifold, equipped with the symplectic structure σ, the restriction of the imaginary part of the Hermitian inner product $\langle\ ,\ \rangle$ on H. Then again $\sigma\in\Omega^2(\mathbf{H}_\omega)$ is a closed 2-form which is non degenerate in the sense that $\sigma^\vee:H_\omega\to H'_\omega=L(H_\omega,\mathbb{R})$ is injective. Let

$$H^*_\omega:=\sigma^\vee(H_\omega)=\sigma(H_\omega,\quad)=\operatorname{Re}\langle H_\omega,\quad\rangle\subset H'_\omega=L(\mathbf{H}_\omega,\mathbb{R})$$

denote the *analytic dual* of H_ω, equipped with the topology induced by the isomorphism with H_ω.

49.18. Remark. All the results leading to the smooth moment mapping can now be carried over to the real analytic setting with *no* changes in the proofs. So all statements from (49.12) to (49.16) are valid in the real analytic situation. We summarize this in one more result:

49.19. Theorem. *Consider the injective linear continuous G-equivariant mapping $i:H_\omega\to\mathbf{H}_\infty$. Then for the smooth moment mapping $\mu:H_\infty\to\mathfrak{g}'$ from* (49.16) *the composition $\mu\circ i:H_\omega\to H_\infty\to\mathfrak{g}'$ is real analytic. It is called the real analytic moment mapping.*

Proof. It is immediately clear from (49.10) and the formula (49.15) for the smooth moment mapping, that $\mu\circ i$ is real analytic. □

50. Applications to Perturbation Theory of Operators

The material of this section is mostly due to [Alekseevsky, Kriegl, Losik, Michor, 1997]. We want to show that relatively simple applications of the calculus developed in the first part of this book can reproduce results which partly are even stronger than the best results from [Kato, 1976]. We start with choosing roots of smoothly parameterized polynomials in a smooth way. For more information on this see the reference above. Let

$$P(t)=x^n-a_1(t)x^{n-1}+\dots+(-1)^na_n(t)$$

be a polynomial with all roots real, smoothly parameterized by t near 0 in $\mathbb{R}$. Can we find n smooth functions $x_1(t), \dots, x_n(t)$ of the parameter t defined near 0, which are roots of $P(t)$ for each t? We can reduce the problem to $a_1 = 0$, replacing the variable x by the variable $y = x - a_1(t)/n$. We will say that the curve (1) is smoothly solvable near $t = 0$ if such smooth roots $x_i(t)$ exist.

50.1. Preliminaries. We recall some known facts on polynomials with real coefficients. Let

$$P(x) = x^n - a_1 x^{n-1} + \dots + (-1)^n a_n$$

be a polynomial with real coefficients $a_1, \dots, a_n$ and roots $x_1, \dots, x_n \in \mathbb{C}$. It is known that $a_i = \sigma_i(x_1, \dots, x_n)$, where σ_i $(i = 1, \dots, n)$ are the elementary symmetric functions in n variables:

$$\sigma_i(x_1, \dots, x_n) = \sum_{1 \le j_1 < \dots < j_i \le n} x_{j_1} \dots x_{j_i}.$$

Denote by s_i the Newton polynomials $\sum_{j=1}^n x_j^i$, which are related to the elementary symmetric function by

$$\text{(1)} \quad s_k - s_{k-1}\sigma_1 + s_{k-2}\sigma_2 + \dots + (-1)^{k-1} s_1 \sigma_{k-1} + (-1)^k k \sigma_k = 0 \quad (k \le n).$$

The corresponding mappings are related by a polynomial diffeomorphism ψ^n, given by (1):

$$\begin{aligned} \sigma^n &:= (\sigma_1, \dots, \sigma_n) : \mathbb{R}^n \to \mathbb{R}^n \\ s^n &:= (s_1, \dots, s_n) : \mathbb{R}^n \to \mathbb{R}^n \\ s^n &:= \psi^n \circ \sigma^n. \end{aligned}$$

Note that the Jacobian (the determinant of the derivative) of s^n is $n!$ times the Vandermond determinant: $\det(ds^n(x)) = n! \prod_{i>j}(x_i - x_j) =: n! \operatorname{Van}(x)$, and even the derivative itself $d(s^n)(x)$ equals the Vandermond matrix up to factors i in the i-th row. We also have $\det(d(\psi^n)(x)) = (-1)^{n(n+3)/2} n! = (-1)^{n(n-1)/2} n!$, and consequently $\det(d\sigma^n(x)) = \prod_{i>j}(x_j - x_i)$. We consider the so-called *Bezoutiant*

$$B := \begin{pmatrix} s_0 & s_1 & \dots & s_{n-1} \\ s_1 & s_2 & \dots & s_n \\ \vdots & \vdots & & \vdots \\ s_{n-1} & s_n & \dots & s_{2n-2} \end{pmatrix}.$$

Let B_k be the minor formed by the first k rows and columns of B. From

$$B_k(x) = \begin{pmatrix} 1 & 1 & \dots & 1 \\ x_1 & x_2 & \dots & x_n \\ \vdots & \vdots & & \vdots \\ x_1^{k-1} & x_2^{k-1} & \dots & x_n^{k-1} \end{pmatrix} \cdot \begin{pmatrix} 1 & x_1 & \dots & x_1^{k-1} \\ 1 & x_2 & \dots & x_2^{k-1} \\ \vdots & \vdots & & \vdots \\ 1 & x_n & \dots & x_n^{k-1} \end{pmatrix}$$

it follows that

$$\text{(2)} \quad \Delta_k(x) := \det(B_k(x)) = \sum_{i_1 < i_2 < \dots < i_k} (x_{i_1} - x_{i_2})^2 \dots (x_{i_1} - x_{i_n})^2 \dots (x_{i_{k-1}} - x_{i_k})^2,$$

since for $n \times k$-matrices A one has $\det(AA^\top) = \sum_{i_1 < \dots < i_k} \det(A_{i_1, \dots, i_k})^2$, where $A_{i_1, \dots, i_k}$ is the minor of A with the indicated rows. Since the Δ_k are symmetric we have $\Delta_k = \tilde{\Delta}_k \circ \sigma^n$ for unique polynomials $\tilde{\Delta}_k$, and similarly we shall use $\tilde{B}$.

50.2. Result. [Sylvester, 1853, pp.511], [Procesi, 1978] *The roots of P are all real if and only if the matrix $\tilde{B}(P) \geq 0$. Then we have $\tilde{\Delta}_k(P) := \tilde{\Delta}_k(a_1, \dots, a_n) \geq 0$ for $1 \leq k \leq n$. The rank of $\tilde{B}(P)$ equals the number of distinct roots of P, and its signature equals the number of distinct real roots.*

50.3. Proposition. *Let now P be a smooth curve of polynomials*

$$P(t)(x) = x^n - a_1(t)x^{n-1} + \dots + (-1)^n a_n(t)$$

with all roots real and distinct for $t = 0$. Then P is smoothly solvable near 0.

This is also true in the real analytic case and for higher dimensional parameters, and in the holomorphic case for complex roots.

Proof. The derivative $\frac{d}{dx}P(0)(x)$ does not vanish at any root, since they are distinct. Thus, by the implicit function theorem we have local smooth solutions $x(t)$ of $P(t,x) = P(t)(x) = 0$. $\square$

50.4. Splitting Lemma. *Let P_0 be a polynomial*

$$P_0(x) = x^n - a_1 x^{n-1} + \dots + (-1)^n a_n.$$

If $P_0 = P_1 \cdot P_2$, where P_1 and P_2 are polynomials with no common root. Then for P near P_0 we have $P = P_1(P) \cdot P_2(P)$ for real analytic mappings of monic polynomials $P \mapsto P_1(P)$ and $P \mapsto P_2(P)$, defined for P near P_0, with the given initial values.

Proof. Let the polynomial P_0 be represented as the product

$$P_0 = P_1.P_2 = (x^p - b_1 x^{p-1} + \dots + (-1)^p b_p)(x^q - c_1 x^{q-1} + \dots + (-1)^q c_q).$$

Let x_i for $i = 1, \dots, n$ be the roots of P_0, ordered in such a way that for $i = 1, \dots, p$ we get the roots of P_1, and for $i = p+1, \dots, p+q = n$ we get those of P_2. Then $(a_1, \dots, a_n) = \phi^{p,q}(b_1, \dots, b_p, c_1, \dots, c_q)$ for a polynomial mapping $\phi^{p,q}$, and we get

$$\sigma^n = \phi^{p,q} \circ (\sigma^p \times \sigma^q),$$
$$\det(d\sigma^n) = \det(d\phi^{p,q}(b,c)) \det(d\sigma^p) \det(d\sigma^q).$$

From (50.1) we conclude

$$\prod_{1 \leq i < j \leq n} (x_i - x_j) = \det(d\phi^{p,q}(b,c)) \prod_{1 \leq i < j \leq p} (x_i - x_j) \prod_{p+1 \leq i < j \leq n} (x_i - x_j)$$

which in turn implies

$$\det(d\phi^{p,q}(b,c)) = \prod_{1 \leq i \leq p < j \leq n} (x_i - x_j) \neq 0,$$

so that $\phi^{p,q}$ is a real analytic diffeomorphism near (b,c). $\square$

50.5. For a continuous function f defined near 0 in $\mathbb{R}$ let the *multiplicity* or *order of flatness* $m(f)$ at 0 be the supremum of all integers p such that $f(t) = t^p g(t)$ near 0 for a continuous function g. If f is C^n and $m(f) < n$ then $f(t) = t^{m(f)} g(t)$ where now g is $C^{n-m(f)}$ and $g(0) \neq 0$. If f is a continuous function on the space of polynomials, then for a fixed continuous curve P of polynomials we will denote by $m(f)$ the multiplicity at 0 of $t \mapsto f(P(t))$.

The splitting lemma (50.4) shows that for the problem of smooth solvability it is enough to assume that all roots of $P(0)$ are equal.

Proposition. *Suppose that the smooth curve of polynomials*

$$P(t)(x) = x^n + a_2(t)x^{n-2} - \cdots + (-1)^n a_n(t)$$

is smoothly solvable with smooth roots $t \mapsto x_i(t)$, and that all roots of $P(0)$ are equal. Then for $(k = 2, \ldots, n)$

$$m(\tilde{\Delta}_k) \geq k(k-1) \min_{1 \leq i \leq n} m(x_i).$$
$$m(a_k) \geq k \min_{1 \leq i \leq n} m(x_i).$$

This result also holds in the real analytic case and in the smooth case.

Proof. This follows by (50.1.2) for Δ_k and by $a_k(t) = \sigma_k(x_1(t), \ldots, x_n(t))$. □

50.6. Lemma. *Let P be a polynomial of degree n with all roots real. If $a_1 = a_2 = 0$ then all roots of P are equal to zero.*

Proof. From (50.1.1) we have $\sum x_i^2 = s_2(x) = \sigma_1^2(x) - 2\sigma_2(x) = a_1^2 - 2a_2 = 0$. □

50.7. Multiplicity lemma. *For an integer $r \geq 1$ consider a C^{nr} curve of polynomials*

$$P(t)(x) = x^n + a_2(t)x^{n-2} - \cdots + (-1)^n a_n(t)$$

with all roots real. Then the following conditions are equivalent:

(1) *$m(a_k) \geq kr$ for all $2 \leq k \leq n$.*
(2) *$m(\tilde{\Delta}_k) \geq k(k-1)r$ for all $2 \leq k \leq n$.*
(3) *$m(a_2) \geq 2r$.*

Proof. We only have to treat $r > 0$.

(1) implies (2): From (50.1.1) we have $m(\tilde{s}_k) \geq rk$, and from the definition of $\tilde{\Delta}_k = \det(\tilde{B}_k)$ we get (2).

(2) implies (3) since $\tilde{\Delta}_2 = -2na_2$.

(3) implies (1): From $a_2(0) = 0$ and lemma (50.6) it follows that all roots of the polynomial $P(0)$ are equal to zero and, then, $a_3(0) = \cdots = a_n(0) = 0$. Therefore, $m(a_3), \ldots, m(a_n) \geq 1$. Under these conditions, we have $a_2(t) = t^{2r} a_{2,2r}(t)$ and $a_k(t) = t^{m_k} a_{k,m_k}(t)$ for $k = 3, \ldots, n$, where the m_k are positive integers and

$a_{2,2r}, a_{3,m_3}, \dots, a_{n,m_n}$ are continuous functions, and where we may assume that either $m_k = m(a_k) < \infty$ or $m_k \geq kr$.

Suppose now indirectly that for some $k > 2$ we have $m_k = m(a_k) < kr$. Then we put

$$m := \min\left(r, \tfrac{m_3}{3}, \dots, \tfrac{m_n}{n}\right) < r.$$

We consider the following continuous curve of polynomials for $t \geq 0$:

$$\bar{P}_m(t)(x) := x^n + a_{2,2r}(t)t^{2r-2m}x^{n-2} \\ - a_{3,m_3}(t)t^{m_3-3m}x^{n-3} + \dots + (-1)^n a_{n,m_n}(t)t^{m_n-nm}.$$

If $x_1, \dots, x_n$ are the real roots of $P(t)$ then $t^{-m}x_1, \dots, t^{-m}x_n$ are the roots of $\bar{P}_m(t)$, for $t > 0$. So for $t > 0$, $\bar{P}_m(t)$ is a family of polynomials with all roots real. Since by theorem (50.2) the set of polynomials with all roots real is closed, $\bar{P}_m(0)$ is also a polynomial with all roots real.

By lemma (50.6), all roots of the polynomial $\bar{P}_m(0)$ are equal to zero, and for those k with $m_k = km$ we have $nr - m_k \geq kr - m_k \geq 1$, thus a_{k,m_k} is $C^{nr-m_k} \subseteq C^1$ and $a_{k,m_k}(0) = 0$, therefore $m(a_k) > m_k$, a contradiction. □

50.8. Algorithm. Consider a smooth curve of polynomials

$$P(t)(x) = x^n - a_1(t)x^{n-1} + a_2(t)x^{n-2} - \dots + (-1)^n a_n(t)$$

with all roots real. The algorithm has the following steps:

(1) If all roots of $P(0)$ are pairwise different, P is smoothly solvable for t near 0 by (50.3).

(2) If there are distinct roots at $t = 0$ we put them into two subsets which splits $P(t) = P_1(t).P_2(t)$ by the splitting lemma (50.4). We then feed $P_i(t)$ (which have lower degree) into the algorithm.

(3) All roots of $P(0)$ are equal. We first reduce $P(t)$ to the case $a_1(t) = 0$ by replacing the variable x by $y = x - a_1(t)/n$. Then all roots are equal to 0, so $m(a_2) > 0$.

(3a) If $m(a_2)$ is finite then it is even since $\tilde{\Delta}_2 = -2na_2 \geq 0$, $m(a_2) = 2r$, and by the multiplicity lemma (50.7) $a_i(t) = a_{i,ir}(t)t^{ir}$ ($i = 2, \dots, n$) for smooth $a_{i,ir}$. Consider the following smooth curve of polynomials

$$P_r(t)(x) = x^n + a_{2,2r}(t)x^{n-2} - a_{3,3r}(t)x^{n-3} + \dots + (-1)^n a_{n,nr}(t).$$

If $P_r(t)$ is smoothly solvable and $x_k(t)$ are its smooth roots, then $x_k(t)t^r$ are the roots of $P(t)$, and the original curve P is smoothly solvable, too. Since $a_{2,2m}(0) \neq 0$, not all roots of $P_r(0)$ are equal, and we may feed P_r into step 2 of the algorithm.

(3b) If $m(a_2)$ is infinite and $a_2 = 0$, then all roots are 0 by (50.6), and thus the polynomial is solvable.

(3c) But if $m(a_2)$ is infinite and $a_2 \neq 0$, then by the multiplicity lemma (50.7) all $m(a_i)$ for $2 \leq i \leq n$ are infinite. In this case we keep $P(t)$ as factor of the original curve of polynomials with all coefficients infinitely flat at $t = 0$ after forcing $a_1 = 0$. This means that all roots of $P(t)$ meet of infinite order of flatness (see (50.5)) at $t = 0$ for any choice of the roots. This can be seen as follows: If $x(t)$ is any root of $P(t)$ then $y(t) := x(t)/t^r$ is a root of $P_r(t)$, hence by (50.9) bounded, so $x(t) = t^{r-1}.ty(t)$ and $t \mapsto ty(t)$ is continuous at $t = 0$.

This algorithm produces a splitting of the original polynomial

$$P(t) = P^{(\infty)}(t)P^{(s)}(t),$$

where $P^{(\infty)}$ has the property that each root meets another one of infinite order at $t = 0$, and where $P^{(s)}(t)$ is smoothly solvable, and no two roots meet of infinite order at $t = 0$, if they are not equal. Any two choices of smooth roots of $P^{(s)}$ differ by a permutation.

The factor $P^{(\infty)}$ may or may not be smoothly solvable. For a flat function $f \geq 0$ consider:

$$x^4 - (f(t) + t^2)x^2 + t^2 f(t) = (x^2 - f(t)).(x - t)(x + t).$$

Here the algorithm produces this factorization. For $f(t) = g(t)^2$ the polynomial is smoothly solvable. There exist smooth functions f (see (25.3) or [Alekseevsky, Kriegl, Losik, Michor, 1997, 2.4]) such that $x^2 = f(t)$ is not smoothly solvable, in fact not C^2-solvable. Moreover, in loc. cit. one finds a polynomial $x^2 + a_2(t)x - a_3(t)$ with smooth coefficient functions a_2 and a_3 which is not C^1-solvable.

50.9. Lemma. *For a polynomial*

$$P(x) = x^n - a_1(P)x^{n-1} + \cdots + (-1)^n a_n(P)$$

with all roots real, i.e. $\tilde{\Delta}_k(P) = \tilde{\Delta}_k(a_1, \ldots, a_n) \geq 0$ *for* $1 \leq k \leq n$, *let*

$$x_1(P) \leq x_2(P) \leq \cdots \leq x_n(P)$$

be the roots, increasingly ordered.

Then all $x_i : \sigma^n(\mathbb{R}^n) \to \mathbb{R}$ *are continuous.*

Proof. We show first that x_1 is continuous. Let $P_0 \in \sigma^n(\mathbb{R}^n)$ be arbitrary. We have to show that for every $\varepsilon > 0$ there exists some $\delta > 0$ such that for all $|P - P_0| < \delta$ there is a root $x(P)$ of P with $x(P) < x_1(P_0) + \varepsilon$ and for all roots $x(P)$ of P we have $x(P) > x_1(P_0) - \varepsilon$. Without loss of generality we may assume that $x_1(P_0) = 0$.

We use induction on the degree n of P. By the splitting lemma (50.4) for the C^0-case, we may factorize P as $P_1(P) \cdot P_2(P)$, where $P_1(P_0)$ has all roots equal to $x_1 = 0$ and $P_2(P_0)$ has all roots greater than 0, and both polynomials have coefficients which depend real analytically on P. The degree of $P_2(P)$ is now smaller than n, so by induction the roots of $P_2(P)$ are continuous and thus larger than $x_1(P_0) - \varepsilon$ for P near P_0.

Since 0 was the smallest root of P_0 we have to show that for all $\varepsilon > 0$ there exists a $\delta > 0$ such that for $|P - P_0| < \delta$ any root x of $P_1(P)$ satisfies $|x| < \varepsilon$. Suppose there is a root x with $|x| \geq \varepsilon$. Then we get a contradiction as follows, where n_1 is the degree of P_1. From

$$-x^{n_1} = \sum_{k=1}^{n_1} (-1)^k a_k(P_1) x^{n_1-k}$$

we have

$$\varepsilon \leq |x| = \Big|\sum_{k=1}^{n_1} (-1)^k a_k(P_1) x^{1-k}\Big| \leq \sum_{k=1}^{n_1} |a_k(P_1)|\,|x|^{1-k} < \sum_{k=1}^{n_1} \frac{\varepsilon^k}{n_1}\varepsilon^{1-k} = \varepsilon,$$

provided that $n_1|a_k(P_1)| < \varepsilon^k$, which is true for P_1 near P_0, since $a_k(P_0) = 0$. Thus, x_1 is continuous.

Now we factorize $P = (x - x_1(P)) \cdot P_2(P)$, where $P_2(P)$ has roots $x_2(P) \leq \cdots \leq x_n(P)$. By Horner's algorithm ($a_n = b_{n-1}x_1$, $a_{n-1} = b_{n-1} + b_{n-2}x_1$, ..., $a_2 = b_2 + b_1 x_1$, $a_1 = b_1 + x_1$), the coefficients b_k of $P_2(P)$ are continuous, and so we may proceed by induction on the degree of P. Thus, the claim is proved. □

50.10. Theorem. *Consider a smooth curve of polynomials*

$$P(t)(x) = x^n + a_2(t)x^{n-2} - \cdots + (-1)^n a_n(t)$$

with all roots real, for $t \in \mathbb{R}$. Let one of the two following equivalent conditions be satisfied:

(1) *If two of the increasingly ordered continuous roots meet of infinite order somewhere then they are equal everywhere.*
(2) *Let k be maximal with the property that $\tilde{\Delta}_k(P)$ does not vanish identically for all t. Then $\tilde{\Delta}_k(P)$ vanishes nowhere of infinite order.*

Then the roots of P can be chosen smoothly, and any two choices differ by a permutation of the roots.

Proof. The local situation. We claim that for any t_0, without loss of generality $t_0 = 0$, the following conditions are equivalent:

(1) If two of the increasingly ordered continuous roots meet of infinite order at $t = 0$ then their germs at $t = 0$ are equal.
(2) Let k be maximal with the property that the germ at $t = 0$ of $\tilde{\Delta}_k(P)$ is not 0. Then $\tilde{\Delta}_k(P)$ is not infinitely flat at $t = 0$.
(3) The algorithm (50.8) never leads to step (3c).

(3) ⇒ (1) Suppose indirectly that two of the increasingly ordered continuous non-equal roots meet of infinite order at $t = 0$. Then in each application of step (2) these two roots stay with the same factor. After any application of step (3a) these two roots lead to nonequal roots of the modified polynomial which still meet of infinite

order at $t = 0$. They never end up in a factor leading to step (3b) or step (1). So they end up in a factor leading to step (3c).

(1) $\Rightarrow$ (2) Let $x_1(t) \leq \dots \leq x_n(t)$ be the continuous roots of $P(t)$. From (50.1.2) we have

$$(4) \qquad \tilde{\Delta}_k(P(t)) = \sum_{i_1 < i_2 < \dots < i_k} (x_{i_1} - x_{i_2})^2 \dots (x_{i_1} - x_{i_n})^2 \dots (x_{i_{k-1}} - x_{i_k})^2.$$

The germ of $\tilde{\Delta}_k(P)$ is not 0, so the germ of one summand is not 0. If $\tilde{\Delta}_k(P)$ were infinitely flat at $t = 0$, then each summand was infinitely flat, there were two roots among the x_i which met of infinite order, thus by assumption their germs were equal, so this summand vanished.

(2) $\Rightarrow$ (3) Since the leading $\tilde{\Delta}_k(P)$ vanishes only of finite order at zero, P has exactly k different roots off 0. Suppose indirectly that the algorithm (50.8) leads to step (3c). Then $P = P^{(\infty)}P^{(s)}$ for a nontrivial polynomial $P^{(\infty)}$. Let $x_1(t) \leq \dots \leq x_p(t)$ be the roots of $P^{(\infty)}(t)$ and $x_{p+1}(t) \leq \dots \leq x_n(t)$ those of $P^{(s)}$. We know that each x_i meets some x_j of infinite order and does not meet any x_l of infinite order, for $i, j \leq p < l$. Let $k^{(\infty)} > 2$ and $k^{(s)}$ be the number of generically different roots of $P^{(\infty)}$ and $P^{(s)}$, respectively. Then $k = k^{(\infty)} + k^{(s)}$, and an inspection of the formula for $\tilde{\Delta}_k(P)$ above leads to the fact that it must vanish of infinite order at 0, since the only non-vanishing summands involve exactly $k^{(\infty)}$ many generically different roots from $P^{(\infty)}$.

The global situation. From the first part of the proof we see that the algorithm (50.8) allows to choose the roots smoothly in a neighborhood of each point $t \in \mathbb{R}$, and that any two choices differ by a (constant) permutation of the roots. Thus, we may glue the local solutions to a global solution. □

50.11. Theorem. *Consider a curve of polynomials*

$$P(t)(x) = x^n - a_1(t)x^{n-1} + \dots + (-1)^n a_n(t), \quad t \in \mathbb{R},$$

with all roots real, where all a_i are of class C^n. Then there is a differentiable curve $x = (x_1, \dots, x_n) : \mathbb{R} \to \mathbb{R}^n$ whose coefficients parameterize the roots.

That this result cannot be improved to C^1-roots is shown in [Alekseevsky, Kriegl, Losik, Michor, 1996, 2.4].

Proof. First we note that the multiplicity lemma (50.7) remains true in the C^n-case for $n > 2$ and $r = 1$ in the following sense, with the same proof:
If $a_1 = 0$ then the following two conditions are equivalent

(1) *$a_k(t) = t^k a_{k,k}(t)$ for a continuous function $a_{k,k}$, for $2 \leq k \leq n$.*
(2) *$a_2(t) = t^2 a_{2,2}(t)$ for a continuous function $a_{2,2}$.*

In order to prove the theorem itself, we follow one step of the algorithm. First we replace x by $x + \frac{1}{n}a_1(t)$, or assume without loss of generality that $a_1 = 0$. Then we choose a fixed t, say $t = 0$.

If $a_2(0) = 0$ then it vanishes of second order at 0, for if it vanishes only of first order then $\tilde{\Delta}_2(P(t)) = -2na_2(t)$ would change sign at $t = 0$, contrary to the assumption that all roots of $P(t)$ are real, by (50.2). Thus, $a_2(t) = t^2 a_{2,2}(t)$, so by the variant of the multiplicity lemma (50.7) described above we have $a_k(t) = t^k a_{k,k}(t)$ for continuous functions $a_{k,k}$, for $2 \le k \le n$. We consider the following continuous curve of polynomials

$$P_1(t)(x) = x^n + a_{2,2}(t)x^{n-2} - a_{3,3}(t)x^{n-3} \cdots + (-1)^n a_{n,n}(t).$$

with continuous roots $z_1(t) \le \cdots \le z_n(t)$, by (50.9). Then $x_k(t) = z_k(t)t$ are differentiable at 0 and are all roots of P, but note that $x_k(t) = y_k(t)$ for $t \ge 0$, but $x_k(t) = y_{n-k}(t)$ for $t \le 0$, where $y_1(t) \le \cdots \le y_n(t)$ are the ordered roots of $P(t)$. This gives us one choice of differentiable roots near $t = 0$. Any choice is then given by this choice and applying afterwards any permutation of the set $\{1, \dots, n\}$ which keeps the function $k \mapsto z_k(0)$ invariant .

If $a_2(0) \ne 0$ then by the splitting lemma (50.4) for the C^n-case we may factor $P(t) = P_1(t) \dots P_k(t)$, where the $P_i(t)$ have again C^n-coefficients, and where each $P_i(0)$ has all roots equal to c_i, and where the c_i are distinct. By the arguments above, the roots of each P_i can be arranged differentiably. Thus P has differentiable roots $y_k(t)$.

But note that we have to apply a permutation on one side of 0 to the original roots, in the following case: Two roots x_k and x_l meet at zero with $x_k(t) - x_l(t) = tc_{kl}(t)$ with $c_{kl}(0) \ne 0$ (we say that they meet slowly). We may apply to this choice an arbitrary permutation of any two roots which meet with $c_{kl}(0) = 0$ (i.e. at least of second order), and we get thus every differentiable choice near $t = 0$.

Now we show that we can choose the roots differentiable on the whole domain $\mathbb{R}$. We start with the ordered continuous roots $y_1(t) \le \cdots \le y_n(t)$. Then we put

$$x_k(t) = y_{\sigma(t)(k)}(t),$$

where the permutation $\sigma(t)$ is given by

$$\sigma(t) = (1,2)^{\varepsilon_{1,2}(t)} \dots (1,n)^{\varepsilon_{1,n}(t)} (2,3)^{\varepsilon_{2,3}(t)} \dots (n-1,n)^{\varepsilon_{n-1,n}(t)},$$

and where $\varepsilon_{i,j}(t) \in \{0,1\}$ will be specified as follows: On the closed set $S_{i,j}$ of all t where $y_i(t)$ and $y_j(t)$ meet of order at least 2 any choice is good. The complement of $S_{i,j}$ is an at most countable union of open intervals, and in each interval we choose a point, where we put $\varepsilon_{i,j} = 0$. Going right (and left) from this point we change $\varepsilon_{i,j}$ in each point where y_i and y_j meet slowly. These points accumulate only in $S_{i,j}$. □

50.12. Theorem. The real analytic case. *Let P be a real analytic curve of polynomials*

$$P(t)(x) = x^n - a_1(t)x^{n-1} + \cdots + (-1)^n a_n(t), \quad t \in \mathbb{R},$$

with all roots real.

Then P is real analytically solvable, globally on $\mathbb{R}$. All solutions differ by permutations.

By a real analytic curve of polynomials we mean that all $a_i(t)$ are real analytic in t, and real analytically solvable means that we may find $x_i(t)$ for $i = 1, \dots, n$ which are real analytic in t and are roots of $P(t)$ for all t. The local existence part of this theorem is due to [Rellich, 1937, Hilfssatz 2], his proof uses Puiseux-expansions. Our proof is different and more elementary.

Proof. We first show that P is locally real analytically solvable near each point $t_0 \in \mathbb{R}$. It suffices to consider $t_0 = 0$. Using the transformation in the introduction we first assume that $a_1(t) = 0$ for all t. We use induction on the degree n. If $n = 1$ the theorem holds. For $n > 1$ we consider several cases:

The case $a_2(0) \neq 0$. Here not all roots of $P(0)$ are equal and zero, so by the splitting lemma (50.4) we may factor $P(t) = P_1(t).P_2(t)$ for real analytic curves of polynomials of positive degree, which have both all roots real, and we have reduced the problem to lower degree.

The case $a_2(0) = 0$. If $a_2(t) = 0$ for all t, then by (50.6) all roots of $P(t)$ are 0, and we are done. Otherwise $1 \leq m(a_2) < \infty$ for the multiplicity of a_2 at 0, and by (50.6) all roots of $P(0)$ are 0. If $m(a_2) > 0$ is odd, then $\tilde{\Delta}_2(P)(t) = -2na_2(t)$ changes sign at $t = 0$, so by (50.2) not all roots of $P(t)$ are real for t on one side of 0. This contradicts the assumption, so $m(a_2) = 2r$ is even. Then by the multiplicity lemma (50.7) we have $a_i(t) = a_{i,ir}(t)t^{ir}$ $(i = 2, \dots, n)$ for real analytic $a_{i,ir}$, and we may consider the following real analytic curve of polynomials

$$P_r(t)(x) = x^n + a_{2,2r}(t)x^{n-2} - a_{3,3r}(t)x^{n-3} \cdots + (-1)^n a_{n,nr}(t)$$

with all roots real. If $P_r(t)$ is real analytically solvable and $x_k(t)$ are its real analytic roots then $x_k(t)t^r$ are the roots of $P(t)$, and the original curve P is real analytically solvable too. Now $a_{2,2r}(0) \neq 0$ and we are done by the case above.

Claim. Let $x = (x_1, \dots, x_n) : I \to \mathbb{R}^n$ be a real analytic curve of roots of P on an open interval $I \subset \mathbb{R}$. Then any real analytic curve of roots of P on I is of the form $\alpha \circ x$ for some permutation α.

Let $y : I \to \mathbb{R}^n$ be another real analytic curve of roots of P. Let $t_k \to t_0$ be a convergent sequence of distinct points in I. Then $y(t_k) = \alpha_k(x(t_k)) = (x_{\alpha_k 1}, \dots, x_{\alpha_k n})$ for permutations α_k. By choosing a subsequence, we may assume that all α_k are the same permutation α. But then the real analytic curves y and $\alpha \circ x$ coincide on a converging sequence, so they coincide on I and the claim follows.

Now from the local smooth solvability above and the uniqueness of smooth solutions up to permutations we can glue a global smooth solution on the whole of $\mathbb{R}$. □

50.13. Now we consider the following situation: Let $A(t) = (A_{ij}(t))$ be a smooth (real analytic, holomorphic) curve of real (complex) $(n \times n)$-matrices or operators, depending on a real (complex) parameter t near 0. What can we say about the eigenvalues and eigenfunctions of $A(t)$?

In the following theorem (50.14) the condition that $A(t)$ is Hermitian cannot be omitted. Consider the following example of real semisimple (not normal) matrices

$$A(t) := \begin{pmatrix} 2t+t^3 & t \\ -t & 0 \end{pmatrix},$$

$$\lambda_\pm(t) = t + \frac{t^2}{2} \pm t^2\sqrt{1+\tfrac{t^2}{4}}, \quad x_\pm(t) = \begin{pmatrix} 1+\frac{t}{2} \pm t\sqrt{1+\frac{t^2}{4}} \\ -1 \end{pmatrix},$$

where at $t=0$ we do not get a base of eigenvectors.

50.14. Theorem. *Let $A(t) = (A_{ij}(t))$ be a smooth curve of complex Hermitian $(n \times n)$-matrices, depending on a real parameter $t \in \mathbb{R}$, acting on a Hermitian space $V = \mathbb{C}^n$, such that no two of the continuous eigenvalues meet of infinite order at any $t \in \mathbb{R}$ if they are not equal for all t.*

Then the eigenvalues and the eigenvectors can be chosen smoothly in t, on the whole parameter domain $\mathbb{R}$.

Let $A(t) = (A_{ij}(t))$ be a real analytic curve of complex Hermitian $(n \times n)$-matrices, depending on a real parameter $t \in \mathbb{R}$, acting on a Hermitian space $V = \mathbb{C}^n$. Then the eigenvalues and the eigenvectors can be chosen real analytically in t on the whole parameter domain $\mathbb{R}$.

The condition on meeting of eigenvalues permits that some eigenvalues agree for all t — we speak of higher 'generic multiplicity' in this situation.

The real analytic version of this theorem is due to [Rellich, 1940]. Our proof is different.

Proof. We prove the smooth case and indicate the changes for the real analytic case. The proof will use an algorithm.

Note first that by (50.10) (by (50.12) in the real analytic case) the characteristic polynomial

$$\begin{aligned} (1) \qquad P(A(t))(\lambda) &= \det(A(t) - \lambda \mathbb{I}) \\ &= \lambda^n - a_1(t)\lambda^{n-1} + a_2(t)\lambda^{n-2} - \dots + (-1)^n a_n(t) \\ &= \sum_{i=0}^{n} \operatorname{tr}(\Lambda^i A(t))\lambda^{n-i} \end{aligned}$$

is smoothly solvable (real analytically solvable), with smooth (real analytic) roots $\lambda_1(t), \dots, \lambda_n(t)$ on the whole parameter interval.

Case 1: distinct eigenvalues. If $A(0)$ has some eigenvalues distinct, then one can reorder them in such a way that for $i_0 = 0 < 1 \le i_1 < i_2 < \dots < i_k < n = i_{k+1}$ we have

$$\lambda_1(0) = \dots = \lambda_{i_1}(0) < \lambda_{i_1+1}(0) = \dots = \lambda_{i_2}(0) < \dots < \lambda_{i_k+1}(0) = \dots = \lambda_n(0).$$

For t near 0 we still have

$$\lambda_1(t),\dots,\lambda_{i_1}(t) < \lambda_{i_1+1}(t),\dots,\lambda_{i_2}(t) < \dots < \lambda_{i_k+1}(t),\dots,\lambda_n(t).$$

For $j = 1,\dots,k+1$ we consider the subspaces

$$V_t^{(j)} = \bigoplus_{i=i_{j-1}+1}^{i_j} \{v \in V : (A(t) - \lambda_i(t))v = 0\}.$$

Then each $V_t^{(j)}$ runs through a smooth (real analytic) vector subbundle of the trivial bundle $(-\varepsilon,\varepsilon)\times V \to (-\varepsilon,\varepsilon)$, which admits a smooth (real analytic) framing $e_{i_{j-1}+1}(t),\dots,e_{i_j}(t)$. We have $V = \bigoplus_{j=1}^{k+1} V_t^{(j)}$ for each t.
In order to prove this statement, note that

$$V_t^{(j)} = \ker\Big((A(t) - \lambda_{i_{j-1}+1}(t)) \circ \dots \circ (A(t) - \lambda_{i_j}(t))\Big),$$

so $V_t^{(j)}$ is the kernel of a smooth (real analytic) vector bundle homomorphism $B(t)$ of constant rank (even of constant dimension of the kernel), and thus is a smooth (real analytic) vector subbundle. This together with a smooth (real analytic) frame field can be shown as follows: Choose a basis of V, constant in t, such that $A(0)$ is diagonal. Then by the elimination procedure one can construct a basis for the kernel of $B(0)$. For t near 0, the elimination procedure (with the same choices) gives then a basis of the kernel of $B(t)$; the elements of this basis are then smooth (real analytic) in t for t near 0.

From the last result it follows that it suffices to find smooth (real analytic) eigenvectors in each subbundle $V^{(j)}$ separately, expanded in the smooth (real analytic) frame field. But in this frame field the vector subbundle looks again like a constant vector space. So feed each of these parts (A restricted to $V^{(j)}$, as matrix with respect to the frame field) into case 2 below.

Case 2: All eigenvalues at 0 are equal. So suppose that $A(t) : V \to V$ is Hermitian with all eigenvalues at $t = 0$ equal to $\frac{a_1(0)}{n}$, see (1).

Eigenvectors of $A(t)$ are also eigenvectors of $A(t) - \frac{a_1(t)}{n}\mathbb{I}$, so we may replace $A(t)$ by $A(t) - \frac{a_1(t)}{n}\mathbb{I}$ and assume that for the characteristic polynomial (1) we have $a_1 = 0$, or assume without loss that $\lambda_i(0) = 0$ for all i, and so $A(0) = 0$.

If $A(t) = 0$ for all t we choose the eigenvectors constant.

Otherwise, let $A_{ij}(t) = tA_{ij}^{(1)}(t)$. From (1) we see that the characteristic polynomial of the Hermitian matrix $A^{(1)}(t)$ is $P_1(t)$ in the notation of (50.8), thus $m(a_i) \geq i$ for $2 \leq i \leq n$, which also follows from (50.5).

The eigenvalues of $A^{(1)}(t)$ are the roots of $P_1(t)$, which may be chosen in a smooth way, since they again satisfy the condition of theorem (50.10). In the real analytic case we just have to invoke (50.12). Note that eigenvectors of $A^{(1)}$ are also eigenvectors of A. If the eigenvalues are still all equal, we apply the same procedure again, until they are not all equal: we arrive at this situation by the assumption of the theorem in the smooth case, and automatically in the real analytic case. Then we apply case 1.

This algorithm shows that one may choose the eigenvectors $x_i(t)$ of $A_i(t)$ in a smooth (real analytic) way, locally in t. It remains to extend this to the whole parameter interval.

If some eigenvalues coincide locally then on the whole of $\mathbb{R}$, by the assumption. The corresponding eigenspaces then form a smooth (real analytic) vector bundle over $\mathbb{R}$, by case 1, since those eigenvalues, which meet in isolated points are different after application of case 2.

So we we get $V = \bigoplus W_t^{(j)}$ where the $W_t^{(j)}$ are real analytic sub vector bundles of $V \times \mathbb{R}$, whose dimension is the generic multiplicity of the corresponding smooth (real analytic) eigenvalue function. It suffices to find global orthonormal smooth (real analytic) frames for each of these; this exists since the vector bundle is smoothly (real analytically) trivial, by using parallel transport with respect to a smooth (real analytic) Hermitian connection. $\square$

50.15. Example. (see [Rellich, 1937, section 2]) That the last result cannot be improved is shown by the following example which rotates a lot:

$$x_+(t) := \begin{pmatrix} \cos\frac{1}{t} \\ \sin\frac{1}{t} \end{pmatrix}, \quad x_-(t) := \begin{pmatrix} -\sin\frac{1}{t} \\ \cos\frac{1}{t} \end{pmatrix}, \quad \lambda_\pm(t) = \pm e^{-\frac{1}{t^2}},$$
$$A(t) := (x_+(t), x_-(t)) \begin{pmatrix} \lambda_+(t) & 0 \\ 0 & \lambda_-(t) \end{pmatrix} (x_+(t), x_-(t))^{-1}$$
$$= e^{-\frac{1}{t^2}} \begin{pmatrix} \cos\frac{2}{t} & \sin\frac{2}{t} \\ \sin\frac{2}{t} & -\cos\frac{2}{t} \end{pmatrix}.$$

Here $t \mapsto A(t)$ and $t \mapsto \lambda_\pm(t)$ are smooth, whereas the eigenvectors cannot be chosen continuously.

50.16. Theorem. *Let $t \mapsto A(t)$ be a smooth curve of unbounded self-adjoint operators in a Hilbert space with common domain of definition and compact resolvent. Then the eigenvalues of $A(t)$ may be arranged increasingly ordered in such a way that each eigenvalue is continuous, and they can be rearranged in such a way that they become C^1-functions.*

Suppose, moreover, that no two of the continuous eigenvalues meet of infinite order at any $t \in \mathbb{R}$ if they are not equal. Then the eigenvalues and the eigenvectors can be chosen smoothly in t on the whole parameter domain.

If on the other hand $t \mapsto A(t)$ is a real analytic curve of unbounded self-adjoint operators in a Hilbert space with common domain of definition and with compact resolvent. Then the eigenvalues and the eigenvectors can be chosen smoothly in t, on the whole parameter domain.

The real analytic version of this theorem is due to [Rellich, 1940], see also [Kato, 1976, VII, 3.9] the smooth version is due to [Alekseevsky, Kriegl, Losik, Michor, 1996]; the proof follows the lines of the latter paper.

That $A(t)$ is a smooth curve of unbounded operators means the following: There is a dense subspace V of the Hilbert space H such that V is the domain of definition of each $A(t)$ and such that $A(t)^* = A(t)$ with the same domains V, where the adjoint operator $A(t)^*$ is defined by $\langle A(t)u, v\rangle = \langle u, A(t)^*v\rangle$ for all v for which the left hand side is bounded as functional in $u \in V \subset H$. Moreover, we require that $t \mapsto \langle A(t)u, v\rangle$ is smooth for each $u \in V$ and $v \in H$. This implies that $t \mapsto A(t)u$ is smooth $\mathbb{R} \to H$ for each $v \in V$ by (2.3). Similar for the real analytic case, by (7.4).

The first part of the proof will show that $t \mapsto A(t)$ smooth implies that the resolvent $(A(t) - z)^{-1}$ is smooth in t and z jointly, and mainly this is used later in the proof. It is well known and in the proof we will show that if for some (t, z) the resolvent $(A(t) - z)^{-1}$ is compact then for all $t \in \mathbb{R}$ and z in the resolvent set of $A(t)$.

Proof. We shall prove the smooth case and indicate the changes for the real analytic case.

For each t consider the norm $\|u\|_t^2 := \|u\|^2 + \|A(t)u\|^2$ on V. Since $A(t) = A(t)^*$ is closed, $(V, \|\ \ \|_t)$ is also a Hilbert space with inner product $\langle u, v\rangle_t := \langle u, v\rangle + \langle A(t)u, A(t)v\rangle$. All these norms are equivalent since $(V, \|\ \ \|_t + \|\ \ \|_s) \to (V, \|\ \ \|_t)$ is continuous and bijective, so an isomorphism by the open mapping theorem. Then $t \mapsto \langle u, v\rangle_t$ is smooth for fixed $u, v \in V$, and by the multilinear uniform boundedness principle (5.18), the mapping $t \mapsto \langle\ \ ,\ \ \rangle_t$ is smooth and into the space of bounded bilinear forms; in the real analytic case we use (11.14) instead. By the exponential law (3.12) the mapping $(t, u) \mapsto \|u\|_t^2$ is smooth from $\mathbb{R} \times (V, \|\ \ \|_s) \to \mathbb{R}$ for each fixed s. In the real analytic case we use (11.18) instead. Thus, all Hilbert norms $\|\ \ \|_t$ are equivalent, since $\{\|u\|_t : |t| \le K, \|u\|_s \le 1\}$ is bounded by $L_{K,s}$ in $\mathbb{R}$, so $\|u\|_t \le L_{K,s}\|u\|_s$ for all $|t| \le K$. Moreover, each $A(s)$ is a globally defined operator $(V, \|\ \ \|_t) \to H$ with closed graph and is thus bounded, and by using again the (multi)linear uniform boundedness principle (5.18) (or (11.14) in the real analytic case) as above we see that $s \mapsto A(s)$ is smooth (real analytic) $\mathbb{R} \to L((V, \|\ \ \|_t), H)$.

If for some $(t, z) \in \mathbb{R} \times \mathbb{C}$ the bounded operator $A(t) - z : V \to H$ is invertible, then this is true locally and $(t, z) \mapsto (A(t) - z)^{-1} : H \to V$ is smooth since inversion is smooth on Banach spaces.

Since each $A(t)$ is Hermitian the *global resolvent set* $\{(t, z) \in \mathbb{R} \times \mathbb{C} : (A(t) - z) : V \to H \text{ is invertible}\}$ is open, contains $\mathbb{R} \times (\mathbb{C} \setminus \mathbb{R})$, and hence is connected.

Moreover $(A(t) - z)^{-1} : H \to H$ is a compact operator for some (equivalently any) (t, z) if and only if the inclusion $i : V \to H$ is compact, since $i = (A(t) - z)^{-1} \circ (A(t) - z) : V \to H \to H$.

Let us fix a parameter s. We choose a simple smooth curve γ in the resolvent set of $A(s)$ for fixed s.

(1) **Claim.** For t near s, there are C^1-functions $t \mapsto \lambda_i(t) : 1 \le i \le N$ which parameterize all eigenvalues (repeated according to their multiplicity) of $A(t)$ in the interior of γ. If no two of the generically different eigenvalues meet of infinite order they can be chosen smoothly.

By replacing $A(s)$ by $A(s)-z_0$ if necessary we may assume that 0 is not an eigenvalue of $A(s)$. Since the global resolvent set is open, no eigenvalue of $A(t)$ lies on γ or equals 0, for t near s. Since

$$t \mapsto -\frac{1}{2\pi i}\int_\gamma (A(t)-z)^{-1}\,dz =: P(t,\gamma)$$

is a smooth curve of projections (on the direct sum of all eigenspaces corresponding to eigenvalues in the interior of γ) with finite dimensional ranges, the ranks (i.e. dimension of the ranges) must be constant: it is easy to see that the (finite) rank cannot fall locally, and it cannot increase, since the distance in $L(H,H)$ of $P(t)$ to the subset of operators of rank $\le N = \operatorname{rank}(P(s))$ is continuous in t and is either 0 or 1. So for t near s, there are equally many eigenvalues in the interior, and we may call them $\mu_i(t) : 1 \le i \le N$ (repeated with multiplicity). Let us denote by $e_i(t) : 1 \le i \le N$ a corresponding system of eigenvectors of $A(t)$. Then by the residue theorem we have

$$\sum_{i=1}^N \mu_i(t)^p e_i(t)\langle e_i(t),\quad\rangle = -\frac{1}{2\pi i}\int_\gamma z^p (A(t)-z)^{-1}\,dz,$$

which is smooth in t near s, as a curve of operators in $L(H,H)$ of rank N, since 0 is not an eigenvalue.

(2) **Claim.** Let $t \mapsto T(t) \in L(H,H)$ be a smooth curve of operators of rank N in Hilbert space such that $T(0)T(0)(H) = T(0)(H)$. Then $t \mapsto \operatorname{tr}(T(t))$ is smooth (real analytic) (note that this implies T smooth (real analytic) into the space of operators of trace class by (2.3) or (2.14.4), (by (10.3) and (9.4) in the real analytic case) since all bounded linear functionals are of the form $A \mapsto \operatorname{tr}(AB)$ for bounded B, see (52.33), e.g.

Let $F := T(0)(H)$. Then $T(t) = (T_1(t), T_2(t)) : H \to F \oplus F^\perp$ and the image of $T(t)$ is the space

$$\begin{aligned} T(t)(H) &= \{(T_1(t)(x), T_2(t)(x)) : x \in H\} \\ &= \{(T_1(t)(x), T_2(t)(x)) : x \in F\} \text{ for } t \text{ near } 0 \\ &= \{(y, S(t)(y)) : y \in F\}, \text{ where } S(t) := T_2(t)\circ (T_1(t)|F)^{-1}. \end{aligned}$$

Note that $S(t) : F \to F^\perp$ is smooth (real analytic) in t by finite dimensional inversion for $T_1(t)|F : F \to F$. Now

$$\begin{aligned} \operatorname{tr}(T(t)) &= \operatorname{tr}\left(\begin{pmatrix} 1 & 0 \\ -S(t) & 1 \end{pmatrix}\begin{pmatrix} T_1(t)|F & T_1(t)|F^\perp \\ T_2(t)|F & T_2(t)|F^\perp \end{pmatrix}\begin{pmatrix} 1 & 0 \\ S(t) & 1 \end{pmatrix}\right) \\ &= \operatorname{tr}\left(\begin{pmatrix} T_1(t)|F & T_1(t)|F^\perp \\ 0 & -S(t)T_1(t)|F^\perp + T_2(t)|F^\perp \end{pmatrix}\begin{pmatrix} 1 & 0 \\ S(t) & 1 \end{pmatrix}\right) \\ &= \operatorname{tr}\left(\begin{pmatrix} T_1(t)|F & T_1(t)|F^\perp \\ 0 & 0 \end{pmatrix}\begin{pmatrix} 1 & 0 \\ S(t) & 1 \end{pmatrix}\right), \quad \text{since rank} = N \\ &= \operatorname{tr}\begin{pmatrix} T_1(t)|F + (T_1(t)|F^\perp)S(t) & T_1(t)|F^\perp \\ 0 & 0 \end{pmatrix} \\ &= \operatorname{tr}\Big(T_1(t)|F + (T_1(t)|F^\perp)S(t) : F \to F\Big), \end{aligned}$$

which visibly is smooth (real analytic) since F is finite dimensional.

From the claim (2) we now may conclude that

$$\sum_{i=-n}^{m} \lambda_i(t)^p = -\frac{1}{2\pi i} \operatorname{tr} \int_\gamma z^p (A(t)-z)^{-1}\, dz$$

is smooth (real analytic) for t near s.

Thus, the Newton polynomial mapping $s^N(\lambda_{-n}(t), \dots, \lambda_m(t))$ is smooth (real analytic), so also the elementary symmetric polynomial $\sigma^N(\lambda_{-n}(t), \dots, \lambda_m(t))$ is smooth, and thus $\{\mu_i(t) : 1 \le i \le N\}$ is the set of roots of a polynomial with smooth (real analytic) coefficients. By theorem (50.11), there is an arrangement of these roots such that they become differentiable. If no two of the generically different ones meet of infinite order, by theorem (50.10) there is even a smooth arrangement. In the real analytic case, by theorem (50.12) the roots may be arranged in a real analytic way.

To see that in the general smooth case they are even C^1 note that the images of the projections $P(t,\gamma)$ of constant rank for t near s describe the fibers of a smooth vector bundle. The restriction of $A(t)$ to this bundle, viewed in a smooth framing, becomes a smooth curve of symmetric matrices, for which by Rellich's result (50.17) below the eigenvalues can be chosen C^1. This finishes the proof of claim (1).

(3) **Claim.** Let $t \mapsto \lambda_i(t)$ be a differentiable eigenvalue of $A(t)$, defined on some interval. Then

$$|\lambda_i(t_1) - \lambda_i(t_2)| \le (1 + |\lambda_i(t_2)|)(e^{a|t_1-t_2|} - 1)$$

holds for a continuous positive function $a = a(t_1, t_2)$ which is independent of the choice of the eigenvalue.

For fixed t near s take all roots λ_j which meet λ_i at t, order them differentiably near t, and consider the projector $P(t,\gamma)$ onto the joint eigenspaces for only those roots (where γ is a simple smooth curve containing only $\lambda_i(t)$ in its interior, of all the eigenvalues at t). Then the image of $u \mapsto P(u,\gamma)$, for u near t, describes a smooth finite dimensional vector subbundle of $\mathbb{R} \times H$, since its rank is constant. For each u choose an orthonormal system of eigenvectors $v_j(u)$ of $A(u)$ corresponding to these $\lambda_j(u)$. They form a (not necessarily continuous) framing of this bundle. For any sequence $t_k \to t$ there is a subsequence such that each $v_j(t_k) \to w_j(t)$ where $w_j(t)$ is again an orthonormal system of eigenvectors of $A(t)$ for the eigenspace of $\lambda_i(t)$. Now consider

$$\frac{A(t)-\lambda_i(t)}{t_k - t} v_i(t_k) + \frac{A(t_k)-A(t)}{t_k - t} v_i(t_k) - \frac{\lambda_i(t_k)-\lambda_i(t)}{t_k - t} v_i(t_k) = 0,$$

take the inner product of this with $w_i(t)$, note that then the first summand vanishes, and let $t_k \to t$ to obtain

$$\lambda_i'(t) = \langle A'(t) w_i(t), w_i(t)\rangle \text{ for an eigenvector } w_i(t) \text{ of } A(t) \text{ with eigenvalue } \lambda_i(t).$$

This implies, where $V_t = (V, \|\ \ \|_t)$,

$$\begin{aligned}|\lambda_i'(t)| &\leq \|A'(t)\|_{L(V_t,H)} \|w_i(t)\|_{V_t} \|w_i(t)\|_H \\ &= \|A'(t)\|_{L(V_t,H)} \sqrt{\|w_i(t)\|_H^2 + \|A(t)w_i(t)\|_H^2} \\ &= \|A'(t)\|_{L(V_t,H)} \sqrt{1 + \lambda_i(t)^2} \leq a + a|\lambda_i(t)|,\end{aligned}$$

for a constant a which is valid for a compact interval of t's since $t \mapsto \|\ \ \|_t^2$ is smooth on V. By Gronwall's lemma (see e.g. [Dieudonné, 1960,] (10.5.1.3)) this implies claim (3).

By the following arguments we can conclude that all eigenvalues may be numbered as $\lambda_i(t)$ for i in $\mathbb{N}$ or $\mathbb{Z}$ in such a way that they are C^1, or C^∞ under the stronger assumption, or real analytic in the real analytic case, in $t \in \mathbb{R}$. Note first that by claim (3) no eigenvalue can go off to infinity in finite time since it may increase at most exponentially. Let us first number all eigenvalues of $A(0)$ increasingly.

We claim that for one eigenvalue (say $\lambda_0(0)$) there exists a C^1 (or C^∞ or real analytic) extension to all of $\mathbb{R}$; namely the set of all $t \in \mathbb{R}$ with a C^1 (or C^∞ or real analytic) extension of λ_0 on the segment from 0 to t is open and closed. Open follows from claim (1). If this interval does not reach infinity, from claim (3) it follows that $(t, \lambda_0(t))$ has an accumulation point (s, x) at the the end s. Clearly x is an eigenvalue of $A(s)$, and by claim (1) the eigenvalues passing through (s, x) can be arranged C^1 (or C^∞ or real analytic), and thus $\lambda_0(t)$ converges to x and can be extended C^1 (or C^∞ or real analytic) beyond s.

By the same argument we can extend iteratively all eigenvalues C^1 (or C^∞ or real analytic) to all $t \in \mathbb{R}$: if it meets an already chosen one, the proof of (50.11) shows that we may pass through it coherently. In the smooth case look at (50.10) instead, and in the real analytic case look at the proof of (50.12).

Now we start to choose the eigenvectors smoothly, under the stronger assumption in the smooth case, and in the real analytic case. Let us consider again eigenvalues $\{\lambda_i(t) : 1 \leq i \leq N\}$ contained in the interior of a smooth curve γ for t in an open interval I. Then $V_t := P(t, \gamma)(H)$ is the fiber of a smooth (real analytic) vector bundle of dimension N over I. We choose a smooth framing of this bundle, and use then the proof of theorem (50.14) to choose smooth (real analytic) sub vector bundles whose fibers over t are the eigenspaces of the eigenvalues with their generic multiplicity. By the same arguments as in (50.14) we then get global vector sub bundles with fibers the eigenspaces of the eigenvalues with their generic multiplicities, and thus smooth (real analytic) eigenvectors for all eigenvalues. □

50.17. Result. ([Rellich, 1969, page 43], see also [Kato, 1976, II, 6.8]). Let $A(t)$ be a C^1-curve of (finite dimensional) symmetric matrices. Then the eigenvalues can be chosen C^1 in t, on the whole parameter interval.

This result is best possible for the degree of continuous differentiability, as is shown by the example in [Alekseevsky, Kriegl, Losik, Michor, 1996, 7.4]

51. The Nash-Moser Inverse Function Theorem

This section treats the hard implicit function theorem of Nash and Moser following [Hamilton, 1982], in full generality and in condensed form, but with all details. The main difficulty of the proof of the hard implicit function theorem is the following: By trying to use the Newton iteration procedure for a nonlinear partial differential equation one quickly finds out that 'loss of derivatives' occurs and one cannot reach the situation, where the Banach fixed point theorem is directly applicable. Using smoothing operators after each iteration step one can estimate higher derivatives by lower ones and finally apply the fixed point theorem.

The core of this presentation is the following: one proves the theorem in a Fréchet space of exponentially decreasing sequences in a Banach space, where the smoothing operators take a very simple form: essentially just cutting the sequences at some index. The statement carries over to certain direct summands which respect 'bounded losses of derivatives', and one can organize these estimates into the concept of tame mappings and thus apply the result to more general situations. However checking that the mappings and also the inverses of their linearizations in a certain problem are tame mappings (a priori estimates) is usually very difficult. We do not give any applications, in view of our remarks before.

51.1. Remark. Let $f : E \supseteq U \to V \subseteq E$ be a diffeomorphisms. Then differentiation of $f^{-1} \circ f = \mathrm{Id}$ and $f \circ f^{-1} = \mathrm{Id}$ at x and $f(x)$ yields using the chain-rule, that $f'(x)$ is invertible with inverse $(f^{-1})'(f(x))$ and hence $x \mapsto f'(x)^{-1}$ is smooth as well.

The inverse function theorem for Banach spaces assumes the invertibility of the derivative only at one point. Openness of $GL(E)$ in $L(E)$ implies then local invertibility and smoothness of $\mathrm{inv} : GL(E) \to GL(E)$ implies the smoothness of $x \mapsto f'(x)^{-1}$.

Beyond Banach spaces we do not have openness of $GL(E)$ in $L(E)$ as the following example shows.

51.2. Example. Let $E := C^\infty(\mathbb{R}, \mathbb{R})$ and $P : E \to E$ be given by $P(f)(t) := f(t) - t\, f(t)\, f'(t)$. Since multiplication with smooth functions and taking derivatives are continuous linear maps, P is a polynomial of degree 2. Its derivative is given by

$$P'(f)(h)(t) = h(t) - t\, h(t)\, f'(t) - t\, f(t)\, h'(t).$$

In particular, the derivative $P'(0)$ is the identity, hence invertible. However, at the constant functions $f_n = \frac{1}{n}$ the derivative $P'(f_n)$ is not injective, since $h_k(t) := t^k$ are in the kernel: $P'(f_n)(h_k)(t) = t^k - t \cdot 0 \cdot t^k - t \cdot \frac{1}{n} \cdot k \cdot t^{k-1} = t^k \cdot (1 - \frac{k}{n})$.

Let us give an even more natural and geometric example:

51.3. Example. Let M be a compact smooth manifold. For $\mathrm{Diff}(M)$ we have shown that the 1-parameter subgroup of $\mathrm{Diff}(M)$ with initial tangent vector $X \in$

$T_{\mathrm{Id}}\,\mathrm{Diff}(M) = \mathfrak{X}(M)$ is given by the flow Fl^X of X, see (43.1). Thus, the exponential mapping $\mathrm{Exp} : T_{\mathrm{Id}}\,\mathrm{Diff}(M) \to \mathrm{Diff}(M)$ is given by $X \mapsto \mathrm{Fl}^X_1$.

The derivative $T_0 \exp : T_eG = T_0(T_eG) \to T_{\exp(0)}(G) = T_eG$ at 0 of the exponential mapping $\exp : \mathfrak{g} = T_eG \to G$ is given by

$$T_0 \exp(X) := \tfrac{d}{dt}|_{t=0} \exp(tX) = \tfrac{d}{dt}|_{t=0} \mathrm{Fl}^{tX}(1,e) = \tfrac{d}{dt}|_{t=0} \mathrm{Fl}^X(t,e) = X_e.$$

Thus, $T_0 \exp = \mathrm{Id}_{\mathfrak{g}}$. In finite dimensions the inverse function theorem now implies that $\exp : \mathfrak{g} \to G$ is a local diffeomorphism.

What is the corresponding situation for $G = \mathrm{Diff}(M)$? We take the simplest compact manifold (without boundary), namely $M = S^1 = \mathbb{R}/2\pi\,\mathbb{Z}$. Since the natural quotient mapping $p : \mathbb{R} \to \mathbb{R}/2\pi\mathbb{Z} = S^1$ is a covering map we can lift each diffeomorphism $f : S^1 \to S^1$ to a diffeomorphism $\tilde{f} : \mathbb{R} \to \mathbb{R}$. This lift is uniquely determined by its initial value $\tilde{f}(0) \in p^{-1}([0]) = 2\pi\mathbb{Z}$. A smooth mapping $\tilde{f} : \mathbb{R} \to \mathbb{R}$ projects to a smooth mapping $f : S^1 \to S^1$ if and only if $\tilde{f}(t+2\pi) \in \tilde{f}(t) + 2\pi\mathbb{Z}$. Since $2\pi\mathbb{Z}$ is discrete, $f(t+2\pi) - f(t)$ has to be $2\pi n$ for some $n \in \mathbb{Z}$ not depending on t. In order that a diffeomorphism $\tilde{f} : \mathbb{R} \to \mathbb{R}$ factors to a diffeomorphism $f : S^1 \to S^1$ the constant n has to be $+1$ or -1. So we finally obtain an isomorphism $\{f \in \mathrm{Diff}(\mathbb{R}) : f(t+2\pi) - f(t) = \pm 1\}/_{2\pi\mathbb{Z}} \cong \mathrm{Diff}(S^1)$. In particular, we have diffeomorphisms R_θ given by translations with $\theta \in S^1$ (In the picture $S^1 \subseteq \mathbb{C}$ these are just the rotations by with angle θ).

Claim. Let $f \in \mathrm{Diff}(S^1)$ be fixed point free and in the image of exp. Then f is conjugate to some translation R_θ.

We have to construct a diffeomorphism $g : S^1 \to S^1$ such that $f = g^{-1} \circ R_\theta \circ g$. Since $p : \mathbb{R} \to \mathbb{R}/2\pi\mathbb{Z} = S^1$ is a covering map it induces an isomorphism $T_tp : \mathbb{R} \to T_{p(t)}S^1$. In the picture $S^1 \subseteq \mathbb{C}$ this isomorphism is given by $s \mapsto s\,p(t)^\perp$, where $p(t)^\perp$ is the normal vector obtained from $p(t) \in S^1$ via rotation by $\pi/2$. Thus, the vector fields on S^1 can be identified with the smooth functions $S^1 \to \mathbb{R}$ or, by composing with $p : \mathbb{R} \to S^1$ with the 2π-periodic functions $X : \mathbb{R} \to \mathbb{R}$. Let us first remark that the constant vector field $X^\theta \in \mathfrak{X}(S^1)$, $s \mapsto \theta$ has as flow $\mathrm{Fl}^{X^\theta} : (t,\varphi) \mapsto \varphi + t\cdot\theta$. Hence $\exp(X^\theta) = \mathrm{Fl}^{X^\theta}_1 = R_\theta$.

Let $f = \exp(X)$ and suppose $g \circ f = R_\theta \circ g$. Then $g \circ \mathrm{Fl}^X_t = \mathrm{Fl}^{X^\theta}_t \circ g$ for $t = 1$. Let us assume that this is true for all t. Then differentiating at $t = 0$ yields $Tg(X_x) = X^\theta_{g(x)}$ for all $x \in S^1$. If we consider g as diffeomorphism $\mathbb{R} \to \mathbb{R}$ this means that $g'(t) \cdot X(t) = \theta$ for all $t \in \mathbb{R}$. Since f was assumed to be fixed point free the vector field X is nowhere vanishing. Otherwise, there would be a stationary point $x \in S^1$. So the condition on g is equivalent to $g(t) = g(0) + \int_0^t \frac{\theta}{X(s)}\,ds$. We take this as definition of g, where $g(0) := 0$, and where θ will be chosen such that g factors to an (orientation preserving) diffeomorphism on S^1, i.e. $\theta \int_t^{t+2\pi} \frac{ds}{X(s)} = g(t+2\pi) - g(t) = 1$. Since X is 2π-periodic this is true for $\theta = 1/\int_0^{2\pi} \frac{ds}{X(s)}$. Since the flow of a transformed vector field is nothing else but the transformed flow we obtain that $g(\mathrm{Fl}^X(t,x)) = \mathrm{Fl}^{X^\theta}(t, g(x))$, and hence $g \circ f = R_\theta \circ g$. $\square$

In order to show that $\exp : \mathfrak{X}(S^1) \to \operatorname{Diff}(S^1)$ is not locally surjective, it hence suffices to find fixed point free diffeomorphisms f arbitrarily close to the identity which are not conjugate to translations. For this consider the translations $R_{2\pi/n}$ and modify them inside the interval $(0, \frac{2\pi}{n})$ such that the resulting diffeomorphism f satisfies $f(\frac{\pi}{n}) \notin \frac{3\pi}{n} + 2\pi\mathbb{Z}$. Then f^k maps 0 to 2π, and thus the induced diffeomorphism on S^1 has $[0]$ as fixed point. If f would be conjugate to a translation, the same would be true for f^k, hence the translation would have a fixed point and hence would have to be the identity. So f^k must be the identity on S^1, which is impossible, since $f(\frac{\pi}{n}) \notin \frac{3\pi}{n} + 2\pi\mathbb{Z}$.

Let us find out the reason for this break-down of the inverse function theorem. For this we calculate the derivative of exp at the constant vector field $X := X^{2\pi/k}$:

$$\begin{aligned}
\exp'(X)(Y)(x) &= \tfrac{d}{ds}|_{s=0} \exp\big((X+sY)(x)\big) \\
&= \tfrac{d}{ds}|_{s=0} \operatorname{Fl}^{X+sY}(1,x) = \int_0^1 Y(x + \tfrac{2t\pi}{k})\, dt,
\end{aligned}$$

where we have differentiated the defining equation for Fl^{X+sY} to obtain

$$\begin{aligned}
\tfrac{\partial}{\partial t}\tfrac{\partial}{\partial s}|_{s=0} \operatorname{Fl}^{X+sY}(t,x) &= \tfrac{\partial}{\partial s}|_{s=0}\tfrac{\partial}{\partial t} \operatorname{Fl}^{X+sY}(t,x) \\
&= \tfrac{\partial}{\partial s}|_{s=0}(X+sY)(\operatorname{Fl}^{X+sY}(t,x)) \\
&= Y(\operatorname{Fl}^X(t,x)) + X'(\dots) \\
&= Y(x + t\tfrac{2\pi}{k}) + 0,
\end{aligned}$$

and the initial condition $\operatorname{Fl}^{X+sY}(0,x) = x$ gives

$$\tfrac{\partial}{\partial s}|_{s=0} \operatorname{Fl}^{X+sY}(t,x) = \int_0^t Y(x + \tau\frac{2\pi}{k})\, d\tau.$$

If we take $x \mapsto \sin(kx)$ as Y then $\exp'(X)(Y) = 0$, so $\exp'(X)$ is not injective, and since X can be chosen arbitrarily near to 0 we have that exp is not locally injective.

So we may conclude that a necessary assumption for an inverse function theorem beyond Banach spaces is the invertibility of $f'(x)$ not only for one point x but for a whole neighborhood.

For Banach spaces one then uses that $x \mapsto f'(x)^{-1}$ is continuous (or even smooth), which follows directly from the smoothness of $\operatorname{inv} : GL(E) \to GL(E)$, see (51.1). However, for Fréchet spaces the following example shows that inv is not even continuous (for the c^∞-topology).

51.4. Example. Let s be the Fréchet space of all fast falling sequences, i.e. $s := \{(x_k)_k \in \mathbb{R}^{\mathbb{N}} : \|(x_k)_k\|_n := \sup\{(1+k)^n|x_k| : k \in \mathbb{N}\} < \infty \text{ for all } n \in \mathbb{N}\}$. Next we consider a curve $c : \mathbb{R} \to GL(s)$ defined by

$$c(t)((x_k)_k) := ((1-h_0(t))x_0, \dots, (1-h_k(t))x_k, \dots),$$

where $h_k(t) := (1-2^{-k})\,h(kt)$ for an $h \in C^\infty(\mathbb{R},\mathbb{R})$ which will be chosen appropriately.

Then $c(t) \in GL(s)$ provided $h(0) = 0$ and $\operatorname{supp} h$ is compact, since then the factors $1 - h_k(t)$ are equal to 1 for almost all k. The inverse is given by multiplying with $1/(1-h_k(t))$, which exists provided $h(\mathbb{R}) \subseteq [0,1]$.

Let us show next that $\operatorname{inv} \circ c : \mathbb{R} \to GL(s) \subseteq L(s)$ is not even continuous. For this take $x \in s$, and consider

$$t \mapsto c\left(\tfrac{1}{k}\right)^{-1}(x) = (\dots; \frac{1}{1-h_k(\frac{1}{k})} x_k; \dots) = (?, \dots, ?; 2^k x_k; ?, \dots),$$

provided $h(1) = 1$. Let x be defined by $x_k := 2^{-k}$, then $\|c(\frac{1}{k})^{-1}(x) - c(0)^{-1}(x)\|_0 \geq 1 - 2^{-k} \not\to 0$.

It remains to show that $c : \mathbb{R} \to GL(s)$ is continuous or even smooth. Since smoothness of a curve depends only on the bounded sets, and boundedness in $GL(E) \subseteq L(E,E)$ can be tested pointwise because of the uniform boundedness theorem (5.18), it is enough to show that $\mathrm{ev}_x \circ s : \mathbb{R} \to GL(s) \to s$ is smooth. Boundedness in a locally convex space can be tested by the continuous linear functionals, so it would be enough to show that $\lambda \circ \mathrm{ev}_x \circ c : \mathbb{R} \to GL(s) \to s \to \mathbb{R}$ is smooth for all $\lambda \in s^*$. We want to use the particular functionals given by the coordinate projections $\lambda_k : (x_k)_k \mapsto x_k$. These, however, do not generate the bornology, but if $B \subseteq s$ is bounded, then so is $\bigcap_{k\in\mathbb{N}} \lambda_k^{-1}(\lambda_k(B))$. In fact, let B be bounded. Then for every $n \in \mathbb{N}$ there exists a constant C_n such that $(1+k)^n |x_k| \leq C_n$ for all k and all $x = (x_k)_k \in B$. Then every $y \in \lambda_k^{-1}(\lambda_k(x))$ (i.e., $\lambda_k(y) = \lambda_k(x)$) satisfies the same inequality for the given k, and hence $\bigcap_{k\in\mathbb{N}} \lambda_k^{-1}(\lambda_k(B))$ is bounded as well. Obviously, $\lambda_k \circ \mathrm{ev}_x \circ c$ is smooth with derivatives $(\lambda_k \circ \mathrm{ev}_x \circ c)^{(p)}(t) = (1-h_k)^{(p)}(t) x_k$. Let $c^p(t)$ be the sequence with these coordinates. We claim that c^p has values in s and is (locally) bounded. So take an $n \in \mathbb{N}$ and consider

$$\|c^p(t)\|_n = \sup_k (1+k)^n |(1-h_k)^{(p)}(t) x_k|.$$

We have $(1-h_k)^{(p)}(t) = 1^{(p)} - (1-2^{-k}) k^p h^{(p)}(kt)$, and hence this factor is bounded by $1 + k^p \|h^{(p)}\|_\infty$. Since $(1+k^n)(1+\|h^{(p)}\|_\infty k^p)|x_k|$ is by assumption on x bounded we have that $\sup_t \|c^p(t)\|_n < \infty$.

Now it is a general argument, that if we are given locally bounded curves $c^p : \mathbb{R} \to s$ such that $\lambda_k \circ c^0$ is smooth with derivatives $(\lambda_k \circ c^0)^{(p)} = \lambda_k \circ c^p$, then c^0 is smooth with derivatives c^p.

In fact, we consider for $c = c^0$ the following expression

$$\lambda_k\left(\frac{1}{t}\left(\frac{c(t)-c(0)}{t} - c^1(0)\right)\right) = \frac{1}{t}\left(\frac{\lambda_k(c(t)) - \lambda_k(c(0))}{t} - \lambda_k(c^1(0))\right),$$

which is by the classical mean value theorem contained in $\{\frac{1}{2}\lambda_k(c^2(s)) : s \in [0,t]\}$. Thus, taking for B the bounded set $\{\frac{1}{2}c^2(s) : s \in [0,1]\}$, we conclude that $\left((c(t)-c(0))/t - c^1(0)\right)/t$ is contained in the bounded set $\bigcap_{k\in\mathbb{N}} \lambda_k^{-1}(\lambda_k(B))$, and hence $\frac{c(t)-c(0)}{t} \to c^1(0)$. Doing the same for $c = c^k$ shows that c^0 is smooth with derivatives c^k.

From this we conclude that in order to obtain an inverse function theorem we have to assume beside local invertibility of the derivative also that $x \mapsto f'(x)^{-1}$ is smooth. That this is still not enough is shown by the following example:

51.5. Example. Let $E := C^\infty(\mathbb{R}, \mathbb{R})$ and consider the map $\exp_* : E \to E$ given by $\exp_*(f)(t) := \exp(f(t))$. Then one can show that $\exp_*$ is smooth. Its (directional) derivative is given by

$$(\exp_*)'(f)(h)(t) = \tfrac{\partial}{\partial s}|_{s=0} e^{(f+sh)(t)} = h(t)\cdot e^{f(t)},$$

so $(\exp_*)'(f)$ is multiplication by $\exp_*(f)$. The inverse of $(\exp_*)'(f)$ is the multiplication operator with $\frac{1}{\exp_*(f)} = \exp_*(-f)$, and hence $f \mapsto (\exp_*)'(f)^{-1}$ is smooth as well. But the image of $\exp_*$ consists of positive functions only, whereas the curve $c : t \mapsto (s \mapsto 1 - ts)$ is a smooth curve in $E = C^\infty(\mathbb{R}, \mathbb{R})$ through $\exp_*(0) = 1$, and $c(t)$ is not positive for all $t \neq 0$ (take $s := \frac{1}{t}$).

So we will need additional assumptions. The idea of the proof is to use that a Fréchet space is built up from Banach spaces as projective limit, to solve the inverse function theorem for the building blocks, and to try to approximate in that way an inverse to the original function. In order to guarantee that such a process converges, we need (a priori) estimates for the seminorms, and hence we have to fix the basis of seminorms on our spaces.

51.6. Definition. A Fréchet space is called *graded*, if it is provided with a fixed increasing basis of its continuous seminorms. A linear map T between graded Fréchet spaces $(E, (p_k)_k)$ and $(F, (q_k)_k)$ is called *tame of degree d and base b* if

$$\forall n \geq b\ \exists C_n \in \mathbb{R}\ \forall x \in E : q_n(Tx) \leq C_n p_{n+d}(x).$$

Recall that T is continuous if and only if

$$\forall n\ \exists m\ \exists C_n \in \mathbb{R}\ \forall x \in E : q_n(Tx) \leq C_n p_m(x).$$

Two gradings are called *tame equivalent of degree r and base b* if and only if the identity is tame of degree r and base b in both directions.

51.7. Examples. Let M be a compact manifold. Then $C^\infty(M, \mathbb{R})$ is a graded Fréchet space, where we consider as k-th norm the supremum of all derivatives of order less or equal to k. In order that this definition makes sense, we can embed M as closed submanifold into some $\mathbb{R}^n$. Choosing a tubular neighborhood $\mathbb{R}^n \supseteq U \to M$ we obtain an extension operator $p^* : C^\infty(M, \mathbb{R}) \to C^\infty(U, \mathbb{R})$, and on the latter space the operator norms of derivatives $f^k(x)$ for $f \in C^\infty(U, \mathbb{R})$ make sense.

Another way to give sense to the definition is to consider the vector bundle $J^k(M,\mathbb{R})$ of k-jets of functions $f : M \to \mathbb{R}$. Its fiber over $x \in M$ consists of all "Taylor-polynomials" of functions $f \in C^\infty(M,\mathbb{R})$. We obtain an injection of $C^\infty(M,\mathbb{R})$ into the space of sections of $J^k(M,\mathbb{R})$ by associating to $f \in C^\infty(M,\mathbb{R})$ the section having the Taylor-polynomial of f at a point $x \in M$. So it remains to define a norm p_k on the space $C^\infty(M \leftarrow J^k(M,\mathbb{R}))$ of sections. This is just the supremum norm, if we consider some metric on the vector bundle $J^k(M,\mathbb{R}) \to M$.

Another method of choosing seminorms would be to take a finite atlas and a partition of unity subordinated to the charts and use the supremum norms of the derivatives of the chart representations.

A second example of a graded Fréchet space, closely related to the first one, is the space $s(E)$ of fast falling sequences in a Banach space E, i.e.

$$s(E) := \{(x_k)_k \in E^{\mathbb{N}} : \|(x_k)_k\|_n := \sup\{(1+k)^n\|x_k| : k \in \mathbb{N}\} < \infty \text{ for all } n \in \mathbb{N}\}.$$

A modification of this is the space $\Sigma(E)$ of very fast falling sequences in a Banach space E, i.e.

$$\Sigma(E) := \{(x_k)_k \in E^{\mathbb{N}} : \|(x_k)_k\|_n := \sum_{k\in\mathbb{N}} e^{nk}\|x_k\| < \infty \text{ for all } n \in \mathbb{N}\}.$$

51.8. Examples.
(1). Let $T : s(E) \to s(E)$ be the multiplication operator with a polynomial p, i.e., $T((x_k)_k) := (p(k)x_k)_k$.
We claim that T is tame of degree $d := \deg(p)$ and base 0. For this we estimate as follows:

$$\begin{aligned}\|T((x_k)_k)\|_n &= \sup\{(1+k)^n p(k)\|x_k\| : k \in \mathbb{N}\}\\ &\le C_n \sup\{(1+k)^{n+d}\|x_k\| : k \in \mathbb{N}\} = C_n\|(x_k)_k\|_{n+d},\end{aligned}$$

where d is the degree of p and $C_n := \sup\{\frac{|p(k)|}{(1+k)^d} : k \in \mathbb{N}\}$. Note that $C_n < \infty$, since $k \mapsto (1+k)^d$ is not vanishing on $\mathbb{N}$, and the limit of the quotient for $k \to \infty$ is the coefficient of p of degree d.

This shows that $s(E)$ is tamely equivalent to the same space, where the seminorms are replaced by $\sum_k (1+k)^n\|x_k\|$. In fact, the sums are larger than the suprema. Conversely, $\sum_k (1+k)^n\|x_k\| \le \sum_k (1+k)^{-2}(1+k)^{n+2}\|x\|_k \le \left(\sum_k (1+k)^{-2}\right)\|x\|_{n+2}$, showing that the identity in the reverse direction is tame of degree 2 and base 0.

(2). Let $T : \Sigma(E) \to \Sigma(E)$ be the multiplication operator with an exponential function, i.e., $T((x_k)_k) := (a^k x_k)_k$.
We claim that T is tame of some degree and base 0. For this we estimate as follows:

$$\begin{aligned}\|T((x_k)_k)\|_n &= \sum_{k\in\mathbb{N}} e^{nk}a^k\|x_k\| = \sum_{k\in\mathbb{N}} e^{(n+\log(a))k}\|x_k\|\\ &\le \sum_{k\in\mathbb{N}} e^{(n+d)k}\|x_k\| = \|(x_k)_k\|_{n+d},\end{aligned}$$

where d is any integer greater or equal to $\log(a)$. Note however, that T is not well defined on $s(E)$ for $a > 1$, and this is the reason to consider the space $\Sigma(E)$.

Note furthermore, that as before one shows that one could equally well replace the sum by the corresponding supremum in the definition of $\Sigma(E)$, one only has to use that $\sum_{k\in\mathbb{N}} e^{-k} = \frac{1}{1-1/e} < \infty$.

(3). As a similar example we consider a linear differential operator D of degree d, i.e., a local operator (the values Df depend at x only on the germ of f at x) which is locally given in the form $Df = \sum_{|\alpha|\le d} g_\alpha \cdot \frac{\partial^{|\alpha|}}{\partial x^\alpha} f$, with smooth coefficient functions $g_\alpha \in C^\infty(M,\mathbb{R})$ on a compact manifold M.
Then $D : C^\infty(M,\mathbb{R}) \to C^\infty(M,\mathbb{R})$ is tame of degree d and base 0. In fact, by the product rule we can write the k-th derivative of Df as linear combination of partial derivatives of the g_α and derivatives of order up to $k+d$ of f.

(4). Now we give an example of a non-tame linear map. For this consider $T : C^\infty([0,1],\mathbb{R}) \to C^\infty([-1,1],\mathbb{R})$ given by $Tf(t) := f(t^2)$. It was shown in the proof of (25.2) that the image of T consists exactly of the space $C^\infty_{\text{even}}([-1,1],\mathbb{R})$ of even functions. Since $(Tf)^{(n)}(t) = f^{(n)}(t^2)(2t)^n + \sum_{0<2k\le n} c^n_k f^{(n-k)}(t^2)t^{n-2k}$ with some $c^n_k \in \mathbb{Z}$, we have that T is tame of order 0 and degree 0. But the inverse is not tame since $(Tf)^{(2n)}(0)$ is proportional to $f^{(n)}(0)$, hence in order to estimate the n-th derivative of $T^{-1}g$ we need the $2n$-th derivative of g.

51.9. Definition. A graded Fréchet space F is called *tame* if there exists some Banach space E such that F is a tame direct summand in $\Sigma(E)$, i.e. there are tame linear mappings $i : F \to \Sigma(E)$ and $p : \Sigma(E) \to F$ with $p \circ i = \mathrm{Id}_F$.

Our next aim is to show that instead of $\Sigma(E)$ we can equally well use $s(E)$. For this we consider a measured space (X,μ) and a measurable positive weight function $w : X \to \mathbb{R}$ and define

$$\mathcal{L}^1_\Sigma(X,\mu,w) := \Big\{ f \in \mathcal{L}^1(X,\mu) : \|f\|_n := \int_X e^{n\,w(x)}|f(x)|\,d\mu(x) < \infty \Big\}.$$

51.10. Proposition. *Every space $\mathcal{L}^1_\Sigma(X,\mu,w)$ is a tame Fréchet space.*

Proof. Let $X_k := \{x \in X : k \le w(x) < k+1\}$. Then the X_k form a countable disjoint covering of X by measurable sets. Let χ_k be the characteristic function of X_k, and let $R : \mathcal{L}^1_\Sigma(X,\mu,w) \to \Sigma\mathcal{L}^1(X,\mu)$ and $L : \Sigma\mathcal{L}^1(X,\mu) \to \mathcal{L}^1_\Sigma(X,\mu,w)$ be defined by $Rf := (\chi_k \cdot f)_k$ and $L((f_k)_k) := \sum_k \chi_k \cdot f_k$. Then obviously $L \circ R = \mathrm{Id}$. The linear map R is well-defined and tame of degree 0 and base 0, since

$$\begin{aligned}\|Rf\|_n &= \sum_k e^{nk}\|\chi_k f\|_1 = \sum_k \int_{X_k} e^{nk}|f|\,d\mu \le \\ &\le \sum_k \int_{X_k} e^{w(x)n}|f(x)|\,d\mu(x) = \int_X e^{w(x)n}|f(x)|\,d\mu(x) = \|f\|_n.\end{aligned}$$

Finally, L is a well-defined linear map, which is tame of degree 0 and base 0, since

$$\begin{aligned}\|L((f_k)_k)\|_n &= \int_X e^{n\,w(x)}\Big|\sum_k \chi_k f_k(x)\Big|\,d\mu(x) = \sum_k \int_{X_k} e^{n\,w(x)}\,|f_k(x)|\,d\mu(x)\\ &\le \sum_k \int_{X_k} e^{n(k+1)}\,|f_k(x)|\,d\mu(x) \le \sum_k \int_{X_k} e^{n(k+1)}\,|f_k(x)|\,d\mu(x)\\ &= e^n \sum_k e^{nk}\|f_k\|_1 = e^n\|(f_k)_k\|_n. \quad \square\end{aligned}$$

51.11. Corollary. *For every Banach space E the space $s(E)$ is a tame Fréchet space.*

Proof. This result follows immediately from the proposition (51.10) above, if one replaces $\mathcal{L}^1_\Sigma(X,\mu,w)$ by the vector valued function space $\mathcal{L}^1_\Sigma(X,\mu,w;E)$ and similarly the space $\mathcal{L}^1(X,\mu)$ by the Banach space $\mathcal{L}^1(X,\mu;E)$. $\square$

Now let us show the converse direction:

51.12. Proposition. *For every Banach space E the space $\Sigma(E)$ is a tame direct summand of $s(E)$.*

Proof. We define $R : \Sigma(E) \to s(E)$ and $L : s(E) \to \Sigma(E)$ by $R((x_k)_k) := (y_k)_k$, where $y_{[e^k]} := x_k$ and 0 otherwise, and $L((y_k)_k) := (y_{[e^k]})_k$. The map R is well-defined, linear and tame, since $\|(y_k)_k\|_n := \sum_k (1+k)^n\|y_k\| = \sum_j (1+[e^j])^n\|x_j\| \le \sum_j (2e^j)^n\|x_j\| = 2^n\|(x_j)_j\|_n$. The map L is well-defined, linear and tame, since $\|(x_k)_k\|_n := \sum_k e^{kn}\|x_k\| = \sum_k e^{kn}\|y_{[e^k]}\| \le \sum_k (1+[e^k])^n\|y_{[e^k]}\| \le \sum_j (1+j)^n\|y_j\| = \|y\|_n$. Obviously, $L\circ R = \mathrm{Id}$. $\square$

51.13. Definition. A non-linear map $f : E \supseteq U \to F$ between graded Fréchet spaces is called *tame of degree r and base b* if it is continuous and every point in U has a neighborhood V such that

$$\forall n \ge b\ \exists C_n\ \forall x \in V : \|f(x)\|_n \le C_n(1+\|x\|_{n+r}).$$

Remark. Every continuous map from a graded Fréchet space into a Banach space is tame.

For fixed $x_0 \in U$ choose a constant $C > \|f(x_0)\|$ and let $V := \{x : \|f(x)\| < C\}$. Then V is an open neighborhood of x_0, and for all n and all $x \in V$ we have $\|f(x)\|_n = \|f(x)\| \le C \le C(1+\|x\|_n)$.

Every continuous map from a finite dimensional space into a graded Fréchet space is tame.

Choose a compact neighborhood V of x_0. Let $C_n := \max\{\|f(x)\|_n : x \in V\}$. Then $\|f(x)\|_n \le C_n \le C_n(1+\|x\|)$.

It is easily checked that the composite of tame linear maps is tame. In fact

$$\begin{aligned}\|f(g(x))\|_n &\le C(1+\|g(x)\|_{n+r})\\ &\le C(1+C(1+\|x\|_{n+r+s})) \le C(1+\|x\|_{n+r+s})\end{aligned}$$

for all x in an appropriately chosen neighborhood and $n \ge b_f$ and $n+r \ge b_g$.

51.14. Proposition. *The definition of tameness of degree r is coherent with the one for linear maps, but the base may change.*

Proof. Let first f be linear and tame as non-linear map. In particular, we have locally around 0

$$\|f(x)\|_n \le C(1+\|x\|_{n+r}) \text{ for all } n \ge b.$$

If we increase b, we may assume that the 0-neighborhood is of the form $\{x : \|x\|_{b+r} \le \varepsilon\}$ for some $\varepsilon > 0$. For $y \ne 0$ let $x := \frac{\varepsilon}{\|y\|_{b+r}} y$, i.e., $\|x\|_{b+r} = \varepsilon$. Thus, $\|f(x)\|_n \le C(1+\|x\|_{n+r})$. By linearity of f, we get

$$\begin{aligned}\|f(y)\|_n = \|f(x)\|_n \frac{\|y\|_{b+r}}{\varepsilon} &\le C\left(\frac{\|y\|_{b+r}}{\varepsilon} + \|x\|_{n+r}\frac{\|y\|_{b+r}}{\varepsilon}\right)\\ &= C\Big(\frac{\|y\|_{b+r}}{\varepsilon} + \|y\|_{n+r}\Big).\end{aligned}$$

Since $\|y\|_{b+r} \le \|x\|_{n+r}$ for $b \le n$ we get

$$\|f(y)\|_n \le C\Big(\frac{1}{\varepsilon}+1\Big)\|x\|_{n+r}.$$

Conversely, let f be a tame linear map. Then the inequality

$$\|f(x)\|_n \le C\|x\|_{n+r} \le C(1+\|x\|_{n+r}) \text{ for all } n \ge b$$

is true. □

Definition. For functions f of two variables we will define tameness of bi-degree (r,s) and base b if locally

$$\forall n \ge b\ \exists C\ \forall x,y : \|f(x,y)\|_n \le C(1+\|x\|_{n+r}+\|y\|_{n+s});$$

and similar for functions in several variables.

51.15. Lemma. *Let $f : U \times E \to F$ be linear in the second variable and tame of base b and degree (r,s) in a $\|\ \|_{b+r} \times \|\ \|_{b+s}$-neighborhood. Then we have*

$$\forall n \ge b\ \exists C : \|f(x)h\|_n \le C(\|h\|_{n+s} + \|x\|_{n+r}\|h\|_{b+s})$$

for all x in a $\|\ \|_{b+r}$-neighborhood and all h.
If $f : U \times E_1 \times E_2$ is tame of base b and degree (r,s,t) in a $\|\ \|_{b+r} \times \|\ \|_{b+s} \times \|\ \|_{b+t}$-neighborhood. Then we have

$$\|f(x)(h,k)\|_n \le C(\|h\|_{n+s}\|k\|_{b+t} + \|h\|_{b+s}\|k\|_{n+t} + \|x\|_{n+r}\|h\|_{b+s}\|k\|_{b+t})$$

for all x in a $b+r$-neighborhood and all h and k.

Proof. For arbitrary h let $\bar h := \frac{\varepsilon}{\|h\|_{b+s}} h$. Then

$$\|f(x)\bar h\| \le C(1+\|x\|_{n+r}+\|\bar h\|_{n+s}).$$

Therefore

$$\begin{aligned}\|f(x)h\|_n &= \frac{\|h\|_{b+s}}{\varepsilon}\|f(x)\bar{h}\|_n \leq \frac{\|h\|_{b+s}}{\varepsilon} C\left(1+\|x\|_{n+r}+\frac{\varepsilon}{\|h\|_{b+s}}\|h\|_{n+s}\right)\\ &\leq \frac{C\|h\|_{b+s}}{\varepsilon}+\frac{C\|h\|_{b+s}}{\varepsilon}\|x\|_{n+r}+C\|h\|_{n+s}\\ &\leq C\left(\frac{1}{\varepsilon}+1\right)\|h\|_{n+s}+\frac{C}{\varepsilon}\|x\|_{n+r}\|h\|_{b+s}.\end{aligned}$$

The second part is proved analogously. □

51.16. Proposition. Interpolation formula for $\Sigma(E)$.

$$\|x\|_n \cdot \|x\|_m \leq \|x\|_{n-r} \cdot \|x\|_{m+r} \textit{ for } 0 \leq r \leq n \leq m.$$

Proof. Let us first consider the special case, where $n = m$ and $r = 1$. Then

$$\begin{aligned}&\|x\|_{n-1}\cdot\|x\|_{n+1} - \|x\|_n^2 =\\ &= \sum_k e^{(n-1)k}\|x_k\| \sum_l e^{(n+1)l}\|x_l\| - \sum_k e^{nk}\|x_k\| \sum_l e^{nl}\|x_l\|\\ &= \sum_{k=l}(e^{(n-1)k}e^{(n+1)k} - e^{2nk})\|x_k\|^2\\ &\quad + \sum_{k<l}(e^{(n-1)k}e^{(n+1)l} + e^{(n+1)k}e^{(n-1)l} - 2e^{n(k+l)})\|x_k\|\|x_l\|.\end{aligned}$$

In both subsummands the expression in brackets is positive, since

$$\begin{aligned}&e^{(n-1)k}e^{(n+1)l} + e^{(n+1)k}e^{(n-1)l} - 2e^{n(k+l)} =\\ &\qquad = 2e^{n(k+l)}(e^{l-k} + e^{k-l} - 2) = 4e^{n(k+l)}(\cosh(l-k) - 1) \geq 0.\end{aligned}$$

By transitivity, it is enough to show the general case for $r = 1$. Without loss of generality we may assume $x \neq 0$. Then this case is equivalent to

$$\frac{\|x\|_n}{\|x\|_{n-1}} \leq \frac{\|x\|_{m+1}}{\|x\|_m} \text{ for } n \leq m.$$

Again by transitivity it is enough to show this for $m = n$. □

51.17. The Nash-Moser inverse function theorem. *Let E and F be tame Fréchet spaces and let $f : E \supseteq U \to F$ be a tame smooth map. Suppose f' has a tame smooth family Ψ of inverses. Then f is locally bijective, and the inverse of f is a tame smooth map.*

The proof will take the rest of this section.

51.18. Proposition. *Let E and F be tame Fréchet spaces and let $f : E \supseteq U \to F$ be a tame smooth map. Suppose f' has a tame smooth family Ψ of linear left inverses. Then f is locally injective.*

51.19. Proposition. *Let E and F be tame Fréchet spaces and let $f : E \supseteq U \to F$ be a smooth tame map. Suppose f' has a tame smooth family Ψ of linear right inverses. Then f is locally surjective (and locally has a smooth right inverse).*

By a *tame smooth mapping* f we will for the moment understand an infinitely often Gâteaux differentiable map, for which the derivatives $f^{(n)}(x)$ are multilinear and are tame as maps $U \times E^n \to F$.

By a tame smooth family of (one-sided) inverses of f' we understand a family $(\Psi(x))_{x\in U} : F \to E$ of (one-sided) inverses of $(f'(x))_{x\in U}$, which gives a tame smooth map $\Psi^\wedge : U \times F \to E$.

Let us start with some preparatory remarks for the proofs. Contrary to good manners the symbol C will almost never denote the same constant even not in the same inequality. This constant may depend on the index of the norm n but not on any argument of the norms.

For all three proofs we may assume that the initial values are $f : 0 \mapsto 0$ (apply translations in the domain and the codomain).

Claim. We may assume that $E = \Sigma(B)$ and $F = \Sigma(C)$.
First for (51.18). In fact, E and F are direct summands in such spaces $\Sigma(B)$ and $\Sigma(C)$. We extend f to a smooth tame mapping $\tilde f : \Sigma(B) \supseteq \tilde U \to \Sigma(B \times C) \cong \Sigma(B) \times \Sigma(C)$, by setting $\tilde U := p^{-1}(U)$, where $p : \Sigma(B) \to E$ is the retraction, and $\tilde f := (\mathrm{Id} - p, f \circ p)$. Note that $(\mathrm{Id} - p)$ preserves exactly that part which gets annihilated by $f \circ p$. More precisely injectivity of $\tilde f$ implies that of f. In fact, $f(x) = f(y)$ implies $x = p(x)$, $y = p(y)$, and hence $(\mathrm{Id} - p)(x) = 0 = (\mathrm{Id} - p)(y)$, and so $\tilde f(x) = \tilde f(y)$. Since $\tilde f'(\tilde x)(\tilde h) = ((\mathrm{Id} - p)(\tilde h), f'(p(\tilde x)) \cdot p(\tilde h))$, let $\tilde\Psi(\tilde x) := (\mathrm{Id} - p) \circ \mathrm{pr}_1 + \Psi(p(\tilde x)) \circ \mathrm{pr}_2$. Then $\tilde\Psi(\tilde x) \circ \tilde f'(\tilde x) = (\mathrm{Id} - p) \circ (\mathrm{Id} - p) + \Psi(p(\tilde x)) \circ f'(p(\tilde x)) \circ p = \mathrm{Id}$.

Now for (51.19). Here we extend f to a smooth tame mapping $\tilde f : \Sigma(B \times C) \cong \Sigma(B) \times \Sigma(C) \supseteq \tilde U \to \Sigma(C)$, by setting $\tilde U := p^{-1}(U) \times \Sigma(C)$ and $\tilde f := (f \circ p) \oplus (\mathrm{Id} - q)$, where $p : \Sigma(B) \to E$ and $q : \Sigma(C) \to F$ are the retractions. Since $\tilde f'(\tilde x, \tilde y) = f'(p(\tilde x)) \circ p \oplus (\mathrm{Id} - q)$ let $\tilde\Psi(\tilde x, \tilde y) : \Sigma(C) \to \Sigma(B) \times \Sigma(C)$ be defined by $\tilde\Psi(\tilde x, \tilde y)(\tilde k) := (\Psi(p(\tilde x))(q(\tilde k)), (\mathrm{Id} - q)(\tilde k))$, i.e. $\tilde\Psi(\tilde x, \tilde y) := (\Psi(p(\tilde x)) \circ q, (\mathrm{Id} - q))$. Then

$$\begin{aligned}
\tilde f'(\tilde x, \tilde y) \circ \tilde\Psi(\tilde x, \tilde y) &= ((f \circ p)'(\tilde x) \oplus (\mathrm{Id} - q)'(\tilde y)) \circ (\Psi(p(\tilde x)) \circ q, (\mathrm{Id} - q)) \\
&= f'(p(\tilde x)) \circ \underbrace{p \circ \Psi(p(\tilde x))}_{\Psi(p(\tilde x))} \circ q + (\mathrm{Id} - q) \circ (\mathrm{Id} - q) \\
&= q + (\mathrm{Id} - 2q + q^2) = \mathrm{Id}.
\end{aligned}$$

Claim. We may assume that $x \mapsto f(x)$, $(x, h) \mapsto f'(x)h$, $(x, h) \mapsto f''(x)(h, h)$ and $(x, k) \mapsto \Psi(x)k$ satisfy tame estimates of degree $2r$ in x, of degree r in h and 0 in k (for some r) and base 0 on the set $\{\|x\|_0 \le 1\}$.
Consider on $\Sigma(B)$ the linear operators ∇^p which are defined by $(\nabla^p x)_k := e^{pk} x_k$. Then $\|\nabla^p x\|_n = \|x\|_{n+p}$. If f satisfies $\|f(x)\|_n \le C(1 + \|x\|_{n+s})$ on $\|x\|_a \le \delta$

for $n \geq b$ then $\tilde{f} := \nabla^q \circ f \circ \nabla^{-p}$ satisfies $\|\tilde{f}(x)\|_m = \|f(\nabla^{-p}x)\|_{m+q} \leq C(1 + \|\nabla^{-p}x\|_{m+q+s}) = C(1 + \|x\|_{m+q+s-p})$ on $\|x\|_{a-p} \leq \delta$ for $m \geq b - q$.
Choosing q and p sufficiently large, we may assume that f, f', f'', and Ψ satisfy tame estimates of base 0 (choose q large in comparison to b) on $\{x : \|x\|_0 \leq \delta\}$ (choose p large in comparison to a). Furthermore, we may achieve that $(x,k) \mapsto \Psi(x)k$ is tame of order 0 (since by linearity we don't need p for the neighborhood, which is now global, but we have to choose it so that $m+q+s-p \leq m$) in k (but we cannot achieve that this is also true for f'). Now take r sufficiently large such that the degrees are dominated by $2r$ and r, and finally replace f by $x \mapsto f(cx)$ to obtain $\delta = 1$.

Claim. On $\|x\|_{2r} \leq 1$ we have for all $n \geq 0$ a $C_n > 0$ such that

$$\begin{aligned}
\|f(x)\|_n &\leq C_n \|x\|_{n+2r}, \\
\|f'(x)h\|_n &\leq C_n(\|h\|_{n+r} + \|x\|_{n+2r}\|h\|_r), \\
\|f''(x)(h_1,h_2)\|_n &\leq C_n(\|h_1\|_{n+r}\|h_2\|_r + \|h_1\|_r\|h_2\|_{n+r} + \|x\|_{n+2r}\|h_1\|_r\|h_2\|_r), \\
\|\Psi(x)k\|_n &\leq C_n(\|k\|_n + \|x\|_{n+2r}\|k\|_0).
\end{aligned}$$

The 2nd, 3rd and 4th inequality follow from the corresponding tameness and (51.15), since the neighborhood is given by a norm with index higher then base + degree. For the first inequality one would expect $\|f(x)\|_n \leq C(1 + \|x\|_{n+2r})$, but since $f(0) = 0$ one can drop the 1, which follows from integration of the second estimate:

$$\|f(x)\|_n = \left\| f(0) + \int_0^1 f'(tx)x\,dt \right\|_n \leq C(\|x\|_{n+r} + \tfrac{1}{2}\|x\|_{n+2r}\|x\|_r).$$

Since $\|x\|_{n+r} \leq \|x\|_{n+2r}$ and $\|x\|_r \leq \|x\|_{2r} \leq 1$ we are done.

Proof of (51.18). The idea comes from the 1-dimensional situation, where $f(x) = f(y)$ implies by the mean value theorem that there exists an $r \in [x,y] := \{tx + (1-t)y : 0 \leq t \leq 1\}$ with $f'(r) = \frac{f(x)-f(y)}{x-y} = 0$.

51.20. Sublemma. *There exists a $\delta > 0$ such that for $\|x_j\|_{2r} \leq \delta$ we have $\|x_1 - x_0\|_0 \leq C\|f(x_1) - f(x_0)\|_0$. In particular, we have that f is injective on $\{x : \|x\|_{2r} \leq \delta\}$.*

Proof. Using the Taylor formula

$$f(x_1) = f(x_0) + f'(x_0)(x_1 - x_0) + \int_0^1 (1-t)f''(x_0 + t(x_1 - x_0))(x_1 - x_0)^2 dt$$

and $\Psi(x_0) \circ f'(x_0) = \mathrm{Id}$, we obtain that $x_1 - x_0 = \Psi(x_0)(k)$, where

$$k := f(x_1) - f(x_0) - \int_0^1 (1-t)f''(x_0 + t(x_1 - x_0))\,(x_1 - x_0)^2\,dt.$$

For $\|x_j\|_{2r} \leq 1$ we can use the tame estimates of f'' and interpolation to get

$$\begin{aligned}
&\|f''(x_0 + t(x_1 - x_0))(x_1 - x_0)^2\|_n \leq \\
&\leq C\Big(\|x_1 - x_0\|_{n+r}\|x_1 - x_0\|_r + (\|x_1\|_{n+2r} + \|x_0\|_{n+2r})\|x_1 - x_0\|_r^2\Big) \\
&\leq C\Big(\|x_1 - x_0\|_{n+2r}\|x_1 - x_0\|_0 + (\|x_1\|_{n+2r} + \|x_0\|_{n+2r})\|x_1 - x_0\|_{2r}\|x_1 - x_0\|_0\Big) \\
&\leq C\Big((\|x_1\|_{n+2r} + \|x_0\|_{n+2r})\|x_1 - x_0\|_0 + (\|x_1\|_{n+2r} + \|x_0\|_{n+2r})2\delta\|x_1 - x_0\|_0\Big) \\
&\leq C(\|x_1\|_{n+2r} + \|x_0\|_{n+2r})\|x_1 - x_0\|_0.
\end{aligned}$$

Using the tame estimate

$$\|\Psi(x_0)k\|_0 \leq C\|k\|_0(1 + \|x_0\|_{2r}) \leq C\|k\|_0,$$

we thus get

$$\begin{aligned}
\|x_1 - x_0\|_0 = \|\Psi(x_0)k\|_0 &\leq C\|k\|_0 \leq \\
&\leq C\left(\|f(x_1) - f(x_0)\|_0 + \tfrac{1}{2}C(\|x_1\|_{2r} + \|x_0\|_{2r})\|x_1 - x_0\|_0\right) \\
&\leq C\left(\|f(x_1) - f(x_0)\|_0 + \|x_1 - x_0\|_r^2\right) \\
&\leq C\left(\|f(x_1) - f(x_0)\|_0 + \|x_1 - x_0\|_{2r} \cdot \|x_1 - x_0\|_0\right).
\end{aligned}$$

Now use $\|x_1 - x_0\|_{2r} \leq \|x_1\|_{2r} + \|x_0\|_{2r} \leq 2\delta$ to obtain

$$\|x_1 - x_0\|_0 \leq C(\|f(x_1) - f(x_0)\|_0 + 2\delta\|x_1 - x_0\|_0).$$

Taking $\delta < \frac{1}{2C}$ yields the result. □

51.21. Corollary. *Let $\|x_j\|_{2r} \leq \delta$ with δ as before. Then for $n \geq 0$ we have*

$$\|x_1 - x_0\|_n \leq C\Big(\|f(x_1) - f(x_0)\|_n + (\|x_1\|_{n+2r} + \|x_0\|_{n+2r})\,\|f(x_1) - f(x_0)\|_0\Big).$$

Proof. As before we have

$$\|f''(x_0 + t(x_1 - x_0))(x_1 - x_0)^2\|_n \leq C(\|x_1\|_{n+2r} + \|x_0\|_{n+2r})\|x_1 - x_0\|_0.$$

Since Ψ is tame we obtain now

$$\begin{aligned}
\|x_1 - x_0\|_n &= \Big\|\Psi(x_0)\Big(f(x_1) - f(x_0) \\
&\qquad - \int_0^1 (1-t)f''(x_0 + t(x_1 - x_0))(x_1 - x_0)^2\Big)\Big\|_n \\
&\leq \Big\|\Psi(x_0)\Big(f(x_1) - f(x_0)\Big)\Big\|_n + \\
&\qquad + \Big\|\Psi(x_0)\Big(\int_0^1 (1-t)f''(x_0 + t(x_1 - x_0))(x_1 - x_0)^2\Big)\Big\|_n
\end{aligned}$$

$$
\begin{aligned}
&\le C\Big(\|f(x_1)-f(x_0)\|_n + \|x_0\|_{n+2r}\cdot\|f(x_1)-f(x_0)\|_0\Big)+\\
&\quad + C\Big(\Big\|\int_0^1(1-t)f''(x_0+t(x_1-x_0))(x_1-x_0)^2\Big\|_n+\\
&\quad + \|x_0\|_{n+2r}\cdot\Big\|\int_0^1(1-t)f''(x_0+t(x_1-x_0))(x_1-x_0)^2\Big\|_0\Big)\\
&\le C\Big(\|f(x_1)-f(x_0)\|_n + (\|x_1\|_{n+2r}+\|x_0\|_{n+2r})\|f(x_1)-f(x_0)\|_0\\
&\quad + (\|x_1\|_{n+2r}+\|x_0\|_{n+2r})\underbrace{\|x_1-x_0\|_0}_{\le C\|f(x_1)-f(x_0)\|_0}\\
&\quad + (\|x_1\|_{n+2r}+\|x_0\|_{n+2r})\cdot\underbrace{(\|x_1\|_{2r}+\|x_0\|_{2r})}_{\le 2\delta}\underbrace{\|x_1-x_0\|_0}_{C\|f(x_1)-f(x_0)\|_0}\Big)\\
&\le C\Big(\|f(x_1)-f(x_0)\|_n + (\|x_1\|_{n+2r}+\|x_0\|_{n+2r})\,\|f(x_1)-f(x_0)\|_0\Big). \qquad \square
\end{aligned}
$$

Proof of (51.19). As in (51.18) we may assume that the initial condition is $f : 0\mapsto 0$ and that $E=\Sigma(B)$ and $F=\Sigma(C)$.
The idea of the proof is to solve the equation $f(x)=y$ via a differential equation for a curve $t\mapsto x(t)$ whose image under f joins 0 and y affinely. More precisely we consider the parameterization $t\mapsto h(t)\,y$ of the segment $[0,y]$, where $h(t):=1-e^{-ct}$ is a smooth increasing function with $h(0)=0$ and $\lim_{t\to+\infty}h(t)=1$. Differentiation of $f(x(t))=h(t)\,y$ yields $f'(x(t))\cdot x'(t)=h'(t)\,y$ and (if $f'(x)$ is invertible) that $x'(t)=c\,\Psi(x(t))\cdot e^{-ct}y$. Substituting $e^{-ct}\,y=(1-h(t))\,y=y-f(x(t))$ gives

$$x'(t)=c\,\Psi(x(t))\cdot(y-f(x(t))).$$

In Fréchet spaces (like $\Sigma(B)$) we cannot guarantee that this differential equation with initial condition $x(0)=0$ has a solution. The subspaces $B_t:=\{(x_k)_k\in\Sigma(B):x_k=0 \text{ for } k>t\}$ however are Banach spaces (isomorphic to finite products of B), and they are direct summands with the obvious projections. So the idea is to modify the differential equation in such a way that for finite t it factors over B_t and to prove that the solution of the modified equation still converges for $t\to\infty$ to a solution x_∞ of $f(x_\infty)=y$. Since t is a non-discrete parameter we have to consider the spaces B_t as a continuous family of Banach spaces, and so we have to find a family $(\sigma_t)_{t\in\mathbb{R}}$ of projections (called smoothing operators). For this we take a smooth function $\sigma:\mathbb{R}\to[0,1]$ with $\sigma(t)=0$ for $t\le 0$ and $\sigma(t)=1$ for $t\ge 1$. Then we set $\sigma_t(x)(k):=\sigma(t-k)\cdot x(k)$.

We have to show that $\sigma_t\to\mathrm{Id}$, more precisely we

Claim. For $n\ge m$ there exists a $c_{n,m}$ such that $\|\sigma_t x\|_n\le c_{n,m}e^{(n-m)t}\|x\|_m$ and $\|(1-\sigma_t)x\|_m\le c_{n,m}e^{(m-n)t}\|x\|_n$.
Recall that $\|x\|_n:=\sum_k e^{nk}\|x_k\|$. Since $\|(\sigma_t x)_k\|\le\|x_k\|$ for all t and k and

$(\sigma_t x)_k = 0$ for $t \le k$ and $((1-\sigma_t)x)_k = 0$ for $t \ge k+1$ we have

$$\|\sigma_t x\|_n = \sum_k e^{nk}\|(\sigma_t x)_k\|$$
$$\le \sum_{k\le t} e^{nk}\|x_k\| \le \sum_{k\le t} e^{(n-m)k}e^{mk}\|x_k\| \le e^{(n-m)t}\|x\|_m,$$
$$\|(1-\sigma_t)x\|_m \le \sum_{k\ge t-1} e^{mk}\|x_k\| \le \sum_{k\ge t-1} e^{(m-n)k}e^{nk}\|x_k\| \le e^{n-m}\, e^{(m-n)t}\|x\|_k.$$

Now we modify our differential equation by projecting the arguments of the defining function to B_t, i.e.

$$x'(t) = c\,\Psi(\sigma_t(x(t)))\cdot(\sigma_t(y - f(x(t)))) \text{ with } x(0)=0.$$

Thus, our modified differential equation factors for finite t over some Banach space. The following sublemma now provides us with local solutions

51.22. Sublemma. *If a function $f : F \supseteq U \to F$ factors via smooth maps over a Banach space E – i.e., $f = g\circ h$, where $h : F \supseteq U \to W \subseteq E$ and $g : E \supseteq W \to F$ are smooth maps – then the differential equation $y'(t) = f(y(t))$ has locally unique solutions depending continuously (smoothly) on the initial condition $y_0 \in U$.*

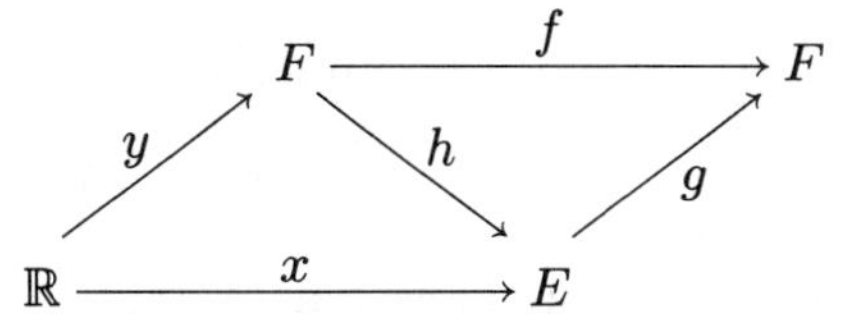

Proof. Suppose y is a solution of the differential equation $y' = f\circ y$ with initial condition $y(0) = y_0$, or equivalently $y(t) = y_0 + \int_0^t f(y(s))\,ds$. The idea is to consider the curve $x := h\circ y$ in the Banach space E. Thus,

$$x(t) = h\Big(y_0 + \int_0^t g(h(y(s)))\,ds\Big) = h\Big(y_0 + \int_0^t g(x(s))\,ds\Big).$$

Now conversely, if x is a solution of this integral equation, then $t\mapsto y(t) := y_0 + \int_0^t g(x(s))\,ds$ is a solution of the original integral equation and hence also of the differential equation, since $x(t) = h(y_0 + \int_0^t g(x(s))\,ds) = h(y(t))$, and so $y(t) = y_0 + \int_0^t g(x(s))\,ds = y_0 + \int_0^t g(h(y(s)))\,ds = y_0 + \int_0^t f(y(s))\,ds$.
In order to show that x exists, we consider the map

$$k : x \mapsto \Big(t\mapsto h\Big(y_0 + \int_0^t g(x(s))\,ds\Big)\Big)$$

and show that it is a contraction.
Since h is smooth we can find a seminorm $\|\ \ \|_q$ on F, a $C>0$ and an $\eta>0$ such that

$$\|h(y_1) - h(y_0)\| \le C\,\|y_1 - y_0\|_q \text{ for all } \|y_j\|_q \le \eta.$$

Furthermore, since g is smooth we find a constant $C > 0$ and $\theta > 0$ such that

$$\|g(x_1) - g(x_0)\|_q \le C\|x_1 - x_0\| \text{ for all } \|x_j\| \le \theta.$$

Since we may assume that $h(0) = 0$, that $\|g(0)\|_q \le C$ and that $\theta \le 1$. So we obtain

$$\|h(y)\| \le C\,\|y\|_q \text{ for all } \|y\|_q \le \eta \text{ and } \|g(x)\|_q \le 2C \text{ for all } \|x\| \le \theta.$$

Let $\tilde U := \{y_0 \in F : \|y_0\|_q \le \delta\}$, let $\tilde V := \{x \in C([0,\varepsilon], E) : \|x(t)\| \le \theta \text{ for all } t\}$, and let $k : F \times C([0,\varepsilon], E) \supseteq \tilde U \times \tilde V \to C([0,\varepsilon], E)$ be given by

$$k(y_0, x)(t) := h\Big(y_0 + \int_0^t g(x(s))\,ds\Big).$$

Then k is continuous with values in $\tilde V$ and is a $C^2\varepsilon$-contraction with respect to x. In fact, $\|y_0 + \int_0^t g(x(s))\,ds\|_q \le \|y_0\|_q + \varepsilon \sup\{\|g(x(s))\|_q : s\} \le \delta + 2C\,\varepsilon \le \eta$ for sufficiently small δ and ε. So $\|k(y_0,x)(t)\| \le C\,\eta \le \theta$ for sufficiently small η. Hence, $k(y_0,x) \in \tilde V$. Furthermore,

$$\begin{aligned}
\|k(y_0,x_1)(t) - k(y_0,x_0)(t)\| &\le C\Big\|\int_0^t g(x_1(s)) - g(x_0(s))\,ds\Big\|_q \\
&\le C\,\varepsilon \sup\{\big\|g(x_1(s)) - g(x_0(s))\big\|_q : s\} \\
&\le C\,\varepsilon\, C \sup\{\|x_1(s) - x_0(s)\| : s\}.
\end{aligned}$$

Thus, by Banach's fixed point theorem, we have a unique solution x of $x = k(y_0, x)$ for every $y_0 \in \tilde U$ and x depends continuously on y_0. As said before, it follows that $y := y_0 + \int_0^t g(x(s))\,ds$ solves the original differential equation. □

Let $k(t) := y - f(x(t))$ be the error we make at time t. For our original differential equation the error was $k(t) = e^{-ct}y$, and hence k was the solution of the homogeneous linear differential equation $k'(t) = -c\,k(t)$. In the modified situation we have

Claim. If $x(t)$ is a solution, $k(t) := y - f(x(t))$, and $g(t)$ is defined by $(f'(\sigma_t \cdot x_t) - f'(x_t))\,x'(t)$ then

$$k'(t) + c\,\sigma_t \cdot k(t) = g(t).$$

From $x'(t) = c\,\Psi(\sigma_t(x(t))) \cdot (\sigma_t(y - f(x(t)))) = c\,\Psi(\sigma_t(x(t))) \cdot (\sigma_t k(t))$ we conclude that $f'(\sigma_t x_t) \cdot x'(t) = c\,\sigma_t k_t$, and by the chain rule $\frac{d}{dt} f(x(t)) = f'(x_t) \cdot x'(t)$ we get

$$k'(t) + c\sigma_t k(t) = (f'(\sigma_t x(t)) - f'(x(t))) \cdot x'(t) =: g(t).$$

In order to estimate that error k, we now consider a general inhomogeneous linear differential equation:

Sublemma. *If the solution k of the differential equation $k'(t) + c\,\sigma_t\, k(t) = g(t)$ exists on $[0,T]$ then for all $p \geq 0$ and $0 < q < c$ we have*

$$\int_0^T e^{qt}\|k(t)\|_p\, dt \leq C\|k(0)\|_{p+q} + C\int_0^T \Big(e^{qt}\|g(t)\|_p + \|g(t)\|_{p+q}\Big)\, dt$$

Proof. If k is a solution of the equation above then the coordinates k_j are solutions of the ordinary inhomogeneous linear differential equation

$$k_j'(t) + c\sigma(t-j)k_j(t) = g_j(t).$$

The solution k_j of that equation can be obtained as usual by solving first the homogeneous equation via separation of variables and applying then the method of variation of the constant. For this recall that for a general 1-dimensional inhomogeneous linear differential equation $k'(t)+\sigma(t)k(t) = g(t)$ of order 1, one integrates the homogeneous equation $\frac{dk_h(t)}{k_h(t)} = -\sigma(t)dt$, i.e., $\log(k_h(t)) = C - \int_0^t \sigma(\tau)d\tau$ or $k_h(t)$ is a multiple of $e^{-\int_0^t \sigma(\tau)d\tau}$. For vector valued equations this is no longer a solution, since $\sigma(t)$ does not commute with $\int_0^t \sigma(\tau)d\tau$ in general, and hence $\frac{d}{dt}e^{-\int_0^t \sigma(\tau)d\tau}$ need not be $-\sigma(t)\cdot e^{-\int_0^t \sigma(\tau)d\tau}$. But in our case, where σ is just a multiplication operator it is still true. For the inhomogeneous equation one makes the ansatz $k(t) := k_h(t)\, C(t)$, which is a solution of the inhomogeneous equation $g(t) = k'(t) + \sigma(t)k(t) = k_h'(t)\, C(t) + k_h(t)\, C'(t) + \sigma(t)\, k_h(t)\, C(t) = -\sigma(t)\, k_h(t)\, C(t) + k_h(t)\, C'(t) + \sigma(t)\, k_h(t)\, C(t) = k_h(t)\, C'(t)$ if and only if $C'(t) = k_h(t)^{-1}\, g(t) = e^{\int_0^t \sigma(\tau)d\tau}\, g(t)$ or $C(t) = C(0) + \int_0^t e^{\int_0^\rho \sigma(\tau)\, g(t)d\tau}\, d\rho$. So the solution of the inhomogeneous equation is given by

$$k(t) = k_h(t)\, C(t) = e^{-\int_0^t \sigma(\tau)d\tau}\Big(C(0) + \int_0^t g(\rho)\, e^{\int_0^\rho \sigma(\tau)d\tau}\, d\rho\Big).$$

In particular we have $k(0) = C(0)$, and using $a(t,s) := e^{-\int_s^t \sigma(\tau)\, d\tau}$ we get $k(t) = a(0,t)k(0) + \int_0^t a(\rho,t)g(\rho)\, d\rho$. We set $a_{j,s,t} := \exp(-c\int_s^t \sigma(\tau-j)d\tau)$. Then

$$k_j(t) = a_{j,0,t}k_j(0) + \int_0^t a_{j,\tau,t}g_j(\tau)d\tau.$$

We claim that $e^{ct}a_{j,s,t} \leq e^c(e^{cs} + e^{cj})$ for $0 \leq s \leq t \leq T$.
For $t \leq j+1$ this follows from $a_{j,s,t} \leq 1$. Let now $t > j+1$. If $s \geq j+1$ then $\sigma(\tau - j) = 1$ for $\tau \geq s$, and so $a_{j,s,t} = e^{-c(t-s)}$ and $e^{ct}a_{j,s,t} = e^{cs}$. If otherwise $s \leq j+1$ then

$$a_{j,s,t} \leq \exp\Big(-c\int_{j+1}^t \sigma(\tau-j)\, d\tau\Big) = e^{-c(t-j-1)}.$$

So $e^{ct}a_{j,s,t} \leq e^{c(j+1)} = e^c e^{cj}$, and the claim holds.

Next we claim that for $0 < q < c$ we have $\int_s^\infty e^{qt} a_{j,s,t} dt \le C(e^{qj} + e^{qs})$. For $j \le s \ge t$ we have $a_{j,s,t} \le Ce^{c(s-t)}$ by the previous claim and

$$\int_s^\infty e^{qt} a_{j,s,t} dt \le Ce^{cs} \int_s^\infty e^{(q-c)t} dt = Ce^{cs} \frac{e^{(q-c)t}}{q-c}\Big|_{t=s}^\infty \le Ce^{qs}$$

using $q < c$. For $s < j$ we split the integral into two parts. From $a_{j,s,t} \le 1$ we conclude that $\int_s^j e^{qt} a_{j,s,t} dt \le \int_s^j e^{qt} dt \le Ce^{qj}$, and from $a_{j,s,t} \le Ce^{c(j-t)}$ (by the previous claim) we conclude that $\int_j^\infty e^{qt} a_{j,s,t} dt \le Ce^{cj} \int_j^\infty e^{(q-c)t} dt \le Ce^{qj}$, which proves the claim.

Now the main claim. We have

$$\begin{aligned} \int_0^T e^{qt} \|k(t)\|_p \, dt &= \int_0^T \sum_j e^{qt} e^{pj} \|k_j(t)\| dt \le \\ &\le \int_0^T \sum_j e^{qt} e^{pj} \Big(a_{j,0,t} \|k_j(0)\| + \int_0^t a_{j,s,t} \|g_j(s)\| ds \Big) dt. \end{aligned}$$

The first summand is bounded by

$$\sum_j e^{pj} \Big(\int_0^T e^{qt} a_{j,0,t} \Big) dt \|k_j(0)\| \le C \sum_j e^{(p+q)j} \|k_j(0)\| \le C \|k(0)\|_{p+q}$$

and the second by

$$\begin{aligned} \int_0^T \sum_j e^{pj} \Big(\int_s^T e^{qt} a_{j,s,t} dt \Big) \|g_j(s)\| ds &\le \int_0^T \sum_j e^{pj} (e^{qj} + q^{qs}) \|g_j(s)\| ds \\ &\le C \int_0^T \|g(s)\|_{p+q} + e^{qs} \|g(s)\|_p ds. \quad \square \end{aligned}$$

We will next show that the domain of definition is not finite and that the limit $\lim_{t\to+\infty} x(t) =: x_\infty$ exists and is a solution of $f(x_\infty) = y$. For this we need the

Sublemma. *If x is a solution and $\|y\|_{2r}$ is sufficiently small then for $n \ge 2r$ and $q \ge 0$ we have*

$$\|x(T)\|_{n+q} \le \int_0^T \|x'(t)\|_{n+q} \, dt \le C_{n,q} e^{qT} \|y\|_n.$$

Proof. Using the inequality for smoothing operators we get for $n \ge 0$ and $q \ge 0$ that

$$\begin{aligned} \|x'(t)\|_{n+q} = \|c\Psi(\sigma_t x(t)) \cdot \sigma_t k(t)\|_{n+q} &\le C \|\sigma_t k_t\|_{n+q} + C \|\sigma_t x(t)\|_{n+q+2r} \|\sigma_t k(t)\|_0 \\ &\le Ce^{qt} \Big(\|k_t\|_n + \|x(t)\|_{n+2r} \|k_t\|_0 \Big). \end{aligned}$$

We will from now on assume that $\|x(t)\|_{2r} \le 1$. Later we will show that this is automatically satisfied. Then

$$\|x'(t)\|_q \le \|k(t)\|_0 \underbrace{(1 + \|x(t)\|_{2r})}_{\le 2} Ce^{qt},$$

and hence

$$\|x(t)\|_q \le \int_0^t \|x'(s)\|_q ds \le C \int_0^t e^{qs} \|k(s)\|_0 \, ds.$$

Using the estimates for k given by the sublemma, we have to estimate norms of

$$g(t) = (f'(\sigma_t x(t)) - f'(x(t))) \cdot x'(t) = -b(x(t), \sigma_t x(t))((1 - \sigma_t)x(t), x'(t)),$$

where $b(x_0, x_1)(h_0, h_1) := \int_0^1 f''(x_0 + t(x_1 - x_0))(h_0, h_1)\, dt$. Obviously, b is smooth, is bilinear with respect to (h_0, h_1), and satisfies a tame estimate of degree $2r$ with respect to x_0, x_1 and degree r in h_0 and h_1 with base 0, i.e.

$$\begin{aligned}\|b(x_0, x_1)(h_0, h_1)\|_n \le C\Big(&\|h_0\|_{n+r}\|h_1\|_r + \|h_0\|_r\|h_1\|_{n+r} \\ &+ (\|x_0\|_{n+2r} + \|x_1\|_{n+2r})\|h_0\|_r\|h_1\|_r\Big).\end{aligned}$$

From this and $\|x_t\|_{2r} \le 1$ we obtain

$$\begin{aligned}\|g(t)\|_n \le C\Big(&\underbrace{\|(1-\sigma_t)x_t\|_{n+r}}_{\le Ce^{-rt}\|x_t\|_{n+2r}} \underbrace{\|x'(t)\|_r}_{\le Ce^{rt}\|k_t\|_0} + \underbrace{\|(1-\sigma_t)x_t\|_r}_{\le Ce^{-(n+r)t}\|x_t\|_{n+2r}} \underbrace{\|x'(t)\|_{n+r}}_{\le Ce^{(n+r)t}\|k_t\|_0} \\ &+ \underbrace{(\|x_t\|_{n+2r} + \|\sigma_t x_t\|_{n+2r})}_{\le \|x_t\|_{n+2r} + C\|x_t\|_{n+2r}} \underbrace{\|(1-\sigma_t)x_t\|_r}_{\le Ce^{-rt}\|x_t\|_{2r}} \underbrace{\|x'(t)\|_r}_{\le Ce^{rt}\|k_t\|_0}\Big) \\ \le C&\|x_t\|_{n+2r}\|k_t\|_0.\end{aligned}$$

In order to estimate $\|x(t)\|_q$, we need the following estimate using the sublemma giving an estimate for the solution $k'(t) + c\sigma(t)k(t) = g(t)$ for $c > 2r$

$$\begin{aligned}&\int_0^T e^{qt}\|k_t\|_p\, dt \le C\Big(\|k(0)\|_{p+q} + \int_0^T \Big(e^{qt} \underbrace{\|g(t)\|_p}_{\le \|x_t\|_{p+2r}\|k_t\|_0} + \underbrace{\|g(t)\|_{p+q}}_{\le \|x_t\|_{p+q+2r}\|k_t\|_0}\Big)\Big) \le \\ &\le C\Big(\|k(0)\|_{p+q} + \int_0^T \underbrace{\Big(e^{qt}C\int_0^t e^{(p+2r)s}\|k_s\|_0\, ds + C\int_0^t e^{(p+q+2r)s}\|k_s\|_0\, ds\Big)}_{\le Ce^{qt}\int_0^t e^{(p+2r)s}\|k_s\|_0\, ds}\|k_t\|_0\, dt\Big) \\ &\le C\Big(\|k(0)\|_{p+q} + \int_0^T e^{qt}\|k_t\|_0\, dt \cdot \int_0^T e^{(p+2r)s}\|k_s\|_0 ds\Big).\end{aligned}$$

This inequality is of recursive nature. In fact, for $p = 0$ and $q = 2r$ it says $K_T := \int_0^T e^{2rt}\|k_t\|_0\, dt \le C\|y\|_{2r} + CK_T^2$ and hence $K_T(1 - CK_T) \le C\|y\|_{2r}$. If $K_T \le \frac{1}{2C}$ then $1 - CK_T \ge \frac{1}{2}$ and hence $K_T \le 2C\|y\|_{2r}$. Thus, $K_T \notin (2C\|y\|_{2r}, \frac{1}{2C}]$.

Therefore, choosing $\|y\|_{2r} < \frac{1}{4C^2}$ makes this a nonempty interval, and continuity of $T \mapsto K_T$ and $K_0 = 0$ shows that

$$\int_0^T e^{2rt}\|k_t\|_0 dt = K_T \le 2C\|y\|_{2r} \text{ for all } \|y\|_{2r} \le \delta.$$

Let us now show that the requirement $\|x(t)\|_{2r} \le 1$ is automatically satisfied. Suppose not, then there is a minimal $t_0 > 0$ with $\|x(t_0)\|_{2r} \ge 1$ since $x(0) = 0$. Thus, for $0 \le t < t_0$ we have $\|x(t)\| < 1$, and hence the above estimates hold on the interval $[0, t_0]$. From $\|x'(t)\|_{2r} \le Ce^{2rt}\|k(t)\|_0$ we obtain by integration that $\|x(t)\|_{2r} \le \int_0^T \|x'(t)\|_{2r} \le \int_0^t Ce^{2rt}\|k(t)\|_0 dt \le 2C\|y\|_{2r}$. Thus, if $\|y\|_{2r} \le \delta$ with $C\delta < 1$ then $\|x(t_0)\|_{2r} < 1$, a contradiction. Note that this shows at the same time that the sublemma is valid for $q = 0$ and $n = 2r$.

Now we proceed to show that

$$\int_0^T e^{qs}\|k_s\|_0\, ds \le C\|y\|_q \text{ for } q \ge 2r.$$

In fact, $q = 2r$ and $q = 2r + 1$ will be sufficient, and hence we have a common C. The above estimate for $p = 0$ gives

$$\int_0^T e^{qt}\|k_s\|_0\, dt \le C\Big(\|\underbrace{k(0)}_{y}\|_q + \int_0^T e^{qt}\|k_t\|_0\, dt \cdot \underbrace{\int_0^T e^{2rs}\|k_s\|_0\, ds}_{\le C\|y\|_{2r} \le C}\Big).$$

Thus

$$(1 - c\|y\|_{2r}) \cdot \int_0^T e^{qt}\|k_s\|_0\, dt \le C\|y\|_q \quad \text{and} \quad \int_0^T e^{qt}\|k_s\|_0\, dt \le C\|y\|_q$$

for $q \ge 2r$.Now for $q \ge p + 2r$

$$\begin{aligned}\int_0^T e^{qt}\|k_s\|_p\, dt &\le C\Big(\|k(0)\|_{p+q} + \overbrace{\int_0^T e^{qt}\|k_t\|_0\, dt}^{\le C\|y\|_q} \cdot \overbrace{\int_0^T e^{(p+2r)s}\|k_s\|_0 ds}^{\le C\|y\|_{p+2r}}\Big)\\ &\le C(\|y\|_{p+q} + \|y\|_{p+q} \cdot \underbrace{\|y\|_{2r}}_{\le 1}) \le C\|y\|_{p+q}.\end{aligned}$$

Now we prove the main claim by induction on $n = p + 2r$. For $p = 0$ we have shown it already. Next for $n + 1 = p + 1 + 2r$: Using the inequality at the very beginning of the proof of the sublemma for n replaced by p and q by $1 + 2r + q$

$$\begin{aligned}\int_0^T \|x'(t)\|_{p+1+2r+q}\, dt &\le \int_0^T Ce^{(1+2r+q)t}\Big(\|k_t\|_p + \|x(t)\|_{p+2r}\|k_t\|_0\Big)\, dt\\ &\le Ce^{qT}\int_0^T \Big(e^{(1+2r)t}\|k(t)\|_p + e^{(1+2r)t}\underbrace{\|x(t)\|_{p+2r}}_{\le C\|y\|_{p+2r}} \cdot \|k_t\|_0\Big) dt\\ &\le Ce^{qT}\Big(C\|y\|_{p+2r+1} + C\|y\|_{p+2r}\|y\|_{2r+1} +\\ &\qquad + \|y\|_{p+2r}C\|y\|_{2r+1}\Big)\\ &\le Ce^{qT}\Big(\|y\|_{p+2r+1} + \|y\|_{p+2r}\|y\|_{2r+1}\Big).\end{aligned}$$

For $\|y\|_{2r} \le \delta \le 1$ we get by interpolation that

$$\|y\|_{p+2r}\|y\|_{2r+1} \le C\|y\|_{p+2r+1}\|y\|_{2r} \le C\|y\|_{p+1+2r},$$

which eliminates the last summand and completes the induction.

Claim. For sufficiently small $\|y\|_{2r}$ the solution x exists globally, $\lim_{t\to+\infty} x(t) =: x_\infty$ exists and solves $f(x_\infty) = y$. Moreover, we have $\|x_\infty\|_n \le c_n\|y\|_n$.
Furthermore, $\|x(t)\|_n \le \int_0^t \|x'(s)\|_n ds \le C\|y\|_n$ for $n > 2r$ and $\|y\|_{2r} \le \delta$ using the main claim for $q = 0$.

Suppose x exists on $[0,\omega)$ with ω chosen maximally. Since $\int_0^T \|x'(t)\|_n dt \le C\|y\|_n$ with C independent on $T < \omega$ we have $\int_0^\omega \|x'(t)\|_n dt \le C\|y\|_n < \infty$, and hence $\lim_{t\to\omega}\int_t^\omega \|x'(\tau)\|d\tau = 0$. Thus, we obtain $\|x(s) - x(t)\|_n \le \int_t^s \|x'(r)\|dr \to 0$ for $s,t\to\omega$. Hence, $x(\omega) := \lim_{t\nearrow\omega} x(t)$ exists and $\|x(\omega)\|_{2r} \le 1$.

Thus, we can extend the solution in a neighborhood of ω, a contradiction to the maximality of ω. Then by the same argument as before $x_\infty := \lim_{t\nearrow\infty} x(t)$ exists and $\|x_\infty\|_{2r} \le 1$ and $\|x_\infty\|_n \le C\|y\|_n$ for all $n \ge 2r$. Since $x'(t) = c\Psi(\sigma(t)x(t))\cdot(\sigma(t)(y - f(x(t))))$ we have that $\lim_{t\to\infty} x'(t) = c\Psi(x_\infty)(y - f(x_\infty))$ exists, and since $\int_0^\infty \|x'(t)\|_n dt \le C\|y\|_n < \infty$ we have that $(x_\infty)(y - f(x_\infty)) = \lim_{t\nearrow\infty} x'(t) = 0$. So we get $y - f(x_\infty) = 0$, hence we have obtained an inverse. □

Proof of the inverse function theorem. By what we have shown so far, we know that f is locally bijective and $\|f^{-1}y\|_n \le C\|y\|_n$ for all $n \ge 2r$. Furthermore,

$$\|f^{-1}y_1 - f^{-1}y_0\|_n \le C\Big(\|y_1 - y_0\|_n + (\|f^{-1}y_1\|_{n+2r} + \|f^{-1}y_0\|_{n+2r})\|y_1 - y_0\|_0\Big),$$

which shows continuity and Lipschitzness of the inverse, and the inverse f^{-1} is tame and locally Lipschitz.

We next show that f^{-1} is Gâteaux-differentiable with derivative as expected, i.e.

$$(f^{-1})'(y)(k) = f'(f^{-1}y)\cdot k.$$

For this let $c(t) := f^{-1}(y + tk)$, $x := c(0) = f^{-1}(y)$ and $\ell := \Psi(x) := f'(f^{-1}(y))^{-1}$. Then c is locally Lipschitz, and we have to show that $\frac{c(t)-c(0)}{t} \to \ell\cdot k$ for $t \to 0$. Now

$$\begin{aligned}
\frac{c(t)-c(0)}{t} - \ell(k) &= (\ell\circ f'(x))\Big(\frac{c(t)-c(0)}{t}\Big) - \ell(k)\\
&= \ell\Big(f'(x)\Big(\frac{c(t)-c(0)}{t}\Big) - \frac{f(c(t)) - f(c(0))}{t}\Big)\\
&= \ell\Big(\int_0^1 \underbrace{f'(x) - f'(x + s(c(t)-x))}_{=:g(t,s)}\, ds\cdot\Big(\frac{c(t)-c(0)}{t}\Big)\Big).
\end{aligned}$$

Since $t\mapsto c(t)$ is locally Lipschitz, the map $(t,s)\mapsto f'(x) - f'(x + s(c(t)-x))$ is locally Lipschitz, and hence in particular continuous. Therefore, $g(t,s) \to g(0,s)$

for $t \to 0$ uniformly on all $s \in [0,1]$. Thus, $\int_0^1 g(t,s)ds \to \int_0^1 g(0,s)ds = \int_0^1 0\,ds = 0$ in $L(E,F)$, and since $\frac{c(t)-c(0)}{t}$ stays bounded, this proves the claim.

Thus, we have for the Gâteaux-derivative of the inverse function the formula

$$(f^{-1})' = \operatorname{inv} \circ f' \circ f^{-1} = \Psi \circ f^{-1}.$$

Since Ψ and f^{-1} are tame, so is $(f^{-1})'$. By induction, using the chain rule for differentiable and for tame maps we conclude that (f^{-1}) is a tame smooth map, since Ψ was assumed to be so. Note that in order to apply the chain-rule it is not enough to have Gâteaux-differentiability, but because of tameness (or local Lipschitzness) of the derivative we have the appropriate type of differentiability automatical. In fact, we have to consider $d(f^{-1}) := ((f^{-1})')^\wedge = \Psi^\wedge \circ (f^{-1} \times \mathrm{Id})$ and apply induction to that. □

Let us finally show that it is enough to assume that Ψ is a tame continuous in order to assure that it is a tame smooth map.

51.23. Lemma. *Let $\Phi : E \supseteq U \to GL(F)$ be a tame smooth map and Ψ, defined by $\Psi(x) := \Phi(x)^{-1}$, be a continuous tame map. Then Ψ is a tame smooth map.*

Proof. For smoothness it is enough to show smoothness along continuous curves, so we may assume that $E = \mathbb{R} = U$, and so Φ is a curve denoted c. Then

$$\frac{c(t)^{-1} - c(s)^{-1}}{t-s} = -\operatorname{comp}(\operatorname{inv}(c(t)), \frac{c(t)-c(s)}{t-s}, \operatorname{inv}(c(s)))$$

and hence is locally bounded, i.e., c is locally Lipschitz. Now let s be fixed. Then

$$t \mapsto -\operatorname{comp}(\operatorname{inv}(c(t+s)), \frac{c(t+s)-c(s)}{t}, \operatorname{inv}(c(s)))$$

is locally Lipschitz, since $t \mapsto \frac{c(t+s)-c(s)}{t} = \int_0^1 c'(s+rt)dr$ is smooth. In particular,

$$\frac{1}{t}\left(\frac{c(t+s)^{-1} - c(s)^{-1}}{t} + \operatorname{inv}(c(s)) \circ c'(s) \circ \operatorname{inv}(c(s))\right)$$

is locally bounded, and hence $\operatorname{inv} \circ c$ is differentiable with derivative

$$(\operatorname{inv} \circ c)'(s) = -\operatorname{inv}(c(s)) \circ c'(s) \circ \operatorname{inv}(c(s)).$$

Thus, $\operatorname{inv} \circ c$ is smooth by induction, and

$$\Psi'(x)(y) = -\Psi(x) \circ \Phi'(x)(y) \circ \Psi(x).$$

Tameness of $\Psi^\wedge$ follows since the differential of $\Psi^\wedge$ is given by

$$\begin{aligned} d\Psi^\wedge(x,h;y,k) &= \Psi'(x)(y)(h) + \Psi(x)'(h)(k) = \Psi'(x)(y)(h) + \Psi(x)(k) \\ &= (-\Psi(x) \circ \Phi'(x)(y) \circ \Psi(x))(h) + \Psi(x)(k) \\ &= \Psi^\wedge(x,k) - \Psi^\wedge(x, \partial_1\Phi^\wedge(x,y)(\Psi^\wedge(x,h))). \quad \square \end{aligned}$$

52. Appendix: Functional Analysis

The aim of this appendix is the following. This book needs prerequisites from functional analysis, in particular about locally convex spaces, which are beyond usual knowledge of non-specialists. We have used as unique reference the book [Jarchow, 1981]. In this appendix we try to sketch these results and to connect them to more widespread knowledge in functional analysis: for this we decided to use [Schaefer, 1971].

52.1. Basic concepts. A *locally convex space* E is a vector space together with a Hausdorff topology such that addition $E \times E \to E$ and scalar multiplication $\mathbb{R} \times E \to E$ (or $\mathbb{C} \times E \to E$) are continuous and 0 has a basis of neighborhoods consisting of (absolutely) convex sets. Equivalently, the topology on E can be described by a system $\mathcal{P}$ of (continuous) seminorms. A *seminorm* $p : E \to \mathbb{R}$ is specified by the following properties: $p(x) \geq 0$, $p(x+y) \leq p(x) + p(y)$, and $p(\lambda x) = |\lambda| p(x)$.

A set B in a locally convex space E is called *bounded* if it is absorbed by each 0-neighborhood, equivalently, if each continuous seminorm is bounded on B. The family of all bounded subsets is called the *bornology* of E. The *bornologification of a locally convex space* is the finest locally convex topology with the same bounded sets, which is treated in detail in (4.2) and (4.4). A locally convex space is called *bornological* if it is stable under the bornologification, see also (4.1). The *ultrabornologification of a locally convex space* is the finest locally convex topology with the same bounded absolutely convex sets for which E_B is a Banach space.

52.2. Result. [Jarchow, 1981, 6.3.2] *&* [Schaefer, 1971, I.1.3] *The Minkowski functional* $q_A : x \mapsto \inf\{t > 0 : x \in t.A\}$ *of a convex absorbing set* A *containing 0 is a convex function.*

A subset A in a vector space is called *absorbing* if $\bigcup\{rA : r > 0\}$ is the whole space.

52.3. Result. [Jarchow, 1981, 6.4.2.(3)] *For an absorbing radial set* U *in a locally convex space* E *the closure is given by* $\{x \in E : q_U(x) \leq 1\}$*, where* q_U *is the Minkowski functional.*

52.4. Result. [Jarchow, 1981, 3.3.1] *Let* X *be a set and let* F *be a Banach space. Then the space* $\ell^\infty(X, F)$ *of all bounded mappings* $X \to F$ *is itself a Banach space, supplied with the supremum norm.*

52.5. Result. [Jarchow, 1981, 3.5.6, p66] *&* [Schaefer, 1971, I.3.6] *A Hausdorff topological vector space E is finite dimensional if and only if it admits a precompact neighborhood of 0.*

A subset K of E is called *precompact* if finitely many translates of any neighborhood of 0 cover K.

52.6. Result. [Jarchow, 1981, 6.7.1, p112] *&* [Schaefer, 1971, II.4.3] *The absolutely convex hull of a precompact set is precompact.*

A set B in a vector space E is called *absolutely convex* if $\lambda x + \mu y \in B$ for $x, y \in B$ and $|\lambda| + |\mu| \leq 1$. By E_B we denote the linear span of B in E, equipped with the Minkowski functional q_B. This is a normed space.

52.7. Result. [Jarchow, 1981, 4.1.4] *&* [Horvath, 1966] *A basis of neighborhoods of 0 of the direct sum $\mathbb{C}^{(\mathbb{N})}$ is given by the sets of the form $\{(z_k)_k \in \mathbb{C}^{(\mathbb{N})} : |z_k| \leq \varepsilon_k$ for all $k\}$ where $\varepsilon_k > 0$.*

The *direct sum* $\bigoplus_i E_i$, also called the *coproduct* $\coprod_i E_i$ of locally convex spaces E_i is the subspace of the cartesian product formed by all points with only finitely many non-vanishing coordinates supplied with the finest locally convex topology for which the inclusions $E_j \to \coprod_i E_i$ are continuous. It solves the universal problem for a coproduct: For continuous linear mappings $f_i : E_i \to F$ into a locally convex space there is a unique continuous linear mapping $f : \coprod_i E_i \to F$ with $f \circ \text{incl}_j = f_j$ for all j. The bounded sets in $\bigoplus_i E_i$ are exactly those which are contained and bounded in a finite subsum. If all spaces E_i are equal to E and the index set is Γ, we write $E^{(\Gamma)}$ for the direct sum.

52.8. Result. [Jarchow, 1981, 4.6.1, 4.6.2, 6.6.9] *&* [Schaefer, 1971, II.6.4 and II.6.5] *Let E be the strict inductive limit of a sequence of locally convex vector spaces E_n. Then every E_n carries the trace topology of E, and every bounded subset of E is contained in some E_n, i.e., the inductive limit is regular.*

Let E be a functor from a small (index) category into the category of all locally convex spaces with continuous linear mappings as morphisms. The *colimit* $\operatorname{colim} E$ of the functor E is the unique (up to isomorphism) locally convex space together with continuous linear mappings $l_i : E(i) \to \operatorname{colim} E$ which solves the following universal problem: Given continuous linear $g_i : E(i) \to F$ into a locally convex space F with $g_j \circ E(f) = g_i$ for each morphism $f : i \to j$ in the index category. Then there exists a unique continuous linear mapping $g : \operatorname{colim} E \to F$ with $g \circ l_i = g_i$ for all i.

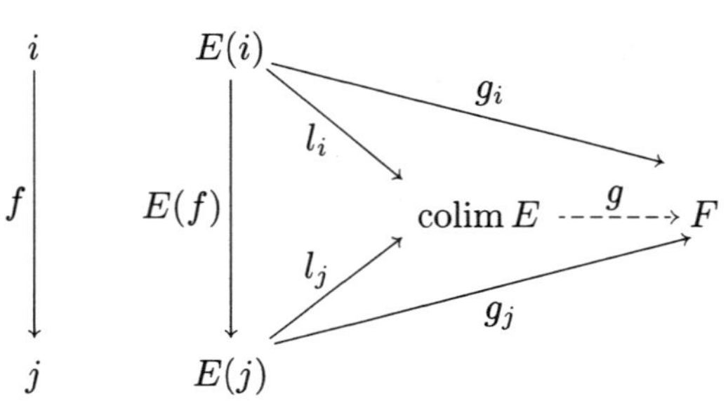

The colimit is given as the locally convex quotient of the direct sum $\coprod_i E(i)$ by the closed linear subspace generated by all elements of the form $\text{incl}_i(x) - (\text{incl}_j \circ E(f))(x)$ for all $x \in E(i)$ and $f : i \to j$ in the index category. Compare [Jarchow, 1981, p.82 & p.110], but we force here inductive limits to be Hausdorff. A *directed set* Γ is a partially ordered set such that for any two elements there is another one that is larger that the two. The *inductive limit* is the colimit of a functor from a directed set (considered as a small category); one writes $\varinjlim_j E_j$ for this. A *strict inductive limit* is the inductive limit of a functor E on the directed set $\mathbb{N}$ such that $E(n < n+1) : E(n) \to E(n+1)$ is the topological embedding of a closed linear subspace.

The dual notions (with the arrows between locally convex spaces reversed) are called the *limit* $\lim E$ of the functor E, and the *projective limit* $\varprojlim_j E_j$ in the case of a directed set. It can be described as the linear subset of the cartesian product $\prod_i E(i)$ consisting of all $(x_i)_i$ with $E(f)(x_i) = x_j$ for all $f : i \to j$ in the index category.

52.9. Result. [Jarchow, 1981, 5.1.4+11.1.6] *&* [Schaefer, 1971, III.5.1, Cor. 1] *Every separately continuous bilinear mapping on Fréchet spaces is continuous.*

A *Fréchet space* is a complete locally convex space with a metrizable topology, equivalently, with a countable base of seminorms. See [Jarchow, 1981, 2.8.1] or [Schaefer, 1971, p.48].

Closed graph and open mapping theorems. These are well known if Banach spaces or even Fréchet spaces are involved. We need a wider class of situations where these theorems hold; those involving webbed spaces. Webbed spaces were introduced for exactly this reason by de Wilde in his thesis, see [de Wilde, 1978]. We do not give their (quite lengthy) definition here, only the results and the permanence properties.

52.10. Result. Closed Graph Theorem. [Jarchow, 1981, 5.4.1] *Any closed linear mapping from an inductive limit of Baire locally convex spaces into a webbed locally convex space is continuous.*

52.11. Result. Open Mapping Theorem. [Jarchow, 1981, 5.5.2] *Any continuous surjective linear mapping from a webbed locally convex space into an inductive limit of Baire locally convex spaces vector spaces is open.*

52.12. Result. *The Fréchet spaces are exactly the webbed spaces with the Baire property.*

This corresponds to [Jarchow, 1981, 5.4.4] by noting that Fréchet spaces are Baire.

52.13. Result. [Jarchow, 1981, 5.3.3] *Projective limits and inductive limits of sequences of webbed spaces are webbed.*

52.14. Result. *The bornologification of a webbed space is webbed.*

This follows from [Jarchow, 1981, 13.3.3 and 5.3.1.(d)] since the bornologification is coarser that the ultrabornologification, [Jarchow, 1981, 13.3.1].

52.15. Definition. [Jarchow, 1981, 6.8] *For a zero neighborhood U in a locally convex vector space E we denote by $\widetilde{E_{(U)}}$ the completed quotient of E with the Minkowski functional of U as norm.*

52.16. Result. Hahn-Banach Theorem. [Jarchow, 1981, 7.3.3] *Let E be a locally convex vector space and let $A \subset E$ be a convex set, and let $x \in E$ be not in the closure of A. Then there exists a continuous linear functional ℓ with $\ell(x)$ not in the closure of $\ell(A)$.*

This is a consequence of the usual Hahn-Banach theorem, [Schaefer, 1971,II.9.2]

52.17. Result. [Jarchow, 1981, 7.2.4] *Let $x \in E$ be a point in a normed space. Then there exists a continuous linear functional $x' \in E^*$ of norm 1 with $x'(x) = \|x\|$.*

This is another consequence of the usual Hahn-Banach theorem, cf. [Schaefer, 1971, II.3.2].

52.18. Result. Bipolar Theorem. [Jarchow, 1981, 8.2.2] *Let E be a locally convex vector space and let $A \subset E$. Then the bipolar A^{oo} in E with respect to the dual pair (E, E^*) is the closed absolutely convex hull of A in E.*

For a duality $\langle \ , \ \rangle$ between vector spaces E and F and a set $A \subseteq E$ the *polar* of A is $A^o := \{y \in F : |\langle x, y\rangle| \leq 1 \text{ for all } x \in A\}$. The *weak topology* $\sigma(E, F)$ is the locally convex topology on E generated by the seminorms $x \mapsto |\langle x, y\rangle|$ for all $y \in F$.

52.19. Result. [Schaefer, 1971, IV.3.2] *A subset of a locally convex vector space is bounded if and only if every continuous linear functional is bounded on it.*

This follows from [Jarchow, 1981, 8.3.4], since the weak topology $\sigma(E, E')$ and the given topology are compatible with the duality, and a subset is bounded for the weak topology, if and only if every continuous linear functional is bounded on it.

52.20. Result. Alaoğlu-Bourbaki Theorem. [Jarchow, 1981, 8.5.2 & 8.5.1.b] *&* [Schaefer, 1971, III.4.3 and II.4.5] *An equicontinuous subset K of E' has compact closure in the topology of uniform convergence on precompact subsets; On K the latter topology coincides with the weak topology $\sigma(E', E)$.*

52.21. Result. [Jarchow, 1981, 8.5.3, p157] *&* [Schaefer, 1971, III.4.7] *Let E be a separable locally convex vector space. Then each equicontinuous subset of E' is metrizable in the weak* topology $\sigma(E', E)$.*

A topological space is called *separable* if it contains a dense countable subset.

52.22. Result. Banach Dieudonné theorem. [Jarchow, 1981, 9.4.3, p182] *&* [Schaefer, 1971, IV.6.3] *On the dual of a metrizable locally convex vector space E the topology of uniform convergence on precompact subsets of E coincides with the so-called equicontinuous weak*-topology which is the final topology induced by the inclusions of the equicontinuous subsets.*

52.23. Result. [Jarchow, 1981, 10.1.4] *In metrizable locally convex spaces the convergent sequences coincide with the Mackey-convergent ones.*

For Mackey convergence see (1.6).

52.24. Result. [Jarchow, 1981, 10.4.3, p202] *&* [Horvath, 1966, p277] *In Schwartz spaces bounded sets are precompact.*

A locally convex space E is called *Schwartz* if each absolutely convex neighborhood U of 0 in E contains another one V such that the induced mapping $E_{(U)} \to E_{(V)}$ maps U into a precompact set.

52.25. Result. Uniform boundedness principle. [Jarchow, 1981, 11.1.1] *(*[Schaefer, 1971, IV.5.2] *for $F = \mathbb{R}$) Let E be a barrelled locally convex vector space and F be a locally convex vector space. Then every pointwise bounded set of continuous linear mappings from E to F is equicontinuous.*

Note that each Fréchet space is barrelled, see [Jarchow, 1981, 11.1.5].

A locally convex space is called *barrelled* if each closed absorbing absolutely convex set is a 0-neighborhood.

52.26. Result. [Jarchow, 1981, 11.5.1, 13.4.5] *&* [Schaefer, 1971, IV.5.5] *Montel spaces are reflexive.*

By a *Montel space* we mean (following [Jarchow, 1981, 11.5]) a locally convex vector space which is barrelled and in which every bounded set is relatively compact. A locally convex space E is called *reflexive* if the canonical embedding of E into the strong dual of the strong dual of E is a topological isomorphism.

52.27. Result. [Jarchow, 1981, 11.6.2, p231] *Fréchet Montel spaces are separable.*

52.28. Result. [Jarchow, 1981, 12.5.8, p266] *In the strong dual of a Fréchet Schwartz space every converging sequence is Mackey converging.*

The *strong dual of a locally convex space E* is the dual E^* of all continuous linear functionals equipped with the topology of uniform convergence on bounded subsets of E.

52.29. Result. *Fréchet Montel spaces have a bornological strong dual.*

Proof. By (52.26) a Fréchet Montel space E is reflexive, thus it's strong dual E'_β is also reflexive by [Jarchow, 1981, 11.4.5.(f)]. So it is barrelled by [Jarchow, 1981,

11.4.2]. By [Jarchow, 1981, 13.4.4] or [Schaefer, 1971, IV.6.6] the strong dual E'_β of a metrizable locally convex vector space E is bornological if and only if it is barrelled and the result follows. □

52.30. Result. [Jarchow, 1981, 13.5.1] *Inductive limits of ultrabornological spaces are ultrabornological.*

Similar to the definition of bornological spaces in (4.1) we define ultrabornological spaces, see [Jarchow, 1981, 13.1.1]. A *bounded completant set* B in a locally convex vector space E is an absolutely convex bounded set B for which the normed space (E_B, q_B) is complete. A locally convex vector space E is called *ultrabornological* if the following equivalent conditions are satisfied:

(1) For any locally convex vector space F a linear mapping $T : E \to F$ is continuous if it is bounded on each bounded completant set. It is sufficient to know this for all Banach spaces F.
(2) A seminorm on E is continuous if it is bounded on each bounded completant set.
(3) An absolutely convex subset is a 0-neighborhood if it absorbs each bounded completant set.

52.31. Result. [Jarchow, 1981, 13.1.2] *Every ultra-bornological space is an inductive limit of Banach spaces.*

In fact, $E = \varinjlim_B E_B$ where B runs through all bounded closed absolutely convex sets in E. Compare with the corresponding result (4.2) for bornological spaces.

52.32. Nuclear Operators. A linear operator $T : E \to F$ between Banach spaces is called *nuclear* or *trace class* if it can be written in the form

$$T(x) = \sum_{j=1}^{\infty} \lambda_j \langle x, x_j \rangle y_j,$$

where $x_j \in E'$, $y_j \in F$ with $\|x_j\| \le 1$, $\|y_j\| \le 1$, and $(\lambda_j)_j \in \ell^1$. The *trace* of T is then given by

$$\operatorname{tr}(T) = \sum_{j=1}^{\infty} \lambda_j \langle y_j, x_j \rangle.$$

The operator T is called *strongly nuclear* if $(\lambda_j)_j \in s$ is rapidly decreasing.

52.33. Result. [Jarchow, 1981, 20.2.6] *The dual of the Banach space of all trace class operators on a Hilbert space consists of all bounded operators. The duality is given by* $\langle T, B \rangle = \operatorname{tr}(TB) = \operatorname{tr}(BT)$.

52.34. Result. [Jarchow, 1981, 21.1.7] *Countable inductive limits of strongly nuclear spaces are again strongly nuclear. Products and subspaces of strongly nuclear spaces are strongly nuclear.*

A locally convex space E is called *nuclear* (or *strongly nuclear*) if each absolutely convex 0-neighborhood U contains another one V such that the induced mapping

$\widetilde{E_{(V)}} \to \widetilde{E_{(U)}}$ is a nuclear operator (or strongly nuclear operator). A locally convex space is (strongly) nuclear if and only if its completion is it, see [Jarchow, 1981, 21.1.2]. Obviously, a nuclear space is a Schwartz space (52.24) since a nuclear operator is compact. Since nuclear operators factor over Hilbert spaces, see [Jarchow, 1981, 19.7.5], each nuclear space admits a basis of seminorms consisting of Hilbert norms, see [Schaefer, 1971, III.7.3].

52.35. Grothendieck-Pietsch criterion. Consider a directed set $\mathcal{P}$ of non-negative real valued sequences $p = (p_n)$ with the property that for each $n \in \mathbb{N}$ there exists a $p \in \mathcal{P}$ with $p_n > 0$. It defines a complete locally convex space (called *Köthe sequence space*)

$$\Lambda(\mathcal{P}) := \{x = (x_n)_n \in \mathbb{K}^{\mathbb{N}} : p(x) := \sum_n p_n |x_n| < \infty \text{ for all } p \in \mathcal{P}\}$$

with the specified seminorms.

Result. [Jarchow, 1981, 21.8.2] *&* [Treves, 1967, p. 530] *The space $\Lambda(\mathcal{P})$ is nuclear if and only if for each $p \in \mathcal{P}$ there is a $q \in \mathcal{P}$ with*

$$\left(\frac{p_n}{q_n}\right)_n \in \ell^1.$$

The space $\Lambda(\mathcal{P})$ is strongly nuclear if and only if for each $p \in \mathcal{P}$ there is a $q \in \mathcal{P}$ with

$$\left(\frac{p_n}{q_n}\right)_n \in \bigcap_{r>0} \ell^r.$$

52.36. Result. [Jarchow, 1981, 21.8.3.b] *$\mathcal{H}(\mathbb{D}^k, \mathbb{C})$ is strongly nuclear for all k.*

Proof. This is an immediate consequence of the Grothendieck-Pietsch criterion (52.35) by considering the power series expansions in the polycylinder $\mathbb{D}^k$ at 0. The set $\mathcal{P}$ consists of $r(n_1, \dots, n_k) := r^{n_1+\dots+n_k}$ for all $0 < r < 1$. □

52.37. Silva spaces. *A locally convex vector space which is an inductive limit of a sequence of Banach spaces with compact connecting mappings is called a Silva space. A Silva space is ultrabornological, webbed, complete, and its strong dual is a Fréchet space. The inductive limit describing the Silva space is regular. A Silva space is Baire if and only if it is finite dimensional. The dual space of a nuclear Silva space is nuclear.*

Proof. Let E be a Silva space. That E is ultrabornological and webbed follows from the permanence properties of ultrabornological spaces (52.30) and of webbed spaces (52.13). The inductive limit describing E is regular and E is complete by [Floret, 1971, 7.4 and 7.5]. The dual E' is a Fréchet space since E has a countable base of bounded sets as a regular inductive limit of Banach spaces. If E is nuclear then the dual is also nuclear by [Jarchow, 1981, 21.5.3].

If E has the Baire property, then it is metrizable by (52.12). But a metrizable Silva space is finite dimensional by [Floret, 1971, 7.7]. □

53. Appendix: Projective Resolutions of Identity on Banach spaces

One of the main tools for getting results for non-separable Banach spaces is that of projective resolutions of identity. The aim is to construct transfinite sequences of complemented subspaces with separable increment and finally reaching the whole space. This works for Banach spaces with enough projections onto closed subspaces. We will give an account on this, following [Orihuela, Valdivia, 1989]. The results in this appendix are used for the construction of smooth partitions of unity in theorem (16.18) and for obtaining smooth realcompactness in example (19.7)

53.1. Definition. Let E be a Banach space, $A \subseteq E$ and $B \subseteq E'$ $\mathbb{Q}$-linear subspaces. Then (A, B) is called *norming pair* if the following two conditions are satisfied:

$$\forall x \in A : \|x\| = \sup\{|\langle x, x^*\rangle| : x^* \in B, \|x^*\| \leq 1\}$$
$$\forall x^* \in B : \|x^*\| = \sup\{|\langle x, x^*\rangle| : x \in A, \|x\| \leq 1\}.$$

53.2. Proposition. *Let (A, B) be a norming pair on a Banach space E. Then*

(1) *$(\bar{A}, \bar{B})$ is a norming pair.*

(2) *Let $A_0 \subseteq A$, $B_0 \subseteq B$, $\omega \leq |A_0| \leq \lambda$, and $\omega \leq |B_0| \leq \lambda$ for some cardinal number λ.*
Then there exists a norming pair $(\tilde{A}, \tilde{B})$ with $A_0 \subseteq \tilde{A} \subseteq A$, $B_0 \subseteq \tilde{B} \subseteq B$, $|\tilde{A}| \leq \lambda$ and $|\tilde{B}| \leq \lambda$.

(3)

$$x \in A,\ y \in B^o \Rightarrow \|x\| \leq \|x + y\|, \textit{ in particular } A \cap B^o = \{0\}$$
$$x^* \in A^o,\ y^* \in B \Rightarrow \|y^*\| \leq \|y^* + x^*\|, \textit{ in particular } A^o \cap B = \{0\}.$$

Proof. (1) Let $x \in \bar{A}$ and $\varepsilon > 0$. Thus there is some $a \in A$ with $\|x - a\| \leq \varepsilon$ and we get

$$\begin{aligned}\|x\| &\leq \|x - a\| + \|a\| \leq \varepsilon + \sup\{|\langle a, x^*\rangle| : x^* \in B, \|x^*\| \leq 1\}\\ &\leq \varepsilon + \sup\{|\langle a - x, x^*\rangle| : x^* \in B, \|x^*\| \leq 1\}\\ &\qquad + \sup\{|\langle x, x^*\rangle| : x^* \in B, \|x^*\| \leq 1\}\\ &\leq \varepsilon + \|a - x\| + \sup\{|\langle x, x^*\rangle| : x^* \in B, \|x^*\| \leq 1\}\\ &\leq 2\varepsilon + \|x\|,\end{aligned}$$

and for $\varepsilon \to 0$ we get the first condition of a norming pair. The second one is shown analogously.

(2) For every $x \in A$ and $y^* \in B$ choose a countable sets $\psi(x) \subseteq B$ and $\varphi(y^*) \subseteq A$ such that

$$\|x\| = \sup\{|\langle x, x^*\rangle| : x^* \in \psi(x)\} \quad \text{and} \quad \|y^*\| = \sup\{|\langle y, y^*\rangle : y \in \varphi(y^*)\}$$

By recursion on n we construct subsets $A_n \subseteq A$ and $B_n \subseteq B$ with $|A_n| \leq \lambda$ and $|B_n| \leq \lambda$:

$$\begin{aligned} B_{n+1} &:= \langle B_n\rangle_{\mathbb{Q}} \cup \{\psi(x) : x \in \langle A_n\rangle_{\mathbb{Q}}\} \\ A_{n+1} &:= \langle A_n\rangle_{\mathbb{Q}} \cup \{\varphi(x^*) : x^* \in \langle B_n\rangle_{\mathbb{Q}}\}. \end{aligned}$$

Finally let $\tilde{A} := \bigcup_{n\in\mathbb{N}} A_n$ and $\tilde{B} := \bigcup_{n\in\mathbb{N}} B_n$. Then $(\tilde{A}, \tilde{B})$ is the required norming pair. In fact for $x \in A_n$ we have that

$$\|x\| = \sup\{|\langle x, x^*\rangle| : x \in \psi(x)\} \leq \sup\{|\langle x, x^*\rangle| : x \in B_{n+1}\} \leq \|x\|$$

Note that $\varphi(\tilde{B}) := \bigcup_{b\in\tilde{B}} \varphi(b) \subseteq \tilde{A}$.

(3) We have

$$\begin{aligned} \|x\| &= \sup\{|\langle x, x^*\rangle| : x^* \in B, \|x^*\| \leq 1\} \\ &= \sup\{|\langle x + y, x^*\rangle| : x^* \in B, \|x^*\| \leq 1\} \\ &\leq \sup\{|\langle x + y, x^*\rangle| : \|x^*\| \leq 1\} = \|x + y\| \end{aligned}$$

and analogously for the second inequality. □

53.3. Proposition. *Let (A, B) be a norming pair on a Banach space E consisting of closed subspaces. It is called* conjugate pair *if one of the following equivalent conditions is satisfied.*

1. *There is a projection $P : E \to E$ with image A, kernel B^o and $\|P\| = 1$;*
2. *$E = A + B^o$;*
3. *$\{0\} = A^o \cap \overline{B}^{\sigma(E',E)}$;*
4. *The canonical mapping $A \hookrightarrow E \cong (E', \sigma(E', E))' \to (B, \sigma(B, E))'$ is onto.*

Proof. We have the following commuting diagram:

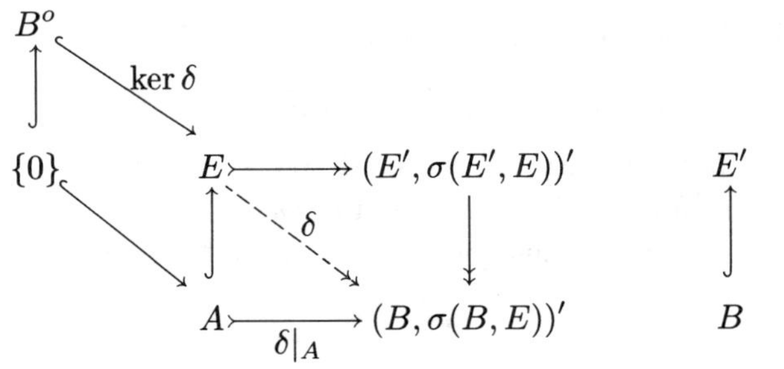

(1)⇒(2) is obvious.

(2)⇔(3) follows immediately from duality.

(2)⇒(4) Let $z \in (B, \sigma(B,E))'$. By Hahn-Banach there is some $x \in E$ with $x|_B = z$. Let $x = a + b$ with $a \in A$ and $b \in B^o$. Then $a|_B = x|_B = z$.

(4)⇒(1) By (4) the mapping $\delta : A \hookrightarrow E \cong (E', \sigma(E',E))' \to (B, \sigma(B,E))'$ is bijective, since $A \cap B^o = \{0\}$, and hence we may define $P(x) := \delta^{-1}(x|_B)$. Then P is the required norm 1 projection, since $\delta : x \mapsto x|_B$ has norm ≤ 1 and δ_A has norm 1 since (A,B) is norming. □

53.4. Corollary. *Let E be a reflexive Banach space. Then any norming pair (A,B) of closed subspaces is a conjugate pair.*

Proof. In fact we then have

$$A^o \cap \overline{B}^{\sigma(E',E)} = A^o \cap \overline{B}^{\|\ \|} = A^o \cap B = \{0\},$$

since the dual of $(E', \sigma(E',E))$ is E and equals E'' the dual of $(E', \|\ \|)$. By [Jarchow, 1981, 8.2.5] convex subsets as B have the same closure in these two topologies. □

53.5. Definition. A *projective generator* φ for a Banach space E is a mapping $\varphi : E' \to 2^E$ for which

(1) $\varphi(x^*)$ is a countable subset of $\{x \in E : \|x\| \leq 1\}$ for all $x^* \in E'$;
(2) $\|x^*\| = \sup\{|\langle x, x^*\rangle| : x \in \varphi(x^*)\}$;
(3) If (A,B) is norming, with $\varphi(B) := \bigcup_{b\in B} \varphi(b) \subseteq A$, then $(\bar{A}, \bar{B})$ is a conjugate pair.

Note that the first two conditions can be always obtained.

We say that the projection P defined by (53.3) for $(\bar{A}, \bar{B})$ is based on the norming pair (A,B), i.e. $P(E) = \bar{A}$ and $\ker(P) = B^o = \bar{B}^o$.

53.6. Corollary. *Every reflexive Banach space has a projective generator φ.*

Proof. Just choose any φ satisfying (53.5.1) and (53.5.2). Then (53.5.3) is by (53.2.1) and (53.4) automatically satisfied. □

53.7. Theorem. *Let φ be a projective generator for a Banach space E. Let $A_0 \subseteq E$ and $B_0 \subseteq E'$ be infinite sets of cardinality at most λ.*

Then there exists a norm 1 projection P based on a norming pair (A,B) with $A_0 \subseteq A$, $B_0 \subseteq B$, $|A| \leq \lambda$, $|B| \leq \lambda$ and $\varphi(B) \subseteq A$.

Proof. By (53.2.3) there is a norming pair (A,B) with

$$A_0 \subseteq A, \quad B_0 \subseteq B, \quad |A| \leq \lambda, \quad |B| \leq \lambda.$$

Note that in the proof of (53.2.3) we used some map φ, and we may take the projective generator for it. Thus we have also $\varphi(B) \subseteq A$. By condition (53.5.3) of the projective generator we thus get that the projection based on (A,B) has the required properties. □

53.8. Proposition. *Every WCD Banach space has a projective generator.*

A Banach space E is called *WCD, weakly countably determined*, if and only if there exists a sequence K_n of weak*-compact subsets of E'' such that for every

$$\forall x \in E\, \forall y \in E'' \setminus E\, \exists n : x \in K_n \text{ and } y \notin K_n.$$

Every WCG Banach space is WCD:
In fact let K be weakly compact (and absolutely convex) such that $\bigcup_{n\in\mathbb{N}} K$ is dense in E. Note that $(E, \sigma(E, E'))$ embeds canonically into $(E'', \sigma(E'', E'))$. Let $K_{n,m} := n\,K + \frac{1}{m}\{x \in E'' : \|x\| \le 1\}$. Then $K_{n,m}$ is weak*-compact, and for any $x \in E$ and $y \in E'' \setminus E$ there exists an $m > 1/\operatorname{dist}(y, E)$ and an n with $\operatorname{dist}(x, n\,K) < \frac{1}{m}$. Hence $x \in K_{n,m}$ and $y \notin E + 1/m\,\{x \in E'' : \|x\| \le 1\} \supseteq K_{n,m}$.
The most important advantage of WCD over WCG Banach spaces are, that they are hereditary with respect to subspaces.
For any finite sequence $n = (n_1, \dots, n_k)$ let

$$C_{n_1,\dots,n_k} := \overline{E \cap K_{n_1} \cap \dots \cap K_{n_k}}^{\sigma(E'',E')}.$$

Then these sets are weak*-compact (since they are contained in K_{n_k}) and if E is not reflexive, then for every $x \in E$ there is a sequence $n : \mathbb{N} \to \mathbb{N}$ such that

$$x \in \bigcap_{k=1}^{\infty} C_{n_1,\dots,n_k} \subseteq E.$$

In fact choose a surjective sequence $n : \mathbb{N} \to \{k : x \in K_k\}$. Then $x \in C_{n_1,\dots,n_k}$ for all k, hence $x \in \bigcap_{k=1}^{\infty} C_{n_1,\dots,n_k}$. If $y \in E'' \setminus E$, then there is some k, such that $y \notin K_{n_k}$ and hence $y \notin C_{n_1,\dots,n_k} \subseteq K_{n_k}$.

Proof of (53.8). Because of (53.6) we may assume that E is not reflexive. For every $x^* \in E'$ we choose a countable set $\varphi(x^*) \subseteq \{x \in E : \|x\| \le 1\}$ such that

$$\|x^*\| = \sup\{|\langle x, x^*\rangle : x \in \varphi(x^*)\} \text{ and}$$
$$\sup\{|\langle x, x^*\rangle| : x \in C_{n_1,\dots,n_k}\} = \sup\{|\langle x, x^*\rangle| : x \in C_{n_1,\dots,n_k} \cap \langle \varphi(x^*)\rangle\}$$

for all finite sequences $(n_1, \dots, n_k)$. We claim that φ is a projective generator:
Let (A, B) be a norming pair with $\varphi(B) \subseteq A$. We use (53.3.3) to show that $(\bar{A}, \bar{B})$ is norming. Assume there is some $0 \ne y^* \in A^o \cap \overline{B}^{\sigma(E',E)}$. Thus we can choose $x_0 \in E$ with $|y^*(x_0)| = 1$ and a net $(y_i^*)_i$ in B that converges to y^* in the Mackey topology $\mu(E', E)$ (of uniform convergence on weakly compact subsets of E). In fact this topology on E' has the same dual E as $\sigma(E', E)$ by the Mackey-Arens theorem [Jarchow, 1981, 8.5.5], and hence the same closure of convex sets by [Jarchow, 1981, 8.2.5]. As before we choose a surjective mapping $n : \mathbb{N} \to \{k : x_0 \in K_k\}$. Then

$$x_0 \in C := \bigcap_{k=1}^{\infty} C_{n_1,\dots,n_k} \subseteq E.$$

and C is weakly compact, hence we find an i_0 such that

$$\sup\{|y^*_{i_0}(x) - y^*(x)| : x \in C\} < \tfrac{1}{2}$$

and in particular we have

$$|y^*_{i_0}(x_0)| \geq |y^*(x_0)| - |y^*_{i_0}(x_0) - y^*(x_0)| > 1 - \tfrac{1}{2} = \tfrac{1}{2}.$$

Since the sets forming the intersection are decreasing, C_{n_1} is $\sigma(E'', E')$-compact and

$$W := \{x^{**} \in E'' : |x^{**}(y^*_{i_0} - y^*)| < \tfrac{1}{2}\}$$

is a $\sigma(E'', E')$-open neighborhood of C there is some $k \in \mathbb{N}$ such that $C_{n_1,\dots,n_k} \subseteq W$, i.e.

$$\sup\{|y^*_{i_0}(x) - y^*(x)| : x \in C_{n_1,\dots,n_k}\} \leq \tfrac{1}{2}.$$

By the definition of φ there is some $y_0 \in C_{n_1,\dots,n_k} \cap \langle\varphi(y^*_{i_0})\rangle$ with $|y^*_{i_0}(y_0)| > 1 - \frac{1}{2}$, thus

$$|y^*(y_0)| \geq |y^*_{i_0}(y_0)| - |y^*_{i_0}(y_0) - y^*(y_0)| > \tfrac{1}{2} - \tfrac{1}{2} = 0.$$

Thus $y^*(y_0) \neq 0$ and $y_0 \in \langle\varphi(B)\rangle \subseteq A$, a contradiction. □

Note that if $P \in L(E)$ is a norm-1 projection with closed image A and kernel B^o, then $P^* \in L(E')$ is a norm-1 projection with image $P^*(E) = \ker P^o = B^{oo} = B$ and kernel $\ker P^* = P(E)^o = A^o$. However not all norm-1 projections onto B can be obtained in this way. Hence we consider the dual of proposition (53.3):

53.9. Proposition. *Let (A, B) be a norming pair on a Banach space E consisting of closed subspaces. It is called dual conjugate pair if one of the following equivalent conditions is satisfied.*

(1) *There is a norm-1 projection $P : E' \to E'$ with image B, kernel A^o;*
(2) $E' = B \oplus A^o$;
(3) $\{0\} = B^o \cap \overline{A}^{\sigma(E'', E')}$;
(4) *The canonical mapping $B \hookrightarrow E' \xrightarrow{(\ )|A} A'$ is onto.*

Proof. This follows by applying (53.3) to the norming pair $(B, A) \subseteq (E', E'')$. □

The dual of definition (53.5) is

53.10. Definition. A dual projective generator ψ for a Banach space E' is a mapping $\psi : E \to 2^{E'}$ for which

(1) $\psi(x)$ is a countable subset of $\{x^* \in E' : \|x^*\| \leq 1\}$ for all $x \in E$;
(2) $\|x\| = \sup\{|\langle x, x^*\rangle| : x^* \in \psi(x)\}$;
(3) If (A, B) is norming, with $\psi(A) := \bigcup_{a \in A} \psi(a) \subseteq B$, then $(\bar{A}, \bar{B})$ satisfies the condition of (53.9).

Note that the first two conditions can be always obtained.

From (53.7) we get:

53.11. Theorem. *Let ψ be a dual projective generator for a Banach space E. Let $A_0 \subseteq E$ and $B_0 \subseteq E'$ be infinite sets of cardinality at most λ.*
Then there exists a norm 1 projection P in E' with $A_0 \subseteq P^*(E'')$, $B_0 \subseteq P(E')$, $|P^*(E'')| \leq \lambda$, $|P(E')| \leq \lambda$.

53.12. Proposition. *A Banach space E is Asplund if and only if there exists a dual projective generator on E.*

Note that if P is a norm-1 projection, then so is P^*. But not all norm-1 projections on the dual are of this form.

Proof. ($\Leftarrow$) Let ψ be a dual projective generator for E. Let A_0 be a separable subspace of E. By (53.11) there is a separable subspace A of E and a norm-1 projection P of E' such that $A_0 \supseteq A$, $P(E')$ is separable and isomorphic with A' via the restriction map. Hence A' is separable and also A_0'. By [Stegall, 1975] E is Asplund.

($\Rightarrow$) Consider the $\|\quad\|$-weak* upper semi-continuous mapping $\phi : X \to 2^{\{x^*:\|x^*\|\leq 1\}}$ given by

$$\phi(x) := \{x^* \in E' : \|x^*\| \leq 1, \langle x, x^*\rangle = \|x\|\}.$$

By the Jayne-Rogers selection theorem [Jayne, Rogers, 1985], see also [Deville, Godefroy, Zizler, 1993, section I.4] there is a map $f : E \to \{x^* \in E' : \|x^*\| \leq 1\}$ with $f(x) \in \phi(x)$ for all $x \in E$ and continuous $f_n : E \to \{x^* : \|x^*\| \leq 1\} \subseteq E'$ with $f_n(x) \to f(x)$ in E' for each $x \in E$. One then shows that

$$\psi(x) := \{f(x), f_1(x), \dots\}$$

defines a dual projective generator, see [Orihuela, Valdivia, 1989]. □

53.13. Definition. Projective Resolution of Identity. Let a "long sequence" of continuous projections $P_\alpha \in L(E,E)$ on a Banach space E for all ordinal numbers $\omega \leq \alpha \leq \operatorname{dens} E$ be given. Recall that $\operatorname{dens}(E)$ is the density of E (a cardinal number, which we identify with the smallest ordinal of same cardinality). Let $E_\alpha := P_\alpha(E)$ and let $R_\alpha := (P_{\alpha+1} - P_\alpha)/(\|P_{\alpha+1} - P_\alpha\|)$ or 0, if $P_{\alpha+1} = P_\alpha$. Then we consider the following properties:

(1) $P_\alpha P_\beta = P_\beta = P_\beta P_\alpha$ for all $\beta \leq \alpha$.
(2) $P_{\operatorname{dens} E} = \operatorname{Id}_E$.
(3) $\operatorname{dens} P_\alpha E \leq \alpha$ for all α.
(4) $\|P_\alpha\| = 1$ for all α.
(5) $\overline{\bigcup_{\beta<\alpha} P_{\beta+1} E} = P_\alpha E$, or equivalently $\overline{\bigcup_{\beta<\alpha} E_\beta} = E_\alpha$ for every limit ordinal $\alpha \leq \operatorname{dens} E$.
(6) For every limit ordinal $\alpha \leq \operatorname{dens} E$ we have $P_\alpha(x) = \lim_{\beta>\alpha} P_\beta(x)$, i.e. $\alpha \mapsto P_\alpha(x)$ is continuous.
(7) $E_{\alpha+1}/E_\alpha$ is separable for all $\omega \leq \alpha < \operatorname{dens} E$.
(8) $(R_\alpha(x))_\alpha \in c_0([\omega, \operatorname{dens} E])$ for all $x \in E$.
(9) $P_\alpha(x) \in \overline{\langle P_\omega(x) \cup \{R_\beta(x) : \omega \leq \beta < \alpha\}\rangle}$.

The family $(P_\alpha)_\alpha$ is called *projective resolution of identity* (*PRI*) if it satisfies (1), (2), (3), (4) and (5).

It is called *separable projective resolution of identity* (*SPRI*) if it satisfies (1), (2), (3), (7), (8) and (9). These are the only properties used in (53.20) and they follow for WCD Banach spaces and for duals of Asplund spaces by (53.15). For $C(K)$ with Valdivia compact K this is not clear, see (53.18) and (53.19). However, we still have (53.21) and in (16.18) we don't use (7), but only (8) and (9) which hold also for PRI, see below.

Remark. Note that from (1) we obtain that $P_\alpha^2 = P_\alpha$ and hence $\|P_\alpha\| \geq 1$, and $E_\alpha := P_\alpha(E)$ is the closed subspace $\{x : P_\alpha(x) = x\}$.

Moreover, $P_\alpha P_\beta = P_\beta = P_\beta P_\alpha$ for $\beta \leq \alpha$ is equivalent to $P_\alpha^2 = P_\alpha$, $P_\beta(E) \subseteq P_\alpha(E)$ and $\ker P_\beta \supseteq \ker P_\alpha$.
($\Rightarrow$) $P_\beta x = P_\alpha P_\beta x \in P_\alpha(E)$ and $P_\alpha x = 0$ implies that $P_\beta x = P_\beta P_\alpha x$.
($\Leftarrow$) For $x \in E$ there is some $y \in E$ with $P_\beta x = P_\alpha y$, hence $P_\alpha P_\beta x = P_\alpha P_\alpha y = P_\alpha y = P_\beta x$. And $P_\beta(1 - P_\alpha)x = 0$, since $(1 - P_\alpha)x \in \ker P_\alpha \subseteq \ker P_\beta$.

Note that $E_{\alpha+1}/E_\alpha \cong (P_{\alpha+1} - P_\alpha)(E)$, since $E_\alpha \to E_{\alpha+1}$ has $P_\alpha|_{E_{\alpha+1}}$ as right inverse, and so $E_{\alpha+1}/E_\alpha \cong \ker(P_\alpha|_{E_{\alpha+1}}) = (1 - P_\alpha)P_{\alpha+1}(E) = (P_{\alpha+1} - P_\alpha)(E)$.

(5) $\Leftarrow$ (9), since for $x \in E_\alpha$ we have $x = P_\alpha(x)$ and $E_\omega \cup \{R_\beta(x) : \beta < \alpha\} \subseteq E_\alpha$ for all α.

(3) $\Leftarrow$ (5) & (7) By transfinite induction we get that for successor ordinals $\alpha = \beta + 1$ we have $\operatorname{dens}(E_\alpha) = \operatorname{dens}(E_\beta) + \operatorname{dens}(E_\alpha/E_\beta) = \operatorname{dens}(E_\beta) \leq \beta \leq \alpha$, since $\operatorname{dens}(E_\alpha/E_\beta) \leq \omega$. For limit ordinals it follows from (5), since $\operatorname{dens}(E_\alpha) = \operatorname{dens}(\bigcup_{\beta<\alpha} E_\beta) = \sup\{\operatorname{dens}(E_\beta) : \beta < \alpha\} \leq \sup\{\beta : \beta < \alpha\} = \alpha$.

(6) $\Leftarrow$ (4) & (1) & (5) For every limit ordinal $0 < \alpha \leq \operatorname{dens} E$ and for all $x \in E$ the net $(P_\beta(x))_{\beta<\alpha}$ converges to $P_\alpha(x)$.
Let first $x \in P_\alpha(E)$ and $\varepsilon > 0$. By (5) there exists a $\gamma < \alpha$ and an $x_\gamma \in P_\gamma(E)$ with $\|x - x_\gamma\| < \varepsilon$. Hence for $\gamma \leq \beta < \alpha$ we have by (1) that $P_\beta(x_\gamma) = P_\alpha(x_\gamma)$ and so

$$\begin{aligned}\|P_\alpha(x) - P_\beta(x)\| &= \|P_\alpha(x - x_\gamma)| + \|P_\alpha(x_\gamma) - P_\beta(x_\gamma)| - P_\beta(x_\gamma - x)\| \\ &\leq (\|P_\alpha\| + \|P_\beta\|)\, \|x - x_\gamma\| < 2\,\varepsilon.\end{aligned}$$

If $x \in E$ is arbitrary, then $P_\alpha(x) \in P_\alpha(E)$, hence by (1)

$$P_\beta(x) = P_\beta(P_\alpha(x)) \to P_\alpha(P_\alpha(x)) = P_\alpha(x) \text{ for } \beta \nearrow \alpha. \quad \square$$

(8) $\Leftarrow$ (1) & (6) Let $\varepsilon > 0$. Then the set $\{\beta : \beta < \alpha, \|R_\beta(x)\| \geq \varepsilon\}$ is finite, since otherwise there would be an increasing sequence (β_n) such that $\|R_{\beta_n}(x)\| \geq \varepsilon$ and since $\|P_{\alpha+1} - P_\alpha\| = \|(1 - P_\alpha)P_{\alpha+1}\| \geq 1$ also $\|(P_{\beta_n+1} - P_{\beta_n})(x)\| \geq \varepsilon$. Let $\beta_\infty := \sup_n \beta_n$. Then $\beta_\infty \leq \alpha$ is a limit ordinal and $P_{\beta_\infty}(x) = \lim_{\beta<\beta_\infty} P_\beta(x)$ according to (6), a contradiction.

(9) $\Leftarrow$ (6) We prove by transfinite induction that $P_\alpha(x)$ is in the closure of the linear span of $\{R_\beta(x) : \omega \leq \beta < \alpha\} \cup P_\omega(x)$.

For $\alpha = \omega$ this is obviously true. Let now $\alpha = \beta + 1$ and assume $P_\beta(x)$ is in the closure of the linear span of $\{R_\gamma(x) : \omega \leq \gamma < \beta\} \cup P_\omega(x)$. Since $P_\alpha(x) = P_\beta(x) + \|P_\alpha - P_\beta\|\, R_\beta(x)$ we get that $P_\alpha(x)$ is in the closure of the linear span of $\{R_\gamma(x) : \omega \leq \gamma < \alpha\} \cup P_\omega(x)$.
Let now α be a limit ordinal and let $P_\beta(x)$ be in the closure of the linear span of $\{R_\gamma(x) : \omega \leq \gamma < \alpha\} \cup P_\omega(x)$ for all $\beta < \alpha$. Then by (6) we get that $P_\alpha(x) = \lim_{\beta<\alpha} P_\beta(x)$ is in this closure as well.

Proposition. *Suppose all complemented subspaces of a Banach space E have PRI then E has a SPRI.*

Proof. We proceed by induction on $\mu := \operatorname{dens} E$. For $\mu = \omega$ nothing is to be shown. Now let $(P_\alpha)_{0\leq\alpha\leq\mu}$ be a PRI of E. For every $\alpha < \mu$ we have $\alpha + 1 < \mu$ and so $\mu_\alpha := \operatorname{dens}((P_{\alpha+1} - P_\alpha)(E)) \leq \operatorname{dens}(P_{\alpha+1}(E)) \leq \alpha < \mu$, hence there is a SPRI $(P^\alpha_\beta)_{0\leq\beta<\mu_\alpha}$ of $(P_{\alpha+1} - P_\alpha)(E)$. Now consider

$$P_{\alpha,\beta} := P_\alpha + P^\alpha_\beta(P_{\alpha+1} - P_\alpha) = (P_\alpha + P^\alpha_\beta(1 - P_\alpha))P_{\alpha+1}$$

for $\omega \leq \alpha < \mu$ and $\omega \leq \beta \leq \mu_\alpha$ with the lexicographical ordering. This is a well-ordering and since the cardinality of μ^2 is μ and $\mu_\alpha < \mu$ it corresponds to the ordinal segment $[\omega, \mu)$. In fact for any limit ordinal $\alpha > \omega$ we have

$$|[\omega, \alpha)| = \sum_{\omega\leq\beta<\alpha} 1 \leq \sum_{\omega\leq\beta<\alpha} |[\omega, \mu_\alpha)| \leq |[\omega, \alpha)|^2 \leq |[\omega, \alpha)|.$$

Obviously the $P_{\alpha,\beta}$ are projections that satisfy (1) and (3).
(1) For $P_{\alpha,\beta}$ with the same α this follows from (1) for $P^\alpha_\beta : R_\alpha(E) \to R_\alpha(E)$: In fact

$$\begin{aligned} P_{\alpha,\beta}\, P_{\alpha,\beta'} &:= \Big(P_\alpha + P^\alpha_\beta(P_{\alpha+1} - P_\alpha)\Big)\Big(P_\alpha + P^\alpha_{\beta'}(P_{\alpha+1} - P_\alpha)\Big) \\ &= P^2_\alpha + P^\alpha_\beta(P_{\alpha+1} - P_\alpha)P_\alpha + P_\alpha P^\alpha_{\beta'}(P_{\alpha+1} - P_\alpha) \\ &\quad + P^\alpha_\beta(P_{\alpha+1} - P_\alpha)P^\alpha_{\beta'}(P_{\alpha+1} - P_\alpha) \\ &= P^2_\alpha + 0 + 0 + P^\alpha_{\min\{\beta,\beta'\}}(P_{\alpha+1} - P_\alpha) \end{aligned}$$

For different α this follows, since $P_{\alpha_1,\beta}E \subseteq P_{\alpha_1+1}E \subseteq P_{\alpha_2}$ and

$$\begin{gathered} P_\alpha E \subseteq P_{\alpha,\beta} \subseteq P_{\alpha+1} \\ \ker P_\alpha \supseteq \ker P_{\alpha,\beta} = \ker(P_\alpha + P^\alpha_\beta(1 - P_\alpha))P_{\alpha+1} \supseteq \ker P_{\alpha+1} \end{gathered}$$

(3) The density of $P_{\alpha,\beta}E$ is less or equal to $\alpha + 1$.
And clearly they satisfy (7) as well, since $R_{\alpha,\beta} = R^\alpha_\beta(P_{\alpha+1} - P_\alpha)$.
(9) Since this is true for the P_α and the P^β_α it follows for $P_{\alpha,\beta}$ as well.
In fact $P_{\alpha,\beta}(x)$ belongs to the closure of the linear span of $P_\alpha(x)$ and the $R_{\alpha,\beta'} = R^\alpha_{\beta'}(P_{\alpha+1} - P_\alpha)(x)$ for $\beta' < \beta$ by the property of the P^α_β. Furthermore $P_\alpha(x)$ belongs

to the closure of the linear span of $R_{\alpha'}(x)$ for $\alpha' < \alpha$ and $P_\omega(x)$ by the property of the P_α and $R_{\alpha'}(x)$ belongs to the closure of the linear span of all $R_\beta^{\alpha'}(R_{\alpha'}x)$ for all $\beta < \operatorname{dens} R_{\alpha'}E$.

(8) For x in the linear span of all $R_{\alpha,\beta}E$ we obviously have that $(R_{\alpha,\beta}(x))_{\alpha,\beta} \in c_0$. In fact for $x := \sum_{i=1}^n \lambda^i R_{\alpha_i,\beta_i}(x_i)$, we have that $R_{\alpha_i,\beta_i}(x) = \lambda^i R_{\alpha_i,\beta_i}(x_i)$ and $R_{\alpha,\beta}(x) = 0$ for all $(\alpha,\beta) \notin \{(\alpha_1,\beta_1),\dots,(\alpha_n,\beta_n)\}$.

$$R_\alpha R_\beta = (P_{\alpha+1} - P_\alpha)(P_{\beta+1} - P_\beta) = (1 - P_\alpha)P_{\alpha+1}P_{\beta+1}(1 - P_\beta) = 0,$$

if $\alpha + 1 \le \beta$ or $\beta + 1 \le \alpha$, since the factors commute. For general x we find by (9) a point $\tilde{x}$ in the linear span of the $R_{\alpha,\beta}x$ with $\|x - \tilde{x}\| < \varepsilon$. Then

$$\{(\alpha,\beta) : \|R_{\alpha,\beta}(x)\| \ge \varepsilon\} \subseteq \{(\alpha_1,\beta_1),\dots,(\alpha_n,\beta_n)\}.$$

Note however that we don't have $\|P_{\alpha,\beta}\| = 1$. □

53.14. Theorem. *Let E be a Banach space with projective generator φ. Then E admits a PRI $(P_\alpha)_\alpha$, where each P_α is based on a norming pair (A_α, B_α) with*

(1) $|A_\alpha| \le \alpha$, $|B_\alpha| \le \alpha$ *for all* α;
(2) $A_\beta \subseteq A_\alpha$ *and* $B_\beta \subseteq B_\alpha$ *for all* $\beta \le \alpha$;
(3) $\bigcup_{\omega\le\beta<\alpha} A_\beta = A_\alpha$ *for all limit ordinals* α;
(4) $\bigcup_{\omega\le\beta<\alpha} B_\beta = B_\alpha$ *for all limit ordinals* α;

Proof. Choose a dense subset $\{x_\alpha : \alpha < \operatorname{dens} E\}$. We construct by transfinite recursion for every ordinal $\alpha \le \operatorname{dens} E$ a norming pair (A_α, B_α) with

$$A_\alpha \supseteq \{x_\beta : \beta < \alpha\}, \quad |A_\alpha| \le \alpha, \quad |B_\alpha| \le \alpha, \quad \varphi(B_\alpha) \subseteq A_\alpha$$
$$A_\beta \subseteq A_\alpha \text{ and } B_\beta \subseteq A_\alpha \text{ for } \beta \le \alpha.$$

For the ordinal ω let $A_0 := \{x_\alpha : \alpha < \omega\}$ and let B_0 be a countable subset of E' such that

$$\|x\| = \sup\{|\langle x, x^*\rangle : x^* \in B_0\} \text{ for all } x \in A_0.$$

By (53.7) there is a norming pair (A_ω, B_ω) with $|A_\omega|, |B_\omega| \le \omega$, $A_\omega \supseteq A_0$, $B_\omega \supseteq B_0$ and $\varphi(B_\omega) \subseteq A_\omega$.

If α is a successor ordinal, i.e. $\alpha = \beta + 1$, then let $A_0 := A_\beta \cup \{x_\beta\}$ and $B_0 := B_\beta$. Again by (53.7) we get a norming pair (A_α, B_α), such that

$$A_0 \subseteq A_\alpha, \quad B_0 \subseteq B_\alpha \subseteq E', \quad |A_\alpha| \le \alpha, \quad |B_\alpha| \le \alpha, \quad \varphi(B_\alpha) \subseteq A_\alpha$$

If α is a limit ordinal, we set

$$A_\alpha := \bigcup_{\beta<\alpha} A_\beta$$
$$B_\alpha := \bigcup_{\beta<\alpha} B_\beta \subseteq E'.$$

Then obviously (A_α, B_α) is a norming pair with $\varphi(B_\alpha) \subseteq A_\alpha$.

Now using the property of the projective generator φ we have that there are norm-1 projections $P_\alpha \in L(E)$ with $P_\alpha(E) = \overline{A_\alpha}$ and $\ker P_\alpha = (\overline{B_\alpha})^o = (B_\alpha)^o$. Hence

$$\text{(53.13.1)} \quad P_\alpha P_\beta = P_\beta = P_\beta P_\alpha \text{ for } \beta \le \alpha$$

$$\text{(53.13.3)} \quad \operatorname{dens} P_\alpha E \le \alpha, \quad \operatorname{dens} P_\alpha^*(E')_\sigma \le \alpha,$$

$$\text{(53.13.5)} \quad P_\alpha(E) = \overline{A_\alpha} = \overline{\bigcup_{\beta<\alpha} P_\beta E}$$

$$\text{(53.13.4)} \quad \|P_\alpha\| = 1$$

and since $\{x_\alpha : \alpha < \operatorname{dens} E\}$ is dense in E we also have (53.13.2). Furthermore we have that B_α is weak*-dense in $P_\alpha^* E'$. □

53.15. Corollary. *WCD and duals of Asplund spaces have SPRI.* □

53.16. Definition. A compact set K is called *Valdivia compact* if there exists some set Γ with $K \subseteq \mathbb{R}^\Gamma$ and $\{x \in K : \operatorname{carr}(x) \text{ is countable}\}$ being dense in K.

53.17. Lemma. *For a Valdivia compact set $K \subseteq \mathbb{R}^\Gamma$ we consider the set $E := \{x \in \mathbb{R}^\Gamma : \operatorname{carr}(x)$ is countable$\}$. Let μ be the density number of $K \cap E$. Then there exists an increasing long sequence of subsets $\Gamma_\alpha \subseteq \Gamma$ for $\omega \le \alpha \le \mu$ satisfying:*

(i) $|\Gamma_\alpha| \le \alpha$;
(ii) $\bigcup_{\beta<\alpha} \Gamma_\beta = \Gamma_\alpha$ *for limit ordinals* α;
(iii) $\Gamma_\mu = \bigcup cup\{\operatorname{carr}(x) : x \in K\}$;

and such that $K_\alpha := Q_{\Gamma_\alpha}(K) \subseteq K$, where $Q_{\Gamma'} : \mathbb{R}^\Gamma \to \mathbb{R}^{\Gamma'} \hookrightarrow \mathbb{R}^\Gamma$, i.e.

$$Q_{\Gamma'}(x)_\gamma := \begin{cases} x_\gamma & \text{for } \gamma \in \Gamma' \\ 0 & \text{for } \gamma \notin \Gamma \setminus \Gamma' \end{cases}.$$

Thus $K_\alpha \subseteq K$ is a retract via Q_{Γ_α}.

Note that for any Valdivia compact set $K \subseteq \mathbb{R}^\Gamma$ we may always replace Γ by $\bigcup\{\operatorname{carr}(x) : x \in K\} = \bigcup\{\operatorname{carr}(x) : x \in K \cap E\}$, and then (iii) says $\Gamma_\mu = \Gamma$.

Proof. The proof is based on the following claim: Let $\Delta \subseteq \Gamma$ be a infinite subset. Then there exists some subset $\tilde{\Delta}$ with $\Delta \subseteq \tilde{\Delta} \subseteq \Gamma$ and $|\Delta| = |\tilde{\Delta}|$ and $Q_{\tilde{\Delta}}(K) \subseteq K$. By induction we construct a sequence $\Delta =: \Delta_0 \subseteq \Delta_1 \subseteq \dots \subseteq \Delta_k \subseteq \dots \subseteq \Gamma$ with $|\Delta_k| = |\Delta_0|$ and $Q_{\Delta_k}(\{x \in K \cap E : \operatorname{carr}(x) \subseteq \Delta_{k+1}\})$ being dense in $Q_{\Delta_k}(K)$:
(k+1) Since $K \cap E$ is dense in K, we have that $Q_{\Delta_k}(K \cap E)$ is dense in $Q_{\Delta_k}(K) \subseteq \mathbb{R}^{\Delta_k} \times \{0\} \subseteq \mathbb{R}^\Gamma$. And since the topology of $\mathbb{R}^{\Delta_k}$ has a basis of cardinality $|\Delta_k|$, there is a subset $D \subseteq K \cap E$ with $|D| \le |\Delta_k|$ and $Q_{\Delta_k}(D)$ dense in $Q_{\Delta_k}(K)$. Let $\Delta_{k+1} := \Delta_k \cup \bigcup_{x\in D} \operatorname{carr}(x)$ then $\Delta_{k+1} \supseteq \Delta_k$ and $|\Delta_{k+1}| = |\Delta_k|$. Furthermore $Q_{\Delta_k}(\{x \in K \cap E : \operatorname{carr}(x) \subseteq \Delta_{k+1}\}) \supseteq Q_{\Delta_k}(D)$ is dense in $Q_{\Delta_k}(K)$.
Now $\tilde{\Delta} := \bigcup_k \Delta_k$ is the required set. In order to show that $Q_{\tilde{\Delta}}(K) \subseteq K$ let $x \in K$ be arbitrary. Since $Q_{\Delta_k}(x)$ is contained in the closure of $Q_{\Delta_k}(\{x_k \in K \cap E :$

$\operatorname{carr}(x_k) \subseteq \Delta_{k+1}\})$ and hence in the closed set $Q_{\Delta_k}(\{x_k \in K : \operatorname{carr}(x_k) \subseteq \tilde{\Gamma}\})$. Thus there is an $x_k \in K$ with $\operatorname{carr}(x_k) \subseteq \tilde{\Gamma}$ and such that x agrees with x_k on Δ_k. Thus $K \ni x_k \to Q_{\tilde{\Delta}}(x)$, since every finite subset of $\tilde{\Delta}$ is contained in some Δ_k and outside $\tilde{\Delta}$ all x_k and $Q_{\tilde{\Delta}}(x)$ are zero. Since K is closed we get $Q_{\tilde{\Delta}}(x) \in K$.

Without loss of generality we may assume that $\mu > \omega$. Let $\{x_\alpha : \omega \le \alpha < \mu\}$ be a dense subset of $K \cap E$. Let $\Gamma_\omega := \operatorname{carr}(x_\omega)$. By transfinite induction we define

$$\Gamma_\alpha := \begin{cases} (\Gamma_\beta \cup \operatorname{carr}(x_\beta))^\sim & \text{for } \alpha = \beta + 1, \\ \bigcup_{\beta<\alpha} \Gamma_\beta & \text{for limit ordinals } \alpha. \end{cases}$$

Then the Γ_α satisfy all the requirements. □

53.18. Corollary. *Let K be Valdivia compact. Then $C(K)$ has a PRI.*

Proof. We choose Γ_α as in (53.17) and set $K_\alpha := Q_{\Gamma_\alpha}(K)$. Let $Q_\alpha := Q_{\Gamma_\alpha}|_K$. Then Q_α is a continuous retraction.

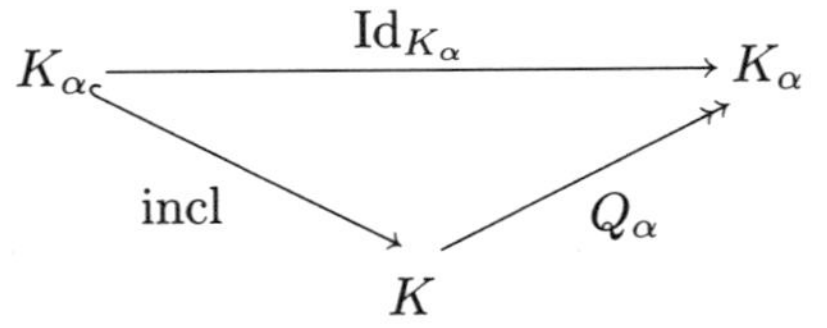

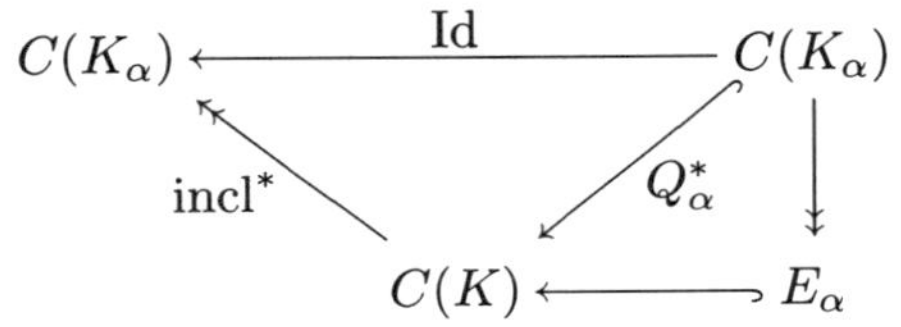

We have $\operatorname{dens}(C(\mathbb{R}^{\Gamma_\alpha})) = |\alpha|$, since we have a base of the topology of this space of that cardinality. Hence $\operatorname{dens}(C(K_\alpha)) \le |\alpha|$. Let $E_\alpha := (Q_\alpha)^*(C(K_\alpha))$. Then E_α is a closed subspace of $C(K)$ and (53.13.3) holds. Furthermore $P_\alpha := Q_\alpha \circ \operatorname{incl}^*_{K_\alpha}$ is a norm-1 projection from $C(K)$ to E_α. The inclusion $\Gamma_\alpha \subseteq \Gamma_\beta$ for $\alpha \le \beta$ implies (53.13.1). To see (53.13.6) and (53.13.5) let $\varepsilon > 0$ and choose a finite covering of K_α by sets

$$U_j := \{x \in \mathbb{R}^{\Gamma_\alpha} : |x_\gamma - x^j_\gamma| < \delta_j \text{ for all } \gamma \in \Delta_j\},$$

where $x^j \in \mathbb{R}^{\Gamma_\alpha}$, $\delta_j > 0$ and $\Delta_j \subseteq \Gamma_\alpha$ is finite and such that for $x', x'' \in U_j \cap K$ we have $|f(x') - f(x'')| < \varepsilon$. Now choose $\alpha_0 < \alpha$ such that $\Gamma_{\alpha_0} \supseteq \Delta_j$ for all of the finitely many j. Since the U_j cover K_α, we have $x \in K_\alpha \cap U_j$ for some j and hence $Q_\beta(x) \in K_\alpha \cap U_j$ for all $\alpha_0 \le \beta < \alpha$. Hence $|f(x) - f(Q_\beta(x))| < \varepsilon$ for all $x \in K_\alpha$ and so $\|P_\alpha(f) - P_\beta(f)\| = \|(1 - P_\beta)P_\alpha(f)\| \le \varepsilon$. Thus we have shown that E has a PRI $(P_\alpha)_\alpha$, with all $E_\alpha \cong C(K_\alpha)$ and $\operatorname{dens}(K_\alpha) \le |\Gamma_\alpha| \le \alpha$. □

53.19. Remark. *The space $C([0,\alpha])$ has a PRI given by*

$$P_\beta(f)(\mu) := \begin{cases} f(\mu) & \text{for } \mu \le \beta \\ f(\beta) & \text{for } \mu \ge \beta \end{cases}.$$

However, there is no PRI on the hyperplane $E := \{f \in C([0,\omega_1]) : f(\omega_1) = 0\}$ *of the space* $C[0,\omega_1]$. *And, in particular,* $C[0,\omega_1]$ *is not WCD.*

Proof. Assume $\{P_\alpha : \omega \le \alpha \le \omega_1\}$ is a PRI on E. Put $\alpha_0 := \omega_0$. We may find $\beta_0 < \omega_1$ with

$$P_{\alpha_0}E \subseteq E_{\beta_0} := \{f \in E : f(\alpha) = 0 \text{ for } \alpha > \beta_0\},$$

because for each f in dense countable subset $D \subseteq P_{\alpha_0}E$ we find a β_f with $f(\alpha) = 0$ for $\alpha \ge \beta_f$. Since E_{β_0} is separable, there is an $\alpha_0 < \alpha_1 < \omega_1$ such that

$$E_{\beta_0} \subseteq P_{\alpha_1}E,$$

in fact $D \subseteq E_{\beta_0}$ is dense and hence for each $f \in D$ and $n \in \mathbb{N}$ there exists an $\alpha_{f,n} < \omega_1$ and $\tilde{f} \in P_{\alpha_{f,n}}E$ such that $\|f - \tilde{f}\| \le 1/n$. Then $\alpha_1 := \sup\{\alpha_{f,n} : n \in \mathbb{N}, f \in D\}$ fulfills the requirements.

Now we proceed by induction. Let $\alpha_\infty := \sup_n \alpha_n$ and $\beta_\infty := \sup_n \beta_n$. Then

$$P_{\alpha_\infty}E = \overline{\bigcup_n P_{\alpha_n}E} = F_{\beta_\infty} := \{f \in E : f(\alpha) = 0 \text{ for } \alpha \ge \beta_\infty\}.$$

But F_{β_∞} is not the image of a norm-1 projection: Suppose P were a norm-1 projection on F_{β_∞}. Let $\pi : E \to C(X)$ be the restriction map, where $X := [0,\beta_\infty]$. It is left inverse to the inclusion ι given by $f \mapsto \tilde{f}$ with $\tilde{f}(\gamma) = 0$ for $\gamma \ge \beta_\infty$. Let $\tilde{P} := \pi \circ P \circ \iota \in L(C(X))$. Then $\tilde{P}$ is a norm-1 projection with image $C_{\beta_\infty}(X) := \{f \in C[0,\beta_\infty] : f(\beta_\infty) = 0\}$. Then $C(X) = \ker(\tilde{P}) \oplus C_{\beta_\infty}(X)$. We pick $0 \ne f_0 \in \ker(\tilde{P})$. Since $f_0 \notin \tilde{P}(C(X)) = C_{\beta_\infty}(X) = \ker(\mathrm{ev}_{\beta_\infty})$, we have $f_0(\beta_\infty) \ne 0$, and without loss of generality we may assume that $f_0(\beta_\infty) = 1$. For $f \in C(X)$ we have that $f - \tilde{P}(f) \in \ker\tilde{P}$ and hence there is a $\lambda_f \in \mathbb{R}$ with $f - \tilde{P}(f) = \lambda_f f_0$. In fact evaluating at β_∞ gives $f(\beta_\infty) - 0 = \lambda_f 1$, hence $\tilde{P}(f) = f - f(\beta_\infty) f_0$. Since β_∞ is a limit point, there is for each $\varepsilon > 0$ a $x_\varepsilon < \beta_\infty$ with $f_0(x_\varepsilon) > 1 - \varepsilon$. Now choose $f_\varepsilon \in C(X)$ with $\|f_\varepsilon\| = 1 = -f_\varepsilon(\beta_\infty) = f_\varepsilon(x_\varepsilon)$. Then

$$\begin{aligned}\|Pf_\varepsilon\|_\infty &= \|f_\varepsilon - f_\varepsilon(\beta_\infty) f_0\|_\infty\\ &\ge |f_\varepsilon(x_\varepsilon) - f_\varepsilon(\beta_\infty) f_0(x_\varepsilon)|\\ &\ge 1 + 1(1-\varepsilon) = 2 - \varepsilon.\end{aligned}$$

Hence $\tilde{P} \ge 2$, a contradiction.

Note however that every separable subspace is contained in a 1-complemented separable subspace. □

53.20. Theorem. [Biström, 1993, 3.16] *If* E *is a realcompact (i.e. non-measurable) Banach space admitting a SPRI, then there is a non-measurable set* Γ *and a injective continuous linear operator* $T : E \to c_0(\Gamma)$.

Proof. We proof by transfinite induction that for every ordinal α with $\alpha \le \mu := \mathrm{dens}(E)$ there is a non-measurable set Γ_α and an injective linear operator T_α :

$E_\alpha := P_\alpha(E) \to c_0(\Gamma_\alpha)$ with $\|T_\alpha\| \le 1$.
Note that if E is separable, then there are $x_n^* \in E'$ with $\|x_n^*\| \le 1$, and which are $\sigma(E', E)$ dense in the unit-ball of E'. Then $T : E \to c_0(\mathbb{N})$, defined by $T(x)_n := \frac{1}{n} x_n^*(x)$, satisfies the requirements: It is obviously a continuous linear mapping into c_0, and it remains to show that it is injective. So let $x \neq 0$. By Hahn-Banach there is a $x^* \in E'$ with $x^*(x) = \|x\|$ and $\|x^*\| \le 1$. Hence there is some n with $|(x_n^* - x^*)(x)| < \|x\|$ and hence $x_n^*(x) \neq 0$.
In particular we have $T_{\omega_0} : E_{\omega_0} \to c_0(\Gamma_{\omega_0})$.
For successor ordinals $\alpha + 1$ we have $E_{\alpha+1} \cong E_\alpha \times (E_{\alpha+1}/E_\alpha) = E_\alpha \times (P_{\alpha+1} - P_\alpha)(E)$. Let $R_\alpha := (P_{\alpha+1} - P_\alpha)/\|P_{\alpha+1} - P_\alpha\|$, let $F := (P_{\alpha+1} - P_\alpha)(E)$ and let $T : F \to c_0$ be the continuous injection for the, by (53.13.7), separable space F with $\|T\| \le 1$. Then we define $\Gamma_{\alpha+1} := \Gamma_\alpha \sqcup \mathbb{N}$ and $T_{\alpha+1} : E_{\alpha+1} \to c_0(\Gamma_{\alpha+1})$ by

$$T_{\alpha+1}(x)_\gamma := \begin{cases} T_\alpha(\frac{P_\alpha(x)}{\|P_\alpha\|})_\gamma & \text{for } \gamma \in G_\alpha \\ T(R_\alpha(x))_\gamma & \text{for } \gamma \in \mathbb{N} \end{cases}.$$

Now let α be a limit ordinal. We set

$$\Gamma_\alpha := \Gamma_\omega \sqcup \bigsqcup_{\omega \le \beta < \alpha} \Gamma_{\beta+1},$$

and define $T_\alpha : E_\alpha := P_\alpha(E) \to c_0(\Gamma_\alpha)$ by

$$T_\alpha(x)_\gamma := \begin{cases} T_\omega(\frac{P_\omega(x)}{\|P_\omega\|}) & \text{for } \gamma \in \Gamma_\omega \\ T_{\beta+1}(R_\beta(x))_\gamma & \text{for } \gamma \in \Gamma_{\beta+1} \end{cases}$$

We show first that $T_\alpha(x) \in c_0(\Gamma_\alpha)$ for all $x \in E$. So let $\varepsilon > 0$. Then the set $\{\beta : \|R_\beta(x)\| \ge \varepsilon,\ \beta < \alpha\}$ is finite by (53.13.8).
Obviously T_α is linear and $\|T_\alpha\| \le 1$. It is also injective: In fact let $T_\alpha(x) = 0$ for some $x \in E_\alpha$. Then $R_\beta(x) = 0$ for all $\beta < \alpha$ and $P_\omega(x) = 0$, hence by $x = P_\alpha(x) = 0$.
As card(E) is non-measurable, also the smaller cardinal dens(E) is non-measurable. Thus the union Γ_α of non-measurable sets over a non-measurable index set is non-measurable. □

53.21. Corollary. *The WCD Banach spaces and the duals of Asplund spaces continuously and linearly inject into some $c_0(\Gamma)$. The same is true for $C(K)$, where K is Valdivia compact.*

For WCG spaces this is due to [Amir, Lindenstrauss, 1968] and for $C(K)$ with K Valdivia compact it is due to [Argyros, Mercourakis, Negrepontis, 1988.]

Proof. For WCD and duals of Asplund spaces this follows using (53.15). For Valdivia compact spaces K one proceeds by induction on dens(K) and uses the PRI constructed in (53.18). The continuous linear injection $C(K) \to c_0(\Gamma)$ is then given as in (53.20) for $\alpha :=$ dens(K), where T_β exists for $\beta < \alpha$, since $E_\beta \cong C(K_\beta)$ with K_β Valdivia compact and dens$(K_\beta) \le \beta < \alpha$. □

53.22. Theorem. [Bartle, Graves, 1952] *Let $T : E \to F$ be a bounded linear surjective mapping between Banach spaces. Then there exists a continuous mapping $S : F \to E$ with $T \circ S = \mathrm{Id}$.*

Proof. By the open mapping theorem there is a constant $M_0 > 0$ such that for all $\|y\| \le 1$ there exists an $x \in T^{-1}(y)$ with $\|y\| \le M_0$. In fact there is an M_0 such that $B_{1/M_0} \subseteq T(B_1)$ or equivalently $B_1 \subseteq T(B_{M_0})$. Let $(f_\gamma)_{\gamma\in\Gamma}$ be a continuous partition of unity on $oF := \{y \in F : \|y\| \le 1\}$ with $\operatorname{diam}(\operatorname{supp}(f_\gamma)) \le 1/2$. Choose $x_\gamma \in T^{-1}(\operatorname{carr}(f_\gamma))$ with $\|x_\gamma\| \le M_0$ and for $\|y\| \le 1$ set

$$S_0 y := \sum_{\gamma\in\Gamma} f_\gamma(y) x_\gamma \quad \text{and recursively}$$
$$S_{n+1} y := S_n y + \frac{1}{a_n} S_n(a_n(y - TS_n y)),$$

where $a_n := 2^{2^n}$.

By induction we show that the continuous mappings $S_n : \{y : \|y\| \le 1\} \to E$ satisfy $\|y - TS_n y\| \le 1/a_n$ and $\|S_n y\| \le M_n := M_0 \cdot \prod_{k=0}^{n-1}(1 + 1/a_k)$.

$(n = 0)$ Obviously $\|S_0 y\| \le \sum_\gamma f_\gamma(y)\|x_\gamma\| \le M_0$ and

$$\|y - TS_0 y\| = \Big\|\sum_\gamma f_\gamma(y)(y - Tx_\gamma)\Big\| \le \sum_{\gamma\in\Gamma_y} f_\gamma(y)\,\|y - Tx_\gamma\| \le \frac{1}{2} = a_0,$$

where $\Gamma_y := \{\gamma \in \Gamma : f_\gamma(y) \ne 0\}$.

$(n + 1)$ For $\|y\| \le 1$ and $y_n := a_n(y - TS_n y)$ we have $\|y_n\| \le 1$ by induction hypothesis. Then

$$\|S_{n+1} y\| \le \|S_n y\| + \frac{1}{a_n}\|S_n y_n\| \le M_n + \frac{1}{a_n} M_n = M_{n+1}.$$

Furthermore

$$\begin{aligned}\|y - TS_{n+1} y\| &= \|y - TS_n y - \frac{1}{a_n} TS_n(a_n(y - TS_n y))\| \\ &\le \frac{1}{a_n}\|y_n - TS_n y_n\| \le \frac{1}{a_n} \cdot \frac{1}{a_n} = \frac{1}{a_{n+1}}.\end{aligned}$$

Now (S_n) is Cauchy with respect to uniform convergence on $\{y : \|y\| \le 1\}$. In fact

$$\|S_{n+1} y - S_n y\| \le \frac{1}{a_n}\|S_n(a_n(y - TS_n y))\| \le \frac{M_n}{a_n} \le \frac{M_\infty}{a_n},$$

where $M_\infty := \lim_n M_n$. Thus $S := \lim_n S_n$ is continuous and $\|y - TSy\| = \lim_n \|y - TS_n y\| = 0$, i.e. $TSy = y$. Now $S : F \to E$ defined by $S(y) := \|y\| S(\frac{y}{\|y\|})$ and $S(0) := 0$ is the claimed continuous section. $\square$

References

Abraham, R., *Lectures of Smale on differential topology*, Lecture Notes, Columbia University, New York, 1962.

Adam, E., *Smoothly real compact spaces*, Diplomarbeit, Universität Wien, 1993.

Adam, E., *Approximation properties and spaces of smooth functions*, doctoral thesis, Universität Wien, Vienna, 1995.

Adam, E.; Biström, P.; Kriegl, A., *Countably evaluating homomorphisms on real function algebras*, Preprint (1995).

Adams, M.; Ratiu, T.; Schmid, R., *The Lie group structure of diffeomorphism groups and invertible Fourier Integral operators*, Infinite dimensional groups with applications (V. Kac, ed.), Springer-Verlag, New York, 1985, pp. 1–69.

Adams, M.; Ratiu, T.; Schmid, R., *A Lie group structure for pseudo differential operators*, Math. Ann. **273** (1986a), 529–551.

Adams, M.; Ratiu, T.; Schmid, R., *A Lie group structure for Fourier integral operators*, Math. Ann. **276** (1986b), 19–41.

Aharoni, I.; Lindenstrauss, J., *Uniform equivalence between Banach spaces*, Bull. Am. Math. Soc. **84** (1978), 281–283.

Alekseevsky, D.; Michor, P.W., *Differential geometry of* $\mathfrak{g}$*-manifolds*, Differ. Geom. Appl. **5** (1995a), 371–403.

Alekseevsky, D.; Michor, P.W., *Differential geometry of Cartan connections*, Publ. Math. **47** (1995b), 349–375.

Alekseevsky, D.; Kriegl, A.; Losik, M.; Michor, P.W., *Choosing roots of polynomials smoothly*, ESI preprint 314 (1997), 24.

Ambrose, W.; Palais, R.S.; Singer, I.M., *Sprays*, An. Acad. Bras. Cienc. **32** (1960), 163–178.

Amir, D.; Lindenstrauss, J., *The structure of weakly compact sets in Banach spaces*, Ann. Math. **88** (1968), 35–46.

Argyros, S.; Mercourakis, S.; Negrepontis, S., *Functional-analytic properties of Corson-compact spaces*, Stud. Math. **89** (1988), 197–229.

Arias-de-Reyna, J., *A real valued homomorphism on algebras of differentiable functions*, Proc. Am. Math. Soc. **104, No. 4** (1988), 1054–1058.

Arnold, V.I., *Sur la géometrie differentielle des groupes de Lie de dimension infinie et ses applications à l'hydrodynamique des fluides parfaits*, Ann. Inst. Fourier **16** (1966a), 319–361.

Arnold, V.I., *An a priori estimate in the theory of hydrodynamic stability*, Russian, Izv. Vyssh. Uchebn. Zaved. Mat. **54,5** (1966b), 3–5.

Arnold, V.I., *Mathematical methods of classical mechanics*, Graduate Texts in Math. 60, Springer-Verlag, New York, Heidelberg, 1978.

Aron, R.M., *Compact polynomials and compact differentiable mappings between Banach spaces*, Sem. P. Lellong 1974/1975, Lecture notes in Math. 524, Springer-Verlag, 1976, pp. 231–222.

Aronszajn, N., *Differentiability of Lipschitzian mappings between Banach spaces*, Stud. Math. **57** (1976), 147–190.

Asplund, E., *Fréchet differentiability of convex functions*, Acta Math. **121** (1968), 31–48.

Atiyah, M., *Complex analytic connections on fiber bundles*, Trans. Am. Math. Soc. **85** (1957), 181–207.

Averbukh, V.I.; Smolyanov, O.G., *The various definitions of the derivative in linear topological spaces*, Russ. Math. Surv. **23** (1968), 67–113.

Balanzat, M.M., *La différential en les espacios métricas affines*, Math. Notae **9** (1949), 29–51.

Balanzat, M.M., *La différentielle d'Hadamard-Fréchet dans les espaces vectoriels topologiques*, C. R. Acad. Sci. Paris **251** (1960), 2459–2461.

Banach, S., *Théorie des opérations linéaires*, Warszawa, 1932.

Banaszczyk, W., *On the existence of exotic Banach Lie groups*, Math. Ann. **264** (1983), 485–493.

Banaszczyk, W., *Characterization of nuclear spaces by means of additive subgroups*, Math. Z. **186** (1984), 125–133.

Banaszczyk, W., *Pontryagin duality for subgroups and quotients of nuclear spaces*, Math. Ann. **273** (1986), 653–664.

Banaszczyk, W., *Additive subgroups of topological vector spaces*, Lecture Notes in Math. 1466, Springer-Verlag, 1991.

Banyaga, A., *Sur la structure du groupe des diffeomorphismes qui preservent une forme symplectique*, Comment. Math. Helv. **53** (1978), 174–227.

Banyaga, A., *On fixed points of symplectic maps*, Invent. Math. **56** (1980), 215–229.

Bartle, R.G.; Graves, L.M., *Mappings between function spaces*, Trans. Am. Math. Soc. **72** (1952), 400–413.

Bastiani, A., *Applications différentiables et variétés différentiables de dimension infinie*, J. Anal. Math. **13** (1964), 1–114.

Baumgärtel, H., *Endlichdimensionale analytische Störungstheorie*, Akademie-Verlag, Berlin, 1972.

Bemelmans, J.; Hildebrandt, St.; von Wahl, W., *Partielle Differentialgleichungen und Variantionsrechnung*, Ein Jahrhundert Mathematik 1890–1990 (Fischer, G.; Hirzebruch, F.; Scharlau, W.; Törnig, W., eds.), Festschrift zum Jubiläum der DMV, Vieweg, Braunschweig, Wiesbaden, 1990, pp. 149–230.

Bessaga, C., *Every infinite-dimensional Hilbert space is diffeomorphic with its unit sphere*, Bull. Pol. Acad. Sci. Math. **14** (1966), 27–31.

Bessaga, C.; Pełczyński, A., *Selected topics in infinite dimensional topology*, Monografie Matematyczne, Tom 58, Polish Scientific Publishers, Warszawa, 1975.

Binz, E., *Ein Differenzierbakeitsbegriff in limitierten Vektorräumen*, Comment. Math. Helv. **41** (1966), 137–156.

Binz, E., *Two natural metrics and their covariant derivatives on a manifold of embeddings*, Monatsh. Math. **89** (1980), 275–288.

Binz, E.; Fischer, H.R., *The manifold of embeddings of a closed manifold*, Proc. Differential geometric methods in theoretical physics, Clausthal 1978, Lecture Notes in Physics 139, Springer-Verlag, 1981.

Biström, P., *The homomorphisms on algebras of functions defined on locally convex spaces and bounding sets*, Doctoral thesis, University of Åbo, 1993.

Biström, P.; Bjon, S.; Lindström, M., *Homomorphisms on some function algebras*, Monatsh. Math. **111** (1990), 5.

Biström, P.; Bjon, S.; Lindström, M., *Function algebras on which homomorphisms are point evaluations on sequences*, Manuscr. Math. **73** (1991), 179–185.

Biström, P.; Bjon, S.; Lindström, M., *Remarks on homomorphisms on certain subalgebras of $C(X)$*, Math. Jap. **37** (1992), 105–109.

Biström, P.; Bjon, S.; Lindström, M., *On C^m-bounding sets*, J. Aust. Math. Soc. **54** (1993), 20–28.

Biström, P.; Jaramillo, J.A., *C^∞-bounding sets and compactness*, Math. Scand. **75** (1994), 82–86.

Biström, P.; Jaramillo, J.A.; Lindström, M., *Algebras of real analytic functions; Homomorphisms and bounding sets*, Stud. Math. **115** (1995), 23–37.

Biström, P.; Lindström, M., *Homomorphisms on $C^\infty(E)$ and C^∞-bounding sets*, Monatsh. Math. **115** (1993a), 257–266.

Biström, P.; Lindström, M., *Characterization of the spectra of certain function algebras*, Arch Math. **60** (1993b), 177–181.

Bochnak, J.; Siciak, J., *Analytic functions in topological vector spaces*, Stud. Math. **39** (1971), 77–112.

Bockstein, M., *Une théorème de séparabilité pour les produits topologiques*, Fundam. Math. **35** (1948), 242–246.

Bolza, O., *Vorlesungen über Variationsrechnung*, Leipzig, 1909; Second edition, New York, Chelsea, 1962.

Boman, J., *Differentiability of a function and of its compositions with functions of one variable*, Math. Scand. **20** (1967), 249–268.

Bonic, R.; Frampton, J., *Differentiable functions on certain Banach spaces*, Bull. Am. Math. Soc. **71** (1965), 393–395.

Bonic, R.; Frampton, J., *Smooth functions on Banach manifolds*, J. Math. Mech. **15** (1966), 877–898.

Bonic, R.; Reis, F., *A characterization of Hilbert space*, An. Acad. Bras. Cienc. **38** (1966), 239–241.

Boothby, W.M., *The transitivity of the automorphisms of certain geometric structures*, Trans. Am. Math. Soc. **137** (1969), 93–100.

Borel, A.; Harish-Chandra, *Arithmetic subgroups of algebraic groups*, Ann. Math. **75** (1962), 485–535.

Borel, A.; Serre, J.P., *Corners and arithmetic groups*, Comment. Math. Helv. **48** (1973), 436–491.

Borwein, J.; Fitzpatrick, S.; Kenderov, P., *Minimal convex uscos and monotone operators on small sets*, Can. J. Math. **43** (1991), 461–476.

Bourbaki, N., *General Topology*, Hermann, Paris, 1966.

Bourguignon, J.-P., *Une stratification de l'espace des structures Riemanniens*, Compos. Math. **30** (1975), 1–41.

Bröcker, T.; Jänich, K., *Einführung in die Differentialtopologie*, Heidelberger Taschenbücher 143, Springer-Verlag, Heidelberg, 1973.

Browder, W., *Diffeomorphisms of 1-connected manifolds*, Trans. Am. Math. Soc. **128** (1967), 155–163.

Brown, R., *Some problems of algebraic topology*, Ph. D. Thesis, Oxford, 1961.

Brown, R., *Ten topologies for $X \times Y$*, Q. J. Math. Oxf. **14** (1963), 303–319.

Brown, R., *Function spaces and product topologies*, Q. J. Math. Oxf. **15** (1964), 238–250.

Buchwalter, H., *Espaces vectoriels bornologiques*, Publ. Dep. Math. Lyon **2** (1965), 1–53.

Cap, A.; Kriegl, A.; Michor, P.W.; Vanžura, J., *The Frölicher-Nijenhuis bracket in non commutative differential geometry*, Acta Math. Univ. Comenianae **62** (1993), 17–49.

Carne, T.K.; Cole, B.; Gamelin, T.W., *A uniform algebra of analytic functions on a Banach space*, Trans. Am. Math. Soc. **314** (1989), 639–659.

Cartan, H.; Eilenberg, S., *Homological Algebra*, Princeton University Press, Princeton, 1956.

Cascales, B.; Orihuela, J., *On compactness in locally convex spaces*, Math. Z. **195** (1987), 365–381.

Cerf, J., *La stratification naturelle des espaces de fonctions différentiables reélles et le théoréme de la pseudoisotopie*, Publ. Math. Inst. Hautes Etud. Sci. **39** (1970).

Cervera, V.; Mascaro, F.; Michor, P.W., *The orbit structure of the action of the diffeomorphism group on the space of immersions*, Differ. Geom. Appl. **1** (1991), 391–401.

Ciesielski, K.; Pol, R., *A weakly Lindelöf function space $C(K)$ without continuous injection into* $c_0(\Gamma)$, Bull. Pol. Acad. Sci. Math. **32** (1984), 681–688.

Cigler, J.; Losert, V.; Michor, P.W., *Banach modules and functors on categories of Banach spaces*, Lecture Notes in Pure and Applied Mathematics 46, Marcel Dekker Inc., 1979, pp. xv+282.

Collier, J.B., *A class of strong differentiability spaces*, Proc. Am. Math. Soc. **53** (1975), 420–422.

Colombeau, J.F., *Différentiation et Bornologie*, Dissertation, Bordeaux, 1973.

Colombeau, J.F., *Infinite dimensional C^∞-mappings with a given sequence of derivatives at a given point*, J. Math. Anal. Appl. **71** (1979), 95–104.

Colombeau, J.F., *Differential Calculus and Holomorphy*, Math. Studies 64, North Holland, Amsterdam, 1982.

Colombeau, J.F.; Meise, R., *C^∞-functions on locally convex and on bornological vector spaces*, Functional analysis, holomorphy, and approximation theory, Proc. Semin., Rio de Janeiro 1978, Lect. Notes Math. 843, Springer-Verlag, 1981, pp. 195–216.

Colombeau, J.F.; Perrot, B., *The (partial d) equation in DFN spaces.*, J. Math. Anal. Appl. **78** (1980), 466–487.

Corson, H.H., *The weak topology of a Banach space*, Trans. Am. Math. Soc. **101** (1961), 1–15.

Cudia, D., *The geometry of Banach spaces. Smoothness*, Trans. Am. Math. Soc. **110** (1964), 284–314.

Daniell, P.J., *The derivative of a functional*, Bull. Am. Math. Soc. **25** (1919), 414–416.

Day, M.M., *Strict convexity and smoothness of normed spaces*, Trans. Am. Math. Soc. **78** (1955), 516–528.

Deville, R.; Godefroy, G.; Zizler, V.E., *The three space problem for smooth partitions of unity and $C(K)$ spaces*, Math. Ann. **288** (1990), 613–625.

Deville, R.; Godefroy, G.; Zizler, V.E., *Smoothness and renormings in Banach spaces*, Pitman Monographs and Surveys in Pure and Applied Mathematics 64, Longman, John Wiley, London, New York, 1993.

De Wilde, M., *Closed graph theorems and webbed spaces*, Pitman, London, 1978.

DeWitt, B.S., *Quantum theory of gravity. I. The canonical theory*, Phys. Rev. **160 (5)** (1967), 1113–1148.

Diestel, J., *Geometry of Banach spaces, selected topics*, Lecture Notes in Math. 485, Springer-Verlag, 1975, pp. New York.

Diestel, J., *Sequences and series in Banach spaces*, Springer Verlag, 1984.

Dieudonné, J., *Sur les espaces uniformes complets*, Ann. Sci. Éc. Norm. Super. **56** (1939), 277–291.

Dieudonné, J.A., *Foundations of modern analysis, I*, Academic Press, New York, London, 1960.

Dieudonné, J., *Sur un theoreme de Glaeser*, J. Anal. Math. **23** (1970), 85–88.

Douady, A., *Le problème des modules pour les sous-espaces analytiques compacts d'un espace analytique donné*, Ann. Inst. Fourier **16** (1966), 1–95.

Dubois-Violette, M.; Michor, P.W., *A common generalization of the Frölicher-Nijenhuis bracket and the Schouten bracket for symmetric multi vector fields*, Indag. Math. New Ser. **6** (1995), 51–66.

Dubois-Violette, M.; Michor, P.W., *More on the Frölicher-Nijenhuis bracket in non commutative differential geometry*, to appear, J. Pure Appl. Algebra (1997).

Ebin, D., *The manifold of Riemannian metrics*, Proc. Symp. Pure Math. 15, Am. Math. Soc., 1970, pp. 11–40.

Ebin, D.G.; Marsden, J.E., *Groups of diffeomorphisms and the motion of an incompressible fluid*, Ann. Math. **92** (1970), 102–163.

Edgar, G.A., *Measurability in Banach spaces*, Indiana Univ. Math. J. **26** (1977), 663–677.

Edgar, G.A., *Measurability in a Banach space, II*, Indiana Univ. Math. J. **28** (1979), 559–579.

Eells, J., *A setting for global analysis*, Bull. Am. Math. Soc. **72** (1966), 571–807.

Eells, J.; Elworthy, K.D., *Open embeddings of certain Banach manifolds*, Ann. Math. **91** (1970), 465–485.

Eichhorn, J., *Gauge theory on open manifolds of bounded geometry*, Int. J. Mod. Phys. A **7** (1992), 3927–3977.

Eichhorn, J., *The manifold structure of maps between open manifolds*, Ann. Global Anal. Geom. **11** (1993), 253–300.

Eichhorn, J., *Spaces of Riemannian metrics on open manifolds*, Preprint (1994).

Ehresmann, C., *Les prolongements d'une variété différentiable, I: Calcul des jets, prolongement principal, II: L'espace des jets d'ordre r de V_n dans V_m*, C. R. Acad. Sci. Paris **233** (1951), 598–600, 777–779.

Ekeland, I.; Lebourg, G., *Generic Fréchet-differentiability and perturbed optimization problems in Banach spaces*, Trans. Am. Math. Soc. **224** (1976), 193–216.

Enflo, P., *Uniform structures and square roots in topological groups II*, Isr. J. Math. **9** (1970), 253–272.

Enflo, P., *On Banach spaces which can be given an equivalent uniformly convex norm*, Isr. J. Math. **13** (1972), 114–146.

Enflo, P.; Lindenstrauss, J.; Pisier, G., *On the 'three space problem'*, Math. Scand. **36** (1975), 199–210.

Engelking, R., *Outline of General Topology*, Polish Scientific Publishers, Warszawa, 1968.

Engelking, R., *General Topology*, revised and completed edition, Heldermann-Verlag, Berlin, 1989.

Epstein, D.B.A., *The simplicity of certain groups of homeomorphisms*, Compos. Math. **22** (1970), 165–173.

Fabian, M.; Godefroy, G., *The dual of every Asplund space admits a projectional resolution of identity*, Stud. Math. **91** (1988), 141–151.

Fantappié, L., *I functionali analitici*, Atti Accad. Naz. Lincei, Mem. **3–11** (1930), 453–683.

Fantappié, L., *Überblick über die Theorie der analytischen Funktionale und ihre Anwendungen*, Jahresber. Dtsch. Math. Ver. **43** (1933), 1–25.

Faure, C.-A., *Sur un théorème de Boman*, C. R. Acad. Sci. Paris **309** (1989), 1003–1006.

Faure, C.-A., *Théorie de la différentiation dans les espaces convenables*, Thése, Université de Genéve, 1991.

Federer, H., *Geometric Measure Theory*, Springer-Verlag, Berlin, 1969.

Fitzpatrick, S., *Metric projections and the differentiability of distance functions*, Bull. Aust. Math. Soc. **22** (1980), 291–312.

Floret, K., *Lokalkonvexe Sequenzen mit kompakten Abbildungen*, J. Reine Angew. Math. **247** (1971), 155–195.

Fréchet, M., *Sur la notion de différentielle*, C. R. Acad. Sci. Paris **152** (1911), 845–847, 1050–1051.

Fréchet, M., *Sur la notion de différentielle dans l'analyse générale*, C. R. Acad. Sci. Paris **180** (1925a), 806–809.

Fréchet, M., *La notion de différentielle dans l'Analyse générale*, Ann. Sci. Ec. Norm. Super. **XLII** (1925b), 293–323.

Fréchet, M., *Sur la notion de différentielle*, J. Math. Pures Appl. **16** (1937), 233–250.

Freed, D.S. ; Groisser, D., *The basic geometry of the manifold of Riemannian metrics and of its quotient by the diffeomorphism group*, Mich. Math. J. **36** (1989), 323–344.

Freed, D.; Uhlenbeck, K., *Instantons and 4-manifolds*, MSRI Publications 1, Springer-Verlag, New York, 1984.

Freedman, M.H., *The topology of four dimensional manifolds*, J. Differ. Geo. **17** (1982), 357–454.

Freifeld, C., *One-parameter subgroups do not fill a neighborhood of the identity in an infinite dimensional Lie (pseudo)-group*, Battelle Rencontres, Lectures in Mathematics and Physics, Benjamin, New York, 1967, pp. 538–543.

Frölicher, A., *Categories cartésiennement fermées engendrées par des monoides*, Cah. Topologie Géom. Differ. **21** (1980), 367–375.

Frölicher, A., *Applications lisses entre espaces et variétés de Fréchet*, C. R. Acad. Sci. Paris **293** (1981), 125–127.

Frölicher, A.; Bucher, W., *Calculus in vector spaces without norm*, Lecture Notes in Math. 30, Springer-Verlag, Heidelberg, 1966.

Frölicher, A.; Gisin, B.; Kriegl, A., *General differentiation theory*, Various publications series No. 35, Aarhus Universitet, 1983, pp. 125–153.

Frölicher, A.; Kriegl, A., *Convergence vector spaces for Analysis*, Convergence Structures 1984, Akademie Verlag, Berlin, 1985, pp. 115–126.

Frölicher, A.; Kriegl, A., *Linear spaces and differentiation theory*, Pure and Applied Mathematics, J. Wiley, Chichester, 1988.

Frölicher, A.; Kriegl, A., *Differentiable extensions of functions*, Differ. Geom. Appl. **3** (1993), 71–90.

Fuks, D.B., *Cohomology of infinite dimensional Lie algebras*, (Russian), Nauka, Moscow, 1984; English, Contemporary Soviet Mathematics, Consultants Bureau (Plenum Press), New York, 1986.

Fulp, R.O., *Connections on the path bundle of a principal fiber bundle*, Differ. Geom. Appl. **4** (1994), 201–237.

Gabriel, P.; Zisman, M., *Fondements de la topologie simpliciale*, Séminaire homotopique, Université de Strasbourg, 1963/64.

Gähler, W., *Grundstrukturen der Analysis I, II*, Birkhäuser/Akademie Verlag, Basel/Berlin, 1977, 1978.

Garrido, I.; Gómez, J.; Jaramillo, J.A., *Homomorphisms on function algebras*, Can. J. Math. **46** (1994), 734–745.

Gâteaux, M.R., *Sur les fonctionnelles continues et les fonctionnelles analytiques*, C. R. Acad. Sci. Paris **157** (1913), 325–327.

Gâteaux, M.R., *Fonctions d'une infinité de variables independantes*, Bull. Soc. Math. Fr. **47** (1919), 70–96.

Gâteaux, M.R., *Sur les fonctionnelles continues et les fonctionnelles analytiques*, Bull. Soc. Math. Fr. **50** (1922), 1–21.

Gelfand, I.M.; Dorfman, I.Y., *Hamiltonian operators and the algebraic structures connected with them*, Funct. Anal. Appl. **13** (1979), 13–30.

Gieraltowska-Kedzierska, M.; Van Vleck, F.S., *Fréchet differentiability of regular locally Lipschitzian functions*, J. Math. Anal. Appl. **159** (1991), 147–157.

Gillman, L.; Jerison, M., *Rings of continuous functions*, Van Nostrand, Princeton, 1960.

Gil-Medrano, O.; Michor, P.W., *The Riemannian manifold of all Riemannian metrics*, Q. J. Math. Oxf. **42** (1991), 183–202.

Gil-Medrano, O.; Michor, P.W.; Neuwirther, M., *Pseudoriemannian metrics on spaces of bilinear structures*, Q. J. Math. Oxf. **43** (1992), 201–221.

Gil-Medrano, O.; Michor, P.W., *Pseudoriemannian metrics on spaces of almost hermitian structures*, to appear, Isr. J. Math. (1994), 11.

Glaeser, G., *Racine carré d'une fonction différentiable*, Ann. Inst. Fourier **13,2** (1963), 203–210.

Godefroy, G.; Pelant, J.; Whitefield, J.H.M.; Zizler, V.E., *Banach space properties of Ciesielski-Pol's $C(K)$-space*, Proc. Am. Math. Soc. **103** (1988), 1087–1093.

Godefroy, G.; Troyanski, S.L; Whitfield, J.H.M; Zizler, V.E., *Smoothness in weakly compactly generated spaces*, J. Funct. Anal. **52** (1983), 344–352.

Godefroy, G.; Troyanski, S.L.; Whitfield, J.H.M; Zizler, V.E., *Locally uniformly rotund renorming and injection into $c_0(\Gamma)$*, Can. Math. Bull. **27** (1984), 494–500.

Godement, R., *Topologie algébrique et théorie des fausceaux*, Hermann, Paris, 1958.

Gómez, J.; Llavona, J.G., *Multiplicative functionals on function algebras*, Rev. Mat. Univ. Complutense Madrid **1** (1988), 19–22.

Grabowski, J., *Free subgroups of diffeomorphism groups*, Fundam. Math. **131** (1988), 103–121.

Grabowski, J., *Derivative of the exponential mapping for infinite dimensional Lie groups*, Ann. Global Anal. Geom. **11** (1993), 213–220.

Grauert, H., *On Levi's problem and the embedding of real analytic manifolds*, Ann. Math. **68** (1958), 460–472.

Greub, W.; Halperin, S.; Vanstone, R., *Connections, Curvature, and Cohomology I, II, III*, Academic Press, New York, London, 1972, 1973, 1976.

Griffiths, P., *On Cartan's method of Lie groups and moving frames as applied to uniqueness and existence questions in differential geometry*, Duke Math. J. **41** (1974), 775–814.

Gromov, M., *Partial differential relations*, Ergebnisse 3. Folge, Band 9, Springer-Verlag, Berlin, Heidelberg, New York, 1986.

Grothendieck, A., *Sur certains espaces de fonctions holomorphes. I*, J. Reine Angew. Math. **192** (1953), 35–64.

Grothendieck, A., *Produits tensoriels topologiques et espaces nucléaires*, Mem. Am. Math. Soc. 16, Am. Math. Soc., 1955.

Gunning, R.C.; Rossi, H., *Analytic functions in several complex variables*, Prentice-Hall, Englewood Cliffs, N.J., 1965.

Günther, M., *Beiträge zur lokalen Lösbarkeit nichtlinearer partieller Differentialgleichungen*, Dissertation, Karl-Marx-Universität, Leipzig, 1988, pp. 150.

Günther, M., *On the perturbation problem associated to isometric embeddings of Riemannian manifolds*, Ann. Global Anal. Geom. **7,1** (1989), 69–77.

Günther, M., *Isometric embeddings of Riemannian manifolds*, Proc. Int. Congr. Math., Kyoto 1990, Vol. II, 1991, pp. 1137–1143.

Hadamard, J., *Leçons sur le calcul des variations*, Hermann, Paris, 1910.

Hadamard, J., *La notion de différentielle dans l'enseignement*, Scripta Univ. Ab. 3, Bib Hierosolymitanarum, Jérusalem, 1923.

Hahn, H., *Über lineare Gleichungssysteme in linearen Räumen*, J. Reine Angew. Math. **157** (1926), 214–229.

Hahn, H., *Reelle Funktionen I* (1932), Leipzig.

Hamilton, R.S., *The inverse function theorem of Nash and Moser*, Bull. Am. Math. Soc. **7** (1982), 65–222.

de la Harpe, P., *Classical Banach-Lie algebras and Banach-Lie groups of operators in Hilbert space*, Lecture Notes in Math. 285, Springer-Verlag, 1972.

Hatakeyama, Y., *Some notes on the groups of automorphisms of contact and symplectic structures*, Tôhoku Math. J. **18** (1966), 338–347.

Hatcher, A., *A proof of a Smale conjecture*, Ann. Math. **117** (1983), 553–607.

Hatcher, A.; Wagoner, J., *Pseudo-isotopies of compact manifolds*, Asterisque **6** (1973).

Haydon, R., *A counterexample to several questions about scattered compact spaces*, Bull. Lond. Math. Soc. **22** (1990), 261–268.

Haydon, R., *Trees and renormings*, Publ. Math. Univ. Pierre Marie Curie **104, No.8** (1991).

Heber, G., *Die Topologie des Konfigurationsraumes der Yang-Mills-Theorie über offenen Mannigfaltigkeiten beschränkter Geometrie*, Dissertation, Universität Greifswald, 1994.

Hewitt, E., *Rings of real valued continuous functions. I*, Trans. Am. Math. Soc. **64** (1948), 45–99.

Hewitt, E., *Linear functionals on spaces of continuous functions*, Fundam. Math. **37** (1950), 161–189.

Hirai, T., *Irreducible unitary representations of the group of diffeomorphisms of a non-compact manifold*, J. Math. Kyoto Univ. **33, 3** (1993), 827–864.

Hirsch, M.W., *Differential topology*, GTM 33, Springer-Verlag, New York, 1976.

Hirzebruch, F., *Neue topologische Methoden in der algebrischen Topologie*, Ergebnisse der Math. 9, 2. edition, Springer-Verlag, Berlin, Göttingen, Heidelberg, 1962.

Hogbe-Nlend, H., *Complétions, tenseurs, nucléarité en bornologie*, J. Math. Pures Appl. (1970).

Hogbe-Nlend, H., *Théorie des Bornologies et Applications*, Lecture Notes in Math. 213, Springer-Verlag, Berlin, 1971.

Hogbe-Nlend, H., *Bornologies and Functional Analysis*, Math. Studies 26, North Holland, Amsterdam, 1977.

Hörmander, L., *The Analysis of linear partial differential operators I*, Grundlehren 256, Springer-Verlag, Berlin, 1983.

Horváth, J., *Topological vector spaces and distributions*, Addison Wesley, Reading, Mass., 1966.

Igusa, K., *What happens to Hatcher and Wagoner's formula for $\pi_0(C(M))$ when the first Postnikoc invariant of M is nontrivial*, Lecture Notes in Math. 1046, Springer-Verlag, New York, Heidelberg, 1984, pp. 104–172.

Imayoshi, Y., *Holomorphic maps of compact Riemannian surfaces into 2-dimensional compact C-hyperbolic manifolds*, Math. Ann. **270** (1985), 403–416.

Imayoshi, Y., *Holomorphic maps of projective algebraic manifolds into compact C-hyperbolic manifolds*, J. Math. Soc. Japan **46** (1994), 289–307.

Jacobowitz, H., *Implicit function theorems and isometric embeddings*, Ann. Math. **95** (1972), 191–225.

James, R.C., *A non-reflexive Banach space isometric with its second conjugate space*, Proc. Natl. Acad. Sci. USA **37** (1951), 174–177.

Jaramillo, J.A., *Multiplicative functionals on algebras of differentiable functions*, To appear in, Arch Math. **58** (1992), 384–387.

Jarchow, H., *Locally convex spaces*, Teubner, Stuttgart, 1981.

Jayne, J.E.; Rogers, C.A., *Borel selectors for upper semicontinuous set valued maps*, Acta Math. **155** (1985), 41–79.

John, K.; Torunczyk, H.; Zizler, V., *Uniformly smooth partitions of unity on superreflexive Banach spaces*, Stud. Math. **70** (1981), 129–137.

John, K.; Zizler, V., *A note on strong differentiability spaces*, Commentat. Math. Univ. Carol. **17** (1976), 127–134.

John, K.; Zizler, V., *On rough norms on Banach spaces*, Commentat. Math. Univ. Carol. **19** (1978), 335–349.

Johnson, W.B.; Lindenstrauss, J., *Some remarks on weakly compactly generated Banach spaces*, Isr. J. Math. **17** (1974), 219–230.

Joris, H., *Une C^∞-application non-immersive qui possède la propriété universelle des immersions.*, Arch Math. **39** (1982), 269–277.

Josefson, B., *Bounding subsets of $\ell^\infty(A)$*, J. Math. Pures. Appl. **57** (1978), 397–421.

Kadec, M.I., *Spaces isomorphic to a locally uniformly convex space*, Izv. Vyssh. Učebn. Zaved. Mat. **13** (1959), 51–57.

Kadec, M.I., *Letters to the editors*, Izv. Vyssh. Učebn. Zaved. Mat. **15** (1961), 139–141.

Kadec, M.I., *Some conditions of the differentiability of the norm of Banach spaces*, Uspechi Mat. Nauk **20** (1965), 183–187.

Kainz, G., *A note on the manifold of immersions and its Riemannian curvature*, Monatsh. Math. **98** (1984), 211–217.

Kainz, G. ; Kriegl, A. ; Michor, P.W., *C^∞-algebras from the functional analytic viewpoint*, J. Pure Appl. Algebra **46** (1987), 89–107.

Kainz, G.; Michor, P.W., *Natural transformations in differential geometry*, Czech. Math. J. **37** (1987), 584–607.

Kashiwara, M.; Kawai, T.; Kimura, T., *Foundations of algebraic analysis*, Princeton Univ. Press, Princeton, 1986.

Katětov, M., *Measures in fully normal spaces*, Fundam. Math. **38** (1951), 73–84.

Kato, T., *Perturbation theory for linear operators*, Grundlehren 132, Springer-Verlag, Berlin, 1976.

Keller, H.H., *Räume stetiger multilinerarer Abbildungen als Limesräume*, Math. Ann. **159** (1965), 259–270.

Keller, H.H., *Differential calculus in locally convex spaces*, Lecture Notes in Math. 417, Springer-Verlag, 1974.

Kenderov, P.S., *The quasi-differentiable functionals are almost everywhere differentiable*, Math. Educ. **2** (1974), 123–126.

Kervaire, M; Milnor, J., *Groups of homotopy spheres, I*, Ann. Math. **77** (1963), 504–537.

Kirillov, A.A., *Elements of the theory of representations*, Springer-Verlag, Berlin, 1976.

Kirillov, A.A., *The orbits of the group of diffeomorphisms of the circle, and local Lie superalgebras*, Funct. Anal. Appl. **15** (1981), 135–136.

Kirillov, A.A., *The orbit method I: geometric quantization; II: Infinite dimensional Lie groups and Lie algebras*, Notes by J. Adams, A. Kleppner, R. Lipsman, J.-Sh. Li, and J. Rosenberg, Department of Math., University of Maryland, 1991.

Knapp, A.W., *Representation theory of semisimple Lie groups*, Princeton University Press, Princeton, 1986.

Kneser, A., *Lehrbuch der Variationsrechnung*, Vieweg, Braunschweig, 1900.

Kodaira, K; Spencer, D.C., *On the deformations of complex analytic structures*, Ann. Math. **67** (1958), 328–460.

Kodaira, K; Nirenberg, L; Spencer, D.C., *On the existence of deformations of complex analytic structures*, Ann. Math. **68** (1958), 450–459.

Kolář, I.; Michor, P.W., *Determination of all natural concomitants of the type of the Frölicher Nijenhuis bracket*, Rend. Circ. Mat. Palermo, Suppl. **16** (1987), 101–108.

Kolář, I.; Michor, P.W.; Slovák, J., *Natural operations in differential geometry*, Springer-Verlag, Berlin, Heidelberg, New York, 1993.

Koppell, N., *Commuting diffeomorphisms*, Proc. Symp. Pure Math. 14 (1970), Am. Math. Soc., 165–184.

Koszul, J.L., *Homologie et Cohomologie des algèbres de Lie*, Bull. Soc. Math. Fr. **78** (1950), 65–127.

Köthe, G., *Dualität in der Funktionentheorie*, J. Reine Angew. Math. **191** (1953), 30–49.

Kriegl, A., *Eine Theorie glatter Mannigfaltigkeiten und Vektorbündel*, Dissertation, Universität Wien, 1980.

Kriegl, A., *Die richtigen Räume für Analysis im Unendlich - Dimensionalen*, Monatsh. Math. **94** (1982), 109–124.

Kriegl, A., *Eine kartesisch abgeschlossene Kategorie glatter Abbildungen zwischen beliebigen lokalkonvexen Vektorräumen*, Monatsh. Math. **95** (1983), 287–309.

Kriegl, A., *A cartesian closed extension of the category of smooth Banach manifolds*, Categorical Topology, Proc. Conference Toledo, Ohio, 1983 (Bentley, L., eds.), Heldermann, Berlin, 1984, pp. 323–336.

Kriegl, A., *Remarks on germs in infinite dimensions*, Acta Math. Univ. Comenianae **66** (1997), 1–18.

Kriegl, A.; Michor, P.W., *A convenient setting for real analytic mappings*, Acta Math. **165** (1990), 105–159.

Kriegl, A.; Michor, P.W., *Aspects of the theory of infinite dimensional manifolds*, Differ. Geom. Appl. **1** (1991), 159–176.

Kriegl, A.; Michor, P.W., *More smoothly real compact spaces*, Proc. Am. Math. Soc. **117** (1993), 467–471.

Kriegl, A.; Michor, P.W., *Regular infinite dimensional Lie groups*, to appear, J. Lie Theory **7,1** (1997), 61–99, ESI Preprint 200.

Kriegl, A.; Michor, P.W. ; Schachermayer, W., *Characters on algebras of smooth functions*, Ann. Global Anal. Geom. **7, No. 2** (1989), 85–92.

Kriegl, A.; Nel, L.D., *A convenient setting for holomorphy*, Cah. Topologie Géom. Differ. **26** (1985), 273–309.

Kriegl, A.; Nel, L.D., *Convenient vector spaces of smooth functions*, Math. Nachr. **147** (1990), 7.

Kuranishi, M., *On the locally complete families of complex analytic structures*, Ann. Math. **75** (1962), 536–577.

Kurzweil, J., *On approximation in real Banach spaces*, Stud. Math. **14** (1954), 214–231.

Ky Fan, M., *Sur quelques notions fondamentales de l'analyse générale*, J. Math. Pures Appl. **21** (1942), 289–368.

Lang, S., *Differential manifolds*, Addison Wesley, 1972.

Lawvere, F.W., *Categorical dynamics*, Lectures given 1967 at the University of Chicago, reprinted in, Topos Theoretical Methods in Geometry (A. Kock, ed.), Aarhus Math. Inst. Var. Publ. Series 30, Aarhus Universitet, 1979.

Lawvere, F.W.; Schanuel, S.H.; Zame, W.R., *On C^∞-function spaces*, Preprint, State University of New York, Buffalo (1981).

Lazet, D., *Sur la différentiabilité des applications analytiques en dimension infinie*, C. R. Acad. Sci. Paris **273** (1971), 155–157.

Leach, E.B.; Whitfield, J.H.M., *Differentiable functions and rough norms on Banach spaces*, Proc. Am. Math. Soc. **33** (1972), 120–126.

Leduc, M., *Densité de certaines familles d'hyperplans tangents*, C. R. Acad. Sci. Paris **270** (1970), 326–328.

Leslie, J., *On a differential structure for the group of diffeomorphisms*, Topology **6** (1967), 264–271.

Leslie, J., *On the group of real analytic diffeomorphisms of a compact real analytic manifold*, Trans. Am. Math. Soc. **274 (2)** (1982), 651–669.

Lévy, M.P., *Leçons d'analyse fonctionnelle*, Gauthier-Villars, Paris, 1922.

Libermann, P.; Marle, C.M., *Symplectic geometry and analytical mechanics*, Mathematics and its applications, D. Reidel, Dordrecht, 1987.

Lindenstrauss, J.; Tzafriri, L., *On the complemented subspace problem,*, Isr. J. Math. **9** (1971), 263–269.

Llavona, J.G., *Approximation of continuously differentiable functions*, North-Holland, 1986.

Long de Foglio, S. Fernandez, *La différentielle au sens d'Hadamard dans les espaces L vectorielles*, Port. Math. **19** (1960), 165–184.

Lovaglia, A.R., *Locally uniformly convex Banach spaces*, Trans. Am. Math. Soc. **78** (1955), 225–238.

Lychagin, V., *On sufficient orbits of a group of contact diffeomorphisms*, Math. USSR Sb. **33** (1977), 223–242.

Mackey, G.W., *Equivalence of a problem in measure theory to a problem in the theory of vector lattices*, Bull. Am. Math. Soc. **50** (1944), 719–722.

Mackey, G.W., *On infinite dimensional spaces*, Trans. Am. Math. Soc. **57** (1945), 155–207.

MacLaughlin, D.P., *Smooth partitions of unity in preduals of WCG spaces*, Math. Z. **211** (1992), 189–194.

Maissen, B., *Lie Gruppen mit Banach Räumen als Parameterräumen*, Acta Math. **108** (1962), 229–270.

Maissen, B., *Über Topologien im Endomorphismenraum eines topologischen Vektorraums*, Math. Ann. **151** (1963), 283–285.

Malgrange, B., *Ideals of differentiable functions*, Oxford Univ. Press, London, 1966.

Margalef Roig, J.; Outerelo Dominguez, E., *Une variedad diferenciable de dimension infinita, seperada y no regular*, Rev. Mat. Hisp.-Amer. **42** (1982), 51–55.

Marinescu, G., *Espaces vectoriels pseudo topologiques et le theorie des distributions*, Deutsche Verlag d. Wiss., Berlin, 1963.

Mather, J.N., *Commutators of diffeomorphisms I*, Comment. Math. Helv. **49** (1974), 512–528.

Mather, J.N., *Commutators of diffeomorphisms II*, Comment. Math. Helv. **50** (1975), 33–40.

Mather, J.N., *A curious remark concerning the geometric transfer map*, Comment. Math. Helv. **59** (1984), 86–110.

Mather, J.N., *Commutators of diffeomorphisms III*, Comment. Math. Helv. **60** (1985), 122–124.

Mattes, J., *On the convenient setting for real analytic mappings*, Monatsh. Math. **116** (1993), 127–141.

Mauhart, M.; Michor, P.W., *Commutators of flows and fields*, Arch. Math. Brno **28** (1992), 228–236.

Mazur, S., *Über konvexe Mengen in linearen normierten Räumen*, Stud. Math. **4** (1933), 70–84.

Mazur, S., *On continuous mappings on cartesian products*, Fundam. Math. **39** (1952), 229–238.

Mazur, S.; Orlicz, W., *Grundlegende Eigenschaften der polynomischen Operationen*, Stud. Math. **5** (1935), 50–68.

Meshkov, V.Z., *Smoothness properties in Banach spaces*, Stud. Math. **63** (1978), 111–123.

Michal, A.D., *Differential calculus in linear topological spaces*, Proc. Natl. Acad. Sci. USA **24** (1938), 340–342.

Michor, P.W., *Manifolds of smooth maps*, Cah. Topologie Géom. Differ. **19** (1978), 47–78.

Michor, P.W., *Manifolds of smooth maps II: The Lie group of diffeomorphisms of a non compact smooth manifold*, Cah. Topologie Géom. Differ. **21** (1980a), 63–86.

Michor, P.W., *Manifolds of smooth maps III: The principal bundle of embeddings of a non compact smooth manifold*, Cah. Topologie Géom. Differ. **21** (1980b), 325–337.

Michor, P.W., *Manifolds of differentiable mappings*, Shiva, Orpington, 1980c.

Michor, P.W., *The manifold of embeddings of a non compact manifold*, Appendix to: E. Binz, H.R. Fischer: The manifold of embeddings of a closed manifold, Proc. Differential geometric methods in theoretical physics, Clausthal 1978, Lecture Notes in Physics 139, Springer-Verlag, 1981.

Michor, P.W., *Manifolds of smooth mappings IV: Theorem of De Rham*, Cah. Topologie Géom. Differ. **24** (1983), 57–86.

Michor, P.W., *A convenient setting for differential geometry and global analysis I, II*, Cah. Topologie Géom. Differ. **25** (1984a), 63–109, 113–178..

Michor, P.W., *Applications of Hamilton's inverse function theorem to manifolds of mappings*, Proc. Convergence structures and applications II, Schwerin 1983, Akademie Verlag, Berlin, 1984b, pp. 159–162.

Michor, P.W., *The cohomology of the diffeomorphism group is a Gelfand-Fuks cohomology*, Rend. Circ. Mat. Palermo, Suppl. **14** (1987a), 235– 246.

Michor, P.W., *Remarks on the Frölicher-Nijenhuis bracket*, Proc. Conference Differential Geometry and its Applications, Brno 1986, D. Reidel, 1987b, pp. 197–220.

Michor, P.W., *Gauge theory for the diffeomorphism group*, Proc. Conf. Differential Geometric Methods in Theoretical Physics, Como 1987, K. Bleuler and M. Werner (eds), Kluwer, Dordrecht, 1988, pp. 345–371.

Michor, P.W., *Knit products of graded Lie algebras and groups*, Rend. Circ. Mat. Palermo, Suppl. **22** (1989), 171–175.

Michor, P.W., *The moment mapping for a unitary representation*, Ann. Global Anal. Geom. **8, No 3** (1990), 299–313.

Michor, P.W., *Gauge theory for fiber bundles*, Monographs and Textbooks in Physical Sciences, Lecture Notes 19, Bibliopolis, Napoli, 1991, pp. 107.

Michor, P.W., *All unitary representations admit moment mappings*, Colloquia Mathematica Societatis János Bolyai, 56. Differential Geometry, Eger (Hungary), 1989 (J. Szenthe, L. Tamássy, eds.), János Bolyai Math. Soc. and Elsevier, Budapest and Amsterdam, 1992, pp. 477–489.

Michor, P.W.; Ratiu, T., *Curvature of the Virasoro-Bott group*, Preprint (1997).

Michor, P.W.; Schichl, H., *No slices on the space of generalized connections*, Preprint (1997).

Michor, P.W.; Vanžura, J., *Characterizing algebras of smooth functions on manifolds*, Commentat. Math. Univ. Carol. **37,3** (1996), 519–521.

Michor, P.W.; Vizman, C., *n-transitivity of certain diffeomorphism groups*, Acta Math. Univ. Comenianae **63** (1994), 4.

Mikusinski, J., *Distributions a valeurs dans le reúnions d'espaces de Banach*, Stud. Math. **19** (1960), 251–285.

Milnor, J., *Topology from the differentiable viewpoint*, University Press of Virginia, Charlottesville, 1965.

Milnor, J., *Remarks on infinite dimensional Lie groups*, Relativity, Groups, and Topology II, Les Houches, 1983, B.S. DeWitt, R. Stora, Eds., Elsevier, Amsterdam, 1984.

Milnor, J.; Stasheff, J., *Characteristic classes*, Annals of Math. Studies 76, Princeton University Press, Princeton, 1974.

Moerdijk, I.; Reyes, G.E., *Models for smooth infinitesimal analysis*, Springer-Verlag, Heidelberg, Berlin, 1991.

Moreau, J.J., *Proximité et dualité dans un espace hilbertien*, Bull. Soc. Math. Fr. **93** (1965), 273–299.

Morrow, J., *The denseness of complete Riemannian metrics*, J. Differ. Geo. **4** (1970), 225–226.

Moser, J., *A new technique for the construction of solutions of nonlinear differential equations*, Proc. Natl. Acad. Sci. USA **47** (1961), 1824–1831.

Moser, J., *A rapidly cenvergent iteration method and non-linear partial differential equations I, II*, Ann. Sc. Norm. Super. Pisa **20** (1966), 265–315, 499–535.

Namioka, I.; Phelps, R.R, *Banach spaces which are Asplund spaces*, Duke Math. J. **42** (1975), 735–749.

Nash, J., *The embedding problem for Riemannian manifolds*, Ann. Math. **63** (1956), 20–63.

Natarajan, L.; Rodrgúez-Carrington, E.; Wolf, J.A., *Differentiable structures for direct limit groups*, Lett. Math. Phys. **23** (1991), 99–109.

Natarajan, L.; Rodrgúez-Carrington, E.; Wolf, J.A., *New classes of infinite dimensional Lie groups*, Proceedings of the 1991 Am. Math. Soc. Summer Research Institute 'Algebraic groups and their generalizations', to appear.

Natarajan, L.; Rodrgúez-Carrington, E.; Wolf, J.A., *Locally convex Lie groups*, Preprint.

Negrepontis, S., *Banach spaces and topology,*, Handbook of set theoretic topology (K. Kunen and J.E. Vaughan, eds.), North-Holland, 1984.

Neuwirther, M., *A family of pseudo metrics on the space of all pseudo metrics*, Diplomarbeit, Universität Wien, 1990.

Neuwirther, M., *Submanifold geometry and Hessians on the Pseudoriemanninan manifold of metrics*, Dr. rer. nat. Dissertation, Universität Wien, 1992.

Neuwirther, M., *Submanifold geometry and Hessians on the Pseudoriemanninan manifold of metrics*, Acta Math. Univ. Comenianae **62** (1993), 51–85.

Nijenhuis, A., *On a class of common properties of some different types of algebras I, II*, Nieuw Arch. Wiskd. **17** (1969), 17–46, 87–108.

Nijenhuis, A.; Richardson, R., *Cohomology and deformations in graded Lie algebras*, Bull. Am. Math. Soc. **72** (1966), 1–29.

Nijenhuis, A.; Richardson, R., *Deformation of Lie algebra structures*, J. Math. Mech. **17** (1967), 89–105.

Nomizu, K.; Ozeki, H., *The existence of complete Riemannian metrics*, Proc. Am. Math. Soc. **12** (1961), 889–891.

Omori, H., *Infinite dimensional Lie transformation groups*, Lecture Notes in Math. 427, Springer-Verlag, Berlin, 1974.

Omori, H., *Theory of infinite dimensional Lie groups*, Kinokuniya, 1978a. (Japanese)

Omori, H., *On Banach Lie groups acting on finite dimensional manifolds*, Tôhoku Math. J. **30** (1978b), 223–250.

Omori, H.; de la Harpe, P., *About interactions between Banach Lie groups and finite dimensional manifolds*, J. Math. Kyoto Univ. **12** (1972), 543–570.

Omori, H.; Maeda, Y.; Yoshioka, A., *On regular Fréchet Lie groups I. Some differential geometric expressions of Fourier integral operators on a Riemannian manifold*, Tokyo J. Math. **3** (1980), 353–390.

Omori, H.; Maeda, Y.; Yoshioka, A., *On regular Fréchet Lie groups II. Composition rules of Fourier Integral operators on a Riemannian manifold*, Tokyo J. Math. **4** (1981a), 221–253.

Omori, H.; Maeda, Y.; Yoshioka, A., *On regular Fréchet Lie groups III*, Tokyo J. Math. **4** (1981b), 255–277.

Omori, H.; Maeda, Y.; Yoshioka, A., *On regular Fréchet Lie groups IV. Definitions and fundamental theorems*, Tokyo J. Math. **5** (1982), 365–398.

Omori, H.; Maeda, Y.; Yoshioka, A., *On regular Fréchet Lie groups V. Several basic properties*, Tokyo J. Math. **6** (1983), 39–64.

Omori, H.; Maeda, Y.; Yoshioka, A.; Kobayashi, O., *On regular Fréchet Lie groups VI. Infinite dimensional Lie groups which appear in general relativity*, Tokyo J. Math. **6** (1983), 217–246.

Onishchik, A.L., *On the classification of fiber spaces*, Sov. Math. Dokl. **2** (1961), 1561–1564.

Onishchik, A.L., *Connections with zero curvature and the de Rham theorem*, Sov. Math. Dokl. **5** (1964), 1654–1657.

Onishchik, A.L., *Some concepts and applications of non-abelian cohomology theory*, Trans. Mosc. Math. Soc. **17** (1967), 49–98.

Orihuela, J.; Valdivia, M., *Projective generators and resolutions of identity in Banach spaces*, Rev. Mat. Univ. Complutense Madrid **2** (1989), 179–199.

Ovsienko, V.Y; Khesin, B.A., *Korteweg–de Vrieß superequations as an Euler equation*, Funct. Anal. Appl. **21** (1987), 329–331.

Palais, R., *Foundations of global non-linear analysis*, Benjamin, New York, 1968.

Palais, R.S.; Terng, C.L., *Critical point theory and submanifold geometry*, Lecture Notes in Mathematics 1353, Springer-Verlag, Berlin, 1988.

Phelps, R.R., *Convex functions, monotone operators and differentiability*, Lecture Notes in Math. 1364, Springer-Verlag, New York, 1989.

Pierpont, J., *Theory of functions of real variables*, Vol. 1, Boston, 1905.

Pisanelli, D., *Applications analytiques en dimension infinie*, C. R. Acad. Sci. Paris **274** (1972a), 760–762.

Pisanelli, D., *Applications analytiques en dimension infinie*, Bull. Sci. Math. **96** (1972b), 181–191.

Preiss, D., *Gâteaux differentiable functions are somewhere Fréchet differentiable*, Rend. Circ. Mat. Pal. **33** (1984), 122–133.

Preiss, D., *Differentiability of Lipschitz functions on Banach spaces*, J. Funct. Anal. **91** (1990), 312–345.

Pressley, A.; Segal, G., *Loop groups*, Oxford Mathematical Monographs, Oxford University Press, 1986.

Procesi, C., *Positive symmetric functions*, Adv. Math. **29** (1978), 219–225.

Quinn, F., *Ends III*, J. Differ. Geo. **17** (1982), 503–521.

Rademacher, H., *Über partielle und totale Differenzierbarkeit von Funktionen mehrerer Variablen und über die Transformation der Doppelintegrale*, Math. Ann. **79** (1919).

Ramadas, T.R., *On the space of maps inducing isomorphic connections*, Ann. Inst. Fourier **32** (1982), 263–276.

Rao, M.M., *Characterisation of Hilbert spaces by smoothness*, Nederl. Akad. v. W. **71** (1967), 132–135.

Ratiu, T.; Schmid, R., *The differentiable structure of three remarkable diffeomorphism groups*, Math. Z. **177** (1981), 81–100.

Rellich, F., *Störungstheorie der Spektralzerlegung, I*, Math. Ann. **113** (1937), 600–619.

Rellich, F., *Störungstheorie der Spektralzerlegung, V*, Math. Ann. **118** (1940), 462–484.

Rellich, F., *Perturbation theory of eigenvalue problems*, Lecture Notes, New York University, 1953; Gordon and Breach, New York, London, Paris, 1969.

Restrepo, G., *Differentiable norms in Banach spaces*, Bull. Am. Math. Soc. **70** (1964), 413–414.

Restrepo, G., *Differentiable norms*, Bol. Soc. Mat. Mex. **10** (1965), 47–55.

Riemann, B., *Ueber die Hypothesen, welche der Geometrie zu Grunde liegen*, Habilitationsvortrag 10. Juni 1854, Gesammelte Werke (H. Weber, ed.), Teubner, Leipzig, 1876.

Schaefer, H.H., *Topological vector spaces*, GTM 3, Springer-Verlag, New York, 1971.

Schichl, H., *On the existence of slice theorems for moduli spaces on fiber bundles*, Doctoral dissertation, Univ. Wien, 1997.

Schlafli, R, *Universal connections*, Invent. Math. **59** (1980), 59–65.

Schmidt, H.-J., *The metric in the superspace of Riemannian metrics and its relation to gravity*, Differential Geometry and its Applications, Brno, Csechoslovakia, 1989, World Scientific, Singapur, 1990, pp. 405–411.

Sebastião e Silva, J., *As funções analíticas e a análise funcional*, Dissertation, 1948.

Sebastião e Silva, J., *As funções analíticas e a análise funcional*, Port. Math. **9** (1950a), 1–130.

Sebastião e Silva, J., *Sobre a topologia dos espaços funcionais analíticos*, Rev. Fac. Cienc. Lisboa, ser. A1 **2** (1950b), 23–102.

Sebastião e Silva, J., *Sui fondamenti della teoria dei funzionali analitici*, Port. Math. **12** (1953), 1–46.

Sebastião e Silva, J., *Le calcul différentielle et intégral dans les espaces localement convexes, réels ou complexes, I*, Atti. Acad. Naz. Lincei **20** (1956a), 743–750.

Sebastião e Silva, J., *Le calcul différentielle et intégral dans les espaces localement convexes, réels ou complexes, II*, Atti. Acad. Naz. Lincei **21** (1956b), 40–46.

Sebastião e Silva, J., *Conceitos de funçã o diferenciável em espaços localmente convexos*, Publ. do Centro Estudos Math. de Lisboa (1957).

Sebastião e Silva, J., *Les espaces à bornés et la notion de fonction différentiable*, Colloque sur l'analyse fonctionnelle, Louvain, 1960, CBRM, 1961, pp. 57–63.

Seeley, R.T., *Extension of C^∞-functions defined in a half space*, Proc. Am. Math. Soc. **15** (1964), 625–626.

Segal, G., *The geometry of the KdV equation*, Int. J. Mod. Phys. A **6** (1991), 2859–2869.

Seip, U., *Kompakt erzeugte Vektorräume und Analysis,*, Lecture Notes in Math. 273, Springer-Verlag, Heidelberg, 1972.

Seip, U., *Differential calculus and cartesian closedness*, Categorical Topology, Mannheim 1975 (Binz, E., Herrlich, H., eds.), Lecture Notes in Math. 540, Springer-Verlag, Heidelberg, 1976, pp. 578–604.

Seip, U., *A convenient setting for differential calculus*, J. Pure Appl. Algebra **14** (1979), 73–100.

Seip, U., *A convenient setting for smooth manifolds*, J. Pure Appl. Algebra **21** (1981), 279–305.

Shirota, *A class of topological spaces*, Osaka Math. J. **4** (1952), 23–40.

Siegl, E., *A free convenient vector space for holomorphic spaces*, Monatsh. Math. **119** (1995), 85–97.

Siegl, E., *Free convenient vector spaces*, doctoral thesis, Universität Wien, Vienna, 1997.

Smale, S., *Diffeomorphisms of the 2-sphere*, Proc. Am. Math. Soc. **10** (1959), 621–626.

Sova, M., *Conditions of differentiability in topological vector spaces*, Czech. Math. J. **16** (1966a), 339–362.

Sova, M., *General theory of differentiability in linear topological spaces,*, Czech. Math. J. **14** (1966b), 485–508.

Steenrod, N.E., *A convenient category for topological spaces*, Mich. Math. J. **14** (1967), 133–152.

Stegall, Ch., *The Radon-Nikodym property in conjugate Banach spaces*, Trans. Am. Math. Soc. **206** (1975), 213–223.

Stegall, Ch., *The duality between Asplund spaces and spaces with the Radon-Nikodym property*, Isr. J. Math. **29** (1978), 408–412.

Stegall, Ch., *The Radon-Nikodym property in conjugate Banach spaces II*, Trans. Am. Math. Soc. **264** (1981), 507–519.

Stein, E.M., *Singular integrals and differentiability properties of functions*, Princeton Univ. Press, Princeton, 1970.

Stolz, O., *Grundz"uge der Differential- und Integralrechnung*, Bd. 1, Leipzig, 1893.

Sullivan, D., *Infinitesimal computations in topology*, Publ. Math. Inst. Hautes Etud. Sci. **47** (1978), 269–331.

Sylvester, J., *On a theory of the syzygetic relations of two rational integral functions, comprising an application to the theory of Sturm's functions, and that of the greatest algebraic common measure*, Philos. Trans. R. Soc. Lond. **143** (1853), 407–548; *Mathematical papers*, Vol. I, At the University Press, Cambridge, 1904, pp. 511ff.

Talagrand, M., *Renormages de quelques $C(K)$*, Isr. J. Math. **54** (1986), 327–334.

Thurston, W., *Foliations and groups of diffeomorphisms*, Bull. Am. Math. Soc. **80** (1974), 304–307.

Toeplitz, O., *Die linearen vollkommenen Räume der Funktionentheorie*, Comment. Math. Helv. **23** (1949), 222–242.

Toruńzcyk, H., *Smooth partitions of unity on some non-separable Banach spaces*, Stud. Math. **46** (1973), 43–51.

Toruńczyk, H., *Characterizing Hilbert space topology*, Fundam. Math. **111** (1981), 247–262.

Toruńczyk, H., *A correction of two papers concerning Hilbert manifolds*, Fundam. Math. **125** (1985), 89–93.

Tougeron, J.C., *Idéaux de fonctions différentiables*, Ergebnisse 71, Springer-Verlag, Heidelberg, 1972.

Treves, F., *Topological vector spaces, distributions, and kernels*, Academic Press, New York, 1967.

Triantafillou, G., *Diffeomorphisms of manifolds with finite fundamental group*, Bull. Am. Math. Soc. **31** (1994), 50–53.

Troyanski, S.L., *Equivalent norms in unseparable B-spaces with an unconditional basis*, Teor. Funkts. Funkts. Anal. Prilozh. **6** (1968), 59–65.

Troyanski, S.L., *On locally convex and differentiable norm in certain non-separable Banach spaces*, Stud. Math. **37** (1971), 173–180.

Turner, E., *Diffeomorphisms of a product of spheres*, Invent. Math. **8** (1969), 69–82.

Tzafriri, L., *Some directions of results in Banach space theory*, Functional analysis, surveys and recent results, North. Holl., 1980, pp. 2–18.

Valdivia, M., *Topics in locally convex spaces*, North Holland, 1982.

Valdivia, M., *Resolution of identity in certain metrizable locally convex spaces*, Rev. R. Acad. Cienc. Exactas Fis. Nat. Madr. **83** (1989), 75–96.

Valdivia, M., *Projective resolution of identity in $C(K)$ spaces*, Arch. Math. **54** (1990), 493–498.

Vanderwerff, J., *Smooth approximations in Banach spaces*, Proc. Am. Math. Soc. **115** (1992), 113–120.

Van Est, W.T. ; Korthagen, T.J., *Nonenlargable Lie algebras*, Indag. Math. **26** (1964), 15–31.

Van Hove, L., *Topologie des espaces fontionnels analytiques et de groupes infinis de transformations*, Bull. Cl. Sci., Acad. R. Belg. **38** (1952), 333–351.

Varadarajan, V.S., *Harmonic analysis on real reductive groups*, Lecture Notes in Mathematics 576, Springer-Verlag, Berlin, Heidelberg, New York, 1977.

Vašák, L., *On one generalization of weakly compactly generated Banach spaces*, Stud. Math. **70** (1981), 11–19.

Veblen, O.; Whitehead, J.H.C., *The foundations of differential geometry*, Cambridge University Press, Cambridge, 1932.

Ver Eecke, P., *Fondements du calcul différentiel*, Presses Universitaires de France, Paris, 1983.

Ver Eecke, P., *Applications du calcul différentiel*, Presses Universitaires de France, Paris, 1985.

Verde-Star, L., *Interpolation and Combinatorial Functions*, Stud. Appl. Math. **79** (1988), 65–92.

Vogt, R., *Convenient categories of topological spaces for Homotopie theory,*, Arch. Math. **22** (1971), 545–555.

Volterra, V., *Sopra le funzioni che dipendone de altre funzioni, Nota I,II,III*, Rend. Acad. Naz. Lincei **4 (3)** (1887), 97–105, 141–146, 153–158.

Waelbroeck, L., *Études spectrale des algèbres complètes*, Mém. Cl. Sci., Collect., Acad. R. Belg. **31** (1960), 142–.

Waelbroeck, L., *Some theorems about bounded structures*, J. Funct. Anal. **1** (1967a), 392–408.

Waelbroeck, L., *Differentiable Mappings in b-spaces*, J. Funct. Anal. **1** (1967b), 409–418.

Warner, F.W., *Foundations of differentiable manifolds and Lie groups*, Scott Foresman, 1971.

Warner, G., *Harmonic analysis on semisimple Lie groups, Volume I*, Springer-Verlag, New York, 1972.

Wegenkittl, K., *Topologien auf Räumen differenzierbarer Funktionen*, Diplomarbeit (1987), Universität Wien, 1–72.

Wegenkittl, K., *The space of isometric immersions into Euclidean space*, Dissertation (1989), Universität Wien, 1–147.

Weil, A., *Théorie des points proches sur les variétés differentielles*, Colloque de topologie et géométrie différentielle, Strasbourg, 1953, pp. 111–117.

Weinstein, A., *Symplectic manifolds and their Lagrangian submanifolds*, Adv. Math. **6** (1971), 329–345.

Weinstein, A., *Lectures on symplectic manifolds*, Reg. Conf. Ser. Math. **29** (1977), Am. Math. Soc..

Wells, J.C., *Differentiable functions on Banach spaces with Lipschitz derivatives*, J. Differ. Geo. **8** (1973), 135–152.

Whitney, H., *Analytic extensions of differentiable functions defined in closed sets*, Trans. Am. Math. Soc. **36** (1934), 63–89.

Whitney, H., *Differentiable Even Functions*, Duke Math. J. **10** (1943), 159–166.

Whitney, H.; Bruhat, F., *Quelques propriétés fondamentales des ensembles analytiques-réels*, Comment. Math. Helv. **33** (1959), 132–160.

Wojtyński, W., *One parameter subgroups and B-C-H formula*, Stud. Math. **111** (1994), 163–185.

Yamamuro, S., *A theory of differentiation in locally convex spaces*, Mem. Am. Math. Soc. **212** (1979).

Yamamuro, S., *Notes on the inverse mapping theorem in locally convex spaces*, Bull. Austr. Math. Soc. **21** (1980), 419–461.

Young, W.H., *The fundamental theorems of differential calculus*, University Press, Cambridge, 1910.

Zhivkov, N.V., *Generic Gâteaux differentiability of directionally differentiable mappings*, Rev. Roum. Math. Pure. Appl. **32** (1987), 179–188.

Živkov, N.V., *Generic Gâteaux differentiability of locally Lipschitzian functions*, Constructive Function Theory, 1981, Sofia, 1983, pp. 590–594..

Zorn, M.A., *Gâteaux-differentiability and essential boundedness*, Duke Math. J. **12** (1945), 579–583.

Index

D

E

F

L

M

S

Selected Titles in This Series

(Continued from the front of this publication)

17 **Le Baron O. Ferguson,** Approximation by polynomials with integral coefficients, 1980
16 **O. Timothy O'Meara,** Symplectic groups, 1978
15 **J. Diestel and J. J. Uhl, Jr.,** Vector measures, 1977
14 **V. Guillemin and S. Sternberg,** Geometric asymptotics, 1977
13 **C. Pearcy, Editor,** Topics in operator theory, 1974
12 **J. R. Isbell,** Uniform spaces, 1964
11 **J. Cronin,** Fixed points and topological degree in nonlinear analysis, 1964
10 **R. Ayoub,** An introduction to the analytic theory of numbers, 1963
9 **Arthur Sard,** Linear approximation, 1963
8 **J. Lehner,** Discontinuous groups and automorphic functions, 1964
7.2 **A. H. Clifford and G. B. Preston,** The algebraic theory of semigroups, Volume II, 1961
7.1 **A. H. Clifford and G. B. Preston,** The algebraic theory of semigroups, Volume I, 1961
6 **C. C. Chevalley,** Introduction to the theory of algebraic functions of one variable, 1951
5 **S. Bergman,** The kernel function and conformal mapping, 1950
4 **O. F. G. Schilling,** The theory of valuations, 1950
3 **M. Marden,** Geometry of polynomials, 1949
2 **N. Jacobson,** The theory of rings, 1943
1 **J. A. Shohat and J. D. Tamarkin,** The problem of moments, 1943